EMS Tracts in Mathematics 22

EMS Tracts in Mathematics

This series includes advanced texts and monographs covering all fields in pure and applied mathematics. *Tracts* will give a reliable introduction and reference to special fields of current research. The books in the series will in most cases be authored monographs, although edited volumes may be published if appropriate. They are addressed to graduate students seeking access to research topics as well as to the experts in the field working at the frontier of research.

For a complete listing see our homepage at www.ems-ph.org.

8 Sergio Albeverio et al., *The Statistical Mechanics of Quantum Lattice Systems*
9 Gebhard Böckle and Richard Pink, *Cohomological Theory of Crystals over Function Fields*
10 Vladimir Turaev, *Homotopy Quantum Field Theory*
11 Hans Triebel, *Bases in Function Spaces, Sampling, Discrepancy, Numerical Integration*
12 Erich Novak and Henryk Woźniakowski, *Tractability of Multivariate Problems. Volume II: Standard Information for Functionals*
13 Laurent Bessières et al., *Geometrisation of 3-Manifolds*
14 Steffen Börm, *Efficient Numerical Methods for Non-local Operators.* $\mathcal{H}^2$*-Matrix Compression, Algorithms and Analysis*
15 Ronald Brown, Philip J. Higgins and Rafael Sivera, *Nonabelian Algebraic Topology. Filtered Spaces, Crossed Complexes, Cubical Homotopy Groupoids*
16 Marek Janicki and Peter Pflug, *Separately Analytical Functions*
17 Anders Björn and Jana Björn, *Nonlinear Potential Theory on Metric Spaces*
18 Erich Novak and Henryk Woźniakowski, *Tractability of Multivariate Problems. Volume III: Standard Information for Operators*
19 Bogdan Bojarski, Vladimir Gutlyanskii, Olli Martio and Vladimir Ryazanov, *Infinitesimal Geometry of Quasiconformal and Bi-Lipschitz Mappings in the Plane*
20 Hans Triebel, *Local Function Spaces, Heat and Navier–Stokes Equations*
21 Kaspar Nipp and Daniel Stoffer, *Invariant Manifolds in Discrete and Continuous Dynamical Systems*
23 Augusto C. Ponce, *Elliptic PDEs, Measures and Capacities. From the Poisson Equation to Nonlinear Thomas–Fermi Problems*
24 Hans Triebel, *Hybrid Function Spaces, Heat and Navier-Stokes Equations*

Patrick Dehornoy

with François Digne
Eddy Godelle
Daan Kramer
Jean Michel

Foundations of Garside Theory

European Mathematical Society

Authors:

Patrick Dehornoy
Laboratoire de Mathématiques Nicolas Oresme, CNRS UMR 6139
Université de Caen
14032 Caen, France

E-mail: patrick.dehornoy@unicaen.fr

with

François Digne
Laboratoire Amiénois de Mathématique
Fondamentale et Appliquée, CNRS UMR 7352
Université de Picardie Jules-Verne
80039 Amiens, France

E-mail: digne@u-picardie.fr

Eddy Godelle
Laboratoire de Mathématiques Nicolas Oresme, CNRS UMR 6139
Université de Caen
14032 Caen, France

E-mail: godelle@math.unicaen.fr

Daan Krammer
Mathematics Institute
University of Warwick
Coventry CV4 7AL, United Kingdom

E-mail: D.Krammer@warwick.ac.uk

Jean Michel
Institut de Mathématiques de Jussieu, CNRS UMR 7586
Université Denis Diderot Paris 7
75205 Paris 13, France

E-mail: jmichel@math.jussieu.fr

2010 Mathematical Subject Classification: Primary: 20Fxx, 20F05, 20F10, 20F36, 20F60, 20F65, 20M05, 20M10; Secondary: 05A05, 18B40, 18G35, 20B30, 20F55, 20L05, 20M50, 20N02, 68Q17

Key words: Group, monoid, category, greedy decomposition, normal decomposition, symmetric normal decomposition, Garside family, Garside map, Garside element, Garside monoid, Garside group, word problem, conjugacy problem, braid group, Artin–Tits group; Deligne–Luzstig variety, self-distributivity, ordered group, Yang–Baxter equation, cell decomposition

ISBN 978-3-03719-139-2

The Swiss National Library lists this publication in The Swiss Book, the Swiss national bibliography, and the detailed bibliographic data are available on the Internet at http://www.helveticat.ch.

Contact address:

European Mathematical Society Publishing House
Seminar for Applied Mathematics
ETH-Zentrum SEW A27
CH-8092 Zürich, Switzerland

Phone: +41 (0)44 632 34 36 | Email: info@ems-ph.org | Homepage: www.ems-ph.org

Printing and binding: Beltz Bad Langensalza GmbH, Bad Langensalza, Germany
∞ Printed on acid free paper

9 8 7 6 5 4 3 2 1

Introduction

A natural, but slowly emerging program. In his PhD thesis prepared under the supervision of Graham Higman and defended in 1965 [124], and in the article that followed [125], F.A. Garside (1915–1988) solved the Conjugacy Problem for Artin's braid group B_n by introducing a submonoid B_n^+ of B_n and a distinguished element Δ_n of B_n^+ that he called fundamental, and showing that every element of B_n can be expressed as a fraction of the form $\Delta_n^m g$, with m an integer and g an element of B_n^+. Moreover, he proved that any two elements of the monoid B_n^+ admit a least common multiple, thus extending to the non-abelian groups B_n some of the standard tools available in a torsion-free abelian group $\mathbb{Z}^n$.

In the beginning of the 1970's, it was soon realized by E. Brieskorn and K. Saito [36] using an algebraic approach, and by P. Deligne [101] using a more geometric approach, that Garside's results extend to all generalized braid groups associated with finite Coxeter groups, that is, all Artin (or, better, Artin–Tits) groups of spherical type.

The next step forward was the possibility of defining, for every element of the braid monoid B_n^+ (and, more generally, of every spherical Artin–Tits monoid) a distinguished decomposition involving the divisors of the fundamental element Δ_n. The point is that, if g is an element of B_n^+, then there exists a (unique) greatest common divisor g_1 for g and Δ_n and, moreover $g \neq 1$ implies $g_1 \neq 1$. Then g_1 is a distinguished fragment of g (the "head" of g); repeating the operation with g' determined by $g = g_1 g'$, we extract the head g_2 of g' and, iterating, we end up with an expression $g_1 \cdots g_p$ of g in terms of divisors of Δ_n. Although F. Garside was very close to such a decomposition when he proved that greatest common divisors exist in B_n^+, the result does not appear in his work explicitly, and it seems that the first instances of such distinguished decompositions, or *normal forms*, go back to the 1980's in independent work by S. Adyan [2], M. El-Rifai and H. Morton [117], and W. Thurston (circulated notes [226], later appearing as Chapter IX in the book [119] by D. Epstein *et al.*). The normal form was soon used to improve Garside's solution of the Conjugacy Problem [117] and, extended from the monoid to the group, to serve as a paradigmatic example in the then emerging theory of automatic groups of J. Cannon, W. Thurston, and others. Sometimes called the *greedy normal form*—or *Garside normal form*, or *Thurston normal form*—it became a standard tool in the investigation of braids and Artin–Tits monoids and groups from a viewpoint of geometric group theory and of theory of representations, essential in particular in D. Krammer's algebraic proof of the linearity of braid groups [162, 163].

In the beginning of the 1990's, it was realized by one of us that some ideas from F. Garside's approach to braid monoids can be applied in a different context to analyze a certain "geometry monoid" M_{LD} that appears in the study of the so-called left-selfdistributivity law $x(yz) = (xy)(xz)$. In particular, the criterion used by F. Garside to establish that the braid monoid B_n^+ is left-cancellative (that is, $gh = gh'$ implies $h = h'$) can be adapted to M_{LD} and a normal form reminiscent of the greedy normal form exists—with the main difference that the pieces of the normal decompositions are not the divisors of some unique element similar to the Garside braid Δ_n, but they are divisors of elements Δ_T that depend on some object T (actually a tree) attached to the element one wishes to decompose. The approach led to results about the exotic left-selfdistributivity law [73] and, more unexpectedly, about braids and their orderability when it turned out that the monoid M_{LD} naturally projects to the (infinite) braid monoid B_∞^+ [72, 75, 77].

At the end of the 1990's, following a suggestion by L. Paris, the idea arose of listing the abstract properties of the monoid B_n^+ and the fundamental braid Δ_n that make the algebraic theory of B_n possible. This resulted in the notions of a *Garside monoid* and a *Garside element* [99]. In a sense, this is just reverse engineering, and establishing the existence of derived normal decompositions with the expected properties essentially means checking that nothing has been forgotten in the definition. However, it soon appeared that a number of new examples are eligible, and, specially after some cleaning of the definitions was completed [80], that the new framework is really more general than the original braid framework. One benefit of the approach is that extending the results often resulted in discovering new improved arguments no longer relying on superfluous assumptions or specific properties. This program turned out to be rather successful and it led to many developments by a number of different authors [7, 10, 12, 18, 19, 20, 57, 56, 68, 122, 129, 138, 139, 171, 170, 180, 198, 211, ...]. Today the study of Garside monoids is still far from complete, and many questions remain open.

However, in the meanwhile, it soon became clear that, although efficient, the framework of Garside monoids, as stabilized in the 1990s, is far from optimal. Essentially, several assumptions, in particular Noetherianity conditions, are superfluous and they just discard further natural examples. Also, excluding nontrivial invertible elements appears as an artificial limiting assumption. More importantly, one of us (DK) in a 2005 preprint subsequently published as [165] and two of us (FD, JM) [109], as well as David Bessis in an independent research [9], realized that normal forms similar to those involved in Garside monoids can be developed and usefully applied in a context of categories, leading to what they naturally called *Garside categories*. By the way, similar structures are already implicit in the 1976 paper [103] by P. Deligne and G. Lusztig, as well as in the above mentioned monoid M_{LD} [75, 77], and in EG's PhD thesis [134].

It was therefore time around 2007 for the development of a new, unifying framework that would include all the previously defined notions, remove all unneeded assumptions, and allow for optimized arguments. This program was developed in particular during a series of workshops and meetings between 2007 and 2012, and it resulted in the current text. As the above description suggests, the emphasis is put on the normal form and its mechanism, and the framework is that of a general category with only one assumption, namely left-cancellativity. Then the central notion is that of a *Garside family*, defined to be any family that gives rise to a normal form of the expected type. Then, of course, every Garside element Δ in a Garside monoid provides an example of a Garside family, namely the set of all divisors of Δ, but many more Garside families may exist—and they do, as we shall see in the text. Note that, in a sense, our current generalization is the ultimate one since, by definition, no further extension may preserve the existence of a greedy normal form. However, different approaches might be developed, either by relaxing the definition of a greedy decomposition (see the Notes at the end of Chapter III) or, more radically, by placing the emphasis on other aspects of Garside groups rather than on normal forms. Typically, several authors, including J. Crisp, J. McCammond and one of us (DK) proposed to view a Garside group mainly as a group acting on a lattice in which certain intervals of the form $[1, \Delta]$ play a distinguished role, thus paving the way for other types of extensions.

Our hope—and our claim—is that the new framework so constructed is quite satisfactory. By this, we mean that most of the properties previously established in more particular contexts can be extended to larger contexts. It is *not* true that all properties of, say, Garside monoids extend to arbitrary categories equipped with a Garside family but, in most cases, addressing the question in an extended framework helps improving the arguments and really capturing the essential features. Typically, almost all known properties of Garside monoids do extend to categories that admit what we call a bounded Garside family, and the proofs cover for free all previously considered notions of Garside categories.

It is clear that a number of future developments will continue to involve particular types of monoids or categories only: we do not claim that our approach is universal... However, we would be happy if the new framework—and the associated terminology—could become a useful reference for further works.

About this text. The aim of the current text is to give a state-of-the-art presentation of this approach. Finding a proper name turned out to be not so obvious. On the one hand, "Garside calculus" would be a natural title, as the greedy normal form and its variations are central in this text: although algorithmic questions are not emphasized, most constructions are effective and the mechanism of the normal form is indeed a sort of calculus. On the other hand, many results, in particular those of structural nature, exploit the normal form but are not reducible to it, making a title like "Garside structures" or "Garside theory" more appropriate. But such a title is certainly too ambitious for what we can offer: no genuine structure theory or no exhaustive classification of, say, Garside families, is to be expected at the moment. What we do here is develop a framework that, we think and hope, can become a good base for a still-to-come theory. Another option could have been "Garside categories", but it will be soon observed that no notion with that name is introduced here: in view of the subsequent developments, a reasonable meaning could be "a cancellative category that admits a Garside map", but a number of variations are still possible, and any particular choice could become quickly obsolete—as is, in some sense, the notion of a Garside group. Finally, we hope that our current title, "Foundations of Garside Theory", reflects the current content in a proper way: the current text is an invitation to further research, and does not aim at being exhaustive—reporting about all previous results involving Garside structures would already be very difficult—but concentrates on what seems to be the core of the subject.

The text in divided into two parts. Part A is devoted to general results and offers a careful treatment of the bases. Here complete proofs are given, and the results are illustrated with a few basic examples. By contrast, Part B consists of essentially independent chapters explaining further examples, or families of examples, that are in general more elaborate. Here some proofs are omitted, and the discussion is centered around what can be called the Garside aspects in the considered structures.

Our general scheme will be to start from an analysis of normal decompositions and then introduce Garside families as the framework guaranteeing the existence of normal decompositions. Then the three main questions we shall address and a chart of the corresponding chapters looks as follows:

- **How** do Garside structures work? (mechanism of normal decomposition)
  - Chapter III (domino rules, geometric aspects)
  - Chapter VII (compatibility with subcategories)
  - Chapter VIII (connection with conjugacy)
- **When** do Garside structures exist? (existence of normal decomposition)
  - Chapter IV (recognizing Garside families)
  - Chapter VI (recognizing Garside germs)
  - Chapter V (recognizing Garside maps)
- **Why** consider Garside structures? (examples and applications)
  - Chapter I (basic examples)
  - Chapter IX (braid groups)
  - Chapter X (Deligne–Lusztig varieties)
  - Chapter XI (selfdistributivity)
  - Chapter XII (ordered groups)
  - Chapter XIII (Yang–Baxter equation)
  - Chapter XIV (four more examples)

Above, and in various places, we use "Garside structure" as a generic and informal way to refer to the various objects occurring with the name "Garside": Garside families, Garside groups, Garside maps, *etc.*

The chapters. To make further reference easy, each chapter in Part A begins with a summary of the main results. At the end of each chapter, exercises are proposed, and a Notes section provides historical references, comments, and questions for further research.

Chapter I is introductory and lists a few examples. The chapter starts with classical examples of Garside monoids, such as free abelian monoids or classical and dual braid monoids, and it continues with some examples of structures that are not Garside monoids, but nevertheless possess a normal form similar to that of Garside monoids, thus providing a motivation for the construction of a new, extended framework.

Chapter II is another introductory chapter in which we fix some terminology and basic results about categories and derived notions, in particular connected with divisibility relations that play an important rôle in the sequel. A few general results about Noetherian categories and groupoids of fractions are established. The final section describes an general method, called reversing, for investigating a presented category. As the question is not central in our current approach (and although it owes much to F.A. Garside's methods), some proofs of this section are deferred to the Appendix at the end of the book.

Chapter III is the one where the theory really starts. Here the notion of a normal decomposition is introduced, as well as the notion of a Garside family, abstractly introduced as a family that guarantees the existence of an associated normal form. The mechanism of the normal form is analyzed, both in the case of a category ("positive case") and in the case of its enveloping groupoid ("signed case"): some simple diagrammatic patterns, the domino rules, are crucial, and their local character directly implies various geometric consequences, in particular a form of automaticity and the Grid Property, a strong convexity statement.

Chapter IV is devoted to obtaining concrete characterizations of Garside families, in other words, conditions that guarantee the existence of normal decompositions. In this chapter, one establishes external characterizations, meaning that we start with a category $\mathcal{C}$ and look for conditions ensuring that a given subfamily $\mathcal{S}$ of $\mathcal{C}$ is a Garside family. Various answers are given, in a general context first, and then in particular contexts where some conditions come for free: typically, if the ambient category $\mathcal{C}$ is Noetherian and admits unique least common right-multiples, then a subfamily $\mathcal{S}$ of $\mathcal{C}$ is a Garside family if and only if it generates $\mathcal{C}$ and is closed under least common right-multiple and right-divisor.

In Chapter V, we investigate particular Garside families that are called bounded. Essentially, a Garside family $\mathcal{S}$ is bounded is there exists a map Δ (an element in the case of a monoid) such that $\mathcal{S}$ consists of the divisors of Δ (in some convenient sense). Not all Garside families are bounded, and, contrary to the existence of a Garside family, the existence of a bounded Garside family is not guaranteed in every category. Here we show that a bounded Garside family is sufficient to prove most of the results previously established for a Garside monoid, including the construction of Δ-normal decompositions, a variant of the symmetric normal decompositions used in groupoids of fractions.

Chapter VI provides what can be called internal (or intrinsic) characterizations of Garside families: here we start with a family $\mathcal{S}$ equipped with a partial product, and we wonder whether there exists a category $\mathcal{C}$ in which $\mathcal{S}$ embeds as a Garside family. The good news is that such characterizations do exist, meaning that, when the conditions are satisfied, all properties of the generated category can be read inside the initial family $\mathcal{S}$. This local approach turns to be useful for constructing examples and, in particular, it can be used to construct a sort of unfolded, torsion-free version of convenient groups, typically braid groups starting from Coxeter groups.

Chapter VII is devoted to subcategories. Here one investigates natural questions such as the following: if $\mathcal{S}$ is a Garside family in a category $\mathcal{C}$ and $\mathcal{C}_1$ is a subcategory of $\mathcal{C}$, then is $\mathcal{S} \cap \mathcal{C}_1$ a Garside family in $\mathcal{C}_1$ and, if so, what is the connection between the associated normal decompositions? Of particular interest are the results involving subgerms, which provide a possibility of reading inside a given Garside family $\mathcal{S}$ the potential properties of the subcategories generated by the subfamilies of $\mathcal{S}$.

In Chapter VIII, we address conjugacy, first in the case of a category equipped with an arbitrary Garside family, and then, mainly, in the case of a category equipped with a bounded Garside family. Here again, most of the results previously established for Garside monoids can be extended, including the cycling, decycling, and sliding transformations which provide a decidability result for the Conjugacy Problem whenever convenient finiteness assumptions are satisfied. We also extend the geometric methods of Bestvina to describe periodic elements in this context.

Part B begins with Chapter IX, devoted to (generalized) braid groups. Here we show how both the reversing approach of Chapter II and the germ approach of Chapter VI can be applied to construct and analyze the classical and dual Artin–Tits monoids. We also mention the braid groups associated with complex reflection groups, as well as several exotic Garside structures on B_n. The applications of Garside structures in the context of braid groups are too many to be described exhaustively, and we just list some of them in the Notes section.

Chapter X is a direct continuation of Chapter IX. It reports about the use of Garside-type methods in the study of Deligne–Lusztig varieties, an ongoing program that aims at establishing by a direct proof some of the consequences of the Broué Conjectures about finite reductive groups. Several questions in this approach directly involve conjugacy in generalized braid groups, and the results of Chapter VIII are then crucial.

Chapter XI is an introduction to the Garside structure hidden in the above mentioned algebraic law $x(yz) = (xy)(xz)$, a typical example where a categorical framework is needed (or, at the least, the framework of Garside monoids is not sufficient). Here a promising contribution of the Garside approach is a natural program possibly leading to the so-called Embedding Conjecture, a deep structural result so far resisting all attempts.

In Chapter XII, we develop an approach to ordered groups based on divisibility properties and Garside elements, resulting in the construction of groups with the property that the associated space of orderings contains isolated points, which answers one of the natural questions of the area. Braid groups are typical examples, but considering what we call triangular presentations leads to a number of different examples.

Chapter XIII is a self-contained introduction to set-theoretic solutions of the Yang–Baxter equation and the associated structure groups, an important family of Garside groups. The exposition is centered on the connection between the RC-law $(xy)(xz) = (yx)(yz)$ and the right-complement operation on the one hand, and what is called the geometric I-structure on the other hand. Here the Garside approach both provides a specially efficient framework and leads to new results.

In Chapter XIV, we present four unrelated topics involving interesting Garside families: divided categories and decompositions categories with two applications, then an extension of the framework of Chapter XIII to more general RC-systems, then what is called the braid group of $\mathbb{Z}^n$, a sort of analog of Artin's braid group in which permutations of $\{1, ..., n\}$ are replaced with linear orderings of $\mathbb{Z}^n$, and, finally, an introduction to groupoids of cell decompositions that arise when the mapping class group approach to braid groups is extended by introducing sort of roots of the generators σ_i.

The final Appendix contains the postponed proofs of some technical statements from Chapter II for which no complete reference exists in literature.

Exercises are proposed at the end of most chapters. Solutions to some of them, as well as a few proofs from the main text that are skipped in the book, can be found at the address

`www.math.unicaen.fr/~garside/Addenda.pdf`

as well as in arXiv:1412.5299.

Thanks. We primarily wish to thank David Bessis, who was part of the crew at an early stage of the project, but then quit it for personal reasons.

Next, we thank all the colleagues and students who participated in the various meetings dedicated to the project and who contributed by their questions and suggestions to this text. A (certainly non-exhaustive) list includes Marc Autord, Serge Bouc, Michel Broué, Matthieu Calvez, Ruth Corran, Pierre-Louis Curien, Matthew Dyer, Jean Fromentin, Volker Gebhardt, Tomas Gobet, Juan González-Meneses, Yves Guiraud, Tatiana Ivanova–Gateva, Jean-Yves Hée, Christophe Hohlweg, Eric Jespers, Eon-Kyung Lee, Sang-Jin Lee, Gilbert Levitt, Jon McCammond, Ivan Marin, Paul-André Melliès, François Métayer, Jan Okniński, Luis Paris, Matthieu Picantin, Wolfgang Rump, Elena Sendroiu, Maya Van Campenhout, Bert Wiest. Also, special thoughts go to Joan Birman and Hugh Morton, whose support and interest in the subject has always been strong.

The project was supported partly by the French ANR grant TheoGar ANR-08-BLAN-0269-02.

Patrick Dehornoy
François Digne
Eddy Godelle
Daan Krammer
Jean Michel

About notation

We use

- $\mathbb{N}$: set of all nonnegative integers
- $\mathbb{Z}$: set of all integers
- $\mathbb{Q}$: set of all rational numbers
- $\mathbb{R}$: set of all real numbers
- $\mathbb{C}$: set of all complex numbers

As much as possible, different letters are used for different types of objects, according to the following list:

- $\mathcal{C}$: category
- $\mathcal{S}, \mathcal{X}$: generic subfamily of a category
- $\mathcal{A}$: atom family in a category
- x, y, z: generic object of a category
- f, g, h: generic element (morphism) in a category, a monoid, or a group
- a, b, c, d, e: special element in a category (atom, endomorphism, etc.)
- ϵ: invertible element in a category
- r, s, t: element of a distinguished subfamily (generating, Garside, ...)
- i, j, k : integer variable (indices of sequences)
- ℓ, m, n, p, q: integer parameter (fixed, for instance, length of a sequence)
- $\mathsf{a}, \mathsf{b}, \mathsf{c}$: constant element of a category or a monoid for concrete examples (a special notation to distinguish from variables)
- u, v, w : path (or word)
- α, β, γ : binary address (in terms and binary trees)
- ϕ, ψ, π : functor

Contents

Part A

General theory

Chapter I
Some examples

Our aim in this text is to unify and better understand a number of a priori unrelated examples and situations occurring in various frameworks. Before developing a general approach, we describe here some basic examples, in particular the seminal example of braids as analyzed by F.A. Garside in the 1960's, but also several other ones. These examples will be often referred to in the sequel. Our description is informal: most proofs are skipped, the emphasis being put on the features that will be developed in the subsequent chapters.

The chapter is organized as follows. We begin in Section 1 with some examples that can be called classical, in that they all have been considered previously and are the initial examples from which earlier unifications grew. In Section 2, we present the now more or less standard notions of a Garside monoid and a Garside group, as they emerged around the end of the 1990's. Finally, in Section 3, we mention other examples which do not enter the framework of Garside monoids and nevertheless enjoy similar properties: as can be expected, such examples will be typical candidates for the extended approach that will be developed subsequently.

1 Classical examples

This section contains a quick introduction to three of the most well known examples of Garside structures, namely the standard (Artin) braid monoids, the dual (Birman–Ko–Lee) braid monoids, and, to start with an even more basic structure, the free abelian monoids.

1.1 Free abelian monoids

Our first example, free abelian monoids, is particularly simple, but it is fundamental as it serves as a paradigm for the sequel: our goal will be to obtain for more complicated structures counterparts of a number of results that are trivial in the case of free abelian monoids.

REFERENCE STRUCTURE #1 **(free abelian group and monoid)**.

• *For $n \geqslant 1$, an element g of $\mathbb{Z}^n$ is viewed as a map from $\{1, ..., n\}$ to $\mathbb{Z}$, and its kth entry is denoted by $g(k)$.*

• *For f, g in $\mathbb{Z}^n$, we define fg by $fg(k) = f(k)+g(k)$ for each k (we use a multiplicative notation for coherence with other examples).*

- *Put* $\mathbb{N}^n = \{g \in \mathbb{Z}^n \mid \forall k\, (g(k) \geqslant 0)\}$.
- *For* $i \leqslant n$, *define* $\mathtt{a}_i$ *in* $\mathbb{Z}^n$ *by* $\mathtt{a}_i(k) = 1$ *for* $k = i$, *and* 0 *otherwise.*
- *Define* Δ_n *in* $\mathbb{Z}^n$ *by* $\Delta_n(k) = 1$ *for every* k.
- *Put* $S_n = \{s \in \mathbb{Z}^n \mid \forall k\, (s(k) \in \{0, 1\})\}$.

Then $\mathbb{Z}^n$ *equipped with the above-defined product is a free abelian group admitting* $\{\mathtt{a}_1, ..., \mathtt{a}_n\}$ *as a basis, that is, it is an abelian group and, if* G *is any abelian group and* f *is any map of* $\{\mathtt{a}_1, ..., \mathtt{a}_n\}$ *to* G, *there exists a (unique) way of extending* f *to a homomorphism of* $\mathbb{Z}^n$ *to* G*—so, in particular, every abelian group generated by at most* n *elements is a quotient of* $\mathbb{Z}^n$. *The group* $\mathbb{Z}^n$ *is generated by* $\mathtt{a}_1, ..., \mathtt{a}_n$ *and presented by the relations* $\mathtt{a}_i\mathtt{a}_j = \mathtt{a}_j\mathtt{a}_i$ *for all* i, j. *As for* $\mathbb{N}^n$, *it is the submonoid of the group* $\mathbb{Z}^n$ *generated by* $\{\mathtt{a}_1, ..., \mathtt{a}_n\}$, *and it is a free abelian monoid based on* $\{\mathtt{a}_1, ..., \mathtt{a}_n\}$.

Now, here is an (easy) result about the group $\mathbb{Z}^n$. For f, g in $\mathbb{N}^n$, say that $f \leqslant g$ is true if $f(i) \leqslant g(i)$ holds for every i in $\{1, ..., n\}$. Note that S_n is $\{s \in \mathbb{N}^n \mid s \leqslant \Delta_n\}$. Then we have

Proposition 1.1 (normal decomposition). *Every element of* $\mathbb{Z}^n$ *admits a unique decomposition of the form* $\Delta_n^d s_1 \cdots s_p$ *with* d *in* $\mathbb{Z}$ *and* $s_1, ..., s_p$ *in* S_n *satisfying* $s_1 \neq \Delta_n$, $s_p \neq 1$, *and, for every* $i < p$,

$$\forall g \in \mathbb{N}^n \setminus \{1\}\ (\, g \leqslant s_{i+1} \Rightarrow s_i g \nleqslant \Delta_n \,). \tag{1.2}$$

Condition (1.2) is a maximality statement: it says that, in the sequence $(s_1, ..., s_p)$, it is impossible to extract a nontrivial fragment g from an entry s_{i+1} and incorporate it into the previous entry s_i without going beyond Δ_n. So each entry s_i is in a sense maximal with respect to the entry that lies at its right.

A direct proof of Proposition 1.1 is easy. However, we can derive the result from the following order property.

Lemma 1.3. *The relation* $\leqslant$ *of Proposition* 1.1 *is a lattice order on* $\mathbb{N}^n$, *and* S_n *is a finite sublattice with* 2^n *elements.*

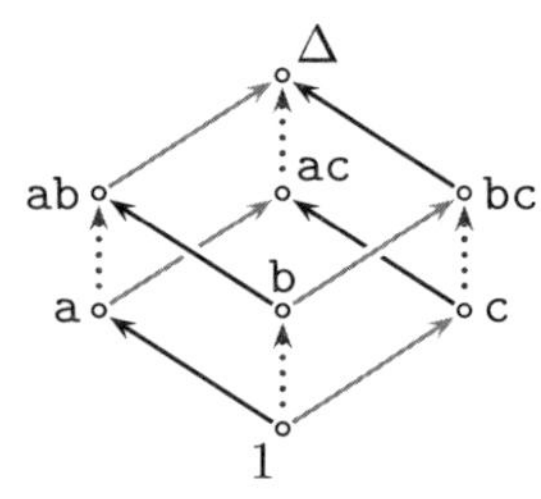

We recall that a *lattice order* is a partial order in which every pair of elements admits a greatest lower bound and a lowest upper bound. The diagram on the right displays the lattice made of the eight elements that are smaller than Δ_3 in the monoid $\mathbb{N}^3$, here a cube (a, b, c stand for $\mathtt{a}_1, \mathtt{a}_2, \mathtt{a}_3$).

Once Lemma 1.3 is available, Proposition 1.1 easily follows: indeed, starting from g, if g is not 1, there exists a maximal element s_1 of S_n that satisfies $s_1 \leqslant g$. The latter relation implies the existence of g' satisfying $g = s_1 g'$. If g' is not 1, we iterate the process with g', finding s_2 and g'', etc. The sequence $(s_1, s_2, ...)$ so obtained then satisfies the maximality condition of (1.2).

Example 1.4. Assume $n = 3$, and write a, b, c for $\mathtt{a}_1, \mathtt{a}_2, \mathtt{a}_3$. With this notation Δ_3 (hereafter written Δ) is abc, and S_3 has eight elements, namely 1, a, b, c, ab, bc, ca, and Δ. The Hasse diagram of the poset $(S_3, \leqslant)$ is the cube displayed on the right of Lemma 1.3.

Let $g = \mathtt{a}^3\mathtt{b}\mathtt{c}^2$, that is, $g = (3, 1, 2)$. The maximal element of S_3 that lies below g is Δ, and we have $g = \Delta \cdot \mathtt{a}^2\mathtt{c}$. The maximal element of S_3 that lies below $\mathtt{a}^2\mathtt{c}$ is $\mathtt{ac}$, and we have $\mathtt{a}^2\mathtt{c} = \mathtt{ac} \cdot \mathtt{a}$. The latter element lies below Δ. So the decomposition of g provided by Proposition 1.1 is $\Delta \cdot \mathtt{ac} \cdot \mathtt{a}$, see Figure 1.

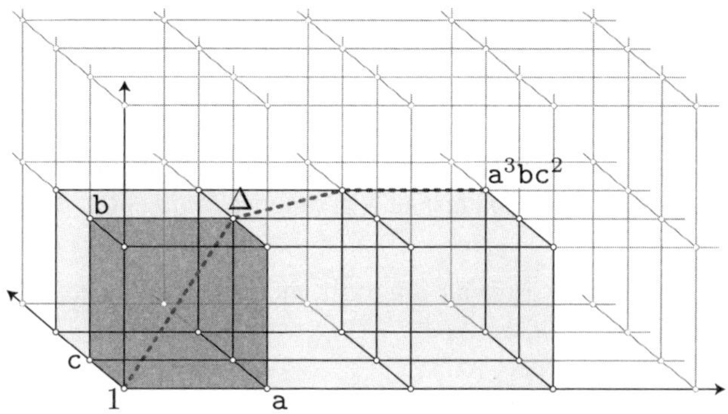

Figure 1. The free abelian monoid $\mathbb{N}^3$; the cube $\{g \in \mathbb{N}^3 \mid g \leqslant \Delta\}$ is in dark grey; among the many possible ways to go from 1 to $\mathtt{a}^3\mathtt{b}\mathtt{c}^2$, the distinguished decomposition of Proposition 1.1 consists in choosing at each step the largest possible element below Δ that lies below the considered element, that is, to remain in the light grey domain.

1.2 Braid groups and monoids

Our second reference structure involves braids. As already mentioned, this fundamental example is the basis we shall build on.

REFERENCE STRUCTURE #2 **(braids)**.

• *For $n \geqslant 1$, denote by B_n the group of all n-strand braids.*

• *For $1 \leqslant i \leqslant n-1$, denote by σ_i the braid that corresponds to a crossing of the $(i+1)$st strand over the ith strand.*

• *Denote by B_n^+ the submonoid of B_n generated by $\sigma_1, ..., \sigma_{n-1}$; its elements are called* positive *braids.*

• *Denote by Δ_n the isotopy class of the geometric braid corresponding to a positive half-turn of the n strands.*

Here are some explanations. We refer to standard textbooks about braids [17, 100, 157] *for details and proofs. See also the Notes at the end of this chapter and Section* XIV.4 *for alternative approaches that are not needed now.*

Let D_n denote the disk of $\mathbb{R}^2$ with diameter $(0,0)$, $(n+1,0)$. An n-strand geometric braid *is a collection of n disjoint curves living in the cylinder $D_n \times [0,1]$, connecting* *the n points $(i, 0, 0)$, $1 \leqslant i \leqslant n$, to the n points $(i, 0, 1)$, $1 \leqslant i \leqslant n$, and such that the intersection with each plane $z = z_0$, $0 \leqslant z_0 \leqslant 1$, consists of n points exactly.*

Two n-strand geometric braids β, β' are said to be isotopic *if there exists a continuous family of geometric braids $(\beta_t)_{t \in [0,1]}$ with $\beta_0 = \beta$ and $\beta_1 = \beta'$ or, equivalently, if there exists a homeomorphism of the cylinder $D_n \times [0, 1]$ into itself that is the identity on the boundary and that deforms β to β'.*

There exists a natural product on geometric braids: if β_1, β_2 are n-strand geometric braids, squeezing the image of β_1 into the cylinder $D_n \times [0, 1/2]$ and the image of β_2 into $D_n \times [1/2, 1]$ yields a new, well-defined geometric braid. This product induces a well defined product on isotopy classes, and gives to the family of all classes the structure of a group: the class of the geometric braid consisting on n straight line segments is the neutral element, and the inverse of the class of a geometric braid is the class of its reflection in the plane $z = 1/2$.

An n-strand braid diagram *is a planar diagram obtained by concatenating one over the other finitely many diagrams of the type*

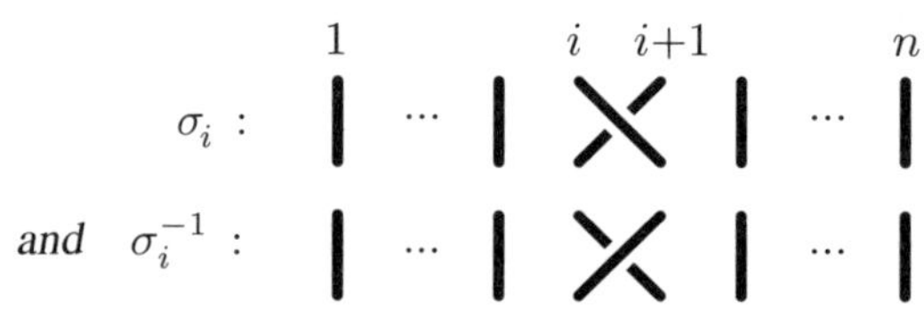

with $1 \leqslant i \leqslant n-1$. An n-strand braid diagram is encoded in a word in the alphabet $\{\sigma_1, \sigma_1^{-1}, \ldots, \sigma_{n-1}, \sigma_{n-1}^{-1}\}$. Provided the 'front–back' information is translated into the segment breaks of the σ_i-diagrams, projection on the plane $y = 0$ induces a surjective map of the isotopy classes of the family of all n-strand geometric braids onto the family of all n-strand braid diagrams: every n-strand geometric braid is isotopic to one whose projection is an n-strand braid diagram. The situation is summarized in the picture below, in which we see a general 4-strand geometric braid, an isotopic normalized version, the corresponding 4-strand braid diagram, and its decomposition in terms of the elementary diagrams σ_i and σ_i^{-1}, here encoded in the braid word $\sigma_1\sigma_3\sigma_2^{-1}\sigma_1^{-1}\sigma_2$:

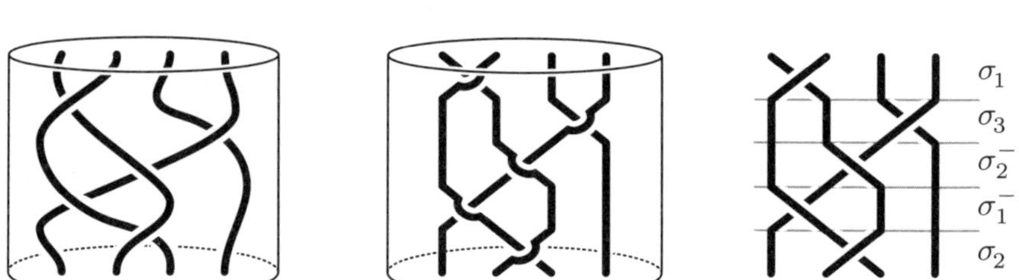

E. Artin proved in [4] that, conversely, two n-strand braid diagrams are the projection of isotopic geometric braids if and only if they are connected by a finite sequence of transformations of the following types:

- *replacing $\sigma_i\sigma_j$ with $\sigma_j\sigma_i$ for some i, j satisfying $|i - j| \geqslant 2$,*
- *replacing $\sigma_i\sigma_j\sigma_i$ with $\sigma_j\sigma_i\sigma_j$ for some i, j satisfying $|i - j| = 1$,*

- *deleting or inserting some factor* $\sigma_i\sigma_i^{-1}$ or $\sigma_i^{-1}\sigma_i$,

in which case the braid diagrams—and the braid words that encode them—are called equivalent. *Geometrically, these relations correspond to the following three types of braid diagrams equivalence:*

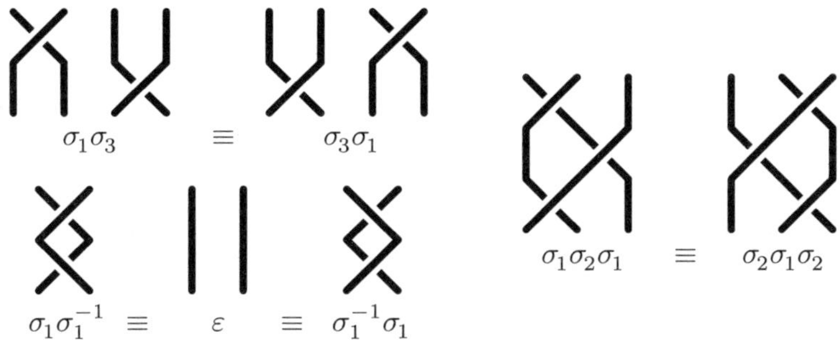

namely moving remote crossings (here $i = 1, j = 3$*), moving adjacent crossings (here* $i = 1, j = 2$*), and removing/adding trivial factors (*ε *stands for the empty word). Clearly the diagrams are projections of isotopic braids; Artin's result says that, conversely, the projection of every isotopy can be decomposed into a finite sequence of transformations of the above types.*

So the group B_n *is both the group of isotopy classes of* n*-strand geometric braids and the group of equivalence classes of* n*-strand braid diagrams. Its elements are called* n-strand braids. *By the above description,* B_n *admits the presentation*

(1.5)
$$\left\langle \sigma_1, \dots, \sigma_{n-1} \;\middle|\; \begin{array}{ll} \sigma_i\sigma_j = \sigma_j\sigma_i & \text{for } |i-j| \geqslant 2 \\ \sigma_i\sigma_j\sigma_i = \sigma_j\sigma_i\sigma_j & \text{for } |i-j| = 1 \end{array} \right\rangle.$$

It follows from the above description that the braids Δ_n *are defined inductively by the relations*

(1.6)
$$\Delta_1 = 1, \quad \Delta_n = \Delta_{n-1}\,\sigma_{n-1}\cdots\sigma_2\sigma_1 \quad \text{for } n \geqslant 2,$$

on the shape of

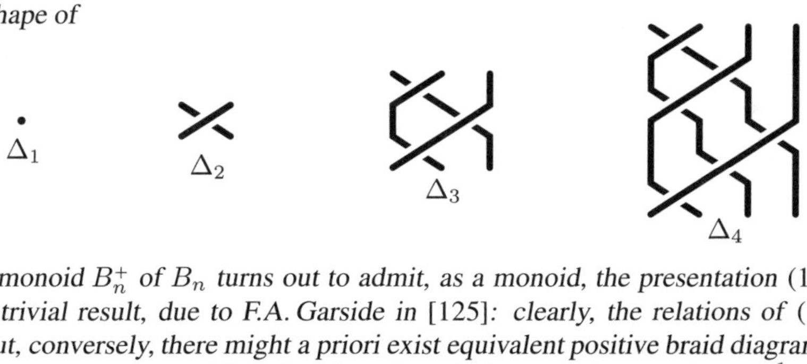

The submonoid B_n^+ *of* B_n *turns out to admit, as a monoid, the presentation* (1.5)*. This is a nontrivial result, due to F.A. Garside in* [125]*: clearly, the relations of* (1.5) *hold in* B_n^+ *but, conversely, there might a priori exist equivalent positive braid diagrams whose equivalence cannot be established without introducing negative crossings* σ_i^{-1}*. What can be proved—see Chapter* IX *below—is that* B_n^+ *and the monoid* $\underline{B}_n^+$ *that admits* (1.5) *as a presentation are isomorphic.*

Every n-strand braid diagram naturally defines a permutation of $\{1, ..., n\}$, namely the permutation that specifies from which position the strand that finishes at position i starts. This permutation is invariant under isotopy, and induces a (non-injective) surjective homomorphism from B_n onto the symmetric group $\mathfrak{S}_n$. The braids whose permutation is the identity are called pure, *and the group of all pure n strand braids is denoted by PB_n. By definition, we then have an exact sequence*

$$1 \longrightarrow PB_n \longrightarrow B_n \longrightarrow \mathfrak{S}_n \longrightarrow 1. \tag{1.7}$$

Here is now a (significant) result about the braid group B_n. For f, g in B_n^+, say that f *left-divides* g, written $f \preccurlyeq g$, if $fg' = g$ holds for some g' lying in B_n^+, and write $\mathrm{Div}(g)$ for the family of all left-divisors of g. Then we have

Proposition 1.8 (normal decomposition). *Every n-strand braid admits a unique decomposition of the form $\Delta_n^d\, s_1 \cdots s_p$ with d in $\mathbb{Z}$ and $s_1, ..., s_p$ in $\mathrm{Div}(\Delta_n)$ satisfying $s_1 \neq \Delta_n$, $s_p \neq 1$, and, for every i,*

$$\forall g \in B_n^+ \setminus \{1\} \; (\, g \preccurlyeq s_{i+1} \;\Rightarrow\; s_i g \not\preccurlyeq \Delta_n \,). \tag{1.9}$$

Note the formal similarity between Proposition 1.1 and Proposition 1.8. In particular, the coordinatewise order $\leqslant$ of Proposition 1.1 is the left-divisibility relation of the monoid $\mathbb{N}^n$, and condition (1.9) is a copy of the maximality condition of (1.2): as the latter, it means that, at each step, one cannot extract a fragment of an entry and incorporate it in the previous entry without going out from the family of left-divisors of Δ_n.

Contrary to Proposition 1.1 about free abelian groups, Proposition 1.8 about braid groups is not trivial, and it is even the key result in the algebraic investigation of braid groups. In particular, it is directly connected with the decidability of the Word and Conjugacy Problems for B_n, as will be seen in Chapters III and VIII.

However, although quite different both in difficulty and interest, Proposition 1.8 can be proved using an argument that is similar to that explained above for Proposition 1.1, namely investigating the submonoid B_n^+ generated by $\sigma_1, ..., \sigma_{n-1}$ and establishing

Lemma 1.10. *The left-divisibility relation on B_n^+ is a lattice order. The family $\mathrm{Div}(\Delta_n)$ of the left-divisors of Δ_n is a finite sublattice with $n!$ elements.*

In the current case, the finite lattice $(\mathrm{Div}(\Delta_n), \preccurlyeq)$ turns out to be isomorphic to what is called the weak order on the symmetric group $\mathfrak{S}_n$, see Figure 2. Geometrically, the elements of $\mathrm{Div}(\Delta_n)$ correspond to the braid diagrams in which all crossings are positive and any two strands cross at most once ('simple braids'). It turns out that the restriction of the surjective map from B_n to $\mathfrak{S}_n$ to the left-divisors of Δ_n in B_n^+ is a bijection. Under this bijection, the braid 1 corresponds to the identity mapping, the braid σ_i corresponds to the transposition $(i, i+1)$, and the braid Δ_n corresponds to the symmetry $i \mapsto n+1-i$.

Once Lemma 1.10 is available, the distinguished decomposition of Proposition 1.8 can be obtained for an element g of B_n^+ by defining s_1 to be the greatest common left-divisor of g and Δ_n, that is, the greatest lower bound of g and Δ_n, and then using induction.

As the left-divisors of Δ_n in B_n^+ correspond to the braid diagrams in which any two strands cross at most once, extracting the maximal left-divisor of a braid g that lies

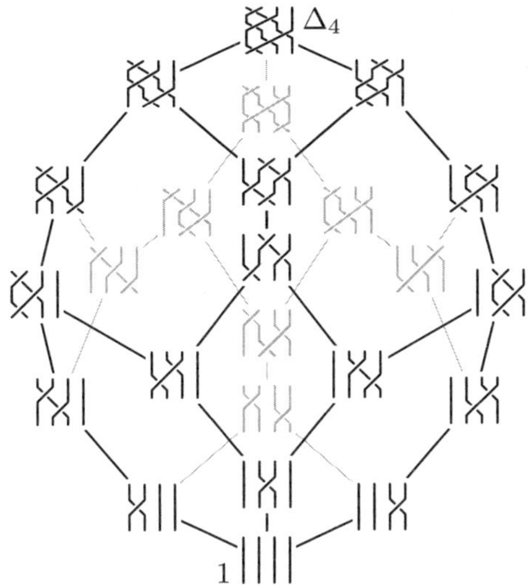

Figure 2. The lattice $(\mathrm{Div}(\Delta_4), \preccurlyeq)$ made by the 24 left-divisors of the braid Δ_4 in the monoid B_4^+: a copy of the permutahedron associated with the symmetric group $\mathfrak{S}_4$ equipped with the weak order; topologically, this is a 2-sphere tessellated by hexagons and squares which correspond to the relations of (1.5).

in $\mathrm{Div}(\Delta_n)$ amounts to pushing the crossings upwards as much as possible until every pair of adjacent strands that later cross have already crossed. The geometric meaning of Proposition 1.8 in B_n^+ is that this strategy leads to a unique well defined result.

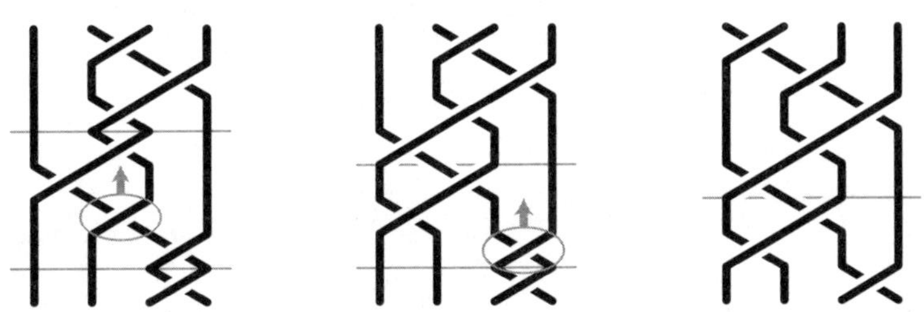

Figure 3. Normalization of the braid $\sigma_2\sigma_3\sigma_2^2\sigma_1\sigma_2\sigma_3^2$ by pushing the crossings upwards in order to maximize the first simple factors and minimize the number of simple factors: here we start with three simple factors, and finish with two only.

Example 1.11. (See Figure 3) Consider $g = \sigma_2\sigma_3\sigma_2\sigma_2\sigma_1\sigma_2\sigma_3\sigma_3$ in B_4^+. Looking to the strands with repeated crossings leads to the decomposition $g = \sigma_2\sigma_3\sigma_2 \cdot \sigma_2\sigma_1\sigma_2\sigma_3 \cdot \sigma_3$. However, the fourth σ_2 crossing can be pushed upwards, becoming σ_1 and leading to the new decomposition $g = \sigma_2\sigma_3\sigma_2\sigma_1 \cdot \sigma_2\sigma_1\sigma_3 \cdot \sigma_3$. But then the second σ_3 crossing can be pushed in turn, leading to the optimal decomposition $g = \sigma_1\sigma_2\sigma_3\sigma_2\sigma_1 \cdot \sigma_2\sigma_1\sigma_3$.

1.3 Dual braid monoids

Here is a third example, whose specificity is to involve the same group as Reference Structure 2, but with a different generating family.

REFERENCE STRUCTURE #3 **(dual braids)**.

- *For $n \geqslant 1$, denote by B_n the group of n-strand braids.*
- *For $1 \leqslant i < j \leqslant n$, denote by $a_{i,j}$ the braid that corresponds to a crossing of the jst strand over the ith strand, and under all possible intermediate strands.*
- *Denote by B_n^{+*} the submonoid of B_n generated by all $a_{i,j}$ with $1 \leqslant i < j \leqslant n$.*
- *Put $\Delta_n^* = a_{1,2} \cdots a_{n-1,n}$.*

A precise definition is $a_{i,j} = \sigma_i \cdots \sigma_{j-2}\sigma_{j-1}\sigma_{j-2}^{-1} \cdots \sigma_i^{-1}$, corresponding to the diagram

Note that $a_{i,i+1}$ coincides with σ_i: the braids $a_{i,j}$ form a redundant family of generators for the group B_n. For $n \geqslant 3$, the braid $a_{1,3}$ does not belong to the monoid B_n^+; therefore, the monoid B_n^{+*}, which is called the n-strand dual braid monoid *properly includes the braid monoid B_n^+ (Reference Structure 2, page 5).*

The family of all braids $a_{i,j}$ enjoys nice invariance properties with respect to cyclic permutations of the indices, which are better visualized when braid diagrams are represented on a cylinder and $a_{i,j}$ is associated with the chord that connects the vertices i and j in a circle with n marked vertices, as suggested below:

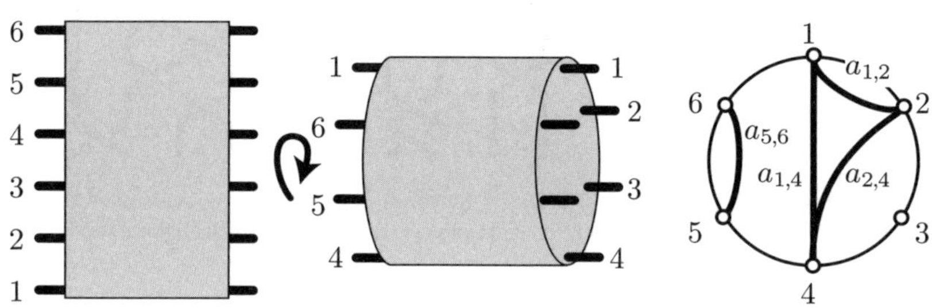

In terms of the $a_{i,j}$, the braids group B_n is presented by the relations

(1.12) $\qquad a_{i,j}a_{i',j'} = a_{i',j'}a_{i,j}$ for $[i, j]$ and $[i', j']$ disjoint or nested,

(1.13) $\qquad a_{i,j}a_{j,k} = a_{j,k}a_{i,k} = a_{i,k}a_{i,j}$ for $1 \leqslant i < j < k \leqslant n$.

In the chord representation, relations of type (1.12) say that the generators associated with non-intersecting chords commute: for instance, in the above picture, we see that $a_{1,4}$

and $a_{5,6}$ commute. On the other hand, relations of type (1.13) say that, for each chord triangle, the product of two adjacent edges taken clockwise does not depend on the edges: in the above picture, the triangle $(1,2,4)$ gives $a_{1,2}a_{2,4} = a_{2,4}a_{1,4} = a_{1,4}a_{1,2}$.

The following result should not be a surprise. Here again, we use $\preccurlyeq$ for the left-divisibility relation of the considered monoid, here B_n^{+*}: for f, g in B_n^{+*}, we say that $f \preccurlyeq g$ holds if $fg' = g$ holds for some g' belonging to B_n^{+*}. Note that the restriction of this relation to B_n^+ is not the left-divisibility relation of B_n^+ as, here, the quotient element is only required to belong to B_n^{+*}.

Proposition 1.14 (normal decomposition). *Every element of B_n admits a unique decomposition of the form $\Delta_n^{*d} s_1 \cdots s_p$ with d in $\mathbb{Z}$ and $s_1, \ldots, s_p$ in $\mathrm{Div}(\Delta_n^*)$ satisfying $s_1 \neq \Delta_n^*$, $s_p \neq 1$, and, for every i,*

$$\forall g \in B_n^{+*} \setminus \{1\} \; (\, g \preccurlyeq s_{i+1} \;\Rightarrow\; s_i g \not\preccurlyeq \Delta_n^* \,). \tag{1.15}$$

Proposition 1.14 is entirely similar to Propositions 1.1 and 1.8, and its proof is directly connected with the following counterpart of Lemma 1.10.

Lemma 1.16. *The left-divisibility relation on B_n^{+*} is a lattice order. The family $\mathrm{Div}(\Delta_n^*)$ of the left-divisors of Δ_n^* is a finite sublattice with $\frac{1}{n+1}\binom{2n}{n}$ elements.*

In the current case, the finite lattice $(\mathrm{Div}(\Delta_n^*), \preccurlyeq)$, whose size is the nth Catalan number, turns out to be isomorphic to lattice of so-called non-crossing partitions of $\{1, \ldots, n\}$ ordered by inclusion, see Figure 4.

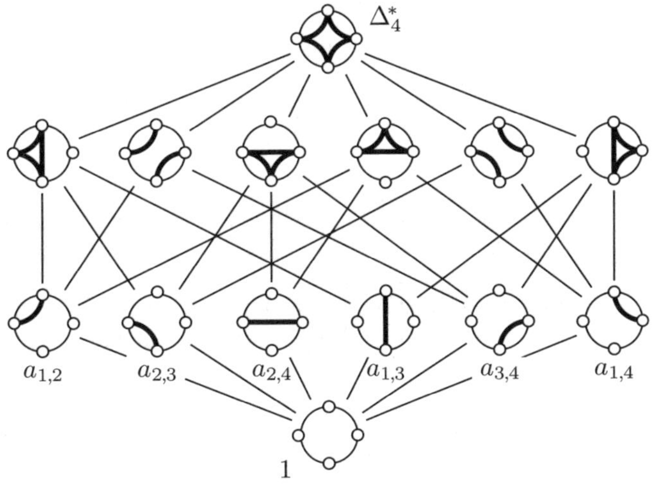

Figure 4. The lattice $(\mathrm{Div}(\Delta_4^*), \preccurlyeq)$ made by the left-divisors of the braid Δ_4^* in the monoid B_4^{+*}: a copy of the inclusion lattice of the 14 non-crossing partitions of $\{1, \ldots, 4\}$.

As can be expected, going from Lemma 1.16 to Proposition 1.14 is similar to going from Lemma 1.10 to Proposition 1.8.

2 Garside monoids and groups

The previous examples share many properties, and it is natural to extract the common basic features that enable one to unify the developments. The notions of a *Garside group* and a *Garside monoid* emerged in this way.

2.1 The notion of a Garside monoid

We introduced above the notion of a left-divisor in a monoid; right-divisors are defined symmetrically: an element f is said to be a *right-divisor* of g in the monoid M if there exists g' in M satisfying $g = g'f$. Under mild assumptions, the left-divisibility relation is a partial ordering, and, when is exists, a corresponding least common upper bound (*resp.* greatest common lower bound) is usually called a least common right-multiple, or right-lcm (*resp.* a greatest common left-divisor, or left-gcd). Left-lcms and right-gcds are defined similarly from the right-divisibility relation.

Definition 2.1 (Garside monoid). A *Garside monoid* is a pair (M, Δ), where M is a monoid and

• M is left- and right-cancellative,

• there exists $\lambda : M \to \mathbb{N}$ satisfying $\lambda(fg) \geqslant \lambda(f) + \lambda(g)$ and $g \neq 1 \Rightarrow \lambda(g) \neq 0$,

• any two elements of M have a left- and a right-lcm and a left- and a right-gcd,

• Δ is a *Garside element* of M, this meaning that the left- and right-divisors of Δ coincide and generate M,

• the family of all divisors of Δ in M is finite.

We also fix a terminology for the following variant sometimes used:

Definition 2.2 (quasi-Garside monoid). A *quasi-Garside monoid* is a pair (M, Δ) that satisfies the conditions of Definition 2.1, except possibly the last one (finiteness of the number of divisors of Δ).

We recall that a group G is said to be a *group of left-fractions* for a monoid M if M is included in G and every element of G can be expressed as $f^{-1}g$ with f, g in M.

Definition 2.3 (Garside and quasi-Garside group). A group G is said to be a *Garside group* (*resp. quasi-Garside group*) if there exists a Garside (*resp.* quasi-Garside) monoid (M, Δ) such that G is a group of left-fractions for M.

In the above context, the terminology "(quasi)-Garside monoid" is frequently used for the monoid alone. Then we mean that M is a (quasi)-Garside monoid if there exists an element Δ of M such that (M, Δ) is a (quasi)-Garside monoid. In this context, it is also common to say that (M, Δ) provides a *Garside structure* for the considered group.

Always using $\preccurlyeq$ for the left-divisibility relation, one can show

Proposition 2.4 (normal decomposition). *Assume that G is a Garside group, group of fractions for the Garside monoid (M, Δ). Then every element of G admits a unique*

decomposition of the form $\Delta^d s_1 \cdots s_p$ *with* d *in* $\mathbb{Z}$ *and* $s_1, \ldots, s_p$ *in* $\mathrm{Div}(\Delta)$ *satisfying* $s_1 \neq \Delta$, $s_p \neq 1$, *and, for every* i,

$$\forall g \in M \backslash \{1\} \; (\, g \preccurlyeq s_{i+1} \; \Rightarrow \; s_i g \not\preccurlyeq \Delta \,). \tag{2.5}$$

Proof (sketch). The first, easy step is to go from elements of the group G to elements of the monoid M. To this aim, one observes that M is invariant under conjugation by Δ as, for every g in M, there exists g' in M satisfying $\Delta g = g'\Delta$. From there, one deduces that every element of the monoid M is a right (and left) divisor of the element Δ^d for d large enough. This implies that every element g of G can be expressed as $\Delta^d g'$ with d in $\mathbb{Z}$ and g' in M. Taking d maximal provides the expected uniqueness result.

From that point on, the problem is to show the existence and uniqueness of a convenient decomposition for the elements of M. So let g belong to M. If g is 1, the empty sequence provides the expected decomposition. Otherwise, as, by assumption, $(\mathrm{Div}(\Delta), \preccurlyeq)$ is a lattice, the elements g and Δ admit a greatest lower bound, that is, a greatest common left-divisor, say s_1. We write $g = s_1 g'$, and, if g' is not 1, we iterate the process and write g' as $s_2 g''$ with s_2 the greatest common left-divisor of g' and Δ, *etc.* As the left-divisors of Δ generate M, the element s_1 cannot be 1, which implies $\lambda(g') < \lambda(g)$, where λ is the function from M to $\mathbb{N}$ whose existence is assumed in the definition of a (quasi)-Garside monoid. This guarantees the termination of the induction after finitely many steps. The maximality of the choice of $s_1, s_2, \ldots$ implies (2.5), and the existence of the expected decomposition follows.

As for uniqueness, there is a subtle point, which is crucial for applications, in particular in terms of the existence of an automatic structure. Indeed, condition (2.5) is a local condition in that it only involves two adjacent entries s_i, s_{i+1} at a time, whereas the greatest common divisor relation used in the construction is a global condition which involves $g, g', \ldots$, whence all of the entries s_i at a time. It is easy to show that, if s_1, s_2 satisfy (2.5), then s_1 must be the greatest common divisor of $s_1 s_2$ and Δ. However, it is *not* obvious that, if (s_1, s_2) and (s_2, s_3) satisfy (2.5), then s_1 must be the greatest common divisor of $s_1 s_2 s_3$ and Δ. Proving this requires to use most of the properties of the Garside element Δ, in particular the assumption that the left- and right-divisors of Δ coincide and the assumption that any two left-divisors of Δ admit a least upper bound with respect to left-divisibility. Typically, the following grouping property

(2.6) If s_1, s_2, s_3 lies in $\mathrm{Div}(\Delta)$ and (s_1, s_2) and (s_2, s_3) satisfy (2.5), then $(s_1, s_2 s_3)$ satisfies (2.5) as well.

holds whenever Δ is a Garside element in a Garside monoid, and it is crucial. □

The framework of Garside groups successfully unifies the examples we have considered so far. Indeed, it is easy to check that, for every n, the free abelian group $\mathbb{Z}^n$ has the structure of a Garside group with respect to the Garside monoid $(\mathbb{N}^n, \Delta_n)$. Similarly, one can check—and this is essentially what F.A. Garside did in [124, 125]—that the braid group B_n has the structure of a Garside group with the respect to the Garside monoid (B_n^+, Δ_n). Finally, one can check—and this is essentially what J. Birman, K.Y. Ko, and S.J. Lee did in [21]—that the braid group B_n has the structure of a Garside group with the respect to the Garside monoid (B_n^{+*}, Δ_n^*). For $n \geqslant 3$, the monoids B_n^+

and B_n^{+*} are not isomorphic and, therefore, the braid group B_n is an example of a group that admits two genuinely different Garside structures, in the sense that they involve different monoids—we shall see that, if (M, Δ) is a Garside monoid, then so is (M, Δ^m) for every positive m, so, for a given group, a Garside structure is (almost) never unique when it exists.

2.2 More examples

A number of Garside monoids have been described in literature, and a number of them will be mentioned in the sequel of this text. For further reference, we introduce here two more examples. The first one is a Garside group.

Example 2.7 (torus-type groups). For $n \geqslant 1$ and $p_1, \dots, p_n \geqslant 2$, denote by $T_{p_1,\dots,p_n}$ (or simply T) the group generated by n elements $\mathtt{a}_1, \dots, \mathtt{a}_n$ with the defining relations

$$\mathtt{a}_1^{p_1} = \mathtt{a}_2^{p_2} = \cdots = \mathtt{a}_n^{p_n}.$$

Denote by $T^+_{p_1,\dots,p_n}$ the submonoid of T generated by $\mathtt{a}_1$, ..., $\mathtt{a}_n$, and define Δ to be the element $\mathtt{a}_1^{p_1}$ of T. Then $(T^+_{p_1,\dots,p_n}, \Delta)$ is a Garside monoid. The lattice of divisors of Δ is especially simple: it consists of n chains of respective lengths $p_1, \dots, p_n$ connecting the minimal element 1 to the maximal element Δ. Calling such monoids and groups "of torus type" is natural owing to the fact that the group T_{p_1,p_2} is the fundamental group of the complement of a (p_1, p_2)-torus knot.

The second one is a quasi-Garside group in which the Garside element has infinitely many divisors.

Example 2.8 (Hurwitz action on a free group). For $n \geqslant 1$, let F_n the free group based on n generators $\mathtt{a}_1, \dots, \mathtt{a}_n$. We define an action of the braid group B_n on F_n^n by

$$(s_1, \dots, s_n) \bullet \sigma_i = (s_1, \dots, s_{i-1}, s_i s_{i+1} s_i^{-1}, s_i, s_{i+2}, \dots, s_n). \tag{2.9}$$

Then denote by O_n the orbit of $(\mathtt{a}_1, \dots, \mathtt{a}_n)$ under the action of B_n, and put

$$S_n = \{g \in F_n \mid \exists (s_1, \dots, s_n) \in O_n \; \exists i \; \exists j \; (i \leqslant j \text{ and } g = s_i \cdots s_j)\}.$$

Now let F_n^+ be the submonoid of F_n generated by S_n, and let $\Delta_n = \mathtt{a}_1 \cdots \mathtt{a}_n$. Then (F_n^+, Δ_n) is a quasi-Garside monoid and F_n is the associated group. In the case $n = 2$, the orbit O_2 looks like

$$\dots \xmapsto{\sigma_1} (\mathtt{Bab}, \mathtt{BAbab}) \xmapsto{\sigma_1} (\mathtt{b}, \mathtt{Bab}) \xmapsto{\sigma_1} (\mathtt{a}, \mathtt{b}) \xmapsto{\sigma_1} (\mathtt{abA}, \mathtt{a}) \xmapsto{\sigma_1} (\mathtt{abaBA}, \mathtt{abA}) \xmapsto{\sigma_1} \dots$$

where we use $\mathtt{a}$ and $\mathtt{b}$ for the generators of F_2 and $\mathtt{A}, \mathtt{B}$ for $\mathtt{a}^{-1}$ and $\mathtt{b}^{-1}$. The family S_2 is a double infinite series

$$\dots, \mathtt{ababABA}, \mathtt{abaBA}, \mathtt{abA}, \mathtt{a}, \mathtt{b}, \mathtt{Bab}, \mathtt{BAbab}, \mathtt{BABabab}, \dots$$

in which the product of any two adjacent entries is ab. For general n, one can interpret F_n as the fundamental group of an n-punctured disk, with $\mathtt{a}_i$ corresponding to a simple loop around the ith puncture. Fix a loop γ with no self-intersection representing Δ_n. Then the elements of S_n are the isotopy classes that contain at least one loop with no self-intersection and whose interior is included in the interior of γ, see Figure 5. The action of n-strand braids on the nth power of F_n is known as the Hurwitz action on F_n.

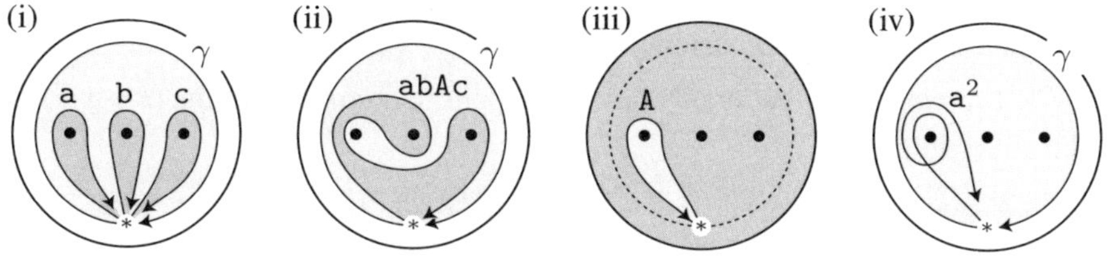

Figure 5. Geometric interpretation of the quasi-Garside structure on a free group F_3 via the action on a 3-punctured disk (see the final Notes of this chapter and Subsection XIV.4.1): (i) three loops realizing the base generators, here denoted by a, b, c, and a loop γ realizing the element Δ_3; (ii) a loop realizing the element abAc: it has no self-intersection and its interior is included in the interior of γ, hence abAc belongs to S_3; (iii) no loop realizing A, that is, $\mathtt{a}^{-1}$, has its interior included in the interior of γ, so A does not belong to S_3; (iv) every loop realizing $\mathtt{a}^2$ has a self-intersection, so $\mathtt{a}^2$ does not belong to S_3.

Remark 2.10. Let us note here for further reference that the Hurwitz action of the n-strand braid group B_n on the nth power of a free group as introduced in (2.9) can be defined by the same formula for the nth power of *any* group G.

We refer to [99] for more examples.

3 Why a further extension?

We now describe four more examples showing that distinguished decompositions similar to those of Proposition 2.4 may exist in cases that do not enter the framework of Garside groups and monoids.

3.1 Infinite braids

First, easy examples arise when we consider certain monoids of infinite type, that is, which do not admit a finite generating family, but are close to, typically direct limits of finite type monoids.

For instance, let I be an arbitrary set and let $\mathbb{Z}^{(I)}$ denote the free abelian group consisting of all I-indexed sequences of integers with finite support, that is, all sequences g such that $g(k) \neq 0$ may hold for finitely many indices k only. In this case, the definition of the Garside element Δ_n makes no longer sense. However, put

$$S_I = \{g \in \mathbb{Z}^{(I)} \mid \forall k \in I\ (g(k) \in \{0,1\})\}.$$

For $I = \{1, ..., n\}$, the set S_I coincides with the divisors of Δ_n. Now, in all cases, the decomposition result of Proposition 1.1 remains true with S_I replacing $\mathrm{Div}(\Delta_n)$.

Here is a similar example involving braids.

REFERENCE STRUCTURE #4 **(infinite braids)**.

• *Denote by B_∞ the group of all braids on infinitely many strands indexed by positive integers (see precise definition below).*

• *For $i \geqslant 1$, denote by σ_i the braid that corresponds to a crossing of the $(i+1)$st strand over the ith strand.*

• *Denote by B_∞^+ the submonoid of B_∞ generated by all σ_i's with $i \geqslant 1$.*

• *Put $S_\infty = \bigcup_{n \geqslant 1} \mathrm{Div}(\Delta_n)$, where Δ_n is the n-strand braid introduced in* Reference Structure 2.

To make the above definitions precise, we observe that, for every n, the identity map of $\{\sigma_1, ..., \sigma_{n-1}\}$ extends to a homomorphism $\iota_{n,n+1}$ of B_n to B_{n+1}: geometrically, this amounts to adding one unbraided strand at the right of an n-strand braid diagram. Then B_∞ is the direct limit of the system so obtained. As $\iota_{n,n+1}$ is easily seen to be injective, we can identify B_n with its image in B_{n+1}. Up to this identification, B_∞ is simply the increasing union of all braid groups B_n and, similarly, B_∞^+ is the union of all braid monoids B_n^+.

In this context, the definition of S_∞ is unambiguous as Δ_n always divides Δ_{n+1} and the equality $\mathrm{Div}(\Delta_n) = \mathrm{Div}(\Delta_{n+1}) \cap B_n^+$ holds for every n.

It is easy to see that the monoid B_∞^+ contains no Garside element. Indeed, B_∞^+ admits no finite generating family since the crossings occurring in a finite subfamily of B_∞^+ can only involve finitely many strands, whereas B_∞^+ contains elements representing crossings of strands with an arbitrarily large index. Hence it is impossible that the divisors of an element of B_∞^+ generate the monoid, one of the requirements in the definition of a Garside element. So, we cannot expect a full counterpart to Proposition 1.8. However, we have

Proposition 3.1 (normal decomposition). *Every braid in B_∞^+ admits a unique decomposition of the form $s_1 \cdots s_p$ with $s_1, ..., s_p$ in S_∞ satisfying $s_p \neq 1$, and, for every i,*

(3.2) $$\forall g \in S_\infty \setminus \{1\} \; (g \preccurlyeq s_{i+1} \;\Rightarrow\; s_i g \notin S_\infty).$$

So, here again, the results remain valid provided we replace the family of divisors of the Garside element with a larger family. These easy observations lead us to addressing the question of finding the most general properties a subfamily S of a monoid has to satisfy in order to guarantee the existence and uniqueness of a distinguished decomposition such as the one of Proposition 2.4. Being the family of all divisors of a Garside element is a sufficient condition, but the examples of $\mathbb{Z}^{(I)}$ with infinite I and of B_∞ show this is not a necessary condition.

3.2 The Klein bottle group

Another easy but interesting example is provided by the fundamental group of the Klein bottle, hereafter called the Klein bottle group.

REFERENCE STRUCTURE #5 (**Klein bottle monoid**).

• *Put $K = \mathbb{Z} \rtimes \mathbb{Z}$, where the second copy of $\mathbb{Z}$ acts nontrivially on the first copy, that is, $(0,1)$ acts by mapping $(p,0)$ to $(-p,0)$ for every p.*

• *Put $\mathtt{a} = (0,1)$ and $\mathtt{b} = (1,0)$, and let K^+ be the submonoid of K generated by $\mathtt{a}$ and $\mathtt{b}$.*

• *Put $\Delta = \mathtt{a}^2$, and $S = \{g \in K \mid 1 \preccurlyeq g \preccurlyeq \Delta\}$, where $f \preccurlyeq g$ is declared to hold if there exists g' in K^+ satisfying $fg' = g$.*

• *Call a word of $\{\mathtt{a},\mathtt{b}\}^*$* canonical *if it is of the form $\mathtt{a}^p\mathtt{b}^q$ with $p, q \geqslant 0$* or *$\mathtt{a}^p\mathtt{b}^q\mathtt{a}$ with $p \geqslant 0$ and $q \geqslant 1$.*

• *For w a word in $\{\mathtt{a},\mathtt{b}\}^*$, we denote by $|w|_\mathtt{a}$ (resp. $|w|_\mathtt{b}$) the total number of occurrences of $\mathtt{a}$ (resp. $\mathtt{b}$) in w.*

The group K can be presented by $\langle \mathtt{a},\mathtt{b} \mid \mathtt{ba} = \mathtt{ab}^{-1}\rangle$, or, equivalently, $\langle \mathtt{a},\mathtt{b} \mid \mathtt{a} = \mathtt{bab}\rangle$. Every element of K admits a unique expression of the form $\mathtt{a}^p\mathtt{b}^q$, with the explicit multiplication

$$\mathtt{a}^{p_1}\mathtt{b}^{q_1} \cdot \mathtt{a}^{p_2}\mathtt{b}^{q_2} = \mathtt{a}^{p_1+p_2}\mathtt{b}^{(-1)^{p_2}q_1+q_2}.$$

We claim that K^+ admits the presentation $\langle \mathtt{a},\mathtt{b} \mid \mathtt{a} = \mathtt{bab}\rangle^+$. Indeed, call the words of the form $\mathtt{a}^p\mathtt{b}^q$ or $\mathtt{a}^p\mathtt{b}^q\mathtt{a}$ canonical. Then the explicit formulas

$$\mathtt{a}^p\mathtt{b}^q \cdot \mathtt{a} = \begin{cases} \mathtt{a}^{p+1} & \text{for } q = 0,\\ \mathtt{a}^p\mathtt{b}^q\mathtt{a} & \text{for } q > 0,\end{cases} \qquad \mathtt{a}^p\mathtt{b}^q \cdot \mathtt{b} = \mathtt{a}^p\mathtt{b}^{q+1},$$

$$\mathtt{a}^p\mathtt{b}^q\mathtt{a} \cdot \mathtt{a} = \mathtt{a}^{p+2}\mathtt{b}^q, \qquad \mathtt{a}^p\mathtt{b}^q\mathtt{a} \cdot \mathtt{b} = \begin{cases} \mathtt{a}^{p+1} & \text{for } q = 1,\\ \mathtt{a}^p\mathtt{b}^{q-1}\mathtt{a} & \text{for } q > 1,\end{cases}$$

show that every word of $\{\mathtt{a},\mathtt{b}\}^*$ is equivalent to a canonical word modulo the relation $\mathtt{a} = \mathtt{bab}$. On the other hand, canonical words receive pairwise distinct evaluations in the group K, and, therefore, the relation $\mathtt{a} = \mathtt{bab}$ forces no equivalence between such words. Hence, the monoid $\langle \mathtt{a},\mathtt{b} \mid \mathtt{a} = \mathtt{bab}\rangle^+$ is isomorphic to K^+, and every element of K^+ has a unique canonical expression.

Note that, as K^+ admits a presentation involving the unique relation $\mathtt{a} = \mathtt{bab}$, whose left and right terms contain one letter $\mathtt{a}$ each, the parameter $|w|_\mathtt{a}$ on $\{\mathtt{a},\mathtt{b}\}^*$ induces a well-defined function on K^+. For g in K^+, we shall naturally denote by $|g|_\mathtt{a}$ the common value of $|w|_\mathtt{a}$ for all words w representing g.

The left-divisibility relation of K^+ is a linear order: for any two elements f, g, at least one of $f \preccurlyeq g$, $g \prec f$ holds. Indeed, the above formulas show that $f \preccurlyeq g$ holds if and only if, in Figure 6, the vertex associated with f lies below the vertex associated with g (that is, $|f|_\mathtt{a} < |g|_\mathtt{a}$ holds), or they lie on the same row and, if $|f|_\mathtt{a}$ is odd (resp. *even), f lies on the left* (resp. *on the right) of g.*

Symmetrically, the formulas

$$\mathtt{a}\cdot\mathtt{a}^p\mathtt{b}^q=\mathtt{a}^{p+1}\mathtt{b}^q,\qquad \mathtt{b}\cdot\mathtt{a}^p\mathtt{b}^q=\begin{cases}\mathtt{a}^p\mathtt{b}^{q+1} & \text{for } p \text{ even},\\ \mathtt{a}^{p-1}\mathtt{ba} & \text{for } p \text{ odd and } q=0,\\ \mathtt{a}^p\mathtt{b}^{q-1} & \text{for } p \text{ odd and } q\geqslant 1,\end{cases}$$

$$\mathtt{a}\cdot\mathtt{a}^p\mathtt{b}^q\mathtt{a}=\mathtt{a}^{p+1}\mathtt{b}^q\mathtt{a},\qquad \mathtt{b}\cdot\mathtt{a}^p\mathtt{b}^q\mathtt{a}=\begin{cases}\mathtt{a}^p\mathtt{b}^{q+1}\mathtt{a} & \text{for } p \text{ even},\\ \mathtt{a}^p\mathtt{b}^{q-1}\mathtt{a} & \text{for } p \text{ odd},\end{cases}$$

imply that the right-divisibility relation of K^+ is also a linear order: $f \widetilde{\preccurlyeq} g$ holds if and only if, in Figure 6, (the vertex associated with) f lies below (the vertex associated with) g, or they lie on the same row and f lies on the left of g. Thus left- and right-divisibility agree on even rows, and disagree on odd rows.

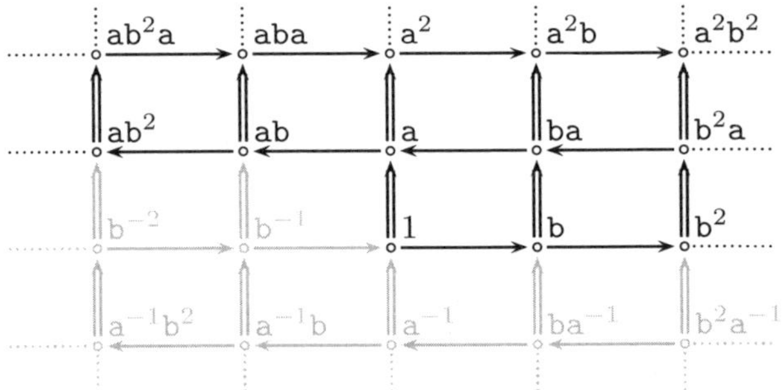

Figure 6. Cayley graph of the Klein bottle monoid K^+ embedded in the Klein bottle group K (in grey). Simple arrows correspond to right-multiplication by $\mathtt{b}$, double arrows to right-multiplication by $\mathtt{a}$. We can see that left-divisibility is a linear ordering. Left-multiplication by $\mathtt{b}$ always corresponds to a horizontal right-oriented translation, whereas left-multiplication by $\mathtt{a}$ corresponds to a vertical bottom-up translation followed by a symmetry with respect to the vertical axis through 1.

The Klein bottle monoid cannot provide a Garside monoid: a generating subset of K^+ must contain at least one element g satisfying $|g|_{\mathtt{a}} = 1$, and every such element has infinitely many left-divisors and, moreover, infinitely many left-divisors making an increasing sequence for left-divisibility. However, checking the following result is easy.

Proposition 3.3 (normal decomposition). *Every element of K admits a unique decomposition of the form $\Delta^d s_1 \cdots s_p$ with d in $\mathbb{Z}$ and $s_1, \ldots, s_p$ in $\mathrm{Div}(\Delta)$ satisfying $s_1 \neq \Delta$, $s_p \neq 1$, and, for every i,*

$$\forall g \in K^+ \setminus \{1\}\ (\, g \preccurlyeq s_{i+1} \ \Rightarrow\ s_i g \not\preccurlyeq \Delta\,). \tag{3.4}$$

As can be expected, Proposition 3.3 can be connected with the fact that the left-divisibility relation on K^+ is a lattice order, actually a quite simple one as it is a linear order. So, in particular, the least upper bound of two elements is simply the larger.

Although very simple and rudimentary, the Klein bottle monoid will provide one useful counter-example, as its divisibility relations satisfy none of the Noetherianity conditions that are valid in classical examples.

3.3 Wreathed free abelian groups

Here is again an example of a monoid that cannot be a Garside monoid as it contains nontrivial invertible elements.

REFERENCE STRUCTURE #6 (**wreathed free abelian group**).

• *For* $n \geqslant 1$ *, define* $\widetilde{\mathbb{Z}^n}$ *to be the set of all pairs* (g, π) *with* g *in* $\mathbb{Z}^n$ *and* π *in the symmetric group* $\mathfrak{S}_n$*; for* (s_1, π_1), (s_2, π_2) *in* $\widetilde{\mathbb{Z}^n}$*, we define* $(s_1, \pi_1) * (s_2, \pi_2) = (s_1(s_2 \circ \pi_1^{-1}), \pi_1\pi_2)$*, here the product of* $\mathbb{Z}^n$ *is written multiplicatively, as in* Reference Structure 1.

• *Put* $\mathtt{a}_i = ((0, ..., 1, ..., 0), \mathrm{id})$*, one* 1 *in position* i*, for* $1 \leqslant i \leqslant n$*; furthermore, put* $\mathtt{s}_i = ((0, ..., 0), (i, i+1))$ *for* $1 \leqslant i < n$*, and put* $\Delta_n = ((1, ..., 1), \mathrm{id})$*.*

• *Denote by* $\widetilde{\mathbb{N}^n}$ *the set of all pairs* (g, π) *in* $\widetilde{\mathbb{Z}^n}$ *satisfying* $\forall k \leqslant n\ (g(k) \geqslant 0)$*.*

• *Denote by* S_n *the set of all pairs* (g, id) *in* $\widetilde{\mathbb{N}^n}$ *satisfying* $\forall k \leqslant n\ (g(k) \in \{0, 1\})$*.*

Then $\widetilde{\mathbb{Z}^n}$ *is a group, namely the wreath product* $\mathbb{Z} \wr \mathfrak{S}_n$*, that is, the semidirect product of the free abelian group* $\mathbb{Z}^n$ (Reference Structure 1) *and the symmetric group* $\mathfrak{S}_n$ *where the latter acts on the coordinates, that is, we have* ${}^{\pi}g(k) = g(\pi^{-1}(k))$*. As for* $\widetilde{\mathbb{N}^n}$*, it is the submonoid of* $\widetilde{\mathbb{Z}^n}$ *generated by the elements* $\mathtt{a}_i$ *and* $\mathtt{s}_i$*. The map* $\iota : \mathbb{Z}^n \to \widetilde{\mathbb{Z}^n}$ *defined by* $\iota(g) = (g, \mathrm{id})$ *is an embedding, as well as its restriction to* $\mathbb{N}^n$*, and we can identify* $\mathbb{Z}^n$ *with its image in* $\widetilde{\mathbb{Z}^n}$*. The group* $\widetilde{\mathbb{Z}^n}$ *is generated by the elements* $\mathtt{a}_i$ *and* $\mathtt{s}_i$*, and it is presented by the relations*

$$
\text{(3.5)} \quad \left\{ \begin{array}{lll}
\mathtt{a}_i\mathtt{a}_j = \mathtt{a}_j\mathtt{a}_i \quad \text{for all } i, j, & & \\
\mathtt{s}_i^2 = 1, \quad \mathtt{s}_i\mathtt{s}_j = \mathtt{s}_j\mathtt{s}_i \text{ for } |i-j| \geqslant 2, & \mathtt{s}_i\mathtt{s}_j\mathtt{s}_i = \mathtt{s}_j\mathtt{s}_i\mathtt{s}_j \text{ for } |i-j| = 1, & \\
\mathtt{s}_i\mathtt{a}_j = \mathtt{a}_j\mathtt{s}_i \text{ for } j \neq i, i+1, & \mathtt{s}_i\mathtt{a}_i = \mathtt{a}_{i+1}\mathtt{s}_i, & \mathtt{s}_i\mathtt{a}_{i+1} = \mathtt{a}_i\mathtt{s}_i,
\end{array} \right.
$$

which also provide a monoid presentation of the monoid $\widetilde{\mathbb{N}^n}$*.*

For (g, π), (g', π') in $\widetilde{\mathbb{Z}^n}$, let us declare $(g, \pi) \leqslant (g', \pi')$ if $\forall k {\leqslant} n\, (g(k) \leqslant g'(k))$ holds. Here is one more decomposition result.

Proposition 3.6 (normal decomposition). *Every element of the group* $\widetilde{\mathbb{Z}^n}$ *admits a unique decomposition of the form* $\Delta_n^d\, s_1 \cdots s_p$ *with* d *in* $\mathbb{Z}$*,* $s_1, ..., s_{p-1}$ *in* S_n*, and* s_p *in* $S_n\mathfrak{S}_n$ *satisfying* $s_1 \neq \Delta_n$*,* $s_p \notin \mathfrak{S}_n$*, and, for every* i*,*

$$
\text{(3.7)} \qquad \forall g \in \widetilde{\mathbb{N}^n}\backslash\{1\}\ (\, g \leqslant s_{i+1} \ \Rightarrow\ s_i g \not\leqslant \Delta_n\,).
$$

It is almost immediate to deduce Proposition 3.6 from Proposition 1.1 using the equality $\widetilde{\mathbb{Z}^n} = \mathbb{Z}^n \rtimes \mathfrak{S}_n$. A direct verification is also possible, based on

Lemma 3.8. *The relation* $\leqslant$ *of Proposition* 3.6 *is a lattice preorder on* $\widetilde{\mathbb{N}^n}$*, and the set* $\{g \in \widetilde{\mathbb{N}^n} \mid g \leqslant \Delta_n\}$ *is a finite sublattice with* $2^n n!$ *elements.*

By a lattice preorder, we mean a preorder, that is, a reflexive and transitive relation, such that any two elements admit a (not necessarily unique) greatest lower bound and a (not necessarily unique) lowest upper bound.

3.4 Ribbon categories

We conclude with one more example. This one is different from the previous ones as there is no group or monoid involved, but instead a groupoid and a category—see Chapter II for basic definitions.

REFERENCE STRUCTURE #7 (**braid ribbons**).

- *For* $n \geqslant 2$ *and* $1 \leqslant i, j < n$, *denote by* $\mathcal{BR}_n(i,j)$ *the family of all braids of* B_n *that contain an* (i,j)*-ribbon.*
- *For* $n \geqslant 2$, *denote by* $\mathcal{BR}_n$ *the groupoid of* n*-strand* braid ribbons, *whose object set is* $\{1, ..., n-1\}$ *and whose family of morphisms with source* i *and target* j *is* $\mathcal{BR}_n(i,j)$.
- *For* $n \geqslant 2$, *denote by* $\mathcal{BR}_n^+$ *the subcategory of* $\mathcal{BR}_n$ *in which the morphisms are required to lie in* B_n^+.
- *For* $1 \leqslant i < n$, *denote by* $\mathcal{S}_n(i)$ *the family of all braids in* B_n^+ *that left-divide* Δ_n *and contain an* (i,j)*-ribbon for some* j.
- *Denote by* $\mathcal{S}_n$ *the union of all families* $\mathcal{S}_n(i)$ *for* $i = 1, ..., n-1$.

We say that a geometric n*-strand braid* β *contains a* ribbon connecting $[i, i+1]$ to $[j, j+1]$, *or simply an* (i,j)*-ribbon, if there exists in the cylinder* $D_n \times [0,1]$ *a surface whose boundary consists of the strands of* β *that start at positions* i *and* $i+1$ *plus the segments* $[(i,0,0),(i+1,0,0)]$ *and* $[(j,0,1),(j+1,0,1)]$, *and that intersects no strand of* β. *A braid is said to contain an* (i,j)*-ribbon if, among the geometric braids that represent it, at least one contains such a ribbon. For instance, we see on the picture*

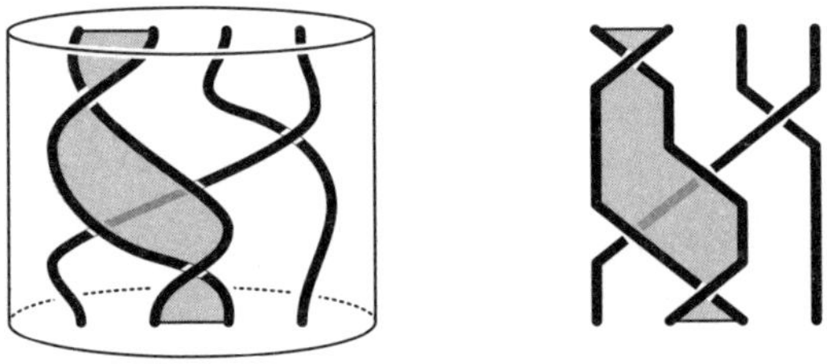

that the braid $\sigma_1\sigma_3\sigma_2^{-1}\sigma_1^{-1}\sigma_2$ *contains a* $(1,2)$*-ribbon, but it contains no* $(2,3)$*-ribbon, although the strand starting at position 2 finishes at position 3 and the strand starting at position* $2{+}1$ *finishes at position* $3{+}1$: *indeed, any surface whose boundary contains these strands is pierced by at least another strand and therefore provides no ribbon.*

The braids β *that contain an* (i,j)*-ribbon turn out to coincide with the braids that satisfy* $\sigma_i\beta = \beta\sigma_j$. *One direction is clear, namely that* $\sigma_i\beta = \beta\sigma_j$ *holds when* β *contains an* (i,j)*-ribbon. Indeed, it suffices to push the additional crossing* σ_i *through the ribbon to establish that* $\sigma_i\beta$ *is isotopic to* $\beta\sigma_j$. *The converse direction is more delicate. First, we observe that the braid* Δ_n *contains an* $(i, n-i)$*-ribbon for every* i, *and, therefore, for every* d, *an* n*-strand braid* β *contains an* (i,j)*-ribbon if and only if the braid* $\Delta_n^d\beta$ *contains one. For every* β *in* B_n, *there exists an integer* d *such that* Δ_n^{2d} *belongs to* B_n^+.

So it suffices to prove that, if β belongs to B_n^+ and $\sigma_i\beta = \beta\sigma_j$ holds, then β contains an (i,j)-ribbon. This is proved using induction on the number of generators σ_i occurring in a decomposition of β. So write $\beta = \sigma_k\beta'$. For $k = i$ and for $|k-i| \geqslant 2$, the braids σ_i and σ_k commute and we deduce $\sigma_i\beta' = \beta'\sigma_j$ by left-cancelling σ_k. For $|k-i| = 1$, we have $\sigma_i\sigma_k\beta' = \sigma_k\beta'\sigma_j$ by assumption, so we have a braid that is a right-multiple both of σ_i and σ_k. It follows that it is a right-multiple of the least common right-multiple of σ_i and σ_k, namely $\sigma_i\sigma_k\sigma_i$, which means that β' is necessarily $\sigma_i\beta''$ for some β'', and we deduce $\sigma_k\beta'' = \beta''\sigma_j$. So, in all cases, we found a shorter braid with the same property and we can apply the induction hypothesis.

As before, we shall use $\preccurlyeq$ for left-divisibility, now in the sense of the category $\mathcal{BR}_n^+$. So, for f, g morphisms of $\mathcal{BR}_n^+$, we have $f \preccurlyeq g$ if there exists a morphism g' of $\mathcal{BR}_n^+$ satisfying $fg' = g$. Then we have

Proposition 3.9 (normal decomposition). *Every n-strand braid ribbon admits a unique decomposition of the form $\Delta_n^d\, s_1 \cdots s_p$ with d in $\mathbb{Z}$ and $s_1, \ldots, s_p$ morphisms of $\mathcal{S}_n$ satisfying $s_1 \neq \Delta_n$, $s_p \neq 1$, and, for every i,*

$$\forall g \in \mathcal{BR}_n^+ \setminus \{1\}\ (\ g \preccurlyeq s_{i+1} \ \Rightarrow\ s_i g \notin \mathcal{S}_n\ ). \tag{3.10}$$

As in the case of Proposition 1.8, the proof of Proposition 3.9 mainly relies on the following order property:

Lemma 3.11. *For $n \geqslant 2$ and $1 \leqslant i < n$, the structure $(\mathcal{S}_n(i), \preccurlyeq)$ is a lattice order.*

By construction, the lattice $(\mathcal{S}_n(i), \preccurlyeq)$ is embedded in the lattice $(\mathrm{Div}(\Delta_n), \preccurlyeq)$ of Lemma 1.10. However, the inclusion is strict in general, as we only allow those braids that admit an (i,j)-ribbon for some j. For instance, $\mathcal{S}_4(1)$, $\mathcal{S}_4(2)$ and $\mathcal{S}_4(3)$ have 12 elements, see Figure 7. Note that, if f, g belong to $\mathcal{S}_n(i)$ and $f \preccurlyeq g$ holds in $\mathcal{BR}_n^+$, the braid g' witnessing $fg' = g$ need not belong to $\mathcal{S}_n(i)$ in general, except in the particular case when f contains an (i,i)-ribbon.

Example 3.12 (braid ribbons). Let us consider the 4-strand braid $g = \sigma_2\sigma_3\sigma_2\sigma_2\sigma_1\sigma_2\sigma_3\sigma_3$ of Example 1.11 again. The braid g contains an $(1,3)$-ribbon, so it belongs to $\mathcal{BR}_4(1)$ and admits a decomposition as in Proposition 3.9. This decomposition is not the decomposition $\sigma_1\sigma_2\sigma_3\sigma_2\sigma_1 \cdot \sigma_2\sigma_1\sigma_3$ provided by Proposition 1.8, as the entries of the latter do not contain ribbons. The $\mathcal{S}_4$-decomposition of g, which turns out to also have length two, is $\sigma_1\sigma_3\sigma_2\sigma_1 \cdot \sigma_2\sigma_3\sigma_2\sigma_1$. Note that the first entry lies in $\mathcal{S}_4(1)$, whereas the second lies in $\mathcal{S}_4(2)$.

The study of the ribbon category $\mathcal{BR}_n$, and in particular Proposition 3.9, proves to be crucial for determining centralizers and normalizers in braid groups. Clearly, Propositions 3.1 and 3.9 are quite similar to Propositions 1.1, 1.8, or 1.14, although they do enter the framework of Garside groups. In particular, no group or monoid is involved in the case of ribbons, where introducing several objects to control composition is necessary.

Thus, the examples presented in this section show that the essential properties of Garside groups as stated in Section 2 extend to a larger class of structures and, in particular, that the assumptions that

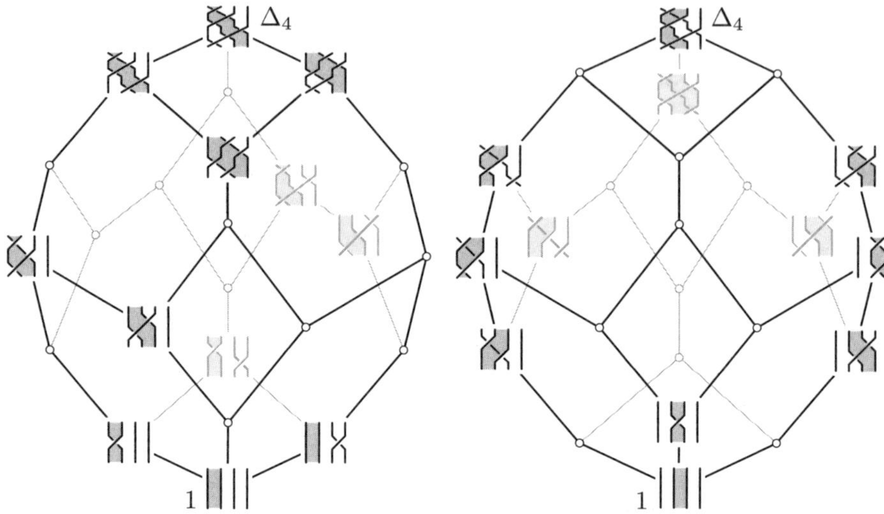

Figure 7. The lattices $S_4(1)$ and $S_4(2)$, here embedded in the lattice $(\mathrm{Div}(\Delta_4), \preccurlyeq)$ of Figure 2: the missing braids contain no ribbon of the expected type. Note that the two lattices are isomorphic; they both contain two hexagonal faces, and six squares.

- the structure is finitely generated,
- the structure is Noetherian (existence of a map λ as in Definition 2.1),
- the structure has no nontrivial invertible elements,
- the multiplication is defined everywhere

are not necessary. It is therefore natural to build a new, extended framework that is relevant for all examples at a time: this is our aim in this text. In one word, the idea will be to go to a category framework, and to introduce a general notion of *Garside family* to play the role of divisors of the element Δ of a Garside monoid. The nice point is that the extended theory is in many respects simpler and more elegant than the old approach.

Exercise

Exercise 1 (braids). Let $\underline{B}_n^+$ be the monoid presented by (1.5) (Reference Structure 2, page 5). Define a braid word to be a word on the alphabet $\{\sigma_1, \sigma_2, ...\}$, and let $\equiv^+$ be the braid word equivalence associated with the relations of (1.5). Inductively define a braid word $\underline{\Delta}_n$ by $\underline{\Delta}_1 = \varepsilon$ and $\underline{\Delta}_n = \underline{\Delta}_{n-1}\sigma_{n-1}\cdots\sigma_2\sigma_1$. (i) Using the relations of (1.5), show that, for every n and every i with $1 \leqslant i \leqslant n-1$, there exists a braid word $w_{i,n}$ satisfying $\sigma_i w_{i,n} \equiv^+ \underline{\Delta}_n$. (ii) Show that $\sigma_i \underline{\Delta}_n \equiv^+ \underline{\Delta}_n \sigma_{n-i}$ holds for $1 \leqslant i \leqslant n-1$. (iii) Prove that, for every ℓ and every sequence $i_1, ..., i_\ell$ in $\{1, ..., n-1\}$, there exists a word w satisfying $\sigma_{i_1}\cdots\sigma_{i_\ell} w \equiv^+ \underline{\Delta}_n^\ell$. (iv) Conclude that any two elements in $\underline{B}_n^+$ admit a common right-multiple.

Notes

Sources and comments. The basic theory of braid groups goes back to E. Artin in [4], and its analysis in term of the braid monoid B_n^+ and the fundamental braid Δ_n is due to F.A. Garside in [124, 125]. The explicit description of the normal form and the geometric interpretation of its computation as illustrated in Figure 3 appears in El-Rifai–Morton [117].

The introduction and analysis of the dual braid monoid B_n^{+*} goes back to Birman–Ko–Lee [21].

The notions of a Garside monoid and a Garside group were introduced by the first author and L. Paris in [99]. The initial definition was slightly more restricted than Definition 2.1 in that it included the additional assumption that the Garside element be the least common multiple of the atoms. The current definition, which became classic in literature, appears in [80].

Torus-type groups are mentioned in [99]. The Garside structure connected with the Hurwitz action on a free group is due to D. Bessis in [10]. The description of the Klein bottle group is folklore, and so is the wreath product of Reference Structure 6, page 19.

Categories of ribbons have been investigated by B. Brink and R.B. Howlett in [37], and by E. Godelle in [134, 135], later extended in [138] and [139].

Further geometric interpretations of braids. Braids will often appear in this text as a seminal example and, although this is needed neither in this introductory chapter nor in most of the subsequent algebraic developments, it will be useful to keep in mind a few further geometrical points involving braids that we briefly mention now—see also Subsection XIV.4.1.

First, a braid can be seen as a continuous function f that assigns to each t in $[0,1]$ a set of n distinct points of D_n, the intersection of the n disjoint curves forming the braid with the slice $D_n \times \{t\}$ of the cylinder $D_n \times [0,1]$. Further, this function f satisfies $f(0) = f(1) = \{1, ..., n\}$. In other words, if $E_n(D_n)$ is the topological space of subsets of D_n of cardinality n, a braid can be viewed as an element of the fundamental group $\pi_1(E_n(D_n), \{1, ..., n\})$.

There is a natural way to construct the space $E_n(D_n)$. First, since the complex plane can be contracted to the disk D_n, it is natural to replace $E_n(D_n)$ by the space $E_n(\mathbb{C})$ of subsets of $\mathbb{C}$ of cardinality n. A way of constructing $E_n(\mathbb{C})$ is by considering the space X_n^{reg} of tuples of points $(x_1, ..., x_n)$ of $\mathbb{C}^n$ with distinct coordinates, and quotienting by the action of the symmetric group $\mathfrak{S}_n$. In other words, we set $X_n^{\mathrm{reg}} = \mathbb{C}^n \setminus \bigcup_{i,j} H_{i,j}$ where $H_{i,j}$ is the hyperplane $x_i = x_j$. There is a natural action of $\mathfrak{S}_n$ on $\mathbb{C}^n$ by permuting the coordinates; then $H_{i,j}$ can be interpreted as the fixed points of the transpositions (i,j) of $\mathfrak{S}_n$. Thus X_n^{reg} is the subspace of $\mathbb{C}^n$ where the action of $\mathfrak{S}_n$ is regular, that is, the images of a point are all distinct. We thus have a regular covering $X_n^{\mathrm{reg}} \to X_n^{\mathrm{reg}}/\mathfrak{S}_n \simeq E_n(\mathbb{C})$ which gives rise to the exact sequence (1.7) where PB_n appears here as $\pi_1(X_n^{\mathrm{reg}}, x)$ and the morphism from B_n to $\mathfrak{S}_n$ is the natural morphism mentioned in Reference Structure 2, page 5.

General results about Garside monoids and groups. The aim of this text is not to develop a complete theory of Garside monoids and Garside groups, but rather to show how to extend the framework. So we shall just briefly mention here a few general results about these structures that will not be developed in the sequel of this text. In some sense, the current work shows that studying Garside monoids and groups is not so crucial, since their definitions include superfluous conditions. However, even with these simplifying conditions, we are still very far from a complete classification, and few general results are known.

The examples provided by classical and dual Artin–Tits monoids (see Chapter IX) show that the family of all Garside monoids is quite large and, even in the case of Garside monoids with two generators, the many examples of Picantin [196] show that an exhaustive classification is difficult. Let us mention $\langle \mathtt{a}, \mathtt{b}, \mathtt{c} \mid \mathtt{acab} = \mathtt{bca}^2, \mathtt{bca}^2\mathtt{c} = \mathtt{cabca}\rangle^+$, which is indeed a Garside monoid but, contrary to all more standard examples, has the property that the lattice of divisors of Δ, shown on the right, is not invariant under top-bottom symmetry. Another similar (but not homogeneous) example is $\langle \mathtt{a}, \mathtt{b}, \mathtt{c} \mid \mathtt{a}^2 = \mathtt{bc}, \mathtt{b}^3 = \mathtt{ca}\rangle^+$, whose group of fractions is the group of the $(3, 4)$-torus knot.

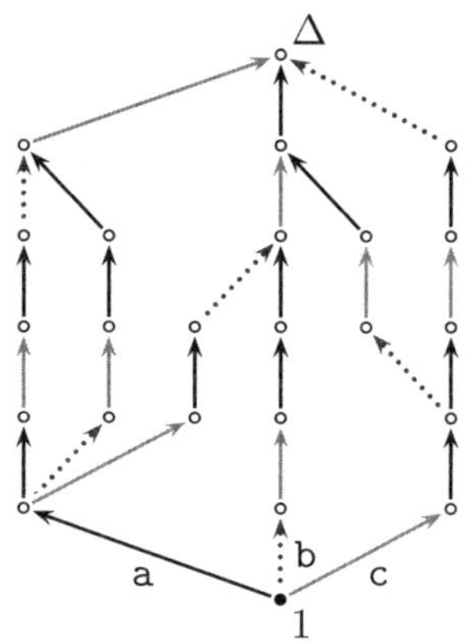

The only case when a (sort of) more complete classification is known is the special case of Garside monoids admitting a presentation with n generators and $\binom{n}{2}$ relations $u = v$ with u and v of length two and at most one relation for each u: as will be seen in Proposition XIII.2.34 (characterization), such Garside monoids are in one-to-one correspondence with the non-degenerate involutive set-theoretic solutions of the Yang–Baxter equation—which however is not a genuine description as a classification of the latter solutions is not known so far. See Question 38 and the discussion at the end of Chapter XIII for hypothetical extensions. Also see [203] for an extension of the above correspondence and a connection with lattice groups.

Some connections between Garside monoids and other families of monoids have been described: for instance, a connection with divisibility monoids [167] is described in Picantin [200]; in another direction, the relation with the factorable groups of C. Bödigheimer and B. Visy [24, 227] was addressed by V. Ozornova in [191]. In another direction, every torus knot group—or even every torus link group—is the group of fractions of various monoids, many of whom are Garside monoids, see Picantin [199]. Actually a knot group is a Garside group if and only if it is a torus knot group. This follows from a more general result from Picantin [195]:

Proposition. *A one-relator group is a Garside group if and only if its center is not trivial.*

The main general structural result about Garside monoids known so far is the following decomposition result, which extends the decomposition of an Artin–Tits group of spherical type in terms of irreducible ones. By definition, a Garside monoid is generated by atoms, which are elements that cannot be expressed as product of at least two non-invertible elements. Say that an element g is *quasi-central* if, writing A for the atom set, we have $gA = Ag$. In a Garside monoid (M, Δ), the Garside element Δ is always quasi-central (and some power Δ^e is central). An Artin–Tits monoid of spherical type has an infinite cyclic center if and only if it is irreducible, that is, the associated Coxeter graph is connected (see Proposition IX.1.38).

Proposition (Picantin [197]). *Every Garside monoid is an iterated Zappa-Szep product of Garside monoids with an infinite cyclic center.*

We recall that a *Zappa-Szep product* (or *bi-crossed product*, or *knit product*) of groups or monoids [230, 221] is the natural extension of a semidirect product in which both groups (or monoids) act on one another, the semidirect product corresponding to the case when one action is trivial, and the direct product to the case when both actions are trivial. Thus the above result reduces the investigation of general Garside monoids and groups to those in which the quasi-center (family of all quasi-central elements) is infinite cyclic. The proof of [197] relies on associating with atom s the element Δ_s that is the least common right-multiple of all elements $s\backslash g$ for g in the monoid, where $s\backslash g$ is the unique element g' such that sg' is the least common right-multiple of s and g, and establishing that the elements Δ_s generate the quasi-center.

As mentioned above, there exists a number of Garside monoids with two generators. However, the associated groups of fractions turn out to be simple.

Question 1. Is every Garside group with two generators either an Artin–Tits group of dihedral type $\langle \mathtt{a}, \mathtt{b} \mid \mathtt{aba}\cdots = \mathtt{bab}\cdots \rangle$, or a torus-type group $\langle \mathtt{a}, \mathtt{b} \mid \mathtt{a}^p = \mathtt{b}^q \rangle$?

At the moment, no example witnessing for a negative answer is known; a positive answer is conjectured in Picantin [196], but no proof is in view.

In another direction, it is known that the braid groups B_n and, more generally, (almost) all Artin–Tits groups of spherical type are linear, that is, admit a faithful representation in the linear group of a finite-dimensional vector space. It is therefore natural to raise

Question 2. Is every Garside group linear?

A negative answer is likely, but it would be desirable to find combinatorial necessary conditions for a Garside group to be linear, more precisely, for it to have a faithful representation over a totally ordered field which realizes the Garside structure.

Chapter II
Preliminaries

The aim of this chapter is to provide some background for the subsequent developments. Section 1 contains a general introduction to basic notions involving categories, in particular category presentations. We should immediately emphasize that, in this text, a category is just viewed as a sort of conditional monoid, in which the product need not be defined everywhere: so, we shall use almost nothing from category theory, just the language of categories.

In Section 2, we define the left- and right-divisibility relations associated with a category, and discuss Noetherianity conditions possibly satisfied by these relations. Most results here are completely standard, as are those of Section 3, where we define the enveloping groupoid of a category and recall Ore's classical theorem about the embedding of a category in a groupoid of fractions.

Finally, the aim of Section 4 is to provide practical criteria for establishing that a presented monoid or category is possibly cancellative and/or admits least common multiples. To this end, we develop a specific method called (right)-reversing, which is directly reminiscent of what F.A. Garside did in the case of braids (using ideas he attributed to G. Higman). This simple combinatorial method is, in a sense, "stupid" in that it exploits no underlying structure as, for instance, the germ method of Chapter VI does, but it turns out to be useful for dealing with a number of the practical examples we are interested in.

In order to reduce the length, some proofs in Sections 3 and 4 have been postponed to the Appendix at the very end of this text.

Main definitions and results

Definition 1.13 **(invertible).** The family of all invertible elements in a category $\mathcal{C}$ is denoted by $\mathcal{C}^\times$.

Notation 1.17 **(relation $=^\times$).** For $\mathcal{C}$ a category and g, g' in $\mathcal{C}$, we write $g =^\times g'$ if there exists ϵ in $\mathcal{C}^\times$ satisfying $g' = g\epsilon$.

Definition 1.21 **(family $\mathcal{S}^\sharp$).** For $\mathcal{C}$ a category and $\mathcal{S} \subseteq \mathcal{C}$, we put $\mathcal{S}^\sharp = \mathcal{S}\mathcal{C}^\times \cup \mathcal{C}^\times$.

Definition 2.1 **(left-divisor, right-multiple).** For f, g in a category $\mathcal{C}$, we say that f is a *left-divisor* of g, or that g is a *right-multiple* of f, denoted by $f \preccurlyeq g$, if there exists g' in $\mathcal{C}$ satisfying $fg' = g$.

Definition 2.9 **(right-lcm, left-gcd).** For h in a left-cancellative category $\mathcal{C}$ and $\mathcal{S}$ included in $\mathcal{C}$, we say that h is a *right-lcm* of $\mathcal{S}$ if h is a right-multiple of all elements of $\mathcal{S}$ and every element of $\mathcal{C}$ that is a right-multiple of all elements of $\mathcal{S}$ is a right-multiple of h. We say h is a *greatest common left-divisor*, or *left-gcd*, of $\mathcal{S}$, if h is a left-divisor of all elements of $\mathcal{S}$ and every element of $\mathcal{C}$ that is a left-divisor of all elements of $\mathcal{S}$ is a left-divisor of h.

Definition 2.26 **(Noetherian).** A subfamily $\mathcal{S}$ of a category $\mathcal{C}$ is called *left-Noetherian* (*resp. right-Noetherian, resp. Noetherian*) if the restriction to $\mathcal{S}$ of the proper left-divisibility relation (*resp.* right-divisibility relation, *resp.* factor relation) is well-founded, that is, every nonempty subfamily has a smallest element.

Definition 2.52 **(atom).** An element g of a left-cancellative category $\mathcal{C}$ is called an *atom* if g is not invertible and every decomposition of g contains at most one non-invertible element.

Proposition 2.58 **(atoms generate).** *Assume that $\mathcal{C}$ is a left-cancellative category that is Noetherian.* (i) *Every subfamily of $\mathcal{C}$ that generates $\mathcal{C}^\times$ and contains at least one element in each $=^\times$-class of atoms generates $\mathcal{C}$.* (ii) *Conversely, every subfamily that generates $\mathcal{C}$ and satisfies $\mathcal{C}^\times\mathcal{S} \subseteq \mathcal{S}^\sharp$ generates $\mathcal{C}^\times$ and contains at least one element in each $=^\times$-class of atoms.*

Definition 3.3 **(enveloping groupoid).** The *enveloping groupoid $\mathcal{E}nv(\mathcal{C})$* of a category $\mathcal{C}$ is the category $\langle\, \mathcal{C} \cup \overline{\mathcal{C}} \mid Rel(\mathcal{C}) \cup Free(\mathcal{C})\,\rangle^+$, where $Rel(\mathcal{C})$ is the family of all relations $fg = h$ with f, g, h in $\mathcal{C}$ and $Free(\mathcal{C})$ is the family of all relations $f\overline{f} = 1_x$ and $\overline{f}f = 1_y$ for f in $\mathcal{C}(x, y)$.

Definition 3.10 **(Ore category).** A category $\mathcal{C}$ is called a *left-Ore* (*resp. right-Ore*) category if it is cancellative and any two elements with the same target (*resp.* source) admit a common left-multiple (*resp.* common right-multiple). A *Ore category* is a category that is both a left-Ore and a right-Ore category.

Proposition 3.11 **(Ore's theorem).** *For $\mathcal{C}$ a category, the following are equivalent:* (i) *There exists an injective functor ι from $\mathcal{C}$ to $\mathcal{E}nv(\mathcal{C})$ and every element of $\mathcal{E}nv(\mathcal{C})$ has the form $\iota(f)^{-1}\,\iota(g)$ for some f, g in $\mathcal{C}$.* (ii) *The category $\mathcal{C}$ is a left-Ore category.*

Proposition 3.21 **(torsion).** *Assume that $\mathcal{C}$ is a right-Ore category that admits right-lcms. Then the torsion elements of $\mathcal{E}nv(\mathcal{C})$ are the elements of the form fhf^{-1} with f in $\mathcal{C}$ and h a torsion element of $\mathcal{C}$.*

Definition 4.1 **(right-complemented presentation).** A category presentation $(\mathcal{S}, \mathcal{R})$ is called *right-complemented* if $\mathcal{R}$ contains no ε-relation (that is, no relation $w = \varepsilon$ with w nonempty), no relation $s... = s...$ with s in $\mathcal{S}$ and, for $s \neq t$ in $\mathcal{S}$, at most one relation $s... = t...$.

Definition 4.14 **(θ-cube condition).** If $(\mathcal{S}, \mathcal{R})$ is a right-complemented presentation, associated with the syntactic right-complement θ, and u, v, w are $\mathcal{S}$-paths, we say that the *sharp θ-cube condition* (*resp.* the *θ-cube condition*) holds for (u, v, w) if (4.15) Either both $\theta_3^*(u, v, w)$ and $\theta_3^*(v, u, w)$ are defined and they are equal (*resp.* they are $\equiv^+_{\mathcal{R}}$-equivalent), or neither is defined, where $\theta_3^*(u, v, w)$ stands for $\theta^*(\theta^*(u, v), \theta^*(u, w))$ with θ^* as in Lemma 4.6. For $\mathcal{X} \subseteq \mathcal{S}^*$, we say that the (sharp) θ-cube condition holds *on $\mathcal{X}$* if it holds for all (u, v, w) in $\mathcal{X}^3$.

Proposition 4.16 **(right-complemented).** *Assume that $(\mathcal{S}, \mathcal{R})$ is a right-complemented presentation associated with the syntactic right-complement θ, and at least one of the* *following conditions is satisfied:* (4.17) *The presentation* $(\mathcal{S}, \mathcal{R})$ *contains only short relations and the sharp* θ*-cube condition holds for every triple of pairwise distinct elements of* $\mathcal{S}$*;* (4.18) *The presentation* $(\mathcal{S}, \mathcal{R})$ *is right-Noetherian and the* θ*-cube condition holds for every triple of pairwise distinct elements of* $\mathcal{S}$*;* (4.19) *The presentation* $(\mathcal{S}, \mathcal{R})$ *is maximal right-triangular. Then:* (i) *The category* $\langle \mathcal{S} \mid \mathcal{R} \rangle^+$ *is left-cancellative.* (ii) *The category* $\langle \mathcal{S} \mid \mathcal{R} \rangle^+$ *admits conditional right-lcms, that is, any two elements of* $\langle \mathcal{S} \mid \mathcal{R} \rangle^+$ *that have a common right-multiple have a right-lcm. More precisely, for all* w, w' *in* $\mathcal{S}^*$*, the elements of* $\langle \mathcal{S} \mid \mathcal{R} \rangle^+$ *represented by* u *and* v *admit a common right-multiple if and only if* $\theta^*(u, v)$ *exists and, in this case,* $u\theta^*(u, v)$ *represents the right-lcm of these elements.* (iii) *Two elements* u, v *of* $\mathcal{S}^*$ *represent the same element of* $\langle \mathcal{S} \mid \mathcal{R} \rangle^+$ *if and only if both* $\theta^*(u, v)$ *and* $\theta^*(v, u)$ *exist and are empty.*

1 The category context

As explained at the end of Chapter I, it will be convenient to place our constructions in a category framework. However, the only purpose of using categories here is to extend the notion of a monoid so as to allow a product that would not be defined everywhere. So, for us, a category is just a variant of a monoid in which objects are used to control which elements can be multiplied: the product of two elements g_1, g_2 exists if and only if the target of g_1 coincides with the source of g_2. The reader who does not like categories may well choose to read this text as a text about monoids and groups exclusively.

This section is a quick introduction to the needed basic notions. First we recall the terminology of categories (Subsection 1.1), and of subcategories (Subsection 1.2). Then, in Subsection 1.3 we gather some remarks about invertible elements. Finally, Subsection 1.4 is devoted to presentations.

1.1 Categories and monoids

We use the standard terminology of categories, as described for instance in [177]. We shall resort to no nontrivial result or construction from category theory and simply use categories as a suitable framework for developing our constructions. An important special case is that of monoids, which remained the only considered one for more than one decade, and which includes some (but not all) of the most interesting examples.

A *category* $\mathcal{C}$ consists of *objects* and *morphisms*. The family of all objects in $\mathcal{C}$ is denoted by $\mathcal{O}bj(\mathcal{C})$, whereas the family of all morphisms, usually denoted by $\mathcal{H}om(\mathcal{C})$, will be simply denoted by $\mathcal{C}$ here: this abuse of notation reflects our viewpoint that a category is a family of elements (morphisms) equipped with a partial product, and that the objects are additional data whose only purpose is to control the existence of the product.

Two objects are attached with each element of a category, namely its *source* and its *target*. The family of all morphisms of $\mathcal{C}$ with source x and target y will be denoted by $\mathcal{C}(x, y)$. We use $\mathcal{C}(x, -)$ for the family of all elements with source x, and similarly $\mathcal{C}(-, y)$ for the family of all elements with target y. For each object x, there is a specific element of $\mathcal{C}(x, x)$ attached with x, namely the *identity* 1_x. Identity elements are also called *trivial*, and the collection of all trivial elements in $\mathcal{C}$ is denoted by $\mathbf{1}_{\mathcal{C}}$. A partial binary operation is defined on elements, which associates with two elements f, g such that the target of f coincides with the source of g a new element denoted by fg, whose source is the source of f, and whose target is the target of g.

Moreover, the following properties are assumed to hold. If fg and gh are defined, then both $f(gh)$ and $(fg)h$, are defined, and one has

(1.1) $$f\,(g\,h) = (f\,g)\,h.$$

Also, for every g in $\mathcal{C}(x, y)$, we have

(1.2) $$1_x\,g = g = g\,1_y.$$

Remark. The product we use is a reversed composition: fg means "f then g". This convention is coherent with the intuition that the elements of a category act on the right, which is convenient here.

If $\mathcal{C}$ is a category, the *complete diagram* of $\mathcal{C}$ is the labeled graph whose vertices are the objects of $\mathcal{C}$ and, for every g in $\mathcal{C}(x, y)$ there is a g-labeled edge from the vertex x to the vertex y. If $\mathcal{S}$ is a generating subfamily of $\mathcal{C}$, that is, every element of $\mathcal{C}$ is a product of elements of $\mathcal{S}$, the *diagram of $\mathcal{C}$ with respect to $\mathcal{S}$* is the subdiagram of the complete diagram of $\mathcal{C}$ in which one only keeps the edges with labels in $\mathcal{S}$.

Example 1.3 (monoid). Assume that M is a monoid, that is, M is a set equipped with an associative product that admits a neutral element (denoted by 1 by default). One associates a category $M^{\bullet}$ with M by defining $\mathcal{O}bj(M^{\bullet})$ to consist of a single object $\bullet$, and the elements of $M^{\bullet}$ to be the elements of M, with $\bullet$ being the source and the target of every element.

Assume that $\mathcal{C}$ is a category, and $\mathcal{S}$ is a subfamily of $\mathcal{C}$ that generates $\mathcal{C}$, that is, every element of $\mathcal{C}$ that is not an identity element is a product of elements of $\mathcal{S}$ (see Subsection 1.4 below). Another (multi)-graph associated with $\mathcal{C}$ is the *Cayley graph* of $\mathcal{C}$ with respect to $\mathcal{S}$: the vertices are the elements of $\mathcal{C}$, and there is a g-labeled arrow from f to h if $h = fg$ holds. See Figure 1 for an example. Note that the diagram of a monoid is a bouquet (only one vertex), whereas its Cayley graph with respect to a generating family is a connected graph.

If $\mathcal{C}, \mathcal{C}'$ are categories, a *functor* from $\mathcal{C}$ to $\mathcal{C}'$ is a map ϕ from $\mathcal{O}bj(\mathcal{C}) \cup \mathcal{C}$ to $\mathcal{O}bj(\mathcal{C}') \cup \mathcal{C}'$ such that ϕ maps objects to objects, elements to elements, and, for all objects x, y and all elements f, g, h,
- $g \in \mathcal{C}(x, y)$ implies $\phi(g) \in \mathcal{C}'(\phi(x), \phi(y))$,
- $h = fg$ implies $\phi(h) = \phi(f)\phi(g)$,
- $\phi(1_x) = 1_{\phi(x)}$ holds.

A functor is called *injective* (*resp. surjective*, *resp. bijective*) if both its restrictions to objects and elements are injective (*resp.* surjective, *resp.* bijective). A bijective functor of a category into itself is also called an *automorphism* of that category.

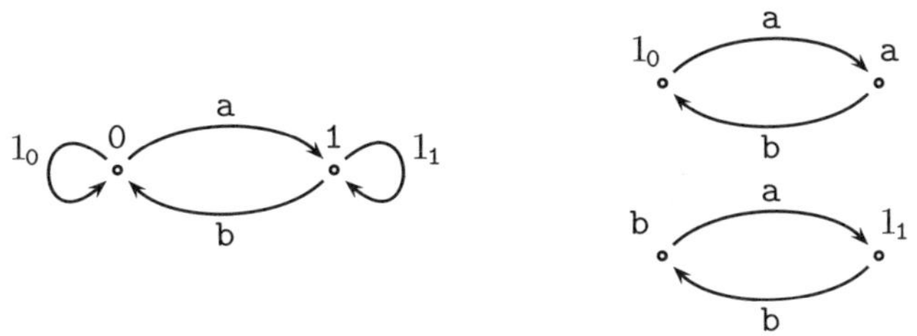

Figure 1. Diagram (left) and Cayley graph with respect to $\{\mathtt{a}, \mathtt{b}\}$ (right) for the category with two objects $\mathtt{0}$, $\mathtt{1}$ and two non-identity elements $\mathtt{a} : \mathtt{0} \to \mathtt{1}$ and $\mathtt{b} : \mathtt{1} \to \mathtt{0}$ satisfying $\mathtt{ab} = 1_0$ and $\mathtt{ba} = 1_1$.

1.2 Subfamilies and subcategories

Whenever the morphisms of a category $\mathcal{C}$ are called the elements of $\mathcal{C}$, it is coherent to say that $\mathcal{S}$ is *included* in $\mathcal{C}$, or, equivalently, that $\mathcal{S}$ is a *subfamily* of $\mathcal{C}$, if $\mathcal{S}$ is any collection of elements of $\mathcal{C}$. However, as the elements of $\mathcal{C}$ come equipped with a source and a target, a more formal definition should include these data.

Definition 1.4 (precategory). A *precategory* is a pair $(\mathcal{O}, \mathcal{S})$ plus two maps, source and target, from $\mathcal{S}$ to $\mathcal{O}$. The elements of $\mathcal{O}$ are called the objects, those of $\mathcal{S}$ are called the elements (or morphisms).

By definition, a category is a precategory, plus a composition map that obeys certain rules. As in the case of a category, we usually identify a precategory with its family of elements, and then denote by $\mathit{Obj}(\mathcal{S})$ the family of objects.

Note that a precategory can be equivalently defined as a (multi)graph, that is, a collection of objects (or vertices), plus, for each pair (x, y) of objects, a family of elements (or edges) with a source x and a target y attached to each element.

If $\mathcal{C}$ is a category, every subfamily $\mathcal{S}$ of $\mathcal{C}$ becomes a precategory when equipped with the restrictions of the source and target maps, and we can then give a formal definition:

Definition 1.5 (subfamily). If $\mathcal{C}$ is a category and $\mathcal{S}$ is included in $\mathcal{C}$, the *subfamily* (associated with) $\mathcal{S}$ is the precategory made of $\mathcal{S}$ together with the restriction of the source and target maps to $\mathcal{S}$.

As already mentioned, in practice we shall not distinguish between a subfamily and its family of elements.

Let us turn to subcategories. A subcategory of a category $\mathcal{C}$ is a category $\mathcal{C}_1$ included in $\mathcal{C}$ and such that the operations of $\mathcal{C}_1$ are induced by those of $\mathcal{C}$:

Definition 1.6 (subcategory). A category $\mathcal{C}_1$ is a *subcategory* of a category $\mathcal{C}$ if we have $\mathit{Obj}(\mathcal{C}_1) \subseteq \mathit{Obj}(\mathcal{C})$ and $\mathcal{C}_1 \subseteq \mathcal{C}$ and, moreover,

(i) For every x in $\mathit{Obj}(\mathcal{C}_1)$, the identity elements 1_x in $\mathcal{C}_1$ and in $\mathcal{C}$ coincide;

(ii) For every g in $\mathcal{C}_1$, the sources and targets of g in $\mathcal{C}_1$ and in $\mathcal{C}$ coincide;

(iii) For all f, g in $\mathcal{C}_1$, the products fg in $\mathcal{C}_1$ and in $\mathcal{C}$ coincide.

In the case of a monoid, we naturally speak of a *submonoid*.

Example 1.7 (subcategory). For every m, the set $m\mathbb{N}$ equipped with addition is a submonoid of the additive monoid $\mathbb{N}$, and so is every set of the form $\sum_i m_i\mathbb{N}$.

Note that, in a proper categorical context, the family of objects of a subcategory may be strictly included in the family of objects of the ambient category.

The following characterization of subcategories is straightforward.

Lemma 1.8. *If $\mathcal{C}$ is a category, then a subfamily $\mathcal{S}$ of $\mathcal{C}$ is a subcategory of $\mathcal{C}$ if and only if $\mathcal{S}$ is closed under identity element and product in $\mathcal{C}$, this meaning that 1_x belongs to $\mathcal{S}$ whenever x belongs to $\mathit{Obj}(\mathcal{S})$ and that gh belongs to $\mathcal{S}$ whenever g and h do and gh is defined in $\mathcal{C}$.*

In the case of a monoid, a subset S of a monoid M is a submonoid of M if and only if it contains 1 and is closed under product. The condition about 1 is not automatic: if a monoid M_1 is included in another monoid M, but the identity elements are not the same, M_1 is *not* a submonoid of M.

Lemma 1.8 implies that, if $\mathcal{C}$ is a category, every intersection of subcategories of $\mathcal{C}$ is a subcategory of $\mathcal{C}$ and, therefore, for every subfamily $\mathcal{S}$ of $\mathcal{C}$, there exists a smallest subcategory of $\mathcal{C}$ that includes $\mathcal{S}$, namely the intersection of all subcategories of $\mathcal{C}$ that include $\mathcal{S}$.

Notation 1.9 (generated subcategory). For $\mathcal{S}$ a subfamily of a category $\mathcal{C}$, we denote by $\mathit{Sub}_{\mathcal{C}}(\mathcal{S})$, or simply $\mathit{Sub}(\mathcal{S})$, the subcategory of $\mathcal{C}$ generated by $\mathcal{S}$.

A concrete description of the subcategory $\mathit{Sub}(\mathcal{S})$ is easy.

Notation 1.10 (product of subfamilies). For $\mathcal{C}$ a category and $\mathcal{S}_1, \mathcal{S}_2$ included in $\mathcal{C}$, we write

$$\mathcal{S}_1\,\mathcal{S}_2 = \{g \in \mathcal{C} \mid \exists g_1 \in \mathcal{S}_1\ \exists g_2 \in \mathcal{S}_2\ (g = g_1 g_2)\}. \tag{1.11}$$

For $\mathcal{S}$ included in $\mathcal{C}$, we write $\mathcal{S}^2$ for $\mathcal{S}\mathcal{S}$, that is, $\{g \in \mathcal{C} \mid \exists g_1, g_2 \in \mathcal{S}\ (g = g_1 g_2)\}$; for $d \geqslant 1$, we denote by $\mathcal{S}^d$ the family of all elements of $\mathcal{C}$ that can be expressed as the product of d elements of $\mathcal{S}$. The notation is extended with the convention that $\mathcal{S}^0$ is the family $\mathbf{1}_{\mathcal{S}}$ of all identity elements 1_x with x in $\mathit{Obj}(\mathcal{S})$.

Lemma 1.12. *For $\mathcal{C}$ a category and $\mathcal{S}$ included in $\mathcal{C}$, the subcategory $\mathit{Sub}(\mathcal{S})$ is $\bigcup_{p \geqslant 0} \mathcal{S}^p$.*

Proof. By Lemma 1.8, $\mathit{Sub}(\mathcal{S})$ must include $\mathbf{1}_{\mathcal{S}}$ and $\mathcal{S}^p$ for every $p \geqslant 1$. Conversely, $\mathbf{1}_{\mathcal{S}} \cup \bigcup_{p \geqslant 1} \mathcal{S}^p$ is closed under identity and product, so, by Lemma 1.8 again, it is a subcategory including $\mathcal{S}$, hence it is $\mathit{Sub}(\mathcal{S})$ as it is included in every subcategory that includes $\mathcal{S}$. □

1.3 Invertible elements

We now fix some terminology and notation for invertible elements in a category. Most examples of monoids and categories we shall consider in this text admit no nontrivial invertible element, that is, the identity elements are the only invertible elements. However, there is no reason for restricting to that particular case, since allowing nontrivial invertible elements will never be a problem provided some mild compatibility conditions explained below are satisfied. As in the case of categories, the reader who is not interested in invertible elements may always forget about nontrivial invertible elements.

The main notion in this short subsection is the closure $\mathcal{S}^\sharp$ of a family $\mathcal{S}$ with respect to right-multiplication by invertible elements (Definition 1.21).

Definition 1.13 (inverse, invertible). If $\mathcal{C}$ is a category and g belongs to $\mathcal{C}(x, y)$, we say that g' is a *left-inverse* (*resp. right-inverse*) for g if g' belongs to $\mathcal{C}(y, x)$ and we have $g'g = 1_y$ (*resp.* $gg' = 1_x$). We say that g' is an *inverse* of g if it is both a left- and a right-inverse of g. An element is called *invertible* if it admits an inverse. The family of all invertible elements in $\mathcal{C}$ is denoted by $\mathcal{C}^\times$; for x, y in $\mathcal{O}bj(\mathcal{C})$, the family of all invertible elements with source x and target y is denoted by $\mathcal{C}^\times(x, y)$.

By construction, $\mathbf{1}_\mathcal{C} \subseteq \mathcal{C}^\times$ always holds: each identity element is its own inverse. Identity elements will be called *trivial* invertible elements and, therefore, a category $\mathcal{C}$ for which the inclusion $\mathbf{1}_\mathcal{C} \subseteq \mathcal{C}^\times$ is an equality will be said to have no nontrivial invertible element. In the particular case of a monoid, this means that the unit 1 is the only invertible element.

Hereafter we shall almost always consider categories that satisfy cancellativity conditions. More precisely, virtually all categories we consider will be assumed to be left-cancellative—however we shall always mention it explicitly.

Definition 1.14 (cancellative). A category $\mathcal{C}$ is called *left-cancellative* (*resp. right-cancellative*) if every relation $fg = fg'$ (*resp.* $gf = g'f$) with f, g, g' in $\mathcal{C}$ implies $g = g'$. We say that $\mathcal{C}$ is *cancellative* if it is both left- and right-cancellative.

One of the (many) advantages of cancellativity assumptions is that they force the above notions of left- and right-inverse to coincide:

Lemma 1.15. *If $\mathcal{C}$ is a left-cancellative category, then, for every g in $\mathcal{C}$, the following conditions are equivalent:*

(i) *The element g has a left-inverse;*

(ii) *The element g has a right-inverse;*

(iii) *There exist f, f' such that fgf' is invertible;*

(iv) *The element g has an inverse, that is, g is invertible.*

Proof. By definition, (iv) implies (i) and (ii), whereas each of (i), (ii) implies (iii). So the point is to prove that (iii) implies (iv). Assume that fgf' has an inverse, say h. Let x be the source of g, and y be that of f. Then we have $fgf'h = 1_y$, hence $fgf'hf = f$, and, therefore $gf'hf = 1_x$ by left-cancelling f. Hence, $f'hf$ is a right-inverse of g. Similarly,

$fgf'h = 1_y$ implies $fgf'hfg = fg$, hence $f'hfg = 1_y$ by left-cancelling fg. Hence, $f'hf$ is also a left-inverse of g, and, finally, it is an inverse of g. □

It is straightforward that a product of invertible elements is again invertible. It follows that, in every (left-cancellative) category $\mathcal{C}$, the family $\mathcal{C}^\times$ equipped with the induced product has the structure of a *groupoid*, that is, it is a category in which every element is invertible.

A direct consequence of Lemma 1.15 is the symmetric result that a product of non-invertible elements is again non-invertible:

Lemma 1.16. *If $\mathcal{C}$ is a left-cancellative category and $g_1, ..., g_p$ are elements of $\mathcal{C}$ such that $g_1 \cdots g_p$ is defined and invertible, then each of $g_1, ..., g_p$ is invertible.*

Proof. By Lemma 1.15(iii), the invertibility of $(g_1 \cdots g_{i-1})\, g_i\, (g_{i+1} \cdots g_p)$ implies that of g_i. □

We shall often be interested in elements that are almost equal in the sense that they are obtained from one another by a left- or right-multiplication by an invertible element.

Notation 1.17 (relations $=^\times$, ${}^\times\!=$, and ${}^\times\!=^\times$). For $\mathcal{C}$ a category and g, g' in $\mathcal{C}$, we write $g =^\times g'$ (*resp.* $g\ {}^\times\!= g'$, $g\ {}^\times\!=^\times g'$) if there exists ϵ in $\mathcal{C}^\times$ (*resp.* ϵ, ϵ' in $\mathcal{C}^\times$) satisfying $g' = g\epsilon$ (*resp.* $\epsilon' g = g'$, *resp.* $\epsilon' g = g'\epsilon$).

By definition, the relations $=^\times$ and ${}^\times\!=$ are included in ${}^\times\!=^\times$, and all are equivalence relations in the ambient category. If the latter contains no nontrivial invertible element, then $=^\times$, ${}^\times\!=$, and ${}^\times\!=^\times$ reduce to equality, and many statements become (slightly) simpler. Many notions that are only defined up to $=^\times$-, ${}^\times\!=$-, or ${}^\times\!=^\times$-equivalence become then unique. One might think of restricting to the case with no nontrivial invertible by going to an appropriate quotient. This need not be always possible: neither $=^\times$ nor ${}^\times\!=^\times$ need be compatible with multiplication in general. However, we have the following connections.

Proposition 1.18 (collapsing invertible elements). *For $\mathcal{C}$ a left-cancellative category, the following are equivalent:*

(i) *The equivalence relation $=^\times$ is compatible with composition, in the sense that, if $g_1 g_2$ is defined and $g'_i =^\times g_i$ holds for $i = 1, 2$, then $g'_1 g'_2$ is defined and $g'_1 g'_2 =^\times g_1 g_2$ holds;*

(ii) *The family $\mathcal{C}^\times(x, y)$ is empty for $x \neq y$ and, for all g, g' sharing the same source, $g\ {}^\times\!=^\times g'$ implies $g =^\times g'$;*

(iii) *The family $\mathcal{C}^\times(x, y)$ is empty for $x \neq y$ and we have*

$$\forall x, y \in \mathcal{O}bj(\mathcal{C})\ \forall g \in \mathcal{C}(x, y)\ \forall \epsilon \in \mathcal{C}^\times(x, x)\ \exists \epsilon' \in \mathcal{C}^\times(y, y)\ (\epsilon g = g\epsilon'). \tag{1.19}$$

When the above conditions are satisfied, the equivalence relation ${}^\times\!=^\times$ is compatible with composition, and the quotient-category $\mathcal{C}/{}^\times\!=^\times$ has no nontrivial invertible element.

The easy proof, which consists in showing that (i) and (ii) both are equivalent to (iii), is left to the reader.

For a category that satisfies the assumptions of Proposition 1.18, one can forget about nontrivial invertible elements at the expense of going to a convenient quotient. However, as the next example shows, this is not always possible.

Example 1.20 (collapsing invertible elements). Consider the wreathed free abelian monoid $\widetilde{\mathbb{N}^n}$ (Reference Structure 6, page 19). The hypotheses of Proposition 1.18 are not satisfied in $\widetilde{\mathbb{N}^n}$. Indeed, $(g, \pi) =^{\times} (g', \pi')$ is then simply $g = g'$. Now we have $\mathtt{s}_i\mathtt{a}_i = \mathtt{a}_{i+1}\mathtt{s}_i$, whence $\mathtt{s}_i\mathtt{a}_i\mathtt{s}_i = \mathtt{a}_{i+1}$. So $\mathtt{a}_i \;{}^{\times}{=}^{\times}\; \mathtt{a}_{i+1}$ holds, whereas $\mathtt{a}_i =^{\times} \mathtt{a}_{i+1}$ fails. Similarly, $\mathtt{s}_i =^{\times}$ id holds, whereas $\mathtt{s}_i\mathtt{a}_i =^{\times} \mathtt{a}_i$ fails.

Note that, in the above example, insisting on identifying ${}^{\times}{=}^{\times}$-equivalent elements would require to identify all elements $\mathtt{a}_i$ and to collapse all elements $\mathtt{s}_i$. The quotient-structure would be at most $\mathbb{Z}$, and we would have lost the whole structure. In such a case, there is no way to collapse invertible elements without collapsing everything and, therefore, it is (much) preferable to stay in the original category and to consider families that are compatible with invertible elements in some convenient sense. To this end, the following terminology will be useful.

Definition 1.21 (family $\mathcal{S}^\sharp$, closed under $=^{\times}$). For $\mathcal{S}$ a subfamily of a left-cancellative category $\mathcal{C}$, we put

$$\mathcal{S}^\sharp = \mathcal{S}\mathcal{C}^{\times} \cup \mathcal{C}^{\times}. \tag{1.22}$$

We say that $\mathcal{S}$ is *closed under* $=^{\times}$, or $=^{\times}$*-closed,* if the conjunction of $g \in \mathcal{S}$ and $g' =^{\times} g$ implies $g' \in \mathcal{S}$.

If $\mathcal{C}$ has no nontrivial invertible element, then the relation $=^{\times}$ is equality, so every subfamily $\mathcal{S}$ of $\mathcal{C}$ is $=^{\times}$-closed and $\mathcal{S}^\sharp$ is $\mathcal{S} \cup \mathbf{1}_{\mathcal{C}}$. In particular, if M is a monoid with no nontrivial element and S is included in M, then $S^\sharp$ is $S \cup \{1\}$. Hereafter, we always reserve the notation $\cdots^\sharp$ for the operation of Definition 1.21. So, for instance, we have $\mathcal{C}^{\times} = (\mathbf{1}_{\mathcal{C}})^\sharp$. Also, the definition easily implies $(\mathcal{S}^\sharp)^\sharp = \mathcal{S}^\sharp$.

Lemma 1.23. *Assume that $\mathcal{C}$ is a left-cancellative category.*

(i) *For every subfamily $\mathcal{S}$ of $\mathcal{C}$, the family $\mathcal{S}^\sharp$ is the smallest subfamily of $\mathcal{C}$ that is $=^{\times}$-closed and includes $\mathcal{S} \cup \mathbf{1}_{\mathcal{C}}$.*

(ii) *Every intersection of $=^{\times}$-closed subfamilies of $\mathcal{C}$ is $=^{\times}$-closed.*

Proof. By definition of $=^{\times}$, a family $\mathcal{S}$ is $=^{\times}$-closed if and only if $\mathcal{S}\mathcal{C}^{\times} \subseteq \mathcal{S}$ holds, hence if and only if $\mathcal{S}\mathcal{C}^{\times} = \mathcal{S}$ since $\mathcal{S}$ is always included in $\mathcal{S}\mathcal{C}^{\times}$. Then all verifications are straightforward. ☐

By definition, an $=^{\times}$-closed family $\mathcal{S}$ is closed under right-multiplication by an invertible element, that is, $\mathcal{S}\mathcal{C}^{\times} \subseteq \mathcal{S}$ holds, but it need not be closed under left-multiplication by an invertible element, that is, $\mathcal{C}^\times\mathcal{S} \subseteq \mathcal{S}$ need not hold. For further reference, we observe that several variants of the latter closure property are equivalent.

Lemma 1.24. *If $\mathcal{C}$ is a left-cancellative category, then, for every subfamily $\mathcal{S}$ of $\mathcal{C}$, the relations* (i) $\mathcal{C}^\times(\mathcal{S}\backslash\mathcal{C}^\times) \subseteq \mathcal{S}^\sharp$, (ii) $\mathcal{C}^\times\mathcal{S} \subseteq \mathcal{S}^\sharp$, *and* (iii) $\mathcal{C}^\times\mathcal{S}^\sharp\mathcal{C}^\times = \mathcal{S}^\sharp$ *are equivalent.*

Proof. As $\mathcal{S}$ is included in $\mathcal{S}^\sharp$ and $\mathbf{1}_\mathcal{C}$ is included in $\mathcal{C}^\times$, (iii) implies (ii), which implies (i). Conversely, assume (i), and let g belong to $\mathcal{C}^\times\mathcal{S}^\sharp\mathcal{C}^\times$, say $g = \epsilon g'\epsilon'$ with ϵ, ϵ' in $\mathcal{C}^\times$ and g' in $\mathcal{S}^\sharp$. If g' is invertible, then so is g, and it belongs to $\mathcal{S}^\sharp$. Otherwise, write $g' = g''\epsilon''$ with g'' in $\mathcal{S}\backslash\mathcal{C}^\times$ and ϵ'' in $\mathcal{C}^\times$. By (i), $\epsilon g''$ belongs to $\mathcal{S}^\sharp$, hence so does g, which is $(\epsilon g'')(\epsilon''\epsilon')$ with $\epsilon''\epsilon'$ invertible. So we have $\mathcal{C}^\times\mathcal{S}^\sharp\mathcal{C}^\times \subseteq \mathcal{S}^\sharp$, and (iii) follows since $\mathcal{S}^\sharp$ is trivially included in $\mathcal{C}^\times\mathcal{S}^\sharp\mathcal{C}^\times$. □

Note that (iii) in Lemma 1.24 means that $\mathcal{S}^\sharp$ is closed under ${}^\times\!\!=^\times$.

Example 1.25 (left-multiplication by invertible). In the monoid $\widetilde{\mathbb{N}^n}$, the family S_n (Reference Structure 6, page 19) satisfies the conditions of Lemma 1.24. Indeed, we first find $S_n^\sharp = S_n\mathfrak{S}_n = \{(g,\pi) \mid \forall i\ (g(i) \in \{0,1\})\}$, a family with $2^n n!$ elements. Then, every permutation leaves S_n globally invariant: permuting a sequence with $0,1$-entries yields a sequence with $0,1$-entries. Hence $\mathfrak{S}_n S_n = S_n\mathfrak{S}_n$ holds in $\widetilde{\mathbb{N}^n}$. We then deduce $\mathfrak{S}_n S_n^\sharp = \mathfrak{S}_n S_n \mathfrak{S}_n = S_n\mathfrak{S}_n = S_n^\sharp$, as expected.

If a category $\mathcal{C}$ contains no nontrivial invertible element, the equivalent conditions of Lemma 1.24 are trivially satisfied by every subfamily of $\mathcal{C}$. In every case, the family $\mathbf{1}_\mathcal{C}$ of all identity elements satisfies the conditions of Lemma 1.24, which in particular says that $\mathcal{C}^\times$, which is $\mathbf{1}_\mathcal{C}^\sharp$, is closed under ${}^\times\!\!=^\times$ (as is obvious directly).

We conclude with easy results about the powers of a family that satisfies the conditions of Lemma 1.24 and about the preservation of invertible elements under automorphisms.

Lemma 1.26. *If $\mathcal{C}$ is a left-cancellative category and $\mathcal{S}$ is a subfamily of $\mathcal{C}$ that satisfies $\mathcal{C}^\times\mathcal{S} \subseteq \mathcal{S}^\sharp$ and includes $\mathbf{1}_\mathcal{C}$, then the relations $(\mathcal{S}^\sharp)^m = \mathcal{S}^{m-1}\mathcal{S}^\sharp = (\mathcal{S}^m)^\sharp$ and $\mathcal{C}^\times\mathcal{S}^m \subseteq (\mathcal{S}^m)^\sharp$ hold for every $m \geqslant 1$.*

Proof. As $\mathcal{S}$ includes $\mathbf{1}_\mathcal{C}$, we have $\mathcal{S}^\sharp = \mathcal{S}\mathcal{C}^\times$. We first prove $(\mathcal{S}^\sharp)^m = \mathcal{S}^{m-1}\mathcal{S}^\sharp$. As $\mathcal{S}$ is included in $\mathcal{S}^\sharp$, we clearly have $\mathcal{S}^{m-1}\mathcal{S}^\sharp \subseteq (\mathcal{S}^\sharp)^m$. For the other inclusion, we use induction on $m \geqslant 1$. The result is obvious for $m = 1$. Assume $m \geqslant 2$. By Lemma 1.24, we have $\mathcal{C}^\times\mathcal{S}^\sharp \subseteq \mathcal{S}^\sharp$, and, using the induction hypothesis, we deduce

$$(\mathcal{S}^\sharp)^m = (\mathcal{S}\mathcal{C}^\times)\mathcal{S}^\sharp(\mathcal{S}^\sharp)^{m-2} = \mathcal{S}(\mathcal{C}^\times\mathcal{S}^\sharp)(\mathcal{S}^\sharp)^{m-2} \subseteq \mathcal{S}(\mathcal{S}^\sharp)^{m-1} \subseteq \mathcal{S}\mathcal{S}^{m-2}\mathcal{S}^\sharp = \mathcal{S}^{m-1}\mathcal{S}^\sharp.$$

Next, as $\mathbf{1}_\mathcal{C}$ is included both in $\mathcal{S}$ and in $\mathcal{S}^m$, we have $\mathcal{S}^{m-1}\mathcal{S}^\sharp = \mathcal{S}^m\mathcal{C}^\times = (\mathcal{S}^m)^\sharp$. Finally, , we obtain $\mathcal{C}^\times\mathcal{S}^m = (\mathcal{C}^\times\mathcal{S})\mathcal{S}^{m-1} \subseteq \mathcal{S}^\sharp\mathcal{S}^{m-1} \subseteq (\mathcal{S}^\sharp)^m = (\mathcal{S}^m)^\sharp$, whence $\mathcal{C}^\times\mathcal{S}^m \subseteq (\mathcal{S}^m)^\sharp$, using the assumption $\mathcal{C}^\times\mathcal{S} \subseteq \mathcal{S}^\sharp$ and the above equalities. □

Lemma 1.27. *If $\mathcal{C}$ is a left-cancellative category, then every automorphism ϕ of $\mathcal{C}$ induces an automorphism of $\mathcal{C}^\times$, and it preserves the relations $=^\times$ and ${}^\times\!\!=^\times$, that is, $g =^\times g'$ (resp. $g \;{}^\times\!\!=^\times g'$) is equivalent to $\phi(g) =^\times \phi(g')$ (resp. $\phi(g) \;{}^\times\!\!=^\times \phi(g')$).*

Proof. First $\epsilon\epsilon^{-1} = 1_x$ implies $\phi(\epsilon)\phi(\epsilon^{-1}) = 1_{\phi(x)}$, so ϕ maps $\mathcal{C}^\times$ into itself. The same holds for ϕ^{-1}, so ϕ induces a permutation of $\mathcal{C}^\times$.

Next $g' = g\epsilon$ with $\epsilon \in \mathcal{C}^\times$ implies $\phi(g') = \phi(g)\phi(\epsilon)$, whence $\phi(g') =^\times \phi(g)$. Conversely, $\phi(g') =^\times \phi(g)$ means $\phi(g') = \phi(g)\epsilon$ for some ϵ in $\mathcal{C}^\times$. As ϕ is bijective on $\mathcal{C}$, we deduce $g' = g\phi^{-1}(\epsilon)$, whence $g' =^\times g$ as ϕ^{-1} maps $\mathcal{C}^\times$ to itself.

The argument is similar for the relation ${}^\times{=}^\times$. □

1.4 Presentations

We now recall basic facts about free categories and presentations of categories. All results are elementary, but the terminology and notation are essential for the sequel.

We begin with free categories. We shall see now that, for every precategory $\mathcal{S}$, there exists a most general category generated by $\mathcal{S}$, hence a *free* category based on $\mathcal{S}$, and that this category can be constructed using $\mathcal{S}$-paths.

Definition 1.28 (path). Assume that $\mathcal{S}$ is a precategory. For $p \geqslant 1$ and x, y in $\mathcal{O}bj(\mathcal{S})$, an $\mathcal{S}$*-path* of length p with source x and target y is a finite sequence $(g_1, ..., g_p)$ of elements of $\mathcal{S}$ such that the source of g_1 is x, the target of g_p is y, and $g_1 \cdots g_p$ is defined. For every x in $\mathcal{O}bj(\mathcal{S})$, one introduces an *empty path* denoted by $()_x$ or ε_x, whose source and target are x, and whose length is declared to be zero. The family of all $\mathcal{S}$-paths of length p is denoted by $\mathcal{S}^{[p]}$; the family of all $\mathcal{S}$-paths is denoted by $\mathcal{S}^*$. The length of a path w is denoted by $\lg(w)$.

In the context of monoids, that is, if there is only one object, it is customary to say $\mathcal{S}$*-word*, or *word in the alphabet* $\mathcal{S}$, rather than $\mathcal{S}$-path. Throughout this text, we shall mostly use the letters w, v, u for words and, by extension, for paths.

There is an obvious way to make $\mathcal{S}^*$ into a category. A source and a target have already been attributed to every path in Definition 1.28. Then we define composition as a *concatenation* of paths (or words).

Notation 1.29 (concatenation). If $(f_1, ..., f_p)$, $(g_1, ..., g_q)$ are $\mathcal{S}$-paths and the target of $(f_1, ..., f_p)$, that is, the target of f_p, coincides with the source of $(g_1, ..., g_q)$, that is, the source of g_1, we put

$$(1.30)\qquad (f_1, ..., f_p)|(g_1, ..., g_q) = (f_1, ..., f_p, g_1, ..., g_q).$$

Moreover, whenever x is the source of g_1 and y is the target of g_q, we put

$$(1.31)\qquad ()_x|(g_1, ..., g_q) = (g_1, ..., g_q) = (g_1, ..., g_q)|()_y.$$

Finally, for each object x, we put $()_x|()_x = ()_x$.

Verifying that $\mathcal{S}^*$ equipped with the concatenation operation is a category is straightforward. Moreover, we can easily formalize the universality of path categories. To this end, we introduce the convenient notion of a (pre)-functor.

Definition 1.32 (prefunctor, functor). (i) If $\mathcal{S}, \mathcal{S}'$ are precategories, a *prefunctor* from $\mathcal{S}$ to $\mathcal{S}'$ is a pair (ϕ_0, ϕ) with $\phi_0 : \mathcal{O}bj(\mathcal{S}) \to \mathcal{O}bj(\mathcal{S}')$, $\phi : \mathcal{S} \to \mathcal{S}'$ such that $\phi(g)$ lies in $\mathcal{S}'(\phi_0(x), \phi_0(y))$ for every g in $\mathcal{S}(x, y)$.

(ii) If $\mathcal{C}, \mathcal{C}'$ are categories, a *functor* from $\mathcal{C}$ to $\mathcal{C}'$ is a prefunctor from $\mathcal{C}$ to $\mathcal{C}'$ satisfying $\phi(1_x) = 1_{\phi_0(x)}$ and $\phi(gh) = \phi(g)\phi(h)$ for all x, g, h.

Proposition 1.33 (free category). *For every precategory $\mathcal{S}$, the category $\mathcal{S}^*$ is freely generated by $\mathcal{S}$, that is, every prefunctor of $\mathcal{S}$ into a category $\mathcal{C}$ extends in a unique way to a functor from $\mathcal{S}^*$ to $\mathcal{C}$.*

Proof (sketch). If ϕ is a prefunctor from $\mathcal{S}$ to a category $\mathcal{C}$, the unique way to extend ϕ to a functor ϕ^* from $\mathcal{S}^*$ consists in putting

$$\phi^*((g_1, \dots, g_p)) = \phi(g_1) \cdots \phi(g_p), \qquad \phi^*(()_x) = 1_{\phi(x)}. \tag{1.34}$$

Verifications are then easy. □

Note that, in the context of Proposition 1.33, the image of a path (or a word) under ϕ^* is the *evaluation* of that path when each element g of $\mathcal{S}$ is given the value $\phi(g)$. By construction, ϕ^* maps $\mathcal{S}^{[p]}$ onto $\mathcal{S}^p$ for each nonnegative integer p.

Proposition 1.33 implies that every category $\mathcal{C}$ generated by a precategory $\mathcal{S}$ is isomorphic to a quotient of $\mathcal{S}^*$. Indeed, using ι for the identity prefunctor on $\mathcal{S}$, we obtain that ι^* is a morphism of $\mathcal{S}^*$ into $\mathcal{C}$. Standard arguments then show that the image of ι^*, which is $\mathcal{C}$ by assumption, is isomorphic to $\mathcal{S}^*/\!\equiv$, where $\equiv$ is the relation $\iota^*(w) = \iota^*(w')$.

Convention 1.35 (path concatenation). In the sequel, we shall mostly skip the concatenation sign $|$, thus writing uv instead of $u|v$ for the concatenation of two $\mathcal{S}$-paths u, v, that is, their product in the free category $\mathcal{S}^*$. However, we shall keep $|$ in the case when the entries of a path are explicitly mentioned. If $\mathcal{S}$ is a precategory, every $\mathcal{S}$-path has a well defined length, and the function of $\mathcal{S}$ into $\mathcal{S}^*$ that maps g to the length one path (g) is injective. Building on this, it is customary to identify g and (g). Then a generic path $(g_1, \dots, g_p)$ of $\mathcal{S}^*$ identifies with the concatenation $g_1|\cdots|g_p$. By keeping the sign $|$ in this case, we avoid the ambiguity between the path $g_1|\cdots|g_p$ and its possible evaluation $g_1 \cdots g_p$ in a category $\mathcal{C}$ generated by $\mathcal{S}$.

If a category $\mathcal{C}$ is a quotient of the free category $\mathcal{S}^*$, there exists a congruence $\equiv$ on $\mathcal{S}^*$ such that $\mathcal{C}$ is isomorphic to the quotient category $\mathcal{S}^*/\!\equiv$. We recall that, if $\mathcal{C}$ is a category, a *congruence* on $\mathcal{C}$ is an equivalence relation $\equiv$ on $\mathcal{C}$ that is compatible with composition, that is, the conjunction of $f \equiv f'$ and $g \equiv g'$ implies $fg \equiv f'g'$. Congruences are the appropriate relations for defining quotients.

In the above context, one can specify the category $\mathcal{C}$, up to isomorphism, by giving a generating precategory $\mathcal{S}$ and describing the corresponding congruence $\equiv$ on $\mathcal{S}^*$ by giving a collection of pairs that generates it, building on the following easy fact.

Definition 1.36 (relation). If $\mathcal{C}$ is a category, a *relation* on $\mathcal{C}$ is a pair (g, h) of elements of $\mathcal{C}$ sharing the same source and the same target.

Lemma 1.37. *If $\mathcal{C}$ is a category and $\mathcal{R}$ is a family of relations on $\mathcal{C}$, there exists a smallest congruence of $\mathcal{C}$ that includes $\mathcal{R}$, namely the reflexive–transitive closure $\equiv^+_{\mathcal{R}}$ of the family $\{(fgh, fg'h) \mid (g, g') \in \mathcal{R}$ or $(g', g) \in \mathcal{R}\}$.*

When the considered category $\mathcal{C}$ is a free category $\mathcal{S}^*$, relations on $\mathcal{C}$ are pairs of $\mathcal{S}$-paths, and Lemma 1.37 says is that two paths are $\equiv^+_{\mathcal{R}}$-equivalent if and only of one can go from the former to the latter by a finite sequence of steps, each of which consists in going

$$\text{from}\quad f_1|\cdots|f_p|g_1|\cdots|g_q|h_1|\cdots|h_m \quad\text{to}\quad f_1|\cdots|f_p|g'_1|\cdots|g'_{q'}|h_1|\cdots|h_m,$$

for some pair $(g_1|\cdots|g_q, g'_1|\cdots|g'_{q'})$ belonging to the symmetric closure of $\mathcal{R}$. Such a sequence is called an *$\mathcal{R}$-derivation*.

Definition 1.38 (presentation). (i) A *category presentation* is a pair $(\mathcal{S}, \mathcal{R})$ where $\mathcal{S}$ is a precategory and $\mathcal{R}$ is a family of relations on $\mathcal{S}^*$. The elements of $\mathcal{S}$ are called the *generators* of the presentation, whereas those of $\mathcal{R}$ are called the *relations*.

(ii) If $(\mathcal{S}, \mathcal{R})$ is a category presentation, the quotient category $\mathcal{S}^*/\equiv^+_{\mathcal{R}}$ is denoted by $\langle \mathcal{S} \mid \mathcal{R}\rangle^+$. Two paths w, w' of $\mathcal{S}^*$ that satisfy $w \equiv^+_{\mathcal{R}} w'$ are called *$\mathcal{R}$-equivalent*. The $\equiv^+_{\mathcal{R}}$-equivalence class of a path w is denoted by $[w]^+_{\mathcal{R}}$ or, simply, $[w]$.

Of course, in the case when $\mathcal{S}$ has only one object, we speak of a *monoid presentation*. In the context of Definition 1.38, one says that a category $\mathcal{C}$ *admits the presentation* $(\mathcal{S}, \mathcal{R})$—or $\langle \mathcal{S} \mid \mathcal{R}\rangle^+$—if $\mathcal{C}$ is isomorphic to $\mathcal{S}^*/\equiv^+_{\mathcal{R}}$, sometimes abusively written $\mathcal{C} = \langle \mathcal{S} \mid \mathcal{R}\rangle^+$. The $+$ sign in the notations $\equiv^+_{\mathcal{R}}$ and $\langle \mathcal{S} \mid \mathcal{R}\rangle^+$ refers to the fact that no inverses are involved for the moment: we consider categories and monoids, not yet groupoids and groups.

Convention 1.39 (relation). In the context of a category presentation $(\mathcal{S}, \mathcal{R})$, it is traditional to write a relation $(f_1|\cdots|f_p, g_1|\cdots|g_q)$, which is a pair of $\mathcal{S}$-paths, as $f_1 \cdots f_p = g_1 \cdots g_q$, because, in the corresponding quotient-structure $\mathcal{S}^*/\equiv^+_{\mathcal{R}}$, the evaluations $f_1 \cdots f_p$ and $g_1 \cdots g_q$ of the considered paths coincide. In this context, it is then coherent to use 1_x (1 in the case of a monoid) rather than ε_x.

Example 1.40 (presentation). For every precategory $\mathcal{S}$, the free category $\mathcal{S}^*$ admits the presentation $\langle \mathcal{S} \mid \emptyset\rangle^+$ and, for every set S, the free monoid S^* admits the presentation $\langle S \mid \emptyset\rangle^+$. On the other hand, the free group based on S, which admits, as a group, the presentation $\langle S \mid \emptyset\rangle$, admits the presentation $\langle S \cup S^{-1} \mid \{ss^{-1} = s^{-1}s = 1 \mid s \in S\}\rangle^+$ when viewed as a monoid.

We conclude with a trivial observation about invertible elements.

Definition 1.41 (ε-relation). A relation $u = v$ is called an *ε-relation* if one of the paths u, v is empty and the other is not.

Lemma 1.42. *If $(\mathcal{S}, \mathcal{R})$ is a category presentation containing no ε-relation, then the category $\langle \mathcal{S} \mid \mathcal{R}\rangle^+$ contains no nontrivial (right)-invertible element.*

Proof. Assume that u, v satisfy $u \not\equiv^+_{\mathcal{R}} \varepsilon_x$ (x the source of u) and $uv \equiv^+_{\mathcal{R}} \varepsilon_x$. By Lemma 1.37, there exist $n \geqslant 1$ and an $\mathcal{R}$-derivation $w_0 = uv, w_1, \ldots, w_n = \varepsilon_x$. Then $w_{n-1} = 1_x$ is a relation of $\mathcal{R}$, and it is an ε-relation. ☐

No simple converse of Lemma 1.42 can be stated. The examples $\langle \mathtt{a} \mid \mathtt{a} = 1\rangle^+$ (which contains no nontrivial invertible element), $\langle \mathtt{a}, \mathtt{b} \mid \mathtt{ab} = 1\rangle^+$ (which contains nontrivial invertible elements), and $\langle \mathtt{a} \mid \mathtt{a}^2 = \mathtt{a}^3 = 1\rangle^+$ (which is a trivial monoid although presentation contains no ε-relation of the form $s = 1$ with s a letter) show that one cannot say more.

2 Divisibility and Noetherianity

In every category—hence in particular in every monoid—there exist natural notions of left- and right-divisibility. As the statement of the decomposition results mentioned in Chapter I suggests, these relations and their properties play a key role in our subsequent approach. Here we fix some notation and establish basic results, in particular about least common multiples, greatest common divisors, and Noetherianity properties.

The section is organized as follows. The left- and right-divisibility relations and some derived notions are defined in Subsection 2.1. Next, we consider in Subsection 2.2 the derived notions of lcms (least common multiples) and gcds (greatest common divisors). From Subsection 2.3 on, we investigate some finiteness conditions called Noetherianity assumptions that guarantee the existence of certain minimal elements. In Subsection 2.4, we introduce height functions, which provide quantitative information about Noetherianity. Finally, in Subsection 2.5, we investigate atoms, which are the elements of height one, and the extent to which atoms must generate the ambient category.

2.1 Divisibility relations

Definition 2.1 (left-divisor, right-multiple). For f, g in a category $\mathcal{C}$, we say that f is a *left-divisor* of g, or, equivalently, that g is a *right-multiple* of f, written $f \preccurlyeq g$, if there exists g' in $\mathcal{C}$ satisfying $fg' = g$. For g in $\mathcal{C}$, we put

(2.2) $$\mathrm{Div}(g) = \{f \in \mathcal{C} \mid f \preccurlyeq g\}.$$

Note that, in the category context, f being a left-divisor of g implies that f and g share the same source.

Example 2.3 (left-divisor). The left-divisibility relation in the monoid $(\mathbb{N}, +)$ coincides with the standard ordering of natural numbers: $f \preccurlyeq g$ holds if there exists $g' \geqslant 0$ satisfying $g = f + g'$, hence if $f \leqslant g$ holds. More generally, in the free abelian monoid $\mathbb{N}^n$ (Reference Structure 1, page 3), $f \preccurlyeq g$ holds if and only if $f(i) \leqslant g(i)$ holds for every i: indeed, $fg' = g$ implies $\forall i\ (f(i) \leqslant f(i) + g'(i) = g(i))$, and $\forall i\ (f(i) \leqslant g(i))$ implies $fg' = g$ for g' determined by $g'(i) = g(i) - f(i)$.

In the case of the Klein bottle monoid K^+ (Reference Structure 5, page 17), the situation is quite special: as we have seen in Subsection I.3.2, the left-divisibility relation of K^+ is a linear ordering, that is, any two elements are comparable.

The basic result about the left-divisibility relation is that it provides a partial preordering, and even a partial ordering in the case of a category that is left-cancellative and contains no nontrivial invertible elements.

Lemma 2.4. (i) *Let $\mathcal{C}$ be a category. The left-divisibility relation of $\mathcal{C}$ is a preorder that is invariant under left-multiplication, that is, $f \preccurlyeq g$ implies $hf \preccurlyeq hg$ for each h whose target is the common source of f and g.*

(ii) *If, moreover, $\mathcal{C}$ is left-cancellative, the equivalence relation associated with the preorder $\preccurlyeq$ is the relation $=^{\times}$: the conjunction of $f \preccurlyeq g$ and $g \preccurlyeq f$ is equivalent to the existence of ϵ in $\mathcal{C}^{\times}$ satisfying $f = g\epsilon$.*

Proof. (i) If g has target y, we have $g = g \cdot 1_y$, hence $\preccurlyeq$ is reflexive. Next, if we have $fg' = g$ and $gh' = h$, we deduce $f(g'h') = (fg')h' = gh' = h$, hence $\preccurlyeq$ is transitive.

Assume now $f \preccurlyeq g$. By definition, we have $fg' = g$ for some g', and we deduce $(hf)g' = h(fg') = hg$, hence $hf \preccurlyeq hg$ whenever hf is defined.

(ii) Assume $fg' = g$ and $gf' = f$. We deduce $f(g'f') = f$ and $g(f'g') = g$. If $\mathcal{C}$ is left-cancellative, this implies $f'g' = 1_x$ and $g'f' = 1_y$, where x is the target of g and y is that of f. So, g' and f' are mutually inverse. □

Note that, in the case of a monoid M, the condition that hf be defined is automatically satisfied, yielding an unrestricted compatibility result: $f \preccurlyeq g$ always implies $hf \preccurlyeq hg$.

Instead of considering left-divisibility, which corresponds to multiplying on the right, we can symmetrically use multiplication on the left. We then obtain a new relation naturally called right-divisibility.

Definition 2.5 (right-divisor, left-multiple). For f, g in a category $\mathcal{C}$, we say that f is a *right-divisor* of g, or, equivalently, that g is a *left-multiple* of f, written $g \succcurlyeq\!\!\!\sim f$, if there exists g' in $\mathcal{C}$ satisfying $g = g'f$. For g in $\mathcal{C}$, we put

(2.6) $$\widetilde{\mathrm{Div}}(g) = \{f \in \mathcal{C} \mid g \succcurlyeq\!\!\!\sim f\}.$$

Example 2.7 (right-divisor). In the wreathed free abelian monoid (Reference Structure 6, page 19), $(g, \pi) \preccurlyeq (g', \pi')$ is equivalent to $\forall i \leqslant n\ (g'(i) \leqslant g(i))$. On the other hand, $(g', \pi') \succcurlyeq\!\!\!\sim (g, \pi)$ holds if and only if there exists a permutation π satisfying $\forall i \leqslant n\ (g'(i) \leqslant g(\pi(i)))$.

Of course, the left- and right-divisibility relations have entirely symmetric properties. However, in this text, left-divisibility will play a more significant role.

We conclude with an obvious observation about preservation of the divisibility relations under automorphisms.

Lemma 2.8. *If $\mathcal{C}$ is a category and ϕ is an automorphism of $\mathcal{C}$, then, for all f, g in $\mathcal{C}$, the relations $f \preccurlyeq g$ and $\phi(f) \preccurlyeq \phi(g)$ are equivalent, and so are $g \succcurlyeq\!\!\!\sim f$ and $\phi(g) \succcurlyeq\!\!\!\sim \phi(f)$.*

Proof. If ϕ is a functor, $f \preccurlyeq g$ implies $\phi(f) \preccurlyeq \phi(g)$. Indeed, $fg' = g$ implies $\phi(f)\phi(g') = \phi(g)$, whence $\phi(f) \preccurlyeq \phi(g)$.

Conversely, assume $\phi(f) \preccurlyeq \phi(g)$. Then we have $\phi(f)h = \phi(g)$ for some h. Whenever ϕ is bijective (or only surjective), this implies $f\phi^{-1}(h) = g$, whence $f \preccurlyeq g$.

The argument is similar for right-divisibility. □

2.2 Lcms and gcds

Once a (pre)-ordering has been introduced, it is natural to consider the associated possible lowest upper bounds and greatest lower bounds. In the case of the (left)-divisibility relation, a specific terminology is usual.

Definition 2.9 (right-lcm, left-gcd). (See Figure 2.) If $\mathcal{C}$ is a left-cancellative category and $\mathcal{X}$ is included in $\mathcal{C}$, an element h of $\mathcal{C}$ is called a *least common right-multiple*, or *right-lcm*, of $\mathcal{X}$, if h is a right-multiple of all elements of $\mathcal{X}$ and every element of $\mathcal{C}$ that is a right-multiple of all elements of $\mathcal{X}$ is a right-multiple of h.

Symmetrically, h is called a *greatest common left-divisor*, or *left-gcd*, of $\mathcal{X}$, if h is a left-divisor of all elements of $\mathcal{S}$ and every element of $\mathcal{C}$ that is a left-divisor of all elements of $\mathcal{X}$ is a left-divisor of h.

If $\mathcal{X}$ consists of two elements f, g, we say "right-lcm of f and g" for "right-lcm of $\{f, g\}$", and "left-gcd of f and g" for "left-gcd of $\{f, g\}$".

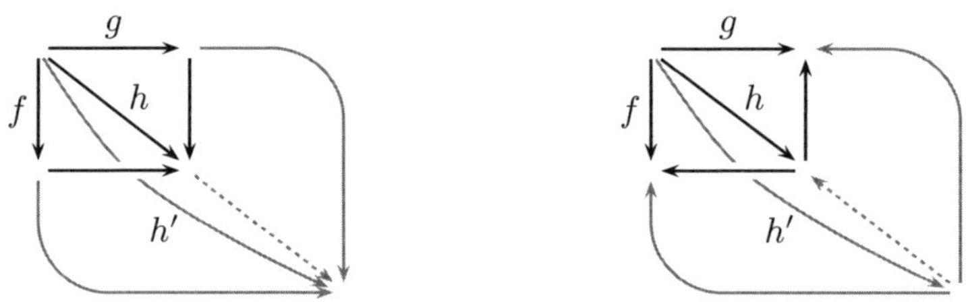

Figure 2. Right-lcm (left) and left-gcd (right) of f and g: a common right-multiple h of f and g that left-divides every common right-multiple h' of f and g (right-lcm), a common left-divisor h of f and g that is a right-multiple of every common left-divisor h' of f and g (left-gcd). Note that a right-lcm and a left-lcm may exist only for elements that share the same source.

The notions of *left-lcm* and *right-gcd* are defined symmetrically using right-divisibility.

As a direct application of Lemma 2.4 we obtain that right-lcms and left-gcds are close to be unique when they exist.

Proposition 2.10 (uniqueness of lcm). *If $\mathcal{C}$ is a left-cancellative category, then, for all f, g in $\mathcal{C}$, any two right-lcms (resp. left-gcds) of f and g are $=^{\times}$-equivalent.*

Proof. Assume that h, h' are right-lcms of f and g. The assumption that h is a right-lcm implies $h \preccurlyeq h'$, whereas the assumption that h' is a right-lcm implies $h' \preccurlyeq h$. By Lemma 2.4, we deduce $h =^{\times} h'$. The argument is similar for left-gcds. □

In the case of a category that is left-cancellative and admits no nontrivial invertible element, Proposition 2.10 is a genuine uniqueness result, and it is natural to introduce

Definition 2.11 (right-complement). If $\mathcal{C}$ is a left-cancellative category with no nontrivial invertible element, then, for f, g in $\mathcal{C}$, we write $\mathrm{lcm}(f, g)$ for the right-lcm of f and g

when it exists, and, in this case, the *right-complement* of f in g, denoted by $f\backslash g$, is the unique element g' satisfying $fg' = \mathrm{lcm}(f, g)$.

Because of their definition, the right-lcm and left-gcd operations share the properties of lowest upper bounds and greatest lower bounds in any poset: associativity, commutativity, idempotency. However, it should be kept in mind that, if there exist nontrivial invertible elements in the considered category, right-lcms and left-gcds are not necessarily unique when they exist. The following characterization of iterated lcms will be used several times in the sequel, and it is good exercise for getting used to lcm computations.

Proposition 2.12 (iterated lcm). (*See Figure* 3.) *If C is a left-cancellative category and f, f_1, f_2, g_1, g_2 are elements of C such that $g_1 f_1$ is a right-lcm of f and g_1, and $g_2 f_2$ is a right-lcm of f_1 and g_2, then $g_1 g_2 f_2$ is a right-lcm of f and $g_1 g_2$.*

Proof. Introduce the elements g'_1, g'_2 satisfying $fg'_1 = g_1 f_1$ and $f_1 g'_2 = g_2 f_2$. Then we have $g_1 g_2 f_2 = g_1 f_1 g'_2 = f g'_1 g'_2$, so $g_1 g_2 f_2$ is a common right-multiple of f and $g_1 g_2$.

Now assume $fg' = (g_1 g_2) f'$. Then we have $fg' = g_1(g_2 f')$. As fg'_1 is a right-lcm of f and g_1, we must have $fg'_1 \preccurlyeq fg'$, that is, there exists h_1 satisfying $fg' = fg'_1 h_1$, hence $g' = g'_1 h_1$ by left-cancelling f. Then we have $g_1 f_1 h_1 = f g'_1 h_1 = fg' = g_1 g_2 f'$, whence $f_1 h_1 = g_2 f'$ by left-cancelling g_1. As $f_1 g'_2$ is a right-lcm of f_1 and g_2, we must have $f_1 g'_2 \preccurlyeq f_1 h_1$, whence $g'_2 \preccurlyeq h_1$. We deduce $g_1 g_2 f_2 = f g'_1 g'_2 \preccurlyeq f g'_1 h_1 = fg'$. Hence $g_1 g_2 f_2$ is a right-lcm of f and $g_1 g_2$. □

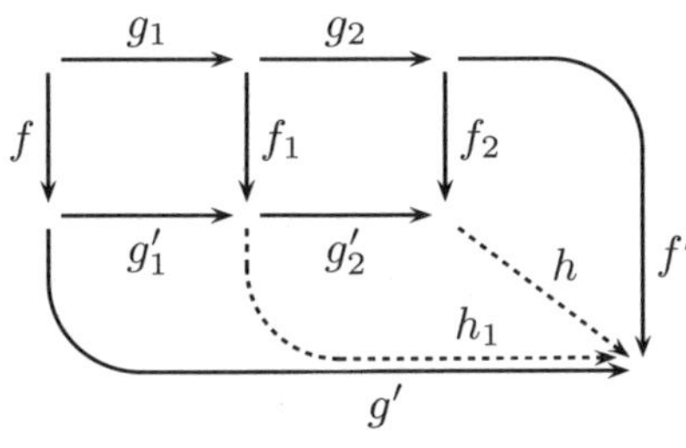

Figure 3. Iterated right-lcm

In the case of a left-cancellative category with no nontrivial invertible element, Proposition 2.12 is equivalent to the associativity of the right-lcm operation, see Exercise 7. Also, it gives simple rules for an iterated right-complement.

Corollary 2.13 (iterated complement). *If C is a left-cancellative category with no nontrivial invertible element, then, for f, g_1, g_2 in C such that f and $g_1 g_2$ admit a right-lcm, we have*

$$f\backslash(g_1 g_2) = (f\backslash g_1) \cdot (g_1\backslash f)\backslash g_2 \quad \text{and} \quad (g_1 g_2)\backslash f = g_2\backslash(g_1\backslash f). \tag{2.14}$$

Proof. Just a translation of Proposition 2.12: with the notation of its proof and of Figure 3, we have $g'_1 = f\backslash g_1$, $f_1 = g_1\backslash f$, $g'_2 = f_1\backslash g_2 = (g_1\backslash f)\backslash g_2$, and $f_2 = g_2\backslash f_1 = g_2\backslash(g_1\backslash f)$. Then Proposition 2.12 says that $g_1 g_2 f_2$ is the right-lcm of f and $g_1 g_2$, so f_2 is $(g_1 g_2)\backslash f$, and $g'_1 g'_2$ is $f\backslash(g_1 g_2)$. □

Applying the previous result in the case when g_2 is a right-complement of g_1 provides both a connection between the right-lcm and the right-complement operations and an algebraic law obeyed by the right-complement operation.

Proposition 2.15 (triple lcm). *If $\mathcal{C}$ is a left-cancellative category with no nontrivial invertible element, then, for f, g, h in $\mathcal{C}$ such that $\{f, g\}$, $\{g, h\}$, $\{f, h\}$, and $\{f, g, h\}$ admit a right-lcm, we have*

$$f\backslash \mathrm{lcm}(g, h) = \mathrm{lcm}(f\backslash g, f\backslash h), \tag{2.16}$$

$$\mathrm{lcm}(g, h)\backslash f = (g\backslash h)\backslash(g\backslash f) = (h\backslash g)\backslash(h\backslash f). \tag{2.17}$$

Proof. Applying Corollary 2.13 with $g_1 = g$ and $g_2 = g\backslash h$ gives

$$f\backslash \mathrm{lcm}(g, h) = f\backslash(g(g\backslash h)) = (f\backslash g) \cdot (g\backslash f)\backslash(g\backslash h), \tag{2.18}$$

$$\mathrm{lcm}(g, h)\backslash f = (g(g\backslash h))\backslash f = (g\backslash h)\backslash(g\backslash f). \tag{2.19}$$

Now $\mathrm{lcm}(g, h)$ is also $\mathrm{lcm}(h, g)$, so exchanging the roles of g and h in (2.19) gives $\mathrm{lcm}(g, h)\backslash f = (h\backslash g)\backslash(h\backslash f)$, thus establishing (2.17). Applying the latter equality to g, h, and f, we find $(g\backslash f)\backslash(g\backslash h) = (f\backslash g)\backslash(g\backslash h)$ and (2.18) implies $f\backslash \mathrm{lcm}(g, h) = (f\backslash g) \cdot (f\backslash g)\backslash(g\backslash h)$. By the definition of the operation $\backslash$, the latter element is $\mathrm{lcm}(f\backslash g, f\backslash h)$, so (2.16) follows. □

The second equality in (2.17) is often called the *right-cyclic law*. It will play a significant role in the study of set theoretic solutions of the Yang–Baxter equation in Chapter XIII and, in a different setting, in the cube condition of Section 4 below.

We continue with connections between the existence of left-gcds and that of right-lcms. It will be useful to fix the following generic terminology.

Definition 2.20 (admit, conditional). We say that a (left-cancellative) category $\mathcal{C}$ *admits common right-multiples* (*resp.* *admits right-lcms*) if any two elements of $\mathcal{C}$ that share the same source admit a common right-multiple (*resp.* a right-lcm). We say that $\mathcal{C}$ *admits conditional right-lcms* if any two elements of $\mathcal{C}$ that admit a common right-multiple admit a right-lcm.

We use a similar terminology with respect to left-gcds, and with respect to their symmetric counterparts involving the right-divisibility relation.

There exist several connections between the existence of lcms and gcds, especially when Noetherianity conditions are satisfied, see below. Here is a result in this direction.

Lemma 2.21. *If $\mathcal{C}$ is a left-cancellative category and every nonempty family of elements of $\mathcal{C}$ sharing the same source admits a left-gcd, then $\mathcal{C}$ admits conditional right-lcms.*

Proof. Assume that f, g are two elements of $\mathcal{C}$ that admit a common right-multiple. Let $\mathcal{S}$ be the family of all common right-multiples of f and g. By assumption, $\mathcal{S}$ is nonempty and it consists of elements whose source is the common source of f and g, hence it admits a left-gcd, say h. By definition, f left-divides every element of $\mathcal{S}$, hence it left-divides h. Similarly, g left-divides h. So h is a common right-multiple of f and g. Moreover, every common right-multiple of f and g lies in $\mathcal{S}$. Hence, by assumption, it is a right-multiple of h. Therefore, h is a right-lcm of f and g. □

Of course, the symmetric version is valid. See Lemma 2.21 for a partial converse, and Exercise 6 for still another connection in the same vein. Here is one more connection, which is slightly more surprising as it involves left- and right-divisibility simultaneously.

Lemma 2.22. *If $\mathcal{C}$ is a cancellative category that admits conditional right-lcms, then any two elements of $\mathcal{C}$ that admit a common left-multiple admit a right-gcd.*

Proof. Let f, g be two elements of $\mathcal{C}$ that admit a common left-multiple, say $f'g = g'f$, hence share the same target. The elements f' and g' admit a common right-multiple, hence they admit a right-lcm, say $f'g'' = g'f''$. By the definition of a right-lcm, there exists h satisfying $f = f''h$ and $g = g''h$. Then h is a common right-divisor of f and g.

Let h_1 be an arbitrary common right-divisor of f and g. There exist f_1, g_1 satisfying $f = f_1h_1$ and $h = g_1h_1$. Then we have $f'g_1h_1 = f'g = g'f = g'f_1h_1$, whence $f'g_1 = g'f_1$. So $f'g_1$ is a common right-multiple of f' and g', hence it is a right-multiple of their right-lcm, that is, there exists h' satisfying $f'g_1 = (f'g'')h'$, whence $g_1 = g''h'$ by left-cancelling f'. We deduce $g''h = g = g_1h_1 = g''h'h_1$, whence $h = h'h_1$ by left-cancelling g''. So h_1 right-divides h, which shows that h is a right-lcm of f and g. □

Remark 2.23. Left-cancellativity and existence of a right-lcm can be seen as two instances of a common condition, namely that, for each pair (s, t) admitting a common right-multiple, there exists a minimal common right-multiple equality $st_0 = ts_0$ through which every equality $st' = ts'$ factors. Applied in the case $s = t$, the condition can be satisfied only for $s_0 = t_0 = 1_y$ (where y is the target of s) and it says that every equality $st' = ss'$ factors through the equality $s1_y = s1_y$, meaning that there exists h satisfying $s' = 1_yh = t'$, indeed an expression of left-cancellativity.

2.3 Noetherianity conditions

A number of the categories and monoids we shall be interested in satisfy some finiteness conditions involving divisibility relations. Generically called Noetherianity conditions, these conditions ensure the existence of certain maximal or minimal elements.

Definition 2.24 (proper divisibility). For f, g in a category $\mathcal{C}$, we say that f is a *proper* left-divisor (*resp.* right-divisor) of g, written $f \prec g$ (*resp.* $f \mathrel{\widetilde{\prec}} g$ or $g \mathrel{\widetilde{\succ}} f$), if we have $fg' = g$ (*resp.* $g = g'f$) for some g' that is not invertible.

If the ambient category is left-cancellative, the relation $\prec$ is antireflexive ($f \prec f$ always fails) and, as the product of two elements can be invertible only if both are invertible, it is transitive. So $\prec$ is a strict partial ordering. Of course, symmetric results hold for proper right-divisibility. However, our default option is to consider left-cancellative categories that need not be right-cancellative, in which case $g \mathrel{\widetilde{\prec}} g$ is not impossible. Also note that, in a non-right-cancellative framework, the existence of a non-invertible element g' satisfying $g = g'f$ does not necessarily rule out the simultaneous existence of an invertible element g'' satisfying $g = g''f$, see Exercise 11.

Definition 2.25 (factor). For f, g in a category $\mathcal{C}$, we say that f is a *factor* of g, written $f \subseteq g$, if there exist g', g'' satisfying $g = g'fg''$; we say that f is a *proper factor* of g, written $f \subset g$, if, moreover, at least one of g', g'' is non-invertible. The family $\{f \in \mathcal{C} \mid f \subseteq g\}$ is denoted by $\mathrm{Fac}(g)$.

The factor relation $\subseteq$ is the smallest transitive relation that includes both the left- and the right-divisibility relations. Note that, even if the ambient category is cancellative, $\subset$ need not be antireflexive: in the Klein bottle monoid $\langle \mathtt{a}, \mathtt{b} \mid \mathtt{a} = \mathtt{bab}\rangle^+$ (Reference Structure 5, page 17), $\mathtt{a} \subset \mathtt{a}$ holds as we have $\mathtt{a} = \mathtt{bab}$ and $\mathtt{b}$ is not invertible.

If R is a binary relation on a family $\mathcal{S}$, a *descending sequence* with respect to R is a (finite or infinite) sequence $(g_0, g_1, ...)$ such that $g_{i+1}\, R\, g_i$ holds for every i. The relation R is called *well-founded* if every nonempty subfamily $\mathcal{X}$ of $\mathcal{S}$ has an R-minimal element, that is, an element g such that $f\, R\, g$ holds for no f in $\mathcal{X}$. By a classic result of Set Theory (and provided a very weak form of the Axiom of Choice called the axiom of Dependent Choices is true), a relation R is well-founded if and only if there exists no infinite descending sequence with respect to R.

Definition 2.26 (Noetherian). (i) A category $\mathcal{C}$ is called *left-Noetherian (resp. right-Noetherian, resp. Noetherian)* if the associated relation $\prec$ *(resp.* $\widetilde{\prec}$, *resp.* $\subset$) is well-founded.

(ii) A category presentation $(\mathcal{S}, \mathcal{R})$ is called *left-Noetherian (resp. right-Noetherian, resp. Noetherian)* if the category $\langle \mathcal{S} \mid \mathcal{R}\rangle^+$ is.

So, a category $\mathcal{C}$ is right-Noetherian if every nonempty subfamily $\mathcal{X}$ of $\mathcal{C}$ contains a $\widetilde{\prec}$-minimal element, that is, an element g such that $f \mathrel{\widetilde{\prec}} g$ holds for no f in $\mathcal{X}$. By the criterion recalled above, a category is left-Noetherian (*resp.* right-Noetherian, *resp.* Noetherian) if and only if there exists no infinite descending sequence in $\mathcal{C}$ with respect to proper left-divisibility (*resp.* proper right-divisibility, *resp.* proper factor relation).

Example 2.27 (Noetherian). Free abelian monoids (Reference Structure 1, page 3) are left-Noetherian, right-Noetherian, and Noetherian. Indeed, for f, g in $\mathbb{N}^n$, the three relations $f \prec g$, $f \mathrel{\widetilde{\prec}} g$, and $f \subset g$ are equivalent to $\forall i\ (f(i) < g(i))$. As there is no infinite descending sequence in $(\mathbb{N}, <)$, the existence of an infinite descending sequence for $\prec$, $\widetilde{\prec}$, or $\subset$ in $\mathbb{N}^n$ is impossible.

By contrast, the Klein bottle monoid K^+ (Reference Structure 5, page 17) is not left-Noetherian, nor is it either right-Noetherian or Noetherian. Indeed, for each i, we have $(\mathtt{a}^{i+1}\mathtt{b})\mathtt{a} = \mathtt{a}^i\mathtt{b}$, whence $\mathtt{a}^{i+1}\mathtt{b} \prec \mathtt{a}^i\mathtt{b}$, and $\mathtt{b}, \mathtt{ab}, \mathtt{a}^2\mathtt{b}, ...$ is an infinite descending sequence in K^+ with respect to proper left-divisibility, so K^+ is not left-Noetherian. Symmetrically, we have $\mathtt{a}(\mathtt{ba}^{i+1}) = \mathtt{ba}^i$, and $\mathtt{b}, \mathtt{ba}, \mathtt{ba}^2, ...$ is an infinite descending sequence in K^+ with respect to proper right-divisibility, so K^+ is not right-Noetherian. Finally, each of the above sequences is descending with respect to $\subset$, and K^+ is not Noetherian.

Before addressing the question of recognizing Noetherianity conditions, we begin with two general results. By definition, right-Noetherianity involves the right-divisibility relation. We observe that this property is equally connected with left-divisibility, at least in a left-cancellative context.

Proposition 2.28 (increasing sequences). *A left-cancellative category $\mathcal{C}$ is right-Noetherian if and only if, for every g in $\mathcal{C}$, every strictly increasing sequence in* $\mathrm{Div}(g)$ *with respect to left-divisibility is finite.*

Proof. Assume that $\mathcal{C}$ is not right-Noetherian. Let $g_0, g_1, \ldots$ be an infinite descending sequence with respect to proper right-divisibility in $\mathcal{C}$. For each i, choose a (necessarily non-invertible) element f_i satisfying $g_{i-1} = f_i g_i$. Then we have $g_0 = f_1 g_1 = (f_1 f_2) g_2 = \ldots$, and the sequence 1_x (x the source of g_0), $f_1, f_1 f_2, \ldots$ is $\prec$-increasing in $\mathrm{Div}(g_0)$.

Conversely, assume that g_0 lies in $\mathcal{C}$ and $h_1 \prec h_2 \prec \ldots$ is a strictly increasing sequence in $\mathrm{Div}(g_0)$. Then, for each i, there exists f_i non-invertible satisfying $h_i f_i = h_{i+1}$. On the other hand, as h_i belongs to $\mathrm{Div}(g_0)$, there exists g_i in $\mathcal{S}$ satisfying $h_i g_i = g_0$. We find $g_0 = h_i g_i = h_{i+1} g_{i+1} = h_i f_i g_{i+1}$. By left-cancelling h_i, we deduce $g_i = f_i g_{i+1}$, hence g_{i+1} is a proper right-divisor of g_i for each i. The sequence $g_0, g_1, \ldots$ witnesses that $\mathcal{C}$ is not right-Noetherian. □

In other words, a left-cancellative category $\mathcal{C}$ is right-Noetherian if and only if every bounded $\prec$-increasing sequence in $\mathcal{C}$ is finite.

In the proof of Proposition 2.28, the assumption that $\mathcal{C}$ is left-cancellative is used only for proving that right-Noetherianity implies the non-existence of bounded increasing sequences, the other direction being valid in any category. The assumption cannot be dropped, see Exercise 12.

The second general result connects the three Noetherianity conditions that have been introduced, in a way that is stronger than what could be a priori expected.

Proposition 2.29 (Noetherian). *A left-cancellative category is Noetherian if and only if it is both left- and right-Noetherian.*

Proof. Assume that $\mathcal{C}$ is a left-cancellative category. One direction is obvious. Indeed, both $\prec$ and $\mathrel{\tilde{\prec}}$ are included in $\subset$, so an infinite descending sequence with respect to $\prec$ or $\mathrel{\tilde{\prec}}$ is an infinite descending sequence with respect to $\subset$. So, if $\mathcal{C}$ is not left-Noetherian, it cannot be Noetherian, and, similarly, if $\mathcal{C}$ is not right-Noetherian, it cannot be Noetherian.

Conversely, assume that $\mathcal{C}$ is left-Noetherian, but not Noetherian. We shall see that $\mathcal{C}$ is not right-Noetherian. Let $f_0, f_1, \ldots$ be an infinite descending sequence with respect to $\subset$. For each i, write $f_i = g_i f_{i+1} h_i$, where at least one of g_i, h_i is non-invertible. Let $f'_i = g_0 \cdots g_{i-1} f_i$. Then we have $f'_i = f'_{i+1} h_i$ for each i, whence $f'_{i+1} \mathrel{\tilde{\preccurlyeq}} f'_i$. The assumption that $\prec$ is well-founded implies the existence of n such that $f'_{i+1} =^{\times} f'_i$ holds for $i \geqslant n$. So h_i is invertible for $i \geqslant n$, and the assumption that $g_i h_i$ is not invertible implies that g_i is not invertible. Let $f''_i = f_i h_{i-1} \cdots h_0$. Then we have $f''_i = g_i f''_{i+1}$ for every i, and $f''_n, f''_{n+1}, \ldots$ is an infinite descending sequence with respect to $\mathrel{\tilde{\prec}}$. Hence $\mathcal{C}$ is not right-Noetherian. □

We turn to the question of recognizing that a category is Noetherian. First, finiteness implies Noetherianity.

Proposition 2.30 (finite implies Noetherian). *Every left-cancellative category with finitely many $=^{\times}$-classes is Noetherian.*

Proof. Assume that $\mathcal{C}$ is left-cancellative and has finitely many $=^\times$-classes. As $\mathcal{C}$ is left-cancellative, $f \prec g$ implies $f \neq^\times g$: the element g' possibly satisfying $g = fg'$ is unique when it exists, and it is either invertible or non-invertible. Therefore, a $\prec$-decreasing sequence must consist of non-$=^\times$-equivalent elements, and the assumption implies that every such sequence is finite. So $\mathcal{C}$ is left-Noetherian.

For the same reason, a $\prec$-increasing sequence consists of non-$=^\times$-equivalent elements, and the assumption implies that every such sequence is finite. By Proposition 2.28, $\mathcal{C}$ must be right-Noetherian, and Proposition 2.29 then implies that $\mathcal{C}$ is Noetherian. □

In particular, every finite left-cancellative category must be Noetherian. Note that the left-cancellativity assumption cannot be dropped: the two-element monoid $\langle \mathtt{a} \mid \mathtt{a}^2 = \mathtt{a}\rangle^+$ is neither left- or right-Noetherian since $\mathtt{a}$ is a proper left- and right-divisor of itself.

The standard criterion for recognizing Noetherianity properties consists in establishing the existence of certain witnesses. Hereafter we denote by **Ord** the collection of all ordinals, see for instance [174]. If $(\mathcal{S}, \mathcal{R})$ is a category presentation, we say that a map λ^* defined on $\mathcal{S}^*$ is $\equiv^+_{\mathcal{R}}$*-invariant* if $w \equiv^+_{\mathcal{R}} w'$ implies $\lambda^*(w) = \lambda^*(w')$.

Definition 2.31 (witness). (i) A left-Noetherianity (*resp.* right-Noetherianity, *resp.* Noetherianity) *witness* for a category $\mathcal{C}$ is a map $\lambda : \mathcal{C} \to \mathbf{Ord}$ such that $f \prec g$ (*resp.* $f \,\widetilde{\succ}\, g$, *resp.* $f \subset g$) implies $\lambda(f) < \lambda(g)$; the witness is called *sharp* if moreover $f \preccurlyeq g$ (*resp.* $f \,\widetilde{\succcurlyeq}\, g$, *resp.* $f \subseteq g$) implies $\lambda(f) \leqslant \lambda(g)$.

(ii) A left-Noetherianity (*resp.* right-Noetherianity, *resp.* Noetherianity) *witness* for a category presentation $(\mathcal{S}, \mathcal{R})$ is an $\equiv^+_{\mathcal{R}}$-invariant map $\lambda^* : \mathcal{S}^* \to \mathbf{Ord}$ satisfying $\lambda^*(w) \leqslant \lambda^*(w|s)$ (*resp.* $\lambda^*(w) \leqslant \lambda^*(s|w)$, *resp.* both inequalities) for all s in $\mathcal{S}$ and w in $\mathcal{S}^*$ such that $s|w$ is a path, the inequality being strict whenever $[s]$ is not invertible in $\langle \mathcal{S} \mid \mathcal{R}\rangle^+$.

Rather than one single example, let us describe a family of examples.

Proposition 2.32 (homogeneous). *Say that a presentation $(\mathcal{S}, \mathcal{R})$ is* homogeneous *if all relations in $\mathcal{R}$ have the form $u = v$ with $\lg(u) = \lg(v)$. Then every homogeneous presentation admits an $\mathbb{N}$-valued Noetherianity witness.*

Proof. The assumption about $\mathcal{R}$ implies that any two $\mathcal{R}$-equivalent paths have the same length, that is, the map $\lg(\cdot)$ is $\equiv^+_{\mathcal{R}}$-invariant. Then $\lg(\cdot)$ provides the expected witness since, for every path w and every s in $\mathcal{S}$, we have $\lg(w) < \lg(w|s)$ and $\lg(w) < \lg(s|w)$ whenever the paths are defined. □

Proposition 2.33 (witness). (i) *A category $\mathcal{C}$ is left-Noetherian (resp. right-Noetherian, resp. Noetherian) if and only if there exists a left-Noetherianity (resp. right-Noetherianity, resp. Noetherianity) witness for $\mathcal{C}$, if and only if there exists a sharp left-Noetherianity (resp. right-Noetherianity, resp. Noetherianity) witness for $\mathcal{C}$.*

(ii) *A category presentation $(\mathcal{S}, \mathcal{R})$ is left-Noetherian (resp. right-Noetherian, resp. Noetherian) if and only if there exists a left-Noetherianity (resp. right-Noetherianity, resp. Noetherianity) witness for $(\mathcal{S}, \mathcal{R})$.*

Proof. (i) A standard result in the theory of ordered sets asserts that a binary relation R on a collection $\mathcal{S}$ has no infinite descending sequence if and only if there exists a map

$\lambda : \mathcal{S} \to \mathbf{Ord}$ such that $x \mathrel{R} y$ implies $\lambda(x) < \lambda(y)$. Applying this to the relations $\prec$, $\widetilde{\prec}$, and $\subset$, respectively, gives the first equivalence result.

In order to prove the second equivalence, it suffices to show that the existence of a witness implies the existence of a sharp one in each case. We consider the case of left-divisibility. Assume that λ is a left-Noetherianity witness for $\mathcal{C}$. For every g in $\mathcal{C}$, define

$$\lambda'(g) = \min\{\lambda(g') \mid g' =^{\times} g\}.$$

As every nonempty family of ordinals has a smallest element, $\lambda'(g)$ is well defined. We claim that λ' is a sharp left-Noetherianity witness for $\mathcal{C}$. Indeed, assume first $f \prec g$, say $fh = g$ with h not invertible. The well-foundedness of the ordinal order guarantees the existence of f' satisfying $f' =^{\times} f$ and $\lambda(f') = \lambda'(f)$ and, similarly, of g' satisfying $g' =^{\times} g$ and $\lambda(g') = \lambda'(g)$. Write $f' = f\epsilon$ and $g' = g\epsilon'$ with ϵ, ϵ' invertible. Then we have $f' \cdot \epsilon^{-1}h\epsilon' = g'$, and $\epsilon^{-1}h\epsilon'$ is not invertible (the fact that a product of invertible elements is invertible requires no cancellativity assumption). We deduce $\lambda'(f) = \lambda(f') < \lambda(g') = \lambda'(g)$, so λ' is a left-Noetherianity witness. Moreover, by definition, $g' =^{\times} g$ implies $\lambda'(g') = \lambda'(g)$. So λ' is sharp.

The arguments for $\widetilde{\prec}$ and $\subset$ are similar, replacing $=^{\times}$ with ${}^{\times}\!=$ and ${}^{\times}\!=^{\times}$, respectively.

(ii) Again we begin with left-Noetherianity. So assume that $\langle \mathcal{S} \mid \mathcal{R} \rangle^{+}$is left-Noetherian. By (i), there exists a sharp left-Noetherianity witness λ on $\mathcal{C}$. Define $\lambda^* : \mathcal{S}^* \to \mathbf{Ord}$ by $\lambda^*(w) = \lambda([w])$. Then λ^* is a left-Noetherianity witness for $(\mathcal{S}, \mathcal{R})$.

Conversely, assume that λ^* is a left-Noetherianity witness for $(\mathcal{S}, \mathcal{R})$. Let $\mathcal{C} = \langle \mathcal{S} \mid \mathcal{R} \rangle^{+}$. Then λ^* induces a well-defined function λ on $\mathcal{C}$. By assumption, $\lambda(f) \leqslant \lambda(fs)$ holds for every f in $\mathcal{C}$ and s in $\mathcal{S}$ such that fs is defined, and $\lambda(f) < \lambda(fs)$ if s is not invertible in $\mathcal{C}$. As $\mathcal{S}$ generates $\mathcal{C}$, this implies $\lambda(f) \leqslant \lambda(fg)$ whenever fg is defined, and $\lambda(f) < \lambda(fg)$ if g is not invertible. So λ is a (sharp) left-Noetherianity witness for $\mathcal{C}$.

The argument is similar for right-Noetherianity and for Noetherianity. □

A Noetherianity witness is to be seen as a weak measure of length for the elements of the considered category. As both relations $\prec$ and $\widetilde{\prec}$ are included in $\subset$, every Noetherianity witness is automatically a left-Noetherianity witness and a right-Noetherianity witness.

In most cases, using transfinite ordinals is not necessary, and it is enough to resort to $\mathbb{N}$-valued witnesses, as in the case of homogeneous presentations—we recall that natural numbers are special ordinals, namely those smaller than Cantor's ω. However, it is good to remember that this need not be always possible, even in a cancellative framework. We refer to Exercise 14 for a typical example, which is a sort of unfolded version of the non-cancellative monoid $\langle \mathtt{a}, \mathtt{b} \mid \mathtt{b} = \mathtt{ba} \rangle^{+}$ where cancellativity is restored by splitting the generator $\mathtt{b}$ into infinitely many copies.

We turn to consequences of one- or two-sided Noetherianity. By definition, Noetherianity assumptions guarantee the existence of elements that are minimal, or maximal, with respect to the divisibility or factor relations. The basic principle is as follows.

Proposition 2.34 (maximal element). *If $\mathcal{C}$ is a left-cancellative category that is right-Noetherian and $\mathcal{X}$ is a nonempty subfamily of $\mathcal{C}$ that is included in* $\mathrm{Div}(g)$ *for some g, then $\mathcal{X}$ admits a $\prec$-maximal element.*

Proof. Let f be an arbitrary element of $\mathcal{X}$, and let x be its source. Starting from $f_0 = f$, we construct a $\prec$-increasing sequence $f_0, f_1, \ldots$ in $\mathcal{X}$. As long as f_i is not $\prec$-maximal in $\mathcal{X}$, we can find f_{i+1} in $\mathcal{X}$ satisfying $f_i \prec f_{i+1} \preccurlyeq g$. By Proposition 2.28 (which is valid as $\mathcal{C}$ is left-cancellative), the assumption that $\mathcal{C}$ is right-Noetherian implies that the construction stops after a finite number m of steps. Then, by construction, f_m is a $\prec$-maximal element of $\mathcal{X}$. □

Under additional closure assumptions, the existence of a $\prec$-maximal element can be strengthened into the existence of a $\preccurlyeq$-greatest element.

Lemma 2.35. *If $\mathcal{C}$ is a left-cancellative category, $\mathcal{X}$ is a subfamily of $\mathcal{C}$ whose elements all share the same source, and any two elements of $\mathcal{X}$ admit a common right-multiple in $\mathcal{X}$, then every $\prec$-maximal element of $\mathcal{X}$ is a $\preccurlyeq$-greatest element of $\mathcal{X}$.*

Proof. Assume that h is $\prec$-maximal in $\mathcal{X}$. Let f belong to $\mathcal{X}$. By assumption, f and h share the same source, so they admit a common right-multiple that lies in $\mathcal{X}$, say $fg' = hf'$. The assumption that h is $\prec$-maximal in $\mathcal{X}$ implies that f' is invertible, and, therefore, f left-divides h. Hence, h is a $\preccurlyeq$-greatest element of $\mathcal{X}$. □

(See Lemma VI.1.27 for an improvement of Lemma 2.35.) Merging Proposition 2.34 and Lemma 2.35, we immediately deduce

Corollary 2.36 (greatest element). *If $\mathcal{C}$ is a left-cancellative category that is right-Noetherian and $\mathcal{X}$ is a nonempty subfamily of $\mathcal{C}$ that is included in* $\mathrm{Div}(g)$ *for some g, then $\mathcal{X}$ admits a $\preccurlyeq$-greatest element.*

As an application of the previous general principle, we obtain the following converse of Lemma 2.21.

Lemma 2.37. *If $\mathcal{C}$ is a left-cancellative category that is right-Noetherian and admits conditional right-lcms, then every nonempty family of elements of $\mathcal{C}$ sharing the same source has a left-gcd.*

Proof. Assume that $\mathcal{X}$ is a nonempty family of elements $\mathcal{C}$ that share the same source x. Let h be an element of $\mathcal{X}$, and let $\mathcal{Y}$ be the family of the elements of $\mathcal{C}$ that left-divide every element of $\mathcal{X}$. By definition, $\mathcal{Y}$ is included in $\mathrm{Div}(h)$, and it is nonempty since it contains 1_x. By Proposition 2.34, $\mathcal{Y}$ has a $\prec$-maximal element, say g. Now, let f be any element of $\mathcal{Y}$. Then h is a common right-multiple of f and g. As $\mathcal{C}$ admits conditional right-lcms, f and g admit a right-lcm, say g'. Let h' be any element of $\mathcal{X}$. As f and g belong to $\mathcal{Y}$, we have $f \preccurlyeq h'$ and $g \preccurlyeq h'$, whence $g' \preccurlyeq h'$. This shows that g' belongs to $\mathcal{Y}$. As, by definition, we have $g \preccurlyeq g'$ and g is $\prec$-maximal in $\mathcal{Y}$, we deduce $g' =^{\times} g$, hence $f \preccurlyeq g$. Hence every element of $\mathcal{Y}$ left-divides g, and g is a left-gcd of $\mathcal{X}$. □

Another application of Noetherianity is the existence, in the most general case, of minimal common multiples, which are a weak form of least common multiples.

Definition 2.38 (minimal common right-multiple). For f, g, h in a category $\mathcal{C}$, we say that h is a *minimal common right-multiple*, or *right-mcm*, of f and g if h is a right-multiple of f and g, and no proper left-divisor of h is a right-multiple of f and g. We say that $\mathcal{C}$ *admits right-mcms* if, for all f, g in $\mathcal{C}$, every common right-multiple of f and g is a right-multiple of some right-mcm of f and g.

Example 2.39 (right-mcm). Let M be the monoid $\langle \mathtt{a}, \mathtt{b} \mid \mathtt{ab} = \mathtt{ba}, \mathtt{a}^2 = \mathtt{b}^2 \rangle^+$. Then M is a cancellative monoid, and any two elements of M admit common left-multiples and common right-multiples. So M satisfies Ore's conditions (see Section 3), and its group of fractions is the direct product $\mathbb{Z} \times \mathbb{Z}_2$. In M, the elements $\mathtt{a}$ and $\mathtt{b}$ admit two right-mcms, namely $\mathtt{ab}$ and $\mathtt{a}^2$, but none is a right-lcm. It is easy to check that M admits right-mcms.

The connection between a right-mcm and a right-lcm is the same as the connection between a minimal element and a minimum: for h to be a right-mcm of f and g, we do not assume that every common multiple of f and g is a multiple of h, but we require that the latter be minimal in the family of common multiples of f and g. So a right-lcm always is a right-mcm and, conversely, if any two right-mcms of two elements f, g are $=^\times$-equivalent, then these right-mcms are right-lcms (see Exercise 5). In particular, every category that admits conditional right-lcms a fortiori admits right-mcms.

Proposition 2.40 (right-mcm). *Every left-cancellative category that is left-Noetherian admits right-mcms.*

Proof. Assume that $\mathcal{C}$ is left-cancellative and left-Noetherian. By assumption, the proper left-divisibility relation has no infinite descending sequence in $\mathcal{C}$, which implies that every nonempty subfamily of $\mathcal{C}$ has a $\prec$-minimal element. Assume that h is a common right-multiple of f and g. If $\mathrm{Mult}(g)$ denotes the family of all right-multiples of g, a $\prec$-minimal element in $\mathrm{Mult}(f) \cap \mathrm{Mult}(g) \cap \mathrm{Div}(h)$ is a common right-multiple of f and g left-dividing h of which no proper left-divisor is a right-multiple of f and g, hence a right-mcm of f and g left-dividing h. $\square$

2.4 Height

We continue our investigation of Noetherianity conditions, and refine the qualitative results of Subsection 2.3 with quantitative results that involve the so-called rank functions associated with the left- and right-divisibility relations and the factor relation. This leads to introducing the smallest Noetherianity witnesses, here called heights.

Our starting point is the following standard result from the theory of relations.

Lemma 2.41. *If R is a binary relation on a family $\mathcal{S}$ and every initial segment $\{f \in \mathcal{S} \mid f\,R\,g\}$ of R is a set, then there exists a unique* rank function *for $(\mathcal{S}, R)$, defined to be a (partial) function $\rho : \mathcal{S} \to \mathbf{Ord}$ such that $\rho(g)$ is defined if and only if the restriction of R to $\{f \in \mathcal{S} \mid f\,R\,g\}$ is well-founded, and ρ obeys the rule*

(2.42) $$\rho(g) = \sup\{\rho(f) + 1 \mid f\,R\,g\}.$$

Note that, if (2.42) is satisfied, then $f\,R\,g$ implies $\rho(f) < \rho(g)$. So, as there exists no infinite descending sequence of ordinals, $\rho(g)$ cannot be defined if $\{f \in \mathcal{S} \mid f\,R\,g\}$ is not well-founded. The converse implication, namely the existence of $\rho(g)$ whenever $\{f \in \mathcal{S} \mid f\,R\,g\}$ is well-founded, is established using ordinal induction. Applying the result in the context of divisibility leads to introducing three partial functions.

Definition 2.43 (height). For $\mathcal{C}$ a category, we define the *left height* ht_{L} (*resp.* the *right height* ht_{R}, *resp.* the *height* ht) of $\mathcal{C}$ to be the rank function associated with proper left-divisibility (*resp.* proper right-divisibility, *resp.* proper factor relation) on $\mathcal{C}$.

By definition, the initial segment determined by an element g with respect to $\prec$ is $\mathrm{Div}(g) \setminus \{g' \mid g' =^{\times} g\}$. The restriction of $\prec$ to $\mathrm{Div}(g) \setminus \{g' \mid g' =^{\times} g\}$ is well-founded if and only if the restriction of $\prec$ to $\mathrm{Div}(g)$ is well-founded, and, therefore, $\mathrm{ht}_{\mathrm{L}}(g)$ is defined if and only if the restriction of $\prec$ to $\mathrm{Div}(g)$ is well-founded. Similarly, $\mathrm{ht}_{\mathrm{R}}(g)$ is defined if and only if the restriction of $\widetilde{\prec}$ to $\widetilde{\mathrm{Div}}(g)$ is well-founded, and $\mathrm{ht}(g)$ is defined if and only if the restriction of $\subset$ to $\mathrm{Fac}(g)$ is well-founded. See Exercise 14 for an example of a monoid where an element has a right height, but no left height and no height.

We begin with general properties of left height.

Lemma 2.44. *Assume that $\mathcal{C}$ is a left-cancellative category.*

(i) *The left height is $=^{\times}$-invariant: if $g' =^{\times} g$ holds and $\mathrm{ht}_{\mathrm{L}}(g)$ is defined, then $\mathrm{ht}_{\mathrm{L}}(g')$ is defined as well and $\mathrm{ht}_{\mathrm{L}}(g') = \mathrm{ht}_{\mathrm{L}}(g)$ holds.*

(ii) *An element g is invertible if and only if $\mathrm{ht}_{\mathrm{L}}(g)$ is defined and equal to 0.*

(iii) *If $\mathrm{ht}_{\mathrm{L}}(fg)$ is defined, so are $\mathrm{ht}_{\mathrm{L}}(f)$ and $\mathrm{ht}_{\mathrm{L}}(g)$ and we have*

$$\mathrm{ht}_{\mathrm{L}}(fg) \geqslant \mathrm{ht}_{\mathrm{L}}(f) + \mathrm{ht}_{\mathrm{L}}(g). \tag{2.45}$$

(iv) *The left height is invariant under every automorphism of $\mathcal{C}$.*

Proof. (i) The relation $g' =^{\times} g$ implies $\mathrm{Div}(g') = \mathrm{Div}(g)$.

(ii) Assume that ϵ is invertible in $\mathcal{C}$. Then the family $\{f \in \mathcal{C} \mid f \prec \epsilon\}$ is empty, and the inductive definition of (2.42) gives $\mathrm{ht}_{\mathrm{L}}(\epsilon) = 0$.

Conversely, assume $\mathrm{ht}_{\mathrm{L}}(g) = 0$, and let x be the source of g. The assumption implies that no element f of $\mathcal{C}$ satisfies $f \prec g$. In particular, $1_x \prec g$ must be false. As $g = 1_x g$ holds, this means that g must be invertible.

(iii) If $f_0, f_1, \cdots$ is an infinite descending sequence in $\mathrm{Div}(f)$ witnessing that $\prec$ is not well-founded, it is also a sequence in $\mathrm{Div}(fg)$. So, if $\mathrm{ht}_{\mathrm{R}}(f)$ is not defined, then $\mathrm{ht}_{\mathrm{R}}(fg)$ is not defined as well. On the other hand, if $g_0, g_1, \ldots$ is an infinite descending sequence in $\mathrm{Div}(g)$ witnessing that $\prec$ is not well-founded, then $fg_0, fg_1, \ldots$ is a sequence in $\mathrm{Div}(fg)$ witnessing that $\prec$ is not well-founded. So, if $\mathrm{ht}_{\mathrm{R}}(g)$ is not defined, then $\mathrm{ht}_{\mathrm{R}}(fg)$ is not defined as well.

We prove (2.45) using induction on $\mathrm{ht}_{\mathrm{L}}(g)$. If g is invertible, then we have $fg =^{\times} f$, whence $\mathrm{ht}_{\mathrm{L}}(fg) = \mathrm{ht}_{\mathrm{L}}(f)$, and $\mathrm{ht}_{\mathrm{L}}(g) = 0$ by (ii). Otherwise, assume $\alpha < \mathrm{ht}_{\mathrm{L}}(g)$. By (2.42), there exists $h \prec g$ satisfying $\mathrm{ht}_{\mathrm{L}}(h) \geqslant \alpha$. The existence of $\mathrm{ht}_{\mathrm{L}}(fg)$ implies that of $\mathrm{ht}_{\mathrm{L}}(fh)$, and the induction hypothesis implies $\mathrm{ht}_{\mathrm{L}}(fh) \geqslant \mathrm{ht}_{\mathrm{L}}(f) + \alpha$. As $fh \prec fg$ holds, we deduce $\mathrm{ht}_{\mathrm{L}}(fg) \geqslant \mathrm{ht}_{\mathrm{L}}(f) + \alpha + 1$. This being true for every $\alpha < \mathrm{ht}_{\mathrm{L}}(g)$, we obtain (2.45).

(iv) Assume that ϕ is an automorphism of $\mathcal{C}$. We prove that $\mathrm{ht}_{\mathrm{L}}(\phi(g)) = \mathrm{ht}_{\mathrm{L}}(g)$ holds for $\mathrm{ht}_{\mathrm{L}}(g) \leqslant \alpha$ using induction on α. Assume $\alpha = 0$. Then, by (ii), g must be invertible, hence so is $\phi(g)$, which implies $\mathrm{ht}_{\mathrm{L}}(\phi(g)) = 0 = \mathrm{ht}_{\mathrm{L}}(g)$. Assume now $\alpha > 0$. Then, by definition, we obtain

$$\mathrm{ht}_{\mathrm{L}}(\phi(g)) = \sup\{\mathrm{ht}_{\mathrm{L}}(h) + 1 \mid h \preccurlyeq \phi(g)\} = \sup\{\mathrm{ht}_{\mathrm{L}}(h) + 1 \mid \phi^{-1}(h) \preccurlyeq g\}.$$

As $\phi^{-1}(h) \preccurlyeq g$ implies $\mathrm{ht}_{\mathrm{L}}(\phi^{-1}(h)) < \mathrm{ht}_{\mathrm{L}}(g) \leqslant \alpha$, the induction hypothesis implies $\mathrm{ht}_{\mathrm{L}}(\phi^{-1}(h)) = \mathrm{ht}_{\mathrm{L}}(h)$ for h in the above family, whence

$$\mathrm{ht}_{\mathrm{L}}(\phi(g)) = \sup\{\mathrm{ht}_{\mathrm{L}}(\phi^{-1}(h)) + 1 \mid \phi^{-1}(h) \preccurlyeq g\} = \sup\{\mathrm{ht}_{\mathrm{L}}(f) + 1 \mid f \preccurlyeq g\} = \mathrm{ht}_{\mathrm{L}}(g),$$

where the second equality holds because ϕ is surjective. ☐

Similar results hold for right height and height. In Lemma 2.44(i), $=^{\times}$ has to be replaced with ${}^{\times}\!=$ in the case of the right height, and with ${}^{\times}\!=^{\times}$ in the case of the height. For Lemma 2.44(ii), the results are the same. As for Lemma 2.44(iii), the result is similar for the right height, with the difference that (2.45) is to be replaced with

(2.46) $$\mathrm{ht}_{\mathrm{R}}(fg) \geqslant \mathrm{ht}_{\mathrm{R}}(g) + \mathrm{ht}_{\mathrm{R}}(f).$$

A category is called *small* if the families of elements and objects are sets (and not proper classes), a very mild condition that is always satisfied in all examples considered in this text. If $\mathcal{C}$ is a small category, then, for every g in $\mathcal{C}$, the families $\mathrm{Div}(g)$, $\widetilde{\mathrm{Div}}(g)$, and $\mathrm{Fac}(g)$ are sets.

Proposition 2.47 (height). *If $\mathcal{C}$ is a small left-cancellative category, then $\mathcal{C}$ is left-Noetherian* (*resp. right-Noetherian, resp. Noetherian*) *if and only if the left height* (*resp. the right height, resp. the height*) *is defined everywhere on $\mathcal{C}$. In this case, the latter is a sharp left-Noetherianity* (*resp. right-Noetherianity, resp. Noetherianity*) *witness for $\mathcal{C}$, and it is the smallest witness: if λ is any witness of the considered type for $\mathcal{C}$, then $\lambda(g) \geqslant \mathrm{ht}_{\mathrm{L}}(g)$* (*resp.* $\mathrm{ht}_{\mathrm{R}}(g)$, *resp.* $\mathrm{ht}(g)$) *holds for every g in $\mathcal{C}$.*

Proof. We consider the case of left-Noetherianity, the other cases are similar. Assume that $\mathcal{C}$ is left-Noetherian. Then the relation $\prec$ is well-founded on $\mathcal{C}$ and, therefore, for every g in $\mathcal{C}$, the restriction of $\prec$ to $\mathrm{Div}(g)$ is well-founded as well. Hence, by Lemma 2.41, $\mathrm{ht}_{\mathrm{L}}(g)$ is defined. Moreover, again by Lemma 2.41, $f \prec g$ implies $\mathrm{ht}_{\mathrm{L}}(f) < \mathrm{ht}_{\mathrm{L}}(g)$, and, by Lemma 2.44(i), $g =^{\times} g'$ implies $\mathrm{ht}_{\mathrm{L}}(g) = \mathrm{ht}_{\mathrm{L}}(g')$. So ht_{L} is a sharp Noetherianity witness for $\mathcal{C}$.

Conversely, assume that $\mathcal{C}$ is not left-Noetherian. Then there exists an infinite descending sequence $g_0 \succ g_1 \succ \cdots$. Then $\mathrm{ht}_{\mathrm{L}}(g_i)$ cannot be defined for every i. Indeed, by Lemma 2.41, $g \succ f$ implies $\mathrm{ht}_{\mathrm{L}}(g) > \mathrm{ht}_{\mathrm{L}}(f)$, and, therefore, $\mathrm{ht}_{\mathrm{L}}(g_0)$, $\mathrm{ht}_{\mathrm{L}}(g_1)$, $\cdots$ would be an infinite descending sequence of ordinals, which is impossible.

Finally, assume that λ is an arbitrary left-Noetherianity witness for $\mathcal{C}$. We prove using induction on the ordinal α that $\mathrm{ht}_{\mathrm{L}}(g) \geqslant \alpha$ implies $\lambda(g) \geqslant \alpha$. So assume $\mathrm{ht}_{\mathrm{L}}(g) \geqslant \alpha$. If α is zero, then $\lambda(g) \geqslant \alpha$ certainly holds. Otherwise, let β be an ordinal smaller than α. By (2.42), there exists f satisfying $f \prec g$ and $\mathrm{ht}_{\mathrm{L}}(f) \geqslant \beta$. By induction hypothesis, we have $\lambda(f) \geqslant \beta$, whence $\lambda(g) \geqslant \beta + 1$. This being true for every $\beta < \alpha$, we deduce $\lambda(g) \geqslant \alpha$. ☐

Owing to Proposition 2.29, it follows from Proposition 2.47 that, if $\mathcal{C}$ is a small left-cancellative category, then the height function is defined everywhere on $\mathcal{C}$ if and only if the left- and right height functions are defined everywhere. Inspecting the proof of Proposition 2.29 gives a local version, namely that, if $\mathcal{C}$ is a left-cancellative category and g is any element of $\mathcal{C}$, then $\mathrm{ht}(g)$ is defined if and only if $\mathrm{ht}_{\mathrm{L}}(g)$ and $\mathrm{ht}_{\mathrm{R}}(g)$ are defined.

We consider now the elements with a finite height. By Lemma 2.44(ii) and its right counterpart, having left height zero is equivalent to having right height zero, and to having height zero. A similar result holds for all finite values.

Proposition 2.48 (finite height). *For every left-cancellative category $\mathcal{C}$ and for all g in $\mathcal{C}$ and n in $\mathbb{N}$, the following conditions are equivalent:*

(i) *Every decomposition of g contains at most n non-invertible entries.*

(ii) *The integer $\mathrm{ht}_{\mathrm{L}}(g)$ is defined and is at most n.*

(iii) *The integer $\mathrm{ht}_{\mathrm{R}}(g)$ is defined and is at most n.*

(iv) *The integer $\mathrm{ht}(g)$ is defined and is at most n.*

Proof. We prove the equivalence of (i) and (ii) using induction on n. For $n = 0$, the equivalence is Lemma 2.44(ii). Assume first that g satisfies (i). Let f be any proper left-divisor of g, say $g = fg'$ with g' non-invertible. If f had a decomposition containing more than $n - 1$ non-invertible elements, g would have a decomposition into more than n non-invertible elements, contradicting the assumption. By the induction hypothesis, we have $\mathrm{ht}_{\mathrm{L}}(f) \leqslant n - 1$. By (2.42), we deduce $\mathrm{ht}_{\mathrm{L}}(g) \leqslant n$. So (i) implies (ii).

Conversely, assume $\mathrm{ht}_{\mathrm{L}}(g) \leqslant n$. Let $g_1|\cdots|g_\ell$ be a decomposition of g. If all elements g_i are invertible, then g is invertible, and (i) holds. Otherwise, let i be the largest index such that g_i is non-invertible, and let $f = g_1 \cdots g_{i-1}$ if $i \geqslant 2$ holds, and $f = 1_x$, x the source of g, otherwise. We have $f \prec g$ whence $\mathrm{ht}_{\mathrm{L}}(f) \leqslant n - 1$ by (2.42) and, by the induction hypothesis, every decomposition of f contains at most $n - 1$ non-invertible entries, so, in particular, there are at most $n - 1$ non-invertible entries among $g_1, \ldots, g_{i-1}$. Hence, there are at most n non-invertible entries among $g_1, \ldots, g_\ell$. So (ii) implies (i).

The equivalence of (i) and (iii), and of (i) and (iv), is established in the same way. □

We deduce the equivalence of several strong forms of Noetherianity.

Corollary 2.49 (finite height). *For every left-cancellative category $\mathcal{C}$, the following conditions are equivalent:*

(i) *For every g in $\mathcal{C}$, there exists a number n such that every decomposition of g contains at most n non-invertible entries,*

(ii) *Every element of $\mathcal{C}$ has a finite left height.*

(iii) *Every element of $\mathcal{C}$ has a finite right height.*

(iv) *Every element of $\mathcal{C}$ has a finite height.*

(v) *The category $\mathcal{C}$ admits an $\mathbb{N}$-valued left-Noetherianity witness.*

(vi) *The category $\mathcal{C}$ admits an $\mathbb{N}$-valued right-Noetherianity witness.*

(vii) *The category $\mathcal{C}$ admits an $\mathbb{N}$-valued Noetherianity witness.*

(viii) *There exists $\lambda : \mathcal{C} \to \mathbb{N}$ satisfying*

$$\lambda(fg) \geqslant \lambda(f) + \lambda(g) \quad \text{and} \quad f \notin \mathcal{C}^{\times} \text{ implies } \lambda(f) \geqslant 1. \tag{2.50}$$

All these conditions are satisfied whenever $\mathcal{C}$ admits a homogeneous presentation.

Proof. Proposition 2.48 directly gives the equivalence of (i)–(iv). Then, owing to Proposition 2.47, (ii) is equivalent to (v) since the left height is a minimal left-Noetherianity witness. Similarly, (iii) is equivalent to (vi), and (iv) is equivalent to (vii). Moreover, (v) and (viii) are equivalent by the $\subset$-counterpart of Lemma 2.44(iii). Finally, as observed in

Proposition 2.32, the existence of a homogeneous presentation guarantees that the length of paths provides an $\mathbb{N}$-valued left-Noetherianity witness. □

Definition 2.51 (strongly Noetherian). A left-cancellative category is called *strongly Noetherian* if it satisfies the equivalent conditions (i)–(viii) of Corollary 2.49. A category presentation $(\mathcal{S}, \mathcal{R})$ defining a left-cancellative category is called *strongly Noetherian* if the category $\langle \mathcal{S} \mid \mathcal{R} \rangle^+$ is strongly Noetherian.

Thus, in particular, every (left-cancellative) category that admits a homogeneous presentation, that is, a presentation in which both terms of each relation have the same length, is strongly Noetherian.

2.5 Atoms

The previous results lead to considering the elements of height one.

Definition 2.52 (atom). An element g of a left-cancellative category $\mathcal{C}$ is called an *atom* if g is not invertible and every decomposition of g in $\mathcal{C}$ contains at most one non-invertible element.

In other words, g is an atom if g is not invertible but there is no equality $g = g_1 g_2$ with g_1, g_2 both not invertible.

Example 2.53 (atom). If $(\mathcal{S}, \mathcal{R})$ is a homogeneous category presentation, then the category $\langle \mathcal{S} \mid \mathcal{R} \rangle^+$ has no nontrivial invertible element, and all elements of $\mathcal{S}$ are atoms. Indeed, by assumption, the various families $\mathcal{S}^p$ are pairwise disjoint. In particular, no element of $\mathcal{S}$ may belong to $\mathcal{S}^p$ with $p \geqslant 2$.

Proposition 2.48 directly implies:

Corollary 2.54 (atom). *An element g of a left-cancellative category $\mathcal{C}$ is an atom if and only if g has left height* 1, *if and only if g has right height* 1, *if and only if g has height* 1.

Owing to the characterization of atoms as elements of height one, the $\subset$-counterpart of Lemma 2.44(i) implies that the family of atoms of a left-cancellative category is closed under ${}^{\times}\!\!=^{\times}$-equivalence. On the other hand, it follows from the inductive definition (2.42) that the values of a rank function always make an initial segment in the family **Ord** (there cannot be any jump). So, by Proposition 2.47, every category that is left- or right-Noetherian and contains at least one non-invertible element must contain an atom. Actually, we can state a more precise result.

Proposition 2.55 (atom). *If $\mathcal{C}$ is a left-cancellative category that is left-Noetherian* (*resp. right-Noetherian*), *then every non-invertible element of $\mathcal{C}$ is left-divisible* (*resp. right-divisible*) *by an atom.*

Proof. Assume that g is a non-invertible element of $\mathcal{C}$. If g is not an atom, there exist non-invertible elements g_1, h_1 satisfying $g = g_1h_1$. If g_1 is not an atom, we similarly find non-invertible elements g_2, h_2 satisfying $g_1 = g_2h_2$. As long as g_i is not an atom, we can repeat the process. Now, by construction, $g, g_1, g_2, \ldots$ is a descending sequence with respect to proper left-divisibility. The assumption that $\mathcal{C}$ is left-Noetherian implies that the process cannot repeat forever, that is, that there exists i such that g_i is an atom. $\square$

It is natural to wonder whether atoms plus invertible elements generate every left- or right-Noetherian category. The following example shows that this need not be the case.

Example 2.56 (atoms not generating). Let M be the monoid with the presentation

$$\langle \mathtt{a}_1, \mathtt{a}_2, \ldots, \mathtt{b}_1, \mathtt{b}_2, \cdots \mid \mathtt{b}_i = \mathtt{b}_{i+1}\mathtt{a}_i \text{ for each } i\rangle^+.$$

Using the toolbox of Section 4.1, it is easy to check that the monoid M is left-cancellative. On the other hand, let $S = \{\mathtt{a}_1, \mathtt{a}_2, \ldots, \mathtt{b}_1, \mathtt{b}_2, \cdots\}$, and let λ be the mapping of S^* to ordinals defined by $\lambda(s_1|\cdots|s_\ell) = \lambda(s_\ell) + \cdots + \lambda(s_1)$, with $\lambda(\mathtt{a}_1) = \lambda(\mathtt{a}_2) = \cdots = 1$ and $\lambda(\mathtt{b}_1) = \lambda(\mathtt{b}_2) = \cdots = \omega$. As we have $\lambda(\mathtt{b}_i) = \omega = 1 + \omega = \lambda(\mathtt{b}_{i+1}\mathtt{a}_i)$ for each i, the map λ induces a well defined map on M. Hence, by Proposition 2.33, M is right-Noetherian. Now the atoms of M are the elements $\mathtt{a}_i$, which do not generate M as $\lambda(g)$ is finite for each g in the submonoid generated by the $\mathtt{a}_i$'s, whereas $\lambda(\mathtt{b}_1)$ is infinite.

The monoid of Example 2.56 is not finitely generated. On the other hand, if we try to modify the presentation by identifying all elements $\mathtt{b}_i$, then we obtain relations $\mathtt{b} = \mathtt{b}\mathtt{a}_i$ that contradict left-cancellativity. We refer to Exercise 16 for an example of a finitely presented monoid that is cancellative, right-Noetherian, but not generated by its atoms and its invertible elements (this monoid however contains nontrivial invertible elements, which leaves the question of finding a similar example with no nontrivial invertible element open).

Such situations cannot happen in the case of Noetherianity, that is, when both left- and right-Noetherianity are satisfied. We begin with a characterization of generating families.

Lemma 2.57. *Assume that $\mathcal{C}$ is a left-cancellative category that is right-Noetherian, and $\mathcal{S}$ is included in $\mathcal{C}$.*

(i) *If every non-invertible element of $\mathcal{C}$ is left-divisible by at least one non-invertible element of $\mathcal{S}$, then $\mathcal{S} \cup \mathcal{C}^\times$ generates $\mathcal{C}$.*

(ii) *Conversely, if $\mathcal{S} \cup \mathcal{C}^\times$ generates $\mathcal{C}$ and $\mathcal{S}$ satisfies $\mathcal{C}^\times\mathcal{S} \subseteq \mathcal{S}^\sharp$, then every non-invertible element of $\mathcal{C}$ is left-divisible by at least one non-invertible element of $\mathcal{S}$.*

Proof. (i) Let g be an element of $\mathcal{C}$. If g is invertible, g belongs to $\mathcal{C}^\times$. Otherwise, by assumption, there exist a non-invertible element s_1 in $\mathcal{S}$ and g' in $\mathcal{C}$ satisfying $g = s_1g'$. If g' is invertible, g belongs to $\mathcal{S}\mathcal{C}^\times$. Otherwise, there exist a non-invertible element s_2 in $\mathcal{S}$ and g'' satisfying $g' = s_2g''$, and so on. By Proposition 2.28, the sequence 1_x, $s_1, s_1s_2, \ldots$, which is increasing with respect to proper left-divisibility and lies in $\mathrm{Div}(g)$, must be finite, yielding p and $g = s_1 \cdots s_p\epsilon$ with $g_1, \ldots, g_p$ in $\mathcal{S}$ and ϵ in $\mathcal{C}^\times$. This proves (i).

(ii) Let g be a non-invertible element of $\mathcal{C}$. Let $s_1|\cdots|s_p$ be a decomposition of g such that g_i lies in $\mathcal{S} \cup \mathcal{C}^\times$ for every i. As g is not invertible, there exists i such that s_i is not invertible. Assume that i is minimal with this property. Then $s_1 \cdots s_{i-1}$ is invertible and, as $\mathcal{C}^\times \mathcal{S} \subseteq \mathcal{S}^\sharp$ holds, there exists s' in $\mathcal{S} \setminus \mathcal{C}^\times$ and ϵ in $\mathcal{C}^\times$ satisfying $s_1 \cdots s_i = s'\epsilon$. Then s' is a non-invertible element of $\mathcal{S}$ left-dividing g. □

Proposition 2.58 (atoms generate). *Assume that $\mathcal{C}$ is a left-cancellative category that is Noetherian.*

(i) *Every subfamily of $\mathcal{C}$ that generates $\mathcal{C}^\times$ and contains at least one element in each $=^\times$-class of atoms generates $\mathcal{C}$.*

(ii) *Conversely, every subfamily $\mathcal{S}$ that generates $\mathcal{C}$ and satisfies $\mathcal{C}^\times \mathcal{S} \subseteq \mathcal{S}^\sharp$ generates $\mathcal{C}^\times$ and contains at least one element in each $=^\times$-class of atoms.*

Proof. (i) Assume that $\mathcal{S}$ generates $\mathcal{C}^\times$ and contains at least one element in each $=^\times$-class of atoms. Let g be a non-invertible element of $\mathcal{C}$. By (the right counterpart of) Proposition 2.55, g is left-divisible by some atom, hence by some element of $\mathcal{S}$. By Lemma 2.57(i), this implies that $\mathcal{S} \cup \mathcal{C}^\times$ generates $\mathcal{C}$. As, by assumption, $\mathcal{S}$ generates $\mathcal{C}^\times$, hence $\mathcal{S} \cup \mathcal{C}^\times$, we deduce that $\mathcal{S}$ generates $\mathcal{C}$.

(ii) Assume that $\mathcal{S}$ generates $\mathcal{C}$ and we have $\mathcal{C}^\times \mathcal{S} \subseteq \mathcal{S}^\sharp$. First, $\mathcal{S}$ must generate $\mathcal{C}^\times$, which is included in $\mathcal{C}$. Now let g be an atom of $\mathcal{C}$. By Lemma 2.57(ii), there must exist a non-invertible element s of $\mathcal{S}$ left-dividing g. As g is an atom, the only possibility is $s =^\times g$. So some atom in the $=^\times$-class of g lies in $\mathcal{S}$. □

Corollary 2.59 (generating subfamily). *If $\mathcal{C}$ is a left-cancellative category that is Noetherian and contains no nontrivial invertible element, then a subfamily of $\mathcal{C}$ generates $\mathcal{C}$ if and only if it contains all atoms of $\mathcal{C}$.*

3 Groupoids of fractions

We now review some basic results about the enveloping groupoid of a category and the special case of a groupoid of fractions as described in a classical theorem of Ore. In general, the constructions are made in the context of monoids and groups, but adapting them to categories introduces no difficulty and remains an easy extension of the embedding of the additive monoid of natural numbers into the group of integers, and of the multiplicative monoid of non-zero integers into the group of rational numbers.

The section comprises three subsections. Subsection 3.1 contains a few general remarks about the enveloping groupoid of a category. In Subsection 3.2, we consider the special case of a groupoid of fractions and establish Ore's theorem, which gives necessary and sufficient conditions for a category to admit a groupoid of fractions. Finally, we show in Subsection 3.3 that the torsion elements in the groupoid of fractions of a category that admits right-lcms are connected with the torsion elements of the category.

3.1 The enveloping groupoid of a category

A groupoid is a category in which all elements are invertible. For instance, a groupoid with one object only is (or identifies with) a group.

We shall see below that, for every category $\mathcal{C}$, there exists a groupoid $\mathcal{G}$, unique up to isomorphism, with the universal property that every functor from $\mathcal{C}$ to a groupoid factors through $\mathcal{G}$. This groupoid is called the enveloping groupoid of the category (the enveloping group of the monoid when there is only one object).

Let us start with a straightforward observation.

Lemma 3.1. *Every category $\mathcal{C}$ admits the presentation $\langle \mathcal{C} \mid Rel(\mathcal{C})\rangle^+$, where $Rel(\mathcal{C})$ is the family of all relations $fg = h$ for f, g, h in $\mathcal{C}$ satisfying $fg = h$.*

Proof. Let ι be the map from $\mathcal{C}$ to itself that sends each object and each element to itself. By definition, the image under ι of every relation of $Rel(\mathcal{C})$ is valid in $\mathcal{C}$, hence ι extends to a functor of $\langle \mathcal{C} \mid Rel(\mathcal{C})\rangle^+$ to $\mathcal{C}$, which, by construction, is surjective and injective. □

We shall define the enveloping groupoid using a presentation similar to that of Lemma 3.1. To this end, we first introduce the notion of a signed path in a category, which extends the notion of a path as introduced in Definition 1.28, and corresponds to a zigzag in which the arrows can be crossed in both directions.

Definition 3.2 (signed path). If $\mathcal{S}$ is a precategory, we denote by $\overline{\mathcal{S}}$ the precategory consisting of a copy $\overline{s}$ for each element s of $\mathcal{S}$, with the convention that the source of $\overline{s}$ is the target of s and the target of $\overline{s}$ is the source of s. A *signed path in $\mathcal{S}$*, or *signed $\mathcal{S}$-path*, is a $(\mathcal{S} \cup \overline{\mathcal{S}})$-path. If w is a signed path, $\overline{w}$ denotes the signed path obtained from w by exchanging s and $\overline{s}$ everywhere and reversing the order of the entries.

So a signed $\mathcal{S}$-path is either an indexed empty sequence $()_x$ with x in $\mathcal{O}bj(\mathcal{S})$, or a nonempty sequence whose entries lie in $\mathcal{S} \cup \overline{\mathcal{S}}$. In this context, it is sometimes convenient to use s^{ε}, with the convention that s^{+1} stands for s and s^{-1} for $\overline{s}$. Then a nonempty signed path takes the form $(s_1^{\varepsilon_1}, ..., s_p^{\varepsilon_p})$ with $s_1, ..., s_p$ in $\mathcal{S}$ and $\varepsilon_1, ..., \varepsilon_p$ in $\{-1, +1\}$. The entries lying in $\mathcal{S}$ are then called *positive*, whereas the entries lying in $\overline{\mathcal{S}}$ are called *negative*. In practice, we write ε_x for $()_x$, and we identify s^{ε} with the length one sequence (s^{ε}). Then $(s_1^{\varepsilon_1}, ..., s_p^{\varepsilon_p})$ identifies with $s_1^{\varepsilon_1}|\cdots|s_p^{\varepsilon_p}$. We recall from Convention 1.35 that, when $\mathcal{S}$ is included in some category $\mathcal{C}$, a clear distinction should be made between a path $s_1|\cdots|s_p$, that is, the sequence $(s_1, ..., s_p)$, and its evaluation in $\mathcal{C}$. Similarly, in the case when $\mathcal{S}$ is included in some groupoid $\mathcal{G}$, a clear distinction should be made between the signed path $s_1^{\varepsilon_1}|\cdots|s_p^{\varepsilon_p}$ and its possible evaluation in $\mathcal{G}$ when $\overline{s}$ is given the value s^{-1}.

Definition 3.3 (enveloping groupoid). The *enveloping groupoid* $\mathcal{E}nv(\mathcal{C})$ of a category $\mathcal{C}$ is the category presented by

(3.4) $$\langle\, \mathcal{C} \cup \overline{\mathcal{C}} \mid Rel(\mathcal{C}) \cup Free(\mathcal{C})\,\rangle^+,$$

where $Free(\mathcal{C})$ is the family of all relations $g\overline{g} = 1_x$ and $\overline{g}g = 1_y$ for g in $\mathcal{C}(x, y)$.

The above terminology makes sense as we have

Proposition 3.5 (enveloping groupoid). *For every category $\mathcal{C}$, the category $\mathcal{E}nv(\mathcal{C})$ is a groupoid, the identity mapping on $\mathcal{C}$ defines a morphism ι of $\mathcal{C}$ to $\mathcal{E}nv(\mathcal{C})$ and, up to isomorphism, $\mathcal{E}nv(\mathcal{C})$ is the only groupoid with the property that each morphism from $\mathcal{C}$ to a groupoid factors through ι.*

Proof (sketch). The argument is straightforward. If ϕ is a morphism of $\mathcal{C}$ to a groupoid $\mathcal{G}$, then the image under ϕ of every relation of $Rel(\mathcal{C})$ must be satisfied in $\mathcal{G}$, and so does the image under ϕ of every relation of $Free(\mathcal{C})$ since $\mathcal{G}$ is a groupoid. Hence ϕ factors through $\mathcal{E}nv(\mathcal{C})$. □

Of course, in the case of a monoid, it is customary to speak of the *enveloping group*. Note that $\mathcal{E}nv(\mathcal{C})$, being defined by a presentation, is defined up to isomorphism only. By definition, the elements of an enveloping groupoid $\mathcal{E}nv(\mathcal{C})$ are equivalence classes of signed paths in $\mathcal{C}$. Let us fix some notation.

Notation 3.6 (relation $\equiv_{\mathcal{C}}^{\pm}$). For $\mathcal{C}$ a category and w, w' signed paths in $\mathcal{C}$, we write $w \equiv_{\mathcal{C}}^{\pm} w'$ if w and w' represent the same element of $\mathcal{E}nv(\mathcal{C})$, that is, they are equivalent with respect to the congruence generated by $Rel(\mathcal{C}) \cup Free(\mathcal{C})$.

In general, the morphism ι of Proposition 3.5 need not be injective, that is, a category need not embed in its enveloping groupoid.

Example 3.7 (not embedding). Let M be the monoid $\langle \mathtt{a}, \mathtt{b}, \mathtt{c} \mid \mathtt{ab} = \mathtt{ac} \rangle^{+}$. In M, the elements $\mathtt{b}$ and $\mathtt{c}$ do not coincide because no relation applies to the length-one words $\mathtt{a}$ and $\mathtt{b}$. However, we can write $\mathtt{b} \equiv_{M}^{\pm} \overline{\mathtt{a}}|\mathtt{a}|\mathtt{b} \equiv_{M}^{\pm} \overline{\mathtt{a}}|\mathtt{a}|\mathtt{c} \equiv_{M}^{\pm} \mathtt{c}$. Therefore we have $\iota(\mathtt{b}) = \iota(\mathtt{c})$, and M does not embed in its enveloping group.

Here are two elementary observations about the possible embeddability of a category in its enveloping groupoid.

Lemma 3.8. (i) *A category $\mathcal{C}$ embeds in its enveloping groupoid if and only if there exists at least one groupoid in which $\mathcal{C}$ embeds.*

(ii) *A category that embeds in its enveloping groupoid is cancellative.*

Proof. (i) Assume that ϕ is an injective morphism from $\mathcal{C}$ to a groupoid. By Proposition 3.5, ϕ factors through the canonical morphism ι of $\mathcal{C}$ to $\mathcal{E}nv(\mathcal{C})$. Then $\iota(g) = \iota(g')$ implies $\phi(g) = \phi(g')$, whence $g = g'$.

(ii) Assume that $\mathcal{C}$ is a category and f, g, g' are elements of $\mathcal{C}$ that satisfy $fg = fg'$. As ι is a morphism, we deduce $\iota(f)\,\iota(g) = \iota(f)\,\iota(g')$ in $\mathcal{E}nv(\mathcal{C})$, whence $\iota(g) = \iota(g')$ because $\mathcal{E}nv(\mathcal{C})$ is a groupoid. If ι is injective, this implies $g = g'$. Hence $\mathcal{C}$ is left-cancellative. The argument is symmetric for right-cancellation. □

However the next example shows that the condition of Lemma 3.8(ii) is not sufficient.

Example 3.9 (not embedding). Consider the monoid

$$M = \langle \mathtt{a}, \mathtt{b}, \mathtt{c}, \mathtt{d}, \mathtt{a}', \mathtt{b}', \mathtt{c}', \mathtt{d}' \mid \mathtt{ac} = \mathtt{a}'\mathtt{c}', \mathtt{ad} = \mathtt{a}'\mathtt{d}', \mathtt{bc} = \mathtt{b}'\mathtt{c}' \rangle^{+}.$$

Using for instance the criterion of Proposition 4.44, one checks that M is cancellative. However, M does not embed in a group. Indeed, $\mathtt{bd} = \mathtt{b}'\mathtt{d}'$ fails in M, whereas

$$\mathtt{b}|\mathtt{d} \equiv^{\pm}_{M} \mathtt{b}|\mathtt{c}|\overline{\mathtt{c}}|\overline{\mathtt{a}}|\mathtt{a}|\mathtt{d} = \mathtt{b}|\mathtt{c}|(\overline{\mathtt{a}|\mathtt{c}})|\mathtt{a}|\mathtt{d} \equiv^{\pm}_{M} \mathtt{b}'|\mathtt{c}'|(\overline{\mathtt{a}'|\mathtt{c}'})|\mathtt{a}'|\mathtt{d}' \equiv^{\pm}_{M} \mathtt{b}'|\mathtt{d}'$$

holds, whence $\iota(\mathtt{bd}) = \iota(\mathtt{b}'\mathtt{d}')$.

3.2 Groupoid of fractions

We leave the general question of characterizing which categories embed in their enveloping groupoid open, and consider now a more restricted case, namely that of categories $\mathcal{C}$ embedding in their enveloping groupoid and such that, in addition, all elements of $\mathcal{E}nv(\mathcal{C})$ all have the form of a fraction. In this case, $\mathcal{E}nv(\mathcal{C})$ is called a groupoid of fractions for $\mathcal{C}$. We shall give below (Proposition 3.11) a simple characterization for this situation.

Definition 3.10 (Ore category). A category $\mathcal{C}$ is said to be a *left-Ore* (*resp. right-Ore*) category if it is cancellative and any two elements with the same target (*resp.* source) admit a common left-multiple (*resp.* common right-multiple). An *Ore category* is a category that is both a left-Ore and a right-Ore category.

Proposition 3.11 (Ore's theorem). *For every category $\mathcal{C}$, the following are equivalent:*

(i) *There exists an injective functor ι from $\mathcal{C}$ to $\mathcal{E}nv(\mathcal{C})$ and every element of $\mathcal{E}nv(\mathcal{C})$ has the form $\iota(f)^{-1}\,\iota(g)$ for some f, g in $\mathcal{C}$.*

(ii) *The category $\mathcal{C}$ is a left-Ore category.*

When (i) *and* (ii) *are met, every element of $\mathcal{E}nv(\mathcal{C})$ is represented by a negative–positive path $\overline{f}|g$ with f, g in $\mathcal{C}$, and two such paths $\overline{f}|g$, $\overline{f'}|g'$ represent the same element of $\mathcal{E}nv(\mathcal{C})$ if and only if they satisfy the relation*

(3.12) $$\exists h, h' \in \mathcal{C}\ (\, hf' = h'f \quad \text{and} \quad hg' = h'g \,),$$

hereafter written $(f, g) \bowtie (f', g')$. Moreover, every presentation of $\mathcal{C}$ as a category is a presentation of $\mathcal{E}nv(\mathcal{C})$ as a groupoid.

The proof, which is a non-commutative generalization of the construction of rational numbers from integers, is postponed to the final Appendix.

We now mention two easy transfer results from a left-Ore category to its groupoid of left-fractions. The first one is that every functor between left-Ore categories extends to the associated groupoids of fractions.

Lemma 3.13. *If $\mathcal{C}_1, \mathcal{C}$ are left-Ore categories and ϕ is a functor from $\mathcal{C}_1$ to $\mathcal{C}$, then there exists a unique way to extend ϕ to a functor $\widehat{\phi}$ from $\mathcal{E}nv(\mathcal{C}_1)$ to $\mathcal{E}nv(\mathcal{C})$, namely putting $\widehat{\phi}(f^{-1}g) = \phi(f)^{-1}\phi(g)$ for f, g in $\mathcal{C}_1$.*

Proof. Assume $f^{-1}g = f'^{-1}g'$ are two fractional expressions for some element h of $\mathcal{E}nv(\mathcal{C}_1)$. Then $(f,g) \bowtie (f',g')$ holds, so there exist h, h' in $\mathcal{C}_1$ satisfying $hf' = h'f$ and $hg' = h'g$. As ϕ is a functor, we have $\phi(h)\phi(f') = \phi(h')\phi(f)$ and $\phi(h)\phi(g') = \phi(h')\phi(g)$, hence $(\phi(f), \phi(g)) \bowtie (\phi(f'), \phi(g'))$. So putting $\widehat{\phi}(f^{-1}g) = \phi(f)^{-1}\phi(g)$ gives a well-defined map from $\mathcal{E}nv(\mathcal{C}_1)$ to $\mathcal{E}nv(\mathcal{C})$. That $\widehat{\phi}$ is a functor is then easy. □

The second observation is that the left-divisibility relation of a left-Ore category naturally extends to its enveloping groupoid.

Definition 3.14 (left-divisibility). If $\mathcal{G}$ is a groupoid of left-fractions for a left-Ore category $\mathcal{C}$ and f, g lie in $\mathcal{G}$, we say that g is a *right-$\mathcal{C}$-multiple* of f, or that f is a *left-$\mathcal{C}$-divisor* of g, written $f \preccurlyeq_{\mathcal{C}} g$, if $f^{-1}g$ belongs to $\iota(\mathcal{C})$.

In other words, we have $f \preccurlyeq_{\mathcal{C}} g$ if there exists g' in $\mathcal{C}$ satisfying $g = f\iota(g')$. Note that $f \preccurlyeq_{\mathcal{C}} g$ implies that f and g share the same source. Of course, $\preccurlyeq_{\mathcal{C}}$ depends on $\mathcal{C}$, and not only on $\mathcal{E}nv(\mathcal{C})$: a groupoid may be a groupoid of fractions for several non-isomorphic left-Ore categories, leading to different left-divisibility relations.

Proposition 3.15 (left-divisibility). *Assume that $\mathcal{G}$ is a groupoid of left-fractions for an Ore category $\mathcal{C}$.*

(i) *The left-divisibility relation $\preccurlyeq_{\mathcal{C}}$ is a partial preordering on $\mathcal{G}$.*

(ii) *For f, g in $\mathcal{C}$, the relation $\iota(f) \preccurlyeq_{\mathcal{C}} \iota(g)$ is equivalent to $f \preccurlyeq g$.*

(iii) *Any two elements of $\mathcal{G}$ sharing the same source admit a common left-$\mathcal{C}$-divisor.*

(iv) *Two elements of $\mathcal{G}$ sharing the same source admit a common right-$\mathcal{C}$-multiple if and only if any two elements of $\mathcal{C}$ sharing the same source admit a common right-multiple.*

Proof. Item (i) is obvious. Item (ii) follows from ι being an embedding. For (iii), assume that f, g belong to $\mathcal{G}$ and share the same source. Then $f^{-1}g$ exists. As $\mathcal{G}$ is a group of left-fractions for $\mathcal{C}$, there exist f', g' in $\mathcal{C}$ satisfying $f^{-1}g = \iota(f')^{-1}\,\iota(g')$. Then we have $f\,\iota(f')^{-1} = g\,\iota(g')^{-1}$. Let h be the latter element. Then we have $f = h\,\iota(f')$ and $g = h\,\iota(g')$, whence $h \preccurlyeq_{\mathcal{C}} f$ and $h \preccurlyeq_{\mathcal{C}} g$. So f and g admit a common left-$\mathcal{C}$-divisor.

Assume that any two elements of $\mathcal{C}$ with the same source admit a common right-multiple. Let f and g be elements of $\mathcal{G}$ with the same source, and let f', g' be elements of $\mathcal{C}$ that satisfy $f^{-1}g = \iota(f')^{-1}\iota(g')$. Let f'', g'' satisfy $f'g'' = g'f''$. Then we obtain

$$f\,\iota(g'') = f\,\iota(f')^{-1}\,\iota(f'g'') = g\,\iota(g')^{-1}\,\iota(g'f'') = g\,\iota(f''),$$

and so f and g admit a common right-$\mathcal{C}$-multiple.

Conversely, if any two elements of $\mathcal{G}$ sharing the same source admit a common right-$\mathcal{C}$-multiple, this applies in particular to elements in the image of ι, which, by (ii), implies that any two elements of $\mathcal{C}$ sharing the same source admit a common right-multiple □

We conclude with a few observations involving subcategories. If $\mathcal{C}$ is a left-Ore category $\mathcal{C}$ and $\mathcal{C}_1$ is a subcategory of $\mathcal{C}$ that is itself a left-Ore category—in which case we simply say that $\mathcal{C}_1$ is a *left-Ore subcategory of* $\mathcal{C}$, the question arises of connecting the groupoids of fractions of $\mathcal{C}_1$ and $\mathcal{C}$. More precisely, two questions arise, namely whether $\mathcal{E}nv(\mathcal{C}_1)$ embeds in $\mathcal{E}nv(\mathcal{C})$ and whether $\mathcal{C}_1$ coincides with $\mathcal{E}nv(\mathcal{C}_1) \cap \mathcal{C}$. Both are answered in Proposition 3.18 below. Before that, we begin with easy preparatory observations.

Definition 3.16 (closure under quotient). A subfamily $\mathcal{S}$ of a left-cancellative category $\mathcal{C}$ is said to be *closed under right-quotient* if the conjunction of $g \in \mathcal{S}$ and $gh \in \mathcal{S}$ implies $h \in \mathcal{S}$. Symmetrically, a subfamily $\mathcal{S}$ of a right-cancellative category $\mathcal{C}$ is said to be *closed under left-quotient* if the conjunction of $h \in \mathcal{S}$ and $gh \in \mathcal{S}$ implies $g \in \mathcal{S}$.

Lemma 3.17. *Assume that $\mathcal{C}$ is a left-Ore category.*

(i) *A subcategory $\mathcal{C}_1$ of $\mathcal{C}$ is a left-Ore subcategory if and only if any two elements of $\mathcal{C}_1$ with the same target admit a common left-multiple in $\mathcal{C}_1$.*

(ii) *If $\mathcal{C}_1$ is closed under left-quotient in $\mathcal{C}$, then $\mathcal{C}_1$ is a left-Ore subcategory of $\mathcal{C}$ if and only if, for all g, h in $\mathcal{C}_1$ with the same target, at least one of the common left-multiples of g and h in $\mathcal{C}$ lies in $\mathcal{C}_1$.*

Proof. (i) Every subcategory of a cancellative category inherits cancellativity, so the only condition that remains to be checked is the existence of common left-multiples, and this is what the condition of the statement ensures.

(ii) If $\mathcal{C}_1$ is closed under left-quotient in $\mathcal{C}$ and h lies in $\mathcal{C}_1$, then every left-multiple of h (in the sense of $\mathcal{C}$) that lies in $\mathcal{C}_1$ is a left-multiple in the sense of $\mathcal{C}_1$: if gh lies in $\mathcal{C}_1$, then g must lie in $\mathcal{C}_1$. □

The embedding problem is then easily solved in the following result:

Proposition 3.18 (left-Ore subcategory). *Assume that $\mathcal{C}_1$ is a left-Ore subcategory of a left-Ore category $\mathcal{C}$.*

(i) *The inclusion of $\mathcal{C}_1$ in $\mathcal{C}$ extends to an embedding of $\mathcal{E}nv(\mathcal{C}_1)$ to $\mathcal{E}nv(\mathcal{C})$.*

(ii) *One has$\mathcal{C}_1 = \mathcal{E}nv(\mathcal{C}_1) \cap \mathcal{C}$ if and only if $\mathcal{C}_1$ is closed under right-quotient in $\mathcal{C}$.*

The proof, which is directly connected with that of Ore's theorem, is given in the Appendix. In the context of Proposition 3.18, we shall usually identify the groupoid $\mathcal{E}nv(\mathcal{C}_1)$ with its image in $\mathcal{E}nv(\mathcal{C})$, that is, drop the letter ι. Let us observe that the conclusion of Proposition 3.18(ii) need not hold when the closure condition is not satisfied.

Example 3.19 (positive elements). Let M be the additive monoid $(\mathbb{N}, +)$ and let M_1 be the submonoid $M \setminus \{1\}$. Then M and M_1 are Ore monoids which both admit $(\mathbb{Z}, +)$ as their group of fractions. Now 1 belongs to $\mathcal{E}nv(M_1) \cap M$, but not to M_1: here M_1 is a proper subset of $\mathcal{E}nv(M_1) \cap M$.

As can be expected, the above results do not extend to the case of a subcategory that is not a left-Ore subcategory. See for instance Exercise 10 for an example of an Ore monoid (actually a Garside monoid) in which some element of the group of left-fractions belongs to the subgroup generated by a subset S, but cannot be expressed as a left-fraction with numerator and denominator in the submonoid generated by S.

Remark 3.20. If a category $\mathcal{C}$ satisfies the Ore conditions on the left and on the right, the enveloping groupoid is both a groupoid of left-fractions and of right-fractions for $\mathcal{C}$. However, there exist cases when the Ore conditions are satisfied on one side only. For instance, consider the monoid $M = \langle \mathtt{a}, \mathtt{b} \mid \mathtt{a} = \mathtt{b}^2\mathtt{ab} \rangle^+$. As can be shown using the techniques of Section 4 below, M is cancellative and any two elements admit a common left-multiple, so that M embeds in a group of left-fractions G that admits, as a group, the same presentation. However, the elements $\mathtt{a}$ and $\mathtt{ba}$ admit no common right-multiple in M, and G is not a group of right-fractions for M: the element $\mathtt{a}^{-1}\mathtt{ba}$ of G cannot be expressed as a right-fraction gh^{-1} with g, h in M.

3.3 Torsion elements in a groupoid of fractions

We now investigate torsion elements. If $\mathcal{C}$ is a category, a *torsion* element in $\mathcal{C}$ is an element g such that some power of g is an identity element, that is, g^m lies in $\mathbf{1}_\mathcal{C}$ for some $m \geqslant 1$. Note that the source and the target of a torsion element must coincide. Here we shall observe that, if an Ore category admits lcms, then there exists a simple connection between the torsion elements of $\mathcal{C}$ and those of $\mathcal{E}nv(\mathcal{C})$.

Proposition 3.21 (torsion). *If $\mathcal{C}$ is a right-Ore category that admits right-lcms, then the torsion elements of $\mathcal{E}nv(\mathcal{C})$ are the elements of the form fhf^{-1} with f in $\mathcal{C}$ and h a torsion element of $\mathcal{C}$.*

Proof. As $(fhf^{-1})^m$ is fh^mf^{-1}, it is clear that, if h is a torsion element of $\mathcal{C}$, then fhf^{-1} is a torsion element of $\mathcal{E}nv(\mathcal{C})$ whenever it is defined.

Conversely, assume that e is an element of $\mathcal{E}nv(\mathcal{C})$ with the same source and target. As $\mathcal{C}$ is right-Ore, there exist f_1, g_1 in $\mathcal{C}$ such that e is $f_1g_1^{-1}$. Then f_1 and g_1 share the same source, hence, by assumption, they admit a right-lcm in $\mathcal{C}$, say $f_1g_2 = g_1f_2$. By construction, the sources of f_2 and g_2 are the common target of f_1 and g_1. So f_2 and g_2 admit a right-lcm, say $f_2g_3 = g_2f_3$. Iterating the argument, we inductively find for every $i \geqslant 1$ two elements f_i, g_i of $\mathcal{C}$ that share the same source and the same target and satisfy $f_ig_{i+1} = g_if_{i+1}$, the latter being the right-lcm of f_i and g_i. Then an iterated use of Proposition 2.12 shows that, for all $p, q \geqslant 1$, we have

$$f_1 \cdots f_p g_{p+1} \cdots g_{p+q} = g_1 \cdots g_q f_{q+1} \cdots f_{p+q},$$

and that the latter is a right-lcm of $f_1 \cdots f_p$ and $g_1 \cdots g_q$. On the other hand, an easy induction on m gives, for every $m \geqslant 1$, the following equalities in $\mathcal{E}nv(\mathcal{C})$:

$$e = (f_1 \cdots f_m)(f_{m+1}g_{m+1}^{-1})(f_1 \cdots f_m)^{-1}, \tag{3.22}$$

$$e^m = (f_1 \cdots f_m)(g_1 \cdots g_m)^{-1}. \tag{3.23}$$

Now, assume $e^m = 1_x$. Put $f = f_1 \cdots f_m$, $g = g_1 \cdots g_m$, and $h = f_{m+1} g_{m+1}^{-1}$. Then (3.23) implies $f = g$, so that f is a right-lcm of f and g. But we saw above that $f g_{m+1} \cdots g_{2m}$ is also a right-lcm of f and g, so the only possibility is that $g_{m+1} \cdots g_{2m}$ is invertible, and, in particular, so is g_{m+1}, and h lies in $\mathcal{C}$. Finally, (3.22) gives $e = f h f^{-1}$, so $e^m = 1_x$ implies $h^m = 1_y$, where y is the common target of f_m and g_m. □

Corollary 3.24 (torsion-free). *If $\mathcal{C}$ is a torsion-free right-Ore category that admits right-lcms, then the enveloping groupoid $\mathcal{E}nv(\mathcal{C})$ is torsion-free.*

A torsion element is necessarily invertible so, in particular, Corollary 3.24 says that the enveloping groupoid of a right-Ore category that admits right-lcms and has no nontrivial invertible element is torsion-free. As, by definition, the quasi-Garside monoids of Definition I.2.2 have no nontrivial invertible element, we deduce

Proposition 3.25 (quasi-Garside). *Every quasi-Garside group is torsion-free.*

The example of the monoid $\langle \mathtt{a}, \mathtt{b} \mid \mathtt{ab} = \mathtt{ba}, \mathtt{a}^2 = \mathtt{b}^2 \rangle^+$ shows that the existence of right-lcms is crucial in Proposition 3.21. Indeed, the latter monoid is right-Ore and has no nontrivial invertible element. However, $\mathtt{ab}^{-1}$ is a nontrivial torsion element in the associated group, since we have $(\mathtt{ab}^{-1})^2 = \mathtt{a}^2\mathtt{b}^{-2} = 1$.

4 Working with presented categories

Almost all developments in the sequel take place in categories or monoids that are left-cancellative. In order to illustrate our constructions with concrete examples that, in general, will be specified using presentations, we need to be able to recognize that our structures are left-cancellative. Several methods are available—in particular, we shall develop a powerful method based on germs in Chapter VI below. However, it is convenient to introduce now a practical method based on the so-called subword (or subpath) reversing approach. The toolbox so provided will be useful for several purposes, in particular proving the existence of least common multiples.

The section is organized as follows. In Subsection 4.1 we state without proof exportable criteria for establishing that a presented category is left-cancellative and admits right-lcms (Proposition 4.16). The rest of the section is devoted to the study of right-reversing, the involved method of proof. Right-reversing is defined in Subsection 4.2, and its termination is (briefly) discussed in Subsection 4.3. Finally, we introduce in Subsection 4.4 a technical condition called completeness and we both show how the latter leads to the results listed in Subsection 4.1 and how to practically establish completeness.

4.1 A toolbox

Here we state (without proof) a simple version of the criterion that will be established in the sequel. This version is sufficient in most cases, and it can be used as a black box.

We recall that, if $(\mathcal{S}, \mathcal{R})$ is a category presentation, then $\mathcal{S}^*$ is the free category of all $\mathcal{S}$-paths and $\equiv_{\mathcal{R}}^+$ is the congruence on $\mathcal{S}^*$ generated by $\mathcal{R}$, so that $\langle \mathcal{S} \mid \mathcal{R} \rangle^+$ is (isomorphic to) $\mathcal{S}^*/\equiv_{\mathcal{R}}^+$. Here we shall consider presentations of a special syntactic form.

Definition 4.1 (right-complemented presentation). A category presentation $(\mathcal{S}, \mathcal{R})$ is called *right-complemented* if $\mathcal{R}$ contains no ε-relation (that is, no relation $w = \varepsilon$ with w nonempty), no relation $s... = s...$ with s in $\mathcal{S}$ and, for $s \neq t$ in $\mathcal{S}$, at most one relation $s... = t...$.

Definition 4.2 (syntactic right-complement). A *syntactic right-complement* on a precategory $\mathcal{S}$ is a partial map θ from $\mathcal{S}^2$ to $\mathcal{S}^*$ such that $\theta(s, s) = \varepsilon_y$ holds for every s in $\mathcal{S}(\cdot, y)$ and, if $\theta(s, t)$ is defined, then s and t share the same source, $\theta(t, s)$ is defined, and both $s|\theta(s, t)$ and $t|\theta(t, s)$ are defined and share the same target; in this case, the list of all relations $s\theta(s, t) = t\theta(t, s)$ with $s \neq t$ in $\mathcal{S}$ is denoted by $\mathcal{R}_\theta$. A syntactic right-complement θ is called *short* if, for all s, t, the path $\theta(s, t)$ has length at most 1 when it is defined, that is, $\theta(s, t)$ belongs to $\mathcal{S}$ or is empty.

The following connection is straightforward:

Lemma 4.3. *A category presentation $(\mathcal{S}, \mathcal{R})$ is right-complemented if and only if there exists a syntactic right-complement θ on $\mathcal{S}$ such that $\mathcal{R}$ is $\mathcal{R}_\theta$.*

In the situation of Lemma 4.3, we naturally say that the presentation $(\mathcal{S}, \mathcal{R})$ is *associated with* the syntactic right-complement θ (which is uniquely determined by the presentation).

Example 4.4 (braids). Consider the Artin presentation (I.1.5) of the braid group B_n (Reference Structure 2, page 5). It contains no ε-relation, no relation $\sigma_i... = \sigma_i...$, and, for $i \neq j$, exactly one relation $\sigma_i... = \sigma_j...$. So the presentation is right-complemented, associated with the syntactic right-complement θ defined by

(4.5)
$$\theta(\sigma_i, \sigma_j) = \begin{cases} \varepsilon & \text{for } i = j, \\ \sigma_j|\sigma_i & \text{for } |i - j| = 1, \\ \sigma_j & \text{for } |i - j| \geqslant 2. \end{cases}$$

As the value of $\theta(\sigma_1, \sigma_2)$ is a length-two word, θ is not short for $n \geqslant 3$.

As, by definition, a right-complemented presentation contains no ε-relation, Lemma 1.42 implies that a category admitting a right-complemented presentation contains no nontrivial invertible element.

On the other hand, if $(\mathcal{S}, \mathcal{R})$ is associated with a syntactic right-complement θ, every pair (s, t) such that $\theta(s, t)$ exists gives rise to a diagram as on the right, which is commutative when evaluated in $\langle \mathcal{S} \mid \mathcal{R} \rangle^+$: the map θ specifies a distinguished common right-multiple for each pair of elements of $\mathcal{S}$.

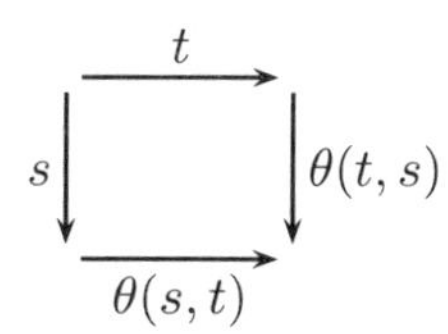

To arrive quickly to the expected criteria, we now state a technical result that will be established later (Lemma 4.32) and, more importantly, that should become more transparent after right-reversing is introduced in Subsection 4.2.

Lemma 4.6. *If $(\mathcal{S},\mathcal{R})$ is a right-complemented presentation associated with the syntactic right-complement θ, there exists a unique minimal extension θ^* of θ to a partial map from $\mathcal{S}^*\times\mathcal{S}^*$ to $\mathcal{S}^*$ that satisfies the rules*

$$\theta^*(s,s)=\varepsilon_y \quad \textit{for } s \textit{ in } \mathcal{S}(\cdot,y), \tag{4.7}$$

$$\theta^*(u_1u_2,v)=\theta^*(u_2,\theta^*(u_1,v)), \tag{4.8}$$

$$\theta^*(u,v_1v_2)=\theta^*(u,v_1)|\theta^*(\theta^*(v_1,u),v_2), \tag{4.9}$$

$$\theta^*(\varepsilon_x,u)=u \quad \textit{and} \quad \theta^*(u,\varepsilon_x)=\varepsilon_y \quad \textit{for } u \textit{ in } \mathcal{S}^*(x,y). \tag{4.10}$$

The map θ^* is such that $\theta^*(u,v)$ exists if and only if $\theta^*(v,u)$ does.

Example 4.11 (braids). Let θ be the braid witness as defined in (4.5). Let us try to evaluate $\theta^*(\sigma_1\sigma_2,\sigma_3\sigma_2)$ by applying the rules above. We find

$$\begin{aligned}
\theta^*(\sigma_1\sigma_2,\sigma_3\sigma_2) &= \theta^*(\sigma_2,\theta^*(\sigma_1,\sigma_3\sigma_2)) && \text{by (4.8) applied to } \theta^*(\sigma_1|\sigma_2,\cdot),\\
&= \theta^*(\sigma_2,\theta^*(\sigma_1,\sigma_3)\theta^*(\theta^*(\sigma_3,\sigma_1),\sigma_2)) && \text{by (4.9) applied to } \theta^*(\cdot,\sigma_3|\sigma_2),\\
&= \theta^*(\sigma_2,\sigma_3\theta^*(\theta^*(\sigma_3,\sigma_1),\sigma_2))) && \text{owing to the value of } \theta(\sigma_1,\sigma_3),\\
&= \theta^*(\sigma_2,\sigma_3\theta^*(\sigma_1,\sigma_2)) && \text{owing to the value of } \theta(\sigma_3,\sigma_1),\\
&= \theta^*(\sigma_2,\sigma_3\sigma_2\sigma_1) && \text{owing to the value of } \theta(\sigma_1,\sigma_2),\\
&= \theta^*(\sigma_2,\sigma_3)\theta^*(\theta^*(\sigma_3,\sigma_2),\sigma_2\sigma_1) && \text{by (4.9) applied to } \theta^*(\cdot,\sigma_3|\sigma_2\sigma_1),\\
&= \sigma_3\sigma_2\theta^*(\theta^*(\sigma_3,\sigma_2),\sigma_2\sigma_1) && \text{owing to the value of } \theta(\sigma_2,\sigma_3),\\
&= \sigma_3\sigma_2\theta^*(\sigma_2\sigma_3,\sigma_2\sigma_1) && \text{owing to the value of } \theta(\sigma_3,\sigma_2),\\
&= \sigma_3\sigma_2\theta^*(\sigma_3,\theta^*(\sigma_2,\sigma_2\sigma_1)) && \text{by (4.8) applied to } \theta^*(\sigma_2|\sigma_3,\cdot),\\
&= \sigma_3\sigma_2\theta^*(\sigma_3,\theta^*(\sigma_2,\sigma_2)\theta^*(\theta^*(\sigma_2,\sigma_2),\sigma_1)) && \text{by (4.9) applied to } \theta^*(\cdot,\sigma_2|\sigma_1),\\
&= \sigma_3\sigma_2\theta^*(\sigma_3,\theta^*(\varepsilon,\sigma_1)) && \text{by (4.7) applied to } \theta^*(\sigma_2,\sigma_2),\\
&= \sigma_3\sigma_2\theta^*(\sigma_3,\sigma_1) && \text{by (4.10) applied to } \theta(\varepsilon,\cdot),\\
&= \sigma_3\sigma_2\sigma_1 && \text{owing to the value of } \theta(\sigma_3,\sigma_1).
\end{aligned}$$

We conclude that $\theta^*(\sigma_1\sigma_2,\sigma_3\sigma_2)$ exists, and is equal to $\sigma_3\sigma_2\sigma_1$.

For the specific purposes of Chapter XII, we also introduce here a (very) particular type of right-complemented presentation.

Definition 4.12 (right-triangular). A right-complemented monoid presentation $(\mathcal{S},\mathcal{R})$ associated with a syntactic right-complement θ is called *right-triangular* if there exists a (finite or infinite) enumeration $s_1, s_2, \ldots$ of $\mathcal{S}$ such that $\theta(s_i,s_{i+1})$ is defined and empty for every i and, for $i+1<j$, either $\theta(s_i,s_j)$ is undefined, or we have

$$\theta(s_i,s_j)=\varepsilon \quad \text{and} \quad \theta(s_j,s_i)=\theta(s_j,s_{j-1})|\cdots|\theta(s_{i+1},s_i) \quad \text{for } i<j; \tag{4.13}$$

then $(\mathcal{R},\mathcal{S})$ is called *maximal* (*resp. minimal*) if $\theta(s_i,s_j)$ is defined for all i,j (*resp.* for $|i-j|\leqslant 1$ only).

A minimal right-triangular presentation is a list of relations of the form $s_i = s_{i+1}w_i$. If a right-triangular presentation is not minimal, (4.13) implies that the relation associated with (s_i, s_j) is a consequence of the above relations $s_i = s_{i+1}w_i$. So the monoid specified by a right-triangular presentation only depends on the minimal right-triangular presentation made of the relations $s_i = s_{i+1}\ldots$. A typical right-triangular presentation is the presentation $(\mathtt{a}, \mathtt{b}, \mathtt{a} = \mathtt{bab})$ of the Klein bottle monoid (Reference Structure 5, page 17).

The announced criteria for categories (and monoids) admitting right-complemented presentations will involve a technical condition that we introduce now.

Definition 4.14 (θ-cube condition). If $(\mathcal{S}, \mathcal{R})$ is a right-complemented presentation, associated with the syntactic right-complement θ, and u, v, w are $\mathcal{S}$-paths, we say that the *sharp θ-cube condition* (*resp.* the *θ-cube condition*) holds for (u, v, w) if

(4.15) Either both $\theta_3^*(u, v, w)$ and $\theta_3^*(v, u, w)$ are defined and they are equal (*resp.* they are $\equiv^+_{\mathcal{R}}$-equivalent), or neither is defined,

where $\theta_3^*(u, v, w)$ stands for $\theta^*(\theta^*(u, v), \theta^*(u, w))$ with θ^* as in Lemma 4.6. For $\mathcal{X} \subseteq \mathcal{S}^*$, we say that the (sharp) θ-cube condition holds *on* $\mathcal{X}$ if it holds for all (u, v, w) in $\mathcal{X}^3$.

As the difference is an equality being replaced with an $\equiv^+_{\mathcal{R}}$-equivalence, the sharp θ-cube condition implies the θ-cube condition. By definition, if θ is a syntactic right-complement, $\theta^*(s, t)$ can be defined only if s and t share the same source, and the inductive definition of θ^* preserves this property. Hence, the θ-cube and sharp θ-cube conditions are vacuously true for every triple of paths (u, v, w) that do not share the same source: in that case, neither $\theta_3^*(u, v, w)$ nor $\theta_3^*(v, u, w)$ may be defined.

Here are the main results for categories with right-complemented presentations:

Proposition 4.16 (right-complemented). *Assume that $(\mathcal{S}, \mathcal{R})$ is a right-complemented presentation associated with the syntactic right-complement θ, and at least one of the following conditions is satisfied:*

(4.17) *The presentation $(\mathcal{S}, \mathcal{R})$ contains only short relations and the sharp θ-cube condition holds for every triple of pairwise distinct elements of $\mathcal{S}$.*

(4.18) *The presentation $(\mathcal{S}, \mathcal{R})$ is right-Noetherian and the θ-cube condition holds for every triple of pairwise distinct elements of $\mathcal{S}$.*

(4.19) *The presentation $(\mathcal{S}, \mathcal{R})$ is maximal right-triangular.*

Then:

(i) *The category $\langle \mathcal{S} \mid \mathcal{R} \rangle^+$ is left-cancellative.*

(ii) The category $\langle \mathcal{S} \mid \mathcal{R} \rangle^+$ admits conditional right-lcms; more precisely, for all w, w' in $\mathcal{S}^*$, the elements of $\langle \mathcal{S} \mid \mathcal{R} \rangle^+$ represented by u and v admit a common right-multiple if and only if $\theta^*(u, v)$ exists and, then, $u\theta^*(u, v)$ represents the right-lcm of these elements.

(iii) Two elements u, v of $\mathcal{S}^*$ represent the same element of $\langle \mathcal{S} \mid \mathcal{R} \rangle^+$ if and only if both $\theta^*(u, v)$ and $\theta^*(v, u)$ exist and are empty.

The results in Proposition 4.16 are special cases of more general results that will be established in Subsection 4.4 below.

Example 4.20 (braids). Consider the Artin presentation (I.1.5). As observed in Example 4.4, this presentation is right-complemented, associated with the syntactic right-complement θ of (4.5). We claim that (4.18) is satisfied. Indeed, the presentation is homogeneous, hence (strongly) Noetherian. So, it suffices to check that the θ-cube condition holds for every triple of pairwise disjoint generators $\sigma_i, \sigma_j, \sigma_k$. For instance, the case $i = 1$, $j = 2$, $k = 3$ amounts to comparing $\theta_3^*(\sigma_1, \sigma_2, \sigma_3)$ and $\theta_3^*(\sigma_2, \sigma_1, \sigma_3)$. In view of Example 4.11, the task may seem extremely boring. Actually it is not so, for two reasons.

First, only a few patterns have to be considered. Indeed, only two types of braid relations exist, so, when considering a triple $(\sigma_i, \sigma_j, \sigma_k)$, what matters is the mutual positions of i, j, k and whether they are neighbors or not. A moment's thought will convince that it is sufficient to consider $(\sigma_1, \sigma_2, \sigma_3)$, $(\sigma_1, \sigma_2, \sigma_4)$, $(\sigma_1, \sigma_3, \sigma_5)$, plus their images under a cyclic permutation. Among these triples, a number are straightforward, namely those that correspond to commutation relations as, if s commutes with all letters of w in the sense that $st = ts$ is a relation of the presentation for every t occurring in w, then we have $\theta^*(w, s) = s$ and $\theta^*(s, w) = w$. It follows that the only nontrivial triples are $(\sigma_1, \sigma_2, \sigma_3)$, $(\sigma_2, \sigma_3, \sigma_1)$, and $(\sigma_3, \sigma_1, \sigma_2)$.

Then, for the latter cases, the diagrammatic technique explained in Subsection 4.2 makes the determination of θ^* much easier than what Example 4.11 suggests. The reader is encouraged to check the values

$$\begin{aligned}
\theta^*(\theta(\sigma_1, \sigma_2), \theta(\sigma_1, \sigma_3)) &= \sigma_3\sigma_2\sigma_1 = \theta^*(\theta(\sigma_2, \sigma_1), \theta(\sigma_2, \sigma_3)),\\
\theta^*(\theta(\sigma_2, \sigma_3), \theta(\sigma_2, \sigma_1)) &= \sigma_1\sigma_2\sigma_3 = \theta^*(\theta(\sigma_3, \sigma_2), \theta(\sigma_3, \sigma_1)),\\
\theta^*(\theta(\sigma_3, \sigma_1), \theta(\sigma_3, \sigma_2)) &= \sigma_2\sigma_1\sigma_3\sigma_2 \equiv^+ \sigma_2\sigma_3\sigma_1\sigma_2 = \theta^*(\theta(\sigma_1, \sigma_3), \theta(\sigma_1, \sigma_2)),
\end{aligned}$$

which complete the proof of (4.18). Then Proposition 4.16 implies that the monoid $\underline{B}_n^+$ presented by (I.1.5) is left-cancellative, and that any two elements of $\underline{B}_n^+$ that have a common right-multiple have a right-lcm. It is not hard to show that any two elements of $\underline{B}_n^+$ admit a common right-multiple (see Exercise 1) and, therefore, we conclude that B_n^+ admits right-lcms. By the way, note that that the sharp θ-cube condition is not satisfied for $n \geqslant 4$ since $\theta_3^*(\sigma_3, \sigma_1, \sigma_2) = \theta_3^*(\sigma_1, \sigma_3, \sigma_2)$ fails.

4.2 Right-reversing: definition

We turn to the proof of Propositions 4.16. As announced, we shall establish more general results and, in particular, no longer restrict to right-complemented presentations. This extension does not make the arguments more complicated and, on the contrary, it helps understanding the situation.

Our main technical tool will be a certain transformation called right-reversing on signed paths (see Definition 3.2). In essence, right-reversing is a strategy for constructing van Kampen diagrams in a context of monoids, that is, equivalently, for finding derivations between words. We shall see that it is especially relevant for investigating right-complemented presentations.

Definition 4.21 (right-reversing). Assume that $(\mathcal{S}, \mathcal{R})$ is a category presentation, and w, w' are signed $\mathcal{S}$-paths. We say that w is *$\mathcal{R}$-right-reversible* to w' *in one step*, written $w \curvearrowright^1_{\mathcal{R}} w'$, if w' is obtained from w either by replacing a length-two subpath $\overline{s}|s$ with ε_y (y the target of s), or by replacing a length-two subpath $\overline{s}|t$ with $v\overline{u}$ such that $sv = tu$ is a relation of $\mathcal{R}$. An *$\mathcal{R}$-right-reversing sequence* is a (finite or infinite) sequence $(w_0, w_1, \cdots)$ such that $w_{i-1} \curvearrowright^1_{\mathcal{R}} w_i$ holds for each i. We say that w is *$\mathcal{R}$-right-reversible to w' in n steps*, written $w \curvearrowright^n_{\mathcal{R}} w'$, if there exists an $\mathcal{R}$-right-reversing sequence of length $n+1$ that goes from w to w'.

When using $\curvearrowright^n_{\mathcal{R}}$ we shall often drop the exponent n, and the subscript $\mathcal{R}$.

Example 4.22 (braids). Consider the presentation (I.1.5) of B_n (Reference Structure 2, page 5). A right-reversing sequence from the signed braid word $\overline{\sigma_2}|\overline{\sigma_1}|\sigma_3|\sigma_2$ is

$$\overline{\sigma_2}|\overline{\sigma_1}|\sigma_3|\sigma_2 \curvearrowright^1 \overline{\sigma_2}|\sigma_3|\overline{\sigma_1}|\sigma_2 \curvearrowright^1 \sigma_3|\sigma_2|\overline{\sigma_3}|\overline{\sigma_2}|\overline{\sigma_1}|\sigma_2 \curvearrowright^1 \sigma_3|\sigma_2|\overline{\sigma_3}|\overline{\sigma_2}|\sigma_2|\sigma_1|\overline{\sigma_2}|\overline{\sigma_1}$$
$$\curvearrowright^1 \sigma_3|\sigma_2|\overline{\sigma_3}|\sigma_1|\overline{\sigma_2}|\overline{\sigma_1}| \curvearrowright^1 \sigma_3|\sigma_2|\sigma_1|\overline{\sigma_3}|\overline{\sigma_2}|\overline{\sigma_1}.$$

So the signed braid word $\overline{\sigma_2}|\overline{\sigma_1}|\sigma_3|\sigma_2$ is right-reversible to $\sigma_3|\sigma_2|\sigma_1|\overline{\sigma_3}|\overline{\sigma_2}|\overline{\sigma_1}$ in five steps, that is, $\overline{\sigma_2}|\overline{\sigma_1}|\sigma_3|\sigma_2| \curvearrowright^5 \sigma_3|\sigma_2|\sigma_1|\overline{\sigma_3}|\overline{\sigma_2}|\overline{\sigma_1}$ holds. The last word is terminal with respect to right-reversing, as it contains no subword of the form $\overline{\sigma_i}|\sigma_j$. Note: using the concatenation symbol $|$ in words, typically braid words, is not usual and may appear clumsy; however, it is useful here to emphasize that right-reversing is a purely syntactic transformation, and *not* a transformation on the braids represented by these words.

If $(\mathcal{S}, \mathcal{R})$ is a right-complemented presentation associated with a syntactic right-complement θ, all relations of $\mathcal{R}$ have the form $s\theta(s,t) = t\theta(t,s)$ with $s \neq t$ in $\mathcal{S}$, so right-reversing consists in either replacing some subpath $\overline{s}|t$ with $s \neq t$ with $\theta(s,t)|\overline{\theta(t,s)}$, or deleting some subpath $\overline{s}|s$, which, with our definition, also means replacing it with $\theta(s,s)|\overline{\theta(s,s)}$ since the latter is an empty path.

It will be convenient—actually essential in order to understand how right-reversing works—to associate with every reversing sequence, say $\vec{w} = (w_0, w_1, ...)$, a rectangular grid diagram $D(\vec{w})$ that we now define. We first consider the special case when the considered presentation $(\mathcal{S}, \mathcal{R})$ is associated with a short syntactic right-complement θ and the initial signed $\mathcal{S}$-path w_0 is negative–positive, meaning that we have $w_0 = \overline{u}|v$ for some positive $\mathcal{S}$-paths u, v. In this case, the diagram $D(\vec{w})$ is a rectangular grid. We begin $D(\vec{w})$ by starting from an initial vertex that represents the common source of u and v and drawing a vertical down-oriented sequence of arrows labeled by the successive entries of u and a horizontal right-oriented sequence of arrows labeled by the successive entries of v. Then we inductively construct $D(\vec{w})$. By definition, w_1 is obtained from w_0 by replacing some subpath $\overline{s}|t$ with $\theta(s,t)|\overline{\theta(t,s)}$, which we first assume to be of length two. In the diagram, $\overline{s}|t$ corresponds to some pattern

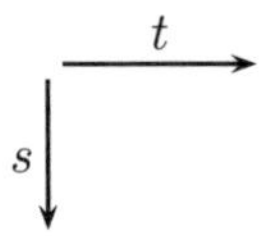

, which we complete to
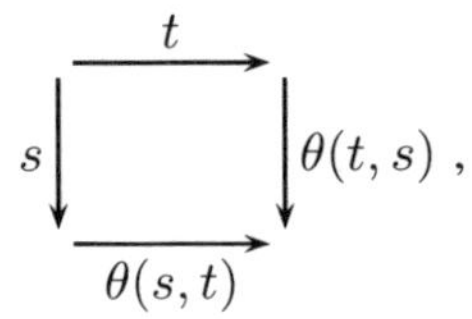

,

and we continue in the same way for w_2, etc., inductively constructing a grid. Special cases occur when empty paths appear: for instance, if $\theta(t,s)$ is empty (hence in particular for $s = t$), we draw
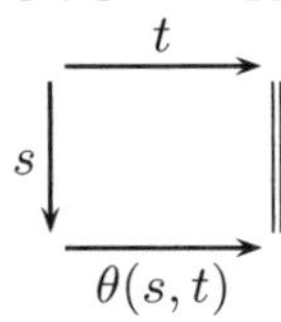

and duplicate all horizontal arrows on the right of t with patterns
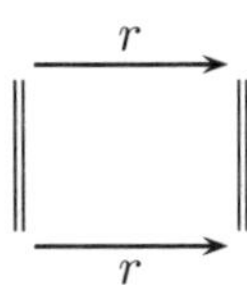

. Similarly, for $\theta(s,t)$ empty, we draw
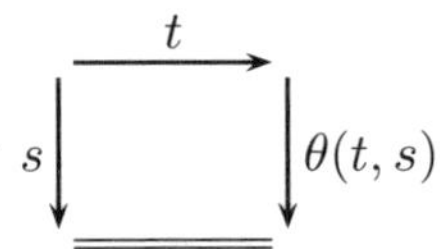

and duplicate the vertical arrows below s using
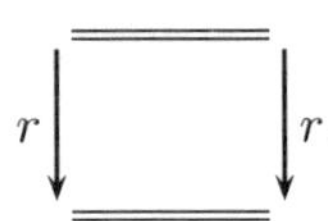

. See an example in Figure 4.

An obvious induction shows that the successive entries of the right-reversing sequence $\vec{w}$ all can be read in the grid from the South-West corner to the North-East corner, using the convention that an s-labeled arrow crossed in the wrong direction (from the target to the source) contributes $\overline{s}$.

Once this is done, extending the construction to more general cases is easy. First, in the case of a reversing sequence $\vec{w}$ whose first entry w_0 is not negative–positive, we draw a similar diagram but start from a staircase that may have more than one stair: details should be clear on Figure 5.

In the case of a right-complement that is not short, the paths $\theta(s,t)$ and $\theta(t,s)$ may have length larger than one. Then we draw the grid as before, but, when necessary, we split the edges of the squares into several shorter arrows, see Figure 6 for an example (in the case of braids).

Finally, in the most general case, that is, for a presentation $(\mathcal{S}, \mathcal{R})$ that need not be right-complemented, the construction of the diagram associated with a reversing sequence is similar: the main difference is that, in the right-complemented case, there is only one way to construct a diagram from a given initial path whereas, in the general case, there may exist several ways since, in the list of relations $\mathcal{R}$, there may exist several relations of the form $s... = t...$, hence several ways of reversing $\overline{s}|t$ and several ways of continuing

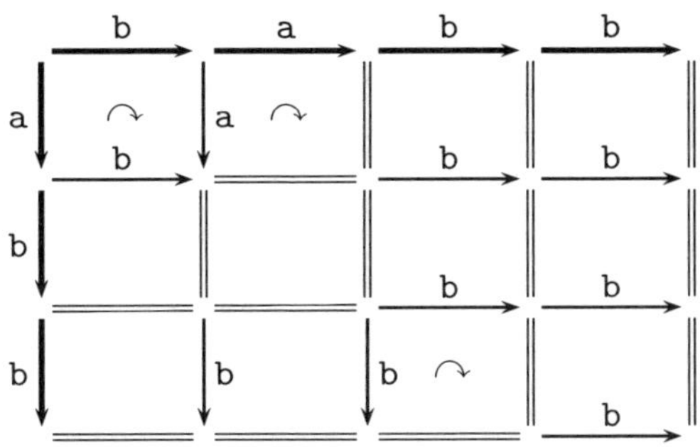

Figure 4. The grid associated with a right-reversing, in a right-complemented case. Here we consider the syntactic right-complement θ on $\{\mathtt{a}, \mathtt{b}\}$ defined by $\theta(\mathtt{a}, \mathtt{a}) = \theta(\mathtt{b}, \mathtt{b}) = \varepsilon$, $\theta(\mathtt{a}, \mathtt{b}) = \mathtt{b}$, and $\theta(\mathtt{b}, \mathtt{a}) = \mathtt{a}$ (with one object only), and start from the negative–positive word $\overline{\mathtt{b}}|\overline{\mathtt{b}}|\overline{\mathtt{a}}|\mathtt{b}|\mathtt{a}|\mathtt{b}|\mathtt{b}$: the initial word is written on the left (negative part, here $\mathtt{a}|\mathtt{b}|\mathtt{b}$) and the top (positive part, here $\mathtt{b}|\mathtt{a}|\mathtt{b}|\mathtt{b}$), and then the grid is constructed using the syntactic right-complement θ to recursively complete the squares. The diagram shows that the initial word right-reverses in three steps to the word $\mathtt{b}$; the diagram has 12 squares, among which the three marked $\curvearrowright$ correspond to reversing steps and the remaining 9 correspond to duplications induced by empty words.

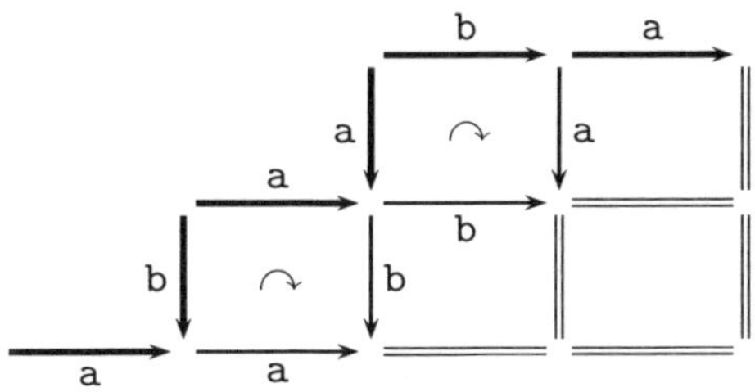

Figure 5. The grid associated with the right-reversing of a signed path that need not be negative–positive. We consider the right-complement from Figure 4, and now start from the word $\mathtt{a}|\overline{\mathtt{b}}|\mathtt{a}|\overline{\mathtt{a}}|\mathtt{b}|\mathtt{a}$: the initial word is written on the top-left boundary (negative edges vertically and positive edges horizontally), and completing the grid to the bottom-right; here there are two reversing steps, and the final word is $\mathtt{a}|\mathtt{a}$.

a diagram in which the pattern 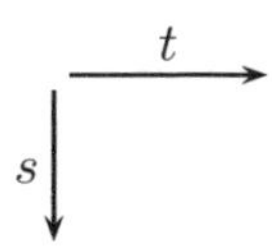 occurs.

We conclude this introduction to right-reversing and the associated diagrams with a first technical result that will be used for many inductions, namely the possibility of splitting a reversing diagram into several reversing subdiagrams, as illustrated in Figure 7.

Lemma 4.23. *Assume that $(\mathcal{S}, \mathcal{R})$ is a category presentation, w, w' are signed $\mathcal{S}$-paths, and $w \curvearrowright_{\mathcal{R}}^{n} w'$ holds. For each decomposition $w = w_1 w_2$, there exist a decomposition $w' = w_1' w_0' w_2'$ and two $\mathcal{S}$-paths u', v' satisfying*

$$w_1 \curvearrowright_{\mathcal{R}}^{n_1} w_1' \overline{u'}, \quad \overline{u'} v' \curvearrowright_{\mathcal{R}}^{n_0} w_0', \quad \textit{and} \quad w_2 \curvearrowright_{\mathcal{R}}^{n_2} v' w_2'$$

with $n_1 + n_0 + n_2 = n$.

Establishing Lemma 4.23 is easy, though a little cumbersome, and we postpone the proof to the Appendix. Applying Lemma 4.23 in the case when w_1 has the form $\overline{u} v_1$ with u, v_1 in $\mathcal{S}^*$ and w_2 belongs to $\mathcal{S}^*$ gives:

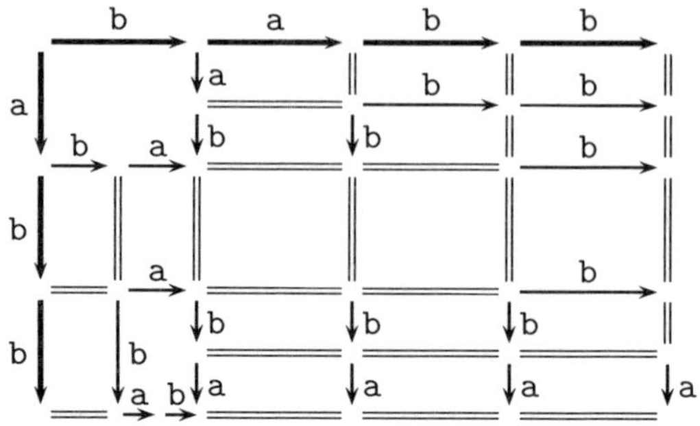

Figure 6. The grid associated with a right-reversing for a syntactic right-complement that is not short. Here we consider the right-complement θ on $\{\mathtt{a}, \mathtt{b}\}$ defined by $\theta(\mathtt{a}, \mathtt{a}) = \theta(\mathtt{b}, \mathtt{b}) = \varepsilon$, $\theta(\mathtt{a}, \mathtt{b}) = \mathtt{b}|\mathtt{a}$, and $\theta(\mathtt{b}, \mathtt{a}) = \mathtt{a}|\mathtt{b}$ (with one object only), and start from the negative–positive word $\overline{\mathtt{b}}|\overline{\mathtt{b}}|\overline{\mathtt{a}}|\mathtt{b}|\mathtt{a}|\mathtt{b}|\mathtt{b}$ (the same as in Figure 4): the difference is that, now, edges of variable size occur, so that the process a priori may never terminate. In the current case, it terminates, and the final word is $\mathtt{a}|\mathtt{b}|\overline{\mathtt{a}}$.

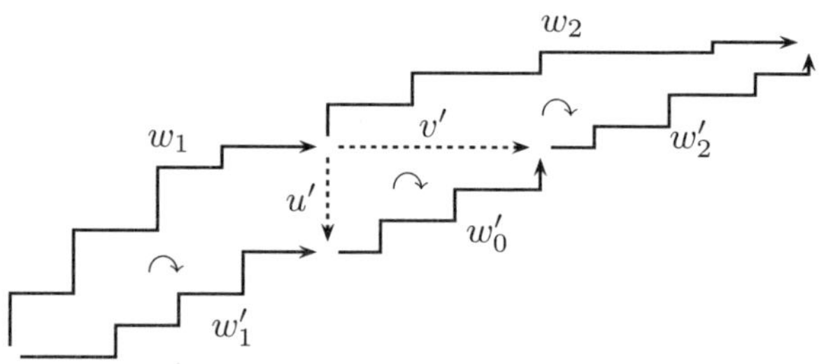

Figure 7. Splitting the right-reversing of a product (general case): every reversing process that goes from the concatenation of two signed paths can be split into three reversing processes.

Lemma 4.24. *If $(\mathcal{S}, \mathcal{R})$ is a category presentation and we have $\overline{u}v_1v_2 \curvearrowright v'\overline{u'}$ with u, v_1, v_2, u', v' in $\mathcal{S}^*$, then there exist a decomposition $v' = v_1'v_2'$ in $\mathcal{S}^*$ and u_1 in $\mathcal{S}^*$ satisfying $\overline{u}v_1 \curvearrowright v_1'\overline{u_1}$ and $\overline{u_1}v_2 \curvearrowright v_2'\overline{u'}$, as shown on the right.*

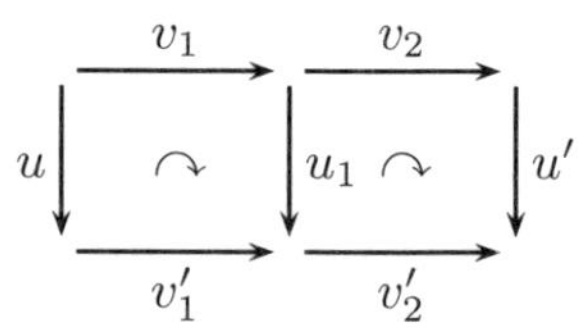

4.3 Right-reversing: termination

Right-reversing is a transformation of paths that pushes the negative entries to the right and the positive entries to the left. By definition, a positive–negative path contains no length-two subpath of the form $\overline{s}|t$ and, therefore, it is *terminal* with respect to right-reversing: no right-reversing sequence of positive length may start from that path. A natural question is whether right-reversing always leads in finitely many steps to a terminal path. Here we gather a few easy observations about this (generally difficult) question.

Definition 4.25 (terminating). If $(\mathcal{S}, \mathcal{R})$ is a category presentation and w is a signed $\mathcal{S}$-path, we say that right-reversing is *terminating* for w in $(\mathcal{S}, \mathcal{R})$ if there is no infinite right-reversing sequence starting from w. We say that right-reversing is *always terminating* in $(\mathcal{S}, \mathcal{R})$ if it is terminating for every signed $\mathcal{S}$-path.

Recognizing whether right-reversing is terminating is often difficult. Here we address the question in a particular case.

Definition 4.26 (short). A relation $u = v$ is called *short* if the length of u and the length of v are at most two. A category presentation $(\mathcal{S}, \mathcal{R})$ is called *short* if all relations of $\mathcal{R}$ are short.

Proposition 4.27 (termination). (i) *If $(\mathcal{S}, \mathcal{R})$ is a category presentation that contains only short relations, which happens in particular if $(\mathcal{S}, \mathcal{R})$ is associated with a short syntactic right-complement, then right-reversing is always terminating in $(\mathcal{S}, \mathcal{R})$. More precisely, every right-reversing sequence starting with a path containing p negative entries and q positive entries has length pq at most.*

(ii) *If, moreover, for all s, t in $\mathcal{S}$ with the same source, there exists at least one relation $s... = t...$ in $\mathcal{R}$, then every right-reversing sequence can be extended so as to finish with a positive–negative path.*

Proof. (i) As every relation in $\mathcal{R}$ is short, applying right-reversing means replacing length-two subpaths with new paths of length at most two. Hence the length cannot increase when right-reversing is performed. If w is an initial path with p negative entries and p positive entries, then every right-reversing diagram for w can be drawn in a $(p{\times}q)$-grid (as is the case in Figures 4 and 5 and as is not the case in Figure 6) and, therefore, it involves at most pq reversing steps.

(ii) The additional assumption guarantees that the only terminal paths are the positive–negative ones: whenever a path is not positive–negative, it contains at least one subpath $\overline{s}|t$, hence it is eligible for at least one right-reversing step. As there is no infinite reversing sequence, every maximal sequence must finish with a positive–negative path. □

The following examples show that, even for simple presentations, termination is no longer guaranteed when paths of length larger than two occur in the presentation.

Example 4.28 (non-terminating). Consider the presentation $(\mathtt{a}, \mathtt{b}, \mathtt{ab} = \mathtt{b}^2\mathtt{a})$ of the Baumslag–Solitar monoid. The presentation is right-complemented, associated with the syntactic right-complement θ defined by $\theta(\mathtt{a}, \mathtt{b}) = \mathtt{b}$ and $\theta(\mathtt{b}, \mathtt{a}) = \mathtt{b}|\mathtt{a}$. Then we find

$$\underline{\overline{\mathtt{a}}|\mathtt{b}|\mathtt{a}} \curvearrowright \mathtt{b}|\overline{\mathtt{a}}|\overline{\mathtt{b}}|\mathtt{a} \curvearrowright \mathtt{b}|\underline{\overline{\mathtt{a}}|\mathtt{b}|\mathtt{a}}|\overline{\mathtt{b}},$$

whence $\overline{\mathtt{a}}|\mathtt{b}|\mathtt{a} \curvearrowright^{2n} \mathtt{b}^n|\overline{\mathtt{a}}|\mathtt{b}|\mathtt{a}|\overline{\mathtt{b}}^n$ for every n. In this case, right-reversing never leads to a terminal word of the form $v\overline{u}$ with u, v in $\{\mathtt{a}, \mathtt{b}\}^*$. The above presentation satisfies (4.18) (finding a right-Noetherianity witness is a little tricky) and, therefore, by Proposition 4.16, the elements $\mathtt{a}$ and $\mathtt{ba}$ admit no common right-multiple.

Similarly, for $(\sigma_1, \sigma_2, \sigma_3, \sigma_1\sigma_2\sigma_1 = \sigma_2\sigma_1\sigma_2, \sigma_2\sigma_3\sigma_2 = \sigma_3\sigma_2\sigma_3, \sigma_1\sigma_3\sigma_1 = \sigma_3\sigma_1\sigma_3)$, a presentation of the Artin–Tits monoid of type $\widetilde{A}_2$ (see Reference Structure 9, page 114 below and Chapter IX), we find

$$\underline{\overline{\sigma_1}|\sigma_2|\sigma_3} \curvearrowright^6 \sigma_2|\sigma_1|\sigma_3|\sigma_2|\sigma_1|\sigma_3|\underline{\overline{\sigma_1}|\sigma_2|\sigma_3}|\overline{\sigma_2}|\overline{\sigma_3}|\overline{\sigma_1}|\overline{\sigma_2}|\overline{\sigma_3}|\overline{\sigma_1},$$

again leading to a non-terminating reversing sequence.

We shall not say more about termination of right-reversing here. In the right-complemented case, there exists at most one maximal right-reversing diagram starting from each initial path (but there are in general many reversing sequences as the reversing steps can be enumerated in different orders), yielding a trichotomy:

(4.29) either right-reversing leads in finitely many steps to a positive–negative path;

(4.30) or one gets stuck at some step with a pattern $\overline{s}|t$ such that $\theta(s,t)$ and $\theta(t,s)$ do not exist;

(4.31) or right-reversing can be extended endlessly reaching neither of the above two cases.

If the syntactic right-complement θ is short, then (4.31) is excluded; if it is everywhere defined (on pairs sharing the same source), (4.30) is excluded.

In the framework now well defined, we can establish Lemma 4.6 that was left pending in Subsection 4.1.

Lemma 4.32. (i) *If $(\mathcal{S},\mathcal{R})$ is a right-complemented category presentation associated with the syntactic right-complement θ, then, for all $\mathcal{S}$-paths u, v, there exist at most one pair of $\mathcal{S}$-paths u', v' satisfying $\overline{u}v \curvearrowright_{\mathcal{R}} v'\overline{u'}$.*

(ii) *Define $\theta^*(u,v)$ to be the above path v' when it exists. Then θ^* is a partial map from $\mathcal{S}^* \times \mathcal{S}^*$ to $\mathcal{S}^*$ that extends θ, it satisfies the induction rules of Lemma 4.6, and, as a family of pairs, it is the least such extension of θ.*

Proof. (i) As $(\mathcal{S},\mathcal{R})$ is right-complemented, for every initial signed $\mathcal{S}$-word w, the maximal right-reversing diagram from w is unique. In case (4.31), there is no final path: then it is impossible that w right-reverses to a positive–negative path, as such a path would be final. In case (4.30), the diagram is finite, there exists a unique final path w', and the latter contains some subpath $\overline{s}|t$ such that $\theta(s,t)$ is not defined, so it is not positive–negative. Finally, in case (4.29), the unique final path is positive–negative. Applying this in the case when w is $\overline{u}v$, we conclude that, in every case, there exists at most one pair of $\mathcal{S}$-paths u', v' satisfying $\overline{u}v \curvearrowright_{\mathcal{R}} v'\overline{u'}$.

(ii) By construction, $w \curvearrowright_{\mathcal{R}} w'$ always implies $\overline{w} \curvearrowright_{\mathcal{R}} \overline{w'}$. So $\overline{u}v \curvearrowright_{\mathcal{R}} v'\overline{u'}$ implies $\overline{v}u \curvearrowright_{\mathcal{R}} u'\overline{v'}$. This shows that, when $\overline{u}v \curvearrowright_{\mathcal{R}} v'\overline{u'}$ holds, the path $\theta^*(v,u)$ exists and equals u'. So $\theta^*(u,v)$ exists if and only if $\theta^*(v,u)$ does.

For all s, t in $\mathcal{S}$ such that $\theta(s,t)$ exists, we have $\overline{s}|t \curvearrowright_{\mathcal{R}} \theta(s,t)|\overline{\theta(t,s)}$, whence $\theta^*(s,t) = \theta(s,t)$. So θ^* extends θ.

Next, $\overline{s}|s \curvearrowright_{\mathcal{R}} \varepsilon_y$ holds for every s in $\mathcal{S}(\cdot, y)$, and, therefore, θ^* satisfies (4.7). Then, when we express the relations of Lemma 4.24 in the right-complemented case, we directly obtain that θ^* satisfies (4.8) and (4.9), as can be read in Figure 8. As for (4.10), it follows from the trivial relations $\varepsilon_x u \curvearrowright_{\mathcal{R}} u$ and $\overline{u}\varepsilon_x \curvearrowright_{\mathcal{R}} \overline{u}$, which hold for every u with source x.

Finally, assume that θ' is an extension of θ to $\mathcal{S}^* \times \mathcal{S}^*$ that satisfies (4.7)–(4.10). Assume that u, v are $\mathcal{S}$-paths and $\theta^*(u,v)$ is defined. Then $\theta^*(v,u)$ is defined as well and we have $\overline{u}v \curvearrowright_{\mathcal{R}}^n \theta^*(u,v)\overline{\theta^*(v,u)}$ for some n. We prove using induction on n that $\theta'(u,v)$ exists as well and coincides with $\theta^*(u,v)$, which will show the minimality and uniqueness of θ^*. For $n = 0$, at least one of the words u, v must be empty, and the result follows from the fact that (4.10) is satisfied by θ^* by construction and by θ' by assumption. For $n = 1$, the paths u and v must have length one, that is, there exist s, t in $\mathcal{S}$ satisfying $u = s$,

$v = t$. Then the result follows from the fact that θ^* extends θ and the assumption that θ' extends θ as well and satisfies (4.7). Assume now $n \geqslant 2$. Then one of the words u, v must have length at least two. Assume v does. Decompose v into $v_1 v_2$. Then we have

$$\overline{u}v \curvearrowright^{n_1}_{\mathcal{R}} \theta^*(u, v_1)\overline{\theta^*(v_1, u)},$$
$$\overline{\theta^*(v_1, u)}v_2 \curvearrowright^{n_2}_{\mathcal{R}} \theta^*(\theta^*(v_1, u), v_2)\overline{\theta^*(v_2, \theta^*(v_1, u))}$$

for some $n_1, n_2 < n$. By the induction hypothesis, we deduce that $\theta'(u, v_1)$, $\theta'(v_1, u)$, $\theta'(\theta'(v_1, u), v_2)$, and $\theta'(v_2, \theta'(v_1, u))$ exist and coincide with their θ^* counterpart. Using the assumption that θ' satisfies (4.8) and (4.9), we deduce that $\theta'(u, v_1 v_2)$ and $\theta'(v_1 v_2, u)$ exist and coincide with $\theta^*(u, v_1 v_2)$ and $\theta^*(v_1 v_2, u)$, respectively. Thus θ' includes θ^* as a family of pairs. □

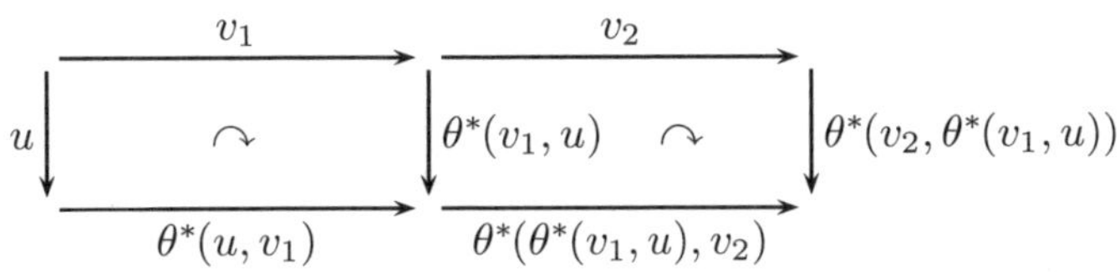

Figure 8. Splitting the right-reversing of a product (complemented case)

So the function θ^* of Lemma 4.6 coincides with that of Lemma 4.32. The double induction in the relations (4.8) and (4.9), which may look mysterious at first, becomes quite natural in the context of reversing diagrams and Figure 8. However, it should be kept in mind that this induction may not terminate, as was seen in Example 4.28.

Even in the case of a right-complemented presentation, right-reversing is not a deterministic procedure: a path may contain several patterns $\overline{s}|t$, leading to several reversing sequences. This ambiguity is however superficial, since Lemma 4.32 shows that, in the right-complemented case, every initial path leads to a unique maximal reversing diagram, and we obtain a deterministic procedure by deciding (for instance) to reverse the leftmost subpath $\overline{s}|t$ of the current path. Then right-reversing becomes an algorithmic procedure.

Algorithm 4.33 (Right-reversing, right-complemented case).

Context: A precategory $\mathcal{S}$, a syntactic right-complement θ on $\mathcal{S}$
Input: A signed $\mathcal{S}$-path w
Output: A positive–negative $\mathcal{S}$-path, or `fail`, or no output

while $\exists i < \lg(w)$ $(w[i] \in \overline{\mathcal{S}}$ and $w[i+1] \in \mathcal{S})$
  put $j := \min\{i \mid w[i] \in \overline{\mathcal{S}}$ and $w[i+1] \in \mathcal{S}\}$
  put $s := \overline{w[j]}$ and $t := w[j+1]$
  if $s = t$ **then**
    remove $\overline{s}|t$ in w
  else
    if $\theta(s, t)$ is defined **then**
      replace $\overline{s}|t$ with $\theta(s, t)|\overline{\theta(t, s)}$ in w at position j
    else
      return `fail`
return w

We recognize the three cases of the trichotomy (4.29)–(4.31): success when there exists a positive–negative path to which the initial path is right-reversible (case of (4.29)), failure when, at some point, one reaches a path with a factor that cannot be reversed (case of (4.30)), non-termination when one never goes out of the **while** loop (case of (4.31)).

4.4 Right-reversing: completeness

Introducing right-reversing is possible for every category presentation. However, this leads to interesting results (such as left-cancellativity) only when some specific condition, called completeness, is satisfied. The intuition behind completeness is that when a presentation is complete for right-reversing the a priori complicated equivalence relation associated with $\mathcal{R}$ can be replaced with the more simple right-reversing relation. As can be expected, this condition follows from the assumptions (4.17)–(4.19) of Proposition 4.16 in the context of a right-complemented presentation.

Before introducing completeness, we start from an easy general property that connects right-reversing with path equivalence.

Proposition 4.34 (reversing implies equivalence). *If $(\mathcal{S}, \mathcal{R})$ is a category presentation, then, for all $\mathcal{S}$-paths u, v, u', v', the relation $\overline{u}v \curvearrowright_{\mathcal{R}} v'\overline{u'}$ implies $uv' \equiv^+_{\mathcal{R}} vu'$. In particular, for u, v with target y,*

$$\overline{u}v \curvearrowright_{\mathcal{R}} \varepsilon_y \quad \textit{implies} \quad u \equiv^+_{\mathcal{R}} v. \tag{4.35}$$

Proof. We use induction on the number of steps n needed to right-reverse $\overline{u}v$ into $v'\overline{u'}$. For $n = 0$, the only possibility is that u or v is empty, in which case we have $u' = u$, $v' = v$, and the result is true. For $n = 1$, the only possibility is that u and v have length one, that is, there exist s and t in $\mathcal{S}$ satisfying $u = s$ and $v = t$. In this case, for $\overline{s}t$ to right-reverse to $v'\overline{u'}$ means that $sv' = tu'$ is a relation of $\mathcal{R}$, and $sv' \equiv^+_{\mathcal{R}} tu'$ holds by definition. For $n \geqslant 2$, at least one of u, v has length at least two. Assume for instance $\lg(v) \geqslant 2$. Then let $v = v_1v_2$ be a decomposition of v with v_1 and v_2 nonempty. Applying Lemma 4.24, we obtain u', v'_1, v'_2 satisfying $v' = v'_1v'_2$, $\overline{u}v_1 \curvearrowright^{n_1}_{\mathcal{R}} v'_1\overline{u_1}$, and $\overline{u_1}v_2 \curvearrowright^{n_2}_{\mathcal{R}} v'_2\overline{u'}$ with $n_1, n_2 < n$. The induction hypothesis implies $uv'_1 \equiv^+_{\mathcal{R}} v_1u_1$ and $u_1v'_2 \equiv^+_{\mathcal{R}} v_2u'$, whence $uv' = uv'_1v'_2 \equiv^+_{\mathcal{R}} v_1u_1v'_2 \equiv^+_{\mathcal{R}} v_1v_2u' = vu'$. □

In the context of a syntactic right-complement, $\overline{u}v \curvearrowright_{\mathcal{R}} \theta^*(u,v)\overline{\theta^*(v,u)}$ holds whenever $\theta^*(u,v)$ is defined, so Proposition 4.34 implies

Corollary 4.36 (reversing implies equivalence). *If $(\mathcal{S}, \mathcal{R})$ is a right-complemented category presentation, associated with the syntactic right-complement θ, then, for all $\mathcal{S}$-paths u, v such that $\theta^*(u, v)$ is defined, we have*

$$u\theta^*(u,v) \equiv^+_{\mathcal{R}} v\theta^*(v,u). \tag{4.37}$$

We now restate the implication of Proposition 4.34 in still another form.

Definition 4.38 (factorable). If $(\mathcal{S}, \mathcal{R})$ is a category presentation, a quadruple $(u, v, \hat{u}, \hat{v})$ of $\mathcal{S}$-paths is said to be *factorable through right-reversing*, or $\curvearrowright$*-factorable*, if there exist $\mathcal{S}$-paths u', v', w' satisfying

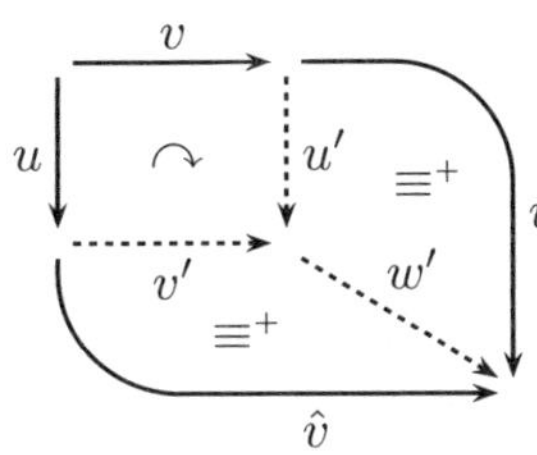

$$\overline{u}v \curvearrowright_{\mathcal{R}} v'\overline{u'}, \quad \hat{u} \equiv^+_{\mathcal{R}} u'w', \quad \text{and} \quad \hat{v} \equiv^+_{\mathcal{R}} v'w',$$

as shown on the right.

Lemma 4.39. *For every category presentation $(\mathcal{S}, \mathcal{R})$, every $\curvearrowright$-factorable quadruple $(u, v, \hat{u}, \hat{v})$ satisfies $u\hat{v} \equiv^+_{\mathcal{R}} v\hat{u}$.*

Proof. Assume that $(u, v, \hat{u}, \hat{v})$ is $\curvearrowright$-factorable. With the notation of Definition 4.38, we have $\overline{u}v \curvearrowright_{\mathcal{R}} v'\overline{u'}$, and Proposition 4.34 implies $uv' \equiv^+_{\mathcal{R}} vu'$. Then we deduce $u\hat{v} \equiv^+_{\mathcal{R}} uv'w \equiv^+_{\mathcal{R}} vu'w \equiv^+_{\mathcal{R}} v\hat{u}$. □

Thus saying that $(u, v, \hat{u}, \hat{v})$ is $\curvearrowright$-factorable means that $u\hat{v} \equiv^+_{\mathcal{R}} v\hat{u}$ holds and that the latter equivalence somehow factors through a right-reversing, whence our terminology.

The notion of completeness. We shall be interested in the case when the converse of Lemma 4.39 is satisfied, that is, when *every* equivalence is $\curvearrowright$-factorable.

Definition 4.40 (complete). Right-reversing is called *complete* for a presentation $(\mathcal{S}, \mathcal{R})$ if every quadruple $(u, v, \hat{u}, \hat{v})$ of $\mathcal{S}$-paths satisfying $u\hat{v} \equiv^+_{\mathcal{R}} v\hat{u}$ is $\curvearrowright$-factorable.

Owing to Lemma 4.39, if right-reversing is complete for $(\mathcal{S}, \mathcal{R})$, then for all $\mathcal{S}$-paths $u, v, \hat{u}, \hat{v}$ the relation $u\hat{v} \equiv^+_{\mathcal{R}} v\hat{u}$ is equivalent to $(u, v, \hat{u}, \hat{v})$ being $\curvearrowright$-factorable.

Example 4.41 (complete). Every category $\mathcal{C}$ admits a presentation for which right-reversing is complete. Indeed, let $\mathcal{R}$ be the family of all relations $f_1 \cdots f_p = g_1 \cdots g_q$ for $p, q \geqslant 1$ and $f_1, ..., g_q$ in $\mathcal{C}$ satisfying $f_1 \cdots f_p = g_1 \cdots g_q$. Then $(\mathcal{C}, \mathcal{R})$ is a presentation of $\mathcal{C}$. Now assume $u\hat{v} \equiv^+_{\mathcal{R}} v\hat{u}$. Then $u\hat{v}$ and $v\hat{u}$ have the same evaluation in $\mathcal{C}$, hence $u\hat{v} = v\hat{u}$ is a relation of $\mathcal{R}$. If both u and v are nonempty, we have $\overline{u}v \curvearrowright_{\mathcal{R}} \hat{v}\overline{\hat{u}}$ by definition, and $u' = \hat{u}$, $v' = \hat{v}$, and w' witness that $(u, v, \hat{u}, \hat{v})$ is $\curvearrowright$-factorable. If u is empty, we put $u' = \varepsilon_y$, where y is the target of v, $v' = v$, and $w = \hat{u}$; then $v \curvearrowright_{\mathcal{R}} v'$ holds and u', v', w' witness that $(u, v, \hat{u}, \hat{v})$ is $\curvearrowright$-factorable. So right-reversing is (trivially) complete for the presentation $(\mathcal{C}, \mathcal{R})$.

On the other hand, right-reversing is not complete for all presentations. Let M be the (stupid!) presented monoid $\langle S \mid R \rangle^+$ with $S = \{\mathtt{a}, \mathtt{b}, \mathtt{c}\}$ and $R = \{\mathtt{a} = \mathtt{b}, \mathtt{b} = \mathtt{c}\}$. In M, which is a free monoid on one generator, we have $\mathtt{a} = \mathtt{c}$, that is, $\mathtt{a} \equiv^+_R \mathtt{c}$ holds in S^*. However, $\overline{\mathtt{a}}|\mathtt{c}$ is right-reversible to no word of the form $v\overline{u}$, since there is no relation $\mathtt{a}... = \mathtt{c}...$ in R. So right-reversing cannot be complete for (S, R). By contrast, it is complete for the (redundant) presentation $\langle \mathtt{a}, \mathtt{b}, \mathtt{c} \mid \mathtt{a} = \mathtt{b}, \mathtt{b} = \mathtt{c}, \mathtt{a} = \mathtt{c} \rangle^+$ of M: in this special case, the failure of completeness amounts to a lack of transitivity.

In this text, we shall only consider completeness in the context of presentations with no ε-relations, in which case completeness takes a more simple form.

Lemma 4.42. *If $(\mathcal{S}, \mathcal{R})$ is a category presentation containing no ε-relation, then right-reversing is complete for $(\mathcal{S}, \mathcal{R})$ if and only if, for all $\mathcal{S}$-paths u, v with target y,*

$$u \equiv^+_{\mathcal{R}} v \quad \textit{implies} \quad \overline{u}v \curvearrowright_{\mathcal{R}} \varepsilon_y. \tag{4.43}$$

Proof. Assume that right-reversing is complete for $(\mathcal{S}, \mathcal{R})$ and $u \equiv^+_{\mathcal{R}} v$ holds. Then we have $u\varepsilon_y \equiv^+_{\mathcal{R}} v\varepsilon_y$, so the quadruple $(u, v, \varepsilon_y, \varepsilon_y)$ is $\curvearrowright$-factorable, so there exist $\mathcal{S}$-paths u', v', w' satisfying $\overline{u}v \curvearrowright_{\mathcal{R}} v'\overline{u'}$, $\varepsilon_y \equiv^+_{\mathcal{R}} u'w'$, and $\varepsilon_y \equiv^+_{\mathcal{R}} v'w'$. As $\mathcal{R}$ contains no ε-relation, the latter two equivalences imply that u' and v' must be empty, so the first relation reads $\overline{u}v \curvearrowright_{\mathcal{R}} \varepsilon_y$.

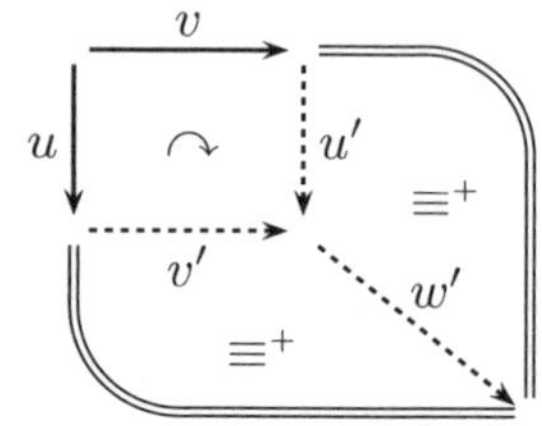

Conversely, assume that $u \equiv^+_{\mathcal{R}} v$ implies $\overline{u}v \curvearrowright_{\mathcal{R}} \varepsilon_y$. Assume $u\hat{v} \equiv^+_{\mathcal{R}} v\hat{u}$. Then, by assumption, $\overline{u\hat{v}}v\hat{u} \curvearrowright_{\mathcal{R}} \varepsilon_y$ holds, where y is the target of $\hat{u}$. By Lemma 4.23, there exist $\mathcal{S}$-paths u', v', w', w'' satisfying $\overline{u}v \curvearrowright_{\mathcal{R}} v'\overline{u'}$, $\overline{u'}\hat{v} \curvearrowright_{\mathcal{R}} \varepsilon_y$, $\overline{\hat{u}}v' \curvearrowright_{\mathcal{R}} \varepsilon_y$, and $\overline{w'}w'' \curvearrowright_{\mathcal{R}} \varepsilon_y$. By Proposition 4.34, we deduce $\hat{u} \equiv^+_{\mathcal{R}} u'w''$, $\hat{v} \equiv^+_{\mathcal{R}} v'w'$, and $w' \equiv^+_{\mathcal{R}} w''$, whence also $\hat{u} \equiv^+_{\mathcal{R}} u'w'$: so $(u, v, \hat{u}, \hat{v})$ is $\curvearrowright$-factorable, and right-reversing is complete for $(\mathcal{S}, \mathcal{R})$. □

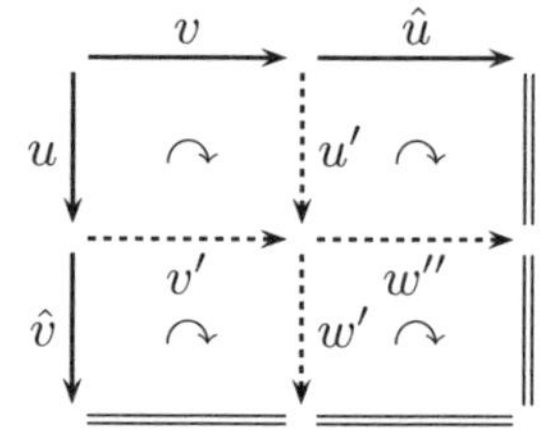

So, when there is no ε-relation, right-reversing is complete for a presentation $(\mathcal{S}, \mathcal{R})$ if $\mathcal{R}$-equivalence can always be detected by right-reversing.

Consequences of completeness. When right-reversing is complete for a category presentation $(\mathcal{S}, \mathcal{R})$, then certain properties of the category $\langle \mathcal{S} \mid \mathcal{R} \rangle^+$ can be recognized easily.

Proposition 4.44 (left-cancellativity). *If $(\mathcal{S}, \mathcal{R})$ is a category presentation for which right-reversing is complete, then the following conditions are equivalent:*

(i) *The category $\langle \mathcal{S} \mid \mathcal{R} \rangle^+$ is left-cancellative.*

(ii) *For every s in $\mathcal{S}$ and every relation $su = sv$ in $\mathcal{R}$ (if any), we have $u \equiv^+_{\mathcal{R}} v$.*

Proof. Assume that $\langle \mathcal{S} \mid \mathcal{R} \rangle^+$ is left-cancellative and $su = sv$ is a relation of $\mathcal{R}$. Then $su \equiv^+_{\mathcal{R}} sv$ holds by definition, hence so does $u \equiv^+_{\mathcal{R}} v$. So (i) implies (ii).

Conversely, assume (ii). In order to prove that $\langle \mathcal{S} \mid \mathcal{R} \rangle^+$ is left-cancellative, it is sufficient to prove that, for all u, v in $\mathcal{S}^*$, if $su \equiv^+_{\mathcal{R}} sv$ holds for some s in $\mathcal{S}$, then $u \equiv^+_{\mathcal{R}} v$ holds as well. So assume $su \equiv^+_{\mathcal{R}} sv$. By the definition of completeness, (s, s, u, v) is $\curvearrowright$-factorable: there exist $\mathcal{S}$-paths u', v', and w' satisfying $\overline{s}s \curvearrowright_{\mathcal{R}} u'\overline{v'}$, $u \equiv^+_{\mathcal{R}} u'w'$, and $v \equiv^+_{\mathcal{R}} v'w'$. Now, by the definition of right-reversing, either $su' = sv'$ is a relation of $\mathcal{R}$, in which case $u' \equiv^+_{\mathcal{R}} v'$ holds by assumption, or u' and v' are empty, and $u' \equiv^+_{\mathcal{R}} v'$ holds as well. Then we find $u \equiv^+_{\mathcal{R}} u'w' \equiv^+_{\mathcal{R}} v'w' \equiv^+_{\mathcal{R}} v$, as claimed. □

Condition (ii) in Proposition 4.44 is vacuously true for a presentation that contains no relation $su = sv$. So we have

Corollary 4.45 (left-cancellative). *If $(\mathcal{S}, \mathcal{R})$ is a category presentation for which right-reversing is complete and $(\mathcal{S}, \mathcal{R})$ contains no relation of the form $su = sv$, then $\langle \mathcal{S} \mid \mathcal{R} \rangle^+$ is left-cancellative.*

We turn to common right-multiples.

Proposition 4.46 (common right-multiple). *If $(\mathcal{S}, \mathcal{R})$ is a category presentation for which right-reversing is complete, then, for all $\mathcal{S}$-paths u, v, the elements of $\langle \mathcal{S} \mid \mathcal{R} \rangle^+$ represented by u and v admit a common right-multiple if and only if there exist $\mathcal{S}$-paths u', v' satisfying $\overline{u}v \curvearrowright_{\mathcal{R}} v'\overline{u'}$. In this case, every common right-multiple is a right-multiple of an element represented by a path uv' with $\overline{u}v \curvearrowright_{\mathcal{R}} v'\overline{u'}$.*

Proof. Let f, g be the elements of $\langle \mathcal{S} \mid \mathcal{R} \rangle^+$ represented by u and v. Assume that h is a common right-multiple of f and g. This means that there exist $\hat{u}, \hat{v}$ in $\mathcal{S}^*$ satisfying $u\hat{v} \equiv^+_{\mathcal{R}} v\hat{u}$ and h is represented by $u\hat{v}$ (and $v\hat{u}$). By the definition of completeness, this implies the existence of u', v', and w satisfying $\overline{u}v \curvearrowright_{\mathcal{R}} v'\overline{u'}$, $\hat{u} \equiv^+_{\mathcal{R}} u'w$, and $\hat{v} \equiv^+_{\mathcal{R}} v'w$. By construction, h is a right-multiple of the element represented by uv'.

Conversely, assume that $\overline{u}v \curvearrowright_{\mathcal{R}} v'\overline{u'}$ holds for some u', v' in $\mathcal{S}^*$. By Proposition 4.34, this implies $uv' \equiv^+_{\mathcal{R}} vu'$, and the element of $\langle \mathcal{S} \mid \mathcal{R} \rangle^+$ represented by uv' and vu' is a common right-multiple of the elements represented by u and v. □

Corollary 4.47 (right-lcm). *If $(\mathcal{S}, \mathcal{R})$ is a right-complemented category presentation for which right-reversing is complete, then any two elements of $\langle \mathcal{S} \mid \mathcal{R} \rangle^+$ that admit a common right-multiple admit a right-lcm. Moreover, for all u, v in $\mathcal{S}^*$, the elements of $\langle \mathcal{S} \mid \mathcal{R} \rangle^+$ represented by u and v admit a common right-multiple if and only if there exist u', v' in $\mathcal{S}^*$ satisfying $\overline{u}v \curvearrowright_{\mathcal{R}} v'\overline{u'}$ and, in this case, their right-lcm is represented by uv'.*

Proof. Assume that the elements f, g of $\langle \mathcal{S} \mid \mathcal{R} \rangle^+$ represented by u and v admit a common right-multiple h. By Proposition 4.46, h is a right-multiple of an element h' represented by uv' and vu', where $\overline{u}v \curvearrowright_{\mathcal{R}} v'\overline{u'}$ holds. The assumption that $(\mathcal{S}, \mathcal{R})$ is right-complemented implies that there exists at most one pair of $\mathcal{S}$-paths u', v' satisfying $\overline{u}v \curvearrowright_{\mathcal{R}} v'\overline{u'}$. So the element h' is unique and does not depend on h. Therefore, it is a right-lcm of f and g. □

Recognizing completeness. Propositions 4.44 and 4.46 would remain useless if we had no practical criteria for recognizing whether right-reversing is complete. Such criteria do exist, as will be explained now.

Definition 4.48 (cube condition). (See Figure 9.) For $(\mathcal{S}, \mathcal{R})$ a category presentation and u, v, w in $\mathcal{S}^*$, we say that the *sharp cube condition* (*resp.* the *cube condition*) holds for (u, v, w) if

For all $u_0, u_1, u_2, v_0, v_1, v_2$ in $\mathcal{S}^*$ satisfying

$$\overline{u}w \curvearrowright_{\mathcal{R}} v_1\overline{u_0}, \quad \overline{w}v \curvearrowright_{\mathcal{R}} v_0\overline{u_1}, \quad \overline{u_0}v_0 \curvearrowright_{\mathcal{R}} v_2\overline{u_2} \quad \text{(if any)},$$

(4.49) there exist u', v', w' in $\mathcal{S}^*$ satisfying

$$\overline{u}v \curvearrowright_{\mathcal{R}} v'\overline{u'}, \quad \overline{u'}u_1 \curvearrowright_{\mathcal{R}} w'\overline{u_2}, \quad \overline{v_1}v' \curvearrowright_{\mathcal{R}} v_2\overline{w'}$$

$$(\textit{resp.} \quad \overline{u}v \curvearrowright_{\mathcal{R}} v'\overline{u'}, \quad u_1u_2 \equiv^+_{\mathcal{R}} u'w', \quad v_1v_2 \equiv^+_{\mathcal{R}} v'w').$$

For $\mathcal{S}'$ included in $\mathcal{S}^*$, we say that the (sharp) cube condition holds *on* $\mathcal{S}'$ if it holds for every triple of elements of $\mathcal{S}'$.

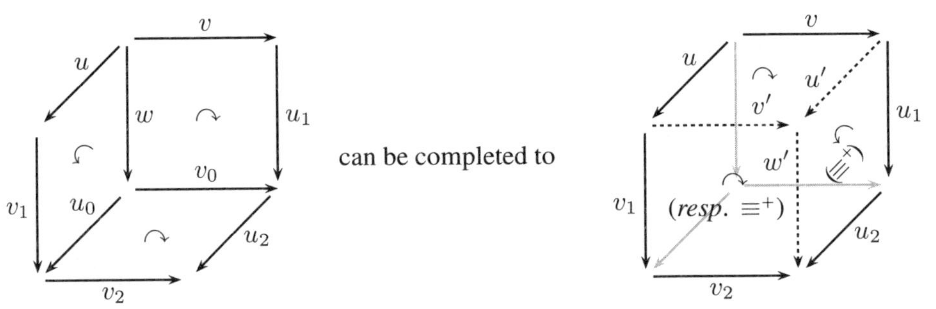

Figure 9. The sharp cube condition (*resp.* the cube condition) for (u, v, w): every half-cube constructed using right-reversing from u, v, w can be completed to a cube using three more right-reversings (sharp cube condition), or one right-reversing and two equivalences (cube condition).

Note that, in (4.49), the premise that $\overline{u}w$ and $\overline{w}v$ are right-reversible to positive–negative paths requires that u, v, and w share the same source. So the (sharp) cube condition vacuously holds for triples of paths not sharing the same source. By Proposition 4.34, $\overline{u'}u_1 \curvearrowright_{\mathcal{R}} w'\overline{u_2}$ implies $u_1u_2 \equiv^+_{\mathcal{R}} u'w'$ and $\overline{v_1}v' \curvearrowright_{\mathcal{R}} v_2\overline{w'}$ implies $v_1v_2 \equiv^+_{\mathcal{R}} v'w'$, so the sharp cube condition implies the cube condition, making the terminology coherent.

Example 4.50 (cube condition). Consider the presentation with two generators $\mathtt{a}, \mathtt{b}$ and two relations $\mathtt{a}^2 = \mathtt{b}^2$, $\mathtt{ab} = \mathtt{ba}$. The cube condition for $\mathtt{a}, \mathtt{a}, \mathtt{b}$ holds, but the sharp cube condition is not. Indeed, the word $\overline{\mathtt{a}}|\mathtt{b}|\overline{\mathtt{b}}|\mathtt{a}$ (that is, $\overline{u}w\overline{w}v$ with $u = v = \mathtt{a}$ and $w = \mathtt{b}$) can be right-reversed in six different ways to a positive–negative word, leading to the six words $\mathtt{b}|\overline{\mathtt{b}}$, $\mathtt{a}|\overline{\mathtt{a}}$, $\mathtt{a}|\mathtt{a}|\overline{\mathtt{b}}|\overline{\mathtt{b}}$, $\mathtt{a}|\mathtt{b}|\overline{\mathtt{a}}|\overline{\mathtt{b}}$, $\mathtt{b}|\mathtt{b}|\overline{\mathtt{a}}|\overline{\mathtt{a}}$, and $\mathtt{b}|\mathtt{a}|\overline{\mathtt{b}}|\overline{\mathtt{a}}$. Now, for each of these six words, the factorization corresponding to the right diagram of Figure 9 exists. For instance, consider $\mathtt{a}|\mathtt{b}|\overline{\mathtt{a}}|\overline{\mathtt{b}}$, which corresponds to $\hat{u} = \mathtt{b}|\mathtt{a}$ and $\hat{v} = \mathtt{a}|\mathtt{b}$: taking $u' = v' = \varepsilon$ and $w' = \mathtt{a}|\mathtt{b}$ works, as we have $\overline{\mathtt{a}}|\mathtt{a} \curvearrowright \varepsilon\overline{\varepsilon}$, $\mathtt{b}|\mathtt{a} \equiv^+ \varepsilon\mathtt{a}|\mathtt{b}$, and $\mathtt{a}|\mathtt{b} \equiv^+ \varepsilon\mathtt{a}|\mathtt{b}$. However, the sharp cube condition fails, as there exist no word w' satisfying $\overline{\mathtt{b}}\varepsilon \curvearrowright \mathtt{a}\overline{w}$ and $\overline{\varepsilon}\mathtt{a} \curvearrowright w\overline{\mathtt{b}}$, see Figure 10. The other verifications are similar.

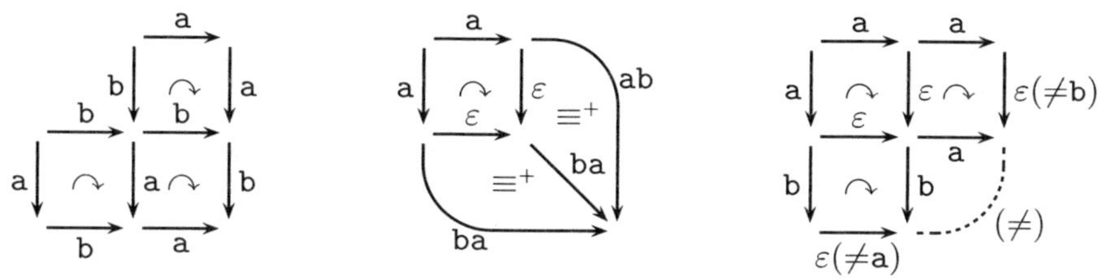

Figure 10. Checking the cube condition for $(\mathtt{a}, \mathtt{a}, \mathtt{b})$ in $(\mathtt{a}, \mathtt{b}, \mathtt{ab} = \mathtt{ba}, \mathtt{a}^2 = \mathtt{b}^2)$: the left diagram corresponds to one of the six ways of right-reversing $\overline{\mathtt{a}}|\mathtt{b}|\overline{\mathtt{b}}|\mathtt{a}$; the central diagram then shows that the corresponding instance of the cube property is satisfied, whereas the right diagram shows that the sharp cube property fails: one can complete the diagram with three squares so that one is a right-reversing and the other two are equivalences, but there is no way to complete the diagram with three right-reversings.

It is easily checked that, if right-reversing is complete for a presentation $(\mathcal{S}, \mathcal{R})$, then the cube condition must hold on $\mathcal{S}^*$, see Exercise 24. What we shall do now is to establish that, conversely, the cube condition guarantees the completeness of right-reversing, the point being that, in good cases, it is enough to check the cube condition for arbitrary triples of paths, but simply for triples of generators. Here is the main result:

Proposition 4.51 (completeness). *Assume that $(\mathcal{S}, \mathcal{R})$ is a category presentation with no ε-relation, and at least one of the following conditions is satisfied:*

(4.52) *The presentation $(\mathcal{S}, \mathcal{R})$ is short and the sharp cube condition holds on $\mathcal{S}$.*

(4.53) *The presentation $(\mathcal{S}, \mathcal{R})$ is right-Noetherian and the cube condition holds on $\mathcal{S}$.*

(4.54) *The presentation $(\mathcal{S}, \mathcal{R})$ is maximal right-triangular.*

Then right-reversing is complete for $(\mathcal{S}, \mathcal{R})$.

Before addressing the proof of Proposition 4.51, let us observe that the latter implies the results of Subsection 4.1. As can be expected, the point is that the θ-cube condition of Subsection 4.1 is directly connected with the current cube condition.

Lemma 4.55. *If the presentation $(\mathcal{S}, \mathcal{R})$ is associated with a syntactic right-complement θ, then, for all $\mathcal{S}$-paths u, v, w, the sharp cube condition (resp. the cube condition) holds on $\{u, v, w\}$ whenever the sharp θ-cube condition (resp. the θ-cube condition) does.*

Proof. Assume that the θ-cube condition holds on $\{u, v, w\}$. We shall show that so does the cube condition. As the assumptions assign symmetric roles to u, v, w, it is enough to consider the cube condition for (u, v, w). So assume $\overline{u}w \curvearrowright_{\mathcal{R}} v_1\overline{u_0}$, $\overline{w}v \curvearrowright_{\mathcal{R}} v_0\overline{u_1}$, and $\overline{u_0}v_0 \curvearrowright_{\mathcal{R}} v_2\overline{u_2}$. The connection of $\curvearrowright_{\mathcal{R}}$ with θ^* implies

$$\begin{aligned} u_0 &= \theta^*(w, u), & u_1 &= \theta^*(v, w), & u_2 &= \theta_3^*(w, v, u), \\ v_0 &= \theta^*(w, v), & v_1 &= \theta^*(u, w), & v_2 &= \theta_3^*(w, u, v), \end{aligned}$$

see Figure 11. As the θ-cube condition holds for (u, v, w), the assumption that $\theta_3^*(w, v, u)$ is defined implies that $\theta_3^*(u, v, w)$ and $\theta_3^*(v, u, w)$ are defined as well, which, in turn, implies that $\theta^*(u, v)$ and $\theta^*(v, u)$ are defined. Put $u' = \theta^*(v, u)$, $v' = \theta^*(u, v)$, $w' =$

$\theta_3^*(u,v,w)$. Then $\overline{u}v \curvearrowright_{\mathcal{R}} v'\overline{u'}$ holds by definition. Next, as $\theta_3^*(v,u,w)$ and $\theta_3^*(u,w,v)$ are defined, we have

$$\overline{u'}u_1 = \overline{\theta^*(v,u)}\theta^*(v,w) \curvearrowright_{\mathcal{R}} \theta_3^*(v,u,w)\overline{\theta_3^*(v,w,u)} \tag{4.56}$$

$$\overline{v_1}v' = \overline{\theta^*(u,w)}\theta^*(u,v) \curvearrowright_{\mathcal{R}} \theta_3^*(u,w,v)\overline{\theta_3^*(u,v,w)}, \tag{4.57}$$

which, by Corollary 4.36, implies

$$u_1\theta_3^*(v,w,u) \equiv_{\mathcal{R}}^+ u'\theta_3^*(v,u,w) \tag{4.58}$$

$$v_1\theta_3^*(u,w,v) \equiv_{\mathcal{R}}^+ v'\theta_3^*(u,v,w) = v'w'. \tag{4.59}$$

As the θ-cube condition holds for (v,w,u), we have $\theta_3^*(v,w,u) \equiv_{\mathcal{R}}^+ \theta_3^*(w,v,u) = u_2$. As the θ-cube condition holds for (u,v,w), we have $\theta_3^*(v,u,w) \equiv_{\mathcal{R}}^+ \theta_3^*(u,v,w) = w'$, so (4.58) implies $u_1u_2 \equiv_{\mathcal{R}}^+ u'w'$. Similarly, as the θ-cube condition holds for (w,u,v), we have $\theta_3^*(u,w,v) \equiv_{\mathcal{R}}^+ \theta_3^*(w,u,v) = v_2$, and (4.59) implies $v_1v_2 \equiv_{\mathcal{R}}^+ v'w'$. Hence the cube condition holds for (u,v,w).

Assume now that the sharp θ-cube condition holds on $\{u,v,w\}$, that is, for (u,v,w), (v,w,u), and (w,u,v). Then the sharp θ-cube condition for (v,w,u) and (u,v,w) implies

$$\theta_3^*(v,w,u) = \theta_3^*(w,v,u) = u_2 \quad \text{and} \quad \theta_3^*(u,v,w) = \theta_3^*(v,u,w),$$

so (4.58) gives $\overline{u'}u_1 \curvearrowright_{\mathcal{R}} w'\overline{u_2}$. Similarly, the sharp θ-cube condition for (v,w,u) implies $\theta_3^*(u,w,v) = \theta_3^*(w,u,v) = v_2$, so (4.59) gives $\overline{v'}u_1 \curvearrowright_{\mathcal{R}} w'\overline{v_2}$. Hence the sharp cube condition holds for (u,v,w). □

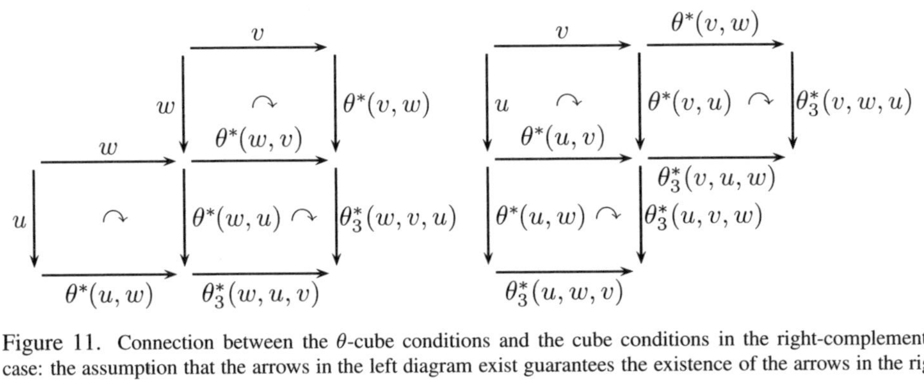

Figure 11. Connection between the θ-cube conditions and the cube conditions in the right-complemented case: the assumption that the arrows in the left diagram exist guarantees the existence of the arrows in the right diagram.

The connection of Lemma 4.55 also goes in the other direction, from the cube condition to the θ-cube condition, but the connection is local in the sharp case only: in the general case, what can be proved is that, if the cube condition holds for all triples of $\mathcal{S}$-paths, then the θ-cube condition holds for all triples of $\mathcal{S}$-paths, see Exercise 25.

Proof of Proposition 4.16 *from Proposition* 4.51. Assume that $(\mathcal{S},\mathcal{R})$ is a right-complemented category, associated with the syntactic right-complement θ. By Lemma 4.55, the (sharp) cube condition holds whenever the (sharp) θ-cube condition does. A priori, there is a difference between the assumptions in Propositions 4.16 and 4.51, namely that, in the former, only triples of pairwise distinct elements are mentioned. However, one easily checks that, if at least two entries in (u, v, w) coincide, then the θ-cube property necessarily holds. Owing to symmetries, it is enough to consider the cases $u = v$ and $u = w$.

So assume that $\theta_3^*(w, u, v)$ is defined and $u = v$ holds. This implies that $\theta^*(w, u)$, and, therefore, $\theta^*(u, w)$, are defined. Let y be the target of u. Then we have $\theta^*(u, v) = \varepsilon_y$, and, by definition, $\theta_3^*(u, v, w)$ is defined, and equal to $\theta^*(u, w)$. And so is $\theta_3^*(v, u, w)$, which is the same expression. Hence, the sharp θ-cube condition holds for (u, v, w).

Assume now that $\theta_3^*(w, u, v)$ is defined and we have $u = w$. This implies that $\theta^*(w, v)$, which is also $\theta^*(u, v)$, and, therefore, $\theta^*(v, w)$, which is $\theta^*(v, u)$, are defined. Let y be the target of u again, and let z be that of $\theta^*(u, v)$. Then we have $\theta^*(u, w) = \varepsilon_y$, whence $\theta_3^*(u, v, w) = \theta^*(\theta^*(u, v), \varepsilon_y) = \varepsilon_z$. On the other hand, as $\theta^*(v, u)$ is defined and has target z, so $\theta_3^*(v, u, w)$ is ε_z, and we obtain $\theta_3^*(u, v, w) = \theta_3^*(v, u, w)$, and the sharp θ-cube condition holds for (u, v, w).

So, in all three cases of Proposition 4.16, the presentation $(\mathcal{S}, \mathcal{R})$ is eligible for Proposition 4.51 and, therefore, right-reversing is complete for $(\mathcal{S}, \mathcal{R})$. Then Corollary 4.45 states that the category $\langle \mathcal{S} \mid \mathcal{R} \rangle^+$ is left-cancellative, and Corollary 4.47 states that $\langle \mathcal{S} \mid \mathcal{R} \rangle^+$ admits conditional right-lcms together with the additional results. Finally, for (iii), we observe that u and v represent the same element of $\langle \mathcal{S} \mid \mathcal{R} \rangle^+$ if the latter element is the right-lcm of the elements represented by u and v. By (ii), this happens if and only if $\theta^*(u)$ and $\theta^*(v)$ exist and are empty. □

Proof of Proposition 4.51. At this point, it only remains to establish Proposition 4.51. The arguments are different in the three cases and, as details require some care, we shall postpone most of the proofs to the final Appendix.

In the case of (4.52), the proof consists in using an induction on the length of paths to stack the elementary cubes provided by the assumption so as to extend the sharp cube condition from $\mathcal{S}$ to $\mathcal{S}^*$, and then deduce the completeness of right-reversing at the end. So the two steps are:

Lemma 4.60. *If $(\mathcal{S}, \mathcal{R})$ is a short category presentation and the sharp cube condition holds on $\mathcal{S}$, then the sharp cube condition holds on $\mathcal{S}^*$.*

Lemma 4.61. *If $(\mathcal{S}, \mathcal{R})$ is a category presentation that contains no ε-relation and the cube condition holds on $\mathcal{S}^*$, then right-reversing is complete for $(\mathcal{S}, \mathcal{R})$.*

In the case of (4.53), the argument, which is is more delicate, relies on several nested inductions involving a right-Noetherianity witness. In this case, one does not establish the cube condition on $\mathcal{S}^*$, but directly prove that the cube condition on $\mathcal{S}$ implies the completeness of right-reversing.

Lemma 4.62. *If $(\mathcal{S}, \mathcal{R})$ is a category presentation that is right-Noetherian, contains no ε-relation, and the cube condition holds on $\mathcal{S}$, then right-reversing is complete for $(\mathcal{S}, \mathcal{R})$.*

Finally, in the (very) special case of a right-triangular presentation, the proof uses an induction on the combinatorial distance and does not involve the cube condition.

Lemma 4.63. *If $(\mathcal{S}, \mathcal{R})$ is a maximal right-triangular presentation, then right-reversing is complete for $(\mathcal{S}, \mathcal{R})$.*

When the proofs of the above lemmas are completed, Proposition 4.51 follows, and therefore all results in this section are established, in particular the recipes of Subsection 4.1.

Remark 4.64 (left-reversing). Everything said so far about right-reversing has a natural counterpart in terms of *left-reversing*, which is the symmetric transformation in which one replaces a pattern of the form $t|\overline{s}$ (s, t in $\mathcal{S}$) with $\overline{u}v$ such that $ut = vs$ is a relation of the considered presentation $(\mathcal{S}, \mathcal{R})$, or one deletes a pattern of the form $s|\overline{s}$ (s in $\mathcal{S}$). Results for left-reversing are entirely similar to those for right-reversing. As in the case of left- and right-divisibility, going from right- to left-reversing just amounts to going from a category to the opposite category. In particular, in the case of a presentation for which left-reversing is complete, the counterparts of Propositions 4.44 and 4.46 provide criteria for right-cancellativity and existence of left-lcms. It should be kept in mind that left-reversing (denoted by $\curvearrowleft$) is not the inverse of right-reversing: for each generator s with target y, the relation $\overline{s}|s \curvearrowright \varepsilon_y$ holds, whereas $\varepsilon_y \curvearrowleft \overline{s}|s$ does *not* hold.

Exercises

Exercise 2 (functor). Assume that $\mathcal{C}$ is a category and ϕ is a functor from $\mathcal{C}$ into itself. Show that, if ϕ is injective (*resp.* surjective) on $\mathcal{C}$, then it is automatically injective (*resp.* surjective) on $\mathcal{O}bj(\mathcal{C})$. [Hint: Use the identity elements.]

Exercise 3 (collapsing invertible elements). (i) Assume that M is a left-cancellative monoid, G is a group, and ϕ is a homomorphism from M to $\mathrm{Aut}(G)$. Show that the semidirect product $M \ltimes_\phi G$ satisfies the conditions (i)–(iii) of Proposition 1.18. [Hint: Observe that GM is included in MG.] (ii) Assume that M is a monoid with no nontrivial invertible element and G is a group with at least two elements. Show that the free product $M * G$ is a monoid that does not satisfy the conditions (i)–(iii) of Proposition 1.18. [Hint: If g is a nontrivial element of M and f is a nontrivial element of G, then fg is an element of $M * G$ that does not belong to MG.]

Exercise 4 (atom). Assume that $\mathcal{C}$ is a left-cancellative category. Show that, for $n \geqslant 1$, every element g of $\mathcal{C}$ satisfying $\mathrm{ht}(g) = n$ admits a decomposition into a product of n atoms.

Exercise 5 (unique right-mcm). Assume that $\mathcal{C}$ is a left-cancellative category that admits right-mcms. (i) Assume thatf, g are elements of $\mathcal{C}$ that admit a common right-multiple and any two right-mcms of f and g are $=^\times$-equivalent. Show that every right-mcm of f and g is a right-lcm of f and g. (ii)

Exercise 6 (right-gcd to right-mcm). Assume that $\mathcal{C}$ is a cancellative category that admits right-gcds, f, g are elements of $\mathcal{C}$ and h is a common right-multiple of f and g. Show that there exists a right-mcm h_0 of f and g such that every common right-multiple of f and g that left-divide h is a right-multiple of h_0.

Exercise 7 (iterated lcm). Assume that $\mathcal{C}$ is a left-cancellative category. (i) Show that, if h is a right-lcm of f and g, then h_0h is a right-lcm of h_0f and h_0g. (ii) Deduce an alternative proof of Proposition 2.12.

Exercise 8 (conditional right-lcm). Assume that $\mathcal{C}$ is a left-cancellative category. (i) Show that every left-gcd of fg_1 and fg_2 (if any) is of the form fg where g is a left-gcd of g_1 and g_2. (ii) Assume moreover that $\mathcal{C}$ admits conditional right-lcms. Show that, if g is a left-gcd of g_1 and g_2 and fg is defined, then fg is a left-gcd of fg_1 and fg_2.

Exercise 9 (left-coprime). Assume that $\mathcal{C}$ is a left-cancellative category. Say that two elements f, g of $\mathcal{C}$ sharing the same source x are *left-coprime* if 1_x is a left-gcd of f and g. Assume that g_1, g_2 are elements of $\mathcal{C}$ sharing the same source, and fg_1 and fg_2 are defined. Consider the properties (i) The elements g_1 and g_2 are left-coprime; (ii) The element f is a left-gcd for fg_1 and fg_2. Show that (ii) implies (i) and that, if $\mathcal{C}$ admits conditional right-lcms, (i) implies (ii). [Hint: Use Exercise 8.]

Exercise 10 (subgroupoid). Let $M = \langle \mathtt{a}, \mathtt{b}, \mathtt{c}, \mathtt{d} \mid \mathtt{ab} = \mathtt{bc} = \mathtt{cd} = \mathtt{da} \rangle^+$ and $\Delta = \mathtt{ab}$. (i) Check that M is a Garside monoid with Garside element Δ. (ii) Let M_1 (*resp.* M_2) be the submonoid of M generated by $\mathtt{a}$ and $\mathtt{c}$ (*resp.* $\mathtt{b}$ and $\mathtt{c}$). Show that M_1 and M_2 are free monoids of rank 2 with intersection reduced to $\{1\}$. (iii) Let G be the group of fractions of M. Show that the intersection of the subgroups of G generated by M_1 and M_2 is not $\{1\}$.

Exercise 11 (weakly right-cancellative). Say that a category $\mathcal{C}$ is weakly right-cancellative if $gh = h$ implies that g is invertible. (i) Observe that a right-cancellative category is weakly right-cancellative; (ii) Assume that $\mathcal{C}$ is a left-cancellative category. Show that $\mathcal{C}$ is weakly right-cancellative if and only if, for all f, g in $\mathcal{C}$, the relation $f \precneqq g$ is equivalent to the conjunction of $f \precnapprox g$ and "$g = g'f$ holds for no g' in $\mathcal{C}^\times$".

Exercise 12 (right-Noetherian). Let $M = \langle \mathtt{a}, \mathtt{b} \mid \mathtt{ba} = \mathtt{b} \rangle^+$. (i) Show that every element of M has a unique expression $\mathtt{a}^p\mathtt{b}^q$ with $p, q \geqslant 0$, and that the map $\lambda : M \to \omega^2$ defined by $\lambda(\mathtt{a}^p\mathtt{b}^q) = \omega \cdot q + p$ is an anti-isomorphism of M to the monoid $(\omega^2, +)$. (ii) Show that $f \neq 1$ implies $\lambda(fg) > \lambda(g)$. (iii) Deduce that M satisfies the condition of Proposition 2.28(ii), but does not satisfy the condition of Proposition 2.28(iv). (iv) Conclude about the necessity of the assumption that $\mathcal{C}$ is left-cancellative in Proposition 2.28.

Exercise 13 (increasing sequences). Assume that $\mathcal{C}$ is a left-cancellative. For $\mathcal{S}$ included in $\mathcal{C}$, put $\mathrm{Div}_{\mathcal{S}}(h) = \{f \in \mathcal{C} \mid \exists g \in \mathcal{S}(fg = h)\}$. Show that the restriction of $\precneqq$ to $\mathcal{S}$ is well-founded (that is, admits no infinite descending sequence) if and only if, for every g in $\mathcal{S}$, every strictly increasing sequence in $\mathrm{Div}_{\mathcal{S}}(g)$ with respect to left-divisibility is finite.

Exercise 14 (lifted omega monoid). (See Figure 12) (i) Let $M_\omega = \langle \mathtt{a}, (\mathtt{b}_i)_{i \in \mathbb{Z}} \mid \mathtt{b}_{i+1} = \mathtt{b}_i\mathtt{a}$ for each $i \rangle^+$. Show that M_ω identifies with the family of all words $\mathtt{a}^p u$ with $p \geqslant 0$ and u a word in the alphabet $\{\mathtt{b}_i \mid i \in \mathbb{Z}\}$ equipped with the product defined by $\mathtt{a}^p u\, \mathtt{a}^q v = \mathtt{a}^{p+q} v$ if u is empty, and $\mathtt{a}^p u\, \mathtt{a}^q v = \mathtt{a}^p u' \mathtt{b}_{i+q} v$ if u is $u'\mathtt{b}_i$. (ii) Show that M_ω is generated by every family $\{\mathtt{a}\} \cup \{\mathtt{b}_i \mid i \leqslant C\}$, where C is any (positive or negative) constant, and that M_ω admits no $\mathbb{N}$-valued right-Noetherianity witness. (iii) Show that M_ω is cancellative. [Hint: Establish explicit formulas for $\mathtt{a} \cdot \mathtt{a}^p u$, $\mathtt{b}_i \cdot \mathtt{a}^p u$, $\mathtt{a}^p u \cdot \mathtt{a}$, and

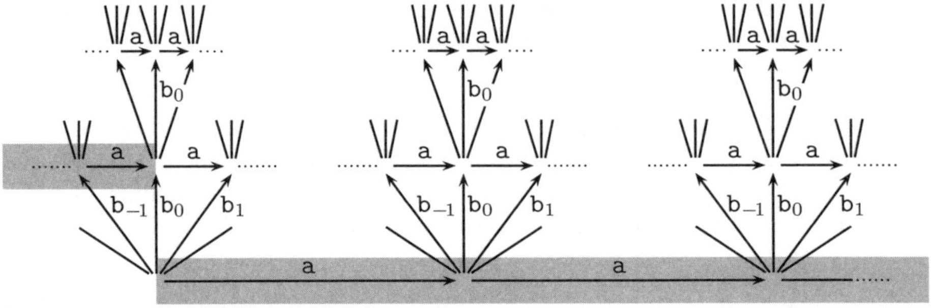

Figure 12. Cayley graph of the monoid of Exercise 14: an $\mathbb{N}$-indexed sequence of infinitely branching trees, with additional horizontal arrows labeled $\mathtt{a}$ and oriented to the right.

$\mathtt{b}_i \cdot \mathtt{a}^p u$ and deduce that the value of g can be recovered from the value of sg, and from the value of gs.] (iv) Show that M_ω is right-Noetherian, but not left-Noetherian. [Hint: Consider λ defined by $\lambda(\mathtt{a}^p u) = \omega \cdot \lg(u) + p$.] Show that the right height of $\mathtt{b}_0$ is not defined, whereas its right height is ω. (v) Let M denote the submonoid of $\mathbb{Z}^2$ consisting of the pairs (x, y) with $y > 0$ plus the pairs $(x, 0)$ with $x \geqslant 0$. Put $\mathtt{a} = (1, 0)$ and $\mathtt{b}_i = (i, 1)$ for $i \in \mathbb{Z}$. Show that M is generated by $\mathtt{a}$ and all elements $\mathtt{b}_i$, and it admits a presentation consisting, in addition to all commutation relations, of the relations $\mathtt{b}_{i+1} = \mathtt{b}_i\mathtt{a}$. Deduce that M is the abelianization of M_ω. Show that, in M, we have $(x, y) \preccurlyeq (x', y')$ if and only if $y < y'$ or $y = y'$ and $x \leqslant x'$. Deduce that M is not right-Noetherian.

Exercise 15 (left-generating). Assume that $\mathcal{C}$ is a left-cancellative category that is right-Noetherian. Say that a subfamily $\mathcal{S}$ of $\mathcal{C}$ *left-generates* (*resp. right-generates*) $\mathcal{C}$ if every non-invertible element of $\mathcal{C}$ admits at least one non-invertible left-divisor (*resp.* right-divisor) belonging to $\mathcal{S}$. (i) Show that $\mathcal{C}$ is right-generated by its atoms. (ii) Show that, if $\mathcal{S}$ is a subfamily of $\mathcal{C}$ that left-generates $\mathcal{C}$, then $\mathcal{S} \cup \mathcal{C}^\times$ generates $\mathcal{C}$. (iii) Conversely, show that, if $\mathcal{S} \cup \mathcal{C}^\times$ generates $\mathcal{C}$ and $\mathcal{C}^\times\mathcal{S} \subseteq \mathcal{S}^\sharp$ holds, then $\mathcal{S}$ left-generates $\mathcal{C}$.

Exercise 16 (atoms not generating). Let $M = \langle \mathtt{a}, \mathtt{b}, \mathtt{e}, \mathtt{E} \mid \mathtt{eE} = \mathtt{Ee} = 1, \mathtt{ea} = \mathtt{ab} \rangle^+$, and let $\mathcal{S}$ consist of the rewrite rules $\mathtt{eE} \to \varepsilon$, $\mathtt{Ee} \to \varepsilon$, $\mathtt{ea} \to \mathtt{ab}$, $\mathtt{Eab} \to \mathtt{a}$. (i) Show that every $\mathcal{S}$-derivation is finite [Hint: Each rule decreases the number of letters $\mathtt{e}, \mathtt{E}$.] (ii) Check that $\mathcal{S}$ is locally confluent, that is, if two rules apply to a word w, leading to two words w', w'', then there exists a word w''' such that w' and w'' rewrite into w''. (iii) Conclude that $\mathcal{S}$ defines a unique normal form for the elements of M. (iv) Show that M is cancellative. [Hint: Show that, for each letter s, the normal form of g can be recovered from that of sg.] (v) Show that M is right-Noetherian, but it is not Noetherian and not generated by atoms and invertible elements. [Hint: The sequence $\mathtt{a}, \mathtt{Ea}, \mathtt{E}^2\mathtt{a}, ...$ is descending with respect to left-divisibility, and $\mathtt{a}$ is not at atom.] (vi) Let M' be the monoid $\langle \mathtt{a}, \mathtt{b}, \mathtt{e} \mid \mathtt{e}^2 = 1, \mathtt{ea} = \mathtt{ab} \rangle^+$. Prove that M' is not left-cancellative. [Hint: $\mathtt{ab}^2 = \mathtt{a}$ holds in M'.]

Exercise 17 (no $\mathbb{N}$-valued Noetherianity witness I). Let M be the monoid presented by $\langle \mathtt{a}_1, \mathtt{a}_2, \cdots \mid \mathtt{a}_1 = \mathtt{a}_2^2 = \mathtt{a}_3^3 = \cdots, \mathtt{a}_i\mathtt{a}_j = \mathtt{a}_j\mathtt{a}_i \text{ for all } i, j \rangle^+$. (i) Show that every element of M has a unique expression as $\mathtt{a}_1^{d_1}\mathtt{a}_2^{d_2} \cdots$ with $d_i \leqslant i$ for every $i \geqslant 2$. (ii) Show that the function λ defined by $\lambda(\prod_{i\geqslant 1} \mathtt{a}_i^{d_i}) = \omega \cdot d_1 + \sum_{i\geqslant 2} d_i$ is a Noetherianity witness for M. (iii) Show that the height of $\mathtt{a}_1$ is infinite.

Exercise 18 (no $\mathbb{N}$-valued Noetherianity witness II). Let M be the monoid presented by $\langle \mathtt{a}, \mathtt{b}, \mathtt{e}, \mathtt{E} \mid \mathtt{eE} = \mathtt{Ee} = 1, \mathtt{eab}^n\mathtt{ae} = \mathtt{ab}^{n+1}\mathtt{a}, n = 0, 1, ...\rangle^+$. (i) Show that the rewrite rules $\mathtt{eE} \to \varepsilon$, $\mathtt{Ee} \to \varepsilon$, $\mathtt{eab}^n\mathtt{ae} \to \mathtt{ab}^{n+1}\mathtt{a}$, $\mathtt{Eab}^{n+1}\mathtt{a} \to \mathtt{ab}^n\mathtt{ae}$, $\mathtt{ab}^{n+1}\mathtt{aE} \to \mathtt{eab}^n\mathtt{a}$ with $n \geqslant 0$ define a unique normal form for the elements of M. [Hint: The rules are length-decreasing, so it suffices to check local confluence, on the shape of Exercise 16.] (ii) Deduce that M is cancellative. (iii) Show that M is left- and right-Noetherian. (iv) Show that the height of $\mathtt{a}^2$ is infinite.

Exercise 19 (compatibility). Assume that $(\mathcal{S}, \mathcal{R})$ is a right-complemented presentation, associated with the syntactic right-complement θ. Prove that right-reversing is complete for $(\mathcal{S}, \mathcal{R})$ if and only if θ^* is compatible with $\equiv_{\mathcal{R}}^+$ in the sense that, if $\theta^*(u, v)$ is defined, then $\theta^*(u', v')$ exists and is $\mathcal{R}$-equivalent to $\theta^*(u, v)$ for all u', v' satisfying $u' \equiv_{\mathcal{R}}^+ u$ and $v' \equiv_{\mathcal{R}}^+ v$.

Exercise 20 (equivalence). Assume that $(\mathcal{S}, \mathcal{R})$ is a category presentation. Say that an element s of $\mathcal{S}$ is $\mathcal{R}$-right-invertible if $sw \equiv_{\mathcal{R}}^+ \varepsilon_x$ (x the source of s) holds for some w in $\mathcal{S}^*$. Show that a category presentation $(\mathcal{S}, \mathcal{R})$ is complete with respect to right-reversing if and only if, for all u, v in $\mathcal{S}^*$, the following are equivalent: (i) u and v are $\mathcal{R}$-equivalent (that is, $u \equiv_{\mathcal{R}}^+ v$ holds), (ii) $\overline{u}v \curvearrowright_{\mathcal{R}} v'\overline{u'}$ holds for some $\mathcal{R}$-equivalent paths u', v' in $\mathcal{S}^*$ all of which entries are $\mathcal{R}$-right-invertible.

Exercise 21 (two generators). Assume that M is $\langle \mathtt{a}, \mathtt{b} \mid \mathtt{a}v = \mathtt{b}u\rangle^+$ and there exists $\lambda : \{\mathtt{a}, \mathtt{b}\}^* \to \mathbb{N}$ that is $\equiv_{\mathtt{a}v=\mathtt{b}u}^+$-invariant and satisfies $\lambda(\mathtt{a}w) > \lambda(w)$ and $\lambda(\mathtt{b}w) > \lambda(w)$ for every w. Show that M is left-cancellative and admits conditional right-lcms. [Hint: Check that Condition (4.18) is satisfied.]

Exercise 22 (complete *vs.* cube, complemented case). Assume that $(\mathcal{S}, \mathcal{R})$ is a right-complemented presentation, associated with the syntactic right-complement θ. Show the equivalence of the following three properties: (i) Right-reversing is complete for $(\mathcal{S}, \mathcal{R})$; (ii) The map θ^* is compatible with $\equiv_{\mathcal{R}}^+$-equivalence in the sense that, if $u' \equiv_{\mathcal{R}}^+ u$ and $v' \equiv_{\mathcal{R}}^+ v$ hold, then $\theta^*(u', v')$ is defined if and only if $\theta(u, v)$ is and, if so, they are $\equiv_{\mathcal{R}}^+$-equivalent; (iii) The θ-cube condition holds for every triple of $\mathcal{S}$-paths. [Hint: For (i) $\Rightarrow$ (ii), if $\theta^*(u, v)$ is defined and $u' \equiv_{\mathcal{R}}^+ u$ and $v' \equiv_{\mathcal{R}}^+ v$ hold, then $(u', v,' \theta^*(u, v), \theta^*(v, u))$ must be $\curvearrowright$-factorable; for (ii) $\Rightarrow$ (iii), compute $\theta^*(w, u\theta^*(u, v))$ and $\theta^*(w, v\theta^*(v, u))$; for (iii) $\Rightarrow$ (i), use Lemma 4.61.]

Exercise 23 (alternative proof). Assume that $(\mathcal{S}, \mathcal{R})$ is a right-complemented presentation associated with the syntactic right-complement θ, that $(\mathcal{S}, \mathcal{R})$ is right-Noetherian, and that the θ-cube condition holds on $\mathcal{S}$. (i) Show that, for all r, s, t in $\mathcal{S}$, the path $\theta^*(r, s\theta(s, t))$ is defined if and only if $\theta^*(r, t\theta(t, s))$ is and, in this case, the relations $\theta^*(r, s\theta(s, t)) \equiv_{\mathcal{R}}^+ \theta^*(r, t\theta(t, s))$ and $\theta^*(s\theta(s, t), r) \equiv_{\mathcal{R}}^+ \theta^*(t\theta(t, s), r)$ hold. (ii) Show that the map θ^* is compatible with $\equiv_{\mathcal{R}}^+$: the conjunction of $u' \equiv_{\mathcal{R}}^+ u$ and $v' \equiv_{\mathcal{R}}^+ v$ implies that $\theta^*(u', v')$ exists if and only if $\theta^*(u, v)$ does and, in this case, they are $\equiv_{\mathcal{R}}^+$-equivalent. [Hint: Show using on $\lambda^*(u\theta^*(u, v))$, where λ^* is a right-Noetherianity witness for $(\mathcal{S}, \mathcal{R})$ that, if $\theta^*(u, v)$ is defined and we have $u' \equiv_{\mathcal{R}}^+ u$ and $v' \equiv_{\mathcal{R}}^+ v$, then $\theta^*(u', v')$ is defined and we have $\theta^*(u', v') \equiv^+_{\mathcal{R}} \theta^*(u, v)$ and $\theta^*(v', u') \equiv^+_{\mathcal{R}} \theta^*(v, u)$.] (iii) Apply Exercise 22 to deduce a new proof of Proposition 4.16 in the right-Noetherian case.

Exercise 24 (cube condition). Assume that $(\mathcal{S}, \mathcal{R})$ is a category presentation (i) Show that the cube condition holds for (u, v, w) if and only if, for all $\hat{u}, \hat{v}$ in $\mathcal{S}^*$ satisfying $\overline{u}w\overline{w}v \curvearrowright_{\mathcal{R}} \hat{v}\overline{\hat{u}}$, the quadruple $(u, v, \hat{u}, \hat{v})$ is $\curvearrowright$-factorable. (ii) Show that, if right-reversing is complete for $(\mathcal{S}, \mathcal{R})$, then the cube condition holds for every triple of $\mathcal{S}$-paths.

Exercise 25 (cube to θ-cube). Assume that $(\mathcal{S}, \mathcal{R})$ is a category presentation associated with a syntactic right-complement θ. (i) Show that, if the sharp cube condition holds on $\{u, v, w\}$, then so does the sharp θ-cube condition. (ii) Show that, if the cube condition holds on $\mathcal{S}^*$, then so does the θ-cube condition. [Hint: First observe that right-reversing must be complete for $(\mathcal{S}, \mathcal{R})$ and then, translating the cube condition for (u, v, w), show that the existence of w' satisfying $\theta^*(u, w)\theta^*_3(w, u, v) \equiv^+_{\mathcal{R}} \theta^*(u, v)w'$ implies that the path $\theta^*(\theta^*_3(w, u, v), \theta^*_3(u, w, v))$ exists and is empty.]

Exercise 26 (1-complete). Say that right-reversing is 1-complete for a category presentation $(\mathcal{S}, \mathcal{R})$ if $u\hat{v} \equiv^+_{\mathcal{R}} v\hat{u}$ implies that $(u, v, \hat{u}, \hat{v})$ is $\curvearrowright$-factorable whenever u, v have length one, that is, lie in $\mathcal{S}$. (i) Prove that, if right-reversing is 1-complete for $(\mathcal{S}, \mathcal{R})$ and $(\mathcal{S}, \mathcal{R})$ is short, then right-reversing is complete for $(\mathcal{S}, \mathcal{R})$. (ii) Same question with the assumption that $(\mathcal{S}, \mathcal{R})$ is right-Noetherian.

Notes

Sources and comments. The definitions of Section 1 are standard. A useful reference is S. MacLane's book [177]. However, we insist once more on the fact that, in this text, categories are only used as conditional monoids, and no specific familiarity with categories is needed.

The definitions of Section 2 and most of the results mentioned there are standard as well. The only less classical point is the emphasis put on the algebraic properties of the least common multiple operation and the derived right-complement operation, which can be found for instance in [80]. Note that, in a categorical terminology, right-lcms and left-gcds are called pushouts (or amalgamated sums) and pullbacks (or fiber products), respectively. The notion of a mimimal common right-multiple (right-mcm, Definition 2.38) is maybe new, but it is a straightforward extension of the notion of a least common right-multiple.

Noetherianity conditions are standard in many domains, with the meaning of forbidding infinite sequences. Our current convention is coherent with the standard definition in terms of non-existence of ascending sequences of ideals in ring theory, and both are related with the possibility of decomposing every element into a product of irreducible elements. Most results of Section 2 may be considered as folklore, yet some examples are maybe new.

All results around Ore's theorem in Section 3 also are standard and can be found, for instance, in the textbook by Clifford and Preston [62]. The only exception is maybe Proposition 3.21 (torsion) and Corollary 3.24 (torsion-free), which can be found in [83].

Right-reversing and its application to proving cancellativity as developed in Section 4 may be seen as an extension of Garside's proof of Theorem H in [124]—and, therefore, it should maybe be attributed to G. Higman, as a foreword in Garside's PhD thesis says "*I am indebted to my supervisor Professor G. Higman for the method of proof of Theorem H and Theorem 4, and I wish to thank him both for this and his generous interest and advise*". This method of proof is reminiscent of the "Kürzungslemma" in Brieskorn–Saito [36], and the monoids eligible for this approach were investigated as "chainable monoids" by R. Corran in [65] and [64], and, in a slightly different setting, by K. Tatsuoka in [225].

Right-reversing itself was first introduced in 1991 in [72, 75] in order to investigate the geometry monoid of left-selfdistributivity that will be described in Chapter XI. The connection with Garside's Theorem H [125] emerged subsequently when a number of presentations proved to be eligible, in particular Artin's presentation of braid groups, see [76]. The extension to presentations that are not right-complemented was developed in [82], see also [89]. We think that the method, which includes all above mentioned methods, is now optimal. The (easy) extension to a category context is probably new. The cancellativity criterion stemming from Proposition 4.51 is efficient for presentations with many relations, whereas Adjan's method explained in Remmers [201] is relevant for presentations with few relations. The completeness criterion of (4.52) appears (in a slightly restricted form) in [80], the one of (4.53) in [76], and the one of (4.54) in [92]. A fourth criterion, not developed here but used in an example of Chapter IX, appears in [78].

Reversing is useful only when it is complete, that is, when it misses no path equivalence. When we start with a right-Noetherian presentation $(\mathcal{S}, \mathcal{R})$ and right-reversing fails to be complete, the cube condition must fail for some triple of generators (r, s, t), and some relation of the form $s\cdots = t\cdots$ is satisfied in $\langle \mathcal{S} \mid \mathcal{R} \rangle^+$, but it is not detected by right-reversing. Adding the latter relation to $(\mathcal{S}, \mathcal{R})$ yields a new presentation $(\mathcal{S}, \mathcal{R}')$ of the same category, for which the (r, s, t) cube condition is satisfied by construction. Repeating the process provides a completion method that produces successive equivalent presentations $(\mathcal{S}, \mathcal{R}')$, $(\mathcal{S}, \mathcal{R}'')$, ... eventually leading in good cases to a presentation for which right-reversing is complete (see [82] for examples).

In terms of algorithmic complexity, Proposition 4.27 implies that reversing has a quadratic time complexity when the closure $\widehat{\mathcal{S}}$ of the initial family of generators $\mathcal{S}$ under reversing is finite—$\widehat{\mathcal{S}}$ being the smallest family of $\mathcal{S}$-paths such that, for all u, v in $\widehat{\mathcal{S}}$, reversing $\overline{u}v$ leads to positive–negative path(s) $v'\overline{u'}$ such that u' and v' belong to $\widehat{\mathcal{S}}$. When $\widehat{\mathcal{S}}$ is infinite, no uniform complexity bound may be given: the standard presentation of the Heisenberg monoid [82] or the presentation of Exercise 122 below lead to a cubic time complexity, whereas the only known upper bound for the complexity of the right-reversing in the of the presentation of the monoid M_{LD} of Definition XI.3.7 is a tower of exponentials of exponential height, see [75, Chapter VIII]. A comparison of reversing as a way of solving the Word Problem and Gröbner–Shirshov bases appears in Autord [5].

We do not claim that the reversing method is a universal tool: in practice, the method is efficient only for complemented presentations, and useful only for proving cancellativity and existence of least common multiples. However, as the latter questions are central in this text and complemented presentations naturally appear in many examples, reversing is specially useful here.

Further questions. All examples of monoids that are Noetherian, but not strongly Noetherian mentioned in the text or in Exercise 17 involve an infinite presentation, leading to

Question 3. Is every finitely generated left-cancellative category that is Noetherian necessarily strongly Noetherian?

We refer to Exercise 18 for an example of a finitely generated (but not finitely presented) cancellative monoid that is Noetherian but not strongly Noetherian.

Chapter III
Normal decompositions

We now enter the core of our subject, namely a certain type of decompositions for the elements of a monoid, a group, or a category, that exist in many different situations, in particular those mentioned in Chapter I. The common unifying principle in these decompositions is that each entry is, in a sense to be made precise, a maximal fragment of what lies after it. The approach we develop here is sort of reverse engineering: we define what we call normal decompositions by their expected properties, and Garside families to be those families that guarantee the existence of normal decompositions. This abstract definition is not suitable for concrete examples, but the quest for more concrete characterizations is postponed to the subsequent Chapters IV to VI. The advantage of this top-down approach is to make it clear that it exhausts the range of possible applications—contrary to the bottom-up approach that once led to Garside groups.

The chapter is organized as follows. In Section 1, we consider the case of monoids and categories and analyze the diagrammatic mechanisms involved in what we call $\mathcal{S}$-normal decompositions. In particular, we show that $\mathcal{S}$-normal decompositions are essentially unique and obey nice construction rules, the domino rule(s) (Proposition 1.53). In Section 2, we similarly consider the case of groups and groupoids and show how to extend, in good cases, $\mathcal{S}$-normal decompositions in a category to what we call symmetric $\mathcal{S}$-normal decompositions in the enveloping groupoid of this category. Again we establish existence and uniqueness results, the main one being that, if $\mathcal{C}$ is an Ore category that admits left-lcms and $\mathcal{S}$ is a Garside family of $\mathcal{C}$, then every element of the enveloping groupoid of $\mathcal{C}$ admits an essentially unique symmetric $\mathcal{S}$-normal decomposition (Proposition 2.20). Finally, in Section 3, we establish various geometric and algorithmic consequences of the existence of $\mathcal{S}$-normal decompositions. We prove that such decompositions are geodesic and we establish the Grid Property (Proposition 3.11), a convexity result that associates with every triangle a certain hexagonal grid in which any two vertices are connected by a geodesic, a result loosely reminiscent of a CAT(0) property. We also show that normal decompositions satisfy a form of the Fellow Traveller Property (Proposition 3.14) and that, when some finiteness assumptions are satisfied, they give rise to an automatic structure and to a practical method for computing homology (Corollary 3.55). Finally, we address the Word Problem (in a category and in its enveloping groupoid) and briefly analyze the complexity of the algorithms deduced from $\mathcal{S}$-normal decompositions (Corollary 3.64).

Main definitions and results

Definition 1.1 (greedy). If $\mathcal{S}$ is a subfamily of a left-cancellative category $\mathcal{C}$, a length-two $\mathcal{C}$-path $g_1|g_2$ in $\mathcal{C}^{[2]}$ is called *$\mathcal{S}$-greedy* if each relation $s \preccurlyeq f g_1 g_2$ with s in $\mathcal{S}$ implies $s \preccurlyeq f g_1$. A path $g_1|\cdots|g_p$ is called *$\mathcal{S}$-greedy* if $g_i|g_{i+1}$ is $\mathcal{S}$-greedy for each $i < p$.

Definition 1.17 **(normal, strict normal).** A path is $\mathcal{S}$-*normal* if it is $\mathcal{S}$-greedy and its entries lie in $\mathcal{S}^\sharp$. An $\mathcal{S}$-normal path is *strict* if no entry is invertible and all entries, except possibly the last one, lie in $\mathcal{S}$.

Definition 1.20 **(deformation).** A path $g'_1|\cdots|g'_q$ in a $\mathcal{C}^\times$-*deformation*, of another path $g_1|\cdots|g_p$ if there exist invertible elements $\epsilon_0, ..., \epsilon_m$, $m = \max(p, q)$, such that ϵ_0, ϵ_m are identity elements and $\epsilon_{i-1}g_i = g'_i\epsilon_i$ holds for $1 \leqslant i \leqslant m$; for $p \neq q$, the shorter path is expanded by identity elements.

Proposition 1.25 **(normal unique).** *If $\mathcal{S}$ is a subfamily of a left-cancellative category $\mathcal{C}$, any two $\mathcal{S}$-normal decompositions of an element of $\mathcal{C}$ (if any) are $\mathcal{C}^\times$-deformations of one another.*

Definition 1.31 **(Garside family).** A subfamily $\mathcal{S}$ of a left-cancellative category $\mathcal{C}$ is called a *Garside family* if every element of $\mathcal{C}$ admits an $\mathcal{S}$-normal decomposition.

Definition 1.38 **(property $\square$).** A family $\mathcal{S}$ satisfies $\square$ if every element of $\mathcal{S}^2$ admits an $\mathcal{S}$-greedy decomposition $s_1|s_2$ with s_1, s_2 in $\mathcal{S}$.

Proposition 1.39 **(recognizing Garside I).** *A subfamily $\mathcal{S}$ of a left-cancellative category $\mathcal{C}$ is a Garside family if and only one of the following equivalent conditions holds:* (1.40) *The family $\mathcal{S}^\sharp$ generates $\mathcal{C}$ and satisfies Property $\square$.* (1.41) *The family $\mathcal{S}^\sharp$ generates $\mathcal{C}$ and every element of $(\mathcal{S}^\sharp)^2$ admits an $\mathcal{S}$-normal decomposition.* (1.42) *The family $\mathcal{S}^\sharp$ generates $\mathcal{C}$, every element of $\mathcal{S}^2$ admits an $\mathcal{S}$-normal decomposition, and $\mathcal{C}^\times\mathcal{S} \subseteq \mathcal{S}^\sharp$ holds.*

Proposition 1.53 **(normal decomposition).** *If $\mathcal{S}$ is a Garside family in a left-cancellative category $\mathcal{C}$, then, for every $\mathcal{S}^\sharp$-path $s_1|\cdots|s_p$, Algorithm* 1.52 *running on $s_1|\cdots|s_p$ returns an $\mathcal{S}$-normal decomposition of $s_1\cdots s_p$. The $\square$-witness is appealed $p(p-1)/2$ times.*

Definition 2.1 **(left-disjoint).** Two elements f, g of a left-cancellative category $\mathcal{C}$ are called *left-disjoint* if f and g share the same source and, for all h, h' in $\mathcal{C}$, the conjunction of $h' \preccurlyeq hf$ and $h' \preccurlyeq hg$ implies $h' \preccurlyeq h$.

Definition 2.9 **(symmetric greedy, symmetric normal).** If $\mathcal{S}$ is a subfamily of a left-cancellative category $\mathcal{C}$, a negative–positive $\mathcal{C}$-path $\overline{f_q}|\cdots|\overline{f_1}|g_1|\cdots|g_p$ is called *symmetric $\mathcal{S}$-greedy* (*resp. symmetric $\mathcal{S}$-normal, resp. strictly symmetric $\mathcal{S}$-normal*) if the paths $f_1|\cdots|f_p$ and $g_1|\cdots|g_q$ are $\mathcal{S}$-greedy (*resp.* $\mathcal{S}$-normal, *resp.* strictly $\mathcal{S}$-normal) and, in addition, f_1 and g_1 are left-disjoint.

Proposition 2.16 **(symmetric normal unique).** *If $\mathcal{S}$ is a subfamily of a left-Ore category $\mathcal{C}$ and $\mathcal{S}^\sharp$ generates $\mathcal{C}$, any two symmetric $\mathcal{S}$-normal decompositions of an element of $\mathcal{E}nv(\mathcal{C})$ are $\mathcal{C}^\times$-deformations of one another.*

Proposition 2.20 **(symmetric normal exist).** *If $\mathcal{S}$ is a Garside family in a left-Ore category $\mathcal{C}$, the following conditions are equivalent:* (i) *Every element of $\mathcal{E}nv(\mathcal{C})$ lying in $\mathcal{C}\mathcal{C}^{-1}$ has a symmetric $\mathcal{S}$-normal decomposition;* (ii) *The category $\mathcal{C}$ admits left-lcms.*

Definition 2.22 **(Garside base).** A subfamily $\mathcal{S}$ of a groupoid $\mathcal{G}$ is a *Garside base* of $\mathcal{G}$ if every element of $\mathcal{G}$ admits a decomposition that is symmetric $\mathcal{S}$-normal with respect to the subcategory of $\mathcal{G}$ generated by $\mathcal{S}$.

Definition 2.29 **(strong Garside).** A Garside family $\mathcal{S}$ in a left-Ore category is called *strong* if, for all s, t in $\mathcal{S}^\sharp$ with the same target, there exist s', t' in $\mathcal{S}^\sharp$ such that $s't$ equals $t's$ and it is a left-lcm of s and t.

Proposition 2.34 **(strong exists).** *If $\mathcal{C}$ is a left-Ore category, then some Garside family of $\mathcal{C}$ is strong if and only if $\mathcal{C}$ viewed a Garside family of $\mathcal{C}$ is strong, if and only if $\mathcal{C}$ admits left-lcms.*

Proposition 2.37 **(symmetric normal, short case).** *If $\mathcal{S}$ is a strong Garside family in a left-Ore category $\mathcal{C}$ that admits left-lcms, then every element of $\mathcal{E}nv(\mathcal{C})$ that can be represented by a positive–negative $\mathcal{S}^\sharp$-path of length ℓ admits a symmetric $\mathcal{S}$-normal decomposition of length at most ℓ.*

Definition 3.6 **(perfect Garside).** A Garside family $\mathcal{S}$ in a left-Ore category $\mathcal{C}$ that admits left-lcms is called *perfect* if there exists a short left-lcm witness $\widetilde{\theta}$ on $\mathcal{S}^\sharp$ such that, for all s, t with the same target, $\widetilde{\theta}(s,t)t$ belongs to $\mathcal{S}^\sharp$.

Proposition 3.11 **(Grid Property).** *If $\mathcal{C}$ is an Ore category that admits unique left-lcms and $\mathcal{S}$ is a perfect Garside family in $\mathcal{C}$, then, for all g_1, g_2, g_3 in $\mathcal{E}nv(\mathcal{C})$, there exists a planar diagram Γ that is the union of three $\mathcal{S}$-grids, admits g_1, g_2, g_3 as extremal vertices, and is such that, for all vertices x, y of Γ, there exists a symmetric $\mathcal{S}$-normal (hence geodesic) path from x to y inside Γ.*

Proposition 3.14 **(fellow traveller).** (i) *If $\mathcal{S}$ is a subfamily of a left-Ore category $\mathcal{C}$ and $\mathcal{S}^\sharp$ generates $\mathcal{C}$, then, for every g in $\mathcal{E}nv(\mathcal{C})$, any two strict symmetric $\mathcal{S}$-normal decompositions of g are 1-fellow travellers.* (ii) *If $\mathcal{S}$ is a strong Garside family in $\mathcal{C}$, then, for every g in $\mathcal{E}nv(\mathcal{C})$ and every s in $\mathcal{S}^\sharp$, any two strict symmetric $\mathcal{S}$-normal decompositions of g and gs are 2-fellow travellers.*

Proposition 3.45 **(resolution).** *If $\mathcal{S}$ is a strong Garside family in a left-cancellative category $\mathcal{C}$ that admits unique left-lcms, the associated complex $(\mathcal{Z}_*, \partial_*)$ is a resolution of the trivial $\mathbb{Z}\mathcal{C}$-module $\mathbb{Z}$ by free $\mathbb{Z}\mathcal{C}$-modules.*

Corollary 3.46 **(homology).** *If $\mathcal{C}$ is an Ore category that admits unique left-lcms and has a finite Garside family, the groupoid $\mathcal{E}nv(\mathcal{C})$ is of FL type. For every n, we have $H_n(\mathcal{E}nv(\mathcal{C}), \mathbb{Z}) = H_n(\mathcal{C}, \mathbb{Z}) = \operatorname{Ker} \mathrm{d}_n / \operatorname{Im} \mathrm{d}_{n+1}$, where d_n is the $\mathbb{Z}$-linear map on $\mathcal{Z}_n$ such that $\mathrm{d}_n \mathcal{C}$ is obtained from $\partial_n \mathcal{C}$ by collapsing all $\mathcal{C}$-coefficients to 1.*

Corollary 3.55 **(dimension).** *If $\mathcal{S}$ is a strong Garside family in a left-Ore category $\mathcal{C}$ that admits unique left-lcms and the height of every element of $\mathcal{S}$ is bounded above by some constant N, the (co)homological dimension of $\mathcal{E}nv(\mathcal{C})$ is at most N.*

Corollary 3.64 **(decidability).** *If $\mathcal{S}$ is a strong Garside family in a left-Ore category $\mathcal{C}$ and there exist a computable $\square$-witness, a computable left-lcm witness on $\mathcal{S}^\sharp$, and a computable $=^\times$-maps for $\mathcal{S}^\sharp$, then the Word Problem for $\mathcal{E}nv(\mathcal{C})$ with respect to $\mathcal{S}^\sharp$ is decidable. Moreover, if $\mathcal{S}^\sharp$ is finite, the Word Problem for $\mathcal{E}nv(\mathcal{C})$ has a quadratic time complexity and a linear space complexity.*

1 Greedy decompositions

We begin with the case of categories (including monoids), and consider the generic question of finding distinguished decompositions for the elements of the category in terms of the elements of some fixed generating subfamily. As announced above, the main intuition is to construct the decomposition by extracting the maximal fragment lying in the considered family, maximal here referring to the left-divisibility relation (Definition II.2.1). Technically, we shall start from an abstract diagrammatic definition, that will be seen to be equivalent subsequently and that makes the inductive constructions easy and natural, in particular with the use of the so-called domino rule.

The main results of this section are Proposition 1.25, which says that an $\mathcal{S}$-normal decomposition is essentially unique when it exists, and Proposition 1.39, which characterizes Garside families, defined to be those families $\mathcal{S}$ such that every element of the ambient category admits an $\mathcal{S}$-normal decomposition.

The section is organized as follows. The notion of an $\mathcal{S}$-greedy path is introduced and studied in Subsection 1.1. Then, in Subsection 1.2, we define the notion of an $\mathcal{S}$-normal decomposition of an element and establish its (near)-uniqueness. In Subsection 1.3, we introduce the notion of a Garside family and establish general properties. Next, in Subsection 1.4, we give alternative definitions of Garside families that are more tractable than the initial definition, and we extract the algorithmic content of the constructions. Finally, in Subsection 1.5, we discuss an additional condition, the second domino rule, which provides an alternative way of determining greedy decompositions.

1.1 The notion of an $\mathcal{S}$-greedy path

As in Chapter II, (almost) all categories we shall consider below will be assumed to be left-cancellative, that is, $fg' = fg$ implies $g = g'$—see Remark 1.16 below for a discussion about removing this assumption. So, assuming that $\mathcal{C}$ is a left-cancellative category and that $\mathcal{S}$ is a fixed subfamily of $\mathcal{C}$, we shall introduce the notion of a $\mathcal{S}$-*greedy* path. The intuition is that a path $g_1|\cdots|g_p$ is $\mathcal{S}$-greedy if each factor g_i contains as much of the $\mathcal{S}$-content of $g_i \cdots g_p$ as possible. However, we do *not* just use the classical definition that a length-two path $g_1|g_2$ is $\mathcal{S}$-greedy if every element of $\mathcal{S}$ that left-divides g_1g_2 has to left-divide g_1. This notion is natural but uneasy to work with when $\mathcal{S}$ is arbitrary, and we shall use a stronger notion. It will be seen subsequently, in Chapter IV, that the two notions coincide when $\mathcal{S}$ has suitable properties.

Definition 1.1 (greedy path). If $\mathcal{S}$ is a subfamily of a left-cancellative category $\mathcal{C}$, a length-two $\mathcal{C}$-path $g_1|g_2$ in $\mathcal{C}^{[2]}$ is called $\mathcal{S}$-*greedy* if each relation $s \preccurlyeq fg_1g_2$ with s in $\mathcal{S}$ implies $s \preccurlyeq fg_1$. A path $g_1|\cdots|g_p$ is called $\mathcal{S}$-*greedy* if $g_i|g_{i+1}$ is $\mathcal{S}$-greedy for each $i < p$.

By definition, a path of length zero or one is always $\mathcal{S}$-greedy. We shall sometimes use "$\mathcal{S}$-greedy sequence" as synonymous with "$\mathcal{S}$-greedy path".

Example 1.2 (greedy). Consider the free abelian monoid $\mathbb{N}^n$ and $S_n = \{g \in \mathbb{N}^n \mid \forall k \leqslant n\ (f(k) \in \{0,1\})\}$ (Reference Structure 1, page 3). For f, g in $\mathbb{N}^n$, as noted in Example II.2.3, the left-divisibility relation $f \preccurlyeq g$ is equivalent to $\forall k \leqslant n\ (f(k) \leqslant g(k))$. In this context, Definition 1.1 amounts to requiring, for each s in S_n,

$$\forall k \leqslant n\ (s(k) \leqslant f(k) + g_1(k) + g_2(k) \Rightarrow s(k) \leqslant f(k) + g_1(k)). \tag{1.3}$$

If $s(k)$ is zero, (1.3) is always true. If $s(k)$ is one, the only case when (1.3) can fail is for $f(k) = g_1(k) = 0$ and $g_2(0) \geqslant 1$. For every k in $\{1, ..., n\}$, there exists an element s in S_n that satisfies $s(k) = 1$ (for instance the element Δ_n), so the conjunction of all implications (1.3) for s in S_n reduces to the unique condition

$$\forall k \leqslant n\ (g_1(k) = 0 \Rightarrow g_2(k) = 0). \tag{1.4}$$

Note that, for any set I, even infinite, the above analysis remains valid in the case of $\mathbb{N}^{(I)}$ and $S_I = \{g \in \mathbb{N}^{(I)} \mid \forall k \in I\ (f(k) \in \{0,1\})\}$.

Diagrams will be used throughout this text. In a diagram, the property that a path $g_1|g_2$ is $\mathcal{S}$-greedy will be indicated by a small arc connecting the arrows associated with g_1 and g_2, as in 

(here we assume that the reference family $\mathcal{S}$ is clear).

The definition of greediness should become immediately more transparent when stated in a diagrammatic way.

Lemma 1.5. *Assume that $\mathcal{C}$ is a left-cancellative category and $\mathcal{S}$ is included in $\mathcal{C}$. Then a $\mathcal{C}$-path $g_1|g_2$ is $\mathcal{S}$-greedy if and only if, for all s in $\mathcal{S}$ and f, h in $\mathcal{C}$,*

$$sh = fg_1g_2 \quad \textit{implies} \quad \exists g \in \mathcal{C}\ (sg = fg_1 \textit{ and } h = gg_2), \tag{1.6}$$

that is, if each commutative diagram 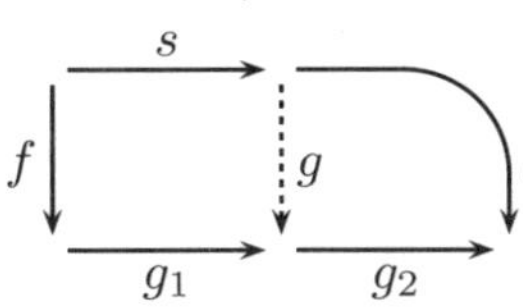

splits as shown.

Proof. Assume that $g_1|g_2$ is $\mathcal{S}$-greedy and $sh = fg_1g_2$ holds, that is, we have a commutative diagram 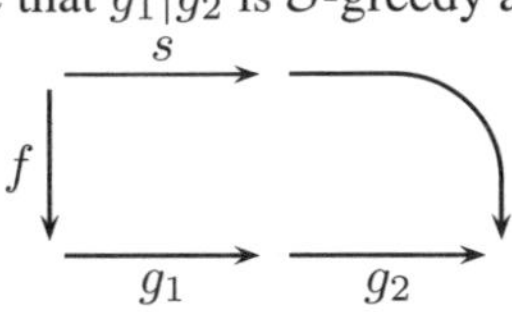

. Then $s \preccurlyeq fg_1g_2$ holds, hence so does $s \preccurlyeq fg_1$ by the definition of $\mathcal{S}$-greediness. So we have $sg = fg_1$ for some g. Moreover, we deduce $sh = fg_1g_2 = sgg_2$, hence $h = gg_2$ by left-cancelling s, which is legal since the ambient category $\mathcal{C}$ is left-cancellative. So (1.6) is true, and the diagram of the lemma splits into two commutative subdiagrams as shown.

Conversely, assume that (1.6) is true, and we have $s \preccurlyeq fg_1g_2$ for some s belonging to $\mathcal{S}$. By the definition of left-divisibility, we have $sh = fg_1g_2$ for some h, whence $sg = fg_1$ by (1.6), and $s \preccurlyeq fg_1$ holds. So $g_1|g_2$ is $\mathcal{S}$-greedy. ∎

A direct consequence is the following connection between greediness and the left- and right-divisibility relations.

Lemma 1.7. *Assume that $\mathcal{C}$ is a left-cancellative category and $\mathcal{S}$ is included in $\mathcal{C}$. If $g_1|g_2$ is $\mathcal{S}$-greedy, then so is $g_1'|g_2'$ for all g_1', g_2' such that g_1' is a left-multiple of g_1 and g_2' is a left-divisor of g_2.*

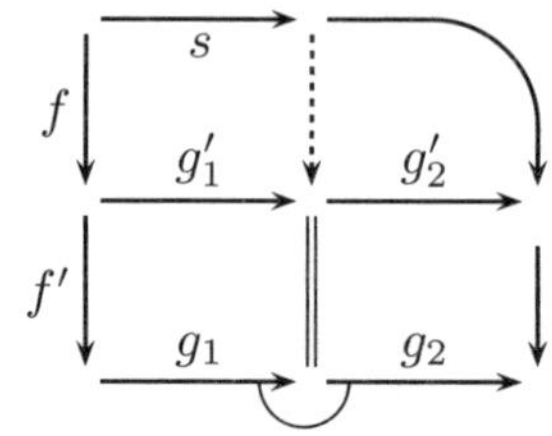

Proof. Assume $s \in \mathcal{S}$ and $s \preccurlyeq f g_1' g_2'$ (see figure on the right). Assuming $g_1' = f' g_1$, we deduce $s \preccurlyeq f f' g_1 g_2$, whence $s \preccurlyeq f f' g_1 = f g_1'$ since $g_1|g_2$ is $\mathcal{S}$-greedy. Hence $g_1'|g_2'$ is $\mathcal{S}$-greedy by Lemma 1.5. □

We continue with the connection between greedy paths and invertible elements.

Lemma 1.8. *Assume that $\mathcal{C}$ is a left-cancellative category, $\mathcal{S}$ is included in $\mathcal{C}$, and $g_1|g_2$ is a $\mathcal{C}$-path.*

(i) *If g_2 is invertible, then $g_1|g_2$ is $\mathcal{S}$-greedy.*

(ii) *If g_1 is invertible and g_2 lies in $\mathcal{S}$, then $g_1|g_2$ is $\mathcal{S}$-greedy (if and) only if g_2 is invertible.*

Proof. (i) If g_2 is invertible, then $g_1 =^{\times} g_1 g_2$ holds, and, therefore, for s in $\mathcal{S}$ and f in $\mathcal{C}$, the relation $s \preccurlyeq f g_1 g_2$ is merely equivalent to $s \preccurlyeq f g_1$. So, in any case, $g_1|g_2$ is greedy.

(ii) Assume that g_1 is invertible and g_2 lies in $\mathcal{C}(x, y)$. Applying Lemma 1.5 to the equality $g_2 \cdot 1_y = g_1^{-1} \cdot g_1 \cdot g_2$ gives the existence of g satisfying $g_2 \cdot g = g_1^{-1} \cdot g_1 = 1_x$. So g is an inverse of g_2. □

Finally, we observe that multiplying by invertible elements preserves greediness.

Lemma 1.9. *Assume that $\mathcal{C}$ is a left-cancellative category, $\mathcal{S}$ is included in $\mathcal{C}$, and we have a commutative diagram as on the right, with $\epsilon_0, \epsilon_1, \epsilon_2$ invertible. Then $g_1|g_2$ is $\mathcal{S}$-greedy if and only if $g_1'|g_2'$ is.*

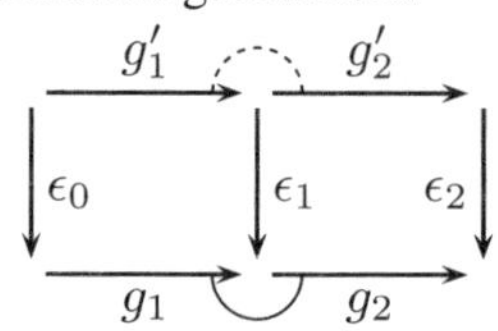

Proof. Assume that $g_1|g_2$ is $\mathcal{S}$-greedy and $s \preccurlyeq f g_1' g_2'$ holds for some s in $\mathcal{S}$. Then we have $s \preccurlyeq f g_1' g_2' \epsilon_2 = (f\epsilon_0) g_1 g_2$, whence $s \preccurlyeq (f\epsilon_0) g_1$ as $g_1|g_2$ is $\mathcal{S}$-greedy. Now we have $(f\epsilon_0) g_1 = f g_1' \epsilon_1$, hence $s \preccurlyeq f g_1' \epsilon_1$, and, therefore, $s \preccurlyeq f g_1'$ since ϵ_1 is invertible. Hence $g_1'|g_2'$ is $\mathcal{S}$-greedy.

The converse implication is similar, using ϵ_i^{-1} in place of ϵ_i. □

In the same vein, we observe now that adding invertible elements in the reference family $\mathcal{S}$ does not change the associated notion of greediness. We recall from Definition II.1.21 that, if $\mathcal{C}$ is a category and $\mathcal{S}$ is included in $\mathcal{C}$, then $\mathcal{S}^\sharp$ is $\mathcal{S}\mathcal{C}^\times \cup \mathcal{C}^\times$, the least subfamily of $\mathcal{C}$ that includes $\mathcal{S}$ and $1_\mathcal{C}$ and is closed under right-multiplication by invertible elements.

Lemma 1.10. *Assume that $\mathcal{C}$ is a left-cancellative category and $\mathcal{S}$ is included in $\mathcal{C}$. Then a $\mathcal{C}$-path is $\mathcal{S}$-greedy if and only if it is $\mathcal{S}^\sharp$-greedy, if and only if it is $\mathcal{C}^\times\mathcal{S}^\sharp$-greedy.*

Proof. The definition implies that, if a $\mathcal{C}$-path is $\mathcal{S}$-greedy, then it is $\mathcal{S}'$-greedy for every $\mathcal{S}'$ included in $\mathcal{S}$. By definition, we have $\mathcal{S} \subseteq \mathcal{S}^\sharp \subseteq \mathcal{C}^\times\mathcal{S}^\sharp$, so being $\mathcal{C}^\times\mathcal{S}^\sharp$-greedy implies being $\mathcal{S}^\sharp$-greedy, which in turn implies being $\mathcal{S}$-greedy.

Conversely, assume that $g_1|\cdots|g_p$ is $\mathcal{S}$-greedy, and let s be an element of $\mathcal{C}^\times\mathcal{S}^\sharp$, say $s = \epsilon_1 s' \epsilon_2$ with ϵ_1, ϵ_2 in $\mathcal{C}^\times$ and s' in $\mathcal{S}$. Assume $s \preccurlyeq f g_i g_{i+1}$. Then we have $s' \preccurlyeq (\epsilon_1^{-1} f) g_i g_{i+1}$. As s' belongs to $\mathcal{S}$ and $g_i|g_{i+1}$ is $\mathcal{S}$-greedy, this implies $s' \preccurlyeq (\epsilon_1^{-1} f) g_i$, say $(\epsilon_1^{-1} f) g_i = s' f'$. This implies in turn $s(\epsilon_2^{-1} f') = \epsilon_1 s' \epsilon_2 \epsilon_2^{-1} f' = \epsilon_1(\epsilon_1^{-1} f) g_i$, whence $s \preccurlyeq f g_i$. So $g_i|g_{i+1}$ is $\mathcal{C}^\times\mathcal{S}^\sharp$-greedy for each i, and $g_1|\cdots|g_p$ is $\mathcal{C}^\times\mathcal{S}^\sharp$-greedy. So the three properties are equivalent. □

Combining the above results, we deduce:

Lemma 1.11. *Assume that $\mathcal{C}$ is a left-cancellative category, $\mathcal{S}$ is included in $\mathcal{C}$, and we have $f_1 f_2 = g_1 g_2$ with f_1, f_2 in $\mathcal{S}^\sharp$ and $g_1|g_2$ $\mathcal{S}$-greedy.*

(i) *If f_1 or f_2 is non-invertible, then g_1 is non-invertible.*

(ii) *If f_1 or f_2 is invertible, then g_2 is invertible.*

Proof. (i) Assume that g_1 is invertible. We have $f_1 \cdot f_2 = g_1 g_2$ with f_1 in $\mathcal{S}^\sharp$ and, by Lemma 1.10, the path $g_1|g_2$ is $\mathcal{S}^\sharp$-greedy, so we deduce $f_1 \preccurlyeq g_1$, and the assumption that g_1 is invertible implies that f_1 is invertible. Then, we can write $f_2 = f_1^{-1} \cdot g_1 g_2$ with f_2 in $\mathcal{S}^\sharp$, whence $f_2 \preccurlyeq f_1^{-1} g_1$ as $g_1|g_2$ is $\mathcal{S}^\sharp$-greedy. This implies that f_2 is invertible, since f_1^{-1} and g_1 are.

(ii) Assume that f_1 is invertible. By assumption, f_2 belongs to $\mathcal{S}^\sharp$ and we have $f_2 = f_1^{-1} g_1 g_2$, whence $f_2 \preccurlyeq f_1^{-1} g_1 g_2$. The assumption that $g_1|g_2$ is $\mathcal{S}$-greedy, hence $\mathcal{S}^\sharp$-greedy by Lemma 1.10, implies $f_2 \preccurlyeq f_1^{-1} g_1$, say $f_1^{-1} g_1 = f_2 g$. Left-multiplying by f_1, we deduce $g_1 = f_1 f_2 g = g_1 g_2 g$, whence $g_2 g \in \mathbf{1}_\mathcal{C}$ by left-cancelling g_1. So g_2 is invertible.

If f_2 is invertible, $f_1 f_2$ belongs to $\mathcal{S}^\sharp$ and it left-divides $g_1 g_2$. As $g_1|g_2$ is $\mathcal{S}^\sharp$-greedy, we deduce that $f_1 f_2$ left-divides g_1 and, as above, that g_2 is invertible. □

We now observe that greediness is preserved when adjacent entries are gathered.

Proposition 1.12 (grouping entries). *Assume that $\mathcal{C}$ is a left-cancellative category and $\mathcal{S}$ is included in $\mathcal{C}$. If a path $g_1|\cdots|g_p$ is $\mathcal{S}$-greedy, then so is every path obtained from $g_1|\cdots|g_p$ by replacing adjacent entries by their product. In particular, $g_1|g_2\cdots g_p$ is $\mathcal{S}$-greedy, that is*

(1.13) *each relation $s \preccurlyeq f g_1 \cdots g_p$ with $s \in \mathcal{S}$ implies $s \preccurlyeq f g_1$.*

Proof. We first prove using decreasing induction on $q \leqslant p$ that, if $g_1|\cdots|g_p$ is $\mathcal{S}$-greedy and s lies in $\mathcal{S}$ and satisfies $s \preccurlyeq f g_1 \cdots g_p$, then, for each q with $1 \leqslant q \leqslant p$, we have $s \preccurlyeq f g_1 \cdots g_q$. For $q = p$, this is the initial assumption. Otherwise, the induction hypothesis gives $s \preccurlyeq f g_1 \cdots g_{q+1}$, that is, $s \preccurlyeq (f g_1 \cdots g_{q-1})\, g_q\, g_{q+1}$. Then the assumption that $g_q|g_{q+1}$ is $\mathcal{S}$-greedy implies $s \preccurlyeq (f g_1 \cdots g_{q-1})\, g_q$, and the induction goes on, see Figure 1.

Now assume that $g_1|\cdots|g_p$ is $\mathcal{S}$-greedy and $1 \leqslant q \leqslant r < p$ hold. Applying the above result to the $\mathcal{S}$-greedy path $g_{q-1}|\cdots|g_r$ shows that every element of $\mathcal{S}$ left-dividing $f g_{q-1}(g_q \cdots g_r)$ left-divides $f g_{q-1}$, which shows that $g_{q-1}|g_q \cdots g_r$ is $\mathcal{S}$-greedy. On the other hand, applying the above result to the $\mathcal{S}$-greedy path $g_q|\cdots|g_{r+1}$ shows that every element of $\mathcal{S}$ left-dividing $f(g_q \cdots g_r) g_{r+1}$ left-divides $f(g_q \cdots g_r)$, which shows that $g_q \cdots g_r|g_{r+1}$ is $\mathcal{S}$-greedy. Hence $g_1|\cdots|g_{q-1}|g_q \cdots g_r|g_{r+1}|\cdots|g_p$ is $\mathcal{S}$-greedy. □

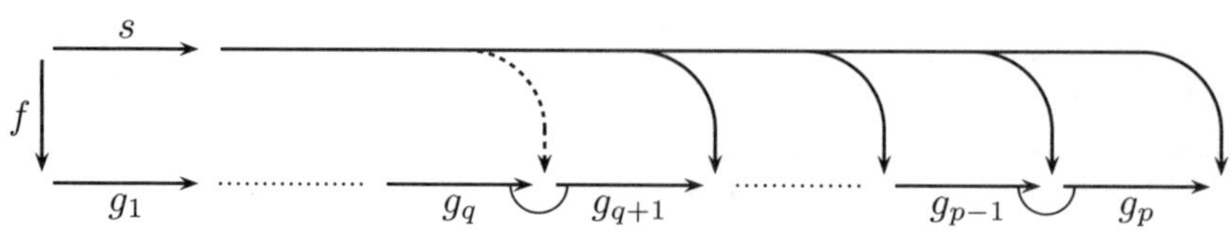

Figure 1. Inductive proof of Proposition 1.12.

We conclude with a result about $\mathcal{S}^m$-greedy paths. We recall from Notation II.1.10 that $\mathcal{S}^m$ denotes the family of all elements of the ambient category that can be expressed as the product of m elements of $\mathcal{S}$.

Proposition 1.14 (power). *If $\mathcal{S}$ is a subfamily of a left-cancellative category $\mathcal{C}$ and $g_1|\cdots|g_p$ is $\mathcal{S}$-greedy, then $g_1\cdots g_m|g_{m+1}\cdots g_p$ is $\mathcal{S}^m$-greedy for $1 \leqslant m \leqslant p$, that is,*

(1.15) *each relation $s \preccurlyeq fg_1\cdots g_p$ with s in $\mathcal{S}^m$ implies $s \preccurlyeq fg_1\cdots g_m$.*

The (easy) proof is left to the reader. Note that Propositions 1.12 and 1.14 imply that, if $g_1|g_2|g_3$ is $\mathcal{S}$-greedy, then $g_1|g_2g_3$ is $\mathcal{S}$-greedy, and $g_1g_2|g_3$ is $\mathcal{S}^2$-greedy.

Remark 1.16. If the ambient category is not left-cancellative, the notion of an $\mathcal{S}$-greedy path still makes sense, but the equivalence of Lemma 1.5 need not be valid in general, and two different notions arise: besides the original notion of Definition 1.1, we have a stronger notion corresponding to the conclusion of Lemma 1.5, namely saying that $g_1|g_2$ is *strongly $\mathcal{S}$-greedy* if, for each s in $\mathcal{S}$ and each equality $sh = fg_1g_2$, there exists g satisfying $sg = fg_1$ and $h = gg_2$. It is then easy to check which results extend to the two variants. In particular, when properly adapted, the assumption that $\mathcal{S}$ is a Garside family still implies the existence of $\mathcal{S}$-greedy decompositions, whereas the uniqueness result of Proposition 1.25 below is valid only for strongly $\mathcal{S}$-greedy paths.

1.2 The notion of an $\mathcal{S}$-normal path

Appealing to $\mathcal{S}$-greedy paths, we now introduce the notion of an $\mathcal{S}$-normal decomposition for an element of a category, and establish its (essential) uniqueness. As usual, we say that a path $g_1|\cdots|g_p$ is a *decomposition* of an element g if $g = g_1\cdots g_p$ holds. Our aim is to use $\mathcal{S}$-greedy paths to obtain distinguished decompositions for the elements of the ambient category. We recall that $\mathcal{S}^\sharp$ stands for $\mathcal{S}\mathcal{C}^\times \cup \mathcal{C}^\times$ (Definition II.1.21).

Definition 1.17 (normal, strict normal). Assume that $\mathcal{C}$ is a left-cancellative category and $\mathcal{S}$ is included in $\mathcal{C}$. A $\mathcal{C}$-path is called *$\mathcal{S}$-normal* if it is $\mathcal{S}$-greedy and every entry lies in $\mathcal{S}^\sharp$. An $\mathcal{S}$-normal path is called *strict* if no entry is invertible and all entries, except possibly the last one, lie in $\mathcal{S}$.

We shall naturally say that $s_1|\cdots|s_p$ is a (strict) $\mathcal{S}$-normal decomposition *for an element g* if $s_1|\cdots|s_p$ is a (strict) $\mathcal{S}$-normal path and $g = s_1\cdots s_p$ holds.

Example 1.18 (normal). Consider again the free abelian monoid $\mathbb{N}^n$ and the family S_n (Reference Structure 1, page 3). We saw in Example 1.2 that a path $g_1|\cdots|g_p$ is S_n-greedy if and only if, for all k and i, the implication $g_i(k) = 0 \Rightarrow g_{i+1}(k) = 0$ is satisfied. For such a path to be S_n-normal means that each entry g_i lies in S_n, that is, $g_i(k)$ is 0 or 1. Then the greediness condition reduces to the condition $g_i(k) \geqslant g_{i+1}(k)$: an S_n-normal path is a path $g_1|\cdots|g_p$ such that, for each k, the sequence of coordinates $(g_1(k), ..., g_p(k))$ is non-increasing.

We can then check that S_n-normal paths are exactly those that satisfy the condition (I.1.2) of Proposition I.1.1: indeed, $s \in S_n$ is equivalent to $s \leqslant \Delta_n$, as both mean $\forall k\ (s(k) \leqslant 1)$. Assume that $s_i|s_{i+1}$ is S_n-normal, and g is a nontrivial element satisfying $g \leqslant s_{i+1}$. The assumption that g is nontrivial implies that $g(k) \geqslant 1$ holds for some k, and, therefore, so does $s_{i+1}(k) \geqslant 1$. As $s_i|s_{i+1}$ is S_n-normal, as seen above, we must have $s_i(k) \geqslant 1$, whence $(s_i g)(k) \geqslant 2$. Hence $s_i g \leqslant \Delta_n$ fails, and (I.1.2) is satisfied.

Conversely, assume that $s_i|s_{i+1}$ satisfies (I.1.2). In order to prove that $s_i|s_{i+1}$ is S_n-normal, it suffices to show that it is impossible to have $s_i(k) = 0$ and $s_{i+1}(k) = 1$ simultaneously, for any k. Now assume $s_{i+1}(k) = 1$. Let $g = \mathtt{a}_k$, that is, $g(j) = 1$ for $j = k$, and $g(j) = 0$ otherwise. Then we have $g \neq 1$ and $g \leqslant s_{i+1}$. Condition (I.1.2) then implies $s_i g \not\leqslant \Delta_n$. By assumption, we have $(s_i g)(j) = s_i(j) \leqslant 1$ for $j \neq k$, so $s_i g \not\leqslant \Delta_n$ necessarily comes from $(s_i g)(k) \geqslant 2$, which is possible only for $s_i(k) = 1$. Hence $s_i|s_{i+1}$ is S_n-normal.

Extending the case of $\mathbb{N}^n$, we shall see below that, if M is a Garside monoid with Garside element Δ, the distinguished decompositions considered in Chapter I coincide with the $\mathrm{Div}(\Delta)$-normal decompositions in the sense of Definition 1.17.

The properties of $\mathcal{S}$-greedy paths established in Subsection 1.1 immediately imply the following behavior of $\mathcal{S}$-normal paths with respect to invertible elements.

Proposition 1.19 (normal *vs.* **invertible).** *Assume that $\mathcal{C}$ is a left-cancellative category and $\mathcal{S}$ is included in $\mathcal{C}$.*

(i) *A path is $\mathcal{S}$-normal if and only if it is $\mathcal{S}^\sharp$-normal.*

(ii) *If $s_1|\cdots|s_p$ is $\mathcal{S}$-normal, then the indices i such that s_i is not invertible make an initial segment of $\{1, ..., p\}$.*

Proof. (i) Assume that $s_1|\cdots|s_p$ is $\mathcal{S}$-normal. By Lemma 1.10, $s_1|\cdots|s_p$ is $\mathcal{S}^\sharp$-greedy. Moreover, by definition, the elements s_i lie in $\mathcal{S}^\sharp$, which coincides with $(\mathcal{S}^\sharp)^\sharp$. Hence $s_1|\cdots|s_p$ is $\mathcal{S}^\sharp$-normal. Conversely, assume that $s_1|\cdots|s_p$ is $\mathcal{S}^\sharp$-normal. Then it is $\mathcal{S}$-greedy by Lemma 1.10, and its entries lie in $(\mathcal{S}^\sharp)^\sharp$, which is $\mathcal{S}^\sharp$. Hence $s_1|\cdots|s_p$ is $\mathcal{S}$-normal.

(ii) Assume that $s_1|\cdots|s_p$ is $\mathcal{S}$-normal. Then $s_1|\cdots|s_p$ is $\mathcal{S}$-greedy, hence, by Lemma 1.10 (or by (i)), $\mathcal{S}^\sharp$-greedy. Then, by Lemma 1.8(ii), if s_i is invertible, then s_{i+1} is invertible too. □

When there is no nontrivial invertible element in $\mathcal{C}$, we have $\mathcal{S}^\sharp = \mathcal{S} \cup 1_\mathcal{C}$, and Proposition 1.19 implies that an $\mathcal{S}$-normal path consists of non-invertible entries in $\mathcal{S}$ possibly followed by identity elements, whereas a strict $\mathcal{S}$-normal path consists of non-invertible entries in $\mathcal{S}$ exclusively.

We now study the uniqueness of $\mathcal{S}$-normal decompositions. The main result is that, when it exists, the $\mathcal{S}$-normal decomposition of an element is essentially unique (Proposition 1.25). However, the possible existence of nontrivial invertible elements may prohibit a genuine uniqueness. In order to state the result, we first introduce the natural notion of a *deformation*.

Definition 1.20 (deformation by invertible elements). (See Figure 2.) Assume that $\mathcal{C}$ is a left-cancellative category. A $\mathcal{C}$-path $f_1|\cdots|f_p$ is said to be a *deformation by invertible elements*, or $\mathcal{C}^\times$*-deformation*, of another $\mathcal{C}$-path $g_1|\cdots|g_q$ if there exist $\epsilon_0, \ldots, \epsilon_m$ in $\mathcal{C}^\times$, $m = \max(p, q)$, such that ϵ_0 and ϵ_m are identity elements and $\epsilon_{i-1}g_i = f_i\epsilon_i$ holds for $1 \leqslant i \leqslant m$, where, for $p \neq q$, the shorter path is expanded by identity elements.

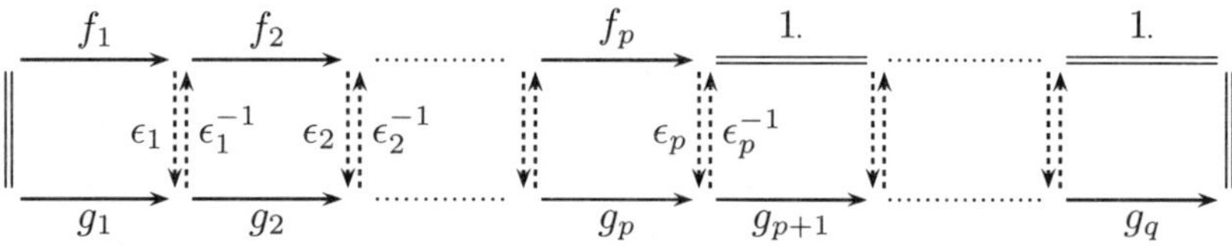

Figure 2. Deformation by invertible elements: invertible elements connect the corresponding entries; if one path is shorter (here we are in the case $p < q$), it is extended by identity elements.

Example 1.21 (deformation by invertible elements). Consider the wreathed free abelian monoid $\widetilde{\mathbb{N}^3}$ (Reference Structure 6, page 19). Then $\mathtt{as_1}|\mathtt{bs_2s_1}|\mathtt{c}|\mathtt{s_2}$ and $\mathtt{a}|\mathtt{a}|\mathtt{as_2s_1}$ are typical deformations by invertible elements of one another, as the diagram

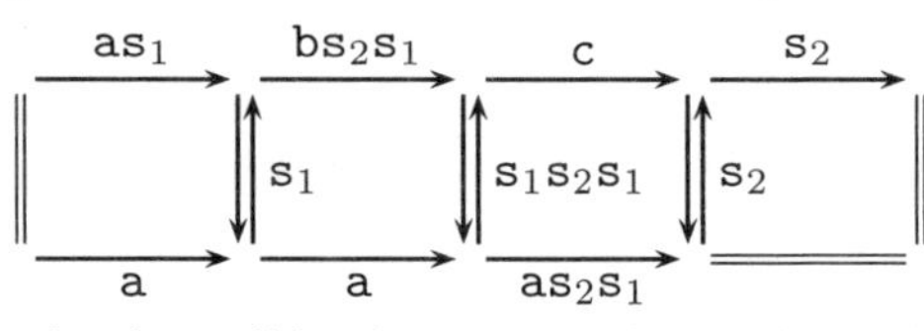

shows (here the connecting invertible elements are involutions, that is, equal to their inverses, but this need not be the case in general).

The definition and Figure 2 make it clear that being a $\mathcal{C}^\times$-deformation is a symmetric relation and that, if $f_1|\cdots|f_p$ is a $\mathcal{C}^\times$-deformation of $g_1|\cdots|g_q$, then

$$f_i \mathrel{{}^\times{=}^\times} g_i \quad \text{and} \quad f_1\cdots f_i =^\times g_1\cdots g_i$$

hold for every i. Conversely, one easily checks that the satisfaction of $f_1\cdots f_i =^\times g_1\cdots g_i$ for each i implies that $f_1|\cdots|f_p$ is a $\mathcal{C}^\times$-deformation of $g_1|\cdots|g_q$ provided the ambient category $\mathcal{C}$ is left-cancellative.

Proposition 1.22 (deformation). *Assume that $\mathcal{C}$ is a left-cancellative category, and $\mathcal{S}$ is a subfamily of $\mathcal{C}$ satisfying $\mathcal{C}^\times\mathcal{S} \subseteq \mathcal{S}^\sharp$. Then, for every g in $\mathcal{C}$, every $\mathcal{C}^\times$-deformation of an $\mathcal{S}$-normal decomposition of g is an $\mathcal{S}$-normal decomposition of g.*

Proof. Assume that $s_1|\cdots|s_p$ is an $\mathcal{S}$-normal decomposition of g, and that $t_1|\cdots|t_q$ is a $\mathcal{C}^\times$-deformation of $s_1|\cdots|s_p$ with witnessing invertible elements ϵ_i as in Definition 1.20 and Figure 2. First, by definition, $t_1|\cdots|t_q$ is also a decomposition of g.

Next, the definition of a $\mathcal{C}^\times$-deformation implies that $t_i \;{}^\times\!\!=^\times s_i$ holds for each i. Then, by Lemma 1.9, the assumption that $s_i|s_{i+1}$ is $\mathcal{S}$-normal, hence in particular $\mathcal{S}$-greedy, implies that $t_i|t_{i+1}$ is $\mathcal{S}$-greedy as well. Moreover, if $q < p$ holds, then $t_q|1_y$ and $1_y|1_y$ are automatically $\mathcal{S}$-greedy. Hence, $t_1|\cdots|t_q$ is an $\mathcal{S}$-greedy path.

Finally, by Lemma II.1.24, which is valid since $\mathcal{C}^\times\mathcal{S} \subseteq \mathcal{S}^\sharp$ is assumed, the relation $s_i \;{}^\times\!\!=^\times t_i$ implies that t_i belongs to $\mathcal{S}^\sharp$ for each i. So $t_1|\cdots|t_q$ is an $\mathcal{S}$-normal path. □

As a first application for the notion of deformation by invertible elements, we revisit the definition of $\mathcal{S}$-normal paths. In Definition 1.17, we introduced a variant called strict. As noted above, the difference is inessential when there exist no nontrivial invertible elements: a strict normal path is one in which final identity elements have been removed. In general, the difference is a genuine one, but the next result shows that, if $\mathcal{C}^\times\mathcal{S} \subseteq \mathcal{S}^\sharp$ holds, the existence of a strict $\mathcal{S}$-normal decomposition is not more demanding than that of a general $\mathcal{S}$-normal decomposition.

Proposition 1.23 (strict normal). *Assume that $\mathcal{C}$ is a left-cancellative category and $\mathcal{S}$ is a subfamily of $\mathcal{C}$ satisfying $\mathcal{C}^\times\mathcal{S} \subseteq \mathcal{S}^\sharp$. Then every non-invertible element of $\mathcal{C}$ that admits an $\mathcal{S}$-normal decomposition with ℓ non-invertible entries admits a strict $\mathcal{S}$-normal decomposition of length ℓ.*

Proof. We shall show that every $\mathcal{S}$-normal path with at least one non-invertible entry can be deformed into a strict $\mathcal{S}$-normal path. So assume that $s_1|\cdots|s_p$ is an $\mathcal{S}$-normal decomposition. Let ℓ be the number of non-invertible entries, which we assume is at least one. By Proposition 1.19, $s_1, \dots, s_\ell$ must be non-invertible, whereas $s_{\ell+1}, \dots, s_p$ are invertible if $p > \ell$ holds.

Put $\epsilon_0 = 1_x$, where x is the source of s_1. The assumption $\mathcal{C}^\times\mathcal{S} \subseteq \mathcal{S}^\sharp$ implies $\mathcal{C}^\times(\mathcal{S} \setminus \mathcal{C}^\times) \subseteq (\mathcal{S}\setminus\mathcal{C}^\times)\mathcal{C}^\times$, because a non-invertible element of $\mathcal{S}^\sharp$ must belong to $(\mathcal{S}\setminus\mathcal{C}^\times)\mathcal{C}^\times$. Using the latter inclusion ℓ times, we find t_i in $\mathcal{S} \setminus \mathcal{C}^\times$ and ϵ_i in $\mathcal{C}^\times$ satisfying $\epsilon_{i-1}s_i = t_i\epsilon_i$ for i increasing from 0 to $\ell - 1$. Finally, put $t_\ell = \epsilon_{\ell-1}s_\ell\cdots s_p$. Then the assumption $\mathcal{C}^\times\mathcal{S} \subseteq \mathcal{S}^\sharp$ implies that t_ℓ belongs to $\mathcal{S}^\sharp$. By construction, the path $t_1|\cdots|t_\ell$ is a $\mathcal{C}^\times$-deformation of the initial path $s_1|\cdots|s_p$. By Proposition 1.22, it is $\mathcal{S}$-normal, and, by construction again, it is strict. □

Example 1.24 (strict normal). Consider the wreathed free abelian monoid $\widetilde{\mathbb{N}^3}$ (Reference Structure 6, page 19) with $S_3 = \{s \in \mathbb{N}^3 \mid \forall k\,(s(k) \in \{0,1\})\}$. Write $\mathtt{a}$ for $(1,0,0)$, $\mathtt{b}$ for $(0,1,0)$, and $\mathtt{c}$ for $(0,0,1)$. Then $\mathtt{as_1}|\mathtt{bs_2s_1}|\mathtt{c}|\mathtt{s_2}$ is a typical S_3-normal path, and it is not strict since $\mathtt{as_1}$ does not belong to S_3. Applying the conjugation relations to push the invertible elements $\mathtt{s}_i$ to the right, we obtain the strict S_3-normal path $\mathtt{a}|\mathtt{a}|\mathtt{as_2s_1}$, another decomposition of the same element of $\widetilde{\mathbb{N}^3}$ which, as seen in Example 1.21, is a deformation of the former and is strictly S_3-normal.

We now address the uniqueness of normal decompositions. The next result completely describes the connection between any two $\mathcal{S}$-normal decompositions of an element.

Proposition 1.25 (normal unique). *If $\mathcal{S}$ is a subfamily of a left-cancellative category $\mathcal{C}$, any two $\mathcal{S}$-normal decompositions of an element of $\mathcal{C}$ are $\mathcal{C}^\times$-deformations of one another.*

Proof. Assume that $s_1|\cdots|s_p$ and $t_1|\cdots|t_q$ are two $\mathcal{S}$-normal decompositions of an element g of $\mathcal{C}$. Let y be the target of g. At the expense of adding factors 1_y at the end of the shorter path if any, we may assume $p = q$: by Lemma 1.8, adding identity elements at the end of an $\mathcal{S}$-normal path yields an $\mathcal{S}$-normal path.

Let $\epsilon_0 = 1_x$, where x is the source of g. By Proposition 1.14, the path $t_1|t_2\cdots t_q$ is $\mathcal{S}$-greedy, hence, by Lemma 1.10, $\mathcal{S}^\sharp$-greedy. Now s_1 belongs to $\mathcal{S}^\sharp$ and left-divides $s_1\cdots s_p$, which is $\epsilon_0 t_1\cdots t_q$, so it must left-divide t_1. In other words, there exists ϵ_1 satisfying $s_1\epsilon_1 = \epsilon_0 t_1$. Left-cancelling s_1 in $s_1\cdots s_p = (s_1\epsilon_1)t_2\cdots t_p$, we deduce $s_2\cdots s_p = \epsilon_1 t_1\cdots t_q$. Now s_2 belongs to $\mathcal{S}^\sharp$ and, by Proposition 1.14 again, $t_2|t_3\cdots t_q$ is $\mathcal{S}$-greedy, so we deduce the existence of ϵ_2 satisfying $s_2\epsilon_2 = \epsilon_1 t_1$, and so on, giving the existence of ϵ_i satisfying $s_i\epsilon_i = \epsilon_{i-1}t_i$ for every $1 \leqslant i \leqslant q$.

Exchanging $s_1|\cdots|s_p$ and $t_1|\cdots|t_q$ and arguing symmetrically from $\epsilon'_0 = 1_x$, we obtain the existence of ϵ'_i satisfying $t_i\epsilon'_i = \epsilon'_{i-1}s_i$ for every $1 \leqslant i \leqslant p$. We deduce, for every i, the equalities $(s_1\cdots s_i)\epsilon_i = t_1\cdots t_i$ and $(t_1\cdots t_i)\epsilon'_i = s_1\cdots s_i$, which imply that ϵ'_i is the inverse of ϵ_i. Hence $s_1|\cdots|s_p$ is a $\mathcal{C}^\times$-deformation of $t_1|\cdots|t_q$. □

When the ambient category has no nontrivial invertible element, Proposition 1.25 is a genuine uniqueness statement: if $s_1|\cdots|s_p$ and $t_1|\cdots|t_q$ are two $\mathcal{S}$-normal decompositions of some element g and $p \leqslant q$ holds, then we have $s_i = t_i$ for $i \leqslant p$, and $t_i = 1_y$ for $p < i \leqslant q$. In the general case, the argument does not work, but we can still obtain a uniqueness result at the expense of considering strict $\mathcal{S}$-normal paths and families $\mathcal{S}$ that are $=^\times$-transverse.

Definition 1.26 (transverse, selector). Assume that $\equiv$ is an equivalence relation on a family $\mathcal{S}$ and $\mathcal{S}' \subseteq \mathcal{S}$ holds. A subfamily $\mathcal{S}''$ of $\mathcal{S}$ is called $\equiv$-*transverse* if distinct elements of $\mathcal{S}''$ are not $\equiv$-equivalent; it is said to be a $\equiv$-*selector* in $\mathcal{S}'$ of $\mathcal{S}$ if $\mathcal{S}''$ in included in $\mathcal{S}'$ and contains exactly one element in each $\equiv$-equivalence class intersecting $\mathcal{S}'$.

By definition, a $\equiv$-selector is $\equiv$-transverse. By the Axiom of Choice, selectors always exist. If $\mathcal{C}$ contains no nontrivial invertible element, every subfamily of $\mathcal{C}$ is $=^\times$-transverse. Proposition 1.25 implies

Corollary 1.27 (transverse). *Assume that $\mathcal{C}$ is a left-cancellative category and $\mathcal{S}$ is a subfamily of $\mathcal{C}$ that is $=^\times$-transverse. Then every non-invertible element of $\mathcal{C}$ admits at most one strict $\mathcal{S}$-normal decomposition.*

Proof. Assume that $f_1|\cdots|f_p$ and $g_1|\cdots|g_q$ are strict $\mathcal{S}$-normal decompositions of some element g. First, p and q are equal by Proposition 1.25 since, by definition, no f_i or g_i may be invertible. Next, an induction on $i < p$ gives $f_i =^{\times} g_i$, whence $f_i = g_i$ since, by assumption, f_i and g_i lie in $\mathcal{S}$. Finally, we have $f_1 \cdots f_{p-1} f_p = g_1 \cdots g_{p-1} g_p$, whence $f_p = g_p$ by left-cancelling $f_1 \cdots f_{p-1}$. □

By contrast, it is easy to see, for instance in the case of the wreathed free abelian monoid $\widetilde{\mathbb{N}^n}$, that, if $\mathcal{S}$ is not $=^{\times}$-transverse, then an element may admit several strict $\mathcal{S}$-normal decompositions.

A direct consequence of Proposition 1.25 is the following invariance result for the number of invertible elements in normal paths.

Corollary 1.28 (normal unique). *If $\mathcal{S}$ is a subfamily of a left-cancellative category $\mathcal{C}$, then, for every element g of $\mathcal{C}$, the number of non-invertible elements in an $\mathcal{S}$-normal decomposition of g (if any) does not depend on the choice of the decomposition.*

Proof. Assume that $s_1|\cdots|s_p$ and $t_1|\cdots|t_q$ are two $\mathcal{S}$-normal decompositions of an element g. By Proposition 1.25, $s_1|\cdots|s_p$ is a $\mathcal{C}^{\times}$-deformation of $t_1|\cdots|t_q$. Therefore, $s_i \,{}^{\times}\!\!=^{\times} t_i$ holds for each i, which implies that s_i is invertible if and only if t_i is. □

Owing to Corollary 1.28, it is natural to introduce the following terminology.

Definition 1.29 ($\mathcal{S}$-length). If $\mathcal{S}$ is a subfamily of a left-cancellative category $\mathcal{C}$ and g is an element of $\mathcal{C}$ that admits at least one $\mathcal{S}$-normal decomposition, then the common number of non-invertible elements in all $\mathcal{S}$-normal decompositions of g is called the $\mathcal{S}$-*length* of g, and denoted by $\|g\|_{\mathcal{S}}$.

So, we have $\|g\|_{\mathcal{S}} = 0$ if and only if g is invertible, and $\|g\|_{\mathcal{S}} = 1$ if and only if g is a non-invertible element of $\mathcal{S}^{\sharp}$. The general results about $\mathcal{S}^m$-greedy paths easily imply an upper bound for the $\mathcal{S}$-length of an element.

Proposition 1.30 (length I). *If $\mathcal{S}$ is a subfamily of a left-cancellative category $\mathcal{C}$, then $\|g\|_{\mathcal{S}} \leqslant m$ holds for every g in $(\mathcal{S}^{\sharp})^m$ that admits at least one $\mathcal{S}$-normal decomposition.*

Proof. Assume that $s_1|\cdots|s_p$ is an $\mathcal{S}$-normal decomposition of g. At the expense of appending identity-entries at the end, we may assume $p \geqslant m$. By Proposition 1.19, $s_1|\cdots|s_p$ is $\mathcal{S}^{\sharp}$-normal, hence, by Proposition 1.14, $s_1 \cdots s_m | s_{m+1} \cdots s_p$ is $(\mathcal{S}^{\sharp})^m$-greedy. As g is assumed to lie in $(\mathcal{S}^{\sharp})^m$ and it left-divides $s_1 \cdots s_p$, we deduce that g left-divides $s_1 \cdots s_m$, which implies that $s_{m+1}, \dots, s_p$ is invertible. Therefore, $\|g\|_{\mathcal{S}} \leqslant m$ holds. □

1.3 The notion of a Garside family

We turn to the existence of $\mathcal{S}$-normal decompositions. A moment's thought shows that such decompositions need not always exist, see Exercise 27(iii). We define a Garside family to be a family $\mathcal{S}$ for which every element of the ambient category admits an $\mathcal{S}$-normal decomposition. In this subsection, we establish a first criterion for recognizing a Garside family. Further results in this direction will be established in Chapters IV and VI.

Definition 1.31 (Garside family). A subfamily $\mathcal{S}$ of a left-cancellative category $\mathcal{C}$ is called a *Garside* family in $\mathcal{C}$ if every element of $\mathcal{C}$ admits at least one $\mathcal{S}$-normal decomposition.

Example 1.32 (Garside family). Let $\mathcal{C}$ be any left-cancellative category. Then $\mathcal{C}$ is a Garside family in itself: for every g in $\mathcal{C}$, the length one path g is a $\mathcal{C}$-normal decomposition of g.

Consider now the free abelian monoid $\mathbb{N}^n$ and $S_n = \{s \in \mathbb{N}^n \mid \forall k\ (s(k) \in \{0,1\})\}$ (Reference Structure 1, page 3). Then S_n is a Garside family in $\mathbb{N}^n$. Indeed, let g be an element of $\mathbb{N}^n$. If g is the identity element, then the empty path is an S_n-normal decomposition for g. Otherwise, let $\ell = \max\{g(k) \mid k = 1, ..., n\}$. For $i \leqslant \ell$, define s_i by $s_i(k) = 1$ if $g(k) \geqslant i$ holds, and $s_i(k) = 0$ otherwise. Then s_i belongs to S_n for every i, and, by construction, $s_1|\cdots|s_\ell$ is a decomposition of g. Moreover, it follows from the characterization of Example 1.18 that $s_{i-1}|s_i$ is S_n-greedy for every i. So $s_1|\cdots|s_\ell$ is an S_n-normal decomposition of g. Hence, by definition, S_n is a Garside family in $\mathbb{N}^n$.

Although trivial, the first example above should be kept in mind, as it shows that the existence of a Garside family is an empty assumption. In the other direction, we refer to Example IV.2.34 for a non-trivial, infinite monoid that has no Garside family except itself.

We now observe that for $\mathcal{S}$ to be a Garside family is actually a property of $\mathcal{S}^\sharp$.

Proposition 1.33 (invariance). *Assume that $\mathcal{C}$ is a left-cancellative category. Then a subfamily $\mathcal{S}$ of $\mathcal{C}$ is a Garside family in $\mathcal{C}$ if and only if $\mathcal{S}^\sharp$ is.*

Proof. Assume that $\mathcal{S}$ is a Garside family in $\mathcal{C}$. Then every element of $\mathcal{C}$ has an $\mathcal{S}$-normal decomposition. By Proposition 1.19, the latter is an $\mathcal{S}^\sharp$-normal decomposition. Hence, $\mathcal{S}^\sharp$ is a Garside family. Conversely, assume that $\mathcal{S}^\sharp$ is a Garside family in $\mathcal{C}$. Then every element of $\mathcal{C}$ has an $\mathcal{S}^\sharp$-normal decomposition, which, by Proposition 1.19, is $\mathcal{S}$-normal as well. Hence, $\mathcal{S}$ is a Garside family. □

Corollary 1.34 (invariance). *Assume that $\mathcal{C}$ is a left-cancellative category.*

(i) *If $\mathcal{S}, \mathcal{S}'$ are included in $\mathcal{C}$ and satisfy $\mathcal{S}^\sharp = \mathcal{S}'^\sharp$, then $\mathcal{S}$ is a Garside family if and only if $\mathcal{S}'$ is.*

(ii) *In particular, if $\mathcal{S}'$ is an $=^\times$-selector in $\mathcal{S}$, then $\mathcal{S}$ is a Garside family if and only if $\mathcal{S}'$ is, if and only if $\mathcal{S}' \setminus \mathbf{1}_\mathcal{C}$ is.*

Proof. Item (i) follows directly from Proposition 1.33. For (ii), we observe that $\mathcal{S}'$ being an $=^\times$-selector in $\mathcal{S}$ implies $\mathcal{S}^\sharp = \mathcal{S}'^\sharp = (\mathcal{S}' \setminus \mathbf{1}_\mathcal{C})^\sharp$, and we apply (i). □

Owing to the above results, we might think of restricting to families of the form $\mathcal{S}^\sharp$, that is, families that contain all identity elements and are closed under right-multiplication by an invertible element. As we shall be interested in Garside families that are as small as possible, padding the families with invertible elements is useless and restrictive. On the contrary, it is often useful to consider Garside families that are $=^\times$-transverse.

Example 1.35 (selector). In the wreathed free abelian monoid $\widetilde{\mathbb{N}^n}$ (Reference Structure 6, page 19), one can check using a construction similar to that of Example 1.32 that the size 2^n family S_n is a Garside family. By Proposition 1.33, the larger family $S_n^\sharp$, which is $S_n\mathfrak{S}_n$, is also a Garside family with $2^n n!$ elements. The $=^\times$-selectors for $S_n^\sharp$ are the families of the form $\{(g, F(g)) \mid g \in S_n\}$ with $F : S_n \to \mathfrak{S}_n$, and each of them is a Garside family in $\widetilde{\mathbb{N}^n}$.

Before entering the investigation of Garside families more precisely, we still state a general transfer result.

Proposition 1.36 (power). *If $\mathcal{S}$ is a Garside family that includes $1_\mathcal{C}$ in a left-cancellative category $\mathcal{C}$, the family $\mathcal{S}^m$ is a Garside family for every $m \geqslant 1$.*

Proof. Let g be an arbitrary element of $\mathcal{C}$. By assumption, g admits an $\mathcal{S}$-normal decomposition, say $s_1|\cdots|s_p$. At the expense of appending identity-entries at the end, we may assume that p is a multiple of m, say $p = qm$. For $i = 1, ..., q$, put $t_i = s_{(i-1)m+1} \cdots s_{im}$. Then, by Proposition 1.14, $t_1|\cdots|t_q$ is an $(\mathcal{S}^\sharp)^m$-greedy decomposition of g. By construction, $t_1, ..., t_q$ lie in $(\mathcal{S}^\sharp)^m$. Anticipating on Proposition 1.39, we have $\mathcal{C}^\times\mathcal{S} \subseteq \mathcal{S}^\sharp$. Hence, by Lemma II.1.26, $t_1, ..., t_q$ also lie in $(\mathcal{S}^m)^\sharp$. Then $t_1|\cdots|t_q$ is a $\mathcal{S}^m$-normal decomposition of g. As every element of $\mathcal{C}$ admits an $\mathcal{S}^m$-normal decomposition, $\mathcal{S}^m$ is a Garside family. ☐

By Proposition 1.33, a subfamily $\mathcal{S}$ of a category $\mathcal{C}$ is a Garside family if and only if $\mathcal{S} \cup 1_\mathcal{C}$ is a Garside family. Applying Proposition 1.36 to the latter gives

Corollary 1.37 (power). *If $\mathcal{S}$ is a Garside family in a left-cancellative category $\mathcal{C}$, the family $\mathcal{S} \cup \mathcal{S}^2 \cup \cdots \cup \mathcal{S}^m$ is a Garside family for every $m \geqslant 1$.*

1.4 Recognizing Garside families

Definition 1.31 is frustrating in that, except in simple cases like free abelian monoids, it gives no practical way of recognizing Garside families and of determining the normal decompositions whose existence is asserted. We shall now establish a first criterion for recognizing a Garside family, and describe an effective method for finding normal decompositions.

Definition 1.38 (property ☐, ☐-witness). Assume that $\mathcal{C}$ is a left-cancellative category and $\mathcal{S}$ is included in $\mathcal{C}$. We say that $\mathcal{S}$ *satisfies Property* ☐ if, for every $s_1|s_2$ in $\mathcal{S}^{[2]}$, there exists an $\mathcal{S}$-greedy decomposition $t_1|t_2$ of s_1s_2 with t_1 and t_2 in $\mathcal{S}$. In this case, a map F that chooses, for every $s_1|s_2$ in $\mathcal{S}^{[2]}$, a pair $t_1|t_2$ as above, is called a ☐*-witness* for $\mathcal{S}$.

For a family $\mathcal{S}$ to satisfy Property $\square$ means that, for each choice of the left and bottom arrows in $\mathcal{S}$ in a diagram as on the right, there exist top and right arrows in $\mathcal{S}$ making an $\mathcal{S}$-greedy path (hence an $\mathcal{S}$-normal path) and making the diagram commute. By definition, $\mathcal{S}$ satisfies Property $\square$ if and only if there exists a $\square$-witness for $\mathcal{S}$.

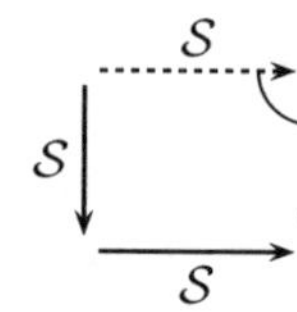

Note that, if $\mathcal{C}$ contains no nontrivial invertible elements, a $\square$-witness is unique when it exists. In the sequel, when investigating a certain subfamily $\mathcal{S}$ of a category, we shall both consider the satisfaction of Property $\square$ by $\mathcal{S}$ and by $\mathcal{S}^\sharp$. These are related conditions, but they need not be equivalent in general.

Proposition 1.39 (recognizing Garside I). *A subfamily $\mathcal{S}$ of a left-cancellative category $\mathcal{C}$ is a Garside family if and only one of the following equivalent conditions holds:*

(1.40) *The family $\mathcal{S}^\sharp$ generates $\mathcal{C}$ and satisfies Property $\square$.*

(1.41) *The family $\mathcal{S}^\sharp$ generates $\mathcal{C}$ and every element of $(\mathcal{S}^\sharp)^2$ admits an $\mathcal{S}$-normal decomposition.*

(1.42) *The family $\mathcal{S}^\sharp$ generates $\mathcal{C}$, every element of $\mathcal{S}^2$ admits an $\mathcal{S}$-normal decomposition, and $\mathcal{C}^\times\mathcal{S} \subseteq \mathcal{S}^\sharp$ holds.*

Before proving Proposition 1.39, let us immediately use it to connect the (quasi)-Garside monoids of Chapter I with Garside families.

Proposition 1.43 (Garside monoid). *Assume that (M, Δ) is a quasi-Garside monoid. Then $\mathrm{Div}(\Delta)$ is a Garside family in M.*

Proof. First, by definition, M is (left)-cancellative, it contains no nontrivial invertible element, and $\mathrm{Div}(\Delta)$, which coincides with $\mathrm{Div}(\Delta)^\sharp$, generates M. So, according to Proposition 1.39, it now suffices to show that, for all s_1, s_2 in $\mathrm{Div}(\Delta)$, the element s_1s_2 admits a $\mathrm{Div}(\Delta)$-normal decomposition. Let t_1 be the left-gcd of s_1s_2 and Δ. By definition, t_1 lies in $\mathrm{Div}(\Delta)$. As t_1 left-divides s_1s_2, we have $s_1s_2 = t_1t_2$ for some t_2, which is unique since M is left-cancellative. As s_1 left-divides s_1s_2 and Δ, it left-divides their left-gcd t_1, that is, $t_1 = s_1g$ holds for some g. We deduce $s_1s_2 = s_1gt_2$, whence $s_2 = gt_2$. As t_2 right-divides s_2, which lies in $\mathrm{Div}(\Delta)$, it lies in $\mathrm{Div}(\Delta)$ as well.

So, in order to prove that $t_1|t_2$ is a $\mathrm{Div}(\Delta)$-normal decomposition of s_1s_2, it only remains to show that $t_1|t_2$ is $\mathrm{Div}(\Delta)$-greedy. Now assume $r \in \mathrm{Div}(\Delta)$ and $r \preccurlyeq ft_1t_2$. As $\mathrm{Div}(\Delta)$ generates M, we can write $f = f_1 \cdots f_p$ with $f_1, ..., f_p$ in $\mathrm{Div}(\Delta)$. By definition, r and f_1 admit a right-lcm, say f_1r_1, see Figure 3. As r and f_1 left-divide Δ, so does f_1r_1. Hence f_1r_1 right-divides Δ, and so does r_1, that is, r_1 lies in $\mathrm{Div}(\Delta)$. Moreover, as ft_1t_2 is a right-multiple of r and of f_1, it is a right-multiple of their right-lcm f_1r_1. Left-cancelling f_1, we deduce $r_1 \preccurlyeq f_2 \cdots f_pt_1t_2$. Repeating the argument with r_1 and f_2, and iterating, we finally find r_p in $\mathrm{Div}(\Delta)$ satisfying $r_p \preccurlyeq t_1t_2$. So r_p left-divides t_1t_2 and Δ, hence it left-divides their left-gcd, which is t_1 by definition. As can be read on the diagram, $r_p \preccurlyeq t_1$ implies $r \preccurlyeq ft_1$, as expected. So $t_1|t_2$ is a $\mathrm{Div}(\Delta)$-normal decomposition of s_1s_2, and $\mathrm{Div}(\Delta)$ is a Garside family in M. $\square$

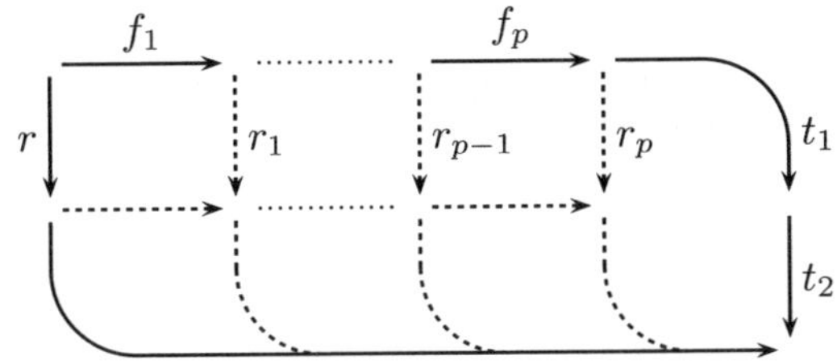

Figure 3. Proof of Proposition 1.43: if (M, Δ) is a Garside monoid, then $\mathrm{Div}(\Delta)$ is a Garside family in M. Here the argument showing that the decomposition $t_1|t_2$ is $\mathrm{Div}(\Delta)$-greedy.

We now turn to the proof of Proposition 1.39. First, we observe that the conditions of the proposition are necessary, which is essentially trivial.

Lemma 1.44. *Assume that $\mathcal{C}$ is a left-cancellative category and $\mathcal{S}$ is included in $\mathcal{C}$. Then* (1.40) *and* (1.41) *are equivalent and, if $\mathcal{S}$ is a Garside family,* (1.40), (1.41), *and* (1.42) *are satisfied.*

Proof. By Proposition 1.30, every element of $(\mathcal{S}^\sharp)^2$ that admits an $\mathcal{S}$-normal decomposition admits an $\mathcal{S}$-normal decomposition of length at most two, hence an $\mathcal{S}$-normal decomposition of length exactly two (possibly with one or two invertible entries). So (1.41) implies (1.40), and, therefore, the latter are equivalent.

Assume that $\mathcal{S}$ is a Garside family in $\mathcal{C}$. Then every element of $\mathcal{C}$ admits an $\mathcal{S}$-normal decomposition, hence, by definition, an expression as a product of elements of $\mathcal{S}^\sharp$. So $\mathcal{S}^\sharp$ generates $\mathcal{C}$. Next, by definition again, every element of $\mathcal{C}$, hence in particular every element of $(\mathcal{S}^\sharp)^2$, admits an $\mathcal{S}$-normal decomposition. So (1.41), hence (1.40) as well, are satisfied. Finally, every element of $\mathcal{S}^2$ admits an $\mathcal{S}$-normal decomposition. Now assume that $\epsilon|s$ belongs to $\mathcal{C}^\times|\mathcal{S}$. By definition, ϵs admits an $\mathcal{S}$-normal decomposition, say $s_1|\cdots|s_p$. Then we have $s \preccurlyeq \epsilon^{-1}s_1(s_2\cdots s_p)$. As s lies in $\mathcal{S}$ and, by Proposition 1.12, $s_1|s_2\cdots s_p$ is $\mathcal{S}$-greedy, we deduce $s \preccurlyeq \epsilon^{-1}s_1$, which implies that $s_2, ..., s_p$ are invertible. Then $s_1\cdots s_p$ belongs to $\mathcal{S}^\sharp$, and we have $\mathcal{C}^\times\mathcal{S} \subseteq \mathcal{S}^\sharp$. So (1.42) is satisfied. □

We now turn to the other direction. The main point is to show that the existence of an $\mathcal{S}$-normal decomposition for every element of $(\mathcal{S}^\sharp)^2$ is sufficient for establishing the existence of an $\mathcal{S}$-normal decomposition for every element. This will be done by using an induction and a simple diagrammatic rule that we call the (first) domino rule, which guarantees that certain length-two paths are greedy whenever some related paths are. Although easy, the result will play a fundamental role in the sequel and it arguably captures the geometric core of the construction of greedy decompositions.

Proposition 1.45 (first domino rule). *Assume that $\mathcal{C}$ is a left-cancellative category, $\mathcal{S}$ is included in $\mathcal{C}$, and we have a commutative diagram with edges in $\mathcal{C}$ as on the right. If $g_1|g_2$ and $g_1'|f$ are $\mathcal{S}$-greedy, then $g_1'|fg_2$ and $g_1'|g_2'$ are $\mathcal{S}$-greedy as well.*

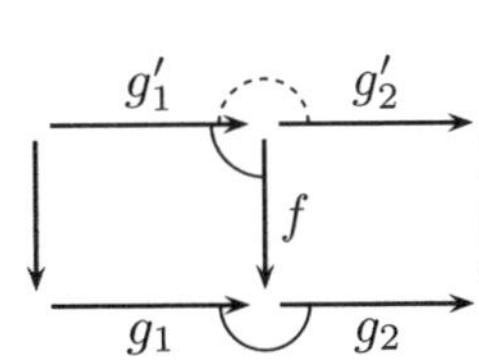

Proof. (See Figure 4.) Assume $s \in \mathcal{S}$ and $s \preccurlyeq f'g_1'fg_2$. Write $g_1'f = f_0g_1$. As the diagram is commutative, we have $s \preccurlyeq f'f_0g_1g_2$. The assumption that $g_1|g_2$ is $\mathcal{S}$-greedy implies $s \preccurlyeq f' f_0 g_1$, hence $s \preccurlyeq f' g_1' f$. Then the assumption that $g_1'|f$ is $\mathcal{S}$-greedy implies $s \preccurlyeq f' g_1'$, and we conclude that $g_1'|f g_2$ is $\mathcal{S}$-greedy. By Lemma 1.7, this implies that $g_1'|g_2'$ is $\mathcal{S}$-greedy as well. $\square$

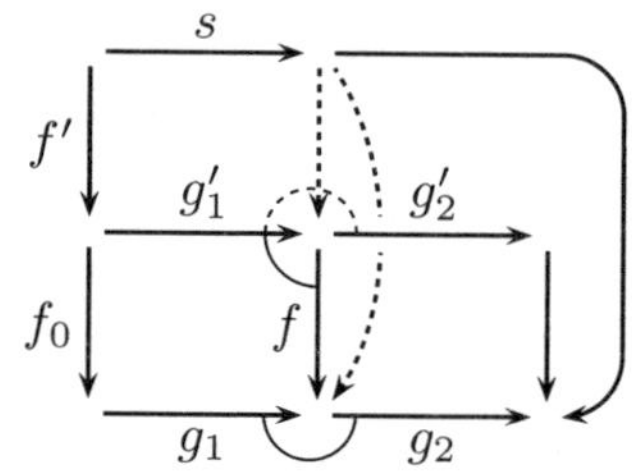

Figure 4. Proof of the first domino rule (Proposition 1.45): If $g_1|g_2$ and $g_1'|f$ are $\mathcal{S}$-greedy, so are $g_1'|fg_2$ and $g_1'|g_2'$.

Lemma 1.46. *If $\mathcal{S}$ is a generating family that satisfies Property $\square$ in a left-cancellative category $\mathcal{C}$, every element of $\mathcal{C} \setminus \mathbf{1}_{\mathcal{C}}$ admits an $\mathcal{S}$-normal decomposition with entries in $\mathcal{S}$.*

Proof. We prove using induction on m that every element g of $\mathcal{C}$ that can be expressed as the product of m elements of $\mathcal{S}$ admits an $\mathcal{S}$-normal decomposition of length m with entries in $\mathcal{S}$. By assumption, the case $m = 0$ is excluded. For $m = 1$, if g lies in $\mathcal{S}$, then the length-one path g is an $\mathcal{S}$-normal decomposition of g. Assume $m \geqslant 2$. Write $g = r_0 f$ with r_0 in $\mathcal{S}$ and f in $\mathcal{S}^{m-1}$. By the induction hypothesis, f admits an $\mathcal{S}$-normal decomposition of length $m-1$, say $s_1|\cdots|s_{m-1}$, with $s_1, \ldots, s_{m-1}$ in $\mathcal{S}$. By Property $\square$, the element $r_0 s_1$ of $\mathcal{S}^2$ admits an $\mathcal{S}$-normal decomposition $s_1'|r_1$ with s_1', r_1 in $\mathcal{S}$. Similarly, by Property $\square$ again, the element $r_1 s_2$ of $(\mathcal{S}^\sharp)^2$ admits an $\mathcal{S}$-normal decomposition $s_2'|r_2$ with s_2', r_2 in $\mathcal{S}$. Continuing in the same way, we recursively find an $\mathcal{S}$-normal decomposition $s_i'|r_i$ of $r_{i-1}s_i$ for $i = 1, \ldots, m-1$, see Figure 5. Then, we have

$$g = r_0 f = r_0 s_1 \cdots s_{m-1} = s_1' r_1 s_2 \cdots s_{m-1}' = \cdots = s_1' \cdots s_{m-1}' r_{m-1}',$$

so $s_1'|\cdots|s_{m-1}'|r_{m-1}$ is a length m decomposition of g, whose entries lie in $\mathcal{S}$ by construction. Moreover, by assumption, $s_i|s_{i+1}$ and $s_i'|r_i$ are $\mathcal{S}$-greedy for every i. Then the first domino rule implies that $s_i'|s_{i+1}'$ is $\mathcal{S}$-greedy as well. As, by assumption, $s_{m-1}'|r_{m-1}$ is $\mathcal{S}$-normal, we deduce that $s_1'|\cdots|s_{m-1}'|r_m$ is an $\mathcal{S}$-normal decomposition of g whose entries lie in $\mathcal{S}$. This completes the induction. $\square$

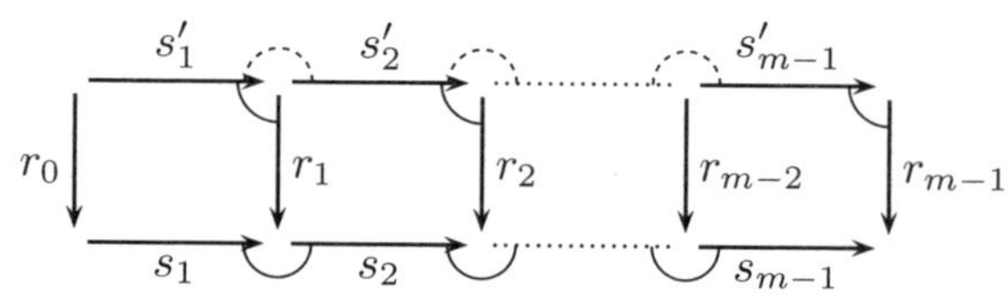

Figure 5. Proof of Lemma 1.46: obtaining an $\mathcal{S}$-normal decomposition for sf starting from an $\mathcal{S}$-normal decomposition $s_1|\cdots|s_{m-1}$ of f and applying $m-1$ times Property $\square$; the first domino rule guarantees that the path $s_1'|\cdots|s_{m-1}'|r_{m-1}$ is $\mathcal{S}$-greedy.

We can now complete the proof of Proposition 1.39.

Proof of Proposition 1.39. Owing to Lemma 1.44, it remains to show that each of (1.40), (1.41), and (1.42) implies that $\mathcal{S}$ is a Garside family in $\mathcal{C}$. Assume first that (1.40) and (1.41) (which are known to be equivalent) are satisfied. Applying Lemma 1.46 to $\mathcal{S}^\sharp$, we see that every element of $\mathcal{C}$ admits an $\mathcal{S}^\sharp$-normal decomposition, hence, by Proposition 1.19, an $\mathcal{S}$-normal decomposition. So $\mathcal{S}$ is a Garside family in $\mathcal{C}$.

Finally, assume that (1.42) is satisfied. Assume that $s_1|s_2$ belongs to $(\mathcal{S}^\sharp)^{[2]}$. By definition, $\mathcal{S}^\sharp$ is $\mathcal{S}\mathcal{C}^\times \cup \mathcal{C}^\times$, which is also $(\mathcal{S} \setminus \mathcal{C}^\times)\mathcal{C}^\times \cup \mathcal{C}^\times$. We consider three cases.

Case 1: s_2 is invertible. By Lemma 1.8, $s_1|s_2$ is $\mathcal{S}$-greedy, and, therefore, it is an $\mathcal{S}$-normal decomposition of s_1s_2.

Case 2: s_1 is invertible. Then s_1s_2 lies in $\mathcal{C}^\times\mathcal{S}^\sharp$, hence, by assumption, in $\mathcal{S}^\sharp$. Therefore s_1s_2 admits a length one $\mathcal{S}$-normal decomposition consisting of itself.

Case 3: s_1 and s_2 lie in $(\mathcal{S} \setminus \mathcal{C}^\times)\mathcal{C}^\times$. Write $s_1 = r_1\epsilon_1$, $s_2 = r_2\epsilon_2$ with r_1, r_2 in $\mathcal{S} \setminus \mathcal{C}^\times$ and ϵ_1, ϵ_2 in $\mathcal{C}^\times$. As r_2 is not invertible, $\epsilon_1 r_2$ lies in $\mathcal{S}^\sharp \setminus \mathcal{C}^\times$, hence, by Lemma II.1.24, in $\mathcal{S}\mathcal{C}^\times$. So we can find r_2' in $\mathcal{S}$ and ϵ_1' in $\mathcal{C}^\times$ satisfying $\epsilon_1 r_2 = r_2'\epsilon_1'$. Then r_1r_2' belongs to $\mathcal{S}^{[2]}$ so, by assumption, it admits an $\mathcal{S}$-normal decomposition, say $t_1|\cdots|t_p$. Then, by Lemma 1.8, $t_1|\cdots|t_p|\epsilon_1'\epsilon_2$ is an $\mathcal{S}$-normal decomposition of $r_1r_2'\epsilon_1'\epsilon_2$, that is, of s_1s_2. So, every element of $(\mathcal{S}^\sharp)^2$ admits an $\mathcal{S}$-normal decomposition. So (1.42) implies (1.41), and therefore it implies that $\mathcal{S}$ is a Garside family in $\mathcal{C}$. □

Remark 1.47. In Proposition 1.39, as many times in the sequel, the condition "$\mathcal{S}^\sharp$ generates $\mathcal{C}$" occurs. Let us observe, once for all, that this condition is equivalent to "$\mathcal{S} \cup \mathcal{C}^\times$ generates $\mathcal{C}$": indeed, $\mathcal{S} \cup \mathcal{C}^\times$ is included in $\mathcal{S}^\sharp$ by definition and, on the other hand, it generates $\mathcal{S}^\sharp$ which, we recall, is $\mathcal{S}\mathcal{C}^\times \cup \mathcal{C}^\times$.

The construction of Lemma 1.46 is effective, and it is worth describing the involved algorithm explicitly. We mention two algorithms, one that computes for s in $\mathcal{S}^\sharp$ an $\mathcal{S}$-normal decomposition of sg starting from an $\mathcal{S}$-normal decomposition of g, and one that computes an $\mathcal{S}$-normal decomposition of g starting from an arbitrary decomposition of g.

Algorithm 1.48 (left-multiplication). (See Figure 6)

Context: A left-cancellative category $\mathcal{C}$, a Garside family $\mathcal{S}$ in $\mathcal{C}$, a □-witness F for $\mathcal{S}^\sharp$
Input: An element s of $\mathcal{S}^\sharp$ and an $\mathcal{S}$-normal decomposition $s_1|\cdots|s_p$ of an element g of $\mathcal{C}$ such that sg exists
Output: An $\mathcal{S}$-normal decomposition of sg

1: put $r_0 := s$
2: **for** i increasing from 1 to p **do**
3: put $(s_i', r_i) := F(r_{i-1}, s_i)$
4: put $s_{p+1}' := r_p$
5: **return** $s_1'|\cdots|s_{p+1}'$

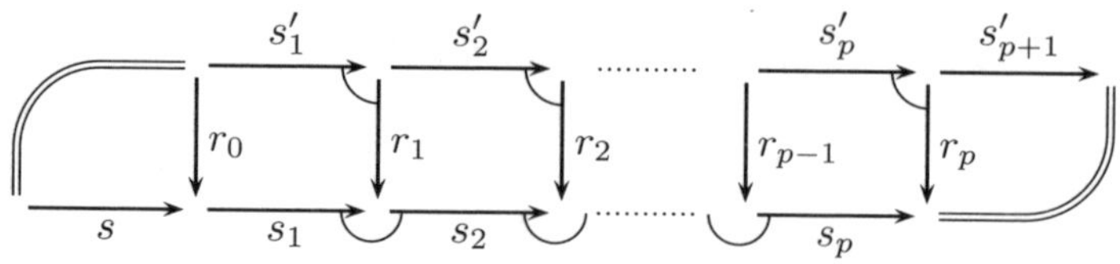

Figure 6. Algorithm 1.48: starting from s in $\mathcal{S}^\sharp$ and an $\mathcal{S}$-normal decomposition of g, it returns an $\mathcal{S}$-normal decomposition of sg.

Proposition 1.49 (left-multiplication). *If $\mathcal{S}$ is a Garside family in a left-cancellative category $\mathcal{C}$ and F is a $\square$-witness for $\mathcal{S}^\sharp$, then Algorithm 1.48 running on an element s of $\mathcal{S}^\sharp$ and an $\mathcal{S}$-normal decomposition $s_1|\cdots|s_p$ of an element g of $\mathcal{C}$ returns an $\mathcal{S}$-normal decomposition of sg. The function F is called p times.*

Moreover, if $s_1|\cdots|s_p$ is strict and F satisfies the condition

(1.50) *if s_1s_2 is not invertible, the first entry of $F(s_1, s_2)$ lies in $\mathcal{S}$,*

then, in the final decomposition $s'_1|\cdots|s'_{p+1}$, either $s'_1|\cdots|s'_{p+1}$, or $s'_1|\cdots|s'_{p-1}|s'_ps'_{p+1}$ is strict. If s is invertible, the latter case always occurs.

Proof. By construction, the diagram of Figure 6 is commutative. By assumption, $s_i|s_{i+1}$ is $\mathcal{S}$-greedy for every i, and, by the definition of a $\square$-witness, (s'_i, r_i) is $\mathcal{S}$-greedy for every i. Then the first domino rule (Proposition 1.45) implies that $s'_i|s'_{i+1}$ is $\mathcal{S}$-normal.

Assume now that $s_1|\cdots|s_p$ is strict. By Lemma 1.11(i), the assumption that s_i is not invertible implies that s'_i is not invertible either, and the assumption that F satisfies (1.50) then implies that s'_i lies in $\mathcal{S}$. If s'_{p+1} is not invertible, then $s'_1|\cdots|s'_{p+1}$ is a strict $\mathcal{S}$-normal path. Otherwise, $s'_ps'_{p+1}$ lies in $\mathcal{S}^\sharp$, and $s'_1|\cdots|s'_{p-1}|s'_ps'_{p+1}$ is strict. If s is invertible, then, by Lemma 1.11(ii), r_1 must be invertible as well, and so are $r_2, r_3, \ldots$ inductively. Hence s'_{p+1}, which is r_p, must be invertible, and we are in the second case of the alternative. $\square$

Observe that the additional condition (1.50) can be satisfied at no extra cost.

Lemma 1.51. *Assume that $\mathcal{C}$ is a left-cancellative category, $\mathcal{S}$ is included in $\mathcal{C}$, and $\mathcal{S}^\sharp$ satisfies Property $\square$. Then there exists a $\square$-witness for $\mathcal{S}^\sharp$ that satisfies* (1.50).

Proof. Assume that $s_1|s_2$ lies in $(\mathcal{S}^\sharp)^{[2]}$ and s_1s_2 is not invertible. Then choose $F(s_1, s_2)$ to be a strict $\mathcal{S}$-normal decomposition of s_1s_2 for $\|s_1s_2\|_\mathcal{S} = 2$, to be a decomposition of s_1s_2 lying in $\mathcal{S}|\mathcal{C}^\times$ for $\|s_1s_2\|_\mathcal{S} = 1$, and to be (s_1, s_2) for $\|s_1s_2\|_\mathcal{S} = 0$. Then F satisfies (1.50). $\square$

The second algorithm is an iteration of Algorithm 1.48.

Algorithm 1.52 (normal decomposition). (See Figure 7)

Context: A left-cancellative category $\mathcal{C}$, a Garside family $\mathcal{S}$ in $\mathcal{C}$, a $\square$-witness F for $\mathcal{S}^\sharp$
Input: An $\mathcal{S}^\sharp$-decomposition $s_1|\cdots|s_p$ of an element g of $\mathcal{C}$
Output: An $\mathcal{S}$-normal decomposition of g

put $s_{0,j} := s_j$ for $1 \leqslant j \leqslant p$
for j decreasing from p to 1 **do**
    for i increasing from 1 to $p - j$ (if any) **do**
        put $(t_{i,j-1}, s_{i,j}) := F(s_{i-1,j}, t_{i,j})$
    put $t_{p-j+1,j-1} := s_{p-j,j}$
return $t_{1,0}|\cdots|t_{p,0}$

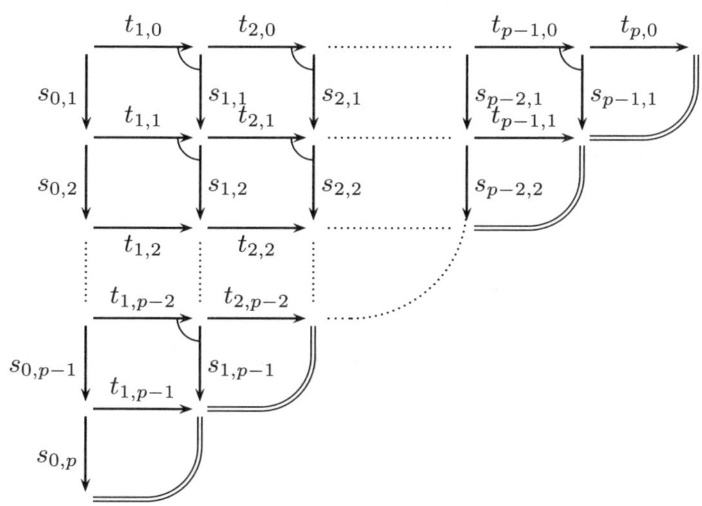

Figure 7. Algorithm 1.52: starting from $s_1|\cdots|s_p$, we fill the triangular grid starting from the bottom and using the $\square$-witness F to append squares from bottom to top, and from left to right.

Proposition 1.53 (normal decomposition). *If $\mathcal{S}$ is a Garside family in a left-cancellative category $\mathcal{C}$, then Algorithm* 1.52 *running on a decomposition $s_1|\cdots|s_p$ of an element g returns an $\mathcal{S}$-normal decomposition of g. The map F is appealed to $p(p-1)/2$ times.*

Proof. By definition, the diagram of Figure 7 can be constructed, as F is defined on every pair of $(\mathcal{S}^\sharp)^{[2]}$. In order to construct the jth row, F is used $p - j - 1$ times, so the complete construction requires $1 + 2 + \cdots + (p-1) = p(p-1)/2$ calls to F.

Then, it inductively follows from the first domino rule or, equivalently, from Proposition 1.49, that every row in the diagram is $\mathcal{S}$-normal. In particular, so is the first row. □

Example 1.54 (normal decomposition). Consider the free abelian monoid $\mathbb{N}^n$ again, with $S_n = \{s \mid \forall k\ (s(k) \leqslant 1)\}$ (Reference Structure 1, page 3). As in Example I.1.4, let $g = (3,1,2) = \mathtt{a}^3\mathtt{b}\mathtt{c}^2$. Starting from the length-one path $\mathtt{c}$, which is an S_n-normal decomposition of $\mathtt{c}$—or from the empty path, which is an S_n-normal decomposition of 1—and left-multiplying by the successive generators, possibly gathered provided one remains in S_n, one finds the (unique) S_n-normal decomposition $\mathtt{abc}|\mathtt{ac}|\mathtt{a}$, as could be expected.

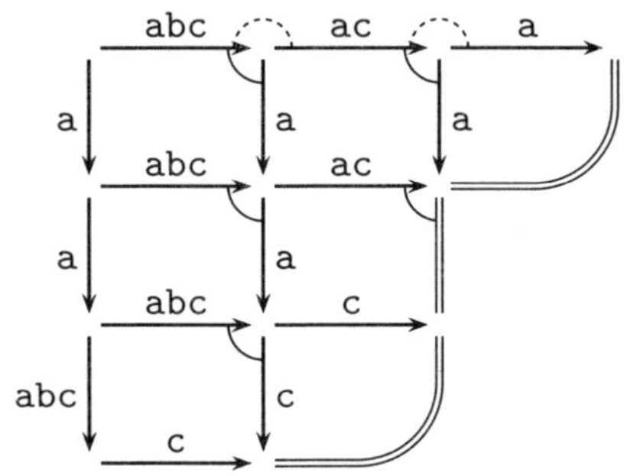

Figure 8. Computing the S-normal decomposition of $\mathtt{a}^3\mathtt{bc}^2$ in the free abelian monoid $\mathbb{N}^3$ using Algorithm 1.52. We may gather adjacent letters provided the product remains in S, so, here, we may start from $\mathtt{a}|\mathtt{a}|\mathtt{abc}|\mathtt{c}$, thus reducing the number of steps.

As a direct application of Proposition 1.49, we find bounds for the $\mathcal{S}$-length.

Corollary 1.55 (length I). *If $\mathcal{S}$ is a Garside family of a left-cancellative category $\mathcal{C}$, then, for every $\mathcal{C}$-path $f|g$, we have*

$$\|g\|_{\mathcal{S}} \leqslant \|fg\|_{\mathcal{S}} \leqslant \|f\|_{\mathcal{S}} + \|g\|_{\mathcal{S}}. \tag{1.56}$$

Proof. We use induction on $\ell = \|f\|_{\mathcal{S}}$. For $\ell = 0$, that is, if f is invertible, the final assertion in Proposition 1.49 implies $\|fg\|_{\mathcal{S}} = \|g\|_{\mathcal{S}}$, which gives (1.56).

Assume now $\ell = 1$, that is, f belongs to $\mathcal{S}^\sharp \setminus \mathcal{C}^\times$. Then Proposition 1.49 says that, if g has a strict $\mathcal{S}$-normal decomposition of length p, then fg has a strict $\mathcal{S}$-normal decomposition of length p or $p+1$. Hence, we have $\|g\|_{\mathcal{S}} \leqslant \|fg\|_{\mathcal{S}} \leqslant 1 + \|g\|_{\mathcal{S}}$.

Assume finally $\ell \geqslant 2$. Then there exists a decomposition $f = sf'$ with s in $\mathcal{S}^\sharp$ and $\|f'\|_{\mathcal{S}} = \ell - 1$. The induction hypothesis gives $\|g\|_{\mathcal{S}} \leqslant \|f'g\|_{\mathcal{S}} \leqslant \ell - 1 + \|g\|_{\mathcal{S}}$, whereas the above argument for $\ell = 1$ gives $\|f'g\|_{\mathcal{S}} \leqslant \|sf'g\|_{\mathcal{S}} \leqslant 1 + \|f'g\|_{\mathcal{S}}$, which together yield (1.56). □

We do not claim that $\|f\|_{\mathcal{S}} \leqslant \|fg\|_{\mathcal{S}}$ needs to hold: see Subsection 1.5 and Example 1.59 below for counter-examples.

1.5 The second domino rule

The first domino rule (Proposition 1.45) provides a way to compute an $\mathcal{S}$-normal decomposition using a left-to-right induction (Proposition 1.49). We now consider a symmetric process based on a right-to-left induction.

Definition 1.57 (second domino rule). Assume that $\mathcal{C}$ is a left-cancellative category and $\mathcal{S}$ is included in $\mathcal{C}$. We say that the *second domino rule* is valid for $\mathcal{S}$ if, when we have a commutative diagram as on the right with edges in $\mathcal{S}^\sharp$ in which $s_1|s_2$ and $t|s_2'$ are $\mathcal{S}$-greedy, then $s_1'|s_2'$ is $\mathcal{S}$-greedy as well.

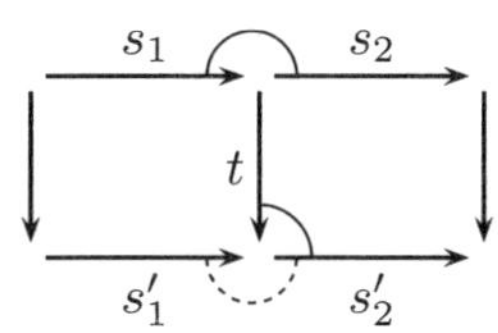

Note that the second domino rule is not an exact counterpart of the first one in that, here, all involved elements are supposed to lie in the reference family $\mathcal{S}^\sharp$, a restriction that is not present in the first domino rule.

Example 1.58 (second domino rule). Let us consider the free abelian monoid $\mathbb{N}^n$ and $S_n = \{s \mid \forall k\ (g(k) \leqslant 1)\}$ once more. We claim that the second domino rule is valid for S_n in $\mathbb{N}^n$. Indeed, assume that $s_1, ..., s_2'$ belong to S_n and satisfy the assumptions of Definition 1.57. Call the left and right vertical arrows t_0 and t_2. According to Example 1.2, what we have to show is that $s_2'(k) = 1$ implies $s_1'(k) = 1$. Now assume $s_2'(k) = 1$. As $t|s_2'$ is S-greedy, we have $t(k) = 1$, whence $(ts_2')(k) = (s_2t_2)(k) = 2$, and $s_2(k) = 1$. As $s_1|s_2$ is S-normal, the latter value implies $s_1(k) = 1$. We deduce $(t_0s_1')(k) = (s_1t_1)(k) = 2$, whence $s_1'(k) = 1$, as expected.

However, there exists no general counterpart to Proposition 1.45 and the second domino rule is not always valid. To see this, we introduce two more reference structures, which will provide useful counterexamples in the sequel.

REFERENCE STRUCTURE #8 **(left-absorbing monoid).**

- *For* $n \geqslant 1$, *put* $L_n = \langle \mathtt{a}, \mathtt{b} \mid \mathtt{ab}^n = \mathtt{b}^{n+1} \rangle^+$.
- *Put* $S_n = \{1, \mathtt{a}, \mathtt{b}, \mathtt{b}^2, ..., \mathtt{b}^{n+1}\}$, *a subset of* L_n *with* $n+3$ *elements.*
- *Put* $\Delta_n = \mathtt{b}^{n+1}$.

The unique relation of the above presentation of L_n *is of the type* $\mathtt{a}\cdots = \mathtt{b}\cdots$, *so it is right-complemented in the sense of Definition II.4.2. Moreover, the two terms have the same length, so, by Proposition* II.2.32 *(homogeneous),* L_n *is (strongly) Noetherian. As there are only two generators, the cube condition vacuously holds on* $\{\mathtt{a}, \mathtt{b}\}$. *Hence, by Proposition* II.4.16 *(right-complemented), the monoid* L_n *is left-cancellative and admits conditional right-lcms. On the other hand,* L_n *is not right-cancellative since we have* $\mathtt{ab}^{n-1} \neq \mathtt{b}^n$ *and* $\mathtt{ab}^n = \mathtt{b}^{n+1}$.

We claim that S_n *is a Garside family in* L_n. *Indeed,* S_n *contains the atoms* $\mathtt{a}$ *and* $\mathtt{b}$, *and, therefore, it generates* L_n. *There exists no nontrivial invertible element in* L_n, *so* $L_n^\times S_n \subseteq S_n^\sharp$ *is trivial. Hence, it suffices to show that every pair of elements of* S_n *admits* *an S_n-normal decomposition. To this end, one can list all length-two S_n-normal paths, namely the paths $\mathtt{a}|\mathtt{b}^p$ with $0 \leqslant p < n$, plus the paths $\mathtt{b}^p|\mathtt{a}$ with $1 \leqslant p \leqslant n$, plus all paths $\Delta_n|g$ with g in S_n. The verification is then easy—an alternative, much quicker argument can be given using Proposition* IV.1.24 *(recognizing Garside* II*) of the next chapter.*

REFERENCE STRUCTURE #9 **(affine braids of type $\widetilde{\mathrm{A}}_2$).**

• *Put* $B^+ = \langle \sigma_1, \sigma_2, \sigma_3 \mid \sigma_1\sigma_2\sigma_1 = \sigma_2\sigma_1\sigma_2, \sigma_2\sigma_3\sigma_2 = \sigma_3\sigma_2\sigma_3, \sigma_3\sigma_1\sigma_3 = \sigma_1\sigma_3\sigma_1\rangle^+$, and $B = \langle \sigma_1, \sigma_2, \sigma_3 \mid \sigma_1\sigma_2\sigma_1 = \sigma_2\sigma_1\sigma_2, \sigma_2\sigma_3\sigma_2 = \sigma_3\sigma_2\sigma_3, \sigma_3\sigma_1\sigma_3 = \sigma_1\sigma_3\sigma_1\rangle$.

• *Put* $E = \{\sigma_1\sigma_2\sigma_3\sigma_2,\ \sigma_2\sigma_3\sigma_1\sigma_3,\ \sigma_3\sigma_1\sigma_3\sigma_1\}$.

• *Define S to be the set of elements that right-divide* (at least) *one of the three elements of E.*

The monoid B^+ is the Artin–Tits monoid of type $\widetilde{\mathrm{A}}_2$, and B is the corresponding Artin–Tits group; both are associated with the (infinite) affine type Coxeter group of type $\widetilde{\mathrm{A}}_2$, see Chapter IX *below. The above presentation of B^+ is right-complemented and homogeneous, hence Noetherian, and one easily checks that the cube condition is satisfied on $\{\sigma_1, \sigma_2, \sigma_3\}$. Hence, by Proposition* II.4.16 *(right-complemented) again, the monoid B^+ is left-cancellative and it admits conditional right-lcms. By the symmetry of the relations, B^+ must be right-cancellative and admit conditional left-lcms. By homogeneity, the family S must be finite since it consists of elements of length at most four; it turns out that S consists of the sixteen elements represented in Figure* 9.

The family S contains the three atoms of B^+, namely σ_1, σ_2, and σ_3. By definition, it is closed under right-divisor, that is, every right-divisor of an element of S belongs to S (by the way, we note that S is not closed under left-divisor: the left-divisor $\sigma_1\sigma_2\sigma_3$ of $\sigma_1\sigma_2\sigma_3\sigma_2$ does not belong to S). Finally, one easily checks using right-reversing that the right-lcm of any two elements of S that admit a common right-multiple in B^+ also lies in S. Anticipating on Corollary IV.2.29 *(recognizing Garside, right-lcm case), we conclude that S is a Garside family in B^+.*

Note that any two elements of the monoid B^+ need not admit a common right-multiple: for instance, σ_1 and $\sigma_2\sigma_3$ do not, since, as seen in Example II.4.20*, the right-reversing of the word $\overline{\sigma_1}|\sigma_2|\sigma_3$ does not terminate. So B^+ is neither a left- or a right-Ore monoid, and the enveloping group of B^+, which is B, is not a group of left- or right-fractions of B^+. It is known that B^+ embeds into B but, for instance, $\sigma_2\sigma_3\sigma_1^{-1}\sigma_2\sigma_3$ is an element of B that can be expressed neither as a left-fraction nor as a right-fraction with respect to B^+.*

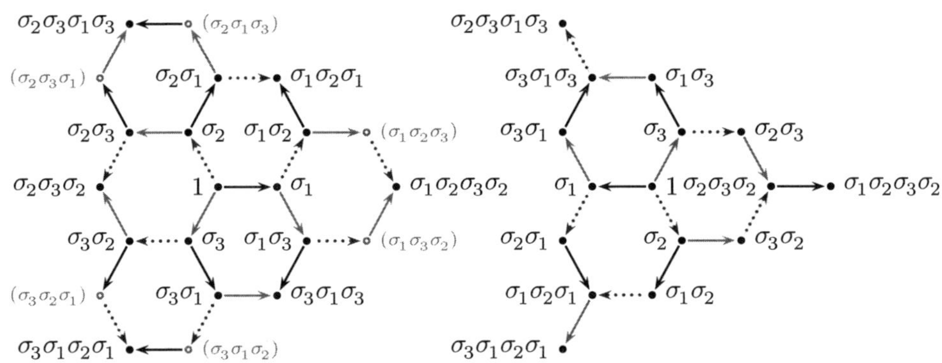

Figure 9. Two representations of the finite Garside family S in the Artin–Tits monoid of type $\widetilde{\mathrm{A}}_2$: on the left, the (usual) Cayley graph of S, where arrows correspond to right-multiplications; on the right, the "co-Cayley" graph with arrows corresponding to left-multiplications; as the family S is closed under right-divisor, but not under left-divisor, some additional vertices not belonging to S appear on the left diagram (the six grey vertices).

Example 1.59 (second domino rule not valid). The second domino rule is not valid for S_n in L_n for $n \geqslant 2$, where L_n and S_n are as in Reference Structure 8. Indeed, the paths $\mathtt{a}|\mathtt{b}$ and $\mathtt{b}^{n+1}|\mathtt{b}$ are S_n-greedy, the diagram on the right is commutative and all edges corresponds to elements of S_n. However $\mathtt{b}|\mathtt{b}$ is not S_n-greedy: the S_n-normal decomposition of $\mathtt{b}^2$ is $\mathtt{b}^2$ itself.

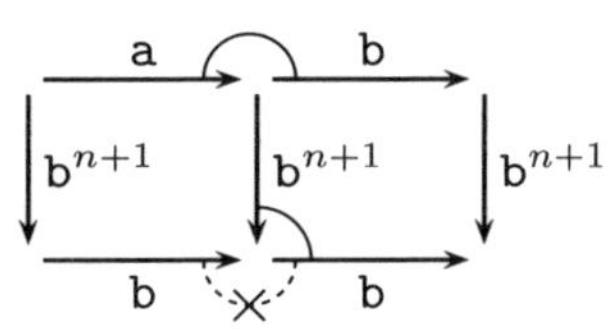

Similarly, the second domino rule is not valid for the Garside family S in the affine braid monoid B^+ of Reference Structure 9. A counter-example is shown on the right: $\sigma_1|\sigma_1\sigma_2$ and $\sigma_1\sigma_2\sigma_2|\sigma_3$ are S-normal, but $\sigma_2|\sigma_3$ is not, since $\sigma_2\sigma_3$ is an element of S. Note that, contrary to L_n, the monoid B^+ is both left- and right-cancellative.

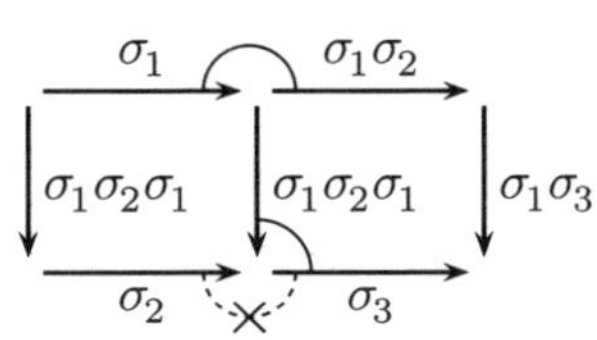

We shall come back to the second domino rule later and establish sufficient conditions for its validity in Propositions IV.1.39 and V.1.52. Here we observe that, when valid, the second domino rule provides a counterpart of Proposition 1.49.

Algorithm 1.60 (right-multiplication). (See Figure 10)

Context: A left-cancellative category $\mathcal{C}$, a Garside family $\mathcal{S}$ of $\mathcal{C}$ for which the second domino rule is valid, a $\square$-witness F for $\mathcal{S}^\sharp$

Input: An element s of $\mathcal{S}^\sharp$ and an $\mathcal{S}$-normal decomposition $s_1|\cdots|s_p$ of an element g of $\mathcal{C}$ such that gs is defined

Output: An $\mathcal{S}$-normal decomposition of gs

```
put r_p := s
for i decreasing from p to 1 do
    put (r_{i-1}, s'_i) := F(s_i, r_i)
put s'_0 := r_0
return s'_0|···|s'_p
```

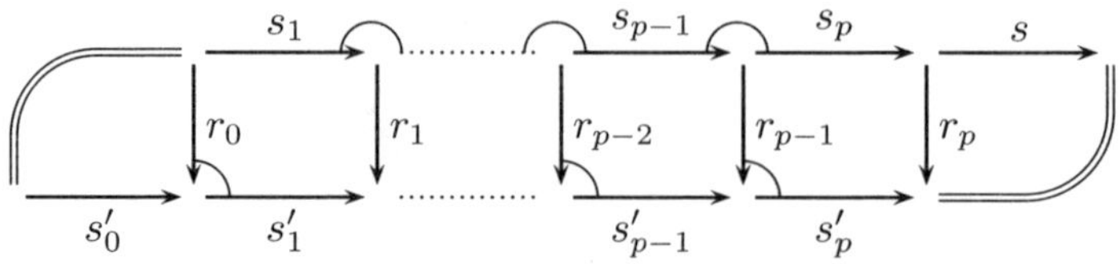

Figure 10. Algorithm 1.60: starting from an $\mathcal{S}$-normal path $s_1|\cdots|s_p$ and an element s of $\mathcal{S}^\sharp$, it returns an $\mathcal{S}$-normal decomposition of $s_1\cdots s_p s$. Attention: the correctness of the method is guaranteed only if the second domino rule is valid for $\mathcal{S}$.

Proposition 1.61 (right-multiplication). *If $\mathcal{S}$ is a Garside family in a left-cancellative category $\mathcal{C}$, if the second domino rule is valid for $\mathcal{S}$, and if F is a $\square$-witness for $\mathcal{S}^\sharp$, then Algorithm* 1.60 *running on an element s of $\mathcal{S}^\sharp$ and an $\mathcal{S}$-normal decomposition $s_1|\cdots|s_p$ of an element g of $\mathcal{C}$ returns an $\mathcal{S}$-normal decomposition of gs. The function F is called p times.*

Proof. As the diagram of Figure 10 is commutative, we have $s'_0\cdots s'_p = r_0 s'_1\cdots s'_p = s_1\cdots s_p r_p = gs$. Applying the second domino rule to each two-square subdiagram of the diagram starting from the right, we see that $s'_0|s'_1|\cdots|s'_p$ is $\mathcal{S}$-greedy. As all entries lie in $\mathcal{S}^\sharp$, this path is $\mathcal{S}$-normal. $\square$

The effect of the second domino rule is to shorten the computation of certain normal decompositions. Assume that $s_1|\cdots|s_p$ and $t_1|\cdots|t_q$ are $\mathcal{S}$-normal paths and $s_p t_1$ is defined. By applying Proposition 1.61, we can compute an $\mathcal{S}$-normal decomposition of the product $s_1\cdots s_p t_1\cdots t_q$ by filling a diagram as in Figure 11. When valid, the second domino rule guarantees that the path consisting of the first q top edges followed by p vertical edges is $\mathcal{S}$-normal, that is, the triangular part of the diagram may be forgotten. The failure of this phenomenon in the context of Example 1.59 is illustrated in Figure 12.

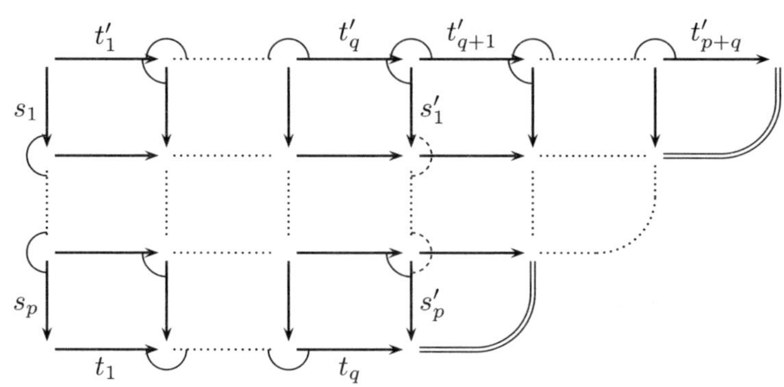

Figure 11. Finding an $\mathcal{S}$-normal decomposition of $s_1\cdots s_p t_1\cdots t_q$ when $s_1|\cdots|s_p$ and $t_1|\cdots|t_q$ are $\mathcal{S}$-normal: using Proposition 1.49, hence the first domino rule only, one determines the $\mathcal{S}$-normal path $t'_1|\cdots|t'_{p+q}$ in $pq + p(p-1)/2$ steps; if the second domino rule is valid, the path $s'_1|\cdots|s'_p$ is already $\mathcal{S}$-normal, and $t'_1|\cdots|t'_q|s'_1|\cdots|s'_p$ is $\mathcal{S}$-normal. Note that, by uniqueness, $t'_{q+1}|\cdots|t'_{p+q}$ must then be a $\mathcal{C}^\times$-deformation of $s'_1|\cdots|s'_p$.

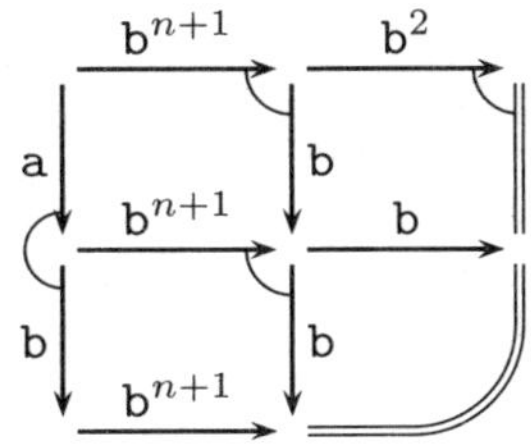

Figure 12. Failure of the second domino rule in the context of Example 1.59: starting with $s_1|s_2 = \mathtt{a}|\mathtt{b}$ and $t_1 = \mathtt{b}^{n+1}$, we obtain the $\mathcal{S}$-normal decomposition $\mathtt{b}^{n+1}|\mathtt{b}^2|1$ of $s_1s_2t_1$ using three normalizing $\square$-tiles, and not two as would be the case if the second domino rule were valid.

Provided some closure assumption is satisfied, another application of the second domino rule is a counterpart of the inequalities (1.56) for the length of a product.

Proposition 1.62 (length II). *If $\mathcal{S}$ is a Garside family of a left-cancellative category $\mathcal{C}$, the second domino rule is valid for $\mathcal{S}$, and every left-divisor of an element of $\mathcal{S}^\sharp$ belongs to $\mathcal{S}^\sharp$, then, for every $\mathcal{C}$-path $f|g$, we have*

$$\max(\|f\|_{\mathcal{S}}, \|g\|_{\mathcal{S}}) \leqslant \|fg\|_{\mathcal{S}} \leqslant \|f\|_{\mathcal{S}} + \|g\|_{\mathcal{S}}, \tag{1.63}$$

and every left- or right-divisor of an element of $\mathcal{S}$-length ℓ has $\mathcal{S}$-length at most ℓ.

Proof. Owing to Corollary 1.55, we only have to prove $\|f\|_{\mathcal{S}} \leqslant \|fg\|_{\mathcal{S}}$. As $\mathcal{S}^\sharp$ generates $\mathcal{C}$, it is sufficient to prove $\|f\|_{\mathcal{S}} \leqslant \|fs\|_{\mathcal{S}}$ for s in $\mathcal{S}^\sharp$. Assume that $s_1|\cdots|s_p$ is a strict $\mathcal{S}$-normal decomposition of f. With the notation of Proposition 1.61 and Figure 10, $r_0|s'_1|\cdots|s'_p$ is an $\mathcal{S}$-normal decomposition of fs. The case $p \leqslant 1$ is trivial, so assume $p \geqslant 2$. If s'_{p-1} was invertible, so would be s'_p, and $r_{p-2}s'_{p-1}s'_p$, which is $s_{p-1}s_ps$, would belong to $\mathcal{S}^\sharp$, and therefore so would do its left-divisor $s_{p-1}s_p$. As $s_{p-1}|s_p$ is $\mathcal{S}$-greedy, we would deduce that s_p is invertible, contrary to the assumption that $s_1|\cdots|s_p$ is strict. So s'_{p-1} cannot be invertible, and the $\mathcal{S}$-length of fs is at least p. $\square$

2 Symmetric normal decompositions

In Section 1, we showed how to construct essentially unique distinguished decompositions for the elements of a left-cancellative category in which a Garside family has been specified. We shall now construct similar distinguished decompositions for the elements of the enveloping groupoid of the category, in the particular case when the considered category is a left-Ore category that admits left-lcms. Let us mention that an alternative, (slightly) different type of distinguished decomposition will be defined, under different assumptions, in Chapter V.

The organization of the current section is parallel to that of Section 1. First, we introduce in Subsection 2.1 the notion of left-disjoint elements, and use it in Subsection 2.2 to define symmetric $\mathcal{S}$-normal decompositions. We show the essential uniqueness of such decompositions in Subsection 2.3, and their existence in left-Ore categories that admit

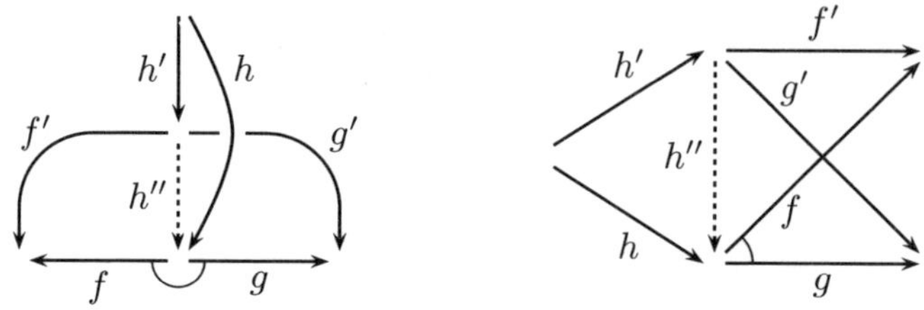

Figure 13. Two equivalent ways of illustrating the left-disjointness of f and g: whenever we have a commutative diagram corresponding to the plain arrows, there exists a factoring dashed arrow. Note that, in the situation of (2.2), if f', g' are the elements satisfying $hf = h'f'$ and $hg = h'g'$, then, by definition, the pair (h, h') witnesses for $(f, g) \bowtie (f', g')$. Attention: a factorization $h = h'h''$ implies $f' = h''f$ and $g' = h''g$ only if the ambient category is left-cancellative; otherwise, the diagram is ambiguous.

left-lcms in Subsection 2.4. Algorithmic questions are addressed in Subsection 2.5. Finally, we briefly discuss in an appendix the extension of the results from left-Ore categories to general cancellative categories.

2.1 Left-disjoint elements

Hereafter our context will be mostly that of a left-Ore category as introduced in Section II.3, that is, a cancellative category in which any two elements admit a common left-multiple. However, all basic definitions and lemmas make sense and work in the context of a general left-cancellative category, and we begin with such a context.

We look for distinguished decompositions for the elements of a groupoid of fractions. As in the elementary case of rational numbers, obtaining unique expressions requires a convenient notion of irreducible fraction.

Definition 2.1 (left-disjoint). (See Figure 13.) Two elements f, g of a left-cancellative category $\mathcal{C}$ are called *left-disjoint* if they have the same source and satisfy

(2.2) $$\forall h, h' \in \mathcal{C}\ ((h' \preccurlyeq hf \text{ and } h' \preccurlyeq hg\ ) \Rightarrow h' \preccurlyeq h).$$

In words: f and g are left-disjoint if every common left-divisor of elements of the form hf and hg must be a left-divisor of h. We shall indicate in diagrams that f and g are left-disjoint using an arc, as in $\xleftarrow{f} \frown \xrightarrow{g}$.

Example 2.3 (left-disjoint). Consider the free abelian monoid $\mathbb{N}^n$ as in Reference Structure 1, page 3. Two elements f, g of $\mathbb{N}^n$ are left-disjoint if and only if, for every i, at least one of $f(i), g(i)$ is zero. Indeed, assume $f(i) \geqslant 1$ and $g(i) \geqslant 1$. Let $h = 1$ and $h' = \mathtt{a}_i$. Then $h' \preccurlyeq hf$ and $h' \preccurlyeq hg$ F are true, but $h' \preccurlyeq 1$ fails, so f, g are not left-disjoint. Conversely, assume $f(i) = 0$. Then $h' \preccurlyeq hf$ implies $h'(i) \leqslant h(i)$. Similarly, $g(i) = 0$

implies $h'(i) \leqslant h(i)$. So, if $f(i)g(i) = 0$ holds for every i, so does $h'(i) \leqslant h(i)$, that is, $h' \preccurlyeq h$ holds. Hence f and g are left-disjoint.

As can be expected, since we aim at introducing irreducible fractions, left-disjointness is related to the non-existence of common divisors.

Definition 2.4 (left-coprime). Two elements f, g of a left-cancellative category $\mathcal{C}$ are called *left-coprime* if f and g share the same source and every common left-divisor of f and g is invertible.

Proposition 2.5 (disjoint *vs.* **coprime).** *Assume that $\mathcal{C}$ is a left-cancellative category.*
(i) *Any two left-disjoint elements are left-coprime.*
(ii) *Conversely, any two left-coprime elements are left-disjoint whenever $\mathcal{C}$ satisfies*

(2.6) *Any two common right-multiples of two elements are right-multiples of some common right-multiple of these elements,*

(iii) *Condition* (2.6) *is satisfied in particular if $\mathcal{C}$ admits left-gcds, and if $\mathcal{C}$ admits conditional right-lcms.*

Proof. (i) Assume that f and g are left-disjoint and h is a common left-divisor of f and g. Let x denote the common source of f and g. Then we have $h \preccurlyeq 1_x f$ and $h \preccurlyeq 1_x g$, whence $h \preccurlyeq 1_x$ by definition. Hence h must be invertible, that is, f and g are left-coprime.

(ii) Assume now that (2.6) is satisfied, and f, g are left-coprime elements of $\mathcal{C}$. Assume $h' \preccurlyeq hf$ and $h' \preccurlyeq hg$, say $hf = h'f'$ and $hg = h'g'$, see Figure 14. By assumption, hf and hg are two common right-multiples of h and h', so, by (2.6), there exists a common right-multiple he of h and h' satisfying $he \preccurlyeq hf$ and $he \preccurlyeq hg$. We deduce $e \preccurlyeq f$ and $e \preccurlyeq g$. As f and g are left-coprime, e must be invertible and, therefore, h', which left-divides he by assumption, also left-divides h. Hence f and g are left-disjoint.

(iii) Assume $f_i \preccurlyeq g_j$ for $i, j = 1, 2$. Assume first that $\mathcal{C}$ admits left-gcds. Then g_1 and g_2 admit a left-gcd, say g. For $i = 1, 2$, the assumption that f_i left-divide g_1 and g_2 implies that f_i left-divides g. So g witnesses for (2.6).

Assume now that $\mathcal{C}$ admits conditional right-lcms. By assumption, f_1 and f_2 admit a common right-multiple, hence they admit a right-lcm, say f. For $j = 1, 2$, the assumption that g_j is a right-multiple of f_1 and f_2 implies that g_j is a right-multiple of f. So f witnesses for (2.6). □

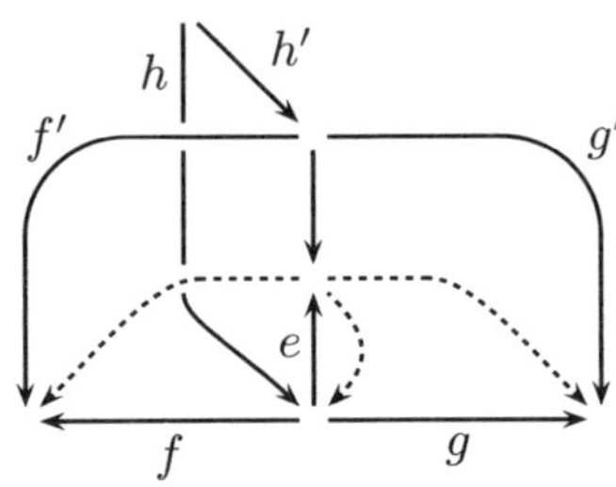

Figure 14. Proof of Proposition 2.5(ii).

Considering the (cancellative) monoid $\langle \mathtt{a}, \mathtt{a}', \mathtt{b}, \mathtt{b}', \mathtt{c}, \mathtt{c}' \mid \mathtt{ab}' = \mathtt{a}'\mathtt{b}, \mathtt{ac}' = \mathtt{a}'\mathtt{c} \rangle^+$, in which $\mathtt{b}$ and $\mathtt{c}$ have no nontrivial common left-divisor, but they are not left-disjoint since we have $\mathsf{a} \preccurlyeq \mathsf{a}'\mathsf{b}$ and $\mathsf{a} \preccurlyeq \mathsf{a}'\mathsf{c}$ but not $\mathsf{a} \preccurlyeq \mathsf{a}'$, we see that some condition about the ambient category is needed to guarantee the converse implication in Proposition 2.5.

On the other hand, when the ambient category is a left-Ore category, left-disjointness has a simple connection with the factorization of fractions.

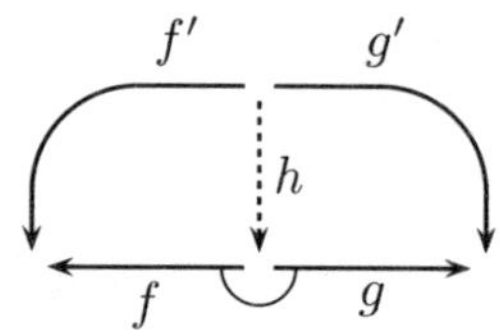

Lemma 2.7. *If $\mathcal{C}$ is a left-Ore category, then, for f, g in $\mathcal{C}$ with the same source, the following conditions are equivalent:*

(i) *The elements f and g are left-disjoint.*

(ii) *For all f', g' in $\mathcal{C}$ satisfying $f^{-1}g = f'^{-1}g'$ in $\mathcal{E}nv(\mathcal{C})$, there exists h satisfying $f' = hf$ and $g' = hg$.*

Proof. Assume that f and g are left-disjoint and $f^{-1}g = f'^{-1}g'$ holds in $\mathcal{E}nv(\mathcal{C})$. By Ore's theorem (Proposition II.3.11), there exist h', h'' witnessing for $(f, g) \bowtie (f', g')$, that is, satisfying $h'f' = h''f$ and $h'g' = h''g$, hence $h' \preccurlyeq h''f$ and $h' \preccurlyeq h''g$. As f and g are left-disjoint, this implies $h' \preccurlyeq h''$, so there exists h satisfying $h'' = h'h$. We deduce $h'f' = h'hf$ and $h'g' = h'hg$, whence $f' = hf$ and $g' = hg$ since $\mathcal{C}$ is left-cancellative. So (i) implies (ii).

Conversely, assume that (ii) holds and we have $h' \preccurlyeq h''f$ and $h' \preccurlyeq h''g$. Write $h''f = h'f'$ and $h''g = h'g'$. Then, in $\mathcal{E}nv(\mathcal{C})$, we have $f^{-1}g = f'^{-1}g'$, and (ii) implies the existence of h satisfying $f' = hf$ and $g' = hg$. We deduce $h''f = h'f' = h'hf$, whence $h'' = h'h$ as $\mathcal{C}$ is right-cancellative. Hence we have $h' \preccurlyeq h''$, and f and g are left-disjoint. So (ii) implies (i). □

Drawn as 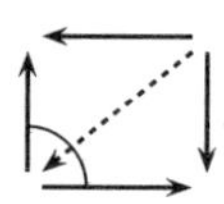, the diagram of Lemma 2.7 is similar to 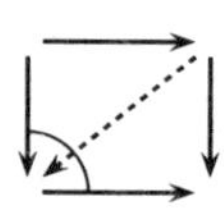, which corresponds to $\mathcal{S}$-greediness: both assert the existence of a factoring arrow for certain pairs of elements with the same endpoints. However, $\mathcal{S}$-greediness refers to $\mathcal{S}$ (the top arrow has to correspond to an element of $\mathcal{S}$), whereas no such restriction holds for left-disjointness.

Finally, in the special case of elements that admit a common right-multiple (always the case in an Ore category), we have a simple connection between left-disjointness and left-lcms.

Lemma 2.8. *Assume that $\mathcal{C}$ is a left-Ore category, and f, g, f', g' are elements of $\mathcal{C}$ satisfying $fg' = gf'$. Then the following are equivalent:*

(i) *The elements f and g are left-disjoint.*

(ii) *The element fg' is a left-lcm of f' and g'.*

Proof. Assume that f and g are left-disjoint, and that $f''g' = g''f'$ is a common left-multiple of f' and g'. In $\mathcal{E}nv(\mathcal{C})$, we have $f''^{-1}g'' = g'f'^{-1} = f^{-1}g$, so, by Lemma 2.7, there must exist h in $\mathcal{C}$ satisfying $f'' = hf$ and $g'' = hg$. This means that $f''g'$ is a left-multiple of fg' and, therefore, fg' is a left-lcm of f' and g'. So (i) implies (ii).

Conversely, assume that fg' is a left-lcm of f' and g'. Let f'', g'' in $\mathcal{C}$ satisfy $f''^{-1}g'' = f^{-1}g$ in $\mathcal{E}nv(\mathcal{C})$. Since $f^{-1}g = g'f'^{-1}$ holds in $\mathcal{E}nv(\mathcal{C})$, we deduce $f''^{-1}g'' = g'f'^{-1}$, whence $f''g' = g''f'$. As fg' is a left-lcm of f' and g', there must exist h in $\mathcal{C}$ satisfying $f''g' = hfg'$, whence $f'' = hf$ and, similarly, $g'' = hg$. By Lemma 2.7, this means that f and g are left-disjoint. So (ii) implies (i). □

2.2 Symmetric normal decompositions

We are now ready to introduce natural distinguished decompositions for the elements of a groupoid of left-fractions by merging the notions of an $\mathcal{S}$-normal path and of left-disjoint elements. If $\mathcal{C}$ is a left-Ore category, then, according to Ore's theorem (Proposition II.3.11), every element of the enveloping groupoid $\mathcal{E}nv(\mathcal{C})$ is a left-fraction, hence it can be expressed by means of two elements of $\mathcal{C}$. If $\mathcal{S}$ is a Garside family in $\mathcal{C}$, every element of $\mathcal{C}$ admits $\mathcal{S}$-normal decompositions, and we deduce decompositions for the elements of $\mathcal{E}nv(\mathcal{C})$ involving two $\mathcal{S}^\sharp$-paths, one for the numerator, one for the denominator. We naturally say that a signed path $\overline{f_q}|\cdots|\overline{f_1}|g_1|\cdots|g_p$ is a *decomposition* for an element g of $\mathcal{E}nv(\mathcal{C})$ if $g = f_q^{-1}\cdots f_1^{-1}g_1\cdots g_p$ holds—we recall that we usually drop the canonical embedding ι of $\mathcal{C}$ into $\mathcal{E}nv(\mathcal{C})$ that is, we identify $\mathcal{C}$ with its image in $\mathcal{E}nv(\mathcal{C})$.

Definition 2.9 (symmetric greedy, symmetric normal). If $\mathcal{S}$ is a subfamily of a left-cancellative category $\mathcal{C}$, a negative–positive $\mathcal{C}$-path $\overline{f_q}|\cdots|\overline{f_1}|g_1|\cdots|g_p$ is called *symmetric $\mathcal{S}$-greedy* (*resp. symmetric $\mathcal{S}$-normal, resp. strict symmetric $\mathcal{S}$-normal*) if the paths $f_1|\cdots|f_q$ and $g_1|\cdots|g_p$ are $\mathcal{S}$-greedy (*resp.* $\mathcal{S}$-normal, *resp.* strict $\mathcal{S}$-normal) and, in addition, f_1 and g_1 are left-disjoint.

It is natural to consider a $\mathcal{C}$-path (hence a positive path) as a positive–negative path in which the negative part is the empty path ε_x, where x is the source of the considered path. Then a positive path is symmetric $\mathcal{S}$-greedy in the sense of Definition 1.1 if and only if it is $\mathcal{S}$-greedy in the sense of Definition 2.9 when viewed as a degenerate negative–positive path. The same holds for symmetric and strict symmetric $\mathcal{S}$-normal paths: indeed, empty paths are strict $\mathcal{S}$-normal and, by very definition, if g is an element of $\mathcal{C}$ with source x, then 1_x and g are left-disjoint since, if f', g', h, and h' satisfy $hf' = h'1_x$ and $hg' = h'g$, the first equality implies $h \preccurlyeq h'$ directly.

Example 2.10 (symmetric greedy, symmetric normal). In the free abelian monoid $\mathbb{N}^n$ with $S_n = \{s \mid \forall k\ (s(k) \leqslant 1)\}$ (Reference Structure 1, page 3), as seen in Example 1.18, a path $s_1|\cdots|s_p$ is S_n-normal if and only if the sequence $(s_1(k), ..., s_p(k))$ is non-increasing for every k. We also observed in Example 2.3 that f and g are left-disjoint if and only if $f(k)g(k) = 0$ holds for every k. Then $\overline{t_q}|\cdots|\overline{t_1}|s_1|\cdots|s_p$ is symmetric S_n-greedy if and only if, for every k, the sequences $(s_1(k), ..., s_p(k))$ and $(t_1(k), ..., t_q(k))$ are non-increasing and at least one of them exclusively consists of zeros.

Then every element g of $\mathbb{Z}^n$ admits a unique strict symmetric S_n-normal decomposition, namely the path $\overline{t_q}|\cdots|\overline{t_1}|s_1|\cdots|s_p$ with $p = \max\{g(k) \mid 1 \leqslant k \leqslant n\}$, $q = \max\{-g(k) \mid 1 \leqslant k \leqslant n\}$, and, for every k with $g(k) < 0$ (*resp.* $g(k) > 0$), we have $t_i(k) = 1$ for $1 \leqslant i \leqslant -g(k)$, $t_i(k) = 0$ otherwise, and $s_i(k) = 0$ for every i (*resp.* $s_i(k) = 1$ for $1 \leqslant i \leqslant g(k)$, $s_i(k) = 0$ otherwise, and $t_i(k) = 0$ for every i).

For instance, writing $\mathtt{a}, \mathtt{b}, \mathtt{c}$ for $\mathtt{a}_1, \mathtt{a}_2, \mathtt{a}_3$, we see that the unique strict symmetric S_3-normal decomposition of $\mathtt{ab}^{-2}\mathtt{c}^2$, that is, of $(1, -2, 2)$, is the length-four path $\overline{\mathtt{b}}|\overline{\mathtt{b}}|\mathtt{ac}|\mathtt{c}$.

The following result is analogous to Proposition 1.12 and enables one to gather the entries of an $\mathcal{S}$-greedy path.

Proposition 2.11 (grouping entries). *Assume that $\mathcal{C}$ is a left-cancellative category, $\mathcal{S}$ is included in $\mathcal{C}$, and $\mathcal{S}^\sharp$ generates $\mathcal{C}$. Then, for all $\mathcal{S}$-greedy paths $f_1|\cdots|f_p$, $g_1|\cdots|g_q$, the following are equivalent:*

(i) *The elements f_1 and g_1 are left-disjoint.*

(ii) *The elements $f_1\cdots f_p$ and $g_1\cdots g_q$ are left-disjoint.*

Proof. Assume (i). Put $f = f_1\cdots f_p$ and $g = g_1\cdots g_q$. In order to establish that f and g are left-disjoint, we shall prove, using induction on m, that, for every h' belonging to $(\mathcal{S}^\sharp)^m$ and every h in $\mathcal{C}$, the conjunction of $h' \preccurlyeq hf$ and $h' \preccurlyeq hg$ implies $h' \preccurlyeq h$. For $m = 0$, the element h' is an identity element, and the result is trivial. Assume $m \geqslant 1$ and write $h' = rh''$ with $r \in \mathcal{S}^\sharp$ and $h'' \in (\mathcal{S}^\sharp)^{m-1}$ (Figure 15). By assumption, we have $rh'' \preccurlyeq hf$, whence a fortiori $r \preccurlyeq hf$, that is, $r \preccurlyeq hf_1\cdots f_p$. By assumption, $f_1|\cdots|f_p$ is $\mathcal{S}$-greedy, hence $\mathcal{S}^\sharp$-greedy by Lemma 1.10, so $r \preccurlyeq hf_1\cdots f_p$ implies $r \preccurlyeq hf_1$. A similar argument gives $r \preccurlyeq hg_1$, and, therefore, the assumption that f_1 and g_1 are left-disjoint implies $r \preccurlyeq h$, that is, there exists h_1 satisfying $h = rh_1$. Then, as $\mathcal{C}$ is left-cancellative, the assumption that h' left-divides hf and hg implies that h'' left-divides h_1f and h_1g. The induction hypothesis then yields $h'' \preccurlyeq h_1$, whence $h' \preccurlyeq h$. So f and g are left-disjoint, and (i) implies (ii).

That (ii) implies (i) is clear: if h' left-divides hf_1 and hg_1, then it a fortiori left-divides hf and hg, and the assumption that f and g are left-disjoint implies that h' left-divides h, hence that f_1 and g_1 are left-disjoint. □

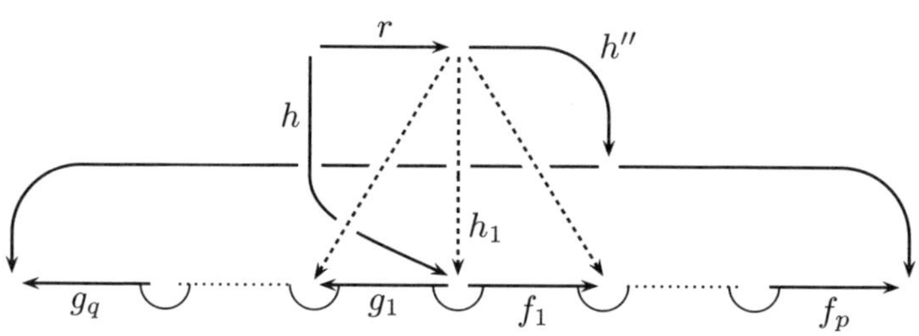

Figure 15. Proof of Proposition 2.11 using an induction on the length of an expression of h.

As an application, we deduce that the existence of a symmetric $\mathcal{S}$-normal decomposition implies the existence of a strict symmetric $\mathcal{S}$-normal one.

Lemma 2.12. *Assume that $\mathcal{C}$ is a left-cancellative category and $\mathcal{S}$ is a subfamily of $\mathcal{C}$ that satisfies $\mathcal{C}^\times\mathcal{S} \subseteq \mathcal{S}^\sharp$. Then, for every symmetric $\mathcal{S}$-normal path $\overline{u}|v$, there exists a strict symmetric $\mathcal{S}$-normal path $\overline{u'}|v'$ satisfying $([u'],[v']) \bowtie ([u],[v])$.*

Proof. Assume that $\overline{u}|v$ is a symmetric $\mathcal{S}$-normal path. By Proposition 1.23, there exist strict $\mathcal{S}$-normal paths u', v' satisfying $[u'] = [u]$ and $[v'] = [v]$. Then the relation $([u'],[v']) \bowtie ([u],[v])$ trivially holds. By Proposition 2.11, the assumption that the first entries of u and v are left-disjoint implies that $[u]$ and $[v]$ are left-disjoint, which in turn implies that the first entries of u' and v' are left-disjoint. Therefore, $\overline{u'}|v'$ is a strict symmetric $\mathcal{S}$-normal path. □

Proposition 2.13 (strict). *Assume that $\mathcal{C}$ is a left-Ore category and $\mathcal{S}$ is a Garside family of $\mathcal{C}$. Then every element of $\mathcal{E}nv(\mathcal{C})$ that admits a symmetric $\mathcal{S}$-normal decomposition admits a strict $\mathcal{S}$-normal decomposition.*

2.3 Uniqueness of symmetric normal decompositions

We now establish the essential uniqueness of symmetric $\mathcal{S}$-normal decompositions when they exist. As in the positive case, the uniqueness statements have to take into account the possible presence of invertible elements in the ambient category. Here is the convenient extension of Definition 1.20.

Definition 2.14 (deformation by invertible elements). (See Figure 16.) A negative–positive $\mathcal{C}$-path $\overline{f_{-q'}}|\cdots|\overline{f_0}|f_1|\cdots|f_{p'}$ in a left-cancellative category $\mathcal{C}$ is called a *deformation by invertible elements*, or *$\mathcal{C}^\times$-deformation*, of another one $\overline{g_{-q}}|\cdots|\overline{g_0}|g_1|\cdots|g_p$ if there exist $\epsilon_{-n}, \dots, \epsilon_m$ in $\mathcal{C}^\times$, $m = \max(p, p') + 1$, $n = \max(q, q')$, such that ϵ_{-n} and ϵ_m are identity elements, $\epsilon_{i-1} g_i = f_i \epsilon_i$ holds for $-n-1 \leqslant i \leqslant m$, where, for $p \neq p'$ or $q \neq q'$, the shorter path is expanded by identity elements.

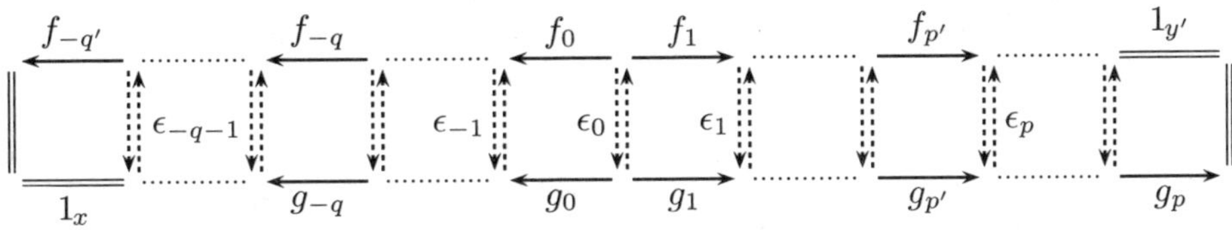

Figure 16. Deformation of a signed path by invertible elements: invertible elements connect the corresponding entries; if some path is shorter than the other (here we have $q < q'$ and $p' < p$), it is extended by identity elements.

Being a deformation is a symmetric relation. As in the positive case, $\mathcal{S}$-normal paths are preserved under deformation.

Proposition 2.15 (deformation). *Assume that $\mathcal{C}$ is a left-cancellative category and $\mathcal{S}$ is a subfamily of $\mathcal{C}$ that satisfies $\mathcal{C}^\times\mathcal{S} \subseteq \mathcal{S}^\sharp$. Then every $\mathcal{C}^\times$-deformation of a symmetric $\mathcal{S}$-normal path is a symmetric $\mathcal{S}$-normal path and the associated pairs are $\bowtie$-equivalent.*

Proof. Assume that $\overline{t_q}|\cdots|\overline{t_1}|s_1|\cdots|s_p$ is symmetric $\mathcal{S}$-normal and $\overline{t'_{q'}}|\cdots|\overline{t'_1}|s'_1|\cdots|s'_{p'}$ is a $\mathcal{C}^\times$-deformation of this path, with witnessing invertible elements ϵ_i as in Definition 2.14 and Figure 16. By definition, $s'_1|\cdots|s'_{p'}$ is a $\mathcal{C}^\times$-deformation of $\epsilon_0 s_1|s_2|\cdots|s_p$, and, similarly, $t'_1|\cdots|t'_{q'}$ is a $\mathcal{C}^\times$-deformation of $\epsilon_0 t_1|t_2|\cdots|t_q$. By Lemma 1.9, the assumption that $t_1|\cdots|t_q$ is $\mathcal{S}$-normal implies that $\epsilon_0 t_1|t_2|\cdots|t_q$ is $\mathcal{S}$-normal as well, so, by Proposition 1.22, we deduce that $t'_1|\cdots|t'_{q'}$ and, similarly, $s'_1|\cdots|s'_{p'}$ are $\mathcal{S}$-normal.

Next, let x be the source of t_1. Then the pair $(\epsilon_0, 1_x)$ witnesses for $(t'_1 \cdots t'_{q'}, s'_1 \cdots s'_{p'}) \bowtie (t_1 \cdots t_q, s_1 \cdots s_p)$. Finally, the assumption that s_1 and t_1 are left-disjoint implies that s'_1 and t'_1 are left-disjoint as well. To see this, put $f = t_1 \cdots t_q$, $g = s_1 \cdots s_p$, and assume $hf = h't'_1$ and $hg = h's'_1$ as in the diagram on the right. Then we deduce $(h'\epsilon_0)t_1 = h(f\epsilon_{-1})$ and $(h'\epsilon_0)s_1 = h(g\epsilon_0)$. As s_1 and t_1 are left-disjoint, we have $h \preccurlyeq h'\epsilon_0$, whence $h \preccurlyeq h'$ since ϵ_0 is invertible, and s'_1 and t'_1 are left-disjoint. □

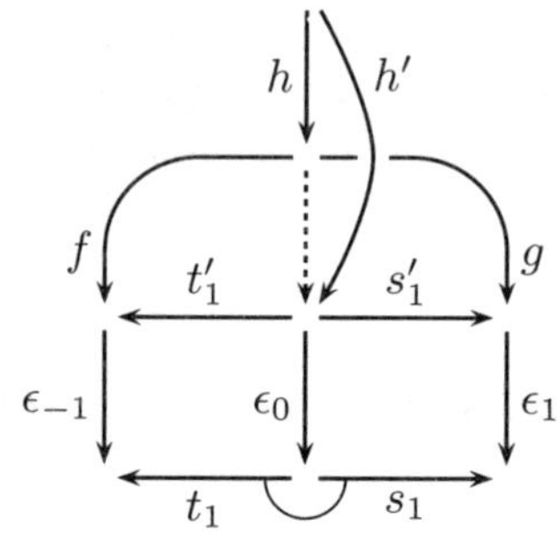

Then the uniqueness of symmetric $\mathcal{S}$-normal decompositions takes the expected form:

Proposition 2.16 (symmetric normal unique I). *Assume that $\mathcal{C}$ is a left-Ore category, $\mathcal{S}$ is included in $\mathcal{C}$, and $\mathcal{S}^\sharp$ generates $\mathcal{C}$ and includes $\mathcal{C}^\times\mathcal{S}$. Then any two symmetric $\mathcal{S}$-normal decompositions of an element of $\mathcal{E}nv(\mathcal{C})$ are $\mathcal{C}^\times$-deformations of one another.*

We shall establish a more general result valid in every left-cancellative category.

Proposition 2.17 (symmetric normal unique II). *Assume that $\mathcal{C}$ is a left-cancellative category, $\mathcal{S}$ is included in $\mathcal{C}$, and $\mathcal{S}^\sharp$ generates $\mathcal{C}$ and includes $\mathcal{C}^\times\mathcal{S}$. If $\overline{u}|v$ and $\overline{u'}|v'$ are symmetric $\mathcal{S}$-normal paths satisfying $([u],[v]) \bowtie ([u'],[v'])$, then $\overline{u}|v$ and $\overline{u'}|v'$ are $\mathcal{C}^\times$-deformations of one another.*

Proposition 2.16 follows from Proposition 2.17 owing to the characterization of equality in $\mathcal{E}nv(\mathcal{C})$ given by Ore's theorem (Proposition II.3.11), so it enough to prove Proposition 2.17. We begin with an auxiliary result.

Lemma 2.18. *Assume that $\mathcal{C}$ is a left-cancellative category, and (f,g), (f',g') are pairs of left-disjoint elements of $\mathcal{C}$ satisfying $(f,g) \bowtie (f',g')$. Then there exists an invertible element ϵ satisfying $f' = \epsilon f$ and $g' = \epsilon g$.*

Proof. Assume $hf' = h'f$ and $hg' = h'g$. The assumption that f and g are left-disjoint implies $h \preccurlyeq h'$, hence the existence of ϵ satisfying $h' = h\epsilon$. Symmetrically, the assumption that f' and g' are left-disjoint implies the existence of ϵ' satisfying $h'\epsilon' = h$. We deduce $h\epsilon\epsilon' = h$, and Lemma II.1.15 implies that ϵ and ϵ' are mutually inverse invertible elements. Then we obtain $hf' = h'f = h\epsilon f$, whence $f' = \epsilon f$ since $\mathcal{C}$ is left-cancellative, and, similarly, $g' = \epsilon g$. □

We can now establish the expected uniqueness result easily.

Proof of Proposition 2.17. By Proposition 2.11, $[u]$ and $[v]$ are left-disjoint, and so are $[u']$ and $[v']$. By assumption, we have $([u],[v]) \bowtie ([u'],[v'])$, hence Lemma 2.18 implies the existence of an invertible element ϵ satisfying $[u'] = \epsilon[u]$ and $[v'] = \epsilon[v]$. Assume $u = t_1|\cdots|t_q$ and $u' = t'_1|\cdots|t'_{q'}$. As $\mathcal{S}^\sharp$ includes $\mathcal{C}^\times\mathcal{S}$, hence $\mathcal{C}^\times\mathcal{S}^\sharp$, the paths $\epsilon t_1|t_2|\cdots|t_q$ and $t'_1|\cdots|t'_{q'}$ are $\mathcal{S}$-normal decompositions of $[u']$. By Proposition 1.25, $t'_1|\cdots|t'_{q'}$ must be a $\mathcal{C}^\times$-deformation of $\epsilon t_1|t_2|\cdots|t_q$. Assuming $v = s_1|\cdots|s_p$ and $v' = s'_1|\cdots|s'_{p'}$, we obtain similarly that $s'_1|\cdots|s'_{p'}$ is a $\mathcal{C}^\times$-deformation of $\epsilon s_1|s_2|\cdots|s_p$. Altogether, this means that $\overline{u'}|v'$, that is, $\overline{t'_{q'+1}}|\cdots|\overline{t'_1}|s'_1|\cdots|s'_{p'+1}$, is a $\mathcal{C}^\times$-deformation of $\overline{t_q}|\cdots|\overline{t_1}|s_1|\cdots|s_p$, that is, of $\overline{u}|v$. □

2.4 Existence of symmetric normal decompositions

We turn to the existence of symmetric normal decompositions. As the results are more complicated for a general left-cancellative category than for a (left)-Ore category, we shall restrict to this case, postponing the general case to an appendix. The main point is that the existence of symmetric normal decompositions requires the existence of left-lcms in the ambient category, but, on the other hand, it does not require more than that.

Lemma 2.19. *Assume that $\mathcal{C}$ is a left-Ore category, $\mathcal{S}$ is a Garside family in $\mathcal{C}$, and f, g are elements of $\mathcal{C}$ with the same target. Then the element gf^{-1} of $\mathcal{E}nv(\mathcal{C})$ admits a symmetric $\mathcal{S}$-normal decomposition if and only if f and g admit a left-lcm in $\mathcal{C}$.*

Proof. Assume that $\overline{t_q}|\cdots|\overline{t_1}|s_1|\cdots|s_p$ is a symmetric $\mathcal{S}$-normal decomposition of gf^{-1}. Let $f' = t_1\cdots t_q$ and $g' = s_1\cdots s_p$. By Proposition 2.11, the elements f' and g' are left-disjoint in $\mathcal{C}$. Hence, by Lemma 2.8, $f'g$, which is also $g'f$, is a left-lcm of f and g.

Conversely, assume that f and g admit a left-lcm, say $g'f = f'g$. By Lemma 2.8, f' and g' are left-disjoint elements of $\mathcal{C}$. Let $t_1|\cdots|t_q$ be an $\mathcal{S}$-normal decomposition of f', and $s_1|\cdots|s_p$ be an $\mathcal{S}$-normal decomposition of g'. By (the trivial part of) Proposition 2.11, t_1 and s_1 are left-disjoint. Then $\overline{t_q}|\cdots|\overline{t_1}|s_1|\cdots|s_p$ is a decomposition of $f'^{-1}g'$, hence of gf^{-1}, and, by construction, it is symmetric $\mathcal{S}$-normal. □

We recall that a category $\mathcal{C}$ is said to admit left-lcms if any two elements of $\mathcal{C}$ with the same target admit a left-lcm. From Lemma 2.19 we immediately deduce:

Proposition 2.20 (symmetric normal exist). *If $\mathcal{S}$ is a Garside family in a left-Ore category $\mathcal{C}$, the following conditions are equivalent:*

(i) *Every element of $\mathcal{E}nv(\mathcal{C})$ lying in $\mathcal{C}\mathcal{C}^{-1}$ has a symmetric $\mathcal{S}$-normal decomposition.*

(ii) *The category $\mathcal{C}$ admits left-lcms.*

The above condition for the existence of symmetric $\mathcal{S}$-normal decompositions does not depend on the particular Garside family $\mathcal{S}$. In the case of an Ore category $\mathcal{C}$, every element of $\mathcal{E}nv(\mathcal{C})$ is a right-fraction, so $\mathcal{C}\mathcal{C}^{-1}$ is all of $\mathcal{E}nv(\mathcal{C})$ and we deduce

Corollary 2.21 (symmetric normal exist). *If $\mathcal{S}$ is a Garside family in an Ore category $\mathcal{C}$, the following conditions are equivalent:*

(i) *Every element of $\mathcal{E}nv(\mathcal{C})$ admits a symmetric $\mathcal{S}$-normal decomposition;*

(ii) *The category $\mathcal{C}$ admits left-lcms.*

Thus, when investigating symmetric normal decompositions, it is natural to consider categories that admit left-lcms. We recall from (the symmetric counterpart of) Lemma II.2.22 that the existence of left-lcms also implies the existence of a left-gcd for elements that admit a common right-multiple. So, in particular, an Ore category that admits left-lcms must also admit left-gcds.

Up to now, the notions of an $\mathcal{S}$-normal decomposition and of a symmetric $\mathcal{S}$-normal decomposition have been defined by referring to some ambient (left-cancellative) category. However, according to Corollary 2.21, symmetric $\mathcal{S}$-normal decompositions provide distinguished decompositions for all elements of the groupoid of left-fractions of the considered category $\mathcal{C}$. It is natural to wonder whether such decompositions can be characterized by referring to $\mathcal{S}$ and the groupoid only. The answer is positive, at least in the case when the family $\mathcal{S}$ generates $\mathcal{C}$. Indeed, in this case, $\mathcal{C}$ is, by assumption, the subcategory of the ambient groupoid $\mathcal{G}$ generated by $\mathcal{S}$ and, as the product of $\mathcal{C}$ is the restriction of the product of $\mathcal{G}$, and the left-divisibility relation of $\mathcal{C}$ is defined in $\mathcal{G}$ by the formula $\exists g' \in \mathcal{C}(fg' = g)$. Therefore, the derived notions of an $\mathcal{S}$-normal path and a symmetric $\mathcal{S}$-normal path are definable from $\mathcal{S}$ in $\mathcal{G}$. This makes it natural to introduce the following counterpart of a Garside family in a groupoid.

Definition 2.22 (Garside base). A subfamily $\mathcal{S}$ of a groupoid $\mathcal{G}$ is a *Garside base* of $\mathcal{G}$ if every element of $\mathcal{G}$ admits a decomposition that is symmetric $\mathcal{S}$-normal with respect to the subcategory of $\mathcal{G}$ generated by $\mathcal{S}$.

Example 2.23 (Garside base). Two Garside bases for Artin's n-strand braid group B_n have been described in Chapter I: the classical Garside base consisting of the divisors of the braid Δ_n in the monoid B_n^+ (Reference Structure 2, page 5), and the dual Garside base consisting of the divisors of the braid Δ_n^* in the monoid B_n^{+*} (Reference Structure 3, page 10). Both are finite, with respectively $n!$ and $\frac{1}{n+1}\binom{2n}{n}$ (the nth Catalan number) elements. The submonoids generated by these Garside bases are B_n^+ and B_n^{+*}, respectively, and the fact that these sets are indeed Garside bases follows from Corollary 2.21, which guarantees the existence of the expected symmetric normal decompositions.

Like the existence of a Garside family, the existence of a Garside base is a vacuous condition: every groupoid $\mathcal{G}$ is a Garside base in itself, since the subcategory of $\mathcal{G}$ generated by $\mathcal{G}$ is $\mathcal{G}$, and every element g of $\mathcal{G}$ admits the length-one decomposition g, which is trivially symmetric $\mathcal{G}$-normal.

As in the cases of Example 2.23, Corollary 2.21 immediately implies

Proposition 2.24 (Garside family to Garside base). *If $\mathcal{C}$ is an Ore category that admits left-lcms, every generating Garside family of $\mathcal{C}$ is a Garside base of the groupoid $\mathcal{E}nv(\mathcal{C})$.*

Restricting to Garside families that generate the ambient category is natural, as, otherwise, the derived notions of $\mathcal{S}$-normality may collapse: for instance, if $\mathcal{C}$ is a groupoid, the empty set $\mathcal{S}$ is a Garside family in $\mathcal{C}$ since the corresponding family $\mathcal{S}^\sharp$ is the whole of $\mathcal{C}$. However, the subcategory of $\mathcal{C}$ generated by $\mathcal{S}$ is $\mathbf{1}_{\mathcal{C}}$, and $\mathcal{S}$ is not a Garside base of $\mathcal{E}nv(\mathcal{C})$ (which coincides with $\mathcal{C}$) whenever $\mathcal{C} \setminus \mathbf{1}_{\mathcal{C}}$ is nonempty: only identity elements can admit symmetric $\mathcal{S}$-normal decompositions in the sense of $\mathbf{1}_{\mathcal{C}}$.

Conversely, it is natural to wonder whether every Garside base necessarily arises in this context, that is, is a Garside family in the subcategory it generates. This is indeed so under mild assumptions, which shows that our approach so far does not restrict generality.

Proposition 2.25 (Garside base to Garside family). *Assume that $\mathcal{G}$ is a groupoid and $\mathcal{S}$ is a Garside base of $\mathcal{G}$ such that the subcategory $\mathcal{C}$ generated by $\mathcal{S}$ contains no nontrivial invertible element. Then $\mathcal{C}$ is a left-Ore category that admits left-lcms and $\mathcal{S}$ is a Garside family in $\mathcal{C}$.*

Proof. First, $\mathcal{C}$ is cancellative, as it is a subcategory of a groupoid. Next, let f, g be two elements of $\mathcal{C}$ sharing the same target. The element gf^{-1} of $\mathcal{G}$ admits a decomposition which is symmetric $\mathcal{S}$-normal, hence of the form $f'^{-1}g'$ with f', g' in $\mathcal{C}$. From $gf^{-1} = f'^{-1}g'$, we deduce $f'g = g'f$ in $\mathcal{G}$, hence in $\mathcal{C}$, showing that f and g admit a common left-multiple in $\mathcal{C}$. Hence, $\mathcal{C}$ is a left-Ore category. Moreover, by Lemma 2.8, the assumption that the above decomposition is symmetric $\mathcal{S}$-normal implies that $f'g$ is a left-lcm of f and g in $\mathcal{C}$, so $\mathcal{C}$ admits left-lcms.

Now let g be an element of $\mathcal{C}(x, \cdot)$. By assumption, g admits a decomposition, say $\overline{t_q}|\cdots|\overline{t_1}|s_1|\cdots|s_p$, that is symmetric $\mathcal{S}$-normal. Therefore, $(1_x, g) \bowtie (t_1 \cdots t_q, s_1 \cdots s_p)$ is satisfied. As $t_1 \cdots t_q$ and $s_1 \cdots s_p$ are left-disjoint by assumption, there must exist h in $\mathcal{C}$ satisfying $1_x = ht_1 \cdots t_q$ (and $g = hs_1 \cdots s_p$). It follows that $t_1, ..., t_q$ are invertible in $\mathcal{C}$ and, therefore, the assumption that $\mathcal{C}$ contains no nontrivial invertible element implies $t_1 = \cdots = t_p = 1_x$. Hence, $s_1|\cdots|s_p$ is an $\mathcal{S}$-normal decomposition of g in $\mathcal{C}$, and $\mathcal{S}$ is a Garside family in $\mathcal{C}$. □

Proposition 2.25 is not an exact converse of Proposition 2.24, as extra assumptions are used in both directions. The counterexamples of Exercise 35 suggest that improving the results is not obvious, at least in a proper categorical context.

Remark 2.26. When starting from a groupoid $\mathcal{G}$, restricting to subfamilies $\mathcal{S}$ such that the subcategory generated by $\mathcal{S}$ contains no nontrivial invertible element is probably needed if one has to expect interesting results. For instance, whenever $\mathcal{S}$ is a generating family of $\mathcal{G}$ that is closed under inverse or, more generally, whenever $\mathcal{S}$ positively generates $\mathcal{G}$, meaning that the subcategory of $\mathcal{G}$ generated by $\mathcal{S}$ is all of $\mathcal{G}$, then $\mathcal{S}$ is a Garside base in $\mathcal{G}$ since the associated family $\mathcal{S}^\sharp$ is $\mathcal{G}$ and every element of $\mathcal{G}$ admits a trivial symmetric $\mathcal{S}$-normal decomposition of length one, an uninteresting situation.

2.5 Computation of symmetric normal decompositions

We now address the question of effectively computing symmetric normal decomposition. Proposition 2.11 immediately gives:

Algorithm 2.27 (symmetric normal decomposition).

Context: A left-Ore category $\mathcal{C}$, a Garside subfamily $\mathcal{S}$ of $\mathcal{C}$, a $\square$-witness F on $\mathcal{S}^\sharp$

Input: A left-fraction $f^{-1}g$ in $\mathcal{E}nv(\mathcal{C})$ such that f and g are left-disjoint

Output: A symmetric $\mathcal{S}$-normal decomposition of $f^{-1}g$

1: compute an $\mathcal{S}$-normal representative $t_1|\cdots|t_q$ of f using Algorithm 1.52

2: compute an $\mathcal{S}$-normal representative $s_1|\cdots|s_p$ of g using Algorithm 1.52

3: **return** $\overline{t_q}|\cdots|\overline{t_1}|s_1|\cdots|s_p$

Algorithm 2.27 is not satisfactory in that it requires to start with an irreducible left-fraction and does not solve the question of finding such a fraction starting from an arbitrary decomposition. We shall begin with the case of elements of the form gf^{-1}, that is, with right-fractions. By Lemma 2.19, finding an irreducible left-fractional decomposition for such an element amounts to determining a left-lcm of f and g.

Definition 2.28 (common multiple witness, lcm witness). Assume that $\mathcal{C}$ is a category and $\mathcal{S}$ is a generating subfamily of $\mathcal{C}$.

(i) A *common right-multiple witness* on $\mathcal{S}$ is a partial map θ from $\mathcal{S} \times \mathcal{S}$ to $\mathcal{S}^*$ such that, for every pair $\{s, t\}$ of elements of $\mathcal{S}$, the elements $\theta(s,t)$ and $\theta(t,s)$ exist if and only if s and t admits a common right-multiple and, in this case, there exists a common right-multiple h of s and t such that $s\theta(s,t)$ and $t\theta(t,s)$ are equal to h. In addition, we assume that $\theta(s,s)$ is equal to ε_x for every s in $\mathcal{C}(\cdot,x)$.

(ii) A *right-lcm witness* is defined similarly, replacing "common right-multiple" with "right-lcm". A *common left-multiple witness* and a *left-lcm witness* $\widetilde{\theta}$ are defined symmetrically, so that $\widetilde{\theta}(s,t)t$ and $\widetilde{\theta}(t,s)s$ are equal to a distinguished common left-multiple.

(iii) A witness θ on $\mathcal{S}$ is *short* if the length of $\theta(s,t)$ is at most one for all s, t, that is, if $\theta(s,t)$ belongs to $\mathcal{S}$ or is empty.

A common right-multiple witness is a map that picks a common right-multiple whenever such one exists or, more exactly, picks what has to be added in order to obtain the chosen common right-multiple. By the Axiom of Choice, all types of witness exist. Note that, by definition, a common right-multiple witness and a right-lcm witness are syntactic right-complements in the sense of Definition II.4.2, whereas a common left-multiple witness and a left-lcm witness are syntactic left-complements. Hence it makes sense to consider the associated right- and left-reversing processes.

Definition 2.29 (strong Garside). A Garside family $\mathcal{S}$ in a left-Ore category is called *strong* if, for all s, t in $\mathcal{S}^\sharp$ with the same target, there exist s', t' in $\mathcal{S}^\sharp$ such that $s't$ equals $t's$ and it is a left-lcm of s and t.

Lemma 2.30. *A Garside family $\mathcal{S}$ in a left-Ore category is strong if and only if there exists a short left-lcm witness on $\mathcal{S}^\sharp$ defined on all pairs (s, t) with the same target.*

Proof. Assume that $\mathcal{S}$ is strong. For all s, t in $\mathcal{S}^\sharp$ with the same target, and possibly using the Axiom of Choice, define $(\widetilde{\theta}(s,t), \widetilde{\theta}(t,s))$ to be a pair (s', t') as in Definition 2.29: then, by definition, $\widetilde{\theta}$ is a short left-lcm witness on $\mathcal{S}^\sharp$. Conversely, if $\widetilde{\theta}$ is such a short left-lcm witness, then, for all s, t in $\mathcal{S}^\sharp$ with the same target, the pair $(\widetilde{\theta}(s,t), \widetilde{\theta}(t,s))$ witnesses that $\mathcal{S}$ is strong. □

Lemma 2.8 connects left-lcm and left-disjointness, and it immediately implies the following alternative definition:

Lemma 2.31. *Assume that $\mathcal{C}$ is a left-Ore category that admits left-lcms. A Garside family $\mathcal{S}$ of $\mathcal{C}$ is strong if and only if, for all s, t with the same target in $\mathcal{S}^\sharp$, there exist s', t' in $\mathcal{S}^\sharp$ satisfying $s't = t's$ and such that s' and t' are left-disjoint.*

Therefore, for a Garside family $\mathcal{S}$ to be strong corresponds to the diagram on the right, meaning that, for every choice of the right and bottoms arrows, there is a choice of the left and top arrows that make the diagram commutative. Note the similarity with the diagram illustrating a □-witness.

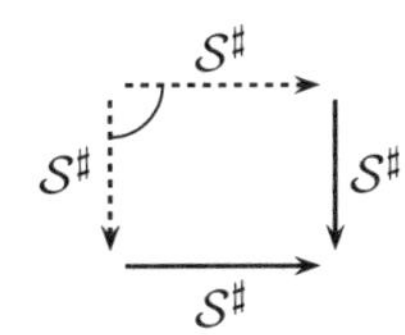

Example 2.32 (strong Garside). In the free abelian monoid $\mathbb{N}^n$, the Garside family S_n is strong: indeed, the monoid $\mathbb{N}^n$ is left-Ore, and it suffices to verify the condition of Lemma 2.31. Now, given s, t in S_n, define s', t' by $s'(k) = t(k) - s(k)$ and $t'(k) = 0$ if $t(k) > s(k)$ holds, and by $s'(k) = 0$, $t'(k) = s(k) - t(k)$ otherwise. Then s' and t' belong to S_n, they are left-disjoint, as seen in Example 2.3, and $s't = t's$ holds by construction.

When a short left-lcm witness $\widetilde{\theta}$ is available, left-lcms of arbitrary elements can be computed easily using a left-reversing process associated with $\widetilde{\theta}$, that is, by constructing a rectangular grid.

Lemma 2.33. *Assume that $\mathcal{C}$ is a right-cancellative category that admits conditional left-lcms, $\mathcal{S}$ is included in $\mathcal{C}$, and $\widetilde{\theta}$ is a short left-lcm witness on $\mathcal{S}^\sharp$. If u, v are $\mathcal{S}^\sharp$-paths with the same target, either $[u]$ and $[v]$ admit a common left-multiple in $\mathcal{C}$ and left-reversing $v\overline{u}$ using $\widetilde{\theta}$ leads to a negative–positive $\mathcal{S}^\sharp$-path $\overline{u'}v'$ such that both $u'v$ and $v'u$ represent a left-lcm of $[u]$ and $[v]$, or they do not and left-reversing $v\overline{u}$ fails.*

Proof. Left-reversing $v\overline{u}$ using $\widetilde{\theta}$ amounts to constructing a rectangular grid based on v (bottom) and u (right), such that the diagonal of each square represents a left-lcm of the bottom and right edges. Owing to the rule for an iterated left-lcm, the left counterpart of Proposition II.2.12 (iterated lcm), every diagonal in the diagram represents a left-lcm of the classes of the corresponding edges and, in particular, the big diagonal $u'v'$, if it exists, represents the left-lcm of $[u]$ and $[v]$. □

A first application of Lemma 2.33 is a characterization of the left-Ore categories that admit a strong Garside family.

Proposition 2.34 (strong exists). *If $\mathcal{C}$ is a left-Ore category, the following are equivalent:*
(i) *Some Garside family of $\mathcal{C}$ is strong.*
(ii) *The category $\mathcal{C}$, viewed as a Garside family in itself, is strong.*
(iii) *The category $\mathcal{C}$ admits left-lcms.*

Proof. The equivalence of (ii) and (iii) follows from the definition. Clearly, (ii) implies (i). Finally, assume that $\mathcal{S}$ is a strong Garside family in $\mathcal{S}$. Then, by definition, any two elements of $\mathcal{S}^\sharp$ with the same target must admit a left-lcm. Now, as $\mathcal{S}^\sharp$ generates $\mathcal{C}$, Lemma 2.33 implies that any two elements of $\mathcal{C}$ with the same target admit a left-lcm, so (i) implies (iii). □

The second application is a practical algorithm for completing Algorithm 2.27 and computing symmetric normal decompositions.

Algorithm 2.35 (symmetric normal, short case I). (See Figure 17.)

Context: A left-Ore category $\mathcal{C}$ admitting left-lcms, a strong Garside family $\mathcal{S}$ of $\mathcal{C}$, a $\square$-witness F and a short left-lcm witness $\widetilde{\theta}$ on $\mathcal{S}^\sharp$
Input: A positive–negative $\mathcal{S}^\sharp$-path $v\overline{u}$
Output: A symmetric $\mathcal{S}$-normal decomposition of $[v\overline{u}]$
1: left-reverse $v\overline{u}$ into a negative–positive $\mathcal{S}^\sharp$-path $\overline{u'}v'$ using $\widetilde{\theta}$
2: compute an $\mathcal{S}$-normal path u'' equivalent to u' using F and Algorithm 1.52
3: compute an $\mathcal{S}$-normal path v'' equivalent to v' using F and Algorithm 1.52
4: **return** $\overline{u''}v''$

The left-reversing transformation mentioned in line 1 is purely syntactic and involves signed paths only, not referring to any specific category: we insist that Definition II.4.21 (right-reversing), which starts with a presentation $(\mathcal{S}, \mathcal{R}_\theta)$, takes place in $\mathcal{S}^*$ but *not* in $\langle \mathcal{S} \mid \mathcal{R}_\theta \rangle^+$. So does its left counterpart considered above.

Example 2.36 (symmetric normal decomposition, short case I). Let us consider the free abelian monoid $\mathbb{N}^3$ based on $\{\mathtt{a}, \mathtt{b}, \mathtt{c}\}$ and the Garside family S consisting of the divisors of $\mathtt{abc}$ (Reference Structure 1, page 3). Let w be the positive–negative path $\mathtt{bc}|\mathtt{ab}|\mathtt{c}|\overline{\mathtt{ac}}|\overline{\mathtt{a}}$. Applying Algorithm 2.35 to $\mathtt{bc}|\mathtt{ab}|\mathtt{c}|\overline{\mathtt{ac}}|\overline{\mathtt{a}}$ means writing the initial word

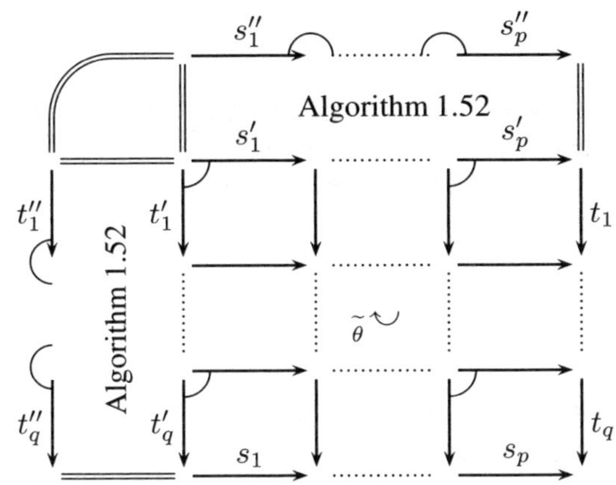

Figure 17. Algorithm 2.35: starting from $v\overline{u}$, here written as $s_1|\cdots|s_p|\overline{t_q}|\cdots|\overline{t_1}$, one first uses $\widetilde{\theta}$ to left-reverse $v\overline{u}$, that is, to construct a rectangular grid starting from the right and the bottom, and, then, one uses Algorithm 1.52 to normalize the numerator $s'_1|\cdots|s'_p$ and the denominator $t'_1|\cdots|t'_q$ of the left-fraction so obtained: the output is the path $\overline{t''_q}|\cdots|\overline{t''_1}|s''_1|\cdots|s''_p$.

as a horizontal-then-vertical path, constructing a rectangular grid using the left-lcm witness (which is unique here since there are no nontrivial invertible element):

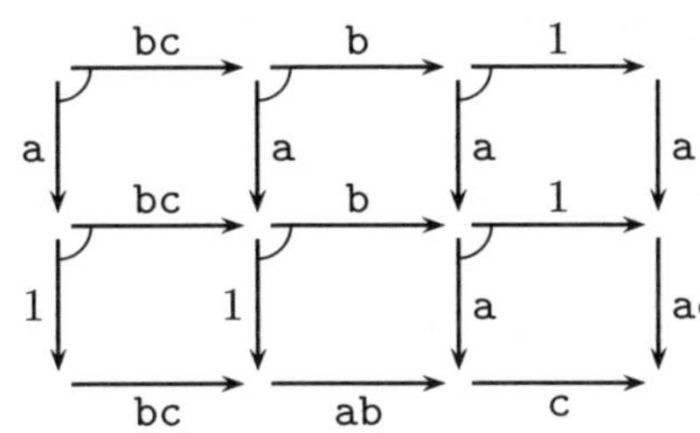

and finally applying the positive normalization algorithm to the left and top edges of the grid: in the current case, the sequences $\mathtt{a}|1$ and $\mathtt{bc}|\mathtt{b}|1$ are S-normal already, so the last step changes nothing. So $\overline{1}|\overline{\mathtt{a}}|\mathtt{bc}|\mathtt{b}|1$ is a symmetric S-normal sequence equivalent to w. Removing the trivial entries gives the strict symmetric S-normal decomposition $\overline{\mathtt{a}}|\mathtt{bc}|\mathtt{b}$.

Proposition 2.37 (symmetric normal, short case I). (i) *If S is a strong Garside family in a left-Ore category $\mathcal{C}$, then every element of $\mathcal{E}nv(\mathcal{C})$ that can be represented by a positive–negative $S^\sharp$-path of length ℓ admits a symmetric S-normal decomposition of length at most ℓ.*

(ii) *Under these assumptions, Algorithm 2.35 running on a path $v\overline{u}$ with $\lg(u) = q$ and $\lg(v) = p$ returns a symmetric S-normal decomposition of $[v\overline{u}]$. The map F is called at most $q(q-1)/2 + p(p-1)/2$ times, and the map $\widetilde{\theta}$ is called at most pq times.*

Proof. Consider (ii) first. By Lemma 2.33, left-reversing $v\overline{u}$ using $\widetilde{\theta}$ leads to a negative–positive $\mathcal{S}^\sharp$-path $\overline{u'}v'$ such that $u'v$ and $v'u$ represent the left-lcm of $[u]$ and $[v]$. Hence, by Lemma 2.8, the elements $[u']$ and $[v']$ are left-disjoint and, by construction, the signed path $\overline{u'}v'$ represents $[v\overline{u}]$. Moreover, we have $\lg(u') \leqslant \lg(u)$ and $\lg(v') \leqslant \lg(v)$ since $\widetilde{\theta}$ is short, and $\widetilde{\theta}$ is called (at most) qp times.

Now, Algorithm 1.52 to u' and v' returns equivalent paths u'' and v'' satisfying $\lg(u'') \leqslant \lg(u')$ and $\lg(v'') \leqslant \lg(v')$. By Proposition 1.53, the map F is called at most $q(q-1)/2+p(p-1)/2$ times. Then $\overline{u''}v''$ also represents $[v\overline{u}]$. By construction, u'' and v'' are $\mathcal{S}$-normal paths, and $[u'']$ and $[v'']$, which are $[u']$ and $[v']$, are left-disjoint. So $\overline{u''}v''$ is symmetric $\mathcal{S}$-normal. This establishes (ii). Point (i) follows since, by construction, $\lg(\overline{u''}v'') \leqslant \lg(v\overline{u})$ holds. □

Remark 2.38. The assumption that the Garside family $\mathcal{S}$ is strong, that is, there exists an everywhere defined short left-lcm witness for $\mathcal{S}^\sharp$, is essential. In every case, there exists a left-lcm witness but, when the latter is not short, the termination of left-reversing is not guaranteed, even when the existence of a left-lcm is guaranteed. In other words, left-reversing associated with a left-lcm witness need not be complete in general, see Exercise 116 in Chapter XII for a counter-example.

In the case of an Ore category, every element of the enveloping groupoid can be expressed as a right-fraction, so Proposition 2.37 implies the existence of a symmetric $\mathcal{S}$-normal decomposition for every element of this groupoid. In order to obtain an algorithmic method, we need to append to Algorithm 1.52 an initial step transforming an arbitrary signed decomposition into an equivalent positive–negative one. The next lemma, which anticipates on more complete results in Chapter IV, says that this is always possible.

Lemma 2.39. *If $\mathcal{S}$ is a Garside family in a left-cancellative category, there exists a short common right-multiple witness on $\mathcal{S}^\sharp$ that is defined for all s, t with the same source.*

Proof. Let s, t be elements of $\mathcal{S}^\sharp$ sharing the same source and admitting a common right-multiple h. Let $r_1|\cdots|r_m$ be an $\mathcal{S}$-normal decomposition of h. Put $h' = r_2\cdots r_m$. By Proposition 1.12, $r_1|h'$ is $\mathcal{S}$-greedy, hence $\mathcal{S}^\sharp$-greedy by Lemma 1.10. Then s lies in $\mathcal{S}^\sharp$ and it left-divides r_1h', hence it must left-divide r_1, say $r_1 = sg$. Similarly, t must left-divide r_1, say $r_1 = tf$. By Corollary 1.55, we have $\|f\|_\mathcal{S} \leqslant \|r_1\|_\mathcal{S} \leqslant 1$ and, similarly, $\|g\|_\mathcal{S} \leqslant \|r_1\|_\mathcal{S} \leqslant 1$, hence f and g must lie in $\mathcal{S}^\sharp$. So, for all s, t in $\mathcal{S}^\sharp$ admitting a common right-multiple, there exist f, g in $\mathcal{S}^\sharp$ satisfying $sg = tf$. Picking such a pair (f, g) for each pair (s, t) gives a short common right-multiple witness. □

Algorithm 2.40 (symmetric normal, short Ore case I).

Context: An Ore category $\mathcal{C}$ admitting left-lcms, a strong Garside family $\mathcal{S}$ of $\mathcal{C}$, a $\square$-witness F, a common right-multiple witness θ on $\mathcal{S}^\sharp$, a short left-lcm witness $\widetilde{\theta}$ on $\mathcal{S}^\sharp$
Input: A signed $\mathcal{S}^\sharp$-path w
Output: A symmetric $\mathcal{S}$-normal decomposition of $[w]$
1: right-reverse w into a positive–negative $\mathcal{S}^\sharp$-path $v\overline{u}$ using θ
2: left-reverse $v\overline{u}$ into a negative–positive $\mathcal{S}^\sharp$-path $\overline{u'}v'$ using $\widetilde{\theta}$
3: compute an $\mathcal{S}$-normal path u'' equivalent to u' using F and Algorithm 1.52
4: compute an $\mathcal{S}$-normal path v'' equivalent to v' using F and Algorithm 1.52
5: **return** $\overline{u''}v''$

Once again, we insist that the right- and left-reversing transformations considered in lines 1 and 2 above are purely syntactic and refer to no category but the free one $\mathcal{S}^{\sharp *}$.

Corollary 2.41 (symmetric normal, short Ore case I). *If $\mathcal{S}$ is a strong Garside family in an Ore category $\mathcal{C}$, then every element of $\mathcal{E}nv(\mathcal{C})$ that can be represented by a signed $\mathcal{S}^\sharp$-path of length ℓ admits a symmetric $\mathcal{S}$-normal decomposition of length at most ℓ. For every signed $\mathcal{S}^\sharp$-path w, Algorithm* 2.40 *running on w returns a symmetric $\mathcal{S}$-normal decomposition of* $[w]$.

Proof. First, by Proposition II.4.27 (termination), the assumption that θ and $\widetilde{\theta}$ are short and everywhere defined (since the existence of left-lcms and of common right-multiples is assumed) guarantees that the associated right- and left-reversing processes terminate. Then the first step of Algorithm 2.40 transforms an arbitrary signed $\mathcal{S}^\sharp$-path w into an equivalent positive–negative $\mathcal{S}^\sharp$-path $v\overline{u}$ satisfying $\lg(v\overline{u}) \leqslant \lg(w)$. The rest follows directly from Proposition 2.37. $\square$

We now describe an alternative way of computing a symmetric normal decomposition. In Algorithm 2.35, the principle is to first left-reverse the initial right-fraction into an equivalent left-fraction and then normalize the numerator and the denominator using Algorithm 1.52. One can equivalently normalize the numerator and the denominator of the initial right-fraction.

Algorithm 2.42 (symmetric normal, short case II). (Figure 18)

Context: A left-Ore category $\mathcal{C}$ admitting left-lcms, a strong Garside family $\mathcal{S}$ of $\mathcal{C}$, a $\square$-witness F, a short left-lcm witness $\widetilde{\theta}$ on $\mathcal{S}^\sharp$
Input: A positive–negative $\mathcal{S}^\sharp$-path $v\overline{u}$
Output: A symmetric $\mathcal{S}$-normal decomposition of $[v\overline{u}]$
compute an $\mathcal{S}$-normal path u' equivalent to u using F and Algorithm 1.52
compute an $\mathcal{S}$-normal path v' equivalent to v using F and Algorithm 1.52
left-reverse $v'\overline{u'}$ into a negative–positive $\mathcal{S}^\sharp$-path $\overline{u''}v''$ using $\widetilde{\theta}$
return $\overline{u''}v''$

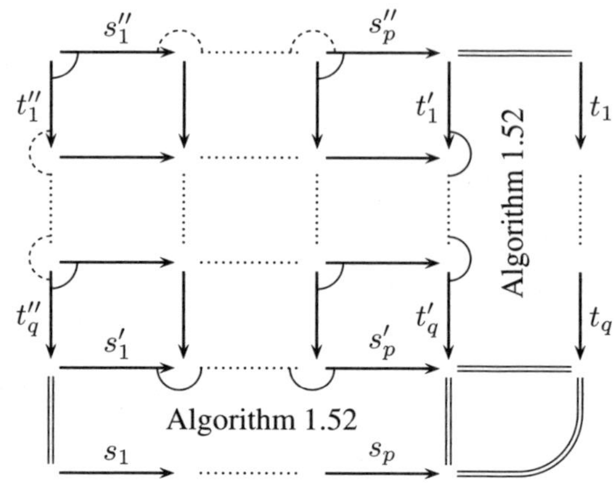

Figure 18. Algorithm 2.42: starting from $\overline{v}u = s_1|\cdots|s_p|\overline{t_q}|\cdots|\overline{t_1}$, we first normalize the numerator and the denominator into $s_1'|\cdots|s_p'$ and $t_1'|\cdots|t_q'$ and then use $\widetilde{\theta}$ to left-reverse the right-fraction $s_1'|\cdots|s_p'|\overline{t_q'}|\cdots|\overline{t_1'}$ into a left-fraction $\overline{t_q''}|\cdots|\overline{t_1''}|s_1''|\cdots|s_p''$ (compare with Figure 17).

Example 2.43 (symmetric normal, short case II). Consider the free abelian monoid $\mathbb{N}^3$ based on $\{\mathtt{a}, \mathtt{b}, \mathtt{c}\}$ and the Garside family S consisting of the divisors of $\mathtt{abc}$ (Reference Structure 1, page 3). Let w be the positive–negative path $\mathtt{bc}|\mathtt{ab}|\mathtt{c}|\overline{\mathtt{ac}}|\overline{\mathtt{a}}$. Applying Algorithm 1.52 to the paths $\mathtt{bc}|\mathtt{ab}|\mathtt{c}$ and $\mathtt{a}|\mathtt{ac}$ yields the S-normal paths $\mathtt{abc}|\mathtt{bc}|1$ and $\mathtt{ac}|\mathtt{a}$. Then, starting from the bottom–right corner, we construct the rectangular grid

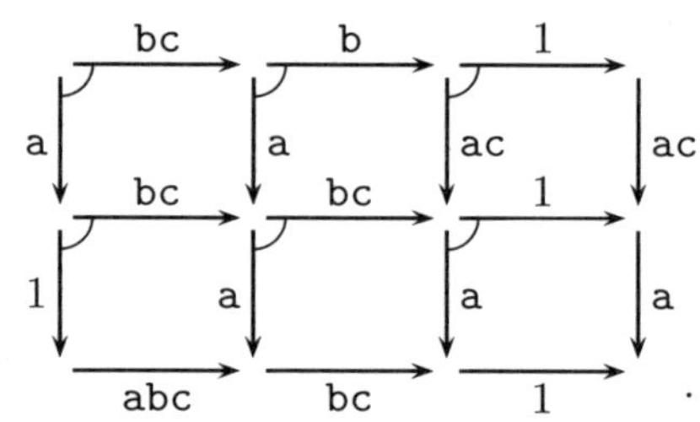

Then, as in Example 2.36, we conclude that $\overline{1}|\overline{\mathtt{a}}|\mathtt{bc}|\mathtt{b}|1$ is a symmetric S-normal path equivalent to w.

Proposition 2.44 (symmetric normal, short case II). *If S is a strong Garside family in a left-Ore category $\mathcal{C}$, then Algorithm 2.42 running on a positive–negative $S^\sharp$-path $v\overline{u}$ with $\lg(u) = q$ and $\lg(v) = p$ returns a symmetric S-normal decomposition of $[v\overline{u}]$. The map F is called at most $q(q-1)/2 + p(p-1)/2$ times, and the map $\widetilde{\theta}$ is called at most pq times.*

To prove this, we shall use a new domino rule similar to those of Section 1.

Proposition 2.45 (third domino rule). *Assume that $\mathcal{C}$ is a left-cancellative category, S is included in $\mathcal{C}$, and we have a commutative diagram with edges in $\mathcal{C}$ as on the right. If $g_1|g_2$ is S-greedy, and f, g_2' are left-disjoint, then $g_1'|g_2'$ is S-greedy as well.*

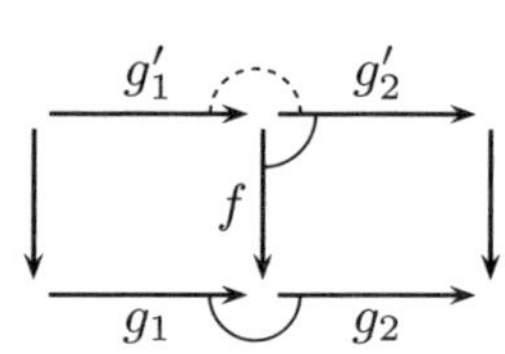

Proof. Assume $s \preccurlyeq f' g'_1 g'_2$ with s in $\mathcal{S}$. Let f_0, f_2 be as in the diagram. Then we have $s \preccurlyeq f' g'_1 g'_2 f_2$, hence $s \preccurlyeq (f' f_0) g_1 g_2$ since the diagram is commutative. The assumption that $g_1 | g_2$ is $\mathcal{S}$-greedy implies $s \preccurlyeq (f' f_0) g_1$, hence $s \preccurlyeq (f' g'_1) f$ using commutativity again. As, by assumption, we have $s \preccurlyeq (f' g'_1) g'_2$, the assumption that f and g'_2 are left-disjoint implies $s \preccurlyeq f' g'_1$. Hence $g'_1 | g'_2$ is $\mathcal{S}$-greedy. □

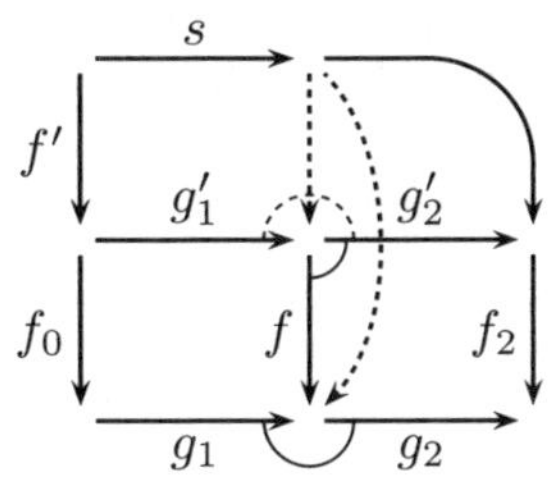

Proof of Proposition 2.44. As in Figure 18, let us write $u = t_1 | \cdots | t_q$ and $v = s_1 | \cdots | s_p$. As $\mathcal{S}$ is a Garside family in $\mathcal{C}$, the $\mathcal{S}^\sharp$-paths $t_1 | \cdots | t_q$ and $s_1 | \cdots | s_p$ are eligible for Algorithm 1.52, which returns equivalent paths $t'_1 | \cdots | t'_q$ and $s'_1 | \cdots | s'_p$, respectively. Moreover, by Proposition 1.53, the map F is called $q(q-1)/2 + p(p-1)/2$ times in the process. Then, as $\widetilde{\theta}$ is a left-lcm witness on $\mathcal{S}^\sharp$, left-reversing $s'_1 | \cdots | s'_p | \overline{t'_q} | \cdots | \overline{t'_1}$ into a negative–positive path $\overline{t''_q} | \cdots | \overline{t''_1} | s''_1 | \cdots | s''_p$ using $\widetilde{\theta}$, that is, completing the grid of Figure 18, requires at most qp calls to $\widetilde{\theta}$.

By construction, s''_1 and t''_1 are left-disjoint. So, to show that $\overline{t''_q} | \cdots | \overline{t''_1} | s''_1 | \cdots | s''_p$ is symmetric $\mathcal{S}$-normal, it is sufficient to show that $s''_1 | \cdots | s''_p$ and $t''_1 | \cdots | t''_q$ are $\mathcal{S}$-normal. This follows from the third domino rule directly: indeed, by assumption, in the diagram of Figure 17, the penultimate right column $t'_1 | \cdots | t'_q$ is $\mathcal{S}$-normal, so the third domino rule inductively implies that every column until the left one $t''_1 | \cdots | t''_q$ is $\mathcal{S}$-normal. Similarly, the penultimate bottom row $s'_1 | \cdots | s'_p$ is $\mathcal{S}$-normal, so the third domino rule inductively implies that every row until the top one $s''_1 | \cdots | s''_p$ is $\mathcal{S}$-normal. □

In Section 1, we established the existence of $\mathcal{S}$-normal decompositions by appealing to Proposition 1.49, an explicit method that, for s in $\mathcal{S}^\sharp$ and g in the ambient category, determines an $\mathcal{S}$-normal decomposition of sg starting from an $\mathcal{S}$-normal decomposition of g. We shall now extend this method to the case when g lies in the enveloping groupoid, that is, determine a symmetric $\mathcal{S}$-normal decomposition of sg starting from a symmetric $\mathcal{S}$-normal decomposition of g.

Algorithm 2.46 (left-multiplication). (See Figure 19.)

Context: A left-Ore category $\mathcal{C}$ admitting left-lcms, a strong Garside subfamily $\mathcal{S}$ of $\mathcal{C}$, a □-witness F, a short left-lcm witness $\widetilde{\theta}$ on $\mathcal{S}^\sharp$

Input: A symmetric $\mathcal{S}^\sharp$-normal decomposition $\overline{t_q} | \cdots | \overline{t_1} | s_1 | \cdots | s_p$ of an element g of $\mathcal{E}nv(\mathcal{C})$ and s in $\mathcal{S}^\sharp$ such that sg is defined

Output: A symmetric $\mathcal{S}$-normal decomposition of sg

put $r_{-q} := s$
for i decreasing from q to 1 **do**
  put $(t'_i, r_{-i-1}) := (\widetilde{\theta}(t_i, r_{-i}), \widetilde{\theta}(r_{-i}, t_i))$
for i increasing from 1 to p **do**
  put $(s'_i, r_i) := F(r_{i-1}, s_i)$
put $s'_{p+1} := r_p$
return $\overline{t'_q} | \cdots | \overline{t'_1} | s'_1 | \cdots | s'_{p+1}$

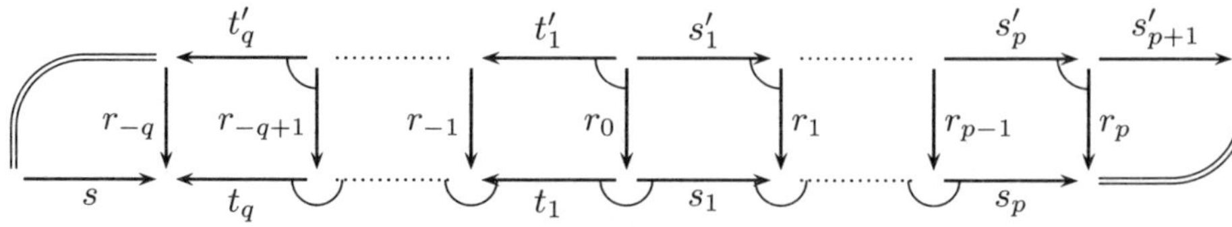

Figure 19. Algorithm 2.46: starting from a symmetric $\mathcal{S}$-normal decomposition $\overline{t_q}|\cdots|\overline{t_1}|s_1|\cdots|s_p$ of g and s in $\mathcal{S}^\sharp$, one obtains a symmetric $\mathcal{S}$-normal decomposition of sg.

Proposition 2.47 (left-multiplication). *If $\mathcal{S}$ is a strong Garside family in a left-Ore category $\mathcal{C}$ that admits left-lcms, Algorithm* 2.46 *running on a symmetric $\mathcal{S}$-normal decomposition of an element g of $\mathcal{E}nv(\mathcal{C})$ and s in $\mathcal{S}^\sharp$ returns a symmetric $\mathcal{S}$-normal decomposition of sg.*

In order to establish Proposition 2.47, we state a new domino rule.

Proposition 2.48 (fourth domino rule). *Assume that $\mathcal{C}$ is a left-cancellative category, $\mathcal{S}$ is included in $\mathcal{C}$, and we have a commutative diagram with edges in $\mathcal{C}$ as on the right. If g_1, g_2 are left-disjoint, and f, g'_2 are left-disjoint, then fg_1, g'_2 are left-disjoint, and g'_1, g'_2 are left-disjoint.*

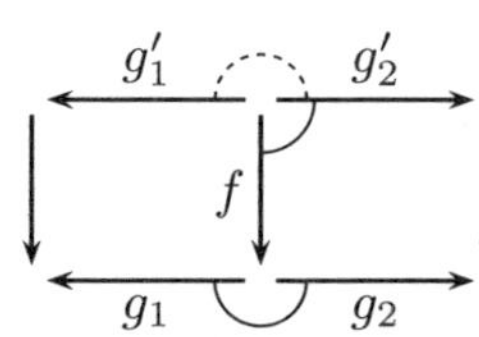

Proof. (See Figure 20.) Assume $h' \preccurlyeq hfg_1$ and $h' \preccurlyeq hg'_2$. Since the diagram is commutative, we deduce $h' \preccurlyeq (hf)g_2$, whence $h' \preccurlyeq hf$ as g_1 and g_2 are left-disjoint. But, now, we have $h' \preccurlyeq hf$ and $h' \preccurlyeq hg'_2$ with f and g'_2 left-disjoint. We deduce $h' \preccurlyeq h$. Hence fg_1 and g'_2 are left-disjoint. As g'_1 left-divides fg_1, this implies that g'_1 and g'_2 are left-disjoint. ☐

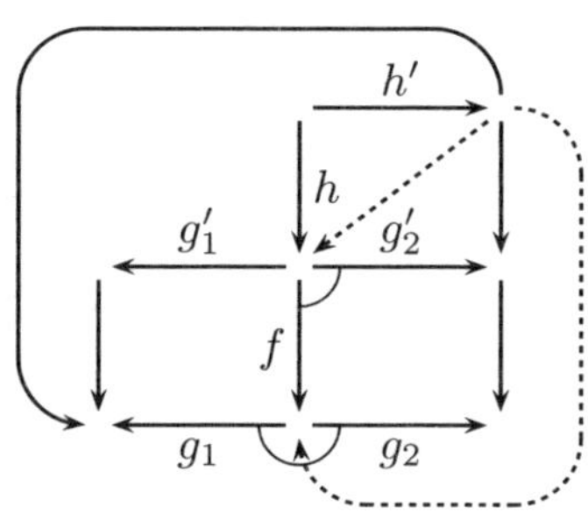

Figure 20. Proof of the fourth domino rule (Proposition 2.48): If g_1, g_2 and f_1, g'_2 are left-disjoint, so are $g_1g'_1$ and g'_2, as well as g'_1 and g'_2.

Proof of Proposition 2.47. Use the notation of Algorithm 2.46 and Figure 19. By construction, the diagram is commutative, so the only point to justify is that the output path $\overline{t'_q}|\cdots|\overline{t'_1}|s'_1|\cdots|s'_{p+1}$ is symmetric $\mathcal{S}$-normal. Now, as $t_1|\cdots|t_q$ is $\mathcal{S}$-normal, the third domino rule implies that $t'_1|\cdots|t'_q$ is $\mathcal{S}$-normal. Next, as t_1 and s_1 are left-disjoint, the fourth domino rule implies that t'_1 and s'_1 are left-disjoint. Then, as $s_1|\cdots|s_p$ is $\mathcal{S}$-normal, the first domino rule implies that $s'_1|\cdots|s'_p$ is $\mathcal{S}$-normal. Finally, $s'_p|s'_{p+1}$ is $\mathcal{S}$-normal by construction. Hence $\overline{t'_q}|\cdots|\overline{t'_1}|s'_1|\cdots|s'_{p+1}$ is symmetric $\mathcal{S}$-normal. ☐

Owing to the symmetry in the definition of symmetric normal paths, it is easy to deduce from Algorithm 2.46 a similar method that, starting from a symmetric $\mathcal{S}$-normal decomposition of an element g of $\mathcal{E}nv(\mathcal{C})$ and an element s of $\mathcal{S}^\sharp$, returns a symmetric $\mathcal{S}$-normal decomposition of gs^{-1}, that is, a right-division algorithm.

Algorithm 2.49 (right-division). (See Figure 21.)

Context: A left-Ore category $\mathcal{C}$ admitting left-lcms, a strong Garside subfamily $\mathcal{S}$ of $\mathcal{C}$, a $\square$-witness F, a short left-lcm witness $\widetilde{\theta}$ on $\mathcal{S}^\sharp$

Input: A symmetric $\mathcal{S}^\sharp$-normal decomposition $\overline{t_q}|\cdots|\overline{t_1}|s_1|\cdots|s_p$ of an element g of $\mathcal{E}nv(\mathcal{C})$ and s in $\mathcal{S}^\sharp$ such that gs^{-1} is defined

Output: A symmetric $\mathcal{S}$-normal decomposition of gs^{-1}

put $r_p := s$
for i decreasing from p to 1 **do**
  put $(s'_i, r_{i-1}) := (\widetilde{\theta}(s_i, r_i), \widetilde{\theta}(r_i, s_i))$
for i increasing from 1 to q **do**
  put $(t'_i, r_{-i}) := F(r_{-i-1}, t_i)$
put $t'_{q+1} := r_{-q}$
return $\overline{t'_{q+1}}|\cdots|\overline{t'_1}|s'_1|\cdots|s'_p$

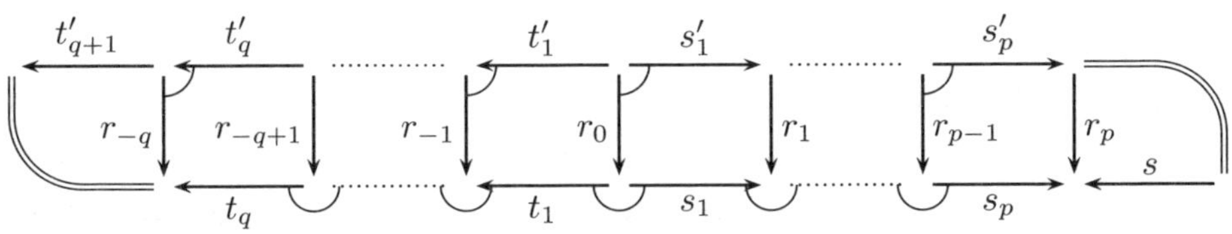

Figure 21. Algorithm 2.49: starting from a symmetric $\mathcal{S}$-normal decomposition $\overline{t_q}|\cdots|\overline{t_1}|s_1|\cdots|s_p$ of g and s in $\mathcal{S}^\sharp$, one obtains a symmetric $\mathcal{S}$-normal decomposition of gs^{-1}.

Corollary 2.50 (right-division). *If $\mathcal{S}$ is a strong Garside family in a left-Ore category $\mathcal{C}$ that admits left-lcms, Algorithm* 2.49 *running on a symmetric $\mathcal{S}$-normal decomposition of an element g of $\mathcal{E}nv(\mathcal{C})$ and s in $\mathcal{S}^\sharp$ returns a symmetric $\mathcal{S}$-normal decomposition of gs^{-1}.*

Appendix: the case of a general left-cancellative category

By Lemma 2.8, the notion of left-disjoint elements is connected with the existence of a left-lcm in a left-Ore category. As seen in the previous two subsections, this connection leads to simple statements on the existence and the computation of symmetric normal decompositions in left-Ore categories that admit left-lcms. However, the mechanism of symmetric normal decompositions does not require such a strong context and most of the results of Subsections 2.4 and 2.5 are valid in all left-cancellative categories, provided the statements are conveniently adapted. Most proofs in this appendix are easy extensions of earlier proofs, and are left to reader.

Definition 2.51 (weak lcm). (See Figure 22.) If f, g, h belong to a category $\mathcal{C}$, we say that h is a *weak left-lcm* of f and g if h is a common left-multiple of f and g, and every common left-multiple $\hat{h}$ of f and g such that h and $\hat{h}$ admit a common left-multiple is a left-multiple of h. We say that $\mathcal{C}$ *admits conditional weak left-lcms* if every common left-multiple of two elements f, g of $\mathcal{C}$ is a left-multiple of some weak left-lcm of f and g.

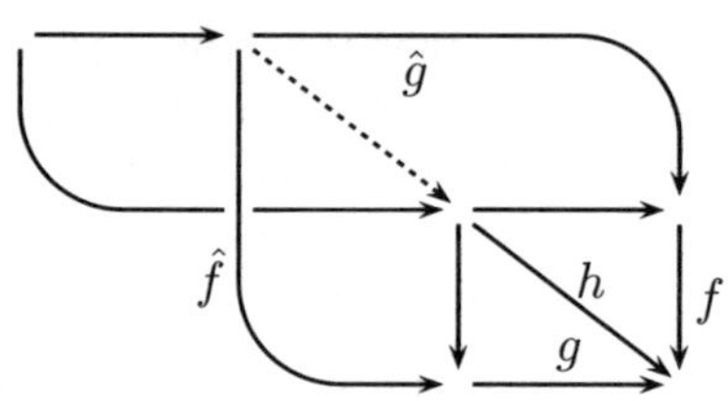

Figure 22. A weak left-lcm of f and g: a common left-multiple h of f and g with the property that every common left-multiple $\hat{f}g$ of f and g such that h and $\hat{f}g$ admit a common left-multiple is a left-multiple of h.

If it exists, a left-lcm is a weak left-lcm: if every common left-multiple of f and g is a left-multiple of h, then a fortiori so is every common left-multiple that satisfies an additional condition. With this notion, the characterization of Lemma 2.8 can be extended.

Lemma 2.52. *For f, g, f', g' in a cancellative category $\mathcal{C}$ satisfying $f'g = g'f$, the following conditions are equivalent:*

(i) *The elements f' and g' are left-disjoint.*

(ii) *The element $f'g$ is a weak left-lcm of f and g.*

The natural extension of Proposition 2.20 is then

Proposition 2.53 (symmetric normal exist). *If $\mathcal{S}$ is a Garside family in a cancellative category $\mathcal{C}$, the following conditions are equivalent:*

(i) *For all f, g in $\mathcal{C}$ admitting a common right-multiple, there exists a symmetric $\mathcal{S}$-normal path $\overline{u}v$ satisfying $(f, g) \bowtie ([u]^+, [v]^+)$.*

(ii) *The category $\mathcal{C}$ admits conditional weak left-lcms.*

We recall that, if w is a path, $[w]^+$ is the element of the category represented by w. In the case when $\mathcal{C}$ is a left-Ore category, Proposition 2.53 is a restatement of Proposition 2.20, since (i) amounts to saying that $\overline{u}v$ is a decomposition of $f^{-1}g$ in $\mathcal{E}nv(\mathcal{C})$ and (ii) amounts to saying that $\mathcal{C}$ admits left-lcms.

In order to extend the mechanism underlying Proposition 2.37, we introduce the following general notion of a strong Garside family.

Definition 2.54 (strong Garside). Assume that $\mathcal{C}$ is a left-cancellative category. A Garside family $\mathcal{S}$ of $\mathcal{C}$ is called *strong* if, whenever s, t lie in $\mathcal{S}^\sharp$ and $fs = gt$ holds (with f, g in $\mathcal{C}$), there exist s', t' in $\mathcal{S}^\sharp$ and h in $\mathcal{C}$ such that s' and t' are left-disjoint and we have $s't = t's$, $f = ht'$, and $g = hs'$.

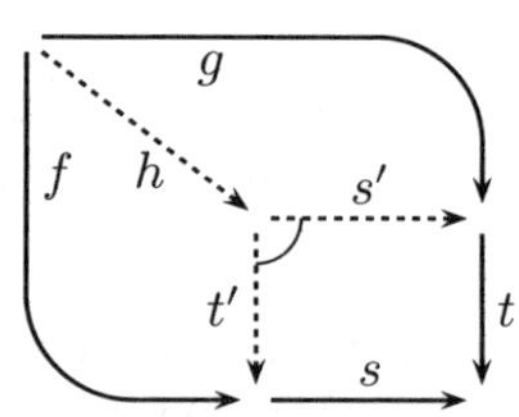

Note than being strong requires nothing for pairs (s, t) that admit no common left-multiple. If $\mathcal{C}$ is a left-Ore category that admits left-lcms, the above definition reduces to the condition of Lemma 2.31, and, therefore, the current definition extends Definition 2.29. As in the left-Ore case, we note, leaving the easy verification to the reader,

Lemma 2.55. *A cancellative category $\mathcal{C}$ admits conditional weak left-lcms if and only if $\mathcal{C}$ is a strong Garside family in itself.*

Then Algorithms 2.35 and 2.42 can be extended easily. We consider the latter.

Proposition 2.56 (symmetric normal, short case III). *If $\mathcal{S}$ is a strong Garside family in a cancellative category $\mathcal{C}$ admitting conditional weak left-lcms, Algorithm 2.42, running on a positive–negative $\mathcal{S}^\sharp$-path $v\overline{u}$ such that $[u]^+$ and $[v]^+$ have common left-multiple $\hat{f}[v]^+ = \hat{g}[u]^+$, returns a symmetric $\mathcal{S}$-normal path $\overline{u''}v''$ satisfying $(f, g) \bowtie ([u'']^+, [v'']^+)$ and $[u''v]^+ = [v''u]^+$; moreover, there exists h satisfying $f = h[u'']^+$ and $g = h[v'']^+$.*

Once again, Proposition 2.56 reduces to Proposition 2.44 in the case of a left-Ore category, as it then says that $\overline{u}v$ is a decomposition of $[v\overline{u}]$ in $\mathcal{E}nv(\mathcal{C})$.

3 Geometric and algorithmic properties

We now establish geometric and algorithmic properties of normal decompositions associated with Garside families. Most results come in two versions: a positive version involving a left-cancellative category $\mathcal{C}$, a Garside family $\mathcal{S}$ of $\mathcal{C}$, and $\mathcal{S}$-normal decompositions, and a signed version involving the groupoid of fractions of a (left)-Ore category $\mathcal{C}$, a strong Garside family $\mathcal{S}$ of $\mathcal{C}$, and symmetric $\mathcal{S}$-normal decompositions. The main results are Proposition 3.11, a convexity property which says that every triangle in the groupoid can be filled with a sort of geodesic planar grid, Proposition 3.20, which says that every group(oid) of fractions of a category with a finite strong Garside family has an automatic structure, and Propositions 3.45 and 3.52, which specify the homology of a groupoid of fractions using a strong Garside family.

The section contains five subsections. In Subsection 3.1 we show that $\mathcal{S}$-normal decompositions are geodesic. Subsection 3.2 deals with the Grid Property alluded to above. In Subsection 3.3, we consider the Fellow Traveller Property, which leads to an automatic structure when finiteness conditions are met. In Subsection 3.4, we use the mechanism of normal decompositions to construct chain complexes and resolutions leading to homology statements in good cases. Finally, we address the Word Problem(s) and its solutions in Subsection 3.5.

3.1 Geodesics

We begin with results asserting that $\mathcal{S}$-normal decompositions are geodesic, that is, shortest among all possible decompositions.

Proposition 3.1 (geodesic, positive case). *If $\mathcal{S}$ is a Garside family in a left-cancellative category $\mathcal{C}$, every strict $\mathcal{S}$-normal path is geodesic among $\mathcal{S}^\sharp$-decompositions.*

Proof. Assume that $s_1|\cdots|s_p$ is a strict $\mathcal{S}$-normal path representing an element g of $\mathcal{C}$ and $t_1|\cdots|t_q$ is an arbitrary $\mathcal{S}^\sharp$-decomposition of g. We claim that $q \geqslant p$ holds. Indeed, Proposition 1.30 implies that every strict $\mathcal{S}$-normal decomposition of g must have length p, and, on the other hand, that g admits a strict $\mathcal{S}$-normal decomposition of length at most q since it admits a $\mathcal{S}^\sharp$-decomposition of length q. Hence, $q \geqslant p$ must hold. □

Proposition 3.2 (geodesic, general case). *If $\mathcal{S}$ is a strong Garside family in a left-Ore category $\mathcal{C}$, every strict symmetric $\mathcal{S}$-normal path is geodesic among $\mathcal{S}^\sharp$-decompositions.*

Proof. Assume that $\overline{t_q}|\cdots|\overline{t_1}|s_1|\cdots|s_p$ is a strict symmetric $\mathcal{S}$-normal path representing an element g of $\mathcal{E}nv(\mathcal{C})$, and let $r_1^{\pm 1}|\cdots|r_\ell^{\pm 1}$ be an arbitrary $\mathcal{S}^\sharp$-decomposition of g. We claim that $\ell \geqslant q + p$ holds.

First, we observe that, for all s, t in $\mathcal{S}^\sharp$ with the same target, the assumption that $\mathcal{S}$ is a strong Garside family implies the existence of s', t' in $\mathcal{S}^\sharp$ satisfying $s't = t's$, hence $ts^{-1} = s'^{-1}t'$ in $\mathcal{E}nv(\mathcal{C})$. Then, starting from $r_1^{\pm 1}|\cdots|r_m^{\pm 1}$ and repeatedly applying the above result to switch the factors (that is, applying a left-reversing transformation), we obtain a new, fractional decomposition of g of the form $\overline{s'_{q'}}|\cdots|\overline{s'_1}|t'_1|\cdots|t'_{p'}$ with $s'_1, \ldots, t'_{p'}$ in $\mathcal{S}^\sharp$ and $q' + p' = \ell$.

Now let $g_- = t_1 \cdots t_q$ and $g_+ = s_1 \cdots s_p$ and, similarly, let $g'_- = s'_1 \cdots s'_{q'}$ and $g'_+ = t'_1 \cdots t'_{p'}$. Then, by construction, $g = g_-^{-1} g_+ = g'_-{}^{-1} g'_+$ hold in $\mathcal{E}nv(\mathcal{C})$. By Proposition 2.11, g_- and g_+ are left-disjoint. By Lemma 2.7, $g'_- = hg_-$ and $g'_+ = hg_+$ hold for some h belonging to $\mathcal{C}$. By Proposition 3.1, we have $q' \geqslant \|g'_-\|_\mathcal{S}$ and $p' \geqslant \|g'_+\|_\mathcal{S}$. On the other hand, by Corollary 1.55, the existence of h implies $\|g'_-\|_\mathcal{S} \geqslant \|g_-\|_\mathcal{S} = q$ and $\|g'_+\|_\mathcal{S} \geqslant \|g_+\|_\mathcal{S} = p$. We deduce $q' \geqslant q$ and $p' \geqslant p$, whence $\ell = q' + p' \geqslant q + p$, as expected. □

If $\mathcal{C}$ is an Ore category (and not only a left-Ore category) and $\mathcal{S}$ is a strong Garside family in $\mathcal{C}$, every element of the groupoid $\mathcal{E}nv(\mathcal{C})$ admits a strict symmetric $\mathcal{S}$-normal decomposition and, therefore, the latter provide geodesic decompositions for all elements of $\mathcal{E}nv(\mathcal{C})$.

3.2 The Grid Property

The Grid Property is a convexity statement that is valid in every groupoid of fractions associated with a category that admits a perfect Garside family, a mild strengthening of the strong Garside families of Section 2 that we shall define below.

We begin with a sort of double domino rule.

Lemma 3.3. *Assume that $\mathcal{C}$ is a left-cancellative category and $\mathcal{S}$ is included in $\mathcal{C}$. Then, for all $f_1, \ldots, f''_2$, $g_1, \ldots, g''_2$, h_1, h_2 in $\mathcal{C}$ such that $f''_1|f''_2$ and $g''_1|g''_2$ are $\mathcal{S}$-greedy, f'_2 and g'_2 are left-disjoint, and the diagram on the right is commutative, the paths $h_1|f'_2$, $h_1|g'_2$, and $h_1|h_2$ are $\mathcal{S}$-greedy.*

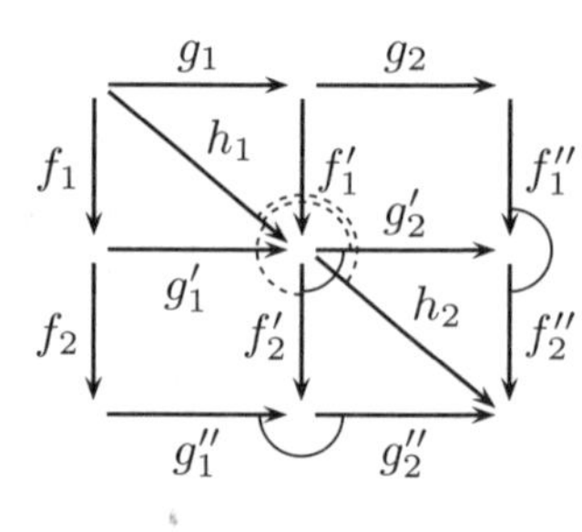

Proof. (See Figure 23.) Assume $s \in \mathcal{S}$ and $h \preccurlyeq fh_1h_2$. As the diagram commutes, $s \preccurlyeq fh_1h_2$ implies $s \preccurlyeq fg_1g_2f_1''f_2''$, whence, as $f_1''|f_2''$ is $\mathcal{S}$-greedy, $s \preccurlyeq fg_1g_2f_1''$, that is, $s \preccurlyeq fh_1g_2'$. A symmetric argument gives $s \preccurlyeq fh_1f_2'$. As f_2' and g_2' are left-disjoint, this implies $s \preccurlyeq fh_1$. Hence $h_1|h_2$ is $\mathcal{S}$-greedy. Lemma 1.7 then implies that $h_1|f_2'$ and $h_1|g_2'$ are $\mathcal{S}$-greedy as well. □

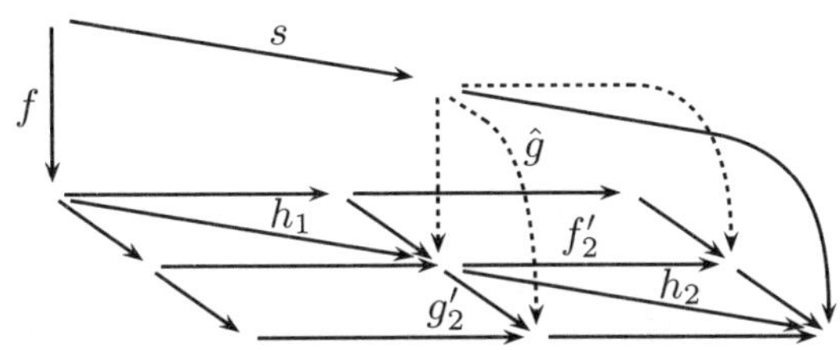

Figure 23. Proof of Lemma 3.3.

Definition 3.4 (grid). Assume that $\mathcal{C}$ is a left-cancellative category, $\mathcal{S}$ is a strong Garside family in $\mathcal{C}$, and $s_1|\cdots|s_p$ and $t_1|\cdots|t_q$ are $\mathcal{S}$-normal paths that share the same target. An *$\mathcal{S}$-grid for $s_1|\cdots|s_p$ and $t_1|\cdots|t_q$* is a rectangular commutative diagram consisting of pq tiles as in Lemma 2.31, with right labels $s_1, ..., s_p$ and bottom labels $t_1, ..., t_q$; the bottom-right vertex of the diagram is called the *extremal* vertex of the grid.

In the sequel, we specify the vertices of a grid using (x, y)-coordinates. By default, the coordinates are supposed to increase along edges, with $(0, 0)$ being the origin of the diagram and (p, q) being the extremal vertex. Then the word "right" (*resp.* "bottom") in Definition 3.4 corresponds to a maximal first (*resp.* second) coordinate.

Lemma 3.5. *Assume that $\mathcal{C}$ is a left-cancellative category and $\mathcal{S}$ is included in $\mathcal{C}$.*

(i) *If $\mathcal{S}$ is a strong Garside family in $\mathcal{C}$, then, for all $\mathcal{S}$-normal paths u, v such that the elements $[u]$ and $[v]$ admit a common left-multiple in $\mathcal{C}$, there exists an $\mathcal{S}$-grid for u and v.*

(ii) *Every path in an $\mathcal{S}$-grid Γ that is diagonal-then-horizontal or diagonal-then-vertical path is $\mathcal{S}$-greedy.*

Proof. (i) The existence of the expected $\mathcal{S}$-grid immediately follows from the construction developed in the proof of Proposition 2.37: starting from u and v, and beginning with the bottom–right corner, we construct the grid inductively using tiles as on the right, which exist as, by assumption, the elements $[u]$ and $[v]$ admit a common left-multiple and $\mathcal{S}$ is a strong Garside family.

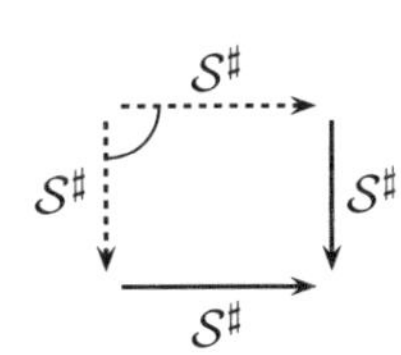

(ii) By the third domino rule (Proposition 2.45), every horizontal or vertical path in Γ corresponds to an $\mathcal{S}$-normal path. By Lemma 3.3, every diagonal path in Γ is $\mathcal{S}$-greedy, and so is every diagonal path followed by a horizontal or a vertical path. □

To obtain the existence of paths that are $\mathcal{S}$-normal, and not only $\mathcal{S}$-greedy, in grids, we add one more assumption—which, we shall see in Chapter V, is satisfied in usual cases.

Definition 3.6 (perfect Garside). A Garside family $\mathcal{S}$ in a left-Ore category $\mathcal{C}$ that admits left-lcms is called *perfect* if there exists a short left-lcm witness $\widetilde{\theta}$ on $\mathcal{S}^\sharp$ such that, for all s, t with the same target, $\widetilde{\theta}(s,t)t$ belongs to $\mathcal{S}^\sharp$.

The difference between strong and perfect is that, in the latter case, we require that the diagonal of the square whose existence is asserted as in the former case represents an element of the reference family $\mathcal{S}^\sharp$, corresponding to the scheme on the right.

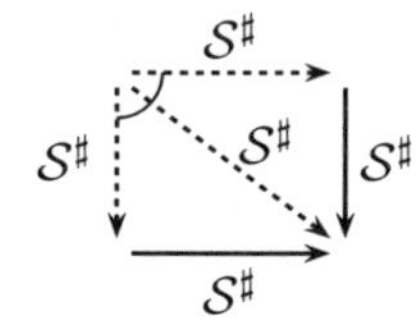

Note that, in an Ore context, a strong Garside family $\mathcal{S}$ is perfect if and only if $\mathcal{S}^\sharp$ is closed under the left-lcm operation, that is, the left-lcm of two elements of $\mathcal{S}^\sharp$ lies in $\mathcal{S}^\sharp$.

Proposition 3.7 (Grid Property, positive case). *If $\mathcal{S}$ is a perfect Garside family in a left-cancellative category $\mathcal{C}$ and Γ is an $\mathcal{S}$-grid, then, for all vertices (i,j), (i',j') in Γ with $i \leqslant i'$ and $j \leqslant j'$, there exists inside Γ an $\mathcal{S}$-normal path from (i,j) to (i',j').*

Proof. The assumptions $i \leqslant i'$ and $j \leqslant j'$ guarantee the existence in Γ of a path from (i,j) to (i',j') of the form diagonal-then-horizontal or diagonal-then-vertical. By Lemma 3.5, such paths are $\mathcal{S}$-greedy. Moreover, the assumption that $\mathcal{S}$ is perfect guarantees that they are $\mathcal{S}$-normal since the diagonals of the squares also correspond to elements of $\mathcal{S}^\sharp$. By Proposition 3.1, such $\mathcal{S}$-normal paths are geodesic, provided possible final invertible edges are incorporated in the last non-invertible entry. □

Example 3.8 (Grid Property, positive case). Consider a free abelian monoid once more (Reference Structure 1, page 3). Starting from the $\mathcal{S}$-normal paths $\mathtt{ab}|\mathtt{a}$ and $\mathtt{ac}|\mathtt{ac}|\mathtt{a}$, we obtain a (unique) $\mathcal{S}$-grid

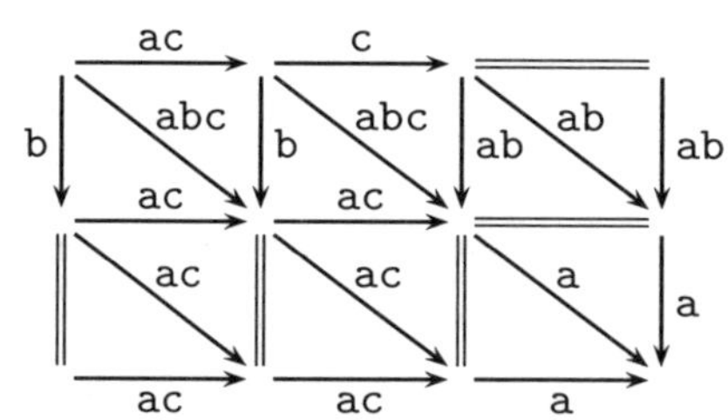

Numbering vertices from the top–left corner, we read for instance that, from the $(0,0)$-vertex to the $(3,2)$-vertex, a geodesic path is $\mathtt{abc}|\mathtt{ac}|\mathtt{a}$, whereas, from the $(1,0)$-vertex to the $(2,2)$-vertex, a geodesic path is $\mathtt{abc}|1$.

For the signed case, the technical core will be a strengtening of Lemma 3.5(ii).

Lemma 3.9. *Assume that $\mathcal{C}$ is a left-cancellative category, $\mathcal{S}$ is included in $\mathcal{C}$, and Γ_1, Γ_2 are $\mathcal{S}$-grids with the same left edge and left-disjoint bottom edges. Let $\mathcal{P}(\Gamma)$ be the collection of all diagonal-then-horizontal and diagonal-then-vertical paths in a grid Γ. Then every negative-positive path in $\Gamma_1 \cup \Gamma_2$ consisting of the inverse of a path of $\mathcal{P}(\Gamma_1)$ followed by a path of $\mathcal{P}(\Gamma_2)$ that contains no pattern "inverse of a diagonal followed by a diagonal" is symmetric $\mathcal{S}$-greedy.*

Proof. Owing to Lemma 3.5(ii), it remains to consider the patterns $\overline{f}|g$ with f in Γ_1 and g in Γ_2. By assumption, the bottom edges of Γ_1 and Γ_2 represent left-disjoint elements of $\mathcal{C}$. The fourth domino rule (Proposition 2.48) then guarantees that all patterns $\overline{f}|g$ of the considered types, that is, all that are not of the type "inverse of diagonal–diagonal", correspond to left-disjoint elements. □

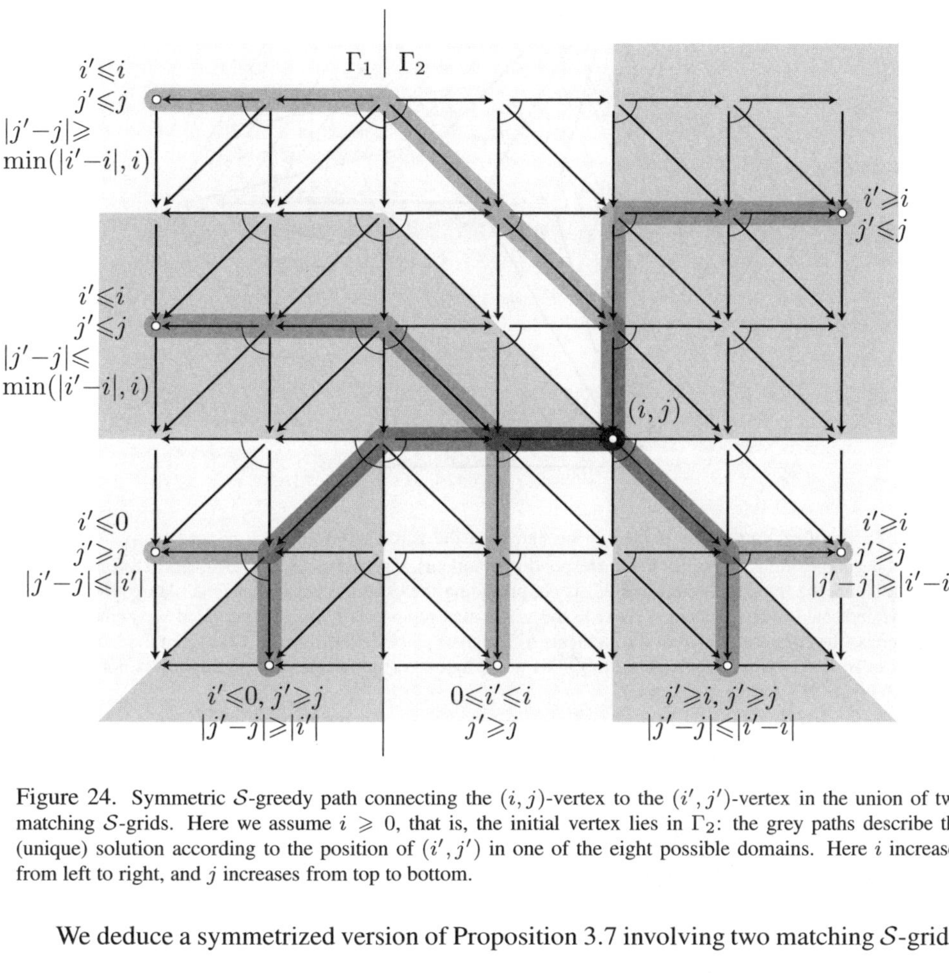

Figure 24. Symmetric $\mathcal{S}$-greedy path connecting the (i, j)-vertex to the (i', j')-vertex in the union of two matching $\mathcal{S}$-grids. Here we assume $i \geqslant 0$, that is, the initial vertex lies in Γ_2: the grey paths describe the (unique) solution according to the position of (i', j') in one of the eight possible domains. Here i increases from left to right, and j increases from top to bottom.

We deduce a symmetrized version of Proposition 3.7 involving two matching $\mathcal{S}$-grids.

Lemma 3.10. *Assume that $\mathcal{C}$ is a left-Ore category, $\mathcal{S}$ is a perfect Garside family in $\mathcal{C}$, and Γ_1, Γ_2 are $\mathcal{S}$-grids with the same left edge and left-disjoint bottom edges. Let Γ be the diagram obtained by gluing along their common edge a symmetric image of Γ_1 and Γ_2. Then, for all vertices M, M' of Γ, there exists inside Γ a symmetric $\mathcal{S}$-normal path from M to M'.*

Proof. As in Proposition 3.7, let us give coordinates to the vertices of Γ, using nonpositive x-coordinates for the vertices of Γ_1 (left) and nonnegative x-coordinates for those of Γ_2 (right), see Figure 24; as above, the y-axis is given a downward orientation. Assume that the coordinates of M and M' are (i, j) and (i', j'), respectively. The figure then displays a symmetric $\mathcal{S}$-normal path from M to M' in all possible cases. □

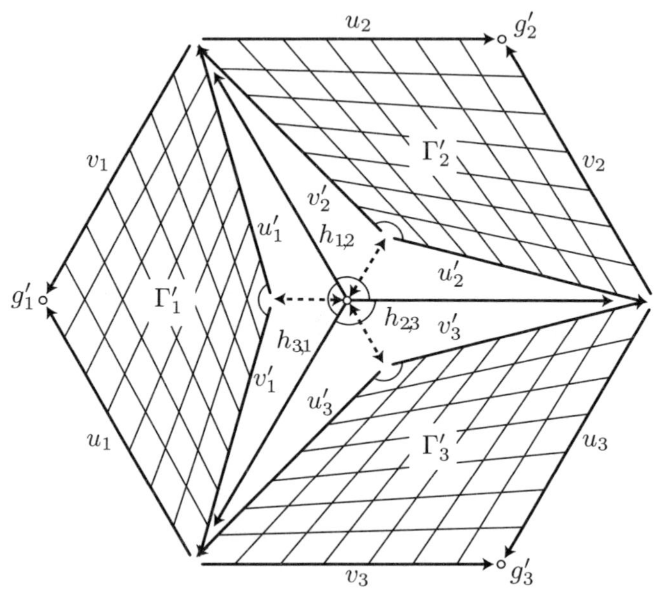

Figure 25. Proof of Proposition 3.11: having extracted the pairwise left-gcds of g'_i and g'_{i+1}, one selects $\mathcal{S}$-normal decompositions u_i, v_i for the corresponding quotients, and one constructs an $\mathcal{S}$-grid Γ'_i for (u_i, v_i). The point is that the three grids necessarily match, providing the expected hexagonal grid. Above (for lack of space), we indicated every element g'_i next to the vertex that represents its target: provided we agree that the central vertex (the common source of the elements g'_i and their pairwise left-lcms $h_{i,j}$) is given the label 1, then the target of g'_i is naturally associated with g'_i—as in a Cayley graph, but the current diagram is not a Cayley graph (see Figure II.1 for the distinction).

We are now ready to state and establish the main result.

Proposition 3.11 (Grid Property). *If $\mathcal{C}$ is an Ore category that admits unique left-lcms and $\mathcal{S}$ is a perfect Garside family in $\mathcal{C}$, then, for all g_1, g_2, g_3 in $\mathcal{E}nv(\mathcal{C})$, there exists a planar diagram Γ that is the union of three $\mathcal{S}$-grids, admits g_1, g_2, g_3 as extremal vertices, and is such that, for all vertices x, y of Γ, there exists a symmetric $\mathcal{S}$-normal (hence geodesic) path from x to y inside Γ.*

Proof. First, there exists h' in $\mathcal{C}$ such that $h'g_1$, $h'g_2$, and $h'g_3$ lie in $\mathcal{C}$. By assumption, $\mathcal{C}$ admits left-lcms, so the left counterpart of Lemma II.2.22 implies that any two elements of $\mathcal{C}$ admit a left-gcd. Let h be the left-gcd of $h'g_1, h'g_2$, and $h'g_3$, and let g'_i be defined by $h'g_i = hg'_i$. Then g'_1, g'_2, g'_3 are three elements of $\mathcal{C}$ with the same source and a trivial left-gcd. Let $h_{i,i+1}$ be a left-gcd of g'_i and g'_{i+1} for $i = 1, 2, 3$ (and indices taken mod 3).

By construction, $h_{i,i+1}$ left-divides g'_i and g'_{i+1}, and we can choose $\mathcal{S}$-normal paths u_i, v_i representing the quotients, namely satisfying

$$g'_i = h_{i-1,i}[u_i] = h_{i,i+1}[v_i],$$

see Figure 25. By Lemma 3.5(i), there exists an $\mathcal{S}$-grid Γ'_i for (u_i, v_i).

The point is that the $\mathcal{S}$-grids $\Gamma'_1, \Gamma'_2, \Gamma'_3$ match in the sense of Lemma 3.10. Indeed, let u'_i and v'_i denote the left and top edges of Γ'_i (when oriented as in the original definition). By construction, the elements $[u'_i]$ and $[v'_i]$ are left-disjoint and satisfy $[u'_i][v_i] = [v'_i][u_i]$. On the other hand, we have $h_{i-1,i}[u_i] = h_{i,i+1}[v_i]$. As $[u'_i]$ and $[v'_i]$ are left-disjoint, there must exist h_i satisfying $h_{i,i+1} = h_i[u'_i]$ and $h_{i-1,i} = h_i[v'_i]$. By construction, h_i left-divides g'_1, g'_2, and g'_3, hence it left-divides their left-gcd, which is trivial by assumption. Hence h_i must be invertible.

Then the assumption that $\mathcal{C}$ admits no non-trivial invertible element implies $[u'_i] = h_{i,i+1} = [v'_{i+1}]$, whence $u'_i = v'_{i+1}$ by uniqueness of the $\mathcal{S}$-normal decomposition, up to possibly adding some identity elements at the end of one path (see Figure 26). Then applying Lemma 3.10 to each of the pairs Γ'_i, Γ'_{i+1} provides symmetric $\mathcal{S}$-normal paths of the expected type in the diagram Γ' which is the union of Γ'_1, Γ'_2, Γ'_3.

By construction, the extremal vertices of Γ' are g'_1, g'_2, g'_3. To obtain the result for g_1, g_2, g_3, it suffices to define Γ to be the image of Γ' under a left-translation by $h^{-1}h'$. ☐

Example 3.12 (Grid Property). Let us come back once more to the free abelian group $\mathbb{Z}^3$ based on $\{\mathtt{a}, \mathtt{b}, \mathtt{c}\}$ with the (perfect) Garside family consisting of the eight divisors of $\mathtt{abc}$. Consider $g_1 = \mathtt{b}$, $g_2 = \mathtt{a}^{-1}\mathtt{b}^{-1}\mathtt{c}^3$, and $g_3 = \mathtt{a}^{-2}\mathtt{bc}^2$. Left-multiplying by $\mathtt{a}^2\mathtt{b}$ yields $g'_1 = \mathtt{a}^2\mathtt{b}^2$, $g'_2 = \mathtt{ac}^3$, $g'_3 = \mathtt{b}^2\mathtt{c}^2$, three elements of $\mathbb{N}^3$ with a trivial left-gcd. With the notation of the proof of Proposition 3.11, we find $h_{1,2} = \mathtt{a}$, $h_{2,3} = \mathtt{c}^2$, and $h_{3,1} = \mathtt{b}^2$, leading to the grid displayed in Figure 26.

3.3 The Fellow Traveller Property

We show now that $\mathcal{S}$-normal decompositions enjoy some geometrical properties which, when suitable finiteness assumptions are realized, imply the existence of an automatic structure. To state the results we shall use a natural extension of the classical Fellow Traveller Property, see for instance [54] (where a groupoid version is even considered) or [119, Theorem 2.3.5] (where the name "fellow traveller" is not used). For $\mathcal{S}$ generating a groupoid $\mathcal{G}$ and f, g in $\mathcal{G}$, we denote by $\mathrm{dist}_{\mathcal{S}}(f, g)$ the $\mathcal{S}$-distance between f and g in $\mathcal{G}$, that is, the minimal length of a signed $\mathcal{S}$-path representing $f^{-1}g$ in $\mathcal{G}$.

Definition 3.13 (fellow traveller). If $\mathcal{S}$ is a subfamily of a left-cancellative category $\mathcal{C}$, then, for $k \geqslant 0$, we say that two signed $\mathcal{S}^\sharp$-paths w, w' are *k-fellow travellers* if, for every i, we have $\mathrm{dist}_{\mathcal{S}^\sharp}([w_{/i}], [w'_{/i}]) \leqslant k$, where $w_{/i}$ is the length i prefix of w for $i \leqslant \lg(w)$ and w otherwise, and $[v]$ is the element of $\mathcal{E}nv(\mathcal{C})$ represented by v.

So two paths are fellow travellers if they remain at uniformly bounded distance from each other when crossed at the same speed.

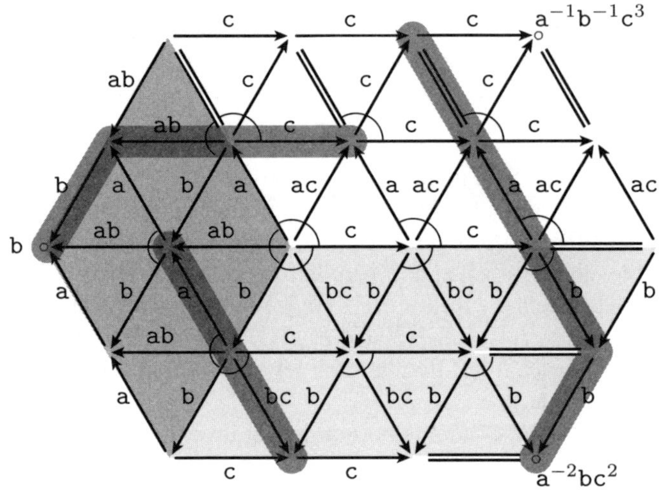

Figure 26. The grid for the triangle $(\mathtt{b}, \mathtt{a}^{-1}\mathtt{b}^{-1}\mathtt{c}^3, \mathtt{a}^{-2}\mathtt{bc}^2)$: the union of three grids for the pairs of normal paths $(\mathtt{ab}|\mathtt{b}, \mathtt{a}|\mathtt{a})$, $(\mathtt{ac}, \mathtt{c}|\mathtt{c}|\mathtt{c})$, and $(\mathtt{c}, \mathtt{b}|\mathtt{b})$ constructed using tiles as in Figure 24. Note that the grids match, but possibly at the expense of adding trivial entries at the end of the normal paths to make the lengths equal. The grey paths are typical geodesic paths connecting two vertices. The point is that the geodesic that connects any two vertices is entirely included in the grid, which is planar.

Proposition 3.14 (fellow traveller). *Assume that $\mathcal{C}$ is a left-Ore category.*

(i) *If $\mathcal{S}$ is included in $\mathcal{C}$ and $\mathcal{S}^\sharp$ generates $\mathcal{C}$, then, for every g in $\mathcal{E}nv(\mathcal{C})$, any two strict symmetric $\mathcal{S}$-normal decompositions of g are 1-fellow travellers.*

(ii) *If $\mathcal{S}$ is a strong Garside family in $\mathcal{C}$, then, for every g in $\mathcal{E}nv(\mathcal{C})$ and every s in $\mathcal{S}^\sharp$, any two strict symmetric $\mathcal{S}$-normal decompositions of g and gs are 2-fellow travellers.*

Proof. (i) The result follows from Proposition 2.16 and the diagram of Figure 16: two strict symmetric $\mathcal{S}$-normal decompositions of an element g are $\mathcal{C}^\times$-deformations of one another. Moreover, being strict, they have the same length and, with the notation of Figure 16, we must have $p = p'$ and $q = q'$. As all invertible elements lie in $\mathcal{S}^\sharp$, the diagram shows that $\overline{t_q}|\cdots|\overline{t_1}|s_1|\cdots|s_p$ and $\overline{t'_{q'}}|\cdots|\overline{t'_1}|s'_1|\cdots|s'_{p'}$ are 1-fellow travellers.

(ii) The result similarly follows from Corollary 2.50. Assume that $\overline{t_q}|\cdots|\overline{t_1}|s_1|\cdots|s_p$ is a strict symmetric $\mathcal{S}$-normal decomposition of gs and Algorithm 2.49 running on $\overline{t_q}|\cdots|\overline{t_1}|s_1|\cdots|s_p$ and s returns $\overline{t'_{q+1}}|\overline{t'_q}|\cdots|\overline{t'_1}|s'_1|\cdots|s'_p$ (we use the notation of Algorithm 2.49). Two cases may happen. If t'_{q+1} is invertible, then $\overline{t'_q t'_{q+1}}|\overline{t'_{q-1}}|\cdots|\overline{t'_1}|s'_1|\cdots|s'_p$ is a strict symmetric $\mathcal{S}$-normal decomposition of g, and, for every i, we have

$$t_q^{-1}\cdots t_i^{-1} = (t'_q t'_{q+1})^{-1} t'^{-1}_{q-1} \cdots t'^{-1}_i \cdot s_{-i+1} \qquad \text{for } q \geqslant i \geqslant 1,$$

$$t_q^{-1}\cdots t_1^{-1} s_1 \cdots s_i = (t'_q t'_{q+1})^{-1} t'^{-1}_{q-1} \cdots t'^{-1}_1 s'_1 \cdots s'_i \cdot s_i \qquad \text{for } 1 \leqslant i \leqslant p.$$

As the final factors s_i lie in $\mathcal{S}^\sharp$, the paths are 1-fellow travellers. On the other hand, if t'_{q+1} is not invertible, $\overline{t'_{q+1}}|\overline{t'_{q'}}|\cdots|\overline{t'_1}|s'_1|\cdots s'_p$ is a strict symmetric $\mathcal{S}$-normal decomposition of g, and, for every i, we have

$$t_q^{-1}\cdots t_i^{-1} = {t'_{q+1}}^{-1}{t'_q}^{-1}\cdots {t'_{i+1}}^{-1}\cdot {t'_i}^{-1}s_{-i+1} \qquad \text{for } q \geqslant i \geqslant 1,$$
$$t_q^{-1}\cdots t_1^{-1}s_1\cdots s_i = {t'_{q+1}}^{-1}{t'_q}^{-1}\cdots {t'_1}^{-1}s'_1\cdots s'_{i-1}\cdot s'_i s_i \qquad \text{for } 1 \leqslant i \leqslant p.$$

As the final factors ${t'_i}^{-1}s_{-i+1}$ and $s'_i s_i$ lie in $(\mathcal{S}^\sharp)^2$, the paths are 2-fellow travellers. □

Similar results hold for symmetric $\mathcal{S}$-normal paths that are not necessarily strict: the difference is that, because of an unbounded number of initial and final invertible entries, one cannot obtain the k-fellow traveller property for some fixed k.

It is now easy to show that symmetric $\mathcal{S}$-normal decompositions are associated with a quasi-automatic structure, "quasi-automatic" here meaning that the involved automata need not be finite. We recall that ε_x is the empty path with source x.

Notation 3.15. If $\mathcal{S}$ is a strong Garside family in a left-Ore category $\mathcal{C}$, we denote by $\mathtt{NF}_\mathcal{S}$ (*resp.* $\mathtt{NF}^\mathrm{s}_\mathcal{S}$) the family of all symmetric $\mathcal{S}$-normal paths (*resp.* strict symmetric $\mathcal{S}$-normal paths).

Proposition 3.16 (quasi-rational I). *Assume that $\mathcal{C}$ is a left-cancellative category and $\mathcal{S}$ is included in $\mathcal{C}$. Let $\mathcal{Q} = \mathcal{S}^\sharp \cup \overline{\mathcal{S}^\sharp}$ and $\widehat{\mathcal{Q}} = \mathcal{Q} \cup \{\varepsilon_x \mid x \in \mathcal{O}bj(\mathcal{C})\}$. Define a partial map $T : \widehat{\mathcal{Q}} \times \mathcal{Q} \to \widehat{\mathcal{Q}}$ by $T(\varepsilon_x, s) = s$ and $T(s,t) = t$ if $s|t$ is symmetric $\mathcal{S}$-greedy. Let T^* be the extension of T to $\widehat{\mathcal{Q}} \times \mathcal{Q}^*$ inductively defined by $T^*(q, \varepsilon_x) = q$, and $T^*(q, w|s) = T(T^*(q,w), s)$ for s in $\mathcal{Q}$. Then a signed $\mathcal{S}$-path w with source x belongs to $\mathtt{NF}_\mathcal{S}$ if and only if $T^*(\varepsilon_x, w)$ is defined.*

Proof. The partial map T is the transition map of an automaton that reads signed $\mathcal{S}^\sharp$-paths and continues as long as it finds no obstruction to being a symmetric $\mathcal{S}$-normal path. The point is that being symmetric $\mathcal{S}$-normal is a local property that only involves pairs of adjacent entries. In particular, for s, t in $\mathcal{Q}$, the relation $T(s,t) = t$ holds if and only if
- s is an empty path ε_x,
- or s, t lie in $\overline{\mathcal{S}^\sharp}$ and $\overline{t}|\overline{s}$ is $\mathcal{S}$-greedy (where, for $s = \overline{r}$ in $\overline{\mathcal{S}^\sharp}$, we put $\overline{s} = r$),
- or s lies in $\overline{\mathcal{S}^\sharp}$, t lies in $\mathcal{S}^\sharp$, and $\overline{s}$ and t are left-disjoint,
- or s, t lie in $\mathcal{S}^\sharp$ and $s|t$ is $\mathcal{S}$-greedy.

Therefore, deciding whether a path is symmetric $\mathcal{S}$-normal only requires to compare two letters at a time. This can be done by memorizing the last previously read letter. According to Definition 2.9, the above definition of the map T guarantees that $T^*(\varepsilon_x, w)$ exists (and is equal to the last letter of w) if and only if w is symmetric $\mathcal{S}$-normal. □

A similar result holds for strict symmetric $\mathcal{S}$-normal paths.

Proposition 3.17 (quasi-rational II). *Assume that $\mathcal{C}$ is a left-cancellative category and $\mathcal{S}$ is included in $\mathcal{C}$. Let $\mathcal{Q} = \mathcal{S}^\sharp \cup \overline{\mathcal{S}^\sharp}$. There exists a family $\widehat{\mathcal{Q}}^\mathrm{s}$ and a partial map T^s : $\widehat{\mathcal{Q}}^\mathrm{s} \times \mathcal{Q} \to \widehat{\mathcal{Q}}^\mathrm{s}$ such that, if T^s is extended into $T^{\mathrm{s}*}$ by $T^{\mathrm{s}*}(q, \varepsilon_x) = q$ and $T^{\mathrm{s}*}(q, w|s) = T(T^{\mathrm{s}*}(q,w), s)$ for s in $\mathcal{Q}$, then a signed $\mathcal{S}$-path w with source x belongs to $\mathtt{NF}^\mathrm{s}_\mathcal{S}$ if and only if $T^{\mathrm{s}*}(\varepsilon_x, w)$ is defined. If $\mathcal{S}^\sharp$ is finite, $\widehat{\mathcal{Q}}^\mathrm{s}$ can be taken finite.*

Proof. The argument is the same as for Proposition 3.16, the difference being that restricting to strict symmetric $\mathcal{S}$-normal paths requires to exclude more cases, namely all paths in which an entry that is neither initial nor final belongs to $\mathcal{S}^\sharp \setminus \mathcal{S}$. This can be done easily, but requires that the information about being the initial or the final letter be recorded in the state q in addition to the last read letter. The state set $\widehat{\mathcal{Q}}^{\mathrm{s}}$ has then to be larger than in Proposition 3.16. However, if $\mathcal{S}^\sharp$ is finite, $\widehat{\mathcal{Q}}^{\mathrm{s}}$ can be constructed to be finite. □

Corollary 3.18 (rational). *If $\mathcal{S}$ is a strong Garside family in a left-cancellative category $\mathcal{C}$ and both $\mathcal{O}bj(\mathcal{C})$ and $\mathcal{S}^\sharp$ are finite, the families $\mathtt{NF}_{\mathcal{S}}$ and $\mathtt{NF}^{\mathrm{s}}_{\mathcal{S}}$ are rational sets, that is, they are recognized by some finite state automaton.*

Proof. If $\mathcal{O}bj(\mathcal{C})$ and $\mathcal{S}^\sharp$ are finite, the families $\widehat{\mathcal{Q}}$ and $\widehat{\mathcal{Q}}^{\mathrm{s}}$ of Propositions 3.16 and 3.17 can be assumed to be finite, and $(\widehat{\mathcal{Q}}, T)$ and $(\widehat{\mathcal{Q}}^{\mathrm{s}}, T^{\mathrm{s}})$ are then finite state automata recognizing the sets $\mathtt{NF}_{\mathcal{S}}$ and $\mathtt{NF}^{\mathrm{s}}_{\mathcal{S}}$, respectively. □

Example 3.19 (rational). Consider a free abelian monoid based on two generators $\mathtt{a}, \mathtt{b}$, with the Garside family $S_2 = \{\mathtt{a}, \mathtt{b}, \mathtt{ab}\}$ (Reference Structure 1, page 3). Then the graph of the automaton defined in Proposition 3.16 is as follows:

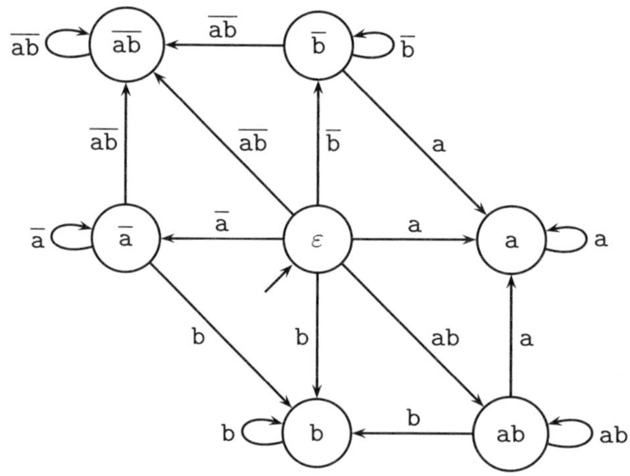

A word in the alphabet $\{\mathtt{a}, \mathtt{b}, \mathtt{ab}, \overline{\mathtt{a}}, \overline{\mathtt{b}}, \overline{\mathtt{ab}}\}$ is symmetric S_2-normal if and only if, starting from the initial state ε, one can read all letters of w successively in the graph above: for instance, $\overline{\mathtt{a}}|\mathtt{b}|\mathtt{b}$ and $\mathtt{ab}|\mathtt{ab}|\mathtt{a}$ are accepted, but neither $\mathtt{a}|\overline{\mathtt{b}}$ nor $\mathtt{a}|\mathtt{ab}$ is.

The existence of a rational family of normal forms satisfying the fellow traveller property is one of the standard definitions for an automatic structure [119, Theorem 2.3.5].

Definition 3.20 (automatic). A groupoid $\mathcal{G}$ is called *automatic* if there exist a finite set $\mathcal{S}$, a rational family of signed $\mathcal{S}$-paths $\mathtt{NF}$, and a constant k, such that every element of $\mathcal{G}$ admits at least one decomposition lying in $\mathtt{NF}$ and, for all f, g in $\mathcal{G}$ such that $f^{-1}g$ lies in $\mathcal{S} \cup \mathbf{1}_{\mathcal{C}}$, any two decompositions of f and g lying in $\mathtt{NF}$ are k-fellow travellers.

Proposition 3.21 (automatic). *If $\mathcal{S}$ is a strong Garside family in an Ore category $\mathcal{C}$ with finitely many objects and $\mathcal{S}^\sharp$ is finite, the groupoid $\mathcal{E}nv(\mathcal{C})$ is automatic.*

Proof. According to Corollary 3.18 and Proposition 3.14, the family of strict symmetric $\mathcal{S}$-normal decompositions enjoys all needed properties. □

In particular, a direct application is

Corollary 3.22 (automatic). *Every Garside group (Definition I.2.3) is automatic.*

Among others, the existence of an automatic structure implies a quadratic isoperimetric inequality with respect to the considered generating family. We shall address the question more systematically in the next subsection.

A natural strengthening of the notion of an automatic group(oid) is the notion of a biautomatic group(oid) where, in addition to the Fellow Traveler Property for the normal decompositions of elements obtained by right-multiplying by an element of the reference generating family $\mathcal{S}$, one requires a symmetric property for left-multiplication by an element of $\mathcal{S}$. For instance, it is known that every Garside group is biautomatic [80]. But we cannot state a general result here: what is missing is a counterpart to Proposition 2.47 connecting the symmetric $\mathcal{S}$-decompositions of g and sg for s in $\mathcal{S}^\sharp$ in general. We shall address the question again in Chapter V, when the Garside family $\mathcal{S}$ is what will be called bounded.

Remark 3.23. The positive case of a Garside family $\mathcal{S}$ in an arbitrary left-cancellative category can be addressed similarly. The counterparts of Propositions 3.16 and 3.17 and, therefore, Corollary 3.18, are then valid. However, the Fellow Traveller Property need not hold in general: for s in $\mathcal{S}$, Proposition 1.49 provides no direct connection between the $\mathcal{S}$-normal decompositions of g and gs. Only when the second domino rule is valid does Proposition 1.61 provide such a connection, implying that the $\mathcal{S}$-normal decompositions of g and gs are 1-fellow travellers.

Combinatorics of normal sequences. Whenever the ambient category is left-cancellative, the fact that $\mathcal{S}$-normal paths are characterized by a local condition and are therefore recognized by (some sort of) automaton implies nice combinatorial properties.

If a category $\mathcal{C}$ admits a finite Garside family $\mathcal{S}$, there exist for every p a finite number of $\mathcal{S}$-normal paths of length p, and *counting* them is a natural task. This amounts to counting the sequences accepted by some finite automaton, and standard results ensure that the associated generating series is rational, controlled by the powers of an adjacency matrix whose entries are (indexed by) the elements of $\mathcal{S}^\sharp$:

Proposition 3.24 (adjacency matrix). *Assume that $\mathcal{S}$ is a Garside family in a left-cancellative category $\mathcal{C}$ and $\mathcal{S}^\sharp$ is finite. Let I_x and M be the $(\mathcal{S}^\sharp{\times}1)$- and $(\mathcal{S}^\sharp{\times}\mathcal{S}^\sharp)$-matrices determined by*

$$(3.25)\qquad (I_x)_t = \begin{cases} 1 & \text{if } t \text{ lies in } \mathcal{C}(x,\cdot),\\ 0 & \text{otherwise,}\end{cases} \qquad M_{s,t} = \begin{cases} 1 & \text{if } s|t \text{ is } \mathcal{S}\text{-normal,}\\ 0 & \text{otherwise.}\end{cases}$$

Then the number $a_{p,x,y}$ of $\mathcal{S}$-normal paths of length p with source x and target y is the 1_y-entry of I_xM^p.

Proof. For s in $\mathcal{S}^\sharp$, let $a_{p,x}(s)$ be the number of $\mathcal{S}$-normal paths of length p with source x and last entry s. We prove $a_{p+1,x}(t) = (I_xM^p)_t$ using induction on p. For $p = 0$, there exists a length-one normal path with source x and last entry t if and only if t has source x, in which case the path is unique. So we have $a_{1,x}(t) = (I_x)_t$. Assume now $p > 0$. Then a length-$(p{+}1)$ path w finishing with t is $\mathcal{S}$-normal if and only if there exists s in $\mathcal{S}^\sharp$ and a length p path w' finishing with s such that $s|t$ is $\mathcal{S}$-normal and w is $w'|t$. Using the induction hypothesis, we deduce

$$a_{p+1,x}(t) = \sum_{s\in\mathcal{S}^\sharp} a_{p,x}(s)(M)_{s,t} = \sum_{s\in\mathcal{S}^\sharp} (I_xM^{p-1})_s(M)_{s,t} = (I_xM^p)_t.$$

Then we have $a_{p,x,y} = a_{p+1,x}(1_y)$ since, for s_p in $\mathcal{S}^\sharp(\cdot,y)$, the path $s_1|\cdots|s_p$ is $\mathcal{S}$-normal if and only if $s_1|\cdots|s_p|1_y$ is. We deduce $a_{p,x,y} = a_{p+1,x}(1_y) = (I_xM^p)_{1_y}$. □

Example 3.26 (adjacency matrix). For $\mathbb{N}^2$ with $S_2 = \{1, \mathtt{a}, \mathtt{b}, \mathtt{ab}\}$ (Reference Structure 1, page 3), the associated $(4{\times}4)$-matrix is $M = \begin{pmatrix} 1 & 0 & 0 & 0\\ 1 & 1 & 0 & 0\\ 1 & 0 & 1 & 0\\ 1 & 1 & 1 & 1\end{pmatrix}$. A straightforward induction shows that the first entry in $(1,1,1,1)M^p$ is $(p+1)^2$, which is the number of normal sequences of length p. By uniqueness of the strict normal decomposition, this number is also the number of elements of $\mathbb{N}^2$ whose S_2-length is at most p.

We refer to the Notes of Chapter IX for more interesting examples.

3.4 The Garside resolution

We shall show now how, if $\mathcal{S}$ is a strong Garside family in a (convenient) cancellative category $\mathcal{C}$, one can use $\mathcal{S}$-normal decompositions to construct a resolution of the free $\mathbb{Z}\mathcal{C}$-module $\mathbb{Z}$ by free $\mathbb{Z}\mathcal{C}$-modules, thus leading to a determination of the homology of $\mathcal{C}$ and of its groupoid of fractions. The construction is a sort of local version of the bar-resolution, and its validity in the current context precisely relies on the third domino rule.

We start from a left-Ore category $\mathcal{C}$ that admits left-lcms. By Ore's theorem, $\mathcal{C}$ embeds in its groupoid of left-fractions $\mathcal{E}nv(\mathcal{C})$. The first observation is that the $\mathbb{Z}$-homology of $\mathcal{E}nv(\mathcal{C})$ coincides with that of $\mathcal{C}$.

Lemma 3.27. [51] *If $\mathcal{C}$ is a left-Ore category and $\mathcal{G}$ is the groupoid of fractions of $\mathcal{C}$, we have $H_*(\mathcal{G},\mathbb{Z}) = H_*(\mathcal{C},\mathbb{Z})$.*

Proof (sketch). Tensorizing by $\mathbb{Z}\mathcal{G}$ over $\mathbb{Z}\mathcal{C}$, one extends every (left) $\mathbb{Z}\mathcal{C}$-module to a $\mathbb{Z}\mathcal{G}$-module. The point is that, under the assumptions, the functor $R \to \mathbb{Z}\mathcal{G} \otimes_{\mathbb{Z}\mathcal{C}} R$ can be proved to be exact. □

From now on, we concentrate on determining the homology of the category $\mathcal{C}$. To this end, we construct for every strong Garside family $\mathcal{S}$ a cubical complex associated with finite subfamilies of $\mathcal{S}$. To avoid redundant cells, we assume that the elements of $\mathcal{S}$ with a given target are ordered. We write $\widetilde{\Delta}_C$ for the left-lcm of a subfamily C of $\mathcal{C}$. We recall that, for f, g in $\mathcal{C}$ admitting a unique left-lcm, f/g, the *left-complement* of g in f, is the (unique) element f' such that $f'g$ is the left-lcm of f and g.

Definition 3.28 (chain complex). Assume that $\mathcal{C}$ is a cancellative category that admits unique conditional left-lcms, $\mathcal{S}$ is a strong Garside family in $\mathcal{C}$, and a linear ordering of $\mathcal{S}(\cdot, y)$ is fixed for each object y. For $n \geqslant 0$, we denote by $\mathcal{S}^{[n]}$ the family of all strictly increasing n-tuples $(s_1, \ldots, s_n)$ in $\mathcal{S}$ such that $s_1, \ldots, s_n$ admit a left-lcm. We denote by $\mathcal{Z}_n$ the free $\mathbb{Z}\mathcal{C}$-module generated by $\mathcal{S}^{[n]}$, and call its elements *n-chains*. The generator of $\mathcal{Z}_n$ associated with an element $(s_1, \ldots, s_n)$ of $\mathcal{S}^{[n]}$ is denoted by $[[s_1, \ldots, s_n]]$, and it is called an *n-cell*. The unique 0-cell is denoted by $[[\,]]$.

As a $\mathbb{Z}$-module, $\mathcal{Z}_n$ is generated by the elements of the form gC with g in $\mathcal{C}$ and C in $\mathcal{S}^{[n]}$; such elements will be called *elementary n-chains*. In order to make a complex out of $\mathcal{Z}_*$, we take a geometrical viewpoint and associate with each n-cell an oriented n-cube reminiscent of a van Kampen diagram, constructed using the left-reversing process of Section II.4. The vertices of the cube are elements of $\mathcal{C}$, whereas the edges are labeled by elements of $\mathcal{S}$. For $s_1 < \cdots < s_n$ in $\mathcal{S}$, the n-cube (associated with) $[[s_1, \ldots, s_n]]$ starts from 1_x, where x is the source of $\widetilde{\Delta}_{s_1,\ldots,s_n}$ and ends at $\widetilde{\Delta}_{s_1,\ldots,s_n}$ (so that the latter left-lcm corresponds to the main diagonal of the cube). To construct the cube, we start with n edges labeled $s_1, \ldots, s_n$ pointing to the final vertex, and we inductively close every pattern

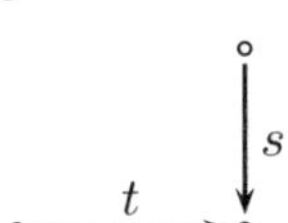

into

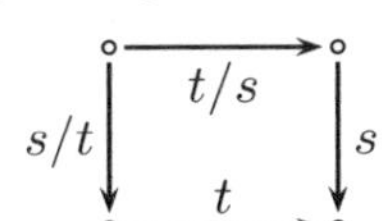

, that is, we construct step by step the left-lcm of the considered n final edges. The construction terminates with 2^n vertices, see Figure 27. The assumption that $\mathcal{S}$ is strong guarantees that all edges in the cube do lie in $\mathcal{S}$. Finally, we associate with the elementary n-chain gC the image of the n-cube (associated with) C under the left translation by g: the cube starts from g instead of starting from 1.

Note that, under the assumptions of Definition 3.28, the left-lcm is unique (when it exists), so there exists a unique left-lcm witness $\widetilde{\theta}$ on $\mathcal{S}$, which is precisely the map $\widetilde{\theta}(s,t) = s/t$. So the above construction of the cube associated with an n-cell can be viewed as closing the initial n final edges under left-reversing with respect to $\widetilde{\theta}$.

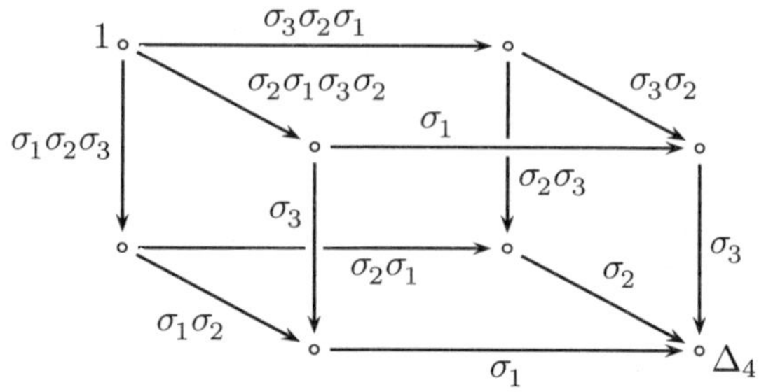

Figure 27. The cell $[[\sigma_1, \sigma_2, \sigma_3]]$ of B_4^+: start from three arrows labeled σ_1, σ_2, σ_3 pointing to the left-lcm, here Δ_4, and construct the cube by completing the faces using the left-complement operation, that is, applying left-reversing associated with the left-lcm witness on $\mathrm{Div}(\Delta_4)$.

We now define boundary maps. The above cube construction should make the idea obvious: for C an n-cell, we define $\partial_n C$ to be the $(n-1)$-chain obtained by enumerating the $(n-1)$-faces of the n-cube associated with C, which are $2n$ in number, with a sign corresponding to their orientation, and taking into account the vertex they start from. In order to handle such enumerations, we extend our notations.

Notation 3.29. (i) For $s_1, ..., s_n$ in $\mathcal{S} \cup \{1_{\mathcal{C}}\}$ (in any order), we write $[[s_1, ..., s_n]]$ for $\varepsilon(\pi)[[s_{\pi(1)}, ..., s_{\pi(n)}]]$ if the elements s_i are not equal to 1, they are pairwise distinct, $(s_{\pi(1)}, ..., s_{\pi(n)})$ is their $<$-increasing enumeration, and $\varepsilon(\pi)$ is the signature of π. In all other cases, $[[s_1, ..., s_n]]$ is defined to be $0_{\mathcal{Z}_n}$.

(ii) For C a cell, say $C = [[s_1, ..., s_n]]$, and s an element of $\mathcal{S}$, we denote by C/s the sequence $(s_1/s, ..., s_n/s)$ and by s/C the element $s/\widetilde{\Delta}_C$. We write C^i (*resp.* $C^{i,j}$) for $[[s_1, ..., \widehat{s_i}, ..., s_n]]$ (*resp.* $[[s_1, ..., \widehat{s_i}, ..., \widehat{s_j}, ..., s_n]]$).

Notation 3.29(i) enables us to freely use the notation $[[s_1, ..., s_n]]$ when $s_1, ..., s_n$ are arbitrary elements of $\mathcal{S} \cup \{1_{\mathcal{C}}\}$ without distinguishing degenerate cells. Note that such an extension is needed in order to define the notation C/s as above since, in general, $s_i \neq s_j$ need not imply $s_i/s \neq s_j/s$, and $s_i < s_j$ need not imply $s_i/s < s_j/s$.

Definition 3.30 (differential). For $n \geqslant 1$ and $C = [[s_1, ..., s_n]]$, we put

$$\partial_n C = \sum_{i=1}^{n} (-1)^i (C^i/s_i) - \sum_{i=1}^{n} (-1)^i (s_i/C^i) C^i, \tag{3.31}$$

and extend ∂_n to $\mathcal{Z}_n$ by $\mathbb{Z}\mathcal{C}$-linearity; we define $\partial_0 : \mathcal{Z}_0 \to \mathbb{Z}$ by $\partial_0[[\,]] = 1$.

So, in low degrees, we find (see Figure 28)

$$\partial_1[[s]] = s[[\,]] - [[\,]], \quad \partial_2[[s,t]] = [[s/t]] + (s/t)[[t]] - [[t/s]] - (t/s)[[s]].$$

Similarly, in the case of the monoid B_4^+, we read on Figure 27 the equality

$$\begin{aligned}\partial_3[[\sigma_1, \sigma_2, \sigma_3]] = &-[[\sigma_1\sigma_2, \sigma_3]] + [[\sigma_2\sigma_1, \sigma_2\sigma_3]] - [[\sigma_1, \sigma_3\sigma_2]]\\ &+ \sigma_3\sigma_2\sigma_1[[\sigma_2, \sigma_3]] - \sigma_2\sigma_1\sigma_3\sigma_2[[\sigma_1, \sigma_3]] + \sigma_1\sigma_2\sigma_3[[\sigma_1, \sigma_2]].\end{aligned}$$

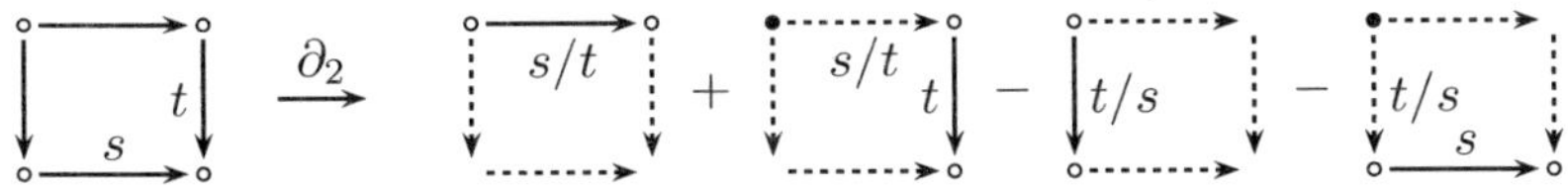

$$\partial_2[[s,t]] \quad = \quad [[s/t]] \;+\; (s/t)[[t]] \;-\; [[t/s]] \;-\; (t/s)[[s]]$$

Figure 28. The boundary operator ∂_2

Lemma 3.32. *The module $(\mathcal{Z}_*, \partial_*)$ is a complex: $\partial_{n-1}\partial_n = 0$ holds for $n \geqslant 1$.*

Proof. First, we have $\partial_1[[s]] = s[[\,]] - [[\,]]$, hence $\partial_0\partial_1[[s]] = s \cdot 1 - 1_x \cdot 1 = 0$ for s in $\mathcal{C}(x,y)$. Assume now $n \geqslant 2$. For $C = [[s_1, \ldots, s_n]]$, we obtain

$$\begin{aligned}
\partial_{n-1}\partial_n C &= \sum\nolimits_i (-1)^i \partial_{n-1}(C^i/s_i) - \sum\nolimits_i (-1)^i (s_i/C^i)\partial_{n-1}C^i \\
&= \sum\nolimits_{i\neq j} (-1)^{i+j+e(i,j)}((C^{i,j}/s_i)/(s_j/s_i)) \\
&\qquad - \sum\nolimits_{i\neq j} (-1)^{i+j+e(j,i)}((s_j/s_i)/(C^{i,j}/s_i))(C^{i,j}/s_i) \\
&\qquad - \sum\nolimits_{i\neq j} (-1)^{i+j+e(j,i)}(s_i/C^i)(C^{i,j}/s_j) \\
&\qquad + \sum\nolimits_{i\neq j} (-1)^{i+j+e(i,j)}(s_i/C^i)(s_j/C^{i,j})C^{i,j},
\end{aligned} \tag{3.33}$$

with $e(i,j) = +1$ for $i < j$, and $e(i,j) = 0$ otherwise. Applying the left-counterpart of (II.2.14) in Corollary II.2.13 (iterated complement), we obtain $(C^{i,j}/s_i)/(s_j/s_i) = C^{i,j}/\widetilde{\Delta}_{s_i,s_j}$, where s_i and s_j play symmetric roles, and the first sum in (3.33) becomes $\sum_{i\neq j}(-1)^{i+j+e(i,j)}C^{i,j}/\widetilde{\Delta}_{s_i,s_j}$, in which each factor $C^{i,j}/\widetilde{\Delta}_{s_i,s_j}$ appears twice, with coefficients $(-1)^{i+j}$ and $(-1)^{i+j+1}$ respectively, so the sum vanishes.

Applying (II.2.14) to s_j, s_i, and $C^{i,j}$ similarly gives $(s_j/s_i)/(C^{i,j}/s_i) = s_j/C^j$. It follows that the second and the third sum in (3.33) contain the same factors, but, as $e(i,j) + e(j,i) = 1$ always holds, the signs are opposite, and the global sum is 0.

Finally, applying (the left counterpart of) (II.2.14) to s_i, s_j, and the left-lcm of $C^{i,j}$ gives $(s_i/C^i)(s_j/C^{i,j}) = \widetilde{\Delta}_{s_i,s_j}/C^{i,j}$, in which s_i and s_j play symmetric roles. So, as in the case of the first sum, every factor in the fourth sum appears twice with opposite signs, and the sum vanishes.

Observe that the case of null factors is not a problem, as we always have $1_y/s = 1_x$ and $s/1_y = s$ for g in $\mathcal{C}(x,y)$, so Formula (3.31) is true for degenerate cells. □

Remark 3.34. The complex $(\mathcal{Z}_*, \partial_*)$ is a cubical complex [51]: instead of completing the computation in the above proof, one could check instead that the module is cubical, from where $\partial_*^2 = 0$ would follow directly. Maybe conceptually more elegant, this approach does not change the needed amount of computation relying on the rules for the left-complement operation.

It is convenient in the sequel to extend the cell notation to the case when the first entry is a path w, thus considering cells of the form $[[w, s_1, \ldots, s_n]]$. Abusing notation, the latter will be denoted by $[[w, C]]$ in the case $C = [[s_1, \ldots, s_n]]$. We recall that, for an $\mathcal{S}$-word w, we use $[w]$ for the element of $\mathcal{C}$ represented by w, that is, the evaluation of w in $\mathcal{C}$.

Definition 3.35 (extended chain). For w an $\mathcal{S}$-path and C in $\mathcal{S}^{[n]}$ sharing the same target, the $(n+1)$-chain $[\![w, C]\!]$ is defined inductively by

$$(3.36)\qquad [\![w, C]\!] = \begin{cases} 0_{\mathcal{Z}_{n+1}} & \text{for } w = \varepsilon_y, \\ [\![v, C/t]\!] + [v]/\widetilde{\Delta}_{C/t}[\![t, C]\!] & \text{for } w = vt \text{ with } t \text{ in } \mathcal{S}. \end{cases}$$

If w has length 1, that is, if v is empty, the inductive clause of (3.36) gives $[\![v(C/t)]\!] = 0$ and $[v]/\widetilde{\Delta}_{C/t} = 1.$, so our current definition of $[\![w, C]\!]$ is compatible with the previous one. Relation (3.36) is coherent with the geometrical intuition that the cell $[\![w, C]\!]$ is associated with a $(n+1)$-parallelotope computing the left-lcm of $[w]$ and $\widetilde{\Delta}_C$ using left-reversing: to compute the left-lcm of $[v]t$ and $\widetilde{\Delta}_C$, we first compute the left-lcm of t and $\widetilde{\Delta}_C$, and then append on the left the left-lcm of $[v]$ and the complement of $\widetilde{\Delta}_C$ in t, which is $\Delta_{C,t}$. The rightmost cell does not start from 1, but from $[v]/\widetilde{\Delta}_{C/t}$, see Figure 29.

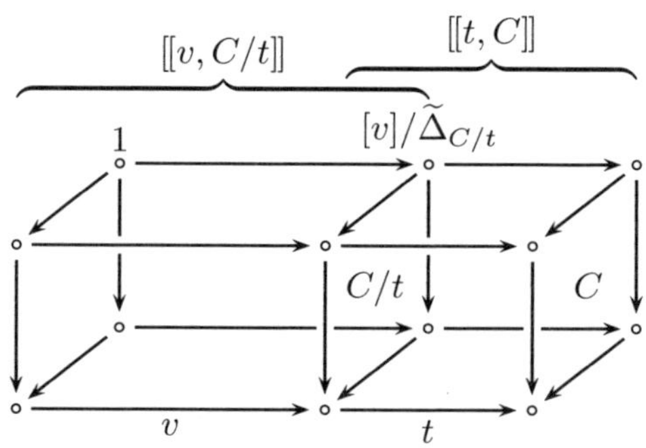

Figure 29. The parallelotope associated with the extended cell $[\![wt, C]\!]$: the concatenation of the parallelotope $[\![v, C/t]\!]$ and the cube $[\![t, C]\!]$ translated by $[v]/\widetilde{\Delta}_{C/t}$.

An easy induction gives, for $s_1, \ldots, s_p$ in $\mathcal{S}$, the simple expression

$$(3.37)\qquad [\![s_1|\cdots|s_p]\!] = \sum_{i=1}^{p} [s_1 \cdots s_{i-1}][\![s_i]\!].$$

Also (3.31) can be rewritten as $\partial_2[\![s, t]\!] = [\![(s/t)\, t]\!] - [\![(t/s)\, s]\!]$, according to the intuition that ∂_2 enumerates the boundary of a square. We shall give below a computational formula that extends (3.31) and describes the boundary of the parallelotope associated with $[\![w, C]\!]$ taking into account the specific role of the w-labeled edge. In order to state the formula, we first extend the definition of the left-complement operation to paths.

Notation 3.38 (operation /). (See Figure 30.) For w an $\mathcal{S}^\sharp$-path and s in $\mathcal{S}^\sharp$ sharing the same target, we write w/s for the path inductively defined by the rules $\varepsilon_y/s = s$ and $(v|t)/s = v/(s/t)|t/s$.

The definition of the path w/s, which may look complicated, is natural as the induction clause corresponds to the formula for an iterated left-lcm: thus, in particular, we always have $[w/s] = [w]/s$, that is, w/s represents the left-complement of the element represented by w and s.

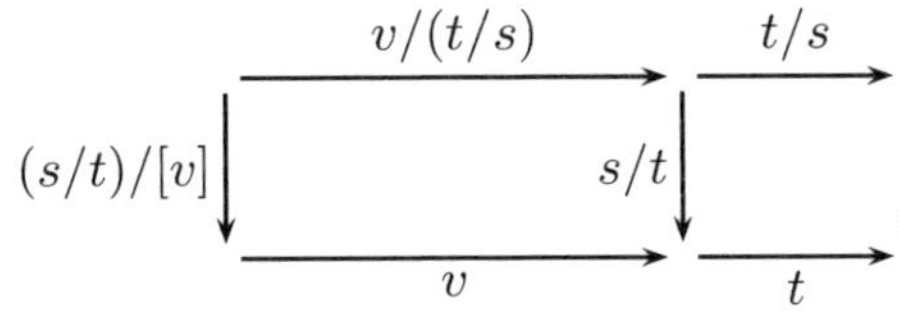

Figure 30. Inductive definition of the path w/s.

Lemma 3.39. *For every path w in $\mathcal{S}^*(x,y)$, we have $\partial_1[\![w]\!] = -[\![\,]\!] + [w]\,[\![\,]\!]$ and, for $n \geqslant 1$ and $C = [\![s_1, ..., s_n]\!]$,*

$$\partial_{n+1}[\![w, C]\!] = -C/[w] - \sum(-1)^i[\![w/s_i, C^i/s_i]\!] + \sum(-1)^i s_i/\widetilde{\Delta}_{[w]\cup C^i}[\![w, C^i]\!] + ([w]/\widetilde{\Delta}_C)C.$$

We skip the proof, which is entirely similar to that of Lemma 3.32, based on the formula for iterated lcm.

Our aim is now to exhibit a contracting homotopy for the complex $(\mathcal{Z}_*, \partial_*)$, that is, a family of $\mathbb{Z}$-linear maps $\sigma_n : \mathcal{Z}_n \to \mathcal{Z}_{n+1}$ satisfying $\partial_{n+1}\sigma_n + \sigma_{n-1}\partial_n = \mathrm{id}_{\mathcal{Z}_n}$ for each degree n. We shall do it using the $\mathcal{S}$-normal form.

Notation 3.40 (normal decomposition). If $\mathcal{S}$ is a Garside family in a category $\mathcal{C}$ that has no nontrivial invertible element, we write $\mathrm{NF}(g)$ for the (unique) strict $\mathcal{S}$-normal decomposition of an element g of $\mathcal{C}$.

The point for our construction is that, when $\mathcal{C}$ is a left-Ore category that admits unique left-lcms, Proposition 2.44—in fact just the third domino rule—implies the equality

$$\mathrm{NF}(g) = \widetilde{\theta}^*(\mathrm{NF}(gs), s), \tag{3.41}$$

that is, the $\mathcal{S}$-normal decomposition of g is obtained from that of gs by left-reversing the signed word $\mathrm{NF}(gs)|\overline{s}$.

The geometric intuition for constructing a contracting homotopy is then simple: as the chain gC represents the cube C with origin translated to g, we shall define $\sigma_n(gC)$ to be an $(n+1)$-parallelotope whose terminal face (the one with edges in $\mathcal{S}$—and not in $\mathcal{S}^*$—that does not contain the initial vertex) is C starting at g. To specify this cell, we have to describe its $n+1$ terminal edges: n of them are the elements of C; the last one must force the main diagonal to be $g\widetilde{\Delta}_C$: the natural choice is to take the normal form of $g\widetilde{\Delta}_C$, which guarantees in addition that all labels in the initial face lie in $1_{\mathcal{C}}$.

Definition 3.42 (section). (See Figure 31.) For $n \geqslant 0$, the $\mathbb{Z}$-linear map σ_n from $\mathcal{Z}_n$ to $\mathcal{Z}_{n+1}$ is defined for g in $\mathcal{C}$ by

$$\sigma_n(gC) = [\![\mathrm{NF}(g\widetilde{\Delta}_C), C]\!]; \tag{3.43}$$

the $\mathbb{Z}$-linear map σ_{-1} from $\mathbb{Z}$ to $\mathcal{Z}_0$ is defined by $\sigma_{-1}(1) = [\![\,]\!]$.

For g in $\mathcal{C}$ and s in $\mathcal{S}$, we have in particular

$$\sigma_0(g[\![\,]\!]) = [\![\mathrm{NF}(g)]\!] \quad \text{and} \quad \sigma_1(g[\![s]\!]) = [\![\mathrm{NF}(gs), s]\!]. \tag{3.44}$$

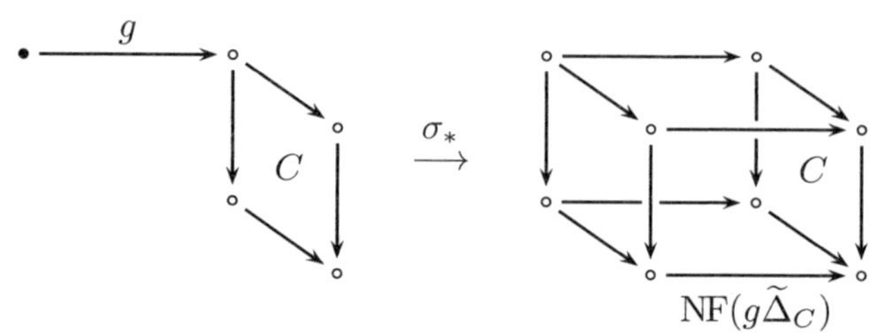

Figure 31. Definition of the contracting homotopy σ_* by filling the parallelotope based on g and C.

Proposition 3.45 (resolution I). *If $\mathcal{S}$ is a strong Garside family in a left-cancellative category $\mathcal{C}$ that admits unique left-lcms, the associated complex $(\mathcal{Z}_*, \partial_*)$ is a resolution of the trivial $\mathbb{Z}\mathcal{C}$-module $\mathbb{Z}$ by free $\mathbb{Z}\mathcal{C}$-modules.*

Proof. We establish the equality $\partial_{n+1}\sigma_n + \sigma_{n-1}\partial_n = \mathrm{id}_{\mathcal{Z}_n}$ for every $n \geqslant 0$. Assume first $n = 0$, and $g \in \mathcal{C}$. Let $w = \mathrm{NF}(g)$. We find $\sigma_0(g[\![\,]\!]) = [\![w]\!]$, hence $\partial_1\sigma_0(g[\![\,]\!]) = \partial_1[\![w]\!] = -[\![\,]\!] + g[\![\,]\!]$, and, on the other hand, $\partial_0(g[\![\,]\!]) = g \cdot 1 = 1$, hence $\sigma_{-1}\partial_0(g[\![\,]\!]) = [\![\,]\!]$, and $(\partial_1\sigma_0 + \sigma_{-1}\partial_0)(g[\![\,]\!]) = g[\![\,]\!]$.

Assume now $n \geqslant 1$. Let $w = \mathrm{NF}(g\widetilde{\Delta}_C)$ with $C = [\![s_1, ..., s_n]\!]$. Applying the definition of σ_n and Lemma 3.39, we obtain

$$\partial_{n+1}\sigma_n(gC) = -C/[w] - \sum(-1)^i[\![(w/s_i), (C^i/s_i)]\!] + \sum(-1)^i(s_i/\widetilde{\Delta}_{[w]\cup C^i})[\![w, C^i]\!] + ([w]/\widetilde{\Delta}_C)C.$$

By construction, each entry s_i right-divides $g\widetilde{\Delta}_C$, that is, of $[w]$, so $C/[w]$ is a degenerate cell of the form $[\![1_x, ..., 1_x]\!]$. At the other end, we find $[w]/\widetilde{\Delta}_C = (g\widetilde{\Delta}_C)/\widetilde{\Delta}_C = g$. Then s_i is a right-divisor of $[w]$, so we have $s_i/\widetilde{\Delta}_{[w]\cup C^i} = 1.$, and so

$$\partial_{n+1}\sigma_n(gC) = -\sum(-1)^i[\![w/s_i, C^i/s_i]\!] + \sum(-1)^i[\![w, C^i]\!] + gC.$$

On the other hand, we have by definition

$$\partial_n(gC) = \sum_i(-1)^i g[\![C^i/s_i]\!] - \sum_i(-1)^i g(s_i/\widetilde{\Delta}_{C^i})[\![C^i]\!].$$

Now we have $g\widetilde{\Delta}_{C^i/s_i} \cdot s_i = g\widetilde{\Delta}_C$, which, by (3.41), implies that the $\mathcal{S}$-normal form of $g\widetilde{\Delta}_{C^i/s_i}$ is w/s_i. Then $g(s_i/\widetilde{\Delta}_{[w]\cup C^i})\widetilde{\Delta}_{C^i}$ is equal to $g\widetilde{\Delta}_C$, and, therefore, its normal form is w. Applying the definition of σ_{n-1}, we deduce

$$\sigma_{n-1}\partial_n(gC) = \sum(-1)^i[\![w/s_i, C^i/s_i]\!] - \sum(-1)^i[\![w, C^i]\!],$$

and, finally, $(\partial_{n+1}\sigma_n + \sigma_{n-1}\partial_n)(gC) = gC$. □

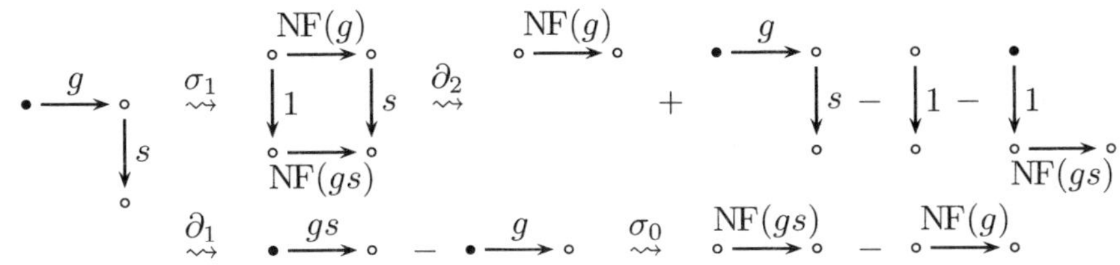

Figure 32. Comparing $\partial_2\sigma_1$ and $\partial_1\sigma_0$: only the trivial left face and $g[\![s]\!]$ remain when the sum is performed.

By definition, $\mathcal{S}^{[n]}$ is a basis for the degree n module $\mathcal{Z}_n$ in the above resolution of $\mathbb{Z}$ by free $\mathbb{Z}\mathcal{C}$-modules. If $\mathcal{S}$ is finite, $\mathcal{S}^{[n]}$ is empty for n larger than the cardinality of $\mathcal{S}$, and the resolution is finite. Using Lemma 3.27 and standard arguments, we deduce:

Corollary 3.46 (homology). *If $\mathcal{C}$ is an Ore category that admits unique left-lcms and has a finite Garside family, the groupoid $\mathcal{E}nv(\mathcal{C})$ is of* FL *type. For every n, we have $H_n(\mathcal{E}nv(\mathcal{C}),\mathbb{Z}) = H_n(\mathcal{C},\mathbb{Z}) = \operatorname{Ker}\mathrm{d}_n/\operatorname{Im}\mathrm{d}_{n+1}$, where d_n is the $\mathbb{Z}$-linear map on $\mathcal{Z}_n$ such that $\mathrm{d}_n C$ is obtained from $\partial_n C$ by collapsing all $\mathcal{C}$-coefficients to* 1.

We recall that a group(oid) is said to be of FL type if $\mathbb{Z}$ admits a finite resolution by free $\mathbb{Z}\mathcal{C}$-modules. Corollary 3.46 applies in particular to every Garside monoid.

Example 3.47 (homology). Consider once more the 3-strand braid group B_3. Then B_3 is a group of fractions for the dual braid monoid B_3^{+*} of Reference Structure 3, page 10, which admits the 4-element family $\mathcal{S} = \{\mathtt{a},\mathtt{b},\mathtt{c},\Delta^*\}$ as a strong Garside family (we write $\mathtt{a},\mathtt{b},\mathtt{c}$ for σ_1, σ_2, and $\sigma_1\sigma_2\sigma_1^{-1}$). We fix the order $\mathtt{a} < \mathtt{b} < \mathtt{c} < \Delta^*$. First, we find $\partial_1[\![\mathtt{a}]\!] = (a-1)[\![\,]\!]$, hence $\mathrm{d}_1[\![\mathtt{a}]\!] = 0$. The result is similar for all 1-cells, so $\operatorname{Ker}\mathrm{d}_1$ is generated by $[\![\mathtt{a}]\!]$, $[\![\mathtt{b}]\!]$, $[\![\mathtt{c}]\!]$, and $[\![\Delta^*]\!]$. Then, we find $\partial_2[\![\mathtt{a},\mathtt{b}]\!] = [\![\mathtt{a}]\!]+\mathtt{a}[\![\mathtt{b}]\!]-[\![\mathtt{c}]\!]-\mathtt{c}[\![\mathtt{a}]\!]$, whence $\mathrm{d}_2[\![\mathtt{a},\mathtt{b}]\!] = [\![\mathtt{b}]\!]-[\![\mathtt{c}]\!]$. Arguing similarly, one finds that $\operatorname{Im}\mathrm{d}_2$ is generated by the images of $[\![\mathtt{a},\mathtt{b}]\!]$, $[\![\mathtt{a},\mathtt{c}]\!]$, and $[\![\mathtt{b},\Delta^*]\!]$, namely $[\![\mathtt{b}]\!]-[\![\mathtt{a}]\!]$, $[\![\mathtt{c}]\!]-[\![\mathtt{b}]\!]$, and $[\![\Delta^*]\!]-[\![\mathtt{b}]\!]-[\![\mathtt{a}]\!]$, and we deduce $H_1(B_3,\mathbb{Z}) = H_1(B_3^{+*},\mathbb{Z}) = \operatorname{Ker}\mathrm{d}_1/\operatorname{Im}\mathrm{d}_2 = \mathbb{Z}$.

Similarly, one checks that $\operatorname{Ker}\mathrm{d}_2$ is generated by $[\![\mathtt{b},\mathtt{c}]\!]+[\![\mathtt{a},\mathtt{b}]\!]-[\![\mathtt{a},\mathtt{c}]\!]$, $[\![\mathtt{a},\Delta^*]\!]-[\![\mathtt{b},\Delta^*]\!]-[\![\mathtt{a},\mathtt{b}]\!]$, and $[\![\mathtt{c},\Delta^*]\!]-[\![\mathtt{a},\Delta^*]\!]+[\![\mathtt{a},c]\!]$. On the other hand, one finds $\mathrm{d}_3[\![\mathtt{a},\mathtt{b},\mathtt{c}]\!] = [\![\mathtt{b},\mathtt{c}]\!]-[\![\mathtt{a},\mathtt{c}]\!]+[\![\mathtt{a},\mathtt{b}]\!]$, and similar equalities for $\mathrm{d}_3[\![\mathtt{a},\mathtt{b},\Delta^*]\!]$ and its counterparts. This shows that $\operatorname{Im}\mathrm{d}_3$ is generated by $[\![\mathtt{b},\mathtt{c}]\!]+[\![\mathtt{a},\mathtt{b}]\!]-[\![\mathtt{a},\mathtt{c}]\!]$, $[\![\mathtt{a},\Delta^*]\!]-[\![\mathtt{b},\Delta^*]\!]-[\![\mathtt{a},\mathtt{b}]\!]$, and $[\![\mathtt{c},\Delta^*]\!]-[\![\mathtt{a},\Delta^*]\!]+[\![\mathtt{a},\mathtt{c}]\!]$, so it coincides with $\operatorname{Ker}\mathrm{d}_2$, and $H_2(B_3,\mathbb{Z})$ is trivial.

In the same way, one checks that $[\![\mathtt{a},\mathtt{b},\mathtt{c}]\!]-[\![\mathtt{a},\mathtt{b},\Delta^*]\!]-[\![\mathtt{b},\mathtt{c},\Delta^*]\!]+[\![\mathtt{a},\mathtt{c},\Delta^*]\!]$ generates $\operatorname{Ker}\mathrm{d}_3$ and that $\mathrm{d}_4[\![\mathtt{a},\mathtt{b},\mathtt{c},\Delta^*]\!] = [\![\mathtt{b},\mathtt{c},\Delta^*]\!]-[\![\mathtt{a},\mathtt{c},\Delta^*]\!]+[\![\mathtt{a},\mathtt{c},\Delta^*]\!]-[\![\mathtt{a},\mathtt{b},\mathtt{c}]\!]$ holds, implying $\operatorname{Im}\mathrm{d}_4 = \operatorname{Ker}\mathrm{d}_3$. So $H_3(B_3,\mathbb{Z})$ is trivial (as will become obvious below).

Remark 3.48. According to Proposition 2.34, the existence of a strong Garside family in a category $\mathcal{C}$ entails that $\mathcal{C}$ admits left-lcms. However, the computation of Proposition 3.45 (as well as that of Proposition 3.52 below) remains valid when $\mathcal{C}$ is only assumed to admit conditional left-lcms whenever $\mathcal{S}$ is "conditionally strong", meaning that, in Definition 2.29, one restricts to pairs s,t that have a common left-multiple.

In general, the resolution of Proposition 3.45 is far from minimal. We conclude this section by briefly explaining how to obtain a shorter resolution by decomposing each n-cube into $n!$ n-simplices. Our aim is to extract from the complex $(\mathcal{Z}_*, \partial_*)$ a subcomplex that is still a resolution of $\mathbb{Z}$. The point is to distinguish those cells that are decreasing with respect to right-divisibility. In order to make the definitions meaningful, we assume in the sequel that the linear order on $\mathcal{S}$ used to enumerate the cells is chosen so that $s < t$ holds whenever t is a proper right-divisor of s.

Definition 3.49 (descending cell). In the context of Definition 3.28, we say that an n-cell $[\![s_1, ..., s_n]\!]$ is *descending* if s_{i+1} is a proper right-divisor of s_i for each i. The submodule of $\mathcal{Z}_n$ generated by descending n-cells is denoted by $\mathcal{Z}'_n$.

According to our intuition that the cell $[\![s_1, ..., s_n]\!]$ is associated with an n-cube representing the computation of the left-lcm of s_1, ...,s_n, a descending n-cell is associated with a special n-cube in which many edges have label 1, and it is accurately associated with an n-simplex, as shown in Figure 33.

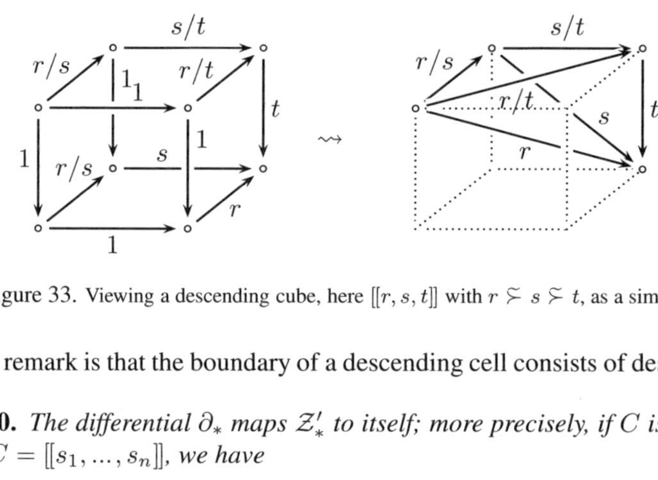

Figure 33. Viewing a descending cube, here $[\![r, s, t]\!]$ with $r \succneqq s \succneqq t$, as a simplex.

The first remark is that the boundary of a descending cell consists of descending cells.

Lemma 3.50. *The differential ∂_* maps $\mathcal{Z}'_*$ to itself; more precisely, if C is a descending n-cell, say $C = [\![s_1, ..., s_n]\!]$, we have*

$$\partial_n[C] = (s_1/s_2)C^1 - \sum\nolimits_{i=2}^{n}(-1)^i C^i + (-1)^n C^n/s_n. \tag{3.51}$$

The proof is an easy verification. So it makes sense to consider the restriction ∂'_* of ∂_* to $\mathcal{Z}'_*$, yielding a new complex.

Proposition 3.52 (resolution II). *If $\mathcal{S}$ is a strong Garside family in a left-cancellative category $\mathcal{C}$ that admits unique left-lcms, the associated complex $(\mathcal{Z}'_*, \partial'_*)$ is a resolution of the trivial $\mathbb{Z}\mathcal{C}$-module $\mathbb{Z}$ by free $\mathbb{Z}\mathcal{C}$-modules.*

In order to prove Proposition 3.52, we shall construct a contracting homotopy. The section σ_* considered in Proposition 3.45 cannot be used, as σ_n does not map $\mathcal{Z}'_n$ to $\mathcal{Z}'_{n+1}$ in general. However, it is easy to construct the desired section by introducing a convenient $\mathbb{Z}\mathcal{C}$-linear map from $\mathcal{Z}_n$ to $\mathcal{Z}'_n$, corresponding to partitioning each n-cube into the union of $n!$ disjoint n-simplexes. Starting from an arbitrary n-cell $[[s_1, \dots, s_n]]$, one can obtain a descending n-cell by taking left-lcm's:

Definition 3.53 (flag cell). For $s_1, \dots, s_n$ in $\mathcal{S}$, we put

$$\langle\!\langle s_1, \dots, s_n \rangle\!\rangle = [[\widetilde{\Delta}_{\{s_1,\dots,s_n\}}, \widetilde{\Delta}_{\{s_2,\dots,s_n\}}, \dots, \widetilde{\Delta}_{\{s_{n-1},s_n\}}, s_n]]. \tag{3.54}$$

If π is a permutation of $\{1, \dots, n\}$, and C is an n-cell, say $C = [[s_1, \dots, s_n]]$, we denote by C^π the cell $\langle\!\langle s_{\pi(1)}, \dots, s_{\pi(n)} \rangle\!\rangle$.

Every cell $\langle\!\langle s_1, \dots, s_n \rangle\!\rangle$ is descending, and it coincides with $[[s_1, \dots, s_n]]$ if and only if the latter is descending. So, for each n-cell C in $\mathcal{Z}_n$, we have a family of $n!$ descending n-cells C^π in $\mathcal{Z}'_n$. The associated simplices constitute a partition of the cube associated with C. For instance, in dimension 2, we have a partition of the square $[[s, t]]$ into the two triangles $\langle\!\langle \widetilde{\Delta}_{s,t}, t \rangle\!\rangle$ and $\langle\!\langle \widetilde{\Delta}_{s,t}, s \rangle\!\rangle$. Similarly, in dimension 3, we have the decomposition of the cube $[[r, s, t]]$ into the six tetrahedra shown in Figure 34.

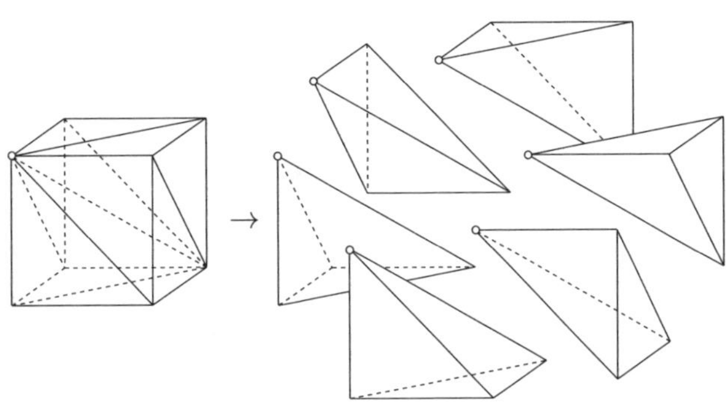

Figure 34. Decomposition of a cube into six tetrahedra.

Proof of Proposition 3.52 (*sketch*). For every n, define a $\mathbb{Z}M$-linear map $f_n : \mathcal{Z}_n \to \mathcal{Z}'_n$ by $f_n(C) = \sum_{\pi \in \mathfrak{S}_n} \varepsilon(\pi) C^\pi$. Then one easily checks that f_n is the identity on $\mathcal{Z}'_n$, and the point is to prove that, for every n-cell C, the equality $\partial_n f_n(C) = f_{n-1}(\partial_n C)$, which, for $n = 3$, essentially amounts to labeling the edges in Figure 34. So f_* provides a retraction from $\mathcal{Z}_*$ to $\mathcal{Z}'_*$. Now, we obtain the expected contracting homotopy by defining $\sigma'_n : \mathcal{Z}'_n \to \mathcal{Z}'_{n+1}$ by $\sigma'_n = f_{n+1}\sigma_n$. It follows that $(\mathcal{Z}'_*, \partial'_*)$ is an exact complex. □

The advantages of restricting to descending cells are clear: first, the length of the resolution, and the dimensions of the modules are drastically reduced; second, the differential is now given by formula (3.51), which has $n+1$ degree n terms only, instead of the $2n$ terms of formula (3.31). In good cases, we deduce an upper bound for the homological dimension.

Corollary 3.55 (dimension). *If $\mathcal{S}$ is a strong Garside family in a left-Ore category $\mathcal{C}$ that admits unique left-lcms and the height of every element of $\mathcal{S}$ is bounded above by some constant N, then the (co)homological dimension of $\mathcal{E}nv(\mathcal{C})$ is at most N.*

Proof. By construction, the maximal length of a nontrivial descending n-cell is the maximal length of a decomposition of an element of $\mathcal{S}$ as a product of nontrivial elements. By Proposition II.2.48 (finite height), this length is bounded by the height (Definition II.2.43). □

Example 3.56 (dimension). In the case of B_3, considered as the group of fractions of the dual braid monoid B_3^{+*} as in Example 3.47, we have a homogeneous presentation and all elements of the Garside family $\{\mathtt{a}, \mathtt{b}, \mathtt{c}, \Delta^*\}$ have height at most 2. So Corollary 3.55 implies that $H_3(B_3, \mathbb{Z})$ is trivial and the homological dimension of B_3 is at most 2.

3.5 Word Problem

We conclude with a few observations about Word Problems and their algorithmic complexity: as $\mathcal{S}$-normal decompositions are essentially unique and effectively computable, they provide solutions.

We recall that the Word Problem for a category $\mathcal{C}$ with respect to a generating system $\mathcal{S}$—which could be more adequately called the *Path Problem*—addresses the existence of an algorithm that, starting with (a proper encoding of) two $\mathcal{S}$-paths u, v, determines whether u and v represent the same element of $\mathcal{C}$. If $\mathcal{C}$ is a groupoid, u and v represent the same element if and only if $\overline{u}v$ represents an identity element and it is customary to state the Word Problem as the one-parameter problem of determining whether a path represents an identity element.

Definition 3.57 ($=^\times$-map). If $\mathcal{S}$ is a subfamily of a left-cancellative category $\mathcal{C}$, an *$=^\times$-map* for $\mathcal{S}^\sharp$ is a partial map E from $\mathcal{S}^\sharp \times \mathcal{S}^\sharp$ to $\mathcal{C}^\times$ such that $E(s,t)$ is defined whenever $s =^\times t$ holds, and one then has $sE(s,t) = t$.

Algorithm 3.58 (Word Problem, positive case).

Context: A left-cancellative category $\mathcal{C}$, a Garside subfamily $\mathcal{S}$ of $\mathcal{C}$, a $\square$-witness F and an $=^\times$-map E for $\mathcal{S}^\sharp$
Input: Two $\mathcal{S}^\sharp$-paths u, v
Output: YES if u and v represent the same element of $\mathcal{C}$, NO otherwise
1: compute an $\mathcal{S}$-normal decomposition $s_1|\cdots|s_p$ of $[u]$ using F and Algorithm 1.52
2: compute an $\mathcal{S}$-normal decomposition $t_1|\cdots|t_q$ of $[v]$ using F and Algorithm 1.52
3: **if** $p < q$ holds **then** put $s_i := 1_y$ for $p < i \leqslant q$ and u in $\mathcal{C}^*(\cdot, y)$
4: **if** $p > q$ holds **then** put $t_i := 1_y$ for $p < i \leqslant q$ and v in $\mathcal{C}^*(\cdot, y)$

put $\epsilon_0 := 1_x$ for u in $\mathcal{C}^*(x, \cdot)$
for i increasing from 1 to $\max(p, q)$ **do**
  if $E(s_i, \epsilon_{i-1}t_i)$ is defined **then**
    put $\epsilon_i := E(s_i, \epsilon_{i-1}t_i)$
  else
    return NO and exit
return YES

Proposition 3.59 (Word Problem, positive case). *Assume that $\mathcal{S}$ is a Garside family in a left-cancellative category $\mathcal{C}$, F is a $\square$-witness for $\mathcal{S}$, and E is an $=^{\times}$-map for $\mathcal{S}^\sharp$. Then Algorithm* 3.58 *solves the Word Problem for $\mathcal{C}$ with respect to $\mathcal{S}^\sharp$.*

Proof. By Proposition 1.53, Algorithm 1.52 always succeeds. By Proposition 1.25, $[u] = [v]$ holds in $\mathcal{C}$ if and only if $t_1|\cdots|t_q$ is a $\mathcal{C}^\times$-deformation of $s_1|\cdots|s_p$, hence if and only if, starting from $\epsilon_0 = 1_x$, there exist invertible elements $\epsilon_1, \epsilon_2, ...$ satisfying $\epsilon_{i-1}t_i = s_i\epsilon_i$ for every i, hence if and only Algorithm 3.58 returns YES. $\square$

We recall that a family is called *decidable* (or *recursive*) if there exists an algorithm that can be implemented on a Turing machine and decides whether an element of the reference structure (specified using a convenient encoding) belongs or not to that family, and that a map is called *computable* (or *recursive*) if there exists an algorithm that can be implemented on a Turing machine and, starting from (encodings of) the arguments, leads to (the encoding of) the value.

Corollary 3.60 (decidability). *If $\mathcal{S}$ is a Garside family in a left-cancellative category $\mathcal{C}$ and there exist computable $\square$- and $=^{\times}$-maps for $\mathcal{S}^\sharp$, the Word Problem for $\mathcal{C}$ with respect to $\mathcal{S}^\sharp$ is decidable. Moreover, if $\mathcal{S}^\sharp$ is finite, the Word Problem for $\mathcal{C}$ has a quadratic time complexity and a linear space complexity.*

Proof. Whenever the maps F and E are computable functions, Algorithm 1.52 is effective. In this case, Proposition 3.59 shows that the Word Problem for $\mathcal{C}$ with respect to $\mathcal{S}$ is decidable.

Assume moreover that $\mathcal{S}^\sharp$ is finite. Then there exist a finite $\square$-witness F and a finite $=^{\times}$-map E for $\mathcal{S}^\sharp$. Then F and E are computable in constant time and space. In this case, the time complexity of Algorithm 1.52 is proportional to the number of calls to F and E. For F, the number is quadratic in the length of the initial paths by Proposition 1.53. As for E, the number is linear in the length of the initial paths. Finally, the space complexity is linear as determining an $\mathcal{S}$-normal decomposition of an element initially specified by a length-ℓ path never requires to store more than ℓ elements of $\mathcal{S}^\sharp$ at a time. $\square$

We now consider left-Ore categories and the Word Problem for their enveloping groupoid. We recall from Section II.4 that left-reversing a signed $\mathcal{S}^\sharp$-path means recursively replacing subpaths of the form $s|\overline{t}$ with equivalent paths $\overline{t'}|s'$: as already explained in Subsection 2.5, this is what a short left-lcm witness enables one to do and, therefore, there is no problem in referring to left-reversing with respect to such a map.

Algorithm 3.61 (Word Problem, general case). (Figure 35)

Context: A left-Ore category $\mathcal{C}$ that admits left-lcms, a strong Garside subfamily $\mathcal{S}$ of $\mathcal{C}$, a $\square$-witness F, a short left-lcm witness $\widetilde{\theta}$ on $\mathcal{S}^\sharp$, and an $=^\times$-map E for $\mathcal{S}^\sharp$

Input: A signed $\mathcal{S}^\sharp$-path w

Output: YES if w represents an identity element in $\mathcal{E}nv(\mathcal{C})$, NO otherwise

1: use $\widetilde{\theta}$ to left-reverse w into a negative–positive path $\overline{v}u$

2: compare $[u]$ and $[v]$ using Algorithm 3.58 and return the answer

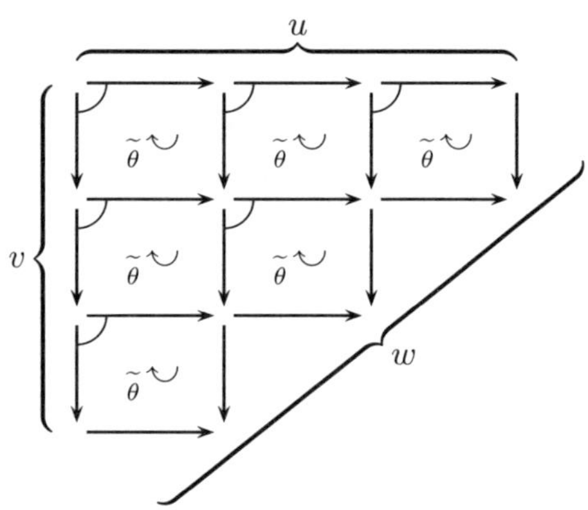

Figure 35. Algorithm 3.61: use the left-lcm witness $\widetilde{\theta}$ to left-reverse w into a negative–positive path.

Proposition 3.62 (Word Problem). *If $\mathcal{S}$ is a strong Garside family in a left-Ore category $\mathcal{C}$ that admits left-lcms, Algorithm* 3.61 *solves the Word Problem for $\mathcal{E}nv(\mathcal{C})$ with respect to $\mathcal{S}^\sharp$.*

Proof. The conjunction of $s' = \widetilde{\theta}(s,t)$ and $t' = \widetilde{\theta}(t,s)$ implies $ts^{-1} = s'^{-1}t'$ in $\mathcal{E}nv(\mathcal{C})$, hence w represents an identity element of $\mathcal{E}nv(\mathcal{C})$ if and only if $\overline{v}u$ does, that is, if u and v represent the same element of $\mathcal{E}nv(\mathcal{C})$. By Proposition II.3.11 (Ore's theorem), $\mathcal{C}$ embeds in $\mathcal{E}nv(\mathcal{C})$, so the latter condition is equivalent to u and v representing the same element of $\mathcal{C}$, hence to a positive answer of Algorithm 1.52 running on u and v. $\square$

Remark 3.63. The signed path $\overline{v}u$ occurring in Algorithm 3.61 need not be symmetric $\mathcal{S}$-normal in general. By Proposition 2.37, this happens if the initial path w is positive–negative. In other cases, we can say nothing: for instance, if s is a non-invertible element of $\mathcal{S}$, Algorithm 3.61 running on $\overline{s}|s$ leads (in zero left-reversing step) to the signed path $\overline{s}|s$, which is not symmetric $\mathcal{S}$-normal (s is not left-disjoint from itself). If we wish to obtain a symmetric $\mathcal{S}$-normal decomposition of the element represented by w, and $\mathcal{C}$ is an Ore (and not only left-Ore) category, we first use common right-multiples to right-reverse w into a positive–negative path, as done in Algorithm 3.61.

Corollary 3.64 (decidability of Word Problem). *If $\mathcal{S}$ is a strong Garside family in a left-Ore category $\mathcal{C}$ and there exist a computable $\square$-witness, a computable left-lcm witness on $\mathcal{S}^\sharp$, and a computable $=^\times$-maps for $\mathcal{S}^\sharp$, the Word Problem for $\mathcal{E}nv(\mathcal{C})$ with respect to $\mathcal{S}^\sharp$ is decidable. Moreover, if $\mathcal{S}^\sharp$ is finite, the Word Problem for $\mathcal{E}nv(\mathcal{C})$ has a quadratic time complexity and a linear space complexity.*

Proof. Whenever the maps $F, \widetilde{\theta}$, and E are computable, Algorithm 3.58 is effective, so Proposition 3.62 shows that the Word Problem for $\mathcal{C}$ with respect to $\mathcal{S}$ is decidable.

If $\mathcal{S}^\sharp$ is finite, then, as in the positive case, the maps F, $\widetilde{\theta}$, and E are finite, hence computable in constant time and space. The time complexity of Algorithm 3.61 is proportional to the number of calls to F, $\widetilde{\theta}$ and E. Assuming that the initial path contains p negative and q positive entries, one calls $\widetilde{\theta}$ at most pq times, and, then, running Algorithm 3.58 on u and v requires at most $O(\max(p,q)^2)$ steps. So the overall time complexity lies in $O((p+q)^2)$, hence is quadratic with respect to $\lg(w)$. As in the positive case, the space complexity is linear, since left-reversing a signed $\mathcal{S}$-path using a short complement does not increase its length. $\square$

Example 3.65 (Word Problem). Consider the braid group B_3 again (Reference Structure 2, page 5), and the Garside family $\{1, \mathtt{a}, \mathtt{b}, \mathtt{ab}, \mathtt{ba}, \Delta\}$ in the braid monoid B_3^+, where Δ stands for $\mathtt{aba}$. One easily computes the table below for the associated square witness and left-lcm witness (which are unique, as there is no nontrivial invertible element):

$F, \widetilde{\theta}$	1	a	b	ab	ba	Δ
1	$1\vert 1$, 1	a$\vert 1$, 1	b$\vert 1$, 1	ab$\vert 1$, 1	ba$\vert 1$, 1	$\Delta\vert 1$, 1
a	a$\vert 1$, a	a$\vert$a, 1	ab$\vert 1$, ba	a$\vert$ab, b	$\Delta\vert 1$, 1	$\Delta\vert$b, 1
b	b$\vert 1$, b	ba$\vert 1$, ab	b$\vert$b, 1	$\Delta\vert 1$, 1	b$\vert$ba, a	$\Delta\vert$a, 1
ab	ab$\vert 1$, ab	$\Delta\vert 1$, ba	ab$\vert$b, a	$\Delta\vert$b, 1	ab$\vert$ba, a	$\Delta\vert$ba, 1
ba	ba$\vert 1$, ba	ba$\vert$a, b	$\Delta\vert 1$, ab	ba$\vert$ab, a	$\Delta\vert$a, 1	$\Delta\vert$ab, 1
Δ	$\Delta\vert 1$, Δ	$\Delta\vert$a, ab	$\Delta\vert$b, ba	$\Delta\vert$ab, b	$\Delta\vert$ba, a	$\Delta\vert\Delta$, 1

For instance, we read that the normal decomposition of $\mathtt{ab}|\mathtt{a}$ is $\Delta|1$, and that the left-lcm of ab and a is $\mathtt{b}\cdot\mathtt{ab}$ and $\mathtt{ab}\cdot\mathtt{a}$. Consider the signed word $\mathtt{b}|\mathtt{a}|\overline{\mathtt{b}}|\overline{\mathtt{b}}|\mathtt{b}|\mathtt{a}$. Applying Step 1 of Algorithm 3.61, that is, left-reversing using $\widetilde{\theta}$, returns the negative–positive word $\overline{\mathtt{a}}|\overline{\mathtt{a}}|1|\mathtt{ba}|\mathtt{b}|\mathtt{a}$. Next, applying Step 2, that is, normalizing, to the positive words $\mathtt{a}|\mathtt{a}$ and $1|\mathtt{ba}|\mathtt{b}|\mathtt{a}$ returns $\mathtt{a}|\mathtt{a}$ and $\Delta|\mathtt{a}|1|1$, which do not coincide up to adding final 1s. We conclude that the initial word does not represent 1 in B_3.

Exercises

Exercise 27 (monoid $\mathbb{N}$). (i) Consider $S = \{n\}$ in the additive monoid $(\mathbb{N}, +)$. Show that the S-greedy paths of length two are the n pairs $0|0, 1|0, ..., n-1|0$, plus all pairs $n_1|n_2$ with $n_1 \geqslant n$. Deduce that, for $n \geqslant 1$, the only S-normal pair is $n|n$. (ii) Assume that S is a finite subset of $\mathbb{N}$ with maximal element n. Show that the S-normal paths of length-two are the pairs $n|q$ with $q \in S$, plus the pairs $p|0$ with $p \in S$ if 0 lies in S. (iii) Show that $\{0, 1, ..., n\}$ is a Garside family in $\mathbb{N}$, but $\{0, 1, 3\}$ is not. [Hint: Show that the $\{0, 1, 3\}$-normal pairs are $0|0$, $1|0$, $3|0$, $3|1$, and $3|3$, and conclude that 2 has no $\{0, 1, 3\}$-normal decomposition.]

Exercise 28 (invertible). Assume that $\mathcal{C}$ is a left-cancellative category and $\mathcal{S}$ is included in $\mathcal{C}$. Show that, if $g_1 \cdots g_p$ belongs to $\mathcal{S}$, then $g_1|\cdots|g_p$ being $\mathcal{S}$-greedy implies that $g_2, ..., g_p$ are invertible.

Exercise 29 (deformation). Assume that $\mathcal{C}$ is a left-cancellative category. Show that a path $g_1|\cdots|g_q$ is a $\mathcal{C}^\times$-deformation of $f_1|\cdots|f_p$ if and only if $g_1 \cdots g_i =^\times f_1 \cdots f_i$ holds for $1 \leqslant i \leqslant \max(p, q)$, the shorter path being extended by identity elements if needed.

Exercise 30 (dropping left-cancellativity). Assume that $\mathcal{C}$ is a category. Say that an element f is *pseudo-invertible* if it admits a *pseudo-inverse*, defined as f' such that there exist g, g' satisfying $gf = g'$ and $g'f' = g$. (i) Show that f is pseudo-invertible if and only if there exists g satisfying $gff' = g$. (ii) Show that, if $\mathcal{C}$ is left-cancellative, then a pseudo-inverse is an inverse. (iii) Declare that two elements g, g' are *pseudo-equivalent* if there exist pseudo-invertible elements f, f' satisfying $gf' = fg'$. Prove that, if $s_1|\cdots|s_p$ and $s'_1|\cdots|s'_{p'}$ are strongly $\mathcal{S}$-normal decompositions of some element g, in the sense of Remark 1.16, and $p \geqslant p'$ holds, then s_i is pseudo-equivalent to s'_i for $i \leqslant p$, and s'_i is pseudo-equivalent to 1_y for $p < i \leqslant p'$.

Exercise 31 (algebraic description of normalization). Assume that $\mathcal{S}$ is a Garside family in a left-cancellative category $\mathcal{C}$ with no nontrivial invertible element. Let F be the map from $(\mathcal{S} \cup \{1\})^{[2]}$ to itself that associates to each pair $s_1|s_2$ the unique $\mathcal{S}$-normal pair $t_1|t_2$ satisfying $t_1 t_2 = s_1 s_2$. (i) Show that, for every $s_1|s_2|s_3$ in $\mathcal{S}^{[3]}$, the $\mathcal{S}$-normal decomposition of $s_1 s_2 s_3$ is the image of $s_1|s_2|s_3$ under $F_2 F_1 F_2$, where F_i means F applied to the ith and $(i+1)$st entries. (ii) Extend the result to $\mathcal{S}^{[p]}$. [Hint: The result is $F_{p-1} \cdots F_2 F_1 \cdot \cdots \cdot F_{p-1} F_{p-2} \cdot F_{p-1}$.] (iii) Show that F is idempotent, that is, satisfies $F^2 = F$, and satisfies $(*)$ $F_2 F_1 F_2 F_1 = F_1 F_2 F_1 F_2 = F_2 F_1 F_2$. (iv) Show that $F_1 F_2 F_1 = F_2 F_1 F_2$ holds whenever the second domino rule is valid for $\mathcal{S}$ in $\mathcal{C}$. (v) Conversely, assume that $\mathcal{S}$ is a subfamily of an arbitrary category $\mathcal{C}$ and F is a partial map from $\mathcal{S}^2$ that is defined on length-two paths, is idempotent, and satisfies $(*)$. Show that F defines unique decompositions for the elements of $\mathcal{C}$.

Exercise 32 (left-disjoint). Assume that $\mathcal{C}$ is a left-cancellative category, and f, g are left-disjoint elements of $\mathcal{C}$. Show that, for every invertible element ϵ such that ϵf is defined, ϵf and ϵg are left-disjoint. [Hint: Use a direct argument or the fourth domino rule.]

Exercise 33 (left-disjoint). Assume that $\mathcal{C}$ is a left-cancellative category, f and g are left-disjoint elements of $\mathcal{C}$, and f left-divides g. Show that f is invertible.

Exercise 34 (normal decomposition). Give a direct argument for deriving Corollary 2.50 from Proposition 2.47 in the case when $\mathcal{S}$ is strong.

Exercise 35 (Garside base). (i) Let $\mathcal{G}$ be the category whose diagram is shown on the right, and let $S = \{\mathtt{a}, \mathtt{b}\}$. Show that $\mathcal{G}$ is a groupoid with nine elements, $\mathcal{S}$ is a Garside base in $\mathcal{G}$, the subcategory $\mathcal{C}$ of $\mathcal{G}$ generated by $\mathcal{S}$ contains no nontrivial invertible element, but $\mathcal{C}$ is not an Ore category. Conclusion?

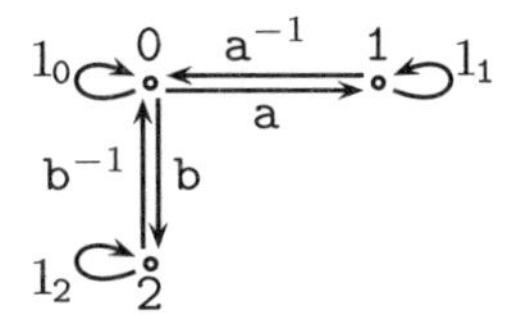

(ii) Let $\mathcal{G}$ be the category whose diagram is on the right, let $\mathcal{S} = \{\epsilon, \mathtt{a}\}$, and let $\mathcal{C}$ be the subcategory of $\mathcal{G}$ generated by $\mathcal{S}$. Show that $\mathcal{G}$ is a groupoid and every element of $\mathcal{G}$ admits a decomposition that is symmetric $\mathcal{S}$-normal in $\mathcal{C}$. Show that $\epsilon\mathtt{a}$ admits a symmetric $\mathcal{S}$-normal decomposition and no $\mathcal{S}$-normal decomposition. Is $\mathcal{S}$ a Garside family in $\mathcal{C}$? Conclusion?

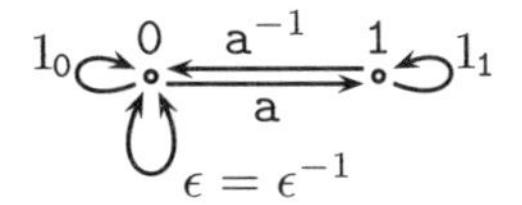

Notes

Sources and comments. The principle of constructing a distinguished decomposition by iteratively extracting the maximal left-divisor of the current remainder appears, in the case of braids, in the PhD thesis of E.A. El-Rifai prepared under the supervision of H. Morton, a part of which appeared as [117]. It also explicitly appears in the notes circulated by W. Thurston [226] around 1988, which have subsequently appeared in an almost unchanged form as Chapter 9 of [119]. The idea is not mentioned in Garside [125], but, according to H. Morton (personal communication), it can be considered implicit in F.A. Garside's PhD thesis [124], at least in the form of extracting the maximal power of Δ that left-divides a given positive braid. It seems that the construction of the symmetric normal decompositions for the elements of the groupoid of fractions goes back to the book by D. Epstein *et al.* [119].

The current development is formally new. The purpose of replacing the standard notion, which defines the normality of $g_1|g_2$ by only considering the left-divisors of g_1g_2, with the stronger notion of Definition 1.1, which involves all divisors of fg_1g_2 and not only of g_1g_2, is to allow for simple general results that almost require no assumption on the reference family $\mathcal{S}$. In particular, uniqueness is essentially obtained for free in this way.

The diagrammatic mechanism of the greedy normal form was certainly known to many authors. Diagrams similar to those used in this text appear for instance in Charney [54, 55]. The (first and second) domino rules appear in [80] and, more explicitly, in D.–Lafont [98] and in [87].

The result that normal and symmetric normal decompositions are geodesic appears in Charney [55], and it is explicitly mentioned for instance in Epstein *et al.* [119]. The Grid Property (Proposition 3.11) seems to be a new result, extending previous observations of D. Krammer in the case of braids. The results about the Fellow Traveler property and the existence of a (sort of) automatic structure is a direct extension of [119], whereas the few remarks about combinatorics follow [85]

The results of Subsection 3.4 come from D.–Lafont [98], using the mechanism of the normal form (precisely the third domino rule) to extend an approach developed by Squier [218] for Artin–Tits groups that heavily depends on the symmetry of the Coxeter relations, see also Squier [217] and Lafont [168]. An alternative approach was developed by R. Charney, J. Meier, and K. Whittlesey in [57], building on Bestvina's paper [15]. Both approaches are essentially equivalent, but the construction of [57] requires stronger assumptions, as only Garside groups and monoids are relevant. Assuming that G is a Garside group, and Δ is a Garside element in some Garside monoid M of which G is a group of fractions, the approach of [57] consists in constructing a finite $K(\pi, 1)$ for G by introducing a flag complex whose 1-skeleton is the fragment of the Cayley graph of G associated with the divisors of Δ. The main point is that this flag complex is contractible. Considering the action of G on the flag complex leads to an explicit free resolution of $\mathbb{Z}$ by $\mathbb{Z}G$-modules, which happens, in the considered cases, to coincide with the one of Proposition 3.52 up to a change of variables analogous to the one involved where one goes from a standard resolution to a bar resolution [43], namely

$$[\![s_1, ..., s_n]\!] \leftrightarrow [s_1/s_2, s_2/s_3, ..., s_{n-1}/s_n, s_n],$$

which is bijective whenever the ambient monoid is right-cancellative. Translating (3.51) then gives for the differential of the new complex the classical bar resolution formula

$$t_1(t_2, ..., t_n) + \sum\nolimits_{i=1}^{n-1} (-1)^i (t_1, ..., t_i t_{i+1}, ..., t_n) + (-1)^n (t_1, ..., t_{n-1}).$$

So the resolution in [57] is a sort of bar resolution restricted to divisors of the Garside element, which makes the nickname "gar resolution" once proposed by D. Bessis—who used "gar nerve" in [7]—quite natural. Note that the usual standard and bar resolutions correspond to the case when the considered Garside family is the whole ambient category. We refer to D.–Lafont [98] for the description of still another (more efficient) resolution based on a well-ordering of the cells in the spirit of Y. Kobayashi's work on the homology of rewriting systems [159].

Exercise 31 is inspired by D. Krammer and led to studying an extension of Coxeter and Artin–Tits groups in which the relations involve words of unequal length [166].

Further questions. At this relatively early point in the text, the mechanism of the normal form is fairly simple and not so many questions arise naturally. With Algorithms 2.46 and 2.49, we gave effective methods for left-multiplying and right-dividing a symmetric normal path by an element of the reference Garside family.

Question 4. Under the assumptions of Proposition 1.49, does there exist effective methods for right-multiplying and left-dividing a symmetric normal path by an element of $\mathcal{S}^\sharp$?

As there exist effective methods for normalizing an arbitrary path, a positive answer is granted. But what we ask for is a method similar in spirit to Algorithms 2.46 and 2.49. We doubt that a general solution exists, but the question is whether the second domino rule might help here. A positive answer will be given in Chapter V below in the special case when the considered Garside family is bounded.

A number of questions potentially arise in connection with the geometrical properties involving normal decompositions, a theme that we do not treat with full details in this text. The Grid Property (Proposition 3.11) seems to be a strong convexity statement, a combinatorial analog of convex sets in real vector spaces or in real hyperbolic spaces.

Question 5. Does the Grid Property characterize (in some sense) Garside groups?

In other words, which properties of Garside group(oid)s follow from the Grid Property? By the way, it would be desirable to exhibit examples of group(oid)s *not* satisfying the Grid Property. Natural candidates could be groups whose Cayley graph includes patterns of the form shown on the right, where the diagonal arrows do not cross.

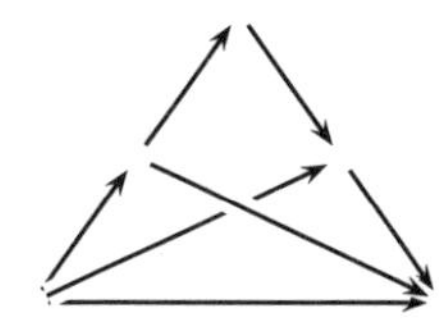

Another related problem is

Question 6. Assume that $\mathcal{C}$ is a cancellative category (with no nontrivial invertible element) and $\mathcal{S}$ is a (perfect) Garside family in $\mathcal{C}$. Does some counterpart of the Grid Property involve right-disjoint decompositions of the form fg^{-1} for the elements of $\mathcal{E}nv(\mathcal{C})$?

The definitions of $\mathcal{S}$-normality and left-disjointness are not symmetric, so the answer is not clear. More generally, very little is known about right-fractional decompositions fg^{-1} in the general case—see Subsection V.3.4 for a discussion in the special context of bounded Garside families.

In the direction of Subsection 3.4, it is natural to raise

Question 7. What is (under suitable assumptions) the connection between the chain complex constructed using the Garside structure in Proposition 3.52 and the Deligne–Salvetti resolution of [101, 208]?

A precise connection with the approach of [57] is established in [98], but it seems that, at the moment, nothing explicit exists for the Deligne–Salvetti complex. See Bessis [11] for further discussions of such questions.

We conclude with a more speculative point:

Question 8. Does an abstract normalization process in the vein of Exercise 31 lead to (interesting) unique normal decompositions beyond the current Garside approach?

Vague as it is, the answer to Question 8 is certainly positive. For instance, we will see in Example IV.2.34 below that the monoid $\langle \mathtt{a}, \mathtt{b} \mid \mathtt{ab} = \mathtt{ba}, \mathtt{a}^2 = \mathtt{b}^2 \rangle^+$ admits no proper Garside family, that is, no Garside family other than the whole monoid and, therefore, no nontrivial normal form can be obtained by means of $\mathcal{S}$-normal decompositions as defined above. However, defining $F(\mathtt{a}|\mathtt{a}) = \mathtt{b}|\mathtt{b}$, $F(\mathtt{b}|\mathtt{a}) = \mathtt{a}|\mathtt{b}$, and $F(w) = w$ for the other pairs leads to a unique normal form on the above monoid. We refer to the PhD theses of A. Hess [146] and V. Ozornova [191] and to the recent preprints [147] and [97] for results in this direction. In particular, it is shown in [97] that Garside normalisation is characterized, among a vast family of normalisation processes called quadratic, by the satisfaction of the first domino rule plus a property called left-weightedness expressing that, for all generators s, t, the first entry of $F(s|t)$ is always a right-multiple of s. Also, it is proved there that Garside normalisation is convergent, meaning that *any* sequence of normalisation steps (applied in any position) must lead in finitely many steps to a normal path, with an explicit upper bound $2^p - p - 1$ for length-p paths. A similar result appears in [147] for a family of normalisations (there called "factorability structures") that is not directly comparable with those of [97]. Better understanding the connection between the two approaches is desirable.

Let us also mention another possible approach, consisting in first introducing by explicit (local) axioms the notion of a *Garside graph* and then defining a Garside group as a group that acts faithfully on a Garside graph, in the sense that there are finitely many orbits of vertices and of edges, and every nontrivial element of the group moves at least one vertex. We refer to Krammer [161] for preliminary developments in this direction; essentially, one should be able to recover results like the domino rules and the Grid Property and, from there, establish the connection with the current approach.

In a completely different direction, one can also wonder whether the current normal decompositions based on a Garside family could be merged with the alternating and rotating normal forms of braids, see Burckel [44], D. [86], and Fromentin [123]: in the latter cases, one keeps the idea of extracting a maximal fragment lying in some families but, here, the involved families consist of (several) submonoids rather than being a Garside family.

Chapter IV
Garside families

We have seen in Chapter III how to construct essentially unique decompositions for the elements of a category $\mathcal{C}$, and of the enveloping groupoid of $\mathcal{C}$ when the latter is a left-Ore category that admits left-lcms, once some subfamily $\mathcal{S}$ of $\mathcal{C}$ has been chosen and it is what we called a Garside family. When doing so, we defined Garside families to be the ones that make the construction possible, but, so far, we did not give practical characterizations of such families. The aim of this chapter is to fill this gap and provide alternative definitions of Garside families (and of greedy paths) that apply in particular to the many examples of Chapter I and make the connection with Garside monoids straightforward.

We shall consider several contexts. In Section 1, we consider the general case, when the ambient category is supposed to satisfy no extra assumption beyond left-cancellativity. The main results are Propositions 1.24 and 1.50, which give several criteria characterizing Garside families, in terms of closure properties and of head functions, respectively. Then, in Section 2, we consider more special contexts, typically when the ambient category satisfies Noetherianity conditions or when the existence of least common multiples is guaranteed. Then some of the conditions characterizing Garside families take a special form or are even automatically satisfied, resulting in more simple criteria: for instance, Noetherianity assumptions give the the existence of a head for free, thus reducing the property of being a Garside family to being closed under right-comultiple and right-divisor (Proposition 2.11), whereas the existence of unique right-lcms reduces the properties of being closed under right-comultiple and right-divisor to the condition that the right-lcm and the right-complement of two elements of the family lie in the family (Proposition 2.40). Finally, we develop in Section 3 some geometrical and algorithmic consequences of the closure properties investigated in Section 1. The main results are Propositions 3.6 and 3.11, where we show that every Garside family naturally gives rise to presentations of the ambient category and, possibly, of its enveloping groupoid, that satisfy quadratic isoperimetric inequalities (Proposition 3.11).

Main definitions and results

Definition 1.1 **(closure I).** (i) A subfamily $\mathcal{S}$ of a left-cancellative category $\mathcal{C}$ is said to be *closed under right-divisor* if every right-divisor of an element of $\mathcal{S}$ is an element of $\mathcal{S}$, that is, if the conjunction of $st' = t$ in $\mathcal{C}$ and $t \in \mathcal{S}$ implies $t' \in \mathcal{S}$. (ii) In the same context, $\mathcal{S}$ is said to be *closed under right-quotient* if the conjunction of $st' = t$ in $\mathcal{C}$ and $s \in \mathcal{S}$ and $t \in \mathcal{S}$ implies $t' \in \mathcal{S}$.

Definition 1.3 **(closure II).** (i) A subfamily $\mathcal{S}$ of a left-cancellative category $\mathcal{C}$ is said to be *closed under right-comultiple* if every common right-multiple of two elements s, t of $\mathcal{S}$

(if any) is a right-multiple of a common right-multiple of s, t lying in $\mathcal{S}$. (ii) In the same context, $\mathcal{S}$ is said to be *closed under right-complement* if, when s, t lie in $\mathcal{S}$ and $sg = tf$ holds in $\mathcal{C}$, there exist s', t' in $\mathcal{S}$ and h in $\mathcal{C}$ satisfying $st' = ts'$, $f = s'h$, and $g = t'h$.

Definition 1.10 **(head).** We say that s is a $\mathcal{S}$-*head* of g if s lies in $\mathcal{S}$, it left-divides g, and every element of $\mathcal{S}$ that left-divides g left-divides s.

Proposition 1.20 **(recognizing greedy).** *Assume $\mathcal{S}$ is a subfamily of a left-cancellative category $\mathcal{C}$ such that $\mathcal{S}$ is closed under right-comultiple and $\mathcal{S}^\sharp$ generates $\mathcal{C}$ and is closed under right-divisor. Then, for every path $s|g$ in $\mathcal{S}^\sharp|\mathcal{C}$, the following are equivalent:* (i) *The path $s|g$ is $\mathcal{S}$-greedy.* (ii) *For every t in $\mathcal{S}$, the relation $t \preccurlyeq sg$ implies $t \preccurlyeq s$, that is, s is an $\mathcal{S}$-head of sg.* (iii) *Every element h of $\mathcal{C}$ satisfying $sh \in \mathcal{S}$ and $h \preccurlyeq g$ is invertible.*

Proposition 1.23 **(Garside closed).** *If $\mathcal{S}$ is a Garside family in a left-cancellative category, then $\mathcal{S}$ is closed under right-comultiple and $\mathcal{S}^\sharp$ is closed under right-comultiple, right-complement, and right-divisor.*

Proposition 1.24 **(recognizing Garside II).** *Assume that $\mathcal{C}$ is a left-cancellative category and $\mathcal{S}$ is a subfamily of $\mathcal{C}$. Then $\mathcal{S}$ is a Garside family if and only if any one of the following conditions is satisfied:* (i) *The family $\mathcal{S}^\sharp$ generates $\mathcal{C}$ and is closed under right-divisor, and every non-invertible element of $\mathcal{C}$ admits an $\mathcal{S}$-head.* (ii) *The family $\mathcal{S}^\sharp$ generates $\mathcal{C}$ and is closed under right-complement, and every non-invertible element of $\mathcal{S}^2$ admits an $\mathcal{S}$-head.* (iii) *The family $\mathcal{S}^\sharp$ generates $\mathcal{C}$ and is closed under right-divisor, $\mathcal{S}$ is closed under right-comultiple, and every non-invertible element of $\mathcal{S}^2$ admits a $\prec$-maximal left-divisor lying in $\mathcal{S}$.*

Proposition 1.35 **(strong and perfect Garside).** *If $\mathcal{S}$ is a Garside family in a cancellative category $\mathcal{C}$ that admits conditional weak left-lcms, the following are equivalent:* (i) *The family $\mathcal{S}$ is a strong* (*resp. perfect*) *Garside family in $\mathcal{C}$.* (ii) *The family $\mathcal{S}^\sharp$ is closed under left-complement* (*resp. this and left-comultiple*).
When the above conditions are met, $\mathcal{S}^\sharp$ is also closed under left-divisor.

Proposition 1.39 **(second domino rule).** *If $\mathcal{S}$ is a Garside family in a cancellative category $\mathcal{C}$ and $\mathcal{S}^\sharp$ is closed under left-comultiple and left-divisor, the second domino rule is valid for $\mathcal{S}$.*

Definition 1.42 **(head function).** If $\mathcal{S}$ is a subfamily of a left-cancellative category $\mathcal{C}$, an $\mathcal{S}$-*head function* is a partial map H from $\mathcal{C}$ to $\mathcal{S}$ such that $\mathrm{Dom}H$ includes $\mathcal{C} \setminus \mathcal{C}^\times$ and $H(g)$ is an $\mathcal{S}$-head of g for every g in $\mathrm{Dom}H$. An $\mathcal{S}$-head function is called *sharp* if $g' =^\times g$ implies $H(g') = H(g)$.

Definition 1.47 **(sharp $\mathcal{H}$-law, $\mathcal{H}$-law).** A (partial) function H from a category $\mathcal{C}$ into itself satisfies the *$\mathcal{H}$-law* (*resp.* the *sharp $\mathcal{H}$-law*) if $H(fg) =^\times H(fH(g))$ (*resp.* $=$) holds whenever the involved expressions are defined.

Proposition 1.50 **(recognizing Garside III).** *Assume that $\mathcal{C}$ is a left-cancellative category and $\mathcal{S}$ is a subfamily $\mathcal{S}$ of $\mathcal{C}$ such that $\mathcal{S}^\sharp$ generates $\mathcal{C}$. Then $\mathcal{S}$ is a Garside family if and only if* (i) *There exists an $\mathcal{S}$-head function that satisfies the sharp $\mathcal{H}$-law.* (ii) *There exists an $\mathcal{S}$-head function that satisfies the $\mathcal{H}$-law.* (iii) *There exists a map $H : \mathcal{C} \setminus \mathcal{C}^\times \to \mathcal{S}$*

that preserves $\preccurlyeq$, satisfies $H(g) \preccurlyeq g$ (resp. $H(g) =^{\times} g$) for every g in $\mathcal{C} \setminus \mathcal{C}^{\times}$ (resp. in $\mathcal{S} \setminus \mathcal{C}^{\times}$), and satisfies the $\mathcal{H}$-law.

Definition 2.1 **(solid).** A subfamily $\mathcal{S}$ of a category $\mathcal{C}$ is said to be *solid* in $\mathcal{C}$ if $\mathcal{S}$ includes $1_{\mathcal{C}}$ and it is closed under right-divisor.

Proposition 2.7 **(recognizing Garside, solid case).** *A solid subfamily $\mathcal{S}$ in a left-cancellative category $\mathcal{C}$ is a Garside family if and only if $\mathcal{S}$ generates $\mathcal{C}$ and one of the following equivalent conditions is satisfied:* (2.8) *Every non-invertible element of $\mathcal{C}$ admits an $\mathcal{S}$-head.* (2.9) *The family $\mathcal{S}$ is closed under right-complement and every non-invertible element of $\mathcal{S}^2$ admits an $\mathcal{S}$-head.* (2.10) *The family $\mathcal{S}$ is closed under right-comultiple and every non-invertible element of $\mathcal{S}^2$ admits a $\prec$-maximal left-divisor lying in $\mathcal{S}$.*

Proposition 2.18 **(solid Garside in right-Noetherian).** *For every solid generating subfamily $\mathcal{S}$ in a left-cancellative category $\mathcal{C}$, the following are equivalent:* (i) *The category $\mathcal{C}$ is right-Noetherian and $\mathcal{S}$ is a Garside family in $\mathcal{C}$.* (ii) *The family $\mathcal{S}$ is locally right-Noetherian and closed under right-comultiple.*

Proposition 2.25 **(recognizing Garside, right-mcm case).** *A subfamily $\mathcal{S}$ of a left-cancellative category $\mathcal{C}$ that is right-Noetherian and admits right-mcms is a Garside family if and only if $\mathcal{S}^{\sharp}$ generates $\mathcal{C}$ and satisfies one of the following equivalent conditions:* (i) *The family $\mathcal{S}^{\sharp}$ is closed under right-mcm and right-divisor.* (ii) *The family $\mathcal{S}^{\sharp}$ is closed under right-mcm and right-complement.*

Corollary 2.29 **(recognizing Garside, right-lcm case).** *A subfamily $\mathcal{S}$ of a left-cancellative category $\mathcal{C}$ that is right-Noetherian and admits conditional right-lcms is a Garside family if and only if $\mathcal{S}^{\sharp}$ generates $\mathcal{C}$ and satisfies one of the equivalent conditions:* (i) *The family $\mathcal{S}^{\sharp}$ is closed under right-lcm and right-divisor.* (ii) *The family $\mathcal{S}^{\sharp}$ is closed under right-lcm and right-complement.*

Corollary 2.32 **(smallest Garside).** *Every strongly Noetherian left-cancellative category $\mathcal{C}$ contains a smallest $=^{\times}$-closed Garside family including $1_{\mathcal{C}}$, namely the closure of the atoms under right-mcm and right-divisor. This Garside family is solid.*

Proposition 2.40 **(recognizing Garside, right-complement case).** *A subfamily $\mathcal{S}$ of a left-cancellative category $\mathcal{C}$ that is right-Noetherian and admits unique conditional right-lcms is a Garside family if and only if $\mathcal{S}^{\sharp}$ generates $\mathcal{C}$ and is closed under right-lcm and right-complement.*

Proposition 2.43 **(finite height).** *If $\mathcal{C}$ is a left-cancellative category and $\mathrm{ht}(s) \leqslant K$ holds for every s in some Garside family $\mathcal{S}$ of $\mathcal{C}$, then $\mathcal{C}$ is strongly Noetherian and $\mathrm{ht}(g) \leqslant K^m$ holds for every g satisfying $\|g\|_{\mathcal{S}} \leqslant m$.*

Proposition 3.6 **(presentation).** *If $\mathcal{S}$ is a Garside family in a left-cancellative category $\mathcal{C}$, then $\mathcal{C}$ admits the presentation $\langle \mathcal{S}^{\sharp} \mid \mathcal{R}_{\mathcal{C}}(\mathcal{S}^{\sharp}) \rangle^{+}$, where $\mathcal{R}_{\mathcal{C}}(\mathcal{S}^{\sharp})$ is the family of all valid relations $rs = t$ with r, s, t in $\mathcal{S}^{\sharp}$.*

Proposition 3.11 **(isoperimetric inequality).** *If $\mathcal{S}$ is a strong Garside family in a left-Ore category $\mathcal{C}$ that admits left-lcms, then $\mathcal{E}nv(\mathcal{C})$ satisfies a quadratic isoperimetric inequality with respect to the presentations $(\mathcal{S}^{\sharp}, \mathcal{R}_{\mathcal{C}}(\mathcal{S}^{\sharp}))$ and $(\mathcal{S} \cup \mathcal{C}^{\times}, \mathcal{R}^{\times}_{\mathcal{C}}(\mathcal{S}))$.*

1 The general case

The aim of this section is to provide new definitions of Garside families that are more easily usable than the original definition of Chapter III and the characterization of Proposition III.1.39. The main technical ingredients here are closure properties, that is, properties asserting that, when some elements s, t lie in the considered family $\mathcal{S}$, then some other elements connected with s and t also lie in $\mathcal{S}$.

The section contains four subsections. In Subsection 1.1, we introduce various closure properties and establish connections between them. In Subsection 1.2, we deduce new characterizations of Garside families in terms of closure properties. In Subsection 1.3, we obtain similar characterizations for strong and perfect Garside families. Finally, in Subsection 1.4, we establish further characterizations of Garside families in terms of what we call head functions.

1.1 Closure properties

As explained above, we shall establish new characterizations of Garside families involving several closure properties. In this subsection, we introduce these properties and establish various connections between them.

We recall that, for f, g in a category $\mathcal{C}$, we say that f is a left- (*resp*. right-) divisor of g or, equivalently, that g is a right- (*resp*. left-) multiple of f, if $g = fg'$ (*resp*. $g = g'f$) holds for some g'. We also recall that $\mathcal{S}^\sharp$ stands for $\mathcal{S}\mathcal{C}^\times \cup \mathcal{C}^\times$.

Definition 1.1 (closure I). (i) A subfamily $\mathcal{S}$ of a left-cancellative category $\mathcal{C}$ is said to be *closed under right-divisor* if every right-divisor of an element of $\mathcal{S}$ is an element of $\mathcal{S}$, that is, if the conjunction of $st' = t$ in $\mathcal{C}$ and $t \in \mathcal{S}$ implies $t' \in \mathcal{S}$.

(ii) In the same context, $\mathcal{S}$ is said to be *closed under right-quotient* if the conjunction of $st' = t$ in $\mathcal{C}$ and $s \in \mathcal{S}$ and $t \in \mathcal{S}$ implies $t' \in \mathcal{S}$.

There exist obvious connections between the above two notions.

Lemma 1.2. *Assume that $\mathcal{C}$ is a left-cancellative category and $\mathcal{S}$ is a subfamily of $\mathcal{C}$.*

(i) *If $\mathcal{S}$ is closed under right-divisor, then it is closed under right-quotient.*

(ii) *If $\mathcal{S}$ generates $\mathcal{C}$ and is closed under right-quotient, then it is closed under right-divisor.*

Proof. Assertion (i) is clear from the definition. For (ii), assume $gs = t$ with t in $\mathcal{S}$. If g is an identity element, we deduce $s = t \in \mathcal{S}$. Otherwise, by assumption, there exist $s_1, ..., s_p$ in $\mathcal{S}$ satisfying $g = s_1 \cdots s_p$. As $\mathcal{S}$ is closed under right-quotient, $gs \in \mathcal{S}$, which is $s_1(s_2 \cdots s_p s) \in \mathcal{S}$, implies $s_2 \cdots s_p s \in \mathcal{S}$, then similarly $s_3 \cdots s_p s \in \mathcal{S}$, and so on until $s \in \mathcal{S}$. Hence $\mathcal{S}$ is closed under right-divisor in $\mathcal{C}$. □

Definition 1.3 (closure II). (See Figure 1.) (i) A subfamily $\mathcal{S}$ of a left-cancellative category $\mathcal{C}$ is said to be *closed under right-comultiple* if every common right-multiple of two elements s, t of $\mathcal{S}$ (if any) is a right-multiple of a common right-multiple of s, t lying in $\mathcal{S}$.

(ii) In the same context, $\mathcal{S}$ is said to be *closed under right-complement* if, when s, t lie in $\mathcal{S}$ and $sg = tf$ holds in $\mathcal{C}$, there exist s', t' in $\mathcal{S}$ and h in $\mathcal{C}$ satisfying $st' = ts'$, $f = s'h$, and $g = t'h$.

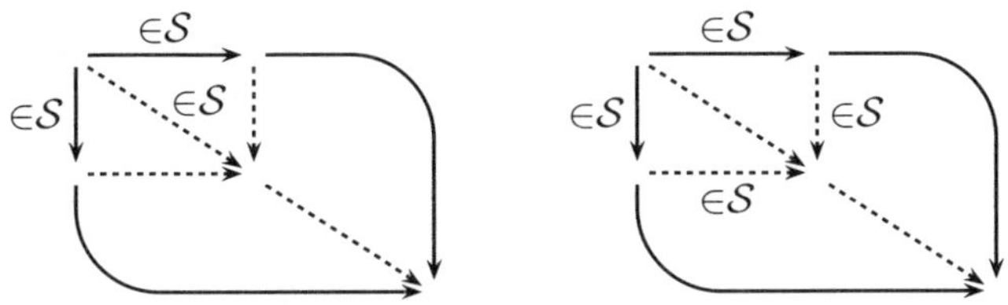

Figure 1. Closure under right-comultiple (left) and right-complement (right): every diagram corresponding to a common right-multiple of elements of $\mathcal{S}$ splits, witnessing for a factorization through a common right-multiple of a special type: in the case of closure under right-comultiple, the diagonal of the square lies in $\mathcal{S}$, whereas, in the case of closure under right-complement, the edges lie in $\mathcal{S}$.

Remark 1.4. The correspondence between the definition of closure under right-comultiple and the diagram of Figure 1 is valid only in the context of a left-cancellative category. Indeed, the diagram splitting amounts to the relation

(1.5) If s, t lie in $\mathcal{S}$ and $sg = tf$ holds in $\mathcal{C}$, there exist f', g', h satisfying $sg' = tf'$, $f = f'h$, $g = g'h$, and such that sg' lies in $\mathcal{S}$,

whereas Definition 1.3(ii) corresponds to

(1.6) If s, t lie in $\mathcal{S}$ and $sg = tf$ holds in $\mathcal{C}$, there exist f', g', h satisfying $sg' = tf'$, $tf = tf'h$, $sg = sg'h$, and such that sg' lies in $\mathcal{S}$.

Clearly, (1.5) implies (1.6), but the converse implication is guaranteed only if one can left-cancel s and t in the last two equalities of (1.5).

Example 1.7 (closure). Assume that (M, Δ) is a Garside monoid (Definition I.2.1). Then the family $\mathrm{Div}(\Delta)$ is closed under right-divisor, right-comultiple, and right-complement. As for closure under right-divisor, it directly follows from the definition: if t is a left-divisor of Δ and t' is a right-divisor of t, then, as t is also a right-divisor of Δ, the element t' is a right-divisor of Δ, hence a left-divisor of Δ.

Next, assume that s, t lie in $\mathrm{Div}(\Delta)$ and $sg = tf$ holds. By the definition of a Garside monoid, s and t admit a right-lcm, say h, hence sg is a right-multiple of h. On the other hand, Δ is a right-comultiple of s and t, hence it is a right comultiple of their right-lcm h. In other words, h lies in $\mathrm{Div}(\Delta)$, and $\mathrm{Div}(\Delta)$ is closed under right-comultiple.

Finally, with the same assumptions, let us write $h = st' = ts'$. As h left-divides Δ and $\mathrm{Div}(\Delta)$ is closed under right-divisor, s' and t' belong to $\mathrm{Div}(\Delta)$. So $\mathrm{Div}(\Delta)$ is closed under right-complement.

The first observation is that the conjunction of closures under right-divisor and right-comultiple implies the closure under right-complement, and even more.

Lemma 1.8. *Assume that $\mathcal{C}$ is a left-cancellative category and $\mathcal{S}$ is a subfamily of $\mathcal{C}$ that is closed under right-divisor and right-comultiple. Then*

(1.9) *If s, t lie in $\mathcal{S}$ and $sg = tf$ holds in $\mathcal{C}$, there exist s', t' in $\mathcal{S}$ and h in $\mathcal{C}$ satisfying $st' = ts'$, $f = s'h$, $g = t'h$ and, in addition, st' lies in $\mathcal{S}$.*

In particular, $\mathcal{S}$ is closed under right-complement.

Proof. Assume $sg = tf$ with s, t in $\mathcal{S}$. As $\mathcal{S}$ is closed under right-comultiple, there exist s', t', h satisfying $st' = ts' \in \mathcal{S}$, $f = s'h$, and $g = t'h$. Now, s' is a right-divisor of ts', which lies in $\mathcal{S}$, hence s' must lie in $\mathcal{S}$ and, similarly, t' must lie in $\mathcal{S}$. ☐

The closure property of (1.9) corresponds to the diagram on the right, which is a fusion of the two diagrams of Figure 1: the common multiple relation factors through a small common multiple relation in which both the edges and the diagonal of the square lie in the considered family $\mathcal{S}$.

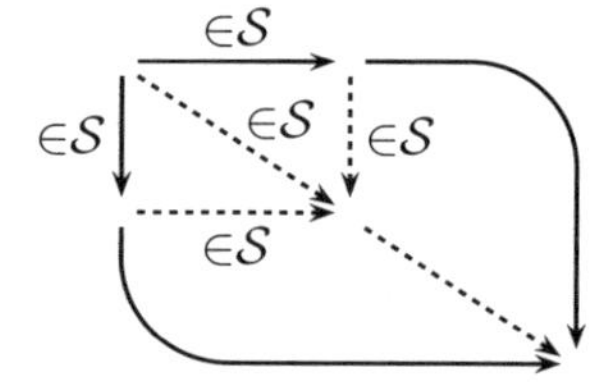

In addition to closure properties, we shall also appeal to the following notion, which corresponds to the intuition of a maximal left-divisor lying in $\mathcal{S}$.

Definition 1.10 (head). If $\mathcal{S}$ is a subfamily of a left-cancellative category $\mathcal{C}$, an element s of $\mathcal{S}$ is called an *$\mathcal{S}$-head* of an element g of $\mathcal{C}$ if s left-divides g and every element of $\mathcal{S}$ left-dividing g left-divides s.

Example 1.11 (head). If (M, Δ) is a Garside monoid, every element g of M admits a $\mathrm{Div}(\Delta)$-head, namely the left-gcd of g and Δ. Indeed, every left-divisor of g that lies in $\mathrm{Div}(\Delta)$ is a common left-divisor of g and Δ, hence a left-divisor of their left-gcd.

There exists a simple connection between the existence of an $\mathcal{S}$-head and closure of $\mathcal{S}$ under right-comultiple.

Lemma 1.12. *If $\mathcal{S}$ is a subfamily of a left-cancellative category $\mathcal{C}$ and every non-invertible element of $\mathcal{C}$ admits an $\mathcal{S}$-head, then $\mathcal{S}$ is closed under right-comultiple.*

Proof. Assume that s, t lie in $\mathcal{S}$ and h is a common right-multiple of s and t. If h is invertible, then so are s and t, and h is a right-multiple of s, which is a common right-multiple of s and t lying in $\mathcal{S}$. Assume now that h is not invertible. Then, by assumption, h admits an $\mathcal{S}$-head, say r. Now, by assumption again, s left-divides h and it lies in $\mathcal{S}$, hence, by definition of an $\mathcal{S}$-head, s left-divides r. The same argument shows that t left-divides r. So h is a right-multiple of r, which is a common right-multiple of s and t lying in $\mathcal{S}$. Hence $\mathcal{S}$ is closed under right-comultiple. ☐

Next, we establish transfer results between $\mathcal{S}$ and $\mathcal{S}^\sharp$.

Lemma 1.13. *Assume that $\mathcal{S}$ is a subfamily of a left-cancellative category $\mathcal{C}$.*

(i) *If $\mathcal{S}$ is closed under right-divisor, then $\mathcal{S}^\sharp$ is closed under right-divisor too.*

(ii) *The family $\mathcal{S}$ is closed under right-comultiple if and only if $\mathcal{S}^\sharp$ is.*

(iii) *If $\mathcal{S}$ is closed under right-complement and $\mathcal{C}^\times\mathcal{S} \subseteq \mathcal{S}$ holds, then $\mathcal{S}^\sharp$ is closed under right-complement too.*

Proof. (i) Assume that $\mathcal{S}$ is closed under right-divisor, t lies in $\mathcal{S}^\sharp$ and t' right-divides t. If t is invertible, then so is t', and therefore t' lies in $\mathcal{S}^\sharp$. Otherwise, we can write $t = s\epsilon$ with s in $\mathcal{S}$ and ϵ in $\mathcal{C}^\times$. The assumption that t' right-divides t implies that $t'\epsilon^{-1}$ right-divides s. Hence $t'\epsilon^{-1}$ lies in $\mathcal{S}$, and therefore t' lies in $\mathcal{S}\mathcal{C}^\times$, hence in $\mathcal{S}^\sharp$. We conclude that $\mathcal{S}^\sharp$ is closed under right-divisor.

The proof of (ii) and (iii) is similar: the closure of $\mathcal{S}^\sharp$ (almost) trivially implies that of $\mathcal{S}$ and, conversely, the case of invertible elements is trivial and the case of elements of $\mathcal{S}\mathcal{C}^\times$ is treated by applying the assumption to the $\mathcal{S}$-components. The (easy) details are left to the reader. ☐

Next, we observe that the technical condition $\mathcal{C}^\times\mathcal{S} \subseteq \mathcal{S}^\sharp$, which is often required, follows from a closure property.

Lemma 1.14. *If $\mathcal{S}$ is a subfamily of a left-cancellative category $\mathcal{C}$ and $\mathcal{S}^\sharp$ is closed under right-complement, then $\mathcal{C}^\times\mathcal{S} \subseteq \mathcal{S}^\sharp$ holds.*

Proof. Assume that ϵ is invertible, t belongs to $\mathcal{S}$, and ϵt is defined. We claim that ϵt belongs to $\mathcal{S}^\sharp$. Indeed, we can write $\epsilon^{-1}(\epsilon t) = t1_y$, where y is the target of t. As $\mathcal{S}^\sharp$ is closed under right-complement, there exist ϵ', t' in $\mathcal{S}^\sharp$ and h in $\mathcal{C}$ satisfying $\epsilon^{-1}t' = t\epsilon'$, $1_y = \epsilon' h$, and $\epsilon t = t'h$. The second equality implies that h (as well as ϵ') is invertible. Then the third equality show that ϵt lies in $\mathcal{S}^\sharp\mathcal{C}^\times$, which is $\mathcal{S}^\sharp$. So we have $\mathcal{C}^\times\mathcal{S}^\sharp \subseteq \mathcal{S}^\sharp$. ☐

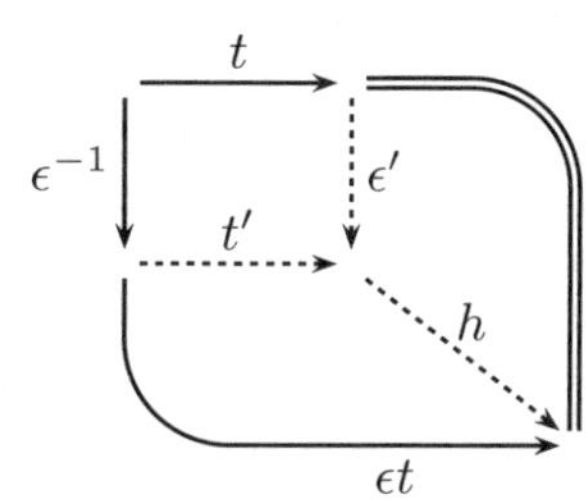

Here comes the main step, which consists in extending the property defining closure under right-complement from a family $\mathcal{S}$ to all families $\mathcal{S}^p$.

Proposition 1.15 (factorization grid). (i) *If $\mathcal{S}$ is a subfamily of a (left-cancellative) category $\mathcal{C}$ and $\mathcal{S}$ is closed under right-complement, then, for every equality $s_1 \cdots s_p g = t_1 \cdots t_q f$ with $s_1, \ldots, s_p$, $t_1, \ldots, t_q$ in $\mathcal{S}$, there exists a commutative diagram as in Figure 2 such that the edges of the squares all correspond to elements of $\mathcal{S}$. In particular, there exist $s'_1, \ldots, s'_p, t'_1, \ldots, t'_q$ in $\mathcal{S}$ and h satisfying*

$$(1.16)\qquad s_1 \cdots s_p t'_1 \cdots t'_q = t_1 \cdots t_q s'_1 \cdots s'_p, \quad f = s'_1 \cdots s'_p h, \quad \textit{and} \quad g = t'_1 \cdots t'_q h.$$

(ii) *If $\mathcal{S}$ satisfies the stronger closure property (1.9), then, in the context of (i), we may assume that the diagonals of the squares in the grid also correspond to elements of $\mathcal{S}$.*

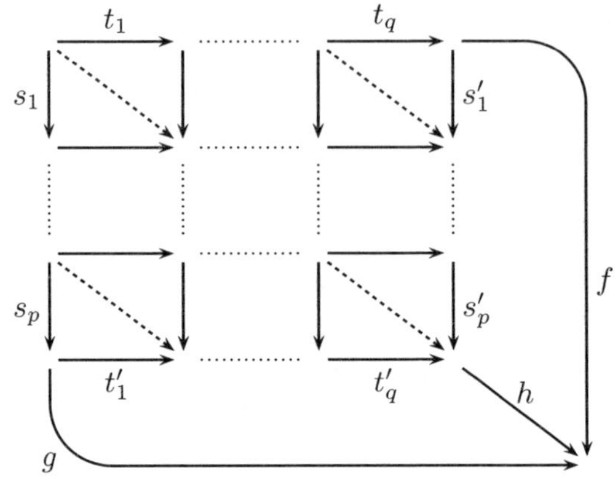

Figure 2. Whenever $\mathcal{S}$ is closed under right-complement, each equality $s_1 \cdots s_p g = t_1 \cdots t_q f$ with $s_1, ..., t_q$ in $\mathcal{S}$ factors through a rectangular grid in which the sides of the squares correspond to elements of $\mathcal{S}$. If $\mathcal{S}$ generates $\mathcal{C}$ and satisfies (1.9), we can assume that the diagonals of the squares also correspond to elements of $\mathcal{S}$.

Proof. (i) We use induction on pq. If p or q is zero, the result is vacuously true. For $p = q = 1$, the result is the assumption that $\mathcal{S}$ is closed under right-complement. Assume now $pq \geqslant 2$, with, say $q \geqslant 2$. Then, by assumption, we have $s_1(s_2 \cdots s_p g) = t_1(t_2 \cdots t_q f)$ with s_1, t_1 in $\mathcal{S}$. As $\mathcal{S}$ is closed under right-complement, there exist $s_{1,1}$ and $t_{1,1}$ in $\mathcal{S}$ and $h_{1,1}$ satisfying $s_1 t_{1,1} = t_1 s_{1,1}$, $s_2 \cdots s_p g = t_{1,1} h_{1,1}$ and $t_2 \cdots t_q f = s_{1,1} h_{1,1}$, see Figure 3.

Next, we have $s_{1,1} h_{1,1} = t_2 \cdots t_q f$ with $s_{1,1}, t_2, ..., t_q$ in $\mathcal{S}$. As $1(q-1) < pq$ holds, the induction hypothesis implies the existence of the expected grid and, in particular, of $t_{2,1}, ..., t_{q,1}, s'_1$ in $\mathcal{S}$ and h_1 making commutative squares and satisfying $f = s'_1 h_1$ and $h_{1,1} = t_{2,1} \cdots t_{q,1} h_1$.

Finally, we have $s_2 \cdots s_p g = t_{1,1} h_{1,1} = t_{1,1} t_{2,1} \cdots t_{q,1} h_1$, with $s_2, ..., s_p, t_{1,1}, ..., t_{q,1}$ in $\mathcal{S}$. As $(p-1)q < pq$ holds, the induction hypothesis implies the existence of $s'_2, ..., s'_q$, $t'_1, ..., t'_q$ in $\mathcal{S}$ and h making commutative squares and satisfying $h_1 = s'_2 \cdots s'_p h$ and $g = t'_1 \cdots t'_q h$. This is the expected result.

(ii) Assume now that $\mathcal{S}$ satisfies (1.9). The inductive argument is the same as for (i). By (1.9), the equality $s_1(s_2 \cdots s_p g) = t_1(t_2 \cdots t_q f)$ implies the existence of $s_{1,1}$ and $t_{1,1}$ in $\mathcal{S}$ and $h_{1,1}$ that satisfy $s_1 t_{1,1} = t_1 s_{1,1}$, $s_2 \cdots s_p g = t_{1,1} h_{1,1}$, and $t_2 \cdots t_q f = s_{1,1} h_{1,1}$ and, in addition, $s_1 t_{1,1} \in \mathcal{S}$. Then the induction proceeds as previously. □

Corollary 1.17 (extension of closure). *If $\mathcal{S}$ is a subfamily of a left-cancellative category $\mathcal{C}$ and $\mathcal{S}$ is closed under right-complement (resp. satisfies (1.9)), then, for every p, the family $\mathcal{S}^p$ is closed under right-complement (resp. satisfies (1.9)). The same holds for the subcategory $\mathcal{S}ub(\mathcal{S})$ generated by $\mathcal{S}$.*

Proof. Applying Proposition 1.15(i) (*resp.* (ii)) with $p = q$ shows that $\mathcal{S}^p$ is closed under right-complement (*resp.* satisfies (1.9)). The result for $\mathcal{S}ub(\mathcal{S})$ follows from the equality $\mathcal{S}ub(\mathcal{S}) = \mathbf{1}_\mathcal{S} \cup \bigcup_p \mathcal{S}^p$, as the property is trivial for identity elements. □

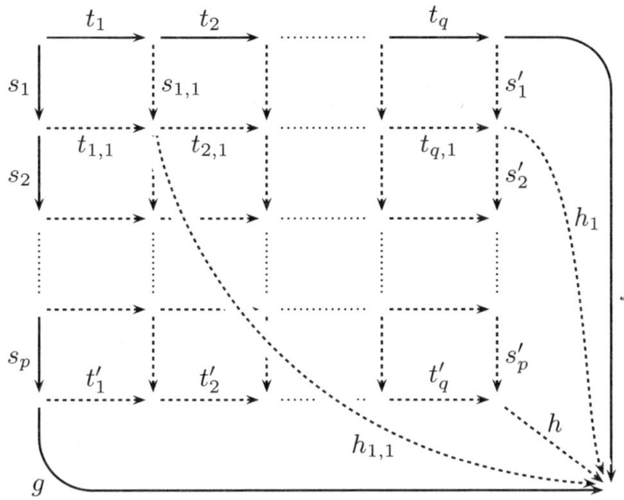

Figure 3. Inductive proof of Proposition 1.15.

We deduce a new connection between closure properties.

Lemma 1.18. *Assume that $\mathcal{C}$ is a left-cancellative category and $\mathcal{S}$ is a subfamily of $\mathcal{C}$ such that $\mathcal{S}^\sharp$ generates $\mathcal{C}$ and is closed under right-complement. Then $\mathcal{S}^\sharp$ is closed under right-divisor.*

Proof. Assume that t belongs to $\mathcal{S}^\sharp$ and g is a right-divisor of t, say $t = sg$. As $\mathcal{S}^\sharp$ generates $\mathcal{C}$, there exists p such that s belongs to $(\mathcal{S}^\sharp)^p$. Applying Proposition 1.15(i) to $\mathcal{S}^\sharp$ and the equality $sg = t1_y$, where y is the target of t, we deduce the existence of s' in $(\mathcal{S}^\sharp)^p$, t' in $\mathcal{S}$, and h satisfying $st' = ts'$, $1_y = s'h$, and $g = t'h$. The second equality implies that h is invertible, and, therefore, the third one implies that g belongs to $\mathcal{S}^\sharp\mathcal{C}^\times$, which is included in $\mathcal{S}^\sharp$. So $\mathcal{S}^\sharp$ is closed under right-divisor. □

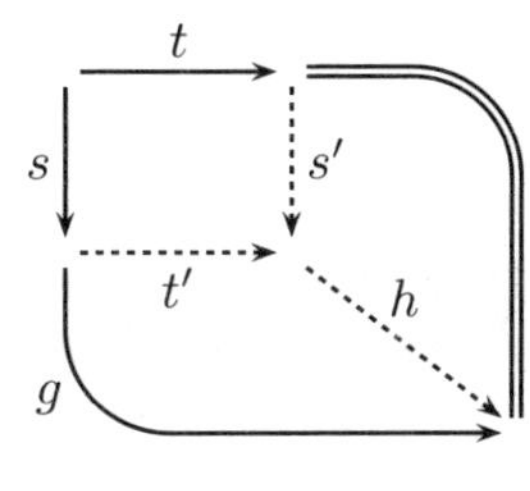

Merging the results, we obtain a simple equivalence:

Lemma 1.19. *If $\mathcal{S}$ is a subfamily of a left-cancellative category $\mathcal{C}$ and $\mathcal{S}^\sharp$ generates $\mathcal{C}$ and is closed under right-comultiple, then $\mathcal{S}^\sharp$ is closed under right-divisor if and only if $\mathcal{S}^\sharp$ is closed under right-complement, if and only if $\mathcal{S}^\sharp$ satisfies* (1.9).

Proof. By Lemma 1.8 and owing to the assumptions on $\mathcal{S}^\sharp$, closure under right-divisor implies (1.9), hence closure under right-complement. Next, by Lemma 1.18, closure under right-complement implies closure under right-divisor, whence (1.9) as above. Finally, (1.9) implies closure under right-complement by definition, whence closure under right-divisor as above. □

At this point, we can establish a first significant result, namely a connection between the current notion of an $\mathcal{S}$-greedy path, as defined in Chapter III, and the classical definition in the literature in terms of maximal left-divisor.

Proposition 1.20 (recognizing greedy). *Assume $\mathcal{S}$ is a subfamily of a left-cancellative category $\mathcal{C}$ such that $\mathcal{S}$ is closed under right-comultiple and $\mathcal{S}^\sharp$ generates $\mathcal{C}$ and is closed under right-divisor. Then, for every path $s|g$ in $\mathcal{S}^\sharp|\mathcal{C}$, the following are equivalent:*
(i) *The path $s|g$ is $\mathcal{S}$-greedy.*
(ii) *For every t in $\mathcal{S}$, the relation $t \preccurlyeq sg$ implies $t \preccurlyeq s$, that is, s is an $\mathcal{S}$-head of sg.*
(iii) *Every element h of $\mathcal{C}$ satisfying $sh \in \mathcal{S}$ and $h \preccurlyeq g$ is invertible.*

Thus, whenever the reference family $\mathcal{S}$ satisfies the closure assumptions of Proposition 1.20, we recover for $\mathcal{S}$-greediness the classical definition in terms of maximal left-divisor lying in $\mathcal{S}$, as it appears for instance in Proposition I.2.4 (normal decomposition).

In fact, we shall establish the following result, which is slightly more precise than Proposition 1.20.

Lemma 1.21. *Assume $\mathcal{S}$ is a subfamily of a left-cancellative category $\mathcal{C}$ and $\mathcal{S}^\sharp$ generates $\mathcal{C}$ For $f|g$ in $\mathcal{C}^{[2]}$, consider the three statements:*
(i) *The path $f|g$ is $\mathcal{S}$-greedy,*
(ii) *For each t in $\mathcal{S}$, the relation $t \preccurlyeq fg$ implies $t \preccurlyeq f$.*
(iii) *Every element h of $\mathcal{C}$ satisfying $fh \in \mathcal{S}$ and $h \preccurlyeq g$ is invertible.*
Then (i) *implies* (ii), *which implies* (iii). *Conversely,* (ii) *implies* (i) *whenever $\mathcal{S}^\sharp$ is closed under right-complement, and* (iii) *implies* (ii) *whenever $\mathcal{S}$ is closed under right-comultiple and f belongs to $\mathcal{S}^\sharp$.*

Proof. That (i) implies (ii) is a direct application of the definition of $\mathcal{S}$-greediness.

Assume that (ii) holds and h satisfies $fh \in \mathcal{S}$ and $h \preccurlyeq g$. Then we have $fh \preccurlyeq fg$ hence $fh \preccurlyeq f$ by (ii). As $\mathcal{C}$ is left-cancellative, we deduce that g left-divides 1_y (y the target of s) hence g is invertible, and (ii) implies (iii).

Assume now that $\mathcal{S}^\sharp$ is closed under right-complement and (ii) holds. Assume $t \in \mathcal{S}$ and $t \preccurlyeq hfg$, say $hfg = t\hat{h}$, see Figure 4 left. As $\mathcal{S}^\sharp$ generates $\mathcal{C}$, the element h lies in $(\mathcal{S}^\sharp)^p$ for some p. Applying Proposition 1.15 to $\mathcal{S}^\sharp$ and the equality $h \cdot (fg) = t \cdot \hat{h}$, we deduce the existence of h' in $(\mathcal{S}^\sharp)^p$ and t' in $\mathcal{S}^\sharp$ satisfying $ht' = th'$ and $t' \preccurlyeq fg$. We claim that $t' \preccurlyeq f$ necessarily holds, from what we can deduce $t \preccurlyeq hf$ and conclude that $f|g$ is $\mathcal{S}$-greedy. Indeed, if t' is invertible, we can write $f = t'(t'^{-1}f)$, and the result is trivial. Otherwise, t' must belong to $\mathcal{S}\mathcal{C}^\times$. So we have $t' = t''\epsilon$ for some t'' in $\mathcal{S}$ and ϵ in $\mathcal{C}^\times$. By construction, we have $t'' \preccurlyeq fg$, hence $t'' \preccurlyeq f$ by (ii). As we have $t' \preccurlyeq t''$, we deduce $t' \preccurlyeq f$, as expected. Therefore, $f|g$ is $\mathcal{S}$-greedy, and (ii) implies (i).

Assume now that $\mathcal{S}$ is closed under right-comultiple, f belongs to $\mathcal{S}^\sharp$, and (iii) holds. Assume $t \in \mathcal{S}$ and $t \preccurlyeq fg$. By Lemma 1.13, $\mathcal{S}^\sharp$ is closed under right-comultiple. Hence, as t and f lie in $\mathcal{S}^\sharp$, there exists a common right-multiple fh of f and t satisfying $fh \preccurlyeq fg$ and $fh \in \mathcal{S}^\sharp$, see Figure 4. By the definition of $\mathcal{S}^\sharp$, there exists h' satisfying $h' =^\times h$ such that fh' either lies in $\mathcal{S}$ or is an identity element. Then we have $t \preccurlyeq fh'$ and $fh' \preccurlyeq fg$, whence $h' \preccurlyeq g$. If $fh' \in \mathcal{S}$ holds, (iii) implies that h' is invertible; if fh' is invertible, then so is h'. Then $t \preccurlyeq fh'$ implies $t \preccurlyeq f$, so (ii) is satisfied, and (iii) implies (ii). □

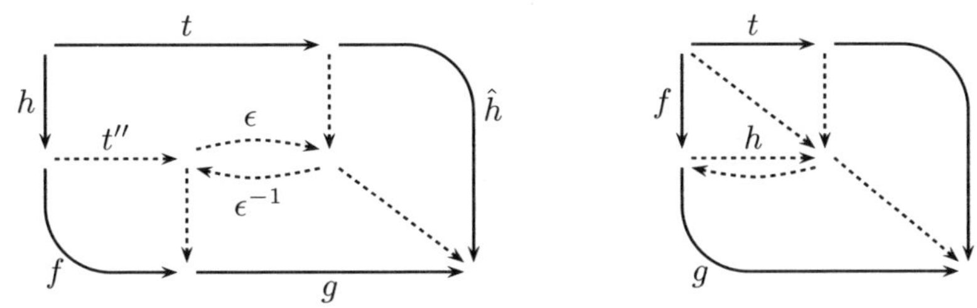

Figure 4. Proof of Lemma 1.21 (ii)⇒(i) on the left, (iii)⇒(ii) on the right.

Proposition 1.20 directly follows from Lemma 1.21 owing to Lemma 1.19, which guarantees that, under the assumptions of the proposition, $\mathcal{S}^\sharp$ is closed under right-divisor if and only if it is closed under right-complement.

1.2 Characterizations of Garside families

We are ready to establish new characterizations of Garside families that are more convenient than those of Chapter III. Note that every such criterion is in fact an existence result, since it provides sufficient conditions for every element of the considered category to admit an $\mathcal{S}$-normal decomposition.

We begin with a preparatory result. We recall that $\mathcal{S}\mathcal{C}$ denotes the subfamily of $\mathcal{C}$ consisting of all elements that can be written as the product of an element of $\mathcal{S}$ and an element of $\mathcal{C}$, that is, that are left-divisible by an element of $\mathcal{S}$.

Lemma 1.22. *If $\mathcal{S}$ is a Garside family in a left-cancellative category $\mathcal{C}$, an element of $\mathcal{C}$ admits an $\mathcal{S}$-head if and only if it belongs to $\mathcal{S}\mathcal{C}$. The latter family includes $\mathcal{C} \setminus \mathcal{C}^\times$.*

Proof. By definition, if s is an $\mathcal{S}$-head for g, then s belongs to $\mathcal{S}$ and g is a right-multiple of s, so g belongs to $\mathcal{S}\mathcal{C}$. So only the elements of $\mathcal{S}\mathcal{C}$ may admit an $\mathcal{S}$-head.

Conversely, assume that g is a non-invertible element of $\mathcal{C}$. By definition, g admits an $\mathcal{S}$-normal decomposition, hence, by Proposition III.1.23 (strict normal), g admits a strict $\mathcal{S}$-normal decomposition, say $s_1|\cdots|s_p$. In the case $p \geqslant 2$, by definition, s_1 belongs to $\mathcal{S}$ and, by Proposition III.1.12 (grouping entries), $s_1|s_2\cdots s_p$ is $\mathcal{S}$-greedy. In the case $p = 1$, by definition again, s_1 is a non-invertible element of $\mathcal{S}^\sharp$, hence we have $s_1 = s\epsilon$ for some s lying in $\mathcal{S}$ and some invertible ϵ. Then $s|\epsilon$ is an $\mathcal{S}$-normal decomposition of g whose first entry lies in $\mathcal{S}$. So, in all cases, we found an $\mathcal{S}$-greedy decomposition $s|g'$ of g with $s \in \mathcal{S}$. Then, by the definition of $\mathcal{S}$-greediness, every element of $\mathcal{S}$ left-dividing g left-divides s, so s is an $\mathcal{S}$-head of g. Hence every non-invertible element of $\mathcal{C}$ admits an $\mathcal{S}$-head. It follows that every such element belongs to $\mathcal{S}\mathcal{C}$, hence that $\mathcal{C} \setminus \mathcal{C}^\times$ is included in $\mathcal{S}\mathcal{C}$.

Finally, assume that g is invertible and belongs to $\mathcal{S}\mathcal{C}$, say $g = sg'$ with s in $\mathcal{S}$. Then s is an $\mathcal{S}$-head for g. Indeed, every element of $\mathcal{S}$ left-dividing g must be invertible, hence it left-divides s (as well as any element of $\mathcal{C}$). □

We are ready to show that every Garside family satisfies several closure properties.

Proposition 1.23 (Garside closed). *If $\mathcal{S}$ is a Garside family in a left-cancellative category, then $\mathcal{S}$ is closed under right-comultiple and $\mathcal{S}^\sharp$ is closed under right-comultiple, right-complement, and right-divisor.*

Proof. Lemma 1.22 says that every non-invertible element g of $\mathcal{C}$ admits an $\mathcal{S}$-head. Hence $\mathcal{S}$ is eligible for Lemma 1.12 and, therefore, it is closed under right-comultiple. By Lemma 1.13, we deduce that $\mathcal{S}^\sharp$ is closed under right-comultiple.

Next, assume that g lies in $\mathcal{S}^\sharp$ and g' right-divides g. By Corollary III.1.55 (length), the inequality $\|g'\|_\mathcal{S} \leqslant \|g\|_\mathcal{S} \leqslant 1$ holds, which implies that g' lies in $\mathcal{S}^\sharp$. So $\mathcal{S}^\sharp$ is closed under right-divisor.

As $\mathcal{S}^\sharp$ is closed under right-comultiple and right-divisor, Lemma 1.8 implies that it is closed under right-complement. □

We show now that, conversely, various combinations of the necessary conditions of Proposition 1.23 (nearly) characterize Garside families.

Proposition 1.24 (recognizing Garside II). *A subfamily $\mathcal{S}$ of a left-cancellative category $\mathcal{C}$ is a Garside family if and only if one of the following equivalent conditions holds:*

(1.25) *The family $\mathcal{S}^\sharp$ generates $\mathcal{C}$ and is closed under right-divisor, and every non-invertible element of $\mathcal{C}$ admits an $\mathcal{S}$-head.*

(1.26) *The family $\mathcal{S}^\sharp$ generates $\mathcal{C}$ and is closed under right-complement, and every non-invertible element of $\mathcal{S}^2$ admits an $\mathcal{S}$-head.*

(1.27) *The family $\mathcal{S}^\sharp$ generates $\mathcal{C}$ and is closed under right-divisor, $\mathcal{S}$ is closed under right-comultiple, and every non-invertible element of $\mathcal{S}^2$ admits a $\prec$-maximal left-divisor lying in $\mathcal{S}$.*

Proof. Thanks to the many preparatory lemmas of Subsection 1.1, there remain not so many facts to establish, and the arguments are simple. Assume first that $\mathcal{S}$ is a Garside family in $\mathcal{C}$. By Proposition III.1.39 (recognizing Garside I), $\mathcal{S}^\sharp$ generates $\mathcal{C}$. Next, by Proposition 1.23, $\mathcal{S}^\sharp$ is closed under right-divisor, right-comultiple, and right-complement. Finally, Lemma 1.22 says that every non-invertible element g of $\mathcal{C}$ admits an $\mathcal{S}$-head. So $\mathcal{S}$ satisfies (1.25)–(1.27).

Conversely, assume that $\mathcal{S}$ satisfies (1.27). First, by Lemma 1.8, $\mathcal{S}^\sharp$ is closed under right-complement. Hence, by Lemma 1.14, $\mathcal{C}^\times\mathcal{S} \subseteq \mathcal{S}^\sharp$ holds. Let s_1, s_2 be elements of $\mathcal{S}$ such that $s_1 s_2$ is defined. By assumption, $s_1 s_2$ admits a $\prec$-maximal left-divisor lying in $\mathcal{S}$, say t_1. Define t_2 by $s_1 s_2 = t_1 t_2$. Then Lemma 1.21 implies that the path $t_1|t_2$ is $\mathcal{S}$-greedy. Moreover, s_1 left-divides t_1, hence, as left-cancelling s_1 is allowed, t_2 right-divides s_2 and, therefore, it belongs to $\mathcal{S}^\sharp$ as the latter is closed under right-divisor. Therefore, $t_1|t_2$ is $\mathcal{S}$-normal, and we proved that $s_1 s_2$ admits an $\mathcal{S}$-normal decomposition. By Proposition III.1.39 (recognizing Garside I), $\mathcal{S}$ is a Garside family in $\mathcal{C}$.

Next, assume that $\mathcal{S}$ satisfies (1.26). By Lemma 1.14, $\mathcal{C}^\times\mathcal{S} \subseteq \mathcal{S}^\sharp$ holds and, by Lemma 1.18, $\mathcal{S}^\sharp$ is closed under right-divisor. Assume that s_1, s_2 lie in $\mathcal{S}$ and $s_1 s_2$ is defined. By assumption, $s_1 s_2$ admits an $\mathcal{S}$-head, say t_1. Define t_2 by $s_1 s_2 = t_1 t_2$. Then the same argument as above shows that $t_1|t_2$ is $\mathcal{S}$-normal and, by Proposition III.1.39 (recognizing Garside I), $\mathcal{S}$ is a Garside family in $\mathcal{C}$.

Finally, assume that $\mathcal{S}$ satisfies (1.25). Then Lemma 1.12 implies that $\mathcal{S}$ is closed under right-comultiple. Therefore, $\mathcal{S}$ satisfies (1.26), hence it is a Garside family by the above argument. □

Example 1.28 (Garside). We established in Proposition III.1.43 (Garside monoid) that, if (M, Δ) is a Garside monoid, then $\mathrm{Div}(\Delta)$ is a Garside family in M. With Proposition 1.24, this is straightforward. Indeed, we saw in Example 1.7 that $\mathrm{Div}(\Delta)$, which coincides with $\mathrm{Div}(\Delta)^\sharp$ as M contains no nontrivial invertible element, is closed under right-divisor. On the other hand, we observed in Example 1.11 that every element of M admits a $\mathrm{Div}(\Delta)$-head. So (1.25) is satisfied, and $\mathrm{Div}(\Delta)$ is a Garside family in M.

Remark 1.29. Comparing (1.25) and (1.26) makes it natural to wonder whether, in (1.25), it is enough to require the existence of an $\mathcal{S}$-head for g in $\mathcal{S}^2$. The answer is negative. Indeed, consider the braid monoid B_3^+ and $S = \{\sigma_1, \sigma_2\}$. Then S generates B_3^+, the set $S^\sharp$, which is $\{1, \sigma_1, \sigma_2\}$, is closed under right-divisor, and σ_i is a S-head of $\sigma_i\sigma_j$ for all i, j. However, S is not a Garside family in B_3^+, since $\sigma_1\sigma_2\sigma_1$ has no S-head.

About (1.27), note that the final condition is a priori weaker than s being an $\mathcal{S}$-head of g: for s to be a $\prec$-maximal left-divisor of g lying in $\mathcal{S}$ is expressed by the conjunction

$$s \preccurlyeq g \text{ and } \forall t \in \mathcal{S}\ (t \preccurlyeq g \Rightarrow s \not\prec t), \tag{1.30}$$

amounting to $s \preccurlyeq g$ and $\forall t \in \mathcal{S}\ (s \preccurlyeq t \preccurlyeq g \Rightarrow t =^\times s)$, whereas s being an $\mathcal{S}$-head of g corresponds to $s \preccurlyeq g$ and $\forall t \in \mathcal{S}\ (t \preccurlyeq g \Rightarrow t \preccurlyeq s)$: the difference is that (1.30) only involves the elements of $\mathcal{S}$ that are right-multiples of s but requires nothing about other elements of $\mathcal{S}$. About (1.27) again, owing to Lemma 1.13, we could replace the condition that $\mathcal{S}$ is closed under right-comultiple with the equivalent condition that $\mathcal{S}^\sharp$ is closed under right-comultiple (making it clear than for $\mathcal{S}$ to be a Garside family is a property of $\mathcal{S}^\sharp$). However, verifying a closure property for $\mathcal{S}$ is simpler than for $\mathcal{S}^\sharp$, so it is natural to keep this formulation in a practical criterion.

By Proposition 1.24, every Garside family is eligible for Proposition 1.20, leading to

Corollary 1.31 (recognizing greedy). *If $\mathcal{S}$ is a Garside family in a left-cancellative category $\mathcal{C}$, then, for all s in $\mathcal{S}$ and g in $\mathcal{C}$ such that sg exists, the following are equivalent:*

(i) *The path $s|g$ is $\mathcal{S}$-greedy.*

(ii) *The element s is an $\mathcal{S}$-head of sg.*

(iii) *The element s is a $\prec$-maximal left-divisor of sg lying in $\mathcal{S}$, that is, every element h of $\mathcal{C}$ satisfying $sh \in \mathcal{S}$ and $h \preccurlyeq g$ is invertible.*

If (i)–(iii) *are satisfied and s is not invertible, we have $\|sg\|_{\mathcal{S}} = \|g\|_{\mathcal{S}} + 1$.*

Proof. Proposition 1.20 gives the equivalence of (i)–(iii) directly. For the last item, let $s_1|\cdots|s_p$ be an $\mathcal{S}$-normal decomposition of g. Then $s|s_1|\cdots|s_p$ is an $\mathcal{S}$-normal decomposition of sg, since the assumption that $s|g$ is $\mathcal{S}$-greedy implies that $s|s_1$ is $\mathcal{S}$-greedy. Moreover, the assumption that sg is not invertible implies that s is not invertible. Then $\|sg\|_{\mathcal{S}}$, which is the number of non-invertible entries in $\{s, s_1, \ldots, s_p\}$, is $\|g\|_{\mathcal{S}} + 1$, since $\|g\|_{\mathcal{S}}$ is the number of non-invertible entries in $\{s_1, \ldots, s_p\}$. □

1.3 Special Garside families

Above we obtained various characterizations of Garside families in terms of closure properties. Here we shall establish similar characterizations for special Garside families, namely strong and perfect Garside families, as well as a sufficient condition for a Garside family to satisfy the second domino rule. These characterizations involve the left counterparts of the closure properties of Definitions 1.1 and 1.3.

Definition 1.32 (closure III). (i) A subfamily $\mathcal{S}$ of a (cancellative) category $\mathcal{C}$ is said to be *closed under left-divisor* if every left-divisor of an element of $\mathcal{S}$ is an element of $\mathcal{S}$.

(ii) (See Figure 5.) In the same context, $\mathcal{S}$ is said to be *closed under left-comultiple* if every common left-multiple of two elements of $\mathcal{S}$ is a left-multiple of some common left-multiple that lies in $\mathcal{S}$.

(iii) In the same context, $\mathcal{S}$ is said to be *closed under left-complement* if, when s, t lie in $\mathcal{S}$ and $ft = gs$ holds in $\mathcal{C}$, there exist s', t' in $\mathcal{S}$ and h in $\mathcal{C}$ satisfying $s't = t's$, $f = hs'$, and $g = ht'$.

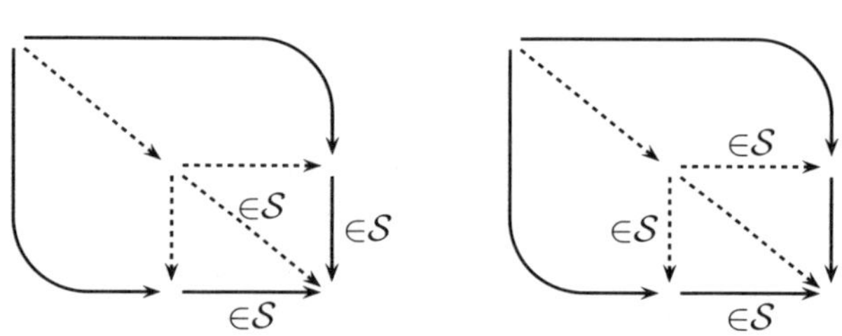

Figure 5. Closure under left-comultiple (left) and left-complement (right): every element that is a common left-multiple of two elements of $\mathcal{S}$ is a left-multiple of a common left-multiple of a certain type.

Example 1.33 (closure). Assume that (M, Δ) is a Garside monoid. Then the set $\mathrm{Div}(\Delta)$ is closed under left-divisor, left-comultiple, and left-complement. First, a left-divisor of a left-divisor of Δ is itself a left-divisor of Δ, so $\mathrm{Div}(\Delta)$ is closed under left-divisor. Next, assume that h is a common left-multiple of two elements s, t of $\mathrm{Div}(\Delta)$. Then h is a left-multiple of the left-lcm of s and t. As s and t right-divide Δ, their left-lcm also right-divides Δ and, therefore, it lies in $\mathrm{Div}(\Delta)$. So $\mathrm{Div}(\Delta)$ is closed under left-comultiple. Moreover, the involved elements s', t' left-divide some right-divisor of Δ, hence they themselves left-divide Δ (since the left- and right-divisors of Δ coincide), and $\mathrm{Div}(\Delta)$ is closed under left-complement.

On the other hand, the following examples show that a general Garside family need not satisfy the closure properties of Definition 1.32.

Example 1.34 (not closed). Consider the left-absorbing monoid L_n and the family S_n with $n \geqslant 2$ (Reference Structure 8, page 113), namely $L_n = \langle \mathtt{a}, \mathtt{b} \mid \mathtt{a}\mathtt{b}^n = \mathtt{b}^{n+1} \rangle$ and $S_n = \{1, \mathtt{a}, \mathtt{b}, \mathtt{b}^2, ..., \mathtt{b}^{n+1}\}$. We saw in Chapter III that L_n is left-cancellative and S_n is a Garside family in L_n—and we can reprove this now easily using Proposition 1.24. The element $\mathtt{ab}$ is a left-divisor of $\mathtt{b}^{n+1}$ that does not belong to S_n, so S_n is not closed under left-divisor. Next, we have $(\mathtt{ab})\,\mathtt{b}^{n-1} = \mathtt{b}^2\,\mathtt{b}^{n-1}$. Now, the only decompositions of $\mathtt{ab}$ in L_n are $1|\mathtt{ab}$, $\mathtt{a}|\mathtt{b}$, and $\mathtt{ab}|1$, whereas those of $\mathtt{b}^2$ are $1|\mathtt{b}^2$, $\mathtt{b}|\mathtt{b}$, and $\mathtt{b}^2|1$. An exhaustive search shows that no combination can witness for closure under left-complement. On the other hand, S_n turns out to be closed under left-comultiple.

Similarly, the 16-element Garside family S in the affine braid monoid of type $\widetilde{A}_2$ (Reference Structure 9, page 114) is neither closed under left-divisor nor under left-comultiple: $\sigma_1\sigma_2\sigma_3\sigma_2$ lies in S, but its left-divisor $\sigma_1\sigma_2\sigma_3$ does not, $\sigma_2\sigma_1$ and $\sigma_3\sigma_1$ lie in S, but their left-lcm $\sigma_2\sigma_3\sigma_2\sigma_1$ does not, so the latter is a common left-multiple of two elements of S that cannot be a left-multiple of a common left-multiple lying in S.

Here come the expected characterizations of strong and perfect Garside families in terms of closure properties. By Proposition III.2.34 (strong exists), strong Garside families exist if and only if the ambient category admits conditional weak left-lcms, and, therefore, it is natural to restrict to such categories.

Proposition 1.35 (strong and perfect Garside). *If S is a Garside family in a cancellative category $\mathcal{C}$ that admits conditional weak left-lcms, the following are equivalent:*

(i) *The family S is a strong (resp. perfect) Garside family in $\mathcal{C}$.*

(ii) *The family $S^\sharp$ is closed under left-complement (resp. this and left-comultiple).*

When the above conditions are met, $S^\sharp$ is also closed under left-divisor.

Proof. First consider the case of strong Garside families. Assume that S is a strong Garside family in $\mathcal{C}$, that s, t lie in $S^\sharp$, and that the equality $ft = gs$ holds. By the definition of a strong Garside family, there exist s' and t' in $S^\sharp$, plus h in $\mathcal{C}$, satisfying $s't = t's$, $f = hs'$, and $g = ht'$. The elements s', t', and h witness that $S^\sharp$ is closed under left-complement, and (i) implies (ii) (here we used neither the assumption that $\mathcal{C}$ admits right-cancellation, nor the assumption that $\mathcal{C}$ admits conditional weak left-lcms.)

Conversely, assume that $\mathcal{S}$ is Garside and $\mathcal{S}^\sharp$ is closed under left-complement. Let s, t be elements of $\mathcal{S}^\sharp$ that admit a common left-multiple, say $ft = gs$, see Figure 6. Since the category $\mathcal{C}$ admits conditional weak left-lcms, ft is a left-multiple of some weak left-lcm of s and t, say $f't = g's$. Next, as s and t belong to $\mathcal{S}^\sharp$, the assumption that $\mathcal{S}^\sharp$ is closed under left-complement implies the existence of s', t' in $\mathcal{S}^\sharp$ and h satisfying $s't = t's$, $f' = hs'$, and $g' = ht'$. Now, by Lemma III.2.52, f' and g' are left-disjoint, which implies that h is invertible, which in turn implies that s' and t' are left-disjoint. Thus we found s' and t' in $\mathcal{S}^\sharp$, left-disjoint, and such that ft is a left-multiple of $s't$. Hence, according to Definition III.2.29, $\mathcal{S}$ is a strong Garside family in $\mathcal{C}$ and (ii) implies (i).

Assume now that (i) and (ii) are satisfied. Then the left counterpart of Lemma 1.18, which is valid as, by assumption, $\mathcal{C}$ is right-cancellative, implies that $\mathcal{S}^\sharp$, which generates $\mathcal{C}$ and is closed under left-complement, is also closed under left-divisor.

For perfect Garside families, the argument is similar. The only difference is that, when we start with a perfect Garside family $\mathcal{S}$ and elements s, t of $\mathcal{S}^\sharp$ that satisfy $ft = gs$, we can assume that the factoring relation $s't = t's$ is such that s', t', but also $s't$, belong to $\mathcal{S}^\sharp$ and, therefore, they simultaneously witness for $\mathcal{S}^\sharp$ being closed under left-complement and under left-comultiple.

Conversely, when we start with a family $\mathcal{S}$ that satisfies (ii) and, again, elements s, t of $\mathcal{S}^\sharp$ that satisfy $ft = gs$, then, by the left counterpart of Lemma 1.8, which is valid as $\mathcal{C}$ is right-cancellative, we can assume the existence of a factoring relation $s't = t's$ that simultaneously witnesses for closure under left-complement and left-comultiple, hence such that s', t', and $s't$, belong to $\mathcal{S}^\sharp$. This shows that $\mathcal{S}$ is a perfect Garside family. □

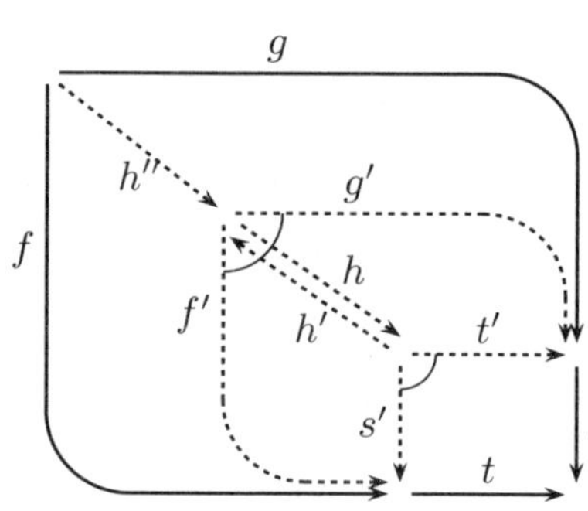

Figure 6. Proof of Proposition 1.35.

Example 1.36 (strong and perfect Garside). Owing to the closure results of Example 1.33, Proposition 1.35 immediately implies that, if (M, Δ) is a Garside monoid, then $\mathrm{Div}(\Delta)$ is a perfect Garside family in M.

In the same vein, we now address the second domino rule (Definition III.1.57) and show that it follows from closure properties. We begin with an alternative form.

Proposition 1.37 (second domino rule, alternative form). *If $\mathcal{S}$ is a Garside family in a left-cancellative category and the second domino rule is valid for $\mathcal{S}$, we have*

(1.38) *For s_1, s_2, s_3 in $\mathcal{S}$, if $s_1|s_2$ is $\mathcal{S}$-normal and $\|s_1s_2s_3\|_{\mathcal{S}} \leqslant 2$ holds, then s_2s_3 lies in $\mathcal{S}^\sharp$.*

Conversely, if $\mathcal{S}$ is a Garside family satisfying (1.38) *and $\mathcal{S}^\sharp$ is closed under left-divisor, the second domino rule is valid for $\mathcal{S}$.*

Proof. Assume that the second domino rule is valid for $\mathcal{S}$ and s_1, s_2, s_3 are elements of $\mathcal{S}$ such that $s_1|s_2$ is $\mathcal{S}$-normal and the inequality $\|s_1s_2s_3\|_{\mathcal{S}} \leqslant 2$ holds. Let $t_1|s'_2$ be an $\mathcal{S}$-normal decomposition of s_2s_3, and $t_0|s'_1$ be an $\mathcal{S}$-normal decomposition of s_1t_1. By Proposition III.1.61, $t_0|s'_1|s'_2$ is an $\mathcal{S}$-normal decomposition of $s_1s_2s_3$. The assumption that $s_1s_2s_3$ has $\mathcal{S}$-length at most 2 implies that s'_2 is invertible. By construction, t_1 lies in $\mathcal{S}^\sharp$, hence so does $t_1s'_2$, which is s_2s_3.

Conversely, assume that $\mathcal{S}$ is a Garside family satisfying (1.38) and $\mathcal{S}^\sharp$ is closed under left-divisor. Consider a domino diagram, namely $s_1, ..., t_2$ in $\mathcal{S}$ satisfying $s_1t_1 = t_0s'_1$ and $s_2t_2 = t_1s'_2$ with $s_1|s_2$ and $t_1|s'_2$ both $\mathcal{S}$-normal. We wish to show that $s'_1|s'_2$ is $\mathcal{S}$-greedy. As $\mathcal{S}$ is a Garside family, by Corollary 1.31, it suffices to show that s'_1 is a maximal left-divisor of $s'_1s'_2$ lying in $\mathcal{S}$.

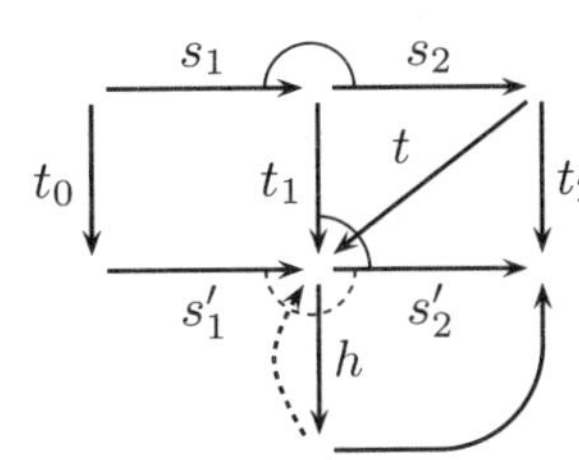

So assume $s'_1h \in \mathcal{S}$ with $h \preccurlyeq s'_2$. As $t_1|s'_2$ is $\mathcal{S}$-greedy, there exists t satisfying $t_1 = s_2t$. Then we have $t_1h = s_2th$. Moreover th left-divides t_2, hence it lies in $\mathcal{S}^\sharp$ as the latter is assumed to be closed under left-divisor. Choose s_3 in $\mathcal{S}$ satisfying $s_3 =^\times th$ (the case $th \in \mathcal{C}^\times$ is trivial). We have

$$\|s_1s_2s_3\|_{\mathcal{S}} = \|s_1s_2th\|_{\mathcal{S}} = \|t_0(s'_1h)\|_{\mathcal{S}} \leqslant 2.$$

Then (1.38) implies $s_2s_3 \in \mathcal{S}^\sharp$, whence $s_2th \in \mathcal{S}^\sharp$, that is, $t_1h \in \mathcal{S}^\sharp$. As $t_1|s'_2$ is $\mathcal{S}$-normal, by Corollary 1.31, the conjunction of $t_1h \in \mathcal{S}^\sharp$ and $h \preccurlyeq s'_2$ implies that h is invertible. Hence, s'_1 is a maximal left-divisor of $s'_1s'_2$ lying in $\mathcal{S}$, and $s'_1|s'_2$ is $\mathcal{S}$-normal. We conclude that the second domino rule is valid for $\mathcal{S}$. □

Proposition 1.39 (second domino rule). *If $\mathcal{S}$ is a Garside family in a cancellative category $\mathcal{C}$ and $\mathcal{S}^\sharp$ is closed under left-comultiple and left-divisor, the second domino rule is valid for $\mathcal{S}$.*

Proof. We show that $\mathcal{S}$ is eligible for Proposition 1.37. To this end, it suffices to check that (1.38) holds. So assume that s_1, s_2, s_3 lie in $\mathcal{S}$, the path $s_1|s_2$ is $\mathcal{S}$-normal, and the inequality $\|s_1s_2s_3\|_{\mathcal{S}} \leqslant 2$ holds. Let $t_1|t_2$ be an $\mathcal{S}$-normal decomposition of $s_1s_2s_3$.

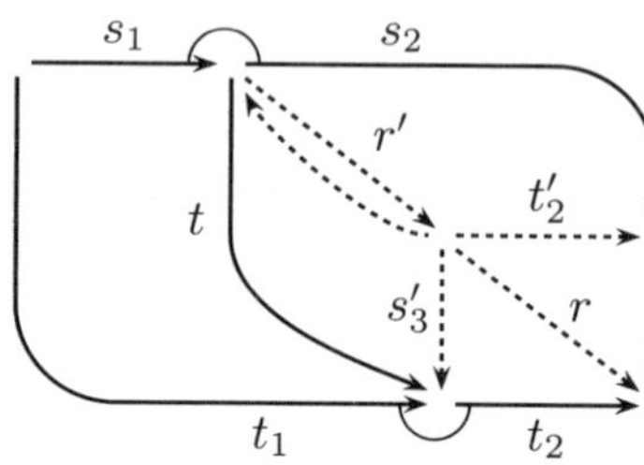

By assumption, we have $s_1 \preccurlyeq t_1t_2$ and $s_1 \in \mathcal{S}$, hence $s_1 \preccurlyeq t_1$, that is, there exists t satisfying $t_1 = s_1t$. We deduce $s_1tt_2 = t_1t_2 = s_1s_2s_3$, hence $tt_2 = s_2s_3$. As t_2 and s_3 belong to $\mathcal{S}^\sharp$ and $\mathcal{S}^\sharp$ is closed under left-comultiple, there exists r in $\mathcal{S}^\sharp$ such that r is a common left-multiple of s_3 and t_2, say $r = t's_3 = s't_2$, and s_2s_3 is a left-multiple of r, say $s_2s_3 = r'r$.

Then we have $r'r = r't's_3 = s_2s_3$, hence $s_2 = r't'$ by right-cancelling s_3. Hence r' is a left-divisor of s_2, which belongs to $\mathcal{S}^\sharp$, and, therefore, r' belongs to $\mathcal{S}^\sharp$ as the latter is closed under left-divisor. On the other hand, we also have $r'r = r's't_2 = tt_2$, hence $t = r's'$ and, therefore, $t_1 = s_1t = s_1r's'$. So s_1r' left-divides t_1, which belongs to $\mathcal{S}^\sharp$ and, therefore, s_1r' belongs to $\mathcal{S}^\sharp$. By Corollary 1.31, as $s_1|s_2$ is $\mathcal{S}$-normal, hence $\mathcal{S}^\sharp$-normal, the conjunction of $s_1r' \in \mathcal{S}^\sharp$ and $r' \preccurlyeq s_2$ implies that r' is invertible. Hence s_2s_3, which is $r'r$, belongs to $\mathcal{C}^\times\mathcal{S}^\sharp$, hence to $\mathcal{S}^\sharp$ since $\mathcal{S}$ is a Garside family and therefore satisfies $\mathcal{C}^\times\mathcal{S} \subseteq \mathcal{S}^\sharp$. So (1.38) holds in $\mathcal{S}$. Hence, by Proposition 1.37, the second domino rule is valid for $\mathcal{S}$. □

1.4 Head functions

We establish further characterizations of Garside families, this time in terms of the global properties of $\mathcal{S}$-head functions, defined to be those maps that associate with every non-invertible element g of the ambient category an $\mathcal{S}$-head of g. We begin with some easy observations about $\mathcal{S}$-heads in general.

Lemma 1.40. *Assume that $\mathcal{C}$ is a left-cancellative category and $\mathcal{S}$ is included in $\mathcal{C}$. If s and s' are $\mathcal{S}$-heads of $=^\times$-equivalent elements g and g', respectively, then $s =^\times s'$ holds, and even $s = s'$ if $\mathcal{S}$ is $=^\times$-transverse.*

Proof. Assume $g' =^\times g$. As s is an $\mathcal{S}$-head of g, we have $s \preccurlyeq g$, whence $s \preccurlyeq g'$, and $s \preccurlyeq s'$ since s' is an $\mathcal{S}$-head of g'. A symmetric argument gives $s' \preccurlyeq s$, whence $s' =^\times s$. If $\mathcal{S}$ is $=^\times$-transverse, it contains at most one element in each $=^\times$-equivalence class, so $s' =^\times s$ implies $s' = s$ as, by assumption, s and s' belong to $\mathcal{S}$. □

Lemma 1.41. *Assume that $\mathcal{S}$ is a subfamily of a left-cancellative category $\mathcal{C}$. For s in $\mathcal{S}$ and g in $\mathcal{C}$, consider the following three statements:*

(i) *The element s is an $\mathcal{S}$-head of g.*

(ii) *There exists g' in $\mathcal{C}$ such that $s|g'$ is an $\mathcal{S}$-greedy decomposition of g.*

(iii) *There exists an $\mathcal{S}$-normal decomposition of g whose first entry is s.*

Then (iii) *implies* (ii), *which implies* (i). *If $\mathcal{S}^\sharp$ is closed under right-complement,* (i) *and* (ii) *are equivalent, and, if $\mathcal{S}$ is a Garside family,* (i), (ii), *and* (iii) *are equivalent.*

Proof. The fact that (ii) implies (i) and that, if $\mathcal{S}^\sharp$ is closed under right-complement, then (i) implies (ii), was proved in Lemma 1.21.

Assume now that $s|s_2\cdots s_p$ is an $\mathcal{S}$-normal decomposition of g. Let $g' = s_2\cdots s_p$ in the case $p \geqslant 2$, and let $g' = 1_y$ otherwise (y the target of g). By Proposition III.1.12 (grouping entries), $s|g'$ is $\mathcal{S}$-greedy in each case, and $g = sg'$ holds. So (iii) implies (ii).

Conversely, assume that $\mathcal{S}$ is Garside and $s|g'$ is an $\mathcal{S}$-greedy decomposition of g. By definition, g' admits an $\mathcal{S}$-normal decomposition, say $s_2|\cdots|s_p$. By Lemma III.1.7, the assumption that $s|g'$ is $\mathcal{S}$-greedy implies that $s|s_2$ is $\mathcal{S}$-greedy as well, so $s|s_2|\cdots|s_p$ is an $\mathcal{S}$-normal decomposition of g. So (ii) implies (iii) in this case. □

We now consider the functions that associate with every element g an $\mathcal{S}$-head of g, when the latter exists. In view of Lemma 1.22, the maximal domain for such a function is the (proper or unproper) subfamily $\mathcal{S}\mathcal{C}$ of $\mathcal{C}$. As every element of $\mathcal{S} \cap \mathcal{C}^{\times}(x, \cdot)$ is an $\mathcal{S}$-head for every element of $\mathcal{C}^{\times}(x, \cdot)$, the $\mathcal{S}$-heads of invertible elements provide no real information and the interesting part in a head function lies in its restriction to $\mathcal{C} \setminus \mathcal{C}^{\times}$.

Definition 1.42 (head function). If $\mathcal{S}$ is a subfamily of a left-cancellative category $\mathcal{C}$, an $\mathcal{S}$-*head function* is a partial map H from $\mathcal{C}$ to $\mathcal{S}$ such that $\mathrm{Dom}(H)$ includes $\mathcal{C} \setminus \mathcal{C}^{\times}$ and $H(g)$ is an $\mathcal{S}$-head of g for every g in $\mathrm{Dom}(H)$. An $\mathcal{S}$-head function is called *sharp* if $g' =^{\times} g$ implies $H(g') = H(g)$.

As for the uniqueness of head functions, Lemma 1.40 immediately implies:

Proposition 1.43 (head function unique). (i) *If $\mathcal{S}$ is a subfamily of a left-cancellative category $\mathcal{C}$ and H, H' are $\mathcal{S}$-head functions, then the relation $H(g) =^{\times} H'(g)$ holds whenever both sides are defined. If, moreover, $\mathcal{S}$ is $=^{\times}$-transverse, then $H(g) = H'(g)$ holds.*

(ii) *Conversely, if $\mathcal{S}$ is a subfamily of a left-cancellative category $\mathcal{C}$ and H is an $\mathcal{S}$-head function, then every partial function H' from $\mathcal{C}$ to $\mathcal{S}$ satisfying* $\mathrm{Dom}(H') = \mathrm{Dom}(H)$ *and $H'(g) =^{\times} H(g)$ for every g in* $\mathrm{Dom}(H)$ *is an $\mathcal{S}$-head function.*

As for the existence of head functions, if $\mathcal{C}$ is a left-cancellative category and $\mathcal{S}$ is a Garside family in $\mathcal{S}$, then, by Lemma 1.22, every element of $\mathcal{S}\mathcal{C}$ admits an $\mathcal{S}$-head and, therefore, there exists an $\mathcal{S}$-head function. We can say more.

Proposition 1.44 (sharp exist). *If $\mathcal{S}$ is a Garside family of a left-cancellative category $\mathcal{C}$, then there exists a sharp $\mathcal{S}$-head function defined on $\mathcal{S}\mathcal{C}$.*

Proof. Assume that F is a choice function on $\mathcal{S}$ with respect to $=^{\times}$, that is, F is a map from $\mathcal{S}$ to itself that picks one element in each $=^{\times}$-equivalence class. Let $\mathcal{S}_1$ be the image of $\mathcal{S}$ under F. Then $\mathcal{S}_1$ is an $=^{\times}$-selector in $\mathcal{S}$ and we have $\mathcal{S}_1 \subseteq \mathcal{S} \subseteq \mathcal{S}_1^{\sharp} = \mathcal{S}^{\sharp}$. Hence, by Proposition III.1.33 (invariance), $\mathcal{S}_1$ is a Garside family in $\mathcal{C}$. Let H be an $\mathcal{S}_1$-head function defined on $\mathcal{S}\mathcal{C}$. As $\mathcal{S}_1$ is included in $\mathcal{S}$, the function H is also an $\mathcal{S}$-head function. Now, by Lemma 1.40, $g' =^{\times} g$ implies $H(g') =^{\times} H(g)$, hence $H(g) = H(g')$ since $\mathcal{S}_1$ is $=^{\times}$-transverse. So H is sharp. □

Head functions satisfy various algebraic relations, which will eventually lead to new characterizations of Garside families.

Lemma 1.45. *If $\mathcal{S}$ is a Garside family in a left-cancellative category $\mathcal{C}$, every $\mathcal{S}$-head function H satisfies*

$$\text{(1.46)}\qquad \text{(i) } H(g) \preccurlyeq g, \quad \text{(ii) } g \in \mathcal{S} \Rightarrow H(g) =^{\times} g, \quad \text{(iii) } f \preccurlyeq g \Rightarrow H(f) \preccurlyeq H(g).$$

Conversely, if $\mathcal{S}$ is a subfamily of $\mathcal{C}$ and H is a partial function from $\mathcal{C}$ to $\mathcal{S}$ whose domain includes $\mathcal{C} \setminus \mathcal{C}^{\times}$ and which satisfies (1.46), *then H is an $\mathcal{S}$-head function.*

Proof. Assume that $\mathcal{S}$ is a Garside family in $\mathcal{C}$ and H is an $\mathcal{S}$-head function. If $H(g)$ is defined, then, by definition, $H(g)$ is an $\mathcal{S}$-head of g, hence $H(g) \preccurlyeq g$ holds. Moreover, if g belongs to $\mathcal{S}$, as g is a left-divisor of itself, we must have $g \preccurlyeq H(g)$, whence $H(g) =^{\times} g$. Finally, if f left-divides g and $H(f)$ is defined, then $H(f)$ is an element of $\mathcal{S}$ that left-divides g, hence it left-divides $H(g)$, which is an $\mathcal{S}$-head of g. So (1.46)(iii) is satisfied.

Assume now that H is a partial function from $\mathcal{C}$ to $\mathcal{S}$ whose domain includes $\mathcal{C} \setminus \mathcal{C}^{\times}$ and which satisfies (1.46). Assume that $H(g)$ is defined. By assumption, $H(g)$ lies in $\mathcal{S}$ and it left-divides g by (1.46)(i). Next, assume that s is a non-invertible element of $\mathcal{S}$ that left-divides g. By assumption, $H(s)$ is defined and, by (1.46)(ii), the relation $H(s) =^{\times} s$ holds. Moreover, by (1.46)(iii), $s \preccurlyeq g$ implies $H(s) \preccurlyeq H(g)$, whence $s \preccurlyeq H(g)$ because of the relation $s =^{\times} H(s)$. On the other hand, if s lies in $\mathcal{S} \cap \mathcal{C}^{\times}$, then $s \preccurlyeq H(g)$ trivially holds. Hence, $s \preccurlyeq H(g)$ holds for every s in $\mathcal{S}$ left-dividing g, and $H(g)$ is an $\mathcal{S}$-head of g. So H is an $\mathcal{S}$-head function. □

Beyond the previous basic relations, head functions satisfy one more important rule.

Definition 1.47 ($\mathcal{H}$-law). A partial function H from a category $\mathcal{C}$ to itself is said to obey the *$\mathcal{H}$-law* (*resp.* the *sharp $\mathcal{H}$-law*) if the relation

$$\text{(1.48)}\qquad H(fg) =^{\times} H(fH(g)) \qquad (\textit{resp. } =)$$

holds whenever the involved expressions are defined.

Proposition 1.49 ($\mathcal{H}$-law). *If $\mathcal{S}$ is a Garside family in a left-cancellative category $\mathcal{C}$, every $\mathcal{S}$-head function H obeys the $\mathcal{H}$-law, and every sharp $\mathcal{S}$-head function obeys the sharp $\mathcal{H}$-law.*

Proof. Assume that $H(g)$, $H(fg)$, and $H(fH(g))$ are defined. On the one hand, by (1.46), we have $H(g) \preccurlyeq g$, hence $fH(g) \preccurlyeq fg$, hence $H(fH(g)) \preccurlyeq H(fg)$. On the other hand, by Lemma 1.41, there exists an $\mathcal{S}$-normal decomposition of g whose first entry is $H(g)$, say $H(g)|s_2|\cdots|s_p$. We obtain $H(fg) \preccurlyeq fg = fH(g)s_2\cdots s_p$. As $H(fg)$ lies in $\mathcal{S}$ and $H(g)|s_2|\cdots|s_p$ is $\mathcal{S}$-greedy, we deduce $H(fg) \preccurlyeq fH(g)$, whence $H(fg) \preccurlyeq H(fH(g))$ since $H(fg)$ lies in $\mathcal{S}$. So H obeys the $\mathcal{H}$-law.

If H is sharp, the assumption that $H(fg)$ and $H(fH(g))$ belong to $\mathcal{S}$ and are $=^{\times}$-equivalent implies $H(fg) = H(fH(g))$, and the sharp $\mathcal{H}$-law is obeyed. □

We deduce new characterizations of Garside families in terms of head functions.

Proposition 1.50 (recognizing Garside III). *A subfamily $\mathcal{S}$ of a left-cancellative category $\mathcal{C}$ is a Garside family if and only if $\mathcal{S}^\sharp$ generates $\mathcal{C}$ and one of the following equivalent conditions is satisfied:*

(1.51) *There exists an $\mathcal{S}$-head function with domain $\mathcal{S}\mathcal{C}$ obeying the sharp $\mathcal{H}$-law.*

(1.52) *There exists an $\mathcal{S}$-head function obeying the $\mathcal{H}$-law.*

(1.53) *There exists a partial map H from $\mathcal{C}$ to $\mathcal{S}$ whose domain includes $\mathcal{C} \setminus \mathcal{C}^\times$, and which satisfies* (1.46) *and obeys the $\mathcal{H}$-law.*

(1.54) *We have $\mathcal{C}^\times\mathcal{S}^\sharp \subseteq \mathcal{S}^\sharp$ and there exists a partial map H from $\mathcal{C}$ to $\mathcal{S}$ whose domain is $\mathcal{S}\mathcal{C}$, and which satisfies* (1.46) *and obeys the $\mathcal{H}$-law.*

When (1.53) *or* (1.54) *is satisfied, H is an $\mathcal{S}$-head function.*

Proof of Proposition 1.50. Assume that $\mathcal{S}$ is a Garside family. By Proposition III.1.39 (recognizing Garside I), $\mathcal{S}^\sharp$ generates $\mathcal{C}$ and $\mathcal{C}^\times\mathcal{S}^\sharp \subseteq \mathcal{S}^\sharp$ holds. Next, by Proposition 1.44, there exists a sharp $\mathcal{S}$-head function defined on $\mathcal{S}\mathcal{C}$. By Proposition 1.49, the latter satisfies the sharp $\mathcal{H}$-law. So (1.51) is satisfied, and therefore so is (1.52). Moreover, by Lemma 1.45, an $\mathcal{S}$-head function satisfies (1.46), so (1.53) and (1.54) hold as well.

Conversely, it is clear that (1.51) implies (1.52). Moreover, owing to the last statement in Lemma 1.45, a function whose domain includes $\mathcal{C} \setminus \mathcal{C}^\times$ and which satisfies (1.46) must be an $\mathcal{S}$-head function, so (1.53) implies (1.52). On the other hand, if $\mathcal{S}^\sharp$ generates $\mathcal{C}$ and $\mathcal{C}^\times\mathcal{S}^\sharp \subseteq \mathcal{S}^\sharp$ holds, then $\mathcal{C} \setminus \mathcal{C}^\times$ is included in $\mathcal{S}\mathcal{C}$: indeed, in this case, if g is non-invertible, it must be left-divisible by an element of the form ϵs with $\epsilon \in \mathcal{C}^\times$ and $s \in \mathcal{S} \setminus \mathcal{C}^\times$, hence by an element of $\mathcal{S} \setminus \mathcal{C}^\times$, because $\mathcal{C}^\times\mathcal{S}^\sharp$ is included in $\mathcal{S}^\sharp$. Hence (1.54) also implies (1.52), and we are left with proving that (1.52) implies that $\mathcal{S}$ is a Garside family.

Assume that $\mathcal{S}$ satisfies (1.52). By assumption, $\mathcal{S}^\sharp$ generates $\mathcal{C}$ and every non-invertible element of $\mathcal{C}$ admits an $\mathcal{S}$-head. So, in order to show that $\mathcal{S}$ satisfies (1.25) and apply Proposition 1.24, it suffices to show that $\mathcal{S}^\sharp$ is closed under right-divisor.

So assume that t belongs to $\mathcal{S}^\sharp$ and g right-divides t, say $t = t'g$. If g is invertible, it belongs to $\mathcal{S}^\sharp$ by definition. Assume now that g, and therefore t, are not invertible. Then we have $t =^\times s$ for some non-invertible s lying in $\mathcal{S}$. By assumption, $H(t)$ and $H(s)$ are defined and, by (1.46)(iii), we have $H(t) =^\times H(s)$, and $H(s) =^\times s$ by (1.46)(ii), whence $H(t) =^\times s$ by (iii). We deduce $H(t) =^\times t$, that is, $H(t'g) =^\times t'g$. Then the $\mathcal{H}$-law implies $H(t'g) =^\times H(t'H(g))$, whereas (1.46)(i) gives $H(t'H(g)) \preccurlyeq t'H(g)$, so we find

$$t'g =^\times H(t'g) =^\times H(t'H(g)) \preccurlyeq t'H(g),$$

hence $t'g \preccurlyeq t'H(g)$, and $g \preccurlyeq H(g)$ by left-cancelling t'. As $H(g) \preccurlyeq g$ always holds, we deduce $g =^\times H(g)$. As, by definition, $H(g)$ lies in $\mathcal{S}$, it follows that g lies in $\mathcal{S}^\sharp$. So $\mathcal{S}^\sharp$ is closed under right-divisor. Then, by Proposition 1.24, $\mathcal{S}$ is a Garside family in $\mathcal{C}$. □

Further variants are possible, see Exercise 39 for another combination of conditions.

2 Special contexts

In Section 1 we established various characterizations of general Garside families in the context of an arbitrary left-cancellative category. When we consider special families, typically families that satisfy closure properties, or when the ambient category turns out to satisfy additional properties, typically Noetherianity conditions or existence of least common multiples, some of the conditions defining Garside families take special forms, or even are automatically satisfied, and we obtain new, simpler characterizations. In this section, we list results in this direction.

The section contains four subsections. Subsection 2.1 deals with the case of what we call solid families. Subsection 2.2 deals with right-Noetherian categories. Subsection 2.3 deals with categories that admit minimal common right-multiples, an important special case being that of categories that admit conditional right-lcms. Finally, we address in Subsection 2.4 categories that admit unique right-lcms.

2.1 Solid families

Up to now we considered general Garside families. Here we shall now adapt the results to the case of Garside families called *solid*, which satisfy some additional closure conditions. The interest of considering such families is that several statements take a more simple form. Solid Garside families will appear naturally when germs are studied in Chapter VI.

Definition 2.1 (solid). A subfamily $\mathcal{S}$ of a category $\mathcal{C}$ is said to be *solid* in $\mathcal{C}$ if $\mathcal{S}$ includes $\mathbf{1}_\mathcal{C}$ and is closed under right-divisor.

As the terminology suggests, a family that is not solid can be seen as having holes: some identity elements or some right-divisors are missing.

Lemma 2.2. (i) *If $\mathcal{S}$ is a solid subfamily in a left-cancellative category $\mathcal{C}$, then $\mathcal{S}$ includes $\mathcal{C}^\times$, and we have $\mathcal{C}^\times\mathcal{S} = \mathcal{S}$ and $\mathcal{S}^\sharp = \mathcal{S}\mathcal{C}^\times$.*

(ii) *A family of the form $\mathcal{S}^\sharp$ is solid if and only if it is closed under right-divisor.*

Proof. (i) Assume $\epsilon \in \mathcal{C}^\times(\cdot, y)$. By assumption, 1_y belongs to $\mathcal{S}$. As we have $\epsilon^{-1}\epsilon = 1_y$ and $\mathcal{S}$ is closed under right-divisor, ϵ must lie in $\mathcal{S}$. Next, $\mathbf{1}_\mathcal{C}$ is included in $\mathcal{C}^\times$, so $\mathcal{C}^\times\mathcal{S} \supseteq \mathcal{S}$ always holds. In the other direction, assume $\epsilon \in \mathcal{C}^\times$, $s \in \mathcal{S}$, and ϵs exists. Then we have $s = \epsilon^{-1}(\epsilon s)$, so ϵs, which right-divides an element of $\mathcal{S}$, lies in $\mathcal{S}$.

(ii) By definition, $\mathcal{S}^\sharp$ always includes $\mathbf{1}_\mathcal{C}$. □

It turns out that many Garside families are solid:

Lemma 2.3. *If $\mathcal{S}$ is a Garside family in a left-cancellative category $\mathcal{C}$, then $\mathcal{S}^\sharp$ is a solid Garside family. If $\mathcal{C}$ contains no nontrivial invertible element, every Garside family that includes $1_\mathcal{C}$ is solid.*

Proof. Assume that $\mathcal{S}$ is a Garside family in $\mathcal{C}$. By Proposition III.1.33 (invariance), $\mathcal{S}^\sharp$ is a Garside family as well. By definition, $1_\mathcal{C}$ is included in $\mathcal{S}^\sharp$, and, by Proposition 1.24, $\mathcal{S}^\sharp$ is closed under right-divisor. Hence $\mathcal{S}^\sharp$ is solid.

In the case $\mathcal{C}^\times = 1_\mathcal{C}$, we have $\mathcal{S}^\sharp = \mathcal{S} \cup 1_\mathcal{C}$, whence $\mathcal{S}^\sharp = \mathcal{S}$ whenever $\mathcal{S}$ includes $1_\mathcal{C}$, and we apply the above result. □

By Lemma 2.3, every Garside family $\mathcal{S}$ that satisfies $\mathcal{S}^\sharp = \mathcal{S}$ is solid. The converse implication need not be true, see Exercise 48. The assumption that a Garside family $\mathcal{S}$ is solid implies that several properties involving $\mathcal{S}^\sharp$ in the general case are true for $\mathcal{S}$.

Proposition 2.4 (solid Garside). *If $\mathcal{S}$ is a solid Garside family in a left-cancellative category $\mathcal{C}$, then $\mathcal{S}$ generates $\mathcal{C}$, and is closed under right-divisor, right-comultiple, and right-complement.*

Proof. As $\mathcal{S}$ is a Garside family, $\mathcal{S} \cup \mathcal{C}^\times$ generates $\mathcal{C}$. Now, by Lemma 2.2, $\mathcal{C}^\times$ is included in $\mathcal{S}$, so $\mathcal{S} \cup \mathcal{C}^\times$ coincides with $\mathcal{S}$.

That $\mathcal{S}$ is closed under right-divisor follows from the definition of being solid. Next, Lemma 1.12 implies that $\mathcal{S}$ is closed under right-comultiple, and, then, Lemma 1.8 implies that $\mathcal{S}$ is closed under right-complement. □

Another property of the families $\mathcal{S}^\sharp$ shared by all solid Garside families is the existence of $\mathcal{S}$-normal decompositions with entries in $\mathcal{S}$. The key technical point is the following easy observation.

Lemma 2.5. *If $\mathcal{S}$ is a Garside family in a left-cancellative category $\mathcal{C}$ and $\mathcal{S}$ is closed under right-divisor, then $\mathcal{S}$ satisfies Property □.*

Proof. Let $s_1|s_2$ lie in $\mathcal{S}^{[2]}$. By Lemma 1.22, the assumption that $\mathcal{S}$ is a Garside family and s_1s_2 lies in $\mathcal{S}\mathcal{C}$ implies that s_1s_2 admits an $\mathcal{S}$-head, say t_1. Let t_2 be determined by $s_1s_2 = t_1t_2$. By Lemma 1.41, the path $t_1|t_2$ is $\mathcal{S}$-greedy. Moreover, as s_1 lies in $\mathcal{S}$ and left-divides s_1s_2, it must left-divide t_1, so the equality $t_1 = s_1s$ holds for some s. Then we have $s_1s_2 = t_1t_2 = s_1st_2$, whence $s_2 = st_2$ by left-cancelling s_1. So t_2 right-divides s_2, which belongs to $\mathcal{S}$, hence t_2 belongs to $\mathcal{S}$, as the latter is closed under right-divisor. Thus, $t_1|t_2$ is an $\mathcal{S}$-normal decomposition of s_1s_2 whose entries lie in $\mathcal{S}$. □

One can wonder whether, conversely, every generating family $\mathcal{S}$ satisfying Property □ must be solid: the answer is negative, see Exercise 49.

Proposition 2.6 (normal exist, solid case). *If $\mathcal{S}$ is a solid Garside family in a left-cancellative category $\mathcal{C}$, every element of $\mathcal{C}$ admits an $\mathcal{S}$-normal decomposition with entries in $\mathcal{S}$.*

Proof. The argument is the same as for Proposition III.1.39 (recognizing Garside I), replacing $\mathcal{S}^\sharp$ with $\mathcal{S}$ everywhere. The main step consists in showing the counterpart of Proposition III.1.49 (left-multiplication), namely that, if $s_1|\cdots|s_p$ is an $\mathcal{S}$-normal decomposition of s with entries in $\mathcal{S}$ and g is an element of $\mathcal{S}$ such that gs is defined, then an $\mathcal{S}$-normal decomposition of gs with entries in $\mathcal{S}$ is obtained inductively using the assumption that $\mathcal{S}$ satisfies Property $\square$ and the first domino rule. $\square$

We cannot say more, in particular about the length of the decomposition involved in Proposition 2.6: if $\mathcal{S}$ is solid but does not coincide with $\mathcal{S}^\sharp$, then an element of $\mathcal{S}^\sharp \setminus \mathcal{S}$ has $\mathcal{S}$-length one, but it has no $\mathcal{S}$-normal decomposition of length one with entry in $\mathcal{S}$.

We now adapt the characterizations of Garside families established in Subsection 1.2.

Proposition 2.7 (recognizing Garside, solid case). *A solid subfamily $\mathcal{S}$ in a left-cancellative category $\mathcal{C}$ is a Garside family if and only if $\mathcal{S}$ generates $\mathcal{C}$ and one of the following equivalent conditions is satisfied:*

(2.8) *Every non-invertible element of $\mathcal{C}$ admits an $\mathcal{S}$-head.*

(2.9) *The family $\mathcal{S}$ is closed under right-complement and every non-invertible element of $\mathcal{S}^2$ admits an $\mathcal{S}$-head.*

(2.10) *The family $\mathcal{S}$ is closed under right-comultiple and every non-invertible element of $\mathcal{S}^2$ admits a $\prec$-maximal left-divisor lying in $\mathcal{S}$.*

Proof. Assume that $\mathcal{S}$ is a solid Garside family in $\mathcal{C}$. Then, by Proposition 2.4, $\mathcal{S}$ generates $\mathcal{C}$ and, by Proposition 1.24, it satisfies (1.25)–(1.27). The latter include (2.8) and (2.10). Moreover, by Lemma 1.8, $\mathcal{S}$ is closed under right-complement, and, by Proposition 1.24, every non-invertible element of $\mathcal{S}^2$ admits an $\mathcal{S}$-head, so $\mathcal{S}$ satisfies (2.9).

Conversely, if $\mathcal{S}$ is solid, generates $\mathcal{C}$, and satisfies (2.8), it satisfies (1.25) so, by Proposition 1.24, it is a Garside family.

Next, if $\mathcal{S}$ is solid, generates $\mathcal{C}$, and satisfies (2.9), then, by Lemma 2.2, we have $\mathcal{C}^\times\mathcal{S} \subseteq \mathcal{S}$ and then, by Lemma 1.13(iii), $\mathcal{S}^\sharp$ is closed under right-complement. So $\mathcal{S}$ satisfies Condition (1.27) and, by Proposition 1.24 again, it is a Garside family.

Finally, if $\mathcal{S}$ is solid, generates $\mathcal{C}$, and satisfies (2.9), it directly satisfies (1.27) and, once more by Proposition 1.24, it is a Garside family. $\square$

2.2 Right-Noetherian categories

When the ambient category is right-Noetherian, the existence of $\prec$-maximal elements is automatic, and recognizing Garside families is easier than in the general case. A typical result in this direction is the following, which is a consequence of Proposition 2.18 below:

Proposition 2.11 (recognizing Garside, solid right-Noetherian case). *A solid subfamily of a left-cancellative category $\mathcal{C}$ that is right-Noetherian is a Garside family if and only if it generates $\mathcal{C}$ and is closed under right-comultiple.*

In terms of subfamilies that are not necessarily solid, we deduce

Corollary 2.12 (recognizing Garside, Noetherian case). *A subfamily $\mathcal{S}$ of a left-cancellative category $\mathcal{C}$ that is right-Noetherian is a Garside family if and only if $\mathcal{S}^\sharp$ generates $\mathcal{C}$ and satisfies one of the following equivalent conditions:*

(i) *The family $\mathcal{S}^\sharp$ is closed under right-comultiple and right-divisor.*

(ii) *The family $\mathcal{S}^\sharp$ is closed under right-comultiple and right-complement.*

Proof. If $\mathcal{S}$ is a Garside family, then $\mathcal{S}^\sharp$ generates $\mathcal{C}$ and, by Proposition 1.23, it is closed under right-comultiple, right-complement, and right-divisor.

Conversely, assume that $\mathcal{S}^\sharp$ generates $\mathcal{C}$ and satisfies (i). By Lemma 2.2(ii), $\mathcal{S}^\sharp$ is solid, and Proposition 2.11 implies that $\mathcal{S}^\sharp$ is a Garside family. By Proposition III.1.33 (invariance), this implies that $\mathcal{S}$ is a Garside family too. Finally, by Lemma 1.19 and owing to the other assumptions, (i) is equivalent to (ii). □

We recall from Lemma 1.13 that $\mathcal{S}^\sharp$ is closed under right-comultiple if and only if $\mathcal{S}$ is, so some variants of Corollary 2.12 could be stated.

Example 2.13 (recognizing Garside, Noetherian case). If (M, Δ) is a Garside monoid, then, by definition, M is right-Noetherian. So, in order to establish that $\mathrm{Div}(\Delta)$ is a Garside family in M, it is actually enough to check that $\mathrm{Div}(\Delta)$ is closed under right-comultiple and right-divisor, two immediate consequences of the properties of Δ.

We shall prove in Proposition 2.18 below a result that is stronger than Proposition 2.11. Indeed, the Noetherianity assumption can be weakened to a local condition involving only the considered family $\mathcal{S}$.

If $\mathcal{S}$ is a subfamily of a (left-cancellative) category $\mathcal{C}$—more generally, if $\mathcal{S}$ is a precategory equipped with a partial product—we can introduce a local version of divisibility.

Definition 2.14 ($\mathcal{S}$-divisor). If $\mathcal{S}$ is a subfamily of a category $\mathcal{C}$ and s, t lie in $\mathcal{S}$, we say that s is a *right-$\mathcal{S}$-divisor* of t, written $s \mathrel{\widetilde{\preccurlyeq}}_{\mathcal{S}} t$, or equivalently $t \mathrel{\widetilde{\succcurlyeq}}_{\mathcal{S}} s$, if $t = t's$ holds for some t' in $\mathcal{S}$. We say that $s \mathrel{\widetilde{\prec}}_{\mathcal{S}} t$, or $t \mathrel{\widetilde{\succ}}_{\mathcal{S}} s$, holds if $t = t's$ holds for some t' in $\mathcal{S} \setminus \mathcal{S}^\times$, where $\mathcal{S}^\times$ is the family of all invertible elements of $\mathcal{S}$ whose inverse lies in $\mathcal{S}$.

Naturally, there exist symmetric relations $s \preccurlyeq_{\mathcal{S}} t$ and $s \prec_{\mathcal{S}} t$ defined by $st' = t$ with t' in $\mathcal{S}$ and in $\mathcal{S} \setminus \mathcal{S}^\times$ respectively but, in this section, we shall use the right version only. See Exercise 53 for a few results about the left version (whose properties may be different when the assumptions about $\mathcal{C}$ and $\mathcal{S}$ are not symmetric).

Note that $\mathrel{\widetilde{\preccurlyeq}}_{\mathcal{S}}$ need not be transitive in general, and that, in a framework that is not necessarily right-cancellative, there may be more than one element t' satisfying $t = t's$: in particular, there may be simultaneously an invertible and a non-invertible one. Clearly, $s \mathrel{\widetilde{\preccurlyeq}}_{\mathcal{S}} t$ implies $s \mathrel{\widetilde{\preccurlyeq}} t$, but the converse need not be true, unless $\mathcal{S}$ is closed under left-quotient. The inclusion of $\mathrel{\widetilde{\prec}}_{\mathcal{S}}$ in $\mathrel{\widetilde{\prec}}$ is not true in general either.

Lemma 2.15. *If $\mathcal{S}$ is a subfamily of a left-cancellative category $\mathcal{C}$ and $\mathcal{S}^{\times} = \mathcal{C}^{\times} \cap \mathcal{S}$ holds, then $s \mathrel{\widetilde{\prec}}_{\mathcal{S}} t$ implies $s \mathrel{\widetilde{\prec}} t$. This holds in particular if $\mathcal{S}$ is solid.*

Proof. The assumption $\mathcal{S}^{\times} = \mathcal{C}^{\times} \cap \mathcal{S}$ implies that an element of $\mathcal{S} \setminus \mathcal{S}^{\times}$ cannot be invertible in $\mathcal{C}$. So, in this case, $t = t's$ with $t' \in \mathcal{S} \setminus \mathcal{S}^{\times}$ implies $t \mathrel{\widetilde{\succ}} s$. If $\mathcal{S}$ is solid, then, by Lemma 2.2, $\mathcal{S}$ includes all of $\mathcal{C}^{\times}$, so every invertible element of $\mathcal{S}$ has its inverse in $\mathcal{S}$, and we have $\mathcal{S}^{\times} = \mathcal{C}^{\times} = \mathcal{C}^{\times} \cap \mathcal{S}$. □

With the local versions of divisibility naturally come local versions of Noetherianity.

Definition 2.16 (locally right-Noetherian). A subfamily $\mathcal{S}$ of a category is called *locally right-Noetherian* if the relation $\mathrel{\widetilde{\prec}}_{\mathcal{S}}$ is well-founded, that is, every nonempty subfamily of $\mathcal{S}$ has a $\mathrel{\widetilde{\prec}}_{\mathcal{S}}$-minimal element.

Of course there is a symmetric notion of *local left-Noetherianity* involving $\prec_{\mathcal{S}}$.

If $\mathcal{C}$ is a right-Noetherian category, then, by Lemma 2.15, every subfamily $\mathcal{S}$ of $\mathcal{C}$ satisfying $\mathcal{S}^{\times} = \mathcal{C}^{\times} \cap \mathcal{S}$ is locally right-Noetherian, since an infinite descending sequence for $\mathrel{\widetilde{\prec}}_{\mathcal{S}}$ would be a descending sequence for $\mathrel{\widetilde{\prec}}$. But local Noetherianity is a priori doubly weaker than Noetherianity, as it says nothing about descending sequences outside $\mathcal{S}$, and about descending sequences inside $\mathcal{S}$ but with quotients outside $\mathcal{S}$. According to the general scheme explained in Subsection II.2.3, a family $\mathcal{S}$ is locally right-Noetherian if and only if there exists no infinite descending sequence for $\mathrel{\widetilde{\prec}}_{\mathcal{S}}$, if and only if there exists an ordinal-valued map λ on $\mathcal{S}$ such that $s \mathrel{\widetilde{\prec}}_{\mathcal{S}} t$ implies $\lambda(s) < \lambda(t)$. In the case when $\mathcal{S}$ is finite, there exists a criterion that is usually simple to check on examples:

Lemma 2.17. *Assume that $\mathcal{S}$ is a finite subfamily of a category $\mathcal{C}$. Then $\mathcal{S}$ is locally right-Noetherian if and only if $\mathrel{\widetilde{\prec}}_{\mathcal{S}}$ has no cycle.*

Proof. Repeating a cycle for $\mathrel{\widetilde{\prec}}_{\mathcal{S}}$, that is, a finite sequence $s_1, ..., s_p$ of elements of $\mathcal{S}$ satisfying $s_i \mathrel{\widetilde{\succ}}_{\mathcal{S}} s_{i+1}$ for every i and $s_p \mathrel{\widetilde{\succ}}_{\mathcal{S}} s_1$ provides an infinite descending sequence. Conversely, if $\mathcal{S}$ has n elements, every sequence of length larger than n contains at least one entry twice, thus giving a cycle if it is $\mathrel{\widetilde{\prec}}_{\mathcal{S}}$-decreasing. □

Here is the main result we shall establish:

Proposition 2.18 (solid Garside in right-Noetherian). *For every solid generating subfamily $\mathcal{S}$ in a left-cancellative category $\mathcal{C}$, the following are equivalent:*

(i) *The category $\mathcal{C}$ is right-Noetherian and $\mathcal{S}$ is a Garside family in $\mathcal{C}$.*

(ii) *The family $\mathcal{S}$ is locally right-Noetherian and closed under right-comultiple.*

So, whenever $\mathcal{S}$ is locally right-Noetherian, we can forget about the existence of the head when looking whether $\mathcal{S}$ is a Garside family, and the ambient category is automatically right-Noetherian.

Proof of Proposition 2.11 *from Proposition* 2.18. Assume that $\mathcal{S}$ is a solid Garside family in $\mathcal{C}$. Then $\mathcal{S}$ generates $\mathcal{C}$ and is closed under right-comultiple by Proposition 2.4.

Conversely, assume that $\mathcal{S}$ is solid, generates $\mathcal{C}$, and is closed under right-comultiple. As $\mathcal{C}$ is right-Noetherian and $\mathcal{S}$ is solid, $\mathcal{S}$ is locally right-Noetherian since, by Lemma 2.15, a descending sequence for $\precnsim_{\mathcal{S}}$ would be a descending sequence for $\precnsim$. Then Proposition 2.18 implies that $\mathcal{S}$ is a Garside family. □

Proposition 2.18 follows from a series of technical results.

First, we observe that, in the situation of Proposition 2.18, the local right-Noetherianity of $\mathcal{S}$ implies a global right-Noetherianity result involving $\mathcal{S}$, and even $\mathcal{S}^\sharp$.

Lemma 2.19. *Assume that $\mathcal{S}$ is a solid generating subfamily in a left-cancellative category $\mathcal{C}$ and $\mathcal{S}$ is locally right-Noetherian. Then the restriction of $\precnsim$ to $\mathcal{S}^\sharp$ is well-founded.*

Proof. First we observe that, as $\mathcal{S}$ is solid, Lemma 2.2 implies $\mathcal{S}^\sharp = \mathcal{S}\mathcal{C}^\times$ and $\mathcal{C}^\times\mathcal{S}^\sharp = \mathcal{S}^\sharp$. Assume that $\mathcal{X}$ is a nonempty subfamily of $\mathcal{S}^\sharp$. Put

$$\mathcal{Y} = \mathcal{C}\mathcal{X}\mathcal{C}^\times \cap \mathcal{S} = \{s \in \mathcal{S} \mid \exists g{\in}\mathcal{C}\, \exists t{\in}\mathcal{X}\, \exists \epsilon{\in}\mathcal{C}^\times (s = gt\epsilon)\}.$$

For every t in $\mathcal{X}$, there exists ϵ in $\mathcal{C}^\times$ such that $t\epsilon$ lies in $\mathcal{S}$, so $\mathcal{Y}$ is nonempty and, therefore, it has a $\precnsim_{\mathcal{S}}$-minimal element, say s (we recall that this means that $s' \precnsim_{\mathcal{S}} s$ holds for no element s'). Write $s = gt\epsilon$ with t in $\mathcal{X}$ and ϵ in $\mathcal{C}^\times$. We claim that t is $\precnsim$-minimal in $\mathcal{S}^\sharp$. Indeed, assume $t = ht'$ with $h \notin \mathcal{C}^\times$ and $t' \in \mathcal{X}$. Then we have $s = ght'\epsilon$. As $\mathcal{S}^\sharp$ generates $\mathcal{C}$ and $\mathcal{C}^\times\mathcal{S}^\sharp \subseteq \mathcal{S}^\sharp$ holds, there exist $p \geqslant 1$ and $t_1, ..., t_p$ in $\mathcal{S}^\sharp \setminus \mathcal{C}^\times$ satisfying $gh = t_1 \cdots t_p$. Write $t_i = s_i\epsilon_i$ with s_i in $\mathcal{S} \setminus \mathcal{C}^\times$ and ϵ_i in $\mathcal{C}^\times$, and put $g' = \epsilon_1 s_2 \epsilon_2 \cdots s_p \epsilon_p$ and $s' = g't'\epsilon$. Then we have $s = s_1\epsilon_1 \cdots s_p\epsilon_p t'\epsilon = s_1 g't'\epsilon = s_1 s'$. As s' right-divides s and $\mathcal{S}$ is solid, s' lies in $\mathcal{S}$, and the decomposition $g'|t'|\epsilon$ witnesses that s' lies in $\mathcal{Y}$. Now s_1 is non-invertible, so we have $s' \precnsim_{\mathcal{S}} s$, which contradicts the choice of s. Thus the existence of t' is impossible, and t is $\precnsim$-minimal in $\mathcal{S}^\sharp$. □

Next, we show that the conditions of Proposition 2.18 force $\mathcal{S}$ to be a Garside family.

Lemma 2.20. *If $\mathcal{S}$ is a solid generating subfamily in a left-cancellative category $\mathcal{C}$ and $\mathcal{S}$ is locally right-Noetherian and closed under right-comultiple, then $\mathcal{S}$ is a Garside family.*

Proof. Owing to (1.27) in Proposition 1.24, it suffices to prove that every element of $\mathcal{S}^2$ admits a $\prec$-maximal left-divisor lying in $\mathcal{S}$. So assume $g = s_1 s_2$ with s_1, s_2 in $\mathcal{S}$. Put

$$\mathcal{X} = \{r \in \mathcal{S} \mid s_1 r \in \mathcal{S} \text{ and } r \preccurlyeq s_2\}.$$

If y is the target of s_1, then 1_y lies in $\mathcal{X}$, which is nonempty. Towards a contradiction, assume that $\mathcal{X}$ has no $\prec$-maximal element. Then there exists an infinite $\prec$-increasing sequence in $\mathcal{X}$, say $r_1 \prec r_2 \prec \cdots$. By assumption, $r_i \preccurlyeq s_2$ holds for each i, so we have $s_2 = r_i t_i$ for some t_i, which must lie in $\mathcal{S}$ since $\mathcal{S}$ is closed under right-divisor and s_2 lies in $\mathcal{S}$. Now, by left-cancellativity, $r_i s' = r_{i+1}$ implies $t_i = s' t_{i+1}$, and therefore $r_i \prec r_{i+1}$

implies $t_i \succneqq t_{i+1}$. Thus $t_1, t_2, \ldots$ is a $\precsim$-decreasing sequence in $\mathcal{S}$, and its existence would contradict Lemma 2.19. Hence $\mathcal{X}$ has a $\prec$-maximal element, say r. Then it follows that $s_1 r$, which belongs to $\mathcal{S}$ and left-divides g by definition, is a $\prec$-maximal left-divisor of g lying in $\mathcal{S}$. By Proposition 1.24, $\mathcal{S}$ is a Garside family. $\square$

The last preparatory result says that a right-Noetherianity condition on $\mathcal{S}^\sharp$ extends to the whole ambient category when $\mathcal{S}$ is a Garside family.

Lemma 2.21. *Assume that $\mathcal{S}$ is a Garside family in a left-cancellative category $\mathcal{C}$ and the restriction of $\precsim$ to $\mathcal{S}^\sharp$ is well-founded. Then $\mathcal{C}$ is right-Noetherian.*

Proof. Assume that $\mathcal{X}$ is a nonempty subfamily of $\mathcal{C}$, and let m be the minimal value of $\|g\|_\mathcal{S}$ for g in $\mathcal{X}$. If m is zero, that is, if $\mathcal{X}$ contains at least one invertible element, then every such element is $\precsim$-minimal, and we are done. Otherwise, let $\mathcal{X}_m$ be the family of all elements t in $\mathcal{S}^\sharp$ satisfying

$$\exists s_1, \ldots, s_{m-1} \in \mathcal{S}^\sharp \ (s_1|\cdots|s_{m-1}|t \text{ is } \mathcal{S}\text{-normal and } s_1\cdots s_{m-1}t \in \mathcal{X}).$$

By definition, $\mathcal{X}_m$ is nonempty, so it admits a $\precsim$-minimal element, say t_m. Then let $\mathcal{X}_{m-1}$ be the family of all elements t in $\mathcal{S}^\sharp$ satisfying

$$\exists s_1, \ldots, s_{m-2} \in \mathcal{S}^\sharp \ (s_1|\cdots|s_{m-2}|t|t_m \text{ is } \mathcal{S}\text{-normal and } s_1\cdots s_{m-2}tt_m \in \mathcal{X}).$$

Again $\mathcal{X}_{m-1}$ is nonempty, so it admits a $\precsim$-minimal element, say t_{m-1}. We continue in the same way, choosing t_i to be $\precsim$-minimal in $\mathcal{X}_i$ for i decreasing from m to 1. In particular, $\mathcal{X}_1$ is the family of all elements t in $\mathcal{S}^\sharp$ such that $t|t_2|\cdots|t_m$ is $\mathcal{S}$-normal and $tt_2\cdots t_m$ lies in $\mathcal{X}$. Let $g = t_1\cdots t_m$. By construction, g lies in $\mathcal{X}$, and we claim it is $\precsim$-minimal in $\mathcal{X}$.

Indeed, assume that g' belongs to $\mathcal{X}$ and right-divides g, say $g = fg'$. The point is to show that f is invertible. As $\mathcal{S}$ is a Garside family, $\mathcal{S}^\sharp$ generates $\mathcal{C}$, and there exists p such that f belongs to $(\mathcal{S}^\sharp)^p$. As g' right-divides g, we have $\|g'\|_\mathcal{S} \leqslant m$, so g' admits an $\mathcal{S}$-normal decomposition of length m, say $s_1|\cdots|s_m$. Applying Proposition III.1.49 (left multiplication) p times, we find an $\mathcal{S}$-normal decomposition $s'_1|\cdots|s'_m|s'_{m+1}|\cdots|s'_{m+p}$ of fg', that is, of g (see Figure 7). By construction, the vertical arrows represent elements $f_0 = f, f_1, \ldots, f_r$ of $(\mathcal{S}^\sharp)^p$. By Proposition III.1.25 (normal unique), $s_1|\cdots|s_m$ and $s'_1|\cdots|s'_m|s'_{m+1}|\cdots|s'_{m+p}$ must be $\mathcal{C}^\times$-deformations of one another, that is, there exist invertible elements $\epsilon_1, \ldots, \epsilon_{m+p-1}$ making the diagram of Figure 7 commutative. The diagram shows that $\epsilon_m f_m$ is an identity element, and we have $t_m = (\epsilon_{m-1}f_{m-1})s_m$, whence $s_m \precsim t_m$. On the other hand, s_m belongs to $\mathcal{X}_m$ since it is the mth entry in an $\mathcal{S}$-normal decomposition of length m for an element of $\mathcal{X}$, namely g. By the choice of t_m, the relation $s_m \precnsim t_m$ is impossible, so $\epsilon_{m-1}f_{m-1}$, hence f_{m-1}, must be invertible.

Put $t'_{m-1} = s_{m-1}f_{m-1}^{-1}\epsilon_{m-1}^{-1}$, an element of $\mathcal{S}^\sharp$. By Proposition III.1.22 (deformation), $s_1|\cdots|s_{m-2}|t'_{m-1}|t_m$ is an $\mathcal{S}$-normal decomposition of g. Then we have $t_{m-1} = (\epsilon_{m-2}f_{m-2})t'_{m-1}$, whence $t'_{m-1} \precsim t_{m-1}$. On the other hand, t'_{m-1} belongs to $\mathcal{X}_{m-1}$ since it is the $(m-1)$st entry in an $\mathcal{S}$-normal decomposition of length m finishing with t_m for an element of $\mathcal{X}$, namely g. By the choice of t_{m-1}, the relation $t'_{m-1} \precnsim t_{m-1}$ is impossible, so $\epsilon_{m-2}f_{m-2}$, hence f_{m-2}, must be invertible.

Continuing in this way, we conclude after m steps that f_0, that is, f, must be invertible, which establishes that t is $\precsim$-minimal in $\mathcal{X}$. $\square$

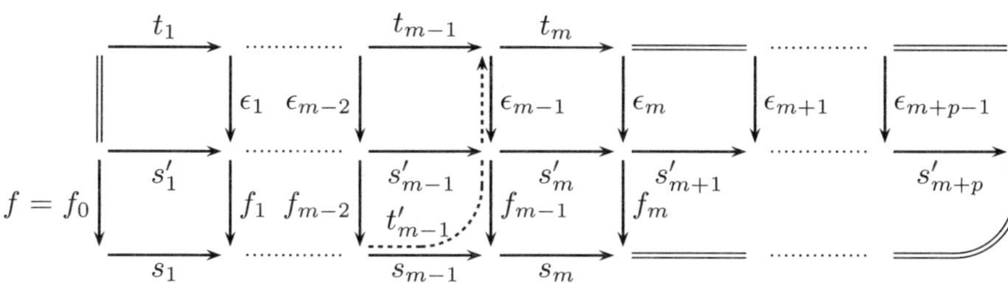

Figure 7. Proof of Lemma 2.21

We can now complete our argument.

Proof of Proposition 2.18. Assume that $\mathcal{C}$ is right-Noetherian and $\mathcal{S}$ is a solid Garside family in $\mathcal{C}$. By Lemma 2.15, the relation $\precneqq_{\mathcal{S}}$ is included in $\precneqq$, so the well-foundedness of $\precneqq$ implies that of $\precneqq_{\mathcal{S}}$ and $\mathcal{S}$ is locally right-Noetherian. On the other hand, by Proposition 1.23, $\mathcal{S}$ is closed under right-comultiple. So (i) implies (ii).

Conversely, assume that $\mathcal{S}$ generates $\mathcal{C}$ and is solid, closed under right-comultiple, and locally right-Noetherian. First, Lemma 2.20 implies that $\mathcal{S}$ is a Garside family. Next, Lemma 2.19 implies that the restriction of $\precneqq$ to $\mathcal{S}^\sharp$ is well-founded. Finally, Lemma 2.21 implies that $\mathcal{C}$ is right-Noetherian. So (ii) implies (i). $\square$

Proposition 2.18 involves solid subfamilies. Remembering that $\mathcal{S}^\sharp$ is solid for every Garside family $\mathcal{S}$, we deduce an analogous statement for general subfamilies.

Corollary 2.22 (Garside in right-Noetherian). *If $\mathcal{S}$ is a subfamily of a left-cancellative category $\mathcal{C}$ and $\mathcal{S}^\sharp$ generates $\mathcal{C}$, the following are equivalent:*

(i) *The category $\mathcal{C}$ is right-Noetherian and $\mathcal{S}$ is a Garside family in $\mathcal{C}$.*

(ii) *The family $\mathcal{S}$ is closed under right-comultiple and $\mathcal{S}^\sharp$ is locally right-Noetherian and closed under right-divisor.*

(iii) *The family $\mathcal{S}$ is closed under right-comultiple and $\mathcal{S}^\sharp$ is locally right-Noetherian and closed under right-complement.*

Proof. Assume (i). By Proposition III.1.33 (invariance) and Lemma 2.3, $\mathcal{S}^\sharp$ is a solid Garside family. Hence, by Propositions 2.18 and 1.23, $\mathcal{S}^\sharp$ is locally right-Noetherian and closed under right-comultiple, right-divisor, and right-complement. By Lemma 1.13, it follows that $\mathcal{S}$ is closed under right-comultiple as well. So (i) implies (ii) and (iii).

Next, (ii) and (iii) are equivalent: if $\mathcal{S}$ is closed under right-comultiple, then so is $\mathcal{S}^\sharp$ by Lemma 1.13 and, therefore, by Lemma 1.19, since $\mathcal{S}^\sharp$ generates $\mathcal{C}$, the family $\mathcal{S}^\sharp$ is closed under right-divisor if and only if it is closed under right-complement.

Finally, assume (ii) and (iii). Then $\mathcal{S}^\sharp$ is a solid generating subfamily of $\mathcal{C}$ that is locally right-Noetherian and closed under right-comultiple. By Proposition 2.18, $\mathcal{C}$ is right-Noetherian and $\mathcal{S}^\sharp$ is a Garside family in $\mathcal{C}$. Then, by Proposition III.1.33 (invariance), $\mathcal{S}$ is a Garside family in $\mathcal{C}$ as well. $\square$

The right-Noetherianity assumption is crucial in the above results: otherwise, the conditions of Proposition 2.11 do not characterize Garside families.

Example 2.23 (Klein bottle monoid). In the Klein bottle monoid K^+ (Reference Structure 5, page 17), consider $S = \{g \in K^+ \mid |g|_{\mathtt{a}} \leqslant 1\}$, where $|g|_{\mathtt{a}}$ is the number of letters $\mathtt{a}$ in an expression of g (we recall that the value does not depend on the expression). As S contains $\mathtt{a}$ and $\mathtt{b}$, it generates K^+. Next, S is closed under right-comultiple, since a common right-multiple of f and g is a right-multiple of $\max_{\preccurlyeq}(f, g)$, which is one of f, g and which is a right-multiple of f and g. Finally, S is closed under right-divisor, since $g' \precneqq g$ implies $|g'|_{\mathtt{a}} \leqslant |g|_{\mathtt{a}}$. However, S is not a Garside family in K^+, as the element $\mathtt{a}^2$ admits no S-head: every element of S left-divides $\mathtt{a}^2$, and S has no maximal element.

2.3 Categories that admit right-mcms

New features appear when the ambient category $\mathcal{C}$, in addition to being right-Noetherian, admits right-mcms, which is the case in particular when $\mathcal{C}$ is Noetherian, and when $\mathcal{C}$ admits right-lcms. The main result is then that, for every subfamily $\mathcal{S}$ of $\mathcal{C}$, there exists a smallest Garside family that includes $\mathcal{S}$ (Corollary 2.32).

We recall from Definitions II.2.9 and II.2.38 that a category $\mathcal{C}$ is said to admit right-mcms (*resp.* to admit conditional right-lcms) if every common right-multiple of two elements is necessarily a right-multiple of some right-mcm (*resp.* right-lcm) of these elements. Note that, if h is a right-lcm of f and g, then the family of all right-lcms of f and g is the $=^{\times}$-equivalence class of h. Our starting point is the following easy observation.

Lemma 2.24. *Assume that $\mathcal{C}$ is a left-cancellative category that admits right-mcms. Then, for every subfamily $\mathcal{S}$ of $\mathcal{C}$, the following are equivalent:*

(i) *The family $\mathcal{S}$ is closed under right-comultiple.*

(ii) *The family $\mathcal{S}^\sharp$ is closed under right-mcm, that is, if s and t belong to $\mathcal{S}^\sharp$, then so does every right-mcm of s and t.*

Proof. Assume that $\mathcal{S}$ is closed under right-comultiple. By Lemma 1.13, $\mathcal{S}^\sharp$ is closed under right-comultiple as well. Assume that s, t belong to $\mathcal{S}^\sharp$, and $sg = tf$ is a right-mcm of s and t. As $\mathcal{S}^\sharp$ is closed under right-comultiple, there exist f', g', h satisfying $sg' = tf' \in \mathcal{S}^\sharp$, $f = f'h$, and $g = g'h$. The assumption that sg is a right-mcm of s and t implies that h is invertible. Hence sg, which is $sg'h$, belongs to $\mathcal{S}^\sharp$. So (i) implies (ii).

Conversely, assume that $\mathcal{S}^\sharp$ is closed under right-mcm. Assume that s, t belong to $\mathcal{S}^\sharp$, and $sg = tf$ holds. By assumption, there exists a right-mcm of s and t, say $sg' = tf'$, that left-divides sg, that is, there exist h satisfying $f = f'h$ and $g = g'h$. By assumption, sg' belongs to $\mathcal{S}^\sharp$, so (f', g', h) witnesses that $\mathcal{S}^\sharp$ is closed under right-comultiple. By Lemma 1.13, so is $\mathcal{S}$, and, therefore, (ii) implies (i). □

Owing to Lemma 2.24, Corollary 2.12 directly implies:

Proposition 2.25 (recognizing Garside, right-mcm case). *A subfamily $\mathcal{S}$ of a left-cancellative category $\mathcal{C}$ that is right-Noetherian and admits right-mcms is a Garside family if and only if $\mathcal{S}^\sharp$ generates $\mathcal{C}$ and satisfies one of the following equivalent conditions:*

(i) *The family $\mathcal{S}^\sharp$ is closed under right-mcm and right-divisor.*

(ii) *The family $\mathcal{S}^\sharp$ is closed under right-mcm and right-complement.*

By Proposition II.2.40 (right-mcm), every left-Noetherian category admits right-mcms, so Proposition 2.25 implies in turn

Corollary 2.26 (recognizing Garside, Noetherian case). *A subfamily $\mathcal{S}$ of a left-cancellative category $\mathcal{C}$ that is Noetherian is a Garside family if and only if $\mathcal{S}^\sharp$ generates $\mathcal{C}$ and satisfies one of the following equivalent conditions:*

(i) *The family $\mathcal{S}^\sharp$ is closed under right-mcm and right-divisor.*

(ii) *The family $\mathcal{S}^\sharp$ is closed under right-mcm and right-complement.*

In the more special case of a category that admits conditional right-lcms, the results become even simpler.

Definition 2.27 (closure under right-lcm). Assume that $\mathcal{C}$ is a left-cancellative category that admits conditional right-lcms. A subfamily $\mathcal{S}$ of $\mathcal{C}$ is said to be *closed* (*resp. weakly closed*) *under right-lcm* if, for all s, t in $\mathcal{S}$ admitting a common right-multiple, every (*resp.* at least one) right-lcm of s and t lies in $\mathcal{S}$.

As the notions of right-mcm and right-lcm coincide in a category that admits conditional right-lcms, we can easily establish an analog of Lemma 2.24.

Lemma 2.28. *Assume that $\mathcal{C}$ is a left-cancellative category that admits conditional right-lcms. Then, for every subfamily $\mathcal{S}$ of $\mathcal{C}$, the following are equivalent:*

(i) *The family $\mathcal{S}$ is closed under right-comultiple.*

(ii) *The family $\mathcal{S}$ is weakly closed under right-lcm.*

(iii) *The family $\mathcal{S}^\sharp$ is closed under right-lcm.*

Proof. The equivalence of (i) and (ii) follows from the definition of a right-lcm. Next, by Lemma 1.13(ii), $\mathcal{S}$ is closed under right-comultiple if and only if $\mathcal{S}^\sharp$ is, hence, by the equivalence of (i) and (ii), if and only $\mathcal{S}^\sharp$ is weakly closed under right-lcm. Now, as every element that is $=^\times$-equivalent to an element of $\mathcal{S}^\sharp$ must belong to $\mathcal{S}^\sharp$, the family $\mathcal{S}^\sharp$ is weakly closed under right-lcm if and only if it is closed under right-lcm. ☐

Proposition 2.25 then implies:

Corollary 2.29 (recognizing Garside, right-lcm case). *A subfamily $\mathcal{S}$ of a left-cancellative category $\mathcal{C}$ that is right-Noetherian and admits conditional right-lcms is a Garside family if and only if $\mathcal{S}^\sharp$ generates $\mathcal{C}$ and one of the following equivalent conditions holds:*

(i) *The family $\mathcal{S}^\sharp$ is closed under right-lcm and right-divisor.*

(ii) *The family $\mathcal{S}^\sharp$ is closed under right-lcm and right-complement.*

(iii) *The family $\mathcal{S}$ is weakly closed under right-lcm and $\mathcal{S}^\sharp$ is closed under right-divisor.*

(iv) *The family $\mathcal{S}$ is weakly closed under right-lcm and $\mathcal{S}^\sharp$ is closed under right-complement.*

Proof. Proposition 2.25 gives the equivalence with (i) and (ii) directly. Lemma 2.28 gives the equivalence of (i) and (iii) on the one hand, and that of (ii) and (iv) on the other hand. ☐

Many examples are eligible for the above criteria. As already mentioned, so is the family of divisors of Δ in a Garside monoid (M, Δ), since $\mathrm{Div}(\Delta)$ is, by its very definition, closed under right-lcm and right-divisor. Here is a less classical example.

Example 2.30 (affine braids). Consider the affine braid monoid B^+ of type $\widetilde{A}_2$ and the sixteen-element family S of (Reference Structure 9, page 114). As mentioned in Chapter III—and as will be more systematically discussed in Chapter IX—the monoid B^+ is not a Garside monoid; it admits conditional right-lcms, but does not admit right-lcms since, for instance, the elements $\sigma_1, \sigma_2, \sigma_3$ have no common right-multiple. Now the family S generates the monoid, it is closed under right-divisor by definition, and it turns out to be closed under right-lcm: indeed, using right-reversing, one can check (which is easy owing to the symmetries) that all pairs of elements of S that do not admit a right-lcm in S actually admit no common right-multiple in the monoid because the associated right-reversing has a repeating pattern, so does not terminate. Hence, by Corollary 2.29, S is a Garside family in B^+, and every element of B^+ admits an S-normal decomposition.

We refer to Exercise 45 for an example of a finite Garside family in a monoid that admits right-mcms but does not admit conditional right-lcms, and Exercise 46 for an example in a monoid that is right-Noetherian but not Noetherian.

Of course, all above characterizations of Garside families admit counterparts in the case of solid families, see Exercise 50.

An interesting consequence of the definition of Garside families in terms of closure is the existence of smallest Garside families.

Proposition 2.31 (smallest Garside). *Assume that $\mathcal{C}$ is a left-cancellative category that is right-Noetherian and admits right-mcms. Then, for every subfamily $\mathcal{S}$ of $\mathcal{C}$ such that $\mathcal{S} \cup \mathcal{C}^\times$ generates $\mathcal{C}$, there exists a smallest $=^\times$-closed Garside family that includes $\mathcal{S}$.*

Proof. Starting from $\mathcal{S}_0 = \mathcal{S} \cup \mathcal{C}^\times$, we inductively construct a sequence of subfamilies of $\mathcal{C}$ by defining $\mathcal{S}_{2i+1}$ to be the family of all right-mcms of two elements of $\mathcal{S}_{2i}$ and $\mathcal{S}_{2i+2}$ to be the family of all right-divisors of an element of $\mathcal{S}_{2i+1}$. Let $\widehat{\mathcal{S}} = \bigcup_{i \geqslant 0} \mathcal{S}_i$. The family $(\mathcal{S}_i)_{i \geqslant 0}$ is increasing: every element t is a right-lcm of t and t, hence $\mathcal{S}_{2i} \subseteq \mathcal{S}_{2i+1}$ holds, and every element t is a right-divisor of itself, hence $\mathcal{S}_{2i+1} \subseteq \mathcal{S}_{2i+2}$ holds. Then $\widehat{\mathcal{S}}$ is a subfamily of $\mathcal{C}$ that includes $\mathcal{S}$ and $\mathcal{C}^\times$, hence it generates $\mathcal{C}$. Moreover, $\widehat{\mathcal{S}}$ is closed under right-mcm: if s, t belong to $\widehat{\mathcal{S}}$ and h is a right-mcm of s and t, there exist i such that s and t belong to $\mathcal{S}_{2i}$ and, therefore, h belongs to $\mathcal{S}_{2i+1}$, hence to $\widehat{\mathcal{S}}$. Similarly, $\widehat{\mathcal{S}}$ is closed under right-divisor: if t belongs to $\widehat{\mathcal{S}}$ and t' is a right-divisor of t, there exists i such that t belongs to $\mathcal{S}_{2i+1}$ and, therefore, h belongs to $\mathcal{S}_{2i+2}$, hence to $\widehat{\mathcal{S}}$. Finally, $\widehat{\mathcal{S}}$ is closed under right-multiplication by an invertible element: if t belongs to $\widehat{\mathcal{S}}$ and $t' =^\times t$ holds, t' is a right-mcm of t and t, hence it belongs to $\widehat{\mathcal{S}}$ as well. Hence, by Proposition 2.25, $\widehat{\mathcal{S}}$ is an $=^\times$-closed Garside family in $\mathcal{C}$.

On the other hand, by Proposition 2.25 again, every $=^\times$-closed Garside family $\widehat{\mathcal{S}}'$ that includes $\widehat{\mathcal{S}}$ includes $\mathcal{S} \cup \mathbf{1}_\mathcal{C}$ and is closed under right-mcm and right-divisor, hence it inductively includes each family $\mathcal{S}_i$, and the union $\widehat{\mathcal{S}}$ of these families. ☐

When we consider Garside families that need not be $=^\times$-closed, the possibly multiple choices for the representatives of the $=^\times$-equivalence classes make it difficult to expect a similar result, and the Axiom of Choice is of no help here.

In the case of a category that is strongly Noetherian, that is, admits an $\mathbb{N}$-valued Noetherianity witness, we can say a little more.

Corollary 2.32 (smallest Garside). *Every strongly Noetherian left-cancellative category $\mathcal{C}$ contains a smallest $=^\times$-closed Garside family including $\mathbf{1}_\mathcal{C}$, namely the closure of the atoms under right-mcm and right-divisor. This Garside family is solid.*

Proof. Let $\mathcal{A}$ be the atom family of $\mathcal{C}$ and let $\mathcal{S}$ be the smallest $=^\times$-closed Garside family that includes $\mathcal{A} \cup \mathbf{1}_\mathcal{C}$, that is, the family obtained from $\mathcal{A} \cup \mathbf{1}_\mathcal{C}$ by closing under right-mcm and right-divisor. By Proposition 2.31, $\mathcal{S}$ is a Garside family. Now assume that $\mathcal{S}'$ is an $=^\times$-closed Garside family of $\mathcal{C}$. By Proposition II.2.58(i) (atoms generate), $\mathcal{S}'$ includes an $=^\times$-selector in $\mathcal{A}$, hence it includes $\mathcal{A}$ and, by Proposition 2.31 again, it includes $\mathcal{S}$.

Finally, by Proposition II.2.58(ii), the family $\mathcal{S}$ generates $\mathcal{C}$, and, by definition, it is closed under right-divisor, hence it is solid. ☐

In practice, Proposition 2.31 leads to a (partly) effective algorithm (see Subsection 2.4). In particular, if the closure $\widehat{\mathcal{S}}$ happens to be finite, then the sequence $(\mathcal{S}_i)_{i \geqslant 0}$ of the proof of Proposition 2.31 has to be eventually constant and one has $\widehat{\mathcal{S}} = \mathcal{S}_i$ where i is the first integer for which $\mathcal{S}_{i+2} = \mathcal{S}_i$ holds.

Example 2.33 (smallest Garside). Consider the free abelian monoid $\mathbb{N}^n$ (Reference Structure 1, page 3). Then $\mathbb{N}^n$ is strongly Noetherian with atoms $\mathtt{a}_1, ..., \mathtt{a}_n$. In the process of Proposition 2.31, we start with $\mathcal{S}_0 = \{\mathtt{a}_1, ..., \mathtt{a}_n\}$. Then $\mathcal{S}_1$ is the closure of $\mathcal{S}_0$ under right-mcm, here $\mathcal{S}_0$ plus all elements $\mathtt{a}_i\mathtt{a}_j$ with $i \neq j$. Then $\mathcal{S}_2$ is the closure of $\mathcal{S}_1$ under right-divisor, in this case $\mathcal{S}_1 \cup \{1\}$. Next, $\mathcal{S}_3$ consists of all elements that are products of at most 4 atoms, and, inductively, $\mathcal{S}_{2i+1}$ consists of all elements that are products of at most 2^i atoms. The process stops for $2^i \geqslant n$, yielding the smallest Garside family in $\mathbb{N}^n$, which consists of the 2^n products of atoms, that is, the family $\mathrm{Div}(\Delta_n)$ illustrated in Lemma I.1.3.

Now, let M be the partially abelian monoid $\langle \mathtt{a}, \mathtt{b}, \mathtt{c} \mid \mathtt{ab} = \mathtt{ba}, \mathtt{bc} = \mathtt{cb}\rangle^+$ (a particular case of the right-angled Artin–Tits monoids of Exercise 102). Arguing as above, one sees that the smallest Garside family in M consists of the 6 elements displayed on the right; note that this Garside family contains no unique maximal element.

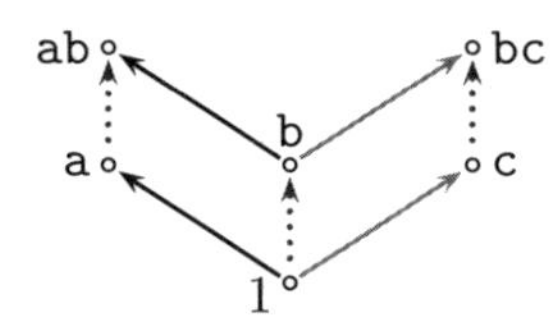

Here are two more examples. The first one shows that, even in the context of Corollary 2.32, the smallest Garside family need not be proper.

Example 2.34 (no proper Garside). Let $M = \langle \mathtt{a}, \mathtt{b} \mid \mathtt{ab} = \mathtt{ba}, \mathtt{a}^2 = \mathtt{b}^2\rangle^+$. Then M has no nontrivial invertible element, and it is strongly Noetherian, as it admits a homogeneous presentation. Its atoms are a and b, which admit two right-mcms, namely ab and $\mathtt{a}^2$. More generally, the elements $\mathtt{a}^m$ and $\mathtt{a}^{m-1}\mathtt{b}$ admit two right-mcms, namely $\mathtt{a}^{m+1}$ and $\mathtt{a}^m\mathtt{b}$, for every m. If S is a Garside family in M, it must contain a and b and be closed under right-mcm. An immediate induction using the above values shows that S must contain $\mathtt{a}^m$ and $\mathtt{a}^{m-1}\mathtt{b}$ for every positive m. Hence S includes all of M, except possibly 1. So M and $M \setminus \{1\}$ are the only Garside families in M. The case of M can be compared with the monoid of Exercise 45, where a finite Garside family exists: both presentations are similar but, in the latter case, the generator a has been split into a and a′, which admit no common right multiple.

The second example shows the necessity of the Noetherianity assumption for the existence of a smallest Garside family.

Example 2.35 (no smallest Garside). Consider the Klein bottle monoid K^+ (Reference Structure 5, page 17), and the subset $\{\mathtt{a}, \mathtt{b}\}$, which generates K^+. We claim that there exists no smallest Garside family including $\{\mathtt{a}, \mathtt{b}\}$. Indeed, for $m \geqslant 0$, consider the family $S_m = \{g \in K^+ \mid g \preccurlyeq \mathtt{ab}^m\mathtt{a}\}$. Then S_m (enumerated in $\preccurlyeq$-increasing order) consists of all elements $\mathtt{b}^p$, plus the elements $\mathtt{b}^p\mathtt{a}$, plus the elements $\mathtt{ab}^p$, plus the elements $\mathtt{ab}^p\mathtt{a}$ with $p \geqslant m$. As the left-divisibility relation of K^+ is a linear order, S^m is closed under right-lcm and every element g of K^+ admits an S_m-head, namely $\min_{\preccurlyeq}(g, \mathtt{ab}^m\mathtt{a})$. On the other hand, the left- and right-divisors of $\mathtt{ab}^m\mathtt{a}$ coincide, so, by Proposition 1.24, S_m is a Garside family in K^+ for every m.

Assume that S is a Garside family including $\{\mathtt{a}, \mathtt{b}\}$ in K^+. Then S is closed under right-divisor, and therefore contains all right-divisors of a, namely all elements $\mathtt{b}^p$ and $\mathtt{ab}^p$

with $p \geqslant 0$. Consider $\mathtt{ba}$. If it does not belong to S, it must admit an S-head s, which by definition satisfies $s \prec \mathtt{ba}$, hence must be either of the form $\mathtt{b}^p$ with $p \geqslant 0$, or of the form $\mathtt{b}^p\mathtt{a}$ with $p \geqslant 1$. Both are impossible: $s = \mathtt{b}^p$ is impossible because $\mathtt{b}^{p+1}$ would be a larger left-divisor of $\mathtt{ba}$ lying in S, and $s = \mathtt{b}^p\mathtt{a}$ is impossible because this would require $\mathtt{b}^p\mathtt{a} \in S$, and, as S is closed under right-divisor, imply $\mathtt{ba} \in S$, contrary to our assumption. So $\mathtt{ba}$ must lie in S. The same argument applies to every element $\mathtt{b}^p\mathtt{a}$, and S must include $\{g \in K^+ \mid |g|_{\mathtt{a}} \leqslant 1\}$. We saw in Example 2.23 that the latter family is not a Garside family, hence the inclusion is proper, and S contains at least one element g with $|g|_{\mathtt{a}} \geqslant 2$. As S is closed under right-divisor, it contains at least one element of the form $\mathtt{ab}^p\mathtt{a}$. Hence S must include at least one of the families S_m. Now, by construction, we have $S_{m+1} \subsetneqq S_m$, and none of the Garside families S_m can be minimal.

See Exercise 46 for an example of a monoid that is right-Noetherian, but not Noetherian, and admits no smallest Garside family.

Existence of lcms. Up to now we concentrated on the question of recognizing Garside families in categories that admit, say, conditional right-lcms. We conclude the section with results in a different direction, namely recognizing the existence of lcms when a Garside family is given. The results here are similar to what was done with Noetherianity in Proposition 2.18: if $\mathcal{S}$ is a Garside family, exactly as it suffices that $\mathcal{S}$ be locally right-Noetherian to ensure that the whole ambient category is right-Noetherian, it suffices that $\mathcal{S}$ locally admits right-lcms to ensure that the ambient category admits right-lcms. So, as in the case of right-Noetherianity, the existence of right-lcms can be decided locally, inside the Garside family $\mathcal{S}$.

We recall from Definition 2.14 that, if $\mathcal{S}$ is a subfamily of a left-cancellative category $\mathcal{C}$ and s, t lie in $\mathcal{S}$, we say that s is a *left-$\mathcal{S}$-divisor* of t, written $s \preccurlyeq_{\mathcal{S}} t$, if $st' = t$ holds for some t' lying in $\mathcal{S}$. We naturally introduce the derived notion of a *right-$\mathcal{S}$-lcm* to be a least upper bounds w.r.t. $\preccurlyeq_{\mathcal{S}}$.

Lemma 2.36. *Assume that $\mathcal{S}$ is a solid Garside family in a left-cancellative category $\mathcal{C}$ and s, t are elements of $\mathcal{S}$ with the same source.*

(i) *If s and t admit a right-$\mathcal{S}$-lcm, the latter is a right-lcm of s and t.*

(ii) *Conversely, if s and t admit a right-lcm, they admit one that lies in $\mathcal{S}$, and the latter is then a right-$\mathcal{S}$-lcm of s and t.*

Proof. (i) Assume that r is a right-$\mathcal{S}$-lcm of s and t. First, $\preccurlyeq_{\mathcal{S}}$ is included in $\preccurlyeq$, so r is a common right-multiple of s and t. Assume that h is a common right-multiple of s and t in $\mathcal{C}$. By Proposition 1.23 (Garside closed), $\mathcal{S}$ is closed under right-divisor and right-comultiple, hence, by Lemma 1.8, $\mathcal{S}$ satisfies (1.9), so there exist s', t', r' in $\mathcal{S}$ satisfying $r' = st' = ts' \preccurlyeq h$. Then r' is a common right-$\mathcal{S}$-multiple of s and t, so the assumption on r implies $r \preccurlyeq_{\mathcal{S}} r'$, whence $r \preccurlyeq h$. Hence r is a right-lcm of s and t.

(ii) Assume that h is a right-lcm of s and t in $\mathcal{C}$, say $h = sg = tf$. By Proposition 1.23 and Lemma 1.8 again, $\mathcal{S}$ satisfies (1.9), so there exist s', t', r in $\mathcal{S}$ satisfying $r = st' = ts' \preccurlyeq h$. The assumption that h is a right-lcm of s and t implies $r =^{\times} h$ and, therefore, r is also a right-lcm of s and t. On the other hand, by construction, r is a common right-$\mathcal{S}$-multiple of s and t, and a least one as a $\preccurlyeq_{\mathcal{S}}$-upper bound must be a $\preccurlyeq$-upper bound. ☐

Lemma 2.37. *Assume that $\mathcal{S}$ is a solid Garside family in a left-cancellative category $\mathcal{C}$. Then the following are equivalent:*

(i) *Any two elements of $\mathcal{S}$ with a common right-multiple admit a right-lcm.*

(ii) *Any two elements of $\mathcal{C}$ with a common right-multiple admit a right-lcm.*

Proof. Since $\mathcal{S}$ is included in $\mathcal{C}$, (i) is a specialization, hence a consequence, of (ii).

Assume (i). By Lemma 2.36, two elements of $\mathcal{S}$ that admit a right-lcm admit one that lies in $\mathcal{S}$. Moreover, by assumption, $\mathcal{S}$ is closed under right-divisor. So there exists an $\mathcal{S}$-valued right-lcm witness θ on $\mathcal{S}$, that is, a map from $\mathcal{S} \times \mathcal{S}$ to $\mathcal{S}$ such that, for every pair $\{s, t\}$ in $\mathcal{S}$ such that s and t admit a common right-multiple, we have $s\theta(s,t) = t\theta(t,s)$ and $s\theta(s,t)$ is a right-lcm of s and t. Assume that f, g belong to $\mathcal{C}$ and admit a common right-multiple. As $\mathcal{S}$ generates $\mathcal{C}$, we can find $\mathcal{S}$-paths $s_1|\cdots|s_p$ and $t_1|\cdots|t_q$ that are decompositions of f and g, respectively. Using θ we right-reverse the signed $\mathcal{S}$-path $\overline{s_p}|\cdots|\overline{s_1}|t_1|\cdots|t_q$, that is, we construct a rectangular grid based on $s_1|\cdots|s_p$ and $t_1|\cdots|t_q$ in which each square corresponds to a right-lcm of the left and top edges, as in Figure 2. The assumption that s and t admit a common right-multiple implies that each pair of elements of $\mathcal{S}$ occurring in the grid admits a common right-multiple, so that the construction of the grid can be completed, and repeated calls to Proposition II.2.12 (iterated lcm) imply that, if $s'_1|\cdots|s'_p$ and $t'_1|\cdots|t'_q$ are the paths occurring on the right and the bottom of the grid, then $ft'_1\cdots t'_q$, which is also $gs'_1\cdots s'_p$, is a right-lcm of s and t. □

By merging Lemmas 2.36 and 2.37, we obtain

Proposition 2.38 (existence of lcm). *If $\mathcal{S}$ is a solid Garside family in a left-cancellative category $\mathcal{C}$, the following are equivalent:*

(i) *Any two elements of $\mathcal{S}$ with a common right-$\mathcal{S}$-multiple admit a right-$\mathcal{S}$-lcm.*

(ii) *The category $\mathcal{C}$ admits conditional right-lcms, that is, any two elements of $\mathcal{C}$ with a common right-multiple admit a right-lcm.*

Of course, a variant of the above result not assuming that the considered Garside family is solid can be obtained by replacing $\mathcal{S}$ with $\mathcal{S}^\sharp$.

2.4 Categories with unique right-lcms

As shown in Example 2.33, the proof of Proposition 2.31 gives an effective method for constructing a smallest Garside family. However, the method is not really effective, as it requires finding all right-divisors of the elements of a family. Except in some very simple cases, there is no practical method to do that. By contrast, we shall now describe a new criterion that, in a more restricted context, provides a really effective method.

Hereafter, we consider the case of a left-cancellative category $\mathcal{C}$ that contains no nontrivial invertible elements and admits conditional right-lcms. In this case, a right-lcm is unique when it exists; conversely, no invertible element may exist when right-lcms are unique. We can therefore equivalently say that $\mathcal{C}$ admit *unique conditional right-lcms*. Then $\mathcal{S}^\sharp = \mathcal{S} \cup 1_\mathcal{C}$ holds for every subfamily $\mathcal{S}$, and right-lcm is a well-defined (partial) binary operation. Associated comes a right-complement operation (Definition II.2.11):

for f, g in $\mathcal{C}$ admitting a common right-multiple, the right-complement $f\backslash g$ is the (unique) element g' such that fg' is the right-lcm of f and g. For instance, in the free abelian monoid $\mathbb{N}^n$ based on $\{\mathtt{a}_1, ..., \mathtt{a}_n\}$ (Reference Structure 1, page 3), the right-lcm of $\mathtt{a}_i$ and $\mathtt{a}_j$ is $\mathtt{a}_i\mathtt{a}_j$ for $i \neq j$, and is $\mathtt{a}_i$ for $i = j$, and we find $\mathtt{a}_i\backslash\mathtt{a}_j = \mathtt{a}_j$ for $i \neq j$ and $\mathtt{a}_i\backslash\mathtt{a}_j = 1$ for $i = j$. Similarly, in the case of braids (Reference Structure 2, page 5), unique right-lcms also exist, and we find $\sigma_i\backslash\sigma_j = \sigma_j\sigma_i$ for $|i-j| = 1$, $\sigma_i\backslash\sigma_j = \sigma_j$ for $|i-j| \geqslant 2$, and $\sigma_i\backslash\sigma_j = 1$ for $i = j$.

Lemma 2.39. *For every subfamily $\mathcal{S}$ of a left-cancellative category $\mathcal{C}$ that is right-Noetherian and admits unique conditional right-lcms, the following are equivalent:*

(i) *The family $\mathcal{S}^\sharp$ is closed under right-complement* (*in the sense of Definition* 1.3).

(ii) *The family $\mathcal{S}^\sharp$ is closed under $\backslash$, that is, if s and t belong to $\mathcal{S}^\sharp$, then so does $s\backslash t$ when defined, that is, when s and t admit a common right-multiple.*

The proof consists in just applying the definitions and it is left to the reader. Owing to Lemma 2.39, Corollary 2.29 and Lemma 2.28 directly imply:

Proposition 2.40 (recognizing Garside, right-complement case). *A subfamily $\mathcal{S}$ of a left-cancellative category $\mathcal{C}$ that is right-Noetherian and admits unique conditional right-lcms is a Garside family if and only if $\mathcal{S}^\sharp$ generates $\mathcal{C}$ and is closed under right-lcm and right-complement.*

Corollary 2.41 (smallest Garside). *For every generating subfamily $\mathcal{S}$ of a left-cancellative category $\mathcal{C}$ that is right-Noetherian and admits unique conditional right-lcms, the smallest Garside $=^\times$-closed family that includes $\mathcal{S}$ is the closure of $\mathcal{S}$ under the right-lcm and right-complement operations.*

The difference between this description of the smallest Garside family and the one of Proposition 2.31 is that the right-lcm and the right-complement are functional operations (at most one result for each choice of the arguments). Moreover, these operations can be computed easily whenever one is given a presentation $(\mathcal{S}, \mathcal{R})$ of the ambient category for which right-reversing is complete. Indeed, Proposition II.4.46 (common right-multiple) implies that, for all $\mathcal{S}$-paths u, v that represent elements admitting a common right-multiple, there exist a unique pair of $\mathcal{S}$-paths u', v' such that $\overline{u}v$ is $\mathcal{R}$-reversible to $v'\overline{u'}$, in which case uv' (and vu') represent the right-lcm of $[u]$ and $[v]$, whereas v' represents $[v]\backslash[u]$ and u' represents $[u]\backslash[v]$. So closing under right-lcm and right-complement essentially means, at the level of $\mathcal{S}$-paths, closing under right-reversing.

Example 2.42 (smallest Garside). Consider the braid monoid B_4^+ (Reference Structure 2, page 5). The atoms are $\sigma_1, \sigma_2, \sigma_3$. To find the smallest Garside family of B_4^+, we start with $S_0 = \{\sigma_1, \sigma_2, \sigma_3\}$ and define S_{i+1} to be S_i completed with the right-lcms (*resp.* right-complements) of elements of S_i when i is odd (*resp.* even). One finds

$S_1 = S_0 \cup \{\sigma_1\sigma_2\sigma_1, \sigma_2\sigma_3\sigma_2, \sigma_1\sigma_3\}$,

$S_2 = S_1 \cup \{1, \sigma_1\sigma_2, \sigma_2\sigma_1, \sigma_2\sigma_3, \sigma_3\sigma_2, \sigma_1\sigma_2\sigma_3, \sigma_3\sigma_2\sigma_1,$

$$\begin{aligned}
&\qquad\qquad\qquad\qquad \sigma_2\sigma_1\sigma_3\sigma_2, \sigma_2\sigma_1\sigma_3\sigma_2\sigma_1, \sigma_1\sigma_2\sigma_3\sigma_1\sigma_2\},\\
S_3 &= S_2 \cup \{\Delta_4, \sigma_2\sigma_3\sigma_1, \sigma_1\sigma_3\sigma_2\sigma_1\sigma_3, \sigma_1\sigma_2\sigma_1\sigma_3, \sigma_2\sigma_3\sigma_2\sigma_1, \sigma_1\sigma_2\sigma_3\sigma_2, \sigma_1\sigma_3\sigma_2\sigma_1\},\\
S_4 &= S_3 \cup \{\sigma_1\sigma_3\sigma_2\},\\
S_5 &= S_6 = S_4.
\end{aligned}$$

Hence, the smallest Garside family S in B_4^+ that contains 1, is the collection of the 24 divisors of Δ_4. Removing the saturation condition amounts to removing 1, so the smallest Garside family in B_4^+ is the collection of the 23 nontrivial divisors of Δ_4.

2.5 Finite height

We saw in Subsection 2.2 that local conditions, namely conditions involving the elements of the considered Garside family only, guarantee right-Noetherianity. It is natural to address a similar question in the case of strong Noetherianity, that is, when every element has a finite height. We shall establish now a partial result, in the case when the height is uniformly bounded on the considered Garside family.

We recall from Proposition II.2.48 (finite height) that an element g has height at most m if every decomposition of g contains at most m non-invertible entries.

Proposition 2.43 (finite height). *If $\mathcal{C}$ is a left-cancellative category and the inequality $\mathrm{ht}(s) \leqslant K$ holds for every s in some Garside family $\mathcal{S}$ of $\mathcal{C}$, then $\mathcal{C}$ is strongly Noetherian and the inequality $\mathrm{ht}(g) \leqslant K^p$ holds for every element g of $\mathcal{C}$ satisfying $\|g\|_{\mathcal{S}} \leqslant p$.*

Proposition 2.43 follows from a stronger technical result—we shall see in Chapter XI a situation where only the latter is relevant.

Lemma 2.44. *Assume that $\mathcal{C}$ is a left-cancellative category, $\mathcal{S}$ is a Garside family of $\mathcal{C}$, and there exists a map $K : \mathit{Obj}(\mathcal{C}) \to \mathbb{N}$ satisfying $\forall g \in \mathcal{S}(\cdot, y)\ (\mathrm{ht}(g) \leqslant K(y))$ and such that $\mathcal{C}(y, y') \neq \emptyset$ implies $K(y) \leqslant K(y')$. Then $\mathrm{ht}(g) \leqslant K(y)^p$ holds for every g in $\mathcal{C}(\cdot, y)$ satisfying $\|g\|_{\mathcal{S}} \leqslant p$.*

Proof. As a preliminary remark, we observe that $\mathrm{ht}(g) \leqslant K(y)$ for every g in $\mathcal{S}^\sharp(\cdot, y)$. Indeed, by (the counterpart of) Lemma II.2.44(i), $g' =^{\times} g$ implies $\mathrm{ht}(g') = \mathrm{ht}(g)$.

Assume that g belongs to $\mathcal{C}(\cdot, y)$ and satisfies $\|g\|_{\mathcal{S}} \leqslant p$. Then g lies in $(\mathcal{S}^\sharp)^p(\cdot, y)$. By assumption, g has an $\mathcal{S}$-normal decomposition of length p, say $s_1|\cdots|s_p$. Let $t_1|\cdots|t_q$ be an arbitrary decomposition of g. As the equality $s_1 \cdots s_p = t_1 \cdots t_q$ holds, Proposition 1.15, which is relevant by Proposition 1.24, guarantees the existence of a rectangular grid as in Figure 8, where the edges and diagonals of the squares lie in $\mathcal{S}^\sharp$. Let $N(j)$ be the number of non-invertible elements on the jth row of the diagram. We claim that the inequality $N(j) \leqslant K(y)^{p-j}$ is satisfied.

Indeed, let $t_{1,j}, \ldots, t_{q,j}$ be the elements of $\mathcal{S}^\sharp$ that appear on the jth row of the diagram and, similarly, let $s_{i,1}, \ldots, s_{i,p}$ be those appearing on the jth column. We use induction on j decreasing from p to 0. For $j = p$, all elements $t_{i,j}$ left-divide 1_y, so they all must be invertible, and $N(p) = 0 \leqslant K(y)^0$ holds.

Assume that $t_{i,j}$ is non-invertible in the jth row. In that row, $t_{i,j}$ is followed by a certain number of invertible entries, say n (possibly zero). Let us evaluate how many of the entries $t_{i,j-1}, \ldots, t_{i+n,j-1}$ may be non-invertible. Locally, the diagram looks like

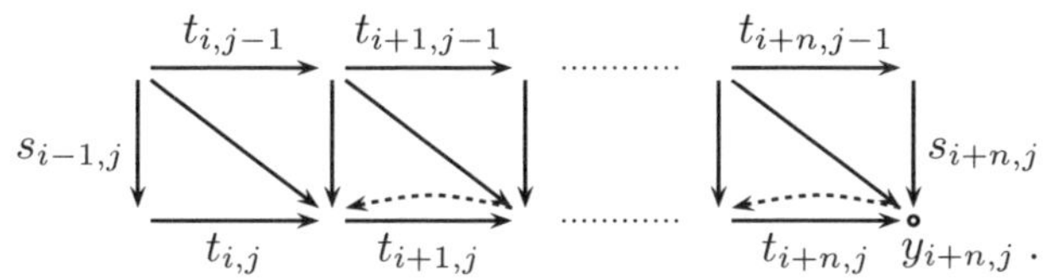

By assumption, $s_{i-1,j}t_{i,j}$ belongs to $\mathcal{S}^\sharp$ and $t_{i+1,j}, \ldots, t_{i+n,j}$ are invertible, so $s_{i-1,j}$ $t_{i,j} \cdots t_{i+n,j}$, which is also $t_{i,j-1} \cdots t_{i+n,j-1} s_{i+n,j}$, belongs to $\mathcal{S}^\sharp$. It follows that, among $t_{i,j-1}, \ldots, t_{i+n,j-1}$ (and $s_{i+n,j}$), there are at most $K(y_{i+n,j})$ non-invertible entries, where $y_{i+n,j}$ is the common target of $t_{i+n,j}$ and $s_{i+n,j}$. By construction, there exists a path from $y_{i+n,j}$ to y in the diagram, so $\mathcal{C}(y_{i+n,j}, y)$ is nonempty and, therefore, we have $K(y_{i+n,j}) \leqslant K(y)$. Thus every non-invertible arrow in the jth row gives rise to at most $K(y)$ non-invertible arrows in the above row. Actually, the number is even at most $K(y)-1$, unless $s_{i+n,j}$ is invertible, in which case the number of non-invertible elements at the right of $s_{i+n,j}$ is the same on the jth and $(j-1)$th rows. The block of invertible elements possibly starting with $t_{1,j}$ together with the vertical arrow $s_{1,j}$ may result in $K(y)$ more non-invertible elements on the $(j-1)$th row. Finally, an easy bookkeeping argument shows that, in all cases, the number of non-invertible elements on the $(j-1)$th row is at most $(N(j)+1)(K(y)-1)+1$, hence, using the induction hypothesis, at most $K(y)^{p-j+1}$. For $j = 0$, we obtain $N(0) \leqslant K(y)^p$, which is the expected result.

Finally, as $\mathcal{S}$ is a Garside family, every element g of $\mathcal{C}$ satisfies $\|g\|_\mathcal{S} \leqslant p$ for some p, and, therefore, $\mathrm{ht}(g)$ is finite for every element of $\mathcal{C}$. □

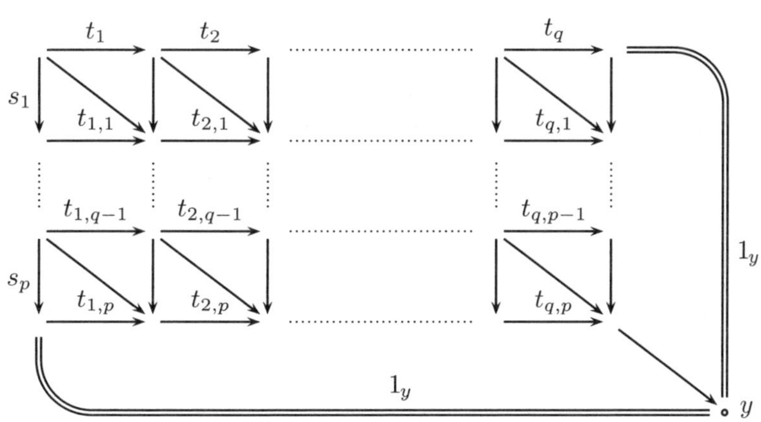

Figure 8. Proof of Proposition 2.43: get an upper bound on the number of non-invertible elements on each row of the diagram, starting from the bottom.

Proof of Proposition 2.43. The result is a direct consequence of Lemma 2.44 since, if $\mathrm{ht}(s) \leqslant K$ holds for every s in $\mathcal{S}$, then conditions (i) and (ii) of Lemma 2.44 are satisfied for the constant function with value K. □

Specializing, we obtain

Corollary 2.45 (finite height). *Every left-cancellative category $\mathcal{C}$ that admits a finite Garside family whose elements have finite height is strongly Noetherian.*

3 Geometric and algorithmic applications

We return to the general context of arbitrary left-cancellative categories, and use the closure properties introduced in Section 1 to derive new geometric and algorithmic results completing those of Section III.3. The main point is the existence of presentations that involve short relations and give rise to quadratic isoperimetric inequalities.

The section is organized as follows. In Subsection 3.1, we show that every Garside family provides distinguished presentations of the ambient category, and we observe that the latter satisfy quadratic isoperimetric inequalities. In Subsection 3.2, we deduce new solutions for the Word Problem for a category or its enveloping groupoid using the previous presentations and right-reversing (Section II.4) rather than appealing to normal decompositions. Finally, in Subsection 3.3, we discuss the specific case of categories that admit unique right-lcms.

3.1 Presentations

We show that every Garside family in a left-cancellative category provides a simple explicit presentation. The point is the following direct consequence of Proposition 1.15.

Lemma 3.1. *Assume that $\mathcal{C}$ is a left-cancellative category and $\mathcal{S}$ is a generating subfamily that is closed under right-complement. Then each equality $s_1 \cdots s_p = t_1 \cdots t_q$ with $s_1, ..., t_q$ in $\mathcal{S}$ satisfied in $\mathcal{C}$ is the consequence of*

(i) *pq relations of the form $st' = ts'$ with s, t, s', t' in $\mathcal{S}$, and*

(ii) *$p + q$ relations of the form $\epsilon = \epsilon'\epsilon''$ with $\epsilon, \epsilon', \epsilon''$ in $\mathcal{C}^\times \setminus \mathbf{1}_\mathcal{C}$.*

If $\mathcal{S}$ is closed under right-divisor and right-comultiple, we may moreover assume that the relations $st' = ts'$ involved in (i) *satisfy $st' \in \mathcal{S}$.*

Proof. Assume $s_1 \cdots s_p = t_1 \cdots t_q$ with $s_1, ..., t_q$ in $\mathcal{S}$. Applying Proposition 1.15(i) with $f = g = 1_y$, where y is the target of s_p and t_q, we obtain a commutative diagram as in Figure 9, in which the edges of the squares all represent elements of $\mathcal{S}$ and, therefore, each square corresponds to an equality of the form $st' = ts'$ with s, t, s', t' in $\mathcal{S}$.

Moreover, as we have $1_y = s'_1 \cdots s'_p = t'_1 \cdots t'_q$, each of $s'_1, ..., s'_p, t'_1, ..., t'_q$, and h is invertible. Therefore, the right and bottom parts of the diagram split into triangular pieces with invertible labels that correspond to relations of the form $\epsilon = \epsilon'\epsilon''$ with $\epsilon, \epsilon', \epsilon''$ in $\mathcal{C}^\times$.

If $\mathcal{S}$ is closed under right-divisor and right-comultiple, Proposition 1.15(ii) says that we may assume that the diagonal of each rectangle represent an element of $\mathcal{S}$, hence that, in the corresponding relation $st' = ts'$, the diagonal st' lies in $\mathcal{S}$. □

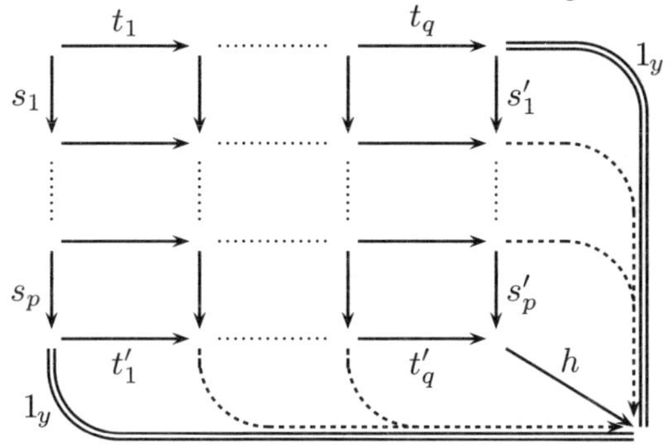

Figure 9. Factorization of the equality $s_1 \cdots s_p = t_1 \cdots t_q$ by means of pq relations of the type $st' = ts'$ with s, t, s', t' in $\mathcal{S}$, plus $p + q$ relations $\epsilon = \epsilon'\epsilon''$ with $\epsilon, \epsilon', \epsilon''$ invertible.

Definition 3.2 (families $\mathcal{R}_\mathcal{C}(\mathcal{S})$ and $\mathcal{R}^\times_\mathcal{C}(\mathcal{S})$). For $\mathcal{C}$ a left-cancellative category and $\mathcal{S}$ a subfamily of $\mathcal{C}$, we define $\mathcal{R}_\mathcal{C}(\mathcal{S})$ to be the family of all relations $r = st$ with r, s, t in $\mathcal{S}$ that are satisfied in $\mathcal{C}$, and $\mathcal{R}^\times_\mathcal{C}(\mathcal{S})$ to be the family of all relations

(i) $r\epsilon = st$ with r, s, t in $\mathcal{S} \setminus \mathcal{C}^\times$ and ϵ in $\mathcal{C}^\times$,

(ii) $\epsilon s = s'\epsilon'$ with s, s' in $\mathcal{S} \setminus \mathcal{C}^\times$ and ϵ, ϵ' in $\mathcal{C}^\times$,

(iii) $\epsilon = \epsilon'\epsilon''$ with $\epsilon, \epsilon', \epsilon''$ in $\mathcal{C}^\times$,

that are satisfied in $\mathcal{C}$.

Note that $\mathcal{R}^\times_\mathcal{C}(\mathcal{S})$ contains as a particular case of type (ii) all relations of the form $\epsilon s = s'$ and $s = s'\epsilon'$ with s, s' in $\mathcal{S} \setminus \mathcal{C}^\times$ and ϵ, ϵ' in $\mathcal{C}^\times$. We first consider the case of solid Garside families.

Lemma 3.3. *Assume that $\mathcal{C}$ is a left-cancellative category and $\mathcal{S}$ is a solid subfamily of $\mathcal{C}$ that generates $\mathcal{C}$ and is closed under right-comultiple. Then $\mathcal{C}$ admits the presentation $\langle \mathcal{S} \mid \mathcal{R}_\mathcal{C}(\mathcal{S}) \rangle^+$, and each equality $s_1 \cdots s_p = t_1 \cdots t_q$ with $s_1, ..., t_q$ in $\mathcal{S}$ satisfied in $\mathcal{C}$ is the consequence of $2pq + p + q$ relations of $\mathcal{R}_\mathcal{C}(\mathcal{S})$.*

Proof. First, by Proposition 1.23, $\mathcal{S}$ is closed under right-divisor, right-comultiple, and right-complement and, therefore, it is eligible for Lemma 3.1. Then every relation $s|t' = t|s'$ of Lemma 3.1(i) with s, t, s', t' and st' in $\mathcal{S}$ is the consequence of two relations of $\mathcal{R}_\mathcal{C}(\mathcal{S})$, namely $s|t' = r$ and $t|s' = r$ with $r = st'$. On the other hand, every relation $r = st$ of Lemma 3.1(ii) is either a free groupoid relation $1_x s = s$, $s 1_y = s$, or $ss^{-1} = 1_x$, that is implicit in every presentation, or a relation involving three elements of $\mathcal{C}^\times \setminus \mathbf{1}_\mathcal{C}$, which belongs to $\mathcal{R}_\mathcal{C}(\mathcal{S})$ since, by Lemma 2.2, $\mathcal{C}^\times \setminus \mathbf{1}_\mathcal{C}$ is included in $\mathcal{S}$.

By Lemma 3.1, every equality $s_1 \cdots s_p = t_1 \cdots t_q$ with $s_1, ..., t_q$ in $\mathcal{S}$ follows from pq relations of type (i) and $p + q$ relations of type (ii), hence from $2pq + p + q$ relations of $\mathcal{R}_\mathcal{C}(\mathcal{S})$. □

Proposition 3.4 (presentation, solid case). *If $\mathcal{S}$ is a solid Garside family in a left-cancellative category $\mathcal{C}$, then $\mathcal{C}$ admits the presentation $\langle \mathcal{S} \mid \mathcal{R}_\mathcal{C}(\mathcal{S}) \rangle^+$.*

Proof. By Proposition 2.4, the assumption that $\mathcal{S}$ is solid implies that $\mathcal{S}$ generates $\mathcal{C}$ and is closed under right-divisor. As $\mathcal{S}$ is Garside family, $\mathcal{S}$ is closed under right-comultiple. Then we apply Lemma 3.3. □

In other words, whenever $\mathcal{S}$ is a solid Garside family, the family of all valid relations $r = st$ with r, s, t in $\mathcal{S}$ constitutes a presentation of the ambient category from $\mathcal{S}$.

We now address arbitrary Garside families using the fact that, if $\mathcal{S}$ is a Garside family, then $\mathcal{S}^\sharp$ is a solid Garside family. We begin with a comparison of $\mathcal{R}_\mathcal{C}(\mathcal{S}^\sharp)$ and $\mathcal{R}^\times_\mathcal{C}(\mathcal{S})$.

Lemma 3.5. *If $\mathcal{S}$ is a subfamily of a left-cancellative category $\mathcal{C}$, every relation of $\mathcal{R}^\times_\mathcal{C}(\mathcal{S})$ is the consequence of at most two relations of $\mathcal{R}_\mathcal{C}(\mathcal{S}^\sharp)$, and every relation of $\mathcal{R}_\mathcal{C}(\mathcal{S}^\sharp)$ is the consequence of at most five relations of $\mathcal{R}^\times_\mathcal{C}(\mathcal{S})$.*

Proof. Every relation of $\mathcal{R}^\times_\mathcal{C}(\mathcal{S})$ of type (i) or (ii) has the form $st' = ts'$ with s, t, s', t' in $\mathcal{S}^\sharp$ such that $r = st'$ lies in $\mathcal{S}^\sharp$, hence it is the consequence of the two corresponding relations $st' = r$ and $ts' = r$, which lie in $\mathcal{R}^\times_\mathcal{C}(\mathcal{S})$. On the other hand, every relation of $\mathcal{R}^\times_\mathcal{C}(\mathcal{S})$ of type (iii) is a relation of $\mathcal{R}_\mathcal{C}(\mathcal{S}^\sharp)$.

Conversely, let t_1, t_2, t_3 belong to $\mathcal{S}^\sharp$ and $t_1t_2 = t_3$ be a relation of $\mathcal{R}_\mathcal{C}(\mathcal{S}^\sharp)$, that is, $t_1t_2 = t_3$ holds in $\mathcal{C}$. Assume first that t_1 and t_2 are invertible. Then t_3 is invertible as well, and $t_1t_2 = t_3$ is a relation of $\mathcal{R}^\times_\mathcal{C}(\mathcal{S})$ of type (iii).

Assume now that t_1 is invertible and t_2 is not. Then t_3 is not invertible either, so t_2 and t_3 lie in $(\mathcal{S} \setminus \mathcal{C}^\times)\mathcal{C}^\times$. Hence, for $i = 2, 3$, we have $t_i = s_i\epsilon_i$ for some s_i in $\mathcal{S} \setminus \mathcal{C}^\times$ and ϵ_i in $\mathcal{C}^\times$. Then the equality $t_1t_2 = t_3$ follows from the path derivation

$$t_1|t_2 \equiv_{\text{(ii)}} t_1|s_2|\epsilon_2 \equiv_{\text{(ii)}} s_3|\epsilon_3\epsilon_2^{-1}|\epsilon_2 \equiv_{\text{(iii)}} s_3|\epsilon_3 \equiv_{\text{(ii)}} t_3,$$

which involves three relations of $\mathcal{R}^\times_\mathcal{C}(\mathcal{S})$ of type (ii) and one of type (iii).

Assume now that t_2 is invertible and t_1 is not. Then t_3 is not invertible, and t_1 and t_3 lie in $(\mathcal{S} \setminus \mathcal{C}^\times)\mathcal{C}^\times$. As above, write $t_1 = s_1\epsilon_1$ with s_1 in $\mathcal{S} \setminus \mathcal{C}^\times$ and ϵ_1 in $\mathcal{C}^\times$. Then the equality $t_1t_2 = t_3$ follows from the path derivation

$$t_1|t_2 \equiv_{\text{(ii)}} s_1|\epsilon_1|t_2 \equiv_{\text{(iii)}} s_1|\epsilon_1t_2 \equiv_{\text{(ii)}} t_3,$$

which involves two relations of $\mathcal{R}^\times_\mathcal{C}(\mathcal{S})$ of type (ii) and one of type (iii).

Assume finally that neither t_1 nor t_2 is invertible. Then t_3 is not invertible. For $i = 1, 3$ write $t_i = s_i\epsilon_i$ with s_i in $\mathcal{S} \setminus \mathcal{C}^\times$ and ϵ_i in $\mathcal{C}^\times$. By assumption, $\mathcal{C}^\times\mathcal{S}$ is included in $\mathcal{S}^\sharp$, so we must have $\epsilon_1t_2 = t'_2\epsilon'_1$ for some t'_2 in $\mathcal{S} \setminus \mathcal{C}^\times$ and ϵ'_1 in $\mathcal{C}^\times$. Then the equality $t_1t_2 = t_3$ follows from the path derivation

$$t_1|t_2 \equiv_{\text{(ii)}} s_1|\epsilon_1|t_2 \equiv_{\text{(ii)}} s_1|t'_2|\epsilon'_1 \equiv_{\text{(i)}} s_3|\epsilon_3{\epsilon'_1}^{-1}|\epsilon'_1 \equiv_{\text{(iii)}} s_3|\epsilon_3 \equiv_{\text{(ii)}} t_3,$$

which involves one relation of $\mathcal{R}_\mathcal{C}(\mathcal{S})$ of type (i), three of type (ii), and one of type (iii). Thus, every relation of $\mathcal{R}_\mathcal{C}(\mathcal{S}^\sharp)$ follows from at most five relations of $\mathcal{R}_\mathcal{C}(\mathcal{S})$. □

We can now establish the general result.

Proposition 3.6 (presentation). *If $\mathcal{S}$ is a Garside family in a left-cancellative category $\mathcal{C}$, then $\mathcal{C}$ admits the presentations $\langle \mathcal{S}^\sharp \mid \mathcal{R}_\mathcal{C}(\mathcal{S}^\sharp)\rangle^+$ and $\langle \mathcal{S} \cup \mathcal{C}^\times \mid \mathcal{R}^\times_\mathcal{C}(\mathcal{S})\rangle^+$.*

Proof. By Lemma 2.3, $\mathcal{S}^\sharp$ is a solid Garside family. By Proposition 1.24, $\mathcal{S}^\sharp$ is closed under right-comultiple, and, therefore, it is eligible for Proposition 3.4, so that $\mathcal{C}$ admits the presentation $\langle \mathcal{S}^\sharp \mid \mathcal{R}_\mathcal{C}(\mathcal{S}^\sharp)\rangle^+$. On the other hand, $\mathcal{S}^\sharp$ generates $\mathcal{C}$ and, by Lemma 3.5, the families $\mathcal{R}_\mathcal{C}(\mathcal{S}^\sharp)$ and $\mathcal{R}^\times_\mathcal{C}(\mathcal{S})$ generate the same congruence. So $\langle \mathcal{S}^\sharp \mid \mathcal{R}^\times_\mathcal{C}(\mathcal{S})\rangle^+$ is also a presentation of $\mathcal{C}$. □

When the ambient category $\mathcal{C}$ contains nontrivial invertible elements, the presentation $(\mathcal{S}^\sharp, \mathcal{R}_\mathcal{C}(\mathcal{S}^\sharp))$ is likely to be redundant, and $(\mathcal{S} \cup \mathcal{C}^\times, \mathcal{R}^\times_\mathcal{C}(\mathcal{S}))$ is more compact. Note that $(\mathcal{S}^\sharp, \mathcal{R}_\mathcal{C}(\mathcal{S}^\sharp))$ and $(\mathcal{S} \cup \mathcal{C}^\times, \mathcal{R}^\times_\mathcal{C}(\mathcal{S}))$ are finite if $\mathcal{S}$ and $\mathcal{C}^\times$ are finite.

Example 3.7 (presentation). Consider the braid monoid B_3^+ (Reference Structure 2, page 5) with the solid Garside family $\mathcal{S} = \{1, \mathtt{a}, \mathtt{b}, \mathtt{ab}, \mathtt{ba}, \Delta\}$, where Δ stands for $\mathtt{aba}$. Then the family $\mathcal{R}_\mathcal{C}(\mathcal{S})$ consists of the six relations $\mathtt{a}|\mathtt{b} = \mathtt{ab}$, $\mathtt{b}|\mathtt{a} = \mathtt{ba}$, $\mathtt{ab}|\mathtt{a} = \Delta$, $\mathtt{ba}|\mathtt{b} = \Delta$, $\mathtt{a}|\mathtt{ba} = \Delta$, and $\mathtt{b}|\mathtt{ab} = \Delta$, plus all relations $1|s = s$ and $s|1 = s$.

Consider now the wreathed free abelian monoid $\widetilde{\mathbb{N}^n}$ (Reference Structure 6, page 19) with the Garside family S_n. Then S_n has 2^n elements and $(\widetilde{\mathbb{N}^n})^\times$, which is isomorphic to $\mathfrak{S}_n$, has $n!$ elements. The family $\mathcal{R}_{\widetilde{\mathbb{N}^n}}(S_n^\times)$, which includes the complete diagram of the cube S_n, plus the complete table of the group $\mathfrak{S}_n$, plus all semicommutation relations between the latter, is redundant. Restricting to atoms of S_n and extracting a generating family for $\mathfrak{S}_n$ leads to the more compact presentation (I.3.5).

We conclude with the enveloping groupoid. By Proposition II.3.11 (Ore's theorem), if $\mathcal{C}$ is a left-Ore category, every presentation of $\mathcal{C}$ is automatically a presentation of the enveloping groupoid $\mathcal{E}nv(\mathcal{C})$. So Proposition 3.6 directly implies

Corollary 3.8 (presentation). *If $\mathcal{S}$ is a strong Garside family in a left-Ore category $\mathcal{C}$, then the groupoid $\mathcal{E}nv(\mathcal{C})$ admits the presentations $\langle \mathcal{S}^\sharp \mid \mathcal{R}_\mathcal{C}(\mathcal{S}^\sharp)\rangle$ and $\langle \mathcal{S} \cup \mathcal{C}^\times \mid \mathcal{R}^\times_\mathcal{C}(\mathcal{S})\rangle$.*

A direct application of the rectangular grids of Figure 9 is the satisfaction of quadratic isoperimetric inequalities.

Definition 3.9 (isoperimetric inequality). Assume that $\mathcal{C}$ is a category, $(\mathcal{S}, \mathcal{R})$ is a presentation of $\mathcal{C}$, and F is a function of $\mathbb{N}$ to itself. We say that $\mathcal{C}$ satisfies an *F-isoperimetric inequality* with respect to $(\mathcal{S}, \mathcal{R})$ if there exist a constant K such that, for all $\mathcal{S}$-paths u, v representing the same element of $\mathcal{C}$ and satisfying $\lg(u) + \lg(v) \leqslant n$ with $n \geqslant K$, there exists an $\mathcal{R}$-derivation of length at most $F(n)$ from u to v.

In practice, only the order of magnitude of growth rate of the function F matters, and one speaks of linear, quadratic, cubic... isoperimetric inequality. If the category $\mathcal{C}$ is a groupoid, it is sufficient to consider one word w that represents an identity element.

Every category $\mathcal{C}$ satisfies a linear isoperimetric inequality with respect to the exhaustive presentation $(\mathcal{C}, \mathcal{R})$ with $\mathcal{R}$ consisting of all relations $fg = h$ that are satisfied in $\mathcal{C}$, so the notion is meaningful mostly for finite presentations. It is known [119] that switching from one finite presentation to another one does not change the order of magnitude of a minimal function F for which an isoperimetric inequality is satisfied.

As the presentations $(\mathcal{S}^\sharp, \mathcal{R}_\mathcal{C}(\mathcal{S}^\sharp))$ and $(\mathcal{S} \cup \mathcal{C}^\times, \mathcal{R}^\times_\mathcal{C}(\mathcal{S}))$ associated with a Garside family are eligible for Proposition 3.4, we immediately obtain:

Proposition 3.10 (isoperimetric inequality, positive case). *If $\mathcal{S}$ is a Garside family in a left-cancellative category $\mathcal{C}$, then $\mathcal{C}$ satisfies a quadratic isoperimetric inequality with respect to the presentations $(\mathcal{S}^\sharp, \mathcal{R}_\mathcal{C}(\mathcal{S}^\sharp))$ and $(\mathcal{S} \cup \mathcal{C}^\times, \mathcal{R}_\mathcal{C}^\times(\mathcal{S}))$.*

In the case of a left-Ore category, the result extends to the enveloping groupoid.

Proposition 3.11 (isoperimetric inequality, general case). *If $\mathcal{S}$ is a strong Garside family in a left-Ore category $\mathcal{C}$, then $\mathcal{E}nv(\mathcal{C})$ satisfies a quadratic isoperimetric inequality with respect to the presentations $(\mathcal{S}^\sharp, \mathcal{R}_\mathcal{C}(\mathcal{S}^\sharp))$ and $(\mathcal{S} \cup \mathcal{C}^\times, \mathcal{R}_\mathcal{C}^\times(\mathcal{S}))$.*

Proof. Assume that w is a signed $\mathcal{S}^\sharp$-path of length n that represents an identity element in $\mathcal{E}nv(\mathcal{C})$. Owing to Proposition 3.10, it suffices to show that, using at most $O(n^2)$ relations of $\mathcal{R}_\mathcal{C}(\mathcal{S}^\sharp)$, we can transform w into an equivalent negative–positive path $\overline{u}v$ of the same length. Indeed, the assumption that w represents an identity element implies that u and v represent the same element of $\mathcal{C}$, and, by Proposition 3.10, we know that u can transformed into v using $O(n^2)$ relations, hence $\overline{u}v$ can be transformed into an empty path using $O(n^2)$ relations of $\mathcal{R}_\mathcal{C}(\mathcal{S}^\sharp)$.

Now, exactly as in Algorithm III.3.61 (Word Problem, general case), we can start from the signed path w and, calling p (*resp.* q) the number of negative (*resp.* positive) entries in w, and apply at most pq relations of the form $s't = t's$ with s, t, s', t' in $\mathcal{S}^\sharp$ to transform w into a negative–positive path of the expected type (we recall that the existence of a strong Garside family in $\mathcal{C}$ implies that $\mathcal{C}$ admits left-lcms, so the above relations can be chosen so that $s't$ is a left-lcm of s and t). By Lemma 3.4, each of the above relations, which are of the type considered in Lemma 3.1, can be decomposed into at most five relations of $\mathcal{R}_\mathcal{C}(\mathcal{S}^\sharp)$.

Going from $(\mathcal{S}^\sharp, \mathcal{R}_\mathcal{C}(\mathcal{S}^\sharp))$ to $(\mathcal{S} \cup \mathcal{C}^\times, \mathcal{R}_\mathcal{C}^\times(\mathcal{S}))$ is then easy. □

3.2 Word Problem

In Section III.3, we described algorithms that solve the Word Problem for a category and, possibly, its groupoid of fractions using (symmetric) normal decompositions. Using the presentations of Proposition 3.6, we obtain alternative algorithms that can be conveniently described using the reversing method of Section II.4.

We recall that the reversing method is useful only when some technical condition, the completeness condition of Definition II.4.40, is satisfied. It turns out that, in general, right-reversing is not complete for the presentation $(\mathcal{S}^\sharp, \mathcal{R}_\mathcal{C}(\mathcal{S}^\sharp))$, see Exercise 56. However, in this case, the failure of completeness is an easily fixed lack of transitivity.

Definition 3.12 (families $\widehat{\mathcal{R}}_\mathcal{C}(\mathcal{S})$ and $\widehat{\mathcal{R}}_\mathcal{C}^\times(\mathcal{S})$). For $\mathcal{C}$ a left-cancellative category and $\mathcal{S}$ included in $\mathcal{C}$, we define $\widehat{\mathcal{R}}_\mathcal{C}(\mathcal{S})$ to be the family of all relations $st' = ts'$ with s, t, s', t', and st' in $\mathcal{S}$ that are satisfied in $\mathcal{C}$, and $\widehat{\mathcal{R}}_\mathcal{C}^\times(\mathcal{S})$ to be the family of all relations obtained from relations of $\widehat{\mathcal{R}}_{\mathcal{C}}(\mathcal{S})$ by choosing an expression for every element of $\mathcal{S}^\sharp \setminus \mathcal{C}^\times$ as a product of an element of $\mathcal{S}$ and an element of $\mathcal{C}^\times$ and replacing elements of $\mathcal{S}^\sharp$ with their decompositions in terms of $\mathcal{S}$ and $\mathcal{C}^\times$.

In other words, $\widehat{\mathcal{R}}_{\mathcal{C}}(\mathcal{S})$ is the family of all relations $st' = ts'$ such that there exists r in $\mathcal{S}$ such that both $r = st'$ and $r = ts'$ are relations of $\mathcal{R}_{\mathcal{C}}(\mathcal{S})$. So, by definition, $\mathcal{R}_{\mathcal{C}}(\mathcal{S})$ is included in $\widehat{\mathcal{R}}_{\mathcal{C}}(\mathcal{S})$, whereas every relation of $\widehat{\mathcal{R}}_{\mathcal{C}}(\mathcal{S})$ follows from at most two relations of $\mathcal{R}_{\mathcal{C}}(\mathcal{S})$. Hence, $(\mathcal{S}, \mathcal{R}_{\mathcal{C}}(\mathcal{S}))$ and $(\mathcal{S}, \widehat{\mathcal{R}}_{\mathcal{C}}(\mathcal{S}))$ are equivalent presentations.

Lemma 3.13. *If $\mathcal{S}$ is a Garside family in a left-cancellative category $\mathcal{C}$, right-reversing is complete for the presentations $(\mathcal{S}^\sharp, \widehat{\mathcal{R}}_{\mathcal{C}}(\mathcal{S}))$ and $(\mathcal{S} \cup \mathcal{C}^\times, \widehat{\mathcal{R}}^{\times}_{\mathcal{C}}(\mathcal{S}))$.*

Proof. The result directly follows from Proposition 1.15. Indeed, assume that $u, v, \hat{u}, \hat{v}$ are $\mathcal{S}^\sharp$-paths and $u\hat{v}$ and $v\hat{u}$ represent the same element of $\mathcal{C}$. Assume $u = s_1|\cdots|s_p$, $v = t_1|\cdots|t_q$, and let $\hat{f} = [\hat{u}]$, $\hat{g} = [\hat{v}]$. Then, in $\mathcal{C}$, we have $s_1 \cdots s_p \hat{g} = t_1 \cdots t_q \hat{f}$. As $\mathcal{S}^\sharp$ is a solid Garside family, it satisfies the hypotheses of Proposition 1.15(ii), and the latter implies the existence of a grid as in Figure 2, with edges and diagonals of the squares lying in $\mathcal{S}^\sharp$. Using the notation of the figure, put $u' = s'_1|\cdots|s'_p$ and $v' = t'_1|\cdots|t'_q$, and let w be an $\mathcal{S}^\sharp$-path representing the element h. Then, by definition, the path $\overline{u}v$ is right-reversible to $v'\overline{u'}$ with respect to $\widehat{\mathcal{R}}_{\mathcal{C}}(\mathcal{S})$, whereas $\hat{u}$ is $\widehat{\mathcal{R}}_{\mathcal{C}}(\mathcal{S})$-equivalent to $u'w$ and $\hat{v}$ is $\widehat{\mathcal{R}}_{\mathcal{C}}(\mathcal{S})$-equivalent to $v'w$. Thus u', v', w witness for the expected instance of completeness.

The translation for $(\mathcal{S} \cup \mathcal{C}^\times, \widehat{\mathcal{R}}^{\times}_{\mathcal{C}}(\mathcal{S}))$ is easy. □

Then right-reversing provides a new way of deciding whether two paths represent the same element in the ambient category.

Algorithm 3.14 (Word Problem, positive case).

Context: A left-cancellative category $\mathcal{C}$, a Garside subfamily $\mathcal{S}$ of $\mathcal{C}$ such that $\widehat{\mathcal{R}}_{\mathcal{C}}(\mathcal{S}^\sharp)$ is finite, a decision test for the membership and Word Problems of $\mathcal{C}^\times$
Input: Two $\mathcal{S}^\sharp$-paths u, v
Output: YES if u and v represent the same element of $\mathcal{C}$, NO otherwise

```
compute all positive–negative paths w_1, ..., w_N into which ū v is right-reversible with
    respect to R̂_C(S)
for i increasing from 1 to N do
    if w_i belongs to (C^×)* and [w_i] belongs to 1_C then
        return YES and exit
return NO
```

Proposition 3.15 (Word Problem, positive case). *If $\mathcal{S}$ is a Garside family in a left-cancellative category $\mathcal{C}$ and $\widehat{\mathcal{R}}_{\mathcal{C}}(\mathcal{S}^\sharp)$ is finite, then Algorithm* 3.14 *solves the Word Problem for $\mathcal{C}$ with respect to $\mathcal{S}^\sharp$.*

Proof. First, as all relations of $\widehat{\mathcal{R}}_{\mathcal{C}}(\mathcal{S})$ involve words of length at most two, every $\widehat{\mathcal{R}}_{\mathcal{C}}(\mathcal{S})$-reversing terminates in finitely many steps, and the computation of the paths w_i is feasible. So Algorithm 3.14 always returns YES or NO.

Assume $[u] = [v]$. Let y be the target of u (and v). By the definition of completeness, there exist u', v', w such that $\overline{u}v$ is $\widehat{\mathcal{R}}_\mathcal{C}(\mathcal{S})$-right-reversible to $v'\overline{u'}$, and $1_y = [u'][w] = [v'][w]$ holds. Hence, $[u']$, $[v']$, and $[w]$ consist of invertible entries, and $v'\overline{u'}$ represents an identity element. In this case, Algorithm 3.14 returns YES. Conversely, if Algorithm 3.14 returns YES, we have $[u][v'] = [v][u']$ and $[u'] = [v'] \in \mathcal{C}^\times$, whence $[u] = [v]$. □

Corollary 3.16 (decidability). *If $\mathcal{S}$ is a Garside family in a left-cancellative category $\mathcal{C}$ and $\widehat{\mathcal{R}}_\mathcal{C}(\mathcal{S}^\sharp)$ is finite, the Word Problem for $\mathcal{C}$ with respect to $\mathcal{S}^\sharp$ is decidable.*

Note that the assumptions of Corollary 3.16 are slightly different from those of Corollary III.3.60. Whenever $\widehat{\mathcal{R}}_\mathcal{C}(\mathcal{S}^\sharp)$ is finite, Algorithm 3.14 is effective. However, the presentation $(\mathcal{S}^\sharp, \widehat{\mathcal{R}}_\mathcal{C}(\mathcal{S}^\sharp))$ need not be right-complemented: for each pair of elements s, t in $\mathcal{S}^\sharp$, there are in general several relations of the form $s... = t...$ in $\widehat{\mathcal{R}}_\mathcal{C}(\mathcal{S}^\sharp)$ and, therefore, there are in general several ways of reversing a signed word $\overline{u}v$. If $\widehat{\mathcal{R}}_\mathcal{C}(\mathcal{S}^\sharp)$ comprises m relations, there are at most m possibilities for each reversing step, so, as the length cannot increase, if we start with a path of length ℓ, the only upper bound for the number N of positive–negative paths that can be reached using right-reversing is m^ℓ.

We now turn to the Word Problem for the enveloping groupoid of a left-Ore category. As recalled in the proof of Proposition 3.11 above, whenever $\mathcal{S}$ is a strong Garside family in a left-Ore category, we can use the relations of $\mathcal{R}_\mathcal{C}(\mathcal{S}^\sharp)$ to transform an arbitrary signed $\mathcal{S}^\sharp$-path into an equivalent negative–positive path $\overline{u}v$, and then apply any comparison algorithm to the paths u and v. We can then repeat Algorithm III.3.61 replacing Algorithm III.3.58 with Algorithm 3.14:

Algorithm 3.17 (Word Problem, general case).

Context: A left-Ore category $\mathcal{C}$ that admits left-lcms, a strong Garside subfamily $\mathcal{S}$ of $\mathcal{C}$ such that $\widehat{\mathcal{R}}_\mathcal{C}(\mathcal{S}^\sharp)$ is finite, a short left-lcm witness $\widetilde{\theta}$ for $\mathcal{S}^\sharp$
Input: A signed $\mathcal{S}^\sharp$-path w
Output: YES if w represents an identity element in $\mathcal{E}nv(\mathcal{C})$, NO otherwise
1: left-reverse w into a negative–positive path $\overline{v}u$ using $\widetilde{\theta}$
2: compare $[u]$ and $[v]$ using Algorithm 3.14

Proposition 3.18 (Word Problem). *If $\mathcal{S}$ is a strong Garside family in a left-Ore category $\mathcal{C}$ that admits left-lcms and $\widehat{\mathcal{R}}_\mathcal{C}(\mathcal{S}^\sharp)$ is finite, Algorithm* 3.17 *solves the Word Problem for $\mathcal{E}nv(\mathcal{C})$ with respect to $\mathcal{S}^\sharp$.*

Corollary 3.19 (decidability). *If $\mathcal{S}$ is a strong Garside family in a left-Ore category $\mathcal{C}$ that admits left-lcms and $\widehat{\mathcal{R}}_\mathcal{C}(\mathcal{S}^\sharp)$ is finite, Then the Word Problem for $\mathcal{E}nv(\mathcal{C})$ with respect to $\mathcal{S}^\sharp$ is decidable.*

If the Garside family turns out to be perfect (Definition III.3.6), that is, when $\mathcal{S}^\sharp$ is closed under left-lcm, the description of Algorithm 3.17 becomes more symmetric: in this case, the conjunction of $s' = \widetilde{\theta}(s, t)$ and $t' = \widetilde{\theta}(t, s)$ implies that $st' = t's$ is a relation of $\widehat{\mathcal{R}}_\mathcal{C}(\mathcal{S}^\sharp)$, so that left-$\widetilde{\theta}$-reversing coincides with left-$\widehat{\mathcal{R}}_\mathcal{C}(\mathcal{S}^\sharp)$-reversing. In this case, Algorithm 3.17 corresponds to left-reversing the signed word w into a negative–positive path and then right-reversing the latter, both times using the set of relations $\widehat{\mathcal{R}}_\mathcal{C}(\mathcal{S}^\sharp)$.

Example 3.20 (Word Problem). As in Example III.3.65, consider the braid group B_3 and start with the initial word $\mathtt{b}|\mathtt{a}|\overline{\mathtt{b}}|\overline{\mathtt{b}}|\mathtt{b}|\mathtt{a}$. The first step in Algorithm 3.17 is the same as in Algorithm III.3.61, namely left-reversing the given word using the left-lcm witness $\widetilde{\theta}$, thus leading to the negative–positive word $\overline{\mathtt{a}}|\overline{\mathtt{a}}|1|\mathtt{ba}|\mathtt{b}|\mathtt{a}$. Step 2 now is different: instead of normalizing the numerator and denominator of the above fractional word, we compare them using right-reversing. In the current case, the positive words to be compared are $\mathtt{a}|\mathtt{a}$ and $\mathtt{ba}|\mathtt{b}|\mathtt{a}$. Using the presentation of Example 3.7, we check that right-reversing transforms $\overline{\mathtt{a}}|\overline{\mathtt{a}}|1|\mathtt{ba}|\mathtt{b}|\mathtt{a}$ into the word $\mathtt{ba}|\mathtt{ab}|\overline{\mathtt{ba}}$. The latter is nonempty, so we conclude again that the initial word does not represent 1 in B_3.

3.3 The case of categories with lcms

As already observed, although their theoretical complexity is quadratic in good cases, Algorithms 3.14 and 3.17 need not be efficient when compared with the methods of Chapter III based on normal decompositions: reversing is non-deterministic in general, and exhaustively enumerating all paths into which a signed path can be reversed is tedious and inefficient.

However, there exists particular cases when the method is interesting in practice, namely when reversing is deterministic, that is, when there exists at most one way of reversing a given path. By Definition II.4.2, this corresponds to the case of right-complemented presentations, which, by Corollary II.4.47 (right-lcm) can happen only in the case of categories that admit conditional right-lcms. We shall see now that, in the latter case, reversing does provide efficient algorithms.

It follows from their definition that, except in very simple cases (like the first one in Example 3.7), the presentations $(\mathcal{S}^\sharp, \widehat{\mathcal{R}}_\mathcal{C}(\mathcal{S}))$ and $(\mathcal{S}\cup\mathcal{C}^\times, \widehat{\mathcal{R}}^\times_\mathcal{C}(\mathcal{S}))$ are not right-complemented, as, for all s, t in $\mathcal{S}^\sharp$, there may in general exist several common right-multiples of s and t that lie in $\mathcal{S}^\sharp$. However, whenever right-lcms exist, there exist alternative presentations that are more economical. For simplicity we only consider the case of categories with no nontrivial invertible element. We recall that a left-cancellative category $\mathcal{C}$ is said to admit conditional right-lcms if any two elements of $\mathcal{C}$ that have a common right-multiple have a right-lcm. Next, from Definition III.2.28, we recall that, in such a case, a right-lcm witness on a subfamily $\mathcal{S}$ of $\mathcal{C}$ is a partial map θ from $\mathcal{S} \times \mathcal{S}$ to $\mathcal{S}^*$ such that, for all s, t in $\mathcal{S}$ admitting a common right-multiple, the paths $s|\theta(s,t)$ and $t|\theta(t,s)$ both represent some right-lcm of s and t, the same one for s and t. In a category that admits no nontrivial invertible element, right-lcms are unique when they exist, but right-lcm witnesses need not be unique as an element of the category may admit several $\mathcal{S}$-decompositions.

Proposition 3.21 (right-lcm witness). *Assume that $\mathcal{S}$ is a generating family of a left-cancellative category $\mathcal{C}$ that has no nontrivial invertible element. Assume moreover that $\mathcal{C}$ is right-Noetherian, or that $\mathcal{S}$ is closed under right-complement. Then the following conditions are equivalent:*

(i) *The category $\mathcal{C}$ admits conditional right-lcms.*

(ii) *The category $\mathcal{C}$ admits a right-complemented presentation $(\mathcal{S}, \mathcal{R})$ for which right-reversing is complete.*

(iii) *For every right-lcm witness θ on $\mathcal{S}$, the pair $(\mathcal{S}, \mathcal{R}_\theta)$ is a presentation of $\mathcal{C}$ for which right-reversing is complete.*

Proof. By Corollary II.4.47 (right-lcm), we know that (ii) implies (i), and (iii) implies (ii) since, by definition, a right-lcm witness is a syntactic right-complement, and the derived presentation $(\mathcal{S}, \mathcal{R}_\theta)$ is right-complemented. So the point is to prove that (i) implies (iii) when at least one of the assumptions of the proposition is satisfied. So, let θ be a right-lcm witness on $\mathcal{S}$. We write $\curvearrowright$ for right-reversing associated with $\mathcal{R}_\theta$.

Assume first that $\mathcal{C}$ is right-Noetherian. Let λ be a right-Noetherianity witness on $\mathcal{C}$. We shall prove, using induction on the ordinal α, the statement

$(\mathcal{H}_\alpha)$ For all $\mathcal{S}$-paths $u, v, \hat{u}, \hat{v}$ satisfying $[u\hat{v}] = [v\hat{u}]$ and $\lambda([u\hat{v}]) \leqslant \alpha$, there exist $\mathcal{S}$-paths u', v', w satisfying $\overline{u}v \curvearrowright v'\overline{u'}$, $[\hat{u}] = [u'w]$, and $[\hat{v}] = [v'w]$.

For $\alpha = 0$, the only possibility is that $u, v, \hat{u}, \hat{v}$ are empty, and the result is true, with u', v', and w empty. So assume $\alpha > 0$, and let $u, v, \hat{u}, \hat{v}$ satisfying $[u\hat{v}] = [v\hat{u}]$ and $\lambda([u\hat{v}]) \leqslant \alpha$. If u is empty, the result is true with $u' = w = u$ and $v' = v$, and similarly if v is empty. So assume that neither u nor v is empty. Write $u = s|u_0$ and $v = t|v_0$ with s and t in $\mathcal{S}$, see Figure 10. By assumption, $[u\hat{v}]$ is a common right-multiple of s and t, hence it is a right-multiple of their right-lcm $[s\theta(s,t)]$. By definition, $s\theta(s,t) = t\theta(t,s)$ is a relation of $\mathcal{R}_\theta$, so $\overline{s}|t \curvearrowright \theta(s,t)|\overline{\theta(t,s)}$ is satisfied. As $[u\hat{v}]$ is a right-multiple of $[s\theta(s,t)]$, hence $[\theta(s,t)]$ is a right-multiple of $[u_0]$ since $\mathcal{C}$ is left-cancellative, so there exists an $\mathcal{S}$-path w_0 satisfying $[u_0\hat{v}] = [\theta(s,t)w_0]$ and, for symmetric reasons, $[v_0\hat{u}] = [\theta(t,s)w_0]$. By assumption, we have $[s\theta(s,t)w_0] \leqslant \alpha$, hence $\lambda([\theta(s,t)w_0]) < \alpha$, and, similarly, $\lambda([\theta(t,s)w_0]) < \alpha$. Then the induction hypothesis guarantees the existence of $\mathcal{S}$-paths u_0', v_0', u_1', v_1', w_1, and w_2 satisfying $\overline{u_0}\theta(s,t) \curvearrowright v_1'\overline{u_0'}$, $\overline{\theta(t,s)}v_0 \curvearrowright v_0'\overline{u_1'}$, $[\hat{u}] = [u_1'w_1]$, $[w_0] = [v_0'w_1] = [u_0'w_2]$, and $[\hat{v}] = [v_1'w_2]$. Finally, as we have $\lambda([w_0]) \leqslant \lambda([\theta(t,s)w_0]) < \alpha$, the induction hypothesis guarantees the existence of $\mathcal{S}$-paths u_2', v_2', and w satisfying $\overline{u_0'}v_0' \curvearrowright v_2'\overline{u_2'}$, $[w_1] = [u_2'w]$, and $[w_2] = [v_2'w]$. Put $u' = u_1'u_2'$ and $v' = v_1'v_2'$. Then we have $\overline{u}v \curvearrowright v'\overline{u'}$, $[\hat{u}] = [u_1'w_1] = [u_1'u_2'w] = [u'w]$, and $[\hat{v}] = [v_1'w_2] = [v_1'v_2'w] = [v'w]$. So $(\mathcal{H}_\alpha)$ is satisfied for every α.

We then easily conclude. First, $(\mathcal{S}, \mathcal{R}_\theta)$ is a presentation of $\mathcal{C}$. Indeed, assume that u, v are $\mathcal{S}$-paths that represent the same element of $\mathcal{C}$. Then applying $(\mathcal{H}_\alpha)$ with $\hat{u}$ and $\hat{v}$ empty and $\alpha = \lambda([u])$ gives the existence of u', v', w satisfying $\overline{u}v \curvearrowright v'\overline{u'}$ and $1_y = [u'w] = [v'w]$ where y is the target of u. As $\mathcal{C}$ is assumed to have no nontrivial invertible element, the only possibility is that u', v', and w be empty, which, by Proposition II.4.34 (reversing implies equivalence), implies that u and v are $\mathcal{R}_\theta$-equivalent.

Finally, once we know that $(\mathcal{S}, \mathcal{R}_\theta)$ is a presentation of $\mathcal{C}$, the relation $[u] = [v]$ is equivalent to $u \equiv^+_{\mathcal{R}_\theta} v$, and then the property expressed in $(\mathcal{H}_\alpha)$ is exactly the one defining the completeness of right-reversing for $(\mathcal{S}, \mathcal{R}_\theta)$.

Assume now that $\mathcal{S}$ is closed under right-complement. As we are in the case of unique right-lcms, Lemma 2.39 guarantees that, for all s, t in $\mathcal{S}$ having a common right-multiple, there exist s', t' in $\mathcal{S} \cup \mathbf{1}_\mathcal{C}$ such that st' and $t's$ both are the right-lcm of s and t. We then prove, using induction on ℓ, the statement

$(\mathcal{H}'_\ell)$ For all $\mathcal{S}$-paths $u, v, \hat{u}, \hat{v}$ satisfying $[u\hat{v}] = [v\hat{u}]$ and $\lg(u) + \lg(v) \leqslant \ell$, there exist $\mathcal{S}$-paths u', v', w satisfying $\overline{u}v \curvearrowright v'\overline{u'}$, $[\hat{u}] = [u'w]$, and $[\hat{v}] = [v'w]$.

The argument is entirely similar to the one for the right-Noetherian case, the only difference being the parameter used to control the induction. So the conclusion is the same, and, in both cases, (i) implies (iii). □

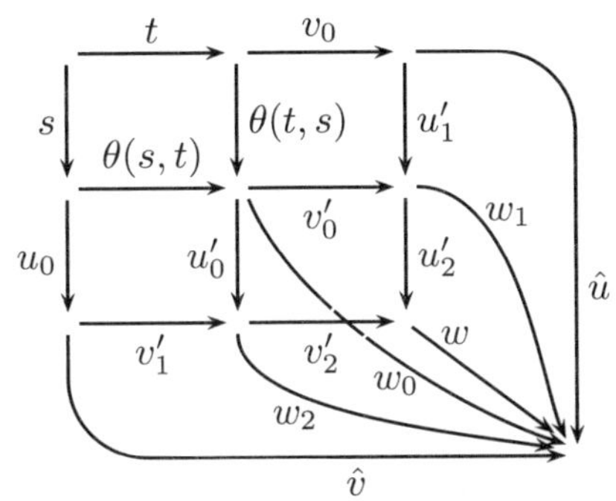

Figure 10. Inductive proof of Proposition 3.21: the argument is the same in the right-Noetherian case and in the case of $\mathcal{S}$ closed under right-complement, with the only difference that the induction parameter is $\lambda([u\hat{v}])$ in the first case, and $\lg(u) + \lg(v)$ in the second case.

Example 3.22 (lcm-witness). Consider the braid monoid B_3^+ once more. There are two ways of applying Proposition 3.21 for defining a presentation for which right-reversing is complete. Let us first take for $\mathcal{S}$ the Garside family $\{1, \mathtt{a}, \mathtt{b}, \mathtt{ab}, \mathtt{ba}, \Delta\}$, which is closed under right-complement. Then we may choose the right-lcm witness θ on $\mathcal{S}$ so that the associated presentation is the one of Example 3.7 (the choice is not unique since, for instance, we may pick for, say, $\theta(\mathtt{a}, \mathtt{b})$ either the length-one word $\mathtt{ba}$ or the length two word $\mathtt{b}|\mathtt{a}$)—thus retrieving the framework of Subsection 3.2. What Proposition 3.21 says is that right-reversing is necessarily complete in this case (this however was already implicit in Proposition 3.15).

Now, B_3^+ is right-Noetherian (and even strongly Noetherian, since it admits a homogeneous presentation), and $\mathcal{A} = \{\mathtt{a}, \mathtt{b}\}$ is a generating family. In terms of $\mathcal{A}$, the right-lcm witness, say θ', is unique: for instance, the only possibility here is $\theta'(\mathtt{a}, \mathtt{b}) = \mathtt{b}|\mathtt{a}$. The resulting presentation is the Artin presentation with the unique relation $\mathtt{aba} = \mathtt{bab}$. Proposition 3.15 guarantees that right-reversing is complete also for this presentation—but this requires to first know that B_3^+ is left-cancellative and admits conditional right-lcms, so this is *not* an alternative to the direct approach based on the cube condition and Proposition II.4.16 (right-complemented).

The assumption that there exists no nontrivial invertible element is crucial in Proposition 3.21: for instance, in the wreathed free abelian monoid $\widetilde{\mathbb{N}}^2$, writing $\mathtt{a}$ and $\mathtt{b}$ for the generators of $\mathbb{N}^2$ and $\mathtt{s}$ for the order-two element, lcms exists, but no right-lcm witness can provide the relation $\mathtt{s}^2 = 1$.

When Proposition 3.21 is relevant, right-reversing is a deterministic process, and the algorithms of Subsection 3.2 become efficient—and extremely simple. Then, by the definition of completeness, right-reversing starting from a negative–positive path $\overline{u}v$ is terminating if and only the elements $[u]$ and $[v]$ admit a common right-multiple, hence always if the ambient category admits right-lcms.

Algorithm 3.23 (Word Problem, positive case II).

Context: A left-cancellative category $\mathcal{C}$ that admits unique right-lcms, a presentation $(\mathcal{S}, \mathcal{R})$ of $\mathcal{C}$ associated with a syntactic right-complement θ for which right-reversing is complete
Input: Two $\mathcal{S}$-paths u, v
Output: YES if u and v represent the same element of $\mathcal{C}$, NO otherwise

```
right-reverse ūv into a positive–negative path v'ū' using θ
if u' and v' are empty then
    return YES
else
    return NO
```

Proposition 3.24 (Word Problem, positive case II). *If $\mathcal{C}$ is a left-cancellative category $\mathcal{C}$ that admits unique right-lcms and $(\mathcal{S}, \mathcal{R})$ is a right-complemented presentation of $\mathcal{C}$ for which right-reversing is complete, then Algorithm* 3.23 *solves the Word Problem for $\mathcal{C}$ with respect to $\mathcal{S}$.*

Proof. The only potential problem is the termination of right-reversing, since we do not assume that the involved right-complement is short. Now, as right-reversing is complete, we know that right-reversing a negative–positive path $\overline{u}v$ terminates if and only if the elements of $\mathcal{C}$ represented by u and v admit a common right-multiple. By assumption, this is always the case in $\mathcal{C}$. □

For the general case of a left-Ore category with left-lcms, we can derive a new algorithm by using left-reversing to transform an initial signed path w into an equivalent negative–positive path $\overline{u}v$, and then apply Algorithm 3.23. However, in the case when right-lcms exist, we can also forget about left-reversing and left-lcms, and instead use a double right-reversing.

Algorithm 3.25 (Word Problem, general case II). See Figure 11

Context: A right-Ore category $\mathcal{C}$ that admits unique right-lcms, a presentation $(\mathcal{S}, \mathcal{R})$ of $\mathcal{C}$ associated with a syntactic right-complement θ for which right-reversing is complete
Input: A signed $\mathcal{S}$-paths w
Output: YES if w represent an identity element of $\mathcal{E}nv(\mathcal{C})$, NO otherwise

```
if w has different source and target then
    return NO
else
    right-reverse w into a positive–negative path vū using θ
    right-reverse ūv into a positive–negative path v'ū' using θ
    if u' and v' are empty then
        return YES
    else
        return NO
```

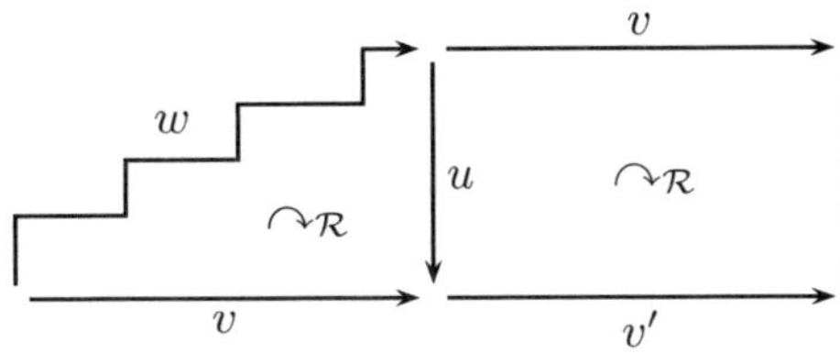

Figure 11. Solution of the Word Problem by means of a double right-reversing: right-reverse w into $v\overline{u}$, copy v, and right-reverse $\overline{u}v$. The initial path w represents an identity element if and only if the second reversing ends with an empty path.

Proposition 3.26 (Word Problem II). *If $\mathcal{C}$ is a right-Ore category that admits unique right-lcms and $(\mathcal{S}, \mathcal{R})$ is a right-complemented presentation of $\mathcal{C}$ for which right-reversing is complete, Algorithm* 3.25 *solves the Word Problem for $\mathcal{E}nv(\mathcal{C})$ with respect to $\mathcal{S}$.*

Proof. The assumption that $\mathcal{C}$ admits right-lcms guarantees that right-$\mathcal{R}$-reversing terminates. Starting with a path w and using the notation of Algorithm 3.25, we see that w represents an identity element in $\mathcal{E}nv(\mathcal{C})$ if and only if $v\overline{u}$ does, hence if and only if u and v represent the same element of $\mathcal{C}$, hence if and only if $\overline{u}v$ reverses to an empty path. □

Example 3.27 (Word Problem II). Return once more to the braid group B_3 and to the signed word $\mathtt{b}|\mathtt{a}|\overline{\mathtt{b}}|\overline{\mathtt{b}}|\mathtt{b}|\mathtt{a}$. Using Algorithm 3.25 with the family of generators $\{\mathtt{a}, \mathtt{b}\}$ leads to first right-reversing the initial word into $\mathtt{a}|\mathtt{a}|\mathtt{a}|\mathtt{b}|\overline{\mathtt{a}}|\overline{\mathtt{b}}$; then we exchange the denominator and the numerator, obtaining the negative–positive word $\overline{\mathtt{a}}|\overline{\mathtt{b}}|\mathtt{a}|\mathtt{a}|\mathtt{a}|\mathtt{b}$, which we right-reverse in turn, obtaining $\mathtt{a}|\mathtt{b}$. The latter word is nonempty, so we conclude that the initial word does not represent 1 in B_3.

3.3.1 Computing upper and lower bounds. As seen in Proposition II.3.15 (left-divisibility), when $\mathcal{C}$ is an Ore category, the left-divisibility relation of $\mathcal{C}$ naturally extends into a relation on the groupoid of fractions $\mathcal{E}nv(\mathcal{C})$, namely the relation, written $\preccurlyeq_{\mathcal{C}}$ or simply $\preccurlyeq$, such that $f \preccurlyeq g$ holds if there exists h in $\mathcal{C}$ satisfying $fh = g$. This relation is a partial preordering on $\mathcal{E}nv(\mathcal{C})$, and a partial ordering if $\mathcal{C}$ admits no nontrivial invertible element. It directly follows from the definition that, if f, g are elements of $\mathcal{C}$, then a least common upper bound (*resp.* a greatest common lower bound) of f and g with respect to $\preccurlyeq$ is (when it exists) a right-lcm (*resp.* a left-gcd) of f and g.

Lemma 3.28. *Assume that $\mathcal{C}$ is an Ore category. Then any two elements of $\mathcal{E}nv(\mathcal{C})$ with the same source admit a least common upper bound (resp. a greatest common lower bound) with respect to $\preccurlyeq$ if and only if any two elements of $\mathcal{C}$ with the same source admit a right-lcm (resp. a left-gcd).*

We skip the easy proof. Then a natural computation problem arises whenever the above bounds exist.

The solution is easy.

Algorithm 3.29 (least common upper bound).

Context: A strong Garside family $\mathcal{S}$ in an Ore category $\mathcal{C}$ that admits unique left- and right-lcms, a right-lcm witness θ and a left-lcm witness $\widetilde{\theta}$ on $\mathcal{S} \cup \mathbf{1}_{\mathcal{C}}$
Input: Two signed $\mathcal{S}$-paths w_1, w_2 with the same source
Output: A signed $\mathcal{S}$-path representing the least common upper bound of $[w_1]$ and $[w_2]$
1: left-reverse $\overline{w_1}w_2$ into a negative–positive path $\overline{u}v$ using $\widetilde{\theta}$
2: right-reverse $\overline{u}v$ into a positive–negative path $v'\overline{u'}$ using θ
3: **return** w_1v'

The solution for greatest lower bound is entirely symmetric.

Algorithm 3.30 (greatest lower bound).

Context: A strong Garside family $\mathcal{S}$ in an Ore category $\mathcal{C}$ that admits unique left- and right-lcms, a right-lcm witness θ and a left-lcm witness $\widetilde{\theta}$ on $\mathcal{S} \cup \mathbf{1}_{\mathcal{C}}$
Input: Two signed $\mathcal{S}$-paths w_1, w_2 with the same source
Output: A signed $\mathcal{S}$-path representing the greatest lower bound of $[w_1]$ and $[w_2]$
1: right-reverse $\overline{w_1}w_2$ into a positive–negative path $u\overline{v}$ using θ
2: left-reverse $u\overline{v}$ into a negative–positive path $\overline{v'}u'$ using $\widetilde{\theta}$
3: **return** $w_1\overline{v'}$

Proposition 3.31 (bounds). *If $\mathcal{S}$ is a strong Garside family in an Ore category $\mathcal{C}$ that admits unique left- and right-lcms, Algorithm* 3.29 (*resp. Algorithm* 3.30) *running on two signed $\mathcal{S}$-paths w_1 and w_2 returns the least upper bound* (*resp. the greatest lower bound*) *of $[w_1]$ and $[w_2]$ with respect to $\preccurlyeq$.*

Proof. First consider Algorithm 3.29. The assumption that $\mathcal{C}$ admits left- and right-lcms guarantees that the considered reversing processes terminate. Step 1 produces two $\mathcal{S}^\sharp$-paths u, v such that, in $\mathcal{E}nv(\mathcal{C})$, we have $[\overline{u}v] = [\overline{w_1}w_2]$, hence $[w_1] = g[u]$ and $[w_2] = g[v]$ for some g, namely the common class of the signed paths $w_1\overline{u}$ and $w_2\overline{v}$. By Corollary II.4.47 (right-lcm), the right-reversing in line 2 produces two $\mathcal{S}$-paths u', v' such that uv' and vu' both represent the right-lcm of $[u]$ and $[v]$, hence their least upper bound with respect to $\preccurlyeq$. As the partial ordering $\preccurlyeq$ is invariant under left-multiplication, the result follows.

The claim for Algorithm 3.30 follows in an analogous way, noting that, for all f, g in $\mathcal{E}nv(\mathcal{C})$ and h in $\mathcal{C}$, the conditions $fh = g$ and $f^{-1} = hg^{-1}$ are equivalent. □

When we restrict Algorithms 3.29 and 3.30 to positive paths, we obtain algorithms that determine right-lcms and left-gcds. In this case, Algorithm 3.29 is a mere right-reversing, since the left-reversing step is trivial: the initial path $\overline{w_1}w_2$ is directly negative–positive. Algorithm 3.30, however, provides a path that represents the left-gcd but, in general, it need not be a positive path, although it must be equivalent to a positive path. In order to obtain a positive output, a third reversing step has to be added.

Algorithm 3.32 (left-gcd).

Context: A strong Garside family $\mathcal{S}$ in an Ore category $\mathcal{C}$ that admits unique left- and right-lcms, a right-lcm witness θ and a left-lcm witness $\widetilde{\theta}$ on $\mathcal{S} \cup \mathbf{1}_{\mathcal{C}}$
Input: Two $\mathcal{S}$-paths u, v with the same source
Output: An $\mathcal{S}$-path that represents the left-gcd of $[u]$ and $[v]$ in $\mathcal{C}$
1: right-reverse $\overline{u}v$ into a positive–negative path $v'\overline{u'}$ using θ
2: left-reverse $v'\overline{u'}$ into a negative–positive path $\overline{u''}|v''$ using $\widetilde{\theta}$
3: left-reverse $u\overline{u''}$ into a positive path w using $\widetilde{\theta}$
4: **return** w

Proposition 3.33 (left-gcd). *If $\mathcal{S}$ is a strong Garside family in an Ore-category that admits unique left- and right-lcms, Algorithm 3.32 running on two $\mathcal{S}$-paths u, v returns a positive $\mathcal{S}$-path that represents the left-gcd of $[u]$ and $[v]$.*

Proof. It follows from Proposition 3.31 that, for u'' as computed in line 2, the path $u\,\overline{u''}$ represents the left-gcd of $[u]$ and $[v]$. As u and v are both positive, this path $u\,\overline{u''}$ must be equivalent to some positive path, say u_0. Then the signed path $u\overline{u''}\,\overline{u_0}$ represents an identity element in $\mathcal{E}nv(\mathcal{C})$, so the positive paths u and u_0u'' are equivalent. Since, by the counterpart of Proposition 3.21, left-reversing is complete for the considered presentation, the path $u\overline{u''}\,\overline{u_0}$ is left-reversible to an empty path, that is, the left-reversing grid constructed from $u\,\overline{u''}\,\overline{u_0}$ has empty arrows everywhere on the left and on the top. This implies in particular that the subgrid corresponding to left-reversing $u\,\overline{u''}$ has empty arrows on the left, that is, with our current notation, that the word w computed in line 3 is positive (and equivalent to u_0). □

Example 3.34 (left-gcd). Let us consider for the last time the monoid B_3^+ and the elements represented by $u = \mathtt{a}|\mathtt{b}|\mathtt{b}$ and $v = \mathtt{b}|\mathtt{a}|\mathtt{b}|\mathtt{b}$ in the alphabet $\mathcal{A} = \{\mathtt{a}, \mathtt{b}\}$. Right-reversing $\overline{u}v$ using the right-lcm witness leads to the positive–negative word $\mathtt{ab}|\overline{\mathtt{a}}$. Then left-reversing the latter word using the left-lcm witness leads to the negative–positive word $\overline{\mathtt{b}}|\mathtt{ab}$. Finally, left-reversing $\mathtt{a}|\mathtt{b}|\mathtt{b}|\overline{\mathtt{b}}$ gives the positive word $\mathtt{a}|\mathtt{b}$, and we conclude that the left-gcd of $[u]$ and $[v]$ in B_3^+ is $\mathtt{ab}$. Note that we used above $\mathcal{S}$-words and the associated lcm witness (where $\mathcal{S}$ is the 6-element Garside family), but we could instead use $\mathcal{A}$-words and the initial right-complement associated with the Artin presentation: the uniqueness of the final result guarantees that the words obtained after each step are pairwise equivalent.

Exercises

Exercise 36 (closure of power). Assume that $\mathcal{C}$ is a left-cancellative category and $\mathcal{S}$ is a subfamily of $\mathcal{C}$ that is closed under right-complement and right-comultiple, and satisfies $\mathcal{S}^\sharp = \mathcal{S}$. Show that $\mathcal{S}^p$ is closed under right-comultiple for each positive p.

Exercise 37 (squarefree words). Assume that A is a nonempty set. Let S be the family of all squarefree words in A^*, that is, the words that admit no factor of the form u^2. Show that S is a Garside family in the free monoid A^*. [Hint: Show that every word admits a longest prefix that is squarefree.]

Exercise 38 (multiplication by invertible). Assume that $\mathcal{C}$ is a cancellative category, and $\mathcal{S}$ is a subfamily of $\mathcal{C}$ that is closed under left-divisor and contains at least one element with source x for each object x. Show that $\mathcal{S}^\sharp = \mathcal{S}$ holds.

Exercise 39 (recognizing Garside). Assume that $\mathcal{C}$ is a left-cancellative category. Show that a subfamily $\mathcal{S}$ of $\mathcal{C}$ is a Garside family if and only if $\mathcal{S}^\sharp$ generates $\mathcal{C}$ and is closed under right-complement, and there exists $H : \mathcal{S}^2 \to \mathcal{S}$ that satisfies (1.46)(i) and (1.46)(iii). Show that H is then the restriction of an $\mathcal{S}$-head function to $\mathcal{S}^2 \setminus \mathcal{C}^\times$.

Exercise 40 (head *vs.* lcm). Assume that $\mathcal{C}$ is a left-cancellative category, $\mathcal{S}$ is included in $\mathcal{C}$, and g belongs to $\mathcal{C} \setminus \mathcal{C}^\times$. Show that s is an $\mathcal{S}$-head of g if and only if it is a right-lcm of $\mathrm{Div}(g) \cap \mathcal{S}$.

Exercise 41 (closed under right-comultiple). Assume that $\mathcal{C}$ is a left-cancellative category, $\mathcal{S}$ is a subfamily of $\mathcal{C}$, and there exists $H : \mathcal{C} \setminus \mathcal{C}^\times \to \mathcal{S}$ satisfying (1.46). Show that $\mathcal{S}$ is closed under right-comultiple.

Exercise 42 (power). Assume that $\mathcal{S}$ is a Garside family in a left-cancellative category $\mathcal{C}$. Show that, if $g_1|\cdots|g_p$ is an $\mathcal{S}^m$-normal decomposition of g and $s_{i,1}|\cdots|s_{i,m}$ is an $\mathcal{S}$-normal decomposition of g_i for every i, then $s_{1,1}|\cdots|s_{1,m}|s_{2,1}|\cdots|s_{2,m}|\cdots|s_{p,1}|\cdots|s_{p,m}$ is an $\mathcal{S}$-normal decomposition of g.

Exercise 43 (mcm of subfamily). Assume that $\mathcal{C}$ is a left-cancellative category and $\mathcal{S}$ is a subfamily in $\mathcal{C}$ such that $\mathcal{S}^\sharp$ generates $\mathcal{C}$ and, for all s, t in $\mathcal{S}^\sharp$, any common right-multiple of s and t is a right-multiple of some right-mcm of s and t. Assume moreover that $\mathcal{C}$ is right-Noetherian, or that $\mathcal{S}^\sharp$ is closed under right-complement. Show that $\mathcal{C}$ admits right-mcms.

Exercise 44 (not right-Noetherian). Let M be the submonoid of $\mathbb{Z}^2$ consisting of the pairs (x, y) with $y > 0$ plus the pairs $(x, 0)$ with $x \geqslant 0$. (i) Show that M is neither left- nor right-Noetherian. (ii) Let $S = \{\mathtt{a}^p \mid p \geqslant 0\} \cup \{\mathtt{b}_i \mid i \geqslant 0\}$. Show that S is a Garside family in M. (iii) Show that the S-head of (x, y) is (x, y) if it lies in S and is $(0, 1)$ otherwise. Show that the S-normal decomposition of (x, y) is $(\mathtt{b}_0, \ldots, \mathtt{b}_0, \mathtt{a}^x)$, y times $\mathtt{b}_0$, for $x > 0$, and is $(\mathtt{b}_0, \ldots, \mathtt{b}_0, \mathtt{b}_{|x|})$, $y - 1$ times $\mathtt{b}_0$, for $x \leqslant 0$.

Exercise 45 (no conditional right-lcm). Let M be the monoid generated by $\mathtt{a}, \mathtt{b}, \mathtt{a}', \mathtt{b}'$ subject to the relations $\mathtt{ab}=\mathtt{ba}$, $\mathtt{a}'\mathtt{b}'=\mathtt{b}'\mathtt{a}'$, $\mathtt{aa}'=\mathtt{bb}'$, $\mathtt{a}'\mathtt{a}=\mathtt{b}'\mathtt{b}$. (i) Show that the cube condition is satisfied on $\{\mathtt{a}, \mathtt{b}, \mathtt{a}', \mathtt{b}'\}$, and that right- and left-reversing are complete for the above presentation. (ii) Show that M is cancellative and admits right-mcms.

(iii) Show that a and b admit two right-mcms in M, but they admit no right-lcm. (iv) Let $S = \{\mathtt{a}, \mathtt{b}, \mathtt{a}', \mathtt{b}', \mathtt{ab}, \mathtt{a}'\mathtt{b}', \mathtt{aa}', \mathtt{a}'\mathtt{a}\}$, see diagram on the side. Show that S is closed under right-mcm, and deduce that $\mathcal{S}$ is a Garside family in M. (v) Show that the (unique) strict S-normal decomposition of the element $\mathtt{a}^2\mathtt{b}'\mathtt{a}'^2$ is $\mathtt{ab}|\mathtt{a}'\mathtt{b}'|\mathtt{b}'$.

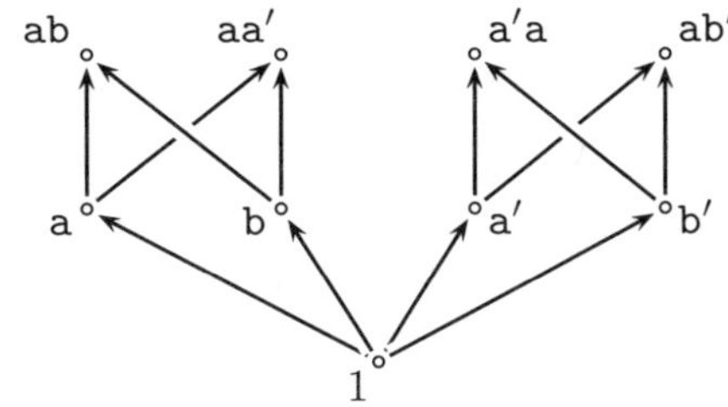

Exercise 46 (lifted omega monoid). Let M_ω be the lifted omega monoid of Exercise 14, which is right-Noetherian. Let $S = \{\mathtt{a}^p \mid p \geqslant 0\} \cup \{\mathtt{b}_i \mid i \leqslant 0\}$. (i) Show that S is closed under right-divisor. (ii) Show that, whenever s and t have a common right-multiple, then $s \preccurlyeq t$ or $t \preccurlyeq s$ is satisfied so that s or t is a right-lcm of s and t. Deduce that M_ω admits conditional right-lcms and that S is closed under right-lcm. (iii) Show that S is a Garside family in M_ω, and so is every family of the form $\{\mathtt{a}^p \mid p \geqslant 0\}$ and $\{\mathtt{b}_i \mid i \in I\}$ with I unbounded on the left. (iv) Show that M_ω admits no minimal Garside family.

Exercise 47 (solid). Assume that $\mathcal{C}$ is a left-cancellative category and $\mathcal{S}$ is a generating subfamily of $\mathcal{C}$. (i) Show that $\mathcal{S}$ is solid in $\mathcal{C}$ if and only if $\mathcal{S}$ includes $1_\mathcal{C}$ and it is closed under right-quotient. (ii) Assume moreover that $\mathcal{S}$ is closed under right-divisor. Show that $\mathcal{S}$ includes $\mathcal{C}^\times \setminus 1_\mathcal{C}$, that $\epsilon \in \mathcal{S} \cap \mathcal{C}^\times$ implies $\epsilon^{-1} \in \mathcal{S}$, and that $\mathcal{C}^\times\mathcal{S} = \mathcal{S}$ holds, but that $\mathcal{S}$ need not be solid.

Exercise 48 (solid). Let M be the monoid $\langle \mathtt{a}, \mathtt{e} \mid \mathtt{ea} = \mathtt{a}, \mathtt{e}^2 = 1\rangle^+$. (i) Show that every element of M has a unique expression of the form $\mathtt{a}^p\mathtt{e}^q$ with $p \geqslant 0$ and $q \in \{0, 1\}$, and that M is left-cancellative. (ii) Let $S = \{1, \mathtt{a}, \mathtt{e}\}$. Show that S is a solid Garside family in M, but that $S = S^\sharp$ does not hold.

Exercise 49 (not solid). Let $M = \langle \mathtt{a}, \mathtt{e} \mid \mathtt{ea} = \mathtt{ae}, \mathtt{e}^2 = 1\rangle^+$, and $S = \{\mathtt{a}, \mathtt{e}\}$. (i) Show that M is left-cancellative. [Hint: M is $\mathbb{N} \times \mathbb{Z}/2\mathbb{Z}$] (ii) Show that S is a Garside family of M, but S is not solid in M. [Hint: ea right-divides a, but does not belong to S.]

Exercise 50 (recognizing Garside, right-lcm solid case). Assume that $\mathcal{S}$ is a solid subfamily in a left-cancellative category $\mathcal{C}$ that is right-Noetherian and admits conditional right-lcms. Show that $\mathcal{S}$ is a Garside family in $\mathcal{C}$ if and only if $\mathcal{S}$ generates $\mathcal{C}$ and it is weakly closed under right-lcm.

Exercise 51 (right-complement). Assume that $\mathcal{C}$ is a left-cancellative category that admits unique conditional right-lcms. Show that, for all f, g, h in $\mathcal{C}$, the relation $f \preccurlyeq gh$ is equivalent to $g\backslash f \preccurlyeq h$.

Exercise 52 (local right-divisibility). Assume that $\mathcal{C}$ is a left-cancellative category and $\mathcal{S}$ is a generating subfamily of $\mathcal{C}$ that is closed under right-divisor. (i) Show that the transitive closure of $\widetilde{\preccurlyeq}_\mathcal{S}$ is the restriction of $\widetilde{\preccurlyeq}$ to $\mathcal{S}$. (ii) Show that the transitive closure of $\widetilde{\prec}_\mathcal{S}$ is almost the restriction of $\widetilde{\prec}$ to $\mathcal{S}$, in the following sense: if $s \mathrel{\widetilde{\prec}} t$ holds, there exists $s' \mathrel{{}^\times\!\!=} s$ satisfying $s' \mathrel{\widetilde{\prec}}{}^*_\mathcal{S}\, t$.

Exercise 53 (local left-divisibility). Assume that $\mathcal{S}$ is a subfamily of a left-cancellative category $\mathcal{C}$. (i) Show that $s \preccurlyeq_\mathcal{S} t$ implies $s \preccurlyeq t$, and that $s \preccurlyeq_\mathcal{S} t$ is equivalent to $s \preccurlyeq t$

whenever $\mathcal{S}$ is closed under right-quotient in $\mathcal{C}$. (ii) Show that, if $\mathcal{S}^\times = \mathcal{C}^\times \cap \mathcal{S}$ holds, then $s \prec_{\mathcal{S}} t$ implies $s \prec t$. (iii) Show that, if $\mathcal{S}$ is closed under right-divisor, then $\preccurlyeq_{\mathcal{S}}$ is the restriction of $\preccurlyeq$ to $\mathcal{S}$ and, if $\mathcal{S}^\times = \mathcal{C}^\times \cap \mathcal{S}$ holds, $\prec_{\mathcal{S}}$ is the restriction of $\prec$ to $\mathcal{S}$.

Exercise 54 (finite implies left-Noetherian). Assume that $\mathcal{C}$ is a left-cancellative category. (i) Show that every subfamily $\mathcal{S}$ of $\mathcal{C}$ such that $=^\times_{\mathcal{S}}$ has finitely many classes is locally left-Noetherian. (ii) Is there a similar result for right-Noetherianity?

Exercise 55 (locally right-Noetherian). Assume that $\mathcal{C}$ is a left-cancellative category and $\mathcal{S}$ is a subfamily of $\mathcal{C}$. (i) Prove that $\mathcal{S}$ is locally right-Noetherian if and only if, for every s in $\mathcal{S}^\sharp$, every $\prec_{\mathcal{S}}$-increasing sequence in $\mathrm{Div}_{\mathcal{S}}(s)$ is finite. (ii) Assume that $\mathcal{S}$ is locally right-Noetherian and closed under right-divisor. Show that $\mathcal{S}^\sharp$ is locally right-Noetherian. [Hint: For $\mathcal{X} \subseteq \mathcal{S}^\sharp$ introduce $\mathcal{X}' = \{s \in \mathcal{S} \mid \exists \epsilon, \epsilon' \in \mathcal{C}^\times\,(\epsilon s \epsilon' \in \mathcal{X})\}$, and construct a $\widetilde{\prec}$-minimal element in $\mathcal{X}$ from a $\widetilde{\prec}$-minimal element in $\mathcal{X}'$.]

Exercise 56 (not complete). Consider the monoid $\mathbb{N}^2$ with $\mathtt{a} = (1,0)$ and $\mathtt{b} = (0,1)$. Put $\Delta = \mathtt{ab}$, and let S be the Garside family $\{1, \mathtt{a}, \mathtt{b}, \Delta\}$. Show that right-reversing is not complete for the presentation $(S, \mathcal{R}_{\mathbb{N}^2}(S))$, that is, $(\{\mathtt{a}, \mathtt{b}, \mathtt{ab}\}, \{\mathtt{ab} = \Delta, \mathtt{ba} = \Delta\})$. [Hint: Show that $\overline{\mathtt{a}}\mathtt{b}$ is only reversible to itself.]

Notes

Sources and comments. The results in this chapter are mostly new, at least in their current exposition, and constitute one of the contributions of this text. An abridged version of the main results appears in D.–Digne–Michel [93]. The algorithmic aspects are addressed in D.–Gebhardt [95]. Of course, many results are mild extensions of previously known results, some of them quite standard. The key point here is the importance of closure conditions, leading to the general philosophy that a Garside family is characterized by two types of properties, namely various (connected) closure properties plus a maximality property (the existence of a head), the latter coming for free in a Noetherian context. In other words, $\mathcal{S}$-normal decompositions essentially exist if and only if the reference family $\mathcal{S}$ satisfies the above two types of properties.

Closure under right-comultiple is crucial, and it captures what is really significant in the existence of least common right-multiples: when lcms exist, closure under right-comultiple is equivalent to closure under right-lcm, but, in any case, even when right-lcms do not exist, closure under right-comultiple is sufficient to obtain most of the standard consequences of closure under right-lcm.

Among the few really new results established here, those involving Noetherianity are probably the most interesting. In particular, the result that local Noetherianity implies global Noetherianity in a Garside context (Proposition 2.18) may appear as unexpected.

Further questions. The left-absorbing monoid L_n (Reference Structure 8, page 113) shows that a Garside family need not be closed under left-divisor in general. Now, in this case, the closure of the Garside family S_n under left-divisor is again a Garside family.

Question 9. If $\mathcal{S}$ is a Garside family in a left-cancellative category, is the closure of $\mathcal{S}$ under left-divisor again a Garside family?

A general answer does not seem obvious. In the same vein, we raise

Question 10. If $\mathcal{S}$ is a solid Garside family in a left-Ore category $\mathcal{C}$ that admits left-lcms, is there any connection between $\mathcal{S}$ being closed under left-complement, and $\mathcal{S}^\sharp$ being closed under left-complement?

An equivalence would allow for a definition of a strong Garside family involving $\mathcal{S}$ rather than $\mathcal{S}^\sharp$; we have no clue about an implication in either direction; the condition $\mathcal{S}\mathcal{C}^\times \subseteq \mathcal{C}^\times\mathcal{S}$, which amounts to $\mathcal{S}\mathcal{C}^\times \subseteq \mathcal{C}^\times\mathcal{S}$ when $\mathcal{S}$ is a Garside family, might be relevant.

The current definition of $\mathcal{S}$-normal decompositions and all derived notions are not invariant under left–right symmetry: what we consider here corresponds to what is sometimes called the *left greedy normal form*. Of course, entirely symmetric results could be stated for the counterpart of a right greedy normal form, with a maximal tail replacing a maximal head. However, questions arise when one sticks to the current definition and tries to reverse some of the assumptions. Typically, the results of Subsection 2.2 assume that the ambient category is right-Noetherian, a property of the right-divisibility relation.

Question 11. Do the results of Subsection 2.2 remain valid when left-Noetherianity replaces right-Noetherianity?

About Noetherianity again, we established in Proposition 2.43 that, if every element in a Garside family has a finite height bounded by some absolute constant K, then every element in the ambient category has finite height, that is, the latter is strongly Noetherian.

Question 12. If a left-cancellative category $\mathcal{C}$ admits a Garside family whose elements have a finite height, does every element of $\mathcal{C}$ have a finite height?

On the other hand, in the case when every element in some Garside family $\mathcal{S}$ has a finite height bounded by some absolute constant K, then what was proved in Lemma 2.44 is that the height of every element in $\mathcal{S}^m$ is bounded by K^m. However, all known examples witness much smaller bounds, compatible with a positive answer to

Question 13. If a left-cancellative category $\mathcal{C}$ admits a Garside family $\mathcal{S}$ and $\mathrm{ht}(s) \leqslant K$ holds for every s in $\mathcal{S}$, is the height of every element in $\mathcal{S}^m$ bounded above by a linear function of m?

In another vein, we saw in Example 2.34 that the monoid $\langle \mathtt{a}, \mathtt{b} \mid \mathtt{ab} = \mathtt{ba}, \mathtt{a}^2 = \mathtt{b}^2 \rangle^+$ admits no proper Garside family: every Garside family must coincide with the whole monoid (with 1 possibly removed). In this trivial case, the reason is that the elements $\mathtt{a}$ and $\mathtt{b}$ admit no right-lcm. When right-lcms exist, no such example is known, and the following problem, which had already been considered in a different terminology in the first times of Garside monoids, remains open:

Question 14. Assume that $\mathcal{C}$ is a left-cancellative category that is Noetherian and admits right-lcms (*resp.* this and is finitely generated). Does $\mathcal{C}$ admit admits a proper (*resp.* finite) Garside family?

Chapter V
Bounded Garside families

So far we considered arbitrary Garside families, and we observed that such families trivially exist in every category. In this chapter, we consider Garside families $\mathcal{S}$ of a restricted type, namely those satisfying the condition that, for all objects x, y of the ambient category, the elements of $\mathcal{S}$ with source x admit a least common right-multiple and, symmetrically, the elements of $\mathcal{S}$ with target y admit a least common left-multiple. Typical examples are the Garside family of simple braids in the monoid B_n^+, which admits the braid Δ_n as a right- and a left-lcm and, more generally, the Garside family $\mathrm{Div}(\Delta)$ in a Garside monoid (M, Δ). What we do here is to show that most of the classical properties known in the latter case extend to categories that are bounded in the above sense.

The chapter is organized as follows. In Section 1, we introduce the notion of a right-bounded Garside family, which only appeals to left-divisibility and common right-multiples. We associate with every right-bounded Garside family a certain functor ϕ_Δ that plays a key technical role (Corollary 1.34). We then show that every category that admits a right-bounded Garside family admits common right-multiples (Proposition 1.46) and that, under mild closure assumptions, a right-bounded Garside family is entirely determined by the associated right-bounding map and that the latter can be simply axiomatized as what we call a right-Garside map.

In Section 2, we introduce the notion of a bounded Garside family, a stronger version that involves both left- and right-divisibility, and provides the most suitable framework. The main result is that, in a cancellative category, the assumption that a Garside family is bounded by a map Δ guarantees that the associated functor ϕ_Δ is an automorphism of the ambient category (Proposition 2.17). Here naturally appears the notion of a Garside map, which provides a simple connection with earlier literature.

Finally, in Section 3, we show that the assumption that a Garside family $\mathcal{S}$ is bounded by a map Δ enables one to rephrase the existence of $\mathcal{S}$-normal decompositions in terms of what we call Delta-normal decompositions, which are similar but give a specific role to the elements of the form $\Delta(x)$. More significantly, we show that every category that admits a bounded Garside family—or, equivalently, a Garside map—must be an Ore category, and we construct an essentially unique decomposition for every element of the associated groupoid of fractions (Proposition 3.18), thus obtaining an alternative to the symmetric normal decompositions of Section III.2.

Main definitions and results

Definition 1.1 **(right-bounded).** If $\mathcal{S}$ is a subfamily of a left-cancellative category $\mathcal{C}$ and Δ is a map from $\mathcal{O}bj(\mathcal{C})$ to $\mathcal{C}$, we say that $\mathcal{S}$ is *right-bounded* by Δ if, for each object x of $\mathcal{C}$, every element of $\mathcal{S}(x, \cdot)$ left-divides $\Delta(x)$ and, moreover, $\Delta(x)$ lies in $\mathcal{S}(x, \cdot)$.

Proposition 1.4 **(finite right-bounded).** *If $\mathcal{C}$ is a left-cancellative category that admits common right-multiples, every Garside family $\mathcal{S}$ of $\mathcal{C}$ such that $\mathcal{S}(x, \cdot)$ is finite for every x is right-bounded.*

Definition 1.15 **(right-Garside map).** If $\mathcal{C}$ is a left-cancellative category, a *right-Garside map* in $\mathcal{C}$ is a map Δ from $\mathcal{O}bj(\mathcal{C})$ to $\mathcal{C}$ that satisfies the following conditions: (1.16) For every x, the source of $\Delta(x)$ is x, (1.17) The family $\mathcal{D}iv(\Delta)$ generates $\mathcal{C}$, (1.18) The family $\widetilde{\mathcal{D}iv}(\Delta)$ is included in $\mathcal{D}iv(\Delta)$, (1.19) For every g in $\mathcal{C}(x, \cdot)$, the elements g and $\Delta(x)$ admit a left-gcd.

Proposition 1.20 **(right-Garside map).** *Assume that $\mathcal{C}$ is a left-cancellative category.* (i) *If $\mathcal{S}$ is a Garside family that is right-bounded by a map Δ and $\mathcal{S}^\sharp$ is closed under left-divisor, then Δ is a right-Garside map and $\mathcal{S}^\sharp$ coincides with $\mathcal{D}iv(\Delta)$.* (ii) *Conversely, if Δ is a right-Garside map in $\mathcal{C}$, then $\mathcal{D}iv(\Delta)$ is a Garside family that is right-bounded by Δ and closed under left-divisor.*

Corollary 1.34 **(functor ϕ_Δ).** *If $\mathcal{C}$ is a left-cancellative category and $\mathcal{S}$ is a Garside family of $\mathcal{C}$ that is right-bounded by a map Δ, there exists a unique functor ϕ_Δ from $\mathcal{C}$ into itself that extends ∂_Δ^2. It is determined by $\Delta(x) \in \mathcal{C}(x, \phi_\Delta(x))$ and $g\,\Delta(y) = \Delta(x)\,\phi_\Delta(g)$.*

Proposition 1.46 **(common right-multiple).** *If $\mathcal{C}$ is a left-cancellative category and* (i) *there exists a map Δ from $\mathcal{O}bj(\mathcal{C})$ to $\mathcal{C}$ that satisfies* (1.16), (1.17), *and* (1.18), *or* (ii) *there exists a right-bounded Garside family in $\mathcal{C}$, then any two elements of $\mathcal{C}$ with the same source admit a common right-multiple.*

Definition 2.3 **(bounded).** A Garside family $\mathcal{S}$ in a left-cancellative category $\mathcal{C}$ is said to be *bounded* by a map Δ from $\mathcal{O}bj(\mathcal{C})$ to $\mathcal{C}$ if $\mathcal{S}$ is right-bounded by Δ and, in addition, for every y, there exists x satisfying $\forall s \in \mathcal{S}^\sharp(\cdot, y)\ \exists r \in \mathcal{S}^\sharp(x, \cdot)\ (rs = \Delta(x))$.

Proposition 2.6 **(finite bounded).** *If $\mathcal{C}$ is a cancellative category and $\mathcal{S}$ is a Garside family of $\mathcal{C}$ such that $\mathcal{S}^\sharp$ is finite and $\mathcal{S}$ is right-bounded by a target-injective map Δ, then $\mathcal{S}$ is bounded by Δ and the functor ϕ_Δ is a finite order automorphism.*

Corollary 2.14 **(common left-multiple).** *Every left-cancellative category that admits a bounded Garside family admits common left-multiples.*

Proposition 2.17 **(automorphism).** *If $\mathcal{S}$ is a Garside family in a left-cancellative category $\mathcal{C}$ and $\mathcal{S}$ is right-bounded by a map Δ, the following are equivalent:* (i) *The category $\mathcal{C}$ is cancellative, $\mathcal{S}$ is bounded by Δ, and Δ is target-injective.* (ii) *The functor ϕ_Δ is an automorphism of $\mathcal{C}$. When* (i) *and* (ii) *hold, we have $\widetilde{\mathcal{D}iv}(\Delta) = \mathcal{S}^\sharp = \mathcal{D}iv(\Delta)$, the family $\mathcal{S}^\sharp$ is closed under left-divisor, ∂_Δ and the restriction of ϕ_Δ to $\mathcal{S}^\sharp$ are permutations of $\mathcal{S}^\sharp$, and $(\phi_\Delta^{-1}, \partial_\Delta^{-1})$ is the unique witness for $(\mathcal{S}, \Delta)$.*

Definition 2.19 **(Garside map).** If $\mathcal{C}$ is a left-cancellative category, a *Garside map* in $\mathcal{C}$ is a map Δ from $\mathcal{O}bj(\mathcal{C})$ to $\mathcal{C}$ that satisfies the following conditions: (2.20) For every x in $\mathcal{O}bj(\mathcal{C})$, the source of $\Delta(x)$ is x, (2.21) The map Δ is target-injective, (2.22) The family $\mathcal{D}iv(\Delta)$ generates $\mathcal{C}$, (2.23) The families $\widetilde{\mathcal{D}iv}(\Delta)$ and $\mathcal{D}iv(\Delta)$ coincide, (2.24) For every g in $\mathcal{C}(x, \cdot)$, the elements g and $\Delta(x)$ admit a left-gcd.

Proposition 2.32 **(Garside map).** *If Δ is a Garside map in a cancellative category $\mathcal{C}$:* (i) *The category $\mathcal{C}$ is an Ore category.* (ii) *For s in $\mathcal{D}iv(\Delta)$ and g in $\mathcal{C}$, the pair $s|g$ is*

$\mathcal{D}iv(\Delta)$-greedy if and only if the elements ∂s and g are left-coprime. (iii) *The functor ϕ_Δ is an automorphism of $\mathcal{C}$ and it preserves $\mathcal{D}iv(\Delta)$-normality.* (iv) *For $m \geqslant 1$, $\Delta^{[m]}$ is a Garside map and $\mathcal{D}iv(\Delta^{[m]}) = (\mathcal{D}iv(\Delta))^m$ holds.* (v) *The second domino rule is valid for $\mathcal{D}iv(\Delta)$ and* (2.33) $\max(\sup_\Delta(f), \sup_\Delta(g)) \leqslant \sup_\Delta(fg) \leqslant \sup_\Delta(f) + \sup_\Delta(g)$ *holds for all $f|g$ in $\mathcal{C}^{[2]}$.*

Proposition 2.35 **(lcms and gcds).** *If a cancellative category $\mathcal{C}$ admits a Garside map:* (i) *The category $\mathcal{C}$ admits left-gcds and left-lcms.* (ii) *If $\mathcal{C}$ is left-Noetherian, then $\mathcal{C}$ admits right-lcms and right-gcds.* (iii) *If $\mathcal{C}$ is left-Noetherian and has no nontrivial invertible element, $(\mathcal{C}, \preccurlyeq)$ and $(\mathcal{C}, \widetilde{\succcurlyeq})$ are lattices.*

Definition 3.3 **(delta-normal).** If $\mathcal{S}$ is a Garside family in a left-cancellative category $\mathcal{C}$ that is right-bounded by a map Δ, an $(\mathcal{S}, \Delta)$*-normal path of infimum* m $(m \geqslant 0)$ is a pair $(\Delta^{[m]}(x), s_1|\cdots|s_\ell)$ such that $\Delta(x)|\Delta(\phi_\Delta(x))|\cdots|\Delta(\phi_\Delta^{m-1}(x))|s_1|\cdots|s_\ell$ is $\mathcal{S}$-normal and s_1 is not delta-like (for $\ell \geqslant 1$). We write $\Delta^{[m]}(x)\|s_1|\cdots|s_\ell$, or $\Delta^{[m]}\|s_1|\cdots|s_\ell$, for such a pair. If $\mathcal{S}^\sharp = \mathcal{D}iv(\Delta)$ holds, we say Δ*-normal* for $(\mathcal{S}, \Delta)$-normal.

Definition 3.17 **(delta-normal).** If Δ is a Garside map in a cancellative category $\mathcal{C}$, we say that an element g of $\mathcal{E}nv(\mathcal{C})(x, \cdot)$ *admits the Δ-normal decomposition* $\Delta^{(m)}\|s_1|\cdots|s_\ell$ $(m \in \mathbb{Z})$ if $s_1|\cdots|s_\ell$ is Δ-normal, s_1 is not delta-like, and $g = \iota(\Delta^{[m]}(x))\,\iota(s_1)\cdots\iota(s_\ell)$ holds in the case $m \geqslant 0$ and $g = \iota(\widetilde{\Delta}_{[-m]}(x))^{-1}\,\iota(s_1)\cdots\iota(s_\ell)$ holds in the case $m < 0$.

Proposition 3.18 **(delta-normal).** *If Δ is a Garside map in a cancellative category $\mathcal{C}$, every element of $\mathcal{E}nv(\mathcal{C})$ admits a Δ-normal decomposition. The exponent of Δ is uniquely determined and the $\mathcal{D}iv(\Delta)$-entries are unique up to $\mathcal{C}^\times$-deformation.*

Proposition 3.24 **(interval).** *If Δ is a Garside map in a cancellative category $\mathcal{C}$ and g lies in $\mathcal{E}nv(\mathcal{C})(x, \cdot)$, then $\inf_\Delta(g)$ is the greatest integer p satisfying $\Delta^{[p]}(x) \preccurlyeq g$, and $\sup_\Delta(g)$ is the least integer q satisfying $g \preccurlyeq \Delta^{[q]}(x)$.*

Proposition 3.26 **(inverse).** *If Δ is a Garside map in a cancellative category $\mathcal{C}$ and $\Delta^{(m)}\|s_1|\cdots|s_\ell$ is a Δ-normal decomposition of an element g of $\mathcal{E}nv(\mathcal{C})$, then $\Delta^{(-m-\ell)}\|\partial(\phi^{-m-\ell}s_\ell)|\cdots|\partial(\phi^{-m-1}s_1)$ is a Δ-normal decomposition of g^{-1}.*

1 Right-bounded Garside families

In this section, we consider a weaker notion of boundedness that only involves the left-divisibility relation. With this one-sided version, we can already prove several important results, in particular the existence of the functor ϕ_Δ and the existence of common right-multiples in the ambient category.

The section comprises five subsections. The notion of a right-bounded Garside family is introduced in Subsection 1.1. Next, in Subsection 1.2, we axiomatize the right-bounding maps that occur in the context of right-bounded Garside families and show how to rephrase the results in terms of the right-Garside maps so introduced. Then, we show in Subsection 1.3 that a certain functor ϕ_Δ naturally appears in connection with every right-bounded Garside family, and in Subsection 1.4 that a power of a right-bounded Garside family is still a right-bounded Garside family. Finally, we discuss in Subsection 1.5 the validity of the second domino rule in the context of right-bounded Garside families.

1.1 The notion of a right-bounded Garside family

Our starting point is the following simple and natural notion. We recall that, if $\mathcal{C}$ is a category and x is an object of $\mathcal{C}$, then $\mathcal{C}(x,\cdot)$ is the family of all elements of $\mathcal{C}$ admitting the source x and, similarly, if $\mathcal{S}$ is a subfamily of $\mathcal{C}$, then $\mathcal{S}(x,\cdot)$ is the family of all elements of $\mathcal{S}$ admitting the source x.

Definition 1.1 (right-bounded). If $\mathcal{S}$ is a subfamily of a left-cancellative category $\mathcal{C}$ and Δ is a map from $\mathcal{O}bj(\mathcal{C})$ to $\mathcal{C}$, we say that $\mathcal{S}$ is *right-bounded* by Δ if, for each object x of $\mathcal{C}$, every element of $\mathcal{S}(x,\cdot)$ left-divides $\Delta(x)$ and, moreover, $\Delta(x)$ lies in $\mathcal{S}(x,\cdot)$.

Example 1.2 (right-bounded). A number of the Garside families mentioned so far are right-bounded. For instance, in the case of the free abelian monoid $\mathbb{N}^I$ with base $(\mathtt{a}_i)_{i\in I}$, consider the Garside family S consisting of all elements $\prod_{i\in J}\mathtt{a}_i$ with $J\subseteq I$. If I is finite, $\prod_i \mathtt{a}_i$ belongs to S and every element of S left-divides it, so S is right-bounded by $\prod_i \mathtt{a}_i$. On the other hand, if I is infinite, the monoid contains infinitely many atoms, whereas an element can be (left)-divisible by finitely many elements only, so S cannot be right-bounded by any element. Hence S is right-bounded if and only if I is finite.

More generally, if (M,Δ) is a Garside or quasi-Garside monoid (Definitions I.2.1 and I.2.2), the set $\mathrm{Div}(\Delta)$ is a Garside family in M and, by definition, Δ is a right-lcm for its left-divisors, hence $\mathrm{Div}(\Delta)$ is right-bounded by Δ.

Except B_∞^+, which is similar to $\mathbb{N}^{(I)}$ with I infinite, all examples of Section I.3 correspond to right-bounded Garside families: the Garside family S of the Klein bottle monoid (Reference Structure 5, page 17) is right-bounded by the element $\mathtt{a}^2$, the family S_n of the wreathed free abelian group (Reference Structure 6, page 19) is right-bounded by the element Δ_n, and the family $\mathcal{S}_n$ in the ribbon category $\mathcal{BR}_n$ (Reference Structure 7, page 20) is right-bounded by the constant map from $\{1,\ldots,n-1\}$ to B_n with value Δ_n.

Similarly, in the left-absorbing monoid $L_n=\langle \mathtt{a},\mathtt{b}\mid \mathtt{a}\mathtt{b}^n=\mathtt{b}^{n+1}\rangle^+$ (Reference Structure 8, page 113), the Garside family $S_n=\{\mathtt{a},\mathtt{b},\mathtt{b}^2,\ldots,\mathtt{b}^{n+1}\}$ is right-bounded by $\mathtt{b}^{n+1}$.

By contrast, in the affine braid monoid of type $\widetilde{A}_2$ (Reference Structure 9, page 114), the sixteen-element Garside family S is not right-bounded: every element of S is a right-divisor of (at least) one of the three maximal elements $\sigma_1\sigma_2\sigma_3\sigma_2$, $\sigma_2\sigma_3\sigma_1\sigma_3$, $\sigma_3\sigma_1\sigma_2\sigma_1$, but the three play similar roles and none is a right-multiple of the other two. The same holds for the six-element Garside family in the partially commutative monoid of Example IV.2.33: both $\mathtt{ab}$ and $\mathtt{bc}$ are maximal elements, but none is a multiple of the other and the family is not right-bounded.

Finally, in the lifted omega monoid M_ω of Exercises 14 and 46, the Garside family S is not right-bounded, as $\mathtt{a}$ and $\mathtt{b}_0$, which both lie in S, have no common right-multiple.

The first observation is that a right-bounding map is (almost) unique when it exists.

Proposition 1.3 (uniqueness). *If a subfamily $\mathcal{S}$ of a left-cancellative category is right-bounded by two maps Δ and Δ', then $\Delta'(x) =^{\times} \Delta(x)$ holds for each object x.*

Proof. The assumption that $\Delta(x)$ lies in $\mathcal{S}$ and that $\mathcal{S}$ is right-bounded by Δ' implies $\Delta(x) \preccurlyeq \Delta'(x)$ and, symmetrically, the assumption that $\Delta'(x)$ lies in $\mathcal{S}$ and that $\mathcal{S}$ is right-bounded by Δ implies $\Delta'(x) \preccurlyeq \Delta(x)$, whence $\Delta(x) =^{\times} \Delta'(x)$. □

Another way of stating Definition 1.1 is to say that, for every object x, the element $\Delta(x)$ is a right-lcm for the family $\mathcal{S}(x, \cdot)$. Indeed, we require that $\Delta(x)$ is a right-multiple of every element of $\mathcal{S}(x, \cdot)$ and then demanding that $\Delta(x)$ belongs to $\mathcal{S}(x, \cdot)$ amounts to demanding that it left-divides every common right-multiple of $\mathcal{S}(x, \cdot)$. It follows that, if the existence of common right-multiples is guaranteed, then finiteness is sufficient to ensure right-boundedness:

Proposition 1.4 (finite right-bounded). *If $\mathcal{C}$ is a left-cancellative category that admits common right-multiples, every Garside family $\mathcal{S}$ of $\mathcal{C}$ such that $\mathcal{S}(x, \cdot)$ is finite for every object x is right-bounded.*

Proof. Assume that $\mathcal{S}$ is a Garside family as in the statement, x is an object of $\mathcal{C}$, and s belong to $\mathcal{S}(x, \cdot)$. Define $\Delta(x)$ to be a $\prec$-maximal element in $\mathcal{S}(x, \cdot)$: such an element necessarily exists since $\mathcal{S}(x, \cdot)$ is finite. By assumption, s and $\Delta(x)$ admit a common right-multiple in $\mathcal{C}$. Hence, as s and $\Delta(x)$ belong to $\mathcal{S}$ and $\mathcal{S}$ is a Garside family, hence, by Proposition IV.1.23 (Garside closed), it is closed under right-comultiple, s and $\Delta(x)$ admit a common right-multiple t that lies in $\mathcal{S}$. Now, by construction, $\Delta(x) \prec t$ is impossible. Therefore we have $t =^{\times} \Delta(x)$, whence $s \preccurlyeq \Delta(x)$. So $\mathcal{S}(x, \cdot)$ is a subfamily of $\mathrm{Div}(\Delta(x))$, and $\mathcal{S}$ is right-bounded by Δ. □

The example of a free monoid based on a finite set $\mathcal{S}$ with more than one element shows that the assumption about common right-multiples cannot be removed: in this case, $\mathcal{S}$ is a finite Garside family, but it is not right-bounded. Actually, we shall see in Proposition 1.46 below that the existence of a right-bounded Garside family always implies the existence of common right-multiples.

We now observe that being right-bounded is invariant under right-multiplying by invertible elements. We recall that $\mathcal{S}^{\sharp}$ stands for $\mathcal{S}\mathcal{C}^{\times} \cup \mathcal{C}^{\times}$.

Lemma 1.5. *Assume that $\mathcal{C}$ is a left-cancellative category and Δ is a map from $\mathcal{O}bj(\mathcal{C})$ to $\mathcal{C}$. For $\mathcal{S}$ included in $\mathcal{C}$, consider the statements:*

(i) *The family $\mathcal{S}$ is right-bounded by Δ.*

(ii) *The family $\mathcal{S}^{\sharp}$ is right-bounded by Δ.*

Then (i) *implies* (ii), *and* (ii) *implies* (i) *whenever $\Delta(x)$ lies in $\mathcal{S}$ for every object x.*

Proof. Assume that $\mathcal{S}$ is right-bounded by Δ and s belongs to $\mathcal{S}^\sharp(x,\cdot)$. If s belongs to $\mathcal{C}^\times(x,\cdot)$, then s left-divides every element of $\mathcal{C}(x,\cdot)$, hence in particular $\Delta(x)$. Otherwise, we have $s = t\epsilon$ for some t in $\mathcal{S}(x,\cdot)$ and ϵ in $\mathcal{C}^\times$. By assumption, t left-divides $\Delta(x)$, hence so does s. As $\Delta(x)$ belongs to $\mathcal{S}(x)$, hence to $\mathcal{S}^\sharp(x)$, we conclude that $\mathcal{S}^\sharp$ is right-bounded by Δ, and (i) implies (ii).

Conversely, if $\mathcal{S}^\sharp$ is right-bounded by Δ, as $\mathcal{S}$ is included in $\mathcal{S}^\sharp$, every element of $\mathcal{S}(x,\cdot)$ left-divides $\Delta(x)$, so, if $\Delta(x)$ lies in $\mathcal{S}$, the latter is right-bounded by Δ. □

Notation 1.6 (families $\mathcal{D}iv(\Delta)$ and $\widetilde{\mathcal{D}iv}(\Delta)$). For $\mathcal{C}$ a category and Δ a map from $\mathcal{O}bj(\mathcal{C})$ to $\mathcal{C}$, we denote by $\mathcal{D}iv(\Delta)$ (*resp.* $\widetilde{\mathcal{D}iv}(\Delta)$) the family of all elements of $\mathcal{C}$ that left-divide (*resp.* right-divide) at least one element $\Delta(x)$.

So, by definition, we have

$$\mathcal{D}iv(\Delta) = \{\, g \in \mathcal{C} \mid \exists x{\in}\mathcal{O}bj(\mathcal{C})\ \exists h{\in}\mathcal{C}\ (gh = \Delta(x))\,\},$$

and, symmetrically, $\widetilde{\mathcal{D}iv}(\Delta) = \{\, g \in \mathcal{C} \mid \exists x{\in}\mathcal{O}bj(\mathcal{C})\ \exists f{\in}\mathcal{C}\ (fg = \Delta(x))\,\}$. Note that, if an element g belongs to $\mathcal{D}iv(\Delta)$, then the object x such that g left-divides $\Delta(x)$ must be the source of g. On the other hand, for g in $\widetilde{\mathcal{D}iv}(\Delta)$, nothing guarantees the uniqueness of the object x such that g right-divides $\Delta(x)$ in general. Also note that, whenever Δ is a map from $\mathcal{O}bj(\mathcal{C})$ to $\mathcal{C}$ satisfying $\Delta(x) \in \mathcal{C}(x,\cdot)$ for every object, the family $\mathcal{D}iv(\Delta)$ is right-bounded by Δ. Then we have the following connections between a family that is right-bounded by a map Δ and the families $\mathcal{D}iv(\Delta)$ and $\widetilde{\mathcal{D}iv}(\Delta)$.

Lemma 1.7. *Assume that $\mathcal{C}$ is a left-cancellative category and $\mathcal{S}$ is a subfamily of $\mathcal{C}$ that is right-bounded by a map Δ. Then the relation*

$$\mathcal{S} \subseteq \mathcal{D}iv(\Delta) \tag{1.8}$$

holds, the inclusion being an equality if and only if $\mathcal{S}$ is closed under left-divisor.

Moreover, if $\mathcal{S}$ is closed under right-divisor, we have

$$\widetilde{\mathcal{D}iv}(\Delta) \subseteq \mathcal{S}. \tag{1.9}$$

Proof. By definition, $\mathcal{S}$ being right-bounded by Δ implies (1.8). Next, $\mathcal{D}iv(\Delta)$ is closed under left-divisor by definition, so $\mathcal{S} = \mathcal{D}iv(\Delta)$ requires that $\mathcal{S}$ be closed under left-divisor. Conversely, if $\mathcal{S}$ is closed under left-divisor, then, as $\Delta(x)$ belongs to $\mathcal{S}$, every left-divisor of an element $\Delta(x)$ belongs to $\mathcal{S}$, whence $\mathcal{D}iv(\Delta) \subseteq \mathcal{S}$, and $\mathcal{S} = \mathcal{D}iv(\Delta)$ owing to (1.8).

Symmetrically, if $\mathcal{S}$ is closed under right-divisor, then, as $\Delta(x)$ belongs to $\mathcal{S}$, every right-divisor of an element $\Delta(x)$ belongs to $\mathcal{S}$, and we deduce $\widetilde{\mathcal{D}iv}(\Delta) \subseteq \mathcal{S}$. □

In the case of a Garside family, we deduce:

Proposition 1.10 (inclusions). *If $\mathcal{C}$ is a left-cancellative category and $\mathcal{S}$ is a Garside family of $\mathcal{C}$ that is right-bounded by a map Δ, then the relation*

$$\widetilde{\mathcal{D}iv}(\Delta) \subseteq \mathcal{S}^\sharp \subseteq \mathcal{D}iv(\Delta) \tag{1.11}$$

holds, and the right inclusion is an equality if and only if $\mathcal{S}^\sharp$ is closed under left-divisor.

Proof. By Lemma 1.5, $\mathcal{S}^\sharp$ is right-bounded by Δ, and, by Proposition IV.1.23 (Garside closed), $\mathcal{S}^\sharp$ is closed under right-divisor. Applying Lemma 1.7 to $\mathcal{S}^\sharp$ gives the result. □

The example of the left-absorbing monoid L_n with the Garside family $\mathcal{S}_n$ (Reference Structure 8, page 113) shows that both inclusions in (1.11) may be strict: $\mathtt{a}$ is an element of $\mathcal{S}_n \setminus \widetilde{\mathrm{Div}}(\mathtt{b}^{n+1})$, and $\mathtt{ab}$ is an element of $\mathrm{Div}(\mathtt{b}^{n+1}) \setminus \mathcal{S}_n$ for $n \geqslant 2$. Note that, if a Garside family $\mathcal{S}$ is closed under left-divisor, then, by Lemma IV.1.13, $\mathcal{S}^\sharp$ is also closed under left-divisor, and, therefore, we obtain $\mathcal{S}^\sharp = \mathcal{D}iv(\Delta)$.

On the other hand, in the case of a finite Garside family, the situation is very simple whenever the ambient category is cancellative.

Corollary 1.12 (inclusions, finite case). *Assume that $\mathcal{C}$ is a cancellative category, $\mathcal{S}$ is a Garside family of $\mathcal{C}$ such that $\mathcal{S}^\sharp$ is finite, $\mathcal{S}$ is right-bounded by a map Δ and, for $x \neq y$, the targets of $\Delta(x)$ and $\Delta(y)$ are distinct. Then we have $\widetilde{\mathcal{D}iv}(\Delta) = \mathcal{S}^\sharp = \mathcal{D}iv(\Delta)$.*

Proof. For every s in $\mathcal{D}iv(\Delta)(x, \cdot)$, there exists a unique element ∂s satisfying $s\partial s = \Delta(x)$. By construction, ∂s belongs to $\widetilde{\mathcal{D}iv}(\Delta)$, so ∂ maps $\mathcal{D}iv(\Delta)$ to $\widetilde{\mathcal{D}iv}(\Delta)$. Assume $\partial s = \partial t$ with s in $\mathcal{D}iv(\Delta)(x, \cdot)$ and t in $\mathcal{D}iv(\Delta)(y, \cdot)$. As we have $s\partial s = \Delta(x)$ and $t\partial t = \Delta(y)$, the targets of $\Delta(x)$ and $\Delta(y)$ must coincide, implying $x = y$ by assumption. Then we have $s\partial s = t\partial s = \Delta(s)$, implying $s = t$ using right-cancellativity. So ∂ is injective. We deduce that $\mathcal{D}iv(\Delta)$ is finite and $\#\widetilde{\mathcal{D}iv}(\Delta) \geqslant \#\mathcal{D}iv(\Delta)$ holds, so the inclusions of (1.11) must be equalities. □

1.2 Right-Garside maps

Proposition 1.10 implies constraints for the maps Δ that can be a right-bound for a Garside family: for instance, (1.11) implies that we must have $\widetilde{\mathcal{D}iv}(\Delta) \subseteq \mathcal{D}iv(\Delta)$, that is, every right-divisor of Δ must be a left-divisor of Δ. It turns out that these constraints can be described exhaustively, leading to the notion of a *right-Garside map* and providing an alternative context for the investigation of right-bounded Garside families, more exactly of those right-bounded Garside families $\mathcal{S}$ such that $\mathcal{S}^\sharp$ is closed under left-divisor.

Lemma 1.13. *Assume that $\mathcal{C}$ is a left-cancellative category and $\mathcal{S}$ is a Garside family that is right-bounded by a map Δ.*

(i) *For every x, the source of $\Delta(x)$ is x.*

(ii) *The family $\mathcal{D}iv(\Delta)$ generates $\mathcal{C}$.*

(iii) *The family $\widetilde{\mathcal{D}iv}(\Delta)$ is included in $\mathcal{D}iv(\Delta)$.*

Proof. Assertion (i) holds by the definition of a right-bounding map. Next, the assumption that $\mathcal{S}$ is a Garside family implies that $\mathcal{S}^\sharp$ generates $\mathcal{C}$, hence so does $\mathcal{D}iv(\Delta)$, which includes $\mathcal{S}^\sharp$ by (1.11). So (ii) is satisfied. Finally, again by (1.11) and as already noted, $\widetilde{\mathcal{D}iv}(\Delta)$ must be included in $\mathcal{D}iv(\Delta)$, so (iii) is satisfied. □

Next, there exists a simple connection between heads and left-gcds with elements $\Delta(x)$.

Lemma 1.14. *Assume that $\mathcal{C}$ is a left-cancellative category and $\mathcal{S}$ is a subfamily of $\mathcal{C}$ that is right-bounded by Δ. For g in $\mathcal{C}(x,\cdot)$ and s in $\mathcal{S}$, consider*

(i) *The element s is a left-gcd of g and $\Delta(x)$,*

(ii) *The element s is an $\mathcal{S}$-head of g.*

Then (i) *implies* (ii) *and, if $\mathcal{S}^\sharp$ is closed under left-divisor,* (ii) *implies* (i).

Proof. Assume that s is a left-gcd of g and $\Delta(x)$. First s left-divides g. Next, let t be an element of $\mathcal{S}$ that left-divides g. Then t must belong to $\mathcal{S}(x,\cdot)$, hence we have $t \preccurlyeq \Delta(x)$, whence $t \preccurlyeq s$ since s is a left-gcd of g and $\Delta(x)$. Hence s is an $\mathcal{S}$-head of g, and (i) implies (ii).

Assume now that $\mathcal{S}^\sharp$ is closed under left-divisor, g lies in $\mathcal{C}(x,\cdot)$, and s is an $\mathcal{S}$-head of g. Then s belongs to $\mathcal{S}(x,\cdot)$, hence, by assumption, we have $s \preccurlyeq \Delta(x)$, so s is a common left-divisor of g and $\Delta(x)$. Now let h be an arbitrary common left-divisor of g and $\Delta(x)$. As it left-divides $\Delta(x)$, which belongs to $\mathcal{S}$, hence to $\mathcal{S}^\sharp$, the element h must belong to $\mathcal{S}^\sharp$, and, therefore, as it left-divides g, it left-divides every $\mathcal{S}$-head of g, hence in particular s. So s is a left-gcd of g and $\Delta(x)$. So (ii) implies (i) in this case. □

We deduce a new constraint for a map Δ to be a right-bounding map, namely that, for every element g with source x, there exists a left-gcd for g and $\Delta(x)$. Closure under left-divisor cannot be skipped in Lemma 1.14, even in the case of a Garside family: as observed in Example 1.2, in the left-absorbing monoid L_n (Reference Structure 8, page 113), the Garside family S_n is right-bounded by Δ_n, but the (unique) S_n-head of $\mathtt{ab}$ is $\mathtt{a}$, whereas the (unique) left-gcd of $\mathtt{ab}$ and Δ_n is $\mathtt{ab}$.

It turns out that the above listed constraints also are sufficient for Δ to be a right-bounding map, and it is therefore natural to introduce a specific terminology.

Definition 1.15 (right-Garside map). If $\mathcal{C}$ is a left-cancellative category, a *right-Garside map* in $\mathcal{C}$ is a map Δ from $\mathcal{O}bj(\mathcal{C})$ to $\mathcal{C}$ that satisfies the following conditions:

(1.16) For every x, the source of $\Delta(x)$ is x.

(1.17) The family $\mathcal{D}iv(\Delta)$ generates $\mathcal{C}$.

(1.18) The family $\widetilde{\mathcal{D}iv}(\Delta)$ is included in $\mathcal{D}iv(\Delta)$.

(1.19) For every g in $\mathcal{C}(x,\cdot)$, the elements g and $\Delta(x)$ admit a left-gcd.

Right-bounded Garside families and right-Garside maps are connected as follows:

Proposition 1.20 (right-Garside map). *Assume that $\mathcal{C}$ is a left-cancellative category.*

(i) *If $\mathcal{S}$ is a Garside family of $\mathcal{C}$ that is right-bounded by a map Δ and $\mathcal{S}^\sharp$ is closed under left-divisor, then Δ is a right-Garside map and $\mathcal{S}^\sharp$ coincides with $\mathcal{D}iv(\Delta)$.*

(ii) *Conversely, if Δ is a right-Garside map in $\mathcal{C}$, then $\mathcal{D}iv(\Delta)$ is a Garside family that is right-bounded by Δ and closed under left-divisor.*

Proof. (i) Assume that $\mathcal{S}$ is a Garside family that is right-bounded by Δ and $\mathcal{S}^\sharp$ is closed under left-divisor. By Lemma 1.13, Δ satisfies (1.16)–(1.18). Let g belong to $\mathcal{C}(x,\cdot)$. If g is non-invertible, then g admits an $\mathcal{S}$-head, hence, by Lemma 1.14, g and $\Delta(x)$ admit a left-gcd. If g is invertible, then g itself is a left-gcd of g and $\Delta(x)$. So, in every case, g and $\Delta(x)$ admit a left-gcd, and Δ satisfies (1.19). Hence, by definition, Δ is a right-Garside map. Moreover, by Lemma 1.7, $\mathcal{S}^\sharp$ coincides with $\mathcal{D}iv(\Delta)$.

(ii) Assume that Δ is a right-Garside map. First, by (1.17), $\mathcal{D}iv(\Delta)$ generates $\mathcal{C}$. Next, assume that s lies in $\mathcal{D}iv(\Delta)(x,\cdot)$ and s' right-divides s, say $s = rs'$. By assumption, s left-divides $\Delta(x)$, say $\Delta(x) = st$. We deduce $\Delta(x) = rs't$, whence $s't \in \widetilde{\mathcal{D}iv}(\Delta)$, and $s't \in \mathcal{D}iv(\Delta)$ owing to (1.18). This in turn implies $s' \in \mathcal{D}iv(\Delta)$. Hence $\mathcal{D}iv(\Delta)$ is closed under right-divisor. Finally, let g be an element of $\mathcal{C}(x,\cdot)$. By assumption, g and $\Delta(x)$ admit a left-gcd, say s. By Lemma 1.14, s is a $\mathcal{D}iv(\Delta)$-head of g. So every element of $\mathcal{C}$ admits a $\mathcal{D}iv(\Delta)$-head. Hence, by Proposition IV.1.24 (recognizing Garside II), $\mathcal{D}iv(\Delta)$ is a Garside family in $\mathcal{C}$. Moreover, every element of $\mathcal{D}iv(\Delta)(x,\cdot)$ left-divides $\Delta(x)$, hence $\mathcal{D}iv(\Delta)$ is right-bounded by Δ. On the other hand, $\mathcal{D}iv(\Delta)$ is closed under left-divisor by definition (and it coincides with $\mathcal{D}iv(\Delta)^\sharp$). □

Note that, in Proposition 1.20, (ii) is an exact converse of (i). Proposition 1.20 implies that, if $\mathcal{C}$ is a left-cancellative category, a subfamily $\mathcal{S}$ of $\mathcal{C}$ is a right-bounded Garside family such that $\mathcal{S}^\sharp$ is closed under left-divisor if and only if there exists a right-Garside map Δ such that $\mathcal{S}^\sharp$ coincides with $\mathcal{D}iv(\Delta)$. So, we have two equivalent frameworks, namely that of a Garside family $\mathcal{S}$ that is right-bounded and such that $\mathcal{S}^\sharp$ is closed under left-divisor, and that of a right-Garside map. All subsequent results can therefore be equivalently stated both in terms of right-bounded Garside families and in terms of right-Garside maps. Below we shall mostly use the terminology of right-Garside maps, which allows for shorter statements, but it should be remembered that each statement of the form "... *if Δ is a right-Garside map ..., then $\mathcal{D}iv(\Delta)$ ...*" can be rephrased as "... *if $\mathcal{S}$ is a Garside family that is right-bounded and $\mathcal{S}^\sharp$ is closed under left-divisor ..., then $\mathcal{S}^\sharp$ ...* ".

In the case of a monoid, there is only one object and we shall naturally speak of a right-Garside *element* rather than of a right-Garside map. As the notion is important, we explicitly state the definition.

Definition 1.21 (right-Garside element). If M is a left-cancellative monoid, a *right-Garside element* in M is an element Δ of M that satisfies the following conditions:

(1.22) The family $\mathrm{Div}(\Delta)$ generates M.

(1.23) The family $\widetilde{\mathrm{Div}}(\Delta)$ is included in $\mathrm{Div}(\Delta)$.

(1.24) For every g in M, the elements g and Δ admit a left-gcd.

Proposition 1.20 directly implies

Proposition 1.25 (right-Garside element). *Assume that M is a left-cancellative monoid.*

(i) *If S is a Garside family that right-bounded by an element Δ and $S^\sharp$ is closed under left-divisor, then Δ is a right-Garside element and $S^\sharp$ coincides with* $\mathrm{Div}(\Delta)$.

(ii) *Conversely, if Δ is a right-Garside element in M, then* $\mathrm{Div}(\Delta)$ *is a Garside family of M that is right-bounded by Δ and is closed under left-divisor.*

Remark 1.26. If any two elements with the same source admit a left-gcd, (1.19) becomes trivial, and a (necessary and) sufficient condition for Δ to be a right-Garside map is that it satisfies (1.16)–(1.18). However, the latter conditions are not sufficient in general. Indeed, let for instance M be the monoid $\langle \mathtt{a}, \mathtt{b} \mid \mathtt{ab} = \mathtt{ba}, \mathtt{a}^2 = \mathtt{b}^2 \rangle^+$, and let $\Delta = \mathtt{ab}$. Then M is cancellative, and Δ satisfies satisfies (1.16)–(1.18). However, we saw in Example IV.2.34 that M and $M \setminus \{1\}$ are the only Garside families in M. So, there is no right-bounded Garside family in M and, in particular, $\mathrm{Div}(\Delta)$ cannot be a right-bounded Garside family in M. Note that, in M, the elements Δ and $\mathtt{a}^2$ admit no left-gcd.

1.3 The functor ϕ_Δ

We associate with every right-Garside map Δ a functor ϕ_Δ of the ambient category into itself. This functor will play a crucial rôle in all subsequent developments—especially when it is an automorphism, see Section 2—and its existence is the main technical difference between right-bounded and arbitrary Garside families.

Most of the developments in this subsection and the next one only require that Δ satisfies the conditions (1.16), (1.17), and (1.18). Owing to Lemma 1.13, this implies that the results are valid for every right-bounding map independently of whether the involved family $\mathcal{S}^\sharp$ is closed under left-divisor or not.

Definition 1.27 (duality map ∂_Δ). If $\mathcal{C}$ is a left-cancellative category and Δ is a map from $\mathcal{O}bj(\mathcal{C})$ to $\mathcal{C}$ that satisfies (1.16), (1.17), and (1.18), then, for s in $\mathcal{D}iv(\Delta)(x, \cdot)$, we write $\partial_\Delta(s)$, or simply ∂s, for the unique element t of $\widetilde{\mathcal{D}iv}(\Delta)$ satisfying $st = \Delta(x)$.

By definition, ∂_Δ is a map from $\mathcal{D}iv(\Delta)$ to $\widetilde{\mathcal{D}iv}(\Delta)$, and, for g in $\mathcal{D}iv(\Delta)$, we have $s\,\partial_\Delta(s) = \Delta(x)$. Definition 1.27 makes sense because the assumption that $\mathcal{C}$ is left-cancellative guarantees the uniqueness of the element t and, by definition, every element of $\mathcal{D}iv(\Delta)(x, \cdot)$ left-divides $\Delta(x)$. Then ∂s belongs to $\widetilde{\mathcal{D}iv}(\Delta)(y, x')$, where y is the target of s and x' is the target of $\Delta(x)$.

The inclusion of (1.18) implies that ∂s belongs to $\mathcal{D}iv(\Delta)$. So, ∂ maps $\mathcal{D}iv(\Delta)$ into itself, and it makes sense to iterate it: for every $m \geqslant 1$, we obtain a well-defined map ∂^m from $\mathcal{D}iv(\Delta)$ to itself. We shall see now that ∂^2 can be extended into a functor from the ambient category into itself.

Proposition 1.28 (functor ϕ_Δ). *If $\mathcal{C}$ is a left-cancellative category and Δ is a map from $\mathcal{O}bj(\mathcal{C})$ to $\mathcal{C}$ that satisfies (1.16), (1.17), and (1.18), then there exists a unique functor ϕ_Δ from $\mathcal{C}$ into itself that extends ∂_Δ^2. For x in $\mathcal{O}bj(\mathcal{C})$ and g in $\mathcal{C}(x, y)$, the values $\phi_\Delta(x)$ and $\phi_\Delta(g)$ are determined by*

$$\Delta(x) \in \mathcal{C}(x, \phi_\Delta(x)) \quad \text{and} \quad g\,\Delta(y) = \Delta(x)\,\phi_\Delta(g). \tag{1.29}$$

Moreover, we have

$$\phi_\Delta \circ \Delta = \Delta \circ \phi_\Delta \quad \text{and} \quad \phi_\Delta \circ \partial_\Delta = \partial_\Delta \circ \phi_\Delta, \tag{1.30}$$

and ϕ_Δ maps both $\mathcal{D}iv(\Delta)$ and $\widetilde{\mathcal{D}iv}(\Delta)$ into themselves.

Proof. Assume that ϕ is a functor from $\mathcal{C}$ to itself that extends ∂^2. Let x be an object of $\mathcal{C}$. By definition, 1_x belongs to $\mathcal{D}iv(\Delta)$ and we have $\partial(1_x) = \Delta(x)$, hence $1_{\phi(x)} = \phi(1_x) = \partial^2(1_x) = \partial(\Delta(x)) = 1_y$, where y is the target of $\Delta(x)$, whence $\phi(x) = y$.

Next, assume that s belongs to $\mathcal{D}iv(\Delta)(x, y)$. Then, by assumption, ∂s belongs to $\mathcal{D}iv(\Delta)(y, \cdot)$, and we find

$$s\,\Delta(y) = s\,\partial s\,\partial^2 s = \Delta(x)\,\partial^2 s, \tag{1.31}$$

which implies

$$s\,\Delta(y) = \Delta(x)\,\phi(s). \tag{1.32}$$

Let g be an arbitrary element of $\mathcal{C}(x, y)$. By (1.17), $\mathcal{D}iv(\Delta)$ generates $\mathcal{C}$, so we have $g = s_1 \cdots s_p$ for some $s_1, \ldots, s_p$ lying in $\mathcal{D}iv(\Delta)$. Calling y_i the target of s_i and applying (1.32) p times, we find

$$\begin{aligned} g\,\Delta(y) &= s_1 \cdots s_p\,\Delta(y_p) = s_1 \cdots s_{p-1}\,\Delta(y_{p-1})\,\phi(s_p) \\ &= \cdots = s_1\,\Delta(y_1)\,\phi(s_2) \cdots \phi(s_p) = \Delta(x)\,\phi(s_1) \cdots \phi(s_p) = \Delta(x)\,\phi(g). \end{aligned}$$

So (1.32) must hold for every g in $\mathcal{C}$. As $\mathcal{C}$ is left-cancellative, this implies the uniqueness of $\phi(g)$. So, if there exists a functor ϕ extending ∂^2, it is unique and determined on elements by (1.32).

As for the existence, (1.31) shows that

$$\Delta(x) \preccurlyeq s\Delta(y) \tag{1.33}$$

holds for every s in $\mathcal{D}iv(\Delta)(x, y)$. We claim that (1.33) actually holds for every element g of $\mathcal{C}$. Indeed, assume $g = s_1 \cdots s_p$ with $s_1, \ldots, s_p$ in $\mathcal{D}iv(\Delta)$, and let again y_i be the target of s_i. Applying (1.33) p times, we find

$$\Delta(x) \preccurlyeq s_1\Delta(y_1) \preccurlyeq s_1 s_2\Delta(y_2) \preccurlyeq \cdots \preccurlyeq s_1 s_2 \cdots s_p\,\Delta(y_p) = g\,\Delta(y).$$

Let us define ϕ_Δ by the condition that $\phi_\Delta(x)$ is the target of $\Delta(x)$ for x in $\mathcal{O}bj(\mathcal{C})$ and that $\phi_\Delta(g)$ is given by (1.29) for g in $\mathcal{C}(x, y)$. Then ϕ_Δ is well-defined as (1.33) holds. We claim that ϕ_Δ is the desired functor.

First, assume that g lies in $\mathcal{C}(x, y)$. Then, as (1.29) holds by definition, the source of $\phi_\Delta(g)$ is the target of $\Delta(x)$, hence is $\phi_\Delta(x)$, whereas its target is the target of $\Delta(y)$, hence it is $\phi_\Delta(y)$.

Next, we claim that ϕ_Δ is a functor of $\mathcal{C}$ to itself. Indeed, assume $fg = h$, with $f \in \mathcal{C}(x, y)$ and g in $\mathcal{C}(y, z)$. Then, applying (1.29) three times, we find the equalities

$$\Delta(x)\,\phi_\Delta(h) = h\,\Delta(z) = f\,g\,\Delta(z) = f\,\Delta(y)\,\phi_\Delta(g) = \Delta(x)\,\phi_\Delta(f)\,\phi_\Delta(g),$$

whence $\phi_\Delta(h) = \phi_\Delta(f)\,\phi_\Delta(g)$ by left-cancelling $\Delta(x)$. Furthermore, applying (1.29) to 1_x gives $\Delta(x) = \Delta(x)\phi_\Delta(1_x)$, whence $\phi_\Delta(1_x) = 1_{\phi_\Delta(x)}$ by left-cancelling $\Delta(x)$.

Finally, ϕ_Δ extends ∂^2 as (1.31) and (1.29) give, for every s in $\mathcal{D}iv(\Delta)(x, y)$, the equality $\Delta(x)\partial^2 s = \Delta(x)\phi_\Delta(s)$, whence $\phi_\Delta(s) = \partial^2 s$ by left-cancelling $\Delta(x)$.

So ϕ_Δ is a functor with the expected properties.

As for (1.30), first, by definition, the source of $\Delta(x)$ is x and its target is $\phi_\Delta(x)$, hence applying (1.29) with $g = \Delta(x)$ yields $\Delta(x)\Delta(\phi_\Delta(x)) = \Delta(x)\phi_\Delta(\Delta(x))$, hence $\Delta(\phi_\Delta(x)) = \phi_\Delta(\Delta(x))$ after left-cancelling $\Delta(x)$. On the other hand, for s in $\mathcal{D}iv(\Delta)$, we clearly have $\partial(\phi_\Delta(s)) = \phi_\Delta(\partial s) = \partial^3 s$.

Assume now $s \in \mathcal{D}iv(\Delta)(x, \cdot)$. By definition, there exists t satisfying $st = \Delta(x)$. This implies $\phi_\Delta(s)\phi_\Delta(t) = \phi_\Delta(\Delta(x)) = \Delta(\phi_\Delta(x))$, the last equality by (1.30). We deduce $\phi_\Delta(s) \in \mathcal{D}iv(\Delta)$. The argument is similar for $\widetilde{\mathcal{D}iv}(\Delta)$. □

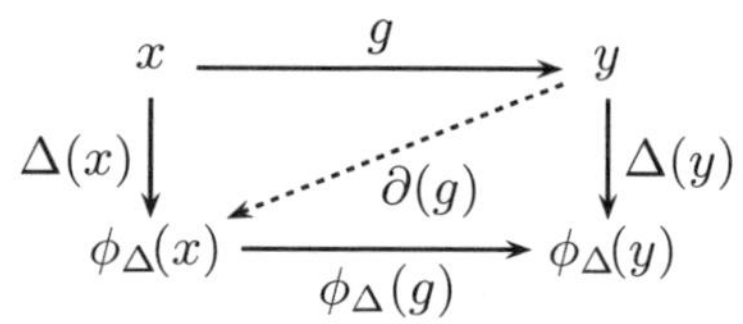

A short way to state (1.29) is to say that the map Δ is a natural transformation from the identity functor on $\mathcal{C}$ to ϕ_Δ, see the diagram on the right. Note that, for g in $\mathcal{D}iv(\Delta)$, the diagonal arrow splits the diagram into two commutative triangles.

We recall that, if ϕ, ϕ' are two functors from $\mathcal{C}$ to itself, a map F from $\mathcal{O}bj(\mathcal{C})$ to $\mathcal{C}$ is a *natural transformation* from ϕ to ϕ' if, for every g in $\mathcal{C}(x, y)$, we have the equality $F(x)\phi'(g) = \phi(g)F(y)$.

In the case of a monoid, the functor ϕ_Δ is an endomorphism, and it corresponds to conjugating by the element Δ since (1.29) takes then the form $g\,\Delta = \Delta\,\phi_\Delta(g)$.

Merging Proposition 1.28 with Lemma 1.13, we deduce:

Corollary 1.34 (functor ϕ_Δ). *If $\mathcal{C}$ is a left-cancellative category and $\mathcal{S}$ is a Garside family of $\mathcal{C}$ that is right-bounded by a map Δ, there exists a unique functor ϕ_Δ from $\mathcal{C}$ into itself that extends ∂_Δ^2. It is determined by* (1.29) *and satisfies* (1.30).

The possible injectivity and surjectivity of the functor ϕ_Δ will play an important rôle in the sequel, and we therefore introduce a specific terminology.

Definition 1.35 (target-injective). If $\mathcal{C}$ is a category, a map Δ from $\mathcal{O}bj(\mathcal{C})$ to $\mathcal{C}$ is called *target-injective* if $x \neq x'$ implies that $\Delta(x)$ and $\Delta(x')$ have different targets.

In view of Proposition 1.28, saying that Δ is target-injective means that the functor ϕ_Δ is injective on objects. On the one hand, it is easy to describe right-bounded Garside families that are right-bounded by maps that are not target-injective (see Example 2.16 below), but, on the other hand, most natural examples involve target-injective maps.

In the framework of right-bounded Garside families, there is a simple connection between the injectivity of ϕ_Δ and right-cancellation in the ambient category.

Proposition 1.36 (right-cancellativity I). *Let $\mathcal{C}$ be a left-cancellative category and Δ be a map from $\mathcal{O}bj(\mathcal{C})$ to $\mathcal{C}$ that satisfies* (1.16), (1.17), *and* (1.18). *Consider the statements:*

(i) *The category $\mathcal{C}$ is right-cancellative.*

(ii) *The functor ϕ_Δ is injective on $\mathcal{C}$.*

Then (ii) *implies* (i) *and, if Δ is target-injective,* (i) *implies* (ii).

Proof. Assume that ϕ_Δ is injective on $\mathcal{C}$ and we have $gs = g's$ with g in $\mathcal{C}(x,y)$ and s in $\mathcal{D}iv(\Delta)$. Then g' must belong to $\mathcal{C}(x,y)$, and we deduce $g\,s\,\partial s = g'\,s\,\partial s$, that is, $g\Delta(y) = g'\Delta(y)$. Then (1.29) gives $\Delta(x)\,\phi_\Delta(g) = g\,\Delta(y) = g'\,\Delta(y) = \Delta(x)\,\phi_\Delta(g')$, whence $\phi_\Delta(g) = \phi_\Delta(g')$ by left-cancelling $\Delta(x)$. As ϕ_Δ is injective on $\mathcal{C}$, we deduce $g = g'$. Since, by assumption, $\mathcal{D}iv(\Delta)$ generates $\mathcal{C}$, this is enough to conclude that $\mathcal{C}$ is right-cancellative. Hence (ii) implies (i).

Conversely, assume that Δ is target-injective and $\mathcal{C}$ is right-cancellative. Let g, g' be elements of $\mathcal{C}$ satisfying $\phi_\Delta(g) = \phi_\Delta(g')$. Assume that g belongs to $\mathcal{C}(x,y)$ and g' belongs to $\mathcal{C}(x',y')$. By construction, $\phi_\Delta(g)$ lies in $\mathcal{C}(\phi_\Delta(x), \phi_\Delta(y))$ and $\phi_\Delta(g')$ lies in $\mathcal{C}(\phi_\Delta(x'), \phi_\Delta(y'))$. So the assumption $\phi_\Delta(g) = \phi_\Delta(g')$ implies $\phi_\Delta(x) = \phi_\Delta(x')$ and $\phi_\Delta(y) = \phi_\Delta(y')$. As Δ is target-injective, we deduce $x' = x$ and $y' = y$. Next, using (1.29) again, we obtain $g\,\Delta(y) = \Delta(x)\,\phi_\Delta(g) = \Delta(x)\,\phi_\Delta(g') = g'\,\Delta(y)$, whence $g = g'$ by right-cancelling $\Delta(y)$. So (i) implies (ii) in this case. □

Using Lemma 1.13, we can restate Proposition 1.36 in terms of Garside families:

Corollary 1.37 (right-cancellativity I). *Let $\mathcal{C}$ be a left-cancellative category and $\mathcal{S}$ be a Garside family of $\mathcal{C}$ that is right-bounded by a map Δ. Consider the statements:*

(i) *The category $\mathcal{C}$ is right-cancellative.*

(ii) *The functor ϕ_Δ is injective on $\mathcal{C}$.*

Then (ii) *implies* (i) *and, if Δ is target-injective,* (i) *implies* (ii).

Example 2.16 shows that the assumption that Δ is target-injective cannot be skipped in Proposition 1.36. On the other hand, note that, if ϕ_Δ is injective on $\mathcal{C}$ (or simply on $\mathcal{D}iv(\Delta)$), then it is necessarily injective on objects, that is, Δ is target-injective. Indeed, as ϕ_Δ is a functor, $\phi_\Delta(x) = \phi_\Delta(x')$ implies $\phi_\Delta(1_x) = \phi_\Delta(1_{x'})$, so we deduce $1_x = 1_{x'}$ and $x = x'$ whenever ϕ_Δ is injective on $\mathbf{1}_\mathcal{C}$.

1.4 Powers of a right-bounded Garside family

By Proposition III.1.36 (power), if $\mathcal{S}$ is a Garside family that includes $\mathbf{1}_\mathcal{C}$ in a left-cancellative category $\mathcal{C}$, then, for every $m \geqslant 2$, the family $\mathcal{S}^m$ is also a Garside family in $\mathcal{C}$ (we recall that the assumption about $\mathcal{S}$ including $\mathbf{1}_\mathcal{C}$ is innocuous: if $\mathcal{S}$ is a Garside family, $\mathcal{S} \cup \mathbf{1}_\mathcal{C}$ is also a Garside family). In this subsection, we establish that, if $\mathcal{S}$ is right-bounded, then $\mathcal{S}^m$ is right-bounded as well, and we describe an explicit right-bounding map for $\mathcal{S}^m$ starting from one for $\mathcal{S}$. A direct consequence is that right-bounded Garside families can exist only in categories that admit common right-multiples.

In the case of a monoid, the construction involves the powers Δ^m of the element Δ. In the general case of a category, Δ maps objects to elements, and taking powers does not make sense. However, it is easy to introduce a convenient notion of iteration.

Definition 1.38 (iterate $\Delta^{[m]}$). Assume that $\mathcal{C}$ is a left-cancellative category and Δ is a map from $\mathcal{O}bj(\mathcal{C})$ to $\mathcal{C}$ that satisfies (1.16), (1.17), and (1.18). For $m \geqslant 0$, the *mth iterate* of Δ is the map $\Delta^{[m]}$ from $\mathcal{O}bj(\mathcal{C})$ to $\mathcal{C}$ inductively specified by

(1.39) $$\Delta^{[0]}(x) = 1_x, \quad \text{and} \quad \Delta^{[m]}(x) = \Delta(x)\,\phi_\Delta(\Delta^{[m-1]}(x)) \text{ for } m \geqslant 1.$$

Since ϕ_Δ is a functor, we have $\phi_\Delta(1_x) = 1_{\phi_\Delta(x)}$, whence $\Delta^{[1]} = \Delta$. Then we obtain

$$\Delta^{[2]}(x) = \Delta(x)\phi_\Delta(\Delta(x)) = \Delta(x)\,\Delta(\phi_\Delta(x)),$$

the equality following from (1.30). An immediate induction gives for $m \geqslant 1$

$$\Delta^{[m]}(x) = \Delta(x)\,\Delta(\phi_\Delta(x)) \cdots \Delta(\phi_\Delta^{m-1}(x)). \tag{1.40}$$

We begin with a few computational formulas involving the maps $\Delta^{[m]}$.

Lemma 1.41. *If $\mathcal{C}$ is a left-cancellative category and Δ is a map from $\mathcal{O}bj(\mathcal{C})$ to $\mathcal{C}$ that satisfies* (1.16), (1.17), *and* (1.18), *then, for every object x and every $m \geqslant 0$, the element $\Delta^{[m]}(x)$ lies in $(\mathcal{D}iv(\Delta))^m$ and, for every g in $\mathcal{C}(x,y)$, we have*

$$g\,\Delta^{[m]}(y) = \Delta^{[m]}(x)\,\phi_\Delta^m(g). \tag{1.42}$$

Proof. For $m = 0$, the element $\Delta^{[m]}(x)$ is 1_x, and (1.42) reduces to $g\,1_y = 1_x\,g$. Assume $m \geqslant 1$. First, $\Delta(y)$ belongs to $\mathcal{D}iv(\Delta)$ for every y, so (1.40) makes it clear that $\Delta^{[m]}(x)$ lies in $(\mathcal{D}iv(\Delta))^m$. Then we obtain

$$\begin{aligned}
g\,\Delta^{[m]}(y) &= g\,\Delta(y)\,\phi_\Delta(\Delta^{[m-1]}(y)) && \text{by the definition of } \Delta^{[m]}(y),\\
&= \Delta(x)\,\phi_\Delta(g)\,\phi_\Delta(\Delta^{[m-1]}(y)) && \text{by (1.29)},\\
&= \Delta(x)\,\phi_\Delta(g\,\Delta^{[m-1]}(y)) && \text{as } \phi_\Delta \text{ is a functor},\\
&= \Delta(x)\,\phi_\Delta(\Delta^{[m-1]}(x)\,\phi_\Delta^{m-1}(g)) && \text{by the induction hypothesis},\\
&= \Delta(x)\,\phi_\Delta(\Delta^{[m-1]}(x))\,\phi_\Delta^{m}(g) && \text{as } \phi_\Delta \text{ is a functor},\\
&= \Delta^{[m]}(x)\,\phi_\Delta^{m}(g) && \text{by the definition of } \Delta^{[m]}(x).
\end{aligned}$$

□

Lemma 1.43. *If $\mathcal{C}$ is a left-cancellative category and Δ is a map from $\mathcal{O}bj(\mathcal{C})$ to $\mathcal{C}$ that satisfies* (1.16), (1.17), *and* (1.18), *then, for every $m \geqslant 1$, every element of $\mathcal{D}iv(\Delta)^m(x,\cdot)$ left-divides $\Delta^{[m]}(x)$.*

Proof. We show using induction on $m \geqslant 1$ the relation $g \preccurlyeq \Delta^{[m]}(x)$ for every g in $\mathcal{D}iv(\Delta)^m(x,\cdot)$. For $m = 1$, this is the definition of $\mathcal{D}iv(\Delta)$. Assume $m \geqslant 2$. Write $g = sg'$ with s in $\mathcal{D}iv(\Delta)$ and g' in $\mathcal{D}iv(\Delta)^{m-1}$. Let y be the source of g', which is also the target of s. Then we obtain

$$\begin{aligned}
g = s\,g' &\preccurlyeq s\,\Delta^{[m-1]}(y) && \text{by the induction hypothesis},\\
&= \Delta^{[m-1]}(x)\,\phi_\Delta^{m-1}(s) && \text{by (1.42)},\\
&\preccurlyeq \Delta^{[m-1]}(x)\,\phi_\Delta^{m-1}(\Delta(x)) && \text{as } s \preccurlyeq \Delta(x) \text{ holds and } \phi_\Delta^{m-1} \text{ is a functor},\\
&= \Delta(x)\,\Delta^{[m-1]}(\phi_\Delta(x)) && \text{by (1.42), as } \phi_\Delta(x) \text{ is the target of } \Delta(x)\\
&= \Delta(x)\,\phi_\Delta(\Delta^{[m-1]}(x)) && \text{by (1.32) applied } m-1 \text{ times},\\
&= \Delta^{[m]}(x) && \text{by the definition.}
\end{aligned}$$

□

Proposition 1.44 (power). *If $\mathcal{C}$ is a left-cancellative category and Δ is a map from $\mathcal{O}bj(\mathcal{C})$ to $\mathcal{C}$ that satisfies* (1.16), (1.17), *and* (1.18), *then, for every $m \geqslant 1$, the map $\Delta^{[m]}$ satisfies* (1.16), (1.17), *and* (1.18), *the family $\mathcal{D}iv(\Delta)^m$ is right-bounded by $\Delta^{[m]}$, and we have then $\phi_{\Delta^{[m]}} = \phi_\Delta^m$.*

Proof. Let $m \geqslant 1$. By construction, the source of $\Delta^{[m]}(x)$ is x, so $\Delta^{[m]}$ satisfies (1.16). Next, $\mathcal{D}iv(\Delta)$ is included in $\mathcal{D}iv(\Delta^{[m]})$, hence $\mathcal{D}iv(\Delta^{[m]})$ generates $\mathcal{C}$, so $\Delta^{[m]}$ satisfies (1.17). Then, assume that g lies in $\widetilde{\mathcal{D}iv}(\Delta^{[m]})(y, \cdot)$, that is, $fg = \Delta^{[m]}(x)$ for some x and some f in $\mathcal{C}(x, y)$. By Lemma 1.41, we have $f\Delta^{[m]}(y) = \Delta^{[m]}(x)\phi_\Delta^m(f)$, that is, $f\Delta^{[m]}(y) = fg\phi_\Delta^m(f)$. By left-cancelling f, we deduce $g\phi_\Delta^m(f) = \Delta^{[m]}(y)$, so g belongs to $\mathcal{D}iv(\Delta^{[m]})$. Hence, $\Delta^{[m]}$ satisfies (1.18).

Then Lemma 1.41 shows that $\Delta^{[m]}(x)$ lies in $(\mathcal{D}iv(\Delta))^m$ for every object x, and Lemma 1.43 shows that every element of $(\mathcal{D}iv(\Delta))^m(x, \cdot)$ left-divides $\Delta^{[m]}(x)$. This implies that $(\mathcal{D}iv(\Delta))^m$ is right-bounded by $\Delta^{[m]}$.

Finally, (1.42) shows that ϕ_Δ^m satisfies the defining relations of $\phi_{\Delta^{[m]}}$, so, by uniqueness, they must coincide. □

Returning to the language of Garside families, we deduce:

Corollary 1.45 (power). *If $\mathcal{C}$ is a left-cancellative category and $\mathcal{S}$ is a Garside family of $\mathcal{C}$ that includes $1_\mathcal{C}$ and is right-bounded by a map Δ, then, for every $m \geqslant 1$, the Garside family $\mathcal{S}^m$ is right-bounded by $\Delta^{[m]}$, and we have $\phi_{\Delta^{[m]}} = \phi_\Delta^m$.*

(We recall that assuming that $1_\mathcal{C}$ is included in $\mathcal{S}$ is needed to guarantee that $\mathcal{S}$ is included in $\mathcal{S}^m$; otherwise, we should use $\mathcal{S} \cup \mathcal{S}^2 \cup \cdots \cup \mathcal{S}^m$ instead.)

A consequence of the above results is the existence of common right-multiples.

Proposition 1.46 (common right-multiple). *If $\mathcal{C}$ is a left-cancellative category and*

(i) *there exists a map Δ from $\mathcal{O}bj(\mathcal{C})$ to $\mathcal{C}$ that satisfies* (1.16), (1.17), *and* (1.18), *or*

(ii) *there exists a right-bounded Garside family in $\mathcal{C}$,*

then any two elements of $\mathcal{C}$ with the same source admit a common right-multiple.

Proof. Proposition 1.44 implies that any two elements of $\mathcal{C}(x, \cdot)$ left-divide $\Delta^{[m]}(x)$ for every m that is large enough. □

Merging with Proposition 1.4, we deduce:

Corollary 1.47 (finite right-bounded). *if $\mathcal{S}$ is a finite Garside family in a left-cancellative category $\mathcal{C}$, the following are equivalent:*

(i) *The family $\mathcal{S}$ is right-bounded.*

(ii) *The category $\mathcal{C}$ admits common right-multiples.*

In the case of a monoid, Proposition 1.46 takes the form:

Corollary 1.48 (common right-multiple). *If M is a left-cancellative monoid and*

(i) *there exists an element Δ of M such that* $\mathrm{Div}(\Delta)$ *generates M and every right-divisor of Δ left-divides Δ, or*

(ii) *there exists a right-bounded Garside family in M,*

then any two elements of M admit a common right-multiple.

We can then wonder whether the existence of a right-bounded Garside family implies the existence of right-lcms. The following example shows that this is not the case.

Example 1.49 (no right-lcm). Let $\mathcal{C}$ be the category with six objects and eight atoms whose diagram with respect to $\{\mathtt{a}, ..., \mathtt{f}\}$ is shown below and that is commutative, that is, we have $\mathtt{bc} = \mathtt{ad}$, etc. The category $\mathcal{C}$ is left-cancellative, as an exhaustive inspection of its finite multiplication table shows, and $\mathcal{C}$ is a finite Garside family in itself.

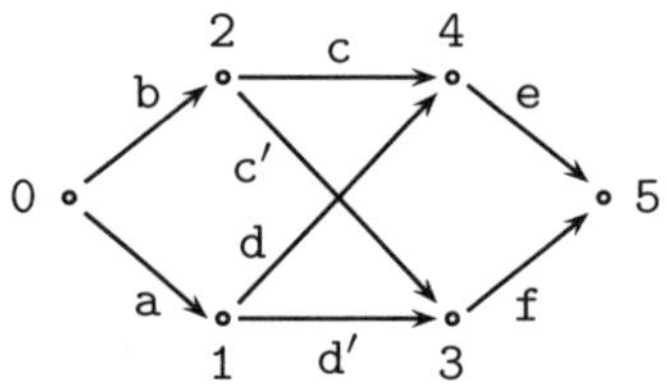

Let Δ be the map that sends every object x to the unique element of $\mathcal{C}$ that connects x to 5: for instance, we have $\Delta(\mathtt{0}) = \mathtt{ade}\,(= \mathtt{ad'f} = \mathtt{bce} = \mathtt{bc'f})$, $\Delta(\mathtt{1}) = \mathtt{de}\,(= \mathtt{d'f}), ..., \Delta(\mathtt{5}) = 1_\mathtt{5}$. Then $\mathcal{C}$ is right-bounded by Δ. However, $\mathtt{a}$ and $\mathtt{b}$ admit no right-lcm as both $\mathtt{ad}$ and $\mathtt{ad'}$ are right-mcms of $\mathtt{a}$ and $\mathtt{b}$. Symmetrically, $\mathtt{e}$ and $\mathtt{f}$ admit no left-lcm.

1.5 Preservation of normality

Here we address the question of whether the ϕ_Δ-image of a normal path necessarily is a normal path. We show in particular that, when this condition is satisfied, the second domino rule (Definition III.1.57) must be valid.

Definition 1.50 (preserving normality). If $\mathcal{S}$ is a subfamily of a left-cancellative category $\mathcal{C}$, we say that a functor ϕ from $\mathcal{C}$ to itself *preserves $\mathcal{S}$-normality* if $\phi(s_1)|\phi(s_2)$ is $\mathcal{S}$-normal whenever $s_1|s_2$ is.

If Δ is a right-Garside map, various conditions imply that the functor ϕ_Δ preserves normality (this meaning $\mathcal{D}iv(\Delta)$-normality, if there is no ambiguity), see in particular Proposition 2.18 below and Exercise 61. On the other hand, simple examples exist where normality is not preserved.

Example 1.51 (not preserving normality). In the left-absorbing monoid L_n (Reference Structure 8, page 113) with $\Delta = \mathtt{b}^{n+1}$, the associated functor ϕ_Δ does not preserve normality. Indeed, we find $\phi_\Delta(\mathtt{a}) = \phi_\Delta(\mathtt{b}) = \mathtt{b}$. Now $\mathtt{a}|\mathtt{b}$ is $\mathrm{Div}(\Delta)$-normal, whereas $\phi_\Delta(\mathtt{a})|\phi_\Delta(\mathtt{b})$, that is $\mathtt{b}|\mathtt{b}$, is not $\mathrm{Div}(\Delta)$-normal.

The main result we shall prove is

Proposition 1.52 (second domino rule). *If Δ is a right-Garside map in a left-cancellative category $\mathcal{C}$ and ϕ_Δ preserves normality, the second domino rule is valid for $\mathcal{D}iv(\Delta)$.*

In order to establish Proposition 1.52, we need a criterion for recognizing greediness in the context of a right-bounded Garside family. The following one will be used several times, and it is of independent interest. We recall that two elements f, g with the same source are called left-coprime if every common left-divisor of f and g is invertible.

Proposition 1.53 (recognizing greedy). *Assume that $\mathcal{C}$ is a left-cancellative category and $\mathcal{S}$ is a Garside family of $\mathcal{C}$ that is right-bounded by Δ. For s in $\mathcal{S}^\sharp$ and g in $\mathcal{C}$, consider*

(i) *The path $s|g$ is $\mathcal{S}$-greedy.*

(ii) *The elements ∂s and g are left-coprime.*

Then (ii) *implies* (i) *and, if $\mathcal{S}^\sharp$ is closed under left-divisor, then* (i) *implies* (ii).

Proof. Assume that ∂s and g are left-coprime, and h is an element of $\mathcal{C}$ satisfying $sh \in \mathcal{S}$ and $h \preccurlyeq g$. Let x be the source of s. As sh belongs to $\mathcal{S}$, we have $sh \preccurlyeq \Delta(x) = s\partial s$, whence $h \preccurlyeq \partial s$ by left-cancelling s. As ∂s and g are left-coprime, h must be invertible. By Proposition IV.1.20 (recognizing greedy), $s|g$ is $\mathcal{S}$-greedy. So (ii) implies (i).

Conversely, assume that $s|g$ is $\mathcal{S}$-greedy and h is a common left-divisor of ∂s and g. Then we have $sh \preccurlyeq s\,\partial s = \Delta(x)$, where x is the source of s. By assumption, $\Delta(x)$ belongs to $\mathcal{S}$, hence, as $\mathcal{S}^\sharp$ is closed under left-divisor, sh belongs to $\mathcal{S}^\sharp$. So there exists h' satisfying $h' =^\times h$ and $sh' \in \mathcal{S}$. Then $h \preccurlyeq g$ implies $h' \preccurlyeq g$, and, by Proposition IV.1.20 (recognizing greedy), h', hence h, is invertible, and (i) implies (ii). □

Owing to Proposition 1.20, Proposition 1.53 can be restated as

Corollary 1.54 (recognizing greedy). *If Δ is a right-Garside map in a left-cancellative category $\mathcal{C}$, then, for s in $\mathcal{D}iv(\Delta)$ and g in $\mathcal{C}$, the path $s|g$ is $\mathcal{D}iv(\Delta)$-greedy if and only if ∂s and g are left-coprime.*

We can now establish Proposition 1.52.

Proof of Proposition 1.52. Put $\mathcal{S} = \mathcal{D}iv(\Delta)$, and assume that s_1, s_2, s_1', s_2', and t_0, t_1, t_2 lie in $\mathcal{S}$ and satisfy $s_1t_1 = t_0s_1'$ and $s_2t_2 = t_1s_2'$, with $s_1|s_2$ and $t_1|s_2'$ both $\mathcal{S}$-normal.

We wish to show that $s_1'|s_2'$ is $\mathcal{D}iv(\Delta)$-greedy. Let x_i be the source of t_i for $i = 0, 1, 2$. Let us introduce the elements ∂t_i and $\phi_\Delta(s_i)$ as shown on the right. We claim that the diagram is commutative. Indeed, applying (1.29), we find $t_0\, s_1'\, \partial t_1 = s_1\, t_1\, \partial t_1 = s_1\, \Delta(x_1) = \Delta(x_0)\, \phi_\Delta(s_1) = t_0\, \partial t_0\, \phi_\Delta(s_1)$, hence $s_1'\, \partial t_1 = \partial t_0\, \phi_\Delta(s_1)$ by left-cancelling t_0. By a similar argument, we obtain $s_2'\, \partial t_2 = \partial t_1\, \phi_\Delta(g)$.

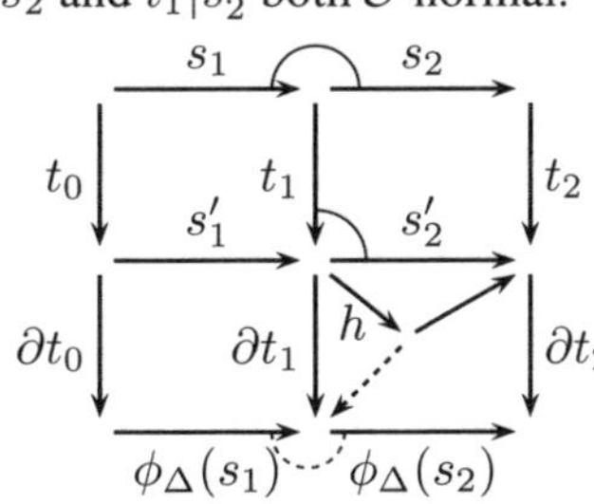

Aiming at applying the criterion of Proposition IV.1.20(iii) (recognizing greedy), assume that $s_1'h$ belongs to $\mathcal{S}$ and h left-divides s_2'. We wish to prove that h is invertible. Now $s_1'h$ left-divides $s_1'\, s_2'\, \partial t_2$, which is also $\partial t_0\, \phi_\Delta(s_1)\, \phi_\Delta(s_2)$. The assumption that ϕ_Δ preserves normality implies that $\phi_\Delta(s_1)|\phi_\Delta(s_2)$ is $\mathcal{D}iv(\Delta)$-greedy, and, therefore,

$s_1'h \preccurlyeq \partial t_0\, \phi_\Delta(s_1)\, \phi_\Delta(s_2)$ implies $s_1'h \preccurlyeq \partial t_0\, \phi_\Delta(s_1)$. As the bottom left square of the diagram is commutative, we deduce $s_1'h \preccurlyeq s_1'\, \partial t_1$, whence $h \preccurlyeq \partial t_1$ by left-cancelling s_1'. So h left-divides both ∂t_1 and s_2'. By Corollary 1.54, the assumption that $t_1|s_2'$ is Δ-normal implies that ∂t_1 and s_2' are left-coprime. Hence h is invertible and, by Proposition IV.1.20(iii) (recognizing greedy), $s_1'|s_2'$ is $\mathcal{D}iv(\Delta)$-greedy. □

Once the second domino rule is known to be valid, the results of Subsection III.1.5 apply and Proposition III.1.62 (length) is valid. It will be convenient to fix a specific notation for the length with respect to a family $\mathcal{D}iv(\Delta)$.

Definition 1.55 (supremum). If Δ is a right-Garside map in a left-cancellative category $\mathcal{C}$ and g belongs to $\mathcal{C}$, the Δ-*supremum* of g, written $\sup_\Delta(g)$, is the $\mathcal{D}iv(\Delta)$-length of g.

Corollary 1.56 (supremum). *If Δ is a right-Garside map in a left-cancellative category $\mathcal{C}$ and ϕ_Δ preserves normality, then, for all $f|g$ in $\mathcal{C}^{[2]}$, we have*

$$\max(\sup_\Delta(f), \sup_\Delta(g)) \leqslant \sup_\Delta(fg) \leqslant \sup_\Delta(f) + \sup_\Delta(g). \tag{1.57}$$

An application of Corollary 1.56 is the following transfer result.

Proposition 1.58 (iterate). *If Δ is a right-Garside map in a left-cancellative category $\mathcal{C}$ and ϕ_Δ preserves normality, then, for each $m \geqslant 2$, the map $\Delta^{[m]}$ is a right-Garside map and we have $\mathcal{D}iv(\Delta^{[m]}) = (\mathcal{D}iv(\Delta))^m$.*

Proof. Assume $m \geqslant 2$. By Lemma 1.43, every element of $(\mathcal{D}iv(\Delta))^m(x, \cdot)$ left-divides $\Delta^{[m]}(x)$. Conversely, $g \preccurlyeq \Delta^{[m]}(x)$ implies $\sup_\Delta(g) \leqslant \sup_\Delta(\Delta^{[m]}(x))$ by Corollary 1.56, whence $\sup_\Delta(g) \leqslant m$. Hence, $g \preccurlyeq \Delta^{[m]}$ implies $g \in (\mathcal{D}iv(\Delta))^m$, and so the equality $\mathcal{D}iv(\Delta^{[m]}) = (\mathcal{D}iv(\Delta))^m$ holds.

As Δ is a right-Garside map, the family $\mathcal{D}iv(\Delta)$ is a Garside family of $\mathcal{C}$ that includes $1_\mathcal{C}$. By Proposition III.1.36 (power), its mth power, which we have seen coincides with $\mathcal{D}iv(\Delta^{[m]})$, is also a Garside family and, by definition, it is right-bounded by $\Delta^{[m]}$ and is closed under left-divisor. By Proposition 1.20, $\Delta^{[m]}$ is a right-Garside map. □

In the context of Proposition 1.58, if we do not assume that ϕ_Δ preserves normality, $\mathcal{D}iv(\Delta)$ is a Garside family that is right-bounded by Δ, and one easily deduces that $\mathcal{D}iv(\Delta)^m$ is a Garside family that is right-bounded by $\Delta^{[m]}$. However, we cannot deduce that $\Delta^{[m]}$ is a right-Garside map as, unless Corollary 1.56 is valid, nothing guarantees that $\mathcal{D}iv(\Delta)^m$ is closed under left-multiplication.

We conclude with another application of preservation of normality. We saw in Proposition 1.36 that, if a Garside family is right-bounded by a map Δ, the injectivity of ϕ_Δ on objects and elements implies that the ambient category is right-cancellative. When ϕ_Δ preserves normality, the injectivity assumption on elements can be weakened.

Proposition 1.59 (right-cancellative II). *If $\mathcal{S}$ is a Garside family in a left-cancellative category $\mathcal{C}$ that is right-bounded by a target-injective map Δ, and ϕ_Δ preserves $\mathcal{S}$-normality and is surjective on $\mathcal{C}^\times$, then the following are equivalent:*

(i) *The category $\mathcal{C}$ is right-cancellative.*

(ii) *The functor ϕ_Δ is injective on $\mathcal{C}$.*

(iii) *The functor ϕ_Δ is injective on $\mathcal{S}^\sharp$.*

In Proposition 1.59, (i) and (ii) are equivalent by Proposition 1.36, and (ii) obviously implies (iii). So the point is to prove that (iii) implies (ii). The argument consists in showing that $\phi_\Delta(s') =^\times \phi_\Delta(s)$ implies $s =^\times s'$ for s, s' in $\mathcal{S}^\sharp$, and then establishing, using induction on p, that $\phi_\Delta(g) = \phi_\Delta(g')$ implies $g = g'$ for all g, g' satisfying $\max(\|g\|_\mathcal{S}, \|g'\|_\mathcal{S}) \leqslant p$. We skip the details here.

2 Bounded Garside families

The definition of a right-bounded Garside family $\mathcal{S}$ requires that the elements of $\mathcal{S}(x, \cdot)$ all left-divide a certain element $\Delta(x)$, thus it only involves left-divisibility. We could also consider a symmetric condition and require that, for each object y, the elements of $\mathcal{S}(\cdot, y)$ all right-divide some prescribed element depending on y. Things become interesting when the two conditions are merged and the same bounding map occurs on both sides. This is the situation investigated in this section. The main technical result is that, provided the ambient category $\mathcal{C}$ is cancellative (on both sides), these boundedness conditions imply, and actually are equivalent to the functor ϕ_Δ being an automorphism of $\mathcal{C}$.

The section comprises five subsections. In Subsection 2.1 we explain how to define a convenient notion of two-sided bounded Garside family and gather direct consequences of the definition. In Subsection 2.2, we establish that bounded Garside families are preserved under power and deduce that their existence implies the existence of common left-multiples. In Subsection 2.3, we investigate bounded Garside families in the context of cancellative categories, with the key result that, in this case, the associated functor ϕ_Δ is an automorphism. In Subsection 2.4, we introduce the notion of a Garside map, a refinement of a right-Garside map, and show how to rephrase most results in this context. Finally, in Subsection 2.5, we discuss the existence of lcms and gcds in a category that admits a Garside map. The main result is that left-gcds and left-lcms always exist, and so do right-lcms and right-gcds whenever the ambient category is left-Noetherian.

2.1 The notion of a bounded Garside family

The notion of a right-bounded Garside family was introduced in Section 1 using the condition that all elements of $\mathcal{S}(x, \cdot)$ left-divide some distinguished element $\Delta(x)$. We now consider a two-sided version in which the elements of $\mathcal{S}$ have to both left-divide and right-divide some element $\Delta(x)$. Several formulations are possible, and to obtain the convenient one, we first restate the definition of a right-bounded Garside family.

Lemma 2.1. *A Garside family $\mathcal{S}$ in a left-cancellative category $\mathcal{C}$ is right-bounded by a map Δ if and only if, for every object x, the element $\Delta(x)$ lies in $\mathcal{S}$ and there exists an object y satisfying*

$$\forall s \in \mathcal{S}^\sharp(x, \cdot)\ \exists t \in \mathcal{S}^\sharp(\cdot, y)\ (st = \Delta(x)). \tag{2.2}$$

Proof. Assume that $\mathcal{S}$ is right-bounded by Δ and x belongs to $\mathit{Obj}(\mathcal{C})$. Let $y = \phi_\Delta(x)$ and $s \in \mathcal{S}^\sharp(x, \cdot)$. Then there exists s' in $\mathcal{S}(x, \cdot)$ satisfying $s' =^\times s$. By assumption, we have $s' \preccurlyeq \Delta(x)$, hence $s \preccurlyeq \Delta(x)$, that is, $st = \Delta(x)$ for some t. By construction, t belongs to $\mathcal{C}(\cdot, y)$ and it lies in $\mathcal{S}^\sharp$ as $\Delta(x)$ does and $\mathcal{S}^\sharp$ is closed under right-divisor. So (2.2) is satisfied. Conversely, if (2.2) holds, then every element of $\mathcal{S}(x, \cdot)$ left-divides $\Delta(x)$, and $\mathcal{S}$ is right-bounded by Δ. □

Relation (2.2) is the one we shall reverse in order to define bounded Garside families.

Definition 2.3 (bounded). A Garside family $\mathcal{S}$ in a left-cancellative category $\mathcal{C}$ is said to be *bounded* by a map Δ from $\mathcal{O}bj(\mathcal{C})$ to $\mathcal{C}$ if $\mathcal{S}$ is right-bounded by Δ and, in addition, for every object y, there exists an object x satisfying

(2.4) $$\forall s \in \mathcal{S}^\sharp(\cdot, y)\ \exists r \in \mathcal{S}^\sharp(x, \cdot)\ (rs = \Delta(x)).$$

Example 2.5 (bounded). In the braid monoid B_n^+ (Reference Structure 2, page 5), the Garside family of simple braids is right-bounded by Δ_n. It is actually bounded by Δ_n: every simple n-strand braid is a right-divisor of Δ_n, so (2.4) is satisfied.

More generally, if (M, Δ) is a Garside or quasi-Garside monoid, the Garside family $\mathrm{Div}(\Delta)$ is bounded by Δ since every left-divisor of Δ is also a right-divisor of Δ. Note that the condition about objects vanishes in the case of a monoid.

On the other hand, the Garside family S_n in the left-absorbing monoid L_n (Reference Structure 8, page 113) is right-bounded, but not bounded by $\mathtt{b}^{n+1}$: indeed, $\mathtt{a}$ lies in S_n, but no element r of S_n (or of L_n) satisfies $r\mathtt{a} = \mathtt{b}^{n+1}$.

Finally, the category $\mathcal{C}$ of Example 1.49, viewed as a Garside family in itself, is right-bounded by the map Δ, but it is not bounded by Δ. For instance, $\mathtt{bd}$ belongs to $S(\mathtt{0}, \mathtt{4})$, and the only element by which $\mathtt{bd}$ can be left-multiplied is $1_\mathtt{0}$. As no object x satisfies $\Delta(x) = \mathtt{bd}$, (2.4) cannot be satisfied and S is not bounded by Δ. However, if $\nabla(y)$ denotes the unique element of $\mathcal{C}$ that maps $\mathtt{0}$ to y, then, for every object y, we have $\forall s \in \mathcal{S}^\sharp(\cdot, y)\ \exists r \in \mathcal{S}^\sharp(\mathtt{0})\ (rs = \nabla(y))$, and, in an obvious sense, S is left-bounded by ∇. But this is not (2.4) because $\nabla(y)$ is not $\Delta(x)$ for any x: in Definition 2.3, we insist that the same map Δ witnesses for right- and left-boundedness.

As the above example shows, being right-bounded does not imply being bounded. However, the two notions coincide in particular cases, notably in the cancellative finite case.

Proposition 2.6 (finite bounded). *If $\mathcal{C}$ is a cancellative category and $\mathcal{S}$ is a Garside family of $\mathcal{C}$ such that $\mathcal{S}^\sharp$ is finite and $\mathcal{S}$ is right-bounded by a target-injective map Δ, then $\mathcal{S}$ is bounded by Δ and the functor ϕ_Δ is a finite order automorphism.*

Proof. First, Corollary 1.12 implies $\mathcal{S}^\sharp = \mathcal{D}iv(\Delta) = \widetilde{\mathcal{D}iv}(\Delta)$.

Next, by Proposition 1.28, ∂_Δ maps the finite set $\mathcal{S}^\sharp$ into itself and, by Proposition 1.36, ϕ_Δ, which is ∂_Δ^2, is injective on $\mathcal{C}$, hence in particular on $\mathcal{S}^\sharp$. So ∂_Δ is injective, hence bijective on $\mathcal{S}^\sharp$. This means that, for every s in $\mathcal{S}^\sharp(\cdot, y)$, there exists r in $\mathcal{S}^\sharp$ satisfying $rs = \Delta(x)$, where x is the source of r. Now, as ϕ_Δ is injective on objects, x must be the unique object satisfying $\phi_\Delta(x) = y$. So (2.4) holds, and $\mathcal{S}$ is bounded by Δ.

Finally, the assumption that $\mathcal{S}^\sharp$ is finite implies that $\mathcal{O}bj(\mathcal{C})$ is finite. As ϕ_Δ is injective on $\mathcal{O}bj(\mathcal{C})$ and on $\mathcal{S}^\sharp$, there must exist an exponent d such that ϕ_Δ^d is the identity on $\mathcal{O}bj(\mathcal{C})$ and on $\mathcal{S}^\sharp$, hence on $\mathcal{C}$. Thus, in particular, ϕ_Δ must be an automorphism. □

In the case of a cancellative monoid, the target-injectivity condition vanishes and Proposition 2.6 says that every right-bounded Garside family $\mathcal{S}$ such that $\mathcal{S}^\sharp$ is finite is bounded. Right-cancellativity cannot be skipped in Proposition 2.6: as seen in Example 2.5, the finite right-bounded family S_n in the left-absorbing monoid L_n (Reference Structure 8, page 113) is not bounded.

We turn to properties of bounded Garside families. If a Garside family is bounded by a map Δ, it is in particular right-bounded by Δ and, therefore, it inherits all properties established in Section 1. We now list additional properties.

Lemma 2.7. *Assume that $\mathcal{S}$ is a Garside family in a left-cancellative category $\mathcal{C}$ and $\mathcal{S}$ is bounded by a map Δ.*

(i) *We have $\widetilde{\mathcal{D}iv}(\Delta) = \mathcal{S}^\sharp \subseteq \mathcal{D}iv(\Delta)$.*

(ii) *For every object y, the element $\Delta(x)$ is a left-lcm for $\mathcal{S}^\sharp(\cdot, y)$, where x is any object for which* (2.4) *holds.*

(iii) *The functor ϕ_Δ is surjective on $\mathcal{O}bj(\mathcal{C})$.*

(iv) *The map ∂_Δ induces a surjection from $\mathcal{S}^\sharp$ onto itself.*

(v) *The functor ϕ_Δ is surjective on $\mathcal{C}$ and induces a surjection from $\mathcal{S}^\sharp$ onto itself.*

Proof. (i) By (2.4), every element of $\mathcal{S}^\sharp$ right-divides some element $\Delta(x)$, so $\mathcal{S}^\sharp$ is included in $\widetilde{\mathcal{D}iv}(\Delta)$. Owing to (1.11), this forces $\widetilde{\mathcal{D}iv}(\Delta) = \mathcal{S}^\sharp \subseteq \mathcal{D}iv(\Delta)$.

(ii) By assumption, $\Delta(x)$ belongs to $\mathcal{S}(\cdot, y)$, hence a fortiori to $\mathcal{S}^\sharp(\cdot, y)$. Now, for $s \in \mathcal{S}^\sharp(\cdot, y)$, (2.4) says that s right-divides $\Delta(x)$, so the latter is a left-lcm for $\mathcal{S}^\sharp(\cdot, y)$.

(iii) Let y be an object of $\mathcal{C}$. By definition, there exists an object x satisfying (2.4). Applying the latter relation to 1_y, an element of $\mathcal{S}^\sharp(y, y)$, we deduce the existence of r in $\mathcal{S}^\sharp(x, y)$ satisfying $r1_y = \Delta(x)$, hence $y = \phi_\Delta(x)$. Hence ϕ_Δ is surjective on $\mathcal{O}bj(\mathcal{C})$.

(iv) By definition, ∂_Δ maps $\mathcal{D}iv(\Delta)$ to $\widetilde{\mathcal{D}iv}(\Delta)$. Owing to (1.11), it induces a map from $\mathcal{S}^\sharp$ to $\widetilde{\mathcal{D}iv}(\Delta)$, which by (i) is $\mathcal{S}^\sharp$. By (2.4), every element of $\mathcal{S}^\sharp$ belongs to the image of this map.

(v) By definition, the restriction of ϕ_Δ to $\mathcal{S}^\sharp$ is the square of the restriction of ∂_Δ to $\mathcal{S}^\sharp$. Hence the sujectivity of the latter implies the surjectivity of the former. It follows that $\mathcal{S}^\sharp$ is included in the image of ϕ_Δ. As ϕ_Δ is a functor and $\mathcal{S}^\sharp$ generates $\mathcal{C}$, it follows that the image of ϕ_Δ includes all of $\mathcal{C}$. □

Remark 2.8. Condition (2.4) implies both $\mathcal{S} \subseteq \widetilde{\mathcal{D}iv}(\Delta)$ and $\mathcal{S}^\sharp \subseteq \widetilde{\mathcal{D}iv}(\Delta)$ (which, contrary to $\mathcal{S} \subseteq \mathcal{D}iv(\Delta)$ and $\mathcal{S}^\sharp \subseteq \mathcal{D}iv(\Delta)$, need not be equivalent conditions), but it is a priori stronger. Defining boundedness by any of the above conditions and not requiring that the elements of $\mathcal{S}^\sharp(\cdot, y)$ all right-divide the same element $\Delta(x)$ seems to be a useless assumption. Note that, if the ambient category is not right-cancellative, (2.4) says nothing about the uniqueness of g and (2.2) says nothing about the injectivity of ϕ_Δ on objects.

2.2 Powers of a bounded Garside family

We proved in Subsection 1.4 that, if $\mathcal{S}$ is a Garside family that is right-bounded by Δ, then, for every $m \geqslant 1$, the power $\mathcal{S}^m$ is a Garside family that is right-bounded by $\Delta^{[m]}$. We shall now establish a similar transfer result for a bounded Garside family.

Definition 2.9 (witness). If $\mathcal{S}$ is a Garside family in a left-cancellative category $\mathcal{C}$ and $\mathcal{S}$ is bounded by a map Δ, a *witness* for $(\mathcal{S}, \Delta)$ is a pair $(\tilde{\phi}, \tilde{\partial})$ with $\tilde{\phi} : \mathcal{O}bj(\mathcal{C}) \to \mathcal{O}bj(\mathcal{C})$, $\tilde{\partial} : \mathcal{S}^\sharp \to \mathcal{S}^\sharp$ that satisfies, for every object y,

$$\forall s \in \mathcal{S}^\sharp(\cdot, y)\ (\tilde{\partial} s\, s = \Delta(\tilde{\phi}(y))). \tag{2.10}$$

The definition of a bounded Garside family says that, for every object y, there exists an object x, and, for every element s, there exists an element r satisfying some properties. Fixing a witness just means choosing such x and r, so every bounded Garside family admits (at least one) witness.

The following result is a weak converse of Proposition 1.28.

Lemma 2.11. *Assume that $\mathcal{S}$ is a Garside family in a left-cancellative category $\mathcal{C}$ and $\mathcal{S}$ is bounded by a map Δ. Let $(\tilde{\phi}, \tilde{\partial})$ be a witness for $(\mathcal{S}, \Delta)$. Then, for every g in $(\mathcal{S}^\sharp)^m(x, y)$, there exists g' in $(\mathcal{S}^\sharp)^m(\tilde{\phi}(x), \tilde{\phi}(y))$ satisfying*

$$g'\, \Delta(\tilde{\phi}(y)) = \Delta(\tilde{\phi}(x))\, g. \tag{2.12}$$

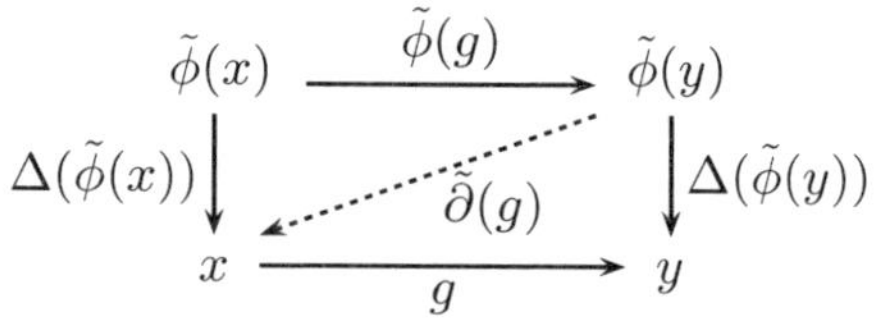

Proof. The argument, which corresponds to the diagram on the right, is similar to the proof of Proposition 1.28, with the difference that we do not claim that $\tilde{\phi}$ is a functor. We begin with $m = 1$. For s in $\mathcal{S}^\sharp$, we put $\tilde{\phi}(s) = \tilde{\partial}^2 s$. Then, for s in $\mathcal{S}^\sharp(x, y)$, we obtain

$$\tilde{\phi}(s)\, \Delta(\tilde{\phi}(y)) = \tilde{\phi}(s)\, \tilde{\partial} s\, s = \tilde{\partial}(\tilde{\partial} s)\, \tilde{\partial}(s)\, s = \Delta(\tilde{\phi}(x))\, s,$$

and (2.12) is satisfied for $s' = \tilde{\phi}(s)$.

Assume now that g lies in $(\mathcal{S}^\sharp)^m(x, y)$. We fix a decomposition of g into a product of elements of $\mathcal{S}^\sharp$, say $g = s_1 \cdots s_m$, and put $g' = \tilde{\phi}(s_1) \cdots \tilde{\phi}(s_m)$. Applying (2.12) in the case $m = 1$ for $s_1, \ldots, s_q$, and denoting by y_i the target of s_i, we obtain

$$\begin{aligned} g'\, \Delta(\tilde{\phi}(y)) &= \tilde{\phi}(s_1) \cdots \tilde{\phi}(s_m)\, \Delta(\tilde{\phi}(y_m)) \\ &= \tilde{\phi}(s_1) \cdots \tilde{\phi}(s_{m-1})\, \Delta(\tilde{\phi}(y_{m-1}))\, s_m \\ &= \tilde{\phi}(s_1) \cdots \tilde{\phi}(s_{m-2})\, \Delta(\tilde{\phi}(y_{m-2}))\, s_{m-1}\, s_m \\ &= \ldots = \Delta(\tilde{\phi}(x))\, s_1 \cdots s_m = \Delta(\tilde{\phi}(x))\, g. \end{aligned}$$

which establishes (2.12) since g' belongs to $(\mathcal{S}^\sharp)^m(\tilde{\phi}(x), \tilde{\phi}(y))$. □

Proposition 2.13 (power). *If $\mathcal{C}$ is a left-cancellative category and $\mathcal{S}$ is a Garside family of $\mathcal{C}$ that includes $1_\mathcal{C}$ and is bounded by a map Δ, then, for every $m \geqslant 1$, the Garside family $\mathcal{S}^m$ is bounded by $\Delta^{[m]}$. Moreover, if $(\tilde{\phi}, \tilde{\partial})$ is a witness for $(\mathcal{S}, \Delta)$, then $(\mathcal{S}, \Delta^{[m]})$ admits a witness of the form $(\tilde{\phi}^m, \cdot)$.*

Proof. Let $(\tilde{\phi}, \tilde{\partial})$ be a witness for $(\mathcal{S}, \Delta)$. We use induction on $m \geqslant 1$. For $m = 1$, the result is the assumption. Assume $m \geqslant 2$. By Proposition 1.44, $\mathcal{S}^m$ is a Garside family in $\mathcal{C}$ and it is right-bounded by $\Delta^{[m]}$. It remains to establish (2.4). So let y be an object and g be an element of $(\mathcal{S}^m)^\sharp(\cdot, y)$. By construction, g belongs to $(\mathcal{S}^\sharp)^m$. Write $g = sg'$ with s in $\mathcal{S}^\sharp$ and g' in $(\mathcal{S}^\sharp)^{m-1}$, see Figure 1. Let z be the target of s. By assumption, we have $\tilde{\partial} s\, s = \Delta(\tilde{\phi}(z))$. On the other hand, by the induction hypothesis, $(\mathcal{S}^\sharp)^{m-1}$ is bounded by $\Delta^{[m-1]}$ with a witness of the form $(\tilde{\phi}^{m-1}, \cdot)$ so, as g' belongs to $(\mathcal{S}^\sharp)^{m-1}(z, y)$, there exists h in $(\mathcal{S}^\sharp)^{m-1}(\tilde{\phi}^{m-1}(y), z)$ satisfying $hg' = \Delta^{[m-1]}(\tilde{\phi}^{m-1}(y))$.

We now apply Lemma 2.11 to the element h of $(\mathcal{S}^\sharp)^{m-1}(\tilde{\phi}^{m-1}(y), z)$, obtaining an element f' in $(\mathcal{S}^\sharp)^{m-1}(\tilde{\phi}^m(y), \tilde{\phi}(z))$ that satisfies $f'\, \Delta(\tilde{\phi}(z)) = \Delta(\tilde{\phi}(\tilde{\phi}^{m-1}(y)))\, h$. Put $f = f'\, \tilde{\partial}(s)$. Then f belongs to $(\mathcal{S}^\sharp)^m$ by construction, and the commutativity of the diagram of Figure 1 gives $f\, g = \Delta(\tilde{\phi}^m(y))\, \Delta^{[m-1]}(\tilde{\phi}^{m-1}(y)) = \Delta^{[m]}(\tilde{\phi}^m(y))$. This is the expected result, as we have $\mathcal{C}^\times \mathcal{S} \subseteq \mathcal{S}^\sharp$, whence $(\mathcal{S}^\sharp)^m = (\mathcal{S}^m)^\sharp$ by Lemma II.1.26. ∎

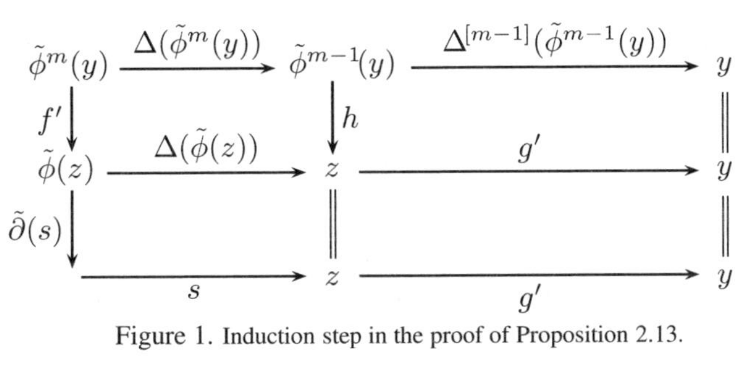

Figure 1. Induction step in the proof of Proposition 2.13.

In Proposition 2.13, explicitly defining a witness for $(\mathcal{S}, \Delta^{[m]})$ would be easy, but this is not useful since that witness would not be not canonical. Using Lemma 2.7(ii), we deduce a new constraint for categories admitting a bounded Garside family.

Corollary 2.14 (common left-multiple). *Every left-cancellative category that admits a bounded Garside family admits common left-multiples. More precisely, if $\mathcal{S}$ is a Garside family that is bounded by Δ, any two elements of $(\mathcal{S}^\sharp)^m(\cdot, y)$ admit a common left-multiple of the form $\Delta^{[m]}(\cdot)$.*

2.3 The case of a cancellative category

When we assume that the ambient category is cancellative on both sides and, in addition, that the functor ϕ_Δ is injective on objects, new results appear, the most important being that the functor ϕ_Δ is an automorphism, which in turn has a number of consequences.

We recall that a map Δ is called *target-injective* if the associated functor ϕ_Δ is injective on objects. In the context of bounded Garside families, bounding maps are often target-injective, in particular due to the following connection.

Proposition 2.15 (target-injective). *If $\mathcal{C}$ is a left-cancellative category with finitely many objects, a map bounding a Garside family of $\mathcal{C}$ (if any) is necessarily target-injective.*

Proof. By Lemma 2.7, ϕ_Δ gives a surjective map from $\mathcal{O}bj(\mathcal{C})$ to itself. If $\mathcal{O}bj(\mathcal{C})$ is finite, this implies that ϕ_Δ is injective as well. □

It is easy to see that Proposition 2.15 does not extend to arbitrary categories.

Example 2.16 (not target-injective). Let $\mathcal{C}$ be the category with infinitely many objects $0, 1, \ldots$, generated by elements $\mathtt{a}_1, \mathtt{a}_2, \ldots$, and whose diagram with respect to $\{\mathtt{a}_1, \mathtt{a}_2, \ldots\}$ is

$$0 \xleftarrow{\ \mathtt{a}_1\ } 1 \xleftarrow{\ \mathtt{a}_2\ } 2 \xleftarrow{\ \mathtt{a}_3\ } 3 \cdots\cdots .$$

The elements of $\mathcal{C}$ are all finite products $\mathtt{a}_i\mathtt{a}_{i-1}\cdots\mathtt{a}_j$ with $j \leqslant i$. Let $\mathcal{S}$ be $\{\mathtt{a}_1, \mathtt{a}_2, \ldots\}$. The category $\mathcal{C}$ is cancellative, and $\mathcal{S}$ is a Garside family in $\mathcal{C}$. Let $\Delta : \mathcal{O}bj(\mathcal{C}) \to \mathcal{C}$ be defined by $\Delta(k) = \mathtt{a}_k$ for $k \geqslant 1$ and $\Delta(0) = 1_0$. Then $\mathcal{S}$ is bounded by Δ, with a witness $(\tilde{\phi}, \tilde{\partial})$ for $(\mathcal{S}, \Delta)$ given by $\tilde{\phi}(k) = k+1$, $\tilde{\partial}(1_k) = \mathtt{a}_{k+1}$, and $\tilde{\partial}(\mathtt{a}_k) = 1_k$ for every k. However, Δ is not target-injective, as we have $\phi_\Delta(1) = \phi_\Delta(0) = 0$. Note that ϕ_Δ is not injective on elements of $\mathcal{C}$ either, as we have $\phi_\Delta(\mathtt{a}_k) = \mathtt{a}_{k-1}$ for $k > 1$ and $\phi_\Delta(\mathtt{a}_1) = 1_0 = \phi_\Delta(1_0)$.

We observed in Proposition 1.36 that, in the context of right-bounded Garside families, there exists a simple connection between the injectivity of ϕ_Δ and right-cancellativity in the ambient category. In the case of bounded Garside families, we can say more.

Proposition 2.17 (automorphism). *If $\mathcal{S}$ is a Garside family in a left-cancellative category $\mathcal{C}$ and $\mathcal{S}$ is right-bounded by a map Δ, the following are equivalent:*

(i) *The category $\mathcal{C}$ is cancellative, $\mathcal{S}$ is bounded by Δ, and Δ is target-injective.*

(ii) *The functor ϕ_Δ is an automorphism of $\mathcal{C}$.*

When (i) *and* (ii) *hold, we have $\widetilde{\mathcal{D}iv}(\Delta) = \mathcal{S}^\sharp = \mathcal{D}iv(\Delta)$, the family $\mathcal{S}^\sharp$ is closed under left-divisor, ∂_Δ and the restriction of ϕ_Δ to $\mathcal{S}^\sharp$ are permutations of $\mathcal{S}^\sharp$, and $(\phi_\Delta^{-1}, \partial_\Delta^{-1})$ is the unique witness for $(\mathcal{S}, \Delta)$.*

Proof. Assume (i). By assumption, Δ is target-injective, hence ϕ_Δ is injective on $\mathcal{O}bj(\mathcal{C})$; by Lemma 2.7(iii), ϕ_Δ is surjective on $\mathcal{O}bj(\mathcal{C})$, so ϕ_Δ is bijective on objects. Next, by

Proposition 1.36, ϕ_Δ is injective on $\mathcal{C}$ and, by Lemma 2.7(v), ϕ_Δ is surjective on $\mathcal{C}$, so ϕ_Δ is bijective on elements as well, and it is an automorphism of $\mathcal{C}$. So (i) implies (ii).

Conversely, assume that ϕ_Δ is an automorphism of $\mathcal{C}$. Then, by Proposition 1.36, the category $\mathcal{C}$ must be right-cancellative since ϕ_Δ is injective on $\mathcal{C}$, and the map Δ must be target-injective since ϕ_Δ is injective on $\mathcal{O}bj(\mathcal{C})$.

Let s be an element of $\mathcal{D}iv(\Delta)(x,\cdot)$. Then $s \preccurlyeq \Delta(x)$ implies $\phi_\Delta^{-1}(s) \preccurlyeq \phi_\Delta^{-1}(\Delta(x))$, hence $\phi_\Delta^{-1}(s) \preccurlyeq \Delta(\phi_\Delta^{-1}(x))$ since, by (1.30), ϕ_Δ and ϕ_Δ^{-1} commute with Δ. So $\phi_\Delta^{-1}(s)$ lies in $\mathcal{D}iv(\Delta)$. By construction, ∂_Δ maps $\mathcal{D}iv(\Delta)$ to $\widetilde{\mathcal{D}iv}(\Delta)$ so, a fortiori, ϕ_Δ maps $\mathcal{D}iv(\Delta)$ to $\widetilde{\mathcal{D}iv}(\Delta)$. As $\phi_\Delta^{-1}(s)$ lies in $\mathcal{D}iv(\Delta)$, we deduce that s, which is $\phi_\Delta(\phi_\Delta^{-1}(s))$, belongs to $\widetilde{\mathcal{D}iv}(\Delta)$. Thus we have $\mathcal{D}iv(\Delta) \subseteq \widetilde{\mathcal{D}iv}(\Delta)$, whence $\widetilde{\mathcal{D}iv}(\Delta) = \mathcal{S}^\sharp = \mathcal{D}iv(\Delta)$ owing to (1.11). As $\mathcal{D}iv(\Delta)$ is closed under left-divisor by definition, so is $\mathcal{S}^\sharp$.

It follows that ∂_Δ is a map from $\mathcal{S}^\sharp$ to itself. This map is injective because its square ϕ_Δ is. We claim that ∂_Δ is also surjective, hence it is a permutation of $\mathcal{S}^\sharp$. Indeed, assume $s \in \mathcal{S}^\sharp(\cdot, y)$. As $\mathcal{S}^\sharp$ coincides with $\widetilde{\mathcal{D}iv}(\Delta)$, there exists x satisfying $\Delta(x) \succcurlyeq s$. So there exists r satisfying $rs = \Delta(x)$. Then r belongs to $\mathcal{D}iv(\Delta)$ and, by construction, we have $\partial r = s$. In addition, we note that y is the target of $\Delta(x)$, whence the equalities $\phi_\Delta(x) = y$, hence $x = \phi_\Delta^{-1}(y)$, and, similarly, $r = \partial^{-1}(s)$. As $\mathcal{S}^\sharp$ coincides with $\mathcal{D}iv(\Delta)$, we have shown that $\partial_\Delta^{-1}(s)\, s = \Delta(\phi_\Delta^{-1}(y))$ holds for every s in $\mathcal{S}^\sharp(\cdot, y)$, showing that $\mathcal{S}$ is bounded by Δ and $(\phi_\Delta^{-1}, \partial_\Delta^{-1})$ is a witness for $(\mathcal{S}, \Delta)$. So (ii) implies (i).

There remain two points. First, as the restriction of ϕ_Δ to $\mathcal{S}^\sharp$ is the square of ∂_Δ and the latter is a permutation of $\mathcal{S}^\sharp$, so is the former.

Finally, assume that $(\tilde{\phi}, \tilde{\partial})$ is a witness for $(\mathcal{S}, \Delta)$. By definition, $\phi_\Delta(\tilde{\phi}(y)) = y$ holds for every object y, which forces $\tilde{\phi}(y) = \phi_\Delta^{-1}(y)$. Similarly, $\tilde{\partial} s\, s = \Delta(x)$ holds for every s in $\mathcal{S}^\sharp(\cdot, \phi_\Delta(x))$. This implies $\partial_\Delta(\tilde{\partial}(s)) = s$, whence $\tilde{\partial}(s) = \partial_\Delta^{-1}(s)$. □

Proposition 2.17 explains why most examples of Garside families are closed under left-divisor: to obtain a Garside family $\mathcal{S}$ such that $\mathcal{S}^\sharp$ is not closed under left-divisor, we must either use a Garside family that is not bounded, or use a category with infinitely many objects (so as to escape Proposition 2.15), or use a non-right-cancellative category, as in the case of the left-absorbing monoid L_n.

When Proposition 2.17 is applicable, that is, when ϕ_Δ is an automorphism, ϕ_Δ and ϕ_Δ^{-1} automatically preserve all properties and relations that are definable from the product. So, in particular, ϕ_Δ and ϕ_Δ^{-1} induce permutations of $\mathcal{C}^\times$, and they preserve the relations $=^\times$ and ${}^\times\!\!=^\times$ (Lemma II.1.27), as well as the left- and right-divisibility relations (Lemma II.2.8) and the derived notions, lcms, gcds, etc. Due to its importance, we explicitly state the following consequence:

Proposition 2.18 (preserving normality). *If $\mathcal{S}$ is a Garside family in a cancellative category $\mathcal{C}$ and $\mathcal{S}$ is bounded by a target-injective map Δ, then ϕ_Δ preserves normality.*

Proof. Let $s_1|s_2$ be an $\mathcal{S}$-normal pair. By Lemma 2.7(iv), ϕ_Δ maps $\mathcal{S}^\sharp$ to itself, so $\phi_\Delta(s_1)$ and $\phi_\Delta(s_2)$ lie in $\mathcal{S}^\sharp$. Assume $s \preccurlyeq \phi_\Delta(s_1)\, \phi_\Delta(s_2)$ with $s \in \mathcal{S}$. By Lemma II.2.8, we deduce $\phi_\Delta^{-1}(s) \preccurlyeq s_1 s_2$, whence $\phi_\Delta^{-1}(s) \preccurlyeq s_1$ as ϕ_Δ^{-1} maps $\mathcal{S}^\sharp$ into itself and $s_1|s_2$, which is $\mathcal{S}$-normal by assumption, is also $\mathcal{S}^\sharp$-normal by Lemma III.1.10. Reapplying ϕ_Δ, we deduce $s \preccurlyeq \phi_\Delta(s_1)$, hence, by Corollary IV.1.31, $\phi_\Delta(s_1)|\phi_\Delta(s_2)$ is $\mathcal{S}$-normal. □

2.4 Garside maps

As in Subsection 1.2, we can change our point of view and describe Garside families that are bounded by a target-injective map in terms of the associated bounding functions. A nice point is that no special assumption is needed here: in the case of right-bounded Garside families, we had to restrict to families $\mathcal{S}$ such that $\mathcal{S}^\sharp$ is closed under left-divisor, whereas, here, by Proposition 2.17, all needed closure conditions are automatic.

Definition 2.19 (Garside map). If $\mathcal{C}$ is a left-cancellative category, a *Garside map* in $\mathcal{C}$ is a map Δ from $\mathcal{O}bj(\mathcal{C})$ to $\mathcal{C}$ that satisfies the following conditions:

(2.20) For every x in $\mathcal{O}bj(\mathcal{C})$, the source of $\Delta(x)$ is x.

(2.21) The map Δ is target-injective.

(2.22) The family $\mathcal{D}iv(\Delta)$ generates $\mathcal{C}$.

(2.23) The families $\widetilde{\mathcal{D}iv}(\Delta)$ and $\mathcal{D}iv(\Delta)$ coincide.

(2.24) For every g in $\mathcal{C}(x, \cdot)$, the elements g and $\Delta(x)$ admit a left-gcd.

Thus a Garside map is a right-Garside map Δ that satisfies two more conditions, namely that Δ is target-injective and that $\widetilde{\mathcal{D}iv}(\Delta)$ coincides with $\mathcal{D}iv(\Delta)$, instead of just being included in it. In the spirit of Proposition 1.20, we have now:

Proposition 2.25 (Garside map). *Assume that $\mathcal{C}$ is a left-cancellative category.*

(i) *If $\mathcal{S}$ is a Garside family that is bounded by a target-injective map Δ and $\mathcal{S}^\sharp$ is closed under left-divisor, then Δ is a Garside map and $\mathcal{S}^\sharp$ coincides with $\mathcal{D}iv(\Delta)$.*

(ii) *Conversely, if Δ is a Garside map in $\mathcal{C}$, then $\mathcal{D}iv(\Delta)$ is a Garside family that is bounded by Δ.*

Proof. (i) Assume that $\mathcal{S}$ is a Garside family that is bounded by a target-injective map Δ and $\mathcal{S}^\sharp$ is closed under left-divisor. By Proposition 1.20, Δ is a right-Garside map. As $\mathcal{S}$ is bounded by Δ, Lemma 2.7 implies $\widetilde{\mathcal{D}iv}(\Delta) = \mathcal{S}^\sharp$ and, as $\mathcal{S}^\sharp$ is closed under left-divisor, Lemma 1.7 implies $\mathcal{S}^\sharp = \mathcal{D}iv(\Delta)$. So $\widetilde{\mathcal{D}iv}(\Delta)=\mathcal{D}iv(\Delta)$ holds, and Δ is a Garside map.

(ii) Assume that Δ is a Garside map in $\mathcal{C}$. Put $\mathcal{S} = \mathcal{D}iv(\Delta)$. Then Δ is a right-Garside map, so, by Proposition 1.20, $\mathcal{S}$ is a Garside family that is right-bounded by Δ and $\mathcal{S}^\sharp$, which is $\mathcal{S}$, is closed under left-divisor. It remains to show that $\mathcal{S}$ is bounded by Δ, that is, for each object y, to find an object x satisfying (2.4). So, let y belong to $\mathcal{O}bj(\mathcal{C})$ and s belong to $\mathcal{S}(\cdot, y)$. By assumption, $\mathcal{S}$ coincides with $\widetilde{\mathcal{D}iv}(\Delta)$, so there exists x (a priori depending on s) such that s right-divides $\Delta(x)$, say $rs = \Delta(x)$. By construction, the target of $\Delta(x)$ is y, implying $\phi_\Delta(x) = y$, and the assumption that Δ is target-injective implies that x does not depend on s. Then r belongs to $\mathcal{S}(x, \cdot)$, and (2.4) is satisfied. So $\mathcal{S}$ is bounded by Δ. □

If the ambient category is cancellative, the result takes a slightly simpler form as, by Proposition 2.17, the condition about left-divisors can be forgotten.

Corollary 2.26 (Garside map). *Assume that $\mathcal{C}$ is a cancellative category.*

(i) *If $\mathcal{S}$ is a Garside family that is bounded by a target-injective map Δ, then Δ is a Garside map and $\mathcal{S}^\sharp$ coincides with $\mathcal{D}iv(\Delta)$.*

(ii) *Conversely, if Δ is a Garside map in $\mathcal{C}$, then $\mathcal{D}iv(\Delta)$ is a Garside family that is bounded by Δ.*

Because of their practical importance, we now restate the definition in the particular case of a monoid. As there is only one object, the target-injectivity condition vanishes.

Definition 2.27 (Garside element). If M is a left-cancellative monoid, a *Garside element* in M is an element Δ of M that satisfies the following conditions:

(2.28) The family $\mathrm{Div}(\Delta)$ generates M.

(2.29) The families $\mathrm{Div}(\Delta)$ and $\widetilde{\mathrm{Div}}(\Delta)$ coincide.

(2.30) For every g in M, the elements g and Δ admit a left-gcd.

Let us immediately note the compatibility between the current notion of a Garside element as defined above and the one commonly used in literature:

Proposition 2.31 (compatibility). *Assume that M is a cancellative monoid that admits left-gcds. Then an element of M is a Garside element in the sense of Definition I.2.2 (Garside monoid) if and only if it is a Garside element in the sense of Definition 2.27.*

Proof. Whenever any two elements of the ambient monoid M admit a left-gcd, (2.30) is automatically satisfied, so an element Δ of M is a Garside element in the sense of Definition 2.27 if and only if the left- and right-divisors of Δ coincide and they generate M, which is Definition I.2.2. □

From now on, cancellative categories equipped with a Garside map, that is, equivalently, with a Garside family bounded by a target-injective map, will be our preferred context. The general philosophy is that all properties previously established for Garside monoids should extend to such categories. For further reference, we summarize the main properties established so far.

Proposition 2.32 (Garside map). *If Δ is a Garside map in a cancellative category $\mathcal{C}$:*

(i) *The category $\mathcal{C}$ is an Ore category.*

(ii) *For s in $\mathcal{D}iv(\Delta)$ and g in $\mathcal{C}$, the pair $s|g$ is $\mathcal{D}iv(\Delta)$-greedy if and only if the elements ∂s and g are left-coprime.*

(iii) *The functor ϕ_Δ is an automorphism of $\mathcal{C}$ and it preserves $\mathcal{D}iv(\Delta)$-normality.*

(iv) *For $m \geqslant 1$, $\Delta^{[m]}$ is a Garside map and $\mathcal{D}iv(\Delta^{[m]}) = (\mathcal{D}iv(\Delta))^m$ holds.*

(v) *The second domino rule is valid for $\mathcal{D}iv(\Delta)$ and, for all $f|g$ in $\mathcal{C}^{[2]}$, we have*

$$\max(\sup_\Delta(f), \sup_\Delta(g)) \leqslant \sup_\Delta(fg) \leqslant \sup_\Delta(f) + \sup_\Delta(g). \tag{2.33}$$

Proof. Put $\mathcal{S} = \mathcal{D}iv(\Delta)$, so that $\mathcal{S}$ is a Garside family that is bounded by Δ. Point (i) follows from Proposition 1.46 and Corollary 2.14, whereas (ii) follows from Corollary 1.54. Then Proposition 2.17 says that ϕ_Δ is an automorphism, and Proposition 2.18 that it preserves $\mathcal{S}$-normality, so (iii) is valid.

Next, Proposition 1.58, which applies since ϕ_Δ preserves normality, guarantees that, for every m, the map $\Delta^{[m]}$ is a right-Garside map and we have $\mathcal{D}iv(\Delta^{[m]}) = (\mathcal{D}iv(\Delta))^m$ so, in particular, $(\mathcal{D}iv(\Delta))^m$ is closed under left-divisor. On the other hand, Proposition 2.13 implies that $(\mathcal{D}iv(\Delta))^m$ is bounded by $\Delta^{[m]}$. Finally, $\Delta^{[m]}$ is target-injective, as $\phi_{\Delta^{[m]}}$ is ϕ_Δ^m and, therefore, it is injective on $\mathcal{O}bj(\mathcal{C})$ whenever ϕ_Δ is. Then, by Proposition 2.25, $\Delta^{[m]}$ is a Garside map, and (iv) holds.

Finally, owing to (iii), Proposition 1.52 and Corollary 1.56 give (v). □

When we consider Garside maps in categories that are left-cancellative, but not necessarily cancellative, some results from Proposition 2.32 remain valid, namely (i), (ii), and the fact that ϕ_Δ is an automorphism, but the results of (iv) and (v) are valid only if we assume that ϕ_Δ preserves normality, which is not automatic. It is natural to wonder whether the latter assumption is weaker than cancellativity: the next result shows that the gap is small since adding the injectivity of ϕ_Δ on $(\mathcal{D}iv(\Delta))^2$ is then sufficient to imply right-cancellativity.

Proposition 2.34 (right-cancellativity III). *If $\mathcal{C}$ is a left-cancellative category and Δ is a Garside map of $\mathcal{C}$ such that ϕ_Δ preserves normality, then $\mathcal{C}$ is right-cancellative if and only if ϕ_Δ is injective on $(\mathcal{D}iv(\Delta))^2$.*

If $\mathcal{C}$ is cancellative, ϕ_Δ is an automorphism, hence injective on $\mathcal{C}$ and a fortiori on $(\mathcal{D}iv(\Delta))^2$, so the condition of Proposition 2.34 is trivially necessary, and the point is to prove that it is sufficient. To this end, on first observes that ϕ_Δ must induce a permutation of $\mathcal{C}^\times$ and one then applies Proposition 1.59. We skip the details.

2.5 Existence of lcms and gcds

As every (left-cancellative) category is a Garside family in itself, the existence of a Garside family cannot imply any structural consequence. Things are different with the existence of a (right)-bounded Garside family: for instance, we have seen that it implies the existence of common right-multiples and, in the bounded case, of common left-multiples. We now address the existence of lcms and gcds. Example 1.49 shows that the existence of a right-bounded Garside family does not imply the existence of right- or left-lcms. By contrast, we shall see now that the existence of a bounded Garside family does imply the existence of left-lcms and left-gcds, and it implies that of right-lcms right-gcds whenever the ambient category is left-Noetherian. This explains why categories with a bounded Garside family but no right-lcm must be somehow complicated.

Proposition 2.35 (lcms and gcds). *If a cancellative category $\mathcal{C}$ admits a Garside map:*
(i) *The category $\mathcal{C}$ admits left-gcds and left-lcms.*
(ii) *If $\mathcal{C}$ is left-Noetherian, then $\mathcal{C}$ admits right-lcms and right-gcds.*
(iii) *If $\mathcal{C}$ is left-Noetherian and has no nontrivial invertible element, $(\mathcal{C}, \preccurlyeq)$ and $(\mathcal{C}, \widetilde{\succcurlyeq})$ are lattices.*

We split the proof into several steps, starting with left-gcds. We saw in Lemma 1.14 that, if Δ is a right-Garside map, then, for every g with source x, the elements g and $\Delta(x)$ must admit a left-gcd. The argument can be extended to more elements.

Lemma 2.36. (i) *In a left-cancellative category that admits a right-Garside map Δ, any two elements sharing the same source and such that at least one of them lies in $\widetilde{\mathcal{D}iv}(\Delta)$ admit a left-gcd.*
(ii) *A cancellative category that admits a Garside map admits left-gcds.*

Proof. (i) Assume that f, g share the same source and f right-divides $\Delta(x)$, say $sf = \Delta(x)$. By definition, $\Delta(x)$ and sg admits a left-gcd, say h. By construction, s left-divides $\Delta(x)$ and sg, so it left-divides h, say $h = sh'$. So sh' is a left-gcd of sf and sg, which implies that h' is a left-gcd of f and g since the ambient category is left-cancellative (see Exercise 8(i) in Chapter II).

(ii) If $\mathcal{C}$ is a cancellative category and Δ is a Garside map in $\mathcal{C}$, then, by Proposition 2.32(iv), so is $\Delta^{[m]}$ for $m \geqslant 1$. By Corollary 2.14, every element of $\mathcal{C}$ right-divides some element $\Delta^{[m]}(\cdot)$ with $m \geqslant 1$, and we apply (i). □

Lemma 2.37. *A cancellative category that admits a Garside map admits left-lcms.*

Proof. Assume that $\mathcal{C}$ is a cancellative category and Δ is a Garside map in $\mathcal{C}$. Assume first that s and t are two elements of $\mathcal{D}iv(\Delta)$ sharing the same target. Let y be the source of t, and x be $\phi_\Delta^{-1}(y)$. Then we have $\Delta(x)t = \phi_\Delta^{-1}(t)\Delta(\cdot)$, so s right-divides $\Delta(x)t$, say $\Delta(x)t = gs$. Let t_1 be a $\mathcal{D}iv(\Delta)$-head of g, let t' be defined by $g = t_1t'$, and let $s' = \partial t_1$. By construction, $s't = t's$ holds. Moreover, by Corollary 1.54, the assumption that $t_1|t'$ is $\mathcal{D}iv(\Delta)$-greedy implies that s' and t' are left-coprime. By Lemma 2.36, the category $\mathcal{C}$ admits left-gcds and, therefore, by Proposition III.2.5 (left-disjoint *vs.* left-coprime), the fact that s' and t' are left-coprime implies that they are left-disjoint; by Lemma III.2.8, this implies that $s't$ is a left-lcm of s and t. Moreover, s' lies in $\mathcal{D}iv(\Delta)$ by construction, and so does t', since the Δ-length of g is at most that of its right-multiple $\Delta(x)t$, which is at most 2. Hence, we proved that any two elements s, t of $\mathcal{D}iv(\Delta)$ sharing the same target admit a left-lcm of the form $t's = s't$ where s' and t' lie in $\mathcal{D}iv(\Delta)$.

It then follows from (the left counterpart of) Proposition II.2.12 (iterated lcm) and an easy induction on $m \geqslant 1$ that any two elements s, t of $\mathcal{D}iv(\Delta)^{[m]}$ sharing the same target admit a left-lcm of the form $t's = s't$, where s' and t' lie in $\mathcal{D}iv(\Delta)^{[m]}$. As, by assumption, $\mathcal{D}iv(\Delta)$ generates $\mathcal{C}$, we conclude that any two elements of $\mathcal{C}$ with the same target admit a left-lcm. □

Once the existence of left-gcds is granted, one easily deduces that of right-lcms provided some Noetherianity condition is satisfied and, from there, that of right-gcds.

Lemma 2.38. (i) *A left-cancellative category that is left-Noetherian and admits left-gcds admits conditional right-lcms.*

(ii) *In a cancellative category that admits conditional right-lcms, any two elements of $\mathcal{C}$ that admit a common left-multiple admit a right-gcd.*

We skip the proof, which has nothing to do with Garside families and which just exploits the definitions: in (i), the right-lcm arises as a $\prec$-minimal left-gcd of common right-multiples, whereas, in (ii), the right-gcd arises as the right-complement of the right-lcm of the left-complements in a common left-multiple.

Proof of Proposition 2.35. (i) The result directly follows from Lemmas 2.36 and 2.37.

(ii) Assume that $\mathcal{C}$ is left-Noetherian. By Lemma 2.38(i), the existence of left-gcds implies that of conditional right-lcms. Now, by Proposition 1.46, any two elements of $\mathcal{C}$ with the same source admit a common right-multiple. Hence, any two elements of $\mathcal{C}$ with the same source admit a right-lcm. On the other hand, by Corollary 2.14, any two elements of $\mathcal{C}$ with the same target admit a common left-multiple. By Lemma 2.38(ii), they must admit a right-gcd.

(iii) Assume in addition that $\mathcal{C}$ has no nontrivial invertible element. Then the left- and right-divisibility preorderings are partial orderings, and the fact that $(\mathcal{C}, \preccurlyeq)$ and $(\mathcal{C}, \succcurlyeq)$ are lattices is a reformulation of the fact that left- and right-lcms and gcds (which are unique in this case) are suprema and infima with respect to these orderings. □

In the context of left-Ore categories that admit left-lcms, we introduced the notions of a strong and a perfect Garside family (Definitions III.2.29 and III.3.6). A direct application of Proposition 2.35 is:

Corollary 2.39 (perfect). *If Δ is a Garside map in a cancellative category $\mathcal{C}$, then the Garside family $\mathcal{D}iv(\Delta)$ is perfect.*

Proof. Let s, t be two elements of $\mathcal{D}iv(\Delta)$ sharing the same target y. By Proposition 2.35, s and t admit a left-lcm, say $h = ft = gs$. Let $x = \phi_\Delta^{-1}(y)$. Then we have $\widetilde{\partial} s\, s = \widetilde{\partial} t\, t = \Delta(x)$, and $\Delta(x)$ is a common left-multiple of s and t, hence a left-multiple of their left-lcm h. In other words, h lies in $\widetilde{\mathcal{D}iv}(\Delta)$, which is also $\mathcal{D}iv(\Delta)$. Moreover, f and g left-divide h, hence they lie in $\mathcal{D}iv(\Delta)$ as well. Then f and g are left-disjoint by Lemma III.2.8, and (f, g, h) witness that $\mathcal{D}iv(\Delta)$ is perfect. □

Returning to Noetherianity, we can look for conditions involving a Garside map Δ guaranteeing that the ambient category is, say, right-Noetherian. By Proposition IV.2.18 (solid Garside in right-Noetherian), a necessary and sufficient condition is that the family $\mathcal{D}iv(\Delta)$ be locally right-Noetherian, leading to the following criterion:

Proposition 2.40 (right-Noetherian). *If Δ is a right-Garside map in a left-cancellative category $\mathcal{C}$ and, for every x, we have $\mathrm{ht}(\Delta(x)) < \infty$ (resp. $\leqslant K$ for some constant K), then $\mathcal{C}$ is right-Noetherian (resp. strongly Noetherian).*

Proof. Assume first that $\mathcal{C}$ is not right-Noetherian. By Proposition IV.2.18 (solid Garside in right-Noetherian), the Garside family $\mathcal{D}iv(\Delta)$ cannot be locally right-Noetherian. So there exists an infinite descending sequence $s_0, s_1,...$ with respect to $\widetilde{\prec}$ inside $\mathcal{D}iv(\Delta)$. Let t_i satisfy $t_i s_{i+1} = s_i$. Then we have $s_0 = t_0 s_1 = t_0 t_1 s_2 = \cdots$ and, calling x the source of s_0, we find $\mathrm{ht}(t_0) < \mathrm{ht}(t_0) + \mathrm{ht}(t_1) < \cdots < \mathrm{ht}(s_0) \leqslant \mathrm{ht}(\Delta(x))$. Hence $\Delta(x)$ cannot have a finite height.

On the other hand, if $\mathrm{ht}(\Delta(x)) \leqslant K$ holds for every object x, then a fortiori $\mathrm{ht}(s) \leqslant K$ holds for every s in the Garside family $\mathcal{D}iv(\Delta)$ and Proposition IV.2.43 (finite height) implies that every element of $\mathcal{C}$ has a finite height. □

In the particular case of a monoid, we deduce

Corollary 2.41 (quasi-Garside monoid). *If M is a cancellative monoid, then the following conditions are equivalent:*

(i) *The monoid M has no nontrivial invertible element and Δ is a Garside element in M that has a finite height.*

(ii) *The pair (M, Δ) is a quasi-Garside monoid.*

Proof. That (ii) implies (i) is clear from the definition of a quasi-Garside monoid.

Assume (i). Then Proposition 2.40 implies that $\mathcal{C}$ is strongly Noetherian, that is, there exists $\lambda : M \to \mathbb{N}$ satisfying $\lambda(fg) \geqslant \lambda(f) + \lambda(g)$ and $g \neq 1 \Rightarrow \lambda(g) \neq 0$. Next, by Proposition 2.35, M admits left- and right-lcms and gcds. Then Proposition 2.31 says that Δ is a Garside element in M in the sense of Definition I.2.2 and, therefore, (M, Δ) is a quasi-Garside monoid. So (i) implies (ii). □

Thus, when one restricts to monoids that contain no nontrivial invertible element and are strongly Noetherian, the context of bounded Garside families essentially coincides with that of quasi-Garside monoids.

3 Delta-normal decompositions

By definition, every Garside family gives rise to distinguished decompositions for the elements of the ambient category and, possibly, of its enveloping groupoid. We show now how to adapt the results in the special case of (right)-bounded Garside categories, or, equivalently, categories equipped with (right)-Garside maps. The principle is to use the elements $\Delta(x)$ as much as possible.

In Subsection 3.1, we address the positive case, that is, the elements of the ambient category, and construct new distinguished decompositions, called delta-normal. Next, in Subsection 3.2, we similarly address the signed case, that is, the elements of the enveloping groupoid. In Subsection 3.3, we compare the delta-normal decompositions with the symmetric normal decompositions of Chapter III. Finally, we discuss in Subsection 3.4 the counterpart of delta-normal decompositions in which, instead of considering maximal left-divisors, one symmetrically consider maximal right-divisors. The results are similar but, interestingly, an additional assumption about the considered Garside map is needed to guarantee the existence of such decompositions.

3.1 The positive case

Our aim is to construct new distinguished decompositions giving a specific role to the bounding map, when it exists. We start from the following simple observation: if a Garside family $\mathcal{S}$ is right-bounded by a map Δ, then $\mathcal{S}$-normal sequences necessarily begin with terms that are $=^{\times}$-equivalent to elements $\Delta(x)$. Thus, the elements $\Delta(x)$, which are $\preccurlyeq$-maximal elements in $\mathcal{S}^{\sharp}$, necessarily occur at the beginning of normal sequences, reminiscent of invertible elements, which are $\preccurlyeq$-minimal elements in $\mathcal{S}^{\sharp}$ and necessarily occur at the end of $\mathcal{S}$-normal sequences.

Definition 3.1 (delta-like). If $\mathcal{S}$ is a Garside family that is right-bounded by a map Δ, an element g of $\mathcal{C}(x, \cdot)$ is called *delta-like* if it is $=^{\times}$-equivalent to $\Delta(x)$.

It can be easily checked that the notion of a delta-like element depends on the Garside family $\mathcal{S}$ only, and not on the choice of the map Δ, see Exercise 58.

Lemma 3.2. *Assume that $\mathcal{C}$ is a left-cancellative category and $\mathcal{S}$ is a Garside family of $\mathcal{C}$ that is right-bounded by Δ.*

(i) *The family of delta-like elements is closed under ${}^{\times}\!\!=^{\times}$-equivalence.*

(ii) *If $s_1|s_2$ is $\mathcal{S}$-normal and s_2 is delta-like, then so is s_1.*

(iii) *If s is delta-like, then $s|g$ is $\mathcal{S}$-greedy for every g.*

(iv) *In every $\mathcal{S}$-normal decomposition of an element g of $\mathcal{C}$, the delta-like entries (if any) occur first, and their number only depends on g.*

Proof. (i) Assume $s\epsilon = \epsilon'\Delta(y)$ with $\epsilon, \epsilon' \in \mathcal{C}^{\times}$. Let x be the source of s and ϵ'. By (1.29), we have $\epsilon'\,\Delta(y) = \Delta(x)\,\phi_\Delta(\epsilon')$. As ϕ_Δ is a functor, $\phi_\Delta(\epsilon')$ is invertible with inverse $\phi_\Delta(\epsilon'^{-1})$, and we deduce $s = \Delta(x)\,\phi_\Delta(\epsilon')\,\epsilon^{-1}$, whence $s =^{\times} \Delta(x)$.

(ii) Assume $s_2 = \Delta(y)\epsilon$ with $\epsilon \in \mathcal{C}^{\times}$. Let x be the source of s_1. Using (1.29), we find $s_1s_2 = s_1\Delta(y)\epsilon = \Delta(x)\phi_\Delta(s_1)\epsilon$, whence $\Delta(x) \preccurlyeq s_1s_2$. As $\Delta(x)$ lies in $\mathcal{S}$ and $s_1|s_2$ is $\mathcal{S}$-normal, we deduce $\Delta(x) \preccurlyeq s_1$. As $s_1 \preccurlyeq \Delta(x)$ follows from the assumption that s_1 belongs to $\mathcal{S}$ and $\mathcal{S}$ is right-bounded by Δ, we deduce $s_1 =^{\times} \Delta(x)$.

(iii) By Lemma III.1.9, we may assume $s = \Delta(x)$ for some object x. Assume that sg is defined, t belongs to $\mathcal{S}$, and $t \preccurlyeq sg$ holds. By definition, we have $t \preccurlyeq \Delta(x)$, that is, $t \preccurlyeq s$. By Proposition IV.1.20 (recognizing greedy), this implies that $s|g$ is $\mathcal{S}$-greedy.

(iv) That the delta-like entries occur first follows from (ii) directly. If $s_1|\cdots|s_p$ and $t_1|\cdots|t_q$ are two $\mathcal{S}$-normal decompositions of g, Proposition III.1.25 (normal unique) implies $s_i \;{}^{\times}\!\!=^{\times}\; t_i$ for each i (at the expense of extending the shorter sequence with identity elements if needed). Then, by (i), s_i is delta-like if and only if t_i is. □

Definition 3.3 (delta-normal). If $\mathcal{S}$ is a Garside family that is right-bounded by a map Δ in a left-cancellative category $\mathcal{C}$, an $(\mathcal{S}, \Delta)$*-normal path of infimum* m ($m \geqslant 0$) is a pair $(\Delta^{[m]}(x), s_1|\cdots|s_\ell)$ such that $\Delta(x)|\Delta(\phi_\Delta(x))|\cdots|\Delta(\phi_\Delta^{m-1}(x))|s_1|\cdots|s_\ell$ is $\mathcal{S}$-normal and s_1 is not delta-like (for $\ell \geqslant 1$). We write $\Delta^{[m]}(x)\|s_1|\cdots|s_\ell$, or $\Delta^{[m]}\|s_1|\cdots|s_\ell$, for such a pair. If $\mathcal{S}^{\sharp} = \mathcal{D}iv(\Delta)$ holds, we say Δ*-normal* for $(\mathcal{S}, \Delta)$-normal.

Thus, an $(\mathcal{S}, \Delta)$-normal path of infimum m begins with m entries of the form $\Delta(y)$, hence Δ repeated m times in the case a monoid. Of course, we say that $\Delta^{[m]}(x)\|s_1|\cdots|s_\ell$ is an $(\mathcal{S}, \Delta)$-*normal decomposition* for g if $g = \Delta^{[m]}(x)\, s_1 \cdots s_\ell$ holds.

Proposition 3.4 (delta-normal). *If $\mathcal{S}$ is a Garside family of $\mathcal{C}$ that is right-bounded by a map Δ in a left-cancellative category $\mathcal{C}$, every element of $\mathcal{C}$ admits an $(\mathcal{S}, \Delta)$-normal decomposition; the infimum is uniquely determined and the $\mathcal{S}$-entries are unique up to $\mathcal{C}^\times$-deformation.*

Proof. We show, using induction on $m \geqslant 0$, that every element that admits an $\mathcal{S}$-decomposition with m delta-like entries admits an $(\mathcal{S}, \Delta)$-normal decomposition. For $m = 0$, every $\mathcal{S}$-normal decomposition is $(\mathcal{S}, \Delta)$-normal and the result is trivial. Assume $m \geqslant 1$, and let g be an element of $\mathcal{C}(x, \cdot)$ that admits an $\mathcal{S}$-decomposition with m delta-like entries. Let $s_1|\cdots|s_p$ be an arbitrary $\mathcal{S}$-normal decomposition of g. By assumption, s_1 is delta-like, so we have $s_1 = \Delta(x)\, \epsilon$ for some invertible element ϵ. Let $g' = \epsilon\, s_2 \cdots s_p$. As $\mathcal{C}^\times\mathcal{S}$ is included in $\mathcal{S}^\sharp$, there exist invertible elements $\epsilon_1 = \epsilon, \epsilon_2, ..., \epsilon_p$ and $g'_2, ..., g'_p$ in $\mathcal{S} \cup \mathcal{C}^\times$ satisfying $\epsilon_{i-1}\, s_i = g'_i\, \epsilon_i$. By construction, $g'_i \;{}^\times\!\!=^\times\; s_i$ holds for each i, so, by Lemma III.1.9, $g'_2|\cdots|g'_p|\epsilon_p$ is an $\mathcal{S}$-normal decomposition of g' and, by Lemma 3.2, it has $m - 1$ delta-like entries. By the induction hypothesis, g' has an $(\mathcal{S}, \Delta)$-normal decomposition. Appending $\Delta(x)$ at the beginning of the latter gives an $(\mathcal{S}, \Delta)$-normal decomposition of g.

For uniqueness, Lemma 3.2(iv) implies that the number of delta-like entries in an $\mathcal{S}$-normal decomposition of an element g does not depend on the choice of the bounding map, so the infimum is uniquely determined. The uniqueness of $\mathcal{S}$-entries then follows from Proposition III.1.25 (normal unique). □

Definition 3.5 (infimum). If $\mathcal{S}$ is a Garside family of $\mathcal{C}$ that is right-bounded by a map Δ in a left-cancellative category $\mathcal{C}$, the $\mathcal{S}$-*infimum* of an element g of $\mathcal{C}$, written $\inf_{\mathcal{S}}(g)$, is the exponent of Δ in an $(\mathcal{S}, \Delta)$-normal decomposition of g. For $\mathcal{S}^\sharp = \mathcal{D}iv(\Delta)$, we write $\inf_\Delta$ for $\inf_{\mathcal{D}iv(\Delta)}$.

We recall from Definition III.1.29 that $\|g\|_{\mathcal{S}}$ is the number of non-invertible entries in an $\mathcal{S}$-normal decomposition of g. Then, by definition, the inequality

$$\inf_{\mathcal{S}}(g) \leqslant \|g\|_{\mathcal{S}} \tag{3.6}$$

always holds: $\|g\|_{\mathcal{S}} - \inf_{\mathcal{S}}(g)$ is the number of "really nontrivial" entries in a $\mathcal{S}$-normal decomposition of g, namely those that are neither invertible nor delta-like. We recall from Definition 1.55 that, when $\mathcal{S}^\sharp = \mathcal{D}iv(\Delta)$ holds, we write $\sup_\Delta(g)$ for $\|g\|_{\mathcal{S}}$. In this case, (3.6) takes the even more natural form $\inf_\Delta(g) \leqslant \sup_\Delta(g)$.

Computation rules for $(\mathcal{S}, \Delta)$-normal decompositions are simple. We saw in Proposition III.1.49 (left-multiplication) how to compute, for s in $\mathcal{S}^\sharp$, an $\mathcal{S}$-normal decomposition for sg starting from an $\mathcal{S}$-normal decomposition of g. We now state a similar result for $(\mathcal{S}, \Delta)$-normal decompositions.

Convention 3.7 (omitting source). To make reading easier, in the sequel (in this chapter, as well as in subsequent chapters, specially Chapters VIII and XIV), we shall often write $\Delta^{[m]}(\cdot)$, or even $\Delta^{[m]}$, for $\Delta^{[m]}(x)$, when there is no need to specify the source x explicitly. Also we often write ϕ for ϕ_Δ.

Algorithm 3.8 (left-multiplication, positive case). (See Figure 2)

Context: A Garside family $\mathcal{S}$ that is right-bounded by a map Δ in a left-cancellative category $\mathcal{C}$, a $\square$-witness F for $\mathcal{S}^\sharp$, an $=^\times$-test E in $\mathcal{C}$
Input: An $(\mathcal{S},\Delta)$-normal decomposition $\Delta^{[m]}\|s_1|\cdots|s_\ell$ of an element g of $\mathcal{C}$ and an element s of $\mathcal{S}^\sharp$ such that sg is defined
Output: An $(\mathcal{S},\Delta)$-normal decomposition of sg

put $r_0 := \phi^m(s)$
for i increasing from 1 to ℓ **do**
  put $(s'_i, r_i) := F(r_{i-1}, s_i)$
if $E(\Delta(\cdot), s'_1)$ is defined **then** (case when s'_1 is Δ-like)
  put $s'_2 := E(\Delta(\cdot), s'_1)s'_2$
  return $\Delta^{[m+1]}\|s'_2|\cdots|s'_\ell|r_\ell$
else (case when s'_1 is not Δ-like)
  return $\Delta^{[m]}\|s'_1|\cdots|s'_\ell|r_\ell$

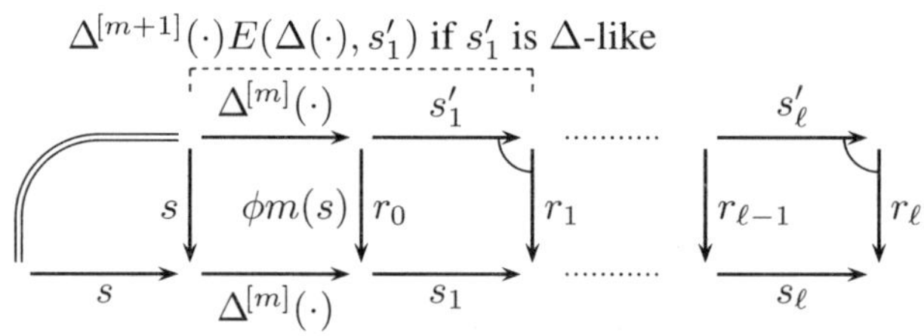

Figure 2. Computing an $(\mathcal{S},\Delta)$-normal decomposition of sg from an $(\mathcal{S},\Delta)$-normal decomposition of g: first use ϕ^m_Δ to cross the m initial $\Delta(\cdot)$ entries, and then apply the standard procedure, making sure that s'_1 is chosen to be $\Delta(\cdot)$ if it is delta-like; if this happens, we merge it in the Δ block.

Proposition 3.9 (left-multiplication, positive case). *Assume that $\mathcal{C}$ is a left-cancellative category, $\mathcal{S}$ is a Garside family in $\mathcal{C}$ that is right-bounded by a map Δ, and F is a $\square$-witness for $\mathcal{S}^\sharp$, and E is an $=^\times$-test E in $\mathcal{C}$. Then Algorithm* 3.8 *running on an $(\mathcal{S},\Delta)$-normal decomposition of g and s returns an $(\mathcal{S},\Delta)$-normal decomposition of sg.*

Proof. Assume that $\Delta^{[m]}\|s_1|\cdots|s_\ell$ is an $(\mathcal{S},\Delta)$-normal decomposition of g and s belongs to $\mathcal{S}^\sharp$. By Lemma 1.41, we have $s\Delta^{[m]} = \Delta^{[m]}\phi^m(s)$, whence $sg = \Delta^{[m]}\,\phi^m(s)\,s_1\cdots s_\ell$. Starting from $r_0 = \phi^m(s)$ and filling the diagram of Figure 2 using the method of Proposition III.1.49 (left-multiplication) gives an $\mathcal{S}$-normal decomposition $s'_1|\cdots|s'_\ell|r_\ell$ for $\phi^m(s)\,s_1\cdots s_\ell$, and we deduce $sg = \Delta^{[m]}\,s'_1\cdots s'_\ell r_\ell$. It only remains to discuss whether $\Delta^{[m]}\|s'_1|\cdots|s'_\ell$ is $(\mathcal{S},\Delta)$-normal. If s'_1 is not delta-like, the sequence is $(\mathcal{S},\Delta)$-normal. Otherwise, s'_1 must be a Δ element, and we can incorporate it to $\Delta^{[m]}$ to obtain $\Delta^{[m+1]}$.

We claim that, in this case, s'_2 cannot be delta-like. Indeed, assume $\Delta \preccurlyeq s'_2$. This implies $\Delta^{[2]} \preccurlyeq s'_1 s'_2$, hence a fortiori $\Delta^{[2]} \preccurlyeq r_0 s_1 \cdots s_\ell$. Using (1.29), we find $\Delta^{[2]} = r_0\,\partial r_0\,\Delta = r_0\,\Delta\,\phi(\partial r_0)$, whence $r_0\,\Delta \preccurlyeq r_0 s_1\cdots s_\ell$, and $\Delta \preccurlyeq s_1\cdots s_\ell$. As $s_1|\cdots|s_\ell$ is $\mathcal{S}$-normal, s_1 must be delta-like, contradicting the assumption that $\Delta^{[m]}\|s_1|\cdots|s_\ell$ is $(\mathcal{S},\Delta)$-normal. So s'_2 cannot be delta-like, and $\Delta^{[m+1]}\|s'_2|\cdots|s'_\ell|r_\ell$ is $(\mathcal{S},\Delta)$-normal. $\square$

Corollary 3.10 (infimum). *If $\mathcal{S}$ is a right-bounded Garside family in a left-cancellative category $\mathcal{C}$, then, for all $f|g$ in $\mathcal{C}^{[2]}$, we have*

$$\inf_{\mathcal{S}}(g) \leqslant \inf_{\mathcal{S}}(fg) \leqslant \sup_{\mathcal{S}}(f) + \inf_{\mathcal{S}}(g). \tag{3.11}$$

Proof. For s in $\mathcal{S}^\sharp$, Proposition 3.9 gives the inequalities

$$\inf_{\mathcal{S}}(g) \leqslant \inf_{\mathcal{S}}(sg) \leqslant 1 + \inf_{\mathcal{S}}(g).$$

If $\sup_{\mathcal{S}}(f)$ is d, then f admits a decomposition as the product of d elements of $\mathcal{S}^\sharp$, and d applications of the above inequalities give (3.11). □

3.2 The general case

We now extend the $(\mathcal{S}, \Delta)$-normal decompositions of Proposition 3.4 to the enveloping groupoid of the involved category.

In this subsection, we always consider cancellative categories. Owing to Corollary 2.26, Garside families that are bounded by a target-injective map are equivalent to Garside maps. From now on, we always adopt the latter framework (which provides slightly simpler statements). According to Definition 3.3, we say Δ-*normal* for $(\mathcal{D}iv(\Delta), \Delta)$-normal.

We saw in Proposition 2.32 that, if Δ is a Garside map in a cancellative category $\mathcal{C}$, then $\mathcal{C}$ is necessarily an Ore category. Hence, by Ore's theorem (Proposition II.3.11), $\mathcal{C}$ embeds in its enveloping groupoid $\mathcal{E}nv(\mathcal{C})$, which is both a groupoid of left and of right fractions for $\mathcal{C}$. In order to extend the definition of Δ-normal decompositions from $\mathcal{C}$ to $\mathcal{E}nv(\mathcal{C})$, the first step is to analyze the relation $\bowtie$ involved in the embedding of $\mathcal{C}$ into $\mathcal{E}nv(\mathcal{C})$.

Lemma 3.12. *If Δ is a Garside map in a cancellative category $\mathcal{C}$, then, for all f, g in $\mathcal{C}$ sharing the same source, there exists a unique triple (x, m, g') with x an object of $\mathcal{C}$, m a nonnegative integer, and g' an element of $\mathcal{C}(x, \cdot)$ satisfying*

$$(\Delta^{[m]}(x), g') \bowtie (f, g) \quad \textit{and} \quad \Delta(x) \not\preccurlyeq g'. \tag{3.13}$$

Proof. Assume that f, g are elements of $\mathcal{C}$ sharing the same source. Let

$$A = \{m \geqslant 0 \mid \exists x{\in}\mathcal{O}bj(\mathcal{C})\ \exists g' \in \mathcal{C}(x, \cdot)\ ((\Delta^{[m]}(x), g') \bowtie (f, g))\}.$$

First, we claim that A is nonempty. Indeed, let y be the target of f. By Corollary 2.14, there exist x and m such that f right-divides $\Delta^{[m]}(x)$, say $\Delta^{[m]}(x) = hf$ with h in $\mathcal{C}(x, \cdot)$. Then we have $\phi_\Delta^m(x) = y$, whence $x = \phi_\Delta^{-m}(y)$. Then the pair $(1_x, h)$ witnesses for $(\Delta^{[m]}(x), hg) \bowtie (f, g)$, and, therefore, m belongs to A.

Let m be the least element of A, and let x and g' satisfy $(\Delta^{[m]}(x), g') \bowtie (f, g)$. We claim that $\Delta(x)$ does not left-divide g'. Indeed, assume $g' = \Delta(x)\, g''$. Put $x' = \phi_\Delta(x)$. Then g'' belongs to $\mathcal{C}(x', \cdot)$, and the pair $(\Delta(x), 1_{x'})$ witnesses for $(\Delta^{[m]}(x), g') \bowtie (\Delta^{[m-1]}(x'), g'')$. By transitivity, $(\Delta^{[m-1]}(x'), g'') \bowtie (f, g)$ holds as well, which shows that $m - 1$ belongs to A and contradicts the choice of m. Thus, at this point, we obtained a triple (x, m, g') satisfying (3.13).

It remains to prove uniqueness. Assume that (x', m', g'') satisfies (3.13). Then m' belongs to A, and we must have $m' \geqslant m$. By construction, we have $\Delta^{[m]}(x) = \widetilde{\Delta}_{[m]}(y)$ and $\Delta^{[m']}(x') = \widetilde{\Delta}_{[m']}(y)$, when y is the target of f, hence of $\Delta^{[m]}(x)$ and $\Delta^{[m']}(x')$. The bijectivity of ϕ_Δ on $\mathcal{O}bj(\mathcal{C})$ implies $x' = \phi_\Delta^{-(m'-m)}(x)$, whence

$$\Delta^{[m']}(x') = \Delta^{[m'-m]}(x')\,\Delta^{[m]}(x). \tag{3.14}$$

Then we have $(\Delta^{[m']}(x'), g'') \bowtie (\Delta^{[m]}(x), g')$, as both pairs are $\bowtie$-equivalent to (f, g). So (3.14) implies $g'' = \Delta^{[m'-m]}(x')\, g'$, and the assumption that (x', m', g'') satisfies (3.13), hence that $\Delta(x')$ does not left-divide g'', implies $m' - m = 0$. As we have $x' = \phi_\Delta^{-(m'-m)}(x)$ and $g'' = \Delta^{[m'-m]}(x')\, g'$, we deduce $x' = x$ and $g'' = g'$. □

We now introduce convenient negative powers of a Garside map.

Notation 3.15 ($\widetilde{\Delta}$, $\widetilde{\Delta}_{[m]}$). If Δ is a Garside map in a cancellative category $\mathcal{C}$ and y is an object of $\mathcal{C}$, we put $\widetilde{\Delta}(y) = \Delta(\phi_\Delta^{-1}(y))$ and, more generally, for $m \geqslant 0$,

$$\widetilde{\Delta}_{[m]}(y) = \Delta^{[m]}(\phi_\Delta^{-m}(y)). \tag{3.16}$$

Note that, by definition, the *target* of $\widetilde{\Delta}_{[m]}(y)$ is y (whereas the source of $\Delta^{[m]}(x)$ is x). Then we extend Δ-normal paths to negative values of the exponent.

Definition 3.17 (delta-normal). If Δ is a Garside map in a cancellative category $\mathcal{C}$, we say that an element g of $\mathcal{E}nv(\mathcal{C})(x, \cdot)$ *admits the Δ-normal decomposition* $\Delta^{(m)} \| s_1 | \cdots | s_\ell$ ($m \in \mathbb{Z}$) if $s_1 | \cdots | s_\ell$ is Δ-normal, s_1 is not delta-like, and we have

$$g = \begin{cases} \iota(\Delta^{[m]}(x))\, \iota(s_1) \cdots \iota(s_\ell), & \text{in the case } m \geqslant 0, \\ \iota(\widetilde{\Delta}_{[-m]}(x))^{-1}\, \iota(s_1) \cdots \iota(s_\ell), & \text{in the case } m < 0. \end{cases}$$

We recall that ι denotes the embedding of $\mathcal{C}$ into $\mathcal{E}nv(\mathcal{C})$, which is often dropped. If the first entry m is nonnegative, Definition 3.17 coincides with Definition 3.3, and it corresponds to a Δ-normal decomposition that begins with m factors of the form $\Delta(\cdot)$. If m is negative, $\Delta^{(m)}(x)$ means $(\widetilde{\Delta}_{[-m]}(x))^{-1}$, so it does not belong to the category $\mathcal{C}$, and the decomposition corresponds to a left fraction whose denominator is a product of $|m|$ factors of the form $\Delta(\cdot)$. One should pay attention to the possible ambiguity resulting from the fact the the exponent -1 is used both to denote the inverse (reciprocal) of a bijective map, and the inverse of an invertible element in a category.

Lemma 3.12 gives the expected existence and uniqueness of Δ-normal decompositions:

Proposition 3.18 (delta-normal). *If Δ is a Garside map in a cancellative category $\mathcal{C}$, every element of $\mathcal{E}nv(\mathcal{C})$ admits a Δ-normal decomposition. The exponent of Δ is uniquely determined and the $\mathcal{D}iv(\Delta)$-entries are unique up to $\mathcal{C}^\times$-deformation.*

Proof. Let h belong to $\mathcal{E}nv(\mathcal{C})$. If h belongs to $\iota(\mathcal{C})$, then the existence follows from Proposition 3.4. Otherwise, h can be written as $\iota(f)^{-1}\iota(g)$ with (f, g) in $\mathcal{C} \times \mathcal{C}$ and $f \not\preccurlyeq g$. Then Lemma 3.12 shows that (f, g) is $\bowtie$-equivalent to a pair $(\Delta^{[m]}(x), g')$ with $m \geqslant 1$ and g' not left-divisible by $\Delta(x)$, for g' in $\mathcal{C}(x, \cdot)$. Then, by construction, another expression of h is $\iota(\Delta^{[m]}(x))^{-1}\iota(g')$. Let $s_1|\cdots|s_\ell$ be a Δ-normal decomposition of g'. The assumption that $\Delta(x)$ does not divide g' is equivalent to s_1 not being delta-like. Thus, in this case, $\Delta^{(-m)}(x)\|s_1|\cdots|s_\ell$ is a Δ-normal decomposition of h.

As for uniqueness, the above two cases are disjoint. In the first case, uniqueness follows from Proposition 3.4. In the second case, it follows from Lemma 3.12, since every $(\mathcal{S}, \Delta)$-normal decomposition for h gives a pair $(\Delta^{[m]}(x), g')$ satisfying (3.13). Finally, we apply the uniqueness result of Proposition III.1.25 (normal unique) for the $\mathcal{S}$-normal decompositions of g'. □

As in the positive case, simple algorithmic rules enable one to compute Δ-normal decompositions in enveloping groupoids. As in Proposition 3.9, we write ϕ for ϕ_Δ and $\Delta(\cdot)$ for a Δ element with unspecified source. We drop the embedding ι, and use ∂ for ∂_Δ and $\widetilde{\partial}$ for ∂_Δ^{-1}. The question is, starting with a Δ-normal decomposition of an element g of $\mathcal{E}nv(\mathcal{C})$ and s in $\mathcal{D}iv(\Delta)$, to determine a Δ-normal decomposition of sg and $s^{-1}g$.

Algorithm 3.19 (left-multiplication). (See Figure 3)

Context: A Garside family $\mathcal{S}$ that is bounded by a map Δ in a cancellative category $\mathcal{C}$, a $\square$-witness F for $\mathcal{S}^\sharp$, an $=^\times$-test E in $\mathcal{C}$

Input: A Δ-normal decomposition $\Delta^{[m]}\|s_1|\cdots|s_\ell$ of an element g of $\mathcal{E}nv(\mathcal{C})$ and an element s of $\mathcal{S}^\sharp$ such that sg is defined

Output: A Δ-normal decomposition of sg

put $r_0 := \phi^m(s)$
for i increasing from 1 to ℓ **do**
  put $(s'_i, r_i) := F(r_{i-1}, s_i)$
if $E(\Delta(\cdot), s'_1)$ is defined **then** (case when s'_1 is Δ-like)
  put $s'_2 := E(\Delta(\cdot), s'_1)s'_2$
  return $\Delta^{[m+1]}\|s'_2|\cdots|s'_\ell|r_\ell$
else (case when s'_1 is not Δ-like)
  return $\Delta^{[m]}\|s'_1|\cdots|s'_\ell|r_\ell$

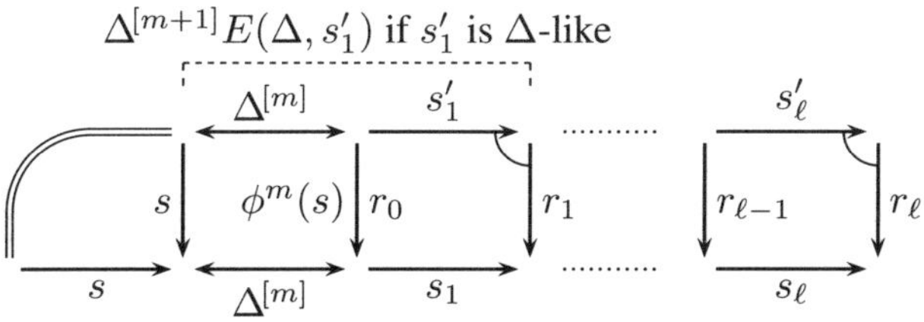

Figure 3. Computing a Δ-normal decomposition for sg from a Δ-normal decomposition of g: Algorithm 3.19 is identical to Algorithm 3.8, the only difference being that the orientation of the left two horizontal arrows now depends on the sign of m.

Proposition 3.20 (left-multiplication). *If Δ is a Garside map in a cancellative category $\mathcal{C}$, F is a $\square$-witness for $\mathcal{S}^\sharp$, and E is an $=^\times$-test E in $\mathcal{C}$, then Algorithm 3.19 running on a Δ-normal decomposition of g and s returns a Δ-normal decomposition of sg.*

Proof. The argument is similar to that for Proposition 3.9, with the difference that the parameter m may now be negative. However, the equality $s\Delta^{(m)}(\cdot) = \Delta^{(m)}(\cdot)(s)\phi^m(s)$ always holds, so, in any case, the constructed path is a decomposition of sg. It remains to see that this path is Δ-normal. If $m \geqslant 0$ holds, this was done for Proposition 3.9. If m is negative, the only problem is when s'_1 is delta-like, say $s'_1 = \Delta(\cdot)\epsilon$. Then the factor $\Delta(\cdot)$ cancels the first $\Delta(\cdot)^{-1}$ in $\Delta^{(|m|)}(\cdot)^{-1}$, and $\epsilon s'_2$ belongs to $\mathcal{C}^\times\mathcal{D}iv(\Delta)$, which is $\mathcal{D}iv(\Delta)$. By Lemma III.1.9, $\epsilon s'_2|s'_3|\cdots|s'_\ell$ is Δ-normal. Finally, as noted in the proof of Proposition 3.9, it is impossible for s'_2 to be delta-like, and no cascade may occur. $\square$

Left-dividing a Δ-normal decomposition is equally easy.

Algorithm 3.21 (left-division). (See Figure 4)

Context: A Garside map Δ in a cancellative category $\mathcal{C}$, a $\square$-witness F for $\mathcal{S}^\sharp$, an $=^\times$-test E in $\mathcal{C}$

Input: A Δ-normal decomposition $\Delta^{[m]}\|s_1|\cdots|s_\ell$ of an element g of $\mathcal{E}nv(\mathcal{C})$ and an element s of $\mathcal{S}^\sharp$ such that $s^{-1}g$ is defined

Output: A Δ-normal decomposition of $s^{-1}g$

put $r_0 := \phi^m(\widetilde{\partial} s)$
for i increasing from 1 to ℓ **do**
  put $(s'_i, r_i) := F(r_{i-1}, s_i)$
if $E(\Delta, s'_1)$ is defined **then** (case when s'_1 is Δ-like)
  put $s'_2 := E(\Delta, s'_1)s'_2$
  return $\Delta^{[m+1]}\|s'_2|\cdots|s'_\ell|r_\ell$
else (case when s'_1 is not Δ-like)
  return $\Delta^{[m]}\|s'_1|\cdots|s'_\ell|r_\ell$

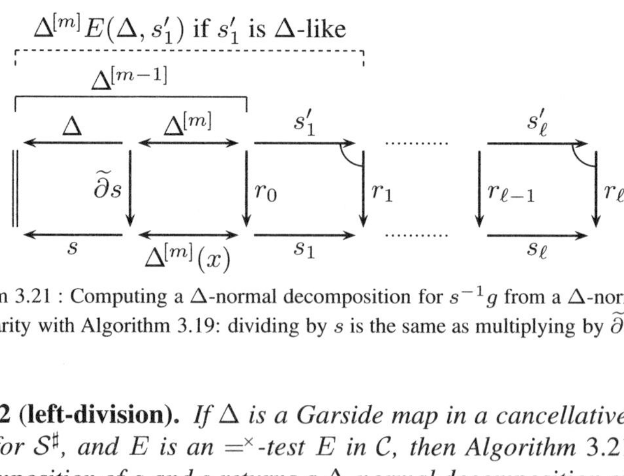

Figure 4. Algorithm 3.21 : Computing a Δ-normal decomposition for $s^{-1}g$ from a Δ-normal decomposition of g. Note the similarity with Algorithm 3.19: dividing by s is the same as multiplying by $\widetilde{\partial} s$.

Proposition 3.22 (left-division). *If Δ is a Garside map in a cancellative category $\mathcal{C}$, F is a $\square$-witness for $\mathcal{S}^\sharp$, and E is an $=^\times$-test E in $\mathcal{C}$, then Algorithm 3.21 running on a Δ-normal decomposition of g and s returns a Δ-normal decomposition of $s^{-1}g$.*

Proof. Let x be the source of $\partial_\Delta^{-1}(s)$, here denoted by $\widetilde{\partial}s$. By definition, we have $\widetilde{\partial}s\, s = \Delta(x)$, whence $s^{-1} = \Delta(x)^{-1}\widetilde{\partial}s$, so left-dividing by s in $\mathcal{E}nv(\mathcal{C})$ amounts to first left-multiplying by $\widetilde{\partial}s$ and then left-dividing by $\Delta(x)$. This is what Algorithm 3.21 does, thus providing a Δ-normal decomposition $s_1'|\cdots|s_\ell'|r_\ell$ of $\phi^m(\widetilde{\partial}s)s_1\cdots s_\ell$ and possibly merging s_1' with $\Delta^{(m)}(\cdot)$ if s_1' is delta-like. Then the final multiplication by $\Delta(x)^{-1}$ simply amounts to diminishing the exponent of Δ by one. □

Next, the Δ-supremum and Δ-infimum functions can be extended to the enveloping groupoid.

Definition 3.23 (infimum, supremum, canonical length). If Δ is a Garside map in a cancellative category $\mathcal{C}$, then, for g in $\mathcal{E}nv(\mathcal{C})$ admitting a Δ-normal decomposition $\Delta^{[m]}\|s_1|\cdots|s_\ell$ with s_ℓ non-invertible, we define the *Δ-infimum* $\inf_\Delta(g)$, the *Δ-supremum* $\sup_\Delta(g)$, and the *canonical Δ-length* $\ell_\Delta(g)$ of g by

$$\inf_\Delta(g) = m, \quad \sup_\Delta(g) = m + \ell, \quad \text{and} \quad \ell_\Delta(g) = \ell.$$

If g admits a Δ-normal decomposition $\Delta^{[m]}$ or $\Delta^{[m]}\|s_1$ with s_1 invertible, we put

$$\inf_\Delta(g) = \sup_\Delta(g) = m \quad \text{and} \quad \ell_\Delta(g) = 0.$$

By construction, for g in $\mathcal{C}$, the values of $\inf_\Delta(\iota(g))$ and $\sup_\Delta(\iota(g))$ as evaluated in $\mathcal{E}nv(\mathcal{C})$ coincide with the previously defined values of $\inf_\Delta(g)$ and $\sup_\Delta(g)$.

We recall from Definition II.3.14 that, if $\mathcal{C}$ is a left-Ore category, and for g, h in $\mathcal{E}nv(\mathcal{C})$, we write $g \preccurlyeq_\mathcal{C} h$ for $g^{-1}h \in \mathcal{C}$. By Proposition II.3.15 (left-divisibility), the relation $\preccurlyeq_\mathcal{C}$ is a partial ordering on $\mathcal{E}nv(\mathcal{C})$ that extends the left-divisibility relation of $\mathcal{C}$. So there is no ambiguity in writing $\preccurlyeq$ for $\preccurlyeq_\mathcal{C}$ when the choice of $\mathcal{C}$ is clear. Then there exist simple connections between the functions $\inf_\Delta$ and $\sup_\Delta$ and the relation $\preccurlyeq$.

Proposition 3.24 (interval). *If Δ is a Garside map in a cancellative category $\mathcal{C}$ and g lies in $\mathcal{E}nv(\mathcal{C})(x, \cdot)$, then* $\inf_\Delta(g)$ *is the greatest integer p satisfying $\Delta^{[p]}(x) \preccurlyeq g$, and* $\sup_\Delta(g)$ *is the least integer q satisfying $g \preccurlyeq \Delta^{[q]}(x)$.*

Proof. Let $\Delta^{(m)}\|s_1|\cdots|s_\ell$ be a Δ-normal decomposition of g with s_ℓ not invertible. Then, by definition, we have $g = \Delta^{(m)}(x)\, s_1\cdots s_\ell$, whence $\Delta^{[m]}(x) \preccurlyeq g$.

On the other hand, we claim that $\Delta^{(m+1)}(x) \preccurlyeq g$ cannot hold. Indeed, assume $g = \Delta^{(m+1)}(x)\, g$. Then we deduce $s_1\cdots s_\ell = \Delta(\cdot)\, g$, whence $\Delta(\cdot) \preccurlyeq s_1\cdots s_\ell$ and, as $s_1|\cdots|s_\ell$ is Δ-greedy, $\Delta(\cdot) \preccurlyeq s_1$, contradicting the definition of an Δ-normal sequence which demands that s_1 be not delta-like. So $\inf_\Delta(g)$ is the maximal p satisfying $\Delta^{(p)}(x) \preccurlyeq g$.

Next, as $s_1, ..., s_\ell$ belong to $\mathcal{D}iv(\Delta)$, we have $s_1\cdots s_\ell \preccurlyeq \Delta^{(\ell)}(\cdot)$ by Lemma 1.43, whence $g \preccurlyeq \Delta^{(m+\ell)}(\cdot)$.

Finally, we claim that $g \preccurlyeq \Delta^{(m+\ell-1)}(\cdot)$ does not hold. Otherwise, we would have $s_1\cdots s_\ell \preccurlyeq \Delta^{(\ell-1)}(\cdot)$. Now, as $\mathcal{C}$ is cancellative and Δ is a Garside map, Corollary 1.56

would imply $\|s_1 \cdots s_\ell\|_\Delta \leqslant \|\Delta^{(\ell-1)}(\cdot)\|_\Delta = \ell - 1$. As the assumption that $s_1|\cdots|s_\ell$ is Δ-normal and s_ℓ is not invertible implies $\|s_1 \cdots s_\ell\|_\Delta = \ell$, the inequality $\|s_1 \cdots s_\ell\|_\Delta \leqslant \ell - 1$ is impossible, and so is $g \preccurlyeq \Delta^{(m+\ell-1)}(\cdot)$. So $\sup_\Delta(g)$ is the least q satisfying $g \preccurlyeq \Delta^{(q)}(x)$.

The particular case when g admits a Δ-normal decomposition $\Delta^{[m]}\|s_1$ with s_1 invertible can be treated similarly. □

Remark 3.25. The assumption that the Garside family consists of all left-divisors of Δ is crucial in Proposition 3.24. In the left-absorbing monoid L_n with $S_n = \{\mathtt{a}, \mathtt{b}, ..., \mathtt{b}^{n+1}\}$ and $\Delta = \mathtt{b}^{n+1}$ (Reference Structure 8, page 113), we have $\sup_{S_n}(\mathtt{ab}) = 2$, whereas $\mathtt{ab} \preccurlyeq \Delta$ holds. Of course, Proposition 3.24 is not relevant here, as S_n is right-bounded but not bounded by Δ, and L_n is not right-cancellative.

We complete the analysis of Δ-normal decompositions with an explicit formula connecting the decompositions of an element and its inverse.

Proposition 3.26 (inverse). *If Δ is a Garside map in a cancellative category $\mathcal{C}$ and $\Delta^{(m)}\|s_1|\cdots|s_\ell$ is a Δ-normal decomposition of an element g of $\mathcal{E}nv(\mathcal{C})$, then $\Delta^{(-m-\ell)}\|\partial(\phi^{-m-\ell}s_\ell)|\cdots|\partial(\phi^{-m-1}s_1)$ is a Δ-normal decomposition of g^{-1}.*

Proof. We first check that the sequence mentioned in the statement is Δ-normal, and then we show that it corresponds to a decomposition of g^{-1}.

So first let $i < \ell$. By assumption, $s_i|s_{i+1}$ is Δ-normal. By Proposition 2.18, ϕ_Δ preserves normality, so $\phi^{-m-i}s_i|\phi^{-m-i}s_{i+1}$ is Δ-normal as well. By Proposition 1.53, this implies that $\partial(\phi^{-m-i}s_i)$ and $\phi^{-m-i}s_{i+1}$ are left-coprime. The latter element is $\phi(\phi^{-m-i-1}s_{i+1})$, hence it is $\partial(\partial(\phi^{-m-i-1}s_{i+1}))$. Reading the above coprimeness result in a symmetric way and applying Proposition 1.53 again, we deduce that the path $\partial(\phi^{-m-i-1}s_{i+1})|\partial(\phi^{-m-i}s_i)$ is $\mathcal{D}iv(\Delta)$-greedy, hence Δ-normal. So

$$\partial(\phi^{-m-\ell}s_\ell)|\cdots|\partial(\phi^{-m-1}s_1)$$

is Δ-normal. Moreover, the assumption that s_1 is not delta-like implies that $\phi^{-m-1}s_1$ is not delta-like either, and, therefore, $\partial(\phi^{-m-1}s_1)$ is not invertible. On the other hand, the assumption that s_ℓ is not invertible implies that $\phi^{-m-\ell}s_\ell$ is not invertible either, and, therefore, $\partial(\phi^{-m-\ell}s_\ell)$ is not delta-like. Hence $\Delta^{(-m-\ell)}\|\partial(\phi^{-m-\ell}s_\ell)|\cdots|\partial(\phi^{-m-1}s_1)$ is Δ-normal.

It remains to show that the above Δ-normal path is a decomposition of g^{-1}. Assume $g \in \mathcal{C}(x, y)$. Using the fact that, by (1.32), ϕ and ∂ commute, and using (1.29) repeatedly, we can push the factor $\Delta^{(m)}$ to the left to obtain

$$\begin{aligned}(3.27)\qquad \Delta^{(-m-\ell)}(\cdot)\,\partial(\phi^{-m-\ell}s_\ell)\cdots\partial(\phi^{-m-1}s_1)\cdot\Delta^{(m)}(\cdot)\,s_1\cdots s_\ell \\ = \Delta^{(-m-\ell)}(\cdot)\,\Delta^{(m)}(\cdot)\,\partial(\phi^{-\ell}s_\ell)\cdots\partial(\phi^{-1}s_1)\,s_1\cdots s_\ell,\end{aligned}$$

and our goal is to prove that this element is 1_y. Let us call the right term of (3.27) $E(m, s_1, ..., s_\ell)$. We shall prove using induction on $\ell \geqslant 0$ that an expression of the form $E(m, s_1, ..., s_\ell)$ equals 1_y.

For $\ell = 0$, what remains for $E(m)$ is $\Delta^{(-m)}(\cdot)\,\Delta^{(m)}(\cdot)$. For $m \geqslant 0$, we find $g = \Delta^{[m]}(x)$, and the value of E is $\widetilde{\Delta}_{[m]}(y)^{-1}\,\Delta^{[m]}(x)$, which is 1_y by definition. For $m < 0$, we find $g = \widetilde{\Delta}_{[|m|]}(x)^{-1}$, and the value of E is $\Delta^{[|m|]}(y)\,\widetilde{\Delta}_{[|m|]}(x)^{-1}$, which is 1_y again.

Assume now $\ell \geqslant 1$. The above result enables us to gather the first two entries of E into $\Delta^{(-\ell)}(\cdot)$. Then $\partial(\phi^{-1}s_1)\,s_1$, which is also $\widetilde{\partial}s_1\,s_1$, equals $\Delta(z)$, where z is the source of $\widetilde{\partial}s_1$. But then we can push this $\Delta(\cdot)$-factor to the left through the $\partial\phi^{-i}(s_i)$ factors with the effect of diminishing the exponents of ϕ by one. In this way, $E(m, s_1, ..., s_\ell)$ becomes $\Delta^{(-\ell)}(\cdot)\,\Delta(\cdot)\,\partial(\phi^{-\ell+1}s_\ell)\cdots\partial(\phi^{-1}s_2)\,s_2\cdots s_\ell$, which is $E(1, s_2, ..., s_\ell)$. By induction hypothesis, its value is 1_y. □

Corollary 3.28 (inverse). *If Δ is a Garside map in a cancellative category $\mathcal{C}$ and g lies in $\mathcal{E}nv(\mathcal{C})(x,\cdot)$, then the following equalities are satisfied:*

(3.29) $\quad \inf_\Delta(g^{-1}) = -\sup_\Delta(g),\ \sup_\Delta(g^{-1}) = -\inf_\Delta(g),$ *and* $\ell_\Delta(g^{-1}) = \ell_\Delta(g).$

Proof. The values can be read on the explicit Δ-normal decomposition of g^{-1} provided by Proposition 3.26. □

With the help of Proposition 3.24 and Corollary 3.28, we deduce various inequalities involving $\inf_\Delta$, $\sup_\Delta$, and ℓ_Δ.

Proposition 3.30 (inequalities). *If Δ is a Garside map in a cancellative category $\mathcal{C}$ and g, h lie in $\mathcal{E}nv(\mathcal{C})$ and gh is defined, then the following relations are satisfied:*

$$\inf_\Delta(g) + \inf_\Delta(h) \leqslant \inf_\Delta(gh) \leqslant \inf_\Delta(g) + \sup_\Delta(h), \tag{3.31}$$

$$\sup_\Delta(g) + \inf_\Delta(h) \leqslant \sup_\Delta(gh) \leqslant \sup_\Delta(g) + \sup_\Delta(h), \tag{3.32}$$

$$|\ell_\Delta(g) - \ell_\Delta(h)| \leqslant \ell_\Delta(gh) \leqslant \ell_\Delta(g) + \ell_\Delta(h). \tag{3.33}$$

Proof. The assertions are obvious if g or h is invertible. Otherwise, assume that the source of g is x and that $\Delta^{[m]}\|g_1|\cdots|g_p$ and $\Delta^{[n]}\|h_1|\cdots|h_q$ are strict Δ-normal decompositions of g and h, respectively. We find

$$gh = \Delta^{[m+n]}(x)\phi_\Delta^n(g_1)\cdots\phi_\Delta^n(g_p)h_1\cdots h_q, \tag{3.34}$$

whence $\Delta^{[m+n]}(x) \preccurlyeq gh$, and, by Proposition 3.24, $m + n \leqslant \inf_\Delta(gh)$, which is the left part of (3.31). Next, (3.34) also implies $gh \preccurlyeq \Delta^{[m+n+p+q]}(x)$, whence, by Proposition 3.24, $\sup_\Delta(gh) \leqslant m + n + p + q$, which is the right part of (3.32).

Applying the above inequalities to gh and h^{-1}, we obtain

$$\inf_\Delta(gh) + \inf_\Delta(h^{-1}) \leqslant \inf_\Delta(g) \quad \text{and} \quad \sup_\Delta(g) \leqslant \sup_\Delta(gh) + \sup_\Delta(h^{-1}).$$

Using the values of $\inf_\Delta(h^{-1})$ and $\sup_\Delta(h^{-1})$ provided by Corollary 3.28 and transferring factors, we deduce the inequalities

$$\inf_\Delta(gh) \leqslant \inf_\Delta(g) + \sup_\Delta(h) \quad \text{and} \quad \sup_\Delta(g) + \inf_\Delta(h) \leqslant \sup_\Delta(gh),$$

which complete (3.31) and (3.32).

Subtracting the inequalities gives $\ell_\Delta(g) - \ell_\Delta(h) \leqslant \ell_\Delta(gh) \leqslant \ell_\Delta(g) + \ell_\Delta(h)$, whence (3.33), the absolute value coming from applying the formula to gh and $(gh)^{-1}$ simultaneously. □

3.3 Symmetric normal decompositions

In Chapter III we showed that, if $\mathcal{C}$ is an Ore category admitting left-lcms and $\mathcal{S}$ is a Garside family in $\mathcal{C}$, then every element of the groupoid $\mathcal{E}nv(\mathcal{C})$ admits a symmetric $\mathcal{S}$-normal decomposition (Corollary III.2.21). By Propositions 2.32 and 2.35, every cancellative category $\mathcal{C}$ that admits a Garside map Δ is an Ore category and admits left-lcms, hence it is eligible for the above results. Thus every element of $\mathcal{E}nv(\mathcal{C})$ possesses both Δ-normal decompositions and symmetric $\mathcal{S}$-normal decompositions. In the positive case (elements of $\mathcal{C}$), the connection follows directly from Proposition 3.4. The following result describes explicitly the connection in the general case.

Proposition 3.35 (symmetric normal *vs.* **Δ-normal).** *Assume that Δ is a Garside map in a cancellative category $\mathcal{C}$ and g is an element of $\mathcal{E}nv(\mathcal{C})$.*

(i) *If $\overline{t_q}|\cdots|\overline{t_1}|s_1|\cdots|s_p$ is a symmetric $\mathcal{D}iv(\Delta)$-normal decomposition of g and t_q is non-invertible, a Δ-normal decomposition of g is $\Delta^{(-q)}\|\widetilde{\partial}^{2q-1}t_q|\cdots|\widetilde{\partial}^3 t_2|\widetilde{\partial}t_1|s_1|\cdots|s_p$.*

(ii) *If $\Delta^{(-m)}\|s_1|\cdots|s_\ell$ is a Δ-normal decomposition of g, a symmetric $\mathcal{D}iv(\Delta)$-normal decomposition of g is*

$$\overline{\partial^{2m-1}s_1}|\cdots|\overline{\partial s_m}|s_{m+1}|\cdots|s_\ell \qquad \textit{for } m < \ell,$$
$$\overline{\partial^{2m-1}s_1}|\cdots|\overline{\partial^{2m-2\ell+1}s_\ell}|\overline{\Delta(\cdot)}|\cdots|\overline{\Delta(\cdot)},\ m-\ell \textit{ times } \Delta, \qquad \textit{for } m \geqslant \ell.$$

Proof. (i) Put $\mathcal{S} = \mathcal{D}iv(\Delta)$. An induction on q gives $t_1\cdots t_q \cdot \partial t_q \cdots \partial^{2q-1}t_1 = \Delta^{[q]}(\cdot)$: the formula is clear for $q = 1$ and, for $q \geqslant 2$, we write

$$\begin{aligned} t_1\cdots t_q \partial t_q \cdots \partial^{2q-1}t_1 &= t_1\cdots t_{q-1}\Delta(\cdot)\partial^3 t_{q-1}\cdots\partial^{2q-1}t_1 \\ &= t_1\cdots t_{q-1}\partial t_{q-1}\cdots\partial^{2q-3}t_1\Delta(\cdot), \end{aligned}$$

and apply the induction hypothesis. We deduce $(t_1\cdots t_q)^{-1} = \partial t_q\cdots\partial^{2q-1}t_1\Delta^{[-q]}(\cdot)$, whence $(t_1\cdots t_q)^{-1} = \Delta^{[-q]}(\cdot)\widetilde{\partial}^{2q-1}t_q\cdots\widetilde{\partial}t_1$ owing to $\partial^r h\Delta^{[-q]}(\cdot) = \Delta^{[-q]}(\cdot)\widetilde{\partial}^{2q-r}h$. Hence the two decompositions represent the same element of $\mathcal{E}nv(\mathcal{C})$. So it remains to check that the second decomposition is Δ-normal.

Now the assumption that t_q is not invertible implies that $\widetilde{\partial}^{2q-1}$ is not delta-like. Next, by Proposition 2.32, the assumption that $t_i|t_{i+1}$ is $\mathcal{S}$-normal for $1 \leqslant i < q$ implies that ∂t_i and t_{i+1} are left-coprime, hence, applying ϕ_Δ^{-2i}, that $\widetilde{\partial}^{2i-1}t_i$ and $\widetilde{\partial}^{2i}t_{i+1}$ are left-coprime, whence, by Proposition 2.32 again, that $\widetilde{\partial}^{2i+1}t_{i+1}|\widetilde{\partial}^{2i-1}t_i$ is $\mathcal{D}iv(\Delta)$-normal. Finally, as already noted in the proof of Proposition 3.37, the existence of left-gcds in $\mathcal{C}$ implies that two elements of $\mathcal{D}iv(\Delta)$ are left-coprime if and only if they are left-disjoint. Hence, the assumption that s_1 and t_1 are left-disjoint implies that they are left-coprime and, therefore, by Proposition 2.32(ii), that $\widetilde{\partial} t_1 | s_1$ is $\mathcal{D}iv(\Delta)$-normal. Hence, as expected, the second decomposition is Δ-normal.

(ii) The argument is entirely similar, going in the other direction: the two paths represent the same element, and the assumption that the first one is Δ-normal implies that the second one is symmetric $\mathcal{D}iv(\Delta)$-normal. □

In Chapter III, we solved the questions of determining symmetric $\mathcal{S}$-normal decompositions of sg and of gs^{-1} from one of g when s lies in $\mathcal{S}^\sharp$, but the symmetric questions of computing a symmetric $\mathcal{S}$-normal decomposition of gs and of $s^{-1}g$ were left open as Question 4. As announced in Chapter III, we shall now solve the question in the case when the considered Garside family $\mathcal{S}$ is bounded by a target-injective map Δ. What makes the solution easy in this case is that left-dividing (or right-multiplying) by an element $\Delta(\cdot)$ is easy, and that left-dividing by an element s of $\mathcal{S}^\sharp$ amounts to first left-dividing by $\Delta(\cdot)$ and then left-multiplying by $\partial(s)$—which we know how to do. As usual, we state the results in terms of Garside maps.

Algorithm 3.36 (left-division). (See Figure 5)

Context: A cancellative category $\mathcal{C}$, a Garside map Δ in $\mathcal{C}$, a □-witness F for $\mathcal{D}iv(\Delta)$

Input: A symmetric $\mathcal{D}iv(\Delta)$-normal decomposition $\overline{t_q}|\cdots|\overline{t_1}|s_1|\cdots|s_p$ of an element g of $\mathcal{E}nv(\mathcal{C})$ and an element s of $\mathcal{S}^\sharp$ such that $s^{-1}g$ is defined

Output: A symmetric $\mathcal{D}iv(\Delta)$-normal decomposition of $s^{-1}g$

put $r_{-q} := s$
for i decreasing from q to 1 **do**
    put $(r_{-i+1}, t'_i) := F(t_i, r_{-i})$
put $r'_0 := \widetilde{\partial}(r_0)$
for i increasing from 1 to p **do**
    put $(s'_i, r'_i) := F(r'_{i-1}, s_i)$
put $t'_0 := \widetilde{\partial}(s'_1)$
return $\overline{t'_q}|\cdots|\overline{t'_1}|\overline{t'_0}|s'_2|\cdots|s'_p|r'_p$

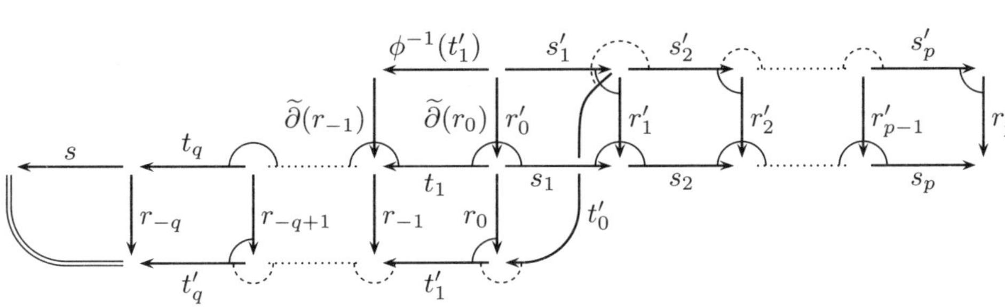

Figure 5. Left-division by an element of $\mathcal{D}iv(\Delta)$ starting from a symmetric $\mathcal{D}iv(\Delta)$-normal path $\overline{t_q}|\cdots|\overline{t_1}|s_1|\cdots|s_p$: the subtlety is the transition from negative to positive entries.

Proposition 3.37 (left-division). *If Δ is a Garside map in a cancellative category $\mathcal{C}$, then Algorithm 3.36 running on a symmetric $\mathcal{D}iv(\Delta)$-normal decomposition of g and s returns a symmetric $\mathcal{D}iv(\Delta)$-normal decomposition of $s^{-1}g$.*

Proof. By construction, the diagram of Figure 5 is commutative, so the returned path is equivalent to $\overline{s}|\overline{t_q}|\cdots|\overline{t_1}|s_1|\cdots|s_p$, hence it is a decomposition of $s^{-1}g$, and its entries are elements of $\mathcal{D}iv(\Delta)$. So the point is to check that the path is symmetric $\mathcal{D}iv(\Delta)$-greedy.

First, the automorphism ϕ_Δ preserves normality by Proposition 2.18, hence, by Proposition 1.52), the second domino rule is valid for $\mathcal{D}iv(\Delta)$. Therefore, the assumption that $t_i|t_{i+1}$ is $\mathcal{D}iv(\Delta)$-normal for $i = 1, ..., q-1$ implies that $t'_i|t'_{i+1}$ is $\mathcal{D}iv(\Delta)$-normal as well. Similarly, as $s_i|s_{i+1}$ is $\mathcal{D}iv(\Delta)$-normal for $i = 1, ..., p-1$, the first domino rule implies that $s'_i|s'_{i+1}$ is $\mathcal{D}iv(\Delta)$-normal as well.

So it remains to check that $t'_0|t'_1$ is $\mathcal{D}iv(\Delta)$-normal and that t'_0 and s'_2 are left-disjoint. Consider $t'_0|t'_1$. Assume that t is an element of $\mathcal{D}iv(\Delta)$ satisfying $t \preccurlyeq \phi_\Delta^{-1}(t'_1)$ and $t \preccurlyeq s'_1$. A fortiori, we have $t \preccurlyeq \phi_\Delta^{-1}(t'_1)\widetilde{\partial} r_{-1}$ and $t \preccurlyeq s'_1 r'_1$, that is, $t \preccurlyeq r'_0 t_1$ and $t \preccurlyeq r'_0 s_1$. By assumption, t_1 and s_1 are left-disjoint, which implies $t \preccurlyeq r'_0$. Next, by assumption, $r_0|t'_1$ is $\mathcal{D}iv(\Delta)$-normal, hence, by Proposition 1.53, ∂r_0 and t'_1 are left-coprime. As ϕ_Δ and, therefore, ϕ_Δ^{-1} is an automorphism, $\phi_\Delta^{-1}(\partial r_0)$ and $\phi_\Delta^{-1}(t'_1)$ are left-coprime as well. Now, by definition, $\phi_\Delta^{-1}(\partial r_0)$ is r'_0. As we have $t \preccurlyeq \phi_\Delta^{-1}(t'_1)$ and $t \preccurlyeq r'_0$, we deduce that t is invertible and, therefore, $\phi_\Delta^{-1}(t'_1)$ and s'_1 left-coprime. Applying the automorphism ϕ_Δ, we deduce that t'_1 and $\phi_\Delta(s'_1)$, that is, t'_1 and $\partial t'_0$, are left-coprime. By Proposition 1.53 again, $t'_0|t'_1$ is $\mathcal{D}iv(\Delta)$-normal.

Finally, again by Proposition 1.53, the assumption that $s'_1|s'_2$ is $\mathcal{D}iv(\Delta)$-normal implies that $\partial s'_1$ and s'_2 are left-coprime: by definition, $\partial s'_1$ is t'_0, so t'_0 and s'_2 are left-coprime. Now, by Proposition 2.35, the category $\mathcal{C}$ admits left-gcds. Hence, by Proposition III.2.5 (left-disjoint *vs.* left-coprime), two elements of $\mathcal{D}iv(\Delta)$ are left-disjoint if and only if they are left-coprime. So we conclude that t'_0 and s'_2 are left-disjoint, and $\overline{t'_q}|\cdots|\overline{t'_1}|\overline{t'_0}|s'_2|\cdots|s'_p|r'_p$ is indeed a symmetric $\mathcal{D}iv(\Delta)$-normal path. □

Example 3.38 (left-division). Consider the element $(\mathtt{ab})^{-1}(\mathtt{ba}^2)$ of B_3^+ once more. Suppose that we found that $\mathtt{ba}|\mathtt{a}$ is a $\mathcal{D}iv(\Delta)$-normal decomposition of $\mathtt{ba}^2$, and we wish to find a (the) $\mathcal{D}iv(\Delta)$-normal decomposition of $(\mathtt{ab})^{-1}(\mathtt{ba}^2)$. Then applying Algorithm 3.36 amounts to completing the diagram on the right, in which we read the expected symmetric normal decomposition $\overline{\mathtt{ab}}|\mathtt{ba}|\mathtt{a}$.

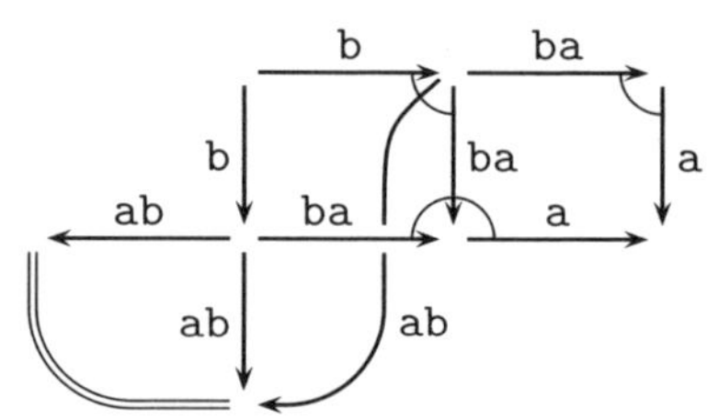

If, in Algorithm 3.36, s is $\Delta(x)$, then each element r_{-i} is of the form $\Delta(\cdot)$, so that t'_i is $\phi_\Delta(t_i)$, hence r'_0 is trivial, and so are all elements r_i with $i \geqslant 0$. We deduce

Corollary 3.39 (left-division by Δ). *If Δ is a Garside map in a cancellative category $\mathcal{C}$ and $\overline{t_q}|\cdots|\overline{t_1}|s_1|\cdots|s_p$ is a symmetric $\mathcal{D}iv(\Delta)$-normal decomposition of an element g of $\mathcal{E}nv(\mathcal{C})(x,\cdot)$, then $\overline{\phi_\Delta(t_q)}|\cdots|\overline{\phi_\Delta(t_1)}|\overline{\partial(s_1)}|s_2|\cdots|s_p$ is a symmetric $\mathcal{D}iv(\Delta)$-normal decomposition of $\Delta(x)^{-1}g$.*

3.4 Co-normal decompositions

If Δ is a Garside map in a cancellative category $\mathcal{C}$, the duality map ∂ associated with Δ and, similarly, the duality maps associated with the iterates $\Delta^{[m]}$, establish a bijective correspondence between the left- and the right-divisors of $\Delta^{[m]}$ that exchanges the roles of left- and right-divisibility. This duality leads to symmetric versions of the normal decompositions considered so far.

Hereafter, we indicate with the sign $\widetilde{\ }$ the symmetric versions of the notions previously introduced in connection with left-divisibility, and use the prefix "co" in terminology. Then, if Δ is a Garside map, using $\widetilde{\Delta}(y)$ for $\Delta(\phi_\Delta^{-1}(y))$ is natural as, by definition, $\widetilde{\Delta}(y)$ is a left-lcm for all elements of $\mathcal{D}iv(\Delta)(\cdot, y)$. So we can say that $\mathcal{D}iv(\Delta)$ is *co-bounded* by $\widetilde{\Delta}$. Similarly, it is natural to use $\widetilde{\partial}$ for ∂^{-1} since, for every g right-dividing $\widetilde{\Delta}(y)$, we have $\partial^{-1}(g)\, g = \widetilde{\Delta}(y)$. The following relations are straightforward:

Lemma 3.40. *If Δ is a Garside map in a cancellative category $\mathcal{C}$, then, for all s, t in $\mathcal{D}iv(\Delta)$, we have $s \preccurlyeq t \Leftrightarrow \partial s \succcurlyeq \partial t$ and $s \succcurlyeq t \Leftrightarrow \widetilde{\partial} s \preccurlyeq \widetilde{\partial} t$.*

The easy verification is left to the reader.

Thus, ∂ and $\widetilde{\partial}$ exchange left- and right-divisibility (changing the orientation) inside each family of the form $\mathrm{Div}(\Delta(x))$. In order to extend the results to arbitrary elements, we appeal to Lemma 1.43, which ensures that every element left-divides a sufficiently large iterate of Δ.

Notation 3.41 (duality maps $\partial^{[m]}$ and $\widetilde{\partial}_{[m]}$). If Δ is a Garside map in a cancellative category $\mathcal{C}$, then, for $m \geqslant 1$, we denote by $\partial^{[m]}$ the duality map associated with $\Delta^{[m]}$, and by $\widetilde{\partial}_{[m]}$ the co-duality map associated with $\widetilde{\Delta}_{[m]}$.

So, for g left-dividing $\Delta^{[m]}(x)$, we have $g\,\partial^{[m]}(g) = \Delta^{[m]}(x)$, and, similarly, for g right-dividing $\widetilde{\Delta}_{[m]}(y)$, we have $\widetilde{\partial}_{[m]}(g)\, g = \widetilde{\Delta}_{[m]}(y)$. In particular, we have $\partial^{[1]} = \partial$ and $\widetilde{\partial}_{[1]} = \widetilde{\partial}$. If g lies in $\mathcal{D}iv(\Delta^{[m]})$, it lies in $\mathcal{D}iv(\Delta^{[m']})$ for $m' \geqslant m$, and one easily checks the equalities (see Exercise 64)

$$\partial^{[m']}(g) = \partial^{[m]}(g)\, \Delta^{[m'-m]}(\phi_\Delta^{m}(x)), \tag{3.42}$$

and, symmetrically, $\widetilde{\partial}_{[m']}(g) = \widetilde{\Delta}_{[m'-m]}(\phi_\Delta^{-m}(y))\, \widetilde{\partial}_{[m]}(g)$ for g in $\widetilde{\mathrm{Div}}(\widetilde{\Delta}_{[m]}(y))$.

Once we consider right-divisibility, it is natural to introduce the symmetric counterparts of the notions of $\mathcal{S}$-greedy sequence, $\mathcal{S}$-normal sequence, and Garside family. If $\mathcal{C}$ is a right-cancellative category and $\mathcal{S}$ is included in $\mathcal{C}$, we say that a length two $\mathcal{C}$-path $g_1|g_2$ is *$\mathcal{S}$-cogreedy* if every relation of the form $g_1 g_2 f \succcurlyeq h$ with h in $\mathcal{S}$ implies $g_2 f \succcurlyeq h$. We similarly introduce the notions of a *$\mathcal{S}$-conormal* path and a *co-Garside family*. Applying the results of Chapter III to the opposite category $\mathcal{C}^{\mathrm{opp}}$ shows that, if $\mathcal{C}$ is a right-cancellative category and $\mathcal{S}$ is a co-Garside family in $\mathcal{C}$, then every element of $\mathcal{C}$ admits a $\mathcal{S}$-conormal decomposition that is unique up to $\mathcal{C}^\times$-deformation. In the current context, it is natural to consider the counterpart of a Garside map.

Definition 3.43 (co-Garside). A *co-Garside map* in a right-cancellative category $\mathcal{C}$ is a map ∇ from $\mathcal{O}bj(\mathcal{C})$ to $\mathcal{C}$ that is a Garside map in the opposite category $\mathcal{C}^{\mathrm{opp}}$.

Thus ∇ is a co-Garside map in $\mathcal{C}$ if (i) for each object y, the target of $\nabla(y)$ is y, (ii) the family $\widetilde{\mathcal{D}iv}(\nabla)$ generates $\mathcal{C}$, (iii) the families $\mathcal{D}iv(\nabla)$ and $\widetilde{\mathcal{D}iv}(\nabla)$ coincide, and (iv) for every g in $\mathcal{C}$ with target y, the elements g and $\nabla(y)$ admit a right-gcd. If ∇ is a co-Garside map, then the family $\widetilde{\mathcal{D}iv}(\nabla)$ is a co-Garside family, and every element of $\mathcal{C}$ admits a $\widetilde{\mathcal{D}iv}(\nabla)$-co-normal decomposition.

The question we address now is whether every Garside map gives rise to a co-Garside map. We have no answer in the most general case, but weak additional assumptions provide a positive answer.

Proposition 3.44 (co-Garside). *If Δ is a Garside map in a cancellative category $\mathcal{C}$, the following conditions are equivalent:*
(i) *The map $\widetilde{\Delta}$ is a co-Garside map in $\mathcal{C}$.*
(ii) *For every g in $\mathcal{C}(\cdot, y)$, the elements g and $\widetilde{\Delta}(y)$ admit a right-gcd.*
(iii) *For every g in $\mathcal{C}(x, \cdot)$, the elements g and $\Delta(x)$ admit a right-lcm.*

We do not know whether every Garside map Δ satisfies the conditions of Proposition 3.44, that is, whether, for every g in $\mathcal{C}(x, \cdot)$, the elements g and $\Delta(x)$ must have a right-lcm. This is so in every category that admits right-lcms, so Proposition 2.35 implies:

Corollary 3.45 (co-Garside). *If Δ is a Garside map in cancellative category $\mathcal{C}$ that is left-Noetherian, then $\widetilde{\Delta}$ is a co-Garside map in $\mathcal{C}$ and every element of $\mathcal{C}$ admits a Δ-conormal decomposition.*

Before proving Proposition 3.44, we begin with an auxiliary duality result.

Lemma 3.46. *If Δ is a Garside map in a cancellative category $\mathcal{C}$ and f, g are elements of $\mathcal{C}(x, \cdot)$, then, for every m satisfying $m \geqslant \max(\sup_\Delta(f), \sup_\Delta(g))$ and for every h in $\mathrm{Div}(\Delta^{[m]}(x))$, the following are equivalent:*
(i) *The element h is a right-lcm of f and g.*
(ii) *The element $\partial^{[m]}h$ is a right-gcd of $\partial^{[m]}f$ and $\partial^{[m]}g$.*

We skip the verification, which simply consists of applying the equivalences of Lemma 3.40 in the definition of a right-lcm.

Proof of Proposition 3.44. First, by definition, the target of $\widetilde{\Delta}(y)$ is y. Next, by definition again, we have $\mathcal{D}iv(\widetilde{\Delta}) = \mathcal{D}iv(\Delta) = \widetilde{\mathcal{D}iv}(\Delta) = \widetilde{\mathcal{D}iv}(\widetilde{\Delta})$, and the latter family generates $\mathcal{C}$. Hence, by definition, $\widetilde{\Delta}$ is a co-Garside map in $\mathcal{C}$ if and only if, for every g in $\mathcal{C}(\cdot, y)$, the elements g and $\widetilde{\Delta}(y)$ have a right-gcd. So (i) and (ii) are equivalent.

Assume now that (i) and (ii) are satisfied, and let g belong to $\mathcal{C}(x, \cdot)$. By Lemma 1.43, there exists m such that g left-divides $\Delta^{[m]}(x)$. The map $\widetilde{\Delta}$ is co-Garside, hence, by the counterpart of Proposition 1.44, so is its iterate $\widetilde{\Delta}_{[m-1]}$. Hence the elements $\partial^{[m]}g$ and $\widetilde{\Delta}_{[m-1]}(\phi^m(x))$ admit a right-gcd. By Lemma 3.46, this implies that $\widetilde{\partial}_{[m]}(\partial^{[m]}g))$ and $\widetilde{\partial}_{[m]}(\widetilde{\Delta}_{[m-1]}(\phi^m(x)))$, that is, g and $\Delta(x)$, admit a right-lcm. So (i) implies (iii).

Conversely, assume (iii). By (1.40), $\Delta^{[m]}(x)$ is a product of factors $\Delta(\cdot)$, so, by Proposition II.2.12 (iterated lcm), for every m and every f in $\mathcal{C}(x, \cdot)$, the elements f and $\Delta^{[m]}(x)$ admit a right-lcm. Let g be an element of $\mathcal{C}(\cdot, y)$. There exists m such that g right-divides $\widetilde{\Delta}_{[m]}(y)$. By the above remark, $\widetilde{\partial}_{[m]}g$ and $\Delta^{[m-1]}(\phi^{-m}(x))$ admit a right-lcm. By Lemma 3.46, this implies that $\partial^{[m]}(\widetilde{\partial}_{[m]}g))$ and $\partial^{[m]}(\Delta^{[m-1]}(\phi^{-m}(y)))$, that is, g and $\widetilde{\Delta}(y)$, admit a right-gcd. So (iii) implies (ii). $\square$

We conclude with an explicit computation of a co-head starting from a map that determines the right-lcm with Δ.

Proposition 3.47 (co-head). *If Δ is a Garside map in a cancellative category $\mathcal{C}$ and $\Delta(x)\widetilde{H}(f)$ is a right-lcm of f and $\Delta(x)$ for every f in $\mathcal{C}(x,\cdot)$, then, for g satisfying $\sup_\Delta(g) \leqslant m$, the element $\partial(\widetilde{H}^{m-1}(\widetilde{\partial}_{[m]}g))$ is a $\widetilde{\Delta}$-co-head of g.*

Proof. First, Proposition II.2.12 (iterated lcm) and (1.40) imply that, for every f in $\mathcal{C}(x,\cdot)$ and every k, a right-lcm of f and $\Delta^{[k]}(x)$ is $\Delta^{[k]}(x)\,\widetilde{H}^k(f)$.

Now assume that g lies in $\mathcal{C}(\cdot,y)$ and satisfies $\sup_\Delta(g) \leqslant m$. By the above remark, a right-lcm of $\widetilde{\partial}_{[m]}g$ and $\Delta^{[m-1]}(\phi^{-m}(y))$ is $\Delta^{[m-1]}(\phi^{-m}(y))\,A^{m-1}(\widetilde{\partial}_{[m]}g)$. Then the proof of Proposition 3.44 shows that $\partial^{[m]}(\Delta^{[m-1]}(\phi^{-m}(y))\,\widetilde{H}^{m-1}(\widetilde{\partial}_{[m]}g))$ is a right-gcd of g and $\widetilde{\Delta}(y)$, hence is a $\widetilde{\Delta}$-co-head of g. Now, by (3.42), the above element is $\partial(\widetilde{H}^{m-1}(\widetilde{\partial}_{[m]}g))$. □

Exercises

Exercise 57 (invertible elements). Assume that $\mathcal{C}$ is a left-cancellative category and $\mathcal{S}$ is a Garside family of $\mathcal{C}$ that is right-bounded by Δ. Show that, for ϵ in $\mathcal{C}^\times(x,y)$, one has $\partial_\Delta(\epsilon) = \epsilon^{-1}\Delta(x)$ and $\phi_\Delta(\epsilon) = \epsilon_1\partial_\Delta(g)$, where s in $\mathcal{S}$ and ϵ_1 in $\mathcal{C}^\times$ satisfy $\epsilon^{-1}\Delta(x) = s\epsilon_1^{-1}$.

Exercise 58 (changing Δ). Assume that $\mathcal{C}$ is a left-cancellative category and $\mathcal{S}$ is a Garside family of $\mathcal{C}$ that is right-bounded by two maps Δ, Δ'. (i) Define $E : \mathcal{O}bj(\mathcal{C}) \to \mathcal{C}^\times$ by $\Delta'(x) = \Delta(x)E(x)$. Show that $\phi_{\Delta'}(g) = E(x)^{-1}\,\phi_\Delta(g)\,E(y)$ holds for every g in $\mathcal{C}(x,y)$. (ii) Show that an element is delta-like with respect to Δ if and only if it is delta-like with respect to Δ'.

Exercise 59 (preserving $\mathcal{D}iv(\Delta)$). Assume that $\mathcal{C}$ is a category, Δ is a map from $\mathcal{O}bj(\mathcal{C})$ to $\mathcal{C}$, and ϕ is a functor from $\mathcal{C}$ to itself that commutes with Δ. Show that ϕ maps $\mathcal{D}iv(\Delta)$ and $\widetilde{\mathcal{D}iv}(\Delta)$ to themselves.

Exercise 60 (preserving normality I). Assume that $\mathcal{C}$ is a cancellative category, $\mathcal{S}$ is a Garside family of $\mathcal{C}$, and ϕ is a functor from $\mathcal{C}$ to itself. (i) Show that, if ϕ induces a permutation of $\mathcal{S}^\sharp$, then ϕ preserves $\mathcal{S}$-normality. (ii) Show that ϕ preserves non-invertibility, that is, $\phi(g)$ is invertible if and only if g is.

Exercise 61 (preserving normality II). Assume that $\mathcal{C}$ is a left-cancellative category and $\mathcal{S}$ is a Garside family of $\mathcal{C}$ that is right-bounded by a map Δ. (i) Show that ϕ_Δ preserves normality if and only if there exists an $\mathcal{S}$-head map H satisfying $H(\phi_\Delta(g)) =^\times \phi_\Delta(H(g))$ for every g in $(\mathcal{S}^\sharp)^2$, if and only if, for each $\mathcal{S}$-head map H, the above relation is satisfied. (ii) Show that a sufficient condition for ϕ_Δ to preserve normality is that ϕ_Δ preserves left-gcds on $\mathcal{S}^\sharp$, that is, if r, s, t belong to $\mathcal{S}^\sharp$ and r is a left-gcd of s and t, then $\phi_\Delta(r)$ is a left-gcd of $\phi_\Delta(s)$ and $\phi_\Delta(t)$.

Exercise 62 (normal decomposition). Assume that $\mathcal{C}$ is a left-cancellative category, Δ is a right-Garside map in $\mathcal{C}$ such that ϕ_Δ preserves normality, and f, g are elements of $\mathcal{C}$ such that fg is defined and $f \preccurlyeq \Delta^{[m]}(\cdot)$ holds, say $ff' = \Delta^{[m]}(\cdot)$ with $m \geqslant 1$. Show that f' and g admit a left-gcd and that, if h is such a left-gcd, then concatenating a $\mathcal{D}iv(\Delta)$-normal decomposition of fh and a $\mathcal{D}iv(\Delta)$-normal decomposition of $h^{-1}g$ yields a $\mathcal{D}iv(\Delta)$-normal decomposition of fg. [Hint: First show that concatenating fh and a $\mathcal{D}iv(\Delta^{[m]})$-normal decomposition of $h^{-1}g$ yields a $\mathcal{D}iv(\Delta^{[m]})$-normal decomposition of fg and apply Exercise 42 in Chapter IV.]

Exercise 63 (no left-gcd). Let $M = (\mathbb{Z} \times \mathbb{N} \times \mathbb{N}) \setminus (\mathbb{Z}_{<0} \times \{0\} \times \{0\})$, $\mathtt{a} = (1, 0, 0)$, $\mathtt{b}_i = (-i, 1, 0)$, and $\mathtt{c}_i = (-i, 0, 1)$ for $i \geqslant 0$.

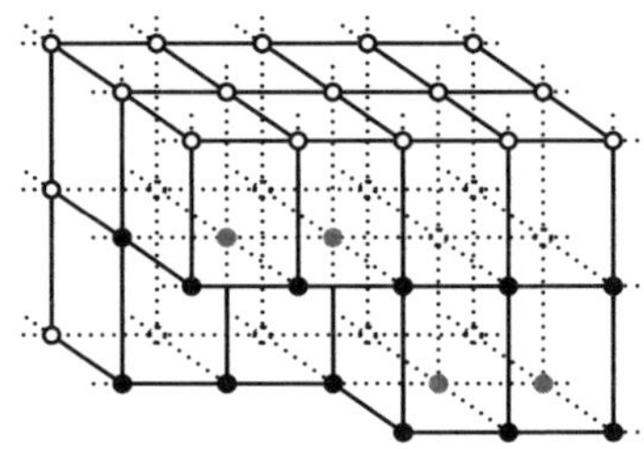

(i) Show that M is a submonoid of the additive monoid $\mathbb{Z} \times \mathbb{N}^2$, and it is presented by the commutation relations plus the relations $\mathtt{b}_i = \mathtt{b}_{i+1}\mathtt{a}$ and $\mathtt{c}_i = \mathtt{c}_{i+1}\mathtt{a}$ for $i \geqslant 0$. (ii) Show that the elements $\mathtt{b}_0$ and $\mathtt{c}_0$ admit no right-lcm. [Hint: Observe that $\mathtt{b}_0\mathtt{c}_i = \mathtt{c}_0\mathtt{c}_i = (-i, 1, 1)$ holds for every i.] (iii) Is $\Delta = (1, 1, 1)$ a Garside element in M? [Hint: Look for a left-gcd of $\mathtt{b}_0^2$ and Δ.]

Exercise 64 (iterated duality). Assume that Δ is a Garside map in a cancellative category $\mathcal{C}$. (i) Show the equality $\partial^{[m']}(g) = \partial^{[m]}(g)\,\Delta^{[m'-m]}(\phi_\Delta^m(x))$ for $m' \geqslant m$ and g in $\mathrm{Div}(\Delta^{[m]}(x))$. (ii) Show the equality $\widetilde{\partial}_{[m']}(g) = \widetilde{\Delta}_{[m'-m]}(\phi_\Delta^{-m}(y))\,\widetilde{\partial}_{[m]}(g)$ for $m' \geqslant m$ and g in $\widetilde{\mathrm{Div}}(\widetilde{\Delta}_{[m]}(y))$.

Notes

Sources and comments. The developments in this chapter are probably new in the current general framework but, of course, they directly extend a number of previously known results involving Garside elements. It seems that the latter have been considered first in D.–Paris [99] (under the additional assumption that the Garside element is the least common multiple of the atoms) and in [80] in the now standard form. The observation that, in a categorical context, a Garside element has to be replaced with a map or, equivalently, with a sequence of elements that locally play the role of a Garside element, which is implicit in [77, Chapter VIII] (see Chapter XI) and in Krammer [165] (see Chapter XIV), appears explicitly in works of Digne and Michel [109], Godelle [139], and Bessis [7]. In particular, the latter source explicitly mentions the construction of the automorphism ϕ_Δ associated with a Garside map Δ as a natural transformation.

The name "*Garside category*" was used in some sources, and it could be used in the current framework to qualify a cancellative category that admits a Garside map, that is, equivalently, that admits a Garside family that is bounded by a target-injective map. We did not find it useful to introduce this terminology here, but it is certainly natural.

The introduction of Δ-normal decompositions goes back at least to the book of D. Epstein *et al.* [119], and it is the one F.A. Garside was closed to in [125]. The extension to the category context is just an exercise, although some fine points like the connection with the symmetric normal decompositions and with co-normal (or "right-normal") decompositions requires some care.

The technical notion of a right-Garside map had not been considered before and probably is of little interest in itself, but it is useful to notice that a number of consequences of the existence of a Garside map, typically the existence of common right-multiples, follow from weaker assumptions. By the way, one could even consider the still weaker assumption that all elements of $\mathcal{S}(x, \cdot)$ admit a common right-multiple, but not demanding that the latter lies in $\mathcal{S}$: it seems that not much can be said starting from that assumption, except when the ambient category is Noetherian, in which case it turns out to be essentially equivalent to being right-bounded.

The main new result in this chapter is perhaps Proposition 2.35 about the existence of gcds and lcms when the ambient category is not assumed to be Noetherian. Corollary 2.41 shows that, when one restricts to monoids that contain no nontrivial invertible element and that are strongly Noetherian, the context of bounded Garside families essentially coincides with that of quasi-Garside monoids. In a sense, this shows that, apart from more or less pathological examples, typically non-Noetherian monoids or categories, no completely new example is to be expected. However, the current approach arguably provides a better understanding of what is crucial and what is not for each specific property, typically the existence of lcms and gcds. Also, we shall see in Chapter XII a number of non-Noetherian examples that would be non-eligible in a restricted framework, although they are not pathological in any sense.

Further questions. For a cancellative category to admit a Garside map imposes constraints, like the existence of left-gcds and left-lcms, but we do not know exactly how strong the condition is. It can be shown that, if $\mathcal{S}$ is any Garside family in a cancellative category $\mathcal{C}$ that admits lcms and gcds, then there exists an extension $\widehat{\mathcal{C}}$ of $\mathcal{C}$ in which $\mathcal{S}$ becomes bounded. The principle is to embed $\mathcal{S}$ into a bounded Garside germ $\underline{\widehat{\mathcal{S}}}$ (see Chapter VI) whose domain is defined by $\widehat{\mathcal{S}}(x, y) = \mathcal{S}(x, y){\times}\{0\} \amalg \mathcal{S}(y, x){\times}\{1\}$, and, with obvious conventions, the multiplication is given by $(s, 0) \bullet (t, 0) = (st, 0)$, $(s, 0) \bullet (t, 1) = (r, 1)$ for $rt = s$, $(s, 1) \bullet (t, 0) = (r, 1)$ for $rs = t$, and $(s, 1) \bullet (t, 1)$ never defined. An amusing verification involving the rules obeyed by the left-gcd and right-lcm shows that the germ $\underline{\widehat{\mathcal{S}}}$ admits an $\mathcal{I}$-function that obeys the $\mathcal{I}$-law and, therefore, it is a Garside germ by Proposition VI.2.8 (recognizing Garside germ I). It is then easy to check that the map $\widehat{\Delta}$ defined by $\widehat{\Delta}(x) = (1_x, 1)$ provides the expected Garside map in the category $\widehat{\mathcal{C}}$ generated by $\underline{\widehat{\mathcal{S}}}$ and that mapping g to $(g, 0)$ embeds $\mathcal{C}$ into $\widehat{\mathcal{C}}$. However, using this abstract construction to deduce new properties of $\mathcal{C}$ remains an open question.

We saw in Example 1.49 that the existence of a right-bounded Garside family does not imply the existence of right-lcms. However, the non-existence of right-lcms in this trivial example comes from the existence of several objects, and from the fact that the family is right-bounded, but not bounded. On the other hand, Proposition 2.35 shows that right-lcms do exist in the case of a bounded Garside family whenever the ambient category is left-Noetherian. This leads to the following questions:

Question 15. Does every right-cancellative monoid that admits a right-bounded Garside family admit right-lcms?

Question 16. Does every cancellative category that admits a bounded Garside family admit right-lcms?

Both questions seem open. For instance, the naive tentative consisting in starting from the presentation $\langle \mathtt{a}, \mathtt{b}, \mathtt{c}, \mathtt{d} \mid \mathtt{ab}{=}\mathtt{bc}{=}\mathtt{ca}, \mathtt{ba}{=}\mathtt{ad}{=}\mathtt{db}\rangle^+$ as on the right fails: indeed, $\mathtt{a}$ and $\mathtt{b}$ admit no right-lcm since both $\mathtt{ab}$ and $\mathtt{ba}$ are distinct right-mcm's, but the divisors of Δ do not make a Garside family since $\mathtt{a}^2\mathtt{b}$ has no unique head, being left-divisible both by $\mathtt{a}^2$ and by $\mathtt{ab}$. See Exercise 63 for another simple example failing to negatively answer Question 15.

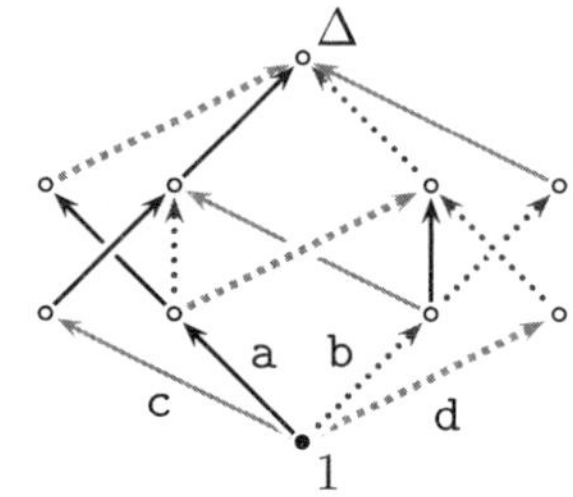

Always in the context of Noetherianity conditions, several questions involve the height of elements (maximal length of a decomposition into non-invertible elements). In particular, Proposition 2.40 leaves the following open:

Question 17. If Δ is a right-Garside map in a left-cancellative category $\mathcal{C}$ and every element $\Delta(x)$ has a finite height, is $\mathcal{C}$ strongly Noetherian?

On the other hand, Question 13 in Chapter IV specializes to

Question 18. If Δ is a Garside map in a cancellative category $\mathcal{C}$ and $\mathrm{ht}(\Delta(x)) \leqslant K$ holds for every x, is $\mathrm{ht}(\Delta^{[m]}(x))$ bounded above by a linear function of m?

We recall that Proposition 2.40 gives for $\mathrm{ht}(\Delta^{[m]}(x))$ the upper bound K^m. The question was addressed for monoids by H. Sibert in [212], but left open: it is proved therein that the answer to Question 18 is positive for the monoid $\langle \mathtt{a}, \mathtt{b} \mid \mathtt{ababa} = \mathtt{b}^2\rangle^+$, which once was a natural candidate for a negative answer. Proving that the monoid is strongly Noetherian is nontrivial, and proving that, for $\Delta = \mathtt{b}^3$, the height of Δ^m is bounded above by a linear function of m resorts on a very tricky argument combining normal and co-normal decompositions. Monoids for which the answer to Question 18 is positive have been called *tame*. The results of Charney, Meier, and Whittlesey [57] about the homology of Garside groups use the assumption that the considered monoid is tame.

We saw in Lemma 3.40 that the duality maps ∂ and $\widetilde{\partial}$, or their iterated versions $\partial^{[m]}$ and $\widetilde{\partial}_{[m]}$, exchange left- and right-divisibility, and one can wonder whether there exists a simple connection between left- and right-Noetherianity when a bounded Garside family exists. There is no clear positive answer, because the formulas of Lemma 3.40 reverse the orientation of inequalities.

Finally, note that the validity of the second domino rule has been established under two different families of conditions. In both cases, the considered Garside family $\mathcal{S}$ is assumed to be closed under left-divisor, but, in Proposition IV.1.39 (domino 2), $\mathcal{S}$ is closed under left-comultiple and the ambient category is cancellative, whereas, in Proposition 1.52, $\mathcal{S}$ is right-bounded by a map Δ such that ϕ_Δ preserves normality.

Question 19. Can one merge the arguments used to establish the second domino rules? Are there underlying common sufficient conditions?

We know of no connection between the latter conditions, but we have no example separating them either.

Chapter VI
Germs

In Chapter IV, we assumed that a category $\mathcal{C}$ is given and investigate necessary and sufficient conditions for a subfamily $\mathcal{S}$ of $\mathcal{C}$ to be a Garside family. In the current chapter, we do not start from a pre-existing category, but consider instead an abstract family $\mathcal{S}$ equipped with a partial product and investigate necessary and sufficient conditions for such a structure, called a germ, to generate a category in which $\mathcal{S}$ embeds as a Garside family. The main result here are Propositions 2.8 and 2.28, which provide simple characterizations of such structures, naturally called Garside germs. These results arguably provide intrinsic axiomatizations of Garside families. As we shall see in several subsequent examples, the germ approach provides a powerful mean for constructing and investigating Garside structures.

The chapter comprises three sections. Section 1 contains basic results about general germs and Garside germs, in particular results showing how a number of global properties of the category can be read inside the germ. Next, we establish in Section 2 the characterizations of Garside germs alluded to above. As can be expected, the criteria take a simpler form (Propositions 2.41 and 2.49) when additional assumptions are satisfied, typically (local versions of) Noetherianity or existence of lcms. We also describe a general scheme for constructing a Garside germ starting from a group with a distinguished family of generators (Proposition 2.69 and its variations). Finally, in Section 3 we introduce the natural notion of a bounded germ, corresponding to the bounded Garside families of Chapter V, and establish a new characterization (Proposition 3.12), as well as a scheme for constructing a Garside germ, this time from a group with a lattice ordering.

Main definitions and results

Definition 1.3 **(germ).** A *germ* is a triple $(\mathcal{S}, \mathbf{1}_\mathcal{S}, \bullet)$ where $\mathcal{S}$ is a precategory, $\mathbf{1}_\mathcal{S}$ is a subfamily of $\mathcal{S}$ consisting of an element 1_x with source and target x for each object x, and $\bullet$ is a partial map from $\mathcal{S}^{[2]}$ into $\mathcal{S}$ that satisfies (1.4) If $s \bullet t$ is defined, the source of $s \bullet t$ is the source of s, and its target is the target of t. (1.5) The relations $1_x \bullet s = s = s \bullet 1_y$ hold for each s in $\mathcal{S}(x, y)$. (1.6) If $r \bullet s$ and $s \bullet t$ are defined, then $(r \bullet s) \bullet t$ is defined if and only if $r \bullet (s \bullet t)$ is, in which case they are equal. A germ is said to be *left-associative* (*resp. right-associative*) if (1.7) if $(r \bullet s) \bullet t$ is defined, then so is $s \bullet t$ (*resp.* (1.8) if $r \bullet (s \bullet t)$ is defined, then so is $r \bullet s$). If $(\mathcal{S}, \mathbf{1}_\mathcal{S}, \bullet)$ is a germ, we denote by $\mathit{Cat}(\mathcal{S}, \mathbf{1}_\mathcal{S}, \bullet)$ the category $\langle \mathcal{S} \mid \mathcal{R}_\bullet \rangle^+$, where $\mathcal{R}_\bullet$ is the family of all relations $st = s \bullet t$ with s, t in $\mathcal{S}$ and $s \bullet t$ defined.

Proposition 1.11 **(germ from Garside).** *If $\mathcal{S}$ is a solid Garside family in a left-cancellative category $\mathcal{C}$, then $\mathcal{S}$ equipped with the induced partial product is a germ $\underline{\mathcal{S}}$ and $\mathcal{C}$ is isomorphic to $\mathit{Cat}(\underline{\mathcal{S}})$.*

Proposition 1.14 **(embedding).** *If $\underline{S}$ is a left-associative germ, the map ι of Notation* 1.12 *is injective and the product of $\mathcal{C}at(\underline{S})$ extends the image of $\bullet$ under ι. Moreover, ιS is a solid subfamily of $\mathcal{C}at(\underline{S})$, and it is closed under left-divisor in $\mathcal{C}at(\underline{S})$ if and only if $\underline{S}$ is right-associative.*

Definition 1.17 **(cancellative).** A germ $\underline{S}$ is called *left-cancellative* (*resp. right-cancellative*) if there exist no triple s, t, t' in $\mathcal{S}$ satisfying $t \neq t'$ and $s \bullet t = s \bullet t'$ (*resp.* $t \bullet s = t' \bullet s$). It is called *cancellative* if it is both left- and right-cancellative.

Definition 1.21 **(Noetherian germ).** A germ $\underline{S}$ is called *left-Noetherian* (*resp. right-Noetherian*) if the relation $\prec_{\mathcal{S}}$ (*resp.* the relation $\widetilde{\prec}_{\mathcal{S}}$) is well-founded. It is called *Noetherian* if it is both left- and right-Noetherian.

Definition 1.23 **(Garside germ).** A germ $\underline{S}$ is a *Garside germ* if there exists a left-cancellative category $\mathcal{C}$ such that $\mathcal{S}$ is a solid Garside family of $\mathcal{C}$.

Definition 2.5 **($\mathfrak{I}$-law, $\mathfrak{J}$-law).** If $\underline{S}$ is a germ and I, J are maps from $\mathcal{S}^{[2]}$ to $\mathcal{S}$, we say that I obeys the $\mathfrak{I}$-*law* if, for every (s_1, s_2, s_3) in $\mathcal{S}^{[3]}$ with $s_1 \bullet s_2$ defined, we have (2.6) $I(s_1, I(s_2, s_3)) =^{\times} I(s_1 \bullet s_2, s_3)$. We say that J obeys the $\mathfrak{J}$-*law* if, for every (s_1, s_2, s_3) in $\mathcal{S}^{[3]}$ with $s_1 \bullet s_2$ defined, we have (2.7) $J(s_1, s_2 \bullet J(s_2, s_3)) =^{\times} s_2 \bullet J(s_1 \bullet s_2, s_3)$. If the counterpart of (2.6) or (2.7) with $=$ replacing $=^{\times}$ is satisfied, we say that I (*resp.* J) obeys the *sharp $\mathfrak{I}$-law* (*resp.* the *sharp $\mathfrak{J}$-law*).

Proposition 2.8 **(recognizing Garside germ I).** *A germ $\underline{S}$ is a Garside germ if and only if it satisfies one of the following equivalent conditions:* (2.9) *$\underline{S}$ is left-associative, left-cancellative, and admits an $\mathfrak{I}$-function obeying the sharp $\mathfrak{I}$-law.* (2.10) *$\underline{S}$ is left-associative, left-cancellative, and admits a $\mathfrak{J}$-function obeying the sharp $\mathfrak{J}$-law.*

Proposition 2.24 **(recognizing Garside IV).** *A solid generating subfamily $\mathcal{S}$ of a left-cancellative category $\mathcal{C}$ is a Garside family if and only if one of the following equivalent conditions is satisfied:* (2.25) *There exists $I : \mathcal{S}^{[2]} \to \mathcal{S}$ satisfying $s_1 \preccurlyeq I(s_1, s_2) \preccurlyeq s_1 s_2$ for all s_1, s_2, and $I(s_1, I(s_2, s_3)) = I(s_1 s_2, s_3)$ for every $s_1|s_2|s_3$ in $\mathcal{S}^{[3]}$ with $s_1 s_2$ in $\mathcal{S}$.* (2.26) *There exists $J : \mathcal{S}^{[2]} \to \mathcal{S}$ satisfying $s_1 J(s_1, s_2) \in \mathcal{S}$ and $J(s_1, s_2) \preccurlyeq s_2$ for all s_1, s_2, and $J(s_1, s_2 J(s_2, s_3)) = s_2 J(s_1 s_2, s_3)$ for every $s_1|s_2|s_3$ in $\mathcal{S}^{[3]}$ with $s_1 s_2$ in $\mathcal{S}$.*

Definition 2.27 **(greatest $\mathfrak{I}$- or $\mathfrak{J}$-function).** An $\mathfrak{I}$-function I for a germ $\underline{S}$ is called a *greatest $\mathfrak{I}$-function* if, for every $s_1|s_2$ in $\mathcal{S}^{[2]}$, the value $I(s_1, s_2)$ is a $\preccurlyeq$-greatest element in $\mathfrak{I}_{\underline{S}}(s_1, s_2)$; same for a *greatest $\mathfrak{J}$-function*, replacing $\mathfrak{I}_{\underline{S}}(s_1, s_2)$ with $\mathfrak{J}_{\underline{S}}(s_1, s_2)$.

Proposition 2.28 **(recognizing Garside germ II).** *A germ $\underline{S}$ is a Garside germ if and only if it satisfies one of the following equivalent conditions:* (2.30) *The germ $\underline{S}$ is left-associative, left-cancellative, and admits a greatest $\mathfrak{J}$-function.* (2.32) *The germ $\underline{S}$ is left-associative, left-cancellative, and admits a greatest $\mathfrak{I}$-function.*

Proposition 2.41 **(recognizing right-Noetherian germ).** *A right-Noetherian germ $\underline{S}$ is a Garside germ if and only if it is left-associative, left-cancellative, and satisfies* (2.40). *In this case, the category $\mathcal{C}at(\underline{S})$ is right-Noetherian.*

Proposition 2.47 **(recognizing Garside germ, atomic case).** *If a germ $\underline{S}$ is associative, left-cancellative, right-Noetherian, and admits conditional right-lcms, and $\mathcal{A}$ is a subfamily of $\mathcal{S}$ such that $\mathcal{A} \cup \mathcal{S}^{\times}$ generates $\underline{S}$ and satisfies $\mathcal{S}^{\times}\mathcal{A} \subseteq \mathcal{A}\mathcal{S}^{\times}$, then $\underline{S}$ is a Garside* *germ whenever it satisfies the condition:* (2.48) *For r in $\mathcal{S}$ and s, s' in $\mathcal{A}$, if $r \bullet s$ and $r \bullet s'$ are defined, then so is $r \bullet t$ for every right-$\underline{\mathcal{S}}$-lcm t of s and s'. If $\underline{\mathcal{S}}$ is Noetherian, the criterion applies in particular when $\mathcal{A}$ is the atom family of $\underline{\mathcal{S}}$.*

Definition 2.52 **(tight).** If Σ positively generates a groupoid $\mathcal{G}$, a $\mathcal{G}$-path $g_1|\cdots|g_p$ is called Σ-*tight* if $\|g_1\cdots g_p\|_\Sigma = \|g_1\|_\Sigma + \cdots + \|g_p\|_\Sigma$ is satisfied.

Definition 2.54 **(Σ-prefix, Σ-suffix).** If Σ positively generates a groupoid $\mathcal{G}$, then for f, g in $\mathcal{G}$ with the same source, we say that f is a Σ-*prefix* of g, written $f \leqslant_\Sigma g$, if $f|f^{-1}g$ is Σ-tight. Symmetrically, we say that h is a Σ-*suffix* of g, written $h \,\widetilde{\leqslant}_\Sigma\, g$, if g, h have the same target and $gh^{-1}|h$ is Σ-tight.

Definition 2.58 **(derived structure).** If a groupoid $\mathcal{G}$ is positively generated by a family Σ and $\mathcal{H}$ is a subfamily of $\mathcal{G}$, the structure *derived from $\mathcal{H}$ and Σ*, denoted by $\mathcal{H}^\Sigma$, is $(\mathcal{H}, \mathbf{1}_\mathcal{H}, \bullet)$, where $\bullet$ is the partial operation on $\mathcal{H}$ such that $f \bullet g = h$ holds if and only if $fg = h$ holds in $\mathcal{G}$ and $f|g$ is Σ-tight.

Proposition 2.66 **(derived Garside II).** *If a groupoid $\mathcal{G}$ is positively generated by a family Σ and $\mathcal{H}$ is a subfamily of $\mathcal{G}$ that is closed under Σ-prefix and Σ-suffix, then $\mathcal{H}^\Sigma$ is a Garside germ whenever the following two conditions are satisfied:* (2.67) *Any two elements of Σ that admit a common $\leqslant_\Sigma$-upper bound in $\mathcal{H}$ admit a least $\leqslant_\Sigma$-upper bound in $\mathcal{H}$.* (2.68) *For f in $\mathcal{H}$ and s, s' in Σ, if $f|s$ and $f|s'$ are Σ-tight, then so is $f|h$ for every least $\leqslant_\Sigma$-upper bound h of s and s'.*

Proposition 2.69 **(derived Garside III).** *If a groupoid $\mathcal{G}$ is positively generated by a family Σ and $\mathcal{H}$ is a subfamily of $\mathcal{G}$ that is closed under Σ-prefix and Σ-suffix and any two elements of $\mathcal{H}$ admit a $\leqslant_\Sigma$-least upper bound, then $\mathcal{H}^\Sigma$ is a Garside germ.*

Definition 3.1 **(right-bounded).** A germ $\underline{\mathcal{S}}$ is called *right-bounded* by a map Δ from $\mathcal{O}bj(\mathcal{S})$ to $\mathcal{S}$ if (i) for every object x, the source of $\Delta(x)$ is x, (ii) for every s in $\mathcal{S}(x, \cdot)$, there exists t in $\mathcal{S}$ satisfying $s \bullet t = \Delta(x)$. If $\underline{\mathcal{S}}$ is left-cancellative, we denote by $\partial_\Delta(s)$, or ∂s, the element t involved in (ii).

Proposition 3.5 **(recognizing right-bounded Garside germ).** *If a germ $\underline{\mathcal{S}}$ is associative, left-cancellative, right-bounded by a map Δ, and admits left-gcds, then $\underline{\mathcal{S}}$ is a Garside germ, and its image in $\mathcal{C}at(\underline{\mathcal{S}})$ is a Garside family that is right-bounded by Δ.*

Definition 3.7 **(bounded germ).** A germ $\underline{\mathcal{S}}$ is called *bounded* by a map Δ from $\mathcal{O}bj(\mathcal{S})$ to $\mathcal{S}$ if it is right-bounded by Δ and, in addition, for every object y, there exists an object x such that, for every s in $\mathcal{S}(\cdot, y)$, there exists r in $\mathcal{S}(x, \cdot)$ satisfying $r \bullet s = \Delta(x)$.

Proposition 3.12 **(recognizing bounded Garside germ).** *For an associative, cancellative, and bounded germ $\underline{\mathcal{S}}$, the following conditions are equivalent:* (i) *the germ $\underline{\mathcal{S}}$ is a Garside germ,* (ii) *the germ $\underline{\mathcal{S}}$ admits left-gcds.*

Proposition 3.13 **(right-cancellativity).** *Assume that $\underline{\mathcal{S}}$ is a germ that is associative, cancellative, with no nontrivial invertible element, and bounded by a map Δ, that Δ is target-injective, that ∂ is a bijection from $\mathcal{S}$ to itself, and that, if $s \bullet t$ is defined, then so is $\partial^{-2}s \bullet \partial^{-2}t$. Then $\mathcal{C}at(\underline{\mathcal{S}})$ is cancellative and ϕ_Δ is an automorphism.*

1 Germs

If $\mathcal{C}$ is a category and $\mathcal{S}$ is a subfamily of $\mathcal{C}$, the restriction of the multiplication to $\mathcal{S}$ is a partial binary operation that gives information about $\mathcal{C}$. This lacunary structure is what we call a *germ*. In this section, we gather basic results about germs and introduce the notion of a Garside germ that will be central in the sequel.

The section is organized as follows. In Subsection 1.1, we introduce the abstract notion of a germ, which is a precategory equipped with a partial product satisfying some weak associativity conditions. In Subsection 1.2, we give a sufficient condition for a germ to embed in the canonically associated category and, when it is so, we establish a few basic results involving derived notions like divisibility or atoms. Finally, in Subsection 1.3, we introduce the notion of a Garside germ and deduce from the results of Chapter IV several transfer results from a Garside germ to the associated category.

1.1 The notion of a germ

If $\mathcal{C}$ is a category and $\mathcal{S}$ is a subfamily of $\mathcal{C}$, then, if s, t lie in $\mathcal{S}$ and st is defined, that is, the target of s is the source of t, the product st may belong or not belong to $\mathcal{S}$. When we restrict to the case when the product belongs to $\mathcal{S}$, we obtain a *partial* map from $\mathcal{S} \times \mathcal{S}$ to $\mathcal{S}$. We observe in this subsection that, when $\mathcal{S}$ is a solid Garside family in $\mathcal{C}$, the whole structure of $\mathcal{C}$ can be recovered from this partial product on $\mathcal{S}$.

Notation 1.1 (partial product $\bullet_{\mathcal{S}}$). If $\mathcal{S}$ is a subfamily of a category $\mathcal{C}$, we denote by $\underline{\mathcal{S}}$ the structure $(\mathcal{S}, \mathbf{1}_{\mathcal{S}}, \bullet)$, where $\bullet$ is the partial map from $\mathcal{S}^{[2]}$ to $\mathcal{S}$ defined by $s \bullet t = st$ whenever st lies in $\mathcal{S}$.

Two typical examples are displayed in Figure 1.

$\bullet$	1	a	b	ab	c	ac	bc	Δ
1	1	a	b	ab	c	ac	bc	Δ
a	a		ab		ac		Δ	
b	b	ab			bc	Δ		
ab	ab				Δ			
c	c	ac	bc	Δ				
ac	ac		Δ					
bc	bc	Δ						
Δ	Δ							

$\bullet$	1	σ_1	σ_2	$\sigma_1\sigma_2$	$\sigma_2\sigma_1$	Δ
1	1	σ_1	σ_2	$\sigma_1\sigma_2$	$\sigma_2\sigma_1$	Δ
σ_1	σ_1		$\sigma_1\sigma_2$		Δ	
σ_2	σ_2	$\sigma_2\sigma_1$		Δ		
$\sigma_1\sigma_2$	$\sigma_1\sigma_2$	Δ				
$\sigma_2\sigma_1$	$\sigma_2\sigma_1$		Δ			
Δ	Δ					

Figure 1. Partial products induced by the ambient monoid multiplication, in the case of the family S_3 in the monoid $\mathbb{N}^3$ (left, Reference Structure 1, page 3, writing $\mathtt{a}, \mathtt{b}, \mathtt{c}$ for $\mathtt{a}_1, \mathtt{a}_2, \mathtt{a}_3$), and in the case of the family $\mathrm{Div}(\Delta_3)$ in the braid monoid B_3^+ (right, Reference Structure 2, page 5, writing Δ for Δ_3).

In the above situation, the partial product necessarily obeys some constraints. We recall that, if $\mathcal{C}$ is a category and $\mathcal{S}$ is included in $\mathcal{C}$, then for x, y in $\mathit{Obj}(\mathcal{C})$ we denote by $\mathcal{S}(x, y)$ the family of all elements of $\mathcal{S}$ with source x and target y.

Lemma 1.2. *Assume that $\mathcal{S}$ is a subfamily of a category $\mathcal{C}$ that includes $\mathbf{1}_\mathcal{C}$. Let $\bullet$ be the partial operation on $\mathcal{S}$ induced by the multiplication of $\mathcal{C}$.*

(i) *If $s \bullet t$ is defined, the source (resp. target) of $s \bullet t$ is that of s (resp. of t).*

(ii) *The relations $1_x \bullet s = s = s \bullet 1_y$ hold for each s in $\mathcal{S}(x, y)$.*

(iii) *If $r \bullet s$ and $s \bullet t$ are defined, then $(r \bullet s) \bullet t$ is defined if and only if $r \bullet (s \bullet t)$ is, in which case they are equal.*

Moreover, if $\mathcal{S}$ is closed under right- (resp. left-) divisor in $\mathcal{C}$, then, if $(r \bullet s) \bullet t$ (resp. if $r \bullet (s \bullet t)$) is defined, then so is $s \bullet t$ (resp. so is $r \bullet s$).

Proof. Points (i) and (ii) follow from $\bullet$ being included in the multiplication of $\mathcal{C}$. As for (iii), it follows from associativity in $\mathcal{C}$: saying that $(r \bullet s) \bullet t$ exists means that the product $(rs)t$ belongs to $\mathcal{S}$, hence so does $r(st)$. As, by assumption, $r \bullet s$ exists, this amounts to $r \bullet (s \bullet t)$ being defined.

For the final point, if $\mathcal{S}$ is closed under right-divisor, the assumption that $(r \bullet s) \bullet t$ is defined implies that $(rs)t$, hence $r(st)$ as well, belong to $\mathcal{S}$, which implies that st belongs to $\mathcal{S}$, hence that $s \bullet t$ is defined. The argument is symmetric when $\mathcal{S}$ is closed under left-divisor. □

In order to investigate the above situation, we shall start from abstract families that obey the rules of Lemma 1.2. To this end, we introduce a specific terminology. We recall from Definition II.1.4 that a precategory consists of two collections $\mathcal{S}$ and $\mathcal{O}bj(\mathcal{S})$ with two maps, source and target, from $\mathcal{S}$ to $\mathcal{O}bj(\mathcal{S})$.

Definition 1.3 (germ). A *germ* is a triple $(\mathcal{S}, \mathbf{1}_\mathcal{S}, \bullet)$, where $\mathcal{S}$ is a precategory, $\mathbf{1}_\mathcal{S}$ is a subfamily of $\mathcal{S}$ consisting of an element 1_x with source and target x for each object x, and $\bullet$ is a partial map from $\mathcal{S}^{[2]}$ into $\mathcal{S}$ that satisfies

(1.4) If $s \bullet t$ is defined, the source of $s \bullet t$ is the source of s, and its target is the target of t,

(1.5) The relations $1_x \bullet s = s = s \bullet 1_y$ hold for each s in $\mathcal{S}(x, y)$,

(1.6) If $r \bullet s$ and $s \bullet t$ are defined, then $(r \bullet s) \bullet t$ is defined if and only if $r \bullet (s \bullet t)$ is, in which case they are equal.

The germ is said to be *left-associative* (*resp. right-associative*) if

(1.7) if $(r \bullet s) \bullet t$ is defined, then so is $s \bullet t$

(1.8) (*resp.* if $r \bullet (s \bullet t)$ is defined, then so is $r \bullet s$).

If $(\mathcal{S}, \mathbf{1}_\mathcal{S}, \bullet)$ is a germ, we denote by $\mathit{Cat}(\mathcal{S}, \mathbf{1}_\mathcal{S}, \bullet)$ the category $\langle \mathcal{S} \mid \mathcal{R}_\bullet \rangle^+$, where $\mathcal{R}_\bullet$ is the family of all relations $st = s \bullet t$ with s, t in $\mathcal{S}$ and $s \bullet t$ defined. In the case of a germ with one object, the associated category is a monoid, and we write $\mathcal{M}on()$ for $\mathit{Cat}()$.

Example 1.9 (germ). The structures of Table 1 are germs as, by definition, the partial products are induced by the composition maps from the monoids $\mathbb{N}^3$ and B_3^+, respectively, so, by Lemma 1.2, (1.4), (1.5), and (1.6) are obeyed. Moreover, as the involved families S_3 and $\mathrm{Div}(\Delta_3)$ are closed under right- and left-divisor, these germs are left- and right-associative.

Here are some notational conventions. First, by default, we shall always use $\bullet$ for the partial product of a germ, and write $r = s \bullet t$ to mean "$s \bullet t$ is defined and equal to r". Next, we shall in general use the notation $\underline{\mathcal{S}}$ for a germ whose domain is $\mathcal{S}$. There would be no big danger in using the same letter for the germ and the underlying domain, but, in this way, the notation makes it visible that a germ is more than a precategory: so, for instance, $\mathcal{C}at(\underline{\mathcal{S}})$ makes sense, whereas $\mathcal{C}at(\mathcal{S})$ does not.

By Lemma 1.2, every subfamily $\mathcal{S}$ of a category $\mathcal{C}$ that includes $\mathbf{1}_{\mathcal{C}}$ gives rise to an induced germ $\underline{\mathcal{S}}$ and, from there, to a derived category $\mathcal{C}at(\underline{\mathcal{S}})$. Then, all relations of $\mathcal{R}_{\bullet}$ are valid in $\mathcal{C}$ by construction, so $\mathcal{C}$ is a quotient of $\mathcal{C}at(\underline{\mathcal{S}})$. In most cases, even if $\mathcal{S}$ generates $\mathcal{C}$, the partial product of $\underline{\mathcal{S}}$ does not determine the product of $\mathcal{C}$, and $\mathcal{C}$ is a proper quotient of $\mathcal{C}at(\underline{\mathcal{S}})$.

Example 1.10 (proper quotient). Assume that $\mathcal{C}$ is a category generated by a family of atoms $\mathcal{A}$ and thus containing no nontrivial invertible element. Then the partial product of the germ $\underline{\mathcal{A} \cup \mathbf{1}_{\mathcal{C}}}$ only consists of the trivial instances listed in (1.5), so the category $\mathcal{C}at(\underline{\mathcal{A} \cup \mathbf{1}_{\mathcal{C}}})$ is a free category based on $\mathcal{A}$. Hence it is not isomorphic to $\mathcal{C}$ when $\mathcal{C}$ is not free.

On the other hand, the following positive result is the origin of our interest in germs. We recall that a subfamily $\mathcal{S}$ of a category $\mathcal{C}$ is called solid if $\mathcal{S}$ includes $\mathbf{1}_{\mathcal{C}}$ and is closed under right-divisor.

Proposition 1.11 (germ from Garside). *If $\mathcal{S}$ is a solid Garside family in a left-cancellative category $\mathcal{C}$, then $\mathcal{S}$ equipped with the induced partial product is a germ $\underline{\mathcal{S}}$ and $\mathcal{C}$ is isomorphic to $\mathcal{C}at(\underline{\mathcal{S}})$.*

Proof. By Proposition IV.1.24 (recognizing Garside II), $\mathcal{S}$ is closed under right-comultiple, so, by Proposition IV.3.4 (presentation, solid case), $\mathcal{C}$ is presented by the relations $r = st$ with r, s, t in $\mathcal{S}$, that is, $\mathcal{C}$ admits a presentation that is precisely that of $\mathcal{C}at(\underline{\mathcal{S}})$. □

In other words, in the case of a solid Garside family $\mathcal{S}$, the induced germ $\underline{\mathcal{S}}$ contains all information needed to determine the ambient category.

1.2 The embedding problem

From now on, our aim is to characterize which germs give rise to Garside families. The first question is whether a germ necessarily embeds in the associated category. Here we show that this need not be in general, but that satisfying (1.7) is a sufficient condition.

Notation 1.12 (relation $\equiv$). For $\underline{S}$ a germ, we denote by $\equiv$ the congruence on S^* generated by the relations of $\mathcal{R}_\bullet$, and by ι the prefunctor from S to $S^*/\equiv$ that is the identity on $\mathcal{O}bj(S)$ and maps s to the $\equiv$-class of (s).

We recall that S^* denotes the free category based on S, that is, the category of all S-paths. Path concatenation is denoted by $|$, and we do not distinguish between an element s of S and the associated length one path. Thus, mixed expressions like $s|w$ with s in S and w in S^* make sense. By definition, the category $\mathcal{C}at(\underline{S})$ is $S^*/\equiv$, and $\equiv$ is the equivalence relation on S^* generated by all pairs

$$(1.13)\qquad (s_1|\cdots|s_i|s_{i+1}|\cdots|s_p \,,\; s_1|\cdots|s_i \bullet s_{i+1}|\cdots|s_p),$$

that is, the pairs in which two adjacent entries are replaced with their $\bullet$-product, assuming that the latter exists.

Proposition 1.14 (embedding). *If $\underline{S}$ is a left-associative germ, the map ι of Notation 1.12 is injective and the product of $\mathcal{C}at(\underline{S})$ extends the image of $\bullet$ under ι. Moreover, ιS is a solid subfamily of $\mathcal{C}at(\underline{S})$, and it is closed under left-divisor in $\mathcal{C}at(\underline{S})$ if and only if $\underline{S}$ is right-associative.*

Proof. We inductively define a partial map Π from S^* to S by

$$(1.15)\qquad \Pi(\varepsilon_x) = 1_x \text{ and } \Pi(s|w) = s \bullet \Pi(w) \text{ if } s \text{ lies in } S \text{ and } s \bullet \Pi(w) \text{ is defined.}$$

We claim that Π induces a well-defined partial map from $\mathcal{C}at(\underline{S})$ to S, more precisely that, if w, w' are $\equiv$-equivalent elements of S^*, then $\Pi(w)$ exists if and only if $\Pi(w')$ does, and in this case they are equal. To prove this, we may assume that $\Pi(w)$ or $\Pi(w')$ is defined and that (w, w') is of the type (1.13). Let $s = \Pi(s_{i+2}|\cdots|s_p)$. The assumption that $\Pi(w)$ or $\Pi(w')$ is defined implies that s is defined. Then (1.15) gives $\Pi(w) = \Pi(s_1|\cdots|s_{i-1}|t)$ whenever $\Pi(w)$ is defined, and $\Pi(w') = \Pi(s_1|\cdots|s_{i-1}|t')$ whenever $\Pi(w')$ is defined, with $t = \Pi(s_i|s_{i+1}|s)$ and $t' = \Pi(s_i \bullet s_{i+1}|s)$, that is,

$$t = s_i \bullet (s_{i+1} \bullet s) \quad \text{and} \quad t' = (s_i \bullet s_{i+1}) \bullet s.$$

So the point is to prove that t is defined if and only if t' is, in which case they are equal. Now, if t is defined, the assumption that $s_i \bullet s_{i+1}$ is defined plus (1.6) imply that t' exists and equals t. Conversely, if t' is defined, (1.7) implies that $s_{i+1} \bullet s$ is defined, and then (1.6) implies that t exists and equals t'.

Assume that s, s' lie in S and $\iota s = \iota s'$ holds, that is, the length-one paths (s) and (s') are $\equiv$-equivalent. The above claim gives $s = \Pi((s)) = \Pi((s')) = s'$, so ι is injective.

Next, assume that s, t belong to S and $s \bullet t$ is defined. We have $s|t \equiv s \bullet t$, which means that the product of ιs and ιt in $\mathcal{C}at(\underline{S})$ is $\iota(s \bullet t)$.

By definition, the objects of $\mathcal{C}at(\underline{S})$ coincide with those of S, so ιS contains all identity elements of $\mathcal{C}at(\underline{S})$. On the other hand, assume that s belongs to S and g is a right-divisor of ιs in $\mathcal{C}at(\underline{S})$. This means that there exist elements $s_1, ..., s_{p+q}$ of S such that ιs

is the $\equiv$-class of $s_1|\cdots|s_{p+q}$ and g is the $\equiv$-class of $s_{p+1}|\cdots|s_{p+q}$. By the claim above, the first relation implies that $\Pi(s_1|\cdots|s_{p+q})$ exists (and equals ιs). By construction, this implies that $\Pi(s_{p+1}|\cdots|s_{p+q})$ exists as well, whence $g = \iota\Pi(s_{p+1}|\cdots|s_{p+q})$, and g belongs to $\iota\mathcal{S}$. Hence, $\iota\mathcal{S}$ is closed under right-divisor in $\mathcal{C}at(\underline{\mathcal{S}})$ and, therefore, it is a solid subfamily.

Assume now that $\underline{\mathcal{S}}$ is right-associative, s belongs to $\mathcal{S}$, and g left-divides ιs in $\mathcal{C}at(\underline{\mathcal{S}})$. As above, there exist $s_1, ..., s_{p+q}$ in $\mathcal{S}$ such that ιs is the $\equiv$-class of $s_1|\cdots|s_{p+q}$ and g is the $\equiv$-class of $s_1|\cdots|s_p$. Hence $\Pi(s_1|\cdots|s_{p+q})$ is defined. Using (1.8) q times to push the brackets to the left, we deduce that $\Pi(s_1|\cdots|s_p)$ is defined as well, so g belongs to $\iota\mathcal{S}$, and $\iota\mathcal{S}$ is closed under left-divisor in $\mathcal{C}at(\underline{\mathcal{S}})$.

Conversely, assume that $\iota\mathcal{S}$ is closed under left-divisor in $\mathcal{C}at(\underline{\mathcal{S}})$ and r, s, t are elements of $\mathcal{S}$ such that $r \bullet (s \bullet t)$ is defined. Then $\iota r\, \iota s$ left-divides $\iota(r \bullet (s \bullet t))$ in $\mathcal{C}at(\underline{\mathcal{S}})$, hence there must exist s' in $\mathcal{S}$ satisfying $\iota r\, \iota s = \iota s'$. By definition, this means that $r|s \equiv s'$ holds. As $\Pi(s')$ exists, the claim above implies that $\Pi(r|s)$ exists as well, which means that $r \bullet s$ is defined. So (1.8) is satisfied and $\underline{\mathcal{S}}$ is right-associative. □

Example 1.16 (not closed). The conclusion of Proposition 1.14 may fail for a germ that is neither left-, nor right-associative. For instance, let $\mathcal{S}$ consist of six elements 1, $\mathtt{a}$, $\mathtt{b}$, $\mathtt{c}$, $\mathtt{d}$, $\mathtt{e}$, all with the same source and target, and $\bullet$ be defined by $1 \bullet x = x \bullet 1 = x$ for each x, plus $\mathtt{a} \bullet \mathtt{b} = \mathtt{c}$ and $\mathtt{c} \bullet \mathtt{d} = \mathtt{e}$. Then $\underline{\mathcal{S}}$ is a germ: indeed, the premises of (1.6) can be satisfied only when at least one of the considered elements is 1, in which case the property follows from (1.5). Now, in the monoid $\mathcal{M}on(\underline{\mathcal{S}})$, we have $\iota\mathtt{e} = \iota\mathtt{c}\,\iota\mathtt{d} = (\iota\mathtt{a}\,\iota\mathtt{b})\,\iota\mathtt{d} = \iota\mathtt{a}\,(\iota\mathtt{b}\,\iota\mathtt{d})$, whereas $\mathtt{b} \bullet \mathtt{d}$ is not defined, and $\iota\mathtt{b}\,\iota\mathtt{d}$ is a right-divisor of an element of $\iota\mathcal{S}$ that does not belong to $\iota\mathcal{S}$.

We refer to Exercise 65 for a similar example where the canonical map ι is not injective (such an example has to be more complicated, as we are to find two expressions that receive different evaluations: to prevent (1.6) to force these expressions to have the same evaluation, we need that the expressions share no common subexpression).

In the context of Proposition 1.14, we shall identify $\mathcal{S}$ with its image in $\mathcal{C}at(\underline{\mathcal{S}})$, that is, drop the canonical injection ι. Before going on, we establish a few consequences of the existence of the function Π used in the proof of Proposition 1.14. To state the results, we introduce local versions adapted to germs of a few general notions.

Definition 1.17 (cancellative). A germ $\underline{\mathcal{S}}$ is called *left-cancellative* (*resp. right-cancellative*) if there exist no triple s, t, t' in $\mathcal{S}$ satisfying $t \neq t'$ and $s \bullet t = s \bullet t'$ (*resp.* $t \bullet s = t' \bullet s$). It is called *cancellative* if it is both left- and right-cancellative.

If a germ $\underline{\mathcal{S}}$ is a category, that is, if the operation $\bullet$ is defined everywhere, (left)-cancellativity as defined above coincides with the usual notion. Whenever $\underline{\mathcal{S}}$ embeds into $\mathcal{C}at(\underline{\mathcal{S}})$, it should be clear that $\mathcal{C}at(\underline{\mathcal{S}})$ can be left-cancellative only if $\underline{\mathcal{S}}$ is left-cancellative but, conversely, the later property is a priori weaker than the former as the left-cancellativity of a germ involves the products that lie inside the germ only.

If $\underline{S}$ is a germ, we have a natural notion of invertible element: an element ϵ of $\mathcal{S}(x,y)$ is called *invertible* if there exists ϵ' in $\mathcal{S}$ satisfying $\epsilon \bullet \epsilon' = 1_x$ and $\epsilon' \bullet \epsilon = 1_y$. We denote by $\mathcal{S}^\times$ the family of all invertible elements of $\underline{S}$.

Definition 1.18 (**$\underline{S}$-divisor, $\underline{S}$-multiple**). Assume that $\underline{S}$ is a germ. For s, t in $\mathcal{S}$, we say that s is a *left-$\underline{S}$-divisor* of t, or that t is a *right-$\underline{S}$-multiple* of s, denoted by $s \preccurlyeq_{\mathcal{S}} t$, if there exists t' in $\mathcal{S}$ satisfying $s \bullet t' = t$. We say that $s \prec_{\mathcal{S}} t$ (*resp.* $s =^{\times}_{\mathcal{S}} t$) holds if there exists t' in $\mathcal{S} \setminus \mathcal{S}^\times$ (*resp.* in $\mathcal{S}^\times$) satisfying $s \bullet t' = t$. Similarly, we say that $s \widetilde{\succcurlyeq}_{\mathcal{S}} t$ (*resp.* $s \widetilde{\succ}_{\mathcal{S}} t$) holds if there exists t' in $\mathcal{S}$ (*resp.* in $\mathcal{S} \setminus \mathcal{S}^\times$) satisfying $t = t' \bullet s$.

In principle, $\preccurlyeq_{\underline{S}}$ would be a better notation than $\preccurlyeq_{\mathcal{S}}$, as the relation involves the partial product on $\mathcal{S}$ and not only $\mathcal{S}$, but the difference is not really visible. Note that the current definition of $\preccurlyeq_{\mathcal{S}}$ and its analogs is compatible with those of Definition IV.2.14: whenever $\underline{S}$ embeds in $\mathcal{C}at(\underline{S})$, the two versions of $s \preccurlyeq_{\mathcal{S}} t$ express that $st' = t$ holds in $\mathcal{C}at(\underline{S})$ for some t' lying in $\mathcal{S}$. So, inasmuch as we consider germs that embed in the associated category, there is no ambiguity in using the same terminology and notation.

Lemma 1.19. *Assume that $\underline{S}$ is a left-associative germ.*

(i) *An element of $\mathcal{S}$ is invertible in $\mathcal{C}at(\underline{S})$ if and only if it is invertible in $\mathcal{S}$, that is, $\mathcal{C}at(\underline{S})^\times = \mathcal{S}^\times$ holds.*

(ii) *For s, t in $\mathcal{S}$, the relation $s \preccurlyeq t$ holds in $\mathcal{C}at(\underline{S})$ if and only if $s \preccurlyeq_{\mathcal{S}} t$ holds.*

(iii) *The relation $\preccurlyeq_{\mathcal{S}}$ is transitive. If $r \bullet s$ and $r \bullet t$ are defined, then $s \preccurlyeq_{\mathcal{S}} t$ implies $r \bullet s \preccurlyeq_{\mathcal{S}} r \bullet t$, and $s =^{\times}_{\mathcal{S}} t$ implies $r \bullet s =^{\times}_{\mathcal{S}} r \bullet t$.*

(iv) *If, in addition, $\underline{S}$ is left-cancellative, then $s =^{\times}_{\mathcal{S}} t$ is equivalent to the conjunction of $s \preccurlyeq_{\mathcal{S}} t$ and $t \preccurlyeq_{\mathcal{S}} s$ and, if $r \bullet s$ and $r \bullet t$ are defined, then $s \preccurlyeq_{\mathcal{S}} t$ is equivalent to $r \bullet s \preccurlyeq_{\mathcal{S}} r \bullet t$, and $s =^{\times}_{\mathcal{S}} t$ is equivalent to $r \bullet s =^{\times}_{\mathcal{S}} r \bullet t$.*

Proof. (i) Assume $\epsilon \in \mathcal{S}$. If $\epsilon \bullet \epsilon' = 1_{\cdot}$ holds in $\mathcal{S}$, then $\epsilon\epsilon' = 1_{\cdot}$ holds in $\mathcal{C}at(\underline{S})$, so invertibility in $\mathcal{S}$ implies invertibility in $\mathcal{C}at(\underline{S})$. In the other direction, assume that $\epsilon\epsilon' = 1_x$ holds in $\mathcal{C}at(\underline{S})$. Then ϵ and 1_x lie in $\mathcal{S}$ and, by Proposition 1.14 (embedding), $\mathcal{S}$ is closed under right-divisor in $\mathcal{C}at(\underline{S})$. Hence, ϵ' must lie in $\mathcal{S}$, and ϵ is invertible in $\mathcal{S}$.

(ii) Assume $s, t \in \mathcal{S}$ and $t = st'$ in $\mathcal{C}at(\underline{S})$. As t belongs to $\mathcal{S}$ and $\mathcal{S}$ is closed under right-divisor in $\mathcal{C}at(\underline{S})$, the element t' must belong to $\mathcal{S}$. So we have $t \equiv s|t'$, whence, applying the map Π of Proposition 1.14, $t = \Pi(t) = \Pi(s|t') = s \bullet t'$. Therefore, $s \preccurlyeq_{\mathcal{S}} t$ is satisfied. The converse implication is straightforward.

(ii) As the relation $\preccurlyeq$ in $\mathcal{C}at(\underline{S})$ is transitive, (i) implies that $\preccurlyeq_{\mathcal{S}}$ is transitive as well (a direct alternative proof from left-associativity is also possible). Now, assume that $r \bullet s$ and $r \bullet t$ are defined, and $t = s \bullet t'$ holds. By (1.6), we deduce $r \bullet t = r \bullet (s \bullet t') = (r \bullet s) \bullet t'$, whence $r \bullet s \preccurlyeq_{\mathcal{S}} r \bullet t$.

(iii) The argument is the same as in Lemma II.2.4: assume $s = t \bullet \epsilon$ and $t = s \bullet \epsilon'$. We deduce $s = (s \bullet \epsilon') \bullet \epsilon$, whence $s = s \bullet (\epsilon' \bullet \epsilon)$ by left-associativity. By left-cancellativity, we deduce $\epsilon' \bullet \epsilon = 1_y$ (y the target of s). So ϵ and ϵ' are invertible, and $s =^{\times}_{\mathcal{S}} t$ holds.

Assume now that $r \bullet s$ and $r \bullet t$ are defined and $r \bullet s \preccurlyeq_{\mathcal{S}} r \bullet t$ holds. So we have $r \bullet t = (r \bullet s) \bullet t'$ for some t'. By left-associativity, we deduce that $s \bullet t'$ is defined and we have $r \bullet t = r \bullet (s \bullet t')$, whence $t = s \bullet t'$ by left-cancellativity. So $s \preccurlyeq_{\mathcal{S}} t$ holds. □

We conclude this preparatory section with a few observations about atoms. Atoms in a category were introduced in Subsection II.2.5 as the elements with no decomposition $g_1|g_2$ with g_1 and g_2 non-invertible. The definition can be mimicked in a germ.

Definition 1.20 (atom). An element s of a germ $\underline{S}$ is called an *atom* if s is not invertible and every decomposition of s in $\underline{S}$ contains at most one non-invertible element.

The connection between the atoms of a germ and those of its category is simple. If $\underline{S}$ is a left-associative germ, the atoms of $\mathcal{C}at(\underline{S})$ are the elements of the form $t\epsilon$ with t an atom of $\underline{S}$ and ϵ an invertible element of $\underline{S}$, see Exercise 67(i). One cannot expect more in general, namely there may exist atoms in the category that do not belong to the germ, see Exercise 67(ii).

As in the category case, atoms need not exist in every germ, but their existence is guaranteed when convenient Noetherianity conditions are satisfied.

Definition 1.21 (Noetherian germ). A germ $\underline{S}$ is called *left-Noetherian* (*resp. right-Noetherian*) if the relation $\prec_S$ (*resp.* the relation $\widetilde{\succ}_S$) is well-founded. It is called *Noetherian* if it is both left- and right-Noetherian.

We recall that a relation R is *well-founded* if every nonempty subfamily of its domain has a minimal element, that is, an element m such that xRm holds for no x. By the standard argument recalled in Chapter II and IV, a germ $\underline{S}$ is left- (*resp.* right-) Noetherian if and only if there exists a left- (*resp.* right-) Noetherianity *witness* for $\underline{S}$, that is, a function $\lambda : S \to \mathbf{Ord}$ such that $s \prec_S t$ (*resp.* $s \widetilde{\succ}_S t$) implies $\lambda(s) < \lambda(t)$. It follows from the already mentioned compatibility that, if a germ $\underline{S}$ embeds in $\mathcal{C}at(\underline{S})$, then the germ $\underline{S}$ is right-Noetherian if and only if the family S is locally right-Noetherian in $\mathcal{C}at(\underline{S})$ in the sense of Definition IV.2.16. Then we obtain the expected result on the existence of atoms:

Proposition 1.22 (atoms generate). *If $\underline{S}$ is a left-associative germ that is Noetherian, the category $\mathcal{C}at(\underline{S})$ is generated by the atoms and the invertible elements of $\underline{S}$.*

Proof. By definition, $\mathcal{C}at(\underline{S})$ is generated by S, so it suffices to show that every element of S is a product of atoms and invertible elements. The argument is the same as for Propositions II.2.55 (atom) and II.2.58 (atoms generate). First, every non-invertible element of S must be left-divisible (in the sense of $\preccurlyeq_S$) by an atom of $\underline{S}$ whenever $\underline{S}$ is left-Noetherian: starting from s, we find $s_1, ..., s_p$ such that s_1 is an atom of $\underline{S}$ and $s = (\cdots(s_1 \bullet s_2)\cdots) \bullet s_p$ holds. As $\underline{S}$ is left-associative, this implies $s_1 \preccurlyeq_S s$. Next, we deduce that every element s of S is a product of atoms and invertible elements using induction on $\lambda(s)$, where λ is a right-Noetherianity witness for S: for $\lambda(s) = 0$, the element s must be invertible; otherwise, we write $s = s_1 \bullet s'$, with s_1 an atom of $\underline{S}$, and apply the induction hypothesis to s', which is legal as $\lambda(s') < \lambda(s)$ holds. $\square$

1.3 Garside germs

The main situation we shall be interested in is that of a germ $\underline{S}$ that embeds in the associated category $\mathcal{C}at(\underline{S})$ and, in addition, is a Garside family in $\mathcal{C}at(\underline{S})$.

Definition 1.23 (Garside germ). A germ $\underline{S}$ is a *Garside germ* if there exists a left-cancellative category $\mathcal{C}$ such that $\mathcal{S}$ is a solid Garside family of $\mathcal{C}$.

Lemma 1.11 says that, if $\mathcal{S}$ is a solid Garside family in some left-cancellative category $\mathcal{C}$, the latter must isomorphic to $\mathcal{C}at(\underline{S})$. So, a germ $\underline{S}$ is a Garside germ if and only if the category $\mathcal{C}at(\underline{S})$ is left-cancellative and $\mathcal{S}$ is a solid Garside family in $\mathcal{C}at(\underline{S})$. In other words, in Definition 1.23, we can always assume that the category $\mathcal{C}$ is $\mathcal{C}at(\underline{S})$. Restricting to Garside families that are solid is natural, because a germ $\underline{S}$ always contains all identity elements of $\mathcal{C}at(\underline{S})$, and it is closed under right-divisor in $\mathcal{C}at(\underline{S})$ whenever it is left-associative, which will always be assumed. On the other hand, considering solid Garside families is not a real restriction as, by Lemma IV.2.3, for every Garside family $\mathcal{S}$, the family $\mathcal{S}^\sharp$ is a solid Garside family that, by Lemma III.1.10, leads to the same normal decompositions as $\mathcal{S}$.

Example 1.24 (Garside germ). The germs of Figure 1 are Garside germs: indeed, they come from Garside families that are closed under right-divisor, so, by Proposition 1.14, the germs embed into the corresponding categories. That their images are Garside families then directly follows from the construction.

By contrast, let $M = \langle \mathtt{a}, \mathtt{b} \mid \mathtt{ab} = \mathtt{ba}, \mathtt{a}^2 = \mathtt{b}^2 \rangle^+$, and $\mathcal{S}$ consist of $1, \mathtt{a}, \mathtt{b}, \mathtt{ab}, \mathtt{a}^2$. The germ $\underline{S}$ induced on $\mathcal{S}$ is shown on the right. It is left-associative and left-cancellative. The monoid $\mathcal{M}on(\underline{S})$ is (isomorphic to) M, as the relations $\mathtt{a}|\mathtt{a} = \mathtt{a}^2 = \mathtt{b}|\mathtt{b}$ and $\mathtt{a}|\mathtt{b} = \mathtt{ab} = \mathtt{b}|\mathtt{a}$ belong to the family $\mathcal{R}_\bullet$. However, we saw in Example IV.2.34 (no proper Garside) that $\mathcal{S}$ is not a Garside family in M. Hence the germ $\underline{S}$ is not a Garside germ.

$\bullet$	1	$\mathtt{a}$	$\mathtt{b}$	$\mathtt{a}^2$	$\mathtt{ab}$
1	1	$\mathtt{a}$	$\mathtt{b}$	$\mathtt{a}^2$	$\mathtt{ab}$
$\mathtt{a}$	$\mathtt{a}$	$\mathtt{a}^2$	$\mathtt{ab}$		
$\mathtt{b}$	$\mathtt{b}$	$\mathtt{ab}$	$\mathtt{a}^2$		
$\mathtt{a}^2$	$\mathtt{a}^2$				
$\mathtt{ab}$	$\mathtt{ab}$				

Before looking for characterizations of Garside germs, we observe that several properties carry over from the germ to the associated category.

Proposition 1.25 (Noetherianity transfers). *If $\underline{S}$ is a Garside germ, the following are equivalent:*

(i) *The germ $\underline{S}$ is right-Noetherian.*

(ii) *The category $\mathcal{C}at(\underline{S})$ is right-Noetherian.*

Proof. Assume (i). By assumption, $\mathcal{S}$ is a solid Garside family in $\mathcal{C}at(\underline{S})$ that is locally right-Noetherian. Then, by Lemmas IV.2.19 and IV.2.21, the category $\mathcal{C}at(\underline{S})$ must be right-Noetherian. So (i) implies (ii).

Conversely, by (the trivial direction in) Proposition IV.2.18 (solid Garside in right-Noetherian), (ii) implies that the family $\mathcal{S}$ is locally right-Noetherian in $\mathcal{C}at(\underline{S})$, hence that the germ $\underline{S}$ is right-Noetherian. So (ii) implies (i). ☐

Having introduced right-$\underline{S}$-multiples, we naturally define (as in Chapter IV) a *right-$\underline{S}$-lcm* to be a common right-$\underline{S}$-multiple that left-$\underline{S}$-divides every common right-$\underline{S}$-multiple. Then the following transfer result is a direct restatement of Proposition IV.2.38 in the language of germs.

Proposition 1.26 (lcms transfer I). *If $\underline{S}$ is a Garside germ, the following are equivalent:*
(i) *Any two elements of S with a common right-$\underline{S}$-multiple admit a right-$\underline{S}$-lcm.*
(ii) *The category $\mathcal{C}at(\underline{S})$ admits conditional right-lcms.*

Thus, properties like right-Noetherianity or the existence of right-lcms in the associated category can be checked by remaining at the level of the germ. Note that no result involves left-Noetherianity or left-lcms: Garside families and Garside germs are oriented notions and there is *no* invariance under a left–right symmetry.

Proposition 1.26 reduces the existence of right-lcms in the ambient category to the existence of local right-lcms in the germ. In some cases, we can reduce the question further. For this we shall appeal to a (local version of) the following refinement of Lemma II.2.35 in which, in order to deduce that a $\prec$-maximal element is a $\preccurlyeq$-greatest element, one concentrates on the elements of some prescribed (small) family, typically atoms. The inductive argument is the same as the one involved in Exercise 23 (alternative proof).

Lemma 1.27. *Assume that $\mathcal{C}$ is a left-cancellative category that is right-Noetherian, $\mathcal{X}$ is a subfamily of $\mathcal{C}(x,\cdot)$ that is closed under left-divisor, $\mathcal{A}$ is a subfamily of $\mathcal{C}$ such that $\mathcal{A}\cup\mathcal{C}^{\times}$ generates $\mathcal{C}$ and satisfies $\mathcal{C}^{\times}\mathcal{A} \subseteq \mathcal{A}\mathcal{C}^{\times}$, and any two elements of $\mathcal{X}$ of the form fa, fb with a, b in $\mathcal{A}$ admit a common right-multiple in $\mathcal{X}$. Then every $\prec$-maximal element of $\mathcal{X}$ is a $\preccurlyeq$-greatest element of $\mathcal{X}$.*

Proof. Assume that m is a $\prec$-maximal element of $\mathcal{X}$ lying in $\mathcal{C}(x, y)$. For f left-dividing m, we write $R(f)$ for the (unique) element satisfying $fR(f) = m$, and we introduce the property $\mathcal{P}(f)$: every right-multiple of f lying in $\mathcal{X}$ left-divides m. Then $\mathcal{P}(m)$ is true since every right-multiple of m must be $=^{\times}$-equivalent to m. On the other hand, if f' is $=^{\times}$-equivalent to f, then $\mathcal{P}(f')$ and $\mathcal{P}(f)$ are equivalent, since f and f' admit the same right-multiples. We shall prove that $\mathcal{P}(f)$ is true for every f left-dividing m using induction on $R(f)$ with respect to $\precnsim$, which, by assumption, is well-founded .

Assume first that $R(f)$ is invertible. Then we have $f =^{\times} m$, whence $\mathcal{P}(f)$ since $\mathcal{P}(m)$ is true. Otherwise, owing to the assumption $\mathcal{C}^{\times}\mathcal{A} \subseteq \mathcal{A}$, some non-invertible element a of $\mathcal{A}$ left-divides $R(f)$, say $R(f) = ah$. Then we find $m = fR(f) = fah$, whence $fa \preccurlyeq m$ and $R(fa) = h$. As a is non-invertible, we have $R(fa) \precnsim R(f)$, so the induction hypothesis implies $\mathcal{P}(fa)$.

Now assume that fg lies in $\mathcal{X}$ (see Figure 2). If g is invertible, we have $fg =^{\times} f$, whence $fg \preccurlyeq m$. Otherwise, as above, some non-invertible element b of $\mathcal{A}$ left-divides g, say $g = bg'$. The assumption that fg lies in $\mathcal{X}$ implies that fb lies in $\mathcal{X}$, as does fa. Hence, by assumption, there exists a common right-multiple c of a and b such that fc lies in $\mathcal{X}$. As $\mathcal{P}(fa)$ and $fa \preccurlyeq fc$ are true, we must have $fc \preccurlyeq m$, whence $fb \preccurlyeq m$. Now, as b is non-invertible, we have $R(fb) \precnsim R(f)$, and the induction hypothesis implies $\mathcal{P}(fb)$. Since $(fb)g'$ belongs to $\mathcal{X}$, we deduce $(fb)g' \preccurlyeq m$, that is, $fg \preccurlyeq m$, and $\mathcal{P}(f)$ is true.

Thus $\mathcal{P}(f)$ is true for every f left-dividing m. In particular, $\mathcal{P}(1_x)$ is true, which means that every element of $\mathcal{X}$ left-divides m, that is, m is a $\preccurlyeq$-greatest element of $\mathcal{X}$. □

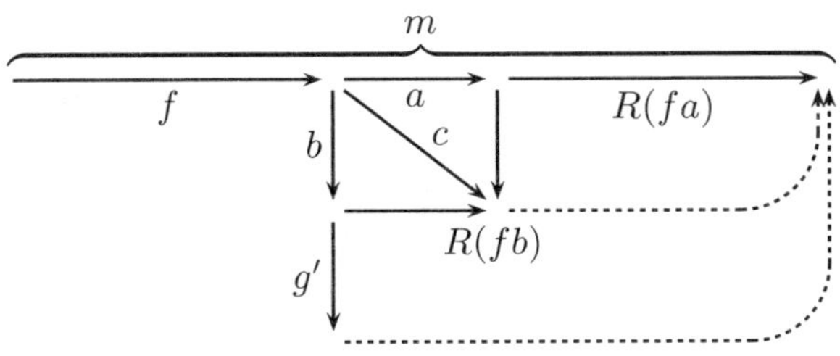

Figure 2. Induction step in the proof of Lemma 1.27. Note the similarity with a right-reversing process and the use of right-Noetherianity to ensure termination.

Note that in the above proof, assuming that the restriction of $\widetilde{\prec}$ to any family of the form $\{g \in \mathcal{C} \mid \exists f{\in}\mathcal{X}\,(fg = m)\}$ is well-founded would be sufficient.

Localizing the above argument to adapt it to a germ context is easy provided the left-divisibility relation of the considered germ resembles enough that of a category. The result can be stated as follows:

Lemma 1.28. *Assume that $\underline{\mathcal{S}}$ is a germ that is associative, left-cancellative, and right-Noetherian, $\mathcal{X}$ is a subfamily of $\mathcal{S}(x, \cdot)$ that is closed under left-$\underline{\mathcal{S}}$-divisor, $\mathcal{A}$ is a subfamily of $\mathcal{S}$ such that $\mathcal{A} \cup \mathcal{S}^\times$ generates $\underline{\mathcal{S}}$ and satisfies $\mathcal{S}^\times\mathcal{A} \subseteq \mathcal{A}\mathcal{S}^\times$, and any two elements of $\mathcal{X}$ of the form fa, fb with a, b in $\mathcal{A}$ admit a common right-multiple in $\mathcal{X}$. Then every $\prec$-maximal element of $\mathcal{X}$ is a $\preccurlyeq$-greatest element of $\mathcal{X}$.*

Proof. The assumptions on $\underline{\mathcal{S}}$ ensure that the properties of $\preccurlyeq_{\mathcal{S}}$ and $=^{\times}_{\mathcal{S}}$ listed in Lemma 1.19 are satisfied in $\underline{\mathcal{S}}$. Inspecting the proof of Lemma 1.27 then shows that all steps remain valid. For instance, if $f' =^{\times}_{\mathcal{S}} f$ holds, say $f' = f \bullet \epsilon$, then we have

$$f \bullet g = (f \bullet (\epsilon \bullet \epsilon^{-1})) \bullet g = ((f \bullet \epsilon) * \epsilon^{-1}) \bullet g = (f \bullet \epsilon) \bullet (\epsilon^{-1} \bullet g),$$

the last equality because $\underline{\mathcal{S}}$ is left-associative, and so every right-$\underline{\mathcal{S}}$-multiple f is also a right-multiple of f'. Similarly, the equality $m = f \bullet (a \bullet h)$ implies $m = (f \bullet a) \bullet h$ because $\underline{\mathcal{S}}$ is right-associative. □

Returning to right-$\underline{\mathcal{S}}$-lcms in germs, we obtain:

Lemma 1.29. *If a germ $\underline{\mathcal{S}}$ is associative, left-cancellative, and right-Noetherian, and $\mathcal{A}$ is a subfamily of $\mathcal{S}$ such that $\mathcal{A} \cup \mathcal{S}^\times$ generates $\underline{\mathcal{S}}$ and satisfies $\mathcal{S}^\times\mathcal{A} \subseteq \mathcal{A}\mathcal{S}^\times$, then the following conditions are equivalent:*

(i) *Any two elements of $\mathcal{A}$ with a common right-$\underline{\mathcal{S}}$-multiple admit a right-$\underline{\mathcal{S}}$-lcm.*

(ii) *Any two elements of $\mathcal{S}$ with a common right-$\underline{\mathcal{S}}$-multiple admit a right-$\underline{\mathcal{S}}$-lcm.*

Proof. Clearly, (ii) implies (i). Assume (i), and let s, t be two elements of $\mathcal{S}$ that admit a common right-$\underline{\mathcal{S}}$-multiple, say r. Let $\mathcal{X}$ be the family of all elements of $\mathcal{S}$ that left-$\underline{\mathcal{S}}$-divide all common right-$\underline{\mathcal{S}}$-multiples of s and t. Then $\mathcal{X}$ is nonempty as it contains s and t. First, as in the proof of Proposition II.2.34 we obtain that $\mathcal{X}$ admits a $\prec_{\mathcal{S}}$-maximal element: indeed, a relation like $f_1 \prec_{\mathcal{S}} f_2 \prec_{\mathcal{S}} \cdots$ in $\mathcal{X}$ (hence in $\mathrm{Div}_{\mathcal{S}}(r)$), would imply $g_1 \widetilde{\succ}_{\mathcal{S}}\, g_2 \widetilde{\succ}_{\mathcal{S}} \cdots$, where g_i is determined by $f_i g_i = r$, and therefore cannot exist since $\underline{\mathcal{S}}$ is right-Noetherian.

Next, $\mathcal{X}$ is closed under left-$\underline{S}$-divisor because $\preccurlyeq_S$ is transitive. Finally, assume that a and b lie in $\mathcal{A}$ and $f \bullet a$ and $f \bullet b$ lie in $\mathcal{X}$. As $f \bullet a$ and $f \bullet b$ admit a common right-$\underline{S}$-multiple, so do a and b, and therefore the latter admit a right-$\underline{S}$-lcm, say c. Every common right-$\underline{S}$-multiple of s and t is a common right-$\underline{S}$-multiple of $f \bullet a$ and $f \bullet b$, hence it can be written $f \bullet g$ with $a \preccurlyeq_S g$ and $b \preccurlyeq_S g$, hence $c \preccurlyeq_S g$, say $c \bullet g' = g$, and the fact that $f \bullet (c \bullet g')$ is defined implies that $f \bullet c$ is since $\underline{S}$ is right-associative. Hence $f \bullet c$ lies in $\mathcal{X}$, and $\mathcal{X}$ satisfies all hypotheses of Lemma 1.28. Therefore, a $\prec_S$-maximal element is a $\preccurlyeq_S$-greatest element. Now, a $\preccurlyeq_S$-greatest element of $\mathcal{X}$ is a right-$\underline{S}$-lcm of s and t. □

Merging the results, we deduce:

Proposition 1.30 (lcms transfer II). *If a Garside germ $\underline{S}$ is Noetherian and associative, then the following conditions are equivalent:*

(i) *Any two atoms of $\mathcal{S}$ with a common right-$\underline{S}$-multiple admit a right-$\underline{S}$-lcm.*

(ii) *The category $\mathcal{C}at(\underline{S})$ admits conditional right-lcms.*

Proof. Let $\mathcal{A}$ be the atom family in $\underline{S}$. By Proposition 1.22, $\underline{S}$ is generated by $\mathcal{A} \cup \mathcal{S}^{\times}$ and, moreover, $\mathcal{S}^{\times}\mathcal{A}$ is included in $\mathcal{A}\mathcal{S}^{\times}$: indeed, if ϵ is invertible and a is an atom, then $\epsilon \bullet a$ is an atom whenever it is defined, as $\epsilon \bullet a = s \bullet t$ implies $a = (\epsilon^{-1} \bullet s) \bullet t$ (because $\underline{S}$ is associative), hence at most one of $\epsilon^{-1} \bullet s$ and t is non-invertible, hence at most one of s, t is non-invertible. Hence $\underline{S}$ and $\mathcal{A}$ are eligible for Lemma 1.29. The result then follows from Proposition 1.26. □

Note that the existence of Garside families that are not closed under left-divisor implies the existence of Garside germs that are not right-associative: for instance, the germs associated with the Garside family S_n in the left-absorbing monoid L_n with $n \geqslant 2$ (Reference Structure 8, page 113) and with the sixteen-element Garside family S in the affine braid monoid B^+ (Reference Structure 9, page 114) are typical finite Garside germs that are not right-associative (the first one is not right-cancellative, whereas the second is.)

2 Recognizing Garside germs

We turn to the main question of the chapter, namely finding characterizations of Garside germs that are simple and easy to use. We establish two main criteria, namely Propositions 2.8 and 2.28. The former involves what we call the $\mathcal{I}$-law and the $\mathcal{J}$-law, and it is reminiscent of characterizing Garside families in terms of the $\mathcal{H}$-law (Proposition IV.1.50). The latter involves $\preccurlyeq_S$-greatest elements, and it is reminiscent of characterizing Garside families in terms of heads (Proposition IV.1.24). The order of the results is dictated by the additional problem that, here, we do not assume that the involved category is left-cancellative, but establish it from the weaker assumption that the germ is left-cancellative.

There are four subsections. In Subsection 2.1, we introduce the $\mathcal{I}$- and $\mathcal{J}$-laws and establish Proposition 2.8. Then, in Subsection 2.2, we investigate greatest $\mathcal{I}$-functions and derive Proposition 2.28. In Subsection 2.3, we consider the specific case of Noetherian germs: as in Chapter IV, some properties become then automatic and we obtain simplified characterizations. Finally, in Subsection 2.4, we show how to construct Garside germs starting from a groupoid with a distinguished family of positive generators.

2.1 The families $\mathfrak{I}$ and $\mathfrak{J}$

The characterizations of Garside germs we shall establish below involve several notions that we first introduce. We recall that, if $\mathcal{S}$ is a precategory, hence in particular if $\mathcal{S}$ is the domain of a germ, $\mathcal{S}^{[p]}$ is the family of all $\mathcal{S}$-paths of length p, that is, the sequences $s_1|\cdots|s_p$ such that s_i lies in $\mathcal{S}$ and the target of s_i is the source of s_{i+1} for every i.

Definition 2.1 (families $\mathfrak{I}$ and $\mathfrak{J}$). For $\underline{\mathcal{S}}$ a germ and $s_1|s_2$ in $\mathcal{S}^{[2]}$, we put

(2.2) $$\mathfrak{I}_{\underline{\mathcal{S}}}(s_1, s_2) = \{t \in \mathcal{S} \mid \exists s \in \mathcal{S}\ (t = s_1 \bullet s \text{ and } s \preccurlyeq_{\mathcal{S}} s_2)\},$$

(2.3) $$\mathfrak{J}_{\underline{\mathcal{S}}}(s_1, s_2) = \{s \in \mathcal{S} \mid s_1 \bullet s \text{ is defined and } s \preccurlyeq_{\mathcal{S}} s_2\}.$$

A map from $\mathcal{S}^{[2]}$ to $\mathcal{S}$ is called an *$\mathfrak{I}$-function* (*resp.* a *$\mathfrak{J}$-function*) if, for every $s_1|s_2$ in $\mathcal{S}^{[2]}$, the value at $s_1|s_2$ lies in $\mathfrak{I}_{\underline{\mathcal{S}}}(s_1, s_2)$ (*resp.* in $\mathfrak{J}_{\underline{\mathcal{S}}}(s_1, s_2)$).

We recall that, in (2.2), writing $t = s_1 \bullet s$ implies that $s_1 \bullet s$ is defined. An element of $\mathfrak{J}_{\underline{\mathcal{S}}}(s_1, s_2)$ is a fragment of s_2 that can be added to s_1 legally, that is, without going out of $\mathcal{S}$. Then s_1 always belongs to $\mathfrak{I}_{\underline{\mathcal{S}}}(s_1, s_2)$ and 1_y always belongs to $\mathfrak{J}_{\underline{\mathcal{S}}}(s_1, s_2)$ (for s_1 in $\mathcal{S}(\cdot, y)$). So, in particular, $\mathfrak{I}_{\underline{\mathcal{S}}}(s_1, s_2)$ and $\mathfrak{J}_{\underline{\mathcal{S}}}(s_1, s_2)$ are never empty.

The connection between $\mathfrak{I}_{\underline{\mathcal{S}}}(s_1, s_2)$ and $\mathfrak{J}_{\underline{\mathcal{S}}}(s_1, s_2)$ is clear: with obvious notation, we have $\mathfrak{I}_{\underline{\mathcal{S}}}(s_1, s_2) = s_1 \bullet \mathfrak{J}_{\underline{\mathcal{S}}}(s_1, s_2)$. It turns out that, depending on cases, using $\mathfrak{I}$ or $\mathfrak{J}$ is more convenient, and it is useful to introduce both notions. Note that, if t belongs to $\mathfrak{I}_{\underline{\mathcal{S}}}(s_1, s_2)$ and $\underline{\mathcal{S}}$ embeds in $\mathcal{C}at(\underline{\mathcal{S}})$, then, in $\mathcal{C}at(\underline{\mathcal{S}})$, we have $t \in \mathcal{S}$ and $t \preccurlyeq s_1 s_2$. However, unless it is known that $\mathcal{C}at(\underline{\mathcal{S}})$ is left-cancellative, the latter relations need not imply $t \in \mathfrak{I}_{\underline{\mathcal{S}}}(s_1, s_2)$ since $s_1 s \preccurlyeq s_1 s_2$ need not imply $s \preccurlyeq s_2$ in general. We insist that the definitions of $\mathfrak{I}$ and $\mathfrak{J}$ entirely sit inside the germ $\underline{\mathcal{S}}$. For further reference, we note that, when $\underline{\mathcal{S}}$ is a Garside germ, $\mathcal{S}$-normal paths are easily described in terms of $\mathfrak{I}$ or $\mathfrak{J}$.

Lemma 2.4. *If $\underline{\mathcal{S}}$ is a Garside germ, the following are equivalent for every $s_1|s_2$ in $\mathcal{S}^{[2]}$:*
(i) *The path $s_1|s_2$ is $\mathcal{S}$-normal.*
(ii) *The element s_1 is $\prec_{\mathcal{S}}$-maximal in $\mathfrak{I}_{\underline{\mathcal{S}}}(s_1, s_2)$.*
(iii) *Every element of $\mathfrak{J}_{\underline{\mathcal{S}}}(s_1, s_2)$ is invertible in $\underline{\mathcal{S}}$.*

Proof. Assume that $s_1|s_2$ is $\mathcal{S}$-normal. Then, by definition, s_1 is an $\mathcal{S}$-head of $s_1 s_2$ in $\mathcal{C}at(\underline{\mathcal{S}})$, hence a greatest left-divisor of $s_1 s_2$ lying in $\mathcal{S}$. This implies that s_1 is $\prec_{\mathcal{S}}$-maximal in $\mathfrak{I}_{\underline{\mathcal{S}}}(s_1, s_2)$. So (i) implies (ii).

Assume now (ii) and $s \in \mathfrak{J}_{\underline{\mathcal{S}}}(s_1, s_2)$. By definition, $s_1 s$ belongs to $\mathfrak{I}_{\underline{\mathcal{S}}}(s_1, s_2)$. The assumption that s_1 is $\prec_{\mathcal{S}}$-maximal in $\mathfrak{I}_{\underline{\mathcal{S}}}(s_1, s_2)$ implies that s is invertible in $\mathcal{C}at(\underline{\mathcal{S}})$, hence in $\mathcal{S}$ by Lemma 1.19. So (ii) implies (iii).

Finally, if $\mathfrak{J}_{\underline{\mathcal{S}}}(s_1, s_2)$ consists of invertible elements solely, s_1 is a $\prec_{\mathcal{S}}$-maximal left-divisor of $s_1 s_2$ lying in $\mathcal{S}$. As $\mathcal{S}$ is a Garside family in $\mathcal{C}at(\underline{\mathcal{S}})$, Corollary IV.1.31 (recognizing greedy) applies, and it says that $s_1|s_2$ is $\mathcal{S}$-normal. So (iii) implies (i). ∎

We shall now characterize Garside germs by the existence of $\mathfrak{I}$- or $\mathfrak{J}$-functions that obey algebraic laws reminiscent of the $\mathcal{H}$-law of Chapter IV.

Definition 2.5 ($\mathfrak{I}$-law, $\mathfrak{J}$-law). If $\underline{S}$ is a germ and I, J are maps from $\mathcal{S}^{[2]}$ to $\mathcal{S}$, we say that I obeys the $\mathfrak{I}$-*law* if, for every $s_1|s_2|s_3$ in $\mathcal{S}^{[3]}$ with $s_1 \bullet s_2$ defined, we have

$$I(s_1, I(s_2, s_3)) =^{\times} I(s_1 \bullet s_2, s_3). \tag{2.6}$$

We say that J obeys the $\mathfrak{J}$-*law* if, for every $s_1|s_2|s_3$ in $\mathcal{S}^{[3]}$ with $s_1 \bullet s_2$ defined, we have

$$J(s_1, s_2 \bullet J(s_2, s_3)) =^{\times} s_2 \bullet J(s_1 \bullet s_2, s_3). \tag{2.7}$$

If the counterpart of (2.6) or (2.7) with $=$ replacing $=^{\times}$ is satisfied, we say that I (*resp.* J) obeys the *sharp* $\mathfrak{I}$-*law* (*resp.* the *sharp* $\mathfrak{J}$-*law*).

Proposition 2.8 (recognizing Garside germ I). *A germ $\underline{S}$ is a Garside germ if and only if it satisfies one of the following equivalent conditions:*

(2.9) *The germ $\underline{S}$ is left-associative, left-cancellative, and admits an $\mathfrak{I}$-function obeying the sharp $\mathfrak{I}$-law.*

(2.10) *The germ $\underline{S}$ is left-associative, left-cancellative, and admits a $\mathfrak{J}$-function obeying the sharp $\mathfrak{J}$-law.*

The proof of Proposition 2.8 will be split into several parts. We insist once more that all conditions in Proposition 2.8 are local in that they only involve the elements of $\mathcal{S}$ and computations taking place inside $\mathcal{S}$. In particular, if $\mathcal{S}$ is finite, all conditions are effectively checkable in finite time.

Example 2.11 (recognizing Garside germ). Let us consider the B_3^+-germ of Figure 1. Checking that the germ is left-associative is a simple verification for each triple (s_1, s_2, s_3). Checking that the germ is left-cancellative means that no value appears twice in a row, a straightforward inspection. Finally, if any one of the following two tables is given,

I	1	σ_1	σ_2	$\sigma_1\sigma_2$	$\sigma_2\sigma_1$	Δ
1	1	σ_1	σ_2	$\sigma_1\sigma_2$	$\sigma_2\sigma_1$	Δ
σ_1	σ_1	σ_1	$\sigma_1\sigma_2$	σ_1	Δ	Δ
σ_2	σ_2	$\sigma_2\sigma_1$	σ_2	Δ	σ_2	Δ
$\sigma_1\sigma_2$	$\sigma_1\sigma_2$	Δ	$\sigma_1\sigma_2$	Δ	$\sigma_1\sigma_2$	Δ
$\sigma_2\sigma_1$	$\sigma_2\sigma_1$	$\sigma_2\sigma_1$	Δ	$\sigma_2\sigma_1$	Δ	Δ
Δ	Δ	Δ	Δ	Δ	Δ	Δ

J	1	σ_1	σ_2	$\sigma_1\sigma_2$	$\sigma_2\sigma_1$	Δ
1	1	σ_1	σ_2	$\sigma_1\sigma_2$	$\sigma_2\sigma_1$	Δ
σ_1	1	1	σ_2	1	$\sigma_2\sigma_1$	$\sigma_2\sigma_1$
σ_2	1	σ_1	1	$\sigma_1\sigma_2$	1	$\sigma_1\sigma_2$
$\sigma_1\sigma_2$	1	σ_1	1	σ_1	1	σ_1
$\sigma_2\sigma_1$	1	1	σ_2	1	σ_2	σ_2
Δ	1	1	1	1	1	1

then checking that the left one satisfies the sharp $\mathfrak{I}$-law or the right one satisfies the sharp $\mathfrak{J}$-law is again a simple verification for each triple $s_1|s_2|s_3$ such that $s_1 \bullet s_2$ is defined.

We first prove that the conditions of Proposition 2.8 are necessary, which is easy. We begin with the equivalence of the two families of conditions.

Lemma 2.12. *If $\mathcal{S}$ is a germ that is left-associative and left-cancellative and $I, J : \mathcal{S}^{[2]} \to \mathcal{S}$ are connected by $I(s_1, s_2) = s_1 \bullet J(s_1, s_2)$ for all s_1, s_2, then I is an $\mathfrak{I}$-function obeying the $\mathfrak{I}$-law (resp. the sharp $\mathfrak{I}$-law) if and only if J is a $\mathfrak{J}$-function obeying the $\mathfrak{J}$-law (resp. the sharp $\mathfrak{J}$-law).*

Proof. First, the assumption that $\mathcal{S}$ is left-cancellative implies that, for every I, there exists at most one associated J and that, when I and J are connected as in the statement, then I is an $\mathfrak{I}$-function for $\mathcal{S}$ if and only if J is a $\mathfrak{J}$-function for $\mathcal{S}$.

Now, assume that I is an $\mathfrak{I}$-function obeying the $\mathfrak{I}$-law, $s_1|s_2|s_3$ lies in $\mathcal{S}^{[3]}$, and $s_1 \bullet s_2$ is defined. By assumption, we have $I(s_1, I(s_2, s_3)) =^{\times} I(s_1 \bullet s_2, s_3)$, which translates into

$$s_1 \bullet J(s_1, s_2 \bullet J(s_2, s_3)) =^{\times} (s_1 \bullet s_2) \bullet J(s_1 \bullet s_2, s_3). \tag{2.13}$$

As $\mathcal{S}$ is left-associative, $s_2 \bullet J(s_1 \bullet s_2, s_3)$ is defined and (2.13) implies

$$s_1 \bullet J(s_1, s_2 \bullet J(s_2, s_3)) =^{\times} s_1 \bullet (s_2 \bullet J(s_1 \bullet s_2, s_3)). \tag{2.14}$$

By Lemma 1.19, we may left-cancel s_1 in (2.14), and what remains is the expected instance of the $\mathfrak{J}$-law. So J obeys the $\mathfrak{J}$-law.

The argument in the case when I obeys the sharp $\mathfrak{I}$-law is similar: now (2.13) and (2.14) are equalities, and, applying the assumption that $\mathcal{S}$ is left-cancellative, we directly deduce the expected instance of the sharp $\mathfrak{J}$-law.

In the other direction, assume that J is a $\mathfrak{J}$-function that obeys the $\mathfrak{J}$-law, $s_1|s_2|s_3$ lies in $\mathcal{S}^{[3]}$, and $s_1 \bullet s_2$ is defined. By the $\mathfrak{J}$-law, $J(s_1, s_2 \bullet J(s_2, s_3))$ and $s_2 \bullet J(s_1 \bullet s_2, s_3)$ are $=^{\times}$-equivalent. By the definition of a $\mathfrak{J}$-function, the expression $s_1 \bullet J(s_1, s_2 \bullet J(s_2, s_3))$ is defined, hence so is $s_1 \bullet (s_2 \bullet J(s_1 \bullet s_2, s_3))$, and we obtain (2.14). Applying (1.6), which is legal since $s_1 \bullet s_2$ is defined, we deduce (2.13), and, from there, the equivalence $I(s_1, I(s_2, s_3)) =^{\times} I(s_1 \bullet s_2, s_3)$, the expected instance of the $\mathfrak{I}$-law. So I obeys the $\mathfrak{I}$-law.

Finally, if J obeys the sharp $\mathfrak{J}$-law, the argument is similar: (2.13) and (2.14) are equalities, and one obtains the expected instance of the sharp $\mathfrak{I}$-law. □

We now observe that the $\mathfrak{I}$-law and the $\mathfrak{J}$-law are closely connected with the $\mathcal{H}$-law of Definition IV.1.47, and deduce the necessity of the conditions of Proposition 2.8.

Lemma 2.15. *Assume that $\underline{\mathcal{S}}$ is a Garside germ. Then $\underline{\mathcal{S}}$ satisfies* (2.9) *and* (2.10).

Proof. Put $\mathcal{C} = \mathcal{C}at(\underline{\mathcal{S}})$. By definition, $\mathcal{S}$ embeds in $\mathcal{C}$, so, by Proposition 1.14, $\underline{\mathcal{S}}$ must be left-associative. Next, by definition again, $\mathcal{C}$ is left-cancellative and, therefore, $\underline{\mathcal{S}}$ must be left-cancellative as, if s, t, t' lie in $\mathcal{S}$ and satisfy $s \bullet t = s \bullet t'$, then $st = st'$ holds in $\mathcal{C}$, implying $t = t'$. So $\underline{\mathcal{S}}$ must be left-cancellative.

Then, owing to Lemma 2.12, it is enough to consider either $\mathfrak{I}$- or $\mathfrak{J}$-functions, and we shall consider the former. By assumption, $\mathcal{S}$ is a Garside family in $\mathcal{C}$. Hence, by Proposition IV.1.44 (sharp exist), there exists a sharp $\mathcal{S}$-head function H. Owing to Lemma IV.1.22, we can assume that H is defined on all of $\mathcal{S}\mathcal{C}$, hence in particular on $\mathcal{S}^2$. Let us now define $I : \mathcal{S}^{[2]} \to \mathcal{S}$ by $I(s_1, s_2) = H(s_1 s_2)$.

First we claim that I is an $\mathcal{I}$-function for $\underline{\mathcal{S}}$. Indeed, assume $s_1|s_2 \in \mathcal{S}^{[2]}$. By definition, we have $s_1 \preccurlyeq H(s_1s_2) \preccurlyeq s_1s_2$ in $\mathcal{C}$, hence $H(s_1s_2) = s_1s$ for some s satisfying $s \preccurlyeq s_2$. As s right-divides $H(s_1s_2)$, which lies in $\mathcal{S}$, and, by assumption, $\mathcal{S}$ is closed under right-divisor, s must lie in $\mathcal{S}$. Hence $H(s_1s_2)$ lies in $\mathcal{I}_{\underline{\mathcal{S}}}(s_1, s_2)$.

Finally, assume that $s_1|s_2|s_3$ lies in $\mathcal{S}$ and $s = s_1 \bullet s_2$ holds. Then the $\mathcal{H}$-law gives $H(s_1H(s_2s_3)) = H(s_1(s_2s_3)) = H(ss_3)$. This translates into $I(s_1, I(s_2s_3)) = I(s_1 \bullet s_2, s_3)$, the expected instance of the sharp $\mathcal{I}$-law. □

We will now prove that the conditions of Proposition 2.8 are sufficient, thus establishing Proposition 2.8 and providing an intrinsic characterization of Garside germs. The principle is obvious, namely using the given $\mathcal{I}$- or $\mathcal{J}$-function to construct a head function on $\mathcal{S}^2$ and then using a characterization of a Garside family in terms of the existence of a head as in Proposition IV.1.24 (recognizing Garside II). However, the argument is more delicate, because we do not know at first that the category $\mathcal{C}at(\underline{\mathcal{S}})$ is left-cancellative and, therefore, eligible for any characterization of Garside families. So what we shall do is to simultaneously construct the head function and prove left-cancellativity. To this end, the main point is to control not only the head $H(g)$ of an element g, but also its tail, defined to be the element g' satisfying $g = H(g)g'$: if the category $\mathcal{C}at(\underline{\mathcal{S}})$ is left-cancellative, this element g' is unique and it should be possible to determine it explicitly. This is what we shall do. For the construction, it is convenient to start here with a $\mathcal{J}$-function.

Lemma 2.16. *Assume that $\underline{\mathcal{S}}$ is a left-associative, left-cancellative germ and J is a $\mathcal{J}$-function for $\underline{\mathcal{S}}$ that obeys the $\mathcal{J}$-law. Define functions*

$$K : \mathcal{S}^{[2]} \to \mathcal{S}, \qquad \underline{H} : \mathcal{S}^* \to \mathcal{S}, \qquad \underline{T} : \mathcal{S}^* \to \mathcal{S}^*$$

by $s_2 = J(s_1, s_2) \bullet K(s_1, s_2)$, $\underline{H}(\varepsilon_x) = 1_x$, $\underline{T}(\varepsilon_x) = \varepsilon_x$ and, for s in $\mathcal{S}$ and w in $\mathcal{S}^*$,

$$\underline{H}(s|w) = s \bullet J(s, \underline{H}(w)) \quad \text{and} \quad \underline{T}(s|w) = K(s, \underline{H}(w))|\underline{T}(w). \tag{2.17}$$

Then, for each w in $\mathcal{S}^*$, we have

$$w \equiv \underline{H}(w)|\underline{T}(w). \tag{2.18}$$

Moreover, $w \equiv w'$ implies $\underline{H}(w) = \underline{H}(w')$ and $\underline{T}(w) \equiv \underline{T}(w')$.

Proof. First, the definition of K makes sense and is unambiguous: indeed, by definition, $J(s_1, s_2) \preccurlyeq_{\mathcal{S}} s_2$ holds, so there exists s in $\mathcal{S}$ satisfying $s_2 = J(s_1, s_2) \bullet s$. Moreover, as $\underline{\mathcal{S}}$ is left-cancellative, the element s is unique.

To prove (2.18), we use induction on the length of w. For $w = \varepsilon_x$, we have $\varepsilon_x \equiv 1_x|\varepsilon_x$. Otherwise, for s in $\mathcal{S}$ and w in $\mathcal{S}^*$, we find

$$\begin{aligned} s|w &\equiv s|\underline{H}(w)|\underline{T}(w) && \text{by the induction hypothesis,} \\ &\equiv s|J(s, \underline{H}(w)) \bullet K(s, \underline{H}(w))|\underline{T}(w) && \text{by the definition of } K, \\ &\equiv s|J(s, \underline{H}(w))|K(s, \underline{H}(w))|\underline{T}(w) && \text{by the definition of } \equiv, \\ &\equiv s \bullet J(s, \underline{H}(w))|K(s, \underline{H}(w))|\underline{T}(w) && \text{as } J(s, \underline{H}(w)) \text{ lies in } \mathcal{J}_{\underline{\mathcal{S}}}(s, \underline{H}(w)), \\ &= \underline{H}(s|w)|\underline{T}(s|w) && \text{by the definition of } \underline{H} \text{ and } \underline{T}. \end{aligned}$$

For the compatibility of $\underline{H}$ and $\underline{T}$ with $\equiv$, owing to the inductive definitions of $\equiv$, $\underline{H}$, and $\underline{T}$, it is sufficient to assume that $s|w$ is a path and $s = s_1 \bullet s_2$ holds with w in $\mathcal{S}^*$ and s_1, s_2 in $\mathcal{S}$, and to establish

$$\underline{H}(s|w) = \underline{H}(s_1|s_2|w) \quad \text{and} \quad \underline{T}(s|w) \equiv \underline{T}(s_1|s_2|w). \tag{2.19}$$

Now, applying (2.7) with $s_3 = \underline{H}(w)$ and the definition of $\underline{H}(s_2|w)$, we find

$$s_2 \bullet J(s, \underline{H}(w)) = J(s_1, \underline{H}(s_2|w)). \tag{2.20}$$

Then, the first relation of (2.19) is satisfied since we can write

$$\begin{aligned}
\underline{H}(s|w) &= (s_1 \bullet s_2) \bullet J(s, \underline{H}(w)) && \text{by the definition of } \underline{H},\\
&= s_1 \bullet (s_2 \bullet J(s, \underline{H}(w))) && \text{by (1.7)},\\
&= s_1 \bullet J(s_1, \underline{H}(s_2|w)) = \underline{H}(s_1|s_2|w) && \text{by (2.20) and the definition of } \underline{H}.
\end{aligned}$$

For the second relation in (2.19), applying the definition of $\underline{H}$, we first find

$$\begin{aligned}
s_2 \bullet J(s_2, \underline{H}(w)) = \underline{H}(s_2|w) &= J(s_1, \underline{H}(s_2|w)) \bullet K(s_1, \underline{H}(s_2|w)) && \text{by definition of } K,\\
&= (s_2 \bullet J(s, \underline{H}(w))) \bullet K(s_1, \underline{H}(s_2|w)) && \text{by (2.20)},\\
&= s_2 \bullet (J(s, \underline{H}(w)) \bullet K(s_1, \underline{H}(s_2|w))) && \text{by (1.7)},
\end{aligned}$$

whence, as $\underline{\mathcal{S}}$ is a left-cancellative germ,

$$J(s_2, \underline{H}(w)) = J(s, \underline{H}(w)) \bullet K(s_1, \underline{H}(s_2|w)). \tag{2.21}$$

We deduce the equalities

$$\begin{aligned}
J(s, \underline{H}(w)) \bullet K(s, \underline{H}(w)) &= \underline{H}(w) && \text{by the definition of } K,\\
&= J(s_2, \underline{H}(w)) \bullet K(s_2, \underline{H}(w)) && \text{by the definition of } K,\\
&= (J(s, \underline{H}(w)) \bullet K(s_1, \underline{H}(s_2|w))) \bullet K(s_2, \underline{H}(w)) && \text{by (2.21)},\\
&= J(s, \underline{H}(w)) \bullet (K(s_1, \underline{H}(s_2|w)) \bullet K(s_2, \underline{H}(w))) && \text{by (1.7)}.
\end{aligned}$$

As $\underline{\mathcal{S}}$ is a left-cancellative germ, we may left-cancel $J(s, \underline{H}(w))$, and we obtain

$$K(s, \underline{H}(w)) = K(s_1, \underline{H}(s_2|w)) \bullet K(s_2, \underline{H}(w)),$$

whence $K(s, \underline{H}(w))|\underline{T}(w) \equiv K(s_1, \underline{H}(s_2|w))|K(s_2, \underline{H}(w))|\underline{T}(w)$. By the definition of $\underline{T}$, this is the second relation in (2.19). $\square$

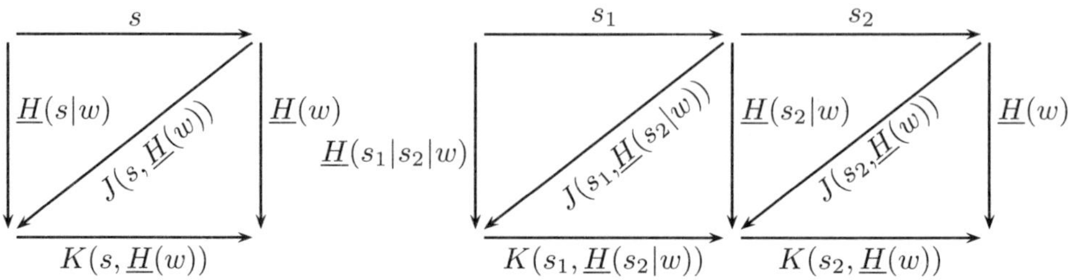

Figure 3. Proof of Lemma 2.16. Attention! As long as the ambient category is not proved to be left-cancellative, the above diagrams should be taken with care.

We can now easily complete the argument.

Proof of Proposition 2.8. Owing to Lemma 2.15 and Lemma 2.12, it suffices to prove now that (2.10) implies that $\underline{\mathcal{S}}$ is a Garside germ. So assume that $\underline{\mathcal{S}}$ is a germ that is left-associative and left-cancellative, and J is a $\mathcal{J}$-function on $\underline{\mathcal{S}}$ that obeys the sharp $\mathcal{J}$-law. Let $\mathcal{C} = \mathcal{C}at(\underline{\mathcal{S}})$. As $\underline{\mathcal{S}}$ is left-associative, Proposition 1.14 implies that $\underline{\mathcal{S}}$ embeds in $\mathcal{C}$ and is solid in $\mathcal{C}$. Now we use the functions K, $\underline{H}$, and $\underline{T}$ of Lemma 2.16. First, we claim that, for each w in $\mathcal{S}^*$ and each s in $\mathcal{S}$ such that the target of s is the source of w, we have

$$w \equiv J(s, \underline{H}(w))|\underline{T}(s|w). \tag{2.22}$$

Indeed, we have $s \bullet J(s, \underline{H}(w)) = \underline{H}(s|w)$ and

$$\begin{aligned} w &\equiv \underline{H}(w)|\underline{T}(w) && \text{by (2.18)},\\ &= J(s, \underline{H}(w)) \bullet K(s, \underline{H}(w))|\underline{T}(w) && \text{by the definition of } K,\\ &\equiv J(s, \underline{H}(w))|K(s, \underline{H}(w))|\underline{T}(w) && \text{by the definition of } \equiv,\\ &= J(s, \underline{H}(w))|\underline{T}(s|w). && \text{by the definition of } \underline{T}(s|w). \end{aligned}$$

Assume now $s|w \equiv s|w'$. First, Lemma 2.16 implies $\underline{H}(s|w) = \underline{H}(s|w')$, that is, $s \bullet J(s, \underline{H}(w)) = s \bullet J(s, \underline{H}(w'))$ owing to the definition of $\underline{H}$. As $\underline{\mathcal{S}}$ is left-cancellative, we may left-cancel s and we deduce $J(s, \underline{H}(w)) = J(s, \underline{H}(w'))$. Then, applying (2.22) twice and Lemma 2.16 again, we find

$$w \equiv J(s, \underline{H}(w))|\underline{T}(s|w) \equiv J(s, \underline{H}(w'))|\underline{T}(s|w') \equiv w',$$

which implies that $\mathcal{C}$ is left-cancellative.

Next, Lemma 2.16 shows that the function $\underline{H}$ induces a well-defined function from $\mathcal{C}$ to $\mathcal{S}$, say H. Then (2.18) implies that $H(s) \preccurlyeq s$ holds for every s in $\mathcal{C}$. On the other hand, assume that r belongs to $\mathcal{S}$, and that $r \preccurlyeq s$ holds in $\mathcal{C}$. This means that there exists w in $\mathcal{S}^*$ such that $r|w$ represents s. By construction, $\underline{H}(r|w) = r \bullet J(r, \underline{H}(w))$ in $\underline{\mathcal{S}}$, which implies $r \preccurlyeq H(s)$ in $\mathcal{C}$. So $H(s)$ is an $\mathcal{S}$-head of s and, therefore, every element of $\mathcal{C}$ admits an $\mathcal{S}$-head.

Finally, by Proposition 1.14, $\mathcal{S}$ is closed under right-divisor in $\mathcal{C}$, which implies that $\mathcal{S}^\sharp$ is also closed under right-divisor: indeed, a right-divisor of an element of $\mathcal{C}^\times$ must lie in $\mathcal{C}^\times$, and, if g right-divides $s\epsilon$ with $s \in \mathcal{S}$ and $\epsilon \in \mathcal{C}^\times$, then $g\epsilon^{-1}$ right-divides s, hence it belongs to $\mathcal{S}$, and therefore g belongs to $\mathcal{S}\mathcal{C}^\times$, hence to $\mathcal{S}^\sharp$. Therefore, $\mathcal{S}$ satisfies (IV.1.25) in $\mathcal{C}$. So, by Proposition IV.1.24, $\mathcal{S}$ is a Garside family in $\mathcal{C}$, and $\underline{\mathcal{S}}$ is a Garside germ. □

Note that Proposition 2.8 provides a new way of establishing that a presented category is left-cancellative (and therefore using Proposition 2.8 for the opposite germ provides a way of establishing right-cancellativity).

Example 2.23 (braids). Starting from $\langle \sigma_1, \sigma_2 \mid \sigma_1\sigma_2\sigma_1 = \sigma_2\sigma_1\sigma_2 \rangle^+$, we would easily obtain the germ of Table 1 by introducing the elements $1, \sigma_1\sigma_2, \sigma_2\sigma_1$, and $\sigma_1\sigma_2\sigma_1$ to put the relations in the form $r = s \bullet t$. Then a systematic verification of the conditions of Proposition 2.8 would provide a new proof of the left-cancellativity of B_3^+—see Subsection 2.4 and Chapter IX. In this case, the germ coincides with the opposite germ, so right-cancellativity follows automatically.

We conclude with an application. The criteria of Proposition 2.8 can also be used in the context of Chapter IV, that is, when one wishes to recognize whether some subfamily $\mathcal{S}$ of a given category $\mathcal{C}$ is a Garside family. Proposition IV.1.50 (recognizing Garside III) provides a criterion involving the $\mathcal{H}$-law on $\mathcal{C} \setminus \mathcal{C}^{\times}$. With the current results, we obtain local versions that involve the $\mathfrak{I}$-law or the $\mathfrak{J}$-law on elements of $\mathcal{S}^2$.

Proposition 2.24 (recognizing Garside IV). *A solid generating subfamily $\mathcal{S}$ of a left-cancellative category $\mathcal{C}$ is a Garside family if and only if one of the following equivalent conditions is satisfied:*

(2.25) *There exists $I : \mathcal{S}^{[2]} \to \mathcal{S}$ satisfying $s_1 \preccurlyeq I(s_1, s_2) \preccurlyeq s_1 s_2$ for all s_1, s_2 and $I(s_1, I(s_2, s_3)) = I(s_1 s_2, s_3)$ for every $s_1|s_2|s_3$ in $\mathcal{S}^{[3]}$ satisfying $s_1 s_2 \in \mathcal{S}$.*

(2.26) *There exists $J : \mathcal{S}^{[2]} \to \mathcal{S}$ satisfying $s_1 J(s_1, s_2) \in \mathcal{S}$ and $J(s_1, s_2) \preccurlyeq s_2$ for all s_1, s_2, and $J(s_1, s_2 J(s_2, s_3)) = s_2 J(s_1 s_2, s_3)$ for every $s_1|s_2|s_3$ in $\mathcal{S}^{[3]}$ satisfying $s_1 s_2 \in \mathcal{S}$.*

Proof. Assume that $\mathcal{S}$ is a solid Garside family in $\mathcal{C}$. By Lemma 1.2, $(\mathcal{S}, \bullet_{\mathcal{S}})$ is a germ, and, by Proposition 1.11, $\mathcal{C}$ is isomorphic to $\mathcal{C}at(\mathcal{S}, \bullet_{\mathcal{S}})$. By definition, $(\mathcal{S}, \bullet_{\mathcal{S}})$ is a Garside germ and, therefore, by Lemma 2.15, there exist functions I and J obeying the expected versions of the sharp $\mathfrak{I}$-law and $\mathfrak{J}$-law. So (2.25) and (2.26) are satisfied.

Conversely, assume that $\mathcal{S}$ satisfies (2.25). Then, by Proposition 1.11, $(\mathcal{S}, \bullet_{\mathcal{S}})$ is a germ and $\mathcal{C}at(\mathcal{S}, \bullet_{\mathcal{S}})$ is isomorphic to $\mathcal{C}$. By Proposition 1.14, the assumption that $\mathcal{S}$ is closed under right-divisor implies that the germ $(\mathcal{S}, \bullet_{\mathcal{S}})$ is left-associative, and the assumption that $\mathcal{C}$ is left-cancellative implies that $(\mathcal{S}, \bullet_{\mathcal{S}})$ is left-cancellative. Finally, the function I whose existence is assumed is an $\mathfrak{I}$-function for $(\mathcal{S}, \bullet_{\mathcal{S}})$, and (2.25) implies that this $\mathfrak{I}$-function obeys the sharp $\mathfrak{I}$-law. Then Proposition 2.8 implies that $(\mathcal{S}, \bullet_{\mathcal{S}})$ is a Garside germ. So, in particular, $\mathcal{S}$ is a Garside family in $\mathcal{C}at(\mathcal{S}, \bullet_{\mathcal{S}})$.

The argument for (2.26) is entirely similar. □

2.2 Greatest $\mathcal{I}$-functions

In Proposition 2.8, we characterized Garside germs by the existence of an $\mathcal{I}$- or a $\mathcal{J}$-function obeying a certain algebraic law. We now establish alternative characterizations that again involve particular $\mathcal{I}$- or $\mathcal{J}$-functions, but in terms of maximality conditions.

Definition 2.27 (greatest $\mathcal{I}$- or $\mathcal{J}$-function). An $\mathcal{I}$-function I for a germ $\underline{S}$ is called a *greatest $\mathcal{I}$-function* if, for every $s_1|s_2$ in $\mathcal{S}^{[2]}$, the value $I(s_1, s_2)$ is a $\preccurlyeq$-greatest element in $\mathcal{I}_{\underline{S}}(s_1, s_2)$; same for a *greatest $\mathcal{J}$-function*, replacing $\mathcal{I}_{\underline{S}}(s_1, s_2)$ with $\mathcal{J}_{\underline{S}}(s_1, s_2)$.

In other words, I is a greatest $\mathcal{I}$-function for $\underline{S}$ if and only if, for every $s_1|s_2$ in $\mathcal{S}^{[2]}$, we have $I(s_1, s_2) = s_1 \bullet s$ for some s satisfying $s \preccurlyeq_{\mathcal{S}} s_2$, and $t \preccurlyeq_{\mathcal{S}} I(s_1, s_2)$ holds for every t in $\mathcal{I}_{\underline{S}}(s_1, s_2)$.

Proposition 2.28 (recognizing Garside germ II). *A germ $\underline{S}$ is a Garside germ if and only if one of the following equivalent conditions is satisfied:*

(2.29) *The germ $\underline{S}$ is left-associative, left-cancellative, and, for every $s_1|s_2$ in $\mathcal{S}^{[2]}$, the family $\mathcal{I}_{\underline{S}}(s_1, s_2)$ admits a $\preccurlyeq_{\mathcal{S}}$-greatest element.*

(2.30) *The germ $\underline{S}$ is left-associative, left-cancellative, and admits a greatest $\mathcal{I}$-function.*

(2.31) *The germ $\underline{S}$ is left-associative, left-cancellative, and, for every $s_1|s_2$ in $\mathcal{S}^{[2]}$, the family $\mathcal{J}_{\underline{S}}(s_1, s_2)$ admits a $\preccurlyeq_{\mathcal{S}}$-greatest element.*

(2.32) *The germ $\underline{S}$ is left-associative, left-cancellative, and admits a greatest $\mathcal{J}$-function.*

Let us immediately note that (2.29) and (2.30) on the one hand, and (2.31) and (2.32) on the other hand, are equivalent: indeed, a greatest function is a function that selects a $\preccurlyeq_{\mathcal{S}}$-greatest element for each argument, and, at the expense of possibly invoking the Axiom of Choice, the existence of the former is equivalent to the existence of the latter. As in the case of Proposition 2.8, proving that the conditions of Proposition 2.28 are satisfied in every Garside germ is (very) easy.

Lemma 2.33. *Every Garside germ satisfies* (2.29)–(2.32).

Proof. Assume that $\underline{S}$ is a Garside germ. By Lemma 2.15, $\underline{S}$ is left-associative and left-cancellative. Then let H be an $\mathcal{S}$-head function defined on $\mathcal{S}Cat(\underline{S})$. Consider the maps $I, J : \mathcal{S}^{[2]} \to \mathcal{S}$ defined by $I(s_1, s_2) = s_1 \bullet J(s_1, s_2) = H(s_1 s_2)$. Then, as in the proof of Lemma 2.15, I is a $\mathcal{I}$-function and J is a $\mathcal{J}$-function for $\underline{S}$.

Moreover, assume that $s_1|s_2$ lies in $\mathcal{S}^{[2]}$ and we have $t = s_1 \bullet s$ with $s \preccurlyeq s_2$. Then we have $t \in \mathcal{S}$ and $t \preccurlyeq s_1 s_2$, whence, as H is an $\mathcal{S}$-head function, $t \preccurlyeq H(s_1 s_2)$, that is,

$t \preccurlyeq I(s_1, s_2)$. Hence $I(s_1, s_2)$ is a $\preccurlyeq_{\mathcal{S}}$-greatest element in $\mathcal{I}_{\underline{\mathcal{S}}}(s_1, s_2)$, and I is a greatest $\mathcal{I}$-function for $\underline{\mathcal{S}}$. The argument is similar for J. □

We shall prove the converse implications by using Proposition 2.8. The main observation is that a greatest $\mathcal{J}$-function necessarily obeys the $\mathcal{J}$-law.

Lemma 2.34. *If a germ $\underline{\mathcal{S}}$ is left-associative and left-cancellative and J is a greatest $\mathcal{J}$-function for $\underline{\mathcal{S}}$, then J obeys the $\mathcal{J}$-law.*

Proof. Assume $s_1|s_2|s_3 \in \mathcal{S}^{[3]}$ with $s_1 \bullet s_2$ defined. Put $t = J(s_1, s_2 \bullet J(s_2, s_3))$ and $t' = s_2 \bullet r'$ with $r' = J(s_1 \bullet s_2, s_3)$. Our aim is to prove that t and t' are $=^{\times}$-equivalent.

First, $s_1 \bullet s_2$ is defined and $s_2 \preccurlyeq_{\mathcal{S}} s_2 \bullet J(s_2, s_3)$ is true, hence the maximality of $J(s_1, s_2 \bullet J(s_2, s_3))$ implies $s_2 \preccurlyeq_{\mathcal{S}} J(s_1, s_2 \bullet J(s_2, s_3))$, that is, $s_2 \preccurlyeq_{\mathcal{S}} t$. Write $t = s_2 \bullet r$. By assumption, t belongs to $\mathcal{J}_{\underline{\mathcal{S}}}(s_1, s_2 \bullet J(s_2, s_3))$, hence we have $t \preccurlyeq_{\mathcal{S}} s_2 \bullet J(s_2, s_3)$, that is, $s_2 \bullet r \preccurlyeq_{\mathcal{S}} s_2 \bullet J(s_2, s_3)$, which implies $r \preccurlyeq_{\mathcal{S}} J(s_2, s_3)$ as $\underline{\mathcal{S}}$ is left-cancellative, whence in turn $r \preccurlyeq_{\mathcal{S}} s_3$ since $J(s_2, s_3)$ belongs to $\mathcal{J}_{\underline{\mathcal{S}}}(s_2, s_3)$. By assumption, $s_1 \bullet t$, that is, $s_1 \bullet (s_2 \bullet r)$, is defined, and so is $s_1 \bullet s_2$. Hence $(s_1 \bullet s_2) \bullet r$ is defined as well, and $r \preccurlyeq_{\mathcal{S}} s_3$ holds. The maximality of $J(s_1 \bullet s_2, s_3)$ gives $r \preccurlyeq_{\mathcal{S}} r'$, whence $t \preccurlyeq_{\mathcal{S}} s_2 \bullet r' = t'$.

For the other direction, the definition of r' implies that $(s_1 \bullet s_2) \bullet r'$ is defined and $r' \preccurlyeq_{\mathcal{S}} s_3$ holds. By left-associativity, the first relation implies that $s_1 \bullet (s_2 \bullet r')$, that is, $s_1 \bullet t'$, is defined. On the other hand, $s_2 \bullet r'$ is defined by assumption and $r' \preccurlyeq_{\mathcal{S}} s_3$ holds, so the maximality of $J(s_2, s_3)$ implies $t' \preccurlyeq_{\mathcal{S}} s_2 \bullet J(s_2, s_3)$. Then the maximality of $J(s_1, s_2 \bullet J(s_2, s_3))$ implies $t' \preccurlyeq_{\mathcal{S}} J(s_1, s_2 \bullet J(s_2, s_3))$, that is, $t' \preccurlyeq_{\mathcal{S}} t$. So $t =^{\times} t'$ is satisfied, which is the desired instance of the $\mathcal{J}$-law. □

Not surprisingly, we have a similar property for greatest $\mathcal{I}$-functions.

Lemma 2.35. *If a germ $\underline{\mathcal{S}}$ is left-associative and left-cancellative and I is a greatest $\mathcal{I}$-function for $\underline{\mathcal{S}}$, then I obeys the $\mathcal{I}$-law.*

Proof. One could mimic the argument used for Lemma 2.34, but the exposition is less convenient, and we shall instead derive the result from Lemma 2.34.

So, assume that I is a greatest $\mathcal{I}$-function for $\mathcal{S}$ and let $J : \mathcal{S}^{[2]} \to \mathcal{S}$ be defined by $I(s_1, s_2) = s_1 \bullet J(s_1, s_2)$. By Lemma 2.12, J is a $\mathcal{J}$-function for $\mathcal{S}$, and the assumption that I is a greatest $\mathcal{I}$-function implies that J is a greatest $\mathcal{J}$-function. Indeed, assume $t \in \mathcal{J}_{\mathcal{S}}(s_1, s_2)$. Then $s_1 \bullet t$ is defined and belongs to $\mathcal{I}_{\mathcal{S}}(s_1, s_2)$, hence $s_1 \bullet t \preccurlyeq_{\mathcal{S}} I(s_1, s_2)$, that is, $s_1 \bullet t \preccurlyeq_{\mathcal{S}} s_1 \bullet J(s_1, s_2)$, whence $t \preccurlyeq_{\mathcal{S}} J(s_1, s_2)$. Then, by Lemma 2.34, J satisfies the $\mathcal{J}$-law. By Lemma 2.12 again, this in turn implies that I satisfies the $\mathcal{I}$-law. □

As in the case of head functions in Chapter IV, we shall now manage to go from the $\mathcal{I}$-law to the sharp $\mathcal{I}$-law, that is, force equality instead of $=^{\times}_{\mathcal{S}}$-equivalence.

Lemma 2.36. *If a germ $\underline{\mathcal{S}}$ is left-associative and left-cancellative, and I is a greatest $\mathcal{I}$-function for $\underline{\mathcal{S}}$, then every function $I' : \mathcal{S}^{[2]} \to \mathcal{S}$ satisfying $I'(s_1, s_2) =^{\times}_{\mathcal{S}} I(s_1, s_2)$ for every $s_1|s_2$ in $\mathcal{S}^{[2]}$ is a greatest $\mathcal{I}$-function for $\underline{\mathcal{S}}$.*

Proof. Let $s_1|s_2$ belong to $\mathcal{S}^{[2]}$. First, we have $s_1 \preccurlyeq_{\mathcal{S}} I(s_1, s_2) \preccurlyeq_{\mathcal{S}} I'(s_1, s_2)$, whence $s_1 \preccurlyeq_{\mathcal{S}} I'(s_1, s_2)$. Write $I(s_1, s_2) = s_1 \bullet s$ and $I'(s_1, s_2) = s_1 \bullet s'$. By definition, we have

$s \preccurlyeq_{\mathcal{S}} s_2$, hence $s_2 = s \bullet r$ for some r, and, by assumption, $s' = s \bullet \epsilon$ for some invertible element ϵ of $\mathcal{S}$. We find $s_2 = (s \bullet (\epsilon \bullet \epsilon^{-1})) \bullet r = ((s \bullet \epsilon) \bullet \epsilon^{-1}) \bullet r = (s \bullet \epsilon) \bullet (\epsilon^{-1} \bullet r)$, whence $s' \preccurlyeq_{\mathcal{S}} s_2$: the second equality comes from (1.6), and the last one from the assumption that $\underline{\mathcal{S}}$ is left-associative. Hence $I'(s_1, s_2)$ belongs to $\mathcal{I}_{\underline{\mathcal{S}}}(s_1, s_2)$.

Now assume $t = s_1 \bullet s$ with $s \preccurlyeq s_2$. Then we have $t \preccurlyeq_{\mathcal{S}} I(s_1, s_2)$ by assumption, hence $t \preccurlyeq_{\mathcal{S}} I'(s_1, s_2)$ by transitivity of $\preccurlyeq_{\mathcal{S}}$. So I' is a greatest $\mathcal{I}$-function for $\underline{\mathcal{S}}$. □

Lemma 2.37. *If a germ $\underline{\mathcal{S}}$ is left-associative, left-cancellative, and admits a greatest $\mathcal{I}$-function, then $\underline{\mathcal{S}}$ admits a greatest $\mathcal{I}$-function obeying the sharp $\mathcal{I}$-law.*

Proof. Let I be a greatest $\mathcal{I}$-function for $\underline{\mathcal{S}}$, and let $\mathcal{S}'$ be an $=^{\times}_{\mathcal{S}}$-selector on $\mathcal{S}$. For $s_1|s_2$ in $\mathcal{S}^{[2]}$, define $I'(s_1, s_2)$ to be the unique element of $\mathcal{S}'$ that is $=^{\times}$-equivalent to $I(s_1, s_2)$. By construction, I' is a function from $\mathcal{S}^{[2]}$ to $\mathcal{S}$ satisfying $I'(s_1, s_2) =^{\times}_{\mathcal{S}} I(s_1, s_2)$ for every $s_1|s_2$ in $\mathcal{S}^{[2]}$, hence, by Lemma 2.36, I' is a greatest $\mathcal{I}$-function for $\underline{\mathcal{S}}$. By Lemma 2.35, I' obeys the $\mathcal{I}$-law, that is, for every $s_1|s_2|s_3$ in $\mathcal{S}^{[3]}$ such that $s_1 \bullet s_2$ is defined, we have

$$I'(s_1, I'(s_2, s_3)) =^{\times} I'(s_1 \bullet s_2, s_3).$$

Now, by the definition of a selector, two elements in the image of the function I' must be equal whenever they are $=^{\times}$-equivalent. So I' obeys the sharp $\mathcal{I}$-law. □

Putting pieces together, we can now complete our argument.

Proof of Proposition 2.28. By Lemma 2.33, the conditions are necessary, and it remains to prove that each of (2.32) and (2.30) implies that $\underline{\mathcal{S}}$ is a Garside germ.

Assume that $\underline{\mathcal{S}}$ is a germ that is left-associative, left-cancellative, and admits a greatest $\mathcal{I}$-function. By Lemma 2.37, $\underline{\mathcal{S}}$ admits an $\mathcal{I}$-function I that satisfies the sharp $\mathcal{I}$-law, so, by Proposition 2.8, $\underline{\mathcal{S}}$ is a Garside germ. Hence, (2.32) implies that $\underline{\mathcal{S}}$ is a Garside germ.

Finally, assume that $\underline{\mathcal{S}}$ is a germ that is left-associative, left-cancellative, and admits a greatest $\mathcal{J}$-function J. As seen in the proof of Lemma 2.35, the function I defined on $\mathcal{S}^{[2]}$ by $I(s_1, s_2) = s_1 \bullet J(s_1, s_2)$ is a greatest $\mathcal{I}$-function for $\underline{\mathcal{S}}$. So (2.30) implies (2.32) and, therefore, it implies that $\underline{\mathcal{S}}$ is a Garside germ. □

Remark 2.38. The above arguments involve both $\mathcal{I}$- and $\mathcal{J}$-functions, and it could appear more elegant to appeal to one type only. However, this is not easy, as the argument for "maximality implies law" is more natural for $\mathcal{J}$-functions, whereas Lemma 2.36 is valid only for $\mathcal{I}$-functions: indeed, unless the considered germ is right-associative, the assumption that $s \bullet t$ is defined and $t' =^{\times}_{\mathcal{S}} t$ holds need not imply that $s \bullet t'$ is defined, so a function that is $=^{\times}_{\mathcal{S}}$-equivalent to a $\mathcal{J}$-function need not be a $\mathcal{J}$-function in general.

2.3 Noetherian germs

The interest in Noetherianity assumptions is the desire to guarantee the existence of minimal (or maximal) elements with respect to divisibility. In germs, we shall use such assumptions to guarantee the existence of a greatest $\mathcal{J}$-function under weak assumptions.

Lemma 2.39. *If $\underline{S}$ is a Garside germ, then:*

(2.40) *For every $s_1|s_2$ in $\mathcal{S}^{[2]}$, any two elements of $\mathcal{J}_{\underline{S}}(s_1, s_2)$ admit a common right-$\underline{S}$-multiple that lies in $\mathcal{J}_{\underline{S}}(s_1, s_2)$.*

Proof. Assume that $\underline{S}$ is a Garside germ. By Proposition 2.28, $\underline{S}$ admits a greatest $\mathcal{J}$-function J. Then, for every $s_1|s_2$ in $\mathcal{S}^{[2]}$, the element $J(s_1, s_2)$ is a right-$\underline{S}$-multiple of every element of $\mathcal{J}_{\underline{S}}(s_1, s_2)$, hence a common right-$\underline{S}$-multiple of any two of them. □

In some contexts, the necessary condition (2.40) turns out to be sufficient:

Proposition 2.41 (recognizing Garside germ, right-Noetherian case). *A right-Noetherian germ $\underline{S}$ is a Garside germ if and only if it is left-associative, left-cancellative, and satisfies* (2.40). *In this case, the category $\mathcal{C}at(\underline{S})$ is right-Noetherian.*

Proof. If $\underline{S}$ is a Garside germ, it is left-associative and left-cancellative by Lemma 2.15, and it satisfies (2.40) by Lemma 2.39. So the conditions are necessary.

Conversely, assume that $\underline{S}$ satisfies the conditions of the statement. Let $s_1|s_2$ belong to $\mathcal{S}^{[2]}$. Put

$$\mathcal{X} = \{t \in \mathcal{S} \mid \exists s \in \mathcal{S}\,(s_1 \bullet s \in \mathcal{S} \text{ and } s \bullet t = s_2)\}.$$

By assumption, $\mathcal{X}$ has a $\widetilde{\prec}_{\mathcal{S}}$-minimal element, say $\hat{t}$, and there exists $\hat{s}$ in $\mathcal{S}$ satisfying $s_1 \bullet \hat{s} \in \mathcal{S}$ and $\hat{s} * \hat{t} = s_2$. By construction, $\hat{s}$ belongs to $\mathcal{J}_{\underline{S}}(s_1, s_2)$. Now let s be an arbitrary element of $\mathcal{J}_{\underline{S}}(s_1, s_2)$. By assumption, $\hat{s}$ and s admit a common right-$\underline{S}$-multiple, say $\hat{s} \bullet s'$, that lies in $\mathcal{J}_{\underline{S}}(s_1, s_2)$. Hence there exists t' satisfying $(\hat{s} \bullet s') \bullet t' = s_2$, whence $\hat{t} = s' \bullet t'$ since $\underline{S}$ is left-associative and left-cancellative. The choice of $\hat{t}$ implies that s' is invertible in $\mathcal{S}$. This means that $\hat{s}$ is a $\preccurlyeq_{\mathcal{S}}$-greatest element in $\mathcal{J}_{\underline{S}}(s_1, s_2)$. By Proposition 2.28, we deduce that $\underline{S}$ is a Garside germ. So the conditions are also sufficient.

Finally, assume that $\underline{S}$ is a Garside germ. Then $\mathcal{C}at(\underline{S})$ is a left-cancellative category, and $\mathcal{S}$ is a solid Garside family of $\mathcal{C}at(\underline{S})$ that is locally right-Noetherian. By Lemma IV.2.21, $\mathcal{C}at(\underline{S})$ must be right-Noetherian. □

We naturally say that a germ $\underline{S}$ *admits right-mcms* if every common right-$\underline{S}$-multiple (if any) of two elements s, s' of $\mathcal{S}$ is a right-$\underline{S}$-multiple of some right-$\underline{S}$-mcm of s and s'. When right-$\underline{S}$-mcms exist, (2.40) can be rewritten in an improved form.

Lemma 2.42. *Assume that $\underline{S}$ is a germ that admits right-mcms, and consider*

(2.43) *For r, s, s' in $\mathcal{S}$, if $r \bullet s$ and $r \bullet s'$ are defined, so is $r \bullet t$ for every right-$\underline{S}$-mcm t of s and s'.*

If $\underline{S}$ satisfies (2.43)*, then it satisfies* (2.40)*. Conversely, if $\underline{S}$ satisfies* (2.40) *and is right-associative or there exists no nontrivial invertible element in $\mathcal{S}$, then it satisfies* (2.43).

Proof. Assume that $s_1|s_2$ belongs to $\mathcal{S}^{[2]}$, and that s, s' lie in $\mathcal{J}_{\underline{S}}(s_1, s_2)$. By assumption, we have $s \preccurlyeq_S s_2$ and $s' \preccurlyeq_S s_2$. As $\underline{S}$ admits right-mcms, there exists a right-$\underline{S}$-mcm t of s and s' that satisfies $t \preccurlyeq_S s_2$. If $\underline{S}$ satisfies (2.43), the assumption that $s_1 \bullet s$ and $s_1 \bullet s'$ are defined implies that $s_1 \bullet t$ is defined, and so t belongs to $\mathcal{J}_S(s_1, s_2)$. Then t is a common right-$\underline{S}$-multiple of s and s' lying in $\mathcal{J}_S(s_1, s_2)$, and (2.40) is satisfied.

Conversely, assume that $\underline{S}$ satisfies (2.40), $r \bullet s$ and $r \bullet s'$ are defined and t is a right-$\underline{S}$-mcm of s and s'. Then, by definition, s and s' belong to $\mathcal{J}_{\underline{S}}(r, t)$. By (2.40), some common right-$\underline{S}$-multiple $\hat{t}$ of s and s' lies in $\mathcal{J}_S(r, t)$. Thus, $r \bullet \hat{t}$ is defined, and $\hat{t} \preccurlyeq_S t$ holds. Write $t = \hat{t} \bullet t'$. As t is a right-$\underline{S}$-mcm of s and s', the element t' must be invertible. By assumption, $r \bullet \hat{t}$, which is $r \bullet (t \bullet t'^{-1})$, is defined. If $\underline{S}$ is right-associative, or if there exists no nontrivial invertible element in $\underline{S}$, we deduce that $r \bullet t$ is defined (in the latter case, we have $t = \hat{t}$). So (2.43) is satisfied. □

Using the fact that Noetherianity implies the existence of right-mcms, we deduce:

Proposition 2.44 (recognizing Garside germ, Noetherian case). *A germ that is left-associative, left-cancellative, Noetherian, and satisfies* (2.43) *is a Garside germ.*

Proof. Assume that $\underline{S}$ satisfies the assumptions of the proposition. As $\underline{S}$ is left-Noetherian, it admits right-mcms. Hence, by Lemma 2.42, it satisfies (2.40) as well. Then, by Proposition 2.41, it is a Garside germ. □

Of course, we shall say that a germ $\underline{S}$ *admits conditional right-lcms* if any two elements of $\mathcal{S}$ that admit a common right-$\underline{S}$-multiple admits a right-$\underline{S}$-lcm. Another condition guaranteeing the existence of right-mcms is the existence of conditional right-lcms.

Corollary 2.45 (recognizing Garside germ, conditional right-lcm case). *A germ $\underline{S}$ that is left-associative, left-cancellative, right-Noetherian, and admits conditional right-lcms is a Garside germ whenever it satisfies:*

(2.46) *For r, s, s' in $\mathcal{S}$, if $r \bullet s$ and $r \bullet s'$ are defined, then so is $r \bullet t$ for every right-$\underline{S}$-lcm t of s and s'.*

Proof. Assume that $\underline{S}$ satisfies the hypotheses of the corollary. The assumption that $\underline{S}$ admits conditional right-lcms implies that it admits right-mcms and, as every right-$\underline{S}$-mcm is a right-$\underline{S}$-lcm, (2.46) implies (2.43), hence (2.40) by Lemma 2.42. Then Proposition 2.41 implies that $\underline{S}$ is a Garside germ. □

We saw at the end of Section 1 how, in a right-Noetherian context, the existence of right-$\underline{S}$-lcms for arbitrary elements follows from that for elements of a generating subfamily, typically atoms. Using the same method, we can refine Corollary 2.45.

Proposition 2.47 (recognizing Garside germ, atomic case). *If a germ $\underline{\mathcal{S}}$ is associative, left-cancellative, right-Noetherian, and admits conditional right-lcms, and $\mathcal{A}$ is a subfamily of $\mathcal{S}$ such that $\mathcal{A} \cup \mathcal{S}^{\times}$ generates $\underline{\mathcal{S}}$ and satisfies $\mathcal{S}^{\times}\mathcal{A} \subseteq \mathcal{A}\mathcal{S}^{\times}$, then $\underline{\mathcal{S}}$ is a Garside germ whenever it satisfies the condition:*

(2.48) *For r in $\mathcal{S}$ and s, s' in $\mathcal{A}$, if $r \bullet s$ and $r \bullet s'$ are defined, then so is $r \bullet t$ for every right-$\underline{\mathcal{S}}$-lcm t of s and s'.*

If $\underline{\mathcal{S}}$ is Noetherian, the criterion applies in particular when $\mathcal{A}$ is the atom family of $\underline{\mathcal{S}}$.

Proof. The argument is similar to the one used for Lemma 1.27 and for the inductions of Sections II.4. So assume that $\underline{\mathcal{S}}$ is a germ that satisfies the hypotheses of the proposition and (2.48). Our aim is to show that (2.46) is valid for all r, s, s', t in $\underline{\mathcal{S}}$.

Assume that $r \bullet s$ and $r \bullet s'$ are defined and t is a right-$\underline{\mathcal{S}}$-lcm of s and s'. We shall prove that $r \bullet t$ is defined, that is, (2.46) holds, using induction on t with respect to $\widetilde{\prec}_{\mathcal{S}}$, which, by assumption, is well-founded. Equivalently, we use induction on $\lambda(t)$, where λ is a fixed sharp Noetherianity witness on $\underline{\mathcal{S}}$. If $\lambda(t) = 0$ holds, then t must be invertible, hence so are s and s', and we have $s =^{\times}_{\mathcal{S}} s' =^{\times}_{\mathcal{S}} t$. As $\underline{\mathcal{S}}$ is right-associative, the assumption that $r \bullet s$ is defined implies that $r \bullet t$ is defined as well.

Assume now $\lambda(t) > 0$. If s' is invertible, we have $t =^{\times} s$ and, as above, the assumption that $r \bullet s$ is defined implies that $r \bullet t$ is defined as well. If s is invertible, the result is the same. Otherwise, as $\mathcal{A} \cup \mathcal{S}^{\times}$ generates $\underline{\mathcal{S}}$ and $\mathcal{S}^{\times}\mathcal{A} \subseteq \mathcal{A}\mathcal{S}^{\times}$ holds, there must exist non-invertible elements s_1, s'_1 in $\mathcal{A}$ left-dividing respectively s and s', say $s = s_1 \bullet s_2$ and $s' = s'_1 \bullet s'_3$. Then the rule for an iterated right-$\underline{\mathcal{S}}$-lcm implies the existence of the diagram of Figure 4, in which each element t_i is a right-$\underline{\mathcal{S}}$-lcm of s_i and s'_i and such that we have $t = t_1 \bullet t_4$. Then t_1 is a right-$\underline{\mathcal{S}}$-lcm of s_1 and s'_1, and $r \bullet s_1$ and $r \bullet s'_1$ are defined because $r \bullet (s_1 \bullet s_2)$ and $r \bullet (s'_1 \bullet s'_2)$ are and $\underline{\mathcal{S}}$ is right-associative. As s_1 and s'_1 lie in $\mathcal{A}$, (2.48) implies that $r \bullet t_1$ is defined. Next, t_2 is a right-$\underline{\mathcal{S}}$-lcm of s_2 and s'_2, and $(r \bullet s_1) \bullet s_2$, which is $r \bullet s$, and $(r \bullet s_1) \bullet s'_2$, which is $r \bullet t_1$, are defined. Moreover, as s_1 is non-invertible, we have $\lambda(t_2) < \lambda(s_1 \bullet t_2) \leqslant \lambda(t_1 \bullet t_4) = \lambda(t)$. Hence, by the induction hypothesis, $(r \bullet s_1) \bullet t_2$, is defined. Mutatis mutandis, we find that $(r \bullet s'_1) \bullet t_3$ is defined. Finally, t_4 is a right-$\underline{\mathcal{S}}$-lcm of s'_4 and s'_4, and we saw above that $(r \bullet t_1) \bullet s_4$, which is $(r \bullet s_1) \bullet t_2$, and $(r \bullet t_1) \bullet s'_4$, which is $(r \bullet s'_1) \bullet t_3$, are defined. As s_1 is not invertible, t_1 is not invertible either, and we have $\lambda(t_4) < \lambda(t_1 \bullet t_4) = \lambda(t)$. Hence, by the induction hypothesis, $(r \bullet t_1) \bullet t_4$ is defined. By left-associativity, it follows that $r \bullet t$ is defined, and (2.46) is satisfied. Then Corollary 2.45 implies that $\underline{\mathcal{S}}$ is a Garside germ.

Finally, assume that $\underline{\mathcal{S}}$ is Noetherian and let $\mathcal{A}$ be the atom family of $\mathcal{S}^{\sharp}$. By Proposition 1.22, $\underline{\mathcal{S}}$ is generated by $\mathcal{A} \cup \mathcal{S}$. Moreover, as observed in the proof of Proposition 1.30, $\mathcal{S}^{\times}\mathcal{A} \subseteq \mathcal{A}\mathcal{S}^{\times}$ holds. Hence $\mathcal{A}$ is eligible for the proposition. □

In another direction, we can consider germs that admit right-lcms, not only conditional right-lcms. Then, provided the germ is associative, there is no need to worry about (2.43).

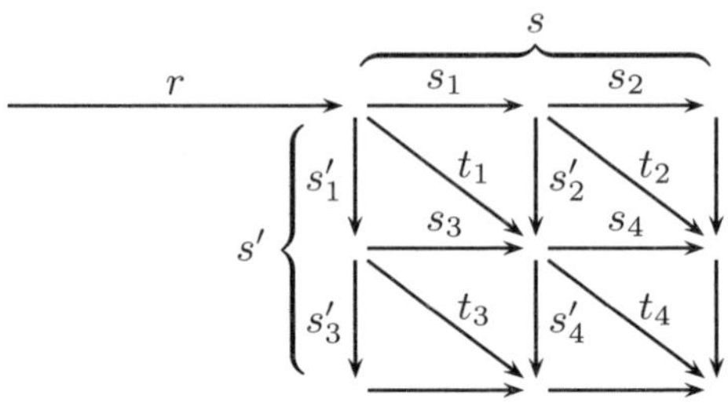

Figure 4. Induction step in the proof of Proposition 2.47. Note the similarity both with a right-reversing process and with Figure 2. Each square represents a right-$\underline{S}$-lcm, and the right-$\underline{S}$-lcm of s and s' corresponds to every path from the top left corner to the bottom right corner.

Proposition 2.49 (recognizing Garside germ, right-lcm case). *A germ that is associative, left-cancellative, right-Noetherian, and admits right-lcms is a Garside germ.*

Proof. Assume that $\underline{S}$ satisfies the hypotheses of the statement. We check that (2.46) is satisfied. So assume that $r \bullet s$ and $r \bullet s'$ are defined and t is a right-$\underline{S}$-lcm of s and s'. Let $\hat{r}$ be a right-$\underline{S}$-lcm of $r \bullet s$ and $r \bullet s'$, which exists by assumption. First, we have $r \preccurlyeq_S r \bullet s \preccurlyeq_S \hat{r}$, whence, by Lemma 1.19, $r \preccurlyeq_S \hat{r}$, so there exists $\hat{s}$ satisfying $\hat{r} = r \bullet \hat{s}$. By Lemma 1.19 again, $r \bullet s \preccurlyeq_S \hat{r}$ implies $s \preccurlyeq_S \hat{s}$ and $r \bullet s' \preccurlyeq_S \hat{r}$ implies $s' \preccurlyeq_S \hat{s}$. So $\hat{s}$ is a common right-$\underline{S}$-multiple of s and s', hence it is a right-$\underline{S}$-multiple of their right-$\underline{S}$-lcm t: we have $\hat{s} = t \bullet \hat{t}$ for some $\hat{t}$. Hence $r \bullet (t \bullet \hat{t})$, which is $r \bullet \hat{s}$, is defined. By right-associativity, this implies that $r \bullet t$ is defined, so (2.46) is true. Then, $\underline{S}$ is a Garside germ by Corollary 2.45. □

2.4 An application: germs derived from a groupoid

We shall now construct an important family of germs. The principle is to start from a group(oid) together with a distinguished generating family, and to derive a germ, leading in turn to a new category and, from there, to a new groupoid. The latter groupoid is a sort of unfolded version of the initial one, as in the seminal examples which start with a Coxeter group and arrive at the (ordinary or dual) braid monoids and groups. As can be expected, the construction is interesting when the derived germ is a Garside germ.

At the technical level, we shall have to consider paths in a groupoid that enjoy a certain length property called *tightness*.

Definition 2.50 (positive generators, length). A subfamily Σ of a groupoid $\mathcal{G}$ is said to *positively generate* $\mathcal{G}$ if every element of $\mathcal{G}$ admits an expression by a Σ-path (no letter in Σ^{-1}). Then, for g in $\mathcal{G}\setminus\{1_{\mathcal{G}}\}$, the Σ-*length* $\|g\|_\Sigma$ is the minimal number ℓ such that g admits an expression by a Σ-path of length ℓ; we complete with $\|1_x\|_\Sigma = 0$ for each x.

If Σ is any family of generators for a groupoid $\mathcal{G}$, then $\Sigma \cup \Sigma^{-1}$ positively generates $\mathcal{G}$. If Σ positively generates a groupoid $\mathcal{G}$, the Σ-length satisfies the triangular inequality

$\|fg\|_\Sigma \leqslant \|f\|_\Sigma + \|g\|_\Sigma$ and, more generally, for every $\mathcal{G}$-path $g_1|\cdots|g_p$,

(2.51) $$\|g_1 \cdots g_p\|_\Sigma \leqslant \|g_1\|_\Sigma + \cdots + \|g_p\|_\Sigma.$$

Definition 2.52 (tight). If Σ positively generates a groupoid $\mathcal{G}$, a $\mathcal{G}$-path $g_1|\cdots|g_p$ is called Σ-*tight* if $\|g_1 \cdots g_p\|_\Sigma = \|g_1\|_\Sigma + \cdots + \|g_p\|_\Sigma$ is satisfied.

Lemma 2.53. *If Σ positively generates a groupoid $\mathcal{G}$, then $g_1|\cdots|g_p$ is Σ-tight if and only if $g_1|\cdots|g_{p-1}$ and $g_1 \cdots g_{p-1}|g_p$ are Σ-tight. Symmetrically, $g_1|\cdots|g_p$ is Σ-tight if and only if $g_2|\cdots|g_p$ and $g_1|g_2 \cdots g_p$ are Σ-tight.*

Proof. To make reading easier, we consider the case of three entries. Assume that $f|g|h$ is Σ-tight. By (2.51), we have $\|fgh\|_\Sigma \leqslant \|fg\|_\Sigma + \|h\|_\Sigma$, whence $\|fg\|_\Sigma \geqslant \|fgh\|_\Sigma - \|h\|_\Sigma = \|f\|_\Sigma + \|g\|_\Sigma$. On the other hand, by (2.51), we have $\|fg\|_\Sigma \leqslant \|f\|_\Sigma + \|g\|_\Sigma$. We deduce $\|fg\|_\Sigma = \|f\|_\Sigma + \|g\|_\Sigma$, and $f|g$ is Σ-tight. Next we have $\|(fg)h\|_\Sigma = \|f\|_\Sigma + \|g\|_\Sigma + \|h\|_\Sigma$, whence $\|(fg)h\|_\Sigma = \|fg\|_\Sigma + \|h\|_\Sigma$ since, as seen above, $f|g$ is Σ-tight. Hence (fg, h) is Σ-tight.

Conversely, assume that $f|g$ and $fg|h$ are Σ-tight. Then we directly obtain $\|fgh\|_\Sigma = \|fg\|_\Sigma + \|h\|_\Sigma = \|f\|_\Sigma + \|g\|_\Sigma + \|h\|_\Sigma$, and $f|g|h$ is Σ-tight.

The argument is similar when gathering final entries instead of initial ones. □

The tightness condition naturally leads to introducing two partial orderings:

Definition 2.54 (Σ-prefix, Σ-suffix). If Σ positively generates a groupoid $\mathcal{G}$, then for f, g in $\mathcal{G}$ with the same source, we say that f is a Σ-*prefix* of g, written $f \leqslant_\Sigma g$, if $f|f^{-1}g$ is Σ-tight. Symmetrically, we say that h is a Σ-*suffix* of g, written $h \,\widetilde{\leqslant}_\Sigma\, g$, if g, h have the same target and $gh^{-1}|h$ is Σ-tight.

Lemma 2.55. *If Σ positively generates a groupoid $\mathcal{G}$, then $\leqslant_\Sigma$ and $\widetilde{\leqslant}_\Sigma$ are partial orders on $\mathcal{G}$ and $1_x \leqslant_\Sigma g$ holds for every g with source x.*

Proof. As $\|1_y\|_\Sigma$ is zero, every path $g|1_y$ is Σ-tight for every g in $\mathcal{G}(\cdot, y)$, so $g \leqslant_\Sigma g$ always holds, and $\leqslant_\Sigma$ is reflexive. Next, as the Σ-length is nonnegative, $f \leqslant_\Sigma g$ implies $\|f\|_\Sigma \leqslant \|g\|_\Sigma$. Now, assume $f \leqslant_\Sigma g$ and $g \leqslant_\Sigma f$. By the previous remark, we must have $\|f\|_\Sigma = \|g\|_\Sigma$, whence $\|f^{-1}g\|_\Sigma = 0$. Hence, $f^{-1}g$ is an identity element, that is, $f = g$ holds. So $\leqslant_\Sigma$ is antisymmetric. Finally, assume $f \leqslant_\Sigma g \leqslant_\Sigma h$. Then $f|f^{-1}g$ and $g|g^{-1}h$, which is $ff^{-1}g|g^{-1}h$, are Σ-tight. By Lemma 2.53, $f|f^{-1}g|g^{-1}h$ is Σ-tight, and then $f|(f^{-1}g)(g^{-1}h)$, which is $f|f^{-1}h$ is Σ-tight. Hence $f \leqslant_\Sigma h$ holds, and $\leqslant_\Sigma$ is transitive. So $\leqslant_\Sigma$ is a partial order on $\mathcal{G}$. Finally, as $\|1_x\|_\Sigma$ is zero, every path $1_x|g$ with x the source of g is Σ-tight, and $1_x \leqslant_\Sigma g$ holds.

The verifications for $\widetilde{\leqslant}_\Sigma$ are entirely similar. □

Example 2.56 (Σ-prefix). Let $\mathcal{G} = (\mathbb{Z}, +)$ and $\Sigma = \{-1, 1\}$. Then Σ is a family of positive generators for $\mathbb{Z}$, and, for every integer p, we have $\|p\|_\Sigma = |p|$. Then $p|q$ is Σ-tight if and only if p and q have the same sign, and p is a Σ-prefix of q if and only if it is a Σ-suffix of q, if and only if p belongs to the interval $[0, q]$.

Consider now $\mathcal{G} = (\mathbb{Z}/n\mathbb{Z}, +)$, with elements $\overline{0}, \overline{1}, ..., \overline{n{-}1}$, and $\Sigma = \{\overline{1}\}$. For $0 \leqslant p < n$, we have $\|\overline{p}\|_\Sigma = p$, so $(\overline{p}, \overline{q})$ is Σ-tight if and only if $p + q < n$ holds, and $\overline{p}$ is a Σ-prefix of $\overline{q}$ for $0 \leqslant q \leqslant p$.

Note that $f|g$ is Σ-tight if and only if f is a Σ-prefix of fg. The partial order $\leqslant_\Sigma$ is weakly compatible with the product in the following sense:

Lemma 2.57. *If Σ positively generates a groupoid $\mathcal{G}$, and f, g are elements of $\mathcal{G}$ such that $f|g$ is Σ-tight and g' is a Σ-prefix of g, then $f|g'$ is Σ-tight and we have $fg' \leqslant_\Sigma fg$.*

Proof. Assume $g' \leqslant_\Sigma g$. Then, by definition, $g'|g'^{-1}g$ is Σ-tight. On the other hand, by assumption, $f|g$, which is $f|g'(g'^{-1}g)$, is Σ-tight. By Lemma 2.53, it follows that $f|g'|g'^{-1}g$ is Σ-tight. First we deduce that $f|g'$ is Σ-tight, that is, $f \leqslant_\Sigma fg'$ holds. Next we deduce that $fg'|g'^{-1}g$ is Σ-tight as well. As $g'^{-1}g$ is also $(fg')^{-1}(fg)$, the latter relation is equivalent to $fg' \leqslant_\Sigma fg$. □

We arrive at our main construction. For $\mathcal{S}$ a subfamily of a category, we use $\mathbf{1}_\mathcal{S}$ for the family of all elements 1_x with x the source or the target of at least one element of $\mathcal{S}$.

Definition 2.58 (derived structure). If a groupoid $\mathcal{G}$ is positively generated by a family Σ and $\mathcal{H}$ is a subfamily of $\mathcal{G}$, the structure *derived from $\mathcal{H}$ and Σ*, denoted by $\mathcal{H}^\Sigma$, is $(\mathcal{H}, \mathbf{1}_\mathcal{H}, \bullet)$, where $\bullet$ is the partial operation on $\mathcal{H}$ such that $f \bullet g = h$ holds if and only if

(2.59) $$fg = h \text{ holds in } \mathcal{G} \text{ and } f|g \text{ is } \Sigma\text{-tight.}$$

So we consider the operation that is induced on $\mathcal{H}$ by the ambient product of $\mathcal{G}$, with the restriction that the products that are not Σ-tight are discarded. We shall be specifically interested in the case when the derived structure is a germ, which is easily characterized.

Lemma 2.60. *If Σ positively generates a groupoid $\mathcal{G}$ and $\mathcal{H}$ is a subfamily of $\mathcal{G}$ that includes $\mathbf{1}_\mathcal{H}$, then $\mathcal{H}^\Sigma$ is a germ. It is cancellative, Noetherian, and contains no nontrivial invertible element.*

Proof. The verifications are easy. First, (1.5) is satisfied since we assume that $\mathbf{1}_\mathcal{H}$ is included in $\mathcal{H}$. Next, assume that f, g, h belong to $\mathcal{H}$ and $f \bullet g$, $g \bullet h$ and $(f \bullet g) \bullet h$ are defined. This means that fg, gh, and $(fg)h$ belong to $\mathcal{H}$ and that the paths $f|g$, $g|h$, and $fg|h$ are Σ-tight in $\mathcal{G}$. By Lemma 2.53, $f|g|h$, and then $f|gh$, which is $f|g \bullet h$, are Σ-tight. As $f(gh)$ belongs to $\mathcal{H}$, we deduce that $f \bullet gh$, that is, $f \bullet (g \bullet h)$, is defined, and it is equal to $f(gh)$. The argument is symmetric in the other direction and, therefore, (1.6) is satisfied. So $\mathcal{H}^\Sigma$ is a germ.

Assume now that f, g, g' belong to $\mathcal{H}$ and $f \bullet g = f \bullet g'$ holds. Then we have $fg = fg'$ in $\mathcal{G}$, whence $g = g'$. So the germ $\mathcal{H}^\Sigma$ is left-cancellative, hence cancellative by a symmetric argument.

Then, assume that f, g lie in $\mathcal{H}$ and we have $g = f \bullet g'$ for some non-invertible g' lying in $\mathcal{H}$. By the definition of $\bullet$, this implies $\|g\|_\Sigma = \|f\|_\Sigma + \|g'\|_\Sigma$, whence $\|f\|_\Sigma < \|g\|_\Sigma$ as, by the definition of the Σ-length, the non-invertibility of g' implies $\|g'\|_\Sigma \geqslant 1$. So the Σ-length is a left-Noetherianity witness for $\mathcal{H}^\Sigma$. By symmetry, it is also a right-Noetherianity witness. So the germ $\mathcal{H}^\Sigma$ is Noetherian.

Finally, assume $\epsilon \bullet \epsilon' = 1_x$ with ϵ, ϵ' in $\mathcal{H}$. Then we must have $\|\epsilon\|_\Sigma + \|\epsilon'\|_\Sigma = \|1_x\|_\Sigma = 0$. The only possibility is $\|\epsilon\|_\Sigma = \|\epsilon'\|_\Sigma = 0$, whence $\epsilon = \epsilon' = 1_x$. □

Example 2.61 (derived germ). Let $\mathcal{G} = (\mathbb{Z}, +)$ and $\Sigma = \{-1, 1\}$. Here are three examples of derived germs according to various choices for $\mathcal{H}$.

$\bullet$	0	1	2	3
0	0	1	2	3
1	1	2	3	
2	2	3		
3	3			

$\bullet$	0	1	2	4
0	0	1	2	4
1	1	2		
2	2		4	
4	4			

$\bullet$	-2	-1	0	1	2
-2			-2		
-1		-2	-1		
0	-2	-1	0	1	2
1			1	2	
2			2		

For $\mathcal{H} = \{0, 1, 2, 4\}$, the germ is not left-associative: $(1 \bullet 1) \bullet 2$ is defined, but $1 \bullet 2$ is not. For $\mathcal{H} = \{-2, -1, 0, 1, 2\}$, the tightness requirement implies that $p \bullet q$ is never defined when p and q have different signs.

Owing to the results of Sections 1 and 2, a germ leads to interesting results only when it is left-associative. As seen in Example 2.61, these properties are not guaranteed for an arbitrary subfamily $\mathcal{H}$, but they turn out to be connected with the closure of the set $\mathcal{H}$ under Σ-suffix and Σ-prefix, where we naturally say that $\mathcal{H}$ is closed under Σ-suffix (*resp.* Σ-prefix) if every Σ-suffix (*resp.* Σ-prefix) of an element of $\mathcal{H}$ lies in $\mathcal{H}$.

Lemma 2.62. *If Σ positively generates a groupoid $\mathcal{G}$ and $\mathcal{H}$ is a subfamily of $\mathcal{G}$ that is closed under Σ-suffix (resp. Σ-prefix), the germ $\mathcal{H}^\Sigma$ is left-associative (resp. right-associative). For f, g in $\mathcal{H}$, the relation $f \preccurlyeq_{\mathcal{H}^\Sigma} g$ is equivalent to f being a Σ-prefix of g in $\mathcal{G}$ (resp. $f \mathrel{\widetilde{\succcurlyeq}}_{\mathcal{H}^\Sigma} g$ is equivalent to f being a Σ-suffix of g in $\mathcal{G}$).*

Proof. Assume that $\mathcal{H}$ is closed under Σ-suffix and f, g, h are elements of $\mathcal{H}$ such that $f \bullet g$ and $(f \bullet g) \bullet h$ are defined. Then fg and $(fg)h$ lie in $\mathcal{H}$ and the paths $f|g$ and $fg|h$ are Σ-tight. By Lemma 2.53, $f|g|h$, and then $f|gh$ are Σ-tight. Hence gh is a Σ-suffix of fgh in $\mathcal{G}$ and, therefore, by assumption, gh belongs to $\mathcal{H}$. By Lemma 2.53 again, the fact that $f|g|h$ is Σ-tight implies that $g|h$ is Σ-tight, and we deduce the equality $gh = g \bullet h$. Thus the germ $\mathcal{H}^\Sigma$ is left-associative.

Assume now that f, g lie in $\mathcal{H}$ and f is a left-divisor of g in the germ $\mathcal{H}^\Sigma$. This means that $f \bullet g' = g$ holds for some g' lying in $\mathcal{H}$. Necessarily g' is $f^{-1}g$, so $f|f^{-1}g$ has to be Σ-tight, which means that f is a Σ-prefix of g. Conversely, assume that f, g lie in $\mathcal{H}$ and f is a Σ-prefix of g. Then $f|f^{-1}g$ is Σ-tight, so $f^{-1}g$ is a Σ-suffix of g. The assumption that $\mathcal{H}$ is closed under Σ-suffix implies that $f^{-1}g$ lies in $\mathcal{H}$, and then $f \bullet f^{-1}g = g$ holds, whence $f \preccurlyeq_{\mathcal{H}^\Sigma} g$.

The arguments for right-associativity and right-divisibility in $\mathcal{H}^\Sigma$ are entirely symmetric, using now the assumption that $\mathcal{H}$ is closed under Σ-prefix. $\square$

We now wonder whether $\mathcal{H}^\Sigma$ is a Garside germ. As $\mathcal{H}^\Sigma$ is Noetherian, it is eligible for Propositions 2.41 and 2.44, and we are led to looking for the satisfaction of (2.43) or (2.46). As the conditions involve the left-divisibility relation of the germ, then, by Lemma 2.62, they can be formulated inside the base groupoid in terms of Σ-prefixes.

Proposition 2.63 (derived Garside I). *If a groupoid $\mathcal{G}$ is positively generated by a family Σ and $\mathcal{H}$ is a subfamily of $\mathcal{G}$ that is closed under Σ-suffix, then $\mathcal{H}^\Sigma$ is a Garside germ whenever the following two conditions are satisfied:*

(2.64) *Any two elements of $\mathcal{H}$ that admit a common $\leqslant_\Sigma$-upper bound admit a least $\leqslant_\Sigma$-upper bound in $\mathcal{H}$,*

(2.65) *For f, g, g' in $\mathcal{H}$, if $f|g$ and $f|g'$ are Σ-tight, then so is $f|h$ for every least $\leqslant_\Sigma$-upper bound h of g and g'.*

Proof. By Lemmas 2.60 and 2.62, the germ $\mathcal{H}^\Sigma$ is left-associative, cancellative, and Noetherian. Hence, by Corollary 2.45, $\mathcal{H}^\Sigma$ is a Garside germ whenever it admits conditional right-lcms and satisfies (2.46). Now, by Lemma 2.62, for g, h in $\mathcal{H}$, the relation $g \preccurlyeq_{\mathcal{H}^\Sigma} h$ is equivalent to $g \leqslant_\Sigma h$. Therefore, h is a right-lcm of g and g' in $\mathcal{H}^\Sigma$ if and only it is a $\leqslant_\Sigma$-least upper bound of g and g'. Hence (2.64) expresses that $\mathcal{H}^\Sigma$ admits conditional right-lcms. Then (2.65) is a direct reformulation of (2.46) owing to the definition of the product in the germ $\mathcal{H}^\Sigma$. $\square$

Appealing to Proposition 2.47 rather than to Corollary 2.45 gives a more economical criterion, whenever the family $\mathcal{H}$ is also closed under Σ-prefix.

Proposition 2.66 (derived Garside II). *If a groupoid $\mathcal{G}$ is positively generated by a family Σ and $\mathcal{H}$ is a subfamily of $\mathcal{G}$ that is closed under Σ-prefix and Σ-suffix, then $\mathcal{H}^\Sigma$ is a Garside germ whenever the following two conditions are satisfied:*

(2.67) *Any two elements of Σ that admit a common $\leqslant_\Sigma$-upper bound in $\mathcal{H}$ admit a least $\leqslant_\Sigma$-upper bound in $\mathcal{H}$,*

(2.68) *For f in $\mathcal{H}$ and s, s' in Σ, if $f|s$ and $f|s'$ are Σ-tight, then so is $f|h$ for every least $\leqslant_\Sigma$-upper bound h of s and s'.*

Proof. We first establish, using induction on ℓ, that all elements g, g' of $\mathcal{H}$ that admit a common $\leqslant_\Sigma$-upper bound gh (that is, $g \bullet h$), in $\mathcal{H}$ satisfying $\|gh\|_\Sigma \leqslant \ell$ admit a least $\leqslant_\Sigma$-upper bound in $\mathcal{H}$. For $\ell = 0$ the result is trivial and, for $\ell \geqslant 1$, we argue using induction on $\|g\|_\Sigma + \|g'\|_\Sigma$. If both g and g' belong to Σ, the result is true by (2.67). Otherwise, assuming $g \notin \Sigma$, we write $g = g_1 \bullet g_2$ with g_1, g_2 in $\mathcal{H}$. As $\mathcal{H}$ is closed under Σ-suffix, (g_2, h) is Σ-tight, and gh is a common $\leqslant_\Sigma$-upper bound of g_1 and g' in $\mathcal{H}$. As we have $\|g_1\|_\Sigma + \|g'\|_\Sigma < \|g\|_\Sigma + \|g'\|_\Sigma$, the induction hypothesis implies that g_1 and g' admit a least $\leqslant_\Sigma$-upper bound, say $g_1 h_1$. Then $g \bullet h$ is a common $\leqslant_\Sigma$-upper bound of $g_1 \bullet (g_2 \bullet h)$

and $g_1 \bullet h_1$, hence $g_2 \bullet h$ is a common $\leqslant_\Sigma$-upper bound of $g_2 \bullet h$ and h_1. By construction, the inequality $\|g_2h\|_\Sigma < \|gh\|_\Sigma$ holds, so the induction hypothesis implies that g_2 and h_1 admit a least $\leqslant_\Sigma$-upper bound in $\mathcal{H}$, say g_2h_2. By Lemma 2.57, we have $gh_2 \leqslant_\Sigma gh$, and gh_2 is a least $\leqslant_\Sigma$-upper bound of g and g'. So (2.67) implies (2.64) and it expresses that the germ $\mathcal{H}^\Sigma$ admits conditional right-lcms.

Now, we observe that the germ $\mathcal{H}^\Sigma$ is eligible for Proposition 2.47, where we take for $\mathcal{A}$ the family Σ: indeed, all assumptions are satisfied (including the condition that the germ is associative on both sides since $\mathcal{H}$ is closed under Σ-prefix and Σ-suffix) and (2.68) is a translation of (2.48). Then the latter proposition gives the result. □

(See Exercise 71 for an alternative argument connecting Propositions 2.66 and 2.63.)

Similarly, if any two elements of the family $\mathcal{H}$ admit a least upper bound with respect to $\leqslant_\Sigma$, Proposition 2.49 provides a simple criterion:

Proposition 2.69 (derived Garside III). *If a groupoid $\mathcal{G}$ is positively generated by a family Σ and $\mathcal{H}$ is a subfamily of $\mathcal{G}$ that is closed under Σ-prefix and Σ-suffix and such that any two elements of $\mathcal{H}$ admit a $\leqslant_\Sigma$-least upper bound, then $\mathcal{H}^\Sigma$ is a Garside germ.*

Proof. By Lemma 2.62, the assumption that $\mathcal{H}$ is closed under Σ-prefix and Σ-suffix implies that the germ $\mathcal{H}^\Sigma$ is associative. Since any two elements of $\mathcal{H}$ admit a $\leqslant_\Sigma$-least upper bound, the germ $\mathcal{H}^\Sigma$ admits right-lcms. Proposition 2.49 then implies that the latter is a Garside germ. □

When Proposition 2.63, 2.66, or 2.69 applies, the results of Section 1 show that the category $\mathcal{C}at(\mathcal{H}^\Sigma)$ is left-cancellative and right-Noetherian, and that it admits conditional right-lcms (and even right-lcms in the case of Proposition 2.69). Note that, if the considered family $\mathcal{H}$ is the whole ambient groupoid $\mathcal{G}$, then the conditions about closure under prefix and suffix becomes trivial. So, for instance, Proposition 2.69 implies

Corollary 2.70 (derived Garside IV). *If a groupoid $\mathcal{G}$ is positively generated by a family Σ and any two elements of $\mathcal{G}$ admit a $\leqslant_\Sigma$-least upper bound, $\mathcal{G}^\Sigma$ is a Garside germ.*

So the point for obtaining a Garside germ is to find a positively generating subfamily Σ of $\mathcal{G}$ such that the partial order $\leqslant_\Sigma$ admits (conditional) least upper bounds.

Example 2.71 (derived germ I). Let $G = (\mathbb{Z}/n\mathbb{Z}, +)$ and $\Sigma = \{\overline{1}\}$. Here G^Σ is $\{\overline{0}, \overline{1}, ..., \overline{n-1}\}$ equipped with the partial operation $\bullet$ defined by $\overline{p} \bullet \overline{q} = \overline{p+q}$ for $p + q < n$, see the table on the right (for $n = 3$). Then, any two elements $\overline{p}, \overline{q}$ of G admit a least $\leqslant_\Sigma$-upper bound, namely $\overline{\max(p, q)}$. So G and Σ are eligible for Corollary 2.70, and the derived germ G^Σ is a Garside germ. The associated monoid is the free monoid $(\mathbb{N}, +)$, and its group of fractions is the free group $(\mathbb{Z}, +)$. So, in this (trivial) case, considering the germ associated with G provides a sort of unfolded, torsion-free version of the initial group.

$\bullet$	$\overline{0}$	$\overline{1}$	$\overline{2}$
$\overline{0}$	$\overline{0}$	$\overline{1}$	$\overline{2}$
$\overline{1}$	$\overline{1}$	$\overline{2}$	
$\overline{2}$	$\overline{2}$		

Here is another example, which will be extended in Chapter IX.

Example 2.72 (derived germ II). Consider the symmetric group $\mathfrak{S}_n$ and, writing σ_i for the transposition $(i, i+1)$, let Σ be $\{\sigma_i \mid 1 \leqslant i < n\}$. Then Σ generates $\mathfrak{S}_n$, and it positively generates $\mathfrak{S}_n$ since it consists of involutions. The Σ-length of a permutation f is the number of inversions of f, that is, the cardinality of the set I_f, defined to be $\{(i, j) \mid i < j \text{ and } f(i) > f(j)\}$. Then $f \leqslant_\Sigma g$ is equivalent to $I_f \subseteq I_g$, see Chapter IX, or for instance [119, Section 9.1]. Then one easily shows that $\leqslant_\Sigma$ provides a lattice structure on $\mathfrak{S}_n$, so, in particular, any two elements admit a $\leqslant_\Sigma$-least upper bound, and (2.67) is satisfied. Moreover, $f|\sigma_i$ is Σ-tight if and only if $f^{-1}(i) < f^{-1}(i+1)$ holds, and it easily follows that, if $f|\sigma_i$ and $f|\sigma_j$ are Σ-tight, then so is $f|\Delta_{i,j}$, where $\Delta_{i,j}$ stands for $\sigma_i\sigma_j$ in the case $|i - j| \geqslant 2$, and for $\sigma_i\sigma_j\sigma_i$ in the case $|i - j| = 1$. So (2.68) is satisfied. By Proposition 2.66, we deduce that $\mathfrak{S}_n^\Sigma$ is a Garside germ. The associated monoid is defined by the family of generators $\mathfrak{S}_n$ and the relations $fg = h$ for $h = fg$ and (f, g) Σ-tight. This list of relations includes the braid relations of (I.1.5) (Reference Structure 2, page 5). Conversely, it is known that any two shortest expressions of a permutation are connected by these relations (see Proposition IX.1.10 below), so the monoid admits the presentation (I.1.5) and, therefore, it is the braid monoid B_n^+.

3 Bounded germs

When a Garside family is bounded by a map Δ, it induces a Garside germ in which the elements $\Delta(x)$ enjoy specific properties. Conversely, if a Garside germ contains elements with such properties, the Garside family it gives rise to is bounded. Quite naturally, we shall say that such a germ is bounded. Here we state a few easy observations involving such bounded germs, in particular a new criterion for recognizing Garside germs. We begin with right-bounded germs (Subsection 3.1), and then continue with bounded germs themselves (Subsection 3.2). Finally, we describe in Subsection 3.3 a general scheme for constructing a bounded Garside germ from a group satisfying convenient lattice conditions.

3.1 Right-bounded germs

As in Chapter V, we begin with a notion of right-boundedness that involves only one side.

Definition 3.1 (right-bounded). A germ $\underline{S}$ is called *right-bounded* by a map Δ from $\mathcal{O}bj(\mathcal{S})$ to $\mathcal{S}$ if

(i) For every object x, the source of $\Delta(x)$ is x.

(ii) For every s in $\mathcal{S}(x, \cdot)$, there exists t in $\mathcal{S}$ satisfying $s \bullet t = \Delta(x)$.

If $\underline{S}$ is left-cancellative, we denote by $\partial_\Delta(s)$, or ∂s, the element t involved in (ii).

Example 3.2 (right-bounded). As the notation suggests, the germs of Figure 1 are right-bounded by the element Δ. More generally, a germ with one object is right-bounded if there exists an element Δ that occurs in every row of the associated table.

Other natural examples arise in the context of the derived germs of Subsection 2.4.

Proposition 3.3 (right-bounded Garside). *Assume that $\mathcal{G}$ is a groupoid, Σ positively generates $\mathcal{G}$, and $\Delta : \mathcal{O}bj(\mathcal{G}) \to \mathcal{G}$ is such that, for every object x, the source of $\Delta(x)$ is x and the family $\mathcal{H}$ of all Σ-prefixes of Δ is closed under Σ-suffix. Assume moreover that $(\mathcal{G}, \Sigma, \mathcal{H})$ is eligible for Proposition 2.63, 2.66, or 2.69. Then $\mathcal{G}^{\mathcal{H}}$ is a Garside germ that is right-bounded by Δ.*

In the above statement, when we speak of the Σ-prefixes of Δ, we naturally mean the family of all Σ-prefixes of elements $\Delta(x)$.

Proof. By the relevant proposition, $\mathcal{G}^{\mathcal{H}}$ is a Garside germ. Now let g belong to $\mathcal{H}(x, \cdot)$. By assumption, g is a Σ-prefix of $\Delta(x)$, so there exists a Σ-suffix h of $\Delta(x)$ such that $g|h$ is Σ-tight and $gh = \Delta(x)$ holds. By assumption, h belongs to $\mathcal{H}$ and, therefore $g \bullet h = \Delta(x)$ holds in the germ $\mathcal{G}^{\mathcal{H}}$. So the latter is right-bounded by Δ. □

For instance, the germs of Examples 2.71 and 2.72 are eligible for Proposition 3.3. They are right-bounded by $\overline{n-1}$ and by the flip permutation $i \mapsto n+1-i$, respectively.

The following result shows that our terminology is coherent with that of Chapter V.

Lemma 3.4. (i) *If $\mathcal{C}$ is a left-cancellative category and $\mathcal{S}$ is a solid Garside family in $\mathcal{C}$ that is right-bounded by a map Δ, then the induced germ $\underline{\mathcal{S}}$ is right-bounded by Δ.*

(ii) *Conversely, if $\underline{\mathcal{S}}$ is a Garside germ that is right-bounded by a map Δ, then the Garside family of $\mathcal{C}at(\underline{\mathcal{S}})$ given by $\mathcal{S}$ is right-bounded by Δ.*

Proof. (i) Assume that s belongs to $\mathcal{S}(x, y)$. By assumption, we have $s \preccurlyeq \Delta(x)$, whence $\Delta(x) = st$ for some t. The assumption that $\mathcal{S}$ is solid, hence in particular closed under right-divisor, implies that t lies in $\mathcal{S}$. Hence, s, t, and $\Delta(x)$ belong to $\mathcal{S}$ and, in the induced germ $\underline{\mathcal{S}}$, we have $s \bullet t = \Delta(x)$. So the germ $\underline{\mathcal{S}}$ is right-bounded by Δ.

(ii) We identify $\mathcal{S}$ with its image in $\mathcal{C}at(\underline{\mathcal{S}})$. Assume that x is an object of $\mathcal{C}at(\underline{\mathcal{S}})$, hence of $\mathcal{S}$, and s is an element of $\mathcal{S}(x, \cdot)$. The existence of t in $\mathcal{S}$ satisfying $s \bullet t = \Delta(x)$ implies $st = \Delta(x)$ in $\mathcal{C}at(\underline{\mathcal{S}})$, so $\mathcal{S}$ is right-bounded by Δ in $\mathcal{C}at(\underline{\mathcal{S}})$. □

We now establish a new characterization of Garside germs in the right-bounded case. We recall from Definition 1.18 that, if $\underline{\mathcal{S}}$ is a germ and s, t lie in $\mathcal{S}$, then s left-$\underline{\mathcal{S}}$-divides t, written $s \preccurlyeq_{\mathcal{S}} t$, if $s \bullet t' = t$ holds for some t' in $\mathcal{S}$. We then have the usual derived notions of a right-$\underline{\mathcal{S}}$-lcm (as in Proposition 1.26) and, symmetrically, *left-$\underline{\mathcal{S}}$-gcd*, that is, a greatest lower bound with respect to $\preccurlyeq_{\mathcal{S}}$. We naturally say that $\underline{\mathcal{S}}$ *admits left-gcds* if any two elements of $\mathcal{S}$ with the same source admit a left-$\underline{\mathcal{S}}$-gcd.

Proposition 3.5 (recognizing right-bounded Garside germ). *If a germ $\underline{\mathcal{S}}$ is associative, left-cancellative, right-bounded by a map Δ, and admits left-gcds, then $\underline{\mathcal{S}}$ is a Garside germ, and its image in $\mathcal{C}at(\underline{\mathcal{S}})$ is a Garside family that is right-bounded by Δ.*

We begin with an auxiliary result.

Lemma 3.6. *Assume that $\underline{\mathcal{S}}$ is a germ that is associative, left-cancellative, and right-bounded by a map Δ. Then, for all s, t in $\mathcal{S}$, the element $s \bullet t$ is defined if and only if the relation $t \preccurlyeq_{\mathcal{S}} \partial_{\Delta} s$ holds.*

Proof. Assume that s lies in $\mathcal{S}(x, y)$ and $s \bullet t$ is defined. Then $s \bullet t$ belongs to $\mathcal{S}(x, \cdot)$ and, therefore, we have $s \bullet t \preccurlyeq_{\mathcal{S}} \Delta(x)$, that is, $s \bullet t \preccurlyeq_{\mathcal{S}} s \bullet \partial_{\Delta} s$. As $\underline{\mathcal{S}}$ is assumed to be left-associative and left-cancellative, Lemma 1.19(iv) implies $t \preccurlyeq_{\mathcal{S}} \partial_{\Delta} s$.

Conversely, assume $t \preccurlyeq_{\mathcal{S}} \partial_{\Delta} s$, say $t \bullet r = \partial_{\Delta} s$. By assumption, $s \bullet \partial_{\Delta} s$, that is, $s \bullet (t \bullet r)$, is defined. By right-associativity, $s \bullet t$ is defined as well. □

Proof of Proposition 3.5. Let $s_1|s_2$ belong to $\mathcal{S}^{[2]}$. Consider the family $\mathfrak{J}_{\underline{\mathcal{S}}}(s_1, s_2)$, that is, by definition, $\{s \in \mathcal{S} \mid s_1 \bullet s \text{ is defined and } s \preccurlyeq_{\mathcal{S}} s_2\}$. By Lemma 3.6, an equivalent definition is $\{s \in \mathcal{S} \mid s \preccurlyeq_{\mathcal{S}} \partial_{\Delta} s_1 \text{ and } s \preccurlyeq_{\mathcal{S}} s_2\}$, that is, $\mathfrak{J}_{\underline{\mathcal{S}}}(s_1, s_2)$ is the family of all common left-$\underline{\mathcal{S}}$-divisors of $\partial_{\Delta} s_1$ and s_2 in $\mathcal{S}$. So the assumption implies that $\mathfrak{J}_{\underline{\mathcal{S}}}(s_1, s_2)$ has a $\preccurlyeq_{\mathcal{S}}$-greatest element, namely any left-$\underline{\mathcal{S}}$-gcd of $\partial_{\Delta} s_1$ and s_2. By Proposition 2.28, we deduce that $\underline{\mathcal{S}}$ is a Garside germ.

Then the category $\mathcal{C}at(\underline{\mathcal{S}})$ is left-cancellative, $\underline{\mathcal{S}}$ embeds in $\mathcal{C}at(\underline{\mathcal{S}})$, and its image is a Garside family in $\mathcal{C}at(\underline{\mathcal{S}})$. The assumption that $\underline{\mathcal{S}}$ is right-bounded by Δ immediately implies that its image in $\mathcal{C}at(\underline{\mathcal{S}})$ is right-bounded by (the image of) Δ. □

3.2 Bounded germs

In order to obtain stronger results and, in particular, to make the conditions of Proposition 3.5 necessary and sufficient, we have to strengthen the assumptions and consider bounded germs.

Definition 3.7 (bounded germ). A germ $\underline{\mathcal{S}}$ is called *bounded* by a map Δ from $\mathcal{O}bj(\mathcal{S})$ to $\mathcal{S}$ if it is right-bounded by Δ and, in addition, for every object y, there exists an object x such that, for every s in $\mathcal{S}(\cdot, y)$, there exists r in $\mathcal{S}(x, \cdot)$ satisfying $r \bullet s = \Delta(x)$.

Once again, natural examples arise in the context of derived germs.

Proposition 3.8 (bounded Garside). *Assume that $\mathcal{G}$ is a groupoid, Σ positively generates $\mathcal{G}$, and $\Delta : \mathcal{O}bj(\mathcal{G}) \to \mathcal{G}$ is such that, for every object x, the source of $\Delta(x)$ is x and the family $\mathcal{H}$ of all Σ-prefixes of Δ coincides with the family of all Σ-suffixes of Δ. If $(\mathcal{G}, \Sigma, \mathcal{H})$ is eligible for Proposition 2.63, or 2.66, or 2.69, then $\underline{\mathcal{G}}^{\mathcal{H}}$ is a Garside germ that is bounded by Δ.*

Proof. Owing to Proposition 3.3, it remains to consider the elements of $\widetilde{\mathcal{D}iv}(\Delta)$. So assume that g belongs to $\mathcal{H}(\cdot, y)$. By assumption, there exists x such that g is a Σ-suffix of $\Delta(x)$, so there exists a Σ-prefix f of $\Delta(x)$ such that $f|g$ is Σ-tight and $fg = \Delta(x)$

holds. By assumption, f belongs to $\mathcal{H}$ and, therefore $f \bullet g = \Delta(x)$ holds in the germ $\mathcal{G}^{\mathcal{H}}$. So the latter is bounded by Δ. ☐

The assumptions of Proposition 3.8 may appear convoluted. The statement is simpler for a group.

Corollary 3.9 (bounded Garside). *Assume that G is a group, Σ positively generates G, and Δ is an element of G such that the family H of all Σ-prefixes of Δ coincides with the family of all Σ-suffixes of Δ. If (G, Σ, H) is eligible for Proposition* 2.63, 2.66, *or* 2.69, *then G^H is a Garside germ that is bounded by Δ.*

The germs of Examples 2.71 and 2.72 are not only right-bounded, but even bounded: this is clear in the case of $\mathbb{Z}/n\mathbb{Z}$, as the group is abelian; in the case of $\mathfrak{S}_n$, this follows from the fact that every permutation is both a Σ-prefix and a Σ-suffix of the flip permutation. We shall come back on this in Chapter IX.

We now list a few general properties of bounded germs and their associated categories.

Lemma 3.10. (i) *If $\mathcal{C}$ is a left-cancellative category and $\mathcal{S}$ is a Garside family in $\mathcal{C}$ that is bounded by a map Δ and satisfies $\mathcal{S}^\sharp = \mathcal{S}$, the induced germ $\underline{\mathcal{S}}$ is associative and bounded by Δ.*

(ii) *Conversely, assume that $\underline{\mathcal{S}}$ is a Garside germ that is associative and bounded by a map Δ. Then the Garside family of $\mathcal{C}at(\underline{\mathcal{S}})$ given by $\mathcal{S}$ is bounded by Δ.*

Proof. (i) By Lemma 3.4, the induced germ $\underline{\mathcal{S}}$ is right-bounded by Δ. By assumption, $\mathcal{S}$ is bounded by Δ and $\mathcal{S} = \mathcal{S}^\sharp$ holds, hence, for every s in $\mathcal{S}(\cdot, y)$, there exists r in $\mathcal{S}(x, \cdot)$ satisfying $rs = \Delta(x)$, whence $r \bullet s = \Delta(x)$, so $\underline{\mathcal{S}}$ is bounded. The assumptions imply that $\mathcal{S}$ is closed under left- and right-divisors in $\mathcal{C}$, hence $\underline{\mathcal{S}}$ is associative.

(ii) By Lemma 3.4 again, the Garside family $\mathcal{S}$ is right-bounded by Δ in $\mathcal{C}at(\underline{\mathcal{S}})$. The assumption that $\underline{\mathcal{S}}$ is associative guarantees that $\mathcal{S}$ is closed under left- and right-divisors in $\mathcal{C}at(\underline{\mathcal{S}})$, hence $\mathcal{S}^\sharp$ coincides with $\mathcal{S}$. Then the assumption that $\underline{\mathcal{S}}$ is bounded by Δ implies that $\mathcal{S}$ satisfies (V.2.4) in $\mathcal{C}at(\underline{\mathcal{S}})$, that is, $\mathcal{S}$ is bounded by Δ in $\mathcal{C}at(\underline{\mathcal{S}})$. ☐

By Proposition V.2.35 (bounded implies gcd), the existence of a bounded Garside family in a category implies the existence of left-gcds. The argument localizes to germs.

Lemma 3.11. *A Garside germ $\underline{\mathcal{S}}$ that is associative, left-cancellative, and bounded admits left-gcds.*

Proof. Assume that $\underline{\mathcal{S}}$ is bounded by Δ. Let s, s' be two elements of $\mathcal{S}$ with the same source. The assumption that $\underline{\mathcal{S}}$ is bounded and right-cancellative implies the existence of x and r satisfying $r \bullet s = \Delta(x)$, whence $s = \partial r$. Then $r|s'$ belongs to $\mathcal{S}^{[2]}$ so Proposition 2.28, which is valid for $\underline{\mathcal{S}}$ since the latter is a Garside germ, implies the existence of a $\preccurlyeq_{\mathcal{S}}$-greatest element t in $\mathcal{J}_{\mathcal{S}}(r, s')$. By definition of the latter family, $t \preccurlyeq s'$ holds and $r \bullet t$ is defined. By Lemma 3.6, the latter relation implies $t \preccurlyeq_{\mathcal{S}} \partial r$, that is, $t \preccurlyeq_{\mathcal{S}} s$. So t is a common left-$\underline{\mathcal{S}}$-divisor of s and s'. Now let t' be an arbitrary common left-$\underline{\mathcal{S}}$-divisor of s and s'. By Lemma 3.6 again (now in the other direction), the assumption $t' \preccurlyeq_{\mathcal{S}} s$, that is, $t' \preccurlyeq_{\mathcal{S}} \partial r$, implies that $r \bullet t'$ is defined. Hence t' belongs to $\mathcal{J}_{\underline{\mathcal{S}}}(r, s')$ and, therefore, $t' \preccurlyeq_{\mathcal{S}} t$ holds. So t is a left-$\underline{\mathcal{S}}$-gcd of s and s', and $\underline{\mathcal{S}}$ admits left-gcds. ☐

We deduce a characterization of Garside germs in the bounded case.

Proposition 3.12 (recognizing bounded Garside germ). *For an associative, cancellative, and bounded germ $\underline{S}$, the following conditions are equivalent:*
(i) *The structure $\underline{S}$ is a Garside germ.*
(ii) *The germ $\underline{S}$ admits left-gcds.*

Proof. Assume that $\underline{S}$ is a Garside germ. As $\underline{S}$ is assumed to be associative and left-cancellative, Lemma 3.11 implies that $\underline{S}$ admits left-gcds. So (i) implies (ii).

Conversely, assume that $\underline{S}$ admits left-gcds. Then Proposition 3.5 implies that $\underline{S}$ is a Garside germ, and (ii) implies (i). □

In the situation of Proposition 3.12, the category $\mathcal{C}at(\underline{S})$ is left-cancellative and $\mathcal{S}$ is a Garside family of $\mathcal{C}at(\underline{S})$ that is bounded by Δ. It is natural to wonder whether further properties of $\mathcal{C}at(\underline{S})$ can be read inside the germ $\underline{S}$, typically right-cancellativity. We saw in Proposition V.1.59 (right-cancellative II) that this is connected with the properties of the functor ϕ_Δ associated with Δ, in particular preservation of normality.

Proposition 3.13 (right-cancellativity). *Assume that $\underline{S}$ is a germ that is associative, cancellative, with no nontrivial invertible element, and bounded by a map Δ, that Δ is target-injective, that ∂ is a bijection from $\mathcal{S}$ to itself, and that, if $s \bullet t$ is defined, then so is $\partial^{-2}s \bullet \partial^{-2}t$. Then $\mathcal{C}at(\underline{S})$ is cancellative and ϕ_Δ is an automorphism.*

Proof. By construction, ϕ_Δ coincides with ∂^2 on $\mathcal{S}$, which is also $\mathcal{S}^\sharp$ since there is no nontrivial invertible element and $1_\mathcal{S}$ is included in $\mathcal{S}$. By assumption, ∂ is injective, hence so is ∂^2, which is the restriction of ϕ_Δ to $\mathcal{S}$. We shall prove that ϕ_Δ preserves $\mathcal{S}$-normality. So assume that $s_1|s_2$ is an $\mathcal{S}$-normal path. We wish to prove that $\phi_\Delta(s_1)|\phi_\Delta(s_2)$ is $\mathcal{S}$-normal as well. Assume that s belongs to $\mathcal{J}_{\underline{S}}(\phi_\Delta(s_1), \phi_\Delta(s_2))$, that is, $\phi_\Delta(s_1) \bullet s$ is defined and we have $s \preccurlyeq_\mathcal{S} \phi_\Delta(s_2)$, hence $\phi_\Delta(s_2) = s \bullet t$ for some t. As $\mathcal{S}$ is bounded by Δ, the map ∂ is surjective, so there exist s' and t' satisfying $s = \partial^2 s'$ and $t = \partial^2 t'$. So, at this point, we have that $\partial^2 s_1 \bullet \partial^2 s'$ is defined, and that $\partial^2 s_2 = \partial^2 s' \bullet \partial^2 t'$ holds. The assumption on ∂^{-2} then implies that $s_1 \bullet s'$ and $s' \bullet t'$ are defined. Then we can write $\partial^2(s' \bullet t') = \phi_\Delta(s't') = \phi_\Delta(s_2) = \partial^2(s_2)$ in $\mathcal{C}at(\underline{S})$, and deduce $s' \bullet t' = s_2$, whence $s' \preccurlyeq_\mathcal{S} s_2$. So s' belongs to $\mathcal{J}_{\underline{S}}(s_1, s_2)$. As $s_1|s_2$ is $\mathcal{S}$-normal, Lemma 2.4 implies that s' must be invertible, hence, under our current assumptions, it is an identity element. Hence so is $\partial^2 s'$, which is s. By Lemma 2.4 again, we deduce that $\phi_\Delta(s_1)|\phi_\Delta(s_2)$ is $\mathcal{S}$-normal. Then all conditions in Proposition V.1.59 (right-cancellative II) are satisfied, and the latter states that $\mathcal{C}at(\underline{S})$ is right-cancellative and that ϕ_Δ is injective on $\mathcal{C}at(\underline{S})$, hence it is an automorphism (as, by assumption, it is surjective). □

Example 3.14 (right-cancellative). Consider the braid germ of Figure 1 right. It is easy to check that ∂^{-2} (which is also ∂^2) is the map that preserves 1 and Δ, and exchanges σ_1 and σ_2, and $\sigma_1\sigma_2$ and $\sigma_2\sigma_1$. Then all assumptions of Proposition 3.13 are satisfied, and we obtain one more way of establishing that the braid monoid B_3^+ is right-cancellative (of course, this is essentially the same argument as in Chapter V).

3.3 An application: germs from lattices

As in Subsection 2.4, we shall now describe a general scheme for associating a (bounded) germ with a group equipped with a partial ordering satisfying convenient (semi)-lattice conditions. The construction is reminiscent of the construction of derived germs in Subsection 2.4, but, instead of using a tightness condition for defining the germ operation, we use an order condition on left cosets. We recall that a poset $(X, \leqslant)$ is called an *inf-semi-lattice* if it admits a least element and any two elements admit a greatest lower bound.

Proposition 3.15 (germ from lattice). *Assume that G is a group, H is a subgroup of G, and $\leqslant$ is an inf-semi-lattice ordering on G/H with least element H and greatest element ΔH. Assume further that*

(3.16) *The relation $fH \leqslant fgH \leqslant fghH$ implies $gH \leqslant ghH$,*

(3.17) *The conjunction of $gH \leqslant ghH$ and $fH \leqslant fghH$ implies $fH \leqslant fgH \leqslant fghH$.*

Let $\bullet$ be the partial operation on G such that $f \bullet g = h$ is true if and only if we have $fg = h$ in G and $fH \leqslant hH$. Write $\underline{G}_{\leqslant}$ for $(G, \bullet, 1)$.

(i) *The structure $\underline{G}_{\leqslant}$ is a Garside germ that is associative, cancellative, and bounded by Δ. The monoid $\mathcal{M}on(\underline{G}_{\leqslant})$ is left-cancellative, its invertible elements are the elements of H, and it admits left-gcds and common right-multiples.*

(ii) *If, moreover, $gH \leqslant hH$ implies $\Delta g \Delta^{-1} H \leqslant \Delta h \Delta^{-1} H$, the monoid $\mathcal{M}on(\underline{G}_{\leqslant})$ is right-cancellative and ϕ_Δ is an automorphism.*

A particular case of Proposition 3.15 is when H is trivial and the lattice ordering exists on the whole group G, in which case the monoid $\mathcal{M}on(\underline{G}_{\leqslant})$ has no nontrivial invertible element. More generally, if H is a normal subgroup of G, then the quotient group G/H is eligible for the construction and we are then in the above particular case.

Proof. (i) First, (1.4) is trivial since there is only one object. Next, $H \leqslant gH$ and $gH \leqslant gH$ hold for every g, so $1 \bullet g$ and $g \bullet 1$ are defined and we have $1 \bullet g = g \bullet 1 = g$. So (1.5) is satisfied in $\underline{G}_{\leqslant}$. Then, assume that $(f \bullet g) \bullet h$ is defined. This means that we have $fH \leqslant fgH$ and $fgH \leqslant fghH$. By (3.16), we deduce $gH \leqslant ghH$ and, directly, $fH \leqslant fghH$, implying that $g \bullet h$ and $f \bullet (g \bullet h)$ are defined. In the other direction, assume that $f \bullet (g \bullet h)$ is defined. Arguing similarly and using (3.17), we deduce that $f \bullet g$ and $(f \bullet g) \bullet h$ are defined. In both cases, the equalities $(f \bullet g) \bullet h = fgh = (f \bullet g) \bullet h$ are then obvious. So (1.6), (1.7), and (1.8) are satisfied in $\underline{G}_{\leqslant}$. Hence the latter is an associative germ.

Next, the germ $\underline{G}_{\leqslant}$ is left-cancellative: $g \bullet h = g \bullet h'$ implies $gh = gh'$, whence $h = h'$. Similarly, $\underline{G}_{\leqslant}$ is right-cancellative as $g \bullet h = g' \bullet h$, which gives $gh = g'h$, implies $g = g'$. Moreover, $\underline{G}_\leqslant$ is bounded by Δ: indeed, for every g in G, we have $gH \leqslant \Delta H = g(g^{-1}\Delta)H$, whence $g \bullet g^{-1}\Delta = \Delta$, and $\Delta g^{-1}H \leqslant \Delta H = (\Delta g^{-1})gH$, whence $\Delta g^{-1} \bullet g = \Delta$.

Then, $\underline{G}_\leqslant$ admits left-gcds: an element g of G is a left-$\underline{G}_\leqslant$-divisor of an element h if there exists h' satisfying $g \bullet h' = h$, which is true if and only if $gH \leqslant hH$ is true, in which case $h' = g^{-1}h$ is convenient. Then the assumption that $(G, \leqslant)$ admits greatest lower bounds implies (actually, is equivalent to) the fact that the germ $\underline{G}_\leqslant$ admits left-gcds. Thus the conditions of Proposition 3.5 are satisfied, and the latter implies that $\underline{G}_\leqslant$ is a Garside germ.

Finally, by Lemma 1.19(i), the invertible elements of $\mathcal{M}on(\underline{G}_\leqslant)$ are the invertible elements g of $\underline{G}_\leqslant$, that is, the elements g of G satisfying $g \bullet h = 1$ for some h. The latter is equivalent to $gH \leqslant H$, whence to $gH = H$, that is, $g \in H$. The other properties follow from $\underline{S}$ being a Garside germ and $\mathcal{C}at(\underline{S})$ admitting a bounded Garside family.

(ii) We show that the assumptions of Proposition 3.13 are satisfied. The map ∂_Δ takes g to $g^{-1}\Delta$, so $g' = \partial_\Delta g$ is equivalent to $g = \Delta g'^{-1}$, hence ∂_Δ is bijective on G. Next, we obtain $\partial_\Delta^{-2} g = \Delta g \Delta^{-1}$. If $g \bullet h$ is defined, we have $gH \leqslant ghH$, whence, by the additional assumption, $\Delta g \Delta^{-1} H \leqslant \Delta gh \Delta^{-1} H$, that is, $\Delta g \Delta^{-1} H \leqslant (\Delta g \Delta^{-1})(\Delta h \Delta^{-1})H$, meaning that $\partial_\Delta^{-2} g \bullet \partial_\Delta^{2} h$ is defined. Then we apply Proposition 3.13. □

Example 3.18 (germ from lattice). Let G be the braid group B_n (Reference Structure 2, page 5) and PB_n be the subgroup of B_n made of all pure braids, that is, the braids whose associated permutation is the identity. We saw in Reference Structure 2, page 5, that PB_n is a normal subgroup of B_n and we have the short exact sequence (I.1.7) $1 \to PB_n \to B_n \to \mathfrak{S}_n \to 1$. Let Σ be the family of all transpositions $(i, i+1)$ in the symmetric group $\mathfrak{S}_n$. As mentioned in Example 2.72, the weak order $\leqslant_\Sigma$ on $\mathfrak{S}_n$ is a lattice ordering on $\mathfrak{S}_n$, that is, on B_n/PB_n. The least element in this order is 1 and the greatest element is the flip permutation Δ (or w_0) that exchanges i and $n+1-i$ for every i. Then the germ constructed from B_n and PB_n using the method of Proposition 3.15 is a sort of extension of the germ of Example 2.72 by the group PB_n: writing π for the projection of B_n onto $\mathfrak{S}_n$, we have that $g \bullet h$ is defined in the current construction if and only if $\pi(g) \leqslant \pi(h)$ holds in $\mathfrak{S}_n$, hence if and only if $\pi(g)|\pi(g)^{-1}\pi(h)$ is Σ-tight, hence if and only if $\pi(g) \bullet \pi(h)$ is defined in the construction of Subsection 2.4. The monoid generated by the current germ is a semi-direct product of B_n^+ by PB_n, with B_n^+ acting by conjugation on PB_n—it is *not* the submonoid of B_n generated by PB_n and B_n^+: for instance, σ_1 does not divides σ_1^2 since the latter is invertible in the monoid whereas the former is not.

The above example is somehow trivial in that the involved subgroup PB_n is normal and we could directly work with the quotient-group B_n/PB_n, that is, $\mathfrak{S}_n$, in which case we obtain the same germ in Example 2.72. We refer to Section XIV.3 for another more novel and interesting example.

Exercises

Exercise 65 (not embedding). Let $\mathcal{S}$ consist of fourteen elements $1, \mathtt{a}, ..., \mathtt{n}$, all with the same source and target, and $\bullet$ be defined by $1 \bullet x = x \bullet 1 = x$ for each x, plus $\mathtt{a} \bullet \mathtt{b} = \mathtt{f}$, $\mathtt{f} \bullet \mathtt{c} = \mathtt{g}$, $\mathtt{d} \bullet \mathtt{e} = \mathtt{h}$, $\mathtt{g} \bullet \mathtt{h} = \mathtt{i}$, $\mathtt{c} \bullet \mathtt{d} = \mathtt{j}$, $\mathtt{b} \bullet \mathtt{j} = \mathtt{k}$, $\mathtt{k} \bullet \mathtt{e} = \mathtt{m}$, and $\mathtt{a} \bullet \mathtt{m} = \mathtt{n}$. (i) Show that $\underline{\mathcal{S}}$ is a germ. (ii) Show that, in $\mathcal{S}$, we have $((\mathtt{a} \bullet \mathtt{b}) \bullet \mathtt{c}) \bullet (\mathtt{d} \bullet \mathtt{e}) = \mathtt{i} \neq \mathtt{n} = \mathtt{a} \bullet ((\mathtt{b} \bullet (\mathtt{c} \bullet \mathtt{d})) \bullet \mathtt{e})$, whereas, in $\mathcal{M}on(\underline{\mathcal{S}})$, we have $\iota \mathtt{i} = \iota \mathtt{n}$. Conclude.

Exercise 66 (multiplying by invertible elements). (i) Show that, if $\underline{\mathcal{S}}$ is a left-associative germ, then $\mathcal{S}$ is closed under left-multiplication by invertible elements in $\mathcal{C}at(\underline{\mathcal{S}})$. (ii) Show that, if $\underline{\mathcal{S}}$ is an associative germ, $s \bullet t$ is defined, and $t' =^{\times}_{\mathcal{S}} t$ holds, then $s \bullet t'$ is defined as well.

Exercise 67 (atoms). (i) Show that, if $\underline{\mathcal{S}}$ is a left-associative germ, the atoms of $\mathcal{C}at(\underline{\mathcal{S}})$ are the elements of the form $t\epsilon$ with t an atom of $\underline{\mathcal{S}}$ and ϵ an invertible element of $\underline{\mathcal{S}}$. (ii) Let $\underline{\mathcal{S}}$ be the germ whose table is shown on the right. Show that the monoid $\mathcal{M}on(\underline{\mathcal{S}})$ admits the presentation $\langle \mathtt{a}, \mathtt{e} \mid \mathtt{ea} = \mathtt{a}, \mathtt{e}^2 = 1 \rangle^+$ (see Exercise 48) and that $\mathtt{a}$ is the only atom of $\underline{\mathcal{S}}$, whereas the atoms of $\mathcal{M}on(\underline{\mathcal{S}})$ are $\mathtt{a}$ and $\mathtt{ae}$. (iii) Show that $\underline{\mathcal{S}}$ is a Garside germ.

$\bullet$	1	a	e
1	1	a	e
a	a		
e	e	a	1

Exercise 68 (families $\mathcal{I}_{\underline{\mathcal{S}}}$ and $\mathcal{J}_{\underline{\mathcal{S}}}$). Assume that $\underline{\mathcal{S}}$ is a left-associative germ. (i) Show that a path $s_1 | s_2$ of $\mathcal{S}^{[2]}$ is $\mathcal{S}$-normal if and only if all elements of $\mathcal{J}_{\underline{\mathcal{S}}}(s_1, s_2)$ are invertible. (ii) Assuming in addition that $\underline{\mathcal{S}}$ is left-cancellative, show that, for $s_1 | s_2$ in $\mathcal{S}^{[2]}$, the family $\mathcal{I}_{\underline{\mathcal{S}}}(s_1, s_2)$ admits common right-multiples if and only if $\mathcal{J}_{\underline{\mathcal{S}}}(s_1, s_2)$ does.

Exercise 69 (positive generators). Assume that Σ is a family of positive generators in a group $\mathcal{G}$ and Σ is closed under inverse, that is, $g \in \Sigma$ implies $g^{-1} \in \Sigma$. (i) Show that $\|g\|_\Sigma = \|g^{-1}\|_\Sigma$ holds for every g in $\mathcal{G}$. (ii) Show that $f^{-1} \leqslant_\Sigma g^{-1}$ is equivalent to $f \,\widetilde{\geqslant}_\Sigma\, g$.

Exercise 70 (minimal upper bound). For $\leqslant$ a partial ordering on a family $\mathcal{S}'$ and f, g, h in $\mathcal{S}'$, say that h is a *minimal upper bound*, or *mub*, for f and g, if $f \leqslant h$ and $g \leqslant h$ holds, but there exists no h' with $h' < h$ satisfying $f \leqslant h'$ and $g \leqslant h'$. Assume that $\mathcal{G}$ is a groupoid, Σ positively generates $\mathcal{G}$, and $\mathcal{H}$ is a subfamily of $\mathcal{G}$ that is closed under Σ-suffix. Show that $\underline{\mathcal{H}}^\Sigma$ is a Garside germ if and only if, for all f, g, g', g'' in $\mathcal{H}$ such that $f \bullet g$ and $f \bullet g'$ are defined and g'' is a $\leqslant_\Sigma$-mub of g and g', the product $f \bullet g''$ is defined.

Exercise 71 (derived germ). Under the assumptions of Proposition 2.66, prove using an induction on ℓ that, if f, g, g' lie in $\mathcal{H}$, if $f \bullet g$, $f \bullet g'$ are defined and lie in $\mathcal{H}$, and g, g' admit a least $\leqslant_\Sigma$-upper bound h satisfying $\|h\|_\Sigma \leqslant \ell$, then $f \bullet h$ is defined. Deduce that (2.68) implies (2.65) and deduce Proposition 2.66 from Proposition 2.63.

Notes

Sources and comments. The notion of a germ and the results of Section 1 first appeared in Bessis, Digne, and Michel [13] under the name “preGarside structure”. It appears also in Digne–Michel [109] and, in a different setting, in Krammer [164]. A lattice-theoretic version of the correspondence between a Garside germ and the associated category via normal decompositions appears in [203]: the role of the germ is played by what the author calls an L-algebra and the role of the category (more exactly its enveloping groupoid) is played by a right l-group, a one-sided version of a lattice group. In this way, the correspondence can be seen as an extension of the classic correspondence between MV-algebras and lattice groups [187, 49].

The characterization of Garside germs given in Proposition 2.8 appears in [109] (with an additional right-associativity condition). The current exposition based on the $\mathcal{J}$-law and the computation of the tail is formally new, as are the results of Subsection 2.2, in particular the result that a greatest $\mathcal{J}$-function has to satisfy the $\mathcal{J}$-law.

The construction of the germ derived from a groupoid as explained in Subsection 2.4 appears for instance in Digne and Michel [109], but it was already known and used earlier in more or less equivalent settings, for instance in Bessis [9, Theorem 0.5.2], which reports that the construction of the germ derived from a group was known in 2000 to one of us (JM): a Garside monoid arises whenever one considers the monoid generated by the set of divisors of an element, with all relations $fg = h$ such that the lengths add, when the divisors of that element on the right and on the left are the same and form a lattice for divisibility [182]. At that time known examples were the Coxeter element of the symmetric group and the longest element of a finite Coxeter group. The approach has been used extensively in the case of braid groups of Coxeter groups and complex reflection groups, where it proved to be especially suitable, see Chapter IX and sources like Digne [106] and [107].

The notion of a bounded Garside germ was considered by D. Bessis, F. Digne and J. Michel (see [13, Proposition 2.23]) and is implicit in Krammer [164], which also contains the principle of the construction of Subsection 3.3. Note that the germ derived from a groupoid in the sense of Subsection 2.4 is automatically bounded when the reference family $\mathcal{H}$ consists of the divisors of some element.

Further questions. At the moment, it is not yet clear that all results are optimal. For instance, right-associativity is superfluous in most cases, but not all. In particular, the following is unclear:

Question 20. Is the right-associativity assumption needed in Lemma 2.42?

Some weak form of right-associativity involving invertible elements is always satisfied in a (left-associative) germ and this might turn to be enough to complete all arguments. On the other hand, two-sided associativity is often guaranteed, in particular in the case of germs derived from a groupoid or a lattice. As the latter are the most interesting examples known so far in the current approach, weakening associativity assumptions is maybe not of real significance.

A less precise but presumably more interesting question would be to connect the current notion of a germ with the "right bricks" introduced by W. Rump in [203]: in the particular case of what is called a modular Garside group, the latter also are fragments that encode the whole structure, here in a lattice-theoretic language and in the spirit of the Mundici correspondence.

Chapter VII
Subcategories

In this chapter, we study subcategories and their connections with Garside families. Typically, starting from a left-cancellative category $\mathcal{C}$ equipped with a Garside family $\mathcal{S}$, we investigate the subcategories $\mathcal{C}_1$ of $\mathcal{C}$ that are compatible with $\mathcal{S}$ in the sense that $\mathcal{S} \cap \mathcal{C}_1$ is a Garside family in $\mathcal{C}_1$ and the $\mathcal{C}_1$-paths that are $(\mathcal{S} \cap \mathcal{C}_1)$-normal coincide with those that are $\mathcal{S}$-normal. We establish various characterizations of such subcategories, as well as results about their intersections. The general philosophy emerging from the results established below is that Garside families and Garside germs behave nicely with respect to subcategories, arguably one more proof of their naturalness.

The chapter is organized as follows. Section 1 contains general background about subcategories and a few results involving specific types subcategories, namely those that are closed under $=^{\times}$, head subcategories, and parabolic subcategories, with in particular useful criteria for recognizing the latter subcategories (Proposition 1.25).

Next, Section 2 is devoted to various forms of compatibility between a subcategory and a Garside family. Here the main result is the existence of local characterizations of compatibility, both in the general case (Propositions 2.10 and 2.14), and in particular cases like that of parabolic subcategories (Corollary 2.22). The connection between compatibility for categories and for their enveloping groupoids is analyzed in Proposition 2.27.

In Section 3, we investigate those subcategories generated by subfamilies of a reference Garside family. The notion of a subgerm arises naturally and the central question is to recognize which properties of the subcategory generated by a subgerm $\underline{\mathcal{S}}_1$ of a germ $\underline{\mathcal{S}}$ can be read inside $\underline{\mathcal{S}}$. Several positive results are established, in particular Proposition 3.14, which compares the category $\mathit{Cat}(\underline{\mathcal{S}}_1)$ with the subcategory of $\mathit{Cat}(\underline{\mathcal{S}})$ generated by $\mathcal{S}_1$, and Proposition 3.16, which gives conditions for $\underline{\mathcal{S}}_1$ to be a Garside germ.

Finally, in Section 4, we consider subcategories arising in connection with a functor, namely subcategories of fixed points and image subcategories. We establish, in particular, an injectivity criterion for a functor involving what we call its correctness (Corollary 4.17), a notion that appears as the appropriate extension of an lcm-homomorphism in contexts where lcms need not exist.

Main definitions and results

Proposition 1.13 **(generated subcategory, right-lcm case).** *Assume that $\mathcal{C}$ is a left-cancellative category that admits conditional right-lcms and $\mathcal{S}_1$ is a subfamily of $\mathcal{C}$ such that $\mathcal{S}_1\mathcal{C}^{\times} \cup \mathcal{C}^{\times}$, denoted by $\mathcal{S}_1^{\sharp}$, is closed under right-diamond in $\mathcal{C}$.* (i) *The subcategory $\mathcal{C}_1$ of $\mathcal{C}$ generated by $\mathcal{S}_1^{\sharp}$ is closed under right-diamond and right-quotient in $\mathcal{C}$, and it admits conditional right-lcms.* (ii) *If $\mathcal{C}$ is right-Noetherian, then so is $\mathcal{C}_1$, and $\mathcal{S}_1$ is a Garside family in $\mathcal{C}_1$.*

Definition 1.19 **(head subcategory).** A subcategory $\mathcal{C}_1$ of a left-cancellative category $\mathcal{C}$ is called a *head subcategory* if $\mathcal{C}_1$ is closed under right-quotient in $\mathcal{C}$ and every element of $\mathcal{C}$ whose source belongs to $\mathcal{O}bj(\mathcal{C}_1)$ admits a $\mathcal{C}_1$-head.

Proposition 1.21 **(head subcategory).** *Assume that $\mathcal{C}_1$ is a subcategory of a left-cancellative category $\mathcal{C}$.* (i) *If $\mathcal{C}_1$ is a head subcategory of $\mathcal{C}$, then $\mathcal{C}_1$ includes $\mathbf{1}_\mathcal{C}$ and is closed under right-quotient and right-comultiple in $\mathcal{C}$.* (ii) *If $\mathcal{C}$ is right-Noetherian, the implication in* (i) *is an equivalence.*

Corollary 1.22 **(head subcategory).** *A subcategory $\mathcal{C}_1$ of a left-cancellative category $\mathcal{C}$ that is right-Noetherian is a head subcategory if and only if $\mathcal{C}_1$ includes $\mathbf{1}_\mathcal{C}$ and is closed under inverse and right-diamond in $\mathcal{C}$, if and only if $\mathcal{C}_1$ is closed under right-quotient and right-comultiple in $\mathcal{C}$.*

Proposition 1.25 **(head on Garside family).** *Assume that $\mathcal{C}$ is a left-cancellative category, $\mathcal{S}$ is a Garside family of $\mathcal{C}$, and $\mathcal{C}_1$ is a subcategory of $\mathcal{C}$ that includes $\mathbf{1}_\mathcal{C}$, is closed under right-quotient in $\mathcal{C}$, and is generated by $\mathcal{S}^\sharp \cap \mathcal{C}_1$. Then $\mathcal{C}_1$ is a head subcategory of $\mathcal{C}$ whenever every element of $\mathcal{S}$ admits a $\mathcal{C}_1$-head that lies in $\mathcal{S}$, and* (i) *The category $\mathcal{C}$ is right-Noetherian, or* (ii) *The family $\mathcal{S}^\sharp$ is closed under left-divisor, the second domino rule is valid for $\mathcal{S}$, and $\mathcal{C}_1$ is compatible with $\mathcal{S}$.*

Definition 1.30 **(parabolic subcategory).** A *parabolic* subcategory of a left-cancellative category $\mathcal{C}$ is a head subcategory of $\mathcal{C}$ that is closed under factor.

Proposition 1.32 **(parabolic subcategory).** *Assume that $\mathcal{C}_1$ is a subcategory of a left-cancellative category $\mathcal{C}$.* (i) *If $\mathcal{C}_1$ is parabolic in $\mathcal{C}$, then $\mathcal{C}_1$ is closed under right-comultiple and factor.* (ii) *If $\mathcal{C}$ is right-Noetherian, the implication of* (i) *is an equivalence.*

Proposition 1.35 **(intersection of parabolic).** *If $\mathcal{C}$ is a left-Noetherian left-cancellative category, every intersection of parabolic subcategories of $\mathcal{C}$ is a parabolic subcategory.*

Definition 2.4 **(compatible subcategory).** A subcategory $\mathcal{C}_1$ of a left-cancellative category $\mathcal{C}$ is said to be *compatible with* a Garside family $\mathcal{S}$ of $\mathcal{C}$ if, putting $\mathcal{S}_1 = \mathcal{S} \cap \mathcal{C}_1$, (2.5) The family $\mathcal{S}_1$ is a Garside family in $\mathcal{C}_1$, and (2.6) A $\mathcal{C}_1$-path is $\mathcal{S}_1$-normal in $\mathcal{C}_1$ if and only if it is $\mathcal{S}$-normal in $\mathcal{C}$.

Proposition 2.10 **(recognizing compatible I).** *If $\mathcal{S}$ is a Garside family in a left-cancellative category $\mathcal{C}$ and $\mathcal{C}_1$ is a subcategory of $\mathcal{C}$ that is closed under right-quotient in $\mathcal{C}$, then $\mathcal{C}_1$ is compatible with $\mathcal{S}$ if and only if the following conditions are satisfied:*(2.11) *Putting $\mathcal{S}_1 = \mathcal{S} \cap \mathcal{C}_1$ and $\mathcal{S}_1^\sharp = \mathcal{S}_1\mathcal{C}_1^\times \cup \mathcal{C}_1^\times$, we have $\mathcal{S}^\sharp \cap \mathcal{C}_1 \subseteq \mathcal{S}_1^\sharp$,* (2.12) *Every element of $\mathcal{C}_1$ admits an $\mathcal{S}$-normal decomposition with entries in $\mathcal{C}_1$.*

Proposition 2.14 **(recognizing compatible II).** *If $\mathcal{S}$ is a Garside family in a left-cancellative category $\mathcal{C}$ and $\mathcal{C}_1$ is a subcategory of $\mathcal{C}$ that is closed under right-quotient in $\mathcal{C}$, then $\mathcal{C}_1$ is compatible with $\mathcal{S}$ if and only if, putting $\mathcal{S}_1 = \mathcal{S} \cap \mathcal{C}_1$ and $\mathcal{S}_1^\sharp = \mathcal{S}_1\mathcal{C}_1^\times \cup \mathcal{C}_1^\times$,* (2.15) *Every element of $\mathcal{C}_1$ admits an $\mathcal{S}$-normal decomposition with entries in $\mathcal{S}_1^\sharp$.*

Proposition 2.21 **(recognizing compatible, $=^\times$-closed case).** *If $\mathcal{S}$ is a Garside family in a left-cancellative category $\mathcal{C}$ and $\mathcal{C}_1$ is a subcategory of $\mathcal{C}$ that is closed under right-quotient and $=^\times$-closed in $\mathcal{C}$, then $\mathcal{C}_1$ is compatible with $\mathcal{S}$ if and only* (2.12) *holds.*

Proposition 2.27 **(strongly compatible).** *If $\mathcal{C}$ is a left-Ore subcategory that admits left-lcms, then, for every Ore subcategory $\mathcal{C}_1$ of $\mathcal{C}$, the following are equivalent:* (i) *The subcategory $\mathcal{C}_1$ is strongly compatible with a strong Garside family $\mathcal{S}$ of $\mathcal{C}$.* (ii) *The subcategory $\mathcal{C}_1$ is closed under quotient and weakly closed under left-lcm.* (iii) *The subcategory $\mathcal{C}_1$ is strongly compatible with every strong Garside family with which it is compatible.*

Definition 3.1 **(subgerm).** A germ $\underline{\mathcal{S}}_1$ is a *subgerm* of a germ $\underline{\mathcal{S}}$ if $\mathcal{S}_1$ is included in $\mathcal{S}$, $\mathcal{O}bj(\underline{\mathcal{S}}_1)$ is included in $\mathcal{O}bj(\underline{\mathcal{S}})$, the source, target, identity, and product maps of $\underline{\mathcal{S}}_1$ are induced by those of $\underline{\mathcal{S}}$, and $\mathrm{Dom}(\bullet_1) = \mathrm{Dom}(\,\bullet\,) \cap \mathcal{S}_1^{[2]}$ holds.

Proposition 3.14 **(Garside subgerm I).** *If $\underline{\mathcal{S}}_1$ is a subgerm of a Garside germ $\underline{\mathcal{S}}$ and $\mathcal{S}_1$ is closed under right-quotient in $\mathcal{S}$, a sufficient condition for $\mathcal{S}ub(\mathcal{S}_1)$ to be isomorphic to $\mathcal{C}at(\underline{\mathcal{S}}_1)$ is that $\mathcal{S}_1$ is closed under right-comultiple in $\mathcal{S}$.*

Corollary 3.18 **(Garside subgerm, Noetherian case).** *If $\underline{\mathcal{S}}_1$ is a right-Noetherian subgerm of a Garside germ $\underline{\mathcal{S}}$ and $\underline{\mathcal{S}}_1$ is closed under right-quotient and right-comultiple in $\underline{\mathcal{S}}$, then $\mathcal{S}ub(\mathcal{S}_1)$ is isomorphic to $\mathcal{C}at(\underline{\mathcal{S}}_1)$ and $\mathcal{S}_1$ is a Garside family in $\mathcal{S}ub(\mathcal{S}_1)$.*

Proposition 4.2 **(fixed points).** *If $\mathcal{S}$ is an $=^{\times}$-transverse Garside family in a left-cancellative category $\mathcal{C}$ and ϕ is an automorphism of $\mathcal{C}$ satisfying $\phi(\mathcal{S}) = \mathcal{S}$, then the subcategory $\mathcal{C}^{\phi}$ is compatible with $\mathcal{S}$; in particular, $\mathcal{S} \cap \mathcal{C}^{\phi}$ is a Garside family in $\mathcal{C}^{\phi}$.*

Definition 4.5 **(correct).** Assume that $\mathcal{C}, \mathcal{C}'$ are left-cancellative categories. (i) For $\mathcal{S} \subseteq \mathcal{C}$, a functor ϕ from $\mathcal{C}$ to $\mathcal{C}'$ is said to be *correct for invertibility on* $\mathcal{S}$ if, when s lies in $\mathcal{S}$ and $\phi(s)$ is invertible in $\mathcal{C}'$, then s is invertible in $\mathcal{C}$ and s^{-1} lies in $\mathcal{S}$. (ii) For $\mathcal{S} \subseteq \mathcal{C}$, a functor ϕ from $\mathcal{C}$ to $\mathcal{C}'$ is said to be *correct for right-comultiples* (*resp. right-complements, resp. right-diamonds*) *on* $\mathcal{S}$ if, when s, t lie to $\mathcal{S}$ and $\phi(s)g = \phi(t)f$ holds in $\mathcal{C}'$ for some g, f, there exist s', t' in $\mathcal{C}$ and h in $\mathcal{C}'$ satisfying $st' = ts'$, $f = \phi(s')h$, and $g = \phi(t')h$, plus $st' \in \mathcal{S}$ (*resp.* plus $s', t' \in \mathcal{S}$, *resp.* plus $s', t', st' \in \mathcal{S}$).

Proposition 4.15 **(correct** *vs.* **divisibility).** *If $\mathcal{C}, \mathcal{C}'$ are left-cancellative categories and ϕ is a functor from $\mathcal{C}$ to $\mathcal{C}'$ that is correct for invertibility and right-complements on a generating subfamily of $\mathcal{C}$, then, for all f, g in $\mathcal{C}$, we have $f \preccurlyeq g \Leftrightarrow \phi(f) \preccurlyeq \phi(g)$.*

Corollary 4.17 **(correct** *vs.* **injectivity).** *If $\mathcal{C}, \mathcal{C}'$ are left-cancellative categories and ϕ is a functor from $\mathcal{C}$ to $\mathcal{C}'$ that is correct for invertibility and right-complements on a generating subfamily of $\mathcal{C}$, then ϕ is injective on $\mathcal{C}$ if and only if it is injective on $\mathcal{C}^{\times}$.*

Proposition 4.22 **(correct** *vs.* **groupoid).** *If $\mathcal{C}, \mathcal{C}'$ are left-Ore categories and ϕ is a functor from $\mathcal{C}$ to $\mathcal{C}'$ that is correct for invertibility and right-complements on a generating family of $\mathcal{C}$ and injective on $\mathcal{C}^{\times}$, then ϕ extends to an injective functor $\phi^{\pm}$ from $\mathcal{E}nv(\mathcal{C})$ to $\mathcal{E}nv(\mathcal{C}')$, and we have then $\phi(\mathcal{C}) = \phi^{\pm}(\mathcal{E}nv(\mathcal{C})) \cap \mathcal{C}'$.*

1 Subcategories

This introductory section is devoted to a few general results about subcategories and various notions of closure. It is organized as follows. In Subsection 1.1, we recall basic facts about subcategories and introduce the closure properties we are interested in. Then, in Subsection 1.2, we discuss the special case of $=^\times$-closed subcategories, which are those subcategories that contain as many invertible elements as possible. Next, in Subsection 1.3, we introduce the notion of a head subcategory, which is a subcategory $\mathcal{C}_1$ such that every element of the ambient category admits a $\mathcal{C}_1$-head, that is, a greatest left-divisor lying in $\mathcal{C}_1$. Finally, in Subsection 1.4, we introduce the notion of a parabolic subcategory.

1.1 Closure properties

We recall from Subsection II.1.2 that a subcategory of a category $\mathcal{C}$ is a category $\mathcal{C}_1$ included in $\mathcal{C}$ and such that the operations of $\mathcal{C}_1$ are induced by those of $\mathcal{C}$. Every subfamily $\mathcal{S}$ of a category $\mathcal{C}$ is included in a smallest subcategory that includes it. Denoted by $\mathcal{S}ub(\mathcal{S})$, the latter is the union of $\mathbf{1}_\mathcal{S}$ and all powers $\mathcal{S}^p$ with $p \geqslant 1$.

We shall now introduce various closure conditions that a subcategory—or, more generally, a subfamily—may satisfy. In view of the subsequent developments, it is convenient to introduce what can be called relative versions in which the reference family need not be the entire ambient category.

Definition 1.1 (closed under identity and product). For $\mathcal{S} \subseteq \mathcal{X} \subseteq \mathcal{C}$ with $\mathcal{C}$ a left-cancellative category, we say that $\mathcal{S}$ is *closed under identity* if 1_x belongs to $\mathcal{S}$ whenever there exists at least one element of $\mathcal{S}$ with source or target x. We say that $\mathcal{S}$ is *closed under product in* $\mathcal{X}$ if st belongs to $\mathcal{S}$ whenever s and t belong to $\mathcal{S}$ and st belongs to $\mathcal{X}$.

Then a subfamily $\mathcal{C}_1$ of $\mathcal{C}$ is a subcategory of $\mathcal{C}$—more exactly, becomes a subcategory when equipped with the operations induced by those of $\mathcal{C}$—if and only if it is closed under identity and product in $\mathcal{C}$. Note that we do not require that a subcategory includes all objects of the initial category.

Definition 1.2 (closed under inverse). For $\mathcal{S} \subseteq \mathcal{X} \subseteq \mathcal{C}$ with $\mathcal{C}$ a left-cancellative category, we say that $\mathcal{S}$ is *closed under inverse in* $\mathcal{X}$ if g^{-1} lies in $\mathcal{S}$ whenever g lies in $\mathcal{S}$, it is invertible, and g^{-1} belongs to $\mathcal{X}$.

The relation between being invertible in the sense of the ambient category and in the sense of a subcategory is then obvious:

Lemma 1.3. *If $\mathcal{C}_1$ is a subcategory of a left-cancellative category $\mathcal{C}$, we have*

$$\mathcal{C}_1^\times \subseteq \mathcal{C}^\times \cap \mathcal{C}_1, \tag{1.4}$$

with equality if and only if $\mathcal{C}_1$ is closed under inverse in $\mathcal{C}$. For every subfamily $\mathcal{S}$ of $\mathcal{C}$, putting $\mathcal{S}_1 = \mathcal{S} \cap \mathcal{C}_1$ and $\mathcal{S}_1^\sharp = \mathcal{S}_1\mathcal{C}_1^\times \cup \mathcal{C}_1^\times$, we have

$$\mathcal{S}_1^\sharp \subseteq \mathcal{S}^\sharp \cap \mathcal{C}_1. \tag{1.5}$$

If $\mathcal{C}$ has no nontrivial invertible element, then so does $\mathcal{C}_1$, and (1.4)–(1.5) *are equalities.*

The proof, which follows directly from the definitions, is left to the reader.

The example of the submonoid $\mathbb{N}$ of $\mathbb{Z}$ viewed as a monoid and, more generally, of every Ore category viewed as a subcategory of its enveloping groupoid, show that (1.4) need not be an equality in general.

We now introduce a stronger closure property.

Definition 1.6 (closed under quotient). For $\mathcal{S} \subseteq \mathcal{X} \subseteq \mathcal{C}$ with $\mathcal{C}$ a left-cancellative category, we say that $\mathcal{S}$ is *closed under right-quotient* (*resp. left-quotient*) *in* $\mathcal{X}$ if, for all f, g, h in $\mathcal{X}$ satisfying $f = gh$, if f and g (*resp.* f and h) belong to $\mathcal{S}$, then h (*resp.* g) belongs to $\mathcal{S}$ too.

Lemma 1.7. *Assume that $\mathcal{C}$ is a left-cancellative category and $\mathcal{C}_1$ is a subcategory of $\mathcal{C}$ that is closed under right-quotient in $\mathcal{C}$.*

(i) *The subcategory $\mathcal{C}_1$ is closed under inverse in $\mathcal{C}$.*

(ii) *For all f, g in $\mathcal{C}_1$, the relation $f \preccurlyeq g$ holds in $\mathcal{C}$ if and only if it does in $\mathcal{C}_1$.*

(iii) *Two elements of $\mathcal{C}_1$ that are left-disjoint in $\mathcal{C}$ are left-disjoint in $\mathcal{C}_1$.*

Proof. (i) Assume that ϵ belongs to $\mathcal{C}^\times \cap \mathcal{C}_1$. Let x be the source of ϵ. Then $\epsilon \in \mathcal{C}_1$ implies $x \in \mathcal{O}bj(\mathcal{C}_1)$, whence $1_x \in \mathcal{C}_1$. Then, in $\mathcal{C}$, we have $\epsilon\epsilon^{-1} = 1_x$. As ϵ and 1_x belong to $\mathcal{C}_1$, the assumption that $\mathcal{C}_1$ is closed under right-quotient implies that ϵ^{-1} lies in $\mathcal{C}_1$.

(ii) In every case, if $f \preccurlyeq g$ holds in $\mathcal{C}_1$, it a fortiori does in $\mathcal{C}$. Conversely, assume that f, g lie in $\mathcal{C}_1$ and $f \preccurlyeq g$ holds in $\mathcal{C}$. Then $fg' = g$ holds for some g' in $\mathcal{C}$. As $\mathcal{C}_1$ is closed under right-quotient in $\mathcal{C}$, the quotient g' belongs to $\mathcal{C}_1$ and, therefore, $f \preccurlyeq g$ holds in $\mathcal{C}_1$.

(iii) Assume that f, g belong to $\mathcal{C}_1$ and they are left-disjoint in $\mathcal{C}$. Assume that h, h' lie in $\mathcal{C}_1$ and satisfy $h' \preccurlyeq hf$ and $h' \preccurlyeq hg$ in $\mathcal{C}_1$. A fortiori, we have $h' \preccurlyeq hf$ and $h' \preccurlyeq hg$ in $\mathcal{C}$, so the assumption that f and g are left-disjoint in $\mathcal{C}$ implies $h' \preccurlyeq h$ in $\mathcal{C}$. By (ii), $h' \preccurlyeq h$ holds in $\mathcal{C}_1$, so, in $\mathcal{C}_1$, the elements f and g are left-disjoint. □

In general, there is no reason why the implication of Lemma 1.7(iii) should be an equivalence: the assumption that two elements f, g of the considered subcategory $\mathcal{C}_1$ are left-disjoint in $\mathcal{C}_1$ says nothing about the elements h, h' of the ambient category $\mathcal{C}$ that possibly satisfy $h' \preccurlyeq hf$ and $h' \preccurlyeq hg$.

There exist connections between the closure properties introduced so far.

Lemma 1.8. *Every subcategory $\mathcal{C}_1$ of a left-cancellative category $\mathcal{C}$ that is closed under inverse and right-complement in $\mathcal{C}$ is closed under right-quotient in $\mathcal{C}$.*

Proof. Assume that f and g lie in $\mathcal{C}_1$ and $g = f\hat{g}$ holds. As $\mathcal{C}_1$ is closed under right-complement in $\mathcal{C}$, there exist f', g' in $\mathcal{C}_1$ and h satisfying $fg' = gf'$, $g'h = \hat{g}$, and $f'h = 1_y$, where y is the target of g. The latter equality implies that f' lies in $\mathcal{C}^\times \cap \mathcal{C}_1$, hence h lies in $\mathcal{C}_1^\times$ since $\mathcal{C}_1$ is closed under inverse. Therefore, $\hat{g}$, which is $g'h$, lies in $\mathcal{C}_1$ as well. So $\mathcal{C}_1$ is closed under right-quotient in $\mathcal{C}$. □

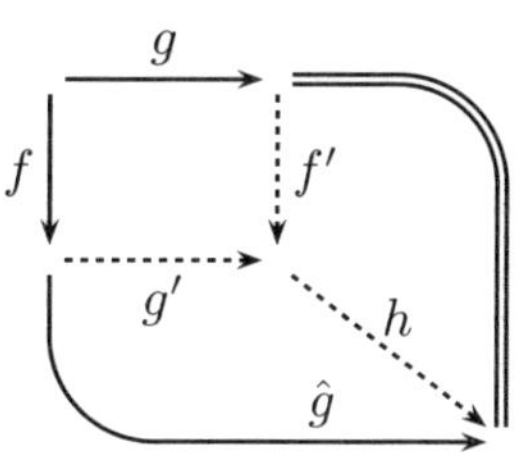

Thus, for a subcategory that is closed under right-complement in the ambient category, closures under inverse and under right-quotient are equivalent conditions.

In the sequel of this chapter, the condition (IV.1.9), which is a common extension of closure under right-comultiple and under right-complement, will appear frequently, and it is convenient to introduce a specific terminology, here in a relative version.

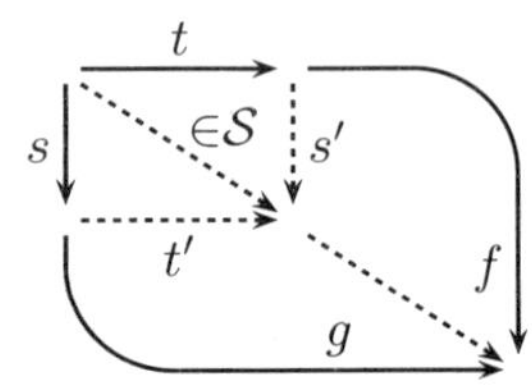

Definition 1.9 (closed under right-diamond). For $\mathcal{S} \subseteq \mathcal{X} \subseteq \mathcal{C}$ with $\mathcal{C}$ a left-cancellative category, we say that $\mathcal{S}$ is *closed under right-diamond* in $\mathcal{X}$ if, when s, t lie in $\mathcal{S}$ and $sg = tf$ holds in $\mathcal{C}$ with sg in $\mathcal{X}$, there exist s', t' in $\mathcal{S}$ and h in $\mathcal{X}$ satisfying $st' = ts'$, $f = s'h$, $g = t'h$ and, moreover, st' lies in $\mathcal{S}$.

By definition, if $\mathcal{S}$ is closed under right-diamond in $\mathcal{X}$, it is closed both under right-comultiple and right-complement in $\mathcal{X}$—with the obvious meaning of this relative version. Conversely, (the relative version of) Lemma IV.1.8 says that, if $\mathcal{S}$ is closed under right-comultiple and right-divisor in $\mathcal{X}$, then it is closed under right-diamond in $\mathcal{X}$. In the case of a subfamily that is a subcategory, we obtain:

Lemma 1.10. *For every subcategory $\mathcal{C}_1$ of a left-cancellative category $\mathcal{C}$, the following conditions are equivalent:*

(i) *The subcategory $\mathcal{C}_1$ is closed under inverse and right-diamond in $\mathcal{C}$.*

(ii) *The subcategory $\mathcal{C}_1$ is closed under right-quotient and right-comultiple in $\mathcal{C}$.*

Proof. Assume (i). As it is closed under right-diamond in $\mathcal{C}$, the subcategory $\mathcal{C}_1$ is closed under right-complement and right-comultiple in $\mathcal{C}$. Then Lemma 1.8 implies that it is also closed under right-quotient in $\mathcal{C}$. So (i) implies (ii).

Conversely, assume (ii). Then, by Lemma 1.7, $\mathcal{C}_1$ is closed under inverse in $\mathcal{C}$. On the other hand, by Lemma IV.1.8, the assumption that it is closed under right-comultiple and right-quotient in $\mathcal{C}$ implies that it is closed under right-diamond in $\mathcal{C}$. ☐

Let us mention an important situation in which a number of subcategories are automatically closed under quotient, and even under divisor.

Proposition 1.11 (balanced presentation). *Call a relation $u = v$* balanced *if the same letters occur in u and v (possibly with different multiplicities), and call a presentation* balanced *if it contains only balanced relations. If a left-cancellative category $\mathcal{C}$ admits a balanced presentation $(\mathcal{S}, \mathcal{R})$, then, for every subfamily $\mathcal{S}_1$ of $\mathcal{S}$, the subcategory $\mathit{Sub}(\mathcal{S}_1)$ is closed under left- and right-divisor and under left- and right-quotient in $\mathcal{C}$.*

Proof. Assume that $\mathcal{S}_1$ is included in $\mathcal{S}$, that g belongs to the subcategory $\mathit{Sub}(\mathcal{S}_1)$, and that h is a factor of g. As g lies in $\mathit{Sub}(\mathcal{S}_1)$, it admits an expression w that is an $\mathcal{S}_1$-path. On the other hand, as h is a factor of g, there exists another expression w' of g such that some subpath v of which is an expression of h. A straightforward induction shows that, if two $\mathcal{S}$-paths w, w' are $\mathcal{R}$-equivalent, then exactly the same letters of $\mathcal{S}$ occur in w and w'. Hence w' must be an $\mathcal{S}_1$-path, and so is a fortiori its subpath v. Then h, which admits an expression by an $\mathcal{S}_1$-path, belongs to $\mathit{Sub}(\mathcal{S}_1)$. So $\mathit{Sub}(\mathcal{S}_1)$ is closed under factor, hence under left- and right-divisor and, a fortiori, under left- and right-quotient. ☐

Example 1.12 (braids). Consider the braid monoid B_n^+ (Reference Structure 2, page 5), and its presentation (I.1.5) in terms of the generators $\sigma_1, ..., \sigma_{n-1}$. The relations are balanced, so Proposition 1.11 states that, for every subset I of $\{1, ..., n-1\}$, the submonoid B_I^+ of B_n^+ generated by $\{\sigma_i \mid i \in I\}$ is closed under left- and right-divisor in B_n^+.

We conclude with a criterion guaranteeing, in good cases, that a subfamily is a Garside family in the subcategory it generates.

Proposition 1.13 (generated subcategory, right-lcm case). *Assume that $\mathcal{C}$ is a left-cancellative category that admits conditional right-lcms and $\mathcal{S}_1$ is a subfamily of $\mathcal{C}$ such that $\mathcal{S}_1\mathcal{C}^\times \cup \mathcal{C}^\times$, denoted by $\mathcal{S}_1^\sharp$, is closed under right-diamond in $\mathcal{C}$.*

(i) *The subcategory $\mathcal{C}_1$ of $\mathcal{C}$ generated by $\mathcal{S}_1^\sharp$ is closed under right-diamond and right-quotient in $\mathcal{C}$, and it admits conditional right-lcms.*

(ii) *If $\mathcal{C}$ is right-Noetherian, then so is $\mathcal{C}_1$, and $\mathcal{S}_1$ is a Garside family in $\mathcal{C}_1$.*

Proof. (i) Assume that, in $\mathcal{C}$, we have an equality $s_1 \cdots s_p g' = t_1 \cdots t_q f'$, with $s_1, ..., s_p$, $t_1, ..., t_q$ in $\mathcal{S}_1^\sharp$. As $\mathcal{S}_1^\sharp$ is closed under right-diamond in $\mathcal{C}$, Proposition IV.1.15 (grid) provides a rectangular diagram with right edges $f'_1, ..., f'_p$ and bottom edges $g'_1, ..., g'_q$, plus h in $\mathcal{C}$, such that all edges and diagonals of the grid belong to $\mathcal{S}_1^\sharp$ and we have $f' = f'_1 \cdots f'_p h$ and $g' = g'_1 \cdots g'_q h$.

Next, as the category $\mathcal{C}$ admits conditional right-lcms, we can construct the above grid so that, for all i, j, the element $s_{i,j+1} t_{i+1,j+1}$, which is also $t_{i+1,j} s_{i+1,j+1}$, is a right-lcm of $s_{i,j+1}$ and $t_{i+1,j}$. Then the rules for an iterated right-lcm imply that $s_1 \cdots s_p g'_1 \cdots g'_q$ is a right-lcm of $s_1 \cdots s_p$ and $t_1 \cdots t_q$. This shows that, whenever two elements f, g of $\mathcal{C}_1$ admit a common right-multiple in $\mathcal{C}$, then some right-lcm of f and g lies in $\mathcal{C}_1$, and therefore, as, by assumption, $\mathcal{C}_1$ includes all of $\mathcal{C}^\times$, every right-lcm of f and g lies in $\mathcal{C}_1$. This implies that $\mathcal{C}_1$ admits conditional right-lcms.

Then, consider the special case when f' is an identity element. Then h must be invertible and, therefore, the equality $g' = g'_1 \cdots g'_q h$ implies that g is an element of $\mathit{Sub}(\mathcal{S}_1^\sharp)$ (here we use the fact that $\mathcal{S}_1^\sharp$, hence $\mathcal{C}_1$, includes all of $\mathcal{C}^\times$). Hence $\mathcal{C}_1$ is closed under right-quotient in $\mathcal{C}$. As it is closed under right-lcm, it is closed under right-diamond in $\mathcal{C}$.

(ii) Assume that $\mathcal{C}$ is right-Noetherian. Then $\mathcal{C}_1$, which is included in $\mathcal{C}$, is a fortiori right-Noetherian. Now $\mathcal{S}_1^\sharp$ is a generating family of $\mathcal{C}_1$ that is closed under right-diamond in $\mathcal{C}$, hence a fortiori in $\mathcal{C}_1$. By definition and Lemma IV.2.28, this implies that $\mathcal{S}_1^\sharp$ is closed under right-complement and right-lcm in $\mathcal{C}_1$, and Corollary IV.2.29 (recognizing Garside, right-lcm case) implies that $\mathcal{S}_1^\sharp$, hence also $\mathcal{S}_1$, is a Garside family in $\mathcal{C}_1$. □

Corollary 1.14 (generated subcategory, right lcm-case). *Under the assumptions of Proposition* 1.13(ii), *$\mathcal{S}_1$ is also a Garside family in the subcategory $\mathit{Sub}(\mathcal{S}_1)$ whenever the latter is closed under right-quotient in $\mathcal{C}_1$.*

Proof. We first observe that, if s_1, s_2 lie in $\mathcal{S}_1$ and $s_1 | s_2$ is $\mathcal{S}_1$-greedy in $\mathcal{C}_1$, then $s_1 | s_2$ is also $\mathcal{S}_1$-greedy in $\mathit{Sub}(\mathcal{S}_1)$. Indeed, assume $s \in \mathcal{S}_1$, $f \in \mathit{Sub}(\mathcal{S}_1)$, and $s \preccurlyeq f s_1 s_2$ in $\mathit{Sub}(\mathcal{S}_1)$. As $\mathit{Sub}(\mathcal{S}_1)$ is included in $\mathcal{C}_1$ and $s_1 | s_2$ is $\mathcal{S}_1$-normal in $\mathcal{C}_1$, there exists f' in $\mathcal{C}_1$ satisfying $s f' = f s_1$. By assumption, s and $f s_1$ belong to $\mathit{Sub}(\mathcal{S}_1)$, so the closure assumption implies that f' belongs to $\mathit{Sub}(\mathcal{S}_1)$ as well. Hence $s \preccurlyeq f s_1$ holds in $\mathit{Sub}(\mathcal{S}_1)$ and $s_1 | s_2$ is $\mathcal{S}_1$-greedy in $\mathit{Sub}(\mathcal{S}_1)$.

Now, let g belong to $\mathcal{S}ub(\mathcal{S}_1)$. Then g belongs to $\mathcal{C}_1$, and, as $\mathcal{S}_1$ is a Garside family in $\mathcal{C}_1$ by Proposition 1.13, g admits an $\mathcal{S}_1^\sharp$-normal decomposition in $\mathcal{C}_1$, say $s_1|\cdots|s_p$. We may assume that $s_1, \ldots, s_{p-1}$ lie in $\mathcal{S}_1$, and that s_{p-1} lies in $\mathcal{S}_1^\sharp$. As g and $s_1 \cdots s_{p-1}$ lie in $\mathcal{S}ub(\mathcal{S}_1)$, the assumption that $\mathcal{S}ub(\mathcal{S}_1)$ is closed under right-quotient in $\mathcal{C}_1$ implies $s_p \in \mathcal{S}ub(\mathcal{S}_1)$. Then, by the result above, $s_1|\cdots|s_p$ is an $\mathcal{S}_1$-normal decomposition of g inside $\mathcal{S}ub(\mathcal{S}_1)$. Every element of $\mathcal{S}ub(\mathcal{S}_1)$ admits an $\mathcal{S}_1$-normal decomposition in $\mathcal{S}ub(\mathcal{S}_1)$, hence $\mathcal{S}_1$ is a Garside family in $\mathcal{S}ub(\mathcal{S}_1)$. □

1.2 Subcategories that are closed under $=^\times$

Various technical problems involving subcategories come from the possible existence of invertible elements of the ambient category that do not belong to the considered subcategory. These problems vanish when we consider subcategories that are $=^\times$-closed, that is, every element $=^\times$-equivalent to an element of the subcategory lies in the subcategory. Here we establish basic properties of such $=^\times$-closed subcategories.

By definition, every category $\mathcal{C}$, viewed as a subfamily of itself, is $=^\times$-closed, and so is the empty subcategory. The point with a subcategory $\mathcal{C}_1$, compared with an arbitrary subfamily, is that the product of two elements of $\mathcal{C}_1$ belongs to $\mathcal{C}_1$. Hence, in order to force $=^\times$-closedness for $\mathcal{C}_1$, it is enough to demand that $\mathcal{C}_1$ contains enough invertible elements, see Exercise 72. For instance, a subcategory of a category $\mathcal{C}$ that includes all of $\mathcal{C}^\times$ is obviously $=^\times$-closed.

Lemma 1.15. *If $\mathcal{C}_1$ is an $=^\times$-closed subcategory of a left-cancellative category $\mathcal{C}$:*

(i) *The subcategory $\mathcal{C}_1$ is closed under inverse in $\mathcal{C}$ and under ${}^\times\!\!=^\times$.*

(ii) *Every $\mathcal{C}^\times$-deformation of a $\mathcal{C}_1$-path is a $\mathcal{C}_1$-path.*

(iii) *For $\mathcal{S} \subseteq \mathcal{C}$, putting $\mathcal{S}_1 = \mathcal{S} \cap \mathcal{C}_1$, we have $\mathcal{S}_1\mathcal{C}_1^\times \cup \mathcal{C}_1^\times = \mathcal{S}^\sharp \cap \mathcal{C}_1$.*

Proof. (i) Let ϵ belongs to $\mathcal{C}^\times \cap \mathcal{C}_1$, say $\epsilon \in \mathcal{C}^\times(x, y)$. Then y belongs to $\mathcal{O}bj(\mathcal{C}_1)$, hence 1_y belongs to $\mathcal{C}_1$. Now $\epsilon^{-1} =^\times 1_y$ holds, so the assumption that $\mathcal{C}_1$ is $=^\times$-closed implies that ϵ^{-1} lies in $\mathcal{C}_1$. We deduce $\mathcal{C}^\times \cap \mathcal{C}_1 \subseteq \mathcal{C}_1^\times$, whence $\mathcal{C}_1^\times = \mathcal{C}^\times \cap \mathcal{C}_1$ by Lemma 1.3.

Next, assume $g \in \mathcal{C}_1(x, \cdot)$. Then 1_x belongs to $\mathcal{C}_1$, and $\epsilon^{-1} =^\times 1_x$ holds. We deduce that ϵ^{-1} lies in $\mathcal{C}_1$, hence that ϵ lies in $\mathcal{C}_1$, since the latter is closed under inverse. It follows that ϵg lies in $\mathcal{C}_1$, hence that $\mathcal{C}_1$ is ${}^\times\!\!=$-closed. As it is $=^\times$-closed, it is ${}^\times\!\!=^\times$-closed, that is, closed under left- and right-multiplication by an invertible element.

(ii) Assume that $f_1|\cdots|f_p$ is a $\mathcal{C}^\times$-deformation of a $\mathcal{C}_1$-path $g_1|\cdots|g_q$. As g_p belongs to $\mathcal{C}_1$, the identity element 1_y, where y is the target of g_p, belongs to $\mathcal{C}_1$, so, if needed, we can always extend the shorter path with identity elements and assume $p = q$. Then, by definition, $f_i \;{}^\times\!\!=^\times g_i$ holds for every i. By (i), this implies that f_i belongs to $\mathcal{C}_1$.

(iii) The inclusion $\mathcal{S}_1\mathcal{C}_1^\times \cup \mathcal{C}_1^\times \subseteq \mathcal{S}^\sharp \cap \mathcal{C}_1$ is obvious. Conversely, assume $g \in \mathcal{S}^\sharp \cap \mathcal{C}_1$. If g is invertible, it belongs to $\mathcal{C}_1^\times$ by (i). Otherwise, assume $g = g'\epsilon$ with $g' \in \mathcal{S}$ and $\epsilon \in \mathcal{C}^\times$. We have $g' =^\times g$, so the assumption that g lies in $\mathcal{C}_1$ implies that g' lies in $\mathcal{C}_1$, hence in $\mathcal{S} \cap \mathcal{C}_1$. On the other hand, we have $\epsilon \;{}^\times\!\!= 1_y$, where y is the target of g, so ϵ lies in $\mathcal{C}_1$, hence in $\mathcal{C}^\times \cap \mathcal{C}_1$, which is $\mathcal{C}_1^\times$ by (i). Therefore, g lies in $\mathcal{S}_1\mathcal{C}_1^\times$, and we deduce the inclusion $\mathcal{S}^\sharp \cap \mathcal{C}_1 \subseteq \mathcal{S}_1\mathcal{C}_1^\times \cup \mathcal{C}_1^\times$. □

We now mention two useful conditions under which a subcategory is $=^{\times}$-closed. The easy proofs are left to the reader.

Lemma 1.16. *Every subcategory that is closed under left- or under right-divisor in a left-cancellative category $\mathcal{C}$ is closed under $=^{\times}$.*

Lemma 1.17. *If $\mathcal{C}, \mathcal{C}'$ are left-cancellative categories, $\phi : \mathcal{C} \to \mathcal{C}'$ is a functor, and $\mathcal{C}'_1$ is a subcategory of $\mathcal{C}'$ that is closed under left-divisor (resp. right-divisor), then so is $\phi^{-1}(\mathcal{C}'_1)$.*

It follows that subcategories that are closed under divisor naturally arise when one considers inverse images under a functor. Indeed combining Lemmas 1.16 and 1.17 gives

Proposition 1.18 (inverse image). *If $\mathcal{C}$ and $\mathcal{C}'$ are left-cancellative categories, ϕ is a functor from $\mathcal{C}$ to $\mathcal{C}'$, and $\mathcal{C}'_1$ is a subcategory of $\mathcal{C}'$ that is closed under right-divisor in $\mathcal{C}'$, then $\phi^{-1}(\mathcal{C}'_1)$ is closed under right-divisor and under $=^{\times}$ in $\mathcal{C}'$.*

Note that, since every factor of an invertible element is invertible, Proposition 1.18 applies in particular when $\mathcal{C}'_1$ is the subcategory $\mathcal{C}'^{\times}$ or one of its subcategories.

1.3 Head subcategories

In Chapter VIII, we shall be interested in particular subcategories called *head subcategories*, which, as the name suggests, are defined by the existence of a head. One of their interests is that every intersection of head subcategories is a head subcategory, which guarantees that every subfamily is included in a smallest head subcategory.

Definition 1.19 (head subcategory). A subcategory $\mathcal{C}_1$ of a left-cancellative category $\mathcal{C}$ is called a *head subcategory* if $\mathcal{C}_1$ is closed under right-quotient in $\mathcal{C}$ and every element of $\mathcal{C}$ whose source belongs to $\mathcal{O}bj(\mathcal{C}_1)$ admits a $\mathcal{C}_1$-head.

We recall from Definition IV.1.10 (head) that a $\mathcal{C}_1$-head for an element g is a greatest left-divisor of g lying in $\mathcal{C}_1$, that is, an element g_1 of $\mathcal{C}_1$ satisfying $g_1 \preccurlyeq g$ and such that the conjunction of $f \in \mathcal{C}_1$ and $f \preccurlyeq g$ implies $f \preccurlyeq g_1$. In the context of Definition 1.19, the assumption that $\mathcal{C}_1$ is closed under right-quotient eliminates the possible ambiguity about the meaning of left-divisibility: by Lemma 1.7, $f \preccurlyeq_{\mathcal{C}_1} h$ and $f \preccurlyeq_{\mathcal{C}} h$ are equivalent.

Example 1.20 (head subcategory). For every m, the submonoid $m\mathbb{N}$ is a head submonoid of the additive monoid $\mathbb{N}$: first, if f, h are multiples of m and $f + g = h$ holds, then g is a multiple of m. On the other hand, for every g in $\mathbb{N}$, there is a unique maximal element h of $m\mathbb{N}$ satisfying $h \leqslant g$.

By contrast, if K^+ is the Klein bottle monoid (Reference Structure 5, page 17), that is, $\langle \mathtt{a}, \mathtt{b} \mid \mathtt{a} = \mathtt{bab} \rangle^+$, the submonoid N of K^+ generated by $\mathtt{b}$ is not a head submonoid of K^+, since $\mathrm{Div}(\mathtt{a}) \cap N$, which is all of N, has no maximal element.

We now establish that, in a Noetherian context, head subcategories are characterized by closure properties.

Proposition 1.21 (head subcategory). *Assume that $\mathcal{C}_1$ is a subcategory of a left-cancellative category $\mathcal{C}$.*

(i) *If $\mathcal{C}_1$ is a head subcategory of $\mathcal{C}$, then $\mathcal{C}_1$ includes $\mathbf{1}_{\mathcal{C}}$ and is closed under right-quotient and right-comultiple in $\mathcal{C}$.*

(ii) *If $\mathcal{C}$ is right-Noetherian, the implication in* (i) *is an equivalence.*

Note that the fact that $\mathcal{C}_1$ includes $\mathbf{1}_{\mathcal{C}}$ implies that the objects of $\mathcal{C}$ and $\mathcal{C}_1$ coincide. Also, we recall that, if $\mathcal{C}_1$ is closed under right-quotient in $\mathcal{C}$, it is closed under right-comultiple if and only it is closed under right-diamond in $\mathcal{C}$.

Proof. (i) Assume that $\mathcal{C}_1$ is a head subcategory of $\mathcal{C}$. Let x be an object of $\mathcal{C}$. Then a $\mathcal{C}_1$-head of 1_x must be an (invertible) element of $\mathcal{C}_1$ whose source is x, so x has to belong to $\mathcal{O}bj(\mathcal{C}_1)$. This in turn implies that 1_x belongs to $\mathcal{C}_1$, so $\mathbf{1}_{\mathcal{C}}$ is included in $\mathcal{C}_1$. Next, $\mathcal{C}_1$ is closed under right-quotient in $\mathcal{C}$ by definition.

Finally, suppose $f, g \in \mathcal{C}_1$ and $f, g \preccurlyeq h$ in $\mathcal{C}$. By assumption, h admits a $\mathcal{C}_1$-head, say h_1. As f and g are elements of $\mathcal{C}_1$ that left-divide h, they must left-divide h_1, so we have $fg_1 = gf_1 = h_1$ for some f_1, g_1. Then f_1 and g_1 belong to $\mathcal{C}_1$ since the latter is closed under right-quotient. Then f_1, g_1, h_1 witness that $\mathcal{C}_1$ is closed under right-diamond, hence under right-comultiple, in $\mathcal{C}$.

(ii) Assume now that $\mathcal{C}$ is right-Noetherian and $\mathcal{C}_1$ includes $\mathbf{1}_{\mathcal{C}}$ and is closed under right-quotient and right-comultiple in $\mathcal{C}$. Let g be an element of $\mathcal{C}(x, \cdot)$. By Proposition II.2.34 (Noetherian implies maximal), the subfamily $\mathrm{Div}(g) \cap \mathcal{C}_1$ of $\mathcal{C}$, which is nonempty as it contains 1_x and is bounded by g, admits a $\prec$-maximal element, say g_1. We claim that g_1 is a $\mathcal{C}_1$-head of g. Indeed, let f be a left-divisor of g in $\mathcal{C}_1$. As g is a right-multiple of g_1 and f, which both lie in $\mathcal{C}_1$, and $\mathcal{C}_1$ is closed under right-comultiple, g is a right-multiple of some common right-multiple g_1' of g_1 and f that lies in $\mathcal{C}_1$. By the definition of g_1, we must have $g_1' =^{\times} g_1$, in $\mathcal{C}$ and in $\mathcal{C}_1$, whence $f \preccurlyeq g_1$. So g_1 is a $\mathcal{C}_1$-head of g, and $\mathcal{C}_1$ is a head subcategory of $\mathcal{C}$. □

Using Lemma 1.10, we deduce

Corollary 1.22 (head subcategory). *A subcategory $\mathcal{C}_1$ of a left-cancellative and right-Noetherian category $\mathcal{C}$ is a head subcategory if and only if $\mathcal{C}_1$ includes $\mathbf{1}_{\mathcal{C}}$ and is closed under inverse and right-diamond in $\mathcal{C}$, if and only if $\mathcal{C}_1$ is closed under right-quotient and right-comultiple in $\mathcal{C}$.*

Specializing even more, we obtain

Corollary 1.23 (head submonoid). *A submonoid M_1 of a left-cancellative monoid M that is right-Noetherian and admits no non-trivial element is a head submonoid of M if and only if M_1 is closed under right-diamond in M, if and only if M_1 is closed under right-quotient and right-comultiple in M.*

Example 1.24 (braids). As in in Example 1.12, for I included in $\{1, ..., n-1\}$, denote by B_I^+ the submonoid of B_n^+ (Reference Structure 2, page 5) generated by the elements σ_i with i in I. We saw in Example 1.12 that B_I^+ is closed under right-quotient in B_n^+. On the other hand, we know that B_n^+ admits right-lcms. An easy induction on the lengths of f and g show that, for f, g in B_I^+, the right-lcm of f and g belongs to B_I^+. So B_I^+ is closed under right-lcm, and therefore under right-comultiple, in B_n^+. Hence B_I^+ is eligible for Corollary 1.23, and the latter says that B_I^+ is is a head submonoid of B_n^+.

As can be expected, the assumption that the whole ambient category is right-Noetherian cannot be removed in Proposition 1.21(ii), and requiring that the considered subcategory be Noetherian does not help: for instance, in the context of Example 1.20, the submonoid generated by $\mathtt{b}$ in K^+ is Noetherian and closed under inverse and right-diamond, but it is not a head submonoid. When no Noetherianity assumption is made, closure conditions are not necessarily sufficient for the existence of a head.

We now turn to different characterizations of head subcategories, involving the existence of a head for particular elements, typically the elements of a Garside family. Here is the result we shall establish—we refer to Exercise 79 for another similar criterion.

Proposition 1.25 (head on Garside family). *Assume that $\mathcal{C}$ is a left-cancellative category, $\mathcal{S}$ is a Garside family of $\mathcal{C}$, and $\mathcal{C}_1$ is a subcategory of $\mathcal{C}$ that includes $1_\mathcal{C}$, is closed under right-quotient in $\mathcal{C}$, and is generated by $\mathcal{S}^\sharp \cap \mathcal{C}_1$. Then $\mathcal{C}_1$ is a head subcategory of $\mathcal{C}$ whenever every element of $\mathcal{S}$ admits a $\mathcal{C}_1$-head that lies in $\mathcal{S}$, and*

(i) *The category $\mathcal{C}$ is right-Noetherian, or*

(ii) *The family $\mathcal{S}^\sharp$ is closed under left-divisor, the second domino rule is valid for $\mathcal{S}$, and $\mathcal{C}_1$ is compatible with $\mathcal{S}$.*

(The condition in (ii) anticipates on Definition 2.4 of the compatibility of a subcategory with a Garside family.) We begin with a technical auxiliary result.

Lemma 1.26. *Assume that $\mathcal{C}$ is a left-cancellative category, $\mathcal{S}$ is a Garside family of $\mathcal{C}$, and $\mathcal{C}_1$ is a subcategory of $\mathcal{C}$ that includes $1_\mathcal{C}$ and is closed under right-quotient in $\mathcal{C}$. Assume moreover that every element of $\mathcal{S}$ admits a $\mathcal{C}_1$-head that lies in $\mathcal{S}$. Then every element of $\mathcal{C}$ admits a $(\mathcal{S}^\sharp \cap \mathcal{C}_1)$-head, and $\mathcal{S}^\sharp \cap \mathcal{C}_1$ is closed under right-complement and right-quotient in $\mathcal{C}$.*

Proof. First, every element of $\mathcal{S}^\sharp$ admits a $\mathcal{C}_1$-head that lies in $\mathcal{S}$, hence in $\mathcal{S} \cap \mathcal{C}_1$. Indeed, if g is $g'\epsilon$ with g' in $\mathcal{S}$ and ϵ in $\mathcal{C}^\times$, a $\mathcal{C}_1$-head of g' is a $\mathcal{C}_1$-head of g, and, if g is invertible with source x, then 1_x is a $\mathcal{C}_1$-head of g since, by assumption, x is an object of $\mathcal{C}_1$.

Next, let g be an arbitrary element of $\mathcal{C}$. Let s be an $\mathcal{S}^\sharp$-head of g, and let s_1 be a $\mathcal{C}_1$-head of s. We claim that s_1 is a $(\mathcal{S}^\sharp \cap \mathcal{C}_1)$-head of g. Indeed, assume $t \in \mathcal{S}^\sharp \cap \mathcal{C}_1$ and $t \preccurlyeq g$. As t lies in $\mathcal{S}^\sharp$, it must left-divide the $\mathcal{S}^\sharp$-head s of g. Now, as t lies in $\mathcal{C}_1$, it must left-divide the $\mathcal{C}_1$-head s_1 of s. So every element of $\mathcal{C}$ admits a $(\mathcal{S}^\sharp \cap \mathcal{C}_1)$-head.

Then, assume that s and t lie in $\mathcal{S}^\sharp \cap \mathcal{C}_1$, and $sg = tf$ holds. Let s_1 be a $(\mathcal{S}^\sharp \cap \mathcal{C}_1)$-head of sg. By definition, s and t must left-divide s_1, say $sg' = tf' = s_1$. Now f' and g' right-divide s_1, an element of $\mathcal{S}^\sharp$, so they belong to $\mathcal{S}^\sharp$ since the latter is closed under right-divisor. On the other hand, s, t, s_1 belong to $\mathcal{C}_1$, so f' and g' belong to $\mathcal{C}_1$ since the latter is closed under right-quotient in $\mathcal{C}$. Hence f' and g' belong to $\mathcal{S}^\sharp \cap \mathcal{C}_1$, and the latter is closed under right-complement (and even right-diamond).

Finally, the proof of Lemma 1.8 implies that $\mathcal{S}^\sharp \cap \mathcal{C}_1$ is closed under right-quotient in $\mathcal{C}$, since it is closed under right-complement and right-multiplication by an invertible element of $\mathcal{C}_1$. □

Proof of Proposition 1.25. Put $\mathcal{S}_1 = \mathcal{S}^\sharp \cap \mathcal{C}_1$. By Lemma 1.26, we can fix a map θ from $\mathcal{C}$ to $\mathcal{S}_1$ that picks, for every element g of $\mathcal{C}$, an $\mathcal{S}_1$-head $\theta(g)$ of that element. Then denote by g' the (unique) element of $\mathcal{C}$ that satisfies $g = \theta(g)g'$, and, for $p \geqslant 1$, put

$$\theta_p(g) = \theta(g)\theta(g')\cdots\theta(g^{(p-1)}) \quad \text{(see Figure 1 below).}$$

We show using induction on $p \geqslant 1$ that

(1.27) For g in $\mathcal{C}$ and $s_1, ..., s_p$ in $\mathcal{S}_1$, the relation $s_1 \cdots s_p \preccurlyeq g$ implies $s_1 \cdots s_p \preccurlyeq \theta_p(g)$.

Assume first $p = 1$. Then $s_1 \preccurlyeq g$ implies $s_1 \preccurlyeq \theta_1(g) = \theta(g)$ since $\theta(g)$ is a $\mathcal{S}_1$-head of g. Assume now $p \geqslant 2$. First, as s_1 lies in $\mathcal{S}^\sharp \cap \mathcal{C}_1$ and left-divides g, it left-divides $\theta(g)$, so there exists t_1 satisfying $s_1 t_1 = \theta(g)$. As s_1 and $\theta(g)$ lie in $\mathcal{S}^\sharp \cap \mathcal{C}_1$ and, by Lemma 1.26, $\mathcal{S}_1$ is closed under right-quotient in $\mathcal{C}$, it follows that t_1 lies in $\mathcal{S}^\sharp \cap \mathcal{C}_1$ as well. Because $\mathcal{S}^\sharp \cap \mathcal{C}_1$ is closed under right-complement in $\mathcal{C}$, we deduce the existence of $s'_2, ..., s'_p$, $t_2, ..., t_p$ in $\mathcal{S}^\sharp \cap \mathcal{C}_1$ making the diagram of Figure 1 commutative. So $s'_2 \cdots s'_p$ left-divides g'. Applying the induction hypothesis to g', we see that $s'_2 \cdots s'_p$ has to left-divide $\theta_{p-1}(g')$, which implies that $s_1 \cdots s_p$ left-divides $\theta(g)\theta_{p-1}(g')$, that is, $\theta_p(g)$.

Assume that the category $\mathcal{C}$ is right-Noetherian. Then the $\preccurlyeq$-increasing sequence $\theta_1(g)$, $\theta_2(g)$, ... is bounded by g, so there must exist q such that $\theta(g^{(i)})$ is invertible for $i \geqslant q$, so that $\theta_p(g) \preccurlyeq \theta_q(g)$ holds for every p. Let f be any element of $\mathcal{C}_1$ that left-divides g. As $\mathcal{S}^\sharp \cap \mathcal{C}_1$ generates $\mathcal{C}_1$, there exists p such that f can be expressed as $s_1 \cdots s_p$ with $s_1, ..., s_p$ in $\mathcal{S}^\sharp \cap \mathcal{C}_1$. By (1.27), f left-divides $\theta_p(g)$, hence $\theta_q(g)$, and the latter is a $\mathcal{C}_1$-head of g.

Assume now that $\mathcal{S}^\sharp$ is closed under left-divisor, the second domino rule is valid for $\mathcal{S}$, and $\mathcal{C}_1$ is compatible with $\mathcal{S}$. Let q be the $\mathcal{S}$-length of g. By Proposition III.1.62 (length II), if f is any left-divisor of g, the $\mathcal{S}$-length of f is at most q. If, in addition f lies in $\mathcal{C}_1$, it admits a strict $(\mathcal{S} \cap \mathcal{C}_1)$-normal decomposition of length at most q. Hence, by (1.27) again, f left-divides $\theta_q(g)$, and the latter is a $\mathcal{C}_1$-head of g. □

We conclude this subsection with intersections of head subcategories.

Proposition 1.28 (intersection of head subcategories). *If $\mathcal{C}$ is a left-cancellative category that is left-Noetherian, every intersection of head subcategories of $\mathcal{C}$ is a head subcategory of $\mathcal{C}$.*

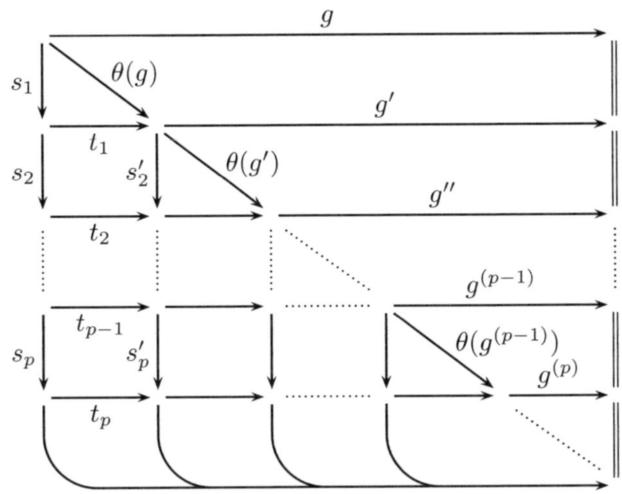

Figure 1. Proof of Proposition 1.25: a sequence of approximations of the expected head is constructed inductively; if the sequence eventually stabilizes, it provides an actual head.

Proof. Assume that $\{\mathcal{C}_i \mid i \in I\}$ is a family of head subcategories of $\mathcal{C}$, and let $\mathcal{C}_* = \bigcap \mathcal{C}_i$. By Proposition 1.21, the families $\mathcal{O}bj(\mathcal{C}_i)$ all coincide with $\mathcal{O}bj(\mathcal{C})$, and $1_\mathcal{C}$ is included in $\mathcal{C}_*$, which is certainly nonempty.

First, assume $\epsilon \in \mathcal{C}^\times \cap \mathcal{C}_*$. Then ϵ belongs to $\mathcal{C}^\times \cap \mathcal{C}_i$ for every i, so the assumption that $\mathcal{C}_i$ is a head subcategory implies that ϵ^{-1} belongs to $\mathcal{C}_i$. Hence ϵ^{-1} belongs to $\mathcal{C}_*$, and $\mathcal{C}_*$ is closed under inverse.

It remains to prove that every element g of $\mathcal{C}$ admits a $\mathcal{C}_*$-head. For simplicity, let us first assume that I is finite with two elements, say $I = \{1, 2\}$, that is, $\mathcal{C}_* = \mathcal{C}_1 \cap \mathcal{C}_2$. We inductively define a sequence $(g_i)_{i \geqslant 0}$ by $g_0 = g$ and the condition that g_i is a $\mathcal{C}_1$-head (*resp.* a $\mathcal{C}_2$-head) of g_{i-1} for i odd (*resp.* even) and positive. As $\mathcal{C}$ is left-Noetherian, the sequence $(g_i)_{i \geqslant 0}$, which is $\preccurlyeq$-decreasing, must be eventually constant: there exists m satisfying $g_m = g_{m+1} = g_{m+2}$, which implies that g_m belongs both to $\mathcal{C}_1$ and to $\mathcal{C}_2$, hence to $\mathcal{C}_*$. Now assume that h lies in $\mathcal{C}_*$ and $h \preccurlyeq g$ holds. We show using induction on i that $h \preccurlyeq g_i$ holds for every i. For $i = 0$, this is the assumption $h \preccurlyeq g$. For i positive and odd (*resp.* even), the induction hypothesis $h \preccurlyeq g_{i-1}$ plus the assumption that h belongs to $\mathcal{C}_1$ (*resp.* $\mathcal{C}_2$) plus the assumption that g_i is a $\mathcal{C}_1$-head (*resp.* a $\mathcal{C}_2$-head) of g_{i-1} imply $h \preccurlyeq g_i$. In particular, we deduce $h \preccurlyeq g_m$, and g_m is a $\mathcal{C}_*$-head of g.

Assume now that I is arbitrary, finite or infinite. The argument is completely similar. However, if I is infinite, we have to appeal to an ordinal induction. The point is that, as above, we need an inductive construction in which each family $\mathcal{C}_i$ appears infinitely many times. So, at the expense of possibly using the Axiom of Choice, we fix a redundant enumeration of the family $\{\mathcal{C}_i \mid i \in I\}$ as a well-ordered sequence $(\mathcal{C}_\alpha)_{\alpha<\theta}$ indexed by ordinals so that, for every i, the indices α satisfying $\mathcal{C}_\alpha = \mathcal{C}_i$ are unbounded in θ. Then we inductively define a sequence $(g_\alpha)_{\alpha<\theta}$ so that $g_0 = g$ holds, $g_{\alpha+1}$ is a $\mathcal{C}_\alpha$-head of g_α for each α, and, for λ limit, g_λ is a $\preccurlyeq$-smallest element of $\{g_\alpha \mid \alpha < \lambda\}$, which exists since the category $\mathcal{C}$ is left-Noetherian.

Let g_θ be a $\preccurlyeq$-smallest element of the sequence $(g_\alpha)_{\alpha<\theta}$, which exists by left-Noetherianity again. Then g_θ lies in $\mathcal{C}_*$: indeed, by construction, there exists $\theta_0 < \theta$ such that $\alpha > \theta_0$ implies $g_\alpha = g_\theta$ and, on the other hand, for each i in I, there exists $\alpha > \theta_0$ such that $\mathcal{C}_\alpha$ equals $\mathcal{C}_i$, and so g_θ lies in $\mathcal{C}_i$.

Finally, assume that h lies in $\mathcal{C}_*$ and $h \preccurlyeq g$ holds. We show using ordinal induction that $h \preccurlyeq g_\alpha$ holds for every α. For $\alpha = 0$, this is the assumption $h \preccurlyeq g$. Next, the assumption that h lies in $\mathcal{C}_*$, hence in $\mathcal{C}_\alpha$ and the induction hypothesis $h \preccurlyeq g_\alpha$ imply $h \preccurlyeq g_{\alpha+1}$. Finally, if λ is a limit ordinal, we have $g_\lambda = \inf_{\alpha<\lambda} g_\alpha$, and the induction hypothesis $h \preccurlyeq g_\alpha$ for $\alpha < \lambda$ implies $h \preccurlyeq g_\lambda$. So we deduce $h \preccurlyeq g_\theta$, and g_θ is a $\mathcal{C}_*$-head of g. □

Note that, even in the case of a finite intersection, the above argument demands that every subcategory be repeated infinitely many times and, therefore, the Noetherianity assumption is needed to guarantee the existence of the expected infima.

Corollary 1.29 (smallest head subcategory). *If $\mathcal{C}$ is a left-cancellative category that is left-Noetherian, then, for every subfamily $\mathcal{S}$ of $\mathcal{C}$, there exists a smallest head subcategory of $\mathcal{C}$ including $\mathcal{S}$.*

Proof. The family of all head subcategories of $\mathcal{C}$ that include $\mathcal{S}$ is nonempty since $\mathcal{C}$ is a head subcategory of itself and it includes $\mathcal{S}$. By Proposition 1.28, the intersection of this nonempty family is a head subcategory of $\mathcal{C}$ that includes $\mathcal{S}$ and, by construction, it is included in every such subcategory. □

1.4 Parabolic subcategories

By merging the previously considered properties, we obtain our last notion, namely that of a parabolic subcategory: the principle is that such a subcategory enjoys all desirable properties and, therefore, we can expect optimal compatibility results.

Definition 1.30 (parabolic subcategory). A *parabolic* subcategory of a left-cancellative category $\mathcal{C}$ is a head subcategory of $\mathcal{C}$ that is closed under factor.

We recall that f is a factor of g is $g'fg'' = g$ holds for some g', g'': so being closed under factor is equivalent to being closed both under left-divisor and right-divisor.

Example 1.31 (parabolic subcategory). We saw in Example 1.20 that, for every m, the submonoid $m\mathbb{N}$ of the monoid $(\mathbb{N}, +)$ is a head submonoid; however, it is not parabolic since (for $m \geqslant 2$) it is not closed under factor: 1 is a divisor of m that does not lie in $m\mathbb{N}$. On the other hand, in the free abelian monoid of rank n based on $\{\mathtt{a}_1, ..., \mathtt{a}_n\}$ (Reference Structure 1, page 3), the submonoid M_I generated by $\{\mathtt{a}_i \mid i \in I\}$ is, for I a nonempty subfamily of $\{1, ..., n\}$, a parabolic submonoid: the M_I-head of an element $\prod_1^n \mathtt{a}_i^{e_i}$ is $\prod_{i\in I} \mathtt{a}_i^{e_i}$, and M_I is closed under left- and right-divisor since, with obvious notation, $e_i + e_i' = 0$ for $i \notin I$ implies $e_i = e_i' = 0$ for $i \notin I$.

More interesting is the submonoid B_I^+ generated by $\{\sigma_i \mid i \in I\}$ in the braid monoid B_n^+ (Reference Structure 2, page 5). We already observed in Example 1.12 that B_I^+ is closed under left- and right-divisor, and in Example 1.24 that B_I^+ is a head submonoid of B_n^+. Hence, B_I^+ is a parabolic submonoid of B_n^+. Actually, every parabolic submonoid of B_n^+ is of the form B_I^+ for some I. Indeed, assume that M is a parabolic submonoid of B_n^+. Let I be the family of all indices i such that at least one element of M admits a decomposition containing σ_i. Then M is included in B_I^+. Now, as M is closed under factor, every element σ_i with i in I must lie in M, and, therefore, M includes the submonoid B_I^+ generated by such elements σ_i. We deduce the equality $M = B_I^+$.

Proposition 1.32 (parabolic subcategory). *Assume that $\mathcal{C}_1$ is a subcategory of a left-cancellative category $\mathcal{C}$.*

(i) *If $\mathcal{C}_1$ is parabolic in $\mathcal{C}$, then $\mathcal{C}_1$ is closed under right-comultiple and factor.*

(ii) *If $\mathcal{C}$ is right-Noetherian, the implication of* (i) *is an equivalence.*

Proof. The results follow directly from Proposition 1.21. For (i), if $\mathcal{C}_1$ is parabolic, it is a head subcategory of $\mathcal{C}$ and Proposition 1.21(i) states that $\mathcal{C}_1$ is closed under right-comultiple; on the other hand, $\mathcal{C}_1$ is closed under factor in $\mathcal{C}$ by definition. For (ii), if $\mathcal{C}$ is closed under right-comultiple and factor, it is closed under right-quotient and, if $\mathcal{C}$ is right-Noetherian, Proposition 1.21(ii) states that $\mathcal{C}_1$ is a head subcategory of $\mathcal{C}$, hence a parabolic subcategory when the assumption that $\mathcal{C}_1$ is closed under factor is added. ☐

As an application of the above characterization, we obtain

Proposition 1.33 (parabolic, bounded case). *Assume that $\mathcal{C}$ is a cancellative right-Noetherian category, Δ is a Garside map in $\mathcal{C}$, and Δ_1 is a partial map from $\mathcal{O}bj(\mathcal{C})$ to $\mathcal{C}$ such that $\Delta_1(x) \preccurlyeq \Delta(x)$ holds whenever $\Delta_1(x)$ is defined. Then the subcategory $\mathcal{C}_1$ of $\mathcal{C}$ generated by $\mathcal{D}iv(\Delta_1)$ is parabolic whenever $\mathcal{D}iv(\Delta_1) = \widetilde{\mathcal{D}iv}(\Delta_1) = \mathcal{D}iv(\Delta) \cap \mathcal{C}_1$ holds.*

Proof. By Proposition 1.32, it suffices to show that $\mathcal{C}_1$ is closed under right-comultiple, right-divisor, and left-divisor in $\mathcal{C}$. Put $\mathcal{S}_1 = \mathcal{D}iv(\Delta_1)$.

We first show that $\mathcal{C}_1$ is closed under right-comultiple in $\mathcal{C}$. So assume that f, g lie in $\mathcal{C}_1$ and there exists an equality $f\hat{g} = g\hat{f}$ in $\mathcal{C}$. Write $f = s_1 \cdots s_p$ and $g = t_1 \cdots t_q$ with $s_1, \ldots, t_q \in \mathcal{S}_1$. As $\mathcal{S}_1$ is included in $\mathcal{D}iv(\Delta)$ and the latter is a Garside family, there exists a rectangular grid as in Proposition IV.1.15 (factorization grid) and Figure IV.2 so that the edges and the diagonals of the squares all lie in $\mathcal{D}iv(\Delta)$. We claim that these edges and diagonals can be chosen to all lie in $\mathcal{S}_1$. Indeed, consider the case of s_1 and t_1: if $s_1 g_1 = t_1 f_1$ holds with s_1, t_1 in $\mathcal{C}(x, \cdot)$, then the left-gcd r_1 of $\Delta_1(x)$ and $s_1 g_1$ is a common right-multiple of s_1 and t_1 that left-divides $s_1 g_1$ and lies in $\mathcal{D}iv(\Delta_1)$. Moreover, the elements s'_1 and t'_1 satisfying $s_1 t'_1 = t_1 s'_1 = r_1$ right-divide r_1, which lies in $\mathcal{D}iv(\Delta_1)$, hence in $\widetilde{\mathcal{D}iv}(\Delta_1)$, so s'_1 and t'_1 lie in $\mathcal{S}_1$. Iteratively constructing the grid in this way taking left-gcd with Δ_1 provides the desired result. Then we found a common right-multiple of f and g that lies in $\mathcal{C}_1$ and divides $f\hat{g}$. Moreover, the quotients also lie in $\mathcal{C}_1$, so $\mathcal{C}_1$ is closed under right-comultiple in $\mathcal{C}$.

Next, we show that $\mathcal{C}_1$ is closed under factor in $\mathcal{C}$. Assume that we have $fg = h$ in $\mathcal{C}$, with $h \in \mathcal{C}_1$. As $\mathcal{D}iv(\Delta)$ generates $\mathcal{C}$, we write $f = s_1 \cdots s_p$, $g = s_{p+1} \cdots s_q$ with $s_1, ..., s_q$ in $\mathcal{D}iv(\Delta)$ and $h = t_1 \cdots t_r$ with $t_1, ..., t_r$ in $\mathcal{D}iv(\Delta_1)$ ($= \mathcal{S}_1$). As above, we construct a rectangular grid where all edges and diagonals lie in $\mathcal{D}iv(\Delta)$. Moreover, by assumption, $\mathcal{C}^\times$ is included in $\mathcal{D}iv(\Delta)$ and $\mathcal{C}^\times \cap \mathcal{C}_1$ is included in $\mathcal{D}iv(\Delta_1)$, so we may assume that the right and bottom edges of the large square are identity elements. Then, starting from the top-right corner and progressing toward the bottom-left corner, we inductively deduce that all edges and diagonals correspond to elements of $\mathcal{S}_1$. Indeed, in each elementary square 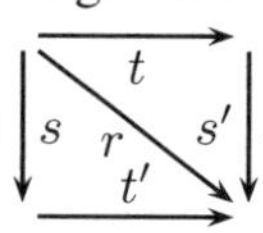, the assumption that t and s' lie in $\mathcal{S}_1$ implies that r lies in $\mathcal{C}_1$, hence in $\mathcal{S}_1$ as we have $\mathcal{S}_1 = \mathcal{D}iv(\Delta) \cap \mathcal{C}_1$, and it implies in turn that s and t' lie in $\mathcal{S}_1$ as we have $\mathcal{D}iv(\Delta_1) = \widetilde{\mathcal{D}iv}(\Delta_1) = \mathcal{D}iv(\Delta) \cap \mathcal{C}_1$. So, in particular, all elements s_i lie in $\mathcal{S}_1$, and both f and g lie in $\mathcal{C}_1$. So $\mathcal{C}_1$ is closed under factor in $\mathcal{C}$ and, by Proposition 1.32, $\mathcal{C}_1$ is a parabolic subcategory of $\mathcal{C}$. □

A typical situation where Proposition 1.33 applies is the submonoid B_I^+ of B_n^+ as in Example 1.31: if Δ_I is the right-lcm of the elements σ_i with i in I, then B_I^+ is generated by the divisors of Δ_I, and $\mathrm{Div}(\Delta_I) = \widetilde{\mathrm{Div}}(\Delta_I) = \mathrm{Div}(\Delta) \cap B_I^+$ is satisfied (in this case, we already observed that B_I^+ is parabolic without appealing to Proposition 1.33).

Remark 1.34. Proposition 1.33 is not optimal: its conclusion remains valid even if the right-Noetherianity assumption is skipped. Indeed, the latter is used in Propositions 1.21 and 1.32 to guarantee the existence of a head. But, in the context of Proposition 1.33, one can obtain a $\mathcal{C}_1$-head for an element g of $\mathcal{C}(x, \cdot)$ satisfying $\sup_\Delta(g) = p$ by taking the left-gcd of g with $\Delta_1^{[p]}$. Moreover, one could show that Δ_1 is a Garside map in $\mathcal{C}_1$.

We conclude this subsection with the closure under intersection of the family of parabolic subcategories.

Proposition 1.35 (intersection of parabolic). *If $\mathcal{C}$ is a left-cancellative and left-Noetherian category, every intersection of parabolic subcategories of $\mathcal{C}$ is a parabolic subcategory of $\mathcal{C}$.*

Proof. Proposition 1.28 states that every intersection of parabolic subcategories is a head subcategory. Moreover, it is obvious that an intersection of subcategories that are closed under factor is closed under factor. □

Then, exactly as in Corollary 1.29, we deduce:

Corollary 1.36 (smallest parabolic). *If $\mathcal{C}$ is a left-cancellative category that is left-Noetherian, then, for every subfamily $\mathcal{S}$ of $\mathcal{C}$, there exists a smallest parabolic category of $\mathcal{C}$ including $\mathcal{S}$.*

2 Compatibility with a Garside family

We now investigate subcategories in the context of a category equipped with a Garside family. A natural notion of compatibility arises when the normal decompositions of an element of the subcategory computed in the ambient category and in the subcategory coincide. The main results of the section provide various necessary and sufficient conditions for this type compatibility to occur.

The section is organized as follows. In Subsection 2.1, we establish a few general relations connecting $\mathcal{S}$- and $(\mathcal{S} \cap \mathcal{C}_1)$-greediness when $\mathcal{C}_1$ is a subcategory. In Subsection 2.2, we introduce a notion of compatibility connecting a subcategory and a Garside family, and give criteria for establishing such a compatibility. In Subsection 2.3, we consider compatibility for special subcategories, typically $=^{\times}$-closed subcategories. Finally, in Subsection 2.4, we consider the signed case and similarly discuss what we call strong compatibility between a subcategory and a strong Garside family.

2.1 Greedy paths

If $\mathcal{C}$ is a left-cancellative category and $\mathcal{C}_1$ is a subcategory of $\mathcal{C}$ that is closed under right-quotient, then, for every subfamily $\mathcal{S}$ of $\mathcal{C}$, there exist connections between the $\mathcal{C}$-paths that are $\mathcal{S}$-greedy and the $\mathcal{C}_1$-paths that are $(\mathcal{S} \cap \mathcal{C}_1)$-greedy. Here we analyze such connections.

Lemma 2.1. *Assume that $\mathcal{C}$ is a left-cancellative category, $\mathcal{S}$ is included in $\mathcal{C}$, and $\mathcal{C}_1$ is a subcategory of $\mathcal{C}$ that is closed under right-quotient in $\mathcal{C}$.*

(i) *Put $\mathcal{S}_1 = \mathcal{S} \cap \mathcal{C}_1$. Every $\mathcal{C}_1$-path that is $\mathcal{S}$-greedy in $\mathcal{C}$ is $\mathcal{S}_1$-greedy in $\mathcal{C}_1$.*

(ii) *Put $\mathcal{S}_1^\sharp = \mathcal{S}_1\mathcal{C}_1^\times \cup \mathcal{C}_1^\times$. If an element g of $\mathcal{C}_1$ admits an $\mathcal{S}$-normal decomposition with entries in $\mathcal{S}_1^\sharp$, every decomposition of g in $\mathcal{C}_1$ that is $\mathcal{S}_1$-normal in $\mathcal{C}_1$ is $\mathcal{S}$-normal in $\mathcal{C}$.*

(iii) *Assume moreover that $\mathcal{C}_1$ is closed under factor in $\mathcal{C}$ and $\mathcal{S}^\sharp$ is closed under right-complement in $\mathcal{C}$. Then a $\mathcal{C}_1$-path is $\mathcal{S}$-greedy in $\mathcal{C}$ if and only if it is $\mathcal{S}_1$-greedy in $\mathcal{C}_1$.*

Proof. (i) Assume that $g_1|g_2$ is $\mathcal{S}$-greedy in $\mathcal{C}$ and g_1, g_2 lie in $\mathcal{C}_1$. Assume that s belongs to $\mathcal{S}_1$ and it left-divides fg_1g_2 in $\mathcal{C}_1$, that is, we have $sf' = fg_1g_2$ for some f' lying in $\mathcal{C}_1$. Then, a fortiori, s left-divides fg_1g_2 in $\mathcal{C}$ so, as $g_1|g_2$ is $\mathcal{S}$-greedy in $\mathcal{C}$, there exists f'' in $\mathcal{C}$ satisfying $sf'' = fg_1$. Now, fg_1 belongs to $\mathcal{C}_1$ since f and g_1 do, and so does s. As $\mathcal{C}_1$ is closed under right-quotient in $\mathcal{C}$, we deduce that f'' belongs to $\mathcal{C}_1$. Hence $g_1|g_2$ is $\mathcal{S}_1$-greedy in $\mathcal{C}_1$.

(ii) Assume that $s_1|\cdots|s_p$ is an $\mathcal{S}$-normal decomposition of g whose entries lie in $\mathcal{S}_1^\sharp$, and let $t_1|\cdots|t_q$ be an arbitrary $\mathcal{S}_1$-normal decomposition of g in $\mathcal{C}_1$. First, by (i), $s_1|\cdots|s_p$ is $\mathcal{S}_1$-greedy in $\mathcal{C}_1$ and, therefore, it is an $\mathcal{S}_1$-normal decomposition of g in $\mathcal{C}_1$. Hence, by Proposition III.1.25 (normal unique), $t_1|\cdots|t_q$ must be a $\mathcal{C}_1^\times$-deformation of $s_1|\cdots|s_p$ in $\mathcal{C}_1$, that is, these sequences are connected by elements $\epsilon_0, ..., \epsilon_{\max(p,q)}$ of $\mathcal{C}_1^\times$. As $\mathcal{C}_1^\times \subseteq \mathcal{C}^\times$ always holds, the elements ϵ_i belong to $\mathcal{C}^\times$, so $t_1|\cdots|t_q$ is also a $\mathcal{C}^\times$-deformation of $s_1|\cdots|s_p$ in $\mathcal{C}$. By Proposition III.1.22 (deformation), this implies that $t_1|\cdots|t_q$ is $\mathcal{S}$-normal in $\mathcal{C}$.

(iii) By (i), every $\mathcal{C}_1$-path that is $\mathcal{S}$-greedy in $\mathcal{C}$ is $\mathcal{S}_1$-greedy in $\mathcal{C}_1$. Conversely, assume that $g_1|\cdots|g_p$ is a $\mathcal{C}_1$-path that is $\mathcal{S}_1$-greedy in $\mathcal{C}_1$. Without loss of generality we may assume $p = 2$. By Lemma IV.1.21, it suffices to prove that $s \preccurlyeq g_1g_2$ implies $s \preccurlyeq g_1$ for every s in $\mathcal{S}$. Now assume $sf = g_1g_2$ in $\mathcal{C}$. As g_1g_2 belongs to $\mathcal{C}_1$ and the latter is closed under factor, s and f must lie in $\mathcal{C}_1$, and s belongs to $\mathcal{S}_1$. The assumption that $g_1|g_2$ is $\mathcal{S}_1$-greedy in $\mathcal{C}_1$ implies the existence of f', in $\mathcal{C}_1$ hence in $\mathcal{C}$, satisfying $sf' = g_1$. So $g_1|g_2$ is $\mathcal{S}$-greedy in $\mathcal{C}$. □

See Exercise 73 for a similar result under the assumption that $\mathcal{C}_1$ is closed under left-quotient in $\mathcal{C}$. As the following example shows, the assumption that $\mathcal{C}_1$ is closed under right-quotient is necessary in Lemma 2.1, even in the case when $\mathcal{S}$ is a Garside family.

Example 2.2 (greedy). Consider the dual braid monoid B_3^{+*} (Reference Structure 3, page 10). We saw that B_3^{+*} admits the presentation $\langle \mathtt{a}, \mathtt{b}, \mathtt{c} \mid \mathtt{ab} = \mathtt{bc} = \mathtt{ca}\rangle^+$, and that $\mathtt{ab}$ is a Garside element in B_3^{+*}, written Δ_3^*. The associated Garside family $\mathcal{S}$ has five elements, namely 1, $\mathtt{a}$, $\mathtt{b}$, $\mathtt{c}$, and Δ_3^*. Let M be the submonoid of B_3^{+*} generated by $\mathtt{a}$ and $\mathtt{b}$. Put $\mathcal{S}_1 = \mathcal{S} \cap M$. Then $\mathcal{S}_1$ consists of 1, $\mathtt{a}$, $\mathtt{b}$, and Δ_3^*. Consider the path $\Delta_3^*|\mathtt{a}$, which belongs to $M^{[2]}$. As $\mathcal{S}$ is bounded by Δ_3^*, the path $\Delta_3^*|\mathtt{a}$ is $\mathcal{S}$-greedy in B_3^{+*}. On the other hand, in M, we have $\mathtt{b}\Delta_3^* = \Delta_3^*\mathtt{a}$, whence $\mathtt{b} \preccurlyeq \Delta_3^*\mathtt{a}$, but $\mathtt{b}$ does not left-divide Δ_3^* since $\mathtt{c}$ is not an element of M. Hence $\Delta_3^*|\mathtt{a}$ is not $\mathcal{S}_1$-greedy in B_3^{+*}.

What is missing in Lemma 2.1 to obtain an optimal connection between the notions of a $\mathcal{S}$-normal path in $\mathcal{C}$ and a $\mathcal{S}_1$-normal path in $\mathcal{C}_1$ is that an entry of an $\mathcal{S}$-normal path that lies in $\mathcal{C}_1$ need not lie in $\mathcal{S}_1^\sharp$, that is, in $\mathcal{S}_1\mathcal{C}_1^\times \cup \mathcal{C}_1^\times$, because relation (1.5) need not be an equality in general. If we introduce the latter as an additional assumption, the statement becomes very simple.

Lemma 2.3. *Assume that $\mathcal{S}$ is a subfamily of a left-cancellative category $\mathcal{C}$ and $\mathcal{C}_1$ is a subcategory of $\mathcal{C}$ that is closed under right-quotient in $\mathcal{C}$ and satisfies $\mathcal{S}^\sharp \cap \mathcal{C}_1 \subseteq \mathcal{S}_1^\sharp$, where we put $\mathcal{S}_1 = \mathcal{S} \cap \mathcal{C}_1$ and $\mathcal{S}_1^\sharp = \mathcal{S}_1\mathcal{C}_1^\times \cup \mathcal{C}_1^\times$.*

(i) *Every $\mathcal{C}_1$-path that is $\mathcal{S}$-normal in $\mathcal{C}$ is $\mathcal{S}_1$-normal in $\mathcal{C}_1$.*

(ii) *If an element g of $\mathcal{C}_1$ admits at least one $\mathcal{S}$-normal decomposition with entries in $\mathcal{C}_1$, then every $\mathcal{S}_1$-normal decomposition of g in $\mathcal{C}_1$ is $\mathcal{S}$-normal in $\mathcal{C}$.*

Proof. (i) Assume that $s_1|\cdots|s_p$ is an $\mathcal{S}$-normal path with entries in $\mathcal{C}_1$. By definition, $s_1|\cdots|s_p$ is $\mathcal{S}$-greedy, hence, by Lemma 2.1, it is $\mathcal{S}_1$-greedy. Moreover, its entries lie in $\mathcal{S}^\sharp \cap \mathcal{C}_1$, hence, by assumption, in $\mathcal{S}_1^\sharp$. So $s_1|\cdots|s_p$ is a $\mathcal{S}_1$-normal path in $\mathcal{C}_1$.

(ii) Owing to the inclusion of $\mathcal{S}^\sharp \cap \mathcal{C}_1$ in $\mathcal{S}_1^\sharp$, the $\mathcal{S}$-normal decomposition of g whose existence is assumed has its entries in $\mathcal{S}_1^\sharp$. Then we apply Lemma 2.1(ii). □

2.2 Compatibility with a Garside family

According to Lemmas 2.1 and 2.3, being normal in the ambient category implies being normal in the subcategory, at least in the case of a subcategory that is closed under right-quotient. We shall be interested in the case when this implication is an equivalence.

Definition 2.4 (compatible subcategory). A subcategory $\mathcal{C}_1$ of a left-cancellative category $\mathcal{C}$ is said to be *compatible with* a Garside family $\mathcal{S}$ of $\mathcal{C}$ if, putting $\mathcal{S}_1 = \mathcal{S} \cap \mathcal{C}_1$,

(2.5) The family $\mathcal{S}_1$ is a Garside family in $\mathcal{C}_1$, and

(2.6) A $\mathcal{C}_1$-path is $\mathcal{S}_1$-normal in $\mathcal{C}_1$ if and only if it is $\mathcal{S}$-normal in $\mathcal{C}$.

Example 2.7 (compatible subcategory). As in Example II.1.7, consider the free (Abelian) monoid $(\mathbb{N}, +)$ and its submonoid $2\mathbb{N}$. Let S_m be the Garside family $\{0, ..., m\}$ of $\mathbb{N}$. First, we have $S_1 \cap 2\mathbb{N} = \{0\}$, so $S_1 \cap 2\mathbb{N}$, which does not generate $2\mathbb{N}$, is not a Garside family in $2\mathbb{N}$, and $2\mathbb{N}$ is not compatible with S_1. However, (2.6) is vacuously satisfied, since the only $2\mathbb{N}$-sequences with entries in $S_1 \cap 2\mathbb{N}$ are the sequences $0|\cdots|0$, which are both S_1-normal in $\mathbb{N}$ and $(S_1 \cap 2\mathbb{N})$-normal in $2\mathbb{N}$. So, (2.6) does not imply (2.5).

Next, assume $m \geqslant 2$. Then we have $S_m \cap 2\mathbb{N} = \{0, 2, ..., 2n\}$ with $n = \lfloor m/2 \rfloor$, so $S_m \cap 2\mathbb{N}$ is a Garside family in $2\mathbb{N}$. Hence (2.5) holds. If m is odd, then $(m-1)|2$ is $(S_m \cap 2\mathbb{N})$-normal in $2\mathbb{N}$, whereas it is not S_m-normal in $\mathbb{N}$, since the S_m-normal decomposition of $m + 1$ in $\mathbb{N}$ is $m|1$. Hence, $2\mathbb{N}$ is not compatible with S_m in this case. This shows that (2.5) does not imply (2.6).

Finally, assume that m is even. Then the $2\mathbb{N}$-sequences that are $(S_m \cap 2\mathbb{N})$-normal in $2\mathbb{N}$ are the sequences $2s_1|\cdots|2s_p$, where $s_1|\cdots|s_p$ is S_n-normal in $\mathbb{N}$, and these are also the S_m-normal sequences in $\mathbb{N}$ whose entries lie in $2\mathbb{N}$. Thus, in this case, (2.6) holds and $2\mathbb{N}$ is compatible with S_m.

See Exercise 74 for further examples. Note that, viewed as a subcategory of itself, a category is always compatible with each of its Garside families.

Lemma 2.8. *Assume that $\mathcal{S}$ is a Garside family in a left-cancellative category $\mathcal{C}$, and $\mathcal{C}_1$ is a subcategory of $\mathcal{C}$ that is compatible with $\mathcal{S}$. Then $\mathcal{S}^\sharp \cap \mathcal{C}_1 = (\mathcal{S} \cap \mathcal{C}_1)\mathcal{C}_1^\times \cup \mathcal{C}_1^\times$ holds.*

Proof. Put $\mathcal{S}_1 = \mathcal{S} \cap \mathcal{C}_1$. We saw in Lemma 1.3 that $\mathcal{S}_1\mathcal{C}_1^\times \cup \mathcal{C}_1^\times \subseteq \mathcal{S}^\sharp \cap \mathcal{C}_1$ always holds. On the other hand, assume $s \in \mathcal{S}^\sharp \cap \mathcal{C}_1$. Then the length-one path s is $\mathcal{S}$-normal, hence $\mathcal{S}_1$-normal by the definition of a compatible subcategory, which, by the definition of an $\mathcal{S}_1$-normal path, implies that its entry lies in $\mathcal{S}_1\mathcal{C}_1^\times \cup \mathcal{C}_1^\times$. We deduce that $\mathcal{S}^\sharp \cap \mathcal{C}_1$ is included in $\mathcal{S}_1\mathcal{C}_1^\times \cup \mathcal{C}_1^\times$. □

Although we are mostly interested in subcategories that are closed under right-quotient in the ambient category, the property is not included in Definition 2.4. The reason is that the property is automatically true provided some weak closure is satisfied.

Lemma 2.9. *If $\mathcal{S}$ is a Garside family in a left-cancellative category $\mathcal{C}$ and $\mathcal{C}_1$ is a subcategory of $\mathcal{C}$ that is compatible with $\mathcal{S}$, then the following statements are equivalent whenever $\mathcal{S} \cap \mathcal{C}_1$ is solid in $\mathcal{C}_1$:*

(i) *The subcategory $\mathcal{C}_1$ is closed under right-quotient in $\mathcal{C}$.*

(ii) *The subfamily $\mathcal{S} \cap \mathcal{C}_1$ is closed under right-quotient in $\mathcal{S}$.*

Proof. Put $\mathcal{S}_1 = \mathcal{S} \cap \mathcal{C}_1$. Assume $t = st'$ with s, t in $\mathcal{S}_1$ and t' in $\mathcal{S}$. Then s, t belong to $\mathcal{C}_1$, hence, if $\mathcal{C}_1$ is closed under right-quotient in $\mathcal{C}$, t' belongs to $\mathcal{C}_1$ as well, hence it belongs to $\mathcal{S}_1$. So (i) implies (ii).

Conversely, assume (ii). Assume that $g = fg'$ holds in $\mathcal{C}$ with f, g in $\mathcal{C}_1$ and g' in $\mathcal{C}$. As $\mathcal{S}_1$ is a solid Garside family in $\mathcal{C}_1$, it generates $\mathcal{C}_1$. We prove $g' \in \mathcal{C}_1$ using induction on the minimal length p of a decomposition of f into a product of elements of $\mathcal{S}_1$. For $p = 0$, that is, if f is an identity element, we have $g' = g \in \mathcal{C}_1$. Assume first $p = 1$, that is, f belongs to $\mathcal{S}_1$. As $\mathcal{S}_1$ is a solid Garside family in $\mathcal{C}$, the element g of $\mathcal{C}_1$ admits in $\mathcal{C}_1$ an $\mathcal{S}_1$-normal decomposition with entries in $\mathcal{S}_1$, say $s_1|\cdots|s_p$. By (2.6), the latter path is $\mathcal{S}$-normal in $\mathcal{C}$. Now f is an element of $\mathcal{S}_1$, hence of $\mathcal{S}$, that left-divides $s_1 \cdots s_p$ in $\mathcal{C}$. As $s_1|\cdots|s_p$ is $\mathcal{S}$-greedy, f must left-divide s_1, say $s_1 = fs_1'$. Now, s_1 and f belong to $\mathcal{S}_1$, so, by (ii), s_1' must belong to $\mathcal{S}_1$. As $\mathcal{C}$ is left-cancellative, we have $g' = s_1's_2 \cdots s_p$, and g' belongs to $\mathcal{C}_1$. Assume finally $p \geqslant 2$. Write $f = sf'$ with s in $\mathcal{S}_1$ and f' admitting an $\mathcal{S}_1$-decomposition of length $p - 1$. Then we have $g = s(f'g')$, and the induction hypothesis implies $f'g' \in \mathcal{C}_1$ first, and then $g' \in \mathcal{C}_1$. □

In Lemma 2.9, the assumption that $\mathcal{S} \cap \mathcal{C}_1$ is solid in $\mathcal{C}_1$ can be skipped at the expense of replacing $\mathcal{S}$ with $\mathcal{S}^\sharp$: if $\mathcal{S} \cap \mathcal{C}_1$ is a Garside family in $\mathcal{C}_1$, then $\mathcal{S}^\sharp \cap \mathcal{C}_1$, which, by Lemma 2.8, is $(\mathcal{S} \cap \mathcal{C}_1)\mathcal{C}_1^\times \cup \mathcal{C}_1^\times$, is a solid Garside family in $\mathcal{C}_1$.

We shall now provide several criteria for recognizing compatibility. The first one is obtained by simply removing from the definition some superfluous conditions.

Proposition 2.10 (recognizing compatible I). *If $\mathcal{S}$ is a Garside family in a left-cancellative category $\mathcal{C}$ and $\mathcal{C}_1$ is a subcategory of $\mathcal{C}$ that is closed under right-quotient in $\mathcal{C}$, then $\mathcal{C}_1$ is compatible with $\mathcal{S}$ if and only if the following conditions are satisfied:*

(2.11) *Putting $\mathcal{S}_1 = \mathcal{S} \cap \mathcal{C}_1$ and $\mathcal{S}_1^\sharp = \mathcal{S}_1\mathcal{C}_1^\times \cup \mathcal{C}_1^\times$, we have $\mathcal{S}^\sharp \cap \mathcal{C}_1 \subseteq \mathcal{S}_1^\sharp$,*

(2.12) *Every element of $\mathcal{C}_1$ admits an $\mathcal{S}$-normal decomposition with entries in $\mathcal{C}_1$.*

Proof. Assume that $\mathcal{C}_1$ is compatible with $\mathcal{S}$. First, by Lemma 2.8, we have $\mathcal{S}^\sharp \cap \mathcal{C}_1 = \mathcal{S}_1^\sharp$, so (2.11) holds. Next, let g belong to $\mathcal{C}_1$. By assumption, $\mathcal{S}_1$ is a Garside family in $\mathcal{C}_1$, so g admits an $\mathcal{S}_1$-normal decomposition in $\mathcal{C}_1$. By (2.6), the latter is $\mathcal{S}$-normal, and, therefore, it is an $\mathcal{S}$-normal decomposition of g whose entries lie in $\mathcal{C}_1$. So (2.12) is satisfied.

Conversely, assume that $\mathcal{C}_1$ satisfies (2.11) and (2.12). Let g belong to $\mathcal{C}_1$. By (2.12), g admits an $\mathcal{S}$-normal decomposition whose entries lie in $\mathcal{C}_1$. By Lemma 2.1, the latter decomposition is $\mathcal{S}_1$-greedy in $\mathcal{C}_1$. Moreover, its entries lie in $\mathcal{S}^\sharp \cap \mathcal{C}_1$, hence in $\mathcal{S}_1^\sharp$ by (2.11), and therefore it is an $\mathcal{S}_1$-normal decomposition of g in $\mathcal{C}_1$. As such a decomposition exists, $\mathcal{S}_1$ is a Garside family in $\mathcal{C}_1$, that is, (2.5) is satisfied.

Next, if $s_1|\cdots|s_p$ is an $\mathcal{S}$-normal $\mathcal{C}_1$-path, then, by the same argument as above, $s_1|\cdots|s_p$ is $\mathcal{S}_1$-normal in $\mathcal{C}_1$. Conversely, assume that $s_1|\cdots|s_p$ is $\mathcal{S}_1$-normal in $\mathcal{C}_1$. As $\mathcal{S}_1$ is a Garside family in $\mathcal{C}_1$, the element $s_1\cdots s_p$ of $\mathcal{C}_1$ admits an $\mathcal{S}_1$-normal decomposition in $\mathcal{C}_1$. Hence, by Lemma 2.3, $s_1|\cdots|s_p$ must be $\mathcal{S}$-normal in $\mathcal{C}$. So (2.6) is satisfied as well, and $\mathcal{C}_1$ is compatible with $\mathcal{S}$. □

As an application, and for further reference, we deduce

Corollary 2.13 (recognizing compatible, inclusion case). *If $\mathcal{S}$ is a Garside family in a left-cancellative category $\mathcal{C}$ and $\mathcal{C}_1$ is a subcategory of $\mathcal{C}$ that is closed under right-quotient in $\mathcal{C}$ and includes $\mathcal{S}$, then $\mathcal{C}_1$ is compatible with $\mathcal{S}$ and $\mathcal{C} = \mathcal{C}_1\mathcal{C}^\times \cup \mathcal{C}^\times$ holds.*

Proof. First, assume $s \in \mathcal{S}^\sharp \cap \mathcal{C}_1$. If s is invertible, it belongs to $\mathcal{C}^\times \cap \mathcal{C}_1$, which is $\mathcal{C}_1^\times$ since $\mathcal{C}_1$ is closed under right-quotient, hence under inverse, in $\mathcal{C}$. Otherwise, we have $s = t\epsilon$ with t in $\mathcal{S}$ and ϵ in $\mathcal{C}^\times$. As s and t belong to $\mathcal{C}_1$, the assumption that $\mathcal{C}_1$ is closed under right-quotient in $\mathcal{C}$ implies that ϵ belongs to $\mathcal{C}_1$, hence to $\mathcal{C}^\times \cap \mathcal{C}_1$, which is $\mathcal{C}_1^\times$. So, in every case, s belongs to $\mathcal{S}\mathcal{C}_1^\times \cup \mathcal{C}_1^\times$, and (2.11) is satisfied.

Let g belong to $\mathcal{C}_1$. Then g admits a strict $\mathcal{S}$-normal decomposition, say $s_1|\cdots|s_p$. By definition, $s_1, ..., s_{p-1}$ belong to $\mathcal{S}$, hence to $\mathcal{C}_1$, so the assumption that $\mathcal{C}_1$ is closed under right-quotient in $\mathcal{C}$ implies that s_q belongs to $\mathcal{C}_1$ too. So (2.12) is satisfied, and $\mathcal{C}_1$ is compatible with $\mathcal{S}$.

Finally, let g be a non-invertible element of $\mathcal{C}$. By Proposition III.1.23 (strict), g admits a strict $\mathcal{S}$-normal decomposition, say $s_1|\cdots|s_p$. By definition, $s_1, ..., s_{p-1}$ belong to $\mathcal{S}$, hence to $\mathcal{C}_1$, and s_p belongs to $\mathcal{S}^\sharp$, hence to $\mathcal{C}_1\mathcal{C}^\times$. So g belongs to $\mathcal{C}_1\mathcal{C}^\times$. □

Recombining the conditions, we obtain a slightly different criterion.

Proposition 2.14 (recognizing compatible II). *If $\mathcal{S}$ is a Garside family in a left-cancellative category $\mathcal{C}$ and $\mathcal{C}_1$ is a subcategory of $\mathcal{C}$ that is closed under right-quotient in $\mathcal{C}$, then $\mathcal{C}_1$ is compatible with $\mathcal{S}$ if and only if, putting $\mathcal{S}_1 = \mathcal{S} \cap \mathcal{C}_1$ and $\mathcal{S}_1^\sharp = \mathcal{S}_1\mathcal{C}_1^\times \cup \mathcal{C}_1^\times$,*

(2.15) *Every element of $\mathcal{C}_1$ admits an $\mathcal{S}$-normal decomposition with entries in $\mathcal{S}_1^\sharp$.*

Proof. Assume that $\mathcal{C}_1$ is compatible with $\mathcal{S}$. Let g belong to $\mathcal{C}_1$. Then, by (2.12), g admits an $\mathcal{S}$-normal decomposition whose entries lie in $\mathcal{C}_1$, hence in $\mathcal{S}^\sharp \cap \mathcal{C}_1$. By (2.11), the latter entries lie in $\mathcal{S}_1^\sharp$, so (2.15) is satisfied.

Conversely, assume that $\mathcal{C}_1$ satisfies (2.15). Let g belong to $\mathcal{S}^\sharp \cap \mathcal{C}_1$. By (2.15), g has an $\mathcal{S}$-normal decomposition $s_1|\cdots|s_p$ whose entries lie in $\mathcal{S}_1^\sharp$. As g lies in $\mathcal{S}^\sharp$, its $\mathcal{S}$-length is at most one, so $s_2, ..., s_p$ must lie in $\mathcal{C}^\times \cap \mathcal{C}_1$, hence in $\mathcal{C}_1^\times$. So g belongs to $\mathcal{S}_1^\sharp\mathcal{C}_1^\times$, hence to $\mathcal{S}_1^\sharp$, and (2.11) is satisfied. On the other hand, as $\mathcal{S}_1^\sharp$ is included in $\mathcal{C}_1$, (2.12) is satisfied. So, by Proposition 2.10, $\mathcal{C}_1$ is compatible with $\mathcal{S}$. □

Using the inductive construction of normal decompositions, one obtains an alternative criterion that only involves the elements of $(\mathcal{S}_1^\sharp)^2$:

Proposition 2.16 (recognizing compatible III). *If $\mathcal{S}$ is a Garside family in a left-cancellative category $\mathcal{C}$ and $\mathcal{C}_1$ is a subcategory of $\mathcal{C}$ that is closed under right-quotient in $\mathcal{C}$, then $\mathcal{C}_1$ is compatible with $\mathcal{S}$ if and only if, when we put $\mathcal{S}_1 = \mathcal{S} \cap \mathcal{C}_1$ and $\mathcal{S}_1^\sharp = \mathcal{S}_1\mathcal{C}_1^\times \cup \mathcal{C}_1^\times$,*

(2.17) *The family $\mathcal{S}_1^\sharp$ generates $\mathcal{C}_1$.*

(2.18) *Every element of $(\mathcal{S}_1^\sharp)^2$ admits an $\mathcal{S}$-normal decomposition with entries in $\mathcal{S}_1^\sharp$.*

Proposition 2.16 follows from an auxiliary result of independent interest:

Lemma 2.19. *Assume that $\mathcal{S}$ is a Garside family in a left-cancellative category $\mathcal{C}$ and $\mathcal{S}'$ is a subfamily of $\mathcal{S}^\sharp$ such that $\mathcal{S}'(\mathcal{S}' \cap \mathcal{C}^\times) \subseteq \mathcal{S}'$ holds and every element of $\mathcal{S}'^2$ admits an $\mathcal{S}$-normal decomposition with entries in $\mathcal{S}'$. Then every element in the subcategory of $\mathcal{C}$ generated by $\mathcal{S}'$ admits an $\mathcal{S}$-normal decomposition with entries in $\mathcal{S}'$.*

We skip the proofs, both of Lemma 2.19 (based on the inductive construction of an $\mathcal{S}$-normal decomposition using left-multiplication) and of Proposition 2.16 (which then easily follows using Proposition 2.14).

We refer to Exercise 82 for still another variant.

2.3 Compatibility, special subcategories

When the considered subcategories satisfy additional closure assumptions, simpler characterizations of compatibility appear.

We begin with $=^\times$-closed subcategories. If $\mathcal{C}_1$ is a subcategory of $\mathcal{C}$ that is compatible with a Garside family $\mathcal{S}$, every element of $\mathcal{C}_1$ admits an $\mathcal{S}$-normal decomposition with entries in $\mathcal{C}_1$, and the latter is then $(\mathcal{S} \cap \mathcal{C}_1)$-normal in $\mathcal{C}_1$. But, besides, there may exist $\mathcal{S}$-normal decompositions with entries not in $\mathcal{C}_1$. This cannot happen if $\mathcal{C}_1$ is $=^\times$-closed.

Lemma 2.20. *If $\mathcal{S}$ is a Garside family in a left-cancellative category $\mathcal{C}$, and $\mathcal{C}_1$ is an $=^\times$-closed subcategory of $\mathcal{C}$ that is closed under right-quotient and is compatible with $\mathcal{S}$, then every $\mathcal{S}$-normal decomposition of an element of $\mathcal{C}_1$ is $(\mathcal{S} \cap \mathcal{C}_1)$-normal in $\mathcal{C}_1$.*

Proof. As usual, put $\mathcal{S}_1 = \mathcal{S} \cap \mathcal{C}_1$. Let g belong to $\mathcal{C}_1$ and $s_1|\cdots|s_p$ be an $\mathcal{S}$-normal decomposition of g. By assumption, $\mathcal{C}_1$ is compatible with $\mathcal{S}$, so $\mathcal{S}_1$ is a Garside family in $\mathcal{C}_1$, hence g has an $\mathcal{S}_1$-normal decomposition $t_1|\cdots|t_q$ in $\mathcal{C}_1$. By (2.6), $t_1|\cdots|t_q$ is $\mathcal{S}$-normal in $\mathcal{C}$, hence, by Proposition III.1.25 (normal unique), $s_1|\cdots|s_p$ is a $\mathcal{C}^\times$-deformation of $t_1|\cdots|t_q$. Then, by Lemma 1.15(i), $s_1|\cdots|s_p$ must be a $\mathcal{C}_1$-path, that is, its entries lie in $\mathcal{C}_1$ and, therefore, it is $\mathcal{S}_1$-normal in $\mathcal{C}_1$. $\square$

We deduce a simplified characterization of compatibility.

Proposition 2.21 (recognizing compatible, $=^{\times}$-closed case). *If $\mathcal{S}$ is a Garside family in a left-cancellative category $\mathcal{C}$ and $\mathcal{C}_1$ is a subcategory of $\mathcal{C}$ that is closed under right-quotient and $=^{\times}$-closed in $\mathcal{C}$, then $\mathcal{C}_1$ is compatible with $\mathcal{S}$ if and only if* (2.12) *holds, that is, if and only if every element of $\mathcal{C}_1$ admits an $\mathcal{S}$-normal decomposition with entries in $\mathcal{C}_1$.*

Proof. By Lemma 1.15, (2.11) is automatically satisfied whenever $\mathcal{C}_1$ is $=^{\times}$-closed. Then the result directly follows from Proposition 2.10. □

We observed above that a subcategory that contains all invertible elements of the ambient category is $=^{\times}$-closed, so Proposition 2.21 applies in particular in the case when $\mathcal{C}_1$ includes $\mathcal{C}^{\times}$. On the other hand, owing to Lemma 1.16, every subcategory that is closed under left- or under right-divisor is closed under $=^{\times}$. When both conditions are met, Proposition 2.21 leads to the best result one can expect:

Corollary 2.22 (compatible, closed under factor case). (i) *If $\mathcal{C}$ is a left-cancellative category, every subcategory of $\mathcal{C}$ that is closed under factor in $\mathcal{C}$ is compatible with every Garside family of $\mathcal{C}$.*

(ii) *The compatibility result of* (i) *applies, in particular, to every parabolic subcategory of $\mathcal{C}$ and, if $\mathcal{C}$ admits a balanced presentation $(\mathcal{S}, \mathcal{R})$, to every subcategory of $\mathcal{C}$ generated by a subfamily of $\mathcal{S}$.*

Proof. (i) Assume that $\mathcal{C}_1$ is a subcategory of $\mathcal{C}$ that is closed under left-divisor and $\mathcal{S}$ is a Garside family in $\mathcal{C}$. By Lemma 1.16, $\mathcal{C}_1$ is $=^{\times}$-closed. So, owing to Proposition 2.21, in order to prove that $\mathcal{C}_1$ is compatible with $\mathcal{S}$, it is enough to show that every element of $\mathcal{C}_1$ admits an $\mathcal{S}$-normal decomposition with entries in $\mathcal{C}_1$. Now let g belong to $\mathcal{C}_1$, and let $s_1|\cdots|s_p$ be an $\mathcal{S}$-normal decomposition of g. By construction, each entry s_i is a factor of g. As $\mathcal{C}_1$ is closed both under left- and right-divisor, it is closed under factor, so every entry s_i belongs to $\mathcal{C}_1$. Then, by Proposition 2.21, the latter is compatible with $\mathcal{S}$.

Owing to the definition of a parabolic subcategory and to Proposition 1.11, (ii) directly follows from (i) since, in both cases, the considered subcategory is closed under factor in the ambient category. □

For instance, in the context of Example 1.12, the submonoid B_I^+ of B_n^+ is parabolic, so Corollary 2.22 implies that B_I^+ is compatible with every Garside family of B_n^+. This applies, in particular, to the divisors of Δ_n, implying that $\mathrm{Div}(\Delta_n) \cap B_I^+$ is a Garside family in B_I^+.

It is natural to wonder whether a converse of Corollary 2.22 might be true, that is, whether every subcategory that is compatible with all Garside families of the ambient category must be closed under factor. The answer is negative, see Exercise 77.

2.4 Compatibility with symmetric decompositions

So far, we considered compatibility between a subcategory $\mathcal{C}_1$ of a left-cancellative category $\mathcal{C}$ and a Garside family $\mathcal{S}$ of $\mathcal{C}$, which involves $\mathcal{S}$-normal decompositions. If $\mathcal{C}$ is a (left)-Ore category, it embeds in a groupoid of left-fractions $\mathcal{E}nv(\mathcal{C})$, and we can similarly address the compatibility of the subgroupoid of $\mathcal{E}nv(\mathcal{C})$ generated by $\mathcal{C}_1$ with $\mathcal{S}$, which involves symmetric $\mathcal{S}$-decompositions. The overall conclusion is that no additional assumption on $\mathcal{S}$ is needed to obtain compatibility with signed decompositions provided the considered subcategory satisfies some appropriate conditions, independently of the Garside family.

We recall from Proposition II.3.18 (left-Ore subcategory) that, if $\mathcal{C}_1$ is a left-Ore subcategory of a left-Ore category $\mathcal{C}$, then the inclusion of $\mathcal{C}_1$ in $\mathcal{C}$ extends to an embedding of $\mathcal{E}nv(\mathcal{C}_1)$ into $\mathcal{E}nv(\mathcal{C})$ and that, moreover, $\mathcal{C}_1 = \mathcal{E}nv(\mathcal{C}_1) \cap \mathcal{C}$ holds if and only if $\mathcal{C}_1$ is closed under right-quotient in $\mathcal{C}$. In this context, we shall usually identify the groupoid $\mathcal{E}nv(\mathcal{C}_1)$ with its image in $\mathcal{E}nv(\mathcal{C})$.

Although the notions will eventually merge, we introduce a strong version of compatibility.

Definition 2.23 (strongly compatible). Assume that $\mathcal{S}$ is a strong Garside family in a left-Ore category $\mathcal{C}$ that admits left-lcms. A subcategory $\mathcal{C}_1$ of $\mathcal{C}$ is said to be *strongly compatible with* $\mathcal{S}$ if, putting $\mathcal{S}_1 = \mathcal{S} \cap \mathcal{C}_1$,

(2.24) The family $\mathcal{S}_1$ is a strong Garside family in $\mathcal{C}_1$, and

(2.25) A signed $\mathcal{C}_1$-path is symmetric $\mathcal{S}_1$-normal in $\mathcal{C}_1$ if and only if it is symmetric $\mathcal{S}$-normal in $\mathcal{C}$.

If a subcategory $\mathcal{C}_1$ is strongly compatible with a Garside family $\mathcal{S}$, it is a fortiori compatible with $\mathcal{S}$: indeed, (2.24) implies that $\mathcal{S}_1$ is a Garside family in $\mathcal{C}_1$, and (2.25) implies in particular that a $\mathcal{C}_1$-path is $\mathcal{S}_1$-normal in $\mathcal{C}_1$ if and only if it is $\mathcal{S}$-normal in $\mathcal{C}$. In the other direction, the next example shows that a subcategory that is compatible with a strong Garside family need not be strongly compatible.

Example 2.26 (not strongly compatible). As in Example II.3.19, let M be the additive monoid $(\mathbb{N}, +)$ and M_1 be $M \setminus \{1\}$. Then M is a strong Garside family in M, and $M \cap M_1$, which is M_1, is a strong Garside family in M_1. Moreover, M_1 is compatible with M viewed as a Garside family in itself: indeed, an M_1-path is M_1-normal if and only if it has length one, if and only if it is M-normal. However, M_1 is not strongly compatible with M: the length-two signed path $\overline{2}|3$ is a strict symmetric M_1-normal path which is not symmetric M-normal, since 2 and 3, which admit 1 (and 2) as a common left-divisor in M, are not left-disjoint in M.

Our aim is to compare compatibility and strong compatibility. We saw above that, as the terminology suggests, every subcategory that is strongly compatible with a strong Garside family $\mathcal{S}$ is necessarily compatible with $\mathcal{S}$. In the other direction, we shall see that the two notions are essentially equivalent, provided one considers subcategories satisfying a convenient closure property not involving $\mathcal{S}$.

We recall from Definition IV.2.27 that a subfamily $\mathcal{S}$ of a category $\mathcal{C}$ is called *weakly closed under left-lcm* if any two elements of $\mathcal{S}$ that admit a left-lcm in $\mathcal{C}$ admit one that lies in $\mathcal{S}$: "weakly" refers here to the fact that we do not demand that every left-lcm lies in $\mathcal{S}$, but only at least one. In the case of a category with no nontrivial invertible element, the left-lcm is unique when it exists and there is no distinction between "closed" and "weakly closed".

Proposition 2.27 (strongly compatible). *If $\mathcal{C}$ is a left-Ore subcategory that admits left-lcms, then, for every Ore subcategory $\mathcal{C}_1$ of $\mathcal{C}$, the following are equivalent:*
(i) *The subcategory $\mathcal{C}_1$ is strongly compatible with a strong Garside family $\mathcal{S}$ of $\mathcal{C}$.*
(ii) *The subcategory $\mathcal{C}_1$ is closed under quotient and weakly closed under left-lcm.*
(iii) *The subcategory $\mathcal{C}_1$ is strongly compatible with every strong Garside family with which it is compatible.*

As every subcategory of a category $\mathcal{C}$ is compatible with at least one Garside family of $\mathcal{C}$, namely $\mathcal{C}$ itself, (iii) obviously implies (i) in Proposition 2.27. Below, we shall prove that (i) implies (ii) and, after establishing an auxiliary result, that (ii) implies (iii).

Proof of (i)$\Rightarrow$(ii) *in Proposition* 2.27. Assume that $\mathcal{S}$ is a strong Garside family in $\mathcal{C}$ and $\mathcal{C}_1$ is strongly compatible with $\mathcal{S}$. Let $\mathcal{S}_1 = \mathcal{S} \cap \mathcal{C}_1$. Assume that $f = hg$ holds in $\mathcal{C}$ with f and g in $\mathcal{C}_1$. By assumption, $\mathcal{S}_1$ is a strong Garside family in $\mathcal{C}_1$, so the element fg^{-1} of $\mathcal{E}nv(\mathcal{C}_1)$ admits a symmetric $\mathcal{S}_1$-normal decomposition $\overline{t_q}|\cdots|\overline{t_1}|s_1|\cdots|s_p$. Moreover, the latter is also a symmetric $\mathcal{S}$-normal decomposition of fg^{-1} since $\mathcal{C}_1$ is strongly compatible with $\mathcal{S}$. Now, if x is the source of h, we have $1_x^{-1}h = (t_1 \cdots t_q)^{-1}(s_1 \cdots s_p)$ in $\mathcal{C}$. As $t_1 \cdots t_q$ and $s_1 \cdots s_p$ are left-disjoint in $\mathcal{C}$, there must exist ϵ in $\mathcal{C}$ satisfying $1_x = \epsilon t_1 \cdots t_q$, which implies that $t_1, \ldots, t_q$ are invertible in $\mathcal{C}$, hence in $\mathcal{C}_1$. So, in $\mathcal{E}nv(\mathcal{C}_1)$, we have $fg^{-1} = (t_1 \cdots t_q)^{-1}(s_1 \cdots s_p) \in \mathcal{C}_1$, that is, h must lie in $\mathcal{C}_1$, and $\mathcal{C}_1$ is closed under left-quotient in $\mathcal{C}$.

Assume now that $gh = f$ holds in $\mathcal{C}$ with g and f in $\mathcal{C}_1$. As $\mathcal{C}_1$ is an Ore category and $\mathcal{S}_1$ is a strong Garside family in $\mathcal{C}_1$, the element $g^{-1}f$ of $\mathcal{E}nv(\mathcal{C}_1)$ admits a symmetric $\mathcal{S}_1$-normal decomposition $\overline{t_q}|\cdots|\overline{t_1}|s_1|\cdots|s_p$. By assumption, the latter is a symmetric $\mathcal{S}$-normal decomposition in $\mathcal{C}$ for $g^{-1}f$, an element of $\mathcal{C}$. By Proposition III.2.16 (symmetric normal unique), the elements $t_1, \ldots, t_q$ must be invertible in $\mathcal{C}$, hence in $\mathcal{C}_1$. So, in $\mathcal{E}nv(\mathcal{C}_1)$, we have $g^{-1}f = (t_1 \cdots t_q)^{-1}(s_1 \cdots s_p) \in \mathcal{C}_1$, that is, h must lie in $\mathcal{C}_1$, and $\mathcal{C}_1$ is closed under right-quotient in $\mathcal{C}$.

Now let f, g be two elements of $\mathcal{C}_1$ sharing the same target. As $\mathcal{C}_1$ is an Ore category and $\mathcal{S}_1$ is a strong Garside family in $\mathcal{C}_1$, the element fg^{-1} of $\mathcal{E}nv(\mathcal{C}_1)$ admits a symmetric $\mathcal{S}_1$-normal decomposition $\overline{t_q}|\cdots|\overline{t_1}|s_1|\cdots|s_p$ in $\mathcal{C}_1$. Let $g' = t_1 \cdots t_q$ and $f' = s_1 \cdots s_p$. Then, in $\mathcal{C}_1$, we have $g'f = f'g$ and the elements f' and g' are left-disjoint. So, by Lemma III.2.8, $g'f$ is a left-lcm of g and f in $\mathcal{C}_1$. On the other hand, as $\mathcal{C}_1$ is strongly compatible with $\mathcal{S}$, the signed path $\overline{t_q}|\cdots|\overline{t_1}|s_1|\cdots|s_p$ is also a symmetric $\mathcal{S}$-normal decomposition of fg^{-1} in $\mathcal{C}$ and, arguing as above but now in $\mathcal{C}$, we deduce that $g'f$ is a left-lcm of g and f in $\mathcal{C}$. Hence at least one left-lcm of f and g in $\mathcal{C}$ lies in $\mathcal{C}_1$, that is, $\mathcal{C}_1$ is weakly closed under left-lcm in $\mathcal{C}$. ☐

For the converse direction in Proposition 2.27, we begin with an auxiliary result. For a subcategory $\mathcal{C}_1$ of a category $\mathcal{C}$ to be weakly closed under left-lcm means that, for all f, g of $\mathcal{C}_1$ sharing the same target, at least one left-lcm of f and g lies in $\mathcal{C}_1$. But, a priori, the latter need not be a left-lcm of f and g in $\mathcal{C}_1$ and the condition says nothing about lcms in $\mathcal{C}_1$. Such distinctions vanish when $\mathcal{C}_1$ is closed under left-quotient.

Lemma 2.28. *Assume that $\mathcal{C}_1$ is a subcategory of a cancellative category $\mathcal{C}$ and $\mathcal{C}_1$ is closed under left-quotient in $\mathcal{C}$ and weakly closed under left-lcm in $\mathcal{C}$.*

(i) *For all f, g in $\mathcal{C}_1$, an element h of $\mathcal{C}_1$ is a left-lcm of f and g in the sense of $\mathcal{C}$ if and only if it is a left-lcm of f and g in the sense of $\mathcal{C}_1$.*

(ii) *If $\mathcal{C}_1$ admits common right-multiples, two elements f and g of $\mathcal{C}_1$ are left-disjoint in $\mathcal{C}$ if and only if they are left-disjoint in $\mathcal{C}_1$.*

Proof. (i) Assume that an element h of $\mathcal{C}_1$ is a left-lcm of f and g in the sense of $\mathcal{C}$. Then there exist f', g' in $\mathcal{C}$ satisfying $h = f'g = g'f$. As $\mathcal{C}_1$ is closed under left-quotient, f' and g' belong to $\mathcal{C}_1$, so h is a left-multiple of f and g in $\mathcal{C}_1$. Moreover, if $f_1 g = g_1 f$ holds in $\mathcal{C}_1$, then it holds in $\mathcal{C}$ and, as h is a left-lcm of f and g in $\mathcal{C}$, we have $f_1 = h_1 f'$ and $g_1 = h_1 g'$ for some h_1 in $\mathcal{C}$. As f' and $\hat{f}$ belong to $\mathcal{C}_1$ and $\mathcal{C}_1$ is closed under left-quotient, h_1 belongs to $\mathcal{C}_1$, which shows that h is a left-lcm of f and g in $\mathcal{C}_1$.

Conversely, assume that h is a left-lcm of f and g in the sense of $\mathcal{C}_1$. As $\mathcal{C}_1$ is weakly closed under left-lcm, some left-lcm h_1 of f and g in $\mathcal{C}$ belongs to $\mathcal{C}_1$. By the above result, h_1 is a left-lcm of f and g in $\mathcal{C}_1$. As $\mathcal{C}_1$ is right-cancellative, the uniqueness of the left-lcm in $\mathcal{C}_1$ implies that $h_1 = \epsilon h$ holds for some ϵ in $\mathcal{C}_1^\times$. Then ϵ belongs to $\mathcal{C}^\times$, so the assumption that h_1 is a left-lcm of f and g in $\mathcal{C}$ implies that h is a left-lcm of f and g in the sense of $\mathcal{C}$.

(ii) Assume that f and g belong to $\mathcal{C}_1$. By Lemma 1.7(iii), the assumption that f and g are left-disjoint in $\mathcal{C}$ implies that they are left-disjoint in $\mathcal{C}_1$.

Conversely, assume that $\mathcal{C}_1$ admits common right-multiples and f and g are left-disjoint in $\mathcal{C}_1$. Then f and g admit a common right-multiple in $\mathcal{C}_1$, say $fg_1 = gf_1$. Then, by Lemma III.2.8, fg_1 is a left-lcm of f_1 and g_1 in $\mathcal{C}_1$. By (i), fg_1 is also a left-lcm of f_1 and g_1 in $\mathcal{C}$ hence, by Lemma III.2.8 now applied in $\mathcal{C}$, the elements f and g are left-disjoint in $\mathcal{C}$. ∎

Proof of (ii)⇒(iii) *in Proposition* 2.27. Assume that $\mathcal{S}$ is a strong Garside family in $\mathcal{C}$ and $\mathcal{C}_1$ is compatible with $\mathcal{S}$. Let $\mathcal{S}_1 = \mathcal{S} \cap \mathcal{C}_1$ and $\mathcal{S}_1^\sharp = \mathcal{S}_1 \mathcal{C}_1^\times \cup \mathcal{C}_1^\times$. Assume that s and t lie in $\mathcal{S}_1^\sharp$ and share the same target. Then, by assumption, there exist s' and t' in $\mathcal{C}_1$ satisfying $s't = t's$ and such that $s't$ is a left-lcm of s and t both in $\mathcal{C}_1$ and in $\mathcal{C}$. On the other hand, as $\mathcal{S}$ is a strong Garside family in $\mathcal{C}$, there exist s'' and t'' in $\mathcal{S}^\sharp$ satisfying $s''t = t''s$ and such that $s''t$ is a left-lcm of s and t in $\mathcal{C}$. The uniqueness of the left-lcm in $\mathcal{C}$ implies the existence of ϵ in $\mathcal{C}^\times$ satisfying $s' = \epsilon s''$ and $t' = \epsilon t''$. So s' and t' lie in $\mathcal{C}^\times \mathcal{S}^\sharp$, hence in $\mathcal{S}^\sharp$, and therefore in $\mathcal{S}^\sharp \cap \mathcal{C}_1$. As $\mathcal{C}_1$ is compatible with $\mathcal{S}$, Proposition 2.10 implies that the latter family is $\mathcal{S}_1^\sharp$. It follows that, in $\mathcal{C}_1$, the Garside family $\mathcal{S}_1$ is a strong Garside family.

Assume now that w is a signed $\mathcal{C}_1$-path, say $w = \overline{t_q}|\cdots|\overline{t_1}|s_1|\cdots|s_p$. Then w is a strict symmetric $\mathcal{S}$-normal path (in $\mathcal{C}$) if and only if $s_1|\cdots|s_p$ and $t_1|\cdots|t_q$ are strict $\mathcal{S}$-normal and s_1 and t_1 are left-disjoint in $\mathcal{C}$. As $\mathcal{C}_1$ is compatible with $\mathcal{S}$, the paths $s_1|\cdots|s_p$ and $t_1|\cdots|t_q$ are strict $\mathcal{S}_1$-normal in $\mathcal{C}_1$. On the other hand, by Lemma 2.28, s_1 and t_1 are left-disjoint in $\mathcal{C}_1$. Hence w is a strict symmetric $\mathcal{S}_1$-normal path in $\mathcal{C}_1$. So, by definition, $\mathcal{C}_1$ is strongly compatible with $\mathcal{S}$. □

When the assumptions of Proposition 2.27 are satisfied, every element in $\mathcal{E}nv(\mathcal{C}_1)$ admits a symmetric $(\mathcal{S} \cap \mathcal{C}_1)$-normal form in $\mathcal{E}nv(\mathcal{C}_1)$, which is also a symmetric $\mathcal{S}$-normal decomposition in $\mathcal{E}nv(\mathcal{C})$. So we can state:

Corollary 2.29 (generated subgroupoid). *Assume that $\mathcal{C}$ is a left-Ore category that admits left-lcms and $\mathcal{C}_1$ is an Ore subcategory of $\mathcal{C}$ that is closed under quotient and weakly closed under left-lcm. Then, for every strong Garside family $\mathcal{S}$ of $\mathcal{C}$, every element of the subgroupoid of $\mathcal{E}nv(\mathcal{C})$ generated by $\mathcal{C}_1$ admits a symmetric $\mathcal{S}$-normal decomposition with entries in $\mathcal{C}_1$.*

In particular, if the ambient category contains no nontrivial invertible element, symmetric normal decompositions of minimal length are unique up to adding identity elements and the result says that the entries in (the) symmetric $\mathcal{S}$-normal decomposition of an element of $\mathcal{C}_1$ must lie in $\mathcal{C}_1$.

3 Subfamilies of a Garside family

We now investigate the subcategories that are generated by the subfamilies of a fixed Garside family $\mathcal{S}$. Then several questions arise, typically whether this subcategory is compatible with $\mathcal{S}$, or which presentation it admits.

The section is organized as follows. In Subsection 3.1, we develop the notion of a subgerm of a germ, which turns out to provide the most convenient framework for addressing the above questions. Next, in Subsection 3.2, we establish transfer results stating that various local closure properties imply global versions in the generated subcategory. Then Subsection 3.3 contains the main results, namely, in the situation when $\underline{\mathcal{S}_1}$ is a subgerm of a Garside germ $\underline{\mathcal{S}}$, local criteria involving $\underline{\mathcal{S}}$ and $\underline{\mathcal{S}_1}$ only and implying various global properties of the subcategory $\mathcal{S}ub(\mathcal{S}_1)$ of $\mathcal{C}at(\underline{\mathcal{S}})$.

3.1 Subgerms

We saw in Chapter VI that, if $\mathcal{S}$ is a solid Garside family in a left-cancellative category $\mathcal{C}$, then the structure $\underline{\mathcal{S}}$ made of $\mathcal{S}$ equipped with the partial product induced by the one of $\mathcal{C}$ is a *germ*, namely it is a precategory equipped with identity elements plus a partial product that satisfies the weak associativity conditions listed in Definition VI.1.3. On the other hand, if we start with an abstract germ $\underline{\mathcal{S}}$ that is at least left-associative, then $\underline{\mathcal{S}}$ embeds in a category $\mathcal{C}at(\underline{\mathcal{S}})$ and it coincides with the germ induced by the product of $\mathcal{C}at(\underline{\mathcal{S}})$.

Whenever $\underline{\mathcal{S}}$ is a germ and $\mathcal{S}_1$ is a subfamily of $\mathcal{S}$, the partial operation on $\mathcal{S}_1$ induced by $\bullet$ on $\mathcal{S}_1$—that is, the operation $\bullet_1$ such that $r \bullet_1 s = t$ holds if r, s, t lie in $\mathcal{S}_1$ and $r \bullet s = t$ holds in $\underline{\mathcal{S}}$—may or may not induce on $\mathcal{S}_1$ the structure of a germ. When it does, we naturally call the resulting structure a subgerm of $\underline{\mathcal{S}}$.

Definition 3.1 (subgerm). A germ $\underline{\mathcal{S}}_1$ is a *subgerm* of a germ $\underline{\mathcal{S}}$ if $\mathcal{S}_1$ is included in $\mathcal{S}$, $\mathcal{O}bj(\underline{\mathcal{S}}_1)$ is included in $\mathcal{O}bj(\underline{\mathcal{S}})$, the source, target, identity, and product of $\underline{\mathcal{S}}_1$ are induced by those of $\underline{\mathcal{S}}$, and the domain of $\bullet_1$ is the intersection with $\mathcal{S}_1^{[2]}$ of the domain of $\bullet$.

Before giving examples, we establish a more practical characterization. Hereafter we shall appeal to the closure properties of Chapter IV and of Section 1 in the context of a germ without redefining them: for instance, if $\underline{\mathcal{S}}$ is a germ, we naturally say that a subfamily $\mathcal{S}_1$ of $\mathcal{S}$ is *closed under product in* $\mathcal{S}$ if $s \bullet t$ belongs to $\mathcal{S}_1$ whenever s and t do and $s \bullet t$ is defined.

Lemma 3.2. *Assume that $\mathcal{S}$ is a Garside family in a left-cancellative category $\mathcal{C}$ and $\mathcal{S}_1$ is a subfamily of $\mathcal{S}$.*

(i) *If $\mathcal{S}ub(\mathcal{S}_1) \cap \mathcal{S} = \mathcal{S}_1$ holds, then $\mathcal{S}_1$ is closed under identity and product in $\mathcal{S}$.*

(ii) *If $\mathcal{S}$ is closed under left- or right-quotient in $\mathcal{C}$, the converse of* (i) *is true, that is, if $\mathcal{S}_1$ is closed under identity and product in $\mathcal{S}$, then $\mathcal{S}ub(\mathcal{S}_1) \cap \mathcal{S} = \mathcal{S}_1$ holds.*

Proof. (i) Assume that x is the source or the target of some element of $\mathcal{S}_1$ and 1_x belongs to $\mathcal{S}$. Then 1_x belongs to $\mathcal{S}ub(\mathcal{S}_1)$, hence to $\mathcal{S}ub(\mathcal{S}_1) \cap \mathcal{S}$ and, therefore, to $\mathcal{S}_1$. So $\mathcal{S}_1$ is closed under identity in $\mathcal{S}$. Next, assume that s, t lie in $\mathcal{S}_1$ and st is defined and it lies in $\mathcal{S}$. Then st belongs to $\mathcal{S}ub(\mathcal{S}_1)$, hence to $\mathcal{S}ub(\mathcal{S}_1) \cap \mathcal{S}$, which is $\mathcal{S}_1$. So $\mathcal{S}_1$ is closed under identity and product in $\mathcal{S}$.

(ii) Assume that $\mathcal{S}$ is closed under right-quotient in $\mathcal{C}$ and t belongs to $\mathcal{S}ub(\mathcal{S}_1) \cap \mathcal{S}$. If t is an identity element, the assumption that $\mathcal{S}_1$ is closed under identity in $\mathcal{S}$ implies $t \in \mathcal{S}_1$. Otherwise, there exist $p \geqslant 1$ and $t_1, ..., t_p$ in $\mathcal{S}_1$ satisfying $t = t_1 \cdots t_p$. As $\mathcal{S}$ is closed under right-quotient in $\mathcal{C}$, the assumption that t_{i-1} lies in $\mathcal{S}_1$ inductively implies that $t_i \cdots t_p$ belongs to $\mathcal{S}$ for i increasing from 2 to $p-1$. Then, as $\mathcal{S}_1$ is closed under product in $\mathcal{S}$, we inductively deduce that $t_i \cdots t_p$ lies in $\mathcal{S}_1$ for i decreasing from $p-1$ to 1. Hence t lies in $\mathcal{S}_1$. The argument is symmetric if $\mathcal{S}$ is closed under left-quotient in $\mathcal{C}$. □

Proposition 3.3 (subgerm). *For $\underline{\mathcal{S}}$ a germ and $\mathcal{S}_1 \subseteq \mathcal{S}$, the following are equivalent:*

(i) *The structure $(\mathcal{S}_1, \mathbf{1}_{\mathcal{S}_1}, \bullet_1)$ is a subgerm of $\underline{\mathcal{S}}$, where $\bullet_1$ is induced by $\bullet$ on $\mathcal{S}_1$.*

(ii) *The family $\mathcal{S}_1$ is closed under identity and product in $\mathcal{S}$.*

Moreover, if $\underline{\mathcal{S}}$ is left- or right-associative, the above conditions are equivalent to

(iii) *The equality $\mathcal{S}ub(\mathcal{S}_1) \cap \mathcal{S} = \mathcal{S}_1$ holds in $\mathcal{C}at(\underline{\mathcal{S}})$.*

In this case, the inclusions of $\mathcal{S}_1$ in $\mathcal{S}$ and of $\mathcal{O}bj(\mathcal{S}_1)$ in $\mathcal{O}bj(\mathcal{S})$ induce a functor ι from $\mathcal{C}at(\underline{\mathcal{S}}_1)$ to the subcategory $\mathcal{S}ub(\mathcal{S}_1)$ of $\mathcal{C}at(\underline{\mathcal{S}})$.

Proof. Assume (i). Put $\underline{\mathcal{S}}_1 = (\mathcal{S}_1, \mathbf{1}_{\mathcal{S}_1}, \bullet_1)$. By the definition of a germ, for every x in $\mathcal{O}bj(\mathcal{S}_1)$, the germ $\underline{\mathcal{S}}_1$ contains an identity element, which, by the definition of a subgerm, must be 1_x, so the latter belongs to $\mathcal{S}_1$. Next, assume that r, s lie in $\mathcal{S}_1$ and $r \bullet s$ is defined. Then, by definition, the product of r and s in $\underline{\mathcal{S}}_1$ must be defined as well, and it must be equal to $r \bullet s$: this is possible only if the latter lies in $\mathcal{S}_1$. So (i) implies (ii).

Conversely, assume (ii). Then Conditions (VI.1.4), (VI.1.5), and (VI.1.6) from the definition of a germ are satisfied in $\underline{\mathcal{S}}_1$. Indeed, for (VI.1.6), if $r \bullet_1 s$, $s \bullet_1 t$, and $r \bullet_1 (s \bullet_1 t)$

are defined, then, by construction, so are $r \bullet s$, $s \bullet t$, and $r \bullet (s \bullet t)$, in which case $(r \bullet s) \bullet t$, which is also $(r \bullet_1 s) \bullet t$, is defined since $\underline{\mathcal{S}}$ is a germ, and the assumption (ii) implies that $(r \bullet_1 s) \bullet t$ belongs to $\mathcal{S}_1$ and, from there, that $(r \bullet_1 s) \bullet_1 t$ is defined. Hence $\underline{\mathcal{S}}_1$ is a germ and, by construction, it is a subgerm of $\underline{\mathcal{S}}$.

Assume now that $\mathcal{S}$ is left- or right-associative. By Proposition VI.1.14 (embedding), $\mathcal{S}$ embeds in $\mathcal{C}at(\underline{\mathcal{S}})$ as a subfamily that is closed under right- or left-divisor. Then, by Lemma 3.2, (ii) is equivalent to $\mathcal{S}ub(\mathcal{S}_1) \cap \mathcal{S} = \mathcal{S}_1$.

Finally, assume that (i)–(iii) are satisfied. By definition, the category $\mathcal{C}at(\underline{\mathcal{S}}_1)$ admits the presentation $\langle \mathcal{S}_1 \mid \mathcal{R}_1 \rangle^+$, where $\mathcal{R}_1$ consists of all relations $r \bullet_1 s = t$, hence it is $\mathcal{S}_1^* /{\equiv_{\bullet_1}}$, where $\equiv_{\bullet_1}$ is the congruence generated by $\mathcal{R}_1$. Similarly, $\mathcal{C}at(\underline{\mathcal{S}})$ is $\mathcal{S}^* /{\equiv_\bullet}$, where $\equiv_\bullet$ is the congruence generated by the family $\mathcal{R}$ of all relations $r \bullet s = t$. For r, s, t in $\mathcal{S}_1$, the relation $r \bullet_1 s = t$ implies $r \bullet s = t$ and, therefore, mapping the $\equiv_{\bullet_1}$-class of an $\mathcal{S}_1$-path to its $\equiv_\bullet$-class (and every object of $\underline{\mathcal{S}}_1$ to itself) defines a functor ι from $\mathcal{C}at(\underline{\mathcal{S}}_1)$ to $\mathcal{C}at(\underline{\mathcal{S}})$. By construction, the image of ι is the subcategory $\mathcal{S}ub(\mathcal{S}_1)$. ☐

So, if $\mathcal{C}$ is a left-cancellative category and $\mathcal{S}$ is a solid Garside family of $\mathcal{C}$, considering subgerms of the germ $\underline{\mathcal{S}}$ and considering subfamilies of $\mathcal{S}$ that are closed under identity and product are equivalent approaches. Note that, in the process of investigating the properties of the subcategory generated by $\mathcal{S}_1$, considering families that satisfy $\mathcal{S}ub(\mathcal{S}_1) \cap \mathcal{S} = \mathcal{S}_1$ is not a restriction since, for every subfamily $\mathcal{S}_1$, if we put $\mathcal{S}'_1 = \mathcal{S}ub(\mathcal{S}_1) \cap \mathcal{S}$, then $\mathcal{S}'_1$ is a subfamily of $\mathcal{S}$ that satisfies both $\mathcal{S}ub(\mathcal{S}'_1) = \mathcal{S}ub(\mathcal{S}_1)$ and $\mathcal{S}ub(\mathcal{S}'_1) \cap \mathcal{S} = \mathcal{S}'_1$.

Example 3.4 (subgerm). Consider the dual braid monoid B_3^{+*} (Reference Structure 3, page 10). Then B_3^{+*} admits the presentation $\langle \mathtt{a}, \mathtt{b}, \mathtt{c} \mid \mathtt{ab} = \mathtt{bc} = \mathtt{ca} \rangle^+$, and $\mathtt{ab}$ is a Garside element, denoted by Δ_3^*. By Proposition VI.1.11 (germ from Garside), the Garside family $\mathrm{Div}(\Delta_3^*)$ gives rise to a 5-element germ $\underline{\mathcal{S}}$, see table on the right, and $\mathcal{M}on(\underline{\mathcal{S}})$ is (isomorphic to) B_3^{+*}.

$\bullet$	1	$\mathtt{a}$	$\mathtt{b}$	$\mathtt{c}$	Δ_3^*
1	1	$\mathtt{a}$	$\mathtt{b}$	$\mathtt{c}$	Δ_3^*
$\mathtt{a}$	$\mathtt{a}$		Δ_3^*		
$\mathtt{b}$	$\mathtt{b}$			Δ_3^*	
$\mathtt{c}$	$\mathtt{c}$	Δ_3^*			
Δ_3^*	Δ_3^*				

Now let $\mathcal{S}_1 = \{1, \mathtt{a}, \mathtt{b}, \Delta_3^*\}$. Then $\mathcal{S}_1$ is closed under identity and product in $\underline{\mathcal{S}}$, so the germ operations of $\underline{\mathcal{S}}$ induce the structure of a subgerm $\underline{\mathcal{S}}_1$ on $\mathcal{S}_1$. By definition, the only nontrivial product in $\underline{\mathcal{S}}_1$ is $\mathtt{a} \bullet_1 \mathtt{b} = \Delta_3^*$, and, therefore, the monoid $\mathcal{M}on(\underline{\mathcal{S}}_1)$ is the monoid $\langle \mathtt{a}, \mathtt{b}, \Delta_3^* \mid \mathtt{ab} = \Delta_3^* \rangle^+$, hence a free monoid based on $\{\mathtt{a}, \mathtt{b}\}$.

$\bullet_1$	1	$\mathtt{a}$	$\mathtt{b}$	Δ_3^*
1	1	$\mathtt{a}$	$\mathtt{b}$	Δ_3^*
$\mathtt{a}$	$\mathtt{a}$		Δ_3^*	
$\mathtt{b}$	$\mathtt{b}$			
Δ_3^*	Δ_3^*			

Observe that, in the above example, $\mathcal{M}on(\underline{\mathcal{S}}_1)$ is *not* a submonoid of $\mathcal{M}on(\underline{\mathcal{S}})$, and that $\mathcal{S}_1$ is a solid Garside family in $\mathcal{M}on(\underline{\mathcal{S}}_1)$, so $\underline{\mathcal{S}}_1$ is a Garside germ.

When $\underline{\mathcal{S}}_1$ is a subgerm of a (Garside) germ $\underline{\mathcal{S}}$—that is, equivalently, when $\mathcal{S}_1$ is a subfamily closed under identity and product of a (solid Garside) family in a left-cancellative category —some properties of $\underline{\mathcal{S}}_1$ are inherited from those of $\underline{\mathcal{S}}$.

Lemma 3.5. *Assume that $\underline{\mathcal{S}}_1$ is a subgerm of a germ $\underline{\mathcal{S}}$.*

(i) *If $\underline{\mathcal{S}}$ is left-associative (resp. right-associative), then so is $\underline{\mathcal{S}}_1$.*

(ii) *If $\underline{\mathcal{S}}$ is left-cancellative (resp. right-cancellative), then so is $\underline{\mathcal{S}}_1$.*

(iii) *The relation $\preccurlyeq_{\mathcal{S}_1}$ is included in $\preccurlyeq_{\mathcal{S}}$.*

(iv) *For every $s_1|s_2$ in $\underline{\mathcal{S}}_1^{[2]}$, we have $\mathcal{J}_{\underline{\mathcal{S}}_1}(s_1, s_2) \subseteq \mathcal{J}_{\underline{\mathcal{S}}}(s_1, s_2) \cap \mathcal{S}_1$.*

Proof. (i) Assume that $\underline{\mathcal{S}}$ is left-associative, and that r, s, t are elements of $\mathcal{S}_1$ such that $(r \bullet_1 s) \bullet_1 t$ is defined. By definition, $(r \bullet s) \bullet t$ is defined, hence so is $s \bullet t$ since $\underline{\mathcal{S}}$ is left-associative, and, therefore, $s \bullet_1 t$ is defined. So $\underline{\mathcal{S}}_1$ is left-associative. The case of right-associativity is similar.

(ii) Assume that $\underline{\mathcal{S}}$ is left-cancellative and $r \bullet_1 s = r \bullet_1 s'$ holds in $\underline{\mathcal{S}}_1$. Then we have $r \bullet s = r \bullet s'$ in $\underline{\mathcal{S}}$, whence $s = s'$ since $\underline{\mathcal{S}}$ is left-cancellative. So $\underline{\mathcal{S}}_1$ is left-cancellative. The case of right-cancellativity is similar.

(iii) Assume $r, s \in \mathcal{S}_1$ and $r \preccurlyeq_{\mathcal{S}_1} s$ holds. By definition, there exists t in $\mathcal{S}_1$ satisfying $r = s \bullet_1 t$. Then t belongs to $\mathcal{S}$ and, in $\mathcal{S}$, we have $r = st$, whence $r \preccurlyeq_{\mathcal{S}} s$.

(iv) Assume $t \in \mathfrak{J}_{\underline{\mathcal{S}}_1}(s_1, s_2)$. By definition, t lies in $\mathcal{S}_1$, $s_1 \bullet t$ is defined in $\underline{\mathcal{S}}_1$, hence in $\underline{\mathcal{S}}$, and $t \preccurlyeq_{\mathcal{S}_1} s_2$ holds. By (iii), the latter implies $t \preccurlyeq_{\mathcal{S}} s_2$, so t lies in $\mathfrak{J}_{\underline{\mathcal{S}}}(s_1, s_2)$. □

More connections appear when the subgerm satisfies closure properties in the ambient germ. We naturally say that a subgerm $\underline{\mathcal{S}}_1$ of a germ $\underline{\mathcal{S}}$ is *closed under right-quotient in* $\underline{\mathcal{S}}$ if the domain $\mathcal{S}_1$ of $\underline{\mathcal{S}}_1$ is closed under right-quotient in the domain $\mathcal{S}$ of $\underline{\mathcal{S}}$.

Lemma 3.6. *Assume that $\underline{\mathcal{S}}$ is a left-cancellative and left-associative germ and $\underline{\mathcal{S}}_1$ is a subgerm of $\underline{\mathcal{S}}$ that is closed under right-quotient in $\underline{\mathcal{S}}$.*

(i) *The relation $\preccurlyeq_{\mathcal{S}_1}$ is the restriction of the relation $\preccurlyeq_{\mathcal{S}}$ to $\mathcal{S}_1$.*

(ii) *An element of $\mathcal{S}_1$ is invertible in $\mathcal{C}at(\underline{\mathcal{S}}_1)$ if and only if it is invertible in $\mathcal{C}at(\underline{\mathcal{S}})$.*

(iii) *For every $g_1|g_2$ in $\mathcal{S}_1^{[2]}$, we have $\mathfrak{J}_{\underline{\mathcal{S}}_1}(g_1, g_2) = \mathfrak{J}_{\underline{\mathcal{S}}}(g_1, g_2) \cap \mathcal{S}_1$.*

Proof. (i) Assume $s, t \in \mathcal{S}_1$. By Lemma 3.5(iii), $s \preccurlyeq_{\mathcal{S}_1} t$ implies $s \preccurlyeq_{\mathcal{S}} t$. Conversely, assume $s \preccurlyeq_{\mathcal{S}} t$. By definition, there exists t' in $\mathcal{S}$ satisfying $s \bullet t' = t$. The assumption that $\mathcal{S}_1$ is closed under right-quotient in $\mathcal{S}$ implies $t' \in \mathcal{S}_1$, so that $s \bullet_1 t' = t$ holds in $\mathcal{S}_1$, implying $s \preccurlyeq_{\mathcal{S}_1} t$. Hence $\preccurlyeq_{\mathcal{S}_1}$ is the restriction of $\preccurlyeq_{\mathcal{S}}$ to $\mathcal{S}_1$.

(ii) By Lemma 3.5(ii), the germ $\underline{\mathcal{S}}_1$ is left-cancellative. Hence, by Lemma VI.1.19, an element of $\mathcal{S}_1$ is invertible in $\mathcal{C}at(\underline{\mathcal{S}}_1)$ if and only if it is invertible in $\mathcal{S}_1$, and, similarly, an element of $\mathcal{S}$, hence in particular an element of $\mathcal{S}_1$, is invertible in $\mathcal{C}at(\underline{\mathcal{S}})$ if and only if it is invertible in $\mathcal{S}$. Now, if ϵ lies in $\mathcal{S}_1(x, y)$ and there exists ϵ' in $\mathcal{S}_1$ satisfying $\epsilon \bullet_1 \epsilon' = 1_x$, then ϵ' is an inverse in $\mathcal{S}$ as well.

On the other hand, if ϵ lies in $\mathcal{S}_1(x, y)$ and there exists ϵ' in $\mathcal{S}$ satisfying $\epsilon \bullet \epsilon' = 1_x$, the assumption that $\mathcal{S}_1$ is closed under identity in $\mathcal{S}$ implies that 1_x lies in $\mathcal{S}_1$ and, then, the assumption that $\mathcal{S}_1$ is closed under right-quotient in $\mathcal{S}$ implies that ϵ' lies in $\mathcal{S}_1$, so it is an inverse of ϵ in $\mathcal{S}_1$.

(iii) Let $s_1|s_2$ belong to $\mathcal{S}_1^{[2]}$. We saw in Lemma 3.5(iv) that $\mathfrak{J}_{\underline{\mathcal{S}}_1}(s_1, s_2)$ is always included in $\mathfrak{J}_{\underline{\mathcal{S}}}(s_1, s_2) \cap \mathcal{S}_1$. Conversely, assume $s \in \mathfrak{J}_{\underline{\mathcal{S}}}(s_1, s_2) \cap \mathcal{S}_1$. By definition, $s_1 \bullet s$ is defined in $\underline{\mathcal{S}}$, hence in $\underline{\mathcal{S}}_1$ since s_1 and s lie in $\mathcal{S}_1$. On the other hand, $s \preccurlyeq_{\mathcal{S}} s_2$ holds, hence so does $s \preccurlyeq_{\mathcal{S}_1} s_2$ by (i). We conclude that s lies in $\mathfrak{J}_{\underline{\mathcal{S}}_1}(s_1, s_2)$. □

A converse of Lemma 3.6(i) is valid, see Exercise 87.

3.2 Transitivity of closure

We now show that, in good cases, the closure properties possibly satisfied by the subcategory $\mathcal{S}ub(\mathcal{S}_1)$ generated by a subgerm $\underline{\mathcal{S}}_1$ of a (Garside) germ $\underline{\mathcal{S}}$ directly rely on the closure properties satisfied by $\underline{\mathcal{S}}_1$ in the germ $\underline{\mathcal{S}}$.

We recall that, if $\mathcal{S}$ is a subfamily of a category $\mathcal{C}$ and s, t lie in $\mathcal{S}$, then $s \preccurlyeq_{\mathcal{S}} t$ stands for $\exists t' \in \mathcal{S}(st' = t)$, that is, s left-divides t and the quotient lies in $\mathcal{S}$. According to our general conventions, if $\mathcal{C}$ is a left-cancellative category, and $\mathcal{S} \subseteq \mathcal{X} \subseteq \mathcal{C}$ holds, we say that $\mathcal{S}$ is *closed under right-comultiple in* $\mathcal{X}$ if, for all s, t in $\mathcal{S}$ and f, g in $\mathcal{X}$ satisfying $sg = tf \in \mathcal{X}$, there exist s', t' in $\mathcal{S}$, and h in $\mathcal{X}$ satisfying $st' = ts' \in \mathcal{S}$, $f = s'h$, and $g = t'h$. When $\mathcal{X}$ is $\mathcal{C}$, this is the notion of Definition IV.1.1 (closure II). The relative version of closure under right-complement is defined similarly. We recall that closure under right-diamond, their common refinement, was introduced above in Definition 1.9.

Like the global versions, the relative versions of closure properties are connected with one another, see Exercise 86. Here we shall consider connections between the local and global versions.

Lemma 3.7. *Assume that $\mathcal{C}$ is a left-cancellative category and $\mathcal{S}$ is a subfamily of $\mathcal{C}$ that is closed under inverse. For $\mathcal{S}_1$ included in $\mathcal{S}$, consider the statements:*

(i) *The family $\mathcal{S}_1$ is closed under inverse in $\mathcal{S}$.*

(ii) *The family $\mathcal{S}_1$ is closed under inverse in $\mathcal{C}$.*

(iii) *The subcategory $\mathcal{S}ub(\mathcal{S}_1)$ is closed under inverse in $\mathcal{C}$.*

Then (i) *and* (ii) *are equivalent,* (ii) *implies* (iii) *whenever $\mathcal{S}_1$ is closed under identity and product in $\mathcal{S}$, and* (iii) *implies* (ii) *whenever $\mathcal{S}_1$ is closed under identity and product in $\mathcal{S}$ and $\mathcal{S}$ is closed under left- or right-quotient in $\mathcal{C}$.*

Proof. Assume (i). Assume that ϵ is invertible in $\mathcal{C}$ and belongs to $\mathcal{S}_1$. Then ϵ belongs to $\mathcal{S} \cap \mathcal{C}^{\times}$, hence, by assumption, ϵ^{-1} belongs to $\mathcal{S}$. Then the assumption that $\mathcal{S}_1$ is closed under inverse in $\mathcal{S}$ implies that ϵ^{-1} belongs to $\mathcal{S}_1$. So (i) implies (ii).

Conversely, assume (ii). Assume that ϵ is invertible in $\mathcal{S}$ and belongs to $\mathcal{S}_1$. Then ϵ is invertible in $\mathcal{C}$ too, so ϵ^{-1} belongs to $\mathcal{S}_1$, and $\mathcal{S}_1$ is closed under inverse in $\mathcal{S}$. So (ii) implies (i).

Assume (ii) with $\mathcal{S}_1$ is closed under identity and product in $\mathcal{S}$. Assume that ϵ is invertible in $\mathcal{C}$ and belongs to $\mathcal{S}ub(\mathcal{S}_1)$. Then either ϵ is an identity element, in which case ϵ^{-1} is equal to ϵ and therefore belongs to $\mathcal{S}ub(\mathcal{S}_1)$, or there exist $p \geqslant 1$ and $\epsilon_1, ..., \epsilon_p$ in $\mathcal{S}_1$ satisfying $\epsilon = \epsilon_1 \cdots \epsilon_p$. As ϵ is invertible, so are $\epsilon_1, ..., \epsilon_p$. As $\mathcal{S}_1$ is closed under inverse in $\mathcal{C}$, the elements $\epsilon_p^{-1}, ..., \epsilon_1^{-1}$ belong to $\mathcal{S}_1$, and, therefore, ϵ^{-1}, which is $\epsilon_p^{-1} \cdots \epsilon_1^{-1}$, belongs to $\mathcal{S}ub(\mathcal{S}_1)$. So $\mathcal{S}ub(\mathcal{S}_1)$ is closed under inverse in $\mathcal{C}$, and (ii) implies (iii).

Finally, assume (iii) with $\mathcal{S}_1$ closed under identity and product in $\mathcal{S}$ and $\mathcal{S}$ closed under left- or right-quotient in $\mathcal{C}$. Let ϵ be an element of $\mathcal{S}_1$ that is invertible in $\mathcal{S}$, that is, ϵ is invertible and ϵ^{-1} belongs to $\mathcal{S}$. As ϵ belongs to $\mathcal{S}ub(\mathcal{S}_1)$ and $\mathcal{S}ub(\mathcal{S}_1)$ is closed under inverse, ϵ^{-1} belongs to $\mathcal{S}ub(\mathcal{S}_1)$, hence to $\mathcal{S}ub(\mathcal{S}_1) \cap \mathcal{S}$, which is $\mathcal{S}_1$ by Lemma 3.2. So $\mathcal{S}_1$ is closed under inverse in $\mathcal{S}$, and (iii) implies (i), hence (ii), in this case. □

Lemma 3.8. *Assume that $\mathcal{C}$ is a left-cancellative category, $\mathcal{S}$ is a subfamily of $\mathcal{C}$ that is closed under right-diamond in $\mathcal{C}$. For $\mathcal{S}_1$ included in $\mathcal{S}$, consider the statements:*

(i) *The family $\mathcal{S}_1$ is closed under right-complement (resp. right-diamond) in $\mathcal{S}$*

(ii) *The family $\mathcal{S}_1$ is closed under right-complement (resp. right-diamond) in $\mathcal{C}$.*

Then (i) *implies* (ii) *whenever $\mathcal{S}$ is closed under right-diamond in $\mathcal{C}$, and* (ii) *implies* (i) *whenever $\mathcal{S}$ is closed under right-quotient in $\mathcal{C}$.*

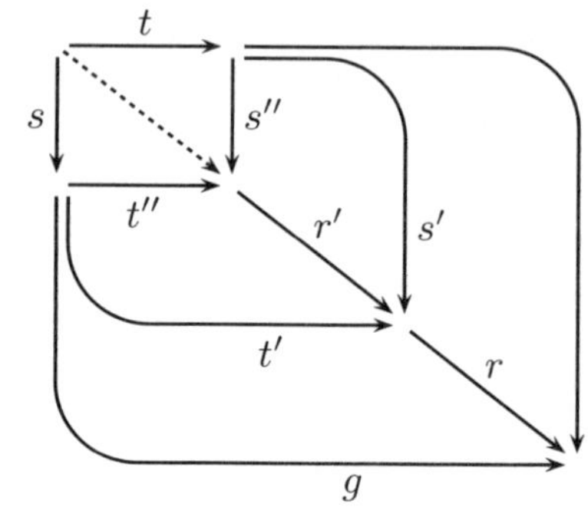

Proof. Assume that $\mathcal{S}_1$ is closed under right-complement in $\mathcal{S}$. Assume $sg = tf$ with s, t in $\mathcal{S}_1$ and f, g in $\mathcal{C}$. Then s and t belong to $\mathcal{S}$, so, as the latter is closed under right-diamond in $\mathcal{C}$, there exist s', t' in $\mathcal{S}$ and r in $\mathcal{C}$ satisfying $st' = ts' \in \mathcal{S}$, $f = s'r$, and $g = t'r$. Next, as $\mathcal{S}_1$ is closed under right-complement in $\mathcal{S}$, there exist s'', t'' in $\mathcal{S}_1$ and r' in $\mathcal{S}$ satisfying $st'' = ts''$, $s' = s''r'$, and $t' = t''r'$. We deduce the equalities $f = s''r'r$ and $g = t''r'r$, and conclude that $\mathcal{S}_1$ is closed under right-complement in $\mathcal{C}$.

The argument is similar for closure under right-diamond, with the only difference that, in this case, st'' can be assumed to belong to $\mathcal{S}_1$. So (i) implies (ii).

Conversely, assume that $\mathcal{S}_1$ is closed under right-complement in $\mathcal{C}$. Assume $sg = tf$ with s, t in $\mathcal{S}_1$ and f, g, and sg in $\mathcal{S}$. By assumption, there exist s', t' in $\mathcal{S}_1$ and r in $\mathcal{C}$ satisfying $st' = ts'$, $f = s'r$, and $g = t'r$. Now s' belongs to $\mathcal{S}_1$, hence to $\mathcal{S}$, and so does f, so the assumption that $\mathcal{S}$ is closed under right-quotient in $\mathcal{C}$ implies that r belongs to $\mathcal{S}$. Therefore, $\mathcal{S}_1$ is closed under right-complement in $\mathcal{S}$.

The argument is again similar for right-diamond, with the difference that st' can then be assumed to belong to $\mathcal{S}_1$. So (ii) implies (i). $\square$

Note that, in the proof that (i) implies (ii) in Lemma 3.8, the assumption that $\mathcal{S}$ is closed under right-complement is not sufficient to deduce that $\mathcal{S}_1$ is closed under right-complement as the assumption on $\mathcal{S}_1$ cannot be applied if st' is not known to lie in $\mathcal{S}$.

The next result gives sufficient local conditions for a subfamily of a solid Garside family to generate a subcategory satisfying closure conditions.

Lemma 3.9. *If $\underline{\mathcal{S}}_1$ is a subgerm of a Garside germ $\underline{\mathcal{S}}$ and $\mathcal{S}_1$ is closed under inverse and right-complement* (*resp. right-diamond*) *in $\mathcal{S}$, then $\mathcal{S}ub(\mathcal{S}_1)$ is closed under right-quotient and right-diamond in $\mathcal{C}at(\underline{\mathcal{S}})$ and $\mathcal{S}_1$ is closed under inverse and right-complement* (*resp. right-diamond*) *in $\mathcal{S}ub(\mathcal{S}_1)$ and in $\mathcal{C}at(\underline{\mathcal{S}})$.*

Proof. Assume that $\mathcal{S}_1$ is closed under inverse and right-complement (*resp.* right-diamond) in $\mathcal{S}$. First, $\mathcal{S}$ is closed under inverse in $\mathcal{C}at(\underline{\mathcal{S}})$ as it is solid, hence it is a fortiori closed under inverse in $\mathcal{S}ub(\mathcal{S}_1)$. Then, by Lemma 3.7, $\mathcal{S}ub(\mathcal{S}_1)$ must be closed under inverse in $\mathcal{C}$, hence a fortiori in $\mathcal{S}ub(\mathcal{S}_1)$.

Next, as it is a solid Garside family, $\mathcal{S}$ is closed under right-diamond in $\mathcal{C}at(\underline{\mathcal{S}})$ by Proposition IV.2.4 (solid Garside). So, by Lemma 3.8, the assumption that $\mathcal{S}_1$ is closed under right-complement (*resp.* right-diamond) in $\mathcal{S}$ implies that $\mathcal{S}_1$ is closed under right-complement (*resp.* right-diamond) in $\mathcal{C}at(\underline{\mathcal{S}})$. Now, Corollary IV.1.17 (extension of closure) implies that $\mathcal{S}ub(\mathcal{S}_1)$ is also closed under right-complement in $\mathcal{C}at(\underline{\mathcal{S}})$, hence under right-diamond, as it is a subcategory of $\mathcal{C}at(\underline{\mathcal{S}})$ and is therefore closed under product.

As for $\mathcal{S}_1$, we already noted above that it is closed under inverse and right-complement (*resp.* right-diamond) in $\mathcal{C}at(\underline{\mathcal{S}})$. So it remains to establish that it is closed in $\mathcal{S}ub(\mathcal{S}_1)$. The result is obvious for inverse as $\mathcal{S}ub(\mathcal{S}_1)$ is included in $\mathcal{C}at(\underline{\mathcal{S}})$. On the other hand, Lemma 1.8 implies that $\mathcal{S}ub(\mathcal{S}_1)$, which is closed under right-complement and inverse in $\mathcal{C}at(\underline{\mathcal{S}})$, is closed under right-quotient in $\mathcal{C}at(\underline{\mathcal{S}})$. By Lemma 3.8 applied with $\mathcal{S}ub(\mathcal{S}_1)$

in place of $\mathcal{S}$, the fact that $\mathcal{S}_1$ is closed under right-complement (*resp.* right-diamond) in $\mathcal{C}at(\underline{\mathcal{S}})$ implies that $\mathcal{S}_1$ is also closed under right-complement (*resp.* right-diamond) in $\mathcal{S}ub(\mathcal{S}_1)$. □

We refer to Exercise 88 for a partial converse of the implication of Lemma 3.9. If we are exclusively interested in the closure of $\mathcal{S}ub(\mathcal{S}_1)$ under right-quotient, it can be shown that $\mathcal{S}_1$ being closed under right-quotient in $\mathcal{S}$ is a necessary condition, see Exercise 89.

3.3 Garside subgerms

When $\underline{\mathcal{S}}_1$ is a subgerm of a Garside germ $\underline{\mathcal{S}}$—that is, equivalently, when $\mathcal{S}$ is a solid Garside family in a left-cancellative category and $\mathcal{S}_1$ is a subfamily of $\mathcal{S}$ that is closed under identity and product—it is natural to wonder which global properties of the subcategory $\mathcal{S}ub(\mathcal{S}_1)$ can be characterized by local properties of $\underline{\mathcal{S}}_1$ inside $\underline{\mathcal{S}}$. We shall now specifically consider three possible properties of $\mathcal{S}ub(\mathcal{S}_1)$:

(3.10) The subcategory $\mathcal{S}ub(\mathcal{S}_1)$ is isomorphic to $\mathcal{C}at(\underline{\mathcal{S}}_1)$.

(3.11) The family $\mathcal{S}_1$ is a Garside family in $\mathcal{S}ub(\mathcal{S}_1)$.

(3.12) The subcategory $\mathcal{S}ub(\mathcal{S}_1)$ is compatible with the Garside family $\mathcal{S}$.

We recall that (3.12) means that an $\mathcal{S}_1$-path is $\mathcal{S}$-normal in $\mathcal{C}at(\underline{\mathcal{S}})$ if and only if it is $\mathcal{S}_1$-normal in $\mathcal{S}ub(\mathcal{S}_1)$. Other properties might be considered, for instance $\mathcal{S}_1$ being a Garside family in $\mathcal{C}at(\underline{\mathcal{S}}_1)$, which differs from (3.11) if (3.10) fails. However, we shall concentrate on the case when (3.10) holds for, otherwise, $\mathcal{C}at(\underline{\mathcal{S}}_1)$ and $\mathcal{C}at(\underline{\mathcal{S}})$ may be extremely different and the germ structure of $\underline{\mathcal{S}}_1$ is of little help in investigating $\mathcal{S}ub(\mathcal{S}_1)$, see Example 3.4 or Exercise 90.

Our first observation is that the properties (3.10)–(3.12), which trivially hold for $\mathcal{S}_1 = \mathcal{S}$, may fail even when $\underline{\mathcal{S}}_1$ is closed under right-quotient in $\underline{\mathcal{S}}$, and that there exist implications between them.

Lemma 3.13. *If $\underline{\mathcal{S}}_1$ is a subgerm of a Garside germ $\underline{\mathcal{S}}$, the implications* (3.12) $\Rightarrow$ (3.11) $\Rightarrow$ (3.10) *hold, but each one of* (3.10), (3.10)$\Rightarrow$(3.11), *and* (3.11)$\Rightarrow$(3.12) *may fail even for $\mathcal{S}_1$ closed under right-quotient in $\mathcal{S}$.*

Proof. First, assume (3.12). By definition, $\mathcal{S}ub(\mathcal{S}_1)$ being compatible with $\mathcal{S}$ implies that $\mathcal{S} \cap \mathcal{S}ub(\mathcal{S}_1)$ is a Garside family in $\mathcal{S}ub(\mathcal{S}_1)$; now, as $\underline{\mathcal{S}}_1$ is a subgerm of $\underline{\mathcal{S}}$, Proposition 3.3 implies the equality $\mathcal{S} \cap \mathcal{S}ub(\mathcal{S}_1) = \mathcal{S}_1$. So (3.12) implies 3.11.

Assume now (3.11). First, $\mathcal{S}_1$ is solid in $\mathcal{S}ub(\mathcal{S}_1)$. Indeed, the assumption that $\underline{\mathcal{S}}_1$ is a subgerm of $\underline{\mathcal{S}}$ implies that $\mathcal{S}_1$ is closed under identity in $\mathcal{S}$, so, for every s in $\mathcal{S}_1(x, y)$, the elements 1_x and 1_y belong to $\mathcal{S}_1$. A straightforward induction then gives the same result for s in $\mathcal{S}ub(\mathcal{S}_1)$. On the other hand, assume that t belongs to $\mathcal{S}_1$ and $t = rs$ holds in $\mathcal{S}ub(\mathcal{S}_1)$. Then s right-divides t in $\mathcal{C}at(\underline{\mathcal{S}})$, so, as t belongs to $\mathcal{S}$ and $\mathcal{S}$ is solid, s belongs to $\mathcal{S}$, hence to $\mathcal{S}ub(\mathcal{S}_1) \cap \mathcal{S}$, which is $\mathcal{S}_1$ by Proposition 3.3. So $\mathcal{S}_1$ is closed under right-divisor in $\mathcal{S}ub(\mathcal{S}_1)$. Hence $\mathcal{S}_1$ is a solid Garside family in $\mathcal{S}ub(\mathcal{S}_1)$. Now, by Proposition IV.3.4 (presentation, solid case), $\mathcal{S}ub(\mathcal{S}_1)$ admits the presentation $\langle \mathcal{S}_1 \mid \mathcal{R}_1 \rangle^+$, where $\mathcal{R}_1$ consists of all relations $rs = t$ with r, s, t in $\mathcal{S}_1$ that are valid in $\mathcal{S}ub(\mathcal{S}_1)$, hence in $\mathcal{C}at(\underline{\mathcal{S}})$: by definition, the latter are the defining relations of $\mathcal{C}at(\underline{\mathcal{S}}_1)$. Therefore, $\mathcal{C}at(\underline{\mathcal{S}}_1)$ is isomorphic to $\mathcal{S}ub(\mathcal{S}_1)$, and (3.10) holds. So (3.11) implies (3.10).

We now construct counter-examples. As for (3.10), we saw in Example 3.4 a subgerm $\underline{\mathcal{S}}_1$ such that $\mathcal{C}at(\underline{\mathcal{S}}_1)$ is not isomorphic to $\mathcal{S}ub(\mathcal{S}_1)$, but, in this case, $\mathcal{S}_1$ is not closed under right-quotient in $\mathcal{S}$, and we construct another one.

Let $\underline{\mathcal{S}}$ be the six-element Garside germ associated with the divisors of Δ_3 in the braid monoid B_3^+, and let $\mathcal{S}_1 = \{1, \mathtt{a}, \mathtt{b}\}$ with $\mathtt{a} = \sigma_1\sigma_2$ and $\mathtt{b} = \sigma_2\sigma_1$. Then $\mathcal{S}_1$ contains 1, and it is closed under product in $\mathcal{S}$: the induced partial product is trivial, that is, the only defined instances are those involving 1. Hence $\mathcal{M}on(\underline{\mathcal{S}}_1)$ is the free monoid based on a and b. On the other hand, in B_3^+, the braids a and b satisfy $\mathtt{a}^3 = \mathtt{b}^3 (= \Delta_3^2)$. The submonoid $\mathcal{S}ub(\mathcal{S}_1)$ is cancellative and Δ_3^2 is a right-lcm of a and b in $\mathcal{S}ub(\mathcal{S}_1)$. Therefore, $\mathcal{S}ub(\mathcal{S}_1)$ is the torus monoid $\langle \mathtt{a}, \mathtt{b} \mid \mathtt{a}^3 = \mathtt{b}^3\rangle^+$, which is not isomorphic to $\mathcal{M}on(\underline{\mathcal{S}}_1)$. So, in this example, (3.10) fails.

Next, let $\underline{\mathcal{S}}$ be the Garside germ associated with the divisors of the Garside element $\mathtt{a}^4$ in the Klein bottle monoid K^+ (Reference Structure 5, page 17)—hence the square of the minimal Garside element $\mathtt{a}^2$—and let $\mathcal{S}_1 = \{g \in K^+ \mid |g|_\mathtt{a} \in \{0, 3\}\}$, where we recall $|g|_\mathtt{a}$ is the number of letters a in any expression of g in terms of a and b. Then $\mathcal{S}_1$ is closed under product in $\underline{\mathcal{S}}$: the products of elements of $\mathcal{S}_1$ that lie in $\mathcal{S}$ all are of the form $f \bullet g$ with $|f|_\mathtt{a} + |g|_\mathtt{a} \leqslant 4$, whence $(|f|_\mathtt{a}, |g|_\mathtt{a}) \in \{(0,0), (0,3), (3,0)\}$, and these belong to $\mathcal{S}_1$. Next, $\mathcal{S}ub(\mathcal{S}_1)$ is $\{g \in K^+ \mid |g|_\mathtt{a} = 0 \pmod 3\}$. The computation rules of K^+ show that $\mathcal{S}ub(\mathcal{S}_1)$ is closed under left- and right-quotient in K^+. Moreover, the map $\mathtt{a} \mapsto \mathtt{a}^3$, $\mathtt{b} \mapsto \mathtt{b}$ induces an isomorphism from K^+ onto $\mathcal{S}ub(K^+)$. As seen in Example IV.2.23, $\{g \in K^+ \mid |g|_\mathtt{a} \leqslant 1\}$ is not a Garside family in K^+, hence $\mathcal{S}_1$ is not a Garside family in $\mathcal{S}ub(\mathcal{S}_1)$. On the other hand, the defining relation of $\mathcal{S}ub(\mathcal{S}_1)$, namely $\mathtt{a}^3 = \mathtt{b} \bullet \mathtt{a}^3 \bullet \mathtt{b}$, holds in the germ $\underline{\mathcal{S}}_1$, so $\mathcal{M}on(\underline{\mathcal{S}}_1)$ is isomorphic to $\mathcal{S}ub(\mathcal{S}_1)$. We see that, in this example, (3.10) holds, but (3.11) fails.

Finally, let $\underline{\mathcal{S}}$ be the four-element germ associated with the divisors of $\mathtt{a}^3$ in the free monoid based on $\{\mathtt{a}\}$, that is, $(\mathbb{N}, +)$, and let $\mathcal{S}_1 = \{1, \mathtt{a}^2\}$. Then $\mathcal{S}$ is a solid Garside family in $\mathcal{M}on(\underline{\mathcal{S}})$, which identifies with $\mathbb{N}$, whereas $\mathcal{S}ub(\mathcal{S}_1)$ identifies with $2\mathbb{N}$. Next, $\mathcal{S}_1$ is closed under right-quotient in $\mathcal{S}$, and it is a solid Garside family in $\mathcal{S}ub(\mathcal{S}_1)$, so (3.11) holds. However, as observed in Example 2.7, $2\mathbb{N}$ is not compatible with $\mathcal{S}$: for instance, $\mathtt{a}^2|\mathtt{a}^2$ is $\mathcal{S}_1$-normal in $\mathcal{S}ub(\mathcal{S}_1)$, but is not $\mathcal{S}$-normal in $\mathcal{C}at(\underline{\mathcal{S}})$, where the $\mathcal{S}$-normal decomposition of $\mathtt{a}^4$ is $\mathtt{a}^3|\mathtt{a}$. So, in this example, (3.11) holds but (3.12) fails. □

What we shall do now is to show that each property (3.10), (3.11), (3.12) is implied by purely local conditions involving $\mathcal{S}_1$ and $\underline{\mathcal{S}}$. We always consider subgerms that are closed under right-quotient in the initial germ.

Proposition 3.14 (Garside subgerm I). *If $\underline{\mathcal{S}}_1$ is a subgerm of a Garside germ $\underline{\mathcal{S}}$ and $\mathcal{S}_1$ is closed under right-quotient in $\mathcal{S}$, a sufficient condition for* (3.10)*, that is, for $\mathcal{S}ub(\mathcal{S}_1)$ to be isomorphic to $\mathcal{C}at(\underline{\mathcal{S}}_1)$, is*

(3.15) *The family $\mathcal{S}_1$ is closed under right-comultiple in $\mathcal{S}$.*

Proof. Assume (3.15). As $\mathcal{S}_1$ is closed under right-quotient in $\mathcal{S}$, (3.15) implies that $\mathcal{S}_1$ is closed under right-diamond in $\mathcal{S}$. On the other hand, as $\underline{\mathcal{S}}$ is a Garside germ, $\mathcal{S}$ is closed under right-divisor (hence right-quotient) and right-diamond in $\mathcal{C}at(\underline{\mathcal{S}})$. By Lemma 3.8, we deduce that $\mathcal{S}_1$ is closed under right-diamond in $\mathcal{C}at(\underline{\mathcal{S}})$. It then follows from the proof of Proposition IV.3.4 (presentation, solid case) that every equality of the form $s_1 \cdots s_p = t_1 \cdots t_q$ holding in $\mathcal{C}$ with $s_1, ..., s_p, t_1, ..., t_q$ in $\mathcal{S}_1$ is the consequence of at most $2pq+p+q$ relations of the form $r = st$ with r, s, t in $\mathcal{S}_1$. By definition, these relations are defining relations of the category $\mathcal{C}at(\underline{\mathcal{S}}_1)$, and therefore the latter make a presentation of $\mathcal{S}ub(\mathcal{S}_1)$. Hence the functor ι of Proposition 3.3 is an isomorphism, and (3.10) is satisfied. □

For (3.11), that is, the question whether $\mathcal{S}_1$ is a Garside family in $\mathcal{S}ub(\mathcal{S}_1)$, which we know implies (3.10), the additional price to pay is small:

Proposition 3.16 (Garside subgerm II). *If $\underline{\mathcal{S}}_1$ is a subgerm of a Garside germ $\underline{\mathcal{S}}$ and $\mathcal{S}_1$ is closed under right-quotient and right-comultiple in $\mathcal{S}$, a necessary and sufficient condition for* (3.11)*, that is, for $\mathcal{S}_1$ to be a Garside family in $\mathcal{S}ub(\mathcal{S}_1)$, is*

(3.17) *For every $s_1|s_2$ in $\mathcal{S}_1^{[2]}$, there exists a $\prec_{\mathcal{S}}$-maximal element in $\mathcal{J}_{\underline{\mathcal{S}}}(s_1, s_2) \cap \mathcal{S}_1$.*

Proof. First, as above, the assumption that $\mathcal{S}_1$ is closed under right-quotient and right-comultiple in $\mathcal{S}$ implies that it is also closed under inverse and right-diamond in $\mathcal{S}$. Hence, by Lemma 3.9, $\mathcal{S}_1$ is closed under inverse and right-diamond in $\mathcal{S}ub(\mathcal{S}_1)$. As $\mathcal{S}_1$ generates $\mathcal{S}ub(\mathcal{S}_1)$, Lemma IV.1.18 implies that $\mathcal{S}_1$ is closed under right-divisor in $\mathcal{S}ub(\mathcal{S}_1)$ and, therefore, $\mathcal{S}_1$ is a solid subfamily in $\mathcal{S}ub(\mathcal{S}_1)$.

Assume that $\mathcal{S}_1$ is a Garside family, hence a solid Garside family, in $\mathcal{S}ub(\mathcal{S}_1)$. By Proposition VI.1.11 (germ from Garside), the germ $\underline{\mathcal{S}}_1'$ induced on $\mathcal{S}_1$ by the product of $\mathcal{S}ub(\mathcal{S}_1)$, is a Garside germ. By Proposition 3.14, $\mathcal{S}ub(\mathcal{S}_1)$ is isomorphic to $\mathcal{C}at(\underline{\mathcal{S}}_1)$, where $\underline{\mathcal{S}}_1$ is the germ induced on $\mathcal{S}_1$ by the product of $\mathcal{C}at(\underline{\mathcal{S}})$. So $\underline{\mathcal{S}}_1$ and $\underline{\mathcal{S}}_1'$ coincide, and $\underline{\mathcal{S}}_1$ is a Garside germ. By Proposition VI.2.28 (recognizing Garside germ II), for every $s_1|s_2$ in $\mathcal{S}_1^{[2]}$, the family $\mathcal{J}_{\underline{\mathcal{S}}_1}(s_1, s_2)$, which is $\mathcal{J}_{\underline{\mathcal{S}}}(s_1, s_2) \cap \mathcal{S}_1$ by Lemma 3.6, admits a $\preccurlyeq_{\mathcal{S}_1}$-greatest element. The latter is a fortiori $\prec_{\mathcal{S}_1}$-maximal, hence $\prec_{\mathcal{S}}$-maximal by Lemma 3.6 again. So (3.17) holds.

Conversely, assume (3.17). We saw above that $\mathcal{S}_1$ is closed under inverse and right-diamond in $\mathcal{S}ub(\mathcal{S}_1)$, of which it is a solid subfamily. Let g belong to $\mathcal{S}_1^2$ in $\mathcal{S}ub(\mathcal{S}_1)$, say $g = s_1 s_2$ with s_1, s_2 in $\mathcal{S}_1$. By assumption, $\mathcal{J}_{\underline{\mathcal{S}}}(s_1, s_2)$ has a $\prec_{\mathcal{S}}$-maximal element, say t. By definition, $s_1 t$ is a left-divisor of g that lies in $\mathcal{S}_1$ and is $\prec_{\mathcal{S}_1}$-maximal: no proper right-multiple of $s_1 t$ may belong to $\mathcal{S}_1$ and left-divide g. So every element of $\mathcal{S}_1^2$ admits a $\prec_{\mathcal{S}_1}$-maximal left-divisor in $\mathcal{S}ub(\mathcal{S}_1)$. By Proposition IV.1.24 (recognizing Garside II), $\mathcal{S}_1$ is a Garside family in $\mathcal{S}ub(\mathcal{S}_1)$. □

Corollary 3.18 (Garside subgerm, Noetherian case). *If $\underline{\mathcal{S}}_1$ is a right-Noetherian subgerm of a Garside germ $\underline{\mathcal{S}}$ and $\underline{\mathcal{S}}_1$ is closed under right-quotient and right-comultiple in $\underline{\mathcal{S}}$, then* (3.10) *and* (3.11) *hold, that is, $\mathcal{S}ub(\mathcal{S}_1)$ is isomorphic to $\mathcal{C}at(\underline{\mathcal{S}}_1)$ and $\mathcal{S}_1$ is a Garside family in $\mathcal{S}ub(\mathcal{S}_1)$.*

Proof. As $\underline{\mathcal{S}}_1$ is right-Noetherian, every bounded subfamily of $\mathcal{S}_1$ admits a $\prec_{\mathcal{S}_1}$-maximal element, so, as $\mathcal{J}_{\underline{\mathcal{S}}}(s_1, s_2) \cap \mathcal{S}_1$ is bounded by s_2, (3.17) is automatically satisfied. □

Corollary 3.18 shows that an example satisfying (3.10) but not (3.11), if any, must be non-Noetherian, thus making the construction in the proof of Lemma 3.13 natural.

Proposition 3.19 (Garside subgerm III). *If $\underline{\mathcal{S}}_1$ is a subgerm of a Garside germ $\underline{\mathcal{S}}$ and $\mathcal{S}_1$ is closed under right-quotient in $\mathcal{S}$, then a necessary and sufficient condition for* (3.12), *that is, for $\mathcal{S}ub(\mathcal{S}_1)$ to be compatible with $\mathcal{S}$, is*

(3.20) *For every $s_1|s_2$ in $\mathcal{S}_1^{[2]}$, at least one $\preccurlyeq_{\mathcal{S}}$-greatest element of $\mathcal{J}_{\underline{\mathcal{S}}}(s_1, s_2)$ lies in $\mathcal{S}_1$.*

In this case, $\mathcal{S}_1$ is a Garside family in $\mathcal{S}ub(\mathcal{S}_1)$, and the latter is isomorphic to $\mathcal{C}at(\underline{\mathcal{S}}_1)$ and closed under right-quotient in $\mathcal{C}at(\underline{\mathcal{S}})$.

Proof. Assume that $\mathcal{S}ub(\mathcal{S}_1)$ is compatible with $\mathcal{S}$ in $\mathcal{C}at(\underline{\mathcal{S}})$, that is, (3.12) holds. By Lemma 3.13, so does (3.10), and $\mathcal{S}_1$ is a Garside family in $\mathcal{S}ub(\mathcal{S}_1)$. Let $s_1|s_2$ belong to $\mathcal{S}_1^{[2]}$. Then there exists an $\mathcal{S}_1$-normal decomposition of s_1s_2 in $\mathcal{S}ub(\mathcal{S}_1)$, say $t_1|t_2$. By construction, we have $t_1 = s_1s$ for some s belonging to $\mathcal{J}_{\underline{\mathcal{S}}_1}(s_1, s_2)$. By assumption, $t_1|t_2$ is also $\mathcal{S}$-normal in $\mathcal{C}at(\underline{\mathcal{S}})$. By Lemma VI.2.4, this means that s is a $\preccurlyeq_{\mathcal{S}}$-greatest element of $\mathcal{J}_{\underline{\mathcal{S}}}(s_1, s_2)$. So (3.20) necessarily holds.

Conversely, assume that (3.20) is satisfied. First, as $\underline{\mathcal{S}}$ is a Garside germ, then, by Proposition VI.2.8 (recognizing Garside germ I), the germ $\underline{\mathcal{S}}$ is left-associative and left-cancellative, hence, by Lemma 3.5, so is $\underline{\mathcal{S}}_1$. Next, for every $s_1|s_2$ in $\mathcal{S}_1^{[2]}$, the family $\mathcal{J}_{\underline{\mathcal{S}}}(s_1, s_2)$ admits a $\preccurlyeq_{\mathcal{S}}$-greatest element.Then, by assumption, there exists such an element that belongs to $\mathcal{S}_1$, and, by Lemma 3.6, it is a $\preccurlyeq_{\mathcal{S}_1}$-greatest element in $\mathcal{J}_{\underline{\mathcal{S}}_1}(s_1, s_2)$. By Proposition VI.2.28 (recognizing Garside germ II), it follows that $\underline{\mathcal{S}}_1$ is a Garside germ.

Assume that $s_1|s_2$ lies in $\mathcal{S}_1^{[2]}$ and is $\mathcal{S}_1$-normal in $\mathcal{C}at(\underline{\mathcal{S}}_1)$. Then, by Lemma VI.2.4, which is eligible since $\mathcal{S}_1$ is a Garside germ, every element of $\mathcal{J}_{\underline{\mathcal{S}}_1}(s_1, s_2)$ is invertible in $\mathcal{C}at(\underline{\mathcal{S}}_1)$. Owing to (3.20), this implies that every element of $\mathcal{J}_{\underline{\mathcal{S}}}(s_1, s_2)$ is invertible in $\mathcal{C}at(\underline{\mathcal{S}})$: if there existed a non-invertible element in $\mathcal{J}_{\underline{\mathcal{S}}}(s_1, s_2)$, a $\prec_{\mathcal{S}}$-maximal element of this family would be non-invertible in $\mathcal{C}at(\underline{\mathcal{S}})$, hence a fortiori in $\mathcal{C}at(\underline{\mathcal{S}}_1)$. Hence, by Lemma VI.2.4 again, $s_1|s_2$ is $\mathcal{S}$-normal in $\mathcal{C}at(\underline{\mathcal{S}})$. Conversely, assume that $s_1|s_2$ lies in $\mathcal{S}_1^{[2]}$ and is $\mathcal{S}$-normal in $\mathcal{C}$. Then, by Lemma VI.2.4 once again, every element of $\mathcal{J}_{\underline{\mathcal{S}}}(s_1, s_2)$ is invertible in $\mathcal{C}$. A fortiori, every element of $\mathcal{J}_{\underline{\mathcal{S}}_1}(s_1, s_2)$, which by Lemma 3.6 is included in $\mathcal{J}_{\underline{\mathcal{S}}}(s_1, s_2)$, is invertible in $\mathcal{C}$, hence, by the same lemma, in $\mathcal{S}_1$. By Lemma VI.2.4, we conclude that $s_1|s_2$ is $\mathcal{S}_1$-normal in $\mathcal{C}at(\underline{\mathcal{S}}_1)$. It follows that an $\mathcal{S}_1$-path is $\mathcal{S}_1$-normal if and only if it is $\mathcal{S}$-normal.

At this point, we know by Proposition 3.3 that $\mathcal{S}_1$ is equal to $\mathcal{S} \cap \mathcal{C}_1$, and we have proved that 2.6 holds. But we are not yet ready to conclude, as we do not know whether $\mathcal{S}_1$ is a Garside family in $\mathcal{S}ub(\mathcal{S}_1)$—and we did not assume that $\mathcal{S}_1$ is closed under right-comultiple in $\mathcal{S}$, so Proposition 3.14 does not apply. Now, assume that f and g belong to $\mathcal{C}at(\underline{\mathcal{S}}_1)$ and $\iota(f) = \iota(g)$ holds. As $\underline{\mathcal{S}}_1$ is a solid Garside family in $\mathcal{C}at(\underline{\mathcal{S}}_1)$, the elements f and g admit $\mathcal{S}_1$-normal decompositions with entries in $\mathcal{S}_1$, say $s_1|\cdots|s_p$ and $t_1|\cdots|t_q$. As seen above, the latter $\mathcal{S}_1$-normal $\mathcal{S}_1$-paths must be $\mathcal{S}$-normal. Now, $s_1|\cdots|s_p$ is an $\mathcal{S}$-normal decomposition of $\iota(f)$ in $\mathcal{C}$, whereas $t_1|\cdots|t_q$ is an $\mathcal{S}$-normal decomposition of $\iota(g)$. Since $\iota(f) = \iota(g)$ holds, $s_1|\cdots|s_p$ is a $\mathcal{C}^\times$-deformation of $t_1|\cdots|t_q$ in $\mathcal{C}$, that is, there exist invertible elements $\epsilon_0, \dots, \epsilon_m$ of $\mathcal{C}$, with ϵ_0, ϵ_m in $1_{\mathcal{C}}$, such that $\epsilon_{i-1}t_i = s_i\epsilon_i$ holds in $\mathcal{C}$ for every i, assuming $m = \max(p, q)$ and $s_i = 1.$ (*resp.* $t_i = 1.$) for $p < i \leqslant m$ (*resp.* $q < i \leqslant m$). We prove, using induction on $i \geqslant 1$, that ϵ_{i-1} and $\epsilon_{i-1}t_i$ belong to $\mathcal{S}_1$. For $i = 1$, the assumption that $\mathcal{S}_1$ is closed under identity in $\mathcal{S}$ implies that ϵ_0 belongs to $\mathcal{S}_1$, and then $\epsilon_0 t_1$, which is t_1, belongs to $\mathcal{S}_1$. Assume $i \geqslant 2$. By induction hypothesis, s_{i-1} and $\epsilon_{i-2}t_{i-1}$ belong to $\mathcal{S}_1$, and $\epsilon_{i-2}t_{i-1} = s_{i-1}\epsilon_{i-1}$ holds in $\mathcal{C}$. As $\mathcal{S}_1$ is closed under right-quotient in $\mathcal{S}$, this implies that ϵ_{i-1} belongs to $\mathcal{S}_1$. As $\mathcal{S}_1$ is closed under inverse in $\mathcal{S}$, we deduce that ϵ_{i-1}^{-1} belongs to $\mathcal{S}_1$, so, as we have $t_i = \epsilon_{i-1}^{-1}(\epsilon_{i-1}t_i)$ and $\mathcal{S}_1$ is closed under right-quotient in $\mathcal{S}$, we deduce that $\epsilon_{i-1}t_i$ belongs to $\mathcal{S}_1$. It follows that, in $\underline{\mathcal{S}}_1$, the $\mathcal{S}_1$-path $s_1|\cdots|s_p$ is a deformation of $t_1|\cdots|t_q$, which implies that, in $\mathcal{C}at(\underline{\mathcal{S}}_1)$, we have $s_1 \cdots s_p = t_1 \cdots t_q$, whence $f = g$. Hence ι is injective, and $\mathcal{C}at(\underline{\mathcal{S}}_1)$ is isomorphic to $\mathcal{S}ub(\mathcal{S}_1)$. As $\mathcal{S}_1$ is a solid Garside family in $\mathcal{C}at(\underline{\mathcal{S}}_1)$, it is a solid Garside family in $\mathcal{S}ub(\mathcal{S}_1)$. Then, by Definition 2.4, the subcategory $\mathcal{S}ub(\mathcal{S}_1)$ is compatible with $\mathcal{S}$, that is, (3.12) holds.

Finally, we already saw above that ι is an isomorphism and that $\underline{\mathcal{S}}_1$ is a Garside germ, and Lemma 2.9 implies that $\mathcal{S}ub(\mathcal{S}_1)$ is closed under right-quotient in $\mathcal{C}at(\underline{\mathcal{S}})$. $\square$

We refer to Exercise 89 for a proof that the assumption that $\mathcal{S}_1$ is closed under right-quotient in $\mathcal{S}$ is necessary for (3.12) to hold.

We shall now conclude by establishing the existence, in good cases, of a smallest compatible subgerm including a subfamily. The result is based on the preservation of compatibility under intersection whenever invertible elements are not a problem.

Definition 3.21 (**$=^{\times}_{\mathcal{S}}$-closed subgerm**). A subgerm $\underline{\mathcal{S}}_1$ of a germ $\underline{\mathcal{S}}$ is called $=^{\times}_{\mathcal{S}}$-*closed* if, for every s in $\mathcal{S}_1$, every s' in $\mathcal{S}$ satisfying $s' =^{\times}_{\mathcal{S}} s$ lies in $\mathcal{S}_1$.

One easily checks that, if $\underline{\mathcal{S}}_1$ is a subgerm of an associative germ $\underline{\mathcal{S}}$, then $\mathcal{S}ub(\mathcal{S}_1)$ is $=^{\times}$-closed in $\mathcal{C}at(\underline{\mathcal{S}})$ if and only if $\underline{\mathcal{S}}_1$ is $=^{\times}_{\mathcal{S}}$-closed in $\mathcal{S}$, see Exercise 91. The interest of considering $=^{\times}_{\mathcal{S}}$-closed subgerms is that, for such subgerms, (3.20) can be rewritten.

Corollary 3.22 (**Garside subgerm, $=^{\times}$-closed case**). *If $\underline{\mathcal{S}}_1$ is subgerm of a Garside germ $\underline{\mathcal{S}}$ and $\mathcal{S}_1$ is closed under right-quotient and $=^{\times}_{\mathcal{S}}$-closed in $\mathcal{S}$, then a necessary and sufficient condition for $\mathcal{S}ub(\mathcal{S}_1)$ to be compatible with $\mathcal{S}$ (3.12) is*

(3.23) *For every $s_1|s_2$ in $\mathcal{S}_1^{[2]}$, every $\preccurlyeq_{\mathcal{S}}$-greatest element of $\mathcal{J}_{\underline{\mathcal{S}}}(s_1, s_2)$ lies in $\mathcal{S}_1$.*

Proof. The only difference with (3.20) is that, in (3.23), "every $\preccurlyeq_{\mathcal{S}}$-greatest element" replaces "at least one $\preccurlyeq_{\mathcal{S}}$-greatest element". Since any two $\preccurlyeq_{\mathcal{S}}$-greatest element in a family $\mathcal{J}_{\underline{\mathcal{S}}}(s_1, s_2)$ always are $=^{\times}_{\mathcal{S}}$-equivalent, the two conditions are equivalent whenever $\underline{\mathcal{S}}_1$ is $=^{\times}_{\mathcal{S}}$-closed. $\square$

The explicit form of the criterion in Corollary 3.22, which only involves universally quantified conditions, makes it clear that every intersection of subgerms eligible for the criterion is again eligible. We deduce:

Proposition 3.24 (**smallest compatible subgerm**). *For every subfamily $\mathcal{X}$ of a Garside germ $\underline{\mathcal{S}}$, there exists a unique smallest $=^{\times}_{\mathcal{S}}$-closed subgerm of $\underline{\mathcal{S}}$ including $\mathcal{X}$ and satisfying* (3.12), *that is, the generated subcategory is compatible with the Garside family $\mathcal{S}$.*

The only fact needed for the proof, already mentioned but not established, is that (3.12) requires that $\mathcal{S}_1$ be closed under right-quotient in $\mathcal{S}$ (Exercise 89).

We conclude this section with a fourth possible property of the subcategory $\mathcal{S}ub(\mathcal{S}_1)$, always in the situation when $\underline{\mathcal{S}}_1$ is a subgerm of a (Garside) germ and $\mathcal{S}_1$ is closed under right-quotient in $\mathcal{S}$, namely

(3.25) The subcategory $\mathcal{S}ub(\mathcal{S}_1)$ is a head subcategory of $\mathcal{C}at(\underline{\mathcal{S}})$.

Definition 3.26 (head in a germ). For $\mathcal{S}_1$ included in a germ $\underline{\mathcal{S}}$ and s in $\mathcal{S}$, and element s_1 of $\mathcal{S}_1$ is called an $\mathcal{S}_1$-*head* of s in $\underline{\mathcal{S}}$ if $t \preccurlyeq_{\mathcal{S}} s_1$ holds for every t in $\mathcal{S}_1$ satisfying $t \preccurlyeq_{\mathcal{S}} s$.

As can be expected, we obtain a criterion connecting (3.25) with the existence of a local $\mathcal{S}_1$-head for the elements of $\mathcal{S}$.

Proposition 3.27 (head subgerm). *Assume that $\underline{\mathcal{S}}_1$ is a subgerm of a right-associative Garside germ $\underline{\mathcal{S}}$ that is closed under right-quotient, and $\underline{\mathcal{S}}$ is right-Noetherian or $\mathcal{S}^\sharp$ is closed under left-divisor. Suppose the second domino rule is valid for $\mathcal{S}$ in $\mathcal{C}at(\underline{\mathcal{S}})$ and* (3.12) *holds. Then a necessary and sufficient condition for $\mathcal{S}ub(\mathcal{S}_1)$ to be a head subcategory of $\mathcal{C}at(\underline{\mathcal{S}})$ is*

(3.28) *Every element of $\mathcal{S}$ admits an $\mathcal{S}_1$-head in $\mathcal{S}$.*

Proof. Assume that $\mathcal{S}ub(\mathcal{S}_1)$ is a head subcategory of $\mathcal{C}at(\underline{\mathcal{S}})$, and let s be an element of $\mathcal{S}$ with source in $\mathcal{O}bj(\mathcal{S}_1)$. By assumption, s admits a $\mathcal{S}ub(\mathcal{S}_1)$-head, say s_1. Since $\underline{\mathcal{S}}$ is right-associative, $\mathcal{S}$ is closed under left-divisor in $\mathcal{C}at(\underline{\mathcal{S}})$. Hence, s_1, which left-divides s, must lie in $\mathcal{S}$. So s_1 lies in $\mathcal{S}ub(\mathcal{S}_1) \cap \mathcal{S}$, hence in $\mathcal{S}_1$ since $\underline{\mathcal{S}}_1$ is a subgerm of $\underline{\mathcal{S}}$. Now assume $t \in \mathcal{S}_1$ and $t \preccurlyeq_{\mathcal{S}} s$. In $\mathcal{C}at(\underline{\mathcal{S}})$, we have $t \preccurlyeq s$, whence $t \preccurlyeq s_1$ since s_1 is a $\mathcal{S}ub(\mathcal{S}_1)$-head of s. As $\mathcal{S}_1$ is closed under right-quotient in $\mathcal{S}$, this implies $t \preccurlyeq_{\mathcal{S}} s_1$, and s_1 is an $\mathcal{S}_1$-head of s. So (3.28) holds.

Conversely, assume that (3.28) holds. We first show that $\mathcal{S}_1$ must be closed under right-complement in $\mathcal{C}at(\underline{\mathcal{S}})$. Indeed, by assumption, $\mathcal{S}$ is a solid Garside family in $\mathcal{C}at(\underline{\mathcal{S}})$, hence it is closed under right-diamond in $\mathcal{C}at(\underline{\mathcal{S}})$ by Proposition IV.2.4 (solid Garside). So, owing to Lemma 3.8, in order to show that $\mathcal{S}_1$ is closed under right-complement in $\mathcal{C}at(\underline{\mathcal{S}})$, it is sufficient to show that $\mathcal{S}_1$ is closed under right-complement in $\mathcal{S}$. Assume that s, t lie in $\mathcal{S}_1$ and $s \bullet g = t \bullet f$ holds in $\mathcal{S}$. Let r be an $\mathcal{S}_1$-head of $s \bullet g$ in $\mathcal{S}$, which exists by assumption. Then we have $s \preccurlyeq_{\mathcal{S}} r$ and $t \preccurlyeq_{\mathcal{S}} r$, by definition of a head, so there exist s', t' in $\mathcal{S}$ satisfying $s \bullet t' = t \bullet s' = r$. Moreover, as s, t and r lie in $\mathcal{S}_1$ and $\mathcal{S}_1$ is closed under right-quotient in $\mathcal{S}$, we must have $t' \in \mathcal{S}_1$ and $s' \in \mathcal{S}_1$. Then we have $s \bullet t' \preccurlyeq_{\mathcal{S}} r \preccurlyeq_{\mathcal{S}} s \bullet g$, whence $t' \preccurlyeq_{\mathcal{S}} g$ and, similarly, $s' \preccurlyeq_{\mathcal{S}} f$ by Lemma VI.1.19, which is valid since, by assumption, $\underline{\mathcal{S}}$ is left-cancellative. So $\mathcal{S}_1$ is closed under right-complement in $\mathcal{C}at(\underline{\mathcal{S}})$.

On the other hand, the assumption that $\mathcal{S}_1$ is closed under right-quotient in $\mathcal{S}$ implies that it is closed under inverse in $\mathcal{S}$, hence in $\mathcal{C}at(\underline{\mathcal{S}})$ by Lemma 3.7. Then, by Lemma 3.9, we deduce that $\mathcal{S}ub(\mathcal{S}_1)$ is closed under right-quotient in $\mathcal{C}at(\underline{\mathcal{S}})$. We can now apply Proposition 1.25 and deduce that $\mathcal{S}ub(\mathcal{S}_1)$ is a head-category whenever one of the assumptions is satisfied. $\square$

4 Subcategories associated with functors

We now consider special subcategories of another type, namely subcategories arising in connection with a functor, with the aim of establishing compatibility results with Garside families. Our main claim here is that the notion of a functor correct for right-comultiples (Definition 4.5) is the appropriate extension of the classical notion of an lcm-homomorphism, as defined in [67], which is relevant even when right-lcms need not exist.

The section contains two subsections. In Subsection 4.1 we consider the case of subcategories that consist of the fixed points of some automorphism. Then, in Subsection 4.2, we investigate subcategories that are the images of functors and discuss conditions involving the functor that imply closure or compatibility results for its image; this is where the above alluded notion of correctness appears.

4.1 Subcategories of fixed points

If M is a braid monoid (Reference Structure 2, page 5) or, more generally, an Artin-Tits monoid of spherical type (see Chapter IX) and ϕ is an automorphism of M, then the submonoid M^ϕ of M consisting of all fixed points of ϕ is an Artin-Tits monoid of spherical type [67]. We may ask whether a similar result holds for the fixed points of every automorphism ϕ in a left-cancellative category $\mathcal{C}$. More precisely, we ask whether the subcategory $\mathcal{C}^\phi$ of all fixed points of ϕ is necessarily compatible with every Garside family of $\mathcal{C}$. There exist situations where the answer is positive, see for instance Exercise 85. However, the following example shows that the answer may also be negative, even when we assume that the considered Garside family is globally invariant under ϕ.

Example 4.1 (not compatible). Let M be the monoid $\langle \mathtt{a}, \mathtt{e} \mid \mathtt{e}^2 = 1, \mathtt{ae} = \mathtt{ea}\rangle^+$, let $S = \{\mathtt{a}, \mathtt{ea}\}$, and let ϕ be defined by $\phi(\mathtt{a}) = \mathtt{ea}$ and $\phi(\mathtt{e}) = \mathtt{e}$. Then M is the direct product of $\mathbb{N}$ and the cyclic group $\langle \mathtt{e} \mid \mathtt{e}^2 = 1\rangle$. It is cancellative, S is a Garside family of M, and ϕ is an automorphism of M satisfying $\phi(S) = S$. Now the submonoid M^ϕ contains 1 and $\mathtt{a}^2$, but $M^\phi \cap S$ is empty. Hence M^ϕ is not compatible with S.

The problem in Example 4.1 is that the considered Garside family contains distinct $=^\times$-equivalent elements. When we forbid this, the result is as can be expected:

Proposition 4.2 (fixed points). *If S is an $=^\times$-transverse Garside family in a left-cancellative category $\mathcal{C}$ and ϕ is an automorphism of $\mathcal{C}$ satisfying $\phi(S) = S$, then the subcategory $\mathcal{C}^\phi$ is compatible with S; in particular, $S \cap \mathcal{C}^\phi$ is a Garside family in $\mathcal{C}^\phi$.*

Proof. First, $\mathcal{C}^\phi$ is closed under inverse: indeed, if ϵ lies in $\mathcal{C}^\times(x, y) \cap \mathcal{C}^\phi$, we must have $\phi(x) = x$, whence $\epsilon\epsilon^{-1} = 1_x = \phi(1_x) = \phi(\epsilon\epsilon^{-1}) = \phi(\epsilon)\phi(\epsilon^{-1}) = \epsilon\phi(\epsilon^{-1})$, and $\epsilon^{-1} = \phi(\epsilon^{-1})$, that is, $\epsilon^{-1} \in \mathcal{C}^\phi$.

Next, assume that s belongs to $\mathcal{S}^\sharp \cap \mathcal{C}^\phi$. If s is invertible, then we saw above that s belongs to $(\mathcal{C}^\phi)^\times$. Otherwise, there exist t in $\mathcal{S}$ and ϵ in $\mathcal{C}^\times$ satisfying $s = t\epsilon$. Then we have $t\epsilon = s = \phi(s) = \phi(t)\phi(\epsilon)$, whence $t =^\times \phi(t)$. As $\mathcal{S}$ is assumed to be $=^\times$-transverse, we deduce $t = \phi(t)$, and $\epsilon = \phi(\epsilon)$. This shows that s lies in $(\mathcal{S} \cap \mathcal{C}^\phi)(\mathcal{C}^\phi)^\times$. It follows that $\mathcal{C}^\phi$ satisfies (2.11).

Finally, assume $g \in \mathcal{C}^\phi \setminus \mathcal{C}^\times$. Let $s_1|\cdots|s_p$ be a strict $\mathcal{S}$-normal decomposition of g. One easily checks that $\phi(s_1)|\cdots|\phi(s_p)$ is an $\mathcal{S}$-normal decomposition of $\phi(g)$, hence of g. Moreover, the assumption that $s_1|\cdots|s_p$ is strict implies that $s_1, ..., s_{p-1}$ lies in $\mathcal{S}$, hence so do $\phi(s_1), ..., \phi(s_{p-1})$, and that every element s_i is non-invertible, hence so is every element $\phi(s_i)$. So $\phi(s_1)|\cdots|\phi(s_p)$ is a strict $\mathcal{S}$-normal decomposition of s. By Corollary III.1.28 (normal unique), the assumption that $\mathcal{S}$ is $=^\times$-transverse implies $\phi(s_i) = s_i$ for $i = 1, ..., p$, that is, every entry s_i lies in $\mathcal{C}^\phi$. For $s \in \mathcal{C}^\phi \cap \mathcal{C}^\times$, the length-one path s is an $\mathcal{S}$-normal decomposition of s with entries in $\mathcal{C}^\phi$, so, in every case, every element of $\mathcal{C}^\phi$ admits an $\mathcal{S}$-normal decomposition with entries in $\mathcal{C}^\phi$. So $\mathcal{C}^\phi$ satisfies (2.12). Hence, by Proposition 2.10, $\mathcal{C}^\phi$ is compatible with $\mathcal{S}$. □

When there is no nontrivial invertible element, every family is $=^\times$-transverse, so Proposition 4.2 implies

Corollary 4.3 (fixed points). *If $\mathcal{S}$ is a Garside family in a left-cancellative category $\mathcal{C}$ containing no nontrivial invertible element and ϕ is an automorphism of $\mathcal{C}$ satisfying $\phi(\mathcal{S}) = \mathcal{S}$, the subcategory $\mathcal{C}^\phi$ is compatible with $\mathcal{S}$; in particular, $\mathcal{S} \cap \mathcal{C}^\phi$ is a Garside family in $\mathcal{C}^\phi$.*

Returning to germs, we consider the specific case of germs generated by atoms and describe the atoms in a fixed point subcategory.

Proposition 4.4 (atoms for fixed points). *If $\underline{\mathcal{S}}$ is a left-Noetherian Garside germ and ϕ is an automorphism of $\mathcal{C}at(\underline{\mathcal{S}})$, the germ $\underline{\mathcal{S}}^\phi$ is generated by its atoms, which are the right-lcms of ϕ-orbits of atoms of $\mathcal{S}$ that admit a common right-multiple and are not properly left-divisible by another such right-lcm.*

Proof. Let $\mathcal{A}$ consist of the right-lcms of orbits of atoms of $\mathcal{S}$ that admit a common right-multiple and are left-divisible by no other such right-lcm. By definition, every element of $\mathcal{A}$ lies in $\mathcal{S}^\phi$, and it is an atom in this structure.

Now, let s belong to $\mathcal{S}^\phi$. We claim that s is left-divisible by some element of $\mathcal{A}$. Indeed, by assumption, some atom t of $\mathcal{C}at(\underline{\mathcal{S}})$ left-divides s. Then, for every i, the element $\phi^i(t)$, which is also an atom as ϕ is an automorphism, left-divides s. Hence s is left-divisible by the right-lcm r of the ϕ-orbit of t, which is either an element of $\mathcal{A}$, or a right-multiple of an element of $\mathcal{A}$. If r belongs to $\mathcal{A}$, we are done. Otherwise, there exists an atom t' such that the right-lcm r' of the ϕ-orbit of t' is a proper left-divisor of r. If r' belongs to $\mathcal{A}$, we are done, otherwise we repeat the process. As $\mathcal{S}$ is left-Noetherian, the process leads in finitely many steps to an element of $\mathcal{A}$. □

4.2 Image subcategories

We now consider subcategories occurring as the image of a functor. A functor necessarily preserves invertibility in the sense that, if $\mathcal{C}, \mathcal{C}'$ are left-cancellative categories and ϕ is a functor from $\mathcal{C}$ to $\mathcal{C}'$, then the image of an invertible element of $\mathcal{C}$ under ϕ is an invertible element of $\mathcal{C}'$. Similarly, ϕ preserves right-multiples: if g is a right-multiple of f in $\mathcal{C}$, then $\phi(g)$ is a right-multiple of $\phi(f)$ in $\mathcal{C}'$. By contrast, no preservation is guaranteed when inverse images are considered (unless ϕ is bijective, hence it is an isomorphism). It turns out that several interesting properties can be established when convenient forms of such reversed preservation are satisfied.

Definition 4.5 (correct). Assume that $\mathcal{C}, \mathcal{C}'$ are left-cancellative categories.

(i) For $\mathcal{S} \subseteq \mathcal{C}$, a functor ϕ from $\mathcal{C}$ to $\mathcal{C}'$ is said to be *correct for invertibility on* $\mathcal{S}$ if, when s lies in $\mathcal{S}$ and $\phi(s)$ is invertible in $\mathcal{C}'$, then s is invertible in $\mathcal{C}$ and s^{-1} lies in $\mathcal{S}$.

(ii) For $\mathcal{S} \subseteq \mathcal{C}$, a functor ϕ from $\mathcal{C}$ to $\mathcal{C}'$ is said to be *correct for right-comultiples* (*resp. right-complements, resp. right-diamonds*) *on* $\mathcal{S}$ if, when s, t lie to $\mathcal{S}$ and $\phi(s)g = \phi(t)f$ holds in $\mathcal{C}'$ for some g, f, there exist s', t' in $\mathcal{C}$ and h in $\mathcal{C}'$ satisfying $st' = ts'$, $f = \phi(s')h$, and $g = \phi(t')h$, plus $st' \in \mathcal{S}$ (*resp.* plus $s', t' \in \mathcal{S}$, *resp.* plus $s', t', st' \in \mathcal{S}$).

Thus, roughly speaking, a functor ϕ is correct for invertibility on $\mathcal{S}$ if every invertible element in $\phi(\mathcal{S})$ comes from an invertible element in $\mathcal{S}$ (so, in particular, ϕ being correct for invertibility on the whole ambient category means that the image of a non-invertible element is non-invertible), and it is correct for right-comultiples on $\mathcal{S}$ if every common right-multiple of elements of $\phi(\mathcal{S})$ comes from a common right-multiple in $\mathcal{S}$, see Figure 2.

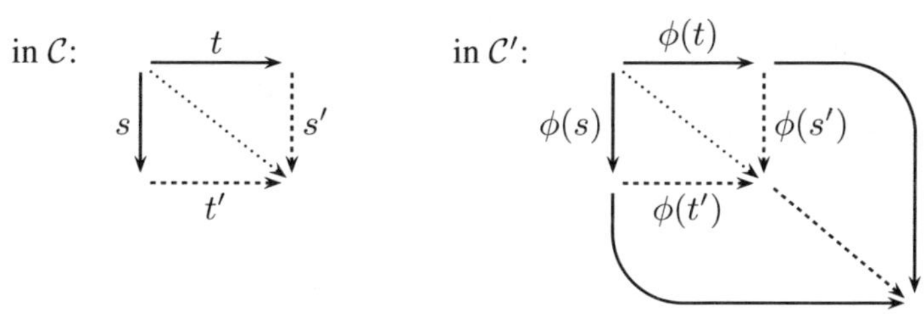

Figure 2. Correctness for right-comultiples on $\mathcal{S}_1$: for s, t in $\mathcal{S}_1$, every common right-multiple of $\phi(s)$ and $\phi(t)$ in $\mathcal{C}$ comes from a right-multiple in $\mathcal{C}_1$. Correctness with right-complements requires that s' and t' lie in $\mathcal{S}_1$, correctness with right-comultiples requires that st' lies in $\mathcal{S}_1$, correctness with right-diamonds requires both simultaneously.

Above we introduced three different notions of correctness on $\mathcal{S}$ involving common right-multiples. We observe that, provided $\mathcal{S}$ satisfies mild closure conditions in $\mathcal{C}$, and, in particular, when $\mathcal{S}$ is all of $\mathcal{C}$, these notions actually coincide.

Lemma 4.6. *Assume that $\mathcal{C}, \mathcal{C}'$ are left-cancellative categories and $\mathcal{S}$ is included in $\mathcal{C}$.*

(i) *A functor from $\mathcal{C}$ to $\mathcal{C}'$ that is correct for right-diamonds on $\mathcal{S}$ is correct for right-comultiples and right-complements on $\mathcal{S}$.*

(ii) *If $\mathcal{S}$ is closed under right-quotient in $\mathcal{C}$, a functor from $\mathcal{C}$ to $\mathcal{C}'$ that is correct for right-comultiples on $\mathcal{S}$ is correct for right-diamonds on $\mathcal{S}$.*

(iii) *If $\mathcal{S}$ is closed under product in $\mathcal{C}$, a functor from $\mathcal{C}$ to $\mathcal{C}'$ that is correct for right-complements on $\mathcal{S}$ is correct for right-diamonds on $\mathcal{S}$.*

Proof. Point (i) is obvious from the definition. For (ii), the assumption that s and st' belong to $\mathcal{S}$ implies that t' belongs to $\mathcal{S}$ whenever $\mathcal{S}$ is closed under right-quotient in $\mathcal{C}$. Finally, for (iii), the assumption that s and t' belong to $\mathcal{S}$ implies that st' belongs to $\mathcal{S}$ whenever $\mathcal{S}$ is closed under product in $\mathcal{C}$. □

As a warm-up exercise, one can observe that correctness properties for an identity-functor just correspond to usual closure properties.

Lemma 4.7. *If $\mathcal{S}$ is any subfamily of a left-cancellative category $\mathcal{C}$, the identity-functor on $\mathcal{C}$ is correct for invertibility (resp. right-comultiples, resp. right-complements, resp. right-diamonds) on $\mathcal{S}$ if and only if $\mathcal{S}$ is closed under inverse (resp. right-comultiple, resp. right-complement, resp. right-diamond) in $\mathcal{C}$.*

More generally, every correctness result implies a closure property for the associated image family. If ϕ is a functor defined on a category $\mathcal{C}$ and $\mathcal{S}$ is a subfamily of $\mathcal{C}$, we naturally write $\phi(\mathcal{S})$ for $\{\phi(s) \mid s \in \mathcal{S}\}$.

Proposition 4.8 (correct implies closed). *If $\mathcal{C}, \mathcal{C}'$ are left-cancellative categories and ϕ is a functor from $\mathcal{C}$ to $\mathcal{C}'$ that is correct for invertibility (resp. right-comultiples, resp. right-complements, resp. right-diamonds) on a subfamily $\mathcal{S}$ of $\mathcal{C}$, then $\phi(\mathcal{S})$ is closed under inverse (resp. right-comultiple, resp. right-complement, resp. right-diamond) in $\mathcal{C}'$.*

Proof. Assume that s' lies in $\phi(\mathcal{S})$ and s' is invertible in $\mathcal{C}'$. Write $s' = \phi(s)$ with s in $\mathcal{C}$. If ϕ is correct for invertibility, s is invertible in $\mathcal{C}$, and s^{-1} lies in $\mathcal{S}$, applying ϕ gives $s'\phi(s^{-1}) \in \mathbf{1}_{\mathcal{C}'}$, whence $s'^{-1} = \phi(s^{-1}) \in \phi(\mathcal{S})$. So $\phi(\mathcal{S})$ is closed under inverse in $\mathcal{C}'$.

Assume now that ϕ is correct for right-comultiples on $\mathcal{S}$. Assume that s, t lie in $\mathcal{S}$ and h is a common right-multiple of $\phi(s)$ and $\phi(t)$. By definition, there exist s', t' in $\mathcal{C}$ satisfying $st' = ts' \in \mathcal{S}$ and such that h is a right-multiple of $\phi(st')$, an element of $\phi(\mathcal{S})$ that is a right-multiple of $\phi(s)$ and $\phi(t)$. So $\phi(\mathcal{S})$ is closed under right-comultiple. The argument is similar for right-complements and right-diamonds. □

The implications of Proposition 4.8 are essentially equivalences when the considered functor is injective (see Exercise 93)—so that, in such a case, introducing a specific terminology would be useless. However, the correctness of ϕ on $\mathcal{S}$ is stronger than the closure of $\phi(\mathcal{S})$ as, in general, there is no reason why an equality $\phi(st') = \phi(ts')$ in $\mathcal{C}'$ should come from an equality $st' = ts'$ in $\mathcal{C}$.

Let us immediately observe that, in the case of a category that admits right-lcms or, at least, right-mcms, correctness corresponds to preservation properties and makes the connection with lcm-homomorphisms (homomorphisms that preserve the right-lcm [67]) explicit.

Proposition 4.9 (lcm-homomorphism). *If $\mathcal{C}, \mathcal{C}'$ are left-cancellative categories, $\mathcal{S}$ is a subfamily of $\mathcal{C}$, and any two elements of $\mathcal{S}$ admit a right-lcm that lies in $\mathcal{S}$, then a functor ϕ from $\mathcal{C}$ to $\mathcal{C}'$ is correct for right-comultiples on $\mathcal{S}$ if and only if, for all r, s, t in $\mathcal{S}$,*

(4.10) *If r is a right-lcm of s and t, then $\phi(r)$ is a right-lcm of $\phi(s)$ and $\phi(t)$.*

Proof. Assume that ϕ is correct for right-comultiples on $\mathcal{S}$, that s, t lie in $\mathcal{S}$, and that r is a right-lcm of s and t that lies in $\mathcal{S}$. As ϕ is a functor, $\phi(r)$ is a common right-multiple of $\phi(s)$ and $\phi(t)$. Assume that h is a common right-multiple of $\phi(s)$ and $\phi(t)$. As ϕ is correct for right-comultiples on $\mathcal{S}$, there exist s', t' in $\mathcal{C}$ satisfying $st' = ts' \in \mathcal{S}$ and $\phi(st') \preccurlyeq h$. As st' is a common right-multiple of s and t, we must have $r \preccurlyeq st'$ in $\mathcal{C}$, whence $\phi(r) \preccurlyeq \phi(st') \preccurlyeq h$ in $\mathcal{C}'$. So $\phi(r)$ is a right-lcm of $\phi(s)$ and $\phi(t)$ in $\mathcal{C}'$. Hence (4.10) is satisfied.

Conversely, assume that (4.10) holds for all r, s, t in $\mathcal{S}$. Assume that s, t lie in $\mathcal{S}$ and h is a common right-multiple of $\phi(s)$ and $\phi(t)$. Let r be a right-lcm of s and t that lies in $\mathcal{S}$. Then $\phi(r)$ is a right-lcm of $\phi(s)$ and $\phi(t)$, so $\phi(r) \preccurlyeq h$ holds. This shows that ϕ is correct for right-comultiples on $\mathcal{S}$. □

If we only assume that the source category $\mathcal{C}$ admits conditional right-lcms, then the direct implication in Proposition 4.9 remains valid, that is, correctness implies (4.10); however, the converse implication is not guaranteed, since it may happen that $\phi(s)$ and $\phi(t)$ admit a common right-multiple whereas s and t do not, and correctness for right-comultiples is a proper extension of lcm preservation. We refer to Exercise 92 for a formulation similar to that of Proposition 4.9 in the case when mcms exist.

On the other hand, if we specialize more and consider categories that admit unique right-lcms, a criterion similar to Proposition 4.9 characterizes correctness for right-complements. We recall that, in such a context, $f\backslash g$ denotes the (unique) element such that $f(f\backslash g)$ is the right-lcm of f and g.

Proposition 4.11 (complement-homomorphism). (i) *If $\mathcal{C}, \mathcal{C}'$ are left-cancellative categories that admit unique right-lcms and $\mathcal{S}$ is a subfamily of $\mathcal{C}$ that is closed under $\backslash$, then a functor ϕ from $\mathcal{C}$ to $\mathcal{C}'$ is correct for right-complements on $\mathcal{S}$ if and only if, for all s, t in $\mathcal{S}$ with the same source, we have*

(4.12) $$\phi(s)\backslash\phi(t) = \phi(s\backslash t).$$

(ii) *If $\mathcal{A}$ is a subfamily of $\mathcal{S}$ that generates $\mathcal{S}$, then* (4.12) *holds for all s, t in $\mathcal{S}$ if and only if it holds for all s, t in $\mathcal{A}$.*

Proof. (i) Assume that ϕ is correct for right-complements on $\mathcal{S}$. Let s, t lie in $\mathcal{S}$ and share the same source. We have $s(s\backslash t) = t(t\backslash s)$ in $\mathcal{C}$, whence $\phi(s)\phi(s\backslash t) = \phi(t)\phi(t\backslash s)$ in $\mathcal{C}'$. By the definition of the operation $\backslash$ in $\mathcal{C}'$, this implies $\phi(s)\backslash\phi(t) \preccurlyeq \phi(s\backslash t)$. On the other hand, as we have $\phi(s)(\phi(s)\backslash\phi(t)) = \phi(t)(\phi(t)\backslash\phi(s))$ in $\mathcal{C}'$ and, as ϕ is correct for right-complements on $\mathcal{S}$, there must exist $\hat{s}, \hat{t}$ in $\mathcal{S}$ and h' in $\mathcal{C}'$ satisfying $s\hat{t} = t\hat{s}$ in $\mathcal{C}$ and $\phi(t)\backslash\phi(s) = \phi(\hat{s})h'$, and $\phi(s)\backslash\phi(t) = \phi(\hat{t})h'$ in $\mathcal{C}'$. The first equality implies $s\backslash t \preccurlyeq \hat{t}$ and $t\backslash s \preccurlyeq \hat{s}$, and we deduce $\phi(s\backslash t) \preccurlyeq \phi(s)\backslash\phi(t)$, whence $\phi(s\backslash t) =^{\times} \phi(s)\backslash\phi(t)$, and $\phi(s\backslash t) = \phi(s)\backslash\phi(t)$ since $\mathcal{C}'$ has no nontrivial invertible element. So (4.12) holds for all s, t in $\mathcal{S}$, hence a fortiori in a subfamily of $\mathcal{S}$.

Conversely, assume first that (4.12) holds for all s, t in $\mathcal{S}$ and, in $\mathcal{C}'$, we have $\phi(s)g = \phi(t)f$ with s, t in $\mathcal{S}$. By definition of $\backslash$ in $\mathcal{C}'$, we must have $f = (\phi(t)\backslash\phi(s))h'$ and $g = (\phi(s)\backslash\phi(t))h'$ for some h' in $\mathcal{C}'$. Then we have $\phi(s)\backslash\phi(t) = \phi(s\backslash t)$ and $\phi(t)\backslash\phi(s) = \phi(t\backslash s)$, so $s\backslash t$ and $t\backslash s$, which belong to $\mathcal{S}$ by assumption, witness for the expected instance of correctness. Hence ϕ is correct for right-complements on $\mathcal{S}$.

(ii) If (4.12) holds for all s, t in a subfamily $\mathcal{A}$ of $\mathcal{S}$ that generates it, it follows from the formulas of Corollary II.2.13 (iterated complement) and from the assumption that ϕ is a functor, hence preserves products, that (4.12) holds for all finite products of elements of $\mathcal{A}$, hence for all elements of $\mathcal{S}$. □

What we do now is to show how correctness can be used to establish nontrivial properties, typically injectivity (Corollary 4.17).

The technical interest of introducing several notions of correctness is that, whereas correctness for right-comultiples is more useful in applications, it does not easily extends from a family to the subcategory generated by that family, but its variants do:

Lemma 4.13. *If $\mathcal{C}, \mathcal{C}'$ are left-cancellative categories and $\mathcal{S}$ is a generating family in $\mathcal{C}$, then every functor from $\mathcal{C}$ to $\mathcal{C}'$ that is correct for invertibility (resp. right-complements, resp. right-diamonds) on $\mathcal{S}$ is correct for invertibility (resp. right-complements, resp. right-diamonds) on $\mathcal{C}$.*

Proof. Assume that g lies in $\mathcal{C}$ and $\phi(g)$ is invertible in $\mathcal{C}'$. Write $g = s_1 \cdots s_p$ with $s_1, \ldots, s_p$ in $\mathcal{S}$. Then we have $\phi(g) = \phi(s_1) \cdots \phi(s_p)$, and the assumption that $\phi(g)$ is invertible implies that $\phi(s_1), \ldots, \phi(s_p)$ are invertible as well. As ϕ is correct for invertibility on $\mathcal{S}$, the element s_i is invertible in $\mathcal{C}$ for each i. Hence so is g.

For right-complements and right-diamonds, the scheme is similar to that used in Chapter IV for Proposition IV.1.15 (grid) and consists in applying the assumption repeatedly to construct a rectangular grid with edges (and diagonals in the case of diamond) that lie in $\phi(\mathcal{S})$ in order to factorize the initial equality, here of the form $\phi(s_1 \cdots s_p)g = \phi(t_1 \cdots t_q)f$ with $s_1, \ldots, t_q$ in $\mathcal{S}$. We skip details (In the case of unique lcms, the construction reduces to the equalities for an iterated complement used at the end of the proof of Proposition 4.11). □

Thus, in order to establish correctness for invertibility and right-complements on the whole source category of a functor, it is enough to establish it on a generating family of that category. Merging the results, we obtain simple criteria for establishing properties of an image category.

Proposition 4.14 (closure of the image). *If $\mathcal{C}, \mathcal{C}'$ are left-cancellative categories and ϕ is a functor from $\mathcal{C}$ to $\mathcal{C}'$ that is correct for invertibility and right-complements on a generating subfamily of $\mathcal{C}$, then $\phi(\mathcal{C})$ is closed under right-quotient and right-diamond in $\mathcal{C}'$.*

Proof. By Lemma 4.13, ϕ is correct for invertibility and right-complements on $\mathcal{C}$. By Lemma 4.6, as $\mathcal{C}$ is closed under product in itself, ϕ is correct for right-diamonds on $\mathcal{C}$. Then Proposition 4.8 implies that $\phi(\mathcal{C})$ is closed under inverse and right-diamond in $\mathcal{C}'$. By Lemma 1.8, this implies in turn that $\phi(\mathcal{C})$ is closed under right-quotient in $\mathcal{C}'$. □

More importantly, it is easy to establish preservation properties of the functor.

Proposition 4.15 (correct *vs.* **divisibility).** *If $\mathcal{C}, \mathcal{C}'$ are left-cancellative categories and ϕ is a functor from $\mathcal{C}$ to $\mathcal{C}'$ that is correct for invertibility and right-complements on a generating subfamily of $\mathcal{C}$, then, for all f, g in $\mathcal{C}$, we have*

$$f \preccurlyeq g \Leftrightarrow \phi(f) \preccurlyeq \phi(g) \quad \text{and} \quad f =^{\times} g \Leftrightarrow \phi(f) =^{\times} \phi(g). \tag{4.16}$$

Proof. First, by Lemma 4.13, ϕ is correct for invertibility and right-complements on $\mathcal{C}$.

As ϕ is a functor, $g = fg'$ implies $\phi(g) = \phi(f)\phi(g')$, so $f \preccurlyeq g$ implies $\phi(f) \preccurlyeq \phi(g)$.

Conversely, assume that $\phi(f) \preccurlyeq \phi(g)$ holds in $\mathcal{C}'$, say $\phi(f)\hat{g} = \phi(g)$. As ϕ is correct for right-complements on $\mathcal{C}$, there exist f', g' in $\mathcal{C}$ and h' in $\mathcal{C}'$ satisfying $fg' = gf'$, $1_{\phi(y)} = \phi(f')h'$, and $\hat{g} = \phi(g')h'$, where y is the target of g. The second equality implies that $\phi(f')$ is invertible in $\mathcal{C}'$. The assumption that ϕ is correct for invertibility implies that f' is invertible in $\mathcal{C}$, and we deduce $f(g'f'^{-1}) = g$ and, from there, $f \preccurlyeq g$ in $\mathcal{C}$.

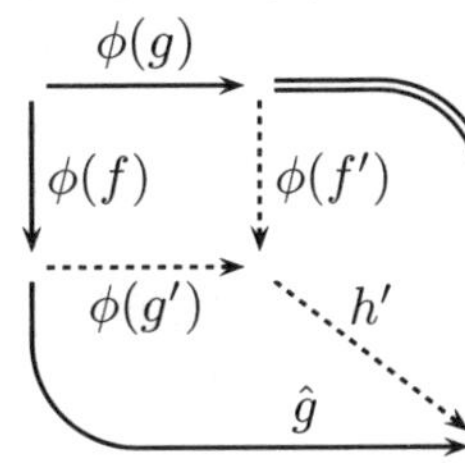

The result for $=^{\times}$ follows as $f =^{\times} g$ is equivalent to the conjunction of $f \preccurlyeq g$ and $g \preccurlyeq f$ in a left-cancellative context. □

In particular, we obtain a simple injectivity criterion:

Corollary 4.17 (correct *vs.* **injectivity).** *If $\mathcal{C}, \mathcal{C}'$ are left-cancellative categories and ϕ is a functor from $\mathcal{C}$ to $\mathcal{C}'$ that is correct for invertibility and right-complements on a generating subfamily of $\mathcal{C}$, then ϕ is injective on $\mathcal{C}$ if and only if it is injective on $\mathcal{C}^{\times}$.*

Proof. Assume that ϕ is injective on $\mathcal{C}^{\times}$ and $\phi(f) = \phi(g)$ holds with $f \in \mathcal{C}(\cdot, x)$. Proposition 4.15 implies $f =^{\times} g$, that is, $f\epsilon = g$ for some ϵ in $\mathcal{C}^{\times}(x, \cdot)$. Applying ϕ, we deduce $\phi(f)\phi(\epsilon) = \phi(g) = \phi(f) = \phi(f)\phi(1_x)$, whence $\phi(\epsilon) = \phi(1_x)$, and $\epsilon = 1_x$ under the specific assumption on ϕ. □

Specializing more and merging with Proposition 4.11, we obtain

Corollary 4.18 (complement *vs.* **injectivity).** *If $\mathcal{C}, \mathcal{C}'$ are left-cancellative categories admitting unique right-lcms and ϕ is a functor from $\mathcal{C}$ to $\mathcal{C}'$ that preserves $\backslash$ on a generating subfamily $\mathcal{A}$ of $\mathcal{C}$, then ϕ is injective on $\mathcal{C}$ if and only if $\phi(s) \in \mathbf{1}_{\mathcal{C}'}$ holds for no s in $\mathcal{A} \setminus \mathbf{1}_{\mathcal{C}}$.*

Proof. If s lies in $\mathcal{A}(x, \cdot) \setminus \mathbf{1}_\mathcal{C}$, then $\phi(s) \in \mathbf{1}_{\mathcal{C}'}$ requires $\phi(s) = \phi(1_x)$ and contradicts the injectivity of ϕ.

Conversely, if $\phi(s) \in \mathbf{1}_{\mathcal{C}'}$ holds for no s in $\mathcal{A} \setminus \mathbf{1}_\mathcal{C}$, then $\phi(g) \in \mathbf{1}_{\mathcal{C}'}$ holds for no g in $\mathcal{C} \setminus \mathbf{1}_\mathcal{C}$ since $\mathcal{A} \setminus \mathbf{1}_\mathcal{C}$ generates $\mathcal{C}$. Therefore, ϕ is correct for invertibility. By Proposition 4.11, ϕ is correct for right-complements on $\mathcal{C}$ since it preserves the operation $\backslash$. Then, as every family $\mathcal{C}^\times(x, \cdot)$ is a singleton, Corollary 4.17 implies that ϕ is injective. □

In Proposition 4.15 and Corollary 4.17, we assume a functor ϕ satisfying correctness assumptions to be given. Before describing a concrete example, let us observe that, at least in the case of unique right-lcms, constructing functors of this kind is easy.

Proposition 4.19 (complement preserving functor). *If $\mathcal{C}, \mathcal{C}'$ are left-cancellative categories admitting unique right-lcms and $\mathcal{C}$ is Noetherian with atom family $\mathcal{A}$, then every map $\phi : \mathcal{A} \to \mathcal{C}'$ satisfying, for all a, b in $\mathcal{A}$ and $c_1|\cdots|c_p$ an $\mathcal{A}$-decomposition of $a\backslash b$,*

$$\phi(a)\backslash\phi(b) = \phi(c_1)\cdots\phi(c_p), \tag{4.20}$$

induces a $\backslash$-preserving functor from $\mathcal{C}$ to $\mathcal{C}'$. The latter is injective whenever $\phi(a) \in \mathbf{1}_{\mathcal{C}'}$ holds for no a in $\mathcal{A}$.

Proof. By Proposition IV.3.21 (right-lcm witness), the category $\mathcal{C}$ admits a presentation by all relations of the form $a\theta(a, b) = b\theta(b, a)$ with a, b in $\mathcal{A}$ and θ a right-lcm witness on $\mathcal{A}$, that is, a map from $\mathcal{A} \times \mathcal{A}$ to the free category $\mathcal{A}^*$ choosing an $\mathcal{A}$-decomposition of $a\backslash b$ for all a, b in $\mathcal{A}$. If (4.20) is satisfied, then the canonical extension ϕ^* of ϕ to $\mathcal{A}^*$ induces a well-defined functor on $\mathcal{C}$ since, by assumption, we have

$$\begin{aligned}\phi^*(a\theta(a, b)) &= \phi(a)\phi^*(\theta(a, b)) = \phi(a)(\phi(a)\backslash\phi(b))\\ &= \phi(b)(\phi(b)\backslash\phi(a)) = \phi(b)\phi^*(\theta(b, a)) = \phi^*(b\theta(b, a)).\end{aligned}$$

Still denoted by ϕ, the induced functor preserves the operation $\backslash$ on $\mathcal{A}$, hence on $\mathcal{C}$, owing to the laws for an iterated complement. Then Corollary 4.18 applies, and it says that ϕ is injective if and only if no atom has a trivial image. □

Example 4.21 (Artin–Tits monoids). Consider the $2n$-strand braid monoid B_{2n}^+ (Reference Structure 2, page 5). Let B^+ be the Artin–Tits monoid of type B_n, that is, the monoid generated by $\sigma_0, \ldots, \sigma_n$, where the elements σ_i with $i > 0$ satisfy the usual braid relations and σ_0 satisfies $\sigma_0\sigma_1\sigma_0\sigma_1 = \sigma_1\sigma_0\sigma_1\sigma_0$ and commutes with $\sigma_2, \ldots, \sigma_n$ (see Chapter IX). Let ϕ be the mapping from $\{\sigma_0, \ldots, \sigma_n\}$ into $\{\sigma_1, \ldots, \sigma_{2n-1}\}$ defined by $\phi(\sigma_0) = \sigma_n$ and $\phi(\sigma_i) = \sigma_{n-i}\sigma_{n+i}$ for $i \geqslant 1$. We claim that (4.20) is satisfied, as an amusing verification shows: the only nontrivial case is the pair (σ_0, σ_1), which is illustrated in the reversing diagram on the right, on which we read the values of $\phi(\sigma_0)\backslash\phi(\sigma_1)$ and $\phi(\sigma_1)\backslash\phi(\sigma_0)$.

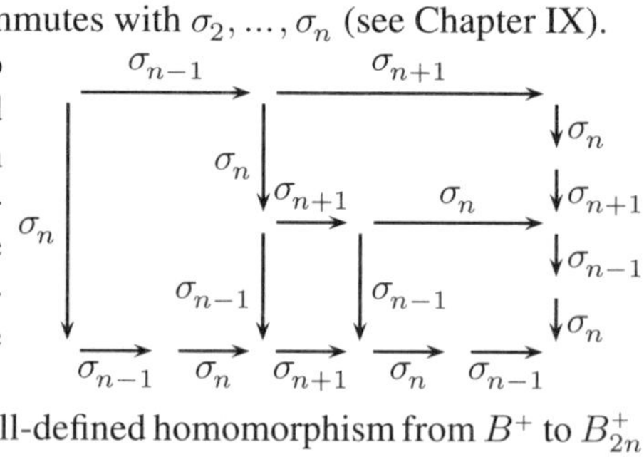

Then Proposition 4.19 says that ϕ induces a well-defined homomorphism from B^+ to B_{2n}^+. Moreover, as $\phi(\sigma_i)$ is trivial for no i, this homomorphism is an embedding: the Artin–Tits monoid of type B_n embeds in the Artin–Tits monoid of type A_{2n-1}.

We refer to Exercise 94 for other similar embeddings.

We conclude with the observation that, in the case of (left-) Ore categories, the injectivity result of Corollary 4.17 extends to the groupoids of fractions.

Proposition 4.22 (correct *vs.* groupoid). *If $\mathcal{C}, \mathcal{C}'$ are left-Ore categories and ϕ is a functor from $\mathcal{C}$ to $\mathcal{C}'$ that is correct for invertibility and right-complements on a generating family of $\mathcal{C}$ and injective on $\mathcal{C}^\times$, then ϕ extends to an injective functor $\phi^\pm$ from $\mathcal{E}nv(\mathcal{C})$ to $\mathcal{E}nv(\mathcal{C}')$, and we have then $\phi(\mathcal{C}) = \phi^\pm(\mathcal{E}nv(\mathcal{C})) \cap \mathcal{C}'$.*

Proof. By Lemma II.3.13, there is a unique way to extend ϕ to a functor $\phi^\pm$ from $\mathcal{E}nv(\mathcal{C})$ to $\mathcal{E}nv(\mathcal{C}')$, namely putting $\phi^\pm(f^{-1}g) = \phi(f)^{-1}\phi(g)$ for f, g in $\mathcal{C}$. Now assume that $\phi^\pm(f^{-1}g)$ is an identity element. By definition, we deduce the equality $\phi(f) = \phi(g)$ in $\mathcal{C}$, whence $f = g$ by Corollary 4.17. Hence $\phi^\pm$ is injective on $\mathcal{E}nv(\mathcal{C})$.

Next, as $\mathcal{C}$ is included in $\mathcal{E}nv(\mathcal{C})$, we have $\phi(\mathcal{C}) \subseteq \phi^\pm(\mathcal{E}nv(\mathcal{C})) \cap \mathcal{C}'$. Conversely, assume $\phi^\pm(f^{-1}g) = h \in \mathcal{C}'$ with f, g in $\mathcal{C}$. By definition, we have $\phi(f)^{-1}\phi(g) = h$, whence $\phi(g) = \phi(f)h$. By Corollary 4.17 again, $\phi(\mathcal{C})$ is closed under right-quotient in $\mathcal{C}'$, so h must belong to $\phi(\mathcal{C})$, that is, we have $h = \phi(h_1)$ for some h_1 in $\mathcal{C}$. We deduce the equality $\phi^\pm(f^{-1}g) = \phi^\pm(h_1)$, whence $f^{-1}g = h_1$ since $\phi^\pm$ is injective. Hence $\phi^\pm(\mathcal{E}nv(\mathcal{C})) \cap \mathcal{C}'$ is included in $\phi(\mathcal{C})$. □

Proposition 4.22 typically applies to the embedding of Artin–Tits monoids described in Example 4.21 (and its analogs of Exercise 94).

Exercises

Exercise 72 ($=^\times$-closed subcategory). (i) Show that a subcategory $\mathcal{C}_1$ of a left-cancellative category $\mathcal{C}$ is $=^\times$-closed if and only if, for each x in $\mathcal{O}bj(\mathcal{C}_1)$, the families $\mathcal{C}^\times(x, \cdot)$ and $\mathcal{C}^\times(\cdot, x)$ are included in $\mathcal{C}_1$. (ii) Deduce that $\mathcal{C}_1$ is $=^\times$-closed if and only if $\mathcal{C}_1^\times$ is a union of connected components of $\mathcal{C}^\times$.

Exercise 73 (greedy paths). Assume that $\mathcal{C}$ is a cancellative category, $\mathcal{S}$ is included in $\mathcal{C}$, and $\mathcal{C}_1$ is a subcategory of $\mathcal{C}$ that is closed under left-quotient. Put $\mathcal{S}_1 = \mathcal{S} \cap \mathcal{C}_1$. Show that every $\mathcal{C}_1$-path that is $\mathcal{S}$-greedy in $\mathcal{C}$ is $\mathcal{S}_1$-greedy in $\mathcal{C}_1$.

Exercise 74 (compatibility with $\mathcal{C}$). Assume that $\mathcal{C}$ is a left-cancellative category. Show that every subcategory of $\mathcal{C}$ that is closed under inverse is compatible with $\mathcal{C}$ viewed as a Garside family in itself.

Exercise 75 (not compatible). Let M be the free abelian monoid generated by $\mathtt{a}$ and $\mathtt{b}$, and let N be the submonoid generated by $\mathtt{a}$ and $\mathtt{ab}$. (i) Show that N is not closed under right-quotient in M. (ii) Let $S = \{1, \mathtt{a}, \mathtt{b}, \mathtt{ab}\}$. Show that N is not compatible with S. (iii) Let $S' = \{\mathtt{a}^p\mathtt{b}^\epsilon \mid p \geqslant 0, i \in \{0, 1\}\}$. Show that N not compatible with S'.

Exercise 76 (not closed under right-quotient). (i) Show that every submonoid $m\mathbb{N}$ of the additive monoid $\mathbb{N}$ is closed under right-quotient, but that $2\mathbb{N}+3\mathbb{N}$ of $\mathbb{N}$ is not. (ii) Let M be the monoid $\mathbb{N} \rtimes (\mathbb{Z}/2\mathbb{Z})^2$, where the generator $\mathtt{a}$ of $\mathbb{N}$ acts on the generators $\mathtt{e}, \mathtt{f}$ of $(\mathbb{Z}/2\mathbb{Z})^2$ by $\mathtt{ae} = \mathtt{fa}$ and $\mathtt{af} = \mathtt{ea}$, and let N be the submonoid of M generated by $\mathtt{a}$ and $\mathtt{e}$. Show that M is left-cancellative, and its elements admit a unique expression of the form $\mathtt{a}^p\mathtt{e}^i\mathtt{f}^j$ with $p \geqslant 0$ and $i, j \in \{0, 1\}$, and that N is $M \setminus \{\mathtt{f}, \mathtt{ef}\}$. (iii) Show that N is not closed under right-quotient in M. (iv) Let $S = \{\mathtt{a}\}$. Show that S is a Garside family in M and determine $S^\sharp$. Show that N is compatible with S. [Hint: Show that $S \cap N$, which is S, is not a Garside family in N.] (v) Show that $S^\sharp \cap N$ is a Garside family in N and N is compatible with $S^\sharp$.

Exercise 77 (not closed under divisor). Let $M = \langle \mathtt{a}, \mathtt{b} \mid \mathtt{ab} = \mathtt{ba}, \mathtt{a}^2 = \mathtt{b}^2 \rangle^+$, and let N be the submonoid of M generated by $\mathtt{a}^2$ and $\mathtt{ab}$. Show that N is compatible with every Garside family S of M, but that M is not closed under left- and right-divisor.

Exercise 78 (head implies closed). Assume that $\mathcal{C}$ is a left-cancellative category, $\mathcal{S}$ is a subfamily of $\mathcal{C}$ that is closed under right-comultiple in $\mathcal{C}$, and $\mathcal{C}_1$ is a subcategory of $\mathcal{C}$. Put $\mathcal{S}_1 = \mathcal{S} \cap \mathcal{C}_1$. (i) Show that, if every element of $\mathcal{S}$ admits a $\mathcal{C}_1$-head that lies in $\mathcal{S}_1$, then $\mathcal{S}_1$ is closed under right-comultiple in $\mathcal{C}$. (ii) Show that, if, moreover, $\mathcal{S}_1$ is closed under right-quotient in $\mathcal{C}$, then $\mathcal{S}_1$ is closed under right-diamond in $\mathcal{C}$.

Exercise 79 (head on generating family). Assume that $\mathcal{C}$ is a left-cancellative category that is right-Noetherian, $\mathcal{C}_1$ is a subcategory of $\mathcal{C}$ that is closed under inverse, and $\mathcal{S}$ is a subfamily of $\mathcal{C}$ such that every element of $\mathcal{S}$ admits a $\mathcal{C}_1$-head that lies in $\mathcal{S}$. Assume moreover that $\mathcal{S}$ is closed under right-comultiple and that $\mathcal{S} \cap \mathcal{C}_1$ generates $\mathcal{C}_1$ and is closed under right-quotient in $\mathcal{C}$. Show that $\mathcal{C}_1$ is a head subcategory of $\mathcal{C}$. [Hint: Apply Exercise 78.]

Exercise 80 (transitivity of compatibility). Assume that $\mathcal{S}$ is a Garside family in a left-cancellative category $\mathcal{C}$, $\mathcal{C}_1$ is a subcategory of $\mathcal{C}$ that is compatible with $\mathcal{S}$, and $\mathcal{C}_2$ is a subcategory of $\mathcal{C}_1$ that is compatible with $\mathcal{S}_1 = \mathcal{S} \cap \mathcal{C}_1$. Show that $\mathcal{C}_2$ is compatible with $\mathcal{S}$.

Exercise 81 (transitivity of head subcategory). Assume that $\mathcal{C}$ is a left-cancellative category, $\mathcal{C}_1$ is a head subcategory of $\mathcal{C}$, and $\mathcal{C}_2$ is a subcategory of $\mathcal{C}_1$. Show that $\mathcal{C}_2$ is a head subcategory of $\mathcal{C}$ if and only if it is a head subcategory of $\mathcal{C}_1$.

Exercise 82 (recognizing compatible IV). Assume that $\mathcal{S}$ is a Garside family in a left-cancellative category $\mathcal{C}$ and $\mathcal{C}_1$ is a subcategory of $\mathcal{C}$ that is closed under right-quotient in $\mathcal{C}$. Show that $\mathcal{C}_1$ is compatible with $\mathcal{S}$ if and only if, putting $\mathcal{S}_1 = \mathcal{S} \cap \mathcal{C}_1$, (i) the family $\mathcal{S}_1$ is a Garside family in $\mathcal{C}_1$, and (ii) a $\mathcal{C}_1$-path is strict $\mathcal{S}_1$-normal in $\mathcal{C}_1$ if and only if it is strict $\mathcal{S}$-normal in $\mathcal{C}$.

Exercise 83 (inverse image). Assume that $\mathcal{C}, \mathcal{C}'$ are left-cancellative categories, ϕ is a functor from $\mathcal{C}$ to $\mathcal{C}'$, and $\mathcal{C}'_1$ is a subcategory of $\mathcal{C}'$ that is closed under left- and right-divisor. Show that the subcategory $\phi^{-1}(\mathcal{C}'_1)$ is compatible with every Garside family of $\mathcal{C}$. (ii) Let B^+ be the Artin–Tits monoid of type B as defined in Example 4.21. Show that the map ϕ defined by $\phi(\sigma_0) = 1$ and $\phi(\sigma_i) = 0$ for $i \geqslant 1$ extends to a homomorphism of B^+

to $\mathbb{N}$, and that the submonoid $N = \{g \in M \mid \phi(g) = 0\}$ of B^+ is compatible with every Garside family of B^+.

Exercise 84 (intersection). Assume that $\mathcal{S}$ is a Garside family in a left-cancellative category $\mathcal{C}$. (i) Let $\mathcal{F}$ be the family of all subcategories of $\mathcal{C}$ that are closed under right-quotient, compatible with $\mathcal{S}$, and $=^\times$-closed. Show that every intersection of elements of $\mathcal{F}$ belongs to $\mathcal{F}$. (ii) Same question when "$=^\times$-closed" is replaced with "including $\mathcal{C}^\times$". (iii) Idem when $\mathcal{C}$ contains no nontrivial invertible element and "$=^\times$-closed" is skipped.

Exercise 85 (fixed points). Assume that $\mathcal{C}$ is a left-cancellative category and $\phi : \mathcal{C} \to \mathcal{C}$ is a functor. Show that the fixed point subcategory $\mathcal{C}^\phi$ is compatible with $\mathcal{C}$, viewed as a Garside family in itself.

Exercise 86 (connection between closure properties). Assume that $\mathcal{S}$ is a subfamily in a left-cancellative category $\mathcal{C}$ and $\mathcal{S}_1$ is a subfamily of $\mathcal{S}$. (i) Show that, if $\mathcal{S}_1$ is closed under product, inverse, and right-complement in $\mathcal{S}$, then $\mathcal{S}_1$ is closed under right-quotient in $\mathcal{S}$. (ii) Assume that $\mathcal{S}_1$ is closed under product and right-complement in $\mathcal{S}$. Show that, if $\mathcal{S}$ is closed under left-divisor in $\mathcal{C}$, then $\mathcal{S}_1$ is closed under right-comultiple in $\mathcal{S}$. Show that, if $\mathcal{S}$ is closed under right-diamond in $\mathcal{C}$, then $\mathcal{S}ub(\mathcal{S}_1)$ is closed under right-comultiple in $\mathcal{S}$. (iii) Show that, if $\mathcal{S}_1$ is closed under identity and product in $\mathcal{S}$, then $\mathcal{S}_1$ is closed under inverse and right-diamond in $\mathcal{S}$ if and only if $\mathcal{S}_1$ is closed under right-quotient and right-comultiple in $\mathcal{S}$.

Exercise 87 (subgerm). Assume that $\underline{\mathcal{S}}$ is a left-cancellative germ and $\underline{\mathcal{S}}_1$ is a subgerm of $\mathcal{S}$ such that the relation $\preccurlyeq_{\mathcal{S}_1}$ is the restriction to $\mathcal{S}_1$ of the relation $\preccurlyeq_{\mathcal{S}}$. Show that $\mathcal{S}_1$ is closed under right-quotient in $\mathcal{S}$.

Exercise 88 (transitivity of closure). Assume that $\underline{\mathcal{S}}_1$ is a subgerm of a Garside germ $\underline{\mathcal{S}}$, the subcategory $\mathcal{S}ub(\mathcal{S}_1)$ is closed under right-quotient and right-diamond in $\mathcal{C}at(\underline{\mathcal{S}})$, and $\mathcal{S}_1$ is closed under inverse and right-complement (*resp.* right-diamond) in $\mathcal{S}ub(\mathcal{S}_1)$. Show that $\mathcal{S}_1$ is closed under inverse and right-complement (*resp.* right-diamond) in $\mathcal{S}$.

Exercise 89 (transfer of closure). Assume that $\mathcal{C}$ is a left-cancellative category, $\mathcal{S}$ is a subfamily of $\mathcal{C}$, and $\mathcal{S}_1$ is a subfamily of $\mathcal{S}$ that is closed under identity and product. (i) Show that, if $\mathcal{S}ub(\mathcal{S}_1)$ is closed under right-quotient in $\mathcal{C}$, then $\mathcal{S}_1$ is closed under right-quotient in $\mathcal{S}$. (ii) Show that, if, moreover, $\mathcal{S}$ is closed under right-quotient in $\mathcal{C}$, then $\mathcal{S}_1$ is closed under right-quotient in $\mathcal{S}ub(\mathcal{S}_1)$.

Exercise 90 (braid subgerm). Let $\underline{\mathcal{S}}$ be the six-element Garside germ associated with the divisors of Δ_3 in the braid monoid B_3^+. (i) Describe the subgerm $\underline{\mathcal{S}}_1$ of $\underline{\mathcal{S}}$ generated by σ_1 and σ_2. Compare $\mathcal{M}on(\underline{\mathcal{S}}_1)$ and $\mathcal{S}ub(\mathcal{S}_1)$ (describe them explicitly). (ii) Same questions with σ_1 and $\sigma_2\sigma_1$. Is $\mathcal{S}_1$ closed under right-quotient in $\underline{\mathcal{S}}$ in this case?

Exercise 91 ($=^\times$-closed). Show that, if $\underline{\mathcal{S}}_1$ is a subgerm of an associative germ $\underline{\mathcal{S}}$, then $\mathcal{S}ub(\mathcal{S}_1)$ is $=^\times$-closed in $\mathcal{C}at(\underline{\mathcal{S}})$ if and only if $\underline{\mathcal{S}}_1$ is $=^\times_{\mathcal{S}}$-closed in $\mathcal{S}$.

Exercise 92 (correct *vs.* mcms). Assume that $\mathcal{C}$ and $\mathcal{C}'$ are left-cancellative categories and $\mathcal{S}$ is included in $\mathcal{C}$. Assume moreover that $\mathcal{C}$ and $\mathcal{C}'$ admit right-mcms and $\mathcal{S}$ is closed under right-mcm. Show that a functor ϕ from $\mathcal{C}$ to $\mathcal{C}'$ is correct for right-comultiples on $\mathcal{S}$ if and only if, for all s, t in $\mathcal{S}$, every right-mcm of $\phi(s)$ and $\phi(t)$ is $=^\times$-equivalent to the image under ϕ of a right-mcm of s and t.

Exercise 93 (correct *vs.* closed). Assume that $\mathcal{C}, \mathcal{C}'$ are left-cancellative categories, $\mathcal{S}$ is included in $\mathcal{C}$, and ϕ is a functor from $\mathcal{C}$ to $\mathcal{C}'$ that is injective on $\mathcal{S}$. (i) Show that ϕ is correct for right-complements (*resp*. right-diamonds) on $\mathcal{S}$ if and only if $\phi(\mathcal{S})$ is closed under right-complement (*resp*. right-diamond) in $\mathcal{C}'$. (ii) Assuming in addition that $\mathcal{S}$ is closed under right-quotient in $\mathcal{C}$, show a similar result for right-comultiples.

Exercise 94 (embeddings of Artin–Tits monoids). (i) Show that the mapping ϕ defined by $\phi(\sigma_0) = \sigma_1^2$ and $\phi(\sigma_i) = \sigma_{i+1}$ for $i \geqslant 1$ induces an embedding of the Artin–Tits monoid of type B_n (see Example 4.21) in the Artin–Tits monoid of type A_{n+1}. (ii) Same question for $\phi : \sigma_1 \mapsto \sigma_1\sigma_3\cdots\sigma_{2\lfloor n/2\rfloor+1}$ and $\sigma_2 \mapsto \sigma_2\sigma_4\cdots\sigma_{2\lfloor (n-1)/2\rfloor}$, defining an embedding of the Artin–Tits of type $\mathrm{I}_2(n)$ into the Artin–Tits monoid of type A_{n-1}.

Notes

Sources and comments. Most of the notions and results in this chapter have never appeared in print, at least in the current form, but they are pretty natural extensions of existing results involving submonoids of Garside monoids, in the vein of the results of Crisp [67] and Godelle [134, 138, 139]. The main interest of the current extension is probably to show that the context of Garside families and Garside germs is well suited and makes most arguments easy and natural. In particular, the notions of an $=^{\times}$-closed subcategory and an $=^{\times}_{\mathcal{S}}$-closed subgerm provide a good solution for avoiding all problems arising from the possible existence of nontrivial invertible elements. Similarly, the notion of a head-category captures the main consequences of Noetherianity in earlier approaches and enables one to skip any such assumption.

The results of Section 2 show that Garside families behave nicely with respect to subcategories, in that weak assumptions are sufficient to guarantee a full compatibility between the associated normal decompositions.

Similarly, the results of Section 3 about subgerms of a Garside germ, which seem to be new, are rather satisfactory in that, in most cases, we obtain either a simple local criterion ensuring that some property or some implication is true, or a counter-example.

About Proposition 4.2 (fixed points), we should note that the fact that the fixed points under an automorphism in an Artin–Tits monoid constitutes still an Artin–Tits monoid appears for the first time in [184, Corollary 4.4].

Finally, the correct functors of Section 4 are a mild extension of the lcm-homomorphisms considered by J. Crisp in [67] and E. Godelle in [137], but they enable one to skip any Noetherianity assumption and to cope with the possible existence of nontrivial invertible elements. The emphasis put on what we call correctness for right-complements and preservation of the operation $\backslash$ and the injectivity criteria of Subsection 4.2 are directly reminiscent of the results of [78].

Further questions. A few natural questions directly inspired by the results of the chapter remain open.

Question 21. Do the results of Propositions 1.25 and 3.27 still hold when neither right-Noetherianity, nor the validity of the second domino rule is assumed?

The question remains open, the problem being the eventual stabilization of the sequence $(\theta_i(g))_{i \geqslant 1}$ involved in the proof of Proposition 1.25. Let us observe that, in the case of the left-absorbing monoid L_n with $n \geqslant 2$ (Reference Structure 8, page 113) and of the submonoid generated by $\mathtt{a}$, the sequence does not stabilize after $\|g\|_{\mathcal{S}}$ steps: for instance, for $n = 2$ and $g = \Delta_2^2$ (of length 2), we find $g = \mathtt{a}^4\mathtt{b}^2$ and the sequence $(\theta_i(g))_{i \geqslant 1}$, here $\mathtt{a}, \mathtt{a}^2, \mathtt{a}^3, \mathtt{a}^4, \mathtt{a}^4, \ldots$ stabilizes after 4 steps only. However, as the monoid L_n is Noetherian, we know that the sequence $(\theta_i(g))_{i \geqslant 1}$ must stabilize for every g, and cannot expect more from this example.

Similarly, it is questionable whether some form of Noetherianity is needed to ensure the results about the intersection of head subcategories or parabolic subcategories:

Question 22. Are the results of Propositions 1.28 and 1.35 and of Corollaries 1.29 and 1.36 still valid when the ambient category $\mathcal{C}$ is not left-Noetherian?

About subgerms, we established in Subsection 3.3 sufficient conditions, in particular for the satisfaction of (3.10), that is, for the subcategory generated by a subgerm to be isomorphic to the category of the subgerm.

Question 23. Is there a converse to Proposition 3.14, that is, if $\underline{\mathcal{S}}_1$ is a subgerm of a Garside germ $\underline{\mathcal{S}}$ and $\mathcal{C}at(\underline{\mathcal{S}}_1)$ and $\mathcal{S}ub(\mathcal{S}_1)$ are isomorphic (and $\mathcal{S}_1$ is closed under right-quotient in $\mathcal{S}$), must $\underline{\mathcal{S}}_1$ must be closed under right-comultiple in $\underline{\mathcal{S}}$?

Always if $\underline{\mathcal{S}}_1$ is a subgerm of a (Garside) germ $\underline{\mathcal{S}}$, by Lemma 3.9, the assumption that $\mathcal{S}_1$ is closed under inverse and right-complement in $\mathcal{S}$ implies that $\mathcal{S}ub(\mathcal{S}_1)$ is closed under right-quotient in $\mathcal{C}$ and, therefore, by the result of Exercise 89, $\mathcal{S}_1$ must be closed under right-quotient in $\mathcal{S}$. We do not know about a possible converse:

Question 24. If $\underline{\mathcal{S}}_1$ is a subgerm of a Garside germ $\mathcal{S}$ and $\mathcal{S}_1$ is closed under right-quotient in $\mathcal{S}$, is the subcategory $\mathcal{S}ub(\mathcal{S}_1)$ necessarily closed under right-quotient in $\mathcal{C}at(\underline{\mathcal{S}})$?

Beyond these technical questions, what remains an essentially open problem is to extend the results from the case of a group of fractions to the general case of an enveloping groupoid. In the case of Artin–Tits monoids and groups, a number of results involving submonoids and subgroups, in particular normalizers and centralizers, can be established in non-spherical cases, typically in what are called the FC and 2-dimensional cases, see Godelle [136, 135, 137]. The general results of the current chapter are weaker and, at the moment, extending the body of work produced in the Artin–Tits case to the framework of an arbitrary left-cancellative category equipped with a Garside family or, more realistically, of a cancellative category equipped with a Garside map, remains out of reach—or, at least, has not yet been done. A real, deep question is whether the formalism of $\mathcal{S}$-normal decompositions and Garside families will enable one to complete such a program, or it will be necessary to introduce radically new tools, for instance normal decompositions of a different type like those of Godelle and Paris [140].

Chapter VIII
Conjugacy

This chapter is devoted to conjugacy in the framework of a left-cancellative category equipped with a Garside family. The existence of distinguished decompositions for the elements of the category and, possibly, its enveloping groupoid, leads to structural results, typically providing distinguished ways for transforming two conjugate elements into one another. Under convenient finiteness conditions, this leads to an effective solution for the Conjugacy Problem, that is, an algorithm that decides whether two elements are conjugate, in a continuation of Garside's solution to the Conjugacy Problem for Artin's braid groups from which the whole current approach emerged.

The chapter contains three sections. In Section 1, we observe that conjugacy naturally gives rise to two related categories, one associated with the general form of conjugacy and one associated with cyclic conjugacy, which is the iteration of the transformation consisting in splitting an element in a product of two factors and exchanging these factors. We show that every Garside family in the base category induces a Garside family in the associated conjugacy categories (Corollary 1.14). We also specifically investigate the case when general and cyclic conjugacies coincide (Proposition 1.24) and, as an application, describe the ribbon category which arises in the study of parabolic submonoids of an Artin–Tits monoid, in particular by characterizing its atoms (Proposition 1.59), thus obtaining a complete description of the normalizer of a parabolic submonoid.

In Section 2, we consider the specific case of a category equipped with a bounded Garside family, that is, equivalently, a Garside map. Then we investigate particular conjugacy transformations called cycling, decycling, and sliding, and we show how to use them to design efficient methods for solving the Conjugacy Problem (Algorithm 2.23 together with Proposition 2.24, Algorithm 2.42 together with Proposition 2.43).

Finally, always in the context of a category equipped with a Garside map, Section 3 is devoted to periodic elements, defined to be those elements that admit a power equal to a power of a Garside element. To this end, we develop geometric methods that extend and adapt to our context results of M. Bestvina. The main result, Proposition 3.34, states that, under suitable assumptions, conjugacy of periodic elements is always a cyclic conjugacy. These results will be used in Chapter X for studying Deligne-Lusztig varieties.

Main definitions and results

Definition 1.1 **(conjugacy category).** For $\mathcal{C}$ a category, we denote by $\mathcal{C}^{\circlearrowleft}$ the subfamily of $\mathcal{C}$ consisting of all elements whose source and target coincide. For e, e' in $\mathcal{C}^{\circlearrowleft}$, we say that an element g of $\mathcal{C}$ *conjugates* e to e' if $eg = ge'$ holds. The *conjugacy category* $\mathit{Conj}\,\mathcal{C}$ of $\mathcal{C}$ is the category whose object family is $\mathcal{C}^{\circlearrowleft}$ and where $\mathit{Conj}\,\mathcal{C}(e, e')$ consists of all triples (e, g, e') such that g belongs to $\mathcal{C}$ and conjugates e to e', the source (*resp.* target) of (e, g, e') being e (*resp.* e'), and the product being defined by $(e, g, e')(e', g', e'') = (e, gg', e'')$. To emphasize the category structure on $\mathit{Conj}\mathcal{C}$, we will write $e \xrightarrow{g} e'$ for (e, g, e'), so that the product formula becomes (1.2) $(e \xrightarrow{g} e')(e' \xrightarrow{g'} e'') = e \xrightarrow{gg'} e''$. For e in $\mathcal{C}^{\circlearrowleft}$, the family $\mathit{Conj}\mathcal{C}(e, e)$ is called the *centralizer* of e in $\mathcal{C}$.

Corollary 1.14 **(Garside in** $\mathit{Conj}\mathcal{C}$**).** *If $\mathcal{S}$ is a Garside family in a left-cancellative category $\mathcal{C}$, then $\{e \xrightarrow{g} \cdot \in \mathit{Conj}\mathcal{C} \mid g \in \mathcal{S}\}$ is a Garside family in $\mathit{Conj}\mathcal{C}$.*

Definition 1.17 **(cyclic conjugacy).** For $\mathcal{C}$ a left-cancellative category, we denote by $\mathit{Cyc}_1\mathcal{C}$ the subfamily of $\mathit{Conj}\mathcal{C}$ consisting of all elements $e \xrightarrow{g} \cdot$ satisfying $g \preccurlyeq e$, and define the *cyclic conjugacy category* $\mathit{Cyc}\mathcal{C}$ to be the subcategory of $\mathit{Conj}\mathcal{C}$ generated by $\mathit{Cyc}_1\mathcal{C}$. For e, e' in $\mathcal{C}^{\circlearrowleft}$, we say that an element g of $\mathcal{C}$ *cyclically conjugates* e to e' if $e \xrightarrow{g} \cdot$ lies in $\mathit{Cyc}\mathcal{C}$ and $e' = e^g$ holds.

Proposition 1.24 **(every conjugate cyclic).** *Assume that $\mathcal{C}$ is a left-cancellative category that is right-Noetherian and admits conditional right-lcms, Δ is a Garside map in $\mathcal{C}$, and e is an element of $\mathcal{C}^{\circlearrowleft}(x, x)$ satisfying $\Delta(x) \preccurlyeq e^n$ for n large enough. Then, for every e' in $\mathcal{C}^{\circlearrowleft}$, one has $\mathit{Cyc}\mathcal{C}(e, e') = \mathit{Conj}\mathcal{C}(e, e')$.*

Definition 2.3 **(cycling).** For e in $\mathcal{G}^{\circlearrowleft}$ satisfying $\inf_\Delta(e) = i$, the *initial factor* $\mathrm{init}(e)$ of e is $H(e\Delta^{[-i]})$, and the *cycling* $\mathrm{cyc}(e)$ of e is $e^{\mathrm{init}(e)}$.

Definition 2.8 **(decycling, final factor).** For e in $\mathcal{G}^{\circlearrowleft}$ admitting the strict Δ-normal decomposition $\Delta^{[i]} \| s_1 | \cdots | s_p$, the *final factor* $\mathrm{fin}(e)$ of e is defined to be s_p, and the *decycling* $\mathrm{dec}(e)$ of e is defined to be $s_p \Delta^{[i]} s_1 \cdots s_{p-1}$.

Corollary 2.12 **(super-summit set).** *In Context* 2.1, *for every e in $\mathcal{G}^{\circlearrowleft}$, the conjugacy class of e contains a well-defined subset* $\mathrm{SSS}(e)$ *on which each one of* $\inf_\Delta$ *and* $\sup_\Delta$ *takes a constant value and such that, for every e' in the conjugacy class of e, we have* $\inf_\Delta(e') \leqslant \inf_\Delta(\mathrm{SSS}(e))$ *and* $\sup_\Delta(e') \geqslant \sup_\Delta(\mathrm{SSS}(e))$. *Furthermore,* $\mathrm{SSS}(e)$ *belongs to the connected component of e in $\mathit{Cyc}\mathcal{G}$.*

Proposition 2.24 **(Conjugacy Problem I).** *If $\mathcal{G}$ is the groupoid of fractions of a Noetherian cancellative category that admits a Garside map and Condition* 2.14 *holds, then Algorithm* 2.23 *solves the Conjugacy Problem for $\mathcal{G}$.*

Definition 2.29 **(prefix, sliding).** For e in $\mathcal{G}^{\circlearrowleft}$, the *prefix* $\mathrm{pr}(e)$ of e is the head of a (any) left-gcd of $\mathrm{init}(e)$ and $\mathrm{init}(e^{-1})$, and the *sliding* $\mathrm{sl}(e)$ of e is the conjugate $e^{\mathrm{pr}(e)}$ of e.

Proposition 2.30 **(sliding circuits).** *In Context* 2.1, *and for every e in $\mathcal{G}^{\circlearrowleft}$:* (i) *We have* $\inf_\Delta(\mathrm{sl}(e)) \geqslant \inf_\Delta(e)$, *and, if there exists e' in the conjugacy class of e satisfying* $\inf_\Delta(e') > \inf_\Delta(e)$, *there exists i satisfying* $\inf_\Delta(\mathrm{sl}^i(e)) > \inf_\Delta(e)$. (ii) *We have* $\sup_\Delta(\mathrm{sl}(e)) \leqslant \sup_\Delta(e)$, *and, if there exists e' in the conjugacy class of e satisfying* $\sup_\Delta(e') < \sup_\Delta(e)$, *there exists i satisfying* $\sup_\Delta(\mathrm{sl}^i(e)) < \sup_\Delta(e)$. (iii) *If $\mathcal{S}/{=^{\times}}$ is finite, sliding is ultimately periodic up to invertible elements in the sense that, for every e in $\mathcal{G}^{\circlearrowleft}$, there exists $i > j \geqslant 0$ satisfying* $\mathrm{sl}^i(e) =^{\times} \mathrm{sl}^j(e)$.

Proposition 2.43 **(Conjugacy Problem II).** *If $\mathcal{G}$ is the groupoid of fractions of a right-Noetherian cancellative category $\mathcal{C}$ that admits a Garside map, and Condition* 2.14 *holds, and $\mathcal{C}^{\times}$ is finite, then Algorithm* 2.42 *solves the Conjugacy Problem for $\mathcal{G}$.*

Definition 3.2 **(periodic).** For $p, q \geqslant 1$, an element e of $\mathcal{G}^{\circlearrowright}$ is called (p,q)*-periodic* if $e^p =^{\times} \Delta^{[q]}$ holds.

Proposition 3.5 **($(p,2)$-periodic).** *In Context* 3.1, *if e is an element of $\mathcal{C}^{\circlearrowright}$ satisfying $e^p =^{\times} \Delta^{[2]}$ for some positive integer p, and putting $r = \lfloor p/2 \rfloor$:* (i) *Some cyclic conjugate d of e satisfies $d^p =^{\times} \Delta^{[2]}$ and $d^r \in \mathcal{D}iv(\Delta)$.* (ii) *Furthermore, if p is even, we have $d^r =^{\times} \Delta$, and, if p is odd, there exists s in $\mathcal{D}iv(\Delta)$ satisfying $d^r s = \Delta$ and $d = s\phi_\Delta(s)\epsilon$, where ϵ is the element of $\mathcal{C}^{\times}$ satisfying $d^p = \Delta^{[2]}\epsilon$.*

Corollary 3.31 **(SSS of periodic).** *In Context* 3.7 *with $\mathcal{C}$ left-Noetherian, if e is an element of $\mathcal{G}^{\circlearrowright}$ that is conjugate to a periodic element, then the super-summit set of e consists of elements of the form $\Delta^{[m]}h$ with m in $\mathbb{Z}$ and h in $\mathcal{D}iv(\Delta)$, and e is conjugate by cyclic conjugacy to such an element.*

Proposition 3.34 **(periodic elements).** *If $\mathcal{C}$ is a Noetherian cancellative category and Δ is a Garside map in $\mathcal{C}$ such that the order of ϕ_Δ is finite:* (i) *Every periodic element e of $\mathcal{C}$ is cyclically conjugate to some element d satisfying $d^p \in \Delta^{[q]}\mathcal{C}^{\times}$ with p and q positive and coprime and such that, for all positive integers p' and q' satisfying $pq' - qp' = 1$, we have $d^{p'} \preccurlyeq \Delta^{[q']}$;* (ii) *For d as in* (i), *the element g satisfying $d^{p'}g = \Delta^{[q']}$ is an element of $\mathcal{D}iv(\Delta)$ whose $=^{\times}$-class is independent of the choice of (p', q'); moreover, $g\Delta^{[-q']}$ is $(p, -qp')$-periodic and $d =^{\times} (g\Delta^{[-q']})^q\Delta^{[qq']}$ holds.*

1 Conjugacy categories

We show how to attach with every left-cancellative category $\mathcal{C}$ two categories $\mathcal{C}onj\mathcal{C}$ and $\mathcal{C}yc\mathcal{C}$ that describe conjugacy in $\mathcal{C}$, and we establish various properties of these categories, showing in particular that every Garside family of $\mathcal{C}$ gives rise to natural Garside families in $\mathcal{C}onj\mathcal{C}$ and $\mathcal{C}yc\mathcal{C}$.

The section contains four subsections. We first consider the case of general conjugacy and introduce the category $\mathcal{C}onj\mathcal{C}$ (Subsection 1.1). Next, we consider a more restricted form of conjugacy, called cyclic conjugacy, and introduce the category $\mathcal{C}yc\mathcal{C}$ (Subsection 1.2). Then, in Subsection 1.3, we consider twisted versions of conjugacy involving an additional automorphism of the base category. Finally, in Subsection 1.4, we describe the category of ribbons as an example of simultaneous conjugacy.

1.1 General conjugacy

We start with a very general framework, namely a category that does not even need to be left-cancellative. However, left-cancellativity will be needed quickly.

Definition 1.1 (conjugacy category, centralizer). For $\mathcal{C}$ a category, we denote by $\mathcal{C}^{\circlearrowleft}$ the subfamily of $\mathcal{C}$ consisting of all elements whose source and target coincide. For e, e' in $\mathcal{C}^{\circlearrowleft}$, we say that an element g of $\mathcal{C}$ *conjugates* e to e' if $eg = ge'$ holds. The *conjugacy category* $Conj\mathcal{C}$ of $\mathcal{C}$ is the category whose object family is $\mathcal{C}^{\circlearrowleft}$ and where $Conj\mathcal{C}(e, e')$ consists of all triples (e, g, e') such that g belongs to $\mathcal{C}$ and conjugates e to e', the source (*resp.* target) of (e, g, e') being e (*resp.* e'), and the product being defined by $(e, g, e')(e', g', e'') = (e, gg', e'')$. To emphasize the category structure on $Conj\mathcal{C}$, we will write $e \xrightarrow{g} e'$ for (e, g, e'), so that the product formula becomes

$$(e \xrightarrow{g} e')(e' \xrightarrow{g'} e'') = e \xrightarrow{gg'} e''. \tag{1.2}$$

For e in $\mathcal{C}^{\circlearrowleft}$, the family $Conj\mathcal{C}(e, e)$ is called the *centralizer* of e in $\mathcal{C}$.

Example 1.3 (conjugacy category). Let M be an abelian monoid (for instance a free abelian monoid Reference Structure 1, page 3). Then $M^{\circlearrowleft}$ coincides with M (as in the case of every monoid), and conjugacy is trivial on M: so the elements of $Conj\,M$ have the form $e \xrightarrow{g} e$, with $(e \xrightarrow{g} e)(e \xrightarrow{g'} e) = e \xrightarrow{gg'} e$. In this case, $Conj\,M$ consists of disjoint copies of M indexed by M.

The elements of $\mathcal{C}^{\circlearrowleft}$ are the endomorphisms of $\mathcal{C}$—whence the choice of e as a preferred letter for elements of $\mathcal{C}^{\circlearrowleft}$. Note that the existence of g satisfying $eg = ge'$ implies that e and e' lie in $\mathcal{C}^{\circlearrowleft}$, so the restriction on the objects of $Conj\mathcal{C}$ is inevitable.

Lemma 1.4. *Definition* 1.1 *is legal, that is, $Conj\mathcal{C}$ is indeed a category. Moreover, the projection π defined on $Conj\mathcal{C}$ by $\pi(e) = x$ for e in $\mathcal{C}(x, x)$ and $\pi(e \xrightarrow{g} e') = g$ is a surjective functor from $Conj\mathcal{C}$ onto $\mathcal{C}$. For e in $\mathcal{C}(x, x)$, we have*

$$\pi(Conj\mathcal{C}(e, \cdot)) = \{g \in \mathcal{C}(x, \cdot) \mid g \preccurlyeq eg\}, \tag{1.5}$$

$$\pi(Conj\mathcal{C}(\cdot, e')) = \{g \in \mathcal{C}(\cdot, x) \mid ge' \succcurlyeq g\}. \tag{1.6}$$

Proof. The product defined by (1.2) is associative, and, for every e in $\mathcal{C}(x, x)$, the element $(e, 1_x, e)$ is an identity element for e. That π is a surjective functor directly follows from the definition.

By definition, $Conj\mathcal{C}(e, \cdot)$ consists of all elements $e \xrightarrow{g} e'$ of $Conj\mathcal{C}$ whose source is e, hence of all triples $e \xrightarrow{g} e'$ satisfying $eg = ge'$. So $\pi(Conj\mathcal{C}(e, \cdot))$ consists of all elements g satisfying $\exists e' \,(eg = ge')$, hence of all elements g satisfying $g \preccurlyeq eg$. The argument is symmetric for $Conj\mathcal{C}(\cdot, e')$. □

To shorten notation, we shall usually write $\pi\mathcal{S}$ for $\pi(\mathcal{S})$. It should be kept in mind that π is not injective in general, and that elements of $Conj\mathcal{C}$ cannot be identified with elements of $\mathcal{C}$, since an element of $\mathcal{C}$ contains no indication of a particular element of $\mathcal{C}^{\circlearrowleft}$ it should acts on.

Remark 1.7. The category $\mathit{Conj}\,\mathcal{C}$ corresponds to conjugating on the right, and it could be called a right-conjugacy category. The opposed category would then correspond to conjugating on the left ("left-conjugacy category").

All subsequent developments involve categories that are (at least) left-cancellative. In such a framework, when an equality $eg = ge'$ holds, the data e and g determine e', allowing for a simplified notation.

Notation 1.8. If $\mathcal{C}$ is a left-cancellative category, then, for e in $\mathcal{C}^{\circlearrowleft}$ and g in $\mathcal{C}$, the unique element e' satisfying $eg = ge'$ is denoted by e^g when it exists, and we write $e \xrightarrow{g} \cdot$ for $e \xrightarrow{g} e^g$ when there is no need to make e^g explicit.

So, in a left-cancellative context, the elements of $\mathit{Conj}\,\mathcal{C}$ can be seen as pairs rather than as triples, a pair $e \xrightarrow{g} \cdot$ in $\mathcal{C}^{\circlearrowleft} \times \mathcal{C}$ being an element of $\mathit{Conj}\,\mathcal{C}$ if and only if there exists e' satisfying $eg = ge'$, that is, if and only if e^g exists. The product of $\mathit{Conj}\,\mathcal{C}$ then takes the form

$$e \xrightarrow{g} e' \xrightarrow{g'} \cdot = e \xrightarrow{gg'} \cdot \quad \text{whenever } eg = ge' \text{ holds,} \tag{1.9}$$

whereas the identity element associated with an element e of $\mathcal{C}^{\circlearrowleft}(x, x)$ is $e \xrightarrow{1_x} \cdot$. A number of properties and derived notions of $\mathit{Conj}\,\mathcal{C}$ directly follow from those of $\mathcal{C}$.

Lemma 1.10. *Assume that $\mathcal{C}$ is a left-cancellative category.*

(i) *The category $\mathit{Conj}\,\mathcal{C}$ is left-cancellative. If $\mathcal{C}$ is cancellative, then so is $\mathit{Conj}\,\mathcal{C}$.*

(ii) *The invertible elements of $\mathit{Conj}\,\mathcal{C}$ are the pairs $e \xrightarrow{\epsilon} \cdot$ with ϵ invertible and $e\epsilon$ defined in $\mathcal{C}$.*

(iii) *For $e \xrightarrow{g} \cdot$ and $d \xrightarrow{h} \cdot$ in $\mathit{Conj}\,\mathcal{C}$, the relation $e \xrightarrow{g} \cdot \preccurlyeq d \xrightarrow{h} \cdot$ (resp. $=^{\times}$, $\prec$) holds in $\mathit{Conj}\,\mathcal{C}$ if and only if $e = d$ and $g \preccurlyeq h$ (resp. $=^{\times}$, $\prec$) hold in $\mathcal{C}$.*

(iv) *If $e \xrightarrow{f} \cdot$ and $e \xrightarrow{g} \cdot$ belong to $\mathit{Conj}\,\mathcal{C}$ and h is a right-lcm of f and g in $\mathcal{C}$, then $e \xrightarrow{h} \cdot$ belongs to $\mathit{Conj}\,\mathcal{C}$ and it is a right-lcm of $e \xrightarrow{f} \cdot$ and $e \xrightarrow{g} \cdot$. If $\mathcal{C}$ admits conditional right-lcms (resp. admits right-lcms), then so does $\mathit{Conj}\,\mathcal{C}$.*

(v) *If $\mathcal{C}$ is right-Noetherian (resp. left-Noetherian), then so is $\mathit{Conj}\,\mathcal{C}$.*

Proof. (i) Assume $e \xrightarrow{f} d \xrightarrow{g} \cdot = e \xrightarrow{f} d' \xrightarrow{g'} \cdot$. By definition, $d = e^f = d'$ holds. Next, applying π, we obtain $fg = fg'$, whence $g = g'$. Hence $\mathit{Conj}\,\mathcal{C}$ is left-cancellative. The argument for right-cancellativity is symmetric.

(ii) The image of an invertible element under the functor π must be invertible, so an invertible element of $\mathit{Conj}\,\mathcal{C}$ necessarily has the form $e \xrightarrow{\epsilon} \cdot$ with ϵ in $\mathcal{C}^{\times}$ and $e\epsilon$ defined. Conversely, assume $e \in \mathcal{C}^{\circlearrowleft}(x, x)$ and $\epsilon \in \mathcal{C}^{\times}(x, \cdot)$. Then $e\epsilon = \epsilon(\epsilon^{-1}e\epsilon)$ holds, so $e \xrightarrow{\epsilon} \cdot$ belongs to $\mathit{Conj}\,\mathcal{C}$, and we have $e \xrightarrow{\epsilon} e^{\epsilon} \xrightarrow{\epsilon^{-1}} \cdot = e \xrightarrow{1_x} \cdot$, so $e \xrightarrow{\epsilon} \cdot$ is invertible in $\mathit{Conj}\,\mathcal{C}$.

(iii) If $e \xrightarrow{g} \cdot \preccurlyeq d \xrightarrow{h} \cdot$ holds in $\mathit{Conj}\,\mathcal{C}$, then $e \xrightarrow{g} \cdot$ and $d \xrightarrow{h} \cdot$ must have the same source, so $e = d$ holds, and applying the projection π to an equality $e \xrightarrow{g} e' \xrightarrow{h'} \cdot = e \xrightarrow{h} \cdot$ in $\mathit{Conj}\,\mathcal{C}$ yields $gh' = h$ in $\mathcal{C}$, whence $g \preccurlyeq h$ in $\mathcal{C}$.

Conversely, assume that $g \preccurlyeq h$ holds in $\mathcal{C}$, say $gh' = h$, and both $e \xrightarrow{g} \cdot$ and $e \xrightarrow{h} \cdot$ belong to $\mathit{Conj}\,\mathcal{C}$. By assumption, we have $eg = ge^g$ and $eh = he^h$, whence $ge^gh' =$

$egh' = eh = he^h = gh'e^h$, and $e^g h' = h'e^h$ by left-cancelling g. So, in $\mathit{Conj}\mathcal{C}$, we have $e \xrightarrow{g} e^g \xrightarrow{h'} \cdot = e \xrightarrow{h} \cdot$, whence $e \xrightarrow{g} \cdot \preccurlyeq e \xrightarrow{h} \cdot$.

Owing to (ii), the argument for $=^\times$ is similar, and the result for $\prec$ then follows.

(iv) By Lemma 1.4, the assumption that $e \xrightarrow{f} \cdot$ and $e \xrightarrow{g} \cdot$ belong to $\mathit{Conj}\mathcal{C}$ implies $f \preccurlyeq ef$ and $g \preccurlyeq eg$, hence a fortiori $f \preccurlyeq eh$ and $g \preccurlyeq eh$. As h is a right-lcm of f and g, we deduce $h \preccurlyeq eh$, which, by Lemma 1.4 again, implies that $e \xrightarrow{h} \cdot$ belongs to $\mathit{Conj}\mathcal{C}$. By (iii), $e \xrightarrow{h} \cdot$ is a right-multiple of $e \xrightarrow{f} \cdot$ and $e \xrightarrow{g} \cdot$. Finally, if $e \xrightarrow{h'} \cdot$ is a right-multiple of $e \xrightarrow{f} \cdot$ and $e \xrightarrow{g} \cdot$, then h' must be a right-multiple of f and g, hence of h', so $e \xrightarrow{h'} \cdot$ is a right-multiple of $e \xrightarrow{h} \cdot$ in $\mathit{Conj}\mathcal{C}$.

Finally, assume that $\mathcal{C}$ admits conditional right-lcms. Two elements of $\mathit{Conj}\mathcal{C}$ with the same source must be of the form $e \xrightarrow{f} \cdot$, $e \xrightarrow{g} \cdot$, with f, g sharing the same source in $\mathcal{C}$. If $e \xrightarrow{f} \cdot$, $e \xrightarrow{g} \cdot$ admit a common right-multiple in $\mathit{Conj}\mathcal{C}$, then f and g admit a common right-multiple, hence a right-lcm, say h, in $\mathcal{C}$. By the previous result, $e \xrightarrow{h} \cdot$ is a right-lcm of $e \xrightarrow{f} \cdot$ and $e \xrightarrow{g} \cdot$ in $\mathit{Conj}\mathcal{C}$. The argument for admitting right-lcms is similar.

(v) Assume that $\mathcal{C}$ is right-Noetherian and $\mathcal{X}$ is a nonempty subfamily of $\mathit{Conj}\mathcal{C}$. Then $\pi\mathcal{X}$ is a nonempty subfamily of $\mathcal{C}$, hence it admits a $\widetilde{\prec}$-minimal element, say g. Let e be such that $e \xrightarrow{g} \cdot$ belongs to $\mathcal{X}$. We claim that $e \xrightarrow{g} \cdot$ is a $\widetilde{\prec}$-minimal element of $\mathcal{X}$. Indeed, assume that $e \xrightarrow{g} \cdot \widetilde{\succ} d \xrightarrow{h} \cdot$ holds. By applying π, we deduce $g \widetilde{\succ} h$. So h cannot belong to $\pi\mathcal{X}$, and $d \xrightarrow{h} \cdot$ cannot belong to $\mathcal{X}$. The argument for left-Noetherianity is similar. □

Lemma 1.10(iii) implies that the subfamily $\pi\mathit{Conj}\mathcal{C}$ of $\mathcal{C}$ and, for each e in $\mathcal{C}^{\circlearrowright}$, the subfamily $\pi\mathit{Conj}\mathcal{C}(e, \cdot)$ are closed under right-multiplication by an invertible element. Also, we have a sort of weak closure under right-quotient for $\mathit{Conj}\mathcal{C}$, in the sense that, if $e \xrightarrow{g} \cdot$ and $e \xrightarrow{h} \cdot$ belong to $\mathit{Conj}\mathcal{C}$ and $gh' = h$ holds in $\mathcal{C}$, then $e^g \xrightarrow{h'} \cdot$ must belong to $\mathit{Conj}\mathcal{C}$—this however is not closure under right-quotient, since we say nothing for elements of $\pi\mathit{Conj}\mathcal{C}$ satisfying no assumption about the elements of $\mathcal{C}^{\circlearrowright}$ they act on.

We now turn to paths in a category $\mathit{Conj}\mathcal{C}$ and to the connection with Garside families.

Notation 1.11 (family $\widehat{\mathcal{S}}$, $\mathit{Conj}\mathcal{C}$-path). (i) If $\mathcal{C}$ is a left-cancellative category and $\mathcal{S}$ is included in $\mathcal{C}$, we put $\widehat{\mathcal{S}} = \{e \xrightarrow{s} \cdot \in \mathit{Conj}\mathcal{C} \mid s \in \mathcal{S}\}$.

(ii) If $\mathcal{C}$ is a left-cancellative category and $e_1 \xrightarrow{s_1} |\cdots| e_p \xrightarrow{s_p} \cdot$ is a $\mathit{Conj}\mathcal{C}$-path, we write $e_1 \xrightarrow{s_1 |\cdots| s_p} \cdot$ for this path.

The convention of Notation 1.11(ii) is legitimate since the data e_1 and $s_1, \ldots, s_p$ determine the path completely: for $i \geqslant 2$, one must have $e_i = e_1^{s_1 \cdots s_{i-1}}$.

Lemma 1.12. *Assume that $\mathcal{C}$ is a left-cancellative category, $\mathcal{S}$ is included in $\mathcal{C}$, $e \xrightarrow{g} \cdot$ is an element of $\mathit{Conj}\mathcal{C}$, and $g_1 | g'$ is an $\mathcal{S}$-greedy decomposition of g.*

(i) *If $e \xrightarrow{g_1 | g'} \cdot$ is a well defined path in $\mathit{Conj}\mathcal{C}$, that is, if e^{g_1} and $e^{g_1 g'}$ are defined, then $e \xrightarrow{g_1 | g'} \cdot$ is $\widehat{\mathcal{S}}$-greedy.*

(ii) *If g_1 belongs to $\mathcal{S}^\sharp$, then e^{g_1} is necessarily defined.*

Proof. (i) Assume $d \xrightarrow{s} \cdot \in \widehat{\mathcal{S}}$ and $d \xrightarrow{s} \cdot \preccurlyeq d \xrightarrow{f} e \xrightarrow{g_1g'} \cdot$ in $\mathcal{C}onj\mathcal{C}$. By projecting using π, we deduce $s \preccurlyeq fg_1g'$ in $\mathcal{C}$, whence $s \preccurlyeq fg_1$ since g_1g' is $\mathcal{S}$-greedy. By assumption, $d \xrightarrow{s} \cdot$ and $d \xrightarrow{fg_1} \cdot$ belong to $\mathcal{C}onj\mathcal{C}$. By Lemma 1.10(iii), $s \preccurlyeq fg_1$ implies $d \xrightarrow{s} \cdot \preccurlyeq d \xrightarrow{fg_1} \cdot$, that is, $d \xrightarrow{s} \cdot \preccurlyeq d \xrightarrow{f} e \xrightarrow{g_1} \cdot$. So $e \xrightarrow{g_1g'} \cdot$ is $\widehat{\mathcal{S}}$-greedy.

(ii) Assume that $e \xrightarrow{g_1g'} \cdot$ belongs to $\mathcal{C}onj\mathcal{C}$ and g_1 belongs to $\mathcal{S}^\sharp$. Then we have $g_1 \preccurlyeq g_1g' \preccurlyeq eg_1g'$, the second relation by Lemma 1.4. As g_1 lies in $\mathcal{S}^\sharp$ and g_1g' is $\mathcal{S}$-greedy, hence $\mathcal{S}^\sharp$-greedy by Lemma III.1.10, we deduce $g_1 \preccurlyeq eg_1$, which, by Lemma 1.4 again, implies that g_1 belongs to $\mathcal{C}onj\mathcal{C}(e, \cdot)$, that is, $e \xrightarrow{g_1} \cdot$ belongs to $\mathcal{C}onj\mathcal{C}$. □

Although technically very easy, Lemma 1.12(ii) is a (small) miracle: in general, the assumption that e^g is defined does not imply that e^{g_1} is defined when g_1 is a left-divisor of g. What Lemma 1.12(ii) says is that e^{g_1} is automatically defined whenever g_1 is an $\mathcal{S}$-head of g, for any family $\mathcal{S}$. This tiny observation is indeed crucial in the connection between conjugacy and Garside calculus. Observe that our current definition of greediness with an additional first factor ("s left-divides fg_1g' implies s left-divides fg_1 for every f") is specially relevant here.

Proposition 1.13 (normal in $\mathcal{C}onj\mathcal{C}$). *If $\mathcal{S}$ is a subfamily of a left-cancellative category $\mathcal{C}$ and $e \xrightarrow{g} \cdot$ belongs to $\mathcal{C}onj\mathcal{C}$, then, for every $\mathcal{S}$-normal decomposition $s_1|\cdots|s_p$ of g (if any), the path $e \xrightarrow{s_1|\cdots|s_p} \cdot$ is well-defined and it is an $\widehat{\mathcal{S}}$-normal decomposition of $e \xrightarrow{g} \cdot$.*

Proof. That $e \xrightarrow{s_1|\cdots|s_p} \cdot$ is well defined follows from Lemma 1.12(ii) using an induction on $p \geqslant 2$: as $s_1|s_2|\cdots|s_p$ is greedy and $e \xrightarrow{s_1(s_2\cdots s_p)} \cdot$ is defined, $e \xrightarrow{s_1} \cdot$ and $e^{s_1} \xrightarrow{s_2|\cdots|s_p} \cdot$ is defined, and we repeat the argument. Then, by definition, $e \xrightarrow{s_1|\cdots|s_p} \cdot$ is a decomposition of $e \xrightarrow{g} \cdot$ and, by Lemma 1.12(i), it is $\widehat{\mathcal{S}}$-greedy. Finally, each entry s_i belongs to $\mathcal{S}^\sharp$, that is, to $\mathcal{D}iv(\Delta) \cup \mathcal{C}^\times$. Then, by Lemma 1.10, the corresponding entry $e^{s_1\cdots s_{i-1}} \xrightarrow{s_i} \cdot$ belongs to $\widehat{\mathcal{S}}(\mathcal{C}onj\mathcal{C})^\times \cup (\mathcal{C}onj\mathcal{C})^\times$. Hence $e \xrightarrow{s_1|\cdots|s_p} \cdot$ is $\widehat{\mathcal{S}}$-normal. □

Corollary 1.14 (Garside in $\mathcal{C}onj\mathcal{C}$). *If $\mathcal{S}$ is a Garside family in a left-cancellative category $\mathcal{C}$, then $\widehat{\mathcal{S}}$ is a Garside family in $\mathcal{C}onj\mathcal{C}$.*

Proof. Let $e \xrightarrow{g} \cdot$ be an arbitrary element of $\mathcal{C}onj\mathcal{C}$. As $\mathcal{S}$ is a Garside family in $\mathcal{C}$, the element g admits an $\mathcal{S}$-normal decomposition $s_1|\cdots|s_p$ in $\mathcal{C}$. Then, by Proposition 1.13, $e \xrightarrow{s_1|\cdots|s_p} \cdot$ is an $\widehat{\mathcal{S}}$-normal decomposition of $e \xrightarrow{g} \cdot$ in $\mathcal{C}onj\mathcal{C}$. So every element of $\mathcal{C}onj\mathcal{C}$ admits an $\widehat{\mathcal{S}}$-normal decomposition. Hence, by definition, $\widehat{\mathcal{S}}$ is a Garside family in $\mathcal{C}onj\mathcal{C}$. □

For special Garside families, we easily obtain similar statements.

Proposition 1.15 (bounded Garside in $\mathcal{C}onj\mathcal{C}$). *Assume that $\mathcal{S}$ is a Garside family in a left-cancellative category $\mathcal{C}$.*

(i) *If $\mathcal{S}$ is right-bounded by a map Δ, then $\widehat{\mathcal{S}}$ is right-bounded by the map $\widehat{\Delta}$ defined by $\widehat{\Delta}(e) = e \xrightarrow{\Delta(x)} \cdot$ for e in $\mathcal{C}^{\circlearrowright}(x, x)$. The associated functor $\phi_{\widehat{\Delta}}$ of $\mathcal{C}onj\mathcal{C}$ is defined by $\phi_{\widehat{\Delta}}(e) = \phi_\Delta(e)$ for e in $\mathcal{C}^{\circlearrowright}$ and $\phi_{\widehat{\Delta}}(e \xrightarrow{g} \cdot) = \phi_\Delta(e) \xrightarrow{\phi_\Delta(g)} \cdot$ for e in $\mathcal{C}onj\mathcal{C}$.*

(ii) *If $\mathcal{C}$ is cancellative, Δ is target-injective (that is, ϕ_Δ is injective on $\mathcal{O}bj(\mathcal{C})$), and $\mathcal{S}$ is bounded by Δ, then $\widehat{\mathcal{S}}$ is bounded by $\widehat{\Delta}$.*

Proof. (i) Let $e \xrightarrow{s} \cdot$ belong to $\widehat{\mathcal{S}}$, with e in $\mathcal{C}^{\circlearrowright}(x, x)$. By definition, s is an element of $\mathcal{S}(x, \cdot)$, hence a left-divisor of $\Delta(x)$ in $\mathcal{C}$. Now $e\Delta(x) = \Delta(x)\phi_\Delta(e)$ holds in $\mathcal{C}$, which shows that $e^{\Delta(x)}$ is defined. So $e \xrightarrow{\Delta(x)} \cdot$ is an element of $\mathcal{C}onj\mathcal{C}$ and, by Lemma 1.10, $e \xrightarrow{s} \cdot \preccurlyeq e \xrightarrow{\Delta(x)} \cdot$ holds in $\mathcal{C}onj\mathcal{C}$. By definition, this is $e \xrightarrow{s} \cdot \preccurlyeq \widehat{\Delta}(e)$, which shows that $\widehat{\mathcal{S}}$ is right-bounded by $\widehat{\Delta}$.

Next, the target of $\widehat{\Delta}(e)$, that is, of $e \xrightarrow{\Delta(x)} \cdot$, is $e^{\Delta(x)}$, that is, $\phi_\Delta(e)$. Moreover, the target of $e \xrightarrow{s} \cdot$ is e^s, and, assuming $s \in \mathcal{C}(\cdot, y)$, hence $e^s \in \mathcal{C}^{\circlearrowright}(y, y)$, we obtain

$$\begin{aligned} e \xrightarrow{s} \cdot \widehat{\Delta}(e^s) &= e \xrightarrow{s} e^s \xrightarrow{\Delta(y)} \cdot = e \xrightarrow{s\Delta(y)} \cdot \\ &= e \xrightarrow{\Delta(x)\phi_\Delta(s)} \cdot = e \xrightarrow{\Delta(x)} e^{\Delta(x)} \xrightarrow{\phi_\Delta(s)} \cdot = \widehat{\Delta}(e)\phi_\Delta(e) \xrightarrow{\phi_\Delta(s)} \cdot. \end{aligned}$$

So, by definition, the functor $\phi_{\widehat{\Delta}}$ is defined on objects by $\phi_{\widehat{\Delta}}(e) = \phi_\Delta(e)$ and on elements by $\phi_{\widehat{\Delta}}(e \xrightarrow{s} \cdot) = \phi_\Delta(e) \xrightarrow{\phi_\Delta(s)} \cdot$.

(ii) By Lemma V.2.7, ϕ_Δ is surjective both on $\mathcal{O}bj(\mathcal{C})$ and on $\mathcal{C}$. Assume $e \in \mathcal{C}^{\circlearrowright}$. Then there exists e' in $\mathcal{C}$ satisfying $\phi_\Delta(e') = e$. The assumption that Δ is target-injective implies that e' belongs to $\mathcal{C}^{\circlearrowright}$. Let $d \xrightarrow{s} \cdot$ be an element of $\widehat{\mathcal{S}}^\sharp(\cdot, e)$. Then, by definition, we have $d^s = e$. As $\mathcal{S}$ is bounded by Δ, there exists r in $\mathcal{S}^\sharp$ satisfying $rs = \Delta(x)$, whence $e'^{rs} = e$. As $\mathcal{C}$ is right-cancellative, the conjunction of $d^s = e$ and $(e'^r)^s = e$ implies $e'^r = d$. Then we obtain $e' \xrightarrow{r} d \xrightarrow{s} \cdot = e' \xrightarrow{\Delta(x)} \cdot$, which is $e' \xrightarrow{r} d \xrightarrow{s} \cdot = \widehat{\Delta}(e')$. This shows that $\widehat{\mathcal{S}}$ is bounded by $\widehat{\Delta}$. □

Remark 1.16. A straightforward generalization of the conjugacy category $\mathcal{C}onj\mathcal{C}$ is the *simultaneous conjugacy category* $\mathcal{C}onj^*\mathcal{C}$, in which objects are nonempty families $(e_i)_{i \in I}$ of elements of $\mathcal{C}^{\circlearrowright}$ sharing the same source and target and elements are triples of the form $(e_i)_{i \in I} \xrightarrow{g} (e'_i)_{i \in I}$ with $e_i g = g e'_i$ for every i in I. See Exercise 95.

1.2 Cyclic conjugacy

A restricted form of conjugacy, called cyclic conjugacy, will be important in applications. As explained in the introduction, cyclic conjugacy is generated by the particular form of conjugacy consisting in repeatedly exchanging the two factors of a decomposition. Specially interesting is the case when any two conjugate elements can be assumed to be cyclically conjugate. For instance, it turns out that any two periodic braids that are conjugate are cyclically conjugate (see Proposition 1.24), a key result for the investigation of braid conjugation.

Definition 1.17 (cyclic conjugacy). For $\mathcal{C}$ a left-cancellative category, we denote by $\mathcal{C}yc_1\mathcal{C}$ the subfamily of $\mathcal{C}onj\mathcal{C}$ consisting of all elements $e \xrightarrow{g} \cdot$ satisfying $g \preccurlyeq e$, and define the *cyclic conjugacy category* $\mathcal{C}yc\mathcal{C}$ to be the subcategory of $\mathcal{C}onj\mathcal{C}$ generated by $\mathcal{C}yc_1\mathcal{C}$. For e, e' in $\mathcal{C}^{\circlearrowleft}$, we say that an element g of $\mathcal{C}$ *cyclically conjugates* e to e' if $e \xrightarrow{g} \cdot$ lies in $\mathcal{C}yc\mathcal{C}$ and $e' = e^g$ holds.

Example 1.18 (cyclic conjugacy). Consider the braid monoid B_n^+ with $n \geqslant 3$ (Reference Structure 2, page 5). Then the elements σ_1 and σ_2 are conjugate in B_n^+, as we have $\sigma_1 \cdot \sigma_2\sigma_1 = \sigma_2\sigma_1 \cdot \sigma_2$. However, σ_1 and σ_2 are not cyclically conjugate, since the only left-divisors of σ_1 are 1 and σ_1, which both conjugate σ_1 to itself.

The category $\mathcal{C}yc\mathcal{C}$ has the same objects as $\mathcal{C}onj\mathcal{C}$ but it contains only the products of elements of the form $e \xrightarrow{g} \cdot$ with $g \preccurlyeq e$. Observe that, for all e in $\mathcal{C}^{\circlearrowleft}$ and g satisfying $g \preccurlyeq e$, the existence of e' satisfying $eg = ge'$ is guaranteed: $e = gg'$ implies $eg = g(g'g)$: whenever $\mathcal{C}$ is left-cancellative (as will always be assumed), cyclically conjugating e means expressing e as gg', switching the entries, and repeating the operation any finite number of times. When $\mathcal{C}$ is cancellative, cyclic conjugacy is a symmetric transformation in that, with obvious definitions, cyclic conjugation from the right coincides with cyclic conjugation from the left. More precisely, $fg = h$ in $\mathcal{C}$ implies $h^f = {}^g h$.

Lemma 1.19. *Assume that $\mathcal{C}$ is a left-cancellative category.*

(i) *The family $\mathcal{C}yc_1\mathcal{C}$ is closed under left- and right-divisor in $\mathcal{C}onj\mathcal{C}$.*

(ii) *If, moreover, $\mathcal{C}$ admits conditional right-lcms, then $\mathcal{C}yc_1\mathcal{C}$ is closed under right-diamond in $\mathcal{C}onj\mathcal{C}$.*

Proof. (i) Assume that $e \xrightarrow{g} \cdot$ belongs to $\mathcal{C}yc_1\mathcal{C}$ and $e \xrightarrow{g} \cdot = e \xrightarrow{f} d \xrightarrow{h} \cdot$ holds in $\mathcal{C}onj\mathcal{C}$. By definition, we have $g = fh$, so the assumption $g \preccurlyeq e$ a fortiori implies $f \preccurlyeq e$. So $e \xrightarrow{f} \cdot$ belongs to $\mathcal{C}yc_1\mathcal{C}$. Next, starting again from $g \preccurlyeq e$, that is, $fh \preccurlyeq e$, we have $fh \preccurlyeq ef$ a fortiori. Now, by assumption, $ef = fd$ holds, hence we obtain $fh \preccurlyeq fd$, whence $h \preccurlyeq d$ by left-cancelling f. So $d \xrightarrow{h} \cdot$ belongs to $\mathcal{C}yc_1\mathcal{C}$ too.

(ii) Assume that $\mathcal{C}$ admits conditional right-lcms. Then, by Lemma 1.10, $\mathcal{C}onj\mathcal{C}$ also admits conditional right-lcms. We claim that $\mathcal{C}yc_1\mathcal{C}$ is closed under right-lcm. Indeed, assume that $e \xrightarrow{f} \cdot$ and $e \xrightarrow{g} \cdot$ belong to $\mathcal{C}yc_1\mathcal{C}$ and $e \xrightarrow{h} \cdot$ is a common right-multiple of $e \xrightarrow{f} \cdot$ and $e \xrightarrow{g} \cdot$ in $\mathcal{C}onj\mathcal{C}$. Then, the assumption that f and g left-divide e implies that h left-divides e. So $e \xrightarrow{h} \cdot$ belongs to $\mathcal{C}yc_1\mathcal{C}$. Hence $\mathcal{C}yc_1\mathcal{C}$ is closed under right-lcm in $\mathcal{C}onj\mathcal{C}$, hence under right-comultiple. As it is closed under right-divisor, it follows that it is also closed under right-diamond. $\square$

Proposition 1.20 (*CycC* **Garside I).** *Assume that $\mathcal{C}$ is a left-cancellative category that admits conditional right-lcms.*

(i) *The subcategory CycC is closed under right-quotient in $Conj\mathcal{C}$.*

(ii) *If, moreover, $\mathcal{C}$ is right-Noetherian, then $Cyc_1\mathcal{C}$ is a Garside family in $Cyc\mathcal{C}$; it is bounded by the map Δ defined by $\Delta(e) = e \xrightarrow{e} \cdot$ for e in $\mathcal{C}^{\circlearrowleft}$; the latter is a Garside map in $Cyc\mathcal{C}$ and the associated functor ϕ_Δ is the identity.*

Proof. (i) We apply Proposition VII.1.13 (generated subcategory, right-lcm case) to the subcategory of $Conj\mathcal{C}$ generated by $Cyc_1\mathcal{C}$, which is $Cyc\mathcal{C}$ by definition. By Lemma 1.10, the category $Conj\mathcal{C}$ admits conditional right-lcms. By Lemma 1.19, $Cyc_1\mathcal{C}$ is closed under right-complement in $Conj\mathcal{C}$ and, moreover, it is closed under right-multiplication by an invertible element since it is closed under left-divisor (or, directly, because $g \preccurlyeq e$ implies $g\epsilon \preccurlyeq e$ whenever ϵ is invertible). So we conclude that $Cyc\mathcal{C}$ is closed under right-quotient in $Conj\mathcal{C}$.

(ii) If $\mathcal{C}$ is right-Noetherian, then Lemma 1.10 says that so is $Conj\mathcal{C}$, and Proposition VII.1.13 says that $Cyc_1\mathcal{C}$ is a Garside family in $Cyc\mathcal{C}$.

Next, it follows from the definition that $Cyc_1\mathcal{C}$ is right-bounded by Δ and, that $g \preccurlyeq e$ implies $e \xrightarrow{g} \cdot \Delta(e^g) = \Delta(e) e \xrightarrow{g} \cdot$, which shows that $\phi_\Delta(e) = e$ holds for every object e and $\phi_\Delta(e \xrightarrow{g} \cdot) = e \xrightarrow{g} \cdot$ holds for every element $e \xrightarrow{g} \cdot$ of $Cyc\mathcal{C}$. So ϕ_Δ is the identity functor. By Proposition V.2.17(i) (automorphism), we deduce that $Cyc_1\mathcal{C}$ is bounded by Δ and Δ is target-injective. Moreover, $Cyc_1\mathcal{C}$ is closed under left-divisor, so Proposition V.2.25(i) (Garside map) implies that Δ is a Garside map in $Cyc\mathcal{C}$. □

Besides $Cyc_1\mathcal{C}$, and under the same hypotheses, another distinguished Garside family arises in $Cyc\mathcal{C}$ whenever a Garside family is given in $\mathcal{C}$. This Garside family is in general smaller than $Cyc_1\mathcal{C}$.

Proposition 1.21 (*CycC* **Garside II).** *Assume that $\mathcal{C}$ is a left-cancellative category that is right-Noetherian and admits conditional right-lcms, and $\mathcal{S}$ is a Garside family in $\mathcal{C}$. Then the family $\{e \xrightarrow{s} \cdot \in Conj\mathcal{C} \mid s \in \mathcal{S}$ and $s \preccurlyeq e\}$ is a Garside family in $Cyc\mathcal{C}$.*

Proof. Put $\widehat{\mathcal{S}}_1 = \{e \xrightarrow{s} \cdot \in \widehat{\mathcal{S}} \mid s \preccurlyeq e\}$, where $\widehat{\mathcal{S}}$ is as in Notation 1.11(i). By Lemma 1.10, the invertible elements of $Conj\mathcal{C}$ are the elements $e \xrightarrow{\epsilon} \cdot$ with ϵ in $\mathcal{C}^\times$, whence $\widehat{\mathcal{S}}^\sharp = \{e \xrightarrow{s} \cdot \in Conj\mathcal{C} \mid s \in \mathcal{S}^\sharp\}$. Moreover, all above invertible elements belong to $Cyc\mathcal{C}$ and $s \preccurlyeq e$ is equivalent to $s\epsilon \preccurlyeq e$ for ϵ invertible, so we deduce $\widehat{\mathcal{S}}_1^\sharp = \{e \xrightarrow{s} \cdot \in \widehat{\mathcal{S}}^\sharp \mid s \preccurlyeq e\}$. The assumption that $\mathcal{S}^\sharp$ generates $\mathcal{C}$ implies that $\widehat{\mathcal{S}}_1^\sharp$ generates $Cyc\mathcal{C}$.

The argument is then similar to that for Proposition 1.20, replacing $Cyc_1\mathcal{C}$ with its subfamily $\widehat{\mathcal{S}}_1^\sharp$. Assume first that $e \xrightarrow{s} \cdot$ belongs to $\widehat{\mathcal{S}}_1^\sharp$ and $d \xrightarrow{t} \cdot$ is a right-divisor of $e \xrightarrow{s} \cdot$ in $Conj\mathcal{C}$. As $\widehat{\mathcal{S}}$ is a Garside family in $Conj\mathcal{C}$, by Proposition IV.1.24 (recognizing Garside II), $\widehat{\mathcal{S}}^\sharp$ is closed under right-divisor, and $d \xrightarrow{t} \cdot$ belongs to $\widehat{\mathcal{S}}^\sharp$. On the other hand, $e \xrightarrow{s} \cdot$ belongs to $Cyc_1\mathcal{C}$, hence so does its right-divisor $d \xrightarrow{t} \cdot$. Hence $d \xrightarrow{t} \cdot$ belongs to $\widehat{\mathcal{S}}_1^\sharp$. So $\widehat{\mathcal{S}}_1^\sharp$ is closed under right-quotient in $Conj\mathcal{C}$.

Next, assume that $e \xrightarrow{s} \cdot$ and $e \xrightarrow{t} \cdot$ belong to $\widehat{\mathcal{S}}_1^\sharp$ and $e \xrightarrow{r} \cdot$ is a common right-multiple of $e \xrightarrow{s} \cdot$ and $e \xrightarrow{t} \cdot$ in $Conj\mathcal{C}$. By Corollary IV.2.29 (recognizing Garside, right-lcm case), r must belong to $\mathcal{S}_1^\sharp$. Moreover, the assumption that s and t left-divide e implies that r

left-divides e. So $e \xrightarrow{r} \cdot$ belongs to $\widehat{\mathcal{S}}_1^\sharp$. Thus, $\widehat{\mathcal{S}}_1^\sharp$ is closed under right-lcm in $\mathcal{C}onj\mathcal{C}$, hence under right-comultiple, and under right-complement since it is closed under right-quotient.

As above, the category $\mathcal{C}onj\mathcal{C}$ is right-Noetherian and admits conditional right-lcms. By applying Proposition VII.1.13 (generated subcategory, right-lcm case), we deduce that $\widehat{\mathcal{S}}_1$ is a Garside family in the subcategory of $\mathcal{C}onj\mathcal{S}$ generated by $\widehat{\mathcal{S}}_1^\sharp$, which is $\mathcal{C}yc\mathcal{C}$. □

Remark 1.22. The assumption that the ambient category admits conditional right-lcms can be (slightly) weakened in the above proofs of Propositions 1.20 and 1.21: what is needed in order to obtain a closure under right-comultiple for $\mathcal{C}yc_1\mathcal{C}$ or $\widehat{\mathcal{S}}_1$ is that, if r is a common right-multiple of s and t and s, t left-divide e, then r is a right-multiple of some common right-multiple r' of s and t that left-divides e. If s and t admit a right-lcm, the latter has the expected property, but the property, which simply says that $\mathrm{Div}(e)$ is closed under right-comultiple, can be true although right-lcms do not exist, see Exercise 96. However, in a category that admits right-mcms, hence in particular in every left-Noetherian category, the property that every family $\mathrm{Div}(s)$ is closed under right-comultiple is equivalent to the existence of conditional right-lcms.

Proposition 1.23 (**$\mathcal{C}yc\mathcal{C}$ closed under left-gcd**). *If $\mathcal{C}$ is a left-cancellative category that is right-Noetherian and admits conditional right-lcms, then $\mathcal{C}yc\mathcal{C}$ is closed under left-gcd in $\mathcal{C}onj\mathcal{C}$ (that is, every left-gcd of two elements of $\mathcal{C}yc\mathcal{C}$ belongs to $\mathcal{C}yc\mathcal{C}$).*

Proof. First, by Lemma 1.10, $\mathcal{C}onj\mathcal{C}$ is right-Noetherian and admits conditional right-lcms, hence it admits left-gcds. Let $e \xrightarrow{f} \cdot$ and $e \xrightarrow{g} \cdot$ belong to $\mathcal{C}yc\mathcal{C}$. Let $e \xrightarrow{f_1|\cdots|f_p} \cdot$ and $e \xrightarrow{g_1|\cdots|g_q} \cdot$ be $\mathcal{C}yc_1\mathcal{C}$-normal decompositions of $e \xrightarrow{f} \cdot$ and $e \xrightarrow{g} \cdot$.

We first prove that, if every common left-divisor of f_1 and g_1 is invertible, then so is every common left-divisor of f and g using induction on $\inf(p, q)$. If $\inf(p, q) = 0$ holds, that is, if one decomposition is empty, then f or g is invertible and the result is clear. Otherwise, assume that h left-divides f and g. The assumption that $e \xrightarrow{f} \cdot$ belongs to $\mathcal{C}yc\mathcal{C}$ and $e \xrightarrow{f_1|\cdots|f_p} \cdot$ is a $\mathcal{C}yc_1\mathcal{C}$- decomposition of $e \xrightarrow{f} \cdot$ implies that f_p left-divides $e^{f_1 \cdots f_{p-1}}$ in $\mathcal{C}$, and we find

$$h \preccurlyeq f_1 \cdots f_p \preccurlyeq f_1 \cdots f_{p-1} e^{f_1 \cdots f_{p-1}} = e f_1 \cdots f_{p-1}.$$

Let eh' be a right-lcm of e and h. Then $h \preccurlyeq e f_1 \cdots f_{p-1}$ implies $h' \preccurlyeq f_1 \cdots f_{p-1}$. By a similar argument, we obtain $h' \preccurlyeq g_1 \cdots g_{q-1}$. Applying the induction hypothesis to $f' = f_1 \cdots f_{p-1}$ and $g' = g_1 \cdots g_{q-1}$, we deduce that h' is invertible, hence that h left-divides e and, therefore, that $e \xrightarrow{h} \cdot$ belongs to $\mathcal{C}yc_1\mathcal{C}$. Now, as, by assumption, $e \xrightarrow{f_1|\cdots|f_p} \cdot$ is $\mathcal{C}yc_1\mathcal{C}$-normal, the assumption that $e \xrightarrow{h} \cdot$ left-divides $e \xrightarrow{f} \cdot$ implies that it left-divides $e \xrightarrow{f_1} \cdot$, and, similarly, $e \xrightarrow{g_1} \cdot$. So, finally, h must be invertible.

We now prove the result. If every common left-divisor of f_1 and g_1 is invertible, then every left-gcd of $e \xrightarrow{f} \cdot$ and $e \xrightarrow{g} \cdot$ in $\mathcal{C}onj\mathcal{C}$ is invertible, hence it lies in $\mathcal{C}yc\mathcal{C}$. Otherwise, let h_1 be a left-gcd of f_1 and g_1, and $f^{(1)}, g^{(1)}$ be defined by $f = h_1 f^{(1)}$, $g = h_1 g^{(1)}$. Similarly, let h_2 be a left-gcd of the first entries in a $\mathcal{C}yc_1\mathcal{C}$-normal decomposition of $e \xrightarrow{f^{(1)}} \cdot$ and $e \xrightarrow{g^{(1)}} \cdot$ and $f^{(2)}$, $g^{(2)}$ be the remainders, etc. Since $\mathcal{C}$ is right-Noetherian, the sequence $h_1, h_1h_2, \ldots$ of increasing left-divisors of f stabilizes at some stage k, meaning that every common left-divisor of the entries of $\mathcal{C}yc\mathcal{C}$-normal decompositions of $f^{(k)}$ and $g^{(k)}$ is invertible. By the first part, every common left-divisor of $f^{(k)}$ and $g^{(k)}$ is invertible, so $h_1 \cdots h_k$ is a left-gcd of f and g in $\mathcal{C}$, and $e \xrightarrow{h_1 \cdots h_k} \cdot$, which belongs to $\mathcal{C}yc\mathcal{C}$ by construction, is a left-gcd of $e \xrightarrow{f} \cdot$ and $e \xrightarrow{g} \cdot$ in $\mathcal{C}onj\mathcal{C}$, hence in $\mathcal{C}yc\mathcal{C}$. □

Finally, we give a quite general context where cyclic conjugacy coincides with conjugacy. Note that, by Proposition V.2.35(ii), if $\mathcal{C}$ is left cancellative, Noetherian, and admits a Garside map, then the assumptions of the next proposition are satisfied.

Proposition 1.24 (every conjugate cyclic). *Assume that $\mathcal{C}$ is a left-cancellative category that is right-Noetherian and admits conditional right-lcms, Δ is a Garside map in $\mathcal{C}$, and e is an element of $\mathcal{C}^{\circlearrowleft}(x,x)$ satisfying $\Delta(x) \preccurlyeq e^n$ for n large enough. Then, for every e' in $\mathcal{C}^{\circlearrowleft}$, one has $\mathcal{C}yc\mathcal{C}(e,e') = \mathcal{C}onj\mathcal{C}(e,e')$.*

Proof. We first note that the property $\exists n \in \mathbb{N}\, (\Delta(x) \preccurlyeq e^n)$ is preserved under conjugacy. Indeed, assume that $\Delta(x) \preccurlyeq e^n$ and e^g is defined with g in $\mathcal{C}(x,y)$. Write $e^n = \Delta(x)d$. Then we deduce $e^{2n} = \Delta(x)d\Delta(x)d = \Delta^{[2]}(x)\phi_\Delta(d)d$, as d must lie in $\mathcal{C}(\phi_\Delta(x),x)$. So we have $\Delta^{[2]}(x) \preccurlyeq e^{2n}$ and, using an easy induction, $\Delta^{[k+1]}(x) \preccurlyeq e^{n(k+1)}$ for every k. Now, there exists k such that g left-divides $\Delta^{[k]}(x)$, say $gg' = \Delta^{[k]}(x)$. Then we have $\Delta^{[k+1]}(x) \preccurlyeq e^{n(k+1)}g$, whence $gg'\Delta(\phi_\Delta^k(x)) \preccurlyeq e^{n(k+1)}g$, and, since $\Delta(y) \preccurlyeq g'\Delta(\phi_\Delta^k(x))$ holds, we deduce $g\Delta(y) \preccurlyeq e^{n(k+1)}g$. This implies $\Delta(y) \preccurlyeq (e^{n(k+1)})^g$, which is also $\Delta(y) \preccurlyeq (e^g)^{n(k+1)}$.

We use this to prove by induction on the right-height of g that $g \in \mathcal{C}onj\mathcal{C}(e,e')$ implies $g \in \mathcal{C}yc\mathcal{C}(e,e')$. This is true if g is invertible. Assume g is non-invertible. Let h be a left-gcd of g and e, which exists since $\mathcal{C}$ admits left-gcds by Lemma II.2.37. Write $g = hg_1$. Then, since $e \xrightarrow{h} \cdot$ belongs to $\mathcal{C}yc\mathcal{C}(e,e^h)$, it is sufficient to prove that $e^h \xrightarrow{g_1} \cdot$, which a priori belongs to $\mathcal{C}onj\mathcal{C}(e^h,e')$, is actually in $\mathcal{C}yc\mathcal{C}(e^h,e')$. If h is not invertible, we are done by induction, since $\mathcal{C}$ is right-Noetherian and e^h satisfies the same condition. Hence it is sufficient to prove that, if g is not invertible, then g and e admit a non-invertible common left-divisor.

We observed above that every element g of $\mathcal{C}onj\mathcal{C}(e,\cdot)$ left-divides some power of e, namely e^{nk} if g left-divides $\Delta^{[k]}(x)$. Hence, it is enough to prove that, if g is a non-invertible element of $\mathcal{C}onj\mathcal{C}(e,\cdot)$ satisfying $g \preccurlyeq e^n$, then g and e admit a non-invertible common left-divisor. We do this by induction on n. Now $g \in \mathcal{C}onj\mathcal{C}(e,\cdot)$ implies $g \preccurlyeq eg$. So, from $g \preccurlyeq e^n$ we deduce $g \preccurlyeq eg_1$ for every left-gcd g_1 of g and e^{n-1}. If g_1 is invertible, we deduce $g \preccurlyeq e$, and g is a non-invertible common left-divisor if g and e. Otherwise, $g_1 \preccurlyeq eg_1$ holds, g_1 is not invertible, and g_1 left-divides e^{n-1}. Then the induction hypothesis implies that g_1 and e, hence a fortiori g and e, admits a non-invertible common left-divisor. □

1.3 Twisted conjugacy

We now consider a twisted version of conjugation involving an automorphism of the ambient category. This will be used in Chapter X for applications to Deligne-Lusztig varieties, but also latter in the current chapter for the automorphism ϕ_Δ associated with a Garside map Δ. We recall that an automorphism of a category $\mathcal{C}$ is a functor of $\mathcal{C}$ into itself that admits an inverse. Note that, if $\mathcal{C}$ is a left-Ore category, every automorphism of $\mathcal{C}$ naturally extends to an automorphism of the enveloping groupoid $\mathcal{E}nv(\mathcal{C})$.

Definition 1.25 (ϕ-conjugacy). For $\mathcal{C}$ a category and ϕ an automorphism of $\mathcal{C}$, we put $\mathcal{C}_\phi^\circlearrowleft = \bigcup_{x \in \mathcal{O}bj(\mathcal{C})} \mathcal{C}(x, \phi(x))$. For e, e' in $\mathcal{C}_\phi^\circlearrowleft$, we say that an element g of $\mathcal{C}$ *ϕ-conjugates* e to e' if $e\phi(g) = ge'$ holds. The *ϕ-conjugacy category* $\mathcal{C}onj_\phi\mathcal{C}$ of $\mathcal{C}$ is the category whose object family is $\mathcal{C}_\phi^\circlearrowleft$ and where $\mathcal{C}onj_\phi\mathcal{C}(e, e')$ consists of all triples $e \xrightarrow{g} e'$ such that g ϕ-conjugates e to e', the source (*resp.* target) of $e \xrightarrow{g} e'$ being e (*resp.* e'), and the product being defined by

$$(e \xrightarrow{g} e')(e' \xrightarrow{g'} e'') = e \xrightarrow{gg'} e''. \tag{1.26}$$

For e in $\mathcal{C}_\phi^\circlearrowleft$, the family $\mathcal{C}onj_\phi\mathcal{C}(e, e)$ is called the *ϕ-centralizer* of e in $\mathcal{C}$.

Of course, ϕ-conjugacy is just conjugacy when ϕ is the identity. Note that the definition of $\mathcal{C}onj_\phi\mathcal{C}(e, e')$ forces the objects of $\mathcal{C}onj_\phi\mathcal{C}$ to belong to some family $\mathcal{C}(x, \phi(x))$. Lemma 1.4 then extends to

Lemma 1.27. *Definition 1.25 is legal, that is, $\mathcal{C}onj_\phi\mathcal{C}$ is indeed a category. Moreover, the map π defined by $\pi(e) = x$ for e in $\mathcal{C}(x, \phi(x))$ and by $\pi(e \xrightarrow{g} e') = g$ on $\mathcal{C}onj_\phi\mathcal{C}$ is a surjective functor from $\mathcal{C}onj_\phi\mathcal{C}$ onto $\mathcal{C}$. For e in $\mathcal{C}(x, \phi(x))$, we have*

$$\pi(\mathcal{C}onj_\phi\mathcal{C}(e, \cdot)) = \{g \in \mathcal{C}(x, \cdot) \mid g \preccurlyeq e\phi(g)\}, \tag{1.28}$$

$$\pi(\mathcal{C}onj_\phi\mathcal{C}(\cdot, e')) = \{g \in \mathcal{C}(\cdot, x) \mid ge' \succcurlyeq \phi(g)\}. \tag{1.29}$$

Proof. Adapting the arguments is easy. For (1.26), the point is that the conjunction of $e\phi(g) = ge'$ and $e'\phi(g') = g'e''$ implies $e\phi(gg') = gg'e''$. ∎

As in the case of ordinary conjugacy, if the ambient category $\mathcal{C}$ is left-cancellative, the data e and g determine e' when $e\phi(g) = ge'$ holds and, therefore, we can use the simplified notation $e \xrightarrow{g} \cdot$ for the corresponding element $e \xrightarrow{g} e'$ of $\mathcal{C}onj_\phi\mathcal{C}$.

It turns out that twisted conjugacy in a category $\mathcal{C}$ can be expressed as an ordinary, non-twisted conjugacy in an extended category reminiscent of a semi-direct product of $\mathcal{C}$ by the cyclic group generated by the considered automorphism.

Definition 1.30 (semi-direct product). For $\mathcal{C}$ a category and ϕ an automorphism of $\mathcal{C}$, the *semi-direct product* $\mathcal{C} \rtimes \langle\phi\rangle$ is the category $\mathcal{C}'$ defined by $\mathcal{O}bj(\mathcal{C}') = \mathcal{O}bj(\mathcal{C})$ and $\mathcal{C}' = \mathcal{C} \times \mathbb{Z}$, the source of (g, m) being that of g, its target being that of $\phi^{-m}(g)$, and the product being given by

$$(g, m)(h, n) = (g\phi^m(h), m + n). \tag{1.31}$$

In practice, the element (g, m) of $\mathcal{C} \rtimes \langle\phi\rangle$ will be denoted by $g\phi^m$; then the formula for the product in $\mathcal{C} \rtimes \langle\phi\rangle$ takes the natural form

(1.32) $$g\phi^m \cdot h\phi^n = g\phi^m(h)\phi^{m+n}.$$

Definition 1.30 is legal: for g in $\mathcal{C}(\cdot, y)$ and h in $\mathcal{C}(z, \cdot)$, the product $g\phi^m(h)$ is defined if and only if the target y of g coincides with the source $\phi^m(z)$ of $\phi^m(h)$, that is, if the target $\phi^{-m}(y)$ of $g\phi^m$ coincides with the source z of $h\phi^n$. Note that the elements $1_x\phi$ actually act like ϕ^{-1}, since every element g of $\mathcal{C}(x, y)$ gives a commutative diagram as on the right.

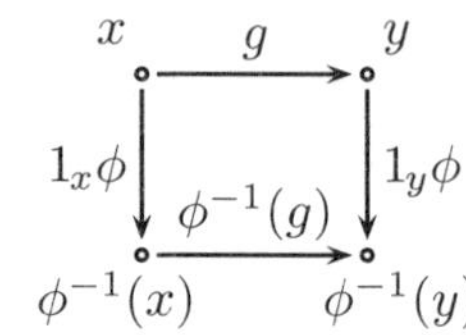

Together with identity on $\mathcal{O}bj(\mathcal{C})$, mapping g to $g\phi^0$ provides an injective functor from $\mathcal{C}$ into $\mathcal{C} \rtimes \langle\phi\rangle$: building on this, we shall hereafter consider $\mathcal{C}$ as included in $\mathcal{C} \rtimes \langle\phi\rangle$ and identify $g\phi^0$ (that is, $(g, 0)$) with g.

Lemma 1.33. *If $\mathcal{C}$ is a left-cancellative category and ϕ is an automorphism of $\mathcal{C}$:*

(i) *The category $\mathcal{C} \rtimes \langle\phi\rangle$ is left-cancellative.*

(ii) *The identity elements of $\mathcal{C} \rtimes \langle\phi\rangle$ are those of $\mathcal{C}$, whereas its invertible elements are those of the form $\epsilon\phi^m$ with ϵ in $\mathcal{C}^\times$, the inverse of $\epsilon\phi^m$ being $\phi^{-m}(\epsilon^{-1})\phi^{-m}$.*

(iii) *For all g, h in $\mathcal{C}$ and m, n in $\mathbb{Z}$, the relation $g\phi^m \preccurlyeq h\phi^n$ holds in $\mathcal{C} \rtimes \langle\phi\rangle$ if and only if $g \preccurlyeq h$ holds in $\mathcal{C}$.*

We skip the straightforward verifications.

Lemma 1.34. *Assume that $\mathcal{C}$ is a left-cancellative category, ϕ is an automorphism of $\mathcal{C}$, and $\mathcal{S}$ is a Garside family of $\mathcal{C}$ that is preserved by ϕ.*

(i) *For all g_1, g' in $\mathcal{C}$ and m in $\mathbb{Z}$, the path $g_1|(g_2\phi^m)$ is $\mathcal{S}$-greedy (resp. $\mathcal{S}$-normal) in $\mathcal{C} \rtimes \langle\phi\rangle$ if and only if $g_1|g_2$ is $\mathcal{S}$-greedy (resp. $\mathcal{S}$-normal) in $\mathcal{C}$.*

(ii) *The family $\mathcal{S}$ is a Garside family in $\mathcal{C} \rtimes \langle\phi\rangle$. If $s_1|\cdots|s_p$ is an $\mathcal{S}$-normal decomposition of s in $\mathcal{C}$, then $s_1|\cdots|s_{p-1}|(s_p\phi^m)$ is an $\mathcal{S}$-normal decomposition of $s\phi^m$ in $\mathcal{C} \rtimes \langle\phi\rangle$.*

Proof. (i) Assume that $g_1|g_2$ is $\mathcal{S}$-greedy, h belongs to $\mathcal{S}$ and, in $\mathcal{C} \rtimes \langle\phi\rangle$, we have $h \preccurlyeq f\phi^n \cdot g_1 \cdot g_2\phi^m$. The latter relation implies $h \preccurlyeq f\phi^n(g_1)\phi^n(g_2)$ in $\mathcal{C}$, whence $\phi^{-n}(h) \preccurlyeq \phi^{-n}(f)g_1g_2$. By assumption, $\phi^{-n}(h)$ lies in $\mathcal{S}$ and $g_1|g_2$ is $\mathcal{S}$-greedy, so we deduce $\phi^{-n}(h) \preccurlyeq \phi^{-n}(f)g_1$, whence $h \preccurlyeq f\phi^n(g_1)$ and $h \preccurlyeq f\phi^n \cdot g_1$ in $\mathcal{C} \rtimes \langle\phi\rangle$. So $g_1|(g_2\phi^m)$ is $\mathcal{S}$-greedy in $\mathcal{C} \rtimes \langle\phi\rangle$. If, moreover, g_1 and g_2 belong to $\mathcal{S}^\sharp$ (in the sense of $\mathcal{C}$), then g_1 and $g_2\phi^m$ belong to $\mathcal{S}^\sharp$ (in the sense of $\mathcal{C} \rtimes \langle\phi\rangle$) and $g_1|(g_2\phi^m)$ is $\mathcal{S}$-normal in $\mathcal{C} \rtimes \langle\phi\rangle$.

The converse implications are similarly easy. Assertion (ii) then follows directly. $\square$

Here is the now expected connection between twisted conjugacy in a category and ordinary conjugacy in a semi-direct extension of that category.

Notation 1.35 (family $\mathcal{S}\phi$). For $\mathcal{C}$ a category, $\mathcal{S}$ included in $\mathcal{C}$, and ϕ an automorphism of $\mathcal{C}$, we denote by $\mathcal{S}\phi$ the subfamily $\{g\phi \mid g \in \mathcal{S}\}$ of $\mathcal{C} \rtimes \langle\phi\rangle$.

Proposition 1.36 (twisted conjugacy). *Assume that $\mathcal{C}$ is a category and ϕ is an automorphism of $\mathcal{C}$. For e in $\mathcal{C}_\phi^\circlearrowleft$, put $\iota(e) = e\phi$ and, for $e \xrightarrow{g} e'$ in $\mathit{Conj}_\phi\mathcal{C}$, put $\iota(e \xrightarrow{g} e') = e\phi \xrightarrow{g} e'\phi$. Then ι is an injective functor from $\mathit{Conj}_\phi\mathcal{C}$ to $\mathit{Conj}(\mathcal{C} \rtimes \langle\phi\rangle)$. The image of $\mathit{Obj}(\mathit{Conj}_\phi\mathcal{C})$ under ι is $\mathcal{C}_\phi^\circlearrowleft\phi$, whereas the image of $\mathit{Conj}_\phi\mathcal{C}$ is the intersection of $\mathit{Conj}(\mathcal{C} \rtimes \langle\phi\rangle)$ with $\mathcal{C}_\phi^\circlearrowleft\phi \times \mathcal{C} \times \mathcal{C}_\phi^\circlearrowleft\phi$.*

Proof. Assume that e belongs to $\mathcal{C}(x, \phi(x))$ and $e \xrightarrow{g} e'$ belongs to $\mathit{Conj}_\phi\mathcal{C}$, that is, we have $e\phi(g) = ge'$. By definition, the source (*resp.* target) of $e \xrightarrow{g} e'$ is e (*resp.* e'), whose image under ι is $e\phi$ (*resp.* $e'\phi$). On the other hand, the image of $e \xrightarrow{g} e'$ under ι is $e\phi \xrightarrow{g} e'\phi$, whose source (*resp.* target) is $e\phi$ (*resp.* $e'\phi$). Next, by definition, the equalities

$$\iota(e) \cdot g = e\phi \cdot g = e\phi(g)\phi = g \cdot e'\phi = g \cdot \iota(e')$$

hold in $\mathcal{C} \rtimes \langle\phi\rangle$, which means that $e\phi \xrightarrow{g} e'\phi$ belongs to $\mathit{Conj}(\mathcal{C} \rtimes \langle\phi\rangle)$. Finally, if $e \xrightarrow{g} e'$ and $e' \xrightarrow{g'} e''$ are elements of $\mathit{Conj}_\phi\mathcal{C}$, we find

$$\begin{aligned}\iota(e \xrightarrow{g} e')\iota(e' \xrightarrow{g'} e'') &= (e\phi \xrightarrow{g} e'\phi)(e'\phi \xrightarrow{g'} e''\phi)\\ &= e\phi \xrightarrow{gg'} e''\phi = \iota(e \xrightarrow{gg'} e'') = \iota((e \xrightarrow{g} e')(e' \xrightarrow{g'} e'')).\end{aligned}$$

So ι is indeed a functor from $\mathit{Conj}_\phi\mathcal{C}$ to $\mathit{Conj}(\mathcal{C} \rtimes \langle\phi\rangle)$.

By construction, the image of $\mathit{Obj}(\mathit{Conj}_\phi\mathcal{C})$ under ι consists of all pairs $e\phi$ with e in $\mathcal{C}_\phi^\circlearrowleft$—note that all such pairs belong to $(\mathcal{C} \rtimes \langle\phi\rangle)^\circlearrowleft$—and the image of $\mathit{Conj}_\phi\mathcal{C}$ consists of all triples $e\phi \xrightarrow{g} e'\phi$ with $e\phi(g) = ge'$, hence of all elements of $\mathit{Conj}(\mathcal{C} \rtimes \langle\phi\rangle)$ that lie in $\mathcal{C}_\phi^\circlearrowleft\phi \times \mathcal{C} \times \mathcal{C}_\phi^\circlearrowleft\phi$. □

Let us denote by $\mathit{Conj}(\mathcal{C}\phi)$ the full subcategory of $\mathit{Conj}(\mathcal{C} \rtimes \langle\phi\rangle)$ obtained by restricting objects to $\mathcal{C}\phi$. By definition, an element of the category $\mathit{Conj}(\mathcal{C} \rtimes \langle\phi\rangle)$ is a triple of the form $e\phi^m \xrightarrow{g\phi^n} e'\phi^{m'}$ satisfying $e\phi^m \cdot g\phi^n = g\phi^n \cdot e'\phi^{m'}$, hence in particular $m = m'$. It follows that $\mathit{Conj}(\mathcal{C}\phi)$ is a union of connected components of $\mathit{Conj}(\mathcal{C} \rtimes \langle\phi\rangle)$ and, therefore, many properties transfer from $\mathit{Conj}(\mathcal{C} \rtimes \langle\phi\rangle)$ to $\mathit{Conj}(\mathcal{C}\phi)$. In particular, we obtain

Proposition 1.37 (Garside in twisted). *Assume that $\mathcal{C}$ is a left-cancellative category, $\mathcal{S}$ is a Garside family in $\mathcal{C}$, and ϕ is an automorphism of $\mathcal{C}$ that preserves $\mathcal{S}$. Then the family of all elements $e \xrightarrow{s} \cdot$ of $\mathit{Conj}_\phi\mathcal{C}$ with s in $\mathcal{S}$ is a Garside family in $\mathit{Conj}_\phi\mathcal{C}$.*

Proof. By Lemma 1.34, $\mathcal{S}$ is a Garside family in $\mathcal{C} \rtimes \langle\phi\rangle$. By Corollary 1.14, the family of all elements $e\phi^m \xrightarrow{s} \cdot$ of $\mathit{Conj}(\mathcal{C} \rtimes \langle\phi\rangle)$ with s in $\mathcal{S}$ is a Garside family in $\mathit{Conj}(\mathcal{C} \rtimes \langle\phi\rangle)$ and, therefore, the family $\widehat{\mathcal{S}}$ of all elements $e\phi \xrightarrow{s} \cdot$ of $\mathit{Conj}(\mathcal{C}\phi)$ with s in $\mathcal{S}$ is a Garside family in $\mathit{Conj}(\mathcal{C}\phi)$. Next, $\mathcal{C}$ is closed under right-quotient in $\mathcal{C}\phi$, hence $\iota(\mathit{Conj}_\phi\mathcal{C})$ is closed under right-quotient in $\mathit{Conj}(\mathcal{C}\phi)$, and it includes $\widehat{\mathcal{S}}$. By Corollary VII.2.13 (closed under right-quotient), $\widehat{\mathcal{S}}$ is a Garside family in $\iota(\mathit{Conj}_\phi\mathcal{C})$. Applying ι^{-1}, we deduce that $\iota^{-1}(\widehat{\mathcal{S}})$, which consists of all $e \xrightarrow{s} \cdot$ in $\mathit{Conj}_\phi\mathcal{C}$ with s in $\mathcal{S}$, is a Garside family in $\mathit{Conj}_\phi\mathcal{C}$. □

The functor ι need not be full in general. However, we have:

Proposition 1.38 (functor ι full). *If ϕ is an automorphism of a category $\mathcal{C}$ and $\mathcal{C}$ is a groupoid, or ϕ has finite order, then the functor ι of Proposition* 1.36 *is full.*

Proof. We have to prove that, for e in $\mathcal{C}$, the conjugation of $e\phi$ by a power of ϕ can be done by an element of $\mathcal{C}$. The equality $(e\phi)^{\phi^{-1}} = (e\phi)^e$ implies that the property is true for negative powers of ϕ, which gives the result when the order of ϕ is finite. If $\mathcal{C}$ is a groupoid, we obtain the result using $(e\phi)^\phi = (e\phi)^{\phi^{-1}(e^{-1})}$. $\square$

To conclude this section, we now consider a twisted version of cyclic conjugacy. Here we just mention a few results without formal statements and proofs. So assume that $\mathcal{C}$ is a left-cancellative category, and ϕ is a finite-order automorphism of the category $\mathcal{C}$. Then, we define $\mathit{Cyc}_\phi\mathcal{C}$ as the subcategory of $\mathit{Conj}_\phi\mathcal{C}$ generated by the elements $e \xrightarrow{g} \cdot$ of $\mathit{Conj}_\phi\mathcal{C}(e, \cdot)$ satisfying $g \preccurlyeq e$, or equivalently, if $\mathcal{C}$ is cancellative, by the elements $\cdot \xrightarrow{g} e'$ of $\mathit{Conj}\mathcal{C}(e, e')$ satisfying $e' \succcurlyeq \phi(g)$. By Proposition 1.38, the functor ι of Proposition 1.36 identifies $\mathit{Cyc}_\phi\mathcal{C}(e, e')$ with the family of elements of $\mathit{Cyc}(\mathcal{C} \rtimes \langle\phi\rangle)(e\phi, e'\phi)$ that lie in $\mathcal{C}$. To simplify notation, we will denote the latter family by $\mathit{Cyc}\mathcal{C}(e\phi, e'\phi)$. If $\mathcal{C}$ is right-Noetherian and admits conditional right-lcms, then the same holds for $\mathcal{C} \rtimes \langle\phi\rangle$. If $\mathcal{S}$ is a Garside family in $\mathcal{C}$ that is preserved by ϕ, translating Proposition 1.21 to the image of ι and then to $\mathit{Cyc}_\phi\mathcal{C}$, we obtain that the family $\{e \xrightarrow{g} \cdot \in \mathit{Conj}_\phi\mathcal{C} \mid g \preccurlyeq e \text{ and } g \in \mathcal{S}\}$ is a Garside family in $\mathit{Cyc}_\phi\mathcal{C}$.

Similarly, Proposition 1.20 says in this context that $\{e \xrightarrow{g} \cdot \in \mathit{Conj}_\phi\mathcal{C} \mid g \preccurlyeq e\}$ is a Garside family in $\mathit{Cyc}_\phi\mathcal{C}$, which is bounded by the Garside map Δ that maps the object e to the element $e \xrightarrow{e} \cdot$ of $\mathit{Cyc}_\phi\mathcal{C}(e, \phi(e))$; the associated functor ϕ_Δ is ϕ.

Then Proposition 1.23 says that, under the assumptions of Proposition 1.20, the subcategory $\mathit{Cyc}_\phi\mathcal{C}$ of $\mathit{Conj}_\phi\mathcal{C}$ is closed under left-gcd.

Finally, Proposition 1.24 says that, if $\mathcal{C}$ is left-cancellative, right-Noetherian, if Δ is a Garside map in $\mathcal{C}$ and ϕ is an automorphism that preserves the family $\mathcal{D}iv(\Delta)$, then, for all e, e' in $\mathcal{C}$ satisfying $\Delta \preccurlyeq (e\phi)^n$ for some n, the equality $\mathit{Cyc}_\phi\mathcal{C}(e, e') = \mathit{Conj}_\phi\mathcal{C}(e, e')$ holds.

For an example, we refer to Figure 1, which represents the connected component of the braid $\beta = \sigma_2\sigma_1\sigma_3\sigma_2\sigma_5$ in the category of twisted cyclic conjugation in the 6-strand braid monoid B_6^+ when ϕ is the flip automorphism that maps σ_i to σ_{6-i} for every i. The braid β, hence also all its conjugates, is a twisted cubic root of Δ_6. By Proposition 1.24 (twisted), cyclic conjugation is the same as (twisted) conjugation for such elements, that is, the equality $\mathit{Cyc}_\phi\mathcal{C}(e, e') = \mathit{Conj}_\phi\mathcal{C}(e, e')$ holds. Note that ϕ is equal to the automorphism ϕ_Δ that stems from the Garside element Δ_6, Then, at the expense of considering that β in the figure represents $\beta\Delta_6$, which is a cubic root of Δ_6^4, we can also see the figure as depicting (non-twisted) conjugacy (identical to cyclic conjugacy) of these elements. With the same notation as above, we have $\mathit{Cyc}\mathcal{C}(e\Delta, e'\Delta) = \mathit{Conj}\mathcal{C}(e\Delta, e'\Delta)$.

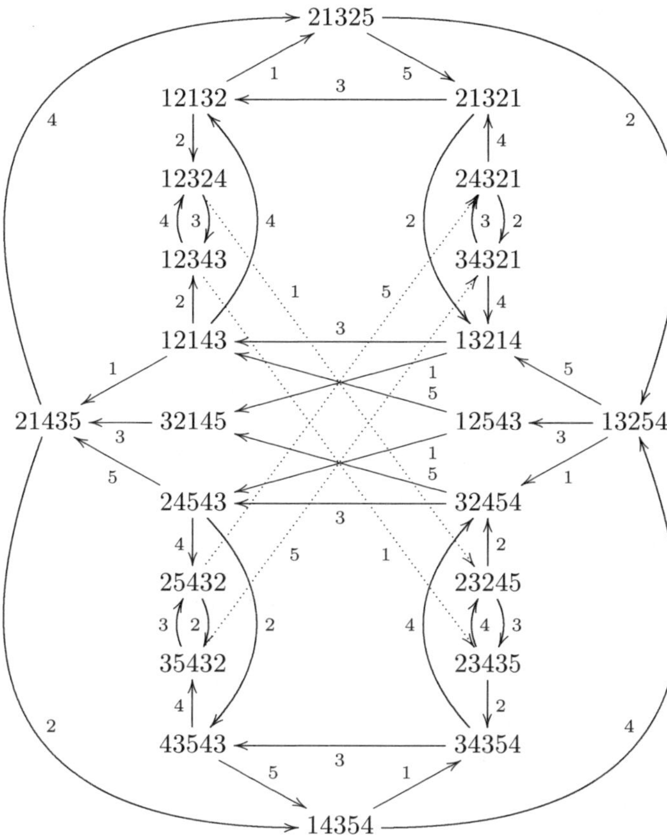

Figure 1. Connected component of $\sigma_2\sigma_1\sigma_3\sigma_2\sigma_5$ in the category of twisted cyclic conjugation in B_6^+ associated with the flip automorphism. The arrows correspond to the atoms of the category, that is, the twisted cyclic conjugacy by $\sigma_1, \ldots, \sigma_5$, here simply denoted by $1, \ldots, 5$. The action of σ_i on the braid β is defined if σ_i left-divides β and maps β to $\sigma_i^{-1}\beta\sigma_{6-i}$.

1.4 An example: ribbon categories

Our aim is to study the simultaneous conjugacy of those submonoids of a reference monoid M that are generated by sets of atoms (in a context where this makes sense). A natural conjugacy category arises, whose objects are the conjugates of some fixed submonoid generated by a set of atoms, and we call it a ribbon category. The seminal example is a monoid of positive braids B_n^+, in which case the geometric notion of a ribbon as explained in Reference Structure 7, page 20 arises.

Our current development takes place in a general context that we describe now.

Context 1.39. • M is a (left- and right-) cancellative monoid that is right-Noetherian, generated by its atoms and its invertible elements, and admits conditional right-lcms;

• $\mathcal{S}$ is a Garside family in M;

• $\mathcal{A}$ is the set of atoms of M;

- $\mathfrak{P}(\mathcal{A})$ is the powerset of $\mathcal{A}$, that is, the collection of all sets of atoms of M;
- for I a set of atoms, M_I is the submonoid of M generated by I.
- $\mathfrak{I}$ is a subset of $\mathfrak{P}(\mathcal{A})$ stable under conjugacy.

By Proposition II.2.58 (atoms generate), if M is Noetherian, then M is automatically generated by its atoms and $M^\times$. Note that, as M_I is generated by a set of atoms, it contains no nontrivial invertible element and I is the complete set of atoms of M_I.

Stable under conjugacy in the above means that whenever I in $\mathfrak{I}$ and g in M are such that, for every s in I, there exists t in M satisfying $sg = gt$, the set $\{s^g \mid s \in I\}$, written I^g, is again in $\mathfrak{I}$. Note that it is not true in general that the conjugate of an atom is an atom. However, this is true if M has a presentation with generating set $\mathcal{A}$ and homogeneous relations, or, equivalently, it admits an additive length function—that is, $\lambda(gh) = \lambda(g) + \lambda(h)$ always holds—with atoms of length 1. An example where this assumption holds is the case of classical or dual Artin–Tits monoids (Reference Structure 2, page 5 and Reference Structure 3, page 10). Then a conjugate of a subset of $\mathcal{A}$ is a subset of $\mathcal{A}$.

Example 1.40 (Coxeter system). Assume that (W, S) is a Coxeter system (see Chapter IX). Let B^+ be the corresponding Artin–Tits monoid and let A be its set of atoms. Then B^+ is cancellative, Noetherian, admits conditional right- (and left-) lcms, and it has a distinguished Garside family, namely the canonical lift $\boldsymbol{W}$ of W in B^+, which consists of the elements whose length with respect to $\mathcal{A}$ is equal to the length of their image in W with respect to S (see Chapter IX). Then taking B^+ for M and $\mathfrak{P}(\mathcal{A})$ for $\mathfrak{I}$ provides a typical example for Context 1.39. The corresponding ribbon category, as defined below, will be used in Chapter X where parabolic Deligne-Lusztig varieties will be associated to elements of the ribbon category and morphisms between these varieties will be associated to the cyclic conjugacy category of the ribbon category.

Definition 1.41 (category $\mathcal{C}onj(M, \mathfrak{P}(\mathcal{A}))$). In Context 1.39, for I, J in $\mathfrak{I}$, we denote by $\mathcal{C}onj(M, \mathfrak{I})(I, J)$ the family of all pairs $I \xrightarrow{g} \cdot$ satisfying $I^g = J$, and by $\mathcal{C}onj(M, \mathfrak{I})$ the category whose objects are the elements of $\mathfrak{I}$ and whose elements are in some family $\mathcal{C}onj(M, \mathfrak{I})(I, J)$, the source (*resp.* target) of $I \xrightarrow{g} J$ being defined to be I (*resp.* J).

Thus $\mathcal{C}onj(M, \mathfrak{I})$ is the family of all connected components of the simultaneous conjugacy category in M of $\mathfrak{I}$. Note that, if $I \xrightarrow{g} \cdot$ lies in $\mathcal{C}onj(M, \mathfrak{I})$, the map $s \mapsto s^g$ is a bijection from I onto I^g, being injective thanks to the right-cancellativity of M. It is easy to extend Corollary 1.14 to our situation to show that the set $\widehat{\mathcal{S}} = \{I \xrightarrow{g} \cdot \mid g \in \mathcal{S}\}$ is a Garside family in $\mathcal{C}onj(M, \mathfrak{I})$ (see Exercise 95).

More conditions will have to be assumed to hold in the sequel. For I included in $\mathfrak{I}$, we shall consider the following conditions:

Condition 1.42 (M_I closed under right-lcm). *The submonoid M_I is weakly closed under right-lcm and closed under right-quotient.*

Remark 1.43. In the case of an Artin–Tits monoid, Condition 1.42 is always satisfied, as can be proved using Corollary IX.1.12 and Proposition IX.1.13. In general, by contrast, Condition 1.42 is true only for some I, as the following example shows: consider the dual braid monoid B_4^{+*} (Reference Structure 3, page 10) in the braid group on four strands, and take for I the set $\{a_{1,3}, a_{2,4}\}$; these two atoms generate a free monoid, and it does not contain the right-lcm Δ_4^* of $a_{1,3}$ and $a_{2,4}$.

When Condition 1.42 is satisfied for a subfamily I of $\mathcal{A}$, then every element g of M has a unique maximal left-divisor in M_I, which we denote by $H_I(g)$: existence is given by Proposition VII.1.21 (head-subcategory) since M_I, being weakly closed under right-lcm, is closed under right-comultiple, whereas uniqueness comes from the fact that two such maximal left-divisors differ by right-multiplication by an invertible element which has to be in M_I since M_I is closed under right-quotient, so is trivial. We denote by $T_I(g)$ the element defined by $g = H_I(g)T_I(g)$.

Definition 1.44 (I-reduced). An element g of M is called *I-reduced* if it is left-divisible by no nontrivial element of M_I, or, equivalently, if $H_I(g) = 1$ holds.

Definition 1.45 (ribbon category). The *ribbon category* $\mathit{Rib}(M, \mathcal{I})$ is the subcategory of $\mathit{Conj}\,(M, \mathcal{I})$ generated by the elements $I \xrightarrow{g} \cdot$ such that g is I-reduced.

Example 1.46 (ribbon category). In the n-strand braid monoid B_n^+ (Reference Structure 2, page 5) with $\mathcal{I} = \{\{\sigma_i\} \mid i < n\}$, an (i, j)-ribbon as defined in Reference Structure 7, page 20 is an element of $\mathit{Rib}(B_n^+, \mathcal{I})$ provided it is not left-divisible by σ_i.

That the above class of elements is closed under composition and right-lcm is the subject of (ii) and (iii) in the next proposition; (i) is a motivation for restricting to I-reduced elements by showing that we lose nothing by doing so.

Proposition 1.47 (composition in $\mathit{Rib}(M, \mathcal{I})$). *In Context* 1.39, *assume that Condition* 1.42 *is satisfied for every I in $\mathcal{I}$.*

(i) *Given I in $\mathcal{I}$ and g in M, the pair $I \xrightarrow{g} \cdot$ lies in $\mathit{Conj}\,(M, \mathcal{I})$ if and only if $H_I(g)$ conjugates I to a subset $I^{H_I(g)}$ of I and $T_I(g)$ conjugates $I^{H_I(g)}$ to a subset I^g of $\mathcal{A}$.*

(ii) *If $I \xrightarrow{g} J$ belongs to $\mathit{Rib}(M, \mathcal{I})$, then $H_I(gh)^g = H_J(h)$ holds for every h in M. In particular, for $I \xrightarrow{g} \cdot$ in $\mathit{Rib}(M, \mathcal{I})$ and $I^g \xrightarrow{h} \cdot$ in $\mathit{Conj}\,(M, \mathcal{I})$, the pair $I \xrightarrow{gh} \cdot$ belongs to $\mathit{Rib}(M, \mathcal{I})$ if and only if $I^g \xrightarrow{h} \cdot$ belongs to $\mathit{Rib}(M, \mathcal{I})$.*

(iii) *If two elements $I \xrightarrow{g} \cdot$ and $I \xrightarrow{g'} \cdot$ of $\mathit{Conj}\,(M, \mathcal{I})$ are I-reduced and admit a right-lcm, then this right-lcm is I-reduced as well.*

Note that, if the right-lcm involved in (iii) above is $I \xrightarrow{h} \cdot$ then, by Lemma 1.10, h must be a right-lcm of g and g' in M.

Proof. (i) We first prove that, if s lies in I and s^g lies in J, then $s^{H_I(g)}$ lies in I. This will establish the result in one direction. The converse is obvious. Now, by assumption, we have $sg = gs'$ for some s' belonging to J. If $s \preccurlyeq g$ holds, let k be the largest integer such that s^k left-divides g, which exists since M is right-Noetherian. We write $g = s^k g'$, whence $sg' = g's'$ by left-cancellation. We have $H_I(g) = s^k H_I(g')$ and we are left with the case when s does not left-divide g.

In that case, since M_I is weakly closed under right-lcm, the elements s and $H_I(g)$ have a right-lcm in M_I that divides sg, hence gs', and can be written $sh = H_I(g)f$, with f non-invertible and h and f in M_I. Left-cancelling s in $sh \preccurlyeq sg$, we obtain that h divides g, so divides $H_I(g)$. Let us write $H_I(g) = hh'$ with h' in M_I; we have $sh = H_I(g)f = hh'f$. By the fifth item of Context 1.39, $h'f$ is an atom, thus h' is invertible, hence equal to 1, and f is an atom. We thus obtain $sH_I(g) = sh = H_I(g)f$, which is the expected result since f is an atom of M_I so it is in I.

(ii) We prove by induction on k that, for $s_1, \dots, s_k$ in I, the relation $s_1 \cdots s_k \preccurlyeq gh$ is equivalent to $(s_1 \cdots s_k)^g \preccurlyeq h$. This will prove (ii) since, the conjugation by g being surjective from I onto J, we can write $H_J(h)$ as $(s_1 s_2 \cdots s_k)^g$ with s_i in I for all i. To start the induction, consider a pair (s, s') with $s' = s^g$ in $I \times J$. Then $s \not\preccurlyeq g$ together with $gs' = sg$ imply that sg is a common multiple of s and g which has to be a right-lcm, since s' is an atom. So $s \preccurlyeq gh$ is equivalent to $gs' \preccurlyeq gh$, that is, $s^g \preccurlyeq h$, whence the result in the case $k = 1$. For a general k, if $s_1 \cdots s_k \preccurlyeq gh$ holds, we use the induction hypothesis to obtain $(s_1 \cdots s_{k-1})^g \preccurlyeq h$. This can be written $gh = s_1 \cdots s_{k-1} g h_1$ for some h_1 in M. We then have $s_k \preccurlyeq gh_1$. By the result for $k = 1$, we deduce $s_k^g \preccurlyeq h_1$, whence the result.

(iii) We claim that a stronger statement is actually true, namely that, if we have $g \preccurlyeq h$ for some g in $\mathcal{C}onj(M, \mathcal{I})(I, .)$ and h in M, then $T_I(g)$ left-divides $T_I(h)$. Then, in the situation of (iii), we obtain that $T_I(h)$ is a common multiple of g and g', which implies $h \preccurlyeq T_I(h)$, whence $H_I(h) = 1$.

Let us prove our claim. Dividing g and h by $H_I(g)$ and replacing I by $I^{H_I(g)}$, we may as well assume $H_I(g) = 1$, since $(I^{H_I(g)})^{T_I(g)} = I^g$ always holds. We write $h = gg_1$ and $J = I^g$. By (ii), we have $H_I(h)^g = H_J(g_1)$, whence

$$H_I(h)g = gH_J(g_1) \preccurlyeq gg_1 = h = H_I(h)T_I(h).$$

Left-cancelling $H_I(h)$, we obtain $g \preccurlyeq T_I(h)$, which is what we want since $g = T_I(g)$ is true. □

The next result shows that $(\widehat{S} \cap \mathcal{R}ib(M, \mathcal{I})) \cup \mathcal{R}ib(M, \mathcal{I})^\times$ generates $\mathcal{R}ib(M, \mathcal{I})$.

Proposition 1.48 (normal decomposition in $\mathcal{R}ib(M, \mathcal{I})$). *In Context 1.39, and provided Condition 1.42 is satisfied for every I in $\mathcal{I}$, all entries in a normal decomposition of an element of $\mathcal{R}ib(M, \mathcal{I})$ in $\mathcal{C}onj(M, \mathcal{I})$ belong to $\mathcal{R}ib(M, \mathcal{I})$.*

Proof. Let $I \xrightarrow{g} \cdot$ belong to $\mathcal{R}ib(M, \mathcal{I})$ and let $I \xrightarrow{g_1 | \cdots | g_q} \cdot$ be a normal decomposition of this element in $\mathcal{C}onj(M, \mathcal{I})$. By Corollary 1.14, generalized as in Exercise 95, $g_1 | \cdots | g_q$ is a normal decomposition in M. Applying Proposition 1.47(ii) with $g_1 \cdots g_{i-1}$ for g and $g_i \cdots g_q$ for h, we obtain $H_{I^{g_1 \cdots g_{i-1}}}(g_i \cdots g_q) = H_I(g)^{g_1 \cdots g_{i-1}} = 1$, whence the result. □

By Proposition 1.47(ii), the subcategory $\mathcal{R}ib(M, \mathcal{I})$ of $\mathcal{C}onj(M, \mathcal{I})$ is closed under right-quotient. Proposition VII.2.10 (recognizing compatible I), whose other assumptions are satisfied by Proposition 1.48, then implies

Corollary 1.49 (Garside in $\mathcal{R}ib(M, \mathcal{I})$). *In Context 1.39, and provided Condition 1.42 is satisfied for every I in $\mathcal{I}$, the set $\widehat{S} \cap \mathcal{R}ib(M, \mathcal{I})$, that is,*

$$\{I \xrightarrow{g} \cdot \in \mathcal{C}onj(M, \mathcal{I}) \mid g \in S \text{ and } H_I(g) = 1\}$$

is a Garside family in $\mathcal{R}ib(M, \mathcal{I})$.

We now show that, under some additional assumptions, the above Garside family in $\mathit{Rib}(M, \mathfrak{I})$ is bounded by a Garside map when the Garside family $\mathcal{S}$ is bounded.

If we do not assume that M_I contains no nontrivial invertible elements, we lose the unicity of H_I, but not its existence. In this case, for g in M we denote by $H_I(g)$ some chosen maximal left-divisor of g lying in M_I.

If $\mathcal{S}$ is bounded by a Garside map Δ, we put $\Delta_I = H_I(\Delta)$.

Lemma 1.50. *In Context* 1.39*, and assuming that Condition* 1.42 *is satisfied for every I in $\mathfrak{I}$, the monoid M is left-Noetherian (thus Noetherian), and $\mathcal{S}$ is bounded by a Garside map Δ, then, for every I in $\mathfrak{I}$, the element Δ_I is a right-Garside element in M_I.*

Proof. The set $\mathcal{S} \cap M_I$ generates M_I, hence also does $\mathcal{D}iv(\Delta) \cap M_I$. As M is left-Noetherian, Proposition V.2.35 (bounded implies gcd) implies that M admits right-lcms. The set $\mathcal{D}iv(\Delta) \cap M_I$ is weakly closed under right-lcm in M since M_I is, and it is closed under right-quotient since M_I and $\mathcal{D}iv(\Delta)$ are both closed under right-quotient. This implies that $\mathcal{D}iv(\Delta) \cap M_I$ is closed under right-complement in M_I, and Corollary IV.2.29 (recognizing Garside, right-lcm case) implies that $\mathcal{D}iv(\Delta) \cap M_I$ is a Garside family in M_I (here we use the fact that M_I has no nontrivial invertible elements).

Now the divisors of Δ which are in M_I are by the definition of H_I the divisors of Δ_I in M_I. Hence, $\mathcal{D}iv(\Delta) \cap M_I$ is closed under left-divisor in M_I and we obtain the result by Proposition V.1.20 (right-Garside map) applied in M_I (here again we use that $M_I^{\times}$ is trivial). □

If M is Noetherian and $\mathcal{S}$ is bounded by a Garside map Δ, we denote by ϕ_I the functor ϕ_{Δ_I} associated to Δ_I as in Proposition V.1.28 (functor ϕ_Δ). We are now interested in the case where ϕ_I is surjective, hence bijective (it is injective since M is right-cancellative). This is equivalent to Δ_I being a Garside element by Proposition V.2.17 (automorphism). We introduce one more property.

Condition 1.51 **(ϕ_I bijective).** *We assume that M is Noetherian, the Garside family $\mathcal{S}$ is bounded by a Garside element Δ, and, for every I in $\mathfrak{I}$, the functor ϕ_I is bijective on M_I.*

By the fifth item of Context 1.39, ϕ_I maps an atom to an atom. Hence its surjectivity on the set I of atoms of M_I is equivalent to its surjectivity on M_I. Since ϕ_I is injective, the above assumption is then equivalent to ϕ_I being bijective on I. Condition 1.51 is satisfied in particular when I is finite. This is the case, for example, if M is a classical or dual Artin–Tits monoid of spherical type, or the semi-direct product of an Artin–Tits monoid of spherical type by an automorphism.

Proposition 1.52 (Garside map in $\mathit{Rib}(M, \mathfrak{I})$). *In Context* 1.39*, and provided Conditions* 1.42 *and* 1.51 *are satisfied for every I in $\mathfrak{I}$, the map Δ defined for I included in $\mathcal{A}$ by $\Delta(I) = I \xrightarrow{\Delta_I^{-1}\Delta} \phi_\Delta(I)$ is a Garside map in the category $\mathit{Rib}(M, \mathfrak{I})$.*

Proof. The bijectivity of ϕ_I and Proposition 1.47(i) show that the conjugation by $T_I(\Delta) = \Delta_I^{-1}\Delta$ is defined on I. Hence $\Delta_I^{-1}\Delta$ is an element of $\widehat{\mathcal{S}} \cap \mathit{Rib}(M, \mathcal{I})$. We first show that every pair $I \xrightarrow{f} \cdot$ in $\widehat{\mathcal{S}} \cap \mathit{Rib}(M, \mathcal{I})$ left-divides $I \xrightarrow{\Delta_I^{-1}\Delta} \cdot$, which is equivalent to $\Delta_I f$ left-dividing Δ. Since Δ_I and f divide Δ, their right-lcm δ left-divides Δ. We claim that δ is equal to $\Delta_I f$. Let us write $\delta = fg$. We have $fg \preccurlyeq \Delta_I f = f\Delta_I^f$, whence $g \preccurlyeq \Delta_I^f$. By definition of $\delta = fg$, we have $\Delta_I \preccurlyeq fg$; from this and $fg \preccurlyeq \Delta$, we deduce $H_I(fg) = \Delta_I$. From Proposition 1.47, applied to the product fg, we obtain $\Delta_I^f = H_I(fg)^f = H_J(g)$. Putting things together, we deduce $g = H_J(g) = \Delta_I^f$, whence $\delta = fg = f\Delta_I^f = \Delta_I f$, which is our claim. We have proved that the collection of elements of the proposition defines a right-Garside map in $\mathit{Rib}(M, \mathcal{I})$. Note for future reference that, since g right-divides Δ, it also left-divides Δ; as g lies in M_J, we conclude that g, which is Δ_I^f, left-divides Δ_J.

To see that $\Delta(I)$ is a Garside map, it remains to show that every element $I \xrightarrow{f} J$ in $\widehat{\mathcal{S}} \cap \mathit{Rib}(M, \mathcal{I})$ right-divides $\Delta^{-1}_{\phi_\Delta^{-1}(J)}\Delta$, which is equivalent to $f\Delta_J$ right-dividing Δ, since ϕ_Δ is an automorphism, hence maps $\Delta_{\phi_\Delta^{-1}(J)}$ to Δ_J. This in turn is equivalent to $f\Delta_J$ left-dividingΔ, since Δ is a Garside element. The result is then a consequence of the following lemma and of the fact that $\Delta_I f$ divides Δ as we have seen in the first part of the proof. □

Lemma 1.53. *In Context* 1.39, *and provided Conditions* 1.42 *and* 1.51 *are satisfied for every I in $\mathcal{I}$, for every pair $I \xrightarrow{f} J$ in $\mathit{Rib}(M, \mathcal{I})$, we have $\Delta_J = \Delta_I^f$, that is, equivalently, $f\Delta_J = \Delta_I f$.*

Proof. It is sufficient to prove the property for $I \xrightarrow{f} J$ in $\widehat{\mathcal{S}} \cap \mathit{Rib}(M, \mathcal{I})$. Since $\Delta_I f$ left-divides Δ, we can write $\Delta = \Delta_I f f_1$ for some f_1. Since Δ conjugates I to $\phi_\Delta(I)$ and Δ_I conjugates I surjectively to itself, which is in turn conjugated by f surjectively to J, we obtain that f_1 conjugates J surjectively to $\phi_\Delta(I)$. In other words, f_1 defines an element $J \xrightarrow{f_1} \phi_\Delta(I)$ of $\mathit{Conj}(M, \mathcal{I})$. This element actually belongs to $\mathit{Rib}(M, \mathcal{I})$: by Proposition 1.47(ii), we have $H_J(f_1) = H_I(ff_1)^f = 1$, the last equality by $H_I(\Delta) = H_I(\Delta_I f f_1) = \Delta_I$. By the first part of the proof of the proposition applied with J and f_1, respectively, instead of I and f, we obtain $\Delta_J^{f_1} \preccurlyeq \Delta_{\phi_\Delta(I)}$. Assume for a contradiction the strict left-divisibility relation $\Delta_I^f \prec \Delta_J$. Putting things together, we obtain $\Delta_I^{ff_1} \prec \Delta_J^{f_1} \preccurlyeq \Delta_{\phi_\Delta(I)}$. But we have $\Delta_I^{ff_1} = \Delta_I^{\Delta_I f f_1} = \Delta_I^{\Delta} = \Delta_{\phi_\Delta(I)}$. Hence we obtain the strict divisibility relation $\Delta_{\phi_\Delta(I)} \prec \Delta_{\phi_\Delta(I)}$, a contradiction. □

We now want to describe the atoms of the ribbon category $\mathit{Rib}(M, \mathcal{I})$. We will do it under the following assumption, which makes sense in Context 1.39.

Condition 1.54. *For every I in $\mathcal{I}$ and every atom s in M, there exists a set J of representatives of the $=^\times$-class of atoms of M that left-divide the right-lcm of s and Δ_I satisfying:*

(i) *The monoid M_J is weakly closed under right-lcm and closed under right-quotient.*

(ii) *The family $\widetilde{\mathrm{Div}}(\Delta_J)$ is included in $\mathrm{Div}(\Delta_J)$ and Φ_J is bijective.*

(iii) *The element Δ_J left-divides the right-lcm of s and Δ_I.*

Note that, by Proposition VII.1.21 (head-subcategory), when condition (i) is satisfied, Δ_J exists as explained below in Remark 1.43. The inclusion condition in (ii) is satisfied in particular if M is Noetherian by Lemma 1.50, which can be applied in Context 1.39, and asserts that then Δ_J is a right-Garside element.

Note also that Condition 1.54 is satisfied in all Artin–Tits monoids of spherical type (see Chapter IX): we have already seen that Conditions 1.42 and 1.51 hold for every subset J of the atoms. As for Condition 1.54(iii), it is satisfied in Artin–Tits monoids of spherical type since, in such a monoid, the set J is equal to $I \cup \{s\}$.

Condition 1.54 is also true in all quasi-Garside dual braid monoids. To show this we first prove two lemmas.

Lemma 1.55. *Assume that M is a quasi-Garside monoid with Garside element Δ such that, for every f in* $\mathrm{Div}(\Delta)$*, the left- and right-divisors of f coincide and such that the square of any atom does not divide Δ.*

(i) *For every f in* $\mathrm{Div}(\Delta)$ *and every decomposition of f as a product $a_1 \cdots a_k$ of atoms, f is a right-lcm of the family $\{a_1, ..., a_k\}$. In particular, f is the right-lcm of the family of all atoms dividing it.*

(ii) *For f and g in* $\mathrm{Div}(\Delta)$*, if fg lies in* $\mathrm{Div}(\Delta)$*, then fg is the right-lcm of f and g.*

(iii) *For f and g in* $\mathrm{Div}(\Delta)$*, if fg and gf lie in* $\mathrm{Div}(\Delta)$*, then we have $fg =^{\times} gf$.*

Proof. Points (ii) and (iii) are immediate consequences of (i). We prove (i) using induction on k. We do not have to distinguish between left- and right-divisibility for elements of $\mathrm{Div}(\Delta)$. Since a_k^2 does not divide Δ, it does not divide the product $a_1 \cdots a_k$. Hence a_k does not divide $a_1 \cdots a_{k-1}$. It follows that every right-lcm of a_k and $a_1 \cdots a_{k-1}$ is a strict right-multiple of $a_1 \cdots a_{k-1}$ that has to divide $a_1 \cdots a_k$, hence is equal to $a_1 \cdots a_k$ up to an invertible element since a_k is an atom. We deduce the result since, by induction, $a_1 \cdots a_{k-1}$ is a right-lcm of $\{a_1, ..., a_{k-1}\}$. □

Lemma 1.56. *Under the assumptions of Lemma* 1.55*, for every δ in* $\mathrm{Div}(\Delta)$*, the monoid generated by the divisors of δ is closed under left- and right-quotient and left- and right-lcm. It is a quasi-Garside monoid with Garside element δ.*

Proof. By definition, the family $\mathrm{Div}(\delta)$ of all divisors of δ is closed under left- and right-quotient and left- and right-lcm in M. Hence $\mathrm{Div}(\delta)$ is closed under right-diamond in $\mathrm{Div}(\Delta)$. By assumption, the ambient monoid M admits conditional right-lcms, it is Noetherian, and it contains no nontrivial invertible element. By Proposition VII.1.13 (generated subcategory, right-lcm case), the submonoid M_1 generated by $\mathrm{Div}(\delta)$ is closed under right-diamond, hence under right-lcm, and under left-lcm as well since we are in a context where left- and right-divisibility coincide. Then $\mathrm{Div}(\delta)$ is a Garside family in M_1. □

In the following result, the notion of dual braid monoids refers to Definitions IX.2.2 and IX.3.11 below. The family of dual braid monoids that are Garside or quasi-Garside monoids contains the dual braid monoids of complex reflection groups (hence in particular of finite Coxeter groups) and of Coxeter groups of type $\widetilde{\mathrm{A}}_n$ and $\widetilde{\mathrm{C}}_n$.

Proposition 1.57 (dual braid monoids). *Assume that M is a dual braid monoid that is a Garside or quasi-Garside monoid. Then Condition* 1.54 *is satisfied in Context* 1.39*, if Condition* 1.42 *is satisfied for every I in $\mathcal{I}$. Moreover, with the notation as in Condition* 1.54*, one has $\Delta_J = \operatorname{lcm}(I, s)$.*

Proof. In a dual braid monoid, there is no non-trivial invertible element and right- and left- divisibility coincide. Hence the monoids considered here satisfy the assumptions of Lemma 1.55. Let I belong to $\mathcal{I}$, let s be an atom, and let J be the set of atoms dividing the right-lcm δ of Δ_I and s. By Lemma 1.55(i), δ is the lcm of J. By Lemma 1.56, M_J is a quasi-Garside monoid with Garside element δ and it is closed under divisor and lcm. Thus Condition 1.54(i) is satisfied. Condition 1.54(ii) is satisfied as well, since the right- and left-divisors of Δ_J coincide. Now Δ_J is a Garside element, so is a multiple of δ. But no atom can divide $\delta^{-1}\Delta_J$, since it would divide δ so that its square would divide Δ, which is false. Hence δ is equal to Δ_J. □

Notation 1.58. Under Condition 1.54, when Δ_J is different from Δ_I (that is, when s does not divide Δ_I) the element $\Delta_I^{-1}\Delta_J$ is denoted by $v(J, I)$.

Since Δ_I^{-1} conjugates I to itself and Δ_J conjugates M_J to itself, the element $v(J, I)$ conjugates I into M_J, hence into J by the fifth item in Context 1.39. Moreover, we have $H_I(v(J, I)) = 1$ since Δ_I left-divides Δ_J. Hence $I \xrightarrow{v(J,I)} \phi_J(I)$ is an element of $\mathit{Rib}(M, \mathcal{I})$.

Proposition 1.59 (atoms of $\mathit{Rib}(M, \mathcal{I})$). *In Context* 1.39*, and provided Condition* 1.42 *is satisfied for every I in $\mathcal{I}$ and Conditions* 1.51 *and* 1.54 *are satisfied:*

(i) *If I lies in $\mathcal{I}$, g lies in M, and we have $H_I(g) = 1$ and there exists $p > 0$ such that the conjugation of Δ_I^p by g is defined in M, then the conjugation of I by g is defined, and $I \xrightarrow{g} \cdot$ lies in $\mathit{Rib}(M, \mathcal{I})$.*

(ii) *Every atom of $\mathit{Rib}(M, \mathcal{I})$ is of the form $I \xrightarrow{v(J,I)} \cdot$ for some I included in $\mathcal{A}$ and some J as in Condition* 1.54*. Conversely, such an element $I \xrightarrow{v(J,I)} \cdot$ is an atom provided $v(J', I) \prec v(J, I)$ holds for no J' as in Condition* 1.54*.*

Proof. (i) Since M is right-Noetherian, it suffices to prove that g is either invertible or left-divisible by some non-invertible element h that belongs to $\mathit{Rib}(M, \mathcal{I})$: indeed, if $g = hg'$ holds with $I \xrightarrow{h} I'$ in $\mathit{Rib}(M, \mathcal{I})$, then, by Proposition 1.47, we have $H_{I'}(g') = 1$ and, as $I^h = I'$ holds, Lemma 1.53 implies $(\Delta_{I'}^p)^{g'} = (\Delta_I^p)^g$ in M, so (i) is equivalent to the same property for g' and, by Noetherianity, the sequence $g, g', g'', ...$ thus constructed terminates with an invertible element. We will prove that, if g is not invertible, then for some J as in 1.54, $v(J, I)$ divides g, which will thus imply (i).

We proceed by decreasing induction on p. We show that, if for $i > 0$ we have $t \preccurlyeq \Delta_I^i g$ for some atom t not in M_I, then $v(J, I)$ left-divides $\Delta_I^{i-1} g$, where J is the set of all atoms dividing the right-lcm of t and Δ_I. By Condition 1.54(iii), Δ_J left-divides the right-lcm of t and Δ_I, hence, from $t \preccurlyeq \Delta^i g$ and $\Delta_I \preccurlyeq \Delta_I^i g$, we deduce $\Delta_J \preccurlyeq \Delta_I^i g$. Since Δ_J is equal to $\Delta_I v(J, I)$, we deduce $v(J, I) \preccurlyeq \Delta_I^{i-1} g$ as expected. The induction starts with $i = p$, taking for t an atom dividing g since then we have $t \preccurlyeq g \preccurlyeq \Delta_I^p g$. Such an atom t, which exists since M is generated by its atoms and its invertible elements does not lie in M_I since $H_I(g) = 1$. Similarly, every atom t' satisfying $t' \preccurlyeq v(J, I)$ does not lie in M_I, hence the induction can go on as long as $i - 1$ is positive.

We deduce (ii) from the proof of (i): every element $I \xrightarrow{g} I^g$ in $\mathit{Rib}(M, \mathfrak{I})$ satisfies the assumption of (i) for $p = \|g\|_S$, whence the result since, in the proof of (i), we have seen that g is a product of some elements $v(J, K)$. □

In this way, we obtained a complete description of the normalizer of a submonoid M_I in a monoid M. Indeed, Proposition 1.59 immediately implies:

Corollary 1.60 (normalizer). *Under the assumptions of Proposition* 1.59, *the normalizer of M_I in M identifies with the family $\mathit{Rib}(M, \mathfrak{I})(I, I)$ in the ribbon category $\mathit{Rib}(M, \mathfrak{I})$.*

Twisted conjugacy. We sketch how ribbons work in the context of twisted conjugacy.

Assume that N is a monoid and $\mathfrak{I}$ is a family of subsets of the atom set of N such that N and $\mathfrak{I}$ satisfy the assumptions of Context 1.39. Let ϕ be an automorphism of N and let $M = N \rtimes \langle\phi\rangle$. Assume that $\mathfrak{I}$ is stable under the action of ϕ (or, equivalently, that $\mathfrak{I}$ is stable under conjugacy in M).

Then, if Conditions 1.42 for every I and 1.54 hold in N, they still hold in M. If we consider just the ribbons thus constructed of the form $I \xrightarrow{g\phi} \cdot$ with g in N, we can consider them as "ribbons for ϕ-twisted conjugacy" in N. For instance, if N is an Artin–Tits monoid with atom set A, we may take $\mathfrak{I} = \mathfrak{P}(\mathcal{A})$.

2 Cycling, sliding, summit sets

In this section, we show how the Conjugacy Problem can be solved in (most of) categories that admit a bounded Garside family, using an approach based on the notions of cycling, decycling, and sliding.

The main result is that there exist a characteristic subset of a conjugacy class, called the *super-summit set*, and an algorithm (repeated application of *cycling* and *decycling*) that maps every element of the conjugacy class to this characteristic subset, resulting (in good cases) in a solution of the Conjugacy Problem (Subsection 2.1). Then we show how to construct a smaller characteristic subset, the set of sliding circuits, obtained by repeated application of another operation called *sliding* (Subsection 2.2).

Throughout this section and the next one, we apply Convention V.3.7 (omitting source), thus writing Δ instead of $\Delta(x)$ when there is no need to specify x explicitly.

2.1 Cycling and decycling

As above, we first fix a common context of all developments in this section.

Context 2.1. • $\mathcal{C}$ is a cancellative category that is right-Noetherian.
• Δ is a Garside map in $\mathcal{C}$; the associated automorphism is ϕ_Δ.
• $\mathcal{G}$ is the groupoid of fractions of $\mathcal{C}$.
• H is a *sharp* head function for $\mathcal{D}iv(\Delta)$; for g in $\mathcal{C}$, we denote by $T(g)$ the element satisfying $g = H(g)T(g)$.

We will write Δ for $\Delta(x)$ when there is no ambiguity on the source x. Since the category is assumed to be cancellative, the functor ϕ_Δ is an automorphism.

Below, we shall consider the categories $\mathcal{C}onj\mathcal{C}$ and $\mathcal{C}yc\mathcal{C}$ of Section 1 (together with their extensions to the enveloping groupoid $\mathcal{G}$). In order to simplify notation, we put

Convention 2.2 ($\mathcal{C}onj\mathcal{C}$)**.** Hereafter, for e, e' in $\mathcal{C}^\circlearrowleft$ and g in $\mathcal{C}$ satisfying $eg = ge'$, that is, for $e \xrightarrow{g} e'$ belonging to $\mathcal{C}onj\mathcal{C}(e, e')$, we will simply say that g belongs to $\mathcal{C}onj\mathcal{C}(e, e')$. We will use the same convention for $\mathcal{C}yc\mathcal{C}$.

We will describe conjugacy in the enveloping groupoid $\mathcal{G}$ and deduce results on the conjugacy in $\mathcal{C}$ as a particular case. For e of $\mathcal{G}^\circlearrowleft$ and g in $\mathcal{C}onj\mathcal{G}(e, \cdot)$, we may write $g^{-1}eg$ for the conjugate e^g. We call *conjugacy class* of e the objects of the connected component of $\mathcal{C}onj\mathcal{G}$ containing e viewed as an object of $\mathcal{C}onj\mathcal{G}$.

If the automorphism ϕ_Δ has finite order n (which is necessarily true if $\mathcal{S}$ is finite), every conjugating map in $\mathcal{C}onj\mathcal{G}(e, \cdot)$ with e in $\mathcal{C}$ gives rise to a map in $\mathcal{C}onj\mathcal{C}(e, \cdot)$. Indeed, for g in $\mathcal{C}onj\mathcal{G}(e, \cdot)$, there exists n such that ϕ_Δ^n is trivial and $g\Delta^{[n]}$ lies in $\mathcal{C}$, so that we have $e^g = \phi_\Delta^n(e^g) = (g\Delta^{[n]})^{-1}e(g\Delta^{[n]}) = e^{g\Delta^{[n]}}$.

We recall that left-divisibility is extended to $\mathcal{G}$ by setting $g \preccurlyeq h$ for $g^{-1}h \in \mathcal{C}$. We use this to define the subcategory $\mathcal{C}yc\mathcal{G}$ of $\mathcal{C}onj\mathcal{G}$ in the same way as for $\mathcal{C}$: it is the subcategory generated by $\{g \in \mathcal{C}onj\mathcal{G}(e, e') \mid g \preccurlyeq e\}$. Then $\mathcal{C}yc\mathcal{C}$ is the full subcategory of $\mathcal{C}yc\mathcal{G}$ whose objects are in $\mathcal{C}$.

We also recall from Proposition V.3.18 (delta-normal) that every element g of $\mathcal{G}$ admits a delta-normal decomposition $\Delta^{[m]}\|s_1|\cdots|s_\ell$, meaning that $g = \Delta^{[m]}s_1\cdots s_\ell$ holds and $s_1|\cdots|s_\ell$ is a $\mathcal{D}iv(\Delta)$-normal decomposition of $\Delta^{[-m]}g$ satisfying $\Delta \not\preccurlyeq s_1$. We then have

$$m = \inf_\Delta(g) = \max\{i \mid \Delta^{[-i]}g \in \mathcal{C}\} = \max\{i \mid \Delta^{[i]} \preccurlyeq g\}.$$

A delta-normal decomposition $\Delta^{[m]}\|s_1|\cdots|s_\ell$ is *strict* if none of $s_1, \dots, s_\ell$ is invertible.

Definition 2.3 (cycling). For e in $\mathcal{G}^\circlearrowleft$ satisfying $\inf_\Delta(e) = i$, the *initial factor* $\mathrm{init}(e)$ of e is defined to be $H(e\Delta^{[-i]})$, and the *cycling* $\mathrm{cyc}(e)$ of e is defined to be $e^{\mathrm{init}(e)}$.

The initial factor $\mathrm{init}(e)$ is a left-gcd of $e\Delta^{[-i]}$ and Δ, and $\inf_\Delta(\mathrm{cyc}(e)) \geqslant \inf_\Delta(e)$ holds. Note that $\mathrm{init}(e)$ lies in $\mathcal{C}onj\mathcal{G}(e, \cdot)$. More precisely

Lemma 2.4. *In Context* 2.1, $\mathrm{init}(e)^{-1}e$ *belongs to* $\mathit{Cyc}\mathcal{G}(\mathrm{cyc}(e), e)$ *for every* e *in* $\mathcal{G}^{\circlearrowleft}$. *Moreover, if* e *lies in* $\mathcal{C}^{\circlearrowleft}$, *then* $\mathrm{init}(e)$ *belongs to* $\mathit{Cyc}\mathcal{C}(e, \mathrm{cyc}(e))$.

By Lemma II.2.41, the assumption that $\mathcal{C}$ is right-Noetherian implies that every element g has admits a right-height $\mathrm{ht}_{\mathrm{R}}(g)$, which is a (finite or infinite) ordinal. When the latter is finite, every $\prec$-increasing sequence of left-divisors of g has length $\mathrm{ht}_{\mathrm{R}}(g)$ at most. Note that $g \preccurlyeq \Delta(x)$ implies $\mathrm{ht}_{\mathrm{R}}(g) \leqslant \mathrm{ht}_{\mathrm{R}}(\Delta(x))$, so the possible finiteness of the right-height for elements $\Delta(x)$ is sufficient to imply its finiteness on $\mathcal{D}iv(\Delta)$.

Proposition 2.5 (cycling and $\inf_\Delta$**).** *In Context* 2.1, *assume that* e *belongs to* $\mathcal{G}^{\circlearrowleft}(x, x)$ *and has a conjugate* e' *in* $\mathcal{G}^{\circlearrowleft}$ *satisfying* $\inf_\Delta(e') > \inf_\Delta(e)$. *Then*

$$\inf_\Delta(\mathrm{cyc}^n(e)) > \inf_\Delta(e) \tag{2.6}$$

holds for n *sufficiently large;* (2.6) *holds for* $n = \mathrm{ht}_{\mathrm{R}}(\Delta(x))$ *if the latter is finite.*

In particular, under the above assumptions, e has a conjugate e' in the same connected component of $\mathit{Cyc}\mathcal{G}$ satisfying $\inf_\Delta(e') > \inf_\Delta(e)$.

Proof. We first show that we can reduce to the case $\inf_\Delta(e) = 0$. Write $e = g\Delta^{[i]}$ with $i = \inf_\Delta(e)$, and consider the translation functor from $\mathit{Conj}\,\mathcal{G}$ to $\mathit{Conj}_{\,\phi_\Delta^{-i}}\mathcal{G}$ which maps the object e to $e\Delta^{[-i]}$ and is the identity on elements. Composing with the functor ι of Proposition 1.36, we map e to $g\phi_\Delta^{-i}$ in the semi-direct product category $\mathcal{G} \rtimes \langle\phi_\Delta^{-i}\rangle$, seen as an object of $\mathit{Conj}(\mathcal{G} \rtimes \langle\phi_\Delta^{-i}\rangle)$. This composition is again the identity on elements. Note that the function $g\phi_\Delta^{-i} \mapsto H(g)$ is a sharp head function in the semi-direct product category. We will still denote it by H. Our identification of e with $g\phi_\Delta^{-i}$ is compatible with cycling, since, by definition, we have $\mathrm{cyc}(e) = e^{H(g)}$ and, in $\mathcal{G} \rtimes \langle\phi_\Delta^{-i}\rangle$, we have $H(g\phi_\Delta^{-i}) = H(g)$ and $\inf_\Delta(g\phi_\Delta^{-i}) = 0$, so that the cycling of $g\phi_\Delta^{-i}$ is also equal to the conjugation by $H(g)$. We thus see that it is sufficient to prove the result for elements with infimum equal to 0.

We assume now $\inf_\Delta(e) = 0$, that is, $e \in \mathcal{C}$ and $\Delta \not\preccurlyeq e$. Let $\Delta\mathcal{C}$ denote the set of elements of $\mathcal{C}$ that are divisible by Δ. The assumption of the proposition is equivalent to the existence of g in $\mathcal{C}$ satisfying $geg^{-1} \in \Delta\mathcal{C}$, since, if there exists such a g in $\mathcal{G}$, we obtain one in $\mathcal{C}$ by multiplying by a suitable power of Δ, using that ϕ_Δ maps $\Delta\mathcal{C}$ onto itself. Since this condition can be rewritten as $ge \succcurlyeq \Delta g$, an easy computation shows that the set of such elements is closed under right-gcd. We may thus assume that g is a minimal such element for right-divisibility; note that $\Delta \not\preccurlyeq e$ implies that g is not invertible.

It will be more convenient to study the composition cyc_Δ of ϕ_Δ^{-1} with cyc rather than cyc itself. We have $\mathrm{cyc}_\Delta(e) = \phi_\Delta^{-1}(\mathrm{cyc}(e)) = \widetilde{\partial}(H(e))e\widetilde{\partial}(H(e))^{-1}$, where we recall $\widetilde{\partial}(H(e))$ is defined by $\widetilde{\partial}(H(e))H(e) = \Delta$. It is clear that $\mathrm{cyc}_\Delta^n(e)$ has the same infimum as $\mathrm{cyc}^n(e)$, since $\mathrm{cyc}_\Delta^n(e) = \phi_\Delta^{-n}(\mathrm{cyc}^n(e))$ holds.

Proposition 2.5 is then an immediate consequence of the following result, by induction on a chain of left-divisors of $H(g)$. □

Lemma 2.7. *With the above notation, define g' in $\mathcal{G}$ by $g = g'\widetilde{\partial}(H(e))$. Then*
(i) *The element g' lies in $\mathcal{C}$.*
(ii) *The element $g'\mathrm{cyc}_\Delta(e)g'^{-1}$ lies in $\Delta\mathcal{C}$.*
(iii) *No strict right-divisor of g' may satisfy* (ii).
(iv) *If $g \neq 1$ holds, we have $H(g') \prec H(g)$.*

Proof. (i) By assumption, we have $ge \succcurlyeq \Delta g$, whence $\Delta \preccurlyeq ge$, then $\Delta \preccurlyeq gH(e)$, then $gH(e) \succcurlyeq \Delta = \widetilde{\partial}(H(f))H(e)$ and, finally, $g \succcurlyeq \widetilde{\partial}(H(f))$.

(ii)–(iii). For h' in $\mathcal{C}$ satisfying $h = h'\widetilde{\partial}(H(f))$, the condition $heh^{-1} \in \Delta\mathcal{C}$ is equivalent to $h'\widetilde{\partial}(H(f))e\widetilde{\partial}(H(f))^{-1}h'^{-1} \in \Delta\mathcal{C}$, hence to $h'\mathrm{cyc}_\Delta(e)h'^{-1} \in \Delta\mathcal{C}$. Applying this with $h' = g'$, we obtain (ii) and applying this with h' a proper right-divisor of g', we obtain (iii) as h would then be a proper right-divisor of g.

(iv). Since g' left-divides g, we have $H(g') \preccurlyeq H(g)$ by (IV.1.46), so it is sufficient to prove that $g \notin \mathcal{C}^\times$ implies $H(g') \neq H(g)$.

If $H(g)$ is equal to $H(g')$, left-dividing by $H(g)$ the equality $g = g'\widetilde{\partial}(H(f))$ yields $T(g) = T(g')\widetilde{\partial}(H(e))$, whence $T(g)H(e) = T(g')\Delta$, from which we obtain $\Delta \preccurlyeq T(g)H(e) \preccurlyeq T(g)e$.

Now, by assumption, there exists f in $\mathcal{C}$ satisfying $ge = \Delta fg$, which may be written $T(g)e = (H(g)^{-1}\Delta)fg$; thus $\Delta \preccurlyeq T(g)e$ implies $\Delta \preccurlyeq (H(g)^{-1}\Delta)fg$, which, by the $\mathcal{H}$-law and (IV.1.46), implies $\Delta \preccurlyeq (H(g)^{-1}\Delta)fH(g) = T(g)eT(g)^{-1}$, contradicting the minimality of g, since $g \notin \mathcal{C}^\times$ implies $g \succ T(g)$. □

Another way of viewing cycling is to choose the strict Δ-normal decomposition of e of the form $\Delta^{[i]}\|s_1|\cdots|s_p$ with $s_j = H(s_j\cdots s_p)$ for $j < p$. We will call this the *H-normal decomposition* of e; then we have $\mathrm{init}(e) = H(\phi_\Delta^{-i}(g_1))$ (equal to $\phi_\Delta^{-i}(g_1)$ if H commutes with ϕ_Δ, which is automatic if $\mathcal{C}$ has no nontrivial invertible element).

Definition 2.8 (decycling, final factor). For e in $\mathcal{G}^{\circlearrowleft}$ admitting the strict Δ-normal decomposition $\Delta^{[i]}\|s_1|\cdots|s_p$, the *final factor* $\mathrm{fin}(e)$ of e is defined to be s_p, and the *decycling* $\mathrm{dec}(e)$ of e is defined to be $s_p\Delta^{[i]}s_1\cdots s_{p-1}$.

With the notation of Definition 2.8, we have

$$\mathrm{dec}(e) = \Delta^{[i]}\phi_\Delta^i(s_p)s_1\cdots s_{p-1} = {}^{\mathrm{fin}(e)}e,$$

where ${}^g e$ denotes the left-conjugation of e by g. So decycling e means left-conjugating e by $\mathrm{fin}(e)$ or, equivalently, right-conjugating it by $\Delta^{[i]}s_1\cdots s_{p-1}$. More precisely, we have

Lemma 2.9. *With the above notation, $\Delta^{[i]}s_1\cdots s_{p-1}$ belongs to $\mathrm{Cyc}\mathcal{G}(e, \mathrm{dec}(e))$, and, if e lies in $\mathcal{C}^{\circlearrowleft}$, then $\mathrm{fin}(e)$ belongs to $\mathrm{Cyc}\mathcal{C}(\mathrm{dec}(e), e)$.*

Decycling is related to cycling: $\phi_\Delta(\mathrm{dec}(e))$ is a conjugate of $\mathrm{cyc}(e^{-1})^{-1}$ by an invertible element. Indeed, for e in $\mathcal{G}$, if $\Delta^{[i]}\|s_1|\cdots|s_p$ is a strict Δ-normal decomposition of e, we have $\sup_\Delta e = i + p = \inf_\Delta e + p = -\inf_\Delta(e^{-1})$. Hence we have $\mathrm{fin}(e) = h^{-1}e$ for some left-gcd h of $\Delta^{[\sup_\Delta(e)-1]}$ and e, hence, by Proposition V.3.26 (inverse), $\mathrm{fin}(e)^{-1}\Delta$ is a left-gcd of $e^{-1}\Delta^{[\sup_\Delta(e)]}$ and Δ, finally, we obtain $\mathrm{fin}(e)^{-1}\Delta =^{\times} \mathrm{init}(e^{-1})$.

We recall that, in Context 2.1, $\mathcal{C}$ is right-Noetherian. It follows that Proposition 2.5 is then valid. From the considerations of the previous two paragraphs we deduce:

Corollary 2.10 (decycling and $\sup_\Delta$). *In Context* 2.1, *assume that* e *lies in* $\mathcal{G}^{\circlearrowright}(x,x)$ *and has a conjugate* e' *satisfying* $\sup_\Delta(e') < \sup_\Delta(e)$. *Then*

$$\sup_\Delta(\mathrm{dec}^n(e)) < \sup_\Delta(e) \tag{2.11}$$

holds for n *sufficiently large;* (2.11) *holds for* $n = \mathrm{ht}_{\mathrm{R}}(\Delta(\phi_\Delta^{-1}(x)))$ *if the latter is finite.*

As cycling does not increase the supremum and decycling does not decrease the infimum, and owing to Proposition 2.5 and Corollary 2.10, an iterated application of cycling and decycling leads to:

Corollary 2.12 (super-summit set). *In Context* 2.1, *for every* e *in* $\mathcal{G}^{\circlearrowright}$, *the conjugacy class of* e *contains a well-defined subset* $\mathrm{SSS}(e)$ *on which each one of* $\inf_\Delta$ *and* $\sup_\Delta$ *takes a constant value, and such that, for every* e' *in the conjugacy class of* e, *we have*

$$\inf_\Delta(e') \leqslant \inf_\Delta(\mathrm{SSS}(e)) \quad \textit{and} \quad \sup_\Delta(e') \geqslant \sup_\Delta(\mathrm{SSS}(e)).$$

Furthermore, $\mathrm{SSS}(e)$ *belongs to the connected component of* e *in* $\mathcal{C}yc\mathcal{G}$.

Definition 2.13 (super-summit set). The subset $\mathrm{SSS}(e)$ involved in Corollary 2.12 is called the *super-summit set* of e. The constant values of $\inf_\Delta$ and $\sup_\Delta$ on $\mathrm{SSS}(e)$ are denoted by $\inf_\Delta(\mathrm{SSS}(e))$ and $\sup_\Delta(\mathrm{SSS}(e))$, respectively.

Note that, by definition, if e belongs to $\mathcal{C}^{\circlearrowright}$, then $\mathrm{SSS}(e)$ is included in $\mathcal{C}^{\circlearrowright}$.

The interest of considering super-summit sets is that, under convenient assumptions, they lead to a solution of the Conjugacy Problem, that is, the question of algorithmically deciding whether two elements of the considered groupoid are conjugate. We now give algorithms providing such a solution. We will have to consider the following conditions involving invertible elements.

Condition 2.14. (i) *The Garside family* $\mathcal{S}$ *is finite up to right-multiplication by* $\mathcal{C}^{\times}$, *that is, the number of* $=^{\times}$*-classes in* $\mathcal{S}$ *is finite.*

(ii) *Distinct* $=^{\times}$*-equivalent elements of* $\mathcal{G}$ *are not conjugate.*

Note that Condition 2.14(ii) is satisfied if $\mathcal{C}$ is a semi-direct product of a category without nontrivial invertible elements by an automorphism as considered in Subsection 1.3 (that condition will be used in the proof of Lemma 2.35).

Owing to Proposition V.2.40 (right-Noetherian) and its symmetric counterpart, Condition 2.14(i) implies that $\mathcal{C}$ is Noetherian and, by Proposition II.2.58 (atom generate), that it is generated by its atoms (which are in $\mathcal{D}iv(\Delta)$) and $\mathcal{C}^{\times}$.

Algorithm 2.15 (element of the super-summit set).

Context: A groupoid $\mathcal{G}$ that is the groupoid of fractions of a cancellative category that is Noetherian and admits a Garside map, an algorithm for cycling, and an algorithm for decycling
Input: An element e of $\mathcal{G}^{\circlearrowright}$
Output: An element of the super-summit set of e

```
put L := {e}
while cyc(e) ∉ L do
    put e := cyc(e)
    put L := L ∪ {e}
while dec(e) ∉ L do
    put e := dec(e)
    put L := L ∪ {e}
return e
```

Proposition 2.16 (element of the super-summit set). *If $\mathcal{G}$ is the groupoid of fractions of a cancellative category that is Noetherian and admits a Garside map, then Algorithm* 2.15 *running on an element e of $\mathcal{G}^{\circlearrowright}$ returns an element of the super-summit set of e.*

Proof. This is a consequence of Proposition 2.5 and Corollary 2.10. □

Example 2.17 (element of the super-summit set). In the 5-strand braid group B_5 (Reference Structure 2, page 5), consider

$$\beta = (\sigma_1\sigma_2\sigma_1\sigma_3\sigma_2\sigma_1\sigma_4)(\sigma_2\sigma_3\sigma_4)(\sigma_4)(\sigma_4)(\sigma_4)(\sigma_4).$$

The given factorization is the delta-normal decomposition of β associated with Δ_5. Repeatedly cycling β gives the following elements, given in delta-normal form:

$$\begin{aligned}
\beta_1 &= (\sigma_1\sigma_2\sigma_1\sigma_3\sigma_2\sigma_1\sigma_4\sigma_3)(\sigma_3)(\sigma_3)(\sigma_3)(\sigma_3\sigma_2\sigma_1),\\
\beta_2 &= (\sigma_1\sigma_2\sigma_1\sigma_3\sigma_2\sigma_1\sigma_4)(\sigma_1)(\sigma_1)(\sigma_1)(\sigma_1\sigma_2\sigma_3\sigma_4),\\
\beta_3 &= \Delta(\sigma_4)(\sigma_4)(\sigma_4)(\sigma_4),\\
\beta_4 &= \Delta(\sigma_1\sigma_4)(\sigma_4)(\sigma_4).
\end{aligned}$$

Cycling becomes constant at the fifth iteration. Decycling β_4 gives $\beta_5 = \Delta(\sigma_1\sigma_4)(\sigma_1\sigma_4)$ and then decycling and cycling of β_5 remain constant. We conclude that β_5 belongs to the super-summit set of (the conjugacy class of) β.

On the other hand, decycling β iteratively gives the following elements

$$\begin{aligned}
\beta'_1 &= (\sigma_1\sigma_2\sigma_1\sigma_3\sigma_4\sigma_3\sigma_2\sigma_1)(\sigma_2\sigma_3\sigma_4)(\sigma_4)(\sigma_4)(\sigma_4),\\
\beta'_2 &= (\sigma_1\sigma_2\sigma_1\sigma_3\sigma_4\sigma_3)(\sigma_1\sigma_2\sigma_3\sigma_2\sigma_1\sigma_4)(\sigma_4)(\sigma_4),\\
\beta'_3 &= (\sigma_1\sigma_2\sigma_1\sigma_3\sigma_4\sigma_3)(\sigma_1\sigma_2\sigma_3\sigma_2\sigma_4)(\sigma_2\sigma_1\sigma_4),\\
\beta'_4 &= (\sigma_1\sigma_2\sigma_1\sigma_4)(\sigma_1\sigma_2\sigma_1\sigma_3\sigma_2\sigma_4\sigma_3\sigma_2)(\sigma_2\sigma_4).
\end{aligned}$$

Decycling becomes constant at the fifth iteration. Cycling β_4' gives β_5. Note that, in order to obtain an element of the super-summit set of β, neither cycling alone nor decycling alone is sufficient.

Algorithm 2.18 (super-summit set).

Context: A groupoid $\mathcal{G}$ that is the groupoid of fractions of a cancellative category that is Noetherian, admits a Garside map, and satisfies Condition 2.14
Input: An element e of $\mathcal{G}^{\circlearrowleft}$
Output: The super-summit set of e

compute an element f of the super-summit set of e using Algorithm 2.15
put $S := \{f\}$
put $S' := \{e' \mid \exists e \in S\, \exists s \in \mathcal{D}iv(\Delta)\, (e' = e^s \text{ and } e' \in \mathrm{SSS}(e'))\}$
while $S' \neq S$ **do**
    put $S := S'$
    put $S' := \{e' \mid \exists e \in S\, \exists s \in \mathcal{D}iv(\Delta)\, (e' = e^s \text{ and } e' \in \mathrm{SSS}(e'))\}$
return S

To prove the correctness of Algorithm 2.18, we begin with an auxiliary result.

Lemma 2.19. *Under the assumptions of Algorithm* 2.18, *if e is an element of $\mathcal{G}$ lying in its super-summit set and if e^f lies in this super-summit set, with f in $\mathcal{C} \cap \mathit{Conj}\mathcal{G}(e, \cdot)$, then, for every head s of f, the element e^s also lies in its super-summit set.*

Proof. Let i be $\inf_\Delta(e) = \inf_\Delta(e^f)$. Let us write $f = sh$ and $e = \Delta^{[i]}e'$. The relation $\Delta^{[i]} \preccurlyeq e^f$ can be written $sh\Delta^{[i]} \preccurlyeq \Delta^{[i]}e'sh$. This implies $\phi_\Delta^i(s) \preccurlyeq e'sh$. Since s is a head of sh, this implies $\phi_\Delta^i(s) \preccurlyeq e's$ which, using $\Delta^{[i]}e' = e$, can be written $s\Delta^{[i]} \preccurlyeq es$, which is in turn equivalent to $\Delta^{[i]} \preccurlyeq e^s$. Since i is the maximum of $\inf_\Delta$ in the conjugacy class, we have $\inf_\Delta(e^s) = i = \inf_\Delta(e)$. Note that, if e lies in its super-summit set, then e^{-1} also lies in its super-summit set. Using then the relations between $\inf_\Delta$ and $\sup_\Delta$ given in Corollary V.3.28 (inverse), we obtain $\sup_\Delta(e^s) = \sup_\Delta(e)$, whence the result. □

Proposition 2.20 (computation of super-summit set). *If $\mathcal{G}$ is the groupoid of fractions of a cancellative category that is Noetherian, admits a Garside map, and satisfies Condition* 2.14, *then Algorithm* 2.18 *running on an element e of $\mathcal{G}^{\circlearrowleft}$ returns the super-summit set of e.*

Proof. First, we claim that, under Condition 2.14, the super-summit set of every element is finite. Indeed, let us choose a set $\mathcal{S}'$ of representatives of the $=^\times$-classes in $\mathcal{S}$. By Condition 2.14(i), the set $\mathcal{S}'$ is finite. Let m and $m+\ell$ be the respective constant values of $\inf_\Delta$ and $\sup_\Delta$ on some super-summit set. Then every element lying in this super-summit set can be written $\Delta^{[m]}t_1 \cdots t_\ell\epsilon$ with t_i in $\mathcal{S}'$ for each i and ϵ in $\mathcal{C}^\times$. By Condition 2.14(ii), there is no other element in the conjugacy class $=^\times$-equivalent to $\Delta^{[m]}t_1 \cdots t_\ell\epsilon$, whence our claim.

As every super-summit set is finite, Algorithm 2.18 terminates.

By Corollary 2.12, a super-summit set is connected in the conjugacy category. Since ϕ_Δ maps a super-summit set onto itself, a super-summit set is connected under conjugation by elements of $\mathcal{C}$. Moreover, Lemma 2.19 shows that conjugating by elements of $\mathcal{D}iv(\Delta)$ as in steps 3 and 6 of Algorithm 2.18 is sufficient to obtain the whole super-summit set. □

Example 2.21 (computation of super-summit set). In the braid group B_5, the super-summit set of $(\sigma_1\sigma_2\sigma_1\sigma_4)(\sigma_4)$ (here written in normal form) has 10 elements, namely

$$(\sigma_1\sigma_2\sigma_1\sigma_4)(\sigma_4), (\sigma_2\sigma_1\sigma_4)(\sigma_1\sigma_4), (\sigma_1\sigma_2\sigma_4)(\sigma_2\sigma_4), (\sigma_1\sigma_3\sigma_4\sigma_3)(\sigma_1), (\sigma_1\sigma_4)(\sigma_1\sigma_2\sigma_4),$$
$$(\sigma_1\sigma_4\sigma_3)(\sigma_1\sigma_3), (\sigma_2\sigma_4)(\sigma_2\sigma_1\sigma_4), (\sigma_1\sigma_3\sigma_4)(\sigma_1\sigma_4), (\sigma_1\sigma_3)(\sigma_1\sigma_3\sigma_4), (\sigma_1\sigma_4)(\sigma_1\sigma_4\sigma_3).$$

Remark 2.22. Algorithm 2.18 is very rough. It can be improved in particular by making some choice of the conjugating elements in steps 3 and 6. We do not enter into details here since we will give a better method in Subsection 2.2 using sliding.

Using Algorithm 2.18, we easily obtain a (first) solution to the Conjugacy Problem.

Algorithm 2.23 (Conjugacy Problem I).

Context: A groupoid $\mathcal{G}$ that is the groupoid of fractions of a cancellative category that is Noetherian, admits a Garside map, and satisfies Condition 2.14
Input: Two elements d, e of $\mathcal{G}^{\circlearrowright}$
Output: YES if d and e are conjugate in $\mathcal{G}$, NO otherwise

```
compute the super-summit set S of e using Algorithm 2.18
compute an element d' of the super-summit set of d using Algorithm 2.15
if d' ∈ S then
    return YES
else
    return NO
```

As the super-summit set characterizes a conjugacy class, the following result is clear.

Proposition 2.24 (Conjugacy Problem I). *If $\mathcal{G}$ is the groupoid of fractions of a Noetherian cancellative category that admits a Garside map and Condition* 2.14 *holds, then Algorithm* 2.23 *solves the Conjugacy Problem for $\mathcal{G}$.*

Corollary 2.25 (decidability of Conjugacy Problem). *If $\mathcal{G}$ is a groupoid satisfying the assumptions of Proposition* 2.24 *and, in addition, the involved family $\mathcal{D}iv(\Delta)$ is finite, the Conjugacy Problem for $\mathcal{G}$ is decidable.*

Proof. The assumption that the Garside family $\mathcal{D}iv(\Delta)$ is finite guarantees that delta-normal decompositions are algorithmically computable, hence so are cycling and decycling, so all underlying procedures used in Algorithm 2.23 are effective. □

We complete the section with two technical results describing the behavior of cycling and decycling with respect to powers and to invertible elements.

Proposition 2.26 (cycling a power). *In Context* 2.1, *for every e in $\mathcal{C}^{\circlearrowright}$ and n in $\mathbb{N}$, the cycling of e^n belongs to $\mathit{CycC}(e,\cdot)$ and the decycling of e^n belongs to $\mathit{CycC}(\cdot,e)$, by which we mean that* $\mathrm{init}(e^n)$ *lies in $\mathit{CycC}(e,\cdot)$ (resp.* $\mathrm{fin}(e^n)$ *lies in $\mathit{CycC}(\cdot,e)$).*

Proof (beginning). By the same considerations as in the beginning of the proof of Proposition 2.5, namely identifying e with the element $e\Delta^{[-i]}\phi_\Delta^{-i}$ of $\mathcal{C}\rtimes\langle\phi_\Delta^{-i}\rangle$ with $i=\inf_\Delta(e)$, we may assume $\inf_\Delta(e)=0$. Let $s_1|\cdots|s_\ell$ be the H-normal decomposition of e^n, that is, $s_i=H(s_i\cdots s_\ell)$ holds for $i<\ell$.

We first prove, using induction on j, that $e^{s_1\cdots s_j}$ lies in $\mathcal{C}$ and s_j in $\mathit{CycC}(e^{s_1\cdots s_{j-1}},\cdot)$. The starting point of the induction is the following technical result. □

Lemma 2.27. *For all e in $\mathcal{C}^{\circlearrowright}$ and n in $\mathbb{N}$, the element $H(e^n)$ belongs to $\mathit{CycC}(e,\cdot)$.*

Proof. For every $i>0$, we have $e^{i-1}\preccurlyeq e^i$, whence $H(e^{i-1})\preccurlyeq H(e^i)$ by (IV.1.46). Define t_i by $H(e^i)=H(e^{i-1})t_i$. We have

$$H(e^{i-1})t_i=H(ee^{i-1})\preccurlyeq eH(e^{i-1}),$$

the last divisibility relation by the $\mathcal{H}$-law. Therefore, $H(e^{i-1})^{-1}eH(e^{i-1})\in\mathcal{C}$ and $t_i\preccurlyeq H(e^{i-1})^{-1}eH(e^{i-1})$ hold, so t_i lies in $\mathit{CycC}(H(e^{i-1})^{-1}eH(e^{i-1}),H(e^i)^{-1}eH(e^i))$, whence the result using $H(e^n)=^{\times}t_1\cdots t_n$. □

Proof of Proposition 2.26 *(end)*. We now do the general step of the induction, assuming that $s_1\cdots s_j$ lies in $\mathit{CycC}(e,\cdot)$. From $e^n=s_1\cdots s_\ell\in\mathit{CycC}(e)$, using that CycC is closed under right-quotient (see remark after 1.21), we obtain that $s_{j+1}\cdots s_\ell$ lies in $\mathit{CycC}(e^{s_1\cdots s_j},\cdot)$. Lemma 2.27 applied to $e^{s_1\cdots s_j}$ gives $H(s_{j+1}\cdots s_\ell s_1\cdots s_j)\in\mathit{CycC}(e^{s_1,\cdots s_j},\cdot)$, since $s_{j+1}\cdots s_\ell s_1\cdots s_j$ is equal to $(e^{s_1,\cdots s_j})^n$. Using that CycC is closed under left-gcd and that s_{j+1} is a left-gcd of $s_{j+1}\cdots s_\ell$ and $H(s_{j+1}\cdots s_\ell s_1\cdots s_j)$, we deduce that s_{j+1} lies in $\mathit{CycC}(e^{s_1\cdots s_j},\cdot)$.

In particular, s_ℓ lies in $\mathit{CycC}(e^{s_1\cdots s_{\ell-1}},e)$, which gives the statement of the proposition about decycling.

For $i=\inf_\Delta(e^n)$, we have $s_1=\cdots=s_i=\Delta$ and, since $\mathrm{init}(e^n)=^{\times}\phi_\Delta^{-i}(s_{i+1})$ holds, the $(i+1)$st step of our induction result gives $\phi_\Delta^i(\mathrm{init}(e^n))\in\mathit{CycC}(e^{\Delta^{[i]}},\cdot)$, which is $\mathit{CycC}(\phi_\Delta^i(e),\cdot)$. We deduce the part of the statement about cycling by applying ϕ_Δ^{-i}. □

Finally, for further reference, it will be useful to know that cycling and decycling are compatible with conjugation by an invertible element.

Lemma 2.28. *For e in $\mathcal{G}$, the cycling (resp. decycling) of a conjugate of e by an element of $\mathcal{C}^{\times}$ is conjugate to the cycling (resp. decycling) of e by an element of $\mathcal{C}^{\times}$.*

Proof. Let $\Delta^{[m]}\|s_1|\cdots|s_\ell$ be the H-normal decomposition of e. Thus, we have $i=\inf_\Delta(e)$ and $s_j=H(s_j\cdots s_\ell)$ for $j<\ell$. Let ϵ be an invertible element of $\mathcal{C}$. Then we find ${}^{\epsilon}e=\Delta^{[m]}\phi_\Delta^i(\epsilon)s_1\cdots s_\ell\epsilon^{-1}$ and $H(\phi_\Delta^i(\epsilon)s_1\cdots s_\ell\epsilon^{-1})=^{\times}\phi_\Delta^i(\epsilon)s_1$. Hence we have $\mathrm{init}({}^{\epsilon}e)=\epsilon\,\mathrm{init}(e)\epsilon'$ for some invertible element ϵ'. We finally obtain the equalities $\mathrm{cyc}({}^{\epsilon}e)=({}^{\epsilon}e)^{\epsilon\,\mathrm{init}(e)\epsilon'}=\mathrm{cyc}(e)^{\epsilon'}$.

Similarly, we find $\mathrm{fin}(\epsilon\Delta^{[m]}s_1\cdots s_\ell\epsilon^{-1}) = \epsilon' s_\ell\epsilon^{-1}$ for some invertible ϵ', since, in the H-normal decomposition of e, the product of the first $r-1$ terms is of the form $\Delta^{[m]}s_1\cdots s_{\ell-1}\epsilon'^{-1}$ for a certain ϵ'. We deduce the equality $\mathrm{dec}({}^{\epsilon}e) = {}^{\epsilon' s_\ell}e = {}^{\epsilon'}\mathrm{dec}(e)$. □

Twisted version. In addition to all previous assumptions, suppose that $\mathcal{C}$ has an automorphism ϕ of finite order satisfying $\phi(\Delta) = \Delta$, so that ϕ commutes with ϕ_Δ. We extend the chosen sharp head function H to $\mathcal{C} \rtimes \langle\phi\rangle$ by $H(e\phi^m) = H(e)$. We define the ϕ-cycling $\mathrm{cyc}_\phi(e)$ of e in $\mathcal{G}$ by $\mathrm{cyc}_\phi(e)\phi = \mathrm{cyc}(e\phi)$, where cyc is the cycling in $\mathcal{C} \rtimes \langle\phi\rangle$. Here $e\phi$ has to be an element of $\mathcal{C} \rtimes \langle\phi\rangle^{\circlearrowleft}$, which means that the target of e is the image by ϕ of the source of e. Similarly, we define the ϕ-decycling by $\mathrm{dec}_\phi(e)\phi = \mathrm{dec}(e\phi)$. Note that the initial factor of $e\phi$ is equal to the initial factor of e, but the final factor of $e\phi$ is $\mathrm{fin}(e)\phi$. We can then translate in this context the above results. In the connected component of an object e in the ϕ-conjugacy category there is a super-summit set where the infimum is maximal and the supremum minimal. This super-summit set is reached from e by a sequence of ϕ-cyclings and ϕ-decyclings.

2.2 Sliding circuits

We now turn to the construction of a new characteristic subset of every conjugacy class that is smaller than the super-summit set. This set, called the set of sliding circuits, is obtained by repeated application of an operation called sliding.

Our setting remains Context 2.1 as in the previous subsection. The starting point is the following easy observation about left-gcds. We recall that the latter necessarily exist in the current context according to Proposition V.2.35 (lefts and gcds). Hereafter, we write "gcd" for left-gcd and use it as a binary operation in formulas, although it is is defined up to right-multiplication by an invertible element only. Now assume that e lies in $\mathcal{G}^{\circlearrowleft}$. Then we can write

$$\begin{aligned}\gcd(\mathrm{init}(e), \mathrm{init}(e^{-1})) &=^{\times} \gcd(\mathrm{init}(e), \mathrm{fin}(e)^{-1}\Delta)\\ &=^{\times} \gcd(e\Delta^{[-\mathrm{inf}_\Delta(e)]}, e^{-1}\Delta^{[\mathrm{sup}_\Delta(e)]}, \Delta) =^{\times} \mathrm{fin}(e)^{-1}H(\mathrm{fin}(e)\mathrm{init}(e)),\end{aligned}$$

where the first relation follows from Proposition V.3.26 (inverse). Note that $\mathrm{init}(e)$ and $\mathrm{fin}(e)^{-1}$ have the same source since, by assumption, e lies in $\mathcal{G}^{\circlearrowleft}$, so the involved gcds are defined.

Definition 2.29 (prefix, sliding). For e in $\mathcal{G}^{\circlearrowleft}$, the *prefix* $\mathrm{pr}(e)$ of e is the head of a (any) left-gcd of $\mathrm{init}(e)$ and $\mathrm{init}(e^{-1})$, and the *sliding* $\mathrm{sl}(e)$ of e is the conjugate $e^{\mathrm{pr}(e)}$ of e.

Although the left-gcd need not be unique in general, the prefix and, therefore, the sliding of an element e of $\mathcal{G}^{\circlearrowleft}$ is well-defined since any two left-gcds are $=^{\times}$-equivalent and, by assumption, H is a sharp head-function, that is, one that takes equal values on

$=^{\times}$-equivalent entries. Note that, as $\mathrm{init}(e)$ and $\mathrm{init}(e^{-1})$ are divisors of Δ, the effect of applying the map H is only to pick the chosen representative in an equivalence class for $=^{\times}$. By definition, the relations $\mathrm{pr}(e) \preccurlyeq \mathrm{init}(e)$ and $\mathrm{fin}(e)\mathrm{pr}(e) \preccurlyeq \Delta$ are satisfied, so the prefix $\mathrm{pr}(e)$ is an element of $\mathit{Conj}\,\mathcal{G}(e, \mathrm{sl}(e))$. The main properties of the sliding are as follows.

Proposition 2.30 (sliding circuits). *In Context* 2.1, *and for every e in* $\mathcal{G}^{\circlearrowleft}$*:*

(i) *We have* $\inf_\Delta(\mathrm{sl}(e)) \geqslant \inf_\Delta(e)$, *and, if there exists* e' *in the conjugacy class of* e *satisfying* $\inf_\Delta(e') > \inf_\Delta(e)$, *there exists* i *satisfying* $\inf_\Delta(\mathrm{sl}^i(e)) > \inf_\Delta(e)$.

(ii) *We have* $\sup_\Delta(\mathrm{sl}(e)) \leqslant \sup_\Delta(e)$, *and, if there exists* e' *in the conjugacy class of* e *satisfying* $\sup_\Delta(e') < \sup_\Delta(e)$, *there exists* i *satisfying* $\sup_\Delta(\mathrm{sl}^i(e)) < \sup_\Delta(e)$.

(iii) *If* $\mathcal{S}/{=^{\times}}$ *is finite, sliding is ultimately periodic up to invertible elements in the sense that, for every e in* $\mathcal{G}^{\circlearrowleft}$, *there exists* $i > j \geqslant 0$ *satisfying* $\mathrm{sl}^i(e) =^{\times} \mathrm{sl}^j(e)$.

Proof (beginning). Let $\Delta^{[m]}\|s_1|\cdots|s_\ell$ be the H-normal decomposition of e. We have

$$\mathrm{sl}(e) = \Delta^{[m]}(\phi_\Delta^m(\mathrm{pr}(e))^{-1}s_1)s_2\cdots(s_\ell\mathrm{pr}(e)),$$

where each of the bracketed terms is a divisor of Δ. Hence $\Delta^{[\inf_\Delta(e)]}$ left-divides $\mathrm{sl}(e)$, which itself left-divides $\Delta^{[\sup_\Delta(e)]}$. We deduce the inequalities $\inf_\Delta(\mathrm{sl}(e)) \geqslant \inf_\Delta(e)$ and $\sup_\Delta(\mathrm{sl}(e)) \leqslant \sup_\Delta(e)$. □

For the other assertions, as in [131, Lemma 4], we first prove

Lemma 2.31. (i) *The relation* $\mathrm{fin}(e)\mathrm{init}(e) \preccurlyeq \Delta$ *is equivalent to* $\sup_\Delta(\mathrm{cyc}(e)) < \sup_\Delta(e)$, *and, when this holds,* $\mathrm{sl}(e)$ *and* $\mathrm{cyc}(e)$ *are conjugate by an element of* $\mathcal{C}^{\times}$.

(ii) *The relation* $\Delta \preccurlyeq \mathrm{fin}(e)\mathrm{init}(e)$ *is equivalent to* $\inf_\Delta(\mathrm{dec}(e)) > \inf_\Delta(e)$, *and, when this holds,* $\mathrm{sl}(e)$ *and* $\phi_\Delta(\mathrm{dec}(e))$ *are conjugate by an element of* $\mathcal{C}^{\times}$.

(iii) *If we are not in case* (i) *or* (ii), *then the three elements* $\mathrm{sl}(e)$, $\mathrm{cyc}(\mathrm{dec}(e))$ *and* $\mathrm{dec}(\mathrm{cyc}(e))$ *are conjugate by elements of* $\mathcal{C}^{\times}$.

Proof. (i) Let again $\Delta^{[m]}\|s_1|\cdots|s_\ell$ be the H-normal decomposition of e, and assume that we have $\mathrm{fin}(e)\mathrm{init}(e) \preccurlyeq \Delta$. Then we find

$$\mathrm{pr}(e) =^{\times} \mathrm{fin}(e)^{-1}H(\mathrm{fin}(e)\mathrm{init}(e)) =^{\times} \mathrm{fin}(e)^{-1}\mathrm{fin}(e)\mathrm{init}(e) = \mathrm{init}(e),$$

thus $\mathrm{sl}(e)$ is conjugate to $\mathrm{cyc}(e)$ by an invertible element. Further $\mathrm{cyc}(e)$, which is $e^{\mathrm{init}(e)}$, hence ${}^{\times}{=}$-equivalent to $\Delta^{[m]}s_2\cdots s_{\ell-1}(s_\ell\,\mathrm{init}(e))$, divides $\Delta^{[\sup_\Delta(e)-1]}$. We will see in the proof of (iii) that $\mathrm{fin}(e)\mathrm{init}(e) \not\preccurlyeq \Delta$ implies $\sup_\Delta(\mathrm{cyc}(e)) = \sup_\Delta(e)$, whence the equivalence.

(ii) Assume $\Delta \preccurlyeq \mathrm{fin}(e)\mathrm{init}(e)$. Then we have $\mathrm{pr}(e) =^{\times} \mathrm{fin}(e)^{-1}\Delta$, thus $\mathrm{sl}(e)$ is conjugate by an invertible element to $\phi_\Delta({}^{\mathrm{fin}(e)}e)$, which is $\phi_\Delta(\mathrm{dec}(e))$, and, since Δ divides $s_2\cdots s_\ell\,\mathrm{init}(e)$, we deduce that $\Delta^{[i+1]}$ divides $\mathrm{dec}(e)$. We will see the converse in the proof of (iii).

(iii) Assume that we are not in case (i) or (ii). Since we have $\mathrm{init}(e) =^{\times} \phi_\Delta^{-i}(s_1)$, we obtain that $\mathrm{cyc}(e)$ is conjugate by an invertible element to g, which is $\Delta^{[m]} s_2 \cdots s_\ell \phi_\Delta^{-i}(s_1)$. Hence, by Lemma 2.28, $\mathrm{dec}(\mathrm{cyc}(e))$ is conjugate to $\mathrm{dec}(g)$ by an invertible element. Now, if $\mathrm{fin}(e)\,\mathrm{init}(e)$, that is, $s_\ell\,\mathrm{init}(e)$, does not divide Δ, then neither does $s_\ell \phi_\Delta^{-i}(s_1)$ and, by Propositions V.1.52 (second domino rule) and III.1.61 (right-multiplication), the last entry g' in the strict normal decomposition of g satisfies $g' =^{\times} T(s_\ell \phi_\Delta^{-i}(s_1))$. In this case, we obtain $\sup_\Delta(\mathrm{cyc}(e)) = \sup_\Delta(g) = \sup_\Delta(e)$. Let $t = H(s_\ell \phi_\Delta^{-i}(s_1))$ and $t' = T(s_\ell \phi_\Delta^{-i}(s_1))$. Then $\mathrm{dec}(g)$ is conjugate to $t' \Delta^{[m]} s_2 \cdots s_{\ell-1} e$ by an invertible element. Since $\mathrm{fin}(e)\mathrm{pr}(e)$ is $=^{\times}$-equivalent to $H(\mathrm{fin}(e)\mathrm{init}(e))$, whence to t, we deduce that $\mathrm{sl}(e)$ is conjugate by an invertible element to $t^{-1} s_\ell e s_\ell^{-1} t$. The latter element is $t^{-1} s_\ell \phi_\Delta^{-i}(s_1) \Delta^{[m]} s_2 \cdots s_{\ell-1} t$, hence it is conjugate by an invertible element to $\mathrm{dec}(\mathrm{cyc}(e))$ since we have $t^{-1} s_\ell \phi_\Delta^{-i}(s_1) = t'$.

Similarly, assume that $s_\ell\,\mathrm{init}(e)$ is not divisible by Δ. By Proposition III.1.45 (first domino rule), we have $\inf_\Delta(\mathrm{dec}(e)) = \inf_\Delta(e)$, and, by the $\mathcal{H}$-law, we have

$$\mathrm{init}(\mathrm{dec}(e)) = H(\mathrm{dec}(e)\Delta^{[-i]}) = H(s_\ell \phi_\Delta^{-i}(s_1) \cdots \phi_\Delta^{-i}(s_\ell)) = H(s_\ell H(\phi_\Delta^{-i}(s_1))).$$

We deduce $\mathrm{init}(\mathrm{dec}(e)) = H(s_\ell\,\mathrm{init}(e))$, whence $\mathrm{cyc}(\mathrm{dec}(e)) = e^g$ where g is the element $s_\ell^{-1} H(s_\ell\,\mathrm{init}(e))$. The latter is equal to $\mathrm{pr}(e)$ up to right-multiplication by an invertible element, whence the result. □

Proof of Proposition 2.30 *(end).* Assume that, for every i, we have

$$\sup_\Delta(\mathrm{sl}^i(e)) = \sup_\Delta(e) \quad \text{and} \quad \inf_\Delta(\mathrm{sl}^i(e)) = \inf_\Delta(e).$$

Then we are in case (iii) of Lemma 2.31 at each iteration of sl. By Lemma 2.28, the transformations cyc and dec are compatible with conjugation by invertible elements, so, up to conjugating by invertible elements, we can replace each iteration of sl by an application of $\mathrm{cyc} \circ \mathrm{dec}$ or of $\mathrm{dec} \circ \mathrm{cyc}$. Also, since $\inf_\Delta$ and $\sup_\Delta$ remain constant for each sequence of transformations cyc and dec applied to e, we are left with an element e' falling in case (iii), that is, $\mathrm{cyc}(\mathrm{dec}(e'))$ and $\mathrm{dec}(\mathrm{cyc}(e'))$ are conjugate by an invertible element. It follows that cyclings and decyclings can be reordered so that $\mathrm{sl}^n(e)$ is equal, up to conjugating by an invertible element, to $\mathrm{cyc}^n(\mathrm{dec}^n(e))$. The latter implies that e lies in its super-summit set.

It follows that, if e does not lie in its super-summit set, some power of sl will diminish the supremum or increase the infimum.

Finally, under the assumption of (iii), there is a finite number of classes modulo right-multiplication by $\mathcal{C}^\times$ of elements with given supremum and infimum. Thus the super-summit set of the element e is finite modulo $\mathcal{C}^\times$. This implies (iii). □

Definition 2.32 (sliding category). For e, e' in $\mathcal{G}^{\circlearrowleft}$, we denote by $\mathcal{C}onj\mathcal{C}(e, e')$ the family of all *positive* elements of $\mathcal{C}onj\mathcal{G}(e, e')$, defined to be those elements that, under Convention 2.2, lie in $\mathcal{C}$. The *sliding category* $\mathrm{SSS}(\mathcal{G})$ of $\mathcal{G}$ is the subcategory of $\mathcal{C}onj\mathcal{G}$ whose objects are the elements of $\mathcal{G}^{\circlearrowleft}$ that have a minimum supremum and a maximum infimum in their conjugacy class and such that, for e, e' in $\mathcal{O}bj(\mathrm{SSS}(\mathcal{G}))$, the set $\mathrm{SSS}(\mathcal{G})(e, e')$ is $\mathcal{C}onj\mathcal{C}(e, e')$.

Note that the existence of elements of $\mathcal{G}^{\circlearrowleft}$ that simultaneously have a minimal supremum and a maximal infimum is guaranteed by Proposition 2.30. Also note that, for every e in $\mathcal{G}^{\circlearrowleft}(x,x)$, the element $\Delta(x)$ lies in $\mathcal{C}onj\mathcal{C}(e, \phi_\Delta(e))$.

Sliding then gives rise to a natural functor from the category $\mathrm{SSS}(\mathcal{G})$ into itself.

Proposition 2.33 (sliding functor). *In Context* 2.1*, define* sl *on the category* $\mathrm{SSS}(\mathcal{G})$ *by mapping* e *to* $\mathrm{sl}(e)$ *for* e *in* $\mathcal{O}bj(\mathrm{SSS}(\mathcal{G}))$ *and* g *to* $\mathrm{pr}(e)^{-1} g\, \mathrm{pr}(e^g)$ *for* g *in* $\mathcal{C}onj\mathcal{C}(e, e^g)$.

(i) *The maps* sl *define a functor from the category* $\mathrm{SSS}(\mathcal{G})$ *to itself.*

(ii) *The functor* sl *preserves left-divisibility, and* $s \in \mathcal{D}iv(\Delta)$ *implies* $\mathrm{sl}(s) \in \mathcal{D}iv(\Delta)$.

(iii) *The functor* sl *preserves left-gcds.*

(iv) *If* $\mathcal{S}/{=^\times}$ *is finite, the functor* sl *is ultimately periodic on elements up to invertible elements, that is, for every* g *in* $\mathrm{SSS}(\mathcal{G})$*, there exists* $i > j \geqslant 0$ *satisfying* $\mathrm{sl}^i(g) =^\times \mathrm{sl}^j(g)$.

Proof. (i) As sl is obviously compatible with composition of elements, the only nontrivial point is to check that, for g in $\mathcal{C}onj\mathcal{C}(e, e^g)$, the element $\mathrm{sl}(g)$ is positive, or, equivalently, that $\mathrm{pr}(e)$ left-divides $g\,\mathrm{pr}(e^g)$ in $\mathcal{C}$. Now we have $\mathrm{pr}(e) =^\times \gcd(\mathrm{init}(e), \mathrm{init}(e^{-1}))$, so it is sufficient to prove $\mathrm{init}(e) \preccurlyeq g\,\mathrm{init}(e^g)$, which, together with the same property for e^{-1}, will give $\mathrm{pr}(e) \preccurlyeq g\,\mathrm{pr}(e^g)$.

We thus have to show the relation

$$\gcd(e\Delta^{[-\inf_\Delta(e)]}, \Delta) \preccurlyeq g\gcd(e^g\Delta^{[-\inf_\Delta(e^g)]}, \Delta)$$

or, equivalently, $\gcd(e\Delta^{[-\inf_\Delta(e)]}, \Delta) \preccurlyeq \gcd(eg\Delta^{[-\inf_\Delta(e^g)]}, g\Delta)$. As we are in $\mathrm{SSS}(\mathcal{G})$, we have $\inf_\Delta(e^g) = \inf_\Delta(e)$, hence it is sufficient to prove $e\Delta^{[-\inf_\Delta(e)]} \preccurlyeq eg\Delta^{[-\inf_\Delta(e)]}$ and $\Delta \preccurlyeq g\Delta$. Now both relations are obvious.

(ii) The argument for left-divisibility is the same as for (i). The second assertion comes from the first and the fact that for e in $\mathcal{G}^{\circlearrowleft}(x,x)$ we have

$$\mathrm{sl}(e, \Delta(x), \phi_\Delta(e)) = (\mathrm{sl}(e), \Delta(y)\epsilon, \phi_\Delta(\mathrm{sl}(e)))$$

where y is the source of $\mathrm{sl}(e)$ and ϵ in $\mathcal{C}^\times$ satisfies $\mathrm{pr}(\phi_\Delta(e)) = \phi_\Delta(\mathrm{pr}(e))\epsilon$.

(iii) We first observe that the subcategory $\mathrm{SSS}(\mathcal{G})$ of $\mathcal{C}onj(\mathcal{G})$ is closed under left-gcd. Indeed, $\inf_\Delta(e^g) \geqslant \inf_\Delta(e)$ is equivalent to $e^g\Delta^{[-\inf_\Delta(e)]} \in \mathcal{C}$, hence also to $g\Delta^{[\inf_\Delta(e)]} \preccurlyeq eg$, and the conjunction of $g\Delta^{[\inf_\Delta(e)]} \preccurlyeq eg$ and $h\Delta^{[\inf_\Delta(e)]} \preccurlyeq eh$ implies $\gcd(g,h)\Delta^{[\inf_\Delta(e)]} \preccurlyeq e\gcd(g,h)$. The same considerations apply to $\sup_\Delta$.

To show (iii), it is sufficient, using (ii), to show that, for g and h in $\mathrm{SSS}(\mathcal{G})(e,\cdot)$, the relation $\gcd(g,h) =^\times 1$ implies $\gcd(\mathrm{sl}(g), \mathrm{sl}(h)) =^\times 1$. So assume $\gcd(g,h) =^\times 1$. As we have $\mathrm{pr}(e)\mathrm{sl}(g) = g\,\mathrm{pr}(e^g)$ and $\mathrm{pr}(e)\mathrm{sl}(h) = h\mathrm{pr}(e^h)$, the relation we have to show is equivalent to $\gcd(g\,\mathrm{pr}(e^g), h\mathrm{pr}(e^h)) =^\times \mathrm{pr}(e)$.

Using again $\mathrm{pr}(e) =^\times \gcd(\mathrm{init}(e), \mathrm{init}(e^{-1}))$, the above relation will be a consequence of the relation $\gcd(g\,\mathrm{init}(e^g), h\,\mathrm{init}(e^h)) =^\times \mathrm{init}(e)$ together with the same relation for e^{-1}. The left-hand side is in turn equal to $\gcd(eg\Delta^{[-m]}, eh\Delta^{[-m]}, g\Delta, h\Delta)$, where we have set $m = \inf_\Delta(e) = \inf_\Delta(e^g) = \inf_\Delta(e^h)$ (the equalities are valid since we are in $\mathrm{SSS}(\mathcal{G})$). Now, by assumption, we have the relations

$$\begin{aligned}\gcd(eg\Delta^{[-m]}, eh\Delta^{[-m]}) &=^\times \gcd(e\Delta^{[-m]}\phi_\Delta^{-i}(g), e\Delta^{[-m]}\phi_\Delta^{-i}(h))\\ &=^\times e\Delta^{[-m]}\phi_\Delta^{-m}(\gcd(g,h)) =^\times e\Delta^{[-m]}.\end{aligned}$$

In the same way, we find the equality $\gcd(g\Delta, h\Delta) =^{\times} \Delta$, whence

$$\gcd(eg\Delta^{[-m]}, eh\Delta^{[-m]}, g\Delta, h\Delta) =^{\times} \gcd(e\Delta^{[-m]}, \Delta) =^{\times} \operatorname{init}(e),$$

and (iii) follows.

(iv) Since $\mathrm{SSS}(\mathcal{G})$ is closed under left-gcd, we have $H(g) =^{\times} \gcd(g, \Delta) \in \mathrm{SSS}(\mathcal{G})$ for every g in $\mathrm{SSS}(\mathcal{G})$. Therefore, in view of an induction on the $\mathcal{S}$-length, it is sufficient to prove the result for g in $\mathcal{S}$. Now, in the latter case, $\mathrm{sl}^i(g)$ lies in $\mathcal{D}iv(\Delta)$ for every i by (ii), and, by Proposition 2.30(iii), the ultimate periodicity is a direct consequence of the finiteness of $\mathcal{S}$ mod $\mathcal{C}^{\times}$. □

Note that, as seen in the proof of Proposition 2.33(ii), the relation $\mathrm{sl}(\Delta) =^{\times} \Delta$ always holds (with the usual abuse of notation: these two elements Δ need not have the same source). Also, note that, when restricted to $\mathrm{SSS}(\mathcal{G})$, the prefix map pr can then be interpreted as a natural transformation between the identity functor and the sliding functor sl.

Gebhardt-González algorithm for computing sliding circuits. We now show how sliding can be used to solve the Conjugacy Problem in good cases, namely when Condition 2.14 is satisfied and moreover $\mathcal{C}^{\times}$ is finite. Recall that Condition 2.14(i) implies that $\mathcal{C}$ is Noetherian and that it is generated by its atoms and $\mathcal{C}^{\times}$.

Definition 2.34 (sliding circuits). In Context 2.1 and assuming that Condition 2.14 is satisfied, the *category of sliding circuits* $\mathrm{SC}(\mathcal{G})$ is the subcategory of $\mathrm{SSS}(\mathcal{G})$ made of all objects and elements that are left fixed by a large enough (divisible enough) power of sl.

By Proposition 2.30, every object of $\mathcal{C}onj\mathcal{G}$ that is fixed by a power of sl belongs to its super-summit set, hence it is an object of $\mathrm{SC}(\mathcal{G})$. The next result implies that (provided Condition 2.14 is satisfied), the category $\mathrm{SC}(\mathcal{G})$ is the intersection of the images of $\mathrm{SSS}(\mathcal{G})$ under sl^i for $i \geqslant 0$.

Lemma 2.35. *In Context* 2.1 *and assuming that Condition* 2.14 *is satisfied, we have:*

(i) *An object or an element of* $\mathrm{SSS}(\mathcal{G})$ *that is fixed up to right-multiplication by an invertible element by a power of* sl *is fixed by that power.*

(ii) *Sliding is ultimately periodic on objects and on elements of* $\mathrm{SSS}(\mathcal{G})$.

Proof. By Proposition 2.33, sliding is ultimately periodic up to right-multiplication by invertible elements on objects and elements. Hence (ii) is a consequence of (i).

Now, if e lies in $\mathcal{O}bj(\mathcal{C}onj\mathcal{G})$ and we have $\mathrm{sl}^k(e) =^{\times} e$, then, since $\mathrm{sl}(e)$ is a conjugate of e, Condition 2.14(ii) implies $\mathrm{sl}^k(e) = e$, whence (i) for objects.

Next, assume $g \in \mathcal{C}onj\mathcal{G}(e, e^g)$ with $\mathrm{sl}^k(e) = e$ and $\mathrm{sl}^k(g) =^{\times} g$. Let us show that this implies $\mathrm{sl}^k(g) = g$. By the definition of sl, there exist elements f and h in $\mathcal{C}yc\mathcal{G}$ satisfying $\mathrm{sl}^k(e) = e^f$, $\mathrm{sl}^k(e^g) = e^{gh}$, and $\mathrm{sl}^k(g) = f^{-1}gh$. Then we have $g =^{\times} \mathrm{sl}^k(g) = f^{-1}gh$, which implies $g^{-1}fg =^{\times} h$, hence $g^{-1}fg = h$ by Condition 2.14(ii) and, finally, $\mathrm{sl}^k(g) = g$, which completes the proof of (i). □

We now describe an algorithm that computes the whole connected component of a given object e in the category of sliding circuits $\mathrm{SC}(\mathcal{G})$. For this, it is sufficient to compute the atoms of $\mathrm{SC}(\mathcal{G})$ with source e.

Lemma 2.36. *Assume that e is an object of $\mathrm{SC}(\mathcal{G})$. If g and h lie in $\mathrm{SC}(\mathcal{G})(e,\cdot)$, then so does $\gcd(g,h)$.*

Proof. By the definition of $\mathrm{SC}(\mathcal{G})$, there exists $i > 0$ satisfying $\mathrm{sl}^i(g) = g$ and $\mathrm{sl}^i(h) = h$. Since sl preserves left-gcds, we deduce $\mathrm{sl}^i(\gcd(g,h)) =^\times \gcd(g,h)$, whence the result by Lemma 2.35(i). □

By Proposition 2.33, we have $\mathrm{sl}(e, \Delta, \phi_\Delta(e)) = (\mathrm{sl}(e), \Delta, \phi_\Delta(\mathrm{sl}(e)))$ and, therefore, Δ lies in $\mathrm{SC}(\mathcal{G})(e,\cdot)$ for every object e of $\mathrm{SC}(\mathcal{G})$. It then follows from Lemma 2.36 that, given an object e of $\mathrm{SC}(\mathcal{G})$ and an atom s of $\mathcal{C}$ with the same source as e, there is a $\preccurlyeq$-minimal element g satisfying $s \preccurlyeq g$ and $g \in \mathrm{SC}(\mathcal{G})(x,\cdot)$ which is unique up to right-multiplication by an invertible element.

Notation 2.37 (element $\widehat{s}$). In the above context, for s an atom of $\mathcal{C}$ with the same source as e, we choose a $\preccurlyeq$-minimal right-multiple $\widehat{s}$ of s that lies in $\mathrm{SC}(\mathcal{G})(e,\cdot)$.

Note that $\widehat{s}$ lies in $\mathcal{D}iv(\Delta)$ for every atom s. The atoms of $\mathrm{SC}(\mathcal{G})(e,\cdot)$ are those elements $\widehat{s}\epsilon$, with $\widehat{s}$ as above and ϵ in $\mathcal{C}^\times$, that are not a proper multiple of another element $\widehat{s'}$ and such that $\widehat{s}\epsilon$ lies in $\mathrm{SC}(\mathcal{G})(e,\cdot)$. In particular, $\widehat{s}$ is an atom if it is not a proper multiple of another element $\widehat{s'}$.

Proposition 2.38 (sliding circuits). *In Context 2.1 with Condition 2.14, assume that e is an object of $\mathrm{SC}(\mathcal{G})$ and s is an atom of $\mathcal{C}$ lying in $\mathcal{C}onj\mathcal{C}(e,\cdot)$. Then*

(i) *If, for every k, we have $\mathrm{sl}^k(s) \notin \mathcal{C}^\times$, then either there exists $k > 0$ satisfying $s \preccurlyeq \mathrm{sl}^k(s)$, and then $\widehat{s} =^\times \mathrm{sl}^{ij}(s)$ holds for i and j minimum satisfying $\mathrm{sl}^{ij}(e) = e$ and $\mathrm{sl}^{(i+1)j}(s) = \mathrm{sl}^{ij}(s)$, or $\widehat{s}$ is not an atom of $\mathrm{SC}(\mathcal{G})$.*

(ii) *If $\mathrm{sl}^k(s)$ is invertible for some integer k, then the relation $s \preccurlyeq \mathrm{pr}(e)$ holds.*

Proof. Let us first consider the case when there is some $k > 0$ satisfying $s \preccurlyeq \mathrm{sl}^k(s)$. Since sl preserves left-divisibility, we have $s \preccurlyeq \mathrm{sl}^k(s) \preccurlyeq \mathrm{sl}^{2k}(s) \preccurlyeq \cdots$, whence $s \preccurlyeq \mathrm{sl}^{ik}(s)$ for every i. Since $\mathcal{C}$ is right-Noetherian, this increasing sequence of left-divisors of Δ must be eventually constant up to invertible elements, hence eventually constant by Condition 2.14. Hence there exists h satisfying $\mathrm{sl}^{hk}(s) = \mathrm{sl}^{lk}(s)$ for every $l \geqslant h$. Taking such an l multiple of the period under sliding of e, we deduce $\mathrm{sl}^{lk}(e) = e$, whence $\mathrm{sl}^{hk}(s) = \mathrm{sl}^{lk}(s) \in \mathrm{SC}(\mathcal{G})(e,\cdot)$, and $\widehat{s} \preccurlyeq \mathrm{sl}^{hk}(s)$. Taking moreover l multiple of the period of $\widehat{s}$ under sliding, from $s \preccurlyeq \widehat{s}$ we deduce $\mathrm{sl}^{hk}(s) = \mathrm{sl}^{lk}(s) \preccurlyeq \mathrm{sl}^{lk}(\widehat{s}) = \widehat{s}$. Putting both together we deduce $\widehat{s} =^\times \mathrm{sl}^{hk}(s)$, so we have also $\widehat{s} =^\times \mathrm{sl}^{ij}(s)$ for every i and j satisfying $\mathrm{sl}^{ij}e = e$ and $\mathrm{sl}^{(i+1)j}(s) = \mathrm{sl}^{ij}(s)$ (such i, j exist, taking for j the length of the sliding circuit of the pair (e, s)).

If, for every $k > 0$, we have both $s \not\preccurlyeq \mathrm{sl}^k(s)$ and $\mathrm{sl}^k(s) \notin \mathcal{C}^\times$, let us take k divisible enough so as to ensure $\mathrm{sl}^k(s) \in \mathrm{SC}(\mathcal{G})(e,\cdot)$ and $\mathrm{sl}^k(\widehat{s}) = \widehat{s}$. From $s \preccurlyeq \widehat{s}$ we deduce $\mathrm{sl}^k(s) \preccurlyeq \mathrm{sl}^k(\widehat{s}) = \widehat{s}$, and $s \not\preccurlyeq \mathrm{sl}^k(s)$ then implies $\mathrm{sl}^k(s) \prec \widehat{s}$. As, by assumption, $\mathrm{sl}^k(s)$ is not invertible, $\widehat{s}$ cannot be an atom.

Finally, assume that an element g of $\mathcal{C}onj\mathcal{C}(e,\cdot)$ is such that, for some k, we have $\mathrm{sl}^k(g) =^{\times} 1$ (we do not assume that g is an atom). We then prove, using induction on k, that $\gcd(g, \mathrm{pr}(e))$ is not 1. This will give the result when g is an atom.

Assume first $k = 1$. Then the equalities $1 =^{\times} \mathrm{sl}(g) = \mathrm{pr}(e)^{-1} g\, \mathrm{pr}(e^g)$ imply $g\, \mathrm{pr}(e^g) =^{\times} \mathrm{pr}(e)$, whence $g \preccurlyeq \mathrm{pr}(e)$.

Assume now $k \geqslant 2$. By the induction hypothesis, $\gcd(\mathrm{sl}(g), \mathrm{pr}(\mathrm{sl}e))$ is not invertible. Applying sl to $\mathrm{pr}(e)$, an element of $\mathcal{C}onj\mathcal{G}(e,\cdot)$, we obtain $\mathrm{sl}(\mathrm{pr}(e)) = \mathrm{pr}(\mathrm{sl}(e))$ by definition. Using the compatibility of sl with left-gcds, we deduce

$$\gcd(\mathrm{sl}(g), \mathrm{pr}(\mathrm{sl}e)) = \gcd(\mathrm{sl}(g), \mathrm{sl}(\mathrm{pr}(e))) = \mathrm{sl}(\gcd(g, \mathrm{pr}(e))).$$

As the left-hand term is not invertible, we deduce that neither is $\gcd(g, \mathrm{pr}(e))$. □

When $\mathcal{C}^{\times}$ is finite, which together with the other assumptions implies that $\mathcal{C}$ has a finite number of atoms, Proposition 2.38 leads to an algorithm for computing the atoms of $\mathrm{SC}(\mathcal{G})(e,\cdot)$.

Algorithm 2.39 (atoms of sliding circuits).

Context: A groupoid $\mathcal{G}$ that is the groupoid of fractions of a cancellative category $\mathcal{C}$ that is right-Noetherian, admits a Garside map, satisfies Condition 2.14, and is such that $\mathcal{C}^{\times}$ is finite, algorithms for cycling and decycling objects and elements of $\mathcal{C}$, the atom set $\mathcal{A}$ of $\mathcal{C}$.

Input: An object e of $\mathrm{SC}(\mathcal{G})$

Output: The atoms of $\mathrm{SC}(\mathcal{G})(e,\cdot)$

put $S := \emptyset$
for s in $\mathcal{A} \cap \mathcal{C}onj\mathcal{C}(e,\cdot)$ **do**
  put $p := (e, s)$ and $L := \{p\}$
  while $\mathrm{sl}(p) \notin L$ **do**
    put $p := \mathrm{sl}(p)$
    put $L := L \cup \{p\}$
  put $(e, \epsilon) :=$ last pair in L whose first term is e
  if $s \preccurlyeq \epsilon$ **then**
    put $S := S \cup \{\epsilon\}$
  if $s \preccurlyeq \mathrm{pr}(e)$ **then**
    put $t :=$ least t satisfying $s \preccurlyeq t \preccurlyeq \mathrm{pr}(e)$ and $e^t \in \mathcal{O}bj(\mathrm{SC}(\mathcal{G}))$
    put $S := S \cup \{t\}$
put $S := S \setminus \{s \in S \mid \exists s' {\in} S\, (s \succnsim s')\}$
for s in S and ϵ in $\mathcal{C}^{\times}$ **do**
  if $e^{s\epsilon} \in \mathcal{O}bj(\mathrm{SC}(\mathcal{G}))$ **then**
    put $S := S \cup \{s\epsilon\}$
return S

Proposition 2.40 (atoms of sliding circuits). *If $\mathcal{G}$ is the groupoid of fractions of a cancellative category $\mathcal{C}$ that is right-Noetherian, admits a Garside map, satisfies Condition* 2.14 *and is such that $\mathcal{C}^\times$ is finite, then Algorithm* 2.39 *running on an object e of* $\mathrm{SC}(\mathcal{G})$ *returns the atoms of* $\mathrm{SC}(\mathcal{G})(e, \cdot)$.

Proof. We know that the atoms of $\mathrm{SC}(\mathcal{G})$ are the elements of the form $\widehat{s}\epsilon$, with the same notation as in 2.37. For each atom $\widehat{s}$ of $\mathcal{C}onj\mathcal{C}(e, \cdot)$, if we are in Case (i) of Proposition 2.38, we can compute parameters i and j as in the statement, and then a candidate $\widehat{s}$ for an atom actually is an atom if s left-divides $\mathrm{sl}^{ij}(s)$. This is exactly what Steps 4 to 8 of Algorithm 2.39 do: take an atom s of $\mathcal{C}onj\mathcal{C}(e, \cdot)$ and apply iteratively sl to the pair (e, s) until it stabilizes to a pair (e, ϵ), which happens by Lemma2.35. We are in Case (i) if ϵ is not invertible, and if s left-divides ϵ, which is checked in Step 9. If we are not in Case (i) of Proposition 2.38 and $\widehat{s}$ is an atom, we are in Case (ii) of *loc. cit.* and then $\widehat{s}$ has to be a divisor of $\mathrm{pr}(e)$ since $\mathrm{pr}(e)$ then lies in $\mathrm{SC}(\mathcal{G})$. So, in order to complete the algorithm, it is sufficient to compute which divisors of $\mathrm{pr}(e)$ are in $\mathrm{SC}(\mathcal{G})$, which can be done since this set of divisors is finite and checking if an element lies in $\mathrm{SC}(\mathcal{G})$ is done by iterating sliding until it stabilizes. We then have to retain amongst all our atom candidates those that are not proper right-multiples of others. This is done in Step 14. Other atoms are $=^\times$-equivalent to these. This is done in Steps 16 and 17. □

Note that the conjunction of Condition 2.14 and the finiteness of $\mathcal{C}^\times$ is equivalent to $\mathcal{D}iv(\Delta)$ being finite.

Once the atoms of a sliding circuit are known, it is easy to compute the full component of an object in $\mathrm{SC}(\mathcal{G})$ and to deduce a new solution to the Conjugacy Problem.

Algorithm 2.41 (sliding circuits).

Context: A groupoid $\mathcal{G}$ that is the groupoid of fractions of a cancellative category $\mathcal{C}$ that is right-Noetherian, admits a Garside map, satisfies Condition 2.14, and is such that $\mathcal{C}^\times$ is finite

Input: An object e of $\mathrm{SC}(\mathcal{G})$

Output: The connected component of e in $\mathrm{SC}(\mathcal{G})$

put $X := \{e\}$
for f in X **do**
    compute the atom set S of $\mathrm{SC}(\mathcal{G})(f, \cdot)$ using Algorithm 2.39
    for s in S **do**
        if $f^s \notin X$ **then**
            put $X := X \cup \{f^s\}$
    for ϵ in $\mathcal{C}^\times \cap \mathcal{C}onj(f, \cdot)$ **do**
        if $f^\epsilon \in \mathrm{SC}(\mathcal{G}) \setminus X$ **then**
            put $X := X \cup \{f^s\}$
until no new element of X has been found
return X

Algorithm 2.42 (Conjugacy Problem II).

Context: A groupoid $\mathcal{G}$ that is the groupoid of fractions of a cancellative category $\mathcal{C}$ that is right-Noetherian, admits a Garside map, satisfies Condition 2.14, and is such that $\mathcal{C}^{\times}$ is finite
Input: Two elements d, e of $\mathcal{G}^{\circlearrowright}$
Output: YES if d and e are conjugate, NO otherwise

```
find d' in a sliding circuit of d by iteratively sliding d until it becomes periodic
find e' in a sliding circuit of e by iteratively sliding e until it becomes periodic
compute X := the connected component of e' in SC(G) using Algorithm 2.41
if d' ∈ X then
    return YES
else
    return NO
```

Proposition 2.43 (Conjugacy Problem II). *If $\mathcal{G}$ is the groupoid of fractions of a right-Noetherian cancellative category $\mathcal{C}$ that admits a Garside map and Condition* 2.14 *holds and $\mathcal{C}^{\times}$ is finite, then Algorithm* 2.42 *solves the Conjugacy Problem for $\mathcal{G}$.*

Proof. First, Algorithm 2.39 computes all atoms of the component of $\mathrm{SC}(\mathcal{G})$ containing a given object. Since the category $\mathrm{SC}(\mathcal{G})$ is generated by its atoms and its invertible elements, conjugating recursively by the atoms, which are in finite number, and then by the invertible elements, which are also in finite number, computes the whole set of sliding circuits of a conjugacy class. This is what Algorithm 2.41 does (it terminates since this set is finite). Now the set of sliding circuits is characteristic of a conjugacy class. Indeed, if e and d are in $\mathrm{SC}(\mathcal{G})$ and conjugate, since sliding is ultimately periodic they are conjugate by an element fixed some power of sl. Hence to know if two elements e and d are conjugate it is sufficient to compute the set$X$ of sliding circuits of the conjugacy class of e, to compute an object d' of the sliding circuits of the conjugacy class of d, and to check if d' is in X. This is exactly what Algorithm 2.42 does. One first obtains e' conjugate to e (*resp.* d' conjugate to d) lying in a sliding circuit by iteratively applying sliding to e (*resp.* d), which is doable by Lemma 2.35. Then Algorithm 2.41 applied to e' gives the full set X of objects of the sliding circuits of the conjugacy class of e. The final test is to check whether d' lies in X. □

Example 2.44 (Conjugacy Problem II). Let us consider the braid β of Example 2.21. The delta-normal decompositions of β and β^{-1}, respectively, are

$$\sigma_2\sigma_1\sigma_4|\sigma_1\sigma_4 \quad \text{and} \quad \Delta^{-2}\|\sigma_2\sigma_1\sigma_3\sigma_2\sigma_1\sigma_4\sigma_3\sigma_2|\sigma_2\sigma_1\sigma_3\sigma_2\sigma_1\sigma_4\sigma_3.$$

Using the fact that Δ^2 is central in the braid monoid, we thus find $\mathrm{init}(\beta) = \sigma_2\sigma_1\sigma_4$ and $\mathrm{init}(\beta^{-1}) = \sigma_2\sigma_1\sigma_3\sigma_2\sigma_1\sigma_4\sigma_3\sigma_2$. The element $\sigma_2\sigma_1$ left-divides $\mathrm{init}(\beta)$ and $\mathrm{init}(\beta^{-1})$.

As $\sigma_2\sigma_1\sigma_4$ does not left-divide $\mathrm{init}(\beta^{-1})$, we deduce that $\mathrm{pr}(\beta)$ is equal to $\sigma_2\sigma_1$. We obtain $\mathrm{sl}(\beta) = \sigma_1\sigma_2\sigma_1\sigma_4\sigma_4$. The computation gives $\mathrm{slsl}(\beta) = \mathrm{sl}(\beta)$. Hence $\beta' = \mathrm{sl}(\beta)$ is an object of $\mathrm{SC}(\mathcal{G})$.

Then Algorithm 2.41 gives that the set of objects of the connected component of $\mathrm{SC}(\mathcal{G})$ containing β' is $\{\beta', \sigma_1\sigma_3\sigma_4\sigma_3\sigma_1\}$. Comparing with the full super-summit set of β as given in Example 2.21 suggests that the sliding approach is much more efficient than the cycling approach.

Left- and right-slidings. To complete this section, we now show that the preferred choice of left-divisors and the resulting normal decompositions does not really matter when defining sliding: choosing right-divisors and the resulting co-normal decompositions as in Subsection V.3.4 leads to isomorphic sliding circuits.

What was developed above will be called *left-sliding*. In order to describe *right-sliding*, a symmetric version, we first need a right-transversal in $\mathcal{D}iv(\Delta)$: we define it as the image of the function $\widetilde{H}$ on $\mathcal{D}iv(\Delta)$ given by $\widetilde{H}(e) = H(\Delta e^{-1})^{-1}\Delta$. The following lemma results from a simple computation.

Lemma 2.45. *For every e in $\mathcal{D}iv(\Delta)$ and every ϵ in $\mathcal{C}^{\times}$, we have $\widetilde{H}(e) \;{}^{\times}\!\!= e$ and, whenever ϵe is defined, $\widetilde{H}(\epsilon e) = \widetilde{H}(e)$.*

Definition 2.46 (right-sliding). For e in $\mathcal{G}^{\circlearrowleft}$, the *suffix* $\widetilde{\mathrm{pr}}(e)$ is defined by

$$\widetilde{\mathrm{pr}}(e) = \widetilde{H}(\widetilde{\mathrm{gcd}}(\Delta^{[-\inf_\Delta(e)]}e, \Delta^{[\sup_\Delta(e)]}e^{-1}, \Delta)),$$

and the *right-sliding* $\widetilde{\mathrm{sl}}(e)$ of e is the left-conjugate ${}^{\widetilde{\mathrm{pr}}(e)}e$ of e.

We extend right-sliding from objects of $\mathcal{C}onj\mathcal{G}$ to elements of $\mathcal{C}onj\mathcal{G}$ by defining, for g in $\mathcal{C}onj\mathcal{C}(e, e^g)$, the element $\widetilde{\mathrm{sl}}(g)$ to be $\widetilde{\mathrm{pr}}(e)\, g\, \widetilde{\mathrm{pr}}(e^g)^{-1}$. A computation similar to that in the proof of Proposition 2.33 shows that $\widetilde{\mathrm{sl}}(g)$ is then well defined. The map $\widetilde{\mathrm{sl}}$ is compatible with composition, so $\widetilde{\mathrm{sl}}$ provides a functor from $\mathrm{SSS}(\mathcal{G})$ to itself.

In the rest of this section, we shall work with the following additional assumption.

Condition 2.47. *The sharp head function H is compatible with left- and right-complement to Δ in the sense that $\Delta = g'g$ implies $H(g') = g' \Leftrightarrow H(g) = g$.*

Condition 2.47 is satisfied in a category that contains no nontrivial invertible element, and also in the case of the semi-direct product of such a category $\mathcal{C}$ by an automorphism ϕ as in Subsection 1.3 when H is extended to $\mathcal{C}\phi$ by $H(g\phi^i) = H(g)$ (see the last paragraph of Subsection 2.2). Condition 2.47 will be used through its following consequence.

Lemma 2.48. *In Context* 2.1, *if Condition* 2.47 *is satisfied, then* $\mathrm{Im}H$ *and* $\mathrm{Im}\widetilde{H}$ *coincide.*

In other words, the left- and right-transversals of $\mathcal{D}iv(\Delta)$ we consider are the same.

Proof. We first show that $\mathrm{Im}\widetilde{H}$ is included in $\mathrm{Im}H$. Let g belong to $\mathrm{Im}\widetilde{H}$. Then we have $g = \widetilde{H}(g) = H(\Delta g^{-1})^{-1}\Delta$, hence $H(\Delta g^{-1})g = \Delta$ whence, by Condition 2.47, $H(g) = g$. Hence g belongs to $\mathrm{Im}H$.

Conversely, assume $g = H(g)$. We want to show $g = \widetilde{H}(g)$. Assume $\Delta = g'g$. By Condition 2.47, we have $H(g') = g'$, hence

$$\widetilde{H}(g) = H(\Delta g^{-1})^{-1}\Delta = H(g')^{-1}\Delta = g'^{-1}\Delta = g. \qquad \square$$

Proposition 2.49 (left- vs. right-sliding). *In Context* 2.1, *if Condition* 2.47 *is satisfied, then, for every* n, *the maps* $\widetilde{\mathrm{sl}}^n$ *from* $\mathrm{Im}(\mathrm{sl}^n)$ *to* $\mathrm{Im}(\widetilde{\mathrm{sl}}^n)$ *and* sl^n *from* $\mathrm{Im}(\widetilde{\mathrm{sl}}^n)$ *to* $\mathrm{Im}(\mathrm{sl}^n)$ *are reciprocal isomorphisms when restricted to* $\mathrm{SSS}(\mathcal{G})$.

Before entering the proof, we inductively define the *k-th iterated prefix* pr_k and the *k-th iterated suffix* $\widetilde{\mathrm{pr}}_k$, by $\mathrm{pr}_1 = \mathrm{pr}$, $\widetilde{\mathrm{pr}}_1 = \widetilde{\mathrm{pr}}$ and $\mathrm{pr}_k(e) = \mathrm{pr}_{k-1}(e)\mathrm{pr}(\mathrm{sl}^{k-1}(e))$ and $\widetilde{\mathrm{pr}}_k(e) = \widetilde{\mathrm{pr}}(\widetilde{\mathrm{sl}}^{k-1}(e))\widetilde{\mathrm{pr}}_{k-1}(e)$ for $k > 1$. Then the equalities

$$\mathrm{sl}^k(e) = e^{\mathrm{pr}_k(e)} \quad \text{and} \quad \widetilde{\mathrm{sl}}^k(e) = {}^{\widetilde{\mathrm{pr}}_k(e)}e$$

hold by definition.

Lemma 2.50. *For every* e *in* $\mathrm{SSS}(\mathcal{C})$ *and every* k, *we have*

$$\widetilde{\mathrm{pr}}_k(e) \preccurlyeq \mathrm{pr}_k(\widetilde{\mathrm{sl}}^k(e)) \quad \textit{and} \quad \widetilde{\mathrm{pr}}_k(\mathrm{sl}^k(e)) \succcurlyeq \mathrm{pr}_k(e).$$

Proof. We shall prove the second property using induction on k. The proof of the first one is similar by reversing the arrows. Let

$$m = \inf_\Delta(e) = \inf_\Delta(\mathrm{sl}(e)) \quad \text{and} \quad m + \ell = \sup_\Delta(e) = \sup_\Delta(\mathrm{sl}(e)).$$

By definition, we have

$$\widetilde{\mathrm{pr}}(e) \mathrel{{}^{\times}{=}} \widetilde{\gcd}(\Delta^{[-m]}e, \Delta^{[m+\ell]}e^{-1}, \Delta) \quad \text{and} \quad \mathrm{pr}(e) =^{\times} \gcd(e\Delta^{[-m]}, e^{-1}\Delta^{[m+\ell]}, \Delta).$$

In particular, $\mathrm{pr}(e)$ left-divides $e\Delta^{[-m]}$, which, conjugating by $\Delta^{[m]}$, implies that the element $\Delta^{[-m]}\mathrm{pr}(e)^{-1}e$ lies in $\mathcal{C}$. Similarly, $\mathrm{pr}(e) \preccurlyeq e^{-1}\Delta^{[m+\ell]}$ implies that the element $\Delta^{[m+\ell]}\mathrm{pr}(e)^{-1}e^{-1}$ lies in $\mathcal{C}$. Now, applying the formula for $\widetilde{\mathrm{pr}}$ to $\mathrm{sl}(e)$ gives

$$\begin{aligned}\widetilde{\mathrm{pr}}(\mathrm{sl}(e)) &\mathrel{{}^{\times}{=}} \widetilde{\gcd}(\Delta^{[-m]}\mathrm{sl}(e), \Delta^{[m+\ell]}\mathrm{sl}(e)^{-1}, \Delta)\\ &\mathrel{{}^{\times}{=}} \widetilde{\gcd}(\Delta^{[-m]}\mathrm{pr}(e)^{-1}e\,\mathrm{pr}(e), \Delta^{[m+\ell]}\mathrm{pr}(e)^{-1}e^{-1}\mathrm{pr}(e), \Delta),\end{aligned}$$

whence $\widetilde{\mathrm{pr}}(\mathrm{sl}(e)) \succcurlyeq \mathrm{pr}(e)$ by applying the above positivity statements and using that $\mathrm{pr}(e)$ is in $\mathcal{S}$, hence right-divides Δ. This completes the proof for $k = 1$.

Assume now $k > 1$. The induction hypothesis applied to $\mathrm{sl}(e)$ instead of to e gives $\widetilde{\mathrm{pr}}_{k-1}(\mathrm{sl}^k(e)) \succcurlyeq \mathrm{pr}_{k-1}(\mathrm{sl}(e))$. The element g satisfying $\widetilde{\mathrm{pr}}_{k-1}(\mathrm{sl}^k(e)) = g\,\mathrm{pr}_{k-1}(\mathrm{sl}(e))$ can be interpreted as an element of $\mathit{Conj}\mathcal{C}(\widetilde{\mathrm{sl}}^{k-1}(\mathrm{sl}^k(e)), \mathrm{sl}(e))$. We then have the commutative diagram

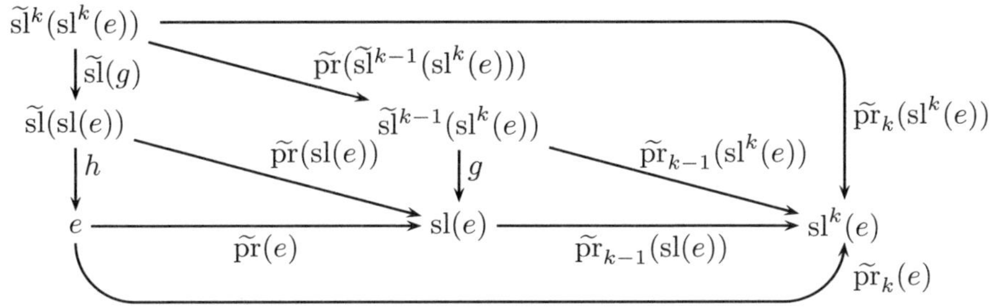

where h comes from the property $\widetilde{\mathrm{pr}}(\mathrm{sl}(e)) \succcurlyeq \mathrm{pr}(e)$. Since all the arrows are given by positive elements, we obtain the expected result. □

Proof of Proposition 2.49. We claim that, for e in $\mathrm{SSS}(\mathcal{G})$ and for every n, we have $\mathrm{sl}^n\widetilde{\mathrm{sl}}^n\mathrm{sl}^n(e) = \mathrm{sl}^n(e)$. By Lemma 2.50, $\widetilde{\mathrm{pr}}_n(\mathrm{sl}^n(e))$ is a left-multiple of $\mathrm{pr}_n(e)$. Assume that g satisfies $g\,\mathrm{pr}_n(e) = \widetilde{\mathrm{pr}}_n(\mathrm{sl}^n(e))$. The commutative diagram of Figure 2 shows that the iterated sliding $\mathrm{sl}^n(g)$ satisfies

$$\mathrm{pr}_n(\widetilde{\mathrm{sl}}^n\mathrm{sl}^n(e))\mathrm{sl}_n(g) = \widetilde{\mathrm{pr}}_n(\mathrm{sl}^n(e)).$$

Thus $\mathrm{pr}_n(\widetilde{\mathrm{sl}}^n\mathrm{sl}^n(e))$ left-divides $\widetilde{\mathrm{pr}}_n(\mathrm{sl}^n(e))$. By Lemma 2.50, $\widetilde{\mathrm{pr}}_n(\mathrm{sl}^n(e))$ left-divides $\mathrm{pr}_n(\widetilde{\mathrm{sl}}^n\mathrm{sl}^n(e))$, hence we must have $\widetilde{\mathrm{pr}}_n(\mathrm{sl}^n(e)) =^{\times} \mathrm{pr}_n(\widetilde{\mathrm{sl}}^n\mathrm{sl}^n(e))$. By Condition 2.47, this implies $\widetilde{\mathrm{pr}}_n(\mathrm{sl}^n(e)) = \mathrm{pr}_n(\widetilde{\mathrm{sl}}^n\mathrm{sl}^n(e))$, hence $\mathrm{sl}^n(g) = 1$, whence our claim.

We have shown that $\mathrm{sl}^n\widetilde{\mathrm{sl}}^n$ is the identity on $\mathrm{sl}^n(\mathrm{SSS}(\mathcal{G}))$. In the same way, $\widetilde{\mathrm{sl}}^n\mathrm{sl}^n$ is the identity on $\widetilde{\mathrm{sl}}^n(\mathrm{SSS}(\mathcal{G}))$. This completes the argument. □

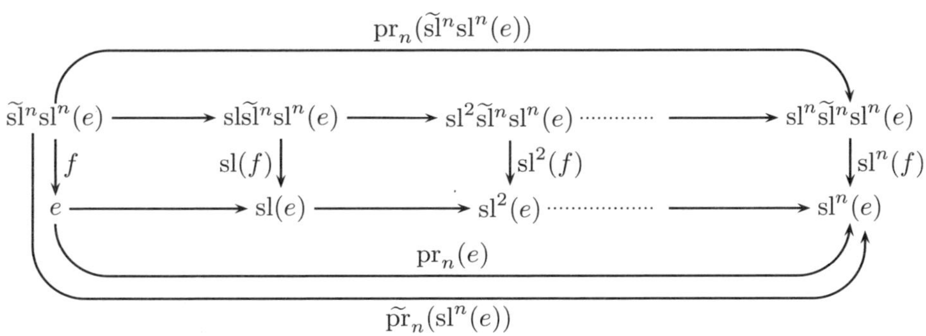

Figure 2. Proof of Proposition 2.49.

The following consequence is straightforward.

Corollary 2.51 (image of sl^n). *In Context* 2.1 *with* 2.47, *for every e in* $\mathrm{SSS}(\mathcal{G})$ *we have*

$$\exists e' \in \mathrm{SSS}(\mathcal{C})\,(e = \mathrm{sl}^n(e')) \;\Leftrightarrow\; \mathrm{sl}^n\widetilde{\mathrm{sl}}^n(e) = e,$$
$$\exists e' \in \mathrm{SSS}(\mathcal{C})\,(e = \widetilde{\mathrm{sl}}^n(e')) \;\Leftrightarrow\; \widetilde{\mathrm{sl}}^n\mathrm{sl}^n(e) = e.$$

Notation 2.52 ($\mathrm{SC}(\mathcal{G})(e)$, $\widetilde{\mathrm{SC}}(\mathcal{G})(e)$)**.** For e belonging to its super-summit set, we denote by $\mathrm{SC}(\mathcal{G})(e)$ the full subcategory of $\mathrm{SC}(\mathcal{G})$ whose objects are in the conjugacy class of e. Similarly, we define $\widetilde{\mathrm{SC}}(\mathcal{G})(e)$ to be the subcategory of $\mathrm{SSS}(\mathcal{C})$ fixed by a large enough (divisible enough) power of $\widetilde{\mathrm{sl}}$ and whose objects are in the conjugacy class of e.

Corollary 2.53 ($\mathrm{SC}(\mathcal{G})(e)$ *vs.* $\widetilde{\mathrm{SC}}(\mathcal{G})(e)$)**.** *In Context* 2.1 *with* 2.47, *for every e belonging to its super-summit set and for n large enough,* sl^n *and* $\widetilde{\mathrm{sl}}^n$ *are mutually inverse isomorphisms between* $\mathrm{SC}(\mathcal{G})(e)$ *and* $\widetilde{\mathrm{SC}}(\mathcal{G})(e)$.

Proof. Since $\mathcal{S}$ is finite modulo $\mathcal{C}^\times$, the categories $\mathrm{SC}(\mathcal{G})(e)$ and $\widetilde{\mathrm{SC}}(\mathcal{G})(e)$ are finite. Hence, for n large enough, the image of sl^n is $\mathrm{SC}(\mathcal{G})(e)$ and the image of $\widetilde{\mathrm{sl}}^n$ is $\widetilde{\mathrm{SC}}(\mathcal{G})(e)$. The result is then a direct consequence of Proposition 2.49. □

Twisted conjugacy. Let us consider again the context of a semi-direct product by an automorphism. For ϕ an automorphism of the category $\mathcal{C}$, we define ϕ-sliding as corresponding to the sliding of elements of the form $e\phi$ in $\mathcal{C} \rtimes \langle\phi\rangle$. As already mentioned, Conditions 2.14 and 2.47 are automatically satisfied in $\mathcal{C} \rtimes \langle\phi\rangle$ if $\mathcal{C}$ has no nontrivial invertible element. Indeed, for the former, if $e\phi^m$ is conjugate to $d\phi^n$, then we must have $\phi^m = \phi^n$ so two conjugate elements cannot differ by an invertible element, and, for the later, if Δ is the product of two elements of $\mathcal{S}(\mathcal{C} \rtimes \langle\phi\rangle)^\times$, it must be a decomposition of the form $\Delta = (g\phi^m)(\phi^{-m}g')$ with g, g' in $\mathcal{S}$; now $H(g\phi^m) = g\phi^m$ implies $\phi^m = \mathrm{id}$, whence $H(\phi^{-m}g') = \phi^{-m}g'$. Thus all results of the previous section hold for $\mathcal{C} \rtimes \langle\phi\rangle$ when $\mathcal{C}$ has no nontrivial invertible element.

3 Conjugacy classes of periodic elements

In this section, we study the conjugacy classes of periodic elements, defined as the roots of a power of Δ. The main result is Proposition 3.34, which precisely describes the conjugacy classes of all periodic elements.

The section contains three subsections. First, we introduce periodic elements and establish results in the special case of roots of $\Delta^{[2]}$ (Subsection 3.1). Next, we develop in Subsection 3.2 a geometric approach involving the subfamily of the considered category consisting of all elements that are not divisible by Δ and viewing this family as a negatively curved space on which the category and its enveloping groupoid act. These methods are then used in Subsection 3.3 to carry out a more complete analysis of periodic elements and of their conjugates.

3.1 Periodic elements

Once again, we fix a common context for the sequel.

Context 3.1. • $\mathcal{C}$ is a Noetherian and cancellative category;
• Δ is a Garside map in $\mathcal{C}$;
• $\mathcal{G}$ is the enveloping groupoid of $\mathcal{C}$;

Definition 3.2 (periodic). For $p, q \geqslant 1$, an element e of $\mathcal{G}^{\circlearrowleft}$ is called (p, q)-*periodic* if the equality $e^p =^{\times} \Delta^{[q]}$ holds.

Note that, by definition, every iteration $\Delta^{[q]}$ of the considered map Δ is periodic, and that, if e is (p, q)-periodic, it is also (np, nq)-periodic for every non-zero integer n.

Example 3.3 (periodic). In a free abelian monoid based on a set I and equipped with the smallest Garside element Δ_I (Reference Structure 1, page 3), the only periodic elements are the powers of Δ_I. By contrast, in the braid monoid B_n^+ equipped with the smallest Garside element Δ_n (Reference Structure 2, page 5), for every permutation $(i_1, i_2, ..., i_{n-1})$ of $\{1, 2, ..., n-1\}$, the element $\sigma_{i_1}\sigma_{i_2}\cdots\sigma_{i_{n-1}}$ is periodic; more precisely we have $(\sigma_{i_1}\sigma_{i_2}\cdots\sigma_{i_n-1})^n = \Delta_n^2$, so these elements are $(n, 2)$-periodic.

Lemma 3.4. *In Context* 3.1, *if, moreover,* (i) ϕ_Δ *has finite order, or* (ii) *there is no non-trivial invertible element in* $\mathcal{C}$, *then, if* e *is* (p, q)*-periodic, the fraction* p/q *depends only on* e *and not on the choice of* p *and* q.

Proof. Assume that e is (p, q)- and (p', q')-periodic. Write $e^p =^{\times} \Delta^{[q]}$ and $e^{p'} =^{\times} \Delta^{[q']}$. If the order of ϕ_Δ is finite, raising these equalities to suitable powers, we can assume that ϕ_Δ^q and $\phi_\Delta^{q'}$ are the identity. In any case, we have $e^{r'p+rp'} =^{\times} \Delta^{[r'q+rq']}$ for all integers r and r' (this is a true equality in case (ii)). In particular, we can choose r and r' such that $r'p+rp'$ is the gcd p'' of p and p'. Raising the equality $e^{p''} =^{\times} \Delta^{[r'q+rq']}$ to the powers p/p'' and p'/p'', using that $\phi_\Delta^{r'q+rq'}$ is the identity in case (i), and identifying the resulting equalities with $e^p =^{\times} \Delta^{[q]}$ and $e^{p'} =^{\times} \Delta^{[q']}$ respectively, we obtain $q = (r'q + rq')p/p''$ and $q' = (r'q + rq')p'/p''$, whence $p/q = p'/q'$. □

If at least one of the assumptions of Lemma 3.4 is satisfied, we call p/q the *period* of e. Note that, if ϕ_Δ has finite order, then a conjugate of a periodic element is periodic with the same period, although the minimal pair (p, q) may change. Indeed, if we have $e^p = \Delta^{[q]}\epsilon$ with ϵ in $\mathcal{C}^\times$ and q large enough to satisfy $\phi_\Delta^p = \mathrm{id}$, then a conjugate e^g satisfies $(e^g)^p = \Delta^{[q]}\epsilon^g =^{\times} \Delta^{[q]}$. We shall see that two periodic elements with the same period are conjugate if and only if they are cyclically conjugate; we will also be able to describe the centralizers of periodic elements. Note that, if ϕ_Δ has finite order d, the centralizer of an element in $\mathcal{G}$ is obtained from its centralizer in $\mathcal{C}$ by multiplying by powers of Δ^d.

We first show by elementary computations that a $(p, 2)$-periodic element of $\mathcal{C}$ is the same, up to cyclic conjugacy, as a $(p/2, 1)$-periodic element when p is even, and obtain a related characterization when p is odd. We will prove similar results for more general fractions in Subsection 3.3 using geometric methods.

Proposition 3.5 (**$(p,2)$-periodic**). *In Context* 3.1, *if e is an element of $\mathcal{C}^{\circlearrowleft}$ satisfying $e^p =^{\times} \Delta^{[2]}$ for some positive integer p, and putting $r = \lfloor p/2 \rfloor$:*

(i) *Some cyclic conjugate d of e satisfies $d^p =^{\times} \Delta^{[2]}$ and $d^r \in \mathcal{D}iv(\Delta)$.*

(ii) *Furthermore, if p is even, we have $d^r =^{\times} \Delta$, and, if p is odd, there exists s in $\mathcal{D}iv(\Delta)$ satisfying $d^r s = \Delta$ and $d = s\phi_\Delta(s)\epsilon$, where ϵ is the element of $\mathcal{C}^{\times}$ satisfying $d^p = \Delta^{[2]}\epsilon$.*

Proof. (i) We prove using increasing induction on i that, for $i \leqslant p/2$, there exists g in $\mathcal{C}yc\mathcal{C}(e,\cdot)$ such that $(e^g)^i$ belongs to $\mathcal{D}iv(\Delta)$ and $(e^g)^p$ belongs to $\Delta^{[2]}\mathcal{C}^{\times}$. We start the induction with $i = 0$, where the result holds trivially with $g = 1$.

Suppose now $i > 0$ and $i + 1 \leqslant p/2$. We assume the result for i and will prove it for $i + 1$. We thus have an element g for step i, thus replacing e by e^g we assume $e^i \in \mathcal{D}iv(\Delta)$ and $e^p \in \Delta^{[2]}\mathcal{C}^{\times}$; we will conclude by finding g in $\mathcal{S}$ satisfying $g \preccurlyeq e$ and $(e^g)^{i+1} \in \mathcal{D}iv(\Delta)$ and $(e^g)^p \in \Delta^{[2]}\mathcal{C}^{\times}$. If e^{i+1} left-divides Δ, we have the desired result with $g = 1$. We may thus assume $\sup_\Delta(e^{i+1}) \geqslant 2$. Since e^{i+1} left-divides $\Delta^{[2]}$, we have actually $\sup_\Delta(e^{i+1}) = 2$ by Proposition III.1.62 (length). Let $e^i g | e'$ be a (strict) normal decomposition of e^{i+1}, where $e^i g$ belongs to $\mathcal{S}$ and e' belongs to $\mathcal{D}iv(\Delta)$. We have

$$e^i g e'(e^i g) \preccurlyeq e^i g e'(e^i g e') = e^{2(i+1)} \preccurlyeq e^p =^{\times} \Delta^{[2]},$$

hence we still have $2 = \sup_\Delta((e^i g)e'(e^i g)) = \sup_\Delta((e^i g)e')$. By Proposition IV.1.37 (second domino rule, alternative form), $e'(e^i g)$ must lie in $\mathcal{D}iv(\Delta)$. Then we deduce $e'(e^i g) = e'((ge')^i)g = (e^g)^{i+1}$, an element of $\mathcal{D}iv(\Delta)$, and $g \preccurlyeq e$ follows.

So g will do if we can show $(e^g)^p \in \Delta^{[2]}\mathcal{C}^{\times}$. Now, e^p is $\Delta^{[2]}\epsilon$ with ϵ invertible, hence e commutes with $\Delta^{[2]}\epsilon$, hence so does e^{i+1}, that is, we have $\phi^2_\Delta(e^{i+1})\epsilon = \epsilon e^{i+1}$ and, equivalently, $\phi^2_\Delta(e^i g)\phi^2_\Delta(e')\epsilon = \epsilon e^i g e'$. Now $\phi^2_\Delta(e^i g) | \phi^2_\Delta(e')\epsilon$ is a (non-necessarily strict) normal decomposition and, since $e^i g | e'$ is also a normal decomposition, Proposition III.1.25 (normal unique) ensures that there exists an invertible element ϵ' satisfying $\phi^2_\Delta(e^i g)\epsilon' = \epsilon e^i g$. Then we have

$$e^i \Delta^{[2]} \phi^2_\Delta(g)\epsilon' = \Delta^{[2]}\phi^2_\Delta(e^i g)\epsilon' = \Delta^{[2]}\epsilon e^i g = e^i \Delta^{[2]}\epsilon g,$$

the last equality because e^i commutes with $\Delta^{[2]}\epsilon$. Cancelling $\Delta^{[2]}$, we find $\phi^2_\Delta(g)\epsilon' = \epsilon g$, whence

$$g(e^g)^p = e^p g = \Delta^{[2]}\epsilon g = \Delta^{[2]}\phi^2_\Delta(g)\epsilon' = g\Delta^{[2]}\epsilon'.$$

Therefore, the result follows by left-cancelling g.

(ii) First, $d^r \in \mathcal{D}iv(\Delta)$ implies the existence of s in $\mathcal{D}iv(\Delta)$ satisfying $d^r s = \Delta$. Assuming $d^p = \Delta^{[2]}\epsilon$ with ϵ invertible, we deduce $d^r s \Delta \epsilon = \Delta^{[2]}\epsilon = d^p$, whence by cancellation $s\Delta\epsilon = d^r d^i$ with $i = 1$ if p is odd, and $i = 0$ if p is even. We deduce the equalities $d^r d^i = s\Delta\epsilon = \Delta\phi_\Delta(s)\epsilon = d^r s\phi_\Delta(s)\epsilon$ and, finally, $s\phi_\Delta(s)\epsilon = d^i$.

If p is odd, we obtain the expected result and, if p is even, we obtain that $s\phi_\Delta(s)$ is invertible, hence so is s and d^r belongs to $\Delta\mathcal{C}^{\times}$. □

Example 3.6 (**$(p, 2)$-periodic**)**.** Let B_n be the braid group on n strands (Reference Structure 2, page 5). The square Δ_n^2 of the Garside element Δ_n is in the center of B_n (in fact, it generates the center), and we have $\phi_\Delta^2 = \mathrm{id}$. If a braid β is a d-th root of Δ_n^2 for some positive integer d, Proposition 3.5 implies that b is cyclically conjugate to a positive braid β_1 whose powers β_1^i are simple braids (that is, positive braids that divide Δ_n in B_n^+) for all $i \leqslant d/2$. Moreover, if d is even, one has $\beta_1^{d/2} = \Delta_n$ and, if d is odd, there exists a (positive) simple braid β' such that $\Delta_n = \beta_1^{(d-1)/2}\beta'$ and $\beta_1 = \beta'\phi_{\Delta_n}(\beta')$.

Let us give a concrete example: take $\beta = \sigma_5\sigma_1\sigma_2\sigma_3\sigma_4\sigma_5$ in B_6. Then β is a 5th root of Δ_6^2, but β^2 does not divide Δ_6. The braid $\beta_1 = \sigma_1\sigma_2\sigma_5\sigma_4\sigma_3\sigma_2$ is cyclically conjugate to β (see Exercise 98) and satisfies $\beta_1^2\beta' = \Delta_6$ with $\beta' = \sigma_5\sigma_4\sigma_3$ and one has $\beta'\phi_{\Delta_6}(\beta') = \beta_1$.

The twisted case: ϕ-periodic elements. Let us apply Proposition 3.5 to a semi-direct product category $\mathcal{C} \rtimes \langle\phi\rangle$ in the case when $\mathcal{C}$ has no nontrivial invertible element and the Garside family $\mathcal{S}$ is invariant under ϕ. Then an element $e\phi$ of $(\mathcal{C}\phi)^{\circlearrowleft}$ is (p, q)-periodic if and only if the equality $(e\phi)^p = \Delta^{[q]}\phi^p$ holds. Proposition 3.5 then shows that, for $q = 2$, the element e must be cyclically ϕ-conjugate to an element e' satisfying $(e'\phi)^r \in \mathcal{S}\phi^r$ and $(e'\phi)^p = \Delta^{[2]}\phi^p$. The latter equality shows that the conjugating element must be fixed under $\phi_\Delta^2\phi^p$.

For $p = 2r$, we have $(e'\phi)^r = \Delta\phi^r$. For $p = 2r + 1$, we have $(e'\phi)^r h = \Delta$ for some h, whence $h = s\phi^{-r}$ for some s in $\mathcal{S}$, and $e'\phi = s\phi(\phi_\Delta\phi^r)(s)$. Further, since $e'\phi$ commutes with $\Delta^{[2]}\phi^p$ and Δ is fixed under ϕ, the equality $(e'\phi)^r s = \Delta\phi^r$ implies that s commutes with $\Delta^{[2]}\phi^p$.

3.2 Geometric methods

In order to extend the above results to arbitrary exponents of Δ, we shall now introduce geometric ideas and consider geodesic walks in some subfamily (or quotient) of the considered groupoid that behaves like a negatively curved space. Our framework remains the same as previously, with the exception that no Noetherianity assumption is included at first. On the other hand, we add the mild assumption that the ambient category is *small*, that is, its families of objects and elements are sets (and not proper classes). This assumption will be used to guarantee the existence of a convenient height function.

Context 3.7. • $\mathcal{C}$ is a small cancellative category.
- Δ is a Garside map in $\mathcal{C}$.
- $\mathcal{G}$ is the enveloping groupoid of $\mathcal{C}$.

We recall from Definition V.3.23 and Proposition V.3.24 (interval) that, for g in $\mathcal{G}$, the parameter $\inf_\Delta(g)$ is the largest integer k satisfying $\Delta^{[k]} \preccurlyeq g$, whereas $\sup_\Delta(g)$ is the least integer k satisfying $g \preccurlyeq \Delta^{[k]}$; the canonical length $\ell_\Delta(g)$ is then $\sup_\Delta(g) - \inf_\Delta(g)$. If g is not of the form $\Delta^{[m]}\epsilon$ with ϵ in $\mathcal{C}^\times$, the canonical length of g is the unique integer ℓ such that g admits a strict Δ-normal decomposition of the form $\Delta^{[m]}\|g_1|\cdots|g_\ell$.

Notation 3.8 (relation $=_\Delta$, family $\mathcal{G}^0$, element g^0). In Context 3.7, we denote by $\mathcal{G}^0$ the family $\{g \in \mathcal{G} \mid \inf_\Delta(g) = 0\}$. For g, g' in $\mathcal{G}$, we write $g =_\Delta g'$ for $\exists q \in \mathbb{Z}\,(g' = g\Delta^{[q]})$, and we put $g^0 = g\Delta^{[-\inf_\Delta(g)]}$.

By the results recalled above, $\mathcal{G}^0$ is also $\mathcal{C} \setminus \Delta\mathcal{C}$, that is, the family of all g in $\mathcal{C}$ that are not left-divisible by Δ. Our starting point is the following straightforward observation.

Lemma 3.9. *In Context* 3.7, *g^0 is, for every g in $\mathcal{G}$, the unique element of $\mathcal{G}^0$ satisfying $g^0 =_\Delta g$.*

Thus $\mathcal{G}^0$ is a family of representatives for the quotient $\mathcal{G}/{=_\Delta}$ (which is also $\mathcal{C}/{=_\Delta}$, since every class contains elements of $\mathcal{C}$), and working with $\mathcal{G}^0$ amounts to working with that quotient. However, it will be convenient here to stick to the current definition, as $\mathcal{G}^0$ being included in $\mathcal{G}$ allows for using the operations and relations of $\mathcal{G}$ without ambiguity. Hereafter, we shall reserve the letters a, b, c for elements of $\mathcal{G}^0$.

The idea will be to view $\mathcal{G}^0$ as a geometric space on which $\mathcal{G}$ acts.

Lemma 3.10. *In Context* 3.7, *put $g \bullet a = (ga)^0$ for g in $\mathcal{G}$ and a in $\mathcal{G}^0$ such that ga is defined. Then $\bullet$ is a left-action of $\mathcal{G}$ on $\mathcal{G}^0$, in the sense that $g \bullet a$ is defined whenever the target of g coincides with the source of a, the source of $g \bullet a$ is the source of g, and we have $1_x \bullet a = a$ for a with source x and $(gh) \bullet a = g \bullet (h \bullet a)$ whenever both are defined.*

Note that, when defined, $g \bullet a$ always has the form $ga\Delta^{[m]}$ for some integer m.

Lemma 3.11. *In Context* 3.7, *the function* dist *defined on $\mathcal{G}^0 \times \mathcal{G}^0$ by*

$$\operatorname{dist}(a,b) = \begin{cases} \infty & \text{if } a \text{ and } b \text{ do not share the same source,} \\ \ell_\Delta(a^{-1}b) & \text{otherwise,} \end{cases}$$

is a quasi-distance, that is, it is symmetric and satisfies the triangle inequality; moreover, dist *is invariant under the action of $\mathcal{G}$ on $\mathcal{G}^0$.*

Proof. Corollary V.3.28 (inverse) gives

$$\operatorname{dist}(b,a) = \ell_\Delta(b^{-1}a) = \ell_\Delta((b^{-1}a)^{-1}) = \ell_\Delta(a^{-1}b) = \operatorname{dist}(a,b).$$

Proposition V.3.30 (inequalities) gives

$$\begin{aligned} \operatorname{dist}(a,c) = \ell_\Delta(a^{-1}c) &= \ell_\Delta((a^{-1}b)(b^{-1}c)) \\ &\leqslant \ell_\Delta(a^{-1}b) + \ell_\Delta(b^{-1}c) = \operatorname{dist}(a,b) + \operatorname{dist}(b,c). \end{aligned}$$

Finally, assume that $g \bullet a$ and $g \bullet b$ are defined. Then there exist m, n satisfying the equalities $g \bullet a = ga\Delta^{[m]}$ and $g \bullet c = gc\Delta^{[n]}$, so we find

$$\begin{aligned} \operatorname{dist}(g \bullet a, g \bullet b) = \ell_\Delta((g \bullet a)^{-1}(g \bullet b)) &= \ell_\Delta((ga\Delta^{[m]})^{-1}(gb\Delta^{[n]})) \\ &= \ell_\Delta(\Delta^{[-m]}a^{-1}b\Delta^{[n]}) = \ell_\Delta(a^{-1}b) = \operatorname{dist}(a,b), \end{aligned}$$

the fourth equality being true because the canonical length is invariant under a left- or a right-multiplication by Δ. ∎

Note that $\mathrm{dist}(1_x, a) = \ell_\Delta(a)$ holds for every a with source x.

Lemma 3.12. *In Context* 3.7, *for all* a, b *in* $\mathcal{G}^0$,
(i) $\mathrm{dist}(a, b) = 0$ *is equivalent to* $a =^\times b$.
(ii) $\mathrm{dist}(a, b) = 1$ *is equivalent to the existence of* s *in* $\mathcal{D}iv(\Delta) \setminus (\mathcal{C}^\times \cup \Delta\mathcal{C}^\times)$ *satisfying either* $b = as$ *or* $a = bs$.

Proof. Assertion (i) is obvious. For (ii), assume $\mathrm{dist}(a, b) = 1$. By definition, we have $\ell_\Delta(a^{-1}b) = 1$, that is, there exist t in $\mathcal{D}iv(\Delta) \setminus (\mathcal{C}^\times \cup \Delta\mathcal{C}^\times)$ and m in $\mathbb{Z}$ satisfying $a^{-1}b = \Delta^{[m]}t$, whence $b = \Delta^{[m]}\phi_\Delta^m(a)t$ and $a = \Delta^{[-m-1]}\phi_\Delta^{-m-1}(b)\phi_\Delta^{-m}(\partial t)$. Now, by assumption, we have $\Delta \not\preccurlyeq b$, that is, $\Delta \not\preccurlyeq \Delta^{[m]}\phi_\Delta^m(a)t$. As $\phi_\Delta^m(t)a$ lies in $\mathcal{C}$, this implies $m \leqslant 0$. Similarly, we have $\Delta \not\preccurlyeq a$, that is, $\Delta \not\preccurlyeq \Delta^{[-m-1]}\phi_\Delta^{-m-1}(b)\phi_\Delta^{-m}(\partial t)$. As $\phi_\Delta^{-m-1}(b)\phi_\Delta^{-m}(\partial t)$ lies in $\mathcal{C}$, this implies $-m-1 \leqslant 0$, that is, $m \geqslant -1$. For $m = 0$, we have $b = as$ with $s = t$, and, for $m = -1$, we have $as = b$ with $s = \partial t$. The converse direction is straightforward. □

We now consider certain geodesics in the space $\mathcal{G}^0$, and first recall the standard notion.

Definition 3.13 (walk, geodesic). In Context 3.7, an *ℓ-step walk* in $\mathcal{G}^0$ is a sequence $(a_0, \ldots, a_\ell)$ satisfying $\mathrm{dist}(a_i, a_{i+1}) \leqslant 1$ for every i; it is called *geodesic* if there exists no ℓ'-step walk from a_0 to a_ℓ with $\ell' < \ell$.

A walk is a concatenation of segments of length at most 1 connecting two endpoints, and a geodesic walk—a *geodesic*, for short—is a walk such that no strictly shorter walk connects the endpoints. Note that all points in a walk must share the same source. If $(a_0, \ldots, a_\ell)$ is a geodesic walk, the triangle inequality (Proposition 3.11) implies that every segment (a_i, a_{i+1}) has length one exactly. Also, it follows from the definition that every subsequence of a geodesic walk is geodesic and that, if $(a_0, \ldots, a_\ell)$ is geodesic, then so is $(a_\ell, \ldots, a_0)$.

Hereafter we shall consider particular geodesics in $\mathcal{G}^0$ that are connected with Δ-normal decompositions in $\mathcal{G}$. In view of Propositions III.3.1 (geodesic, positive case) and III.3.2 (geodesic, general case), a connection is natural. However, the current context is different since the equivalence relation $=_\Delta$ is involved.

Definition 3.14 (normal walk). In Context 3.7, a walk $(a_0, \ldots, a_\ell)$ in $\mathcal{G}^0$ is called *normal* if some strict Δ-normal path $s_1|\cdots|s_n$, then called a *witness* for $(a_0, \ldots, a_\ell)$, satisfies

$$\forall i \; (\, a_i =_\Delta a_0\, s_1 \cdots s_i \,). \tag{3.15}$$

Example 3.16 (normal geodesic). Consider the points $\mathtt{a}$ and $\mathtt{b}^2\mathtt{c}$ in the space $\mathcal{G}^0$ associated with the free abelian monoid based on $\{\mathtt{a}, \mathtt{b}, \mathtt{c}\}$ with $\Delta = \mathtt{abc}$ (Reference Structure 1, page 3). The canonical length of $\mathtt{a}^{-1}\mathtt{b}^2\mathtt{c}$ is 3, so the distance from $\mathtt{a}$ to $\mathtt{b}^2\mathtt{c}$ is 3. Now $(\mathtt{a}, 1, \mathtt{bc}, \mathtt{b}^2\mathtt{c})$ is a normal geodesic from $\mathtt{a}$ to $\mathtt{b}^2\mathtt{c}$, as shows the witness $\mathtt{bc}|\mathtt{bc}|\mathtt{b}$. By contrast, $(\mathtt{a}, \mathtt{ab}, \mathtt{b}, \mathtt{b}^2\mathtt{c})$, which is another geodesic from $\mathtt{a}$ to $\mathtt{b}^2\mathtt{c}$, is not normal since the only candidate-witness, namely $\mathtt{b}|\mathtt{bc}|\mathtt{bc}$, is not normal. Note that the reversed geodesic $(\mathtt{b}^2\mathtt{c}, \mathtt{bc}, 1, \mathtt{a})$ is also normal, since it admits the witness $\mathtt{ac}|\mathtt{a}|\mathtt{a}$. We shall see in Lemma 3.18 below that this is a general fact.

Our first observation is that normal geodesics exist as often as possible, namely whenever the expected endpoints lie at finite distance from one another.

Lemma 3.17. *In Context* 3.7, *for all a, b in $\mathcal{G}^0$, the following are equivalent:*
(i) *The inequalities* $0 < \mathrm{dist}(a,b) < \infty$ *are satisfied.*
(ii) *There exists a geodesic connecting a to b.*
(iii) *There exists a normal geodesic connecting a to b.*

Proof. That (iii) implies (ii) and (ii) implies (i) follows directly from the definition.

Assume now that a and b share the same source and satisfy $\mathrm{dist}(a,b) = \ell \neq 0$. Let $\Delta^{[m]} \| s_1 | \cdots | s_\ell$ be a strict Δ-normal decomposition of $a^{-1}b$. Put $a_0 = a$ and, for $i \leqslant \ell$, put $a_i = (a_0 \phi_\Delta^m(s_1) \cdots \phi_\Delta^m(s_i))^0$. Then we have

$$a_\ell = (a_0 \phi_\Delta^m(s_1) \cdots \phi_\Delta^m(s_\ell))^0 = (a_0 \phi_\Delta^m(s_1) \cdots \phi_\Delta^m(s_\ell) \Delta^{[m]})^0$$
$$= (a \Delta^{[m]} s_1 \cdots s_\ell)^0 = b^0 = b.$$

As ϕ_Δ is an automorphism of $\mathcal{C}$, the path $\phi_\Delta^m(s_1) | \cdots | \phi_\Delta^m(s_\ell)$ is strictly Δ-normal and, by definition, it is a witness for $(a_0, \dots, a_\ell)$. So the latter is a normal walk connecting a to b. Moreover, this walk is geodesic, since ℓ is the distance of a and b. So (i) implies (iii). □

We now establish some general properties of normal geodesics.

Lemma 3.18. *In Context* 3.7,
(i) *A normal walk in $\mathcal{G}^0$ is always geodesic, and it admits a unique witness.*
(ii) *If $(a_0, \dots, a_\ell)$ is a normal geodesic and $g \bullet a_0$ is defined, then $(g \bullet a_0, \dots, g \bullet a_\ell)$ is also normal geodesic.*
(iii) *If $(a_0, \dots, a_\ell)$ is a normal geodesic, then so is $(a_\ell, \dots, a_0)$.*

Proof. (i) Assume that $(a_0, \dots, a_\ell)$ is normal with witness $s_1 | \cdots | s_\ell$. By definition, the canonical length of $s_1 \cdots s_\ell$ is ℓ and we find $\mathrm{dist}(a_0, a_\ell) = \ell_\Delta(a_0^{-1} a_\ell) = \ell$. Hence $(a_0, \dots, a_\ell)$ is geodesic.

Assume that $s'_1 | \cdots | s'_\ell$ is another witness. Then $a_0 s'_1 \cdots s'_i =_\Delta a_0 s_1 \cdots s_i$ holds for each i, whence $s'_1 \cdots s'_i =_\Delta s_1 \cdots s_i \Delta^{[m]}$. By definition, $s'_1 \cdots s'_i$ and $s_1 \cdots s_i$ belong to $\mathcal{G}^0$, so they must be equal. An obvious induction on i then implies $s'_i = s_i$ for every i.

(ii) Assume that $s_1 | \cdots | s_\ell$ is a witness for $(a_0, \dots, a_\ell)$ and $g \bullet a_0$ is defined. Let m be the (unique) integer satisfying $g \bullet a_0 = g a_0 \Delta^{[m]}$. Then we obtain for every i

$$g \bullet a_i =_\Delta g a_i =_\Delta g a_0\, s_1 \cdots s_i = (g \bullet a_0) \Delta^{[-m]}\, s_1 \cdots s_i =_\Delta (g \bullet a_0)\, \phi_\Delta^m(s_1) \cdots \phi_\Delta^m(s_i),$$

which shows that $\phi_\Delta^m(s_1) | \cdots | \phi_\Delta^m(s_\ell)$ is a witness for $(g \bullet a_0, \dots, g \bullet a_\ell)$.

(iii) Assume again that $s_1 | \cdots | s_\ell$ is a witness for $(a_0, \dots, a_\ell)$. For each i, there exists an integer m_i satisfying $a_i = a_0 s_1 \cdots s_i \Delta^{[m_i]}$. We deduce $a_i = a_\ell \Delta^{[-m_\ell]} s_\ell^{-1} \cdots s_{i+1}^{-1} \Delta^{[m_i]}$, whence, using the formula $g^{-1} = \partial g\, \Delta^{[-1]}$ and writing m for m_ℓ,

$$a_i = a_\ell\, \Delta^{[-m]} \partial s_\ell \Delta^{[-1]}\, \partial s_{\ell-1} \Delta^{[-1]} \dots \partial s_{i+1} \Delta^{[-1]}\, \Delta^{[m_i]}$$
$$= a_\ell\, \phi_\Delta^m(\partial s_\ell) \phi_\Delta^{m+1}(\partial s_{\ell-1}) \cdots \phi_\Delta^{m+\ell-i}(\partial s_{i+1})\, \Delta^{[m_i - m + i - \ell]}.$$

By Proposition V.3.26 (inverse), the path $\phi_\Delta^m(\partial s_\ell)|\cdots|\phi_\Delta^{m+\ell-1}(\partial s_1)$ is Δ-normal, and it is strict since the assumption that s_i is neither invertible nor delta-like implies that $\phi_\Delta^n(\partial s_i)$ is neither invertible nor delta-like for every n. Therefore, this path is a witness for $(a_\ell, ..., a_0)$. □

As normal geodesics are defined in terms of normal paths and normality is local in that a path is normal if and only if every length-two subpath is normal, it is natural to expect that normal geodesics admit a local characterization. This is indeed the case.

Proposition 3.19 (normal geodesic local). *In Context* 3.7, *a walk is a normal geodesic in* $\mathcal{G}^0$ *if and only if each of its length-three subsequences is.*

Proof. Assume that $(a_0, ..., a_\ell)$ is a normal geodesic, with witness $s_1|\cdots|s_\ell$. By definition, for each i, we have $a_i = a_0\, s_1 \cdots s_i\, \Delta^{[m_i]}$ for some (unique) integer m_i. Choose i with $0 \leqslant i \leqslant \ell - 2$. Then, writing

$$\begin{aligned} a_{i+1} &= a_0\, s_1 \cdots s_i s_{i+1}\, \Delta^{[m_{i+1}]} \\ &= a_0\, s_1 \cdots s_i \Delta^{[m_i]} \Delta^{[-m_i]} s_{i+1}\, \Delta^{[m_{i+1}]} = a_i \phi_\Delta^{m_i}(s_{i+1})\, \Delta^{[m_{i+1}-m_i]}, \end{aligned}$$

we deduce $a_{i+1} =_\Delta a_i \phi_\Delta^{m_i}(s_{i+1})$ and, similarly, $a_{i+2} =_\Delta a_i \phi_\Delta^{m_i}(s_{i+1})\phi_\Delta^{m_i}(s_{i+2})$. Now, $s_{i+1}|s_{i+2}$ is strictly Δ-normal, hence so is $\phi_\Delta^{m_i}(s_{i+1})|\phi_\Delta^{m_i}(s_{i+2})$. Then the latter is a witness for (a_i, a_{i+1}, a_{i+2}), which is therefore normal, hence geodesic by Lemma 3.18.

Conversely, assume that every length-three subsequence of $(a_0, ..., a_\ell)$ is normal (hence geodesic). We use induction on ℓ. For $\ell \leqslant 2$, there is nothing to prove. Assume $\ell \geqslant 3$. By the induction hypothesis, the walk $(a_0, ..., a_{\ell-1})$ is normal, so it admits a witness, say $s_1|\cdots|s_{\ell-1}$, hence there exist integers m, m' satisfying

$$a_{\ell-2} = a_0 s_1 \cdots s_{\ell-2} \Delta^{[m]} \quad \text{and} \quad a_{\ell-1} = a_0 s_1 \cdots s_{\ell-2} s_{\ell-1} \Delta^{[m']}, \tag{3.20}$$

whence $a_{\ell-1} = a_{\ell-2}\phi_\Delta^m(s_{\ell-1})\Delta^{[m'-m]}$. On the other hand, by assumption, the walk $(a_{\ell-2}, a_{\ell-1}, a_\ell)$ is normal, so it admits a witness, say $t_1|t_2$, hence there exist integers n, n' satisfying $a_{\ell-1} = a_{\ell-2} t_1 \Delta^{[n]}$ and $a_\ell = a_{\ell-2} t_1 t_2 \Delta^{[n']}$. Merging the evaluations of $a_{\ell-1}$ yields $\phi_\Delta^m(s_{\ell-1})\Delta^{[m'-m]} = t_1 \Delta^{[n]}$, whence $\phi_\Delta^m(s_{\ell-1}) =_\Delta t_1$. By definition, t_1 and $s_{\ell-1}$, hence $\phi_\Delta^m(s_{\ell-1})$ as well, lie in $\mathcal{D}iv(\Delta) \setminus (\mathcal{C}^\times \cup \Delta\mathcal{C}^\times)$. It follows that $\phi_\Delta^m(s_{\ell-1}) = t_1$ holds, whence $\Delta^{[m]} t_1 = s_{\ell-1}\Delta^{[m]}$. Let $s_\ell = \phi_\Delta^{-m}(t_2)$, that is, $\Delta^{[m]} t_2 = s_\ell \Delta^{[m]}$. Inserting the value of $a_{\ell-2}$ provided by (3.20) in $a_\ell = a_{\ell-2} t_1 t_2 \Delta^{[n']}$, we obtain

$$\begin{aligned} a_\ell &= a_0 s_1 \cdots s_{\ell-2} \Delta^{[m]} t_1 t_2 \Delta^{[n']} \\ &= a_0 s_1 \cdots s_{\ell-2} s_{\ell-1} \Delta^{[m]} t_2 \Delta^{[n']} = a_0 s_1 \cdots s_{\ell-2} s_{\ell-1} s_\ell \Delta^{[m+n']} \end{aligned}$$

By assumption, $t_1|t_2$ is strictly Δ-normal, hence so is $\phi_\Delta^{-m}(t_1)|\phi_\Delta^{-m}(t_2)$, that is, $s_{\ell-1}|s_\ell$. Therefore, the path $s_1|\cdots|s_\ell$ is strictly Δ-normal and, by construction, it witnesses that $(a_0, ..., a_\ell)$ is a normal walk in $\mathcal{G}^0$. □

We shall now introduce an oriented version of the distance on the space $\mathcal{G}^0$ using the left height function on the category $\mathcal{C}$, which, by Proposition II.2.47 (height) is defined everywhere on $\mathcal{C}$ if and only if $\mathcal{C}$ is left-Noetherian—this is where we use the assumption that $\mathcal{C}$ is small. We recall that the left height function ht_{L} is the ordinal-valued function ht_{L} on $\mathcal{C}$, defined by

$$\mathrm{ht}_{\mathrm{L}}(g) = \begin{cases} 0 & \text{if } g \text{ is invertible,} \\ \sup\{\mathrm{ht}_{\mathrm{L}}(f) + 1 \mid f \prec g\} & \text{otherwise.} \end{cases}$$

Then $f \prec g$ implies $\mathrm{ht}_{\mathrm{L}}(f) < \mathrm{ht}_{\mathrm{L}}(g)$. By Lemma II.2.44, ht_{L} is invariant under ϕ_Δ. Here comes the key observation about left height in the current context.

Lemma 3.21. *In Context* 3.7, *assume that* $\mathcal{C}$ *is left-Noetherian.*

(i) *If* (a, b) *is a geodesic in* $\mathcal{G}^0$, *that is,* $\mathrm{dist}(a, b) = 1$ *implies either* $\mathrm{ht}_{\mathrm{L}}(a) < \mathrm{ht}_{\mathrm{L}}(b)$, *or* $\mathrm{ht}_{\mathrm{L}}(a) > \mathrm{ht}_{\mathrm{L}}(b)$.

(ii) *If* (a, b, c) *is a normal geodesic in* $\mathcal{G}^0$, *then* $\mathrm{ht}_{\mathrm{L}}(a) < \mathrm{ht}_{\mathrm{L}}(b) > \mathrm{ht}_{\mathrm{L}}(c)$ *is impossible.*

Proof. (i) By Lemma 3.12, there exists an element t of $\mathcal{D}iv(\Delta) \setminus (\mathcal{C}^\times \cup \Delta\mathcal{C}^\times)$ satisfying either $b = at$ or $a = bt$. In the first case, we deduce $\mathrm{ht}_{\mathrm{L}}(b) > \mathrm{ht}_{\mathrm{L}}(a)$, in the second case, we have $\mathrm{ht}_{\mathrm{L}}(b) < \mathrm{ht}_{\mathrm{L}}(a)$.

(ii) Let $s_1|s_2$ be a witness for (a, b, c). By Lemma 3.12, there exists t in $\mathcal{D}iv(\Delta) \setminus (\mathcal{C}^\times \cup \Delta\mathcal{C}^\times)$ satisfying either $b = at$ or $a = bt$. This is equivalent to the existence of t_1 in $\mathcal{D}iv(\Delta) \setminus (\mathcal{C}^\times \cup \Delta\mathcal{C}^\times)$ and m in $\{0, -1\}$ satisfying $b = at_1\Delta^{[m]}$. Similarly there exists h_2 in $\mathcal{D}iv(\Delta) \setminus (\mathcal{C}^\times \cup \Delta\mathcal{C}^\times)$ and n in $\{0, -1\}$ satisfying $c = bt_2\Delta^{[n]}$. Combining this with the definition of a witness, we deduce $t_1 = s_1$ and $t_2 = \phi_\Delta^m(s_2)$, and $c = as_1s_2\Delta^{[m+n]}$.

Assume $n = -1$. Then we have $-(m + n) \geqslant 1$, and the latter equality, rewritten as $c\Delta^{[-(m+n)]} = as_1s_2$, implies $\Delta \preccurlyeq as_1s_2$. As $s_1|s_2$ is normal, this implies $\Delta \preccurlyeq as_1$. Hence as_1 does not belong to $\mathcal{G}^0$, whereas, by assumption, $as_1\Delta^{[m]}$, which is b, does. We deduce $m \neq 0$, whence $m = -1$. So $n = -1$ implies $m = -1$.

We claim that, symmetrically, $m = 0$ implies $n = 0$. Indeed, this follows from applying the above result to the reversed geodesic (c, b, a), which is normal by Lemma 3.18(iii). Calling m' and n' the associated parameters, the above result says that $n' = -1$ implies $m' = -1$. Now, by construction, $n' = -1$ is equivalent to $m = 0$ and $m' = -1$ is equivalent to $n = 0$.

We conclude that the conjunction of $m = 0$ and $n = -1$ is impossible. Now, by definition, $m = 0$ corresponds to $\mathrm{ht}_{\mathrm{L}}(a) < \mathrm{ht}_{\mathrm{L}}(b)$, whereas $n = -1$ corresponds to $\mathrm{ht}_{\mathrm{L}}(b) > \mathrm{ht}_{\mathrm{L}}(c)$. ☐

This leads us to considering a new parameter.

Definition 3.22 (oriented distance). In Context 3.7, assume that $\mathcal{C}$ is left-Noetherian. Define $\overrightarrow{\mathrm{dist}}$ from $\mathcal{G}^0 \times \mathcal{G}^0$ to $\mathbf{Ord} \cup \{\infty\}$ by

$$\overrightarrow{\mathrm{dist}}(a, b) = \begin{cases} \infty & \text{if } a \text{ and } b \text{ do not share the same source,} \\ \mathrm{ht}_{\mathrm{L}}((a^{-1}b)^0) & \text{otherwise.} \end{cases}$$

Proposition 3.23. *In Context* 3.7, *if $\mathcal{C}$ is left-Noetherian:*

(i) *For a in $\mathcal{G}^0(x, \cdot)$, we have $\overrightarrow{\mathrm{dist}}(1_x, a) = \mathrm{ht}_{\mathrm{L}}(a)$.*

(ii) *For all a, b, b' in $\mathcal{G}^0$, the relation $b' =^{\times} b$ implies $\overrightarrow{\mathrm{dist}}(a, b') = \overrightarrow{\mathrm{dist}}(a, b)$.*

(iii) *The function $\overrightarrow{\mathrm{dist}}$ is invariant under $\bullet$, that is, $\overrightarrow{\mathrm{dist}}(g \bullet a, g \bullet b) = \overrightarrow{\mathrm{dist}}(a, b)$ holds whenever the points are defined.*

(iv) *If $(b_0, \ldots, b_\ell)$ is a normal geodesic in $\mathcal{G}^0$, then, for every a in $\mathcal{G}^0$ sharing the source of b_0 and every i in $\{1, \ldots, \ell - 1\}$, we have*

$$\overrightarrow{\mathrm{dist}}(a, b_i) < \max(\overrightarrow{\mathrm{dist}}(a, b_0), \overrightarrow{\mathrm{dist}}(a, b_\ell)). \tag{3.24}$$

Of course we declare that $\alpha < \infty$ holds for every ordinal α (alternatively, in the case when $\mathcal{O}bj(\mathcal{C})$ is a set, hence in particular when $\mathcal{C}$ is a small category, we could use an ordinal θ that is larger than $\mathrm{ht}_{\mathrm{L}}(\Delta(x))$ for every object x of $\mathcal{C}$).

Proof. (i) For a in $\mathcal{G}^0$, we have $(1_x^{-1}a)^0 = a^0 = a$, so the result follows from the definition.

(ii) Assume $b' = b\epsilon$ with ϵ invertible. Write $(a^{-1}b)^0 = a^{-1}b\Delta^{[m]}$. Then we have $a^{-1}b' = a^{-1}b\epsilon =_{\Delta} a^{-1}b\epsilon\Delta^{[m]} = a^{-1}b\Delta^{[m]}\phi_{\Delta}^{m}(\epsilon)$, whence $(a^{-1}b')^0 =^{\times} (a^{-1}b)^0$, and $\overrightarrow{\mathrm{dist}}(a, b') = \overrightarrow{\mathrm{dist}}(a, b)$.

(iii) Assume that $g \bullet a$ and $g \bullet b$ are defined (in particular, a and b share the same source). Write $g \bullet a = ga\Delta^{[m]}$ and $g \bullet b = gb\Delta^{[n]}$. Then we find

$$(g \bullet a)^{-1}(g \bullet b) = \Delta^{[-m]}a^{-1}b\Delta^{[n]} = \phi_{\Delta}^{m}(a^{-1}b)\Delta^{[n-m]} =_{\Delta} \phi_{\Delta}^{m}(a^{-1}b),$$

whence, using the invariance of ht_{L} under ϕ_Δ given by Lemma II.2.44(iv),

$$\overrightarrow{\mathrm{dist}}(g \bullet a, g \bullet b) = \mathrm{ht}_{\mathrm{L}}(\phi_{\Delta}^{m}(a^{-1}b)) = \mathrm{ht}_{\mathrm{L}}(a^{-1}b) = \overrightarrow{\mathrm{dist}}(a, b).$$

(iv) Let x be the common source of $b_0, \ldots, b_\ell$ and a be an element of $\mathcal{G}^0$ with source x. By definition, $a^{-1} \bullet b_i$ is defined for every i. Put $c_i = a^{-1} \bullet b_i$. By Lemma 3.18(ii), $(c_0, \ldots, c_\ell)$ is a normal geodesic. Then, using (iii) and (i), we have for each i

$$\overrightarrow{\mathrm{dist}}(a, b_i) = \overrightarrow{\mathrm{dist}}(a^{-1} \bullet a, a^{-1} \bullet b_i) = \overrightarrow{\mathrm{dist}}(1_x, c_i) = \mathrm{ht}_{\mathrm{L}}(c_i).$$

So we are left with establishing $\mathrm{ht}_{\mathrm{L}}(c_i) < \max(\mathrm{ht}_{\mathrm{L}}(c_0), \mathrm{ht}_{\mathrm{L}}(c_\ell))$ for $1 \leqslant i \leqslant \ell - 1$ whenever $(c_0, \ldots, c_\ell)$ is a normal geodesic. This follows from Lemma 3.21(ii), since, by the latter, no point except the endpoints may be a local maximum of the left height. □

Definition 3.25 (center). For $\mathcal{O}$ included in $\mathcal{G}^0$ and b in $\mathcal{G}^0$, put

$$\overrightarrow{\mathrm{dist}}_{\mathcal{O}}(b) = \sup_{a \in \mathcal{O}} \overrightarrow{\mathrm{dist}}(a, b). \tag{3.26}$$

A point of $\mathcal{G}^0$ is said to be a *center* of $\mathcal{O}$ if it achieves the minimum of $\overrightarrow{\mathrm{dist}}_{\mathcal{O}}$ on $\mathcal{G}^0$.

Notation 3.27 (family $\mathcal{S}/{=^{\times}}$). For $\mathcal{S}$ included in $\mathcal{C}$, we denote by $\mathcal{S}/{=^{\times}}$ the family of all $=^{\times}$-classes of elements of $\mathcal{S}$.

By definition, $\mathcal{G}^0$ is included in $\mathcal{C}$, so the notation $\mathcal{O}/{=^{\times}}$ makes sense for $\mathcal{O} \subseteq \mathcal{G}^0$.

Proposition 3.28 (center). *In Context* 3.7 *with $\mathcal{C}$ left-Noetherian and $\mathcal{O}$ included in $\mathcal{G}^0$:*
(i) *There exists at least one center in $\mathcal{O}$.*
(ii) *If all elements of $\mathcal{O}$ share the same source x, every center of $\mathcal{O}$ has source x.*
(iii) *If, moreover, $\mathcal{O}/{=^\times}$ is finite, any two centers c, c' of $\mathcal{O}$ satisfy* $\mathrm{dist}(c, c') \leqslant 1$.

Proof. (i) The supremum of (3.26) is always defined in $\mathbf{Ord}\cup\{\infty\}$ and the well-ordering property of ordinals guarantees that the minimum is achieved by a point of $\mathcal{G}^0$.

(ii) By definition, $\overrightarrow{\mathrm{dist}}_{\mathcal{O}}(b)$ is ∞ if and only if the source of b is not x. In particular, $\overrightarrow{\mathrm{dist}}_{\mathcal{O}}(b) < \infty$ holds for b in $\mathcal{O}$, hence for c a center of $\mathcal{O}$. This implies that such a center admits x as its source.

(iii) Let $\mathcal{O}'$ be a (finite) $=^\times$-selector on $\mathcal{O}$, that is, a subfamily of $\mathcal{O}$ that contains one point in each $=^\times$-class. Then, for every b in $\mathcal{G}^0$, we have

$$\overrightarrow{\mathrm{dist}}_{\mathcal{O}}(b) = \max\nolimits_{a\in\mathcal{O}'} \overrightarrow{\mathrm{dist}}(a, b) \tag{3.29}$$

since, by Proposition 3.23(ii), $b =^\times b'$ implies $\overrightarrow{\mathrm{dist}}(a, b') = \overrightarrow{\mathrm{dist}}(a, b)$. Now, by definition, c and c' share the common source of all elements of $\mathcal{O}$, hence $\mathrm{dist}(c, c') < \infty$ holds. By Lemma 3.17, there exists a normal geodesic, say $(c_0, ..., c_\ell)$, connecting c to c'. If $\ell \geqslant 2$ holds, then, for every point a in $\mathcal{O}'$, (3.24) gives

$$\overrightarrow{\mathrm{dist}}(a, c_1) < \max(\overrightarrow{\mathrm{dist}}(a, c), \overrightarrow{\mathrm{dist}}(a, c')),$$

whence, by taking the maxima over the finite set $\mathcal{O}'$ and using (3.29),

$$\overrightarrow{\mathrm{dist}}_{\mathcal{O}}(c_1) < \overrightarrow{\mathrm{dist}}_{\mathcal{O}}(c) = \overrightarrow{\mathrm{dist}}_{\mathcal{O}}(c').$$

This contradicts the assumption that c and c' achieve the minimum of $\overrightarrow{\mathrm{dist}}_{\mathcal{O}}$ and, therefore, we must have $\ell \leqslant 1$. ☐

3.3 Conjugacy of periodic elements

We can now return to periodic elements of $\mathcal{G}$ and their conjugates and apply the results of Subsection 3.2. We first use the results of that subsection to give a characterization of the elements which have a periodic conjugate and we prove that, in that case, they have a conjugate close to a power of Δ.

Proposition 3.30 (conjugate to periodic). *In Context* 3.7 *with $\mathcal{C}$ left-Noetherian, for e in $\mathcal{G}^{\circlearrowleft}$ and a in $\mathcal{G}^0$ such that ea is defined, define the* e-orbit of a *to be* $\{e^n \bullet a \mid n \in \mathbb{Z}\}$. *Then an element e of $\mathcal{G}^{\circlearrowleft}$ is conjugate to a periodic element if and only if there exists an e-orbit $\mathcal{O}$ such that $\mathcal{O}/{=^\times}$ is finite. In this case, e is conjugate to an element of the form $\Delta^{[m]}h$ with m in $\mathbb{Z}$ and h in $\mathcal{D}iv(\Delta) \setminus \Delta\mathcal{C}^\times$.*

Proof. Assume that e^g is periodic. We note that $(e^g)^p =^\times \Delta^{[q]}$ implies $(e^{g\Delta^{[m]}})^p =^\times \Delta^{[q]}$ for every r, and, therefore $e^{g'}$ is also periodic for every g' satisfying $g' =_\Delta g$, with the same parameters. Hence, some conjugate of e of the form e^a with a in $\mathcal{G}^0$ is periodic. Now, by definition, e^a is periodic if and only if there exists p satisfying $(e^a)^p =^\times \Delta^{[q]}$ for some q. The latter relation can be recast as $e^pa =^\times a\Delta^{[q]}$, and, as $g =^\times g'$ implies $g^0 =^\times g'^0$, it implies $e^p \bullet a = (e^p a)^0 =^\times (a\Delta^{[q]})^0 = a$. So, in this case, the e-orbit of a intersects at most p many $=^\times$-classes.

Conversely, assume that the e-orbit of some a has a finite image in $\mathcal{G}^0/=^\times$. As $\mathcal{G}$ is a groupoid, there must exist p satisfying $e^p \bullet a =^\times a$. This means that there exists an integer q satisfying $e^p a \Delta^{[-q]} =^\times a$, whence $e^p a =^\times a\Delta^{[q]}$, that is, $(e^a)^p =^\times \Delta^{[q]}$. So e^a is periodic.

Assume that we are in the case above, that is, the equality $(e^a)^p =^\times \Delta^{[q]}$ holds. Let x be the source of e and a, and let $\mathcal{O}$ be the e-orbit of a. By Proposition 3.28(i), $\mathcal{O}$ admits at least one center, say c and, by Proposition 3.28(ii), the source of c is x. We claim that $e \bullet c$ is also a center of $\mathcal{O}$. Indeed, $\mathcal{O}$ is invariant under left-translation by e, and, for all a, b in $\mathcal{G}^0$, Proposition 3.23(iii) implies $\overrightarrow{\mathrm{dist}}(e \bullet a, e \bullet b) = \overrightarrow{\mathrm{dist}}(a, b')$. Hence, if c achieves the minimum of $\overrightarrow{\mathrm{dist}}$ on $\mathcal{O}$, so does $e \bullet c$. Now, by assumption, the image of $\mathcal{O}$ in $\mathcal{G}^0/=^\times$ is finite, so Proposition 3.28(iii) implies $\mathrm{dist}(c, e \bullet c) \leqslant 1$. By definition, the last relation reads $\ell_\Delta(c^{-1}(e \bullet c)) \leqslant 1$, which is also $\ell_\Delta(e^c) \leqslant 1$ since ℓ_Δ is invariant under right-multiplication by Δ. So $\ell_\Delta(e^c)$ must be 0 or 1. In the first case, e^c belongs to $\Delta^{[m]}\mathcal{C}^\times$ for some m. In the second case, e^c has an expression of the form $\Delta^{[m]}h$ with h in $\mathcal{D}iv(\Delta) \setminus (\mathcal{C}^\times \cup \Delta\mathcal{C}^\times)$, hence also an expression of the form $h\Delta^{[m]}$ with h in $\mathcal{D}iv(\Delta) \setminus (\mathcal{C}^\times \cup \Delta\mathcal{C}^\times)$. In all cases, e^c has the form $\Delta^{[m]}h$ with h in $\mathcal{D}iv(\Delta) \setminus \Delta\mathcal{C}^\times$. □

Corollary 3.31 (SSS of periodic). *In Context* 3.7 *with $\mathcal{C}$ left-Noetherian, if e is an element of $\mathcal{G}^{\circlearrowleft}$ that is conjugate to a periodic element, then the super-summit set of e consists of elements of the form $\Delta^{[m]}h$ with m in $\mathbb{Z}$ and h in $\mathcal{D}iv(\Delta)$, and e is conjugate by cyclic conjugacy to such an element.*

Proof. By Proposition 3.30, e is conjugate to an element e' of the form $\Delta^{[m]}h$ with m in $\mathbb{Z}$ and h in $\mathcal{D}iv(\Delta) \setminus \Delta\mathcal{C}^\times$. Such an element must lie in its super-summit set. Indeed, , we find $\inf_\Delta(e') = \sup_\Delta(e') = m$ if h is invertible, and $\sup_\Delta(e') = \inf_\Delta(e') + 1 = m + 1$ otherwise; in each case, e' is in its super-summit set since cycling does not change its shape. Furthermore, every element of the super-summit set of e' has the same shape, since the values of $\inf_\Delta$ and $\sup_\Delta$ must be the same. So the super-summit set is as claimed.

As for the last point, it follows from Corollary 2.12, which guarantees that one can go from e to an element of its super-summit set using cyclic conjugacies. □

Proposition 3.32 (cyclically conjugate). *In Context* 3.7 *with $\mathcal{C}$ Noetherian, any two periodic elements of $\mathcal{C}$ that are conjugate are cyclically conjugate.*

Proof. This is a special case of Proposition 1.24. □

In the case when the automorphism ϕ_Δ has a finite order, we can come back to Proposition 3.30 and obtain a more complete description of the periodic elements.

Lemma 3.33. *In Context* 3.7 *with ϕ_Δ of finite order, every conjugate of a periodic element must be a periodic element.*

Proof. Assume that ϕ_Δ^n is the identity and $e^p =^\times \Delta^{[q]}$ holds. For every g such that e^g is defined, we obtain $(e^g)^{pn} = g^{-1}e^{pn}g =^\times g^{-1}\Delta^{[qn]}g = \Delta^{[qn]}$, since $\Delta^{[qn]}$ commutes with every element. □

Proposition 3.34 (periodic elements). *If $\mathcal{C}$ is a Noetherian cancellative category and Δ is a Garside map in $\mathcal{C}$ such that the order of ϕ_Δ is finite:*

(i) *Every periodic element e of $\mathcal{C}$ is cyclically conjugate to some element d satisfying $d^p \in \Delta^{[q]}\mathcal{C}^\times$ with p and q positive and coprime and such that, for all positive integers p' and q' satisfying $pq' - qp' = 1$, we have $d^{p'} \preccurlyeq \Delta^{[q']}$.*

(ii) *For d as in (i), the element g satisfying $d^{p'}g = \Delta^{[q']}$ is an element of $\mathcal{Div}(\Delta)$ whose $=^\times$-class is independent of the choice of (p', q'); moreover, $g\Delta^{[-q']}$ is $(p, -qp')$-periodic and $d =^\times (g\Delta^{[-q']})^q\Delta^{[qq']}$ holds.*

Proof. (i) We use the notions and notation of Subsection 3.2. Since ϕ_Δ has finite order, Lemma 3.33 implies that all conjugates of e are periodic, and, therefore, all orbits of a periodic element have a finite image in $\mathcal{G}^0/{=^\times}$. Let $\mathcal{O}$ be an e-orbit in $\mathcal{G}^0$ and let $\mathcal{Z}$ be an e-orbit in the family of the centers of $\mathcal{O}$. Then $\mathcal{Z}/{=^\times}$ is finite, and every two elements c, c' of $\mathcal{Z}$ satisfy $\operatorname{dist}(c, c') \leqslant 1$. Take c in $\mathcal{Z}$ and set $d = e^c$. Here d is a priori in $\mathcal{G}$; we will prove eventually that it is in fact in $\mathcal{C}$, which will imply, by Proposition 3.32, that it is cyclically conjugate to e.

The set $c^{-1} \bullet \mathcal{Z}$ is a d-orbit in the set of centers of the d-orbit $c^{-1} \bullet \mathcal{O}$ and it consists of all elements $(d^i)^0$ with i in $\mathbb{Z}$. Let p be the cardinal of $\mathcal{Z}/{=^\times}$, and $\mathcal{Z}_0$ be the set $\{(d^i)^0 \mid i = 0, ..., p-1\}$. Thus, we have $d^p =^\times \Delta^{[q]}$ for some integer q. For each i, the elements d^i and d^{p+i} are $=_\Delta$-equivalent to $=^\times$-equivalent elements. Since p/q is the period of d, it is also the period of e, which is positive, since e is in $\mathcal{C}$. Hence $p > 0$ implies $q > 0$. If $p = 1$ holds, $pq' - qp' = 1$ is equivalent to $q' = 1 + kq$ and $p' = kq$ for some k in $\mathbb{Z}$, and (i) holds.

We now assume that d^0 is not invertible in $\mathcal{C}$. Then $\operatorname{dist}((d^i)^0, (d^{i+1})^0)$ is positive and, since $(d^i)^0$ and $(d^{i+1})^0$ are centers, we have $\operatorname{dist}((d^i)^0, (d^{i+1})^0) = 1$. Let x be the source of d. We have $\operatorname{dist}(1_x, (d^i)^0) = 1$, hence all elements $(d^i)^0$ lie in $\mathcal{Div}(\Delta)$. By Lemma 3.12, the relation $\operatorname{dist}((d^i)^0, (d^j)^0) = 1$ forbids $\operatorname{ht}_{\mathrm{L}}((d^i)^0) = \operatorname{ht}_{\mathrm{L}}((d^j)^0)$, so we may label the elements of $\mathcal{Z}_0$ as $c_0 = 1, c_1, ..., c_{p-1}$ so that that $\operatorname{ht}_{\mathrm{L}}(c_j) < \operatorname{ht}_{\mathrm{L}}(c_{j+1})$ holds for every j. Then, by Lemma 3.12 and our assumption on left heights, we have $c_i = c_1 g_1 \cdots g_{i-1}$, where c_i and g_i are non-invertible elements of $\mathcal{Div}(\Delta)$ for $i > 0$. There exists a positive integer p' such that $d^{p'}$ maps c_1 to an invertible element. We then have $d^{p'}c_1 = \Delta^{[q']}\epsilon$ for some integer q' and some invertible element ϵ. Hence, for $i \geqslant 1$, the action of $d^{p'}$, which is $\Delta^{[q']}\epsilon c_1^{-1}$, maps c_i to $\phi_\Delta^{-q'}(\epsilon g_1 \cdots g_{i-1})$, an element whose left height is at most that of c_i, since $\operatorname{ht}_{\mathrm{L}}$ is invariant under ϕ_Δ. Since the elements of $\mathcal{Z}_0$ have distinct left heights, the action of $d^{p'}$ on $\mathcal{Z}_0/{=^\times}$ must map the $=^\times$-class of c_j to that of $c_{j-1 \pmod p}$. We claim that the equality $d^{-jp'} =^\times c_j\Delta^{[-jq']}$ for $1 \leqslant j \leqslant p-1$ holds.

This is true for $j = 1$. By induction on j, we obtain

$$d^{-(j+1)p'} =^{\times} (c_1\epsilon^{-1}\Delta^{[-q']})c_j\Delta^{[-jq']} = c_1\epsilon^{-1}\phi_\Delta^{q'}(c_j)\Delta^{[-(j+1)q']}.$$

If Δ left-divided $c_1\epsilon^{-1}\phi_\Delta^{q'}(c_j)$, we would have $\Delta = c_1\epsilon^{-1}c'$ for some left-divisor c' of $\phi_\Delta^{q'}(c_j)$, but then the left height of c_{j+1}, which is $=^{\times}$-equivalent to $(d^{-(j+1)p'})^0$ would be at most the left height of c_j. As this is impossible, we deduce that $(d^{-(j+1)p'})^0$ is $=^{\times}$-equivalent to $c_1e^{-1}\phi_\Delta^{q'}(c_j)$, and our claim follows.

Now $d^{-p'}$ maps c_{p-1} to an invertible element. This means that $c_1\epsilon^{-1}\phi_\Delta^{q'}(c_{p-1})\Delta^{[-pq']}$ is a power of Δ up to right-multiplication by an invertible element. Since Δ does not divide c_{j-1}, we obtain $c_1\epsilon^{-1}\phi_\Delta^{q'}(c_{p-1}) =^{\times} \Delta$, whence

$$d^{-p'p} =^{\times} c_1\epsilon^{-1}\phi_\Delta^{q'}(c_{p-1})\Delta^{[-pq']} =^{\times} \Delta^{[1-pq']}.$$

From $d^{-p'p} =^{\times} \Delta^{[-qp']}$, we deduce $pq' - qp' = 1$. In particular, p and q are coprime so that they are the q and p of the proposition, and $q' > 0$ holds.

Let us check the other statements of (i). From

$$d = d^{-qp'+pq'} = (c_1\epsilon^{-1}\Delta^{[-q']})^q\Delta^{[qq']},$$

we deduce $d \in \mathcal{C}$. Then the fact that d and e are cyclically conjugate follows from Proposition 3.32. The equality $d^{p'}c_1 = \Delta^{[q']}\epsilon$ implies $d^{p'} \preccurlyeq \Delta^{[q']}$. Moreover, the relation $d^p =^{\times} \Delta^{[q]}$ implies $d^{-p'-kp}\Delta^{[q'+kq]} =^{\times} d^{-p'}\Delta^{[q']} \in \mathcal{C}$ for every k, so that, for every pair of integers (p', q') as in (i), we have $d^{p'} \preccurlyeq \Delta^{[q']}$.

(ii) First, $p = 1$ implies $g = \Delta$ and all assertions of (ii) are easily checked. Assume now $p \neq 1$. In the proof of (i), we have obtained $g = d^{-p'}\Delta^{[q']} = c_1\epsilon^{-1}$, so the $=^{\times}$-class of g is independent of (q', p'). Moreover, we have

$$(g\Delta^{[-q']})^p = d^{-pp'} =^{\times} (\Delta^{[-q]})^{p'} =^{\times} \Delta^{[1-pq']},$$

hence $g\Delta^{[-q']}$ is $(p, 1-dd')$-periodic and we have $g \preccurlyeq \Delta$, that is, g belongs to $\mathcal{D}iv(\Delta)$. We have also $(g\Delta^{[-q']})^q = d^{-qp'} = d^{1-pq'} =^{\times} d\Delta^{[-qq']}$, which completes the proof. □

Remark 3.35. (i) With the notation of Proposition 3.34, let us write $q = ip + k$ with $0 < k < p$. Then the following equalities are satisfied:

$$\begin{aligned} d = (g\Delta^{[-q']})^{ip+k}\Delta^{[q'(ip+k)]} &= (g\Delta^{[-q']})^k(g\Delta^{[-q']})^{ip}\Delta^{[q'(ip+k)]} \\ &=^{\times} (g\Delta^{[-q']})^k\Delta^{[-iqp']}\Delta^{[q'(ip+k)]} = (g\Delta^{[-q']})^k\Delta^{[kq']}\Delta^{[i]}. \end{aligned}$$

Hence we can write $d = h\Delta^{[i]}$ with $h \in \mathcal{D}iv(\Delta)$. Indeed, we have

$$(g\Delta^{[-q']})^k\Delta^{[q'k]} \preccurlyeq (g\Delta^{[-q']})^p\Delta^{[q'p]} =^{\times} \Delta^{[-qp']}\Delta^{[q'p]} = \Delta.$$

So the elements thus obtained have the form mentioned in Proposition 3.30.

(ii) In the case $p \neq 1$ of Proposition 3.34, the element d is (p, q)-periodic and satisfies $\inf_\Delta(d^{p'}) = q' - 1$ and $\sup_\Delta(d^{p'}) = q'$. In particular, $d^{p'}$ is in its own super-summit set.

Conversely, if d is (p,q)-periodic with $p \neq 1$ and p, q', q, p' are as in the theorem, assume that $d^{p'}$ has minimal $\sup_\Delta$ in its conjugacy class. Then $\sup_\Delta(d^{p'}) = q'$ and g defined by $d^{p'}g = \Delta^{[q']}$ lies in $\mathcal{D}iv(\Delta) \setminus \mathcal{C}^\times$ since, as in the proof of Proposition 3.34(ii), the element $g\Delta^{[-q']}$ is $(p, 1-pq')$-periodic. Moreover, g is not delta-like, since the relation $(g\Delta^{[-q']})^p =^\times \Delta^{[1-pq']}$ would imply $p = 1$. Thus $\inf_\Delta(d^{p'})$ equals $q' - 1$ and $d^{p'}$ lies in its super-summit set.

(iii) Note that the proof of Proposition 3.34(ii) from (i) does not use the geometric tools of Subsection 3.2.

Corollary 3.36 (centralizer). *Under the assumptions of Proposition* 3.34*, the centralizer of d in $\mathcal{C}$ coincides with the simultaneous centralizer of $g\Delta^{[-q']}$ and of $\Delta^{[q]}\epsilon$, where ϵ is the invertible element defined by $d^p = \Delta^{[q]}\epsilon$.*

Proof. From $g\Delta^{[-q']} = d^{-p'}$, we deduce that every element that centralizes d also centralizes $g\Delta^{[-q']}$. On the other hand, an element that centralizes d obviously centralizes its power $\Delta^{[q]}\epsilon$.

Conversely, the relations $(g\Delta^{[-q']})^q(\Delta^{[q]}\epsilon)^{q'} = d^{-qp'}d^{pq'} = d$ imply that an element centralizing both $g\Delta^{[-q']}$ and $\Delta^{[q]}\epsilon$ also centralizes d. □

The twisted case. As at the end of the previous subsections, we now consider a category $\mathcal{C} \rtimes \langle\phi\rangle$ and spell out how Proposition 3.34 looks when applied to an element of $\mathcal{C}\phi$.

Proposition 3.37 (periodic elements, twisted case). *Assume that $\mathcal{C}$ is a left-Noetherian cancellative category with no nontrivial invertible element, Δ is a Garside map in $\mathcal{C}$, the order of ϕ_Δ is finite, and ϕ is a finite-order automorphism of $\mathcal{C}$ that commutes with ϕ_Δ.*

(i) *If an element d of $\mathcal{C}$ is such that $d\phi$ is periodic of period p/q with p, q positive and coprime, then d is cyclically ϕ-conjugate to some element $e\phi$ satisfying $(e\phi)^p = \Delta^{[q]}\phi^p$ and $(e\phi)^{p'}\phi^{-p'} \preccurlyeq \Delta^{[q']}$ for all positive integers p' and q' satisfying $pq' - qp' = 1$.*

(ii) *For e as in* (i)*, the element g satisfying $(e\phi)^{p'}g = \Delta^{[q']}\phi^{p'}$ belongs to $\mathcal{C}$ and is independent of the choice of (q', p'). We have $(g\Delta^{[-q']}\phi^{-p'})^p = \Delta^{[1-pq']}\phi^{-p'p}$ and $e = (g\Delta^{[-q']}\phi^{-p'})^q\Delta^{[q'q]}\phi^{qp'}$. Moreover, g is fixed under $\phi_\Delta^q\phi^p$.*

(iii) *For e and g as in* (ii)*, and $\psi = \phi_\Delta^{-q'}\phi^{-p'}$, the centralizer of $e\phi$ in $\mathcal{C}$ coincides with $\mathcal{C}yc\mathcal{C}(g\psi, g\psi)$ in the category $\mathcal{C} \rtimes \langle\psi\rangle$; moreover, $\mathcal{C}yc\mathcal{C}(g\psi)$ can be computed inside the category of fixed points $(\mathcal{C}yc\mathcal{C})^{\phi_\Delta^q\phi^p}$.*

Proof. All assertions of (i) and (ii) but the last one are mere translations of Proposition 3.34, taking into account that the only invertible elements are the powers of ϕ, which allows us to replace the $=^\times$ relations with equalities. In particular, $g\phi^{-p'}$ here corresponds to the element g of Proposition 3.34(ii). This element is fixed under $\phi_\Delta^q\phi^p$ since Δ and e are, the former since ϕ commutes with ϕ_Δ and the latter since $\Delta^{[q]}\phi^p$ is a power of $e\phi$.

(iii) By Corollary 3.36, we know that the centralizer of $e\phi$ coincides with the family of all $\phi_\Delta^q\phi^p$-fixed points in the centralizer of $g\Delta^{[-q']}\phi^{-p'}$. Identifying $g\Delta^{[-q']}\phi^{-p'}$ with the element $g\phi_\Delta^{-q'}\phi^{-p'}$ of the category $\mathcal{C} \rtimes \langle\phi_\Delta^{-q'}\phi^{-p'}\rangle$ and applying Proposition 1.24, we obtain that every element f of the centralizer of $g\phi_\Delta^{-q'}\phi^{-p'}$ lies in the cyclic conjugacy family $\mathcal{C}yc\mathcal{C}(g\psi, g\psi)$. More precisely, the proof of Proposition 1.24 shows that f is a product of elements of the cyclic conjugacy class that are inductively computed as left-gcds, the first being the left-gcd of f and g. If f is fixed under $\phi_\Delta^q\phi^p$, this left-gcd is fixed under $\phi_\Delta^q \phi^p$ since g is, whence the result by induction. Finally, $\mathcal{C}yc\mathcal{C}(g\psi)$ can be evaluated inside $(\mathcal{C}yc\mathcal{C})^{\phi_\Delta^q \phi^p}$, since $g\psi$ commutes with $\phi_\Delta^q \phi^p$. □

We now spell out more explicitly how the above result looks like when $q = 2$ and $\phi_\Delta^2 = \mathrm{id}$ hold. A specific feature of this case is that we can replace the use of Proposition 3.34(i) by that of Proposition 3.5 to prove Proposition 3.37, and we thus obtain, by the proof of Proposition 3.5, an explicit sequence of cyclic ϕ-conjugations from d to e.

Corollary 3.38 (case $q = 2$). *Assume that $\mathcal{C}$ is a left-Noetherian cancellative category with no nontrivial invertible element, Δ is a Garside map in $\mathcal{C}$, and ϕ is a finite-order automorphism of $\mathcal{C}$ that commutes with ϕ_Δ. Assume moreover that the element $d\phi$ of $\mathcal{C}\phi$ satisfies $(d\phi)^p = \Delta^{[2]}\phi^p$.*

(i) *If p is even, say $p = 2r$, there exists e such that $\mathcal{C}yc\mathcal{C}(d\phi, e\phi)$ is nonempty and we have $(e\phi)^r = \Delta\phi^r$. The centralizer of $e\phi$ in $\mathcal{C}$ coincides with $\mathcal{C}yc\mathcal{C}(e\phi, e\phi)$, and it can be computed in the category of fixed points $(\mathcal{C}yc\mathcal{C})^{\phi_\Delta \phi^r}$.*

(ii) *If p is odd, say $p = 2r + 1$, there exists e such that $\mathcal{C}yc\mathcal{C}(d\phi, e\phi)$ is nonempty and we have $(e\phi)^p = \Delta^{[2]}\phi^p$ and $(e\phi)^r \phi^{-r} \preccurlyeq \Delta$. The element g defined by $(e\phi)^r g = \Delta\phi^r$ is such that, in the category $\mathcal{C} \rtimes \langle\psi\rangle$ with $\psi = \phi_\Delta \phi^{-r}$, we have $e\psi^2 = (g\psi)^2$ and $(g\psi)^p = \Delta\psi^p$. The centralizer of $e\phi$ in $\mathcal{C}$ coincides with $\mathcal{C}yc\mathcal{C}(g\psi, g\psi)$, and it can be computed in the category of fixed points $(\mathcal{C}yc\mathcal{C})^{\phi^p \phi_\Delta^{-2}}$.*

Proof. The corollary is a direct application of Proposition 3.37, where, for (i), we take $(p, q', q, p') = (r, 1, 1, r - 1)$, implying $g = d$ and, $(p, q', q, p') = (p, 1, 2, r)$ for (ii). □

Exercises

Exercise 95 (simultaneous conjugacy). Show that g belongs to $\mathcal{C}onj^*\mathcal{C}((e_i)_{i\in I})$ (see Remark 1.16) if and only if $g \preccurlyeq e_i a$ holds for every i. Generalize 1.10 to 1.15 in this context.

Exercise 96 (no right-lcm). Let M be the monoid generated by $\mathtt{a}_0, \mathtt{a}_1, \ldots, \mathtt{b}_0, \mathtt{b}_1, \ldots, \mathtt{c}$ with relations $\mathtt{a}_i = \mathtt{a}_{i+1}\mathtt{c}$, $\mathtt{b}_j = \mathtt{b}_{j+1}\mathtt{c}$, and $\mathtt{a}_i\mathtt{b}_j = \mathtt{b}_i\mathtt{a}_j$ for all i, j. (i) Show that $\mathtt{a}_i\mathtt{b}_j = \mathtt{a}_{i+k}\mathtt{b}_{j+k}\mathtt{c}^{2k}$ holds for every k, and deduce that $\mathtt{a}_i$ and $\mathtt{b}_j$ admit no right-lcm. (ii) Show that, nevertheless, every family $\mathrm{Div}(g)$ is closed under right-comultiple.

Exercise 97 (quasi-distance). (i) In Context 3.7, show that $\ell_\Delta(g) = \sup_\Delta(g^0)$ holds for every g in $\mathcal{G}$. (ii) For g, g' in $\mathcal{G}$, define $\mathrm{dist}(g, g')$ to be ∞ if g, g' do not share the same source, and to be $\ell_\Delta(g^{-1}g')$ otherwise. Show that dist is a quasi-distance on $\mathcal{G}$ that is compatible with $=_\Delta$.

Exercise 98 (conjugacy). Show using Algorithm 2.23 or Algorithm 2.42 that the braids β and β_1 of Example 3.6 are conjugate.

Notes

Sources and comments. The Conjugacy Problem for the braid groups was solved by F.A. Garside in his seminal PhD thesis of 1967 [124] and in the paper [125]. The original solution consists in introducing summit sets, which are distinguished subsets of conjugacy classes, in general larger than the super-summit sets but nevertheless finite and effectively computable. At the time, the normal form was not known, and the summit set was defined to consist of all elements of the form $\Delta^i f$ with maximal i. Cycling, decycling, and their connection with the current normal form were introduced and first investigated around 1988 by E. ElRifai and H. Morton in [117]. More recent contributions by J. Birman, K.Y. Ko, and S.J. Lee appear in [22], by V. Gebhardt in [129] and by J. Birman, V. Gebhardt and J. González-Meneses in [20]. Sliding circuits were introduced by V. Gebhardt and J. González-Meneses in [131] and [132].

Our exposition in Section 1 is close to the one given by F. Digne and J. Michel in [109]. In particular, $\mathcal{C}yc_\phi(\mathcal{C})$ is the particular case for $n = 1$ of the category called $\mathcal{C}_n(\phi_\Delta)$ in Subsection XIV.1.2, and $\mathcal{C}(P_1(\phi))$ in [109, 8.3(ii)]. The study of ribbons as a Garside category was introduced in Digne and Michel [109]. It is also implicit in Godelle [133]. Proposition 1.59(i) is a generalization of Paris [193, Proposition 5.6].

The ribbon category considered in Subsection 1.4 is an extension of the ribbon category occurring in the study of the normalizer of a parabolic submonoid of an Artin–Tits monoid (the submonoid generated by a part of the atoms) as developed by L. Paris and E. Godelle in [193, Section 5], and in Godelle [134], [135], [139]. The description of the atoms in a ribbon category given in Proposition 1.59 extends to the case of Artin–Tits monoids associated to infinite Coxeter groups (and thus do not have a Garside element). The following result can be extracted from the proof of Godelle [135, Theorem 0.5].

Proposition. *Assume that B^+ is an Artin–Tits monoid and let B_I^+ be a submonoid generated by a set I of atoms of B^+. The atoms of B_I^+ are the elements $I \xrightarrow{v(J,I)} \phi_J(I)$, where J is the connected component of $I \cup \{s\}$ in the Coxeter graph of atoms of B^+ for s an atom of B^+ such that J has spherical type.*

The current exposition of Section 2 is an extension to a general categorical framework of the main results in the above mentioned papers. Some of the original arguments were simplified along the way. One of the specific features of the current treatment is that nontrivial invertible elements are not discarded. We are indebted to J. González-Meneses for Lemma 2.27 (private communication). Our treatment of sliding is a slight generalization of the results of Gebhardt and González-Meneses [131], which we mostly follow. In particular, the proof of Lemma 2.50 for $k = 1$ is similar to the proof of [131, Proposition 5]. V. Gebhardt and J. González-Meneses have a more efficient replacement for Step 11 of Algorithm 2.39, based on the notion of *pullback*, a partial inverse to sliding on morphisms. However, we decided to omit a description of it since it would lengthen considerably our exposition.

Let us also mention the *ultra-summit set* as introduced by V. Gebhardt in [129]: in general, the ultra-summit set is a proper subset of the super-summit set, and in turn strictly includes the set of sliding circuits. The operations of *transport*, which in the terminology of Section 2 is the map on morphisms induced by the sliding functor, and the related operation *pullback* were first introduced in [129] for ultra-summit set, and later extended to the set of sliding circuits in [131]. The notion of ultra-summit set allows for especially efficient implementations. However, as it was superseded by sliding circuits, we did not give the details here. All the algorithms described in this chapter are implemented in GAP [183].

Alternative solutions to the Conjugacy Problem for braids based on the geometric structure of braids (Nielsen–Thurston classification) rather than on their algebraic structure might also exist: in the case of B_4, such a solution with a polynomial complexity was recently described by M. Calvez and B. Wiest in [47]. Let us mention that, using the Garside structure, M. Calvez has given in [46] an algorithm for finding the type of a braid in the Nielsen–Thurston classification that is quadratic in the braid length.

The results of Section 3 are directly inspired by M. Bestvina's analysis in [15], which was generalized to the Garside context in work of Charney, Meier, and Whittlesey [57]. In particular, Subsection 3.2 is a solution of the exercise suggested in Bessis [7, 8.1], namely rewriting the results of [15] and [57] in a categorical setting. Proposition 3.32 is already in [20, 4.5, 4.6], in the case when $\mathcal{C}$ has no nontrivial invertible element and the order of ϕ_Δ is finite. Here we allow for an additional generalization, namely the presence of invertible elements in the considered category.

Further questions. An open problem related to the Conjugacy Problem for the braid group B_n is to bound the number of slidings required to reach the ultra-summit set.

Although very efficient implementations of sliding circuits have been made, the theoretical complexity of the method is exponential, and the main open problem of the area so far is to know whether the Conjugacy Problem for braid groups B_n with $n \geqslant 5$ admits a polynomial solution.

Owing to its supposed high complexity, it has been proposed to use the Conjugacy Problem for braids to design cryptographical protocols, see [3], [158], or [74] for a survey. Experiments seem promising. However, a potential weakness of the approach is that, because the braid groups are not amenable, it is virtually impossible to state and prove probabilistic statements involving braids: there is no reasonable probability measure on sets of braids, and, therefore, actually *proving* that a cryptosystem based on braids is secure remains a challenging question.

Another natural problem connected with conjugacy is whether any two dth roots of an element of a (quasi)-Garside group are necessarily conjugate. In the case of braid groups, using the Nielsen–Thurston classification, J. González-Meneses proved in [142] that any two dth roots of an element of B_n^+ are conjugate in B_n^+. This is an open problem in general (see Conjecture X.3.10), in particular for other spherical Artin–Tits groups, though a similar behavior to the case of the braid groups is expected: in this case, defining a Nielsen–Thurston classification—this amounts to defining what would be a reducible element—is an open problem. Note that the question of the uniqueness of dth roots up to conjugacy has an obvious solution in the case of a bi-orderable group, since dth roots are then unique. However, there are few examples of bi-orderable quasi-Garside groups; the free groups are such an example (see Example XII.1.15, Example I.2.8 and the notes of Chapter I referring to the paper [10]).

Part B

Specific examples

Chapter IX
Braids

This chapter is about various Garside structures that occur in connection with Artin's braid groups (Reference Structure 2, page 5, Reference Structure 3, page 10) and some generalizations, Artin–Tits groups and braid groups associated with complex reflection groups. Due to the richness of the subject, we concentrate here on the description of the structures, but renounce to explore their innumerable applications, some of which will be just briefly mentioned in the Notes section.

The chapter contains three sections. In Section 1, we introduce Artin–Tits groups and describe Garside families that have been constructed on them using what is called the classical approach, namely the one based on the Artin generators. We establish the needed background about Coxeter groups, and apply both the reversing approach of Chapter II and the germ approach of Chapter VI to construct the classical Garside structure for the braid group associated with a spherical type Coxeter system (Proposition 1.29 and 1.36).

Then, in Section 2, we use the germ approach again to describe another Garside structure existing on certain Artin–Tits groups, namely the so-called dual Garside structure. Here we concentrate on type A (Proposition 2.11) and on the beautiful connection with noncrossing partitions (Proposition 2.7), mentioning in a more sketchy way the extension to arbitrary finite Coxeter groups of other types. We also show that a number of exotic Garside structures arise in particular in connection with braid orderings, such structures involving Garside elements (in the current meaning) but failing to be Garside monoids (in the restricted sense) due to lack of Noetherianity.

Finally, in Section 3, we show how the dual approach of Section 2 can also be developed for the braid groups of (certain) complex reflection groups, namely those that are finite and well-generated (Propositions 3.20, 3.21, and 3.22).

1 The classical Garside structure on Artin–Tits groups

The classical Garside structure on an Artin–Tits group directly relies on the properties of the underlying Coxeter group, and it can be constructed in (at least) two different ways, namely starting from the presentation and using the technique of subword reversing as described in Section II.4, or starting from the properties of Coxeter groups and using the technique of germs and derived groups as described in Chapter VI.

Here we shall treat both approaches: we first recall basic properties of Coxeter groups and the definition of the associated Artin–Tits groups in Subsection 1.1, then develop the reversing approach in Subsection 1.2, and finally describe the germ approach in Subsection 1.3.

1.1 Coxeter groups

Coxeter groups are groups defined on the shape of the symmetric group $\mathfrak{S}_n$. Here we introduce them based on the notion of a Coxeter system. We recall that an *involution* of a group G is a nontrivial endomorphism of G whose square is the identity.

Definition 1.1 (Coxeter group, Coxeter system). Assume that W is a group and Σ is a set of involutions in W. For s and t in Σ, let $m_{s,t}$ be the order of st if this order is finite, and be ∞ otherwise. We say that (W, Σ) is a *Coxeter system*, and that W is a *Coxeter group*, if W admits the presentation

$$(1.2) \quad \langle \Sigma \mid \{s^2 = 1 \mid s \in \Sigma\} \cup \{(s|t)^{[m_{s,t}]} = (t|s)^{[m_{t,s}]} \mid s, t \in \Sigma \text{ and } m_{s,t} \neq \infty\}\rangle,$$

where $(f|g)^{[m]}$ stands for the word $f|g|f|g|...$ with m letters.

A relation $(s|t)^{[m_{s,t}]} = (t|s)^{[m_{t,s}]}$ as in (1.2) is called a *braid relation*. Note that, in presence of the torsion relations $s^2 = 1$, it is equivalent to $(st)^{m_{s,t}} = 1$. Beware that a Coxeter group may originate from several non-isomorphic Coxeter systems (see Exercise 99). Note that, in Definition 1.1, we do not assume Σ to be finite. Nevertheless, in all examples we will be interested in, Σ will be finite.

Example 1.3 (Coxeter system). Consider the symmetric group $\mathfrak{S}_n$, and let Σ_n be the family of all transpositions $(i, i+1)$ for $i = 1, ..., n-1$. Put $\mathtt{s}_i = (i, i+1)$. Each transposition $\mathtt{s}_i$ has order 2, the relations $\mathtt{s}_i\mathtt{s}_j = \mathtt{s}_j\mathtt{s}_i$ hold for $|i-j| \geqslant 2$, and the relations $\mathtt{s}_i\mathtt{s}_j\mathtt{s}_i = \mathtt{s}_j\mathtt{s}_i\mathtt{s}_j$ hold for $|i-j| = 1$. The group $\mathfrak{S}_n$ is generated by Σ_n and it is well known that the above relations make a presentation of $\mathfrak{S}_n$. Moreover, the order of $\mathtt{s}_i\mathtt{s}_j$ is 2 for $|i-j| \geqslant 2$, and 3 for $|i-j| = 1$. It follows that $(\mathfrak{S}_n, \Sigma_n)$ is a Coxeter system, with the associated coefficients $m_{\mathtt{s},\mathtt{t}}$ given by $m_{\mathtt{s}_i,\mathtt{s}_j} = 2$ for $|i-j| \geqslant 2$, and $m_{\mathtt{s}_i,\mathtt{s}_j} = 3$ for $|i-j| = 1$.

All information needed to encode a Coxeter system is the list of the coefficients $m_{s,t}$. It is customary to record the latter in the form of a square matrix. For the diagonal, we complete with the values $m_{s,s} = 1$, which amounts to completing the presentation (1.2) with the trivial relations $s = s$.

Definition 1.4 (Coxeter matrix). A *Coxeter matrix* is a symmetric matrix $(m_{s,t})$ indexed by a set Σ and whose entries are integers or ∞ that satisfy $m_{s,t} \geqslant 2$ for every pair (s,t) of distinct element of Σ and $m_{s,s} = 1$ for every s in Σ.

For instance, the Coxeter matrix associated with the Coxeter system $(\mathfrak{S}_4, \Sigma_4)$ of Example 1.3 is $\begin{pmatrix} 1 & 3 & 2 \\ 3 & 1 & 3 \\ 2 & 3 & 1 \end{pmatrix}$. We thus have associated a Coxeter matrix to a Coxeter system. The following theorem gives a converse. For a proof see in [27, Chapter V, §4 no 3, Proposition 4].

Proposition 1.5 (Coxeter system). *Let $(m_{s,t})$ be an arbitrary Coxeter matrix indexed by a set Σ, and let W be the group presented by the associated presentation* (1.2). *Then (W, Σ) is a Coxeter system.*

The meaning of this result is that, in the group W presented as above, every element of Σ is different from 1 and, for every pair of elements of Σ, the order of st in W is exactly $m_{s,t}$ (a priori, it could be less).

A way to encode a Coxeter matrix is to draw a *Coxeter graph* with one vertex for each element of Σ, and one (unoriented) edge labeled $m_{s,t}$ between the vertices s and t if $m_{s,t} \neq 2$; the convention is that 3-labeled edges are presented as unlabeled. For instance, the Coxeter graph for $\mathfrak{S}_n$ is

If Γ is a Coxeter graph we will denote by $(W_\Gamma, \Sigma_\Gamma)$ the associated Coxeter system.

We now recall basic results about Coxeter groups. For a detailed exposition, see for instance [27, Chapter IV].

If (W, Σ) is a Coxeter system, since the elements of Σ are involutions, they generate W positively, that is, every element of W can be expressed as a product of elements of Σ. We denote by $\|\cdot\|_\Sigma$ the Σ-length on W (see Definition VI.2.50).

Definition 1.6 (reflection, reduced word). If (W, Σ) is a Coxeter system, the elements of $\bigcup_{g\in W} g\Sigma g^{-1}$ are called the *reflections* of W. A Σ-word is called *reduced* if no shorter word represents the same element of W.

The next result, which associates with every element of a Coxeter group a family of reflections, captures in algebraic terms the essential geometric properties of these groups. According to our general conventions, if w is a Σ-word, we denote by $[w]$ the element of W represented by w. We use $\triangle$ for the symmetric difference in a powerset, that is, the operation defined by $X \triangle Y = (X\backslash Y) \cup (Y\backslash X)$ (corresponding to addition mod 2 on indicator functions).

Lemma 1.7. *Assume that (W, Σ) is a Coxeter system and R is the set of reflections of W.*

(i) *There is a unique map N from Σ^* to the powerset of R satisfying $N(s) = \{s\}$ for s in Σ and $N(u|v) = N(u) \triangle [u]N(v)[u]^{-1}$ for all u and v in Σ^*.*

(ii) *For $w = s_1|\cdots|s_k$ with $s_1, \ldots, s_k$ in Σ, we have*

$$N(w) = \{s_1\} \triangle \{s_1 s_2 s_1\} \triangle \cdots \triangle \{s_1 \cdots s_{k-1} s_k s_{k-1} \cdots s_1\}. \tag{1.8}$$

(iii) *If w is a Σ-word, the value of $N(w)$ depends only on $[w]$.*

(iv) *If w is a reduced Σ-word, the reflections appearing in the right-hand side of* (1.8) *are pairwise distinct, the symmetric difference is a union, and the cardinal of $N(w)$ is the Σ-length of $[w]$.*

Proof. If N satisfying (i) exists, it must satisfy (ii), as can be seen by applying inductively the equality $N(u|v) = N(u) \triangle [u]N(v)[u]^{-1}$. This implies the unicity of N. Conversely, if we take the right-hand side of (1.8) as a definition for N, the equality $N(u|v) = N(u) \triangle [u]N(v)[u]^{-1}$ is satisfied for all words u and v.

To prove (iii), it suffices to show that N is invariant under the defining relations of W, which results from the easy equalities $N(s|s) = \emptyset$ and $N((s|t)^{[m_{s,t}]}) = N((t|s)^{[m_{t,s}]})$ for every pair (s,t) of distinct elements of Σ such that $m_{s,t}$ is finite.

For (iv), assume $w = s_1|\cdots|s_k$. Saying that two reflections occurring in (1.8) coincide means that there exist $i < j$ such that $s_1s_2\cdots s_i\cdots s_2s_1 = s_1s_2\cdots s_j\cdots s_2s_1$ holds in W. But, then, we deduce $s_is_{i+1}\cdots s_j = s_{i+1}s_{i+2}\cdots s_{j-1}$, and k cannot be minimal. □

Owing to Lemma 1.7(iii), we shall naturally define $N(g)$ for g in W to be the common value of $N(w)$ for w representing g.

Lemma 1.9. *Assume that (W,Σ) is a Coxeter system and R is the set of reflections of W.*
(i) *For g, h in W, the relation $\|g\|_\Sigma+\|h\|_\Sigma=\|gh\|_\Sigma$ is equivalent to $N(g) \subseteq N(gh)$.*
(ii) *For every r in R, we have $r \in N(r)$.*
(iii) *For every g in W, we have $N(g) = \{r \in R \mid \|rg\|_\Sigma < \|g\|_\Sigma\}$.*

Proof. Let us prove (i). By Lemma 1.7(iv), $\|gh\|_\Sigma = \|g\|_\Sigma + \|h\|_\Sigma$ holds if and only if the cardinal of $N(gh)$ is the sum of the cardinals of $N(g)$ and $N(h)$. Using the equality $N(gh) = N(g) \vartriangle gN(h)g^{-1}$, this is equivalent to $N(g) \subseteq N(gh)$.

For (ii), assume that r is a reflection, and let $s_1\cdots s_{n-1}s_ns_{n-1}\cdots s_1$ be an expression of r as a conjugate of an element of Σ, with $s_1, ..., s_n$ in Σ and n minimal. Let r_i the ith entry in the right-hand side of (1.8) applied to this decomposition (which has $2n-1$ terms). We claim that $r = r_n$ and $r \neq r_i$ hold for $i \neq n$, which will show that r lies in $N(r)$. Indeed, an easy computation shows that, for $i \leqslant n$, we have $rr_ir = r_{2n-i}$; hence $r_{2n-i} = r$ is equivalent to $r_i = r$. But, by the minimality of n, we have $r_i \neq r$ for $i < n$.

Let us prove (iii). Let $s_1|\cdots|s_k$ be a reduced Σ-word representing g. Consider $r = s_1s_2\cdots s_is_{i-1}\cdots s_1$, an element of $N(g)$. Then we find $rg = s_1\cdots s_{i-1}s_{i+1}\cdots s_k$, hence $\|rg\|_\Sigma < \|g\|_\Sigma$. Conversely, for r in $R \setminus N(g)$ we have $r \notin rN(g)r^{-1}$ whereas, by (ii), r lies in $N(r)$. Then, by Lemma 1.7(ii), r must lie in $N(rg)$ and, by the first part, we deduce the relations $\|g\|_\Sigma = \|r(rg)\|_\Sigma < \|rg\|_\Sigma$. □

Proposition 1.10 (exchange property). *If (W,Σ) is a Coxeter system and $s_1|\cdots|s_k$ is a reduced Σ-word representing an element g of W, then, for every s in Σ satisfying $\|sg\|_\Sigma < \|g\|_\Sigma$, there exists i such that the word $s_1|\cdots|s_{i-1}|s_{i+1}|\cdots|s_k$ represents sg. The latter word is reduced, that is, we have $\|sg\|_\Sigma = \|g\|_\Sigma - 1$.*

Proof. By assumption, one obtains a Σ-minimal expression for g by left-multiplying by s a Σ-minimal expression for sg, which implies $s \in N(g)$. It follows that there exists i satisfying $s = s_1\cdots s_i\cdots s_1$, whence $sg = s_1\cdots s_{i-1}s_{i+1}\cdots s_k$. But, in this case, the word $s|s_1|\cdots|s_{i-1}|s_{i+1}|\cdots|s_k$ represents g and it is reduced since it has length k. It follows that its subword $s_1|\cdots|s_{i-1}|s_{i+1}|\cdots|s_k$ is reduced as well. □

The symmetric result obtained by exchanging left and right applies to gs, as can be seen by applying Proposition 1.10 to sg^{-1}.

We now list some consequences of the exchange property.

Corollary 1.11 (Matsumoto's lemma). *Assume that* (W,Σ) *is a Coxeter system.*

(i) *If* $s_1|\cdots|s_k$ *is a non-reduced* Σ*-word representing an element* g *of* W*, there exist* i *and* j *such that the word* $s_1|\cdots|s_{i-1}|s_{i+1}|\cdots|s_{j-1}|s_{j+1}|\cdots|s_k$ *represents* g*.*

(ii) *If* w *and* w' *are reduced* Σ*-words representing the same element of* W*, one can pass from* w *to* w' *using only braid relations. In particular,* w *and* w' *have same length.*

Proof. (i) Let i be the largest index such that the word $s_i|\cdots|s_k$ is not reduced. Then, by Proposition 1.10, there exist $j > i$ satisfying $s_i\cdots s_k = s_{i+1}\cdots s_{j-1}s_{j+1}\cdots s_k$.

(ii) First, Lemma 1.7(iii) and (iv) implies that two reduced words representing the same element have the same length. We then prove the result using induction on the length k of w and w'. Assume $w = s_1|\cdots|s_k$ and $w' = s'_1|\cdots|s'_k$, and w and w' represent g in W. Then the word $s'_1|s_1|s_2|\cdots|s_k$ represents an element of Σ-length $k-1$. Hence, by Proposition 1.10, there exists $i \leqslant k$ such that the words $s'_1|s_1|\cdots|s_{i-1}$ and $s_1|s_2|\cdots|s_i$ represent the same element (and are reduced). If $i < k$ holds, we are done since, by the induction hypothesis, we can pass using only braid relations from $s_1|\cdots|s_i$ to $s'_1|s_1|\cdots|s_{i-1}$ and then from $s_1|\cdots|s_{i-1}|s_{i+1}|\cdots|s_k$ to $s'_2|\cdots|s'_k$. Otherwise, we have $i = k$ and both $s'_1|s_1|\cdots s_{k-1}$ and $s_1|\cdots|s_k$ represent g. Repeating the same argument with s_1 instead of s'_1, we see that either we are done by the induction hypothesis, or $s_1|s'_1|s_1|\cdots|s_{k-2}$ represents g. Going on, we deduce that either we are done by induction hypothesis, or that g is represented by the word $(s_1|s'_1)^{[k]}$. Going further by one step then shows that w is also represented by the word $(s'_1|s_1)^{[k]}$, and, as this word is reduced, k must be the order m_{s_1,s'_1} of $s_1s'_1$ in W. Applying the induction hypothesis, we can pass using only braid relations from w to $(s_1|s'_1)^{[k]}$ and from $(s'_1|s_1)^{[k]}$ to w'. As $(s_1|s'_1)^{[k]} = (s'_1|s_1)^{[k]}$ is a braid relation, we are done. ☐

Corollary 1.12 (parabolic subgroup). *If* (W,Σ) *is a Coxeter system and* W_I *is the subgroup of* W *generated by a subset* I *of* Σ*, then all reduced* Σ*-words representing elements of* W_I *are* I*-words.*

Proof. Assume $g \in W_I$. By definition, there exists at least one I-word, say w, that represents g. By Matsumoto's lemma, one can obtain a reduced word w' representing g starting from w and deleting letters, so w' is an I-word. Finally, by Matsumoto's lemma again, every reduced expression of g is obtained from w' by applying braid relations, hence it must be an I-word. ☐

From the above corollaries we deduce canonical representatives of left-cosets.

Proposition 1.13 (left-cosets). *If* (W,Σ) *is a Coxeter system and* W_I *is the subgroup of* W *generated by a subset* I *of* Σ*, then, for every* g *in* W*, the following are equivalent:*

(1.14) *The element* g *has no nontrivial* Σ*-suffix in* W_I*.*

(1.15) *The equality* $\|gh\|_\Sigma = \|g\|_\Sigma + \|h\|_\Sigma$ *holds for all* $h \in W_I$*.*

(1.16) *The element* g *has minimal* Σ*-length in its coset* gW_I*.*

Further, if g satisfies the conditions above, it is the unique element of gW_I of minimal Σ-length and for every f in W_I, every Σ-suffix of gf in W_I is a Σ-suffix of f.

The symmetric result (exchanging left and right, suffix and prefix) is valid for right-cosets $W_I g$.

Proof. Relation (1.15) implies that g has minimal Σ-length in its coset gW_I. Conversely, assume that g satisfies (1.16). If h is an element of minimal Σ-length in W_I such that (1.15) does not hold, then h can be written fs with $f \in W_I$ and $s \in I$ satisfying $\|fs\|_\Sigma = \|f\|_\Sigma + \|s\|_\Sigma$ and $\|gfs\|_\Sigma < \|gf\|_\Sigma = \|g\|_\Sigma + \|f\|_\Sigma$. By the exchange property, this means that one gets a Σ-word representing gfs by deleting a generator in a Σ-word of minimal length representing gf. As we have $\|fs\|_\Sigma = \|f\|_\Sigma + \|s\|_\Sigma$, the deletion must occur in g. We then obtain $gfs = g'f$ with $\|g'\|_\Sigma < \|g\|_\Sigma$. This contradicts the minimality of g in gW_I. We thus have shown the equivalence of (1.15) and (1.16).

Next, Conditions (1.14) and (1.16) are equivalent. Indeed, if g has a non-trivial Σ-suffix v in W_I, we then have $g = g_1 h$ with $\|g_1\|_\Sigma = \|g\|_\Sigma - \|v\|_\Sigma$. It follows that g is not an element of minimal Σ-length in gW_I. Hence (1.16) implies (1.14). Conversely, if g' is an element of minimal Σ-length in gW_I, we have $g = g'h$ with h in W_I and $\|g\|_\Sigma = \|g'\|_\Sigma + \|h\|_\Sigma$, by (1.15) applied to g'. It follows that h is a nontrivial Σ-suffix of g in W_I, contradicting (1.14).

Now (1.15) shows that all elements of gW_I have a Σ-length strictly larger than that of g, whence the unicity. Moreover, if an element h of W_I is a Σ-suffix of gf with f in W_I, the equality $\|gf\|_\Sigma = \|gfh^{-1}\|_\Sigma + \|h\|_\Sigma$ holds and, by (1.15), it follows that so do $\|gf\|_\Sigma = \|g\|_\Sigma + \|f\|_\Sigma$ and $\|gfh^{-1}\|_\Sigma = \|g\|_\Sigma + \|fh^{-1}\|_\Sigma$. This implies $\|f\|_\Sigma = \|fh^{-1}\|_\Sigma + \|h\|_\Sigma$, so that h is a Σ-suffix of f. □

Further results arise in the case of a finite Coxeter system, that is, when W is finite.

Lemma 1.17. *If (W, Σ) is a Coxeter system and W is finite, then there exists a unique element δ of maximal Σ-length in W. The Σ-length of δ is the total number of reflections in W, and every element of W is both a Σ-prefix and a Σ-suffix of δ. Furthermore, δ is the unique element of W that admits all elements of Σ as prefixes (or suffixes).*

Proof. Since the group W is finite, it contains an element δ of maximal Σ-length. Then, for every reflection r, we have $\|r\delta\|_\Sigma < \|\delta\|_\Sigma$. By Lemma 1.9(iii), this implies $N(\delta) = R$, whence $\|\delta\|_\Sigma = \#(R)$. For every w in W, we have $N(w) \subseteq N(\delta)$, so Lemma 1.9(i) implies that w is a Σ-prefix of δ. The unicity of δ is clear since another element of the same length being a prefix of δ must be equal to it.

Let now g be an element of W_Γ that admits all elements of Σ as prefixes. Since g is a suffix of δ, there exists an element f in W satisfying $\delta = fg$ and $\|\delta\|_\Sigma = \|f\|_\Sigma + \|g\|_\Sigma$. But no element s of Σ can be a suffix of f since it is a prefix of g. Hence f has to be 1. Exchanging prefixes and suffixes gives the analogous result for an element that admits all elements of Σ as suffixes. □

When the group W is finite, the Coxeter system (resp. graph, resp. matrix) is said to be of *spherical type*. Note that, if a Coxeter graph Γ is not connected, then the group W_Γ is the direct product of the Coxeter groups associated with the connected components of Γ so that, in view of an exhaustive description, it is enough to consider connected graphs.

The complete classification of the connected Coxeter graphs of spherical type is shown in Figure 1 (see [27, Chapter VI,§1 no 1, Theorem 1]). In that figure two different types correspond to non-isomorphic groups. Nevertheless, we shall use also the notation $\mathrm{I}_2(m)$ for $m \geqslant 2$: then types $\mathrm{I}_2(3)$, $\mathrm{I}_2(4)$, $\mathrm{I}_2(6)$ are respectively the same as types A_2, B_2, and G_2. The graph of type $\mathrm{I}_2(2)$ is not connected; the corresponding Coxeter group is the direct product of two groups of order 2. We shall also use the notation $\mathrm{I}_2(\infty)$ for the infinite dihedral group, which is the free product of two groups of order 2.

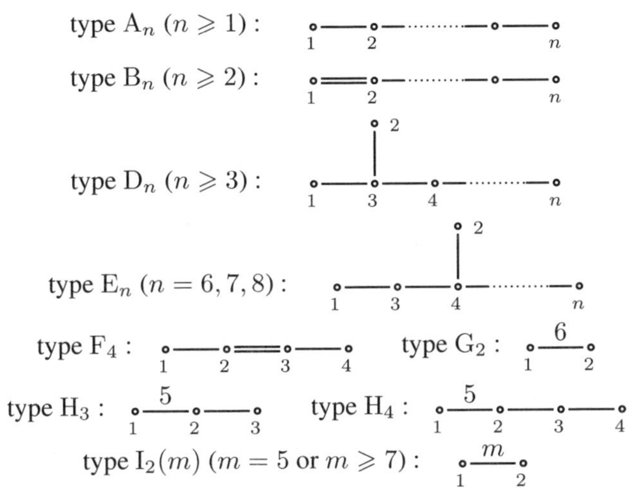

Figure 1. Dynkin classification of the Coxeter graphs such that the associated Coxeter group is finite: no edge means exponent 2, a simple edge means exponent 3, a double edge means exponent 4.

1.2 Artin–Tits groups, reversing approach

When we start with a Coxeter presentation and remove the torsion relations $s_i^2 = 1$, we obtain a new group presentation: this is what is called the *Artin–Tits group* associated with the Coxeter system.

Definition 1.18 (Artin–Tits group, Artin–Tits monoid). If (W, Σ) is a Coxeter system encoded by the Coxeter matrix $(m_{s,t})$, we denote by $\mathcal{P}_{W,\Sigma}$ the presentation $(\Sigma, \mathcal{R})$, where $\mathcal{R}$ is the family of all braid relations

(1.19) $$(s|t)^{[m_{s,t}]} = (t|s)^{[m_{t,s}]} \text{ for } s, t \text{ in } \Sigma \text{ with } m_{s,t} \neq \infty.$$

The *braid group* (*resp. monoid*) of (W, Σ), also called the *Artin–Tits group* (*resp. monoid*) associated with (W, Σ), is the group (*resp.* monoid) presented by $\mathcal{P}_{W,\Sigma}$.

Example 1.20 (Artin–Tits monoid, Artin–Tits group). The Artin–Tits group associated with the Coxeter system $(\mathfrak{S}_n, \Sigma_n)$ of Example 1.3 is the group with presentation

$$\left\langle \sigma_1, \ldots, \sigma_{n-1} \;\middle|\; \begin{array}{ll} \sigma_i\sigma_j = \sigma_j\sigma_i & \text{for } |i-j| \geqslant 2 \\ \sigma_i\sigma_j\sigma_i = \sigma_j\sigma_i\sigma_j & \text{for } |i-j| = 1 \end{array} \right\rangle, \tag{1.21}$$

in which we recognize Artin's braid group B_n (Reference Structure 2, page 5). The associated Artin–Tits monoid is then, of course, Artin's braid monoid B_n^+.

We begin with a few easy remarks.

Lemma 1.22. *Assume that (W, Σ) is a Coxeter system and B (resp. B^+) is the associated Artin–Tits group (resp. monoid).*

(i) *The identity map on Σ induces a surjective (group) morphism π from B to W, and a (monoid) morphism ι from B^+ to B.*

(ii) *Mapping a reduced Σ-word to its image in B^+ provides a well-defined injective map σ of W to B^+ that is a (set theoretical) section of π.*

(iii) *The monoid B^+ has no nontrivial invertible element.*

(iv) *The presentation $\mathcal{P}_{W,\Sigma}$ is homogeneous, the monoid B^+ is Noetherian; every element of B^+ has a finite height, namely the common length of the Σ-words that represent it, and the atoms of B^+ are the elements of Σ.*

Proof. (i) The results are direct consequences of the definitions.

Assertion (ii) follows from Matsumoto's lemma: if two reduced Σ-words have the same image in W, they are connected by braid relations, hence they have the same image in B^+ (and a fortiori in B). By definition, $\pi(\sigma([w])) = [w]$ holds for every reduced Σ-word w, so σ is a section for π and, therefore, it must be injective.

(iii) No braid relation involves an empty word, so a nonempty Σ-word cannot be equivalent to an empty word, hence cannot represent 1.

(iv) Braid relations have the form $u = v$ with u and v of the same length. So the presentation $\mathcal{P}_{W,\Sigma}$ is homogeneous and, according to Proposition II.2.32 (homogeneous), the length provides an $\mathbb{N}$-valued Noetherianity witness. By Proposition II.2.33 (witness), the monoid B^+ is Noetherian, and every element of B^+ has a height which is the common length of the words that represent it. Then B^+ is generated by its atoms, which are the elements of height 1, hence the elements of Σ. □

We shall now explain how to investigate Artin–Tits groups and, first, Artin–Tits monoids, using a direct approach based on the presentation (1.19) and on the reversing technique of Section II.4, which proves to be well suited. The basic (and trivial) observation in the approach is

Lemma 1.23. *For every Coxeter system (W, Σ), the presentation $\mathcal{P}_{W,\Sigma}$ is right- and left-complemented. If $(m_{s,t})$ is the corresponding Coxeter matrix, the involved syntactic right-complement θ and left-complement $\widetilde{\theta}$ are given by*

$$\theta(s,t) = (t|s)^{[m_{s,t}-1]} \qquad \widetilde{\theta}(s,t) = \begin{cases} (t|s)^{[m_{s,t}-1]} & \textit{if } m_{s,t} \textit{ is odd,} \\ (s|t)^{[m_{s,t}-1]} & \textit{if } m_{s,t} \textit{ is even,} \end{cases}$$

and $\theta(s,t)$ and $\widetilde{\theta}(s,t)$ undefined if $m_{s,t}$ is ∞.

Proof. That $\mathcal{P}_{W,\Sigma}$ is right-complemented and associated with θ is straightforward since, for all s, t in Σ, the relation $(t|s)^{[m_{s,t}]} = (s|t)^{[m_{t,s}]}$ can be written as

$$s \cdot (t|s)^{[m_{s,t}-1]} = t \cdot (s|t)^{[m_{t,s}-1]}.$$

The argument is similar on the left, taking into account that $\widetilde{\theta}(s,t)$ must finish with s. □

Lemma 1.23 implies that every Artin–Tits presentation is eligible for the simple, deterministic version of right- and left-reversing—as was already seen in Section II.4, where braid examples were described. In order to take advantage of such tools, we have to know that reversing is complete for the considered presentation. As we are dealing with a homogeneous presentation, this reduces to verifying that the θ-cube condition of Definition II.4.14 is satisfied. So the following technical result is crucial:

Lemma 1.24. *For every Coxeter system (W, Σ), the θ-cube condition is satisfied in $\mathcal{P}_{W,\Sigma}$ for every triple of pairwise distinct elements of Σ.*

Proof. The point is to establish, for r, s, t pairwise distinct in Σ, the cube relations

(1.25)
$$\theta_3^*(r,s,t) \equiv^+ \theta_3^*(s,r,t),\ \theta_3^*(s,t,r) \equiv^+ \theta_3^*(t,s,r), \text{ and } \theta_3^*(t,r,s) \equiv^+ \theta_3^*(r,t,s),$$

where $\theta_3^*(r,s,t)$ stands for $\theta^*(\theta(r,s), \theta(r,t))$ and θ^* is the extension of θ to words described in Lemma II.4.6; the relations mean that either both terms are defined and they are equivalent, or neither is defined. We shall consider all possible values of the triple $(m_{r,s}, m_{s,t}, m_{r,t})$ with entries in $\{2, 3, ..., \infty\}$, which amounts to considering all three-generator Coxeter systems. Not surprisingly, different behaviors appear according to whether the corresponding Coxeter group is finite (first four cases below) or not (remaining cases). When writing "$\geqslant m$" below, we mean "a finite number not smaller than m".

• *Case* $(2,2,\geqslant 2)$: By definition, we have $\theta(r,s) = \theta(t,s) = s$, $\theta(s,r) = r$, $\theta(s,t) = t$, $\theta(r,t) = (t|r)^{[m]}$, $\theta(t,r) = (r|t)^{[m]}$ with $m = m_{r,t} - 1$, and we obtain

$$\theta_3^*(r,s,t) = \theta^*(s, (t|r)^{[m]}) = (t|r)^{[m]} = \theta^*(r,t) = \theta_3^*(s,r,t),$$
$$\theta_3^*(s,t,r) = \theta^*(t,r) = (r|t)^{[m]} = \theta^*(s, (r|t)^{[m]}) = \theta_3^*(t,s,r),$$
$$\theta_3^*(t,r,s) = \theta^*((r|t)^{[m]}, s) = s = \theta^*((t|r)^{[m]}, s) = \theta_3^*(r,t,s),$$

as expected for (1.25).

• *Case* $(2,3,3)$: The verification was already made in Example II.4.20. By definition, we have $\theta(r,s) = s$, $\theta(s,r) = r$, $\theta(s,t) = t|s$, $\theta(t,s) = s|t$, $\theta(r,t) = t|r$, $\theta(t,r) = r|t$, and, as expected, we obtain

$$\theta_3^*(r,s,t) = \theta^*(s, t|r) = t|s|r|t \equiv^+ t|r|s|t = \theta^*(r, t|s) = \theta_3^*(s,r,t),$$
$$\theta_3^*(s,t,r) = \theta^*(t|s, r) = r|t|s = \theta^*(s|t, r|t) = \theta_3^*(t,s,r),$$
$$\theta_3^*(t,r,s) = \theta^*(r|t, s|t) = s|r|t = \theta^*(t|r, s) = \theta_3^*(r,t,s).$$

• *Case* $(2,3,4)$: The verification is entirely similar, now with the values

$$\theta_3^*(r,s,t) = \theta^*(s, t|r|t) = t|s|r|t|r|s|t \equiv^+ t|r|t|s|t|r|t = \theta^*(r, t|s) = \theta_3^*(s,r,t),$$
$$\theta_3^*(s,t,r) = \theta^*(t|s, r) = r|t|s|r|t|r \equiv^+ r|t|r|s|t|r = \theta^*(s|t, r|t|r) = \theta_3^*(t,s,r),$$

$$\theta_3^*(t,r,s) = \theta^*(r|t|r, s|t) = s|t|r|t|s = \theta^*(t|r|t, s) = \theta_3^*(r,t,s).$$

• *Case* $(2,3,5)$: Here the values are

$$\theta_3^*(r,s,t) = \theta^*(s, t|r|t|r) = t|s|r|t|r|t|s|t|r|t|r|s|t$$
$$\equiv^+ t|r|t|r|s|t|r|t|s|r|t|r|t = \theta^*(r, t|s) = \theta_3^*(s,r,t),$$
$$\theta_3^*(s,t,r) = \theta^*(t|s, r) = r|t|s|r|t|r|t|s|t|r|t|r$$
$$\equiv^+ r|t|r|t|s|t|r|t|r|s|t|r = \theta^*(s|t, r|t|r|t) = \theta_3^*(t,s,r),$$
$$\theta_3^*(t,r,s) = \theta^*(r|t|r|t, s|t) = s|t|r|t|s|r|t|r|t|s$$
$$\equiv^+ s|t|r|t|r|s|t|r|t|s = \theta^*(t|r|t|r, s) = \theta_3^*(r,t,s).$$

Note that, in order to establish an equivalence $u \equiv^+ v$ as above, it is *sufficient* to check that reversing $\overline{u}v$ yields an empty word, which is indeed the case.

• *Case* $(\geqslant 3, \geqslant 3, \geqslant 3)$: In this case, as in all subsequent cases, (1.25) is satisfied because none of the involved expressions is defined. The point is that, as already observed in Example II.4.20, reversing the pattern $\overline{r}|s|t$ leads in six steps to a word that includes the initial pattern, see on the right, and, therefore, it cannot lead in finitely many steps to a positive–negative word.

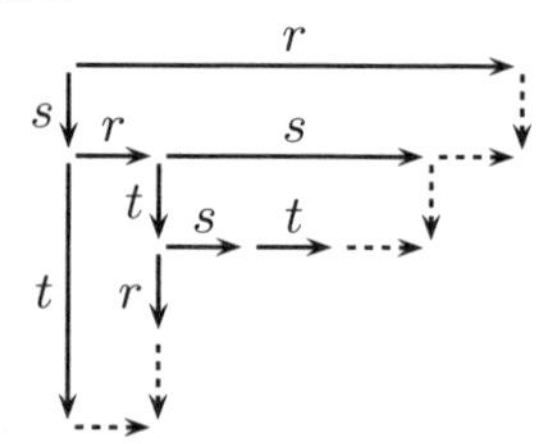

The three indices play symmetric roles, and the same situation holds for every pattern $\overline{s}|t|r$ and the like. It follows that $\theta_3^*(r,s,t)$ is not defined: the latter has the form $\theta^*(s|r|..., t|r|...)$, and reversing $...|\overline{r}|\overline{s}|t|r|...$ cannot terminate, since the latter word includes one of the above forbidden patterns. The same holds for all similar expressions involved in (1.25).

• *Case* $(2, \geqslant 4, \geqslant 4)$: First, we claim that reversing the words $\overline{r}|t|s$ and $\overline{s}|t|r$ does not terminate: as shown on the right, reversing one of these words leads in four steps to a word that includes the other. It follows that $\theta_3^*(r,s,t)$ is not defined:

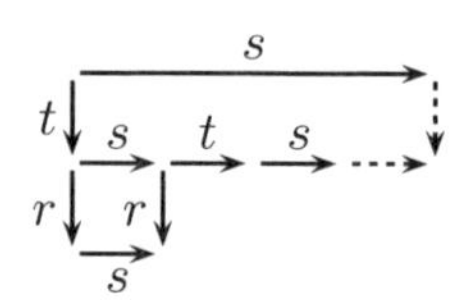

the latter has the form $\theta^*(s, t|r|t|...)$, and reversing $...\overline{s}|t|r|t|...$ cannot terminate since the latter word includes the forbidden pattern $\overline{s}|t|r$. Similarly, $\theta_3^*(s,t,r)$ has the form $\theta^*(t|s|t|..., r)$ and reversing $...|\overline{t}|\overline{s}|\overline{t}|r$ cannot terminate, since the latter word includes the (inverse of the) forbidden pattern $\overline{r}|t|s$, and $\theta_3^*(t,r,s)$ has the form $\theta^*(r|t|r|..., s|t|s|...)$ and reversing $...|\overline{r}|\overline{t}|\overline{r}|s|t|s|...$ cannot terminate since the latter word leads in one step to $...|\overline{r}|\overline{t}|s|\overline{r}|t|s|...$, which includes both forbidden patterns $\overline{r}|t|s$ and (the inverse of) $\overline{s}|t|r$.

• *Case* $(2,3,\geqslant 6)$: This time, reversing $\overline{s}|t|r|t$ does not terminate: as shown on the right, reversing this word leads in ten steps to a word that includes the initial word. It follows again that $\theta_3^*(r,s,t)$ is not defined: the latter has the form $\theta^*(s, t|r|t|r|t|...)$, and reversing

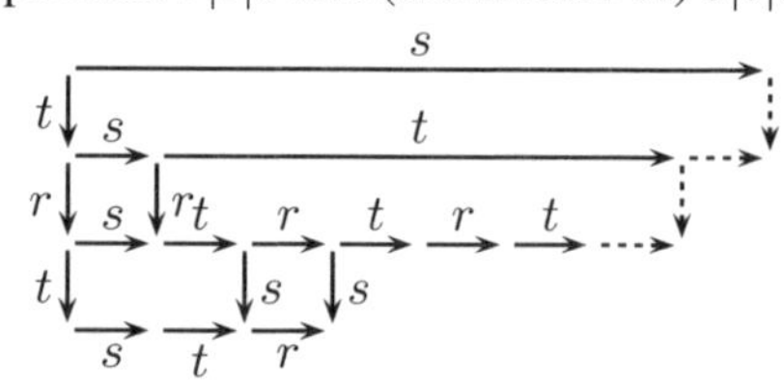

$...\overline{s}|t|r|t|r|t|...$ cannot terminate, since the latter includes the forbidden pattern $\overline{s}|t|r|t$. Similarly, $\theta_3^*(s,t,r)$ is $\theta^*(t|s, r)$ and reversing $\overline{s}|\overline{t}|r$ cannot terminate, since it leads in two steps to $...|r|\overline{s}|t|r|t|r|...$, which includes the forbidden pattern, whereas $\theta_3^*(t,r,s)$ has the form $\theta^*(r|t|r|t|r|..., s|t)$ and reversing $...|\overline{r}|\overline{t}|\overline{r}|\overline{t}|\overline{r}|s|t$ cannot terminate, since the latter word leads in one step to $...|\overline{r}|\overline{t}|\overline{r}|\overline{t}|s|\overline{r}|t$, again including the (inverse of the) forbidden pattern.

• *Case* $(..., ..., \infty)$: Then $\theta(r,t)$ and $\theta(t,r)$ are undefined, hence none of $\theta_3^*(r,s,t)$, $\theta_3^*(r,t,s)$, $\theta_3^*(t,r,s)$, and $\theta_3^*(t,s,r)$ can be defined. If, in addition, $m_{r,s}$ or $m_{s,t}$ is infinite, then all six expressions of (1.25) are undefined, and the cube condition is satisfied. Assume now that $m_{r,s}$ and $m_{s,t}$ are finite, and consider $\theta_3^*(s,r,t)$, that is $\theta^*(\theta(s,r),\theta(s,t))$, hence $\theta^*(r|...,t|...)$: reversing $...|\overline{r}|t|...$ cannot terminate, since the latter word includes the forbidden pattern $\overline{r}|t$. The argument is similar for $\theta_3^*(s,t,r)$.

Thus all possible cases have been considered, and the proof is complete. □

Applying Proposition II.4.16 (right-complemented) and observing that, as the Coxeter relations are palindromic (invariant under left–right symmetry), all results on one side imply their symmetric counterpart, we deduce:

Proposition 1.26. *For every Coxeter system (W,Σ), the associated braid monoid B^+ is cancellative; it admits conditional right- and left-lcms, and left- and right-gcds.*

By Corollary IV.2.41 (smallest Garside), this in turn implies

Corollary 1.27 (smallest Garside). *For every Coxeter system (W,Σ), the associated Artin–Tits monoid admits a smallest Garside family containing 1, namely the closure of the atoms under the right-lcm and right-complement operations, which is also the closure of the atoms under right-division and right-lcm.*

Example 1.28 (dihedral case). For type $\mathrm{I}_2(m)$ with finite m, we have two atoms, say $\mathtt{a}, \mathtt{b}$, and the unique relation $(\mathtt{a}|\mathtt{b})^{[m]} = (\mathtt{b}|\mathtt{a})^{[m]}$. Let Δ be the element represented by $(\mathtt{a}|\mathtt{b})^{[m]}$. Then Δ is the (unique) right-lcm of $\mathtt{a}$ and $\mathtt{b}$. The closure of $\{\mathtt{a},\mathtt{b}\}$ under the right-lcm and right-complement operations is the cardinality $2m$ family $\{1,\mathtt{a},\mathtt{b},\mathtt{ab},\mathtt{ba},...,\Delta\}$ consisting of all elements $(\mathtt{a}|\mathtt{b})^{[k]}$ and $(\mathtt{b}|\mathtt{a})^{[k]}$ with $k \leqslant m$, that is, in other words, the (left or right) divisors of Δ.

For type $\mathrm{I}_2(\infty)$, the monoid is a free monoid based on the two generators $\mathtt{a}, \mathtt{b}$. Then no right-lcm exists, and the smallest Garside family containing 1 is simply $\{1,\mathtt{a},\mathtt{b}\}$.

It is natural to wonder whether the smallest Garside family in an Artin–Tits monoid is bounded. The results of Subsection 1.1 imply that the answer is directly connected with the finiteness of the corresponding Coxeter group. We recall that a Coxeter system (W,Σ) is said to be *of spherical type* if W is finite.

Proposition 1.29 (Garside group I). *Assume that (W,Σ) is a Coxeter system of spherical type and B (resp. B^+) is the associated Artin–Tits group (resp. monoid). Let Δ be the lifting of the longest element of W. Then (B^+,Δ) is a Garside monoid, and B is a Garside group. The element Δ is the right-lcm of Σ, which is the atom set of B^+, and* $\mathrm{Div}(\Delta)$ *is the smallest Garside family of B^+ containing* 1.

Before establishing Proposition 1.29, we establish two auxiliary results.

Lemma 1.30. *For every Coxeter system (W, Σ) with associated Artin–Tits monoid B^+, the image of the lifting σ from W to B^+ is closed under left- and right-divisor in B^+.*

Proof. Assume that g lies in the image of σ, that is, equivalently, we have $g = \sigma(\pi(g))$, and f left-divides g in B^+. By definition, there exist Σ-words u, v satisfying $[u] = f$ and $[uv] = g$. The assumption that g lies in the image of σ, hence that some expression of g is a reduced Σ-word, plus the fact that all expressions of g have the same length imply that uv must be reduced. As right-multiplying by one element σ_i increases the Σ-length by 1 at most, an initial subword of a reduced word must be reduced, so u is a reduced Σ-word. We deduce the equality $f = \sigma(\pi(f))$, so f lies in the image in σ.

The argument for right-divisors is symmetric. □

Lemma 1.31. *If (W, Σ) is a Coxeter system of spherical type and Δ is the lifting of the longest element of W in the associated Artin–Tits monoid B^+, then, for every g in B^+, the following conditions are equivalent:*

(i) *The element g belongs to the image of σ.*

(ii) *The element g is a left-divisor of Δ.*

(iii) *The element g is a right-divisor of Δ.*

Proof. Let δ be the longest element of W, and σ be the lifting from W to B^+. Assume (i). This means that g has an expression u that is a reduced Σ-word. By Lemma 1.17, $[u]$ must be an Σ-prefix of δ in W, that is, $\delta = [uv]$ holds in W for some reduced Σ-word v. Lifting this equality by σ, we obtain $\Delta = g\sigma([v])$ in B^+, which implies (ii). The argument for (iii) is symmetric, using the fact that u must also be an Σ-suffix of δ.

On the other hand, Δ belongs to the image of σ by definition, so, by Lemma 1.30, every left- or right-divisor of Δ must lie in the image of σ, that is, each of (ii) and (iii) implies (i). □

Proof of Proposition 1.29. We claim that Δ is a Garside element in B^+. Indeed, each generator σ_i belongs to the image of σ, hence, by Lemma 1.31, left-divides Δ, so the left-divisors of Δ generate B^+. On the other hand, Lemma 1.31 says that the left- and right-divisors of Δ coincide, since they both coincide with the elements in the image of σ. Finally, as σ is injective, the divisors of Δ in B^+ are in one-to-one correspondence with the elements of W, hence they are finite in number. Hence, by Proposition V.2.31 (compatibility), Δ is a Garside element in B^+, and (B^+, Δ) is a Garside monoid in the sense of Definition I.2.1.

Then, by Corollary V.2.26 (Garside map), the family $\mathrm{Div}(\Delta)$ is a bounded Garside family in B^+ and, by Propositions V.2.32 (Garside map) and II.3.11 (Ore's theorem), the group B is a group of fractions for B^+. Moreover, Δ is the right-lcm of Σ; indeed it is a right-multiple of Σ and, if a left-divisor of Δ is a right-multiple of all atoms, it is in the image of σ by Lemma 1.31, and by Lemma 1.17 its image in W is δ. □

Example 1.32 (braid groups). Proposition 1.29 applies, in particular, to type A_{n-1}, that is, in the case of the braid monoid B_n^+. Then we obtain the results mentioned in Chapter I, namely that the fundamental braid Δ_n is a Garside element in B_n^+ and its divisors make a finite, bounded Garside family. The elements of the latter are the positive braids that can be represented by a braid diagram in which any two strands cross at most once: Δ_n has this property, hence so do all its divisors; conversely, any braid with this property can be multiplied on the left (or on the right) by a positive braid so as to obtain Δ_n.

As a direct application of Algorithm IV.3.25 and Proposition IV.3.26 (Word Problem II), we obtain

Proposition 1.33 (Word Problem). *If (W, Σ) is a Coxeter system of spherical type, then a signed Σ-word w represents 1 in the associated Artin–Tits group B if and only if its double right-reversing ends with the empty word, that is, if there exist positive words u, v satisfying $w \curvearrowright u\overline{v}$ and $\overline{v}u \curvearrowright \varepsilon$. The complexity of the algorithm is quadratic in time and linear in space.*

Example 1.34 (braids). Consider the braid word $\overline{\sigma_1}|\overline{\sigma_1}|\overline{\sigma_2}|\overline{\sigma_2}|\sigma_1|\sigma_1|\sigma_2|\sigma_2$. Right-reversing it leads to the positive–negative word $\sigma_2|\sigma_1|\sigma_1|\sigma_1|\sigma_2|\overline{\sigma_1}|\overline{\sigma_2}|\overline{\sigma_2}|\overline{\sigma_2}|\overline{\sigma_1}$. Exchanging the negative and positive parts and reversing again leads back to the original word. The latter is nonempty, so we conclude that the initial word does not represent 1 in B_n.

1.3 The germ approach

Instead of verifying the cube condition, one can also derive all previous results by starting from the Coxeter group and using the derived germ approach of Subsection VI.2.4. Once the basic properties of Coxeter groups established in Subsection 1.1 are at hand, one obtains a quick and elegant proof of Proposition 1.29 and of Lemmas 1.30 and 1.31.

Starting with a Coxeter system (W, Σ), we shall apply Definition VI.2.58 (derived structure) taking for $(\mathcal{G}, \Sigma)$ the pair (W, Σ) and choosing $\mathcal{H} = W$. By Lemma VI.2.60, the derived germ W^Σ is cancellative and Noetherian, and we can consider the monoid $\mathcal{M}on(W^\Sigma)$ generated by this germ. We first use Proposition VI.2.66 (derived Garside II) to show that this germ is a Garside germ. For r, s in Σ, we shall denote by $\Delta_{r,s}$ the image of the word $(r|s)^{[m_{r,s}]}$ in W.

Proposition 1.35 (Garside germ). *For every Coxeter system (W, Σ), the germ W^Σ is a Garside germ, and the monoid $\mathcal{M}on(W^\Sigma)$ is the associated Artin–Tits monoid.*

Proof. We prove that (W, Σ) is eligible for Proposition VI.2.66 (derived Garside II). We first look at the condition (VI.2.67). Assume that r, s lie in Σ and admit a common $\leqslant_\Sigma$-upper bound. We let $I = \{r, s\}$ and let g be the upper bound. Write $g = fh$, where h has minimal Σ-length in $W_I g$ and f lies in W_I. We have $\|f\|_\Sigma + \|h\|_\Sigma = \|g\|_\Sigma$. Then, by Proposition 1.13, r and s are Σ-prefixes of f. The latter is thus an upper bound of r and s in W_I. Since every element of W_I is equal to a product rsr... or srs..., it follows that rs has finite order and that $f = \Delta_{r,s}$ holds. Thus we found that $\Delta_{r,s}$ is a least $\leqslant_\Sigma$-upper bound of r and s.

Next, we show (VI.2.68). Thus we assume this time that r, s lie in Σ and g lies in W and we have $\|g\|_\Sigma + 1 = \|gr\|_\Sigma = \|gs\|_\Sigma$, and we assume that r, s have a common $\leqslant_\Sigma$-upper bound, which we have seen is $\Delta_{r,s}$. We have to show the equality $\|g\|_\Sigma + \|\Delta_{r,s}\|_\Sigma = \|g\Delta_{r,s}\|_\Sigma$; but this is exactly the fact that (1.14) implies (1.15).

We obtain the last assertion by comparing the presentation of a monoid associated with a germ and the presentation of an Artin–Tits monoid. □

Note that the germ W^{Σ}, seen as a subset of the associated Artin–Tits monoid, is the image of W under the lifting σ defined in Lemma 1.22(ii). If (W, Σ) is of spherical type, then, by Lemma 1.17, the Garside family provided by this germ is equal to $\mathrm{Div}(\Delta)$, hence is bounded by Δ. By the definition of a derived germ, we have recovered Lemmas 1.30 and 1.31, as well as Proposition 1.29:

Proposition 1.36 (Garside group II). *If (W, Σ) is a Coxeter system of spherical type, then the Garside family of the associated Artin–Tits monoid B^+ given by the germ W^{Σ} is bounded by the lift Δ of the longest element of W. In particular, (B^+, Δ) is a Garside monoid and the associated Artin–Tits group B is a Garside group.*

Remark 1.37. Using Proposition II.3.11 (Ore's theorem), we thus obtain an easy proof of the result that the braid monoid associated with a Coxeter system of spherical type embeds in the associated braid group.

Always in the spherical case, we add a few observations about the center and the quasi-center. We recall that an element is called *quasi-central* if it globally commutes with the atom set.

Proposition 1.38 (quasi-center). *If (W, Σ) is a Coxeter system of spherical type and the associated diagram is connected, then every quasi-central element in the associated Artin–Tits monoid B^+ is a power of the lift Δ of the longest element of W.*

Proof. Assume that b is a quasi-central element in B^+. Let s be an element of Σ that left-divides b and let t be an element of Σ that does not commute with s. We claim that t left-divides b. We have $tb = bt'$ for some t' in Σ. From the equality $bt' = tb$ we deduce that s and t left-divide tb. Hence their right-lcm divides tb. This right-lcm is a product $tsts\cdots$ with at least 3 factors, hence, cancelling t, we obtain that st divides b. Let us write b as stb'. We have $b = tb's'$, where s' is the element of Σ satisfying $sb = bs'$. This establishes our claim.

Now, using the connectedness of the Coxeter diagram, we see that all atoms divide b. Since Δ is the right-lcm of the atoms, it has to divide b. Write $b = \Delta b_1$. Then b_1 is quasi-central and we obtain the expected result using induction on the length of b. □

Corollary 1.39 (center). *The center of an irreducible Artin–Tits group of spherical type is cyclic, generated by the smallest central power of Δ.*

Proof. This comes from the facts that Δ has a central power and that any element of the group is equal to a power of Δ times an element of the monoid. □

Let us now briefly consider Coxeter systems (W, Σ) that are not of spherical type. Then one easily checks that the Garside family of the associated braid monoid provided by the germ W^{Σ} is not bounded. Actually, we have a stronger result:

type	spherical	large no ∞	$\widetilde{\mathrm{A}}_2$	$\widetilde{\mathrm{A}}_3$	$\widetilde{\mathrm{A}}_4$	$\widetilde{\mathrm{B}}_3$	$\widetilde{\mathrm{C}}_2$	$\widetilde{\mathrm{C}}_3$
$\#E$	1	$3\binom{\#S}{3}$	3	10	35	14	3	12
$\#S$	$\#W$	$O((\#S)^3)$	16	125	$1,296$	315	24	317

Table 1. Size of the smallest Garside family S in a few Artin–Tits monoids; whenever S is finite, it must consist of all right-divisors of the elements of some finite set E (the "extremal elements").

Proposition 1.40 (not bounded). *The Artin–Tits monoid associated with a Coxeter system not of spherical type admits no right-bounded Garside family.*

Proof. Assume that (W, Σ) is a Coxeter system, B^+ is the associated Artin–Tits monoid, and B^+ admits a Garside family that is right-bounded by an element Δ. Then all atoms of B^+, that is, all elements of Σ, left-divide Δ. First, as Δ has a finite height, Σ must be finite. Next, by Proposition V.1.46 (common right-multiples), any two elements of B^+ admit a common right-multiple, hence a right-lcm. Let Δ_0 be the right-lcm of Σ in B^+, and let δ be the projection of Δ_0 to W. Then, by (VI.2.67), δ is a common $\leqslant_\Sigma$-upper bound of Σ in W. It follows that $N(\delta)$ is the set of all reflections in W, and, therefore, $\|g\|_\Sigma \leqslant \|\delta\|_\Sigma$ holds for every g in W. This implies that W is finite. □

When W is finite, the Garside family $\sigma(W)$ of the associated braid monoid B^+ is the smallest Garside family of B^+: indeed, every Garside family of B^+ must contain the right-lcm Δ of the atoms, hence all right-divisors of Δ. When W is infinite, the situation is different and the following question naturally arises:

Question 25 (smallest Garside). Does every finitely generated Artin–Tits monoid admit a finite Garside family?

For instance, in type $\widetilde{\mathrm{A}}_2$, we have seen in Reference Structure 9, page 114, that the answer is positive, as it is for many similar cases, see Exercises 102 and 103 and Table 1. For the moment, we observe that Question 25 translates into a pure Coxeter problem:

Proposition 1.41 (translation). *For every Coxeter system (W, Σ), the following conditions are equivalent:*

(i) *There exists a finite Garside family in the braid monoid of (W, Σ).*

(ii) *There exists a finite subfamily of W that includes Σ and is closed under least $\leqslant_\Sigma$-upper bound and Σ-suffix.*

Proof. Let B^+ be the braid monoid of (W, Σ). There exists a finite Garside family in B^+ if and only if the smallest Garside family of B^+, namely the closure $\widehat{\Sigma}$ of Σ under right-lcm and right-divisor, is finite. By Proposition 1.35, the copy $\sigma(W)$ of W in B^+ is a Garside family, hence it includes $\widehat{\Sigma}$. Now, by Definition VI.2.58 (derived structure), for all f, g in W, the relation $\sigma(f) \preccurlyeq \sigma(g)$ holds in the germ W^Σ, hence in the monoid B^+, if and only if $f \leqslant_\Sigma g$ holds in W and, symmetrically, $\sigma(f) \succcurlyeq\!\!\!\!\sim \sigma(g)$ holds in B^+ if and only if $f \,\widetilde{\leqslant}_\Sigma\, g$ holds in W. Hence $\widehat{\Sigma}$ is in one-to-one correspondence with the closure of Σ under least $\leqslant_\Sigma$-upper bound and Σ-suffix in W. □

We refer to the final Notes section for the announcement of a positive answer to Question 25.

2 More Garside structures on Artin–Tits groups

A monoid (more generally, a category) admits in general several different Garside families: typically, if $\mathcal{S}$ is a Garside family including $\mathbf{1}_{\mathcal{C}}$ in $\mathcal{C}$, then $\mathcal{S}^2$ and $\mathcal{S}$ are in general distinct Garside families. Beyond that, a group (more generally, a groupoid) may be the group(oid) of fractions of several distinct monoids (categories), leading to possibly unrelated Garside structures. Here we shall see that such a situation occurs for many Artin–Tits groups, which turn out to admit what is known as a *dual Garside structure*.

The section comprises four subsections. Dual germs and dual braid monoids are introduced in full generality in Subsection 2.1, and we state without proof the state-of-the-art description of the cases when the dual germ is a Garside germ (Proposition 2.4). A complete proof of the result in the case of type A_n, that is, for the symmetric group, is given in Subsection 2.2. The case of other finite Coxeter groups is addressed, in a sketchy way, in Subsection 2.3. Finally, we briefly mention in Subsection 2.4 the existence of several other unrelated Garside structures in the (very) special case of B_3.

2.1 The dual braid monoid

In this section, we consider a Coxeter system (W, Σ) with Σ finite. We shall define another derived germ in W that will, in good cases, be a Garside germ and generate another monoid for the Artin–Tits group associated with (W, Σ), now known as a *dual braid monoid.*

The construction of this germ depends on the choice of a Coxeter element, which we introduce now.

Definition 2.1 (Coxeter element). If (W, Σ) is a Coxeter system with Σ finite, a *Coxeter element* is an element of W that can be expressed as the product of all elements of Σ (enumerated in some order).

It is known (see [27, chap. V n° 6.1, prop. 1]) that, if the Coxeter graph has no cycle, then all Coxeter elements are conjugate. This is not true in general.

We shall use Definition VI.2.58 (derived structure) with W as groupoid $\mathcal{G}$ (as in Subsection 1.3) and the set R of *all* reflections of W as generating set (unlike Subsection 1.3 where only the reflections of Σ are used). Next, we choose a Coxeter element c in W, and take for $\mathcal{H}$ of Definition VI.2.58 the set $\mathrm{Pref}(c)$ of all R-prefixes of c.

Definition 2.2 (dual germ, dual monoid). Assume that (W, Σ) is a Coxeter system with Σ finite and c is a Coxeter element in W. Then the germ $\mathrm{Pref}(c)^R$ is called a *dual germ* and the monoid $\mathcal{M}on(\mathrm{Pref}(c)^R)$ is called a *dual braid monoid* for W.

Example 2.3 (dual germ, dual monoid). Consider the type A_3, that is, the symmetric group $\mathfrak{S}_3$ with the generating system $\Sigma = \{\mathtt{s}_1, \mathtt{s}_2\}$ made of the two transpositions $(1, 2)$

and $(2, 3)$. There exist two Coxeter elements, namely $c = s_1s_2$ and $c' = s_2s_1$ (which are conjugate as mentioned above), and the set of reflections R has three elements, namely s_1, s_2, and $s_1s_2s_1$. Choose the Coxeter element c. We find $c = s_1{\cdot}s_2 = s_2{\cdot}s_2s_1s_2 = s_1s_2s_1{\cdot}s_1$, so all three reflections are R-prefixes of c. The associated germ is shown on the right, where s_3 means $s_1s_2s_1$. We see that the corresponding monoid admits the presentation $\langle s_1, s_2, s_3 \mid s_1s_2 = s_2s_3 = s_3s_1\rangle^+$, and recognize the dual braid monoid B_3^{+*} of Reference Structure 3, page 10.

$\bullet$	1	s_1	s_2	s_3	c
1	1	s_1	s_2	s_3	c
s_1	s_1		c		
s_2	s_2			c	
s_3	s_3	c			
c	c				

Note that the dual germ depends in general on the choice of a Coxeter element. Nevertheless, when the Coxeter graph has no cycles, since all Coxeter elements are conjugate in W, all dual germs are isomorphic. This is the case for spherical types.

The next result describes cases where the dual germ is a Garside germ. The affine Coxeter graphs of type $\widetilde{\mathrm{A}}_n$ and $\widetilde{\mathrm{C}}_n$ respectively are the graphs

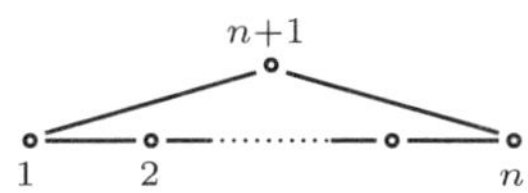

and

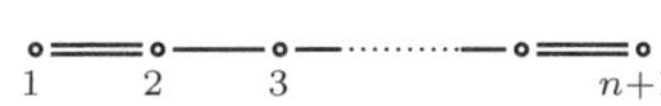

.

Proposition 2.4 (dual germ). *If (W, Σ) is a Coxeter system with Σ finite and c is a Coxeter element in W, then the germ $\mathrm{Pref}(c)^R$ is a bounded Garside germ in the following cases, and in each of them the group obtained is the corresponding Artin–Tits group:*

(i) *If the type of W is spherical and c is any Coxeter element.*

(ii) *If the type of W is $\widetilde{\mathrm{A}}_n$ and c is the product of the generators in consecutive order on the Coxeter diagram.*

(iii) *If the type of W is $\widetilde{\mathrm{C}}_n$ and c is any Coxeter element.*

So, in particular, the dual braid monoid is a Garside or quasi-Garside monoid in the above cases.

The proof of the part of Proposition 2.4 stating that the germ is a bounded Garside germ consists in showing that the germ $\mathrm{Pref}(c)^R$ is eligible for Proposition VI.2.69 (derived Garside III). If we can apply this proposition, since all elements of the germ $\mathrm{Pref}(c)^R$ are prefixes of c, the Garside family thus obtained will be automatically bounded.

Lemma 2.5. *Assume that (W, Σ) is a Coxeter system with Σ finite and c is a Coxeter element in W. Then the germ $\mathrm{Pref}(c)^R$ is closed under R-prefixes and R-suffixes.*

Proof. By definition, $\mathrm{Pref}(c)^R$ is closed under R-prefixes. Now R is closed under conjugacy in W, hence the R-length is invariant under conjugacy, and a R-suffix of an element of W is also a R-prefix of the same element since every product fg in W can also be written $g(g^{-1}fg)$. Whence the lemma. □

So, in order to establish that $\mathrm{Pref}(c)^R$ is a bounded Garside germ, it remains to show that any two elements of $\mathrm{Pref}(c)^R$ admit a $\leqslant_R$-least upper bound. This is the main point of the proof. Then, in order to prove Proposition 2.4, we have to show that the group generated by the germ $\mathrm{Pref}(c)^R$ is indeed the expected Artin–Tits group.

Below, we shall give a complete proof of Proposition 2.4 only for type A_n, that is, when the Coxeter group is a symmetric group, and give the main ideas for the other spherical types following [9] and [31]. For affine types, we refer to [106] and [107]. Let us also mention that the analogous result for complex reflection groups will appear as Proposition 3.20 below.

2.2 The case of the symmetric group

We now prove Proposition 2.4 when W is the symmetric group $\mathfrak{S}_n$. In this case, R is the set of all transpositions and the R-length of an element f is n minus the number of factors in a decomposition of f into a product of disjoint cycles.

Definition 2.6 (noncrossing partition). A partition $(\lambda_1, \dots, \lambda_r)$ of $\{1, \dots, n\}$ is called *noncrossing* if, for every 4-tuple (i, j, k, l) with $1 \leqslant i < j < k < l \leqslant n$, if $\{i, k\}$ is a subset of some part λ_s, then $\{j, l\}$ is not a subset of a part λ_t with $t \neq s$.

Noncrossing partitions can be visualised as follows (see Figure 2): consider an oriented circle and a set $\{1, \dots, n\}$ of n points on that circle numbered 1 to n in cyclic order; a partition $(\lambda_1, \dots, \lambda_r)$ of $\{1, \dots, n\}$ is noncrossing if and only if the convex hulls of any two parts in the disc are equal or disjoint.

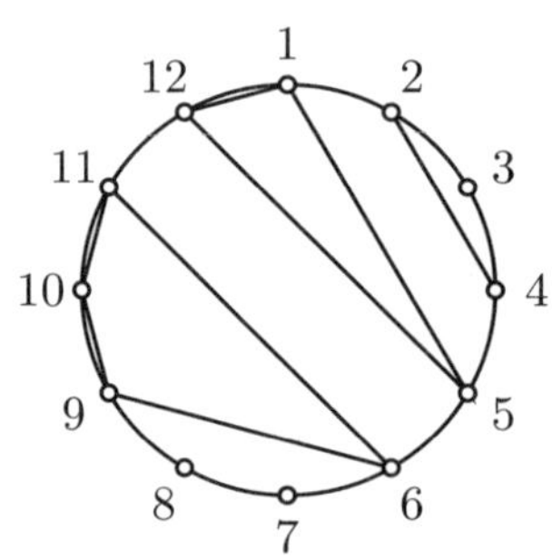

Figure 2. The noncrossing partition λ of $\{1, \dots, 12\}$ given by $\{1, 5, 12\}$, $\{2, 4\}$, $\{3\}$, $\{6, 9, 10, 11\}$, $\{7\}$, and $\{8\}$. The corresponding element of $\mathfrak{S}_{12}$, see Definition 2.8, is $c_\lambda = (1, 5, 12)(2, 4)(6, 9, 10, 11)$.

There is a natural partial order on partitions: a partition λ is *finer* than a partition λ' if every part of λ' is a union of parts of λ.

Noncrossing partitions will allow us to understand the order on $\mathrm{Pref}(c)^R$ when we take the cycle $(1, 2, \dots, n)$ as Coxeter element.

Proposition 2.7 (isomorphism). *Let c be the Coxeter element $(1, 2, \dots, n)$ of $\mathfrak{S}_n$. The map ψ from $\mathfrak{S}_n$ to the set of partitions of $\{1, \dots, n\}$ that sends every element of $\mathfrak{S}_n$ to the partition given by its orbits induces a poset isomorphism from* $\mathrm{Pref}(c)^R$ *to the set of noncrossing partitions of* $\{1, \dots, n\}$.

Before proving Proposition 2.7, we need some definitions and a lemma.

Definition 2.8 (permutation c_λ, cut). (i) For $F \subseteq \{1, \dots, n\}$, say $F = \{i_1, \dots, i_k\}$ with $i_1 < \cdots < i_k$, we denote by c_F the cycle $(i_1, \dots, i_k)$. Then, for $\lambda = (\lambda_1, \dots, \lambda_r)$ a noncrossing partition of $\{1, \dots, n\}$, we put $c_\lambda = c_{\lambda_1} c_{\lambda_2} \cdots c_{\lambda_r}$.

(ii) For $F' \subseteq F \subseteq \{1, \dots, n\}$, say $F = \{i_1, \dots, i_k\}$ with $i_1 < \cdots < i_k$ and $F' = \{i_{j_1}, \dots, i_{j_{k'}}\}$ with $j_1 < j_2 < \cdots < j_{k'}$, we denote by $F' \backslash\!\backslash F$ (*F cut at F'*) the partition of F whose parts are the sets $\{i_{j_l}, i_{j_l+1}, \dots, i_{j_{l+1}-1}\}$, where we make the convention that $j_{k'+1}$ means j_1 and indices of elements i are taken modulo k, see Figure 3.

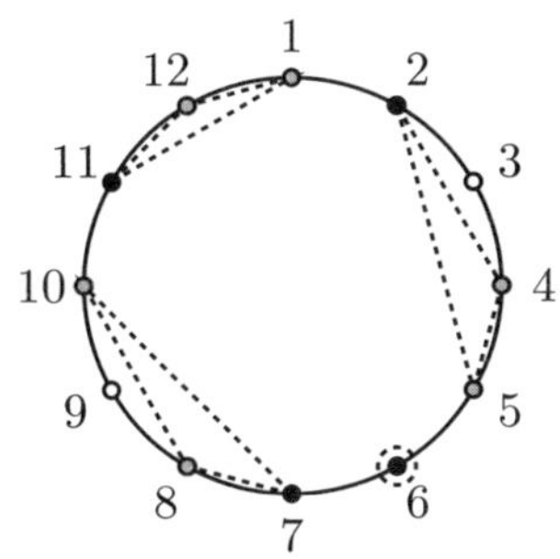

Figure 3. Cutting a subset F, here $\{1, 2, 4, 5, 7, 8, 10, 11, 12\}$, at a subsubset F', here $\{2, 6, 7, 11\}$: black points are the elements of F', grey points are those of $F \setminus F'$, and white points are those not in F; the parts of $F' \backslash\!\backslash F$ are indicated with dashed lines: we attach behind every black point the grey points that lie between this black point and the next one in cyclic order; note that $\{6\}$ also belongs to $F' \backslash\!\backslash F$ (that is why we added the small dashed circle).

Lemma 2.9. (i) *For $F' \subseteq F \subseteq \{1, \dots, n\}$, we have $c_F = c_{F'} c_{F' \backslash\!\backslash F}$.*

(ii) *Assume that λ and λ' are noncrossing partitions of $\{1, \dots, n\}$ and λ' is finer that λ. Then there exists a unique noncrossing partition of $\{1, \dots, n\}$, which we denote by $\lambda' \backslash\!\backslash \lambda$ (λ* cut at *λ'), satisfying $c_{\lambda'} c_{\lambda' \backslash\!\backslash \lambda} = c_\lambda$.*

Proof. We prove (i) by induction on the cardinality of F'. With notation as in Definition 2.8(iii), let $F_1 = \{i_1, \dots, i_{j_2-1}\}$, $F_2 = F \setminus F_1$, and let $F_2' = F' \setminus \{i_1\}$, so that we have $c_{F'} = (i_1, i_{j_2}) c_{F_2'}$. Using the equality $F' \backslash\!\backslash F = \{F_1\} \cup F_2' \backslash\!\backslash F_2$, we obtain

$$c_{F'} c_{F' \backslash\!\backslash F} = c_{F'} c_{F_1} c_{F_2' \backslash\!\backslash F_2} = (i_1, i_{j_2}) c_{F_2'} c_{F_1} c_{F_2' \backslash\!\backslash F_2} = (i_1, i_{j_2}) c_{F_1} c_{F_2'} c_{F_2' \backslash\!\backslash F_2},$$

the last equality since $c_{F_2'}$ commutes to c_{F_1} since $F_2' \cap F_1$ is empty. By the induction hypothesis, we have $c_{F_2} = c_{F_2'} c_{F_2' \backslash\!\backslash F_2}$. This together with the equality

$$c_F = c_{F_1} (i_{j_2-1}, i_{j_2}) c_{F_2} = (i_1, i_{j_2}) c_{F_1} c_{F_2}$$

gives the result.

To prove (ii) we first consider the case when λ has only one part with cardinal at least 2 which we denote by F. Let $\lambda_1', \dots, \lambda_l'$ be the non-trivial parts of λ'. They are all subsets of F. The point is to observe that, since for $i > 1$ the convex hulls of λ_i' and λ_1' do not intersect, each λ_i' lies inside a single connected component of the complement of the convex hull of λ_1' in the convex hull of F, hence each is included in a part of $\lambda_1' \backslash\!\backslash F$. Therefore the result follows from (i) using induction on the number of parts of λ. Uniqueness comes from the equality $c_{\lambda'}c_{\lambda'\backslash\lambda} = c_\lambda$.

The general case of (ii) is easily obtained by applying the above result to all the nontrivial parts of λ. □

Proof of Proposition 2.7. We claim that, if λ is a noncrossing partition and if an element g of $\mathrm{Pref}(c)^R$ satisfies $g \leqslant_R c_\lambda$, then $\psi(g)$ is a noncrossing partition finer than λ, and $g = c_{\psi(g)}$ holds. This will prove the proposition. Indeed, taking for λ the partition with only one part, so that c_λ is c, we see that $\mathrm{Pref}(c)^R$ is the set of all elements c_λ with λ a noncrossing partition and that ψ and $\lambda \mapsto c_\lambda$ are bijections inverse of each other since, for every noncrossing partition λ, we have $\psi(c_\lambda) = \lambda$. This proves also that divisibility in the germ $\mathrm{Pref}(c)^R$ is mapped by ψ on the order relation on partitions.

We prove our claim using decreasing induction on the R-length of g. The starting point of the induction is for $g = c$. In this case, $\psi(g)$ is the partition with only one part. Let g belong to $\mathrm{Pref}(c)^R$; we can assume $g \neq c_\lambda$ so that there exists a transposition (i, j) satisfying $\|(i, j)g\|_R = \|g\|_R + 1$ and $(i, j)g \leqslant_R c_\lambda$. This implies that (i, j) is a prefix of c_λ so that the decomposition of $(i, j)c_\lambda$ into disjoint cycles has one more cycle than the decomposition of c_λ, since the R-length of a product of k cycles in $\mathfrak{S}_n$ is $n - k$. This in turn means that i and j are in the same part of λ, that is, the partition $\psi((i, j))$ is finer than the partition $\psi(y)$. Applying Lemma 2.9(ii), we deduce $g = c_{\psi((i,j))\backslash\lambda}$, which gives the claim. □

Corollary 2.10 (Garside germ). *Let c be the Coxeter element $(1, 2, \dots, n)$ of $\mathfrak{S}_n$. Then the germ $\mathrm{Pref}(c)^R$ is a bounded Garside germ.*

Proof. Since noncrossing partitions and the germ $\mathrm{Pref}(c)^R$ are isomorphic posets and noncrossing partitions make a lattice, we can apply Propositions VI.2.69 (derived Garside III) and Corollary VI.3.9 (bounded Garside), which give the result: Lemma 2.5 guarantees that the Σ-prefixes of c coincide with its Σ-suffixes, so the assumptions of Corollary VI.3.9 are satisfied. □

We now consider the monoid associated with the above Garside germ and its enveloping group.

Proposition 2.11 (braid group). *Let c be the Coxeter element $(1, 2, \dots, n)$ of $\mathfrak{S}_n$. Then the group generated by the germ $\mathrm{Pref}(c)^R$ is isomorphic to the n-strand braid group B_n and the monoid generated by the germ is the dual braid monoid B_n^{+*} of Reference Structure 3, page 10.*

To prove Proposition 2.11, we first define a surjective morphism from B_n to the group generated by the above germ $\mathrm{Pref}(c)^R$. As in Definition 2.8, we write $c_{i,j}$ for $c_{\{i,j\}}$, and s_i for $c_{i,i+1}$.

Lemma 2.12. *Let c be the Coxeter element $(1, 2, \ldots, n)$ of $\mathfrak{S}_n$ and G be the group generated by the germ $\mathrm{Pref}(c)^R$. Then the map $\sigma_i \mapsto s_i$ extends to a surjective morphism from B_n to G.*

Proof. The equalities $c_{i,i+1}c_{i+1,i+2} = c_{i+1,i+2}c_{i,i+2} = c_{i,i+2}c_{i,i+1}$ hold in the germ. We deduce that, in G, we have $s_is_{i+1}s_i = s_{i+1}c_{i,i+2}s_i = s_{i+1}s_is_{i+1}$. We also have $s_is_j = s_js_i$ in the germ for $|i - j| \geqslant 2$. Hence the elements s_i satisfy the braid relations. So the map $\sigma_i \mapsto s_i$ extends to a morphism from B_n to G. By induction on $j - i$ we see that, for $i < j$, we have $c_{i,j} = s_is_{i+1} \cdots s_js_{j-1}^{-1} \cdots s_{i+1}^{-1}s_i^{-1}$. It follows that G is generated by $\{s_1, \ldots, s_{n-1}\}$ and, therefore, the above morphism is surjective. □

To prove that the morphism of Lemma 2.12 is an isomorphism, we first give a presentation of the group G using only R as generating set.

Lemma 2.13. *Let c be the Coxeter element $(1, 2, \ldots, n)$ of $\mathfrak{S}_n$ and G be the group generated and G (resp. M) be the group (resp. monoid) generated by the germ $\mathrm{Pref}(c)^R$. Then M admits the presentation with generators $c_{i,j}$, where $\{i, j\}$ runs over all subsets of cardinality 2 of $\{1, \ldots, n\}$, and relations*

(2.14) $\quad c_{i,j}c_{i',j'} = c_{i',j'}c_{i,j}$ *if $\{i, j\}$ and $\{i', j'\}$ are noncrossing*

(2.15) $\quad c_{i,j}c_{j,k} = c_{j,k}c_{k,i}$ *for $i < j < k$ in the cyclic order.*

The group G admits the same presentation as a group presentation.

Proof. Since M generates G, the last assertion follows from the first one. The monoid M is generated by the elements $c_{i,j}$. Indeed, it is generated by the elements c_λ where λ runs over noncrossing partitions of $\{1, \ldots, n\}$; by definition, c_λ is the product of elements of the form c_F with F a subset of $\{1, \ldots, n\}$ and $c_F = c_{i_1,i_2}c_{i_2,i_3} \cdots c_{i_{k-1},i_k}$ for $F = \{i_1, \ldots, i_k\}$ with $i_1 < \cdots i_k$.

Relations (2.14) and (2.15) hold in the germ. We have to prove that all other relations are consequences of these. By definition, all relations are of the form $f = gh$ where f, g, h are in the germ and $\|f\|_R = \|g\|_R + \|h\|_R$ holds. Writing decompositions of f, g, and h, into products of elements of R, we are reduced to proving that one can pass from any such decomposition of f to any other only by means of relations (2.14) and (2.15). We show that, starting from a fixed decomposition of f, for every s in R left-dividing f, we can obtain a decomposition of f starting with s using only relations (2.14) and (2.15). This gives the result using induction on $\|f\|_R$. We start with a decomposition of f obtained in the following way: we have $f = c_\lambda = c_{\lambda_1} \cdots c_{\lambda_r}$, where $\lambda = (\lambda_1, \ldots, \lambda_r)$ is a noncrossing partition; each part λ_p is of the form $\{i_1, \ldots, i_l\}$ with $i_1 < \cdots < i_l$. We decompose c_{λ_p} as $c_{i_1,i_2}c_{i_2,i_3} \cdots c_{i_{l-1},i_l}$. Since s left-divides f, we have $s = c_{i,j}$ with $i < j$ and i, j in the same part λ_p of λ. Since λ is noncrossing, s commutes with the elements of R that occur in the decompositions of c_{λ_q} for $q \neq p$.

Hence we are left with proving the result for $f = c_{\lambda_p}$. We have $i = i_h$, $j = i_k$ for some indices h and k. Using relation (2.15) repeatedly, we find

$$c_{i,i_{h+1}} \cdots c_{i_{k-1},j} = c_{i,i_{h+1}} c_{i_{h+1},j} c_{i_{h+1},i_{h+2}} \cdots c_{i_{k-2},i_{k-1}}.$$

We have also $c_{i,i_{h+1}} c_{i_{h+1},j} = c_{i,j} c_{i,i_{h+1}}$. Using once more relation (2.15), then the commutation relations (2.14), we can push $c_{i,j}$ to the left and obtain the expected result. □

Proof of Proposition 2.11. Consider in the braid group B_n the elements $a_{i,j}$ defined in Reference Structure 3, page 10. According to (I.1.12), they satisfy the same relations (2.14) and (2.15) as the elements $c_{i,j}$ hence, by Lemma 2.13, there exists a surjective morphism from G to B_n mapping $c_{i,j}$ to $a_{i,j}$. Since this morphism maps s_i to σ_i, it is the inverse of the morphism defined in Lemma 2.12, whence the result. □

Note that Lemma 2.13 is equivalent to the fact that the Hurwitz action of the braid group B_k is transitive on the decompositions of every element of the germ of R-length k into products of atoms (compare with Lemma 3.17). We recall from Remark I.2.10 that the Hurwitz action of B_n on length n sequences with entries in a group G is given by

(2.16) $$\sigma_i \bullet (g_1, \dots, g_n) = (g_1, \dots, g_{i-1}, g_i g_{i+1} g_i^{-1}, g_i, g_{i+1}, \dots, g_n).$$

2.3 The case of finite Coxeter groups

We now sketch the main steps for a proof of Proposition 2.4 in the case when G is a (general) finite Coxeter group, following the approach of Brady and Watt in [30]. We start from the next result, stated for the orthogonal group of a real vector space in [30], but actually valid for an arbitrary linear group.

Lemma 2.17. *Assume that V is a finite-dimensional vector space over a field* k. *For A in* $\mathrm{GL}(V)$, *let $M(A)$ be the image of $A - \mathrm{id}_V$ and let $m(A) = \dim M(A)$ (the number of eigenvalues of A different from* 1 *if A is semi-simple). Then*

(i) *For all A, B in* $\mathrm{GL}(V)$, *we have $m(AB) \leqslant m(A) + m(B)$, and equality occurs if and only if $M(AB) = M(A) \oplus M(B)$ holds.*

(ii) *Write $A \trianglelefteq B$ for $m(A) + m(A^{-1}B) = m(B)$. Then $\trianglelefteq$ defines a partial order on* $\mathrm{GL}(V)$.

Proof (sketch). The key point is the equality $AB - \mathrm{id}_V = (A - \mathrm{id}_V)B + (B - \mathrm{id}_V)$, which implies $M(AB) \subseteq M(A) + M(B)$. □

By contrast, the next result requires a specific context.

Lemma 2.18. *Assume that $G(V)$ is either the orthogonal group of a real vector space, or the unitary group of a complex vector space. Let A belong to $G(V)$. Then the map $B \mapsto M(B)$ is a poset isomorphism between $\{B \in G(V) \mid B \trianglelefteq A\}$ and the poset of subspaces of $M(A)$ ordered by inclusion.*

We assume now that W is a subgroup of $G(V)$ generated by reflections (or complex reflections in the unitary case). Let R be the set of all reflections of W. Then W is generated as a monoid by R.

Lemma 2.19. *For every g in W, we have $\|g\|_R = m(g)$.*

We assume now that V is an n-dimensional real vector space, which is the reflection representation of the irreducible finite Coxeter group W, and that c is a Coxeter element in W. Lemma 2.19 implies $\|c\|_R = n$. Let $\mathrm{Pref}(c)$ be the set of left R-prefixes of c. Lemmas 2.17–2.19 imply that, on $\mathrm{Pref}(c)$, the Σ-prefix relation $\leqslant_R$ identifies with the partial order $\trianglelefteq$ of Lemma 2.17. So, in view of the criteria of Chapter VI, the question reduces to proving that least upper bounds exist for $\trianglelefteq$. Now, the main result of Brady and Watt in [31] may be formulated as

Proposition 2.20 (lattice). *If (W, Σ) is a Coxeter system of spherical type and c is a Coxeter element in W, then the poset $(\mathrm{Pref}(c), \trianglelefteq)$ is a lattice.*

Proof (sketch). Lemma 2.18 suggests an approach that does *not* work: starting with g, h satisfying $g, h \trianglelefteq c$, if we could find f satisfying $M(f) = M(g) \cap M(h)$, then, by the poset isomorphism of Lemma 2.18, we would deduce that f is a greatest lower bound for g and h. However, such an element f does not exist in general. Now, a different approach works. Put $R(g) = \{s \in R \mid s \trianglelefteq g\}$. Then, for all g, h as above, one can show that there exists f satisfying $R(f) = R(g) \cap R(h)$ using a spherical complex structure on the positive roots of W, adapted to the $\trianglelefteq$-relation. We will not say more about the argument here. □

2.4 Exotic Garside structures on B_n

The question of whether there exist further Garside structures on (spherical) Artin–Tits groups beyond the above described classical and dual ones remains open in general. In the (very) special case of the 3-strand braid group B_3, several such exotic Garside structures are known, and we quickly review them here. We also observe that every left-invariant ordering on B_n gives rise to a Garside structure.

Proposition 2.21 (exotic monoids for B_3). *For m in $\mathbb{Z}$, define $B^+_{3,m}$ by*

$$(2.22) \qquad B^+_{3,m} = \begin{cases} \langle \mathtt{a}, \mathtt{b} \mid \mathtt{aba} = \mathtt{ba}^m\mathtt{b} \rangle^+ & \text{for } m \geqslant 0, \\ \langle \mathtt{a}, \mathtt{b} \mid \mathtt{a} = \mathtt{bab}^{|m|-1}\mathtt{ab} \rangle^+ & \text{for } m < 0. \end{cases}$$

Then, for every m, the monoid $B^+_{3,m}$ is an Ore monoid whose group of fractions is the 3-strand braid group B_3, and it admits Δ_3^2 as a Garside element.

Proof. In B_3, put $\mathtt{x} = \sigma_2$ and $\mathtt{y} = \sigma_1\sigma_2^{-\ell+1}$, whence $\sigma_1 = \mathtt{y}\mathtt{x}^{\ell-1}$, with $\ell \in \mathbb{Z}$. The braid relation $\sigma_2\sigma_1\sigma_2 = \sigma_1\sigma_2\sigma_1$ then reads $\mathtt{xyx}^\ell = \mathtt{yx}^\ell\mathtt{yx}^{\ell-1}$, so B_3 admits the presentation $\langle \mathtt{x}, \mathtt{y} \mid \mathtt{xyx} = \mathtt{yx}^\ell\mathtt{y}\rangle$. We deduce that (2.22) is, in every case, a presentation of B_3^+: for $\ell \geqslant 0$, this corresponds to taking $m = \ell$, $\mathtt{a} = \mathtt{x}$, $\mathtt{b} = \mathtt{y}$, that is, $\mathtt{a} = \sigma_2$ and $\mathtt{b} = \sigma_1\sigma_2^{-m+1}$; for $\ell < 0$, this corresponds to taking $m = \ell - 1$, $\mathtt{a} = \mathtt{y}$ and $\mathtt{b} = \mathtt{x}^{-1}$, that is, $\mathtt{a} = \sigma_1\sigma_2^{|m|}$ and $\mathtt{b} = \sigma_2^{-1}$.

The presentations (2.22) are right- and left-complemented. For $m < 0$, the presentation is right-triangular in the sense of Definition II.4.12 and, therefore, by Proposition II.4.51 (completeness), right-reversing is complete for the presentation. Hence $B_{3,m}^+$ is cancellative. Let $\Delta = (\mathtt{ab}^{|m|-1}\mathtt{ab}^{|m|})^2$. A direct verification shows that Δ is central in $B_{3,m}^+$ (as can be expected, since the evaluation of this expression in B_3 is Δ_3^2), and that it is a right-multiple of $\mathtt{a}$ and $\mathtt{b}$. Hence this element is a Garside element in $B_{3,m}^+$. We deduce that $B_{3,m}^+$ is an Ore monoid, hence it embeds in its group of fractions, which is B_3.

For $m = 0$ and $m = 1$, the monoid $B_{3,m}^+$ is (strongly) Noetherian (see Exercise 100), so, by Proposition II.4.51 (completeness) again, right-reversing is complete and, as above, we deduce that $B_{3,m}^+$ is cancellative and admits conditional lcms. For $m \geqslant 2$, the monoid $B_{3,m}^+$ is not Noetherian, and the presentation (2.22) is not eligible for Proposition II.4.51 (completeness). However, since every two-generator right-complemented presentation, it is eligible for [78, Proposition 4.1 and Corollary 4.5] and, again, right-reversing is complete and, therefore, $B_{3,m}^+$ is cancellative and admits conditional lcms. Let $\Delta = (\mathtt{aba}^m)^2$. As above, a direct verification shows that Δ is central in $B_{3,m}^+$ and it is a multiple of $\mathtt{a}$ and $\mathtt{b}$. So it is a Garside element in $B_{3,m}^+$ and, as in the $m < 0$ case, $B_{3,m}^+$ is an Ore monoid, hence it embeds in its group of fractions, which is B_3. □

The monoid $B_{3,1}^+$ is the usual Artin braid monoid B_3^+. The monoid $B_{3,0}^+$, whose presentation is $\langle \mathtt{a}, \mathtt{b} \mid \mathtt{aba} = \mathtt{b}^2\rangle^+$, is a Garside monoid (in the sense of Definition I.2.1), in which the Garside element Δ_3^2 has 8 divisors, forming the lattice shown on the right. The monoid $B_{3,2}^+$ is not Noetherian; an explicit description of the (infinitely many) divisors of Δ_3^2 in $B_{3,2}^+$ is given in Picantin [196, Figure 21].

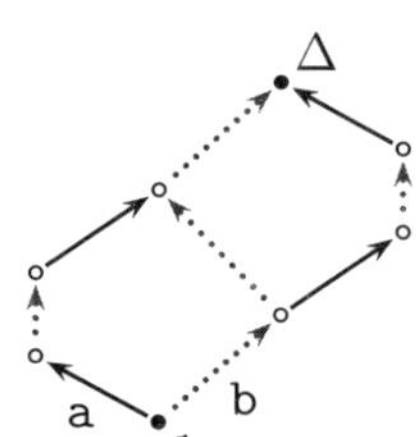

The monoid $B_{3,-1}^+$ is the Dubrovina–Dubrovin monoid $B_3^\oplus$ of Definition XII.3.12 below. Like all other monoids $B_{3,m}^+$ with $m < 0$, it is a monoid of O-type, that is, any two elements are comparable with respect to left-divisibility or, equivalently, the monoid with 1 removed is the positive cone of a left-invariant ordering on its group of fractions, here B_3; see Chapter XII for further details.

More Garside structures on 3-strand braids can be described (the list is not closed).

Example 2.23 (more exotic monoids for B_3**).** Start with $\mathtt{a} = \Delta_3$ in B_3. For $\mathtt{b} = \sigma_1\sigma_2$, the braids σ_1 and σ_2 can be recovered from $\mathtt{a}$ and $\mathtt{b}$ by $\sigma_1 = \mathtt{a}^{-1}\mathtt{b}$ and $\sigma_2 = \mathtt{ba}^{-1}$, leading to the presentation $\langle \mathtt{a}, \mathtt{b} \mid \mathtt{a}^3 = \mathtt{b}^2\rangle$. The associated monoid $\langle \mathtt{a}, \mathtt{b} \mid \mathtt{a}^3 = \mathtt{b}^2\rangle^+$ is indeed the torus-type monoid $T_{3,2}$ of Example I.2.7, which we know is a Garside monoid. The corresponding 5-element lattice of divisors of Δ_3 is shown on the right.

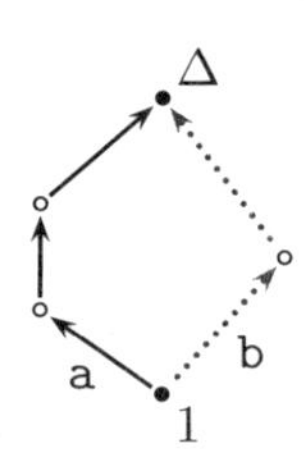

Starting again from $\mathtt{a} = \Delta_3$, consider now $\mathtt{a} = \sigma_2^{-1}$. Then we easily find the presentation $\langle \mathtt{a}, \mathtt{b} \mid \mathtt{a} = \mathtt{babab} \rangle$ for B_3. The corresponding monoid is of O-type, that is, any two elements are comparable with respect to left-divisibility and $\mathtt{a}^2$ (*alias* Δ_3^2), is central; see again Chapter XII for further details.

In all above Garside structures on B_3, in particular in the cases connected with linear orderings, the element Δ_3^2 is always a Garside element of the involved monoid. This is a general result that can be established for all braid groups B_n.

Proposition 2.24 (Garside in ordered case). *If $\leqslant$ is a left-invariant linear order on B_n, and M is the submonoid $\{\beta \in B_n \mid 1 \leqslant \beta\}$ of B_n, then Δ_n^2 belongs either to M or to M^{-1} and, in either case, it is a Garside element in the involved monoid.*

Proof (sketch). That M and M^{-1} are monoids readily follows from the assumption that $\leqslant$ is a left-invariant order on B_n (see Lemma XII.1.3). If Δ_n^2 does not lie in M, that is, if $1 \leqslant \Delta_n^2$ fails, then $\Delta_n^2 < 1$ must hold, implying $1 < \Delta_n^{-2}$, whence $\Delta_n^{-2} \in M$. Next, as Δ_n^2 is central, its left- and right-divisors coincide: indeed $\beta\beta' = \Delta_n^2$ implies $\beta\Delta_n^2 = \Delta_n^2\beta = \beta\beta'\beta$, whence $\beta'\beta = \Delta_n^2$ by left-cancelling β.

Assume from now on $1 < \Delta_n^2$. For $p < q$ in $\mathbb{Z}$, let us write $[p, q]$ for the interval $\{\beta \in B_n \mid \Delta_n^{2p} \leqslant \beta \leqslant \Delta_n^{2q}\}$ of M. We want to show that the left-divisors of Δ_n^2 in M generate M, which amounts to saying that $[0, 1]$ generates M. It is easy to check that the product of p braids in $[0, 1]$ belongs to $[0, p]$, and that the inverse of a braid in $[0, p]$ belongs to $[-p, 0]$. So the point is to show that *every* braid lies in some interval $[p, p+1]$: a priori, there might exist braids outside all these intervals (we do not assume that the order is Archimedean).

Now we claim that the generators σ_i must all lie in the interval $[-1, 1]$, which will clearly imply that each braid lies in some interval, as needed. Let $\delta_n = \sigma_1 \cdots \sigma_{n-1}$. As is well known, we have $\delta_n^n = \Delta_n^2$, so the assumption $1 < \Delta_n$ implies $1 < \Delta_n^2 = \delta_n^n$, hence $1 < \delta_n$, and, therefore, $1 < \delta_n < \delta_n^2 < \cdots < \delta_n^n = \Delta_n^2$.

Assume that we have $\Delta_n^2 \leqslant \sigma_i$ for some i. Let j be any element of $\{1, ..., n-1\}$. All generators σ_j are conjugate under some power of δ_n, so we can find p with $0 \leqslant p \leqslant n-1$ satisfying $\sigma_j = \delta_n^{-p}\sigma_i\delta_n^p$. Then we obtain

$$1 < \delta_n^{n-p} = \delta_n^{-p}\Delta_n^2 \leqslant \delta_n^{-p}\sigma_i \leqslant \delta_n^{-p}\sigma_i\delta_n^p = \sigma_j.$$

So $1 < \sigma_j$ holds for each generator σ_j. It follows that, if a braid β can be represented by a positive braid word that contains at least one letter σ_i, then $\Delta_n^2 \leqslant \beta$ holds. This applies in particular to Δ_n, and we deduce $\Delta_n^2 \leqslant \Delta_n$, which contradicts the assumption $1 < \Delta_n$.

Similarly, assume $\sigma_i \leqslant \Delta_n^{-2}$. Consider again any σ_j. If p is as above, we also have $\sigma_j = \delta_n^{n-p}\sigma_i\delta_n^{p-n}$, since δ_n^n lies in the center of B_n. Then we find

$$\sigma_j = \delta_n^{n-p}\sigma_i\delta_n^{p-n} < \delta_n^{n-p}\sigma_i \leqslant \delta_n^{n-p}\Delta_n^{-2} = \delta_n^{-p} \leqslant 1.$$

This time, $\sigma_j < 1$ holds for each j. As Δ_n is a positive braid, this implies $\Delta_n < 1$, which contradicts the assumption $1 < \Delta_n$ again. So our claim is established, and the proposition follows. □

Example 2.25. For every n, the braid group B_n admits several left-invariant orders, see Chapter XII. For each of them, the associated positive cone (augmented with 1) is a cancellative submonoid of B_n in which Δ_n^2 (or its inverse) is a Garside element. This applies in particular to the positive cone of the Dehornoy order, and to the Dubrovina–Dubrovin monoid $B_n^{\oplus}$ of Definition XII.3.12: the former monoid is not finitely generated, but the second one is. However, $B_n^{\oplus}$ is not a Garside monoid in the sense of Definition I.2.1 since it is (very) far from Noetherian.

3 Braid groups of well-generated complex reflection groups

We now consider an extension of the family of Coxeter groups, the family of complex reflection groups. These groups arise when the construction of Coxeter groups is extended to consider groups generated by pseudo-reflections of finite order in a complex vector space, instead of a real vector space as in the Coxeter case. The main result is that the dual monoid approach can be extended to well-generated reflection groups (Definition 3.7), thus giving rise to bounded Garside germs and, from there, to Garside groups.

3.1 Complex reflection groups

We begin with a quick introduction to reflection groups. A possible reference is [39].

Definition 3.1 (complex reflection). If V is a finite-dimensional vector space over $\mathbb{C}$, a *complex reflection* is an element s of $\mathrm{GL}(V)$ of finite order such that the fixed points of s, that is, $\mathrm{Ker}(s - \mathrm{id}_V)$, form a hyperplane H_s.

If V is of dimension n, a complex reflection s is diagonalizable with $n-1$ eigenvalues equal to 1 and the last one a root of unity ζ different from 1. For $\zeta = -1$, we recover the classical notion of a reflection, which is the only possible complex reflection if we replace V by a vector space over $\mathbb{R}$. To lighten the terminology, we will just say "reflection" for complex reflection in the rest of this section and specify "reflection of order 2" when needed.

Definition 3.2 (reflection group). A *finite complex reflection group* is a finite group W with the property that there exists a finite-dimensional $\mathbb{C}$-vector space V such that W is a subgroup of $\mathrm{GL}(V)$ generated by complex reflections. We then denote by $R(W)$ the set of all reflections in W.

Every finite Coxeter group is a finite complex reflection group, but we shall see below that there exist finite complex reflection groups that are not Coxeter groups, namely V does not admit any real structure preserved by W.

There is an alternative definition of finite complex reflection groups based on algebraic geometry: given a finite linear subgroup W of $\mathrm{GL}(V)$, the quotient variety V/W always exists, but is not necessarily smooth; it is smooth if and only if W is a complex reflection group.

If W is a finite complex reflection group, then V/W is not only smooth, but isomorphic to V again as a variety. This leads to a characterization of complex reflection groups in terms of invariants: the algebra of the variety V (that is the algebra of polynomial functions on V) is $S(V^*)$, the symmetric algebra of the dual vector space; for any choice of basis $x_1, \dots, x_n$ of V^* it is isomorphic to $\mathbb{C}[x_1, \dots, x_n]$. The algebra of the variety V/W is the algebra $S(V^*)^W$ of fixed points under the natural action of W on $S(V^*)$. Thus finite complex reflection groups are characterized by the property that $S(V^*)^W$ is a polynomial algebra $\mathbb{C}[f_1, \dots, f_n]$. The polynomials f_i are not unique, but their degrees d_i (as polynomials in the x_i) are unique: they are called the *reflection degrees* of W. We may choose the polynomials f_i homogeneous of degree d_i, which we will do.

A lot of information on W is encoded in the reflection degrees; for instance, the cardinality of W is $d_1 \cdots d_n$ and the number of reflections of W is $\sum_{i=1}^{i=n}(d_i - 1)$.

Classification of finite complex reflection groups. We say that a reflection group W is *irreducible* if its natural representation on the underlying vector space V is irreducible, that is, equivalently, if V cannot be decomposed as a product $V_1 \times V_2$ such that W is included in $\mathrm{GL}(V_1) \times \mathrm{GL}(V_2)$. Irreducible complex reflection groups have been classified by Shephard and Todd [210] in 1954: they contain an infinite family $G(p, e, n)$ with $e|p$, consisting of the monomial $n \times n$ matrices whose entries are p-th roots of unity and whose product of entries is an e-th root of unity, and 34 exceptional groups which for historical reasons (the original naming of Shephard and Todd) are denoted by G_4 to G_{37}. The irreducible finite Coxeter groups are special cases of irreducible finite complex reflection groups, precisely those whose natural representation can be defined over the reals. They appear in the Shephard–Todd classification as indicated in Table 2.

A_n	B_n	D_n	$\mathrm{I}_2(e)$	H_3	H_4	F_4	E_6	E_7	E_8
$G(1,1,n{+}1)$	$G(2,1,n)$	$G(2,2,n)$	$G(e,e,2)$	G_{23}	G_{30}	G_{28}	G_{35}	G_{36}	G_{37}

Table 2. Correspondence between the Dynkin classification of finite Coxeter groups and the Shephard–Todd classification of complex reflection groups

3.2 Braid groups of complex reflection groups

We now associate with every finite complex reflection group W an analog of the braid group (or Artin–Tits group) associated with every (finite) Coxeter group. To this end, we start from the geometrical viewpoint of hyperplane arrangements.

Assume that W is a finite complex reflection group with underlying vector space V. By a theorem of Steinberg [219] in 1964, the covering $V \to V/W$ is regular outside of the reflecting hyperplanes. This means that we can extend to complex reflection groups the topological definition of braid groups.

Definition 3.3 (braid group). Assume that W is a finite complex reflection group with underlying vector space V. Let V^{reg} be the complement in V of the reflecting hyperplanes. The *braid group* $B(W)$ of W is defined to be the fundamental group $\pi_1(V^{\mathrm{reg}}/W)$; the *pure braid group* $PB(W)$ is defined to be $\pi_1(V^{\mathrm{reg}})$.

The covering $V \to V^{\mathrm{reg}}/W$ gives rise to an exact sequence

$$1 \longrightarrow PB(W) \longrightarrow B(W) \xrightarrow{\pi} W \longrightarrow 1,$$

which makes W a natural quotient of $B(W)$, and generalizes the special case already mentioned in (I.1.7).

As shown by Brieskorn in [34], in the case of a Coxeter group (hence in particular in the case of a symmetric group), the braid group so introduced coincides with the braid group (or Artin–Tits group) considered in Section 1.

Proposition 3.4 (coherence). *If (W, Σ) is a finite Coxeter system, then W can be realized as a real reflection group; the associated braid group $B(W)$ as introduced in Definition* 3.3 *is then isomorphic to the Artin–Tits group as introduced in Definition* 1.18.

Let us return to the case of a general finite complex reflection group W. Then the group $B(W)$ can be generated by *braid reflections*, see for instance [41, Proposition 2.2], where a braid reflection is an element $\boldsymbol{s}$ of $B(W)$ such that $\pi(\boldsymbol{s})$ is a reflection, and such that $\boldsymbol{s}$ is the image in $B(W)$ of a path in V^{reg} of the form shown in Figure 4. Braid reflections above conjugate reflections are conjugate.

Given a basepoint x_0 in V^{reg}, the loop $\boldsymbol{\pi} : t \mapsto e^{2i\pi t}x_0$ from $[0, 1]$ to V^{reg} is central in $PB(W)$. It can actually be shown that it generates the center of this group when W is irreducible, see [108].

Definition 3.5 (regular). If W is a finite complex reflection group with underlying vector space V, an element g of W is called *regular* if it has an eigenvector x in V^{reg}.

If the eigenvalue associated with the eigenvector x of V^{reg} is $e^{2ik\pi/d}$, the path $t \mapsto e^{2ik\pi t/d}x$ becomes a loop in V^{reg}/W; if we denote this loop by $\boldsymbol{\beta}$, we have $\boldsymbol{\beta}^d = \boldsymbol{\pi}^k$ by construction. Then we have the following theorem of Springer about regular elements, see [215, Theorem 4.2]:

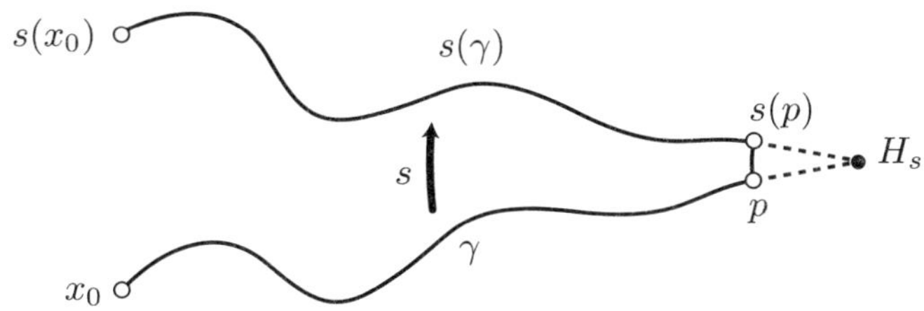

Figure 4. A braid reflection above a reflection s, drawn in V^{reg}. If x_0 in V^{reg} is above the base point in V^{reg}/W, we take any path γ from x_0 to a point p close to the hyperplane H_s, then follow an arc around H_s from p to $s(p)$, and finish by the reflected path $s(\gamma)$. This maps to a loop in V^{reg}/W.

Proposition 3.6 (regular elements). *Assume that W is a finite complex reflection group with underlying vector space V and g is a regular element of W with regular eigenvalue ζ.*

(i) *All other regular elements with regular eigenvalue ζ are conjugate to g.*

(ii) *Let $V_{g,\zeta}$ be the ζ-eigenspace of w. Then $C_W(g)$ acting on $V_{g,\zeta}$ is a complex reflection group, whose reflecting hyperplanes are the traces on $V_{g,\zeta}$ of those of W.*

(iii) *The reflection degrees of $C_W(g)$ are the degrees d_i of W satisfying $\zeta^{d_i} = 1$.*

(iv) *The eigenvalues of g on V are ζ^{1-d_i}.*

3.3 Well-generated complex reflection groups

A finite Coxeter group of dimension n (meaning that the group can be realized as a group of reflections of an n-dimensional real vector space) can be generated by n reflections. In contrast, an irreducible finite complex reflection group of dimension n sometimes needs $n + 1$ reflections to be generated.

Definition 3.7 (well-generated). A finite complex reflection group with underlying vector space of dimension n is called *well-generated* if it can be generated by n reflections.

The irreducible groups that are not well-generated are $G(p, e, n)$ for $p \neq e$ and $e \neq 1$, and $G_7, G_{11}, G_{12}, G_{13}, G_{15}, G_{19}, G_{22}$, and G_{31}.

Hereafter we assume that the fundamental invariants $f_1, \dots, f_n$ have been labeled so as to satisfy $d_1 \leqslant \cdots \leqslant d_n$. The following result can be observed from the classification.

Lemma 3.8. *If W is an irreducible well-generated finite complex reflection group, and $d_1, \dots, d_n$ are the associated degrees (in non-decreasing order), $d_i < d_n$ holds for $i \neq n$.*

It follows by a general regularity criterion [172] that there exists regular elements for the eigenvalue $e^{2i\pi/d_n}$. By Proposition 3.6, such elements form a single conjugacy class.

Definition 3.9 (Coxeter number, Coxeter class). In the context of Lemma 3.8, the maximal degree d_n is denoted by h and is called the *Coxeter number* of W. The conjugacy class of regular elements for the eigenvalue $e^{2i\pi/h}$ is called the *Coxeter class* of W.

If c belongs to the Coxeter class, its $e^{2i\pi/h}$-eigenspace is one-dimensional, and its centralizer is a cyclic group of order h. The following can be shown:

Lemma 3.10. *If W is an irreducible well-generated finite complex reflection group with underlying vector space V of dimension n, every element of the Coxeter class is the product of n reflections, and this is the minimum number of reflections of which it is the product.*

Proof (*sketch*). The fact that n reflections are sufficient is checked case-by-case. In the case where W is real (thus a finite Coxeter group), it results from the fact that the Coxeter class is the conjugacy class of the Coxeter elements defined in Definition 2.1.

On the other hand, Proposition 3.6(iv) implies that an element c of the Coxeter class has no eigenvalue equal to one. Then it follows from Lemma 2.17 that at least n reflections are needed since we have $m(c) = n$. □

Having introduced the Coxeter class, we can now define what will play the role of a dual braid monoid in the general case of a finite complex reflection group.

Definition 3.11. Assume that W is a finite, well-generated complex reflection group. The *dual braid monoid* associated with W is the monoid generated by the germ $\mathrm{Pref}(c)^R$, where c is a chosen element in the Coxeter class of W.

That this definition makes sense is clear, since, by definition, W is generated by R, and even positively generated since R consists of finite order elements. Next, the family $\mathrm{Pref}(c)$ is closed under R-prefix by definition. So Lemmas VI.2.60 and VI.2.62 guarantee that $\mathrm{Pref}(c)^R$ is indeed a cancellative, right-associative germ that is Noetherian. Our main task will be to show that this germ is actually a Garside germ and that the associated group is the braid group $B(W)$, which will be done in Proposition 3.22 below.

3.4 Tunnels

We assume until the end of this section that W is an irreducible well-generated finite complex reflection group. In the proof of Proposition 3.22 as developed by Bessis in [11], the first ingredient is the notion of a tunnel, which we explain now.

For each reflecting hyperplane H for W, let $e_H = |C_W(H)|$ (so that there are $e_H - 1$ reflections with hyperplane H), and let $\alpha_H \in V^*$ be a linear form defining H. Then $\mathrm{Disc} = \prod_H \alpha_H^{e_H}$ is invariant under W, thus is a polynomial in the fundamental invariants $f_1, ..., f_n$, and $\mathrm{Disc} \neq 0$ is the equation which defines V^{reg}/W in the vector space V/W with natural coordinates $f_1, ..., f_n$; the hypersurface $\mathrm{Disc} = 0$ is called the *discriminant*. We have the following result, see [11, 2.6].

Proposition 3.12 (discriminant). *It is possible to choose the invariants f_i so that they satisfy*

$$\mathrm{Disc} = f_n^n + \alpha_2 f_n^{n-2} + \alpha_3 f_n^{n-3} + \cdots + \alpha_n,$$

where α_i is a polynomial in $f_1, ..., f_{n-1}$ weighted homogeneous of degree ih (for the weight d_i assigned to f_i).

Then, the idea for computing the fundamental group $B(W)$ of the complement of the discriminant is to study the intersection of the discriminant with the complex line L_y in V/W obtained by fixing the values $y = (y_1, \dots, y_{n-1})$ of $f_1, \dots, f_{n-1}$. By Proposition 3.12, this intersection is the set of zeros of a degree-n polynomial, thus is a set $\mathrm{LL}(y)$ of points with multiplicities whose total multiplicity is n. To be able to consider loops in various L_y as elements of the same fundamental group $B(W)$, Bessis uses the fact that it is possible to replace a basepoint by any contractible subset $\mathcal{U}$. He proves [11, Lemma 6.2]:

Lemma 3.13. *Let $\mathcal{U} = \bigcup_y \mathcal{U}_y$, where $\mathcal{U}_y$ is the set of all z in L_y satisfying*

$$\Re(z) \neq \Re(x) \text{ for every } x \text{ in } \mathrm{LL}(y), \text{ or}$$
$$\Im(z) > \Im(x) \text{ for every } x \text{ in } \mathrm{LL}(y) \text{ satisfying } \Re(x) = \Re(z).$$

Then $\mathcal{U}$ is contractible.

Using this, it will be possible to use for generating loops in $B(W)$ segments of a special type that Bessis calls tunnels.

Definition 3.14 (tunnel). (See Figure 5.) A *tunnel* of L_y is a horizontal line segment in some L_y that starts and ends in $\mathcal{U}_y$. A tunnel is called *simple* if it crosses only one half-line below some x_i in $\mathrm{LL}(y)$, and crosses that line below every x_i in that line. A tunnel deep enough (with a negative enough imaginary part) and long enough so that it crosses every half-line below all points x_i of that half-line is called the *deep* tunnel of L_y.

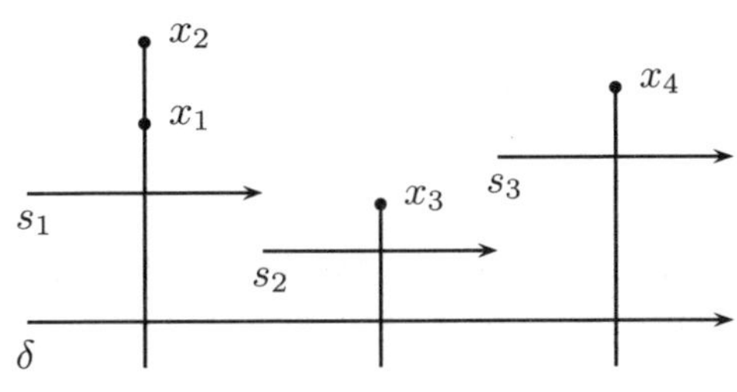

Figure 5. One of the complex lines L_y. We assume that $\mathrm{LL}(y)$ is $\{x_1, x_2, x_3, x_4\}$ with respective multiplicities $2, 1, 1, 2$. The complement of $\mathcal{U}_y$ in L_y consists of 3 half-lines. The simple tunnels s_1, s_2, s_3 are product respectively of $3, 1, 2$ braid reflections. The deep tunnel is $\boldsymbol{\delta} = s_1 s_2 s_3$.

In Definition 3.14, we may use "the" for the deep tunnel, since all such tunnels are clearly homotopic. It is also clear that all the deep tunnels in various lines L_y represent the same element of $B(W)$; but note that this element is nevertheless dependent on our choice of coordinates in V, specific choice of invariants, and choice of basepoint; it turns out that if we change these choices we may obtain every conjugate element in $B(W)$.

Choose for basepoint v in V^{reg}, a $e^{2i\pi/h}v$-eigenvector of a Coxeter element c and let $\boldsymbol{\delta}$ be the h-th root of $\boldsymbol{\pi}$ image of the path $t \mapsto e^{2i\pi t/h}v$. For every i, since f_i is invariant and homogeneous of degree d_i we have

$$f_i(v) = f_i(c(v)) = f_i(e^{2i\pi/h}v) = e^{2i\pi d_i/h} f_i(v),$$

from which we deduce $f_i(v) = 0$ for $i \leqslant n-1$ since then we have $d_i < h$. Thus the loop $\boldsymbol{\delta}$ lives entirely in the line L_{y_0} where y_0 is $(0, \dots, 0)$. The intersection of the discriminant with the line L_{y_0} is given in L_{y_0} by the equation $f_n^n = 0$, thus $\mathrm{LL}(y_0)$ is only one point 0 with multiplicity n. The loop $\boldsymbol{\delta}$ is thus represented by the deep tunnel of L_{y_0}. Hence, all deep tunnels represent $\boldsymbol{\delta}$, and each L_y gives a decomposition of $\boldsymbol{\delta}$ as a product of simple tunnels.

Note the following property (which is not difficult to prove):

Lemma 3.15. *A simple tunnel s included in L_y is the product of m braid reflections, where m is the sum of the multiplicities of the points of* $\mathrm{LL}(y)$ *on the half-line that s crosses.*

3.5 The Lyashko–Looijenga covering and Hurwitz action

Let $Y = \mathbb{C}^{n-1}$, with coordinates identified to $f_1, \dots, f_{n-1}$ and let X_n be the configuration space of multisets of total multiplicity n in $\mathbb{C}$ (also called the punctual Hilbert scheme $\mathbb{C}^{[n]}$). We can see LL as a map from Y to X_n that falls into the subset of points with sum 0; in algebraic terms, this map sends $(f_1, \dots, f_{n-1})$ to $(\alpha_2, \dots, \alpha_n)$ since the multiset $\mathrm{LL}(y)$ of roots of $\mathrm{Disc} = 0$ is determined by the elementary symmetric functions of the roots, which are the α_i. Let X_n^{reg} be the configuration space of subsets of n points of $\mathbb{C}$ with sum 0, that is, the subspace of X_n where each point has multiplicity 1 and the sum of points is 0, and let Y^{reg} be the preimage of X_n^{reg} by LL. Bessis shows [11, 5.3 and 5.6]:

Proposition 3.16 (covering). *The map* $\mathrm{LL} : Y^{\mathrm{reg}} \to X_n^{\mathrm{reg}}$ *is an étale, unramified covering of degree $n! h^n / |W|$.*

It is important for our purposes that the subspace X_n^{gen} of X_n^{reg} consisting of n-tuples of points with distinct real parts be contractible. Thus the ordinary braid group B_n, which is the fundamental group of X_n^{reg}, can be seen as the fundamental group with basepoint X_n^{gen}. Since the fundamental group of a base space acts on the fibers of a covering, we obtain a Galois action of B_n on the preimage Y^{gen} of X_n^{gen}. Note that an element y of Y^{reg} is in Y^{gen} if and only if L_y provides a decomposition of $\boldsymbol{\delta}$ as a product of n simple tunnels which are braid reflections. We have the following result [11, 6.18]:

Lemma 3.17. *The Galois action of B_n induces the Hurwitz action on the decompositions of $\boldsymbol{\delta}$ as a product of n braid reflections.*

The definition of the Hurwitz action was recalled in (2.16). As a corollary, we obtain that the Hurwitz orbit of a decomposition of $\boldsymbol{\delta}$ into simple tunnels is of cardinality less than $n! h^n / |W|$, and there will be only one orbit if we prove that an orbit has this specific cardinality. This last fact turns out to be true, and the main tool to show it is the following proposition, of which unfortunately only a case-by-case proof, using the classification, is known: see [12] for the infinite series. The exceptional groups have been checked using the GAP3 package `CHEVIE` [183].

Proposition 3.18 (decompositions). *For every c in the Coxeter class of W, there exist $n!h^n/|W|$ decompositions of c into a product of n reflections, and the Hurwitz action is transitive on them.*

Using the fact that the Hurwitz action commutes with the projection $B(W) \xrightarrow{\pi} W$, one deduces

Lemma 3.19. *The Hurwitz action of B_n on decompositions of δ into simple tunnels is transitive. Moreover, there is an isomorphism of B_n-sets between such decompositions and decompositions of c as a product of n reflections.*

Actually, more is known by the same case-by-case analysis, with the same references to [12] and to the GAP3 package `CHEVIE` as above:

Proposition 3.20 (Garside germ). *Let R be the set of reflections of W. Let* $\mathrm{Pref}(c)$ *be the set of R-prefixes of c. Then* $\mathrm{Pref}(c)$ *satisfies the assumptions of Proposition* VI.2.69 *(derived Garside* III*), and* $\mathrm{Pref}(c)^R$ *is a bounded Garside germ.*

It follows from Lemma 3.10 that the R-tight decompositions of c are the decompositions of Proposition 3.18.

We should note here an important difference between the real case (the spherical Coxeter case) and the complex case. In the real case, R is a subset of $\mathrm{Pref}(c)$. In the general case, the intersection $\Sigma = R \cap \mathrm{Pref}(c)$ is a proper subset of R, which still generates W. We may as well, and we will, use Σ in Proposition 3.20 instead of R and write the germ derived from $\mathrm{Pref}(c)$ as $\mathrm{Pref}(c)^\Sigma$.

Summarizing what we have so far, we obtain

Proposition 3.21 (tunnels). *Assume that W is an irreducible well-generated finite complex reflection group. Then tunnels equipped with composition form a bounded Garside germ isomorphic to* $\mathrm{Pref}(c)^\Sigma$.

It remains to see that the group associated with the Garside germ $\mathrm{Pref}(c)^\Sigma$ is (isomorphic to) the braid group of W. This is what the next result [11, 8.2] states:

Proposition 3.22 (generated group). *If W is an irreducible well-generated finite complex reflection group and G is the group generated by the germ* $\mathrm{Pref}(c)^\Sigma$*, then the natural morphism from G to $B(W)$ is an isomorphism.*

The proof of Proposition 3.22 uses the transitivity of the Hurwitz action on decompositions of δ and Lemma 3.19 to establish the following presentation for the group derived from $\mathrm{Pref}(c)^\Sigma$.

Lemma 3.23. *A complete set of relations for the group generated by* $\mathrm{Pref}(c)^{\Sigma}$ *is given by* $rr' = r'r''$*, where* r, r', r'' *lie in* Σ *and* rr' *and* $r'r''$ *are* Σ*-prefixes of* c*.*

Compare with Lemma 2.13 in the case of B_n.

In the framework explained, Proposition 3.22 is a consequence of the presentation of $B(W)$ given in [8, Section 3]. The relations given there can be seen to result from the relations of Lemma 3.23 and Hurwitz action.

In his paper [11], Bessis goes on to show that the set of fixed points $(V^{\mathrm{reg}}/W)^{\zeta}$ of V^{reg}/W under a root of unity ζ (for the natural action of $\mathbb{C}$, which in the coordinates f_i is the map $f_i \mapsto \zeta^{d_i} f_i$), if nonempty, is also a $K(\pi, 1)$ space. Along the way, he shows [11, Theorem 12.5] that Springer's theorem has a lift to the braid group when W is well-generated:

Proposition 3.24 (regular braids). *Assume that* W *is an irreducible well-generated finite complex reflection group and* $\boldsymbol{\beta}$ *is an element of* $B(W)$ *satisfying* $\boldsymbol{\beta}^d = \boldsymbol{\pi}$*.*

(i) *All other d-th roots of* $\boldsymbol{\pi}$ *are conjugate to* $\boldsymbol{\beta}$*.*

(ii) *The image* w *of* $\boldsymbol{\beta}$ *in* W *is a regular element for the eigenvalue* $\zeta = e^{2i\pi/d}$.

(iii) *There is a natural map from* $\pi_1((V_{w,\zeta} \cap V^{\mathrm{reg}})/C_W(w))$ *to* $B(W)$*, which is injective, and has for image* $C_{B(W)}(\boldsymbol{\beta})$*.*

Note that, in the above, $(V_{w,\zeta} \cap V^{\mathrm{reg}})/C_W(w)$ is isomorphic to $(V^{\mathrm{reg}}/W)^{\zeta}$.

To conclude, let us mention that, in the case of non-well-generated complex reflection groups, structures of Garside monoids may also exist.

Example 3.25 (2-dimensional complex reflection groups). The braid groups of the 2-dimensional complex reflections groups G_{15}, G_7, G_{11}, G_{19}, and $G(2de, 2e, 2)$ for $d > 1$ are isomorphic to each other (see [6]), and in [45] it is proven that no other braid group of an irreducible complex reflection group is isomorphic to one of these braid groups. For every integer $e \geqslant 1$, these groups admit the presentation $\langle \mathtt{a}, \mathtt{b}, \mathtt{c} \mid \mathtt{abc} = \mathtt{bca}, \mathtt{ca}(\mathtt{b}|\mathtt{c})^{[2e-1]} = (\mathtt{b}|\mathtt{c})^{[2e-1]}\mathtt{ca}\rangle$ [41]. One can check [196] that, for each e, the monoid with this presentation is indeed a Garside monoid associated with $\Delta = \mathtt{ca}(\mathtt{b}|\mathtt{c})^{[2e-1]}$. For $e = 2$ the lattice of the sixteen divisors of Δ is shown on the right. As shown in [6], these braid groups are also isomorphic to the Artin–Tits group of type $\mathrm{A}_1 \times \widetilde{\mathrm{A}}_1$, which gives another structure of quasi-Garside monoid.

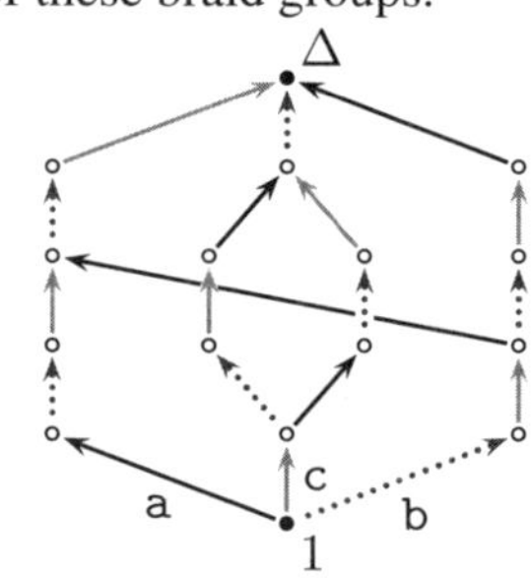

A similar result holds for the braid group of G_{13}, which admits the presentation

$$\langle \mathtt{a}, \mathtt{b}, \mathtt{c} \mid \mathtt{acabc} = \mathtt{bcaba}, \mathtt{bcab} = \mathtt{cabc}, \mathtt{cabca} = \mathtt{abcab}\rangle;$$

in this case, a Garside element is $(\mathtt{abc})^3$, and it admits 90 divisors.

Exercise

Exercise 99 (non-isomorphic). (i) Show that the Coxeter group of type B_3 is isomorphic to the direct product of two Coxeter groups of types A_1 and A_3. (ii) Show that the Artin–

Tits group of type B_3 is *not* isomorphic to the direct product of two Artin–Tits groups of types A_1 and A_3.

Exercise 100 (exotic monoid). Show that the monoid $\langle \mathtt{a}, \mathtt{b} \mid \mathtt{aba} = \mathtt{b}^2 \rangle^+$ (the monoid $B^+_{3,0}$ of Proposition 2.21) is strongly Noetherian. [Hint: Define an $\mathbb{N}$-valued Noetherianity witness by giving weights to a and b.]

Exercise 101 (unfolding). (i) Show that, in Artin's braid group B_n, the elements $\sigma_1\sigma_5\sigma_5\cdots$ and $\sigma_2\sigma_4\sigma_6\cdots$ generate a dihedral group and express the order of the latter in terms of n. (ii) Deduce that the centralizer of Δ_n in B_n is an Artin–Tits group of type B. [Hint: Show that a braid commutes with Δ_n if and only if it is fixed under the flip endomorphism that exchanges σ_i and σ_{n-i} for each i, and deduce that the centralizer of Δ_n is generated by $\sigma_1\sigma_{n-1}$, $\sigma_2\sigma_{n-2}$, etc.]

Exercise 102 (smallest Garside, right-angled type). Assume that B^+ is a right-angled Artin–Tits monoid, that is, B^+ is associated with a Coxeter system (W, Σ) satisfying $m_{s,t} \in \{2, \infty\}$ for all s, t in Σ. (i) For $I \subseteq \Sigma$, denote by Δ_I the right-lcm (here the product) of the elements of I, when it exists (that is, when the elements pairwise commute). Show that the divisors of Δ_I are the elements Δ_J with $J \subseteq I$. (ii) Deduce that the smallest Garside family in B^+ is finite and consists of the elements Δ_I with $I \subseteq \Sigma$.

Exercise 103 (smallest Garside, large type). Assume that B^+ is an Artin–Tits monoid of large type, B^+ is associated with a Coxeter system (W, Σ) satisfying $m_{s,t} \geqslant 3$ for all s, t in Σ. Put $\Sigma_1 = \{s{\in}\Sigma \mid \forall r{\in}\Sigma\, (m_{r,s}{=}\infty)\}$, $\Sigma_2 = \{(s,t){\in}\Sigma^2 \mid \forall r{\in}\Sigma\, (m_{r,s}{+}m_{r,t}{=}\infty)\}$, $\Sigma_3 = \{(r,s,t){\in}\Sigma^3 \mid m_{r,s}{+}m_{r,t}{+}m_{s,t}{<}\infty\}$, and $E = \Sigma_1 \cup \{\Delta_{s,t} \mid (s,t) \in \Sigma_2\} \cup \{r\Delta_{s,t} \mid (r,s,t) \in \Sigma_3\}$ (we write $\Delta_{s,t}$ for the right-lcm of s and t when it exists). (i) Explicitly describe the elements of the closure S of E under right-divisor, and deduce that S is finite. (ii) Show that S is closed under right-lcm and deduce that S is a Garside family in B^+. (iii) Show that S is the smallest Garside family in B^+. (iv) Show that, if Σ has n elements and $m_{s,t} \neq \infty$ holds for all s, t, then E has $3\binom{n}{3}$ elements. (v) Show that, if Σ has n elements and $m_{s,t} = m$ holds for all s, t, then S has $(n + 2m - 5)\binom{n}{2} + n + 1$ elements. Apply to $n = m = 3$.

Notes

Sources and comments. Artin–Tits groups were investigated by J. Tits in the 1960's as "generalized braid groups". They subsequently appeared in the literature under the name "Artin groups". As these groups seem to never appear in Artin's works, calling them "Artin–Tits groups" appears a reasonable compromise.

The reversing approach developed in Subsection 1.2 is close to the one originally developed by F.A. Garside in [124, 125] (in the case of the braid monoids B_n^+). Although perhaps less elegant that the germ approach developed in the subsequent section, this approach allows for a direct and elementary treatment.

The dual monoid for type A_n was first introduced by J. Birman, K.Y. Ko, and S.J. Lee in [21]. The germ approach to this monoid was developed in [13]. The lattice property for the case of the dual monoid in general is a difficult result, of which only a case-by-case proof is known, see [11, 8.14]. Previous to this work, D. Bessis had given a construction for the real case in [9], using case-by-case arguments for the lattice property. The proof of Proposition 2.7 follows [13]. The proof of Proposition 2.11 follows [13] and [9]. There exists a case-free proof for Coxeter groups of spherical type appears in work of Brady and Watt [31].

For type $\widetilde{\mathrm{A}}_n$ the dual germ is a Garside germ for two choices of the Coxeter element only, nevertheless it is a germ for the corresponding Artin–Tits group for all choices of the Coxeter element. It is conjectured that this is the case for all types of Coxeter groups.

It has been announced by J. McCammond [181] that, for Coxeter groups of affine type other than $\widetilde{\mathrm{A}}_n$ or $\widetilde{\mathrm{C}}_n$, the dual germ of Definition 2.2 is never a Garside germ.

D. Bessis proved in [10] that the dual germ is a Garside germ for the universal Coxeter group with a finite number of generators, that is the Coxeter group such that all edges of the Coxeter graph are labeled with ∞ (the associated Artin–Tits group is a free group). See Corran–Picantin [66] for an explicit description of a Garside structure on braid groups associated with complex reflection groups of type (e, e, r).

The source for the Garside structures of Example 3.25 is Picantin [196]. Proposition 2.24 comes from D. and *al.* [100, Proposition II.3.6].

The many uses of the Garside structures of braids. The Garside structure of braid groups and Artin–Tits groups has been used in a number of works that we cannot list here, starting with the already mentioned solutions to the Word and Conjugacy Problems, first in Garside [125] and then in the many developments and extensions already mentioned in the notes of Chapter VIII. Most of them are based on normal decompositions and, therefore, on the Garside structure of braids, as are most of the practical implementations of braid groups. In particular, should braids become important in cryptography, their Garside structure would certainly play a central role.

Among the structural results based on the Garside structure, the most important is maybe the existence of a (bi)-automatic structure established for the braid groups B_n in Thurston [226] and Epstein and *al.* [119], and subsequently extended to all spherical type Artin–Tits groups in Charney [54, 55]. In another direction, the algebraic proof of the linearity of braid groups as established in Krammer [162, 163] and extended to other Artin–Tits groups of spherical type in Digne [105] and Cohen and Wales [63] uses decompositions into divisors of the Garside element. An extension of these methods to general Artin–Tits groups was used by L. Paris in [194] to prove that Artin–Tits monoids inject in their group. In the same vein, the proofs of the $K(\pi, 1)$ property by P. Deligne in [101] for Artin–Tits groups of spherical type and by D. Bessis in [11] for braid groups of complex reflection groups resort to normal decompositions, as does the study of the homology and the $K(\pi, 1)$-property as investigated in Charney, Meier, and Wittlesey [57], Dehornoy and Lafont [98], and Charney and Peifer [58].

The complete description of the normalizers and centralizers in Paris [192], González-Meneses and Wiest [143], and Godelle [134, 135] heavily relies on the existence of a Garside element and the properties of the derived normal decompositions. The connected problem of extracting roots in a braid group was addressed using Δ-normal decompositions in Stychnev [220], then extended in Sibert [211], again using normal forms. So is the recognition of palindromes in Deloup [104].

Even in the topological approach of braid groups as mapping class groups, the algebraic properties of the fundamental braids Δ_n and the existence of Δ-normal decompositions are often useful. Let us for instance mention the recognition of the Nielsen–Thurston type in Lee and Lee [170], the recognition of quasi-positive braids in Orevkov [190], or the study of Kleinian singularities in Brav and Thomas [32]. A topological interpretation of the parameters $\inf_\Delta$ and $\sup_\Delta$ appears in Wiest [229]. We stop here this certainly very incomplete review.

Combinatorics of normal sequences. Every finite Coxeter group gives rise to a finite Garside family in the associated Artin–Tits monoid, and natural counting problems immediately arise, namely counting (strict) normal sequences of a given length. Note that, by Proposition III.3.1 (geodesic, positive case), strict normal sequences are geodesic, so counting strict normal sequences amounts to investigating the metric on the monoid associated with the considered family of generators.

Here we shall only address type A, that is, Artin's braid groups B_n. By Proposition III.3.24 (adjacency matrix), the associated combinatorics is essentially contained in the $(n!\times n!)$-matrix M_n that specifies length-2 normal sequences. By Lemma 1.31, the divisors of Δ_n in B_n^+ are in one-to-one correspondence with the permutations of $\{1, ..., n\}$, and the question arises of characterizing normality in terms of permutations. For f in the symmetric group $\mathfrak{S}_n$, one says that i is a *descent* of f if $f(i) > f(i+1)$ holds.

Lemma. *Let σ be the lifting from $\mathfrak{S}_n$ to B_n^+ as in Lemma* 1.31. *Then, for f, g in $\mathfrak{S}_n$, the pair $\sigma(f)|\sigma(g)$ is Δ_n-normal if and only if every descent of g^{-1} is a descent of f.*

The result follows from the characterization of normality in Proposition V.1.53 (recognizing greedy). We are thus left with the question of investigating the $(\mathfrak{S}_n\times\mathfrak{S}_n)$-matrix M_n whose (f, g) entry is 1 if every descent of g^{-1} is a descent of f, and is 0 otherwise. It turns out that M_n is connected with the descent algebra of Solomon [214], and many interesting questions arise, in particular about the eigenvalues of M_n, see [85]. Let us mention that the size of the matrix M_n, which somehow indicates the size of the automata involved in an automatic structure, can be reduced from $n!$ to the number of partitions of n, see [*loc. cit.*] and Gebhardt [130]. The natural but nontrivial fact that the characteristic polynomial of M_n divides that of M_{n+1} was proved using quasi-symmetric functions by Hivert, Novelli, and Thibon [148]. One of the puzzling open questions in the area involves the spectral radius $\rho(M_n)$ of the matrix M_n (maximal modulus of an eigenvalue), which directly controls the growth of the numbers of Δ_n-length p elements in B_n^+:

Conjecture. *The ratio $\rho(M_{n+1})/n!\rho(M_n)$ tends to* $\log 2$ *when n grows to ∞.*

An application of the counting results of [*loc. cit.*] is the construction of sequences of braids that are finite, but so long that one cannot prove their existence in weak logical subsystems of Peano arithmetic in Carlucci, Dehornoy, and Weiermann [50], a paradoxical situation.

Similar questions arise in the dual case. As shown in Subsection 2.2, the set $\mathrm{Div}(\Delta_n^*)$ in the dual braid monoid B_n^{+*} is in one-to-one correspondence with the set $\mathrm{NC}(n)$ of all noncrossing partitions of $\{1, ..., n\}$. Then the counterpart of the above lemma is

Lemma. *Let σ^* be the map from* $\mathrm{NC}(n)$ *to* B_n^{+*} *stemming from Proposition 2.7. Then, for* λ, μ *in* $\mathrm{NC}(n)$*, the pair* $\sigma^*(\lambda)|\sigma^*(\mu)$ *is* Δ_n^**-normal if and only if the polygons associated with the Kreweras complement of* λ *and with* μ *have no chord in common.*

Above, the polygons associated with a partition are those displayed in Figure 2, and the Kreweras complement of a partition λ is the cut of $\mathbf{1}_n$ at λ, where $\mathbf{1}_n$ is the finest partition of $\{1, ..., n\}$. Then a size $\mathit{Cat}(n) \times \mathit{Cat}(n)$ incidence matrix M_n^* arises and governs the combinatorics of dual normal sequences of braids. Little is known so far, but the number of length two Δ_n^*-normal pairs and the determinant of the matrix M_n^* are computed in Biane and Dehornoy [16] using a connection with free cumulants for a product of independent variables.

Finally, let us mention that very little is known about combinatorics of braids with respect to Artin generators: in the case of the monoid B_n^+, the generating series for the number of length-p braids is easily computed but, even for B_4, it remains unknown whether the generating series for the number of length-p braids with respect to Artin's generators σ_i is rational or not, see Mairesse and Matheus [178].

Further questions. A number of questions involving Artin–Tits groups and braid groups of complex reflection groups remain open, even in the case of basic problems:

Question 26. Are the Word and Conjugacy Problems for an Artin–Tits group decidable?

Many solutions in particular cases are known. Typically, when it exists, a finite bounded Garside structure provides positive answers to Question 26 by means of Corollaries III.3.64 (decidability of Word Problem) and VIII.2.25 (decidability of Conjugacy Problem). By contrast, nothing is clear in the non-spherical case. We addressed in Question 25 the existence of a finite (unbounded) Garside family; at the time of printing, a positive answer has been announced [94, 116]:

Proposition. *Every finitely generated Artin–Tits monoid admits a finite Garside family.*

The proof relies on Proposition 1.41 and on introducing the notion of a low element in a Coxeter group W by using the associated root system Φ [23, Chapter 4]: an element w of W is called *low* if there exists a subset A of the (finite) set of small roots [*loc. cit.*, Section 4.7] such that the right-inversion set $\Phi^+ \cap w(\Phi^-)$ is obtained by taking all roots in the cone (linear combinations with nonnegative coefficients) spanned by A. One can show that the family of all low elements includes Σ, is finite, and is closed under join (that is, least $\leqslant_\Sigma$-upper bound) and Σ-suffix. This requires in particular to analyze what is called the dominance relation on the root system Φ [*loc. cit.*, page 116].

The above result then makes the following question crucial:

Question 27. Can a finite Garside family in the Artin–Tits monoid be used to solve Question 26?

This question is open. A finite Garside family S in an Artin–Tits monoid B^+ gives a solution to the Word Problem for B^+, but the latter is trivial anyway, since one can always enumerate the (finite) equivalence class of a positive word exhaustively. One also obtains a solution for the Conjugacy Problem for B^+ as follows: assume that e, e' lie in B^+ and

$ge = e'g$ holds. As g left-divides eg, the S-head g_1 of g must left-divide the S-head of eg, which, by the $\mathcal{H}$-law, is that of eg_1. This implies $g_1 \preccurlyeq eg_1$, whence the existence of e_1 satisfying $g_1e = e_1g_1$. Repeating the argument, we deduce that two elements e, e' of B^+ are conjugate if, and only if, there exists a finite sequence $e_0, ..., e_p$ with $e_0 = e$ and $e_p = e'$ such that, for each i, there exists g_i in S conjugating e_i to e_{i+1}. As S is finite and all elements e_i must have the same length, the existence of the sequences e_i and g_i as above can be decided effectively.

In the non-spherical type, hence outside the range of Ore's theorem, the problem is to connect the Artin–Tits monoid B^+ with its enveloping group B, typically to control the elements of B by distinguished zigzag-words where positive and negative factors alternate. What Question 27 asks is whether a finite Garside family in B^+ may help.

Let us mention another problem connected with the Word Problem, but maybe easier. If W is an infinite Coxeter group, the (left- or right-) reversing process associated with the presentation (1.19) of the corresponding Artin–Tits monoid does not always terminate and Algorithm IV.3.23 does not solve the Word Problem. However, some variants can be considered and natural questions then appear. Assuming that (S, R) is a positive presentation and w, w' are signed S-words, say that $w \rightsquigarrow w'$ holds if one can transform w into w' by repeatedly right- and left-reversing subwords and replacing positive (*resp.* negative) subwords with equivalent positive (*resp.* negative) words. Say that *extended reversing* solves the Word Problem if $w \rightsquigarrow \varepsilon$ holds for every signed S-word w representing 1. Extended reversing is reminiscent of the Dehn algorithm for hyperbolic groups [119] in that it corresponds to applying the relations of the presentation and the trivial free group relations, but banishing the introduction of factors $\overline{s}|s$ or $s|\overline{s}$.

Question 28. Does extended reversing solve the Word Problem of every Artin–Tits presentation?

By Proposition 1.33, the answer is positive in the spherical case. It is shown in Dehornoy and Godelle [96] that the answer is also positive for Artin–Tits groups of type FC (the closure of spherical type Artin–Tits groups under amalgamated products), and right-angled Artin–Tits groups (those in which all exponents are 2 or ∞).

In another direction, and in view of the preliminary observations of Subsection 2.4, it is natural to raise

Question 29. Do the braid groups B_n admit other Garside structures than the classical and dual ones?

An (extremely) partial answer is given with the exotic Garside structures on B_3 described in Proposition 2.21 (which do not involve Garside monoids as the associated monoids are in general not Noetherian). A more general positive answer is given in Proposition 2.24 when some left-invariant ordering is involved. In the vein of Example 2.23, the following simple question seems to be open

Question 30. Does the submonoid of B_n generated by $\sigma_1, \sigma_1\sigma_2, ..., \sigma_1\sigma_2\cdots\sigma_{n-1}$ admit a finite presentation? Is it a Garside monoid?

In a completely different direction, interesting submonoids of B_n originate in the following result of Sergiescu [209].

Proposition. *Assume that* Γ *is a finite planar connected graph with* n *vertices and no loop of length* 1*, embedded in the oriented plane. Let* $\sigma_1, \dots, \sigma_{N-1}$ *be an enumeration of the edges of* Γ*, and* $\mathcal{R}_\Gamma$ *consist of the relations*

$$(*) \quad \begin{cases} \sigma_i\sigma_j = \sigma_j\sigma_i & \text{if } \sigma_i \text{ and } \sigma_j \text{ are disjoint,} \\ \sigma_i\sigma_j\sigma_i = \sigma_j\sigma_i\sigma_j & \text{if } \sigma_i \text{ and } \sigma_j \text{ have a vertex in common,} \\ \sigma_i\sigma_j\sigma_k\sigma_i = \sigma_j\sigma_k\sigma_i\sigma_j & \text{if } \sigma_i, \sigma_j, \sigma_k \text{ have a vertex in common,} \\ & \qquad \text{and occur in positive order,} \\ \sigma_{i_1}\sigma_{i_2}\cdots\sigma_{i_{p-1}} = \sigma_{i_2}\cdots\sigma_{i_{p-1}}\sigma_{i_p} & \text{if } (\sigma_{i_1}, \dots, \sigma_{i_p}) \text{ is a positive loop with no} \\ & \qquad \text{vertex inside and } i_1 \neq i_2 \text{ and } i_{p-1} \neq i_p. \end{cases}$$

Then $\langle \sigma_1, \dots, \sigma_{N-1} \,|\, \mathcal{R}_\Gamma \rangle$ *is a presentation of the braid group* B_n.

If Γ is a path of length $n-1$, the presentation $(*)$ is the standard Artin presentation of B_n, and the associated monoid is the Artin braid monoid B_n^+. If Γ is a complete graph on n vertices, one obtains the Birman–Ko–Lee presentation, and the associated monoid is the dual braid monoid B_n^{+*}. Considering various graphs provides a number of positive presentations for the braid group B_n and, therefore, a number of monoids that can be called the *Sergiescu monoids*.

Question 31. Which Sergiescu monoids can be given the structure of a Garside monoid?

No general answer is known. We refer to Picantin [196] for partial observations in the case of the graph shown on the right, where a Garside monoid with 589 divisors of the Garside element arises.

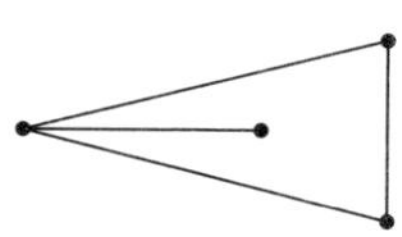

Chapter X
Deligne–Lusztig varieties

This chapter describes an application of the methods of Part A to the representation theory of finite reductive groups and, more specifically, to properties connected with the Broué Abelian Defect Conjecture in the theory of finite groups. In essence, finite reductive groups are central extensions of simple groups in the infinite families of finite simple groups; they constitute a large family that includes all infinite families of finite simple groups except for the alternating groups. The Broué Conjecture in the case of the principal block predicts some deep connection between the representation theory of a finite reductive group and that of its Sylow subgroups. Lacking at the moment a proof of the Broué Conjecture, one can directly address some of its geometric consequences that involve Deligne–Lusztig varieties and are connected with questions about conjugacy and centralizers in braid groups and ribbon categories. This is where Garside theory appears, mainly through results about conjugacy in a category equipped with a bounded Garside family as developed in Chapters V and VIII. Of special interest here is Proposition VIII.1.24, which states that, under convenient assumptions, every conjugation is a cyclic conjugation. Our aim in this chapter is to explain these results and their contribution to a program that could eventually lead to a direct proof of these geometric consequences of the Broué Conjecture. We shall describe mainly the special case of the group $\mathrm{GL}_n(\mathbb{F}_q)$, but in a way where we can hint to what happens for other finite reductive groups.

The chapter comprises four sections, plus an Appendix. First, in Section 1 and 2, we give a quick introduction to the theory of finite reductive groups and their representations, in particular Deligne–Lusztig theory. We then recall some modular representation theory, ending with a geometric version of the Broué Abelian Defect Conjecture in the case of the principal block. Next, for every reductive group, one has an associated Weyl group, hence an Artin–Tits group and a classical Artin–Tits monoid as described in Chapter IX. This monoid admits a distinguished Garside family. We show how to use the results of Chapter VIII to establish some geometric consequences of the Broué Conjecture, first in the case of a torus (Section 3), next in the general case (Section 4). The results are summarized in Propositions 3.12 and 4.10.

In the final Appendix, we prove Proposition A.2, a general result of independent interest about representations of categories with Garside families into bicategories, which generalizes a result of Deligne [102, Théorème 1.5] about representations of spherical braid monoids into a category.

1 Finite reductive groups

We begin with a quick introduction to reductive groups and some of their special subgroups.

1.1 Reductive groups

The basic example of a finite reductive group is the general linear group over a finite field.

Example 1.1 (general linear group). Let $\mathbb{F}_q$ be the finite field of characteristic p with q elements and $\overline{\mathbb{F}}$ be an algebraic closure of $\mathbb{F}_q$. Then the general linear group $\mathrm{GL}_n(\mathbb{F}_q)$ can be seen as the subgroup of the algebraic group $\boldsymbol{G} = \mathrm{GL}_n(\overline{\mathbb{F}})$ consisting of all points over $\mathbb{F}_q$. Now the points over $\mathbb{F}_q$ are also the points that are fixed by the *Frobenius endomorphism* F given by $F(a_{i,j}) = (a_{i,j}^q)$, since the elements of $\mathbb{F}_q$ are the solutions of the equation $x^q = x$ in $\overline{\mathbb{F}}$. So, in this case we have $\mathrm{GL}_n(\mathbb{F}_q) = \boldsymbol{G}^F$.

Finite reductive groups are those groups that similarly arise as subgroups of fixed points in a convenient algebraic group.

Definition 1.2 (reductive group). A *finite reductive group* is a group of the form $\boldsymbol{G}^F$, where $\boldsymbol{G}$ is a connected reductive linear algebraic group and F is a Frobenius endomorphism of $\boldsymbol{G}^F$ (or a similar isogeny in the case of a Ree or a Suzuki group).

In the above definition, reductive algebraic group means that $\boldsymbol{G}$ admits no unipotent radical, that is, it has no normal unipotent subgroup, and a Frobenius endomorphism is one that is defined by the same formula as in GL_n in appropriate local coordinates.

Thus, $\mathrm{GL}_n(\mathbb{F}_q)$ is the typical example of a finite reductive group. Other examples are the other classical groups like the unitary, symplectic, and orthogonal groups over finite fields. There are also the so-called exceptional groups, like the group $E_8(\mathbb{F}_q)$ first constructed by Chevalley. The current unified definition is due to Chevalley and Steinberg. In the sequel, we will denote by $\boldsymbol{G}$ some connected algebraic group and by F a Frobenius endomorphism of $\boldsymbol{G}$. The reader can have in mind that $\boldsymbol{G}$ is GL_n but, unless explicitly stated, the definitions and results are valid for all connected algebraic reductive groups.

1.2 Some important subgroups

Like the general linear group $\mathrm{GL}_n(\mathbb{F}_q)$, every finite reductive group admits some special subgroups.

Definition 1.3 (maximal torus). If $\boldsymbol{G}$ is a linear algebraic group, a *maximal torus* of $\boldsymbol{G}$ is a maximal connected algebraic subgroup consisting of simultaneously diagonalizable elements.

A maximal torus is commutative and it is its own centralizer. All maximal tori are conjugate.

Definition 1.4 (Borel subgroup). If $\boldsymbol{G}$ is a linear algebraic group, a *Borel subgroup* of $\boldsymbol{G}$ is a maximal connected solvable algebraic subgroup.

All Borel subgroups are conjugate. Hereafter we shall fix a maximal torus of $\boldsymbol{G}$, denoted by $\boldsymbol{T}_1$, and a Borel subgroup of $\boldsymbol{G}$, denoted by $\boldsymbol{B}_1$, chosen so that it contains $\boldsymbol{T}_1$ (note that $\boldsymbol{T}_1$ being commutative is included in a Borel subgroup). In the case of a linear group, we can take for $\boldsymbol{T}_1$ the subgroup of diagonal matrices and for $\boldsymbol{B}_1$ the subgroup of upper triangular matrices:

$$\boldsymbol{T}_1 = \begin{pmatrix} * & & 0 \\ & \ddots & \\ 0 & & * \end{pmatrix}, \qquad \boldsymbol{B}_1 = \begin{pmatrix} * & \dots & * \\ & \ddots & \vdots \\ 0 & & * \end{pmatrix}.$$

A general Borel subgroup of GL_n is the stabilizer of a complete flag

$$V_0 \subset V_1 \subset \cdots \subset V_n$$

with $\dim V_i = i$ in $\mathbb{F}^n$.

The quotient $W = N_{\boldsymbol{G}}(\boldsymbol{T}_1)/\boldsymbol{T}_1$ is called the *Weyl group*. It is a Coxeter group. In the case of GL_n, the normalizer $N_{\boldsymbol{G}}(\boldsymbol{T}_1)$ is the group of monomial matrices so that the Weyl group is (isomorphic to) the symmetric group $\mathfrak{S}_n$.

The subgroups containing a Borel subgroup are called *parabolic subgroups*. They need not be reductive, but they always have a *Levi decomposition*: they are the semi-direct product $\boldsymbol{P} = \boldsymbol{V} \rtimes \boldsymbol{L}$ of their unipotent radical by a *Levi subgroup* which is reductive.

Example 1.5. In GL_n, parabolic subgroups containing $\boldsymbol{B}_1$ are block-upper triangular matrices, for which the corresponding block-diagonal matrices is a Levi subgroup:

$$\boldsymbol{P} = \begin{pmatrix} * & \cdots & \cdots & \cdots & \cdots & \cdots & \cdots & * \\ \vdots & & & & & & & \vdots \\ * & \cdots & * & & & & & \\ & & & * & & & & \\ & & & * & * & & & \\ & & & & & * & & \\ & 0 & & & & * & & \vdots \\ & & & & & * & * & * \end{pmatrix}, \quad \boldsymbol{L} = \begin{pmatrix} * & \cdots & * & & & & & \\ \vdots & & \vdots & & & & 0 & \\ * & \cdots & * & & & & & \\ & & & * & * & & & \\ & & & * & * & & & \\ & & & & & * & * & * \\ & 0 & & & & * & * & * \\ & & & & & * & * & * \end{pmatrix}.$$

For every parabolic subgroup $\boldsymbol{P}$, the normalizer $N_{\boldsymbol{G}}(\boldsymbol{P})$ is equal to $\boldsymbol{P}$.

Each F-stable subgroup of $\boldsymbol{G}$ gives rise to a subgroup of $\boldsymbol{G}^F$; we define thus tori, Borel subgroups, parabolic subgroups, etc. of $\boldsymbol{G}^F$. We assume that we have taken $\boldsymbol{T}_1$ and $\boldsymbol{B}_1$ to be F-stable (this is always possible, see Proposition 1.7 for Borel subgroups and maximal tori). This condition is obviously satisfied for our choice in the case of GL_n.

1.3 G^F-conjugacy

An important property of the Frobenius endomorphisms is

Proposition 1.6 (Lang's theorem). *Assume that $\boldsymbol{H}$ is a connected linear algebraic group with a Frobenius endomorphism F. Then the map $x \mapsto x^{-1}F(x)$ is surjective.*

Lang's theorem enables one to study $\boldsymbol{G}^F$-conjugacy. For example one can show:

Proposition 1.7 (**F-stable Borel subgroup**). *There exist F-stable Borel subgroups; they form a single orbit for $\boldsymbol{G}^F$-conjugation.*

Proof. Let us fix a Borel subgroup $\boldsymbol{B}$. Every other Borel subgroup is equal to ${}^g\boldsymbol{B}$ for some g in $\boldsymbol{G}$. It is F-stable if and only if one has ${}^F({}^g\boldsymbol{B}) = {}^g\boldsymbol{B}$ or, equivalently, ${}^{g^{-1}F(g)}({}^F\boldsymbol{B}) = \boldsymbol{B}$. But ${}^F\boldsymbol{B}$ is another Borel subgroup. So we must have ${}^{hF}\boldsymbol{B} = \boldsymbol{B}$ for some h in $\boldsymbol{G}$, and, by Lang's theorem, the equation $g^{-1}F(g) = h$ has a solution. Thus $\boldsymbol{B}_1 = {}^g\boldsymbol{B}$ is an F-stable Borel subgroup.

Now an arbitrary Borel subgroup ${}^g\boldsymbol{B}_1$ is F-stable if and only if $F({}^g\boldsymbol{B}_1) = {}^g\boldsymbol{B}_1$ or, equivalently, $g^{-1}F(g) \in N_{\boldsymbol{G}}(\boldsymbol{B}_1) = \boldsymbol{B}_1$ holds. Since $\boldsymbol{B}_1$ is connected, we may apply Lang's theorem to write $g^{-1}F(g) = b^{-1}F(b)$ with b in $\boldsymbol{B}_1$. Then we deduce $gb^{-1} = F(gb^{-1})$ and ${}^g\boldsymbol{B}_1 = {}^{gb^{-1}}\boldsymbol{B}_1$, and we conclude that ${}^g\boldsymbol{B}_1$ is conjugate to $\boldsymbol{B}_1$ by the F-stable element gb^{-1}. □

Things are different with tori, since a maximal torus need not be equal to its own normalizer, but it is only the connected component of this normalizer, the quotient (the component group) being the Weyl group.

Since every maximal torus of $\boldsymbol{G}$ is included in a Borel subgroup, we see that a maximal torus of a Borel subgroup is a maximal torus of $\boldsymbol{G}$. Arguing as in Proposition 1.7, we obtain that an F-fixed Borel subgroup contains an F-fixed maximal torus. So we obtain the existence of the F-fixed pair $\boldsymbol{T}_1 \subseteq \boldsymbol{B}_1$, as announced.

We now explain the classification of the $\boldsymbol{G}^F$-conjugacy classes of F-fixed maximal tori. To simplify the exposition, except in Subsection 4.2, we consider only *split* Chevalley groups, which means that we assume that the automorphism ϕ of W induced by the Frobenius endomorphism is trivial (we sketch in Subsection 4.2 what happens when this assumption is dropped). With the assumption, it is easy to deduce from Lang's theorem the next result.

Proposition 1.8 (**F-stable torus**). *The $\boldsymbol{G}^F$-class of the F-stable torus $g\boldsymbol{T}_1g^{-1}$ is parameterized by the conjugacy class of the image in W of the element $g^{-1}F(g)$ of $N_{\boldsymbol{G}}(\boldsymbol{T}_1)$.*

If $\boldsymbol{T}$ is an F-stable maximal torus whose $\boldsymbol{G}^F$-class is parameterized by the class of w, we say that $\boldsymbol{T}$ is *of type w* (abusing notation since w is defined up to W-conjugacy). For such a torus the pair $(\boldsymbol{T}, F)$ is conjugate by an element of $\boldsymbol{G}$ to $(\boldsymbol{T}_1, wF)$. We denote by $\boldsymbol{T}_w$ a maximal torus of type w.

In $\mathrm{GL}_n(\mathbb{F}_q)$, the elements that lie in a torus of type w are characterized by the fact that F induces the permutation w on the roots of their characteristic polynomial. If the permutation w is a product of cycles of lengths $n_1, \ldots, n_k$ we have

$$\boldsymbol{T}_w^F \simeq \boldsymbol{T}_1^{wF} \simeq \prod\nolimits_i \mathbb{F}_{q^{n_i}}^{\times}, \quad \text{whence} \quad |\boldsymbol{T}_w^F| = |\boldsymbol{T}_1^{wF}| = \prod_i (q^{n_i} - 1).$$

Note that every degree n polynomial $c_0 + \cdots + c_{n-1}T^{n-1} + T^n$ in $\mathbb{F}_q[T]$ is the characteristic polynomial of some element of GL_n^F, as demonstrated by the companion matrix $\begin{pmatrix} 0 & 1 & & 0 \\ \vdots & \ddots & \ddots & \\ 0 & \cdots & 0 & 1 \\ -c_0 & \cdots & \cdots & -c_{n-1} \end{pmatrix}$ of GL_n^F.

2 Representations

We now turn to complex representations of reductive groups, which will lead us to Deligne–Lusztig varieties and to the Broué Conjecture.

2.1 Complex representations of G^F

For G a finite group, we denote by $\mathrm{Irr}(G)$ the set of all complex irreducible characters of G. A first idea to construct representations of $\boldsymbol{G}^F$ is to use induction from subgroups that resemble it, like a Levi subgroup $\boldsymbol{L}$ isomorphic to a product of smaller reductive groups, that is, in the case of GL_n, to a product $\mathrm{GL}_{n_1} \times \cdots \times \mathrm{GL}_{n_k}$ with $n_1 + \cdots + n_k = n$. However, for χ in $\mathrm{Irr}(\boldsymbol{L}^F)$, the induced representation $\mathrm{Ind}_{\boldsymbol{L}^F}^{\boldsymbol{G}^F}(\chi)$ has too many irreducible components to be understood. A construction which turns out to work better is the Harish-Chandra induction.

Definition 2.1 (cuspidal character). Assume that $\boldsymbol{L}$ is an F-fixed Levi subgroup of an F-fixed parabolic subgroup $\boldsymbol{P} = \boldsymbol{V} \rtimes \boldsymbol{L}$ and χ is a character of $\boldsymbol{L}^F$.

(i) The *Harish-Chandra induction* $R_{\boldsymbol{L}}^{\boldsymbol{G}}(\chi)$ of χ is $\mathrm{Ind}_{\boldsymbol{P}^F}^{\boldsymbol{G}^F}(\tilde{\chi})$, where $\tilde{\chi}$ is the natural extension of χ to $\boldsymbol{P}^F$ through the quotient $1 \to \boldsymbol{V}^F \to \boldsymbol{P}^F \to \boldsymbol{L}^F \to 1$.

(ii) We say that χ is *cuspidal* if it is irreducible and cannot be obtained by Harish-Chandra induction, this meaning that $\langle \chi, R_{\boldsymbol{M}}^{\boldsymbol{L}}(\psi)\rangle_{\boldsymbol{G}^F} = 0$ holds for every F-fixed Levi subgroup M of every F-fixed proper parabolic subgroup of $\boldsymbol{L}$ and every irreducible character ψ of $\boldsymbol{M}^F$.

It can be shown that $R_{\boldsymbol{L}}^{\boldsymbol{G}}(\chi)$ does not depend on the F-fixed parabolic subgroup admitting $\boldsymbol{L}$ as a Levi subgroup, which is why we omit the parabolic subgroup from the notation. When χ is cuspidal, the decomposition of $R_{\boldsymbol{L}}^{\boldsymbol{G}}(\chi)$ into irreducible characters is simple; the multiplicities are the same as in the regular representation of the *relative Weyl group* $N_{\boldsymbol{G}^F}(\boldsymbol{L}, \chi)/\boldsymbol{L}^F$, which is of small size (in particular, of size independent of q).

Thus Harish-Chandra induction reduces the problem of parameterizing $\mathrm{Irr}(\boldsymbol{G}^F)$ to that of parameterizing the cuspidal characters.

2.2 Deligne–Lusztig varieties

We denote by $\mathcal{B}$ the variety of Borel subgroups of $\boldsymbol{G}$. Since $N_{\boldsymbol{G}}(\boldsymbol{B}_1) = \boldsymbol{B}_1$ holds, the variety $\mathcal{B}$ is isomorphic to $\boldsymbol{G}/\boldsymbol{B}_1$.

Definition 2.2 (relative position of Borel subgroups). For w in W, a pair $(\boldsymbol{B}, \boldsymbol{B}')$ in $\mathcal{B} \times \mathcal{B}$ is said to be in *relative position* w if it is $\boldsymbol{G}$-conjugate to the pair $(\boldsymbol{B}_1, {}^w\boldsymbol{B}_1)$. We denote this by $\boldsymbol{B} \xrightarrow{w} \boldsymbol{B}'$ and we denote by $\mathcal{O}(w)$ the subvariety of $\mathcal{B} \times \mathcal{B}$ consisting of all pairs in relative position w.

Example 2.3 (relative position). In the case of GL_n, we saw that Borel subgroups are the stabilizers of complete flags and that the Weyl group is the symmetric group $\mathfrak{S}_n$.

Under this dictionary, a pair of complete flags $((V_i)_i, (V'_i)_i)$ is in relative position w if there exists a basis (e_i) of $\overline{\mathbb{F}}^n$ such that e_i spans V_i/V_{i-1} and $e_{w(i)}$ spans V'_i/V'_{i-1}.

A pair of Borel subgroups has a uniquely defined relative position. This results from the *Bruhat decomposition* $\boldsymbol{G} = \coprod_w \boldsymbol{B}_1 w \boldsymbol{B}_1$. We thus obtain a stratification

$$\mathcal{B} \times \mathcal{B} = \coprod_w \mathcal{O}(w).$$

Let l denote the S-length in W, where S is the generating set of W as a Coxeter group. We denote by w_0 the longest element of W. In $\mathfrak{S}_n$ we have $S = \{(i, i+1) \mid i = 1, \dots, n-1\}$, the S-length is the number of inversions and w_0 is the involution $i \mapsto n+1-i$ (see Example IX.1.3 and Lemma IX.1.17).

The variety $\mathcal{O}(w)$ has dimension $\ell(w_0) + \ell(w)$. For $w = 1$ we obtain that $\mathcal{O}(1)$ is the diagonal in $\mathcal{B} \times \mathcal{B}$. The variety $\mathcal{O}(w_0)$ is the unique open piece in the stratification of $\mathcal{B} \times \mathcal{B}$; it has dimension $\dim(\mathcal{B} \times \mathcal{B}) = 2\ell(w_0)$. In GL_n, the Borel $\boldsymbol{B}_1$ of upper triangular matrices is in relative position w_0 with the Borel subgroup of lower triangular matrices, equal to ${}^{w_0}\boldsymbol{B}_1$.

Definition 2.4 (Deligne–Lusztig variety). For w in W, the *Deligne–Lusztig variety* $\boldsymbol{X}_w$ associated with w is $\{\boldsymbol{B} \mid \boldsymbol{B} \xrightarrow{w} F(\boldsymbol{B})\}$, that is, equivalently, the intersection of $\mathcal{O}(w)$ with the graph of the Frobenius endomorphism.

The dimension of $\boldsymbol{X}_w$ is equal to $\ell(w)$. Let us explain the connection between maximal tori of type w and the variety $\boldsymbol{X}_w$. If $\boldsymbol{T}$ is an F-stable maximal torus and $\boldsymbol{B}$ is a Borel subgroup containing $\boldsymbol{T}$, we have $(\boldsymbol{T}, \boldsymbol{B}) = {}^g(\boldsymbol{T}_1, \boldsymbol{B}_1)$ for some g in $\boldsymbol{G}$. We saw above that $g^{-1}\,{}^F g$ lies in $N(\boldsymbol{T})$ and that (the class of) its image w in the Weyl group is the type of $\boldsymbol{T}$. As we have ${}^F\boldsymbol{B} = {}^F({}^g\boldsymbol{B}_1) = {}^{{}^F g}\boldsymbol{B}_1$, and the latter can be written ${}^{gw}\boldsymbol{B}_1$, we see that $\boldsymbol{B}$ lies in $\boldsymbol{X}_w$.

We recall that, for every prime $\ell \neq p$, a variety $\boldsymbol{X}$ over $\overline{\mathbb{F}}$ has ℓ-adic cohomology groups with compact support $H^i_c(\boldsymbol{X}, \overline{\mathbb{Q}}_\ell)$ which are $\overline{\mathbb{Q}}_\ell$-vector spaces that can be non-trivial only for i in $\{0, \dots, 2\dim \boldsymbol{X}\}$.

The group $\boldsymbol{G}^F$ acts by conjugation on $\boldsymbol{X}_w$; this induces an action on the ℓ-adic cohomology of $\boldsymbol{X}_w$.

Note that $\overline{\mathbb{Q}}_\ell$ and $\mathbb{C}$ are abstractly isomorphic, but an explicit isomorphism depends on the axiom of choice. However, for algebraic numbers like entries in representing matrices, there is no difference between ℓ-adic and complex representations.

Definition 2.5 (Deligne–Lusztig unipotent characters). The *Deligne–Lusztig unipotent characters* $R^{\boldsymbol{G}}_{\boldsymbol{T}_w}(1)$ are the characters defined for g in $\boldsymbol{G}^F$ by

$$R^{\boldsymbol{G}}_{\boldsymbol{T}_w}(1)(g) = \sum_{i=0}^{2\ell(w)} (-1)^i \mathrm{Trace}(g \mid H^i_c(\boldsymbol{X}_w, \overline{\mathbb{Q}}_\ell)).$$

Note that Deligne–Lusztig unipotent characters are *virtual* characters, due to the signs. It can be shown that they depend only on the conjugacy class of w in W.

The notation can seem strange, since there is no torus in the definition, but it is a particular case of a more general construction that involves the torus $\boldsymbol{T}_w$ (see Definition 2.6).

For $w = 1$, the character $R^{\boldsymbol{G}}_{\boldsymbol{T}_w}(1)$ is obtained by Harish-Chandra induction from the trivial character of the torus $\boldsymbol{T}_1$, so that the notations agree: indeed, the variety $\boldsymbol{X}_1$ is the discrete variety $\boldsymbol{G}^F/\boldsymbol{B}_1^F$ of F-stable Borel subgroups; thus its cohomology is only in degree 0 and we have $H^0_c(\boldsymbol{X}_1, \overline{\mathbb{Q}}_\ell) = \overline{\mathbb{Q}}_\ell[\boldsymbol{G}^F/\boldsymbol{B}_1^F]$.

The Deligne–Lusztig unipotent characters are a particular case of the Deligne–Lusztig characters, which we define in general now but will not need further in this chapter.

Definition 2.6 (Deligne–Lusztig character). For θ in $\mathrm{Irr}(\boldsymbol{T}_w^F)$, the Deligne–Lusztig character $R^{\boldsymbol{G}}_{\boldsymbol{T}_w}(\theta)$ *associated with* θ is defined to be $\sum_i (-1)^i H^i_c(\boldsymbol{X}_w, \mathcal{F}_\theta)$, where $\mathcal{F}_\theta$ is a sheaf on $\boldsymbol{X}_w$ that can be associated with θ.

Two main properties of Deligne–Lusztig characters are

Proposition 2.7 (Deligne–Lusztig characters). (i) *For every irreducible representation θ, one has*

$$\langle R^{\boldsymbol{G}}_{\boldsymbol{T}_w}(\theta), R^{\boldsymbol{G}}_{\boldsymbol{T}_w}(\theta)\rangle_{\boldsymbol{G}^F} = |N_{\boldsymbol{G}^F}(\boldsymbol{T}_w, \theta)/\boldsymbol{T}_w^F|.$$

(ii) *Every irreducible representation occurs as a component of some $R^{\boldsymbol{G}}_{\boldsymbol{T}_w}(\theta)$.*

In the case of GL_n, the decomposition of $R^{\boldsymbol{G}}_{\boldsymbol{T}_w}(\theta)$ has a simple description. For example, one has $R^{\boldsymbol{G}}_{\boldsymbol{T}_w}(1) = \sum_{\chi \in \mathrm{Irr}(W)} \chi(w) U_\chi$, where U_χ are irreducible characters. The decomposition is more complicated for other groups.

2.3 Modular representation theory

The algebra $\mathbb{C}\boldsymbol{G}^F$ is semi-simple, hence its blocks, that is, its indecomposable factors, are indexed by the irreducible representations.

When ℓ is a prime dividing the cardinal of $\boldsymbol{G}^F$, the algebra $\overline{\mathbb{F}}_\ell \boldsymbol{G}^F$ is not semi-simple, and the first problem of *modular representation theory* is to understand its blocks. It can be shown that they are the same as the blocks of $\overline{\mathbb{Z}}_\ell \boldsymbol{G}^F$, which are slightly easier to understand since $\overline{\mathbb{Z}}_\ell \boldsymbol{G}^F$ can be compared with $\overline{\mathbb{Q}}_\ell \boldsymbol{G}^F$, which is semisimple.

The *principal* block is the one containing the identity representation. The following conjecture, a particular case of the Broué Conjecture, somehow describes the representations of the principal block when the considered ℓ-Sylow subgroup is abelian.

Conjecture 2.8 (Broué Conjecture for principal block). *If $\boldsymbol{G}$ is a finite reductive group and S is an abelian ℓ-Sylow subgroup of $\boldsymbol{G}^F$, there exists an equivalence of derived categories between the principal block of $\overline{\mathbb{Z}}_\ell \boldsymbol{G}^F$ and that of $\overline{\mathbb{Z}}_\ell N_{\boldsymbol{G}^F}(S)$.*

The rest of the chapter explains some concepts involved in exploring this conjecture for reductive groups. We first describe the abelian Sylow subgroups of reductive groups.

The well known order formula

$$|\mathrm{GL}_n(\mathbb{F}_q)| = \prod_{i=0}^{n-1}(q^n - q^i) = q^{n(n-1)/2} \prod_{d \leqslant n} \Phi_d(q)^{\lfloor n/d \rfloor}$$

generalizes for split reductive groups to

$$|\boldsymbol{G}^F| = q^{\sum_i (d_i - 1)} \prod_i (q^{d_i} - 1) = q^{\sum_i (d_i-1)} \prod_d \Phi_d(q)^{a(d)},$$

where the numbers d_i are the *reflection degrees* of the Weyl group and $a(d)$ is the number of indices i such that $d|d_i$; here we regard W as a reflection group on $\mathbb{Z}^{\dim \boldsymbol{T}_1}$, so there exist $\dim \boldsymbol{T}_1$ reflection degrees satisfying $|W| = \prod_i d_i$. In the case of $\mathfrak{S}_n$, the degrees are $1, 2, \ldots, n$.

Proposition 2.9 (Sylow subgroup). *If ℓ is a prime different from p, the ℓ-Sylow subgroup of $\boldsymbol{G}^F$ is abelian if and only if ℓ does not divide the cardinal of W.*

In the case of GL_n, the condition in Proposition 2.9 reduces to $\ell > n$.

If we have $\ell \neq p$ and $\ell \not| \, |W|$, but still $\ell | \, |\boldsymbol{G}^F|$, then ℓ divides a single factor $\Phi_d(q)^{a(d)}$ of $|\boldsymbol{G}^F|$, and thus determines a number d. It can be shown that there are elements of W whose characteristic polynomial on $\mathbb{Z}^{\dim \boldsymbol{T}_1}$ is a multiple of $\Phi_d(X)^{a(d)}$, which shows that $\boldsymbol{T}_w^F$ contains an ℓ-Sylow subgroup S. For GL_n, we take for w the product of $\lfloor n/d \rfloor$ disjoint d-cycles satisfying $|\boldsymbol{T}_w^F| = (q^d - 1)^{\lfloor n/d \rfloor}(q-1)^{n \bmod d}$.

Definition 2.10 (geometric Sylow subgroup). For S a Sylow subgroup of $\boldsymbol{G}^F$, the minimal connected algebraic subgroup $\boldsymbol{S}$ including S is called the *geometric Sylow subgroup* associated with S.

A geometric Sylow subgroup $\boldsymbol{S}$ is always a subtorus satisfying $|\boldsymbol{S}^F| = \Phi_d(q)^{a(d)}$. Moreover, we have $C_{\boldsymbol{G}}(\boldsymbol{S}) = C_{\boldsymbol{G}}(S)$ and $N_{\boldsymbol{G}}(\boldsymbol{S}) = N_{\boldsymbol{G}}(S)$. As the centralizer of a torus is a Levi subgroup, we have $C_{\boldsymbol{G}}(\boldsymbol{S}) = \boldsymbol{L}$, where $\boldsymbol{L}$ is a Levi subgroup that includes $\boldsymbol{T}_w$. We then also have $N_{\boldsymbol{G}}(\boldsymbol{S}) = N_{\boldsymbol{G}}(\boldsymbol{L})$.

The next result is proved in [40].

Proposition 2.11 (maximal eigenspace). *Let $\boldsymbol{S}$ be a geometric Sylow subgroup and w be such that $\boldsymbol{S}$ is included in a maximal torus of type w. Then w (viewed as a linear transformation) has an $\mathrm{e}^{2i\pi/d}$-eigenspace of maximal dimension $a(d)$, and the centralizer of this space is the Weyl group of the Levi subgroup $C_{\boldsymbol{G}}(\boldsymbol{S})$.*

3 Geometric Broué Conjecture, torus case

We now study a geometric form of the Broué Conjecture, which consists in using the geometric Sylow subgroup $\boldsymbol{S}$ rather than the Sylow subgroup S. More precisely, we shall explain some consequences of this conjecture and we shall present a tentative program that, if completed, would lead to a proof of these consequences. We begin this section with the special case when the Levi subgroup $C_{\boldsymbol{G}}(\boldsymbol{S})$ is a torus $\boldsymbol{T}_w$.

3.1 The geometric approach

By Proposition 2.11, the assumption $C_{\boldsymbol{G}}(\boldsymbol{S}) = \boldsymbol{T}_w$ translates into the fact that w (viewed as a linear transformation) admits an $e^{2i\pi/d}$-eigenspace of maximal dimension $a(d)$ (where d and $a(d)$ are as at the end of Section 2) and, moreover, that the centralizer of this eigenspace is trivial, so there exists an eigenvector for $e^{2i\pi/d}$ outside the reflecting hyperplanes. An element of W with an eigenvector outside the reflecting hyperplanes is called *regular*. It is called d-regular if the associated eigenvalue is $e^{2i\pi/d}$. The d-regular elements have order d and form a single conjugacy class of W, see Proposition IX.3.6 (regular elements). In the case of GL_n, such elements exist exactly for $d|n$ and for $d|n-1$.

In the torus case, the derived equivalence predicted by the Broué Conjecture should be realized "through the cohomology complex" of $\boldsymbol{X}_w$ for an appropriate choice of w in its conjugacy class. More precisely, the Broué Conjecture implies that one should have

Conjecture 3.1 (Broué conjecture for unipotent characters, torus case). *In every conjugacy class of regular elements of W, there exists an element w satisfying*

$$\langle H^i_c(\boldsymbol{X}_w, \overline{\mathbb{Q}}_\ell), H^j_c(\boldsymbol{X}_w, \overline{\mathbb{Q}}_\ell)\rangle_{\boldsymbol{G}^F} = 0 \quad \textit{for } i \neq j.$$

Moreover, there exists an action of $N_{\boldsymbol{G}^F}(S)/C_{\boldsymbol{G}^F}(S)$ on $H^*_c(\boldsymbol{X}_w, \overline{\mathbb{Q}}_\ell)$ which makes its algebra the commuting algebra of this $\boldsymbol{G}^F$-module.

In the above framework, one has $N_{\boldsymbol{G}^F}(S)/C_{\boldsymbol{G}^F}(S) = N_{\boldsymbol{G}^F}(\boldsymbol{T}_w)/\boldsymbol{T}_w \simeq C_W(w)$.

One main problem in Conjecture 3.1 is to construct the action of $C_W(w)$ as $\boldsymbol{G}^F$-endomorphisms of $H^*_c(\boldsymbol{X}_w, \overline{\mathbb{Q}}_\ell)$. The key point is to construct a sufficiently large family of $\boldsymbol{G}^F$-endomorphisms of $H^*_c(\boldsymbol{X}_w, \overline{\mathbb{Q}}_\ell)$. We shall sketch below a two-step program in this direction.

As the element w considered in Conjecture 3.1 is regular, the group $C_W(w)$ is a complex reflection group (in the case of GL_n, it is the group $G(d, 1, \lfloor n/d \rfloor)$, hence isomorphic to a wreath product $\mathbb{Z}/d \wr \mathfrak{S}_{\lfloor n/d \rfloor}$). Being a complex reflection group, it has an associated braid group and one can define cyclotomic Hecke algebras as quotient of the group algebra of this braid group.

Program 3.2. In the context of Conjecture 3.1, construct for a well-chosen representative w of every class of regular elements of W an action of $C_W(w)$ according to the following steps:

1. Define an action of a braid monoid for the braid group of $C_W(w)$ as $\boldsymbol{G}^F$-endomorphisms of $\boldsymbol{X}_w$.

2. Check that this action, on the cohomology, factors through a cyclotomic Hecke algebra for $C_W(w)$.

This would conclude using Tits' theorem which gives an isomorphism of the cyclotomic Hecke algebra and the group algebra of $C_W(w)$.

Here we will address only Step 1 of Program 3.2 since it is for this step that the Garside structure of the Artin–Tits monoid is used. Step 2 has been successfully addressed in particular cases by Broué, Digne, Michel, and Rouquier, see [42], [113], and [111].

3.2 Endomorphisms of Deligne–Lusztig varieties

From now on, our problem, namely Step 1 of Program 3.2, is to construct endomorphisms of $\boldsymbol{X}_w$ when w is a regular element of W. To do this, we shall resort to results from Chapter VIII about the cyclic conjugacy category.

A first attempt would be to directly associate endomorphisms with the elements of W. To this end, we could consider double cosets of the form $\boldsymbol{B}_1 w \boldsymbol{B}_1$. Such double cosets multiply according to the following rule

Lemma 3.3. *For $\ell(w') + \ell(w'') = \ell(w'w'')$, we have $\boldsymbol{B}_1 w' \boldsymbol{B}_1 w'' \boldsymbol{B}_1 = \boldsymbol{B}_1 w'w'' \boldsymbol{B}_1$.*

When $\ell(w') + \ell(w'') = \ell(w'w'')$ fails, $\boldsymbol{B}_1 w' \boldsymbol{B}_1 w'' \boldsymbol{B}_1$ is the union of several double cosets. One deduces

Lemma 3.4. *Assume $\ell(w') + \ell(w'') = \ell(w'w'')$.*

(i) *If we have $\boldsymbol{B} \xrightarrow{w'} \boldsymbol{B}'$ and $\boldsymbol{B}' \xrightarrow{w''} \boldsymbol{B}''$, then we also have $\boldsymbol{B} \xrightarrow{w'w''} \boldsymbol{B}'$.*

(ii) *Conversely, if $\boldsymbol{B} \xrightarrow{w'w''} \boldsymbol{B}'$ holds, there exists a unique Borel subgroup $\boldsymbol{B}'$ satisfying $\boldsymbol{B} \xrightarrow{w'} \boldsymbol{B}'$ and $\boldsymbol{B}' \xrightarrow{w''} \boldsymbol{B}''$.*

Then an idea for constructing $\boldsymbol{G}^F$-endomorphisms of $\boldsymbol{X}_w$ could be to use the following three ingredients:

(i) If we have $w = xy$ with $\ell(w) = \ell(x) + \ell(y)$, then, given $\boldsymbol{B} \xrightarrow{w} F(\boldsymbol{B})$, there exists a unique Borel subgroup $\boldsymbol{B}'$ satisfying $\boldsymbol{B} \xrightarrow{x} \boldsymbol{B}' \xrightarrow{y} F(\boldsymbol{B})$.

(ii) If we have also $\ell(yx) = \ell(y) + \ell(x)$, then, since $\boldsymbol{B}' \xrightarrow{y} F(\boldsymbol{B}) \xrightarrow{x} F(\boldsymbol{B}')$ holds, the subgroup $\boldsymbol{B}'$ belongs to $\boldsymbol{X}_{yx}$, hence $\boldsymbol{B} \mapsto \boldsymbol{B}'$ defines a map D_x from $\boldsymbol{X}_w$ to $\boldsymbol{X}_{yx}$ that commutes with the action of $\boldsymbol{G}^F$ on both varieties.

(iii) If in addition x commutes to w, then D_x is an endomorphism of $\boldsymbol{X}_w$.

However, it is not true in general that every regular element w admits a decomposition $w = xy$ satisfying all conditions listed in (i)–(iii) and the above approach is not sufficient for constructing enough endomorphisms of $\boldsymbol{X}_w$.

In order to solve the problem and construct enough endomorphisms of $\boldsymbol{X}_w$, we shall go out of W and introduce the associated braid monoid B^+ which, as explained in Chapters VI and IX can be seen as an unfolded version of W. In this way, the constraints on the addition of lengths vanish, corresponding to the fact that the partial product of the germ W extends to an everywhere defined product in B^+. The price to pay is that we then have to extend the definition of Deligne–Lusztig varieties $\boldsymbol{X}(w)$, so far defined for w in W, so as to allow w to be a positive braid.

The set S of Coxeter generators positively generates W and, as shown in Chapter IX, W lifts to a subset $\boldsymbol{W}$ which is a germ for the braid monoid B^+. The subset $\boldsymbol{W}$ can be characterized as the set of all braids $\boldsymbol{w}$ whose image w in W has same length as $\boldsymbol{w}$. By definition, for $\boldsymbol{w}, \boldsymbol{w}'$ in $\boldsymbol{W}$ with images w, w' in W, the element $\boldsymbol{w}\boldsymbol{w}'$ lies in $\boldsymbol{W}$ if and only if $\ell(w) + \ell(w') = \ell(ww')$ holds, and then $\boldsymbol{w}\boldsymbol{w}'$ lifts ww'. The Garside element Δ is the lift of the longest element of W—in the case of $\mathfrak{S}_n$ the permutation $\left(\begin{smallmatrix} 1 & 2 & \cdots & n \\ n & n-1 & \cdots & 1 \end{smallmatrix}\right)$.

Every element of B^+ is a product $\boldsymbol{w}_1 \cdots \boldsymbol{w}_r$ with $\boldsymbol{w}_i$ in $\boldsymbol{W}$. We associate a variety

$$\boldsymbol{X}(w_1, \dots, w_r) = \{(\boldsymbol{B}_0, \dots, \boldsymbol{B}_r) \mid \boldsymbol{B}_{i-1} \xrightarrow{w_i} \boldsymbol{B}_i \text{ and } \boldsymbol{B}_r = F(\boldsymbol{B}_0)\}$$

to every sequence w_i of elements of W. The facts that relative positions compose exactly when the partial multiplication is defined in $\boldsymbol{W}$, and that $\boldsymbol{W}$ is a germ for B^+ imply that, up to isomorphism, $\boldsymbol{X}(w_1, \dots, w_r)$ only depends on the product of $\boldsymbol{w}_1, \dots, \boldsymbol{w}_r$ in B.

In order to actually attach a well defined variety with every braid $\boldsymbol{b}$ in B^+, we need more, namely a *unique* isomorphism between two models associated with different decompositions of the same braid $\boldsymbol{b}$. Such a unique isomorphism does exist by the following special case of Proposition A.2 in the Appendix:

Proposition 3.5 (Deligne's theorem). *Every representation of $\boldsymbol{W}$ into a monoidal category extends uniquely to a representation of B^+ into that monoidal category.*

Applying Deligne's theorem to the varieties

$$\mathcal{O}(w_1, \dots, w_r) = \{(\boldsymbol{B}_0, \dots, \boldsymbol{B}_r) \mid \boldsymbol{B}_{i-1} \xrightarrow{w_i} \boldsymbol{B}_i \text{ for all } i\},$$

we conclude that there is a unique morphism

$$\mathcal{O}(w_1, \dots, w_r) \times_{\mathcal{B}} \mathcal{O}(w'_1, \dots, w'_{r'}) \longrightarrow \mathcal{O}(w_1, \dots, w_r, w'_1, \dots, w'_{r'})$$

defined on the fibered product, that is, on pairs of sequences where the last term of the first agrees with the first term of the second, defining a structure of monoidal category. Then the restriction to 1-term sequences is a representation of $\boldsymbol{W}$ into a monoidal category. By Deligne's theorem, it extends to B^+, giving a well-defined variety $\mathcal{O}(\boldsymbol{b})$ attached with every braid $\boldsymbol{b}$ in B^+, which for every decomposition $\boldsymbol{w}_1|\cdots|\boldsymbol{w}_r$ of $\boldsymbol{b}$ as a product of elements of $\boldsymbol{W}$, is isomorphic to $\mathcal{O}(w_1, \dots, w_r)$. We can then extend Definition 2.4 and put:

Definition 3.6 (extended Deligne–Lusztig variety). For $\boldsymbol{b}$ in B^+, the *Deligne–Lusztig variety* $\boldsymbol{X}_{\boldsymbol{b}}$ associated with $\boldsymbol{b}$ is the intersection of $\mathcal{O}(\boldsymbol{b})$ with the graph of the Frobenius endomorphism.

When $\boldsymbol{b}$ is the lift in $\boldsymbol{W}$ of an element w of W, the variety $\boldsymbol{X}_{\boldsymbol{b}}$ is canonically isomorphic to the ordinary Deligne–Lusztig variety $\boldsymbol{X}_w$.

Now, whenever we have an (arbitrary) left-divisor $\boldsymbol{x}$ of $\boldsymbol{b}$ in B^+, mimicking the three-step approach explained after Lemma 3.4, we obtain a well defined morphism $D_{\boldsymbol{x}}$ from $\boldsymbol{X}_{\boldsymbol{b}}$ to $\boldsymbol{X}_{\boldsymbol{x}^{-1}\boldsymbol{b}\boldsymbol{x}}$ commuting with the action of $\boldsymbol{G}^F$ on both varieties. We can therefore state:

Proposition 3.7 (endomorphisms of $\boldsymbol{X}_{\boldsymbol{b}}$). *For every $\boldsymbol{b}$ in B^+, every element in the cyclic conjugacy category $\mathcal{C}yc B^+(\boldsymbol{b}, \boldsymbol{b})$ gives an endomorphism of $\boldsymbol{X}_{\boldsymbol{b}}$.*

We thus obtained a translation of questions about $\boldsymbol{G}^F$-endomorphisms of Deligne–Lusztig varieties, hence about Step 1 in Program 3.2, into questions about braids. The cyclic conjugacy category of B^+ maps naturally to the category of Deligne–Lusztig varieties with $\boldsymbol{G}^F$-morphisms $D_{\boldsymbol{x}}$: an object $\boldsymbol{b}$ maps to $\boldsymbol{X}_{\boldsymbol{b}}$, and a cyclic conjugation $\boldsymbol{xy} \xrightarrow{\boldsymbol{x}} \boldsymbol{yx}$ maps to the morphism $D_{\boldsymbol{x}}$.

3.3 Periodic elements

Our aim now is to build endomorphisms of the varieties $\boldsymbol{X}_{\boldsymbol{b}}$ for every element of the centralizer of $\boldsymbol{b}$ in the monoid B^+, that is the endomorphisms of $\boldsymbol{b}$ in the conjugacy category of B^+, when $\boldsymbol{b}$ is the lift in B^+ of a regular element of W, according to our Program 3.2.

At this point, the endomorphisms $D_{\boldsymbol{x}}$ we built come from centralizers in the cyclic conjugacy category. A priori, a problem is that the cyclic conjugacy category is smaller than the full conjugacy category and, therefore, the associated centralizers may be different. However—and this is the point—it turns out that, for the particular elements in the conjugacy class of regular elements involved in the Broué Conjecture, the considered centralizers coincide, as stated in Proposition 3.9 below.

The starting observation is that the elements $\boldsymbol{b}$ we have to consider are connected with periodic elements of B^+, see [42, Proposition 3.11 and Théorème 3.12]:

Proposition 3.8 (periodic elements). (i) *In the conjugacy class of every d-regular element of W, there exists an element w whose lift $\boldsymbol{w}$ in the Artin–Tits monoid satisfies $\boldsymbol{w}^d = \Delta^2$.*

(ii) *Conversely, for every $\boldsymbol{b}$ in B^+ satisfying $\boldsymbol{b}^d = \Delta^2$, the image of $\boldsymbol{b}$ in W is a regular element of order d.*

In the case $W = \mathfrak{S}_n$ the second item is due to Kerekjarto and Eilenberg, who classified periodic braids. In that case the only numbers d for which Δ^2 has d-th roots are the divisors of n and the divisors of $n-1$.

Now, Proposition VIII.1.24 (every conjugacy cyclic) applied in the monoid B^+ with the Garside family $\mathcal{D}iv(\Delta)$ implies

Proposition 3.9 (every conjugacy cyclic I). *If an element $\boldsymbol{b}$ of B^+ has some power divisible by Δ, then every conjugation from $\boldsymbol{b}$ is a cyclic conjugation.*

Then, for every element w whose lift $\boldsymbol{w}$ is a root of Δ^2, the conjunction of Propositions 3.8 and 3.9 provides us with an endomorphism $D_{\boldsymbol{x}}$ of $\boldsymbol{X}_w$ for every $\boldsymbol{x}$ in B^+ centralizing $\boldsymbol{w}$.

Moreover, the morphisms so obtained $D_{\boldsymbol{x}}$ are "equivalences of etale sites", which implies that they induce isomorphisms of cohomology spaces, hence Proposition 3.9 also implies that there is essentially only one variety to consider whenever the d-th roots of Δ^2 are conjugate. We are thus led to wondering whether the latter property is true.

Conjecture 3.10 (roots conjugate). *For every finite Coxeter group W and every d, any two d-th roots of an element of the associated Artin–Tits monoid B^+ are conjugate in B^+.*

By [142], Conjecture 3.10 is true for every d when W is the symmetric group $\mathfrak{S}_n$. On the other hand, for an arbitrary finite Coxeter group W, the conjecture is true for certain values of d, namely when $2\ell(\Delta)/d$ is the minimal length in the conjugacy class of $e^{2i\pi/d}$-regular elements, as state the first two items of the following result of [176, 145]:

Proposition 3.11 (minimal length). *Assume that W is a finite Coxeter group and w is an element of W that has minimal length in its conjugacy class and does not lie in a proper parabolic subgroup.*

(i) *Every braid in B^+ that lifts w has a power divisible by Δ.*

(ii) *All liftings in B^+ of elements of minimal length in the W-conjugacy class of w are conjugate in B^+.*

(iii) *If $\boldsymbol{w}$ lifts w, the morphism from $C_{B^+}(\boldsymbol{w})$ to $C_W(w)$ is surjective.*

Proposition 3.11(iii) helps to complete Step 1 in Program 3.2: what we want is an action of a monoid for the braid group associated to the reflection group $C_W(w)$ (see Subsection IX.3.2 for a definition). What we have built here is an action of the centralizer of $\boldsymbol{w}$ in B^+.

In the case of $\mathfrak{S}_n$, the results of [13] show that the centralizer of a d-th root of Δ^2 is the braid group of $G(d,1,\lfloor n/d\rfloor)$, so, in this case, we actually obtain an action of the expected type for this braid group. It has been shown more generally by Bessis that for a well-generated complex reflection group, the centralizer in the braid group of a periodic element is the braid group of the centralizer of its image in the reflection group, see Proposition IX.3.24 (regular braids).

Thus, summarizing, we may state:

Proposition 3.12 (step 1 completed I). *For every finite Coxeter group, Step 1 in Program 3.2 is completed for every class of regular elements of W.*

4 Geometric Broué Conjecture, the general case

When $\boldsymbol{S}$ is a general geometric Sylow subgroup of $\boldsymbol{G}$, the centralizer $C_{\boldsymbol{G}}(\boldsymbol{S})$ is a Levi subgroup that can be larger than the torus $\boldsymbol{T}_w$. There is again a geometric version of the Broué Conjecture, but it involves Deligne–Lusztig varieties that are associated with parabolic subgroups. Then, in order to extend the approach of Section 3, we will have to consider a Garside family in a convenient category rather than in a braid monoid. In this context, we will consider the analog of Step 1 in Program 3.2, see Program 4.7.

4.1 The parabolic case

As everywhere above, we consider the case of split Chevalley groups, that is, we assume that the automorphism ϕ of W induced by the Frobenius endomorphism is trivial. If W is the Weyl group of $\boldsymbol{G}$ and S its set of Coxeter generators, for every subset I of S, we denote by $\boldsymbol{P}_I$ the parabolic subgroup containing $\boldsymbol{B}_1$ such that the Weyl group of its Levi subgroup is the subgroup W_I of W generated by I, see Corollary IX.1.12 (parabolic subgroup) and Proposition IX.1.13 (left-cosets). It is known that there is a unique such parabolic subgroup for every I. For every parabolic subgroup $\boldsymbol{P}$, there is a unique subset I of S such that $\boldsymbol{P}$ is conjugate to $\boldsymbol{P}_I$.

Definition 4.1 (type of a parabolic). If $\boldsymbol{P}$ is a parabolic subgroup conjugate to $\boldsymbol{P}_I$, we say that I is the type of $\boldsymbol{P}$.

Example 4.2 (type of a parabolic). If $\boldsymbol{G}$ is GL_n, a subset I of S corresponds to a partition of the set $\{1, ..., n\}$ and the group W_I is the subgroup of $\mathfrak{S}_n$ that permutes the integers within each part of that partition. The corresponding parabolic subgroup $\boldsymbol{P}_I$ is the group of matrices which are block upper-triangular according to that partition.

We want to replace pairs of Borel subgroups by pairs of parabolic subgroups having a common Levi subgroup (note that a Levi subgroup of a Borel subgroup is a maximal torus and that two Borel subgroups contain always a common maximal torus). Such a pair of parabolic subgroups $(\boldsymbol{P}', \boldsymbol{P}'')$ is $\boldsymbol{G}$-conjugate to the pair $(\boldsymbol{P}_J, w\boldsymbol{P}_K w^{-1})$ for some J, K included in S and some w in W satisfying $J = wKw^{-1}$. From now on, we fix I and we will consider only parabolic subgroups of type J, where J is W-conjugate to I. Then we can define the relative position of two parabolic subgroups having a common Levi subgroup:

Definition 4.3 (relative position of parabolic subgroups). Let $\mathcal{J}$ be the set of all subsets J of S that are W-conjugate to I. Let $\mathcal{C}(\mathcal{J})$ be the category with object set $\mathcal{J}$ and whose elements are conjugations of the form $J \xrightarrow{w} K$ with $K = J^w$. For $J \xrightarrow{w} K$ in $\mathcal{C}(\mathcal{J})$, we say that two parabolic subgroups $\boldsymbol{P}'$ and $\boldsymbol{P}''$ are in *relative position* $J \xrightarrow{w} K$ if the pair $(\boldsymbol{P}', \boldsymbol{P}'')$ is conjugate to $(\boldsymbol{P}_J, w\boldsymbol{P}_K w^{-1})$. We write this as $\boldsymbol{P}' \;{}_J\!\xrightarrow{w}_K\; \boldsymbol{P}''$.

Since the relative position is well defined only up to changing w in its double-coset $W_J w W_K$, we may assume that w is J-reduced, or, equivalently, that its lift $\boldsymbol{w}$ is $\boldsymbol{J}$-reduced (see Definition VIII.1.44), where $\boldsymbol{w}$ is the lift of w to the Artin–Tits monoid and $\boldsymbol{J}$ included in $\boldsymbol{S}$ is the lift of J for $J \subseteq S$.

Definition 4.4 (parabolic Deligne–Lusztig variety). For $J \xrightarrow{w} J$ in $\mathcal{C}(\mathcal{J})$, the *parabolic Deligne–Lusztig variety* associated to $J \xrightarrow{w} J$ is

$$\boldsymbol{X}(J \xrightarrow{w} J) = \{\boldsymbol{P} \mid \boldsymbol{P} \;{}_J\!\xrightarrow{w}_J\; F(\boldsymbol{P})\}.$$

Note that, since we have assumed that F acts trivially on W, the group $F(\boldsymbol{P})$ is conjugate to $\boldsymbol{P}_J$ if $\boldsymbol{P}$ is.

We will identify $\mathcal{J}$ with the set of all B^+-conjugates of $\boldsymbol{I}$ that are included in $\boldsymbol{S}$.

If we choose for the monoid M of Definition VIII.1.45 (ribbon category) the braid monoid B^+, the ribbon category $\mathcal{R}ib(B^+, \mathcal{J})$ has a Garside germ whose arrows are in one-to-one correspondence with $\mathcal{C}(\mathcal{J})$, the partial product of the germ being defined when the lengths add.

To an element $\boldsymbol{J} \xrightarrow{\boldsymbol{w}} \boldsymbol{J}$ of $\mathcal{R}ib(B^+, \mathcal{J})^{\circlearrowleft}$ with $\boldsymbol{w} \in \boldsymbol{W}$, we associate the variety $\boldsymbol{X}(J \xrightarrow{w} J)$. Similarly, to an element

$$\boldsymbol{J} \xrightarrow{\boldsymbol{b}} \boldsymbol{J} = \boldsymbol{J} \xrightarrow{\boldsymbol{w}_1} \boldsymbol{J}_1 | \cdots | \boldsymbol{J}_{n-1} \xrightarrow{\boldsymbol{w}_n} \boldsymbol{J}$$

with $\boldsymbol{w}_i \in \boldsymbol{W}$ for each i, we associate the variety

$$\{\boldsymbol{P}, \boldsymbol{P}_1, \dots, \boldsymbol{P}_n \mid \boldsymbol{P} \;{}_J\!\xrightarrow{w}_{I_1}\; \boldsymbol{P}_1 \cdots \boldsymbol{P}_n \;{}_{J_{n-1}}\!\xrightarrow{w_n}_J\; F(\boldsymbol{P})\}.$$

By using the extension of Deligne's theorem stated as Proposition A.2, one can show that there exists a canonical isomorphism between the varieties attached to two decompositions of $\boldsymbol{J} \xrightarrow{\boldsymbol{b}} \boldsymbol{K}$. Then, as above in Definition 3.6, we can attach to each element of $\mathcal{R}ib(B^+, \mathcal{J})^{\circlearrowleft}$ a unique, well-defined parabolic Deligne–Lusztig variety naturally denoted by $\boldsymbol{X}(\boldsymbol{J} \xrightarrow{\boldsymbol{b}} \boldsymbol{J})$.

At this point, the cyclic conjugacy category of the ribbon category occurs: for every left-divisor $\boldsymbol{x}$ of $\boldsymbol{b}$ in $B^+(\mathcal{J})$, there exists a well-defined morphism

$$D_{\boldsymbol{x}} : \boldsymbol{X}(\boldsymbol{J} \xrightarrow{\boldsymbol{b}} \boldsymbol{J}) \longrightarrow \boldsymbol{X}(\boldsymbol{x}^{-1}\boldsymbol{J}\boldsymbol{x} \xrightarrow{\boldsymbol{x}^{-1}\boldsymbol{b}\boldsymbol{x}} \boldsymbol{x}^{-1}\boldsymbol{J}\boldsymbol{x})$$

in the same way as in the torus case; moreover, as in the torus case, the category with objects the parabolic varieties $\boldsymbol{X}(\boldsymbol{J} \xrightarrow{\boldsymbol{b}} \boldsymbol{J})$ for J in $\mathcal{J}$ and elements the compositions of the endomorphisms $D_{\boldsymbol{x}}$ identifies with the cyclic conjugacy category $\mathcal{C}yc(\mathcal{R}ib(B^+, \mathcal{J}))$.

The next result of [111, 145] is the analog for the general case of Proposition 3.8 for the torus case. For $\boldsymbol{J}$ included in $\boldsymbol{S}$, we will write $\Delta^2/\Delta_{\boldsymbol{J}}^2$ for $\Delta_{\boldsymbol{J}}^{-2}\Delta^2$, which makes sense since, in an Artin–Tits group, the element Δ^2 is central. But beware that $\Delta_{\boldsymbol{J}}^{-1}\Delta$ is not an endomorphism in $\mathcal{R}ib(B^+, \mathcal{J})$ unless $\phi_\Delta(\boldsymbol{J}) = \boldsymbol{J}$ holds.

Proposition 4.5 (root of Δ^2/Δ_I^2). *Let W be a finite Coxeter group. Let $\zeta = e^{2i\pi/d}$ and let V_ζ be a subspace of V on which some element w of W acts by ζ. Then, up to W-conjugacy, we have $C_W(V_\zeta) = W_I$ for some I included in S. Moreover, if w is I-reduced, its lift $\boldsymbol{w}$ to the braid monoid satisfies $\boldsymbol{w}^d = \Delta^2/\Delta_{\boldsymbol{I}}^2$.*

(Note that, in the above statement, we must have $wIw^{-1} = I$). The following sort of converse is established in [112]:

Proposition 4.6 (not extendible). *Assume that $\boldsymbol{w}$ lies in B^+ and we have $\boldsymbol{w}^d = \Delta^2/\Delta_{\boldsymbol{I}}^2$ for some $\boldsymbol{I}$ included in $\boldsymbol{S}$.*

(i) *The relation $\boldsymbol{w}\boldsymbol{I}\boldsymbol{w}^{-1} = \boldsymbol{I}$ holds, $\boldsymbol{w}$ defines an element $\boldsymbol{I} \xrightarrow{\boldsymbol{w}} \boldsymbol{I}$ in $\mathit{Rib}(B^+, \mathcal{I})$, and $C_W(V_d)$ is included in W_I, where V_d is the $e^{2i\pi/d}$-eigenspace of w.*

(ii) *Moreover, $\boldsymbol{w}$ is not extendible, in the sense that there are no proper subset $\boldsymbol{J}$ of $\boldsymbol{I}$ and $\boldsymbol{v}$ in $B_{\boldsymbol{I}}^+$ satisfying $(\boldsymbol{v}\boldsymbol{w})^d = \Delta^2/\Delta_{\boldsymbol{J}}^2$, if and only if we have $C_W(V_d) = W_I$ and V_d is a maximal $e^{2i\pi/d}$-eigenspace of W.*

The assumption that $C_{\boldsymbol{G}}(\boldsymbol{S})$ contains $\boldsymbol{T}_w$ translates into the fact that w has a $e^{2i\pi/d}$-eigenspace of maximal dimension $a(d)$, where d and $a(d)$ are as at the end of Section 2, and moreover that the centralizer of this space is W_I where I is the type of a parabolic subgroup having $C_{\boldsymbol{G}}(\boldsymbol{S})$ as a Levi subgroup. Now, Proposition 4.5 guarantees that we can choose w in its conjugacy class such that its lift $\boldsymbol{w}$ is a root of $\Delta^2/\Delta_{\boldsymbol{I}}^2$, and Proposition 4.6 shows that $\boldsymbol{w}$ is not extendible. This is the reason why we are only interested in elements w satisfying the assumptions of Proposition 4.5 and Property (ii) of Proposition 4.6. In this context, the analog of Step 1 in Program 3.2 is

Program 4.7. For every element w satisfying the assumptions of Proposition 4.5 and Property (ii) of Proposition 4.6 construct an action of a braid monoid for the braid group of the complex reflection group $N_W(wW_I)/W_I$ as $\boldsymbol{G}^F$-endomorphisms of $\boldsymbol{X}(I \xrightarrow{w} I)$.

As in the torus case, Proposition VIII.1.24 (every conjugacy cyclic) applies, here in the case of the category $\mathit{Rib}(B^+, \mathcal{I})$ and the Garside family consisting of all elements of the form $\boldsymbol{I} \xrightarrow{\Delta_{\boldsymbol{I}}^{-1}\Delta} \phi_\Delta(\boldsymbol{I})$, and it gives:

Proposition 4.8 (every conjugacy cyclic II). *Assume that $\boldsymbol{I} \xrightarrow{\boldsymbol{w}} \boldsymbol{I}$ is an element of $\mathit{Rib}(B^+, \mathcal{I})$ such that some power of $\boldsymbol{w}$ is divisible by $\Delta_{\boldsymbol{I}}^{-1}\Delta$. Then every conjugate of $\boldsymbol{I} \xrightarrow{\boldsymbol{w}} \boldsymbol{I}$ is a cyclic conjugate: every element $\boldsymbol{I} \xrightarrow{\boldsymbol{x}} \boldsymbol{I}$ that conjugates $\boldsymbol{I} \xrightarrow{\boldsymbol{w}} \boldsymbol{I}$ to itself, that is, every element of $\mathit{Conj}(\mathit{Rib}(B^+, \mathcal{I}))(\boldsymbol{I} \xrightarrow{\boldsymbol{w}} \boldsymbol{I}, \boldsymbol{I} \xrightarrow{\boldsymbol{w}} \boldsymbol{I})$, belongs to $\mathit{Cyc}(\mathit{Rib}(B^+, \mathcal{I}))(\boldsymbol{I} \xrightarrow{\boldsymbol{w}} \boldsymbol{I}, \boldsymbol{I} \xrightarrow{\boldsymbol{w}} \boldsymbol{I})$.*

Now Proposition 4.8 provides us with endomorphisms of $\boldsymbol{X}_w$ for every $\boldsymbol{x}$ as in the proposition. Note than such an element $\boldsymbol{x}$ must lie in $C_{B^+}(\boldsymbol{w})$.

By [173], if $\boldsymbol{w}$ is a root of $\Delta^2/\Delta_{\boldsymbol{I}}^2$ that is not extendible in the sense of Proposition 4.6(ii), the group $N_W(wW_I)/W_I$ is a complex reflection group. Hence what is missing to complete Program 4.7 is a positive answer to the following conjecture analogous to Proposition IX.3.24.

Conjecture 4.9 (centralizer in braid group equal braid group of centralizer II). *If W is a finite Coxeter group and $\boldsymbol{w}$ is a root of $\Delta^2/\Delta_{\boldsymbol{I}}^2$ that is not extendible in the sense of*

Proposition 4.6(ii), *the enveloping group of the monoid*

$$\{\boldsymbol{x} \in C_{B^+}(\boldsymbol{w}) \mid \boldsymbol{x}\boldsymbol{I}\boldsymbol{x}^{-1} = \boldsymbol{I} \text{ and } \boldsymbol{x} \text{ is } \boldsymbol{I}\text{-reduced}\}$$

is the braid group of the complex reflection group $N_W(wW_I)/W_I$.

Very little is known in general on this property. Summarizing the results, we state:

Proposition 4.10 (step 1 completed II). *If W is a finite Coxeter group and w is an element of W satisfying the the assumptions of Proposition* 4.5 *and Property* (ii) *of Proposition* 4.6 *and for which Conjecture* 4.9 *is true, Program* 4.7 *is completed.*

Note that, as in the torus case, as the morphisms D_x induce isomorphisms of cohomology spaces, the variety that has to be considered for a given $\boldsymbol{I}$ and a given d is essentially unique in the case when all d-th roots of $\Delta^2/\Delta_{\boldsymbol{I}}^2$ are conjugate. We are thus led to addressing the question of whether all d-th roots of $\Delta^2/\Delta_{\boldsymbol{I}}^2$ are conjugate. This is conjectured to be true.

4.2 The really general case

Until now, we made the simplifying assumption that the automorphism ϕ induced by the Frobenius endomorphism F on the Weyl group W is trivial.

This is not always the case: for instance, the unitary group is defined by starting with the same algebraic group $\mathrm{GL}_n(\mathbb{F})$, but composing the Frobenius endomorphism of $\mathrm{GL}_n(\mathbb{F}_q)$ with the transpose and inverse maps. In this example, ϕ acts on W by conjugation by the longest element.

Thus, in the very general case, we have to replace conjugacy by ϕ-conjugacy in the sense of Subsection VIII.1.3, namely defining the ϕ-conjugate of w under v to be the element $v^{-1}w\phi(v)$. The $\boldsymbol{G}^F$-classes of tori are then parameterized by ϕ-conjugacy classes.

We obtain morphisms of Deligne–Lusztig varieties in the cyclic ϕ-conjugacy category $\mathcal{C}yc_\phi$ instead of $\mathcal{C}yc$. To handle this case, we have to add ϕ to the category as an invertible element. The results of Chapter VIII are still relevant in this context. But we need results about ϕ-centralizers and, here already, there are no known results similar to those of Bessis, stated in Proposition IX.3.24 (regular braids), used to establish Proposition 3.12 (step 1 completed I).

Appendix : Representations into bicategories

We now establish a general result that extends Deligne's theorem (Proposition 3.5) used above to attach a Deligne–Lusztig variety to a morphism of a ribbon category. Deligne's theorem is about Artin–Tits monoids of spherical type. Our theorem covers the case of non-spherical Artin–Tits monoids.

We follow the terminology of [177, XII.6] for bicategories. By a "representation of a category $\mathcal{C}$ into a bicategory $\mathcal{B}$", we mean a morphism of bicategories between $\mathcal{C}$ viewed as a trivial bicategory into the given bicategory $\mathcal{B}$. This amounts to giving a map T from $\mathcal{O}bj(\mathcal{C})$ to the 0-cells of X, and for g in $\mathcal{C}$ of source x and target y, an element $T(g)$ of $V(T(x), T(y))$ where $V(T(x), T(y))$ is the category whose objects (*resp.* morphisms) are the 1-cells of X with domain $T(x)$ and codomain $T(y)$ (*resp.* the 2-cells between them), together, with for every path $g|h$, an isomorphism $T(g)T(h) \xrightarrow{\sim} T(gh)$ such that the resulting square

(A.1)
$$\begin{array}{ccc} T(f)T(g)T(h) & \xrightarrow{\sim} & T(fg)T(h) \\ \sim\downarrow & & \downarrow\sim \\ T(f)T(gh) & \xrightarrow{\sim} & T(fgh) \end{array}$$

commutes.

We consider now a Garside family $\mathcal{S}$ in $\mathcal{C}$. We define a *representation* of the Garside family $\mathcal{S}$ as satisfying the same property (A.1), except that the above square is restricted to the case when f, g, fg lie in $\mathcal{S}$ and fgh lies in $\mathcal{S}^\sharp$ (which implies that h and gh lie in $\mathcal{S}^\sharp$ since the latter is closed under right-divisor). Here is the main result:

Proposition A.2 (bicategory representation). *If $\mathcal{C}$ is a left-cancellative category that is right-Noetherian and admits conditional right-lcms and $\mathcal{S}$ is a Garside family in $\mathcal{C}$, every representation of $\mathcal{S}$ into a bicategory extends uniquely to a representation of $\mathcal{C}$ into the same bicategory.*

In order to establish Proposition A.2, we will use *strict* decompositions of an arbitrary element g of $\mathcal{C}$ into elements of $\mathcal{S}^\sharp$. By strict decomposition, we mean a decomposition $g_1|\cdots|g_n$, where $g_1, ..., g_{n-1}$ lie in $\mathcal{S} \setminus \mathcal{C}^\times$ and g_n lies in $\mathcal{S}^\sharp \setminus \mathcal{C}^\times$. The proof goes exactly as in [102], in that what must been proven is a simple connectedness property for the set of strict decompositions of an element of $\mathcal{C}$— this generalizes [102, 1.7] and is used in the same way. In his context, Deligne shows more, namely the contractibility of the set of decompositions; on the other hand, our proof, which follows a suggestion by Serge Bouc to use a version of [26, Lemma 6], is simpler and valid in our more general context.

Fix a non-invertible element g in $\mathcal{C}$. We denote by $E(g)$ the set of all strict decompositions of g. Then $E(g)$ is a poset, the order being defined by

$$g_1|\cdots|g_{i-1}|g_i|g_{i+1}|\cdots|g_n > g_1|\cdots|g_{i-1}|g_i'|g_i''|g_{i+1}|\cdots|g_n$$

if $g_i = g_i' g_i''$ holds in $\mathcal{S}$.

We recall the definition of homotopy in a poset E (a translation of the corresponding notion in a simplicial complex isomorphic as a poset to E). A path from x_1 to x_k in E is a sequence $x_1, \dots, x_k$ where each x_i is comparable with x_{i+1}. The composition of paths is defined by concatenation. Homotopy, denoted by $\sim$, is the finest equivalence relation on paths compatible with concatenation and generated by the following three elementary relations: $x|y|z \sim x|z$ if $x \leqslant y \leqslant z$ and both $x|y|x \sim x$ and $y|x|y \sim y$ when $x \leqslant y$ holds. Homotopy classes form a groupoid, as the composition of a path with source x and of the inverse path is homotopic to the constant path at x. For x in E, we denote by $\Pi_1(E, x)$ the fundamental group of E with base point x, which is the group of homotopy classes of loops starting from x.

A poset E is said to be *simply connected* if it is connected (there is a path linking any two elements of E) and if the fundamental group with some (or any) base point is trivial.

Note that a poset with a smallest or largest element x is simply connected since every path $x|y|z|t|\cdots|x$ is homotopic to $x|y|x|z|x|t|x|\cdots|x$ which is homotopic to the trivial loop.

Proposition A.3 (simply connected). *The set $E(g)$ is simply connected.*

Before proving Proposition A.3, we establish a version of a result from [26] on order preserving maps between posets. If E is a poset, we write $E_{\geqslant x}$ for $\{x' \in E \mid x' \geqslant x\}$, which is a simply connected subposet of E since it has a smallest element. If $f : X \to Y$ is an order preserving map, it is compatible with homotopy (it corresponds to a continuous map between simplicial complexes), so it induces a homomorphism f^* from $\Pi_1(X, x)$ to $\Pi_1(Y, f(x))$.

Lemma A.4. *Let $f : X \to Y$ be an order preserving map between two posets. Assume that Y is connected and that, for every y in Y, the poset $f^{-1}(Y_{\geqslant y})$ is connected and nonempty. Then f^* is surjective. If moreover $f^{-1}(Y_{\geqslant y})$ is simply connected for every y, then f^* is an isomorphism.*

Proof. Let us first show that X is connected. Let x, x' belong to X. We choose a path y_0—$\cdots$—y_n in Y from $y_0 = f(x)$ to $y_n = f(x')$. For $i = 0, \dots, n$, we choose x_i in $f^{-1}(Y_{\geqslant y_i})$ with $x_0 = x$ and $x_n = x'$. Two cases are possible. In the case $y_i \geqslant y_{i+1}$, we have $f^{-1}(Y_{\geqslant y_i}) \subseteq f^{-1}(Y_{\geqslant y_{i+1}})$, and there exists a path in $f^{-1}(Y_{\geqslant y_{i+1}})$ from x_i to x_{i+1}; otherwise, $y_i < y_{i+1}$ holds, we have $f^{-1}(Y_{\geqslant y_i}) \supseteq f^{-1}(Y_{\geqslant y_{i+1}})$, and there exists a path in $f^{-1}(Y_{\geqslant y_i})$ from x_i to x_{i+1}. Concatenating these paths gives a path connecting x and x'.

We fix now x_0 in X. Let $y_0 = f(x_0)$. We prove that $f^* : \Pi_1(X, x_0) \to \Pi_1(Y, y_0)$ is surjective. Let y_0—y_1—$\cdots$—y_n with $y_n = y_0$ be a loop in Y. We lift arbitrarily this loop to a loop x_0—$\cdots$—x_n in X as above, where x_i—x_{i+1} stands for a path from x_i to x_{i+1} which is either in $f^{-1}(Y_{\geqslant y_i})$ or in $f^{-1}(Y_{\geqslant y_{i+1}})$. Then the path $f(x_0$—x_1—$\cdots$—$x_n)$ is homotopic to y_0—$\cdots$—y_n. This can be seen by induction: let us assume that the path $f(x_0$—$x_1\cdots$—$x_i)$ is homotopic to y_0—$\cdots$—y_i—$f(x_i)$. Then the same property holds for $i+1$: indeed, we have y_i—$y_{i+1} \sim y_i$—$f(x_i)$—y_{i+1} as they are two paths in a simply connected set which is either $Y_{\geqslant y_i}$ or $Y_{\geqslant y_{i+1}}$. Similarly, we have $f(x_i)$—y_{i+1}—$f(x_{i+1}) \sim$

$f(x_i—x_{i+1})$. Putting things together gives

$$\begin{aligned} y_0—\cdots—y_i—y_{i+1}—f(x_{i+1}) &\sim y_0—y_1—\cdots—y_i—f(x_i)—y_{i+1}—f(x_{i+1}) \\ &\sim f(x_0—\cdots—x_i)—y_{i+1}—f(x_{i+1}) \\ &\sim f(x_0—\cdots—x_i—x_{i+1}). \end{aligned}$$

We now prove that f^* is injective when all $f^{-1}(Y_{\geqslant y})$ are simply connected.

We first show that if $x_0—\cdots—x_n$ and $x'_0—\cdots—x'_n$ are two loops lifting the same loop $y_0 \cdots y_n$, then they are homotopic. Indeed, we obtain, using induction on i, that $x_0—\cdots—x_i—x'_i$ and $x'_0—\cdots—x'_i$ are homotopic paths, using the fact that x_{i-1}, x_i, x'_{i-1} and x'_i are all in the same simply connected subposet, namely either $f^{-1}(Y_{\geqslant y_{i-1}})$ or $f^{-1}(Y_{\geqslant y_i})$ holds.

It remains to prove that we can lift homotopies, which amounts to showing that, if we lift as above two loops which differ by an elementary homotopy, the liftings are homotopic. If $y—y'—y \sim y$ is an elementary homotopy with $y < y'$ (*resp.* $y > y'$), then we have $f^{-1}(Y_{\geqslant y'}) \subseteq f^{-1}(Y_{\geqslant y})$ (*resp.* $f^{-1}(Y_{\geqslant y}) \subseteq f^{-1}(Y_{\geqslant y'})$) and the lifting of $y—y'—y$ constructed as above is in $f^{-1}(Y_{\geqslant y})$ (*resp.* $f^{-1}(Y_{\geqslant y'})$), so is homotopic to the trivial path. If $y < y' < y''$ holds, a lifting of $y—y'—y''$ constructed as above is in $f^{-1}(Y_{\geqslant y})$, so is homotopic to any path in $f^{-1}(Y_{\geqslant y})$ with the same endpoints. □

Proof of Proposition A.3. We argue by contradiction. If the result fails, we choose g in $\mathcal{C}$ minimal for proper right-divisibility such that $E(g)$ is not simply connected.

Let L be the set of all non-invertible elements of $\mathcal{S}$ that left-divide g. For every I included in L, since the category admits conditional right-lcms and is right-Noetherian, the elements of I have a right-lcm. We fix such a right-lcm Δ_I. Let $E_I(g)$ be the set $\{g_1|\cdots|g_n \in E(g) \mid \Delta_I \preccurlyeq g_1\}$. We claim that $E_I(g)$ is simply connected when I is nonempty. This is clear if g lies in $\Delta_I\mathcal{C}^\times$, in which case we have $E_I(g) = \{(g)\}$. Let us assume this is not the case. In the following, if $\Delta_I \preccurlyeq a$ holds, we denote by a^I the element satisfying $a = \Delta_I a^I$. The set $E(g^I)$ is defined since g does not lie in $\Delta_I\mathcal{C}^\times$. We apply Lemma A.4 to the map $f : E_I(g) \to E(g^I)$ defined by

$$g_1|\cdots|g_n \longmapsto \begin{cases} g_2|\cdots|g_n & \text{for } g_1 = \Delta_I, \\ g_1^I|g_2|\cdots|g_n & \text{otherwise.} \end{cases}$$

This map preserves the order and every set $f^{-1}(Y_{\geqslant(g_1|\cdots|g_n)})$ has a least element, namely $\Delta_I|g_1|\cdots|g_n$, so it is simply connected. As, by the minimality of g, the set $E(g^I)$ is simply connected, Lemma A.4 implies that $E_I(g)$ is simply connected.

Let Y be the set of non-empty subsets of L. We now apply Lemma A.4 to the map f from $E(g)$ to Y defined by $g_1|\cdots|g_n \mapsto \{s \in L \mid s \preccurlyeq g_1\}$, where Y is ordered by inclusion. This map is order preserving since $g_1|\cdots|g_n < g'_1|\cdots|g'_n$ implies $g_1 \preccurlyeq g'_1$. We have $f^{-1}(Y_{\geqslant I}) = E_I(g)$, so this set is simply connected. Since Y, having a greatest element, is simply connected, Lemma A.4 implies that $E(g)$ is simply connected, whence the result. □

Proof of Proposition A.2. We have to define $T(g)$ for each g in $\mathcal{C}$ and isomorphisms $T(g)T(h) \xrightarrow{\sim} T(gh)$ for every composable pair (g,h), starting from the analogous data for the Garside family. To each strict decomposition $g_1|\cdots|g_n$ of a non-invertible g we associate $T(g_1)\cdots T(g_n)$. To each elementary homotopy is associated an isomorphisms between these functors. Now, as in [102, Lemma 1.6], the simple connectedness of $E(g)$, as established in Proposition A.3, implies that this set of isomorphisms generates a transitive system so that we can define $T(g)$ as the projective limit of this system and obtain isomorphisms $T(g)T(h) \simeq T(gh)$, as wanted. □

Notes

Sources and comments. The current exposition is based on Digne and Michel [112] and [111]. Many of the ideas are already present in Broué and Michel [42]. Details and more complete information on reductive groups can be found for instance in the books by A. Borel [25], J. Humphreys [150], or T. Springer [216]. For their representation theory, see the books by R. Carter [52] or by F. Digne and J. Michel [110].

The construction of cuspidal characters was first done for GL_n by J.A. Green (1955) using a complicated inductive procedure. I.G. Macdonald (1970) predicted that the cuspidal characters should be associated to some twisted tori, that is, some $\boldsymbol{T}_w$ with $w \neq 1$. Finally, P. Deligne and G. Lusztig (1976) constructed a variety associated to twisted tori such that the cuspidal representations occur in its ℓ-adic cohomology.

What we have looked at is a simplified version of a particular case of the Broué Conjecture as stated in [38]. The general Broué Conjecture is about the cohomology of a non-necessarily constant sheaf on the variety $\boldsymbol{X}_w$ where moreover the base ring is the ring of ℓ-adic integers instead of $\overline{\mathbb{Q}}_\ell$.

Proposition 3.5 was proved by P. Deligne in the paper [102]. This paper was motivated by a construction of A.I. Bondal and M.M. Kapranov and by a question of M. Broué and J. Michel when they were writing the paper [42] and wanted to define Deligne–Lusztig varieties associated to all elements of an Artin–Tits monoid, see [42, 1.6]. Another proof was given recently by S. Gaussent, Y. Guiraud, and P. Malbos in [128]. The generalization (Proposition A.2) we give in the Appendix appeared in [112].

Chapter XI
Left self-distributivity

Left self-distributivity is the algebraic law

$$\text{(LD)} \qquad x(yz) = (xy)(xz).$$

Syntactically, it appears as a variant of the associativity law $x(yz) = (xy)z$ and, indeed, it turns out that these laws have much in common. However, the repetition of the variable x in the right-hand term of the LD law changes the situation radically and makes the study of the LD law much more delicate than that of associativity. In particular, the Word Problem of the LD law, that is, the question of effectively recognizing whether two formal expressions are equivalent modulo the LD law, is not trivial and remained open until 1991. By contrast, the similar question for associativity is trivial: two expressions are equivalent modulo associativity if and only if they become equal when all parentheses are erased. In the early 1990's a rather sophisticated theory was developed in order to investigate the LD law and, in particular, solve the above mentioned Word Problem, see the Notes section at the end of this chapter. It turns out that, at the heart of the theory, a key role is played by a certain monoid M_{LD} and that this monoid has a Garside structure, yet it is not a Garside monoid. This Garside structure is interesting in several respects. First, it provides a natural and nontrivial example of a Garside family that is not associated with a Garside element. Next, this structure can actually be described in terms of a Garside map, but at the expense of switching to a category framework, thus providing another argument in favor of the category approach. Finally, the Garside structure associated with the LD law turns out to be closely connected with that of braids, the latter being a projection of the former. As a consequence, a number of algebraic results involving braids immediately follow from analogous results involving the LD law: from that point of view, Garside's theory of braids is the emerged part of an iceberg, namely the algebraic theory of the LD law.

In this chapter, we shall give a self-contained overview of this theory, with special emphasis on the associated Garside structure. In particular, we discuss what is known as the Embedding Conjecture, a puzzling open question directly connected with the normal decompositions associated with the Garside structure.

The chapter is organized as follows. In Section 1, we describe the general framework of a (right)-Garside sequence associated with a *partial* action of a monoid on a set. This can be seen as a weak variant for the notion of a (right)-Garside element as investigated in Chapter V: a (right)-Garside sequence in a monoid M consists of elements Δ_x that play a role of *local* (right)-Garside elements for the elements of M that act on x. A (right)-Garside element then corresponds to a constant (right)-Garside sequence and an action that is defined everywhere.

In Section 2, we provide some background about the LD law, introducing in particular the notion of an LD-expansion, a partial order on terms that corresponds to applying the

LD law in the expanding direction only. A certain category $\mathcal{LD}_0$ naturally arises, which contains a map Δ_0 that (at the least) resembles a right-Garside map.

In Section 3, we introduce the monoid M_{LD}, the geometry monoid of the LD law, a counterpart of Richard Thompson's group F in the context of self-distributivity, derive the notion of labeled LD-expansion and prove that the associated category $\mathcal{LD}$ has a right-Garside map (Proposition 3.25). This leads us to introducing and discussing the Embedding Conjecture, a natural statement that would make our understanding of the LD law much more complete but, frustratingly, resisted efforts so far. Some forms of this conjecture directly involve the Garside structure on M_{LD} (Propositions 3.27 and 3.33).

Finally, in Section 4, we describe the connection between the category $\mathcal{LD}$ and Artin's braids: formalized in Proposition 4.8, it enables one to recover a number of braid properties from the results of Section 3, see for instance Proposition 4.14.

1 Garside sequences

One of the main benefits of the existence of a (right)-Garside element Δ in a monoid M is the existence of a Δ-normal decomposition for every element of M. There are cases—and the structure studied from Section 2 will be an example—when no Garside element exists, but instead there exists a sequence $(\Delta_x)_{x\in X}$ of elements playing the role of local Garside elements and again leading to a Garside family, hence to distinguished decompositions for the elements of the considered monoid. In this section, we describe a general context for such constructions, in connection with what will be called a partial action.

There are three subsections. In Subsection 1.1 we introduce the notions of a partial action of a monoid on a set and a natural associated category. Next, in Subsection 1.2, we define the notion of a (right)-Garside sequence in a monoid and connect it with a (right)-Garside map in the associated category. Finally, in Subsection 1.3, we show how the existence of a right-Garside sequence leads to results similar to those following from the existence of a right-Garside map.

1.1 Partial actions

The convenient framework here is that of a monoid with a partial action on a set, namely an action that need not be defined everywhere. Several definitions may be figured out. For our current purpose the suitable one is as follows.

Definition 1.1 (partial action). A *partial action* of a monoid M on a set X is a partial function $F : M{\rightarrow}(X{\rightarrow}X)$ with the following properties, writing $x \bullet g$ for $F(g)(x)$,

(1.2) $x \bullet 1 = x$ for every x in X,

(1.3) $(x \bullet g) \bullet h = x \bullet gh$ for all x in X and g, h in M, this meaning that either both terms are defined and they are equal, or neither is defined.

For x in X, we then put $\mathrm{Def}(x) = \{ g \in M \mid x \cdot g \text{ is defined} \}$. The partial action is called *proper* if, moreover,

(1.4) For every finite subset A of M, there exists x in X such that $x \cdot g$ is defined for every g in A.

In terms of the sets $\mathrm{Def}(x)$, (1.2), (1.3), and (1.4), respectively, amount to

$$1 \in \mathrm{Def}(x), \quad gh \in \mathrm{Def}(x) \Leftrightarrow (g \in \mathrm{Def}(x) \text{ and } h \in \mathrm{Def}(x \cdot g)),$$
$$\forall A \text{ finite} \subseteq M \ \exists x \in X \ (A \subseteq \mathrm{Def}(x)).$$

An action of a monoid M in the standard sense, that is, an everywhere defined action, is also a partial action, and, in this case, $\mathrm{Def}(x) = M$ holds for every x in X. A partial action may well be empty, that is, nowhere defined. However, this is impossible for a proper action, which we shall see behaves in many respects as an everywhere defined action.

Example 1.5 (partial action). Consider the braid monoid B_∞^+ (Reference Structure 2, page 5). We obtain a (trivial) partial action of B_∞^+ on $\mathbb{N}$ by putting

(1.6) $$n \cdot g = n \text{ if } g \text{ lies in } B_n^+, \quad \text{and } n \cdot g \text{ undefined otherwise.}$$

Indeed, for every n, the unit braid 1 belongs to B_n^+, and, if g, h belong to B_n^+, so does the product gh. This action is proper, since, for every finite family of positive braids $g_1, ..., g_p$ in B_∞^+, there exists n such that $g_1, ..., g_p$ all lie in B_n, and, therefore, $n \cdot g_i$ is defined for every i. The set $\mathrm{Def}(n)$ is then the submonoid B_n^+ of B_∞^+.

A category naturally arises when a partial action of a monoid is given.

Definition 1.7 (category $\mathcal{C}_F(M,X)$). For F a partial action of a monoid M on a set X, the category *associated with* F, denoted by $\mathcal{C}_F(M,X)$, or simply $\mathcal{C}(M,X)$, is as follows:
- The objects of $\mathcal{C}_F(M,X)$ are the elements of X;
- The elements of $\mathcal{C}_F(M,X)$ are triples (x, g, y) with x in X, g in M such that $x \cdot g$ is defined and $y = x \cdot g$.
- The source (*resp*. target) of (x, g, y) is x (*resp*. y), and the multiplication of (x, g, y) and (y, h, z) gives (x, gh, z).

In practice, we shall use the same convention as in Chapters VIII and X and write $x \xrightarrow{g} y$ for a triple (x, g, y). So, for instance, the category $\mathcal{C}(B_\infty^+, \mathbb{N})$ associated with the proper partial action (1.6) of B_∞^+ on $\mathbb{N}$ is the category whose objects are positive integers and whose elements are triples $n \xrightarrow{g} n$ with g in B_n^+.

The category $\mathcal{C}_F(M,X)$ is a sort of pointed version of the monoid M in which the action on X is made explicit. If the partial action F is proper, then, for every g in M, there exists at least one element x for which $x \cdot g$ is defined and, therefore, mapping $x \xrightarrow{g} y$ to g provides a surjection of $\mathcal{C}_F(M,X)$ onto M ("forgetful surjection").

We begin with a few observations about the categories $\mathcal{C}_F(M,X)$ that are directly reminiscent of those about the conjugacy categories $\mathcal{C}onj\mathcal{C}$ in Chapter VIII. As in Chapter VIII, the first two entries in a triple $x \xrightarrow{g} y$ determine the third one, so we can write $x \xrightarrow{g} \cdot$ with no ambiguity when there is no need to give a name to the last entry.

Lemma 1.8. *Assume that F is a partial action of a monoid M on a set X.*

(i) *If M is left-cancellative, then so is the category $\mathcal{C}_F(M,X)$.*

(ii) *Conversely, if F is proper and $\mathcal{C}_F(M,X)$ is left-cancellative, then so is M.*

The easy verification is left to the reader, as is the one of the next result:

Lemma 1.9. *Assume that F is a partial action of a monoid M on a set X.*

(i) *If $x \bullet g$ is defined, then $y \xrightarrow{h} \cdot \preccurlyeq x \xrightarrow{g} \cdot$ holds in $\mathcal{C}_F(M,X)$ if and only if we have $y = x$ and $h \preccurlyeq g$ in M.*

(ii) *If $x \bullet f$ and $x \bullet g$ are defined, then $x \xrightarrow{h} \cdot$ is a left-gcd of $x \xrightarrow{f} \cdot$ and $x \xrightarrow{g} \cdot$ in $\mathcal{C}_F(M,X)$ if and only if h is a left-gcd of f and g in M.*

We refer to Exercise 105 for further easy preservation results between the monoid M and the associated categories $\mathcal{C}_F(M,X)$.

1.2 Right-Garside sequences

The example of the category $\mathcal{C}(B_\infty^+, \mathbb{N})$ shows the interest of going from a monoid to a category. The monoid B_∞^+ may contain no Garside element, since it has infinitely many atoms and no element may be a multiple of all atoms simultaneously. However, the partial action of (1.6) enables us to restrict to subsets B_n^+ (submonoids in the current case) for which Garside elements exist. So the category context allows one to capture the fact that B_∞^+ has, in some sense, a local Garside structure. We now formalize this intuition in a general context. The natural idea is to consider, for every x in the set X on which the monoid M acts, an element Δ_x playing the role of a Garside element for the subset $\mathrm{Def}(x)$ of M made by those elements that act on x.

Definition 1.10 (right-Garside sequence). If M is a left-cancellative monoid and F is a partial action of M on a set X, a sequence $(\Delta_x)_{x\in X}$ of elements of M is called a *right-Garside sequence with respect to F* if the following properties are satisfied:

(1.11) For every x in X, the element $x \bullet \Delta_x$ is defined.

(1.12) The monoid M is generated by $\bigcup_{x\in X} \mathrm{Div}(\Delta_x)$.

(1.13) The relation $g \preccurlyeq \Delta_x$ implies $\Delta_x \preccurlyeq g\Delta_{x\bullet g}$.

(1.14) For all x and g such that $x \bullet g$ is defined, g and Δ_x admit a left-gcd.

(1.15) For all x, y in X, the set $\mathrm{Div}(\Delta_x) \cap \mathrm{Def}(y)$ is included in $\mathrm{Div}(\Delta_y)$.

Right-Garside elements trivially give rise to right-Garside sequences:

Lemma 1.16. *If F is an everywhere defined partial action of a left-cancellative monoid M on a set X, then a sequence $(\Delta_x)_{x \in X}$ is a right-Garside sequence in M with respect to F if and only if Δ_x is a right-Garside element for every x and its $=^\times$-equivalence does not depend on x.*

Proof. Since F is defined everywhere, (1.14) implies $\mathrm{Div}(\Delta_x) \subseteq \mathrm{Div}(\Delta_y)$ for all x, y. By symmetry, we deduce $\mathrm{Div}(\Delta_x) = \mathrm{Div}(\Delta_y)$, whence $\Delta_x =^\times \Delta_y$. Then, comparing (1.12)–(1.14) with (V.1.22)–(V.1.24) in Chapter V, we immediately see that Δ_x is a Garside element in M.

Conversely, it is equally easy to see that every sequence $(\Delta_x)_{x \in X}$ such that Δ_x is a Garside element for each x and $\Delta_x =^\times \Delta_y$ holds for all x, y satisfies (1.11)–(1.15). □

Of course, we shall be interested here in the case when the action is genuinely partial.

Example 1.17 (right-Garside sequence). Consider the braid monoid B_∞^+ and its action on $\mathbb{N}$ as defined in (1.6). Let Δ_n be the Garside element of B_n^+ as defined in (I.1.6). Then $(\Delta_n)_{n \in \mathbb{N}}$ is a right-Garside sequence in B_∞^+ with respect to the considered action. Everything is easy in this case, as the set $\mathrm{Def}(n)$ is the braid monoid B_n^+, and the latter admits Δ_n as a (right)-Garside element. In particular, (1.15) is satisfied as the divisors of Δ_n have an intrinsic definition that does not depend on n, namely that any two strands cross at most once in any positive diagram that represent them.

For our current purpose, the point is the following connection between a right-Garside sequence in a monoid M and a right-Garside map in the associated category $\mathcal{C}_F(M,X)$.

Proposition 1.18 (right-Garside sequence). *Assume that M is a left-cancellative monoid with a partial action F of M on a set X.*

(i) *If $(\Delta_x)_{x \in X}$ is a right-Garside sequence in M with respect to F, then the map Δ defined on X by $\Delta(x) = x \xrightarrow{\Delta_x} \cdot$ is a right-Garside map in $\mathcal{C}_F(M,X)$ satisfying*

(1.19) *If $x \xrightarrow{g} \cdot$ lies in $\mathcal{D}iv(\Delta)$, then so does $y \xrightarrow{g} \cdot$ whenever defined.*

(ii) *Conversely, if F is proper and Δ is a right-Garside map in $\mathcal{C}_F(M,X)$ satisfying (1.19), then $(\Delta_x)_{x \in X}$ is a right-Garside sequence in M, where, for x in X, we define Δ_x to be the (unique) element of M satisfying $\Delta(x) = x \xrightarrow{\Delta_x} \cdot$.*

Proof. (i) First, Lemma 1.8 implies that $\mathcal{C}_F(M,X)$ is left-cancellative, so it makes sense to speak of a right-Garside family in $\mathcal{C}_F(M,X)$. Next, for every x in X, the source of $\Delta(x)$ is x, so Δ satisfies (V.1.16).

Then, by Lemma 1.9, $y \xrightarrow{g} \cdot$ left-divides $\Delta(x)$ if and only if we have $y = x$ and $h \in \mathrm{Div}(\Delta_x)$. Hence the family $\mathcal{D}iv(\Delta)$ in $\mathcal{C}_F(M,X)$ is the family of all triples $x \xrightarrow{g} \cdot$ with g in $\bigcup_{x \in X} \mathrm{Div}(\Delta_x)$. Now let $x \xrightarrow{g} \cdot$ be an arbitrary element of $\mathcal{C}_F(M,X)$. By assumption,

$\bigcup_{x\in X}\mathrm{Div}(\Delta_x)$ generates M. Hence there exists a decomposition $s_1|\cdots|s_p$ of g such that, for every i, there exists y_i such that s_i left-divides $\Delta(y_i)$. Now, by assumption, $x \bullet g$ is defined, hence so is $x \bullet s_1\cdots s_i$ for every i. Let $x_0 = x$ and $x_i = x \bullet s_1\cdots s_{i-1}$ for $i = 1, ..., p$. By assumption, we have $x_{i-1} \bullet s_i = x_i$ for every i. Moreover, by (1.15), the assumption that g_i left-divides some element Δ_{y_i} implies that it left-divides $\Delta_{x_{i-1}}$ and, therefore, $x_{i-1} \xrightarrow{s_i} x_i$ is an element of $\mathcal{D}iv(\Delta)$. Hence $\mathcal{D}iv(\Delta)$ generates $\mathcal{C}$, and Δ satisfies (V.1.17).

Assume now that $x \xrightarrow{g} \cdot$ right-divides $\Delta(y)$ in $\mathcal{C}_F(M,X)$. This means that there exists y in X and f in M satisfying $y \xrightarrow{f} x \xrightarrow{g} \cdot = y \xrightarrow{\Delta_y} \cdot$. In particular, $fg = \Delta_y$ holds. Then f left-divides Δ_y, hence, by (1.13), we have $\Delta_y \preccurlyeq f\Delta_{y\bullet f}$, that is, $fg \preccurlyeq f\Delta_x$. Left-cancelling f, we deduce $g \preccurlyeq \Delta_x$, whence $x \xrightarrow{g} \cdot \preccurlyeq \Delta(x)$. This shows that Δ satisfies (V.1.18).

Finally, let $x \xrightarrow{g} \cdot$ be an element of $\mathcal{C}_F(M,X)$ with source x. Then, by definition, $x \bullet g$ is defined and, therefore, by (1.14), the elements g and Δ_x admit a left-gcd, say s. Then, by Lemma 1.9(ii), $x \xrightarrow{s} \cdot$ is a left-gcd of $x \xrightarrow{g} \cdot$ and $\Delta(x)$ in $\mathcal{C}_F(M,X)$. So Δ satisfies (V.1.19) and, therefore, it is a right-Garside map in $\mathcal{C}_F(M,X)$.

For (1.19), assume that $x \xrightarrow{g} \cdot$ belongs to $\mathcal{D}iv(\Delta)$ and $y \xrightarrow{g} \cdot$ is another element of $\mathcal{C}_F(M,X)$. By assumption, g left-divides Δ_x and it belongs to $\mathrm{Def}(y)$. Hence, by (1.15), g left-divides Δ_y, and, therefore, $y \xrightarrow{g} \cdot$ belongs to $\mathcal{D}iv(\Delta)$ as well. So (1.19) is satisfied.

(ii) Let x be an element of X. First, by definition, $\Delta(x)$ is an element of $\mathcal{C}_F(M,X)$ whose source is x, hence of the form $x \xrightarrow{g} \cdot$, so the definition of Δ_x makes sense. Then, by construction, $x \bullet \Delta_x$ is defined. Hence $(\Delta_x)_{x\in X}$ satisfies (1.11).

Let g be an element of M. As F is proper, there exists x in X such that $x\bullet g$ is defined. Then $x \xrightarrow{g} \cdot$ is an element of $\mathcal{C}_F(M,X)$. By assumption, $\mathcal{D}iv(\Delta)$ generates the category, so there exists a decomposition $x_1 \xrightarrow{s_1} \cdot|\cdots|x_p \xrightarrow{s_p} \cdot$ of $x \xrightarrow{g} \cdot$ as a product of elements of $\mathcal{D}iv(\Delta)$. Then, by construction, we have $g = s_1\cdots s_p$, and each element s_i belongs to $\bigcup_{x\in X}\mathrm{Div}(\Delta_x)$. So $(\Delta_x)_{x\in X}$ satisfies (1.12).

Next, assume $g \preccurlyeq \Delta_x$. Let h satisfy $gh = \Delta_x$, and let $y = x \bullet g$. In $\mathcal{C}_F(M,X)$, we have $x \xrightarrow{g} y \xrightarrow{h} \phi_\Delta(x) = \Delta(x)$, so $y \xrightarrow{h} \phi_\Delta(x)$ lies in $\widetilde{\mathcal{D}iv}(\Delta)$. By assumption, Δ satisfies (V.1.18), so $y \xrightarrow{h} \phi_\Delta(x)$ must belong to $\mathcal{D}iv(\Delta)$. The only possibility is that $y \xrightarrow{h} \phi_\Delta(x)$ left-divides $\Delta(y)$ in $\mathcal{C}_F(M,X)$, hence that h left-divides Δ_y in M. It follows that gh, which is $\Delta(x)$, left-divides $g\Delta_y$, that is, $(\Delta_x)_{x\in X}$ satisfies (1.13).

Assume now that $x \bullet g$ is defined. Then, in $\mathcal{C}_F(M,X)$, the elements $x \xrightarrow{g} \cdot$ and $\Delta(x)$ admits a left-gcd. By Lemma 1.9(ii), the latter must be of the form $x \xrightarrow{s} \cdot$ where s is a left-gcd of g and Δ_x in M. So $(\Delta_x)_{x\in X}$ satisfies (1.14).

Finally, assume that g belongs to $\mathrm{Div}(\Delta_x) \cap \mathrm{Def}(y)$. Then $x \xrightarrow{g} \cdot$ is an element of $\mathcal{D}iv(\Delta)$. On the other hand, $y \xrightarrow{g} \cdot$ is another element of $\mathcal{C}_F(M,X)$ with middle entry g. By (1.19), this element must lie in $\mathcal{D}iv(\Delta)$ as well, which means that g must left-divide Δ_y. So $(\Delta_x)_{x\in X}$ satisfies (1.15), and it is a right-Garside sequence in M. □

Note that Condition (1.19) is crucial in the above argument: otherwise, there would be no way to compose the elements of $\mathcal{D}iv(\Delta)$.

1.3 Derived notions

Proposition 1.18 shows that, if we have a right-Garside sequence in a monoid M in connection with a partial action F of M on a set X, then we deduce a right-Garside map in the associated category $\mathcal{C}_F(M,X)$, and, therefore, a distinguished Garside family in this category. The interest of the construction is that it also provides a Garside family in the initial monoid M.

Proposition 1.20 (Garside family). *If F is a proper partial action of a left-cancellative monoid M on X and $(\Delta_x)_{x\in X}$ is a right-Garside sequence in M with respect to F, then $\bigcup_{x\in X}\mathrm{Div}(\Delta_x)$ is a Garside family in M that is closed under left-divisor.*

Proof. Put $S = \bigcup_{x\in X}\mathrm{Div}(\Delta_x)$. First, S contains 1 and is closed under left-divisor by very definition. Next, by the definition again, S generates M and it is closed under right-divisor. So, in order to apply Proposition IV.1.50(i) (recognizing Garside III), it suffices to prove that every element g of M has an S-head. Let g be an arbitrary element of M. Since F is assumed to be proper, there exists x in X such that $x \bullet g$ is defined. Let s be a left-gcd of g and Δ_x, which exists by (1.14). Then s left-divides g and belongs to S. Now let t be any element of S that left-divides g. As $t \preccurlyeq g$ holds, $x \bullet t$ is defined, that is, t lies in $\mathrm{Def}(x)$. By (1.15), the assumption that t lies in some set $\mathrm{Div}(\Delta_y)$ implies that it lies in $\mathrm{Div}(\Delta_x)$. Hence t must left-divide s, and s is an S-head of g. So S is a Garside family in M. □

It follows that, under the assumption that a right-Garside sequence exists in M, we obtain for the elements of M normal decompositions with all the properties explained in Chapter V. However, in general, the Garside family provided by Proposition 1.20 need not be (right)-bounded: a typical example will be described in Section 3 below.

In addition to the existence of distinguished decompositions, we saw in Section V.1 that a right-Garside map implies the existence of common right-multiples. Using Proposition 1.18, we easily derive a similar result from the existence of a right-Garside sequence.

Proposition 1.21 (common right-multiple). *Every left-cancellative monoid containing a right-Garside sequence with respect to a proper partial action necessarily admits common right-multiples.*

Proof. Assume that F is a proper partial action of M on some set X and $(\Delta_x)_{x\in X}$ is a right-Garside sequence in M with respect to F. By Proposition 1.18, the category $\mathcal{C}_F(M,X)$ admits a right-Garside map. Hence, by Corollary V.1.48 (common right-multiple), any two elements of $\mathcal{C}_F(M,X)$ admit a common right-multiple. Let f, g belong to M. As F is assumed to be proper, there exists x in X such that $x \bullet f$ and $x \bullet g$ are defined. Then $x \xrightarrow{f} \cdot$ and $x \xrightarrow{g} \cdot$ are elements of $\mathcal{C}_F(M,X)$ that share the same source, hence they admit a common right-multiple, say $x \xrightarrow{h} \cdot$. Then, by construction, h must be a common right-multiple of f and g in M. □

One of the main tools deriving from a (right)-Garside map Δ is the associated functor ϕ_Δ, hereafter written ϕ. In the case of a right-Garside sequence, we similarly obtain a sort of local endomorphism. Once again, we can appeal to the category $\mathcal{C}_F(M,X)$.

Proposition 1.22 (map ϕ_x). *Assume that M is a left-cancellative monoid, F is a proper partial action of M on X, and $(\Delta_x)_{x\in X}$ is a right-Garside sequence in M. Then, for all x in X and g in M such that $x \bullet g$ is defined, $\Delta_x \preccurlyeq g\Delta_{x\bullet g}$ holds and, if we define $\phi_x(g)$ to by $g\,\Delta_{x\bullet g} = \Delta_x\,\phi_x(g)$, then the equality*

(1.23) $$\phi_x(gh) = \phi_x(g)\phi_{x\bullet g}(h).$$

holds whenever $x \bullet gh$ is defined. Moreover, we have $\phi_x(\Delta_x) = \Delta_{\phi(x)}$.

Proof. Let Δ be the right-Garside map on $\mathcal{C}_F(M,X)$ provided by Proposition 1.20. Then, by construction, $\phi_x(g)$ is the unique element of M satisfying

(1.24) $$\phi_\Delta(x \xrightarrow{\;g\;} y) = \phi(x) \xrightarrow{\;\phi_x(g)\;} \phi(y).$$

Then it suffices to translate the results of Proposition V.1.28 (functor ϕ_Δ) for the associated functor ϕ_Δ in terms of ϕ_x. □

Relation (1.23) is illustrated in the commutative diagram shown on the right. Note that, if the action is defined everywhere and the Δ-sequence is constant, ϕ_x does not depend on x and it is an endomorphism of the ambient monoid, but this need not be the case for an arbitrary partial action.

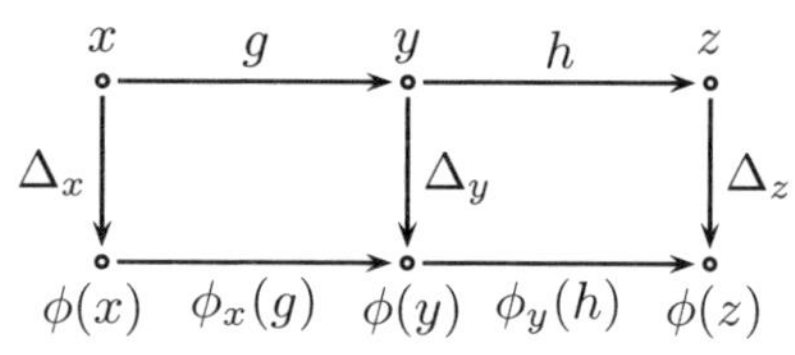

Finally, we conclude with a sort of local version of Lemma 1.16.

Proposition 1.25 (local delta). *Assume that F is a partial action of a left-cancellative monoid M on a set X and x_0 is an element of X such that*

(1.26) *$x_0 \bullet gh$ is defined if and only if $x_0 \bullet g$ and $x_0 \bullet h$ are defined.*

(i) *The set* $\mathrm{Def}(x_0)$ *is a left-cancellative submonoid of M that is closed under left- and right-divisor.*

(ii) *If $(\Delta_x)_{x\in X}$ is a right-Garside sequence in M with respect to F, then Δ_{x_0} is a right-Garside element in the monoid* $\mathrm{Def}(x_0)$.

Proof. (i) Put $M_0 = \mathrm{Def}(x_0)$. First, 1 belongs to M_0 by (1.2). Then (1.26) implies that the product of two elements of M_0 belongs to M_0. Hence, M_0 is a submonoid of M. It is left-cancellative, as a counter-example to left-cancellativity in M_0 would be a counter-example in M. The closure properties are clear since $gh \in M_0$ implies both $g \in M_0$ and $h \in M_0$.

(ii) First, Δ_{x_0} belongs to M_0 by (1.11). Next, let g be an arbitrary element of M_0. By (1.11), we can write $g = s_1 \cdots s_p$ with $s_1, \ldots, s_p$ in $\bigcup_x \mathrm{Div}(\Delta_x)$. Then (1.26) implies that $x_0 \bullet s_i$ is defined for every i, and (1.15) then implies that s_i lies in $\mathrm{Div}(\Delta_{x_0})$. This shows that $\mathrm{Div}(\Delta_{x_0})$, which is included in M_0, is generates M_0. So Δ_{x_0} satisfies (V.1.22) in M_0.

Next, assume $fg = \Delta_{x_0}$. By (1.26), the element g belongs to M_0. As $(\Delta_x)_{x\in X}$ is a right-Garside sequence in M, there exists y such that g left-divides Δ_y, and (1.15) then implies that g left-divides Δ_{x_0}. In other words, $\widetilde{\mathrm{Div}}(\Delta_{x_0})$ is included in $\mathrm{Div}(\Delta_{x_0})$, and Δ_{x_0} satisfies (V.1.23) in M_0.

Finally, assume $g \in M_0$. Then $x_0 \bullet g$ is defined, so, by (1.15), the elements g and Δ_{x_0} admit a left-gcd in M, say s. Assume $g = sg'$ and $\Delta_{x_0} = sh$. First, (1.26) implies that s, g', and h belong to M_0. So s is a common left-divisor of g' and Δ_{x_0} in M_0. Moreover, if t is a common left-divisor of g and Δ_{x_0} in M_0, then it is a common left-divisor in M as well, hence it must left-divide s. So s is a left-gcd of g and Δ_{x_0} in M_0, and Δ_{x_0} satisfies (V.1.24). Hence it is a right-Garside element in M_0. □

We stop with general results here. It should be clear that what was made above for right-Garside elements could be made for Garside elements as well. However, we shall not go into this direction here, as it is not relevant for the sequel of this chapter.

2 LD-expansions and the category $\mathcal{LD}_0$

We now turn to the specific case of left self-distributivity. Our exposition is centered around the notion of an LD-expansion, which corresponds to applying the LD law in the expanding direction. In this way, there arise a poset and, more relevant in our current approach, a category. The main technical result is that this category $\mathcal{LD}_0$ admits a map Δ_0 that is close to be a right-Garside map. However, one of the ingredients for a genuine right-Garside map is missing: as we shall see subsequently, this is one of the many forms of the Embedding Conjecture.

The section is organized as follows. The LD law and the notion of an LD-system are introduced in Subsection 2.1. Next, in Subsection 2.2, we define the partial ordering of terms provided by LD-expansions and sketch a proof of the result that any two LD-equivalent terms admit a common LD-expansion. In Subsection 2.3, we introduce the category $\mathcal{LD}_0$ naturally associated with the partial ordering of LD-expansions and show the existence of a map Δ_0 that is nearly a right-Garside map on $\mathcal{LD}_0$. Finally, in Subsection 2.4, we give an intrinsic characterization of those LD-expansions that left-divide Δ_0 in the category $\mathcal{LD}_0$.

2.1 Free LD-systems

We recall that LD refers to the left self-distributivity law $x(yz) = (xy)(xz)$. An algebraic structure consisting of a set equipped with a binary operation satisfying (LD) will be called an LD-*system*.

Example 2.1 (LD-system). If X is any set and f any map of X to itself, then the operation $*$ defined on X by $x * y = f(y)$ satisfies the LD law, and, therefore, $(X, *)$ is a (trivial) LD-system.

If R is a ring and V is an R-module, then, for t in R, the operation $*$ defined on V by $x * y = (1-t)x + ty$ satisfies the LD law, so $(V, *)$ is an LD-system.

If G is a group, then the conjugacy operation defined on G by $x * y = xyx^{-1}$ satisfies the LD law, so $(G, *)$ is an LD-system.

Up to a few variants, the last two families complete the description of what can be called classical LD-systems. Note that, in both case, the idempotency law $xx = x$ is satisfied. An example of a completely different type involves the braid group B_∞ (Reference Structure 2, page 5). Indeed, let sh be the endomorphism of B_∞ that maps σ_i to σ_{i+1} for every i. Then the operation $*$ defined on B_∞ by $x * y = x\,\mathrm{sh}(y)\,\sigma_1\,\mathrm{sh}(x)^{-1}$ satisfies the LD law, so $(B_\infty, *)$ is an LD-system. As $1 * 1 = \sigma_1$ holds, the LD-system $(B_\infty, *)$ is not idempotent.

In the sequel, we shall investigate particular LD-systems, namely *free* LD-systems. In the case of the LD law as, more generally, in the case of every algebraic law or family of algebraic laws, free systems are the universal objects in the associated variety; they can be described uniformly as quotients of absolutely free structures under convenient congruences.

Definition 2.2 (term). For $n \geqslant 1$, we define $\mathcal{T}_n$ to be the set of all (well-formed) bracketed expressions involving the variables $x_1, \ldots, x_n$, that is, the closure of $\{x_1, \ldots, x_n\}$ under the operation $^\wedge$ defined by $T_1{}^\wedge T_2 = (T_1)(T_2)$. We use $\mathcal{T}$ for the union of all sets $\mathcal{T}_n$. The elements of $\mathcal{T}$ are called *terms*. The *size* $\mu(T)$ of a term T is the number of brackets in T divided by four.

Typical terms are x_1, which has size zero, $x_2{}^\wedge x_1$, which, by definition, is $(x_2)(x_1)$ and has size one, and $x_3{}^\wedge(x_3{}^\wedge x_1)$, which, by definition, is $(x_3)((x_3)(x_1))$ and has size two. It is convenient to think of terms as of rooted binary trees with leaves indexed by variables: the trees associated with the previous terms are 

, 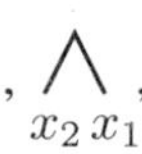

, and 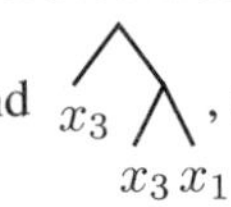

, respectively. The size of a term is the number of internal nodes in the tree associated with T, which is also the number of leaves diminished by one. Then $(\mathcal{T}_n, {}^\wedge)$ is the absolutely free system (or algebra) generated by $x_1, \ldots, x_n$, and every binary system generated by n elements is a quotient of $(\mathcal{T}_n, {}^\wedge)$. So is in particular the free LD-system of rank n.

Definition 2.3 (LD-equivalence). We denote by $=_{\mathrm{LD}}$ the least congruence (that is, equivalence relation compatible with the product) on $(\mathcal{T}_n, {}^\wedge)$ that contains all pairs of the form

$$(T_1{}^\wedge(T_2{}^\wedge T_3), (T_1{}^\wedge T_2){}^\wedge(T_1{}^\wedge T_3)).$$

Two terms T, T' satisfying $T =_{\mathrm{LD}} T'$ are called LD-*equivalent*.

The following result is then standard.

Proposition 2.4 (free LD-system). *For every n, the binary system $(\mathcal{T}_n/{=_{\mathrm{LD}}}, {}^\wedge)$ is a free* LD-*system based on $\{x_1, \ldots, x_n\}$.*

2.2 LD-expansions

The LD-equivalence relation $=_{\text{LD}}$ is a complicated relation, about which some basic questions remain open. In order to investigate it, it is useful to introduce the subrelation that corresponds to applying the LD law in one direction only.

Definition 2.5 (LD-expansion). For two terms T, T', we say that T' is an *atomic* LD-*expansion* of T if T' is obtained from T by replacing a subterm of the form $T_1{}^\wedge(T_2{}^\wedge T_3)$ with the corresponding term $(T_1{}^\wedge T_2)^\wedge(T_1{}^\wedge T_3)$. We say that T' is an LD-*expansion* of T, written $T \leqslant_{\text{LD}} T'$, if there exists a finite sequence of terms $T_0, \dots, T_p$ satisfying $T_0 = T$, $T_p = T'$, and, for every i, the term T_i is an atomic LD-expansion of T_{i-1}.

By definition, being an LD-expansion implies being LD-equivalent, but the converse is not true. For instance, the term $(x^\wedge x)^\wedge(x^\wedge x)$ is an (atomic) LD-expansion of $x^\wedge(x^\wedge x)$, but the latter is not an LD-expansion of the former. However, the relation $=_{\text{LD}}$ is generated by $\leqslant_{\text{LD}}$ in the sense that two terms T, T' are LD-equivalent if and only if there exists a finite zigzag $T_0, T_1, \dots, T_{2p}$ satisfying $T_0 = T$, $T_{2p} = T'$, and $T_{i-1} \leqslant_{\text{LD}} T_i \geqslant_{\text{LD}} T_{i+1}$ for each odd i. The first nontrivial result about LD-equivalence is that zigzags may always be assumed to have length two.

Proposition 2.6 (confluence). *Two terms are* LD-*equivalent if and only if they admit a common* LD-*expansion.*

This result is similar to the property that, if any two elements of a monoid M admit a common right-multiple, then every element in the enveloping group of M can be expressed as a fraction of the form gh^{-1} with g, h in M. Proposition 2.6 plays a fundamental role in the sequel, and we need to explain some elements of its proof.

Definition 2.7 (dilatation). We define a binary operation $^{\wedge\!\!\wedge}$ on terms recursively by

$$T^{\wedge\!\!\wedge}x_i = T^\wedge x_i, \quad T^{\wedge\!\!\wedge}(T_1{}^\wedge T_2) = (T^{\wedge\!\!\wedge}T_1)^\wedge(T^{\wedge\!\!\wedge}T_2). \tag{2.8}$$

Next, for each term T, the term $\phi(T)$ is recursively defined by

$$\phi(x_i) = x_i \quad \phi(T_1{}^\wedge T_2) = \phi(T_1)^{\wedge\!\!\wedge}\phi(T_2). \tag{2.9}$$

An easy induction shows that $T^{\wedge\!\!\wedge}T'$ is the term obtained by distributing T everywhere in T', this meaning replacing each variable x_i with $T^\wedge x_i$. Then $\phi(T)$ is the image of T when $^\wedge$ is replaced with $^{\wedge\!\!\wedge}$ everywhere in the unique expression of T in terms of variables. Examples are given in Figure 1. A straightforward induction shows that $T^{\wedge\!\!\wedge}T'$ is always an LD-expansion of $T^\wedge T'$ and, therefore, that $\phi(T)$ is an LD-expansion of T.

The main step for establishing Proposition 2.6 consists in proving that $\phi(T)$ plays with respect to atomic LD-expansions a role similar to Garside's fundamental braid Δ_n with respect to Artin's generators σ_i in the braid monoid B_∞^+—which makes it natural to call $\phi(T)$ the *fundamental* LD-*expansion* of T.

Lemma 2.10. [77, Lemmas V.3.11 and V.3.12] (i) *The term $\phi(T)$ is an* LD-*expansion of every atomic* LD-*expansion of T.*

(ii) *If T' is an* LD-*expansion of T, then $\phi(T')$ is an* LD-*expansion of $\phi(T)$.*

(iii) *The map ϕ is injective.*

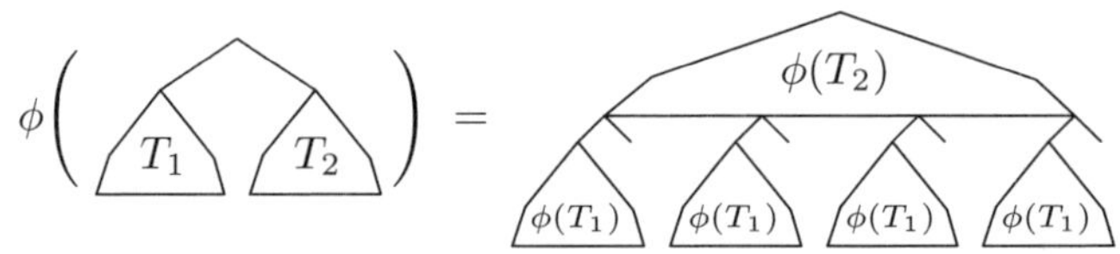

Figure 1. The fundamental LD-expansion $\phi(T)$ of a term T, recursive definition: $\phi(T_1{}^\wedge T_2)$ is obtained by distributing $\phi(T_1)$ everywhere in $\phi(T_2)$.

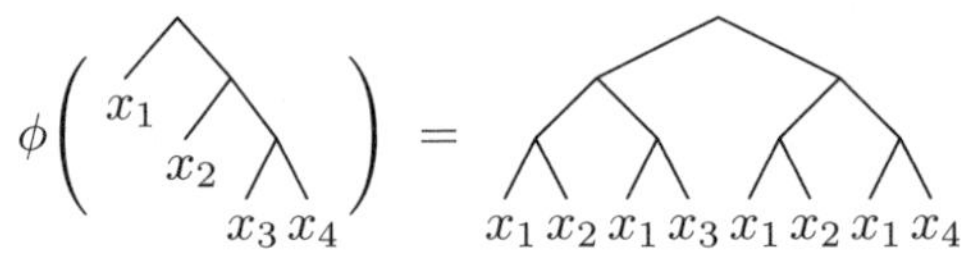

Figure 2. The fundamental LD-expansion $\phi(T)$ of a term T, an example: $T = x_1{}^\wedge(x_2{}^\wedge(x_3{}^\wedge x_4))$ gives $\phi(T) = x_1{}^{\wedge\!\!\!\wedge}(x_2{}^{\wedge\!\!\!\wedge}(x_3{}^{\wedge\!\!\!\wedge}x_4))$, which is realized as shown above.

Proof (*sketch*). In each case, one uses induction on the size of T. Here is a typical case for (i). Assume $T = T_1{}^\wedge(T_2{}^\wedge T_3)$ and $T' = (T_1{}^\wedge T_2){}^\wedge(T_1{}^\wedge T_3)$. Using the fact that $\phi(T) \geqslant_{\mathrm{LD}} T$ always holds, we find

$$\phi(T) = \phi(T_1){}^{\wedge\!\!\!\wedge}\phi(T_2{}^\wedge T_3) \geqslant_{\mathrm{LD}} T_1{}^{\wedge\!\!\!\wedge}(T_2{}^\wedge T_3) = (T_1{}^\wedge T_2){}^\wedge(T_1{}^\wedge T_3) = T'.$$

The case of (ii) is similar.

Let us go into more details for (iii). The aim is to show using induction on the size of T that $\phi(T)$ determines T. The result is obvious if T has size 0. Assume now $T = T_0{}^\wedge T_1$. By construction, the term $\phi(T)$ is obtained by substituting every variable x_i occurring in the term $\phi(T_1)$ with the term $\phi(T_0){}^\wedge x_i$. Hence $\phi(T_0)$ is the $1^{n-1}0$th subterm of $\phi(T)$ (see Definition 3.1 below), where n is the common right height of T and $\phi(T)$. From there, $\phi(T_1)$ can be recovered by replacing the subterms $\phi(T_0){}^\wedge x_i$ of $\phi(T)$ by x_i. Then, by the induction hypothesis, T_0 and T_1, hence T, can be recovered from $\phi(T_1)$ and $\phi(T_0)$. □

Proof of Proposition 2.6 (*sketch*). One uses induction on the size of the involved terms. Once Lemma 2.10 is established, an easy induction on m shows that, if there exists a length-m sequence of atomic LD-expansions connecting T to T', then $\phi^m(T)$ is an LD-expansion of T'. Then a final induction on the length of a zigzag connecting T to T' shows that, if T and T' are LD-equivalent, then $\phi^m(T)$ is an LD-expansion of T' for sufficiently large m, namely for m at least equal to the number of zag's in the zigzag. □

By construction, the relation $\leqslant_{\mathrm{LD}}$ is a partial ordering on the set $\mathcal{T}$ of all terms. Indeed, it is reflexive and transitive by definition, and, as every LD-expansion that is not the identity increases the size of the tree, we may have $T \leqslant_{\mathrm{LD}} T'$ and $T' \leqslant_{\mathrm{LD}} T$ simultaneously only for $T = T'$.

Question 2.11 (lattice). *Is* $(\mathcal{T}, \leqslant_{\mathrm{LD}})$ *an upper semilattice, that is, do every two terms that admit a common upper bound with respect to* $\leqslant_{\mathrm{LD}}$ *admit a least upper bound?*

By Proposition 2.6, two terms are LD-equivalent if and only if they admit a common $\leqslant_{\mathrm{LD}}$-upper bound. So Question 2.11 asks whether any two LD-equivalent terms admit a smallest common LD-expansion. We shall return on it in Section 3 below.

2.3 The category $\mathcal{LD}_0$

We now make the Hasse diagram of the partial ordering $\leqslant_{\text{LD}}$ into a category.

Definition 2.12 (category $\mathcal{LD}_0$). We let $\mathcal{LD}_0$ be the category whose objects are terms, and whose elements are pairs of terms (T, T') satisfying $T \leqslant_{\text{LD}} T'$, the source (*resp.* target) of (T, T') being T (*resp.* T'). For T a term, we put $\Delta_0(T) = (T, \phi(T))$.

Note that the definition of the source and target imposes that the product of (T, T') and (T', T'') necessarily is (T, T'').

Proposition 2.13 (category $\mathcal{LD}_0$). (i) *The category $\mathcal{LD}_0$ is cancellative.*

(ii) *Two elements of $\mathcal{LD}_0$ with the same source have a common right-multiple.*

(iii) *The category $\mathcal{LD}_0$ is strongly Noetherian; the atoms of $\mathcal{LD}_0$ are the pairs (T, T') with T' an atomic* LD*-expansion of T.*

(iv) *The map Δ_0 satisfies* (V.1.16)–(V.1.18) *from the definition of a right-Garside map; if the answer to Question* 2.11 *is positive, Δ_0 is a right-Garside map in $\mathcal{LD}_0$.*

Proof. (i) Cancellativity is obvious from the construction. Indeed, assume $fg = fg'$ in $\mathcal{LD}_0$. This means that there exist terms T, T', T'' satisfying $f = (T, T')$, $g = (T', T'')$, and the only possibility is then $g' = (T', T'')$.

(ii) The existence of common right-multiples is a consequence (actually a restatement) of Proposition 2.6. Indeed, g, g' sharing the same source means that there exist T, T', T'' such that T' and T'' are LD-expansions of T and we have $g = (T, T')$ and $g' = (T, T'')$. By Proposition 2.6, T' and T'' admit a common LD-expansion T''', and then (T, T''') is a common right-multiple of g and g' in $\mathcal{LD}_0$.

(iii) Put $\lambda((T, T')) = \mu(T') - \mu(T)$, where we recall $\mu(T)$ denotes the size of T. Then λ is $\mathbb{N}$-valued since an LD-expansion cannot decrease the size. Moreover, $\mu(T') > \mu(T)$ holds whenever T' is an atomic LD-expansion of T, hence whenever T' is any nontrivial LD-expansion of T. It follows that λ is an $\mathbb{N}$-valued Noetherianity witness for $\mathcal{LD}_0$. Moreover, if T' is an atomic LD-expansion of T, then T' is an immediate successor of T with respect to $\leqslant_{\text{LD}}$, so (T, T') is an atom of $\mathcal{LD}_0$. Conversely, if T' is an LD-expansion of T that is not atomic, there exists T'' such that $(T, T') = (T, T'')(T'', T')$ holds in $\mathcal{LD}_0$ and, therefore, (T, T') is not an atom.

(iv) By definition, Δ_0 is a map of $\mathcal{T}$, that is, of $\mathcal{O}bj(\mathcal{LD}_0)$, to $\mathcal{LD}_0$ and, for every term T, the source of $\Delta_0(T)$ is T. So Δ_0 satisfies Condition (V.1.16).

Next, let (T, T') be any element of $\mathcal{LD}_0$. By definition (or by Noetherianity), there exists at an atomic LD-expansion T'' of T such that (T, T'') left-divides (T, T') in $\mathcal{LD}_0$, that is, $T'' \leqslant_{\text{LD}} T'$ holds. Then, by Lemma 2.10(i), $\phi(T)$ is an LD-expansion of T'', that is, (T, T'') left-divides $\Delta_0(T)$ in $\mathcal{LD}_0$. This shows that every element of $\mathcal{LD}_0$ is left-divisible by an element of $\mathcal{D}iv(\Delta_0)$. By (iii), $\mathcal{LD}_0$ is Noetherian, so, by Lemma II.2.57, this is enough to conclude that $\mathcal{D}iv(\Delta_0)$ generates $\mathcal{LD}_0$. Hence Δ_0 satisfies Condition (V.1.17).

Assume now that g right-divides $\Delta_0(T)$ in $\mathcal{LD}_0$. This means that there exists f satisfying $fg = \Delta_0(T)$, hence there exists an LD-expansion T' of T such that f is (T, T') and g is $(T', \phi(T))$. Then, by Lemma 2.10(ii), $\phi(T')$ is an LD-expansion of $\phi(T)$. Consequently, we have $g = (T', \phi(T)) \preccurlyeq (T', \phi(T')) = \Delta_0(T')$, that is, g belongs to $\mathcal{D}iv(\Delta_0)$. Hence, Δ_0 satisfies Condition (V.1.18).

Finally, assume that the answer to Question 2.11 is positive. Then $\mathcal{LD}_0$ admits conditional right-lcms. As it is Noetherian, it then admits left-gcds. Hence, for every term T and every LD-expansion T' of T, the elements (T, T') and $\Delta_0(T)$ admit a left-gcd. Hence Δ_0 satisfies Condition (V.1.19) and, therefore, it is a right-Garside map in $\mathcal{LD}_0$. $\square$

Remark 2.14. In any case, Δ_0 is not a Garside map in $\mathcal{LD}_0$. Indeed, Δ_0 is target-injective as, according to Lemma 2.10(iii), $T \neq T'$ implies $\phi(T) \neq \phi(T')$. But $\mathcal{D}iv(\Delta_0)$ is certainly not equal to $\widetilde{\mathcal{D}iv}(\Delta_0)$, since the mapping ϕ is not surjective on $\mathcal{T}$: for instance, for $T_0 = x^\wedge(x^\wedge x)$, no term T may satisfy $\phi(T) = T_0$ and, therefore, the pair (T_0, T_0) is an element of $\mathcal{D}iv(\Delta_0)$ that does not lie in $\widetilde{\mathcal{D}iv}(\Delta_0)$.

2.4 Simple LD-expansions

For every term T, the LD-expansions of T that lie below $\phi(T)$, that is, the terms T' that satisfy $T \leqslant_{\mathrm{LD}} T' \leqslant_{\mathrm{LD}} \phi(T)$, turn out to admit a simple intrinsic characterization. Such terms will play an important role in the sequel, namely that of simple braids, that is, braid dividing Δ_n, and we describe them now.

Definition 2.15 (simple LD-expansion). If T, T' are terms, we say that T' is a *simple* LD-expansion of T if $T \leqslant_{\mathrm{LD}} T' \leqslant_{\mathrm{LD}} \phi(T)$ holds.

In order to state the announced characterization of simple LD-expansions, we introduce the covering relation, a geometric notion involving the names and the positions of variables occurring in a term. We recall that a sequence of variables $x_1, x_2, \ldots$ has been fixed and $\mathcal{T}$ denotes the set of all terms built from these variables.

Definition 2.16 (injective, semi-injective, covering). A term T of $\mathcal{T}$ is called *injective* if no variable x_i occurs twice or more in T. It is called *semi-injective* if no variable x_i occurring in T *covers* itself, where x_i *covers* x_j in T if there exists a subterm T' of T such that x_i is the rightmost variable in T' and x_j occurs at some other position in T'.

Example 2.17 (injective, semi-injective, covering). Consider the terms $T = x_1{}^\wedge(x_2{}^\wedge x_3)$, $T' = (x_1{}^\wedge x_2)^\wedge(x_1{}^\wedge x_3)$, and $T'' = ((x_1{}^\wedge x_2)^\wedge x_1)^\wedge((x_1{}^\wedge x_2)^\wedge x_3)$. Then T' is an atomic LD-expansion of T, and T'' is an atomic LD-expansion of T'. Then we have

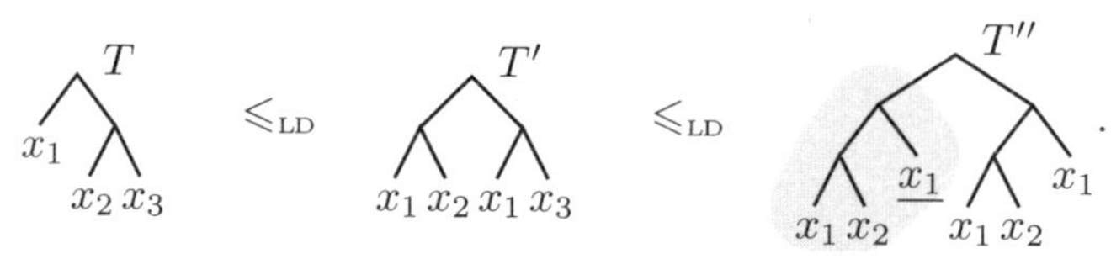

The term T is injective: each variable x_1, x_2, x_3 occurs once, and none is repeated. On the other hand, T' is not injective, as x_1 occurs twice. However T' is semi-injective, as the second occurrence of x_1 does not cover the first one: there is no subterm of T' ending with x_1 and containing another occurrence of x_1. By contrast, T'' is not semi-injective: its left subterm is $(x_1{}^\wedge x_2)^\wedge x_1$ (in grey), ends with x_1 (underlined) and contains another occurrence of x_1. So x_1 covers itself in T''.

Proposition 2.18 (simple LD-expansion). [77, Lemma VIII.5.8] *If T is an injective term and T' is an* LD*-expansion of T, then the following are equivalent:*
(i) *The term T' is a simple* LD*-expansion of T;*
(ii) *The term T' is semi-injective.*

Proof (sketch). The preliminary observation is that an LD-expansion can only add covering: if x_i covers x_j in T, then x_i covers x_j in every LD-expansion of T. Then a relatively simple induction on T shows that $\phi(T)$ is a semi-injective LD-expansion of T, and one easily deduces that (i) implies (ii).

For the other direction, one shows that $\phi(T)$ is a maximal semi-injective LD-expansion of T, that is, no proper LD-expansion of $\phi(T)$ is semi-injective. Then the point is that, if T_0 is a semi-injective term and T_1, T_2 are semi-injective LD-expansions of T_0, then there must exist a semi-injective term T_3 that is an LD-expansion both of T_1 and T_2: one first shows the result when T_1 and T_2 are atomic expansions of T_0 by exhaustively considering all possible cases, and then uses an induction to treat the case of arbitrary expansions. Now, assume that T' is a semi-injective LD-expansion of T. Then $\phi(T)$ is another semi-injective LD-expansion of T, so, by the above result, there must exist a semi-injective term T'' that is an LD-expansion both of T' and $\phi(T)$. Now, as no proper LD-expansion of $\phi(T)$ is semi-injective, the only possibility is that T'' equals $\phi(T)$, that is, $\phi(T)$ is an LD-expansion of T'. Hence T' is a simple LD-expansion of T. □

3 Labeled LD-expansions and the category $\mathcal{LD}$

At this point, we obtained a category $\mathcal{LD}_0$ that describes LD-expansions, but its study remains incomplete in that we do not know about the possible existence of right-lcms in $\mathcal{LD}_0$, or, equivalently, about whether the map Δ_0 is a right-Garside map in $\mathcal{LD}_0$. Our strategy for going further consists in becoming more precise and considering labeled LD-expansions, which are LD-expansions plus explicit descriptions of the way the expansion is performed. This approach relies on introducing a certain monoid M_{LD} whose elements can be used to specify LD-expansions.

The section is organized as follows. In Subsection 3.1, we introduce the operators Σ_α, which correspond to performing an LD-expansion at a specified position in a term. Next, in Subsection 3.2, we study the relations that connect the various operators Σ_α and we are led to introducing a certain monoid M_{LD}. In Subsection 3.3, we define the notion of a labeled LD-expansion, introduce the associated category $\mathcal{LD}$, and show that it admits a natural right-Garside map. Finally, we discuss the Embedding Conjecture, which claims that the categories $\mathcal{LD}_0$ and $\mathcal{LD}$ are isomorphic, in Subsection 3.4.

3.1 The operators Σ_α

By definition, applying the LD law to a term T means selecting some subterm of T and replacing it with a new, LD-equivalent term. When terms are viewed as binary rooted trees, the position of a subterm can be specified by describing the path that connects the root of the tree to the root of the considered subtree, hence by a finite sequence of 0s and 1s, according to the convention that 0 means "forking to the left" and 1 means "forking to the right". Hereafter, we use $\mathbb{A}$ for the set of such sequences, which we call *addresses*, and $\emptyset$ for the empty address, which corresponds to the position of the root in a tree.

Definition 3.1 (subterm, skeleton). For T a term and α an address, we denote by $T_{/\alpha}$ the subterm of T whose root has address α, if it exists, that is, if α is short enough. The *skeleton* of T is the set of all addresses α such that $T_{/\alpha}$ exists.

So, for instance, if T is the tree $x_1{}^\wedge(x_2{}^\wedge x_3)$, we have $T_{/0} = x_1$, $T_{/10} = x_2$, whereas $T_{/00}$ is not defined, and $T_{/\emptyset} = T$ holds, as it holds for every term. The skeleton of T consists of the fives addresses $\emptyset$, 0, 1, 10, and 11.

Definition 3.2 (operator Σ_α). (See Figure 3.) We say that T' is the *Σ_α-expansion of T at α*, written $T' = T \bullet \Sigma_\alpha$, if T' is the atomic LD-expansion of T obtained by applying the LD law in the expanding direction at the position α, that is, replacing the subterm $T_{/\alpha}$, supposed to have the form $T_1{}^\wedge(T_2{}^\wedge T_3)$, with the corresponding term $(T_1{}^\wedge T_2)^\wedge(T_1{}^\wedge T_3)$.

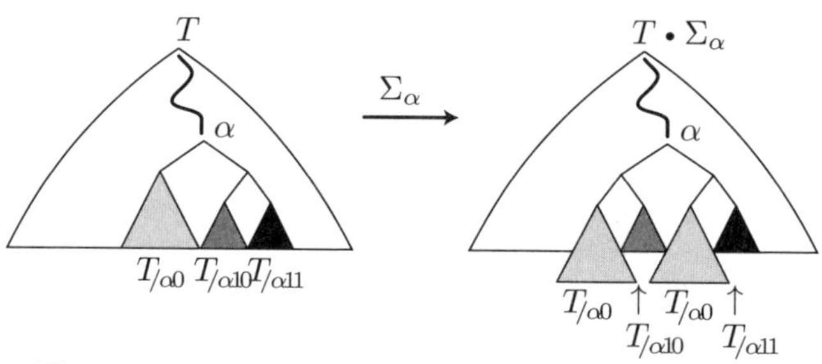

Figure 3. Action of Σ_α on a term T: the LD law is applied to expand T at position α, that is, to replace the subterm $T_{/\alpha}$, which is $T_{/\alpha0}{}^\wedge(T_{/\alpha10}{}^\wedge T_{/\alpha11})$, with $(T_{/\alpha0}{}^\wedge T_{/\alpha10})^\wedge(T_{/\alpha0}{}^\wedge T_{/\alpha11})$; in other words, the light grey subtree is duplicated and distributed to the left of the dark grey and black subtrees.

By construction, every atomic LD-expansion is a Σ_α-expansion for a unique α. Note that $T\bullet\Sigma_\alpha$ need not be defined for all T and α: it is defined if and only if the subterm $T_{/\alpha}$ is defined and can be expressed as $T_1{}^\wedge(T_2{}^\wedge T_3)$. The latter condition amounts (in particular) to $T_{/\alpha10}$ being defined, so $T \bullet \Sigma_\alpha$ is defined if and only if $T_{/\alpha10}$ is.

The idea now is to use the letters Σ_α as labels for LD-expansions. As arbitrary LD-expansions are compositions of finitely many atomic LD-expansions, hence of operators Σ_α, it is natural to use finite sequences of letters Σ_α, that is, words, to label LD-expansions. Hereafter, a word in the alphabet $\{\Sigma_\alpha | \alpha\in\mathbb{A}\}$ will be called a *Σ-word*.

Definition 3.3 (action). For T a term and w a Σ-word, say $w = \Sigma_{\alpha_1}|\cdots|\Sigma_{\alpha_p}$, we define $T \bullet w$ to be $(\cdots((T \bullet \Sigma_{\alpha_1}) \bullet \Sigma_{\alpha_2})\cdots) \bullet \Sigma_{\alpha_p}$ if the latter is defined.

In this way, we obtain a partial action of Σ-words, hence a partial action of the free monoid $\{\Sigma_\alpha \mid \alpha\in\mathbb{A}\}^*$, on $\mathcal{T}$, in the sense of Definition 3.2. This action is directly connected with the notion of an LD-expansion, and a technically significant point is that it is proper and that the domains of definition can be specified precisely. The next result summarizes the needed results. A map $\sigma : \{x_i \mid i \in \mathbb{N}\} \to \mathcal{T}$ is called a *substitution*; if T is a term and σ is a substitution, we denote by T^σ the term obtained from T by applying σ to every occurrence of variable in T, that is, substituting x_i with $\sigma(x_i)$ everywhere.

Lemma 3.4. (i) *A term T' is an* LD*-expansion of a term T if and only if there exists a Σ-word w satisfying $T' = T \bullet w$.*

(ii) *For every Σ-word w, there exists a pair of terms (T_w^-, T_w^+) such that T_w^- is injective and, for every term T, the term $T \bullet w$ is defined if and only if there exists a substitution σ such that T is $(T_w^-)^\sigma$, in which case $T \bullet w$ is $(T_w^+)^\sigma$.*

(iii) *The partial action of $\{\Sigma_\alpha|\alpha\in\mathbb{A}\}^*$ on terms is proper.*

Proof (sketch). Assertion (i) is obvious as, by definition, T' is an LD-expansion of T if and only if there exists a finite sequence of atomic LD-expansions connecting T to T'.

For (ii), one uses induction on the length of w. If w is empty, the result is clear with $T_w^- = T_w^+ = x_1$. Assume now that w has length one, say $w = \Sigma_\alpha$. For $\alpha = \emptyset$, the result is clear with $T_w^- = x_1{}^\wedge(x_2{}^\wedge x_3)$ and $T_w^+ = (x_1{}^\wedge x_2){}^\wedge(x_1{}^\wedge x_3)$. Then we use induction on the length of α. For $\alpha = 0\beta$, we can take $T_\alpha^- = T_\beta^-{}^\wedge x_n$, where n is the size of T_β^-, and, for $\alpha = 1\beta$, we can take $T_\alpha^- = x_1{}^\wedge(T_\beta^-)^\sigma$, where $\sigma(x_i) = x_{i+1}$ for every i. Finally, for $w = uv$, the point is that, because, by the induction hypothesis, T_v^- exists and is injective, there must exist a minimal pair of substitutions (σ, τ) satisfying $(T_u^+)^\sigma = (T_v^-)^\tau$; then defining $T_w^- = (T_u^-)^\sigma$ and $T_w^+ = (T_v^+)^\tau$ gives the result.

For (iii), let $w_1, ..., w_p$ be Σ-words. Because the terms $T_{w_i}^-$ all are injective, there exist substitutions $\sigma_1, ..., \sigma_p$ satisfying $(T_{w_1}^-)^{\sigma_1} = \cdots = (T_{w_p})^{\sigma_p}$. Then $T \bullet w_i$ is defined for every i whenever the skeleton of T includes the common skeleton of $(T_{w_1}^-)^{\sigma_1}, ..., (T_{w_p})^{\sigma_p}$. □

See Exercise 104 for more about the terms T such that $T \bullet w$ is defined.

3.2 The monoid M_{LD}

Various relations connect the actions of the operators Σ_α: different sequences may lead to the same term transformation, in particular some commutation relations hold. We now identify a family of such relations, and introduce the monoid presented by the latter.

Lemma 3.5. *For all addresses α, β, γ, the following pairs have the same action on trees:*

(i) $\Sigma_{\alpha0\beta}|\Sigma_{\alpha1\gamma}$ *and* $\Sigma_{\alpha1\gamma}|\Sigma_{\alpha0\beta}$; *("parallel case")*

(ii) $\Sigma_{\alpha0\beta}|\Sigma_\alpha$ *and* $\Sigma_\alpha|\Sigma_{\alpha00\beta}|\Sigma_{\alpha10\beta}$; *("nested case 1")*

(iii) $\Sigma_{\alpha10\beta}|\Sigma_\alpha$ *and* $\Sigma_\alpha|\Sigma_{\alpha01\beta}$; *("nested case 2")*

(iv) $\Sigma_{\alpha11\beta}|\Sigma_\alpha$ *and* $\Sigma_\alpha|\Sigma_{\alpha11\beta}$; *("nested case 3")*

(v) $\Sigma_\alpha|\Sigma_{\alpha1}|\Sigma_\alpha$ *and* $\Sigma_{\alpha1}|\Sigma_\alpha|\Sigma_{\alpha1}|\Sigma_{\alpha0}$. *("critical case")*

Proof (sketch). The commutation relation of the parallel case is clear, as the transformations involve disjoint subterms. The nested cases are commutation relations as well, but, because one of the involved subtrees is nested in the other, it may be moved and possibly duplicated when the main expansion is performed, so that the nested expansion(s) have different names before and after the main expansion. Finally, the critical case is specific to the LD law, and there is no way to predict it except the verification, see Figure 4. $\square$

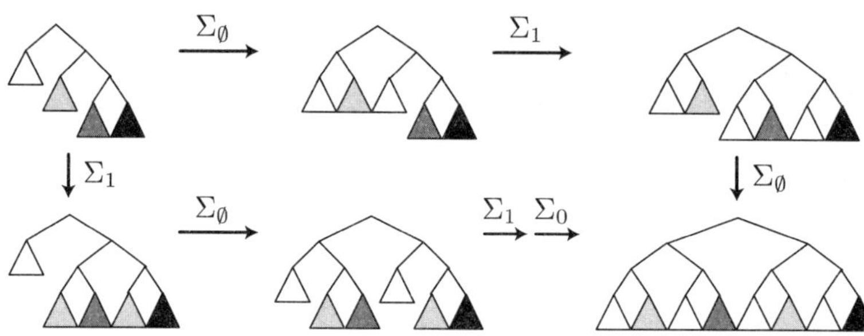

Figure 4. Relations between Σ_α-expansions: the critical case. We read that the action of $\Sigma_\emptyset|\Sigma_1|\Sigma_\emptyset$ and $\Sigma_1|\Sigma_\emptyset|\Sigma_1|\Sigma_0$ coincide.

Question 3.6. *Do the relations of Lemma* 3.5 *exhaust all relations between the operators Σ_α: does every equality $T \bullet w = T \bullet w'$ imply that w and w' are connected by a finite sequence of relations as above?*

We shall come back to the question below. For the moment, we introduce the monoid presented by the above relations.

Definition 3.7 (monoid M_{LD}). We define $\mathcal{R}_{\text{LD}}$ to be the family of all relations of Lemma 3.5, and M_{LD} to be the monoid $\langle\{\Sigma_\alpha|\alpha \in \mathbb{A}\} \mid \mathcal{R}_{\text{LD}}\rangle^+$.

Lemma 3.5 immediately implies

Lemma 3.8. *The partial action of the free monoid $\{\Sigma_\alpha|\alpha{\in}\mathbb{A}\}^*$ on terms induces a well defined proper partial action F of the monoid M_{LD} on $\mathcal{T}$.*

For T a term and g in M_{LD}, we shall naturally denote by $T \bullet g$ the common value of $T \bullet w$ for all Σ-words w that represent g.

We are now left with the technical question of investigating the presented monoid M_{LD}. The next step is to show that the monoid M_{LD} is left-cancellative. To this end, we once more use the method developed in Subsection II.4.

Lemma 3.9. *The presentation $(\{\Sigma_\alpha|\alpha{\in}\mathbb{A}\}, \mathcal{R}_{\text{LD}})$ is right-complemented and (strongly) Noetherian.*

Proof. That $(\{\Sigma_\alpha|\alpha{\in}\mathbb{A}\}, \mathcal{R}_{\text{LD}})$ is right-complemented follows from an inspection of $\mathcal{R}_{\text{LD}}$: there exists no relation of the form $\Sigma_\alpha\cdots = \Sigma_\alpha\cdots$ in $\mathcal{R}_{\text{LD}}$, and, for all distinct α, β, there exists exactly one relation of the form $\Sigma_\alpha\cdots = \Sigma_\beta\cdots$.

As for Noetherianity, we use the partial action of $\{\Sigma_\alpha|\alpha{\in}\mathbb{A}\}^*$ on terms. Let w be a word in the alphabet $\{\Sigma_\alpha|\alpha{\in}\mathbb{A}\}$. As the partial action of $\{\Sigma_\alpha|\alpha{\in}\mathbb{A}\}^*$ on terms is proper, there exists a term T such that $T \bullet w$ is defined. Then, let us put

$$\lambda^*(w) = \min\{\mu(T \bullet w) - \mu(T) \mid T \bullet w \text{ is defined}\}.$$

By definition, λ^* is $\mathbb{N}$-valued. Next we claim that λ^* is $\mathcal{R}_{\text{LD}}$-invariant. Indeed, by Lemma 3.5, if w and w' are $\mathcal{R}_{\text{LD}}$-equivalent Σ-words and T is a term such that $T \bullet w$ is defined, then $T \bullet w'$ is defined as well and it is equal to $T \bullet w'$. It follows that the minimum of $\mu(T \bullet w) - \mu(T)$ for T such that $T \bullet w$ is defined coincides with the minimum of $\mu(T \bullet w') - \mu(T)$ for T such that $T \bullet w'$ is defined, that is, $\lambda^*(w)$ and $\lambda^*(w')$ are equal.

Now let α be an address. Then $\lambda^*(\Sigma_\alpha|w)$ is the minimum of $\mu(T \bullet \Sigma_\alpha|w) - \mu(T)$ for T such that $T \bullet \Sigma_\alpha|w$ is defined. Let T be a term realizing the above minimum. By assumption, $T^\wedge\Sigma_\alpha|w$ is defined, hence, by (1.3), so is $T \bullet \Sigma_\alpha$. Put $T' = T \bullet \Sigma_\alpha$. We find

$$\begin{aligned}
\lambda^*(\Sigma_\alpha|w) &= \mu(T \bullet \Sigma_\alpha|w) - \mu(T)\\
&= \mu(T \bullet \Sigma_\alpha|w) - \mu(T \bullet \Sigma_\alpha) + \mu(T \bullet \Sigma_\alpha) - \mu(T)\\
&= \mu(T' \bullet w) - \mu(T') + \mu(T \bullet \Sigma_\alpha) - \mu(T) \geqslant \lambda^*(w) + 1 > \lambda^*(w).
\end{aligned}$$

Hence λ^* is a right-Noetherianity witness for $(\{\Sigma_\alpha|\alpha{\in}\mathbb{A}\}, \mathcal{R}_{\text{LD}})$. □

Proposition 3.10 (left-cancellativity). *The monoid M_{LD} is left-cancellative, and it admits conditional right-lcms and left-gcds.*

Proof. For the first two properties, owing to Proposition II.4.16 (right-complemented) and to Lemma 3.9, it suffices to show that, for every triple of distinct addresses α, β, γ, the θ-cube condition is true for $\Sigma_\alpha, \Sigma_\beta, \Sigma_\gamma$. This is Proposition VIII.1.9 of [77]. The proof consists in considering all possible mutual positions of the involved addresses. A priori, a high (but finite) number of cases have to be considered, but uniform geometric arguments show that all instances involving the parallel or nested cases in Lemma 3.5 are automatically satisfied. Essentially, the only really nontrivial case is that of a triple $(\alpha, \alpha 1, \alpha 11)$ involving two critical cases. Using θ for the involved syntactic right-complement, we have to check that, if $\theta_3^*(\Sigma_\emptyset, \Sigma_1, \Sigma_{11})$ is defined (which is true), then $\theta_3^*(\Sigma_1, \Sigma_\emptyset, \Sigma_{11})$ is defined as well and is $\mathcal{R}_{\text{LD}}$-equivalent, and similarly when a cyclic rotation is applied. Now, one obtains

$$\begin{aligned}
\theta_3^*(\Sigma_\emptyset, \Sigma_1, \Sigma_{11}) &= \Sigma_{11}|\Sigma_1|\Sigma_\emptyset = \theta_3^*(\Sigma_1, \Sigma_\emptyset, \Sigma_{11}),\\
\theta_3^*(\Sigma_1, \Sigma_{11}, \Sigma_\emptyset) &= \Sigma_\emptyset|\Sigma_1|\Sigma_0|\Sigma_{11}|\Sigma_{01}|\Sigma_{00}\\
&\equiv \Sigma_\emptyset|\Sigma_1|\Sigma_{11}|\Sigma_{10}|\Sigma_0|\Sigma_{01}|\Sigma_{00} = \theta_3^*(\Sigma_{11}, \Sigma_1, \Sigma_\emptyset),\\
\theta_3^*(\Sigma_{11}, \Sigma_\emptyset, \Sigma_1) &= \Sigma_1|\Sigma_\emptyset|\Sigma_{11}|\Sigma_1|\Sigma_{01}|\Sigma_\emptyset\\
&\equiv \Sigma_1|\Sigma_{11}|\Sigma_{10}|\Sigma_\emptyset|\Sigma_1\Sigma_0 = \theta_3^*(\Sigma_\emptyset, \Sigma_{11}, \Sigma_1),
\end{aligned}$$

as expected. Hence, M_{LD} is cancellative and admits conditional right-lcms. The existence of left-gcds then follows using Lemma II.2.37 since M_{LD} is (strongly) Noetherian. □

3.3 The category $\mathcal{LD}$

We are now ready to introduce our main object of interest, namely the refined version of the category $\mathcal{LD}_0$ in which, in addition to an LD-expansion (T, T'), one takes into account an element g of M_{LD} that maps T to T', that is, a distinguished way of expanding T to T'.

Definition 3.11 (category $\mathcal{LD}$). The category $\mathcal{LD}$ of *labeled* LD-*expansions* is the category $\mathcal{C}(M_{\mathrm{LD}}, \mathcal{T})$ associated with the action of M_{LD} on $\mathcal{T}$ in the sense of Definition 1.7.

Thus, the objects of $\mathcal{LD}$ are terms, and the elements are triples $T \xrightarrow{g} T'$ with g in M_{LD} and $T \bullet g = T'$. The source of $T \xrightarrow{g} T'$ is T, and its target is T'. For instance, a typical element of $\mathcal{LD}$ is the triple

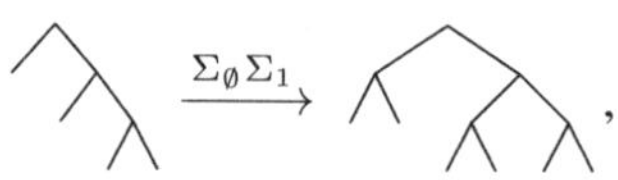

whose source is the term $x^\wedge(x^\wedge(x^\wedge x))$—we adopt the convention that an unspecified variable means some fixed variable x—and whose target is $(x^\wedge x)^\wedge((x^\wedge x)^\wedge(x^\wedge x))$. Two natural projections apply to $\mathcal{LD}$: forgetting the central entry of the triples gives a surjective functor from $\mathcal{LD}$ onto $\mathcal{LD}_0$, whereas forgetting the first and third entries gives a surjective homomorphism from $\mathcal{LD}$ onto the monoid M_{LD}.

Proposition 3.12 (left-cancellativity). *The category $\mathcal{LD}$ is left-cancellative and admits conditional right-lcms and left-gcds.*

Proof. Assume that $T \xrightarrow{g} T' \xrightarrow{h} T'' = T \xrightarrow{g} T' \xrightarrow{h'} T''$ holds in $\mathcal{LD}$. Then, by definition, we have $gh = gh'$ in M_{LD}, whence $h = h'$ by Proposition 3.10, and $T' \xrightarrow{h} T'' = T' \xrightarrow{h'} T''$. Therefore, $\mathcal{LD}$ is left-cancellative.

Assume now that $T \xrightarrow{g} \cdot$ and $T \xrightarrow{g'} \cdot$ admit a common right-multiple in $\mathcal{LD}$. By definition, this common right-multiple is of the form $T \xrightarrow{h} \cdot$, where h is a common right-multiple of g and g'. By Proposition 3.10, g and g' must admit a right-lcm, say g''. Then $T \xrightarrow{g''} \cdot$ is easily seen to be a right-lcm of $T \xrightarrow{g} \cdot$ and $T \xrightarrow{g'} \cdot$ in $\mathcal{LD}$. The argument for left-gcds is similar. $\square$

We shall now prove that the category $\mathcal{LD}$ admits a natural right-Garside map Δ. As can be expected, $\Delta(T)$ will be constructed as a labeled version of $\Delta_0(T)$, that is, of $(T, \phi(T))$. This amounts to fixing some canonical way of expanding a term T into the corresponding term $\phi(T)$. To this end, the natural solution is to follow the recursive definition of the operations $\wedge\!\!\!\wedge$ and ϕ.

For w a Σ-word, we denote by $\mathrm{sh}_0(w)$ the word obtained by replacing each letter Σ_α of w with the corresponding letter $\Sigma_{0\alpha}$, that is, by *shifting* all indices by 0. Similarly, we denote by $\mathrm{sh}_\gamma(w)$ the word obtained by appending γ on the left of each address in w. The

LD-relations of Lemma 3.5 are invariant under shifting: if w and w' represent the same element g of M_{LD}, then, for each γ, the words $\text{sh}_\gamma(w)$ and $\text{sh}_\gamma(w')$ represent the same element, naturally denoted by $\text{sh}_\gamma(g)$, of M_{LD}. By construction, sh_γ is an endomorphism of the monoid M_{LD}. For every g in M_{LD}, the action of $\text{sh}_\gamma(g)$ on a term T corresponds to the action of g on the γ-subterm of T: so, for instance, if $T' = T \bullet g$ holds, then $T'{}^\wedge T_1 = (T^\wedge T_1) \bullet \text{sh}_0(g)$ holds as well, since the 0-subterm of $T^\wedge T_1$ is T, whereas that of $T'{}^\wedge T_1$ is T'.

Definition 3.13 (elements δ_T and Δ_T). For T a term, the elements δ_T and Δ_T of M_{LD} are defined by $\delta_T = \Delta_T = 1$ if T is a variable and, recursively, for $T = T_0{}^\wedge T_1$,

$$\delta_T = \Sigma_\emptyset \cdot \text{sh}_0(\delta_{T_0}) \cdot \text{sh}_1(\delta_{T_1}), \tag{3.14}$$

$$\Delta_T = \text{sh}_0(\Delta_{T_0}) \cdot \text{sh}_1(\Delta_{T_1}) \cdot \delta_{\phi(T_1)}. \tag{3.15}$$

Example 3.16 (elements δ_T and Δ_T). Let $T = x^\wedge(x^\wedge(x^\wedge x))$. Then we have $T_{/0} = x$, whence $\Delta_{T_{/0}} = 1$. Next, $T_{/1}$ is $x^\wedge(x^\wedge x)$, so (3.15) reads $\Delta_T = \text{sh}_1(\Delta_{T_{/1}}) \cdot \delta_{\phi(T_{/1})}$. Then $\phi(T_{/1})$ is $(x^\wedge x)^\wedge(x^\wedge x)$, so applying (3.14),we obtain

$$\delta_{\phi(T_{/1})} = \Sigma_\emptyset \cdot \text{sh}_0(\delta_{x^\wedge x}) \cdot \text{sh}_1(\delta_{x^\wedge x}) = \Sigma_\emptyset \Sigma_0 \Sigma_1.$$

On the other hand, using (3.15) again, we find the equalities

$$\Delta_{T_{/1}} = \text{sh}_0(\Delta_x) \cdot \text{sh}_1(\Delta_{x^\wedge x}) \cdot \delta_{x^\wedge x} = 1 \cdot 1 \cdot \Sigma_\emptyset = \Sigma_\emptyset,$$

and, finally, $\Delta_T = \Sigma_1 \Sigma_\emptyset \Sigma_0 \Sigma_1$. According to the defining relations of the monoid M_{LD}, this element is also $\Sigma_\emptyset \Sigma_1 \Sigma_\emptyset$. Note the compatibility with the examples of Figures 2 and 4.

Lemma 3.17. *For all terms T_0, T, we have*

$$(T_0{}^\wedge T) \bullet \delta_T = T_0{}^{\wedge\!\!\wedge} T, \tag{3.18}$$

$$T \bullet \Delta_T = \phi(T). \tag{3.19}$$

The proof is an easy inductive verification.

According to Lemma 2.10, if T is a term and $T \bullet \Sigma_\alpha$ is defined, the term $\phi(T)$ is an LD-expansion of $T \bullet \Sigma_\alpha$, and the term $\phi(T \bullet \Sigma_\alpha)$ is an LD-expansion of $\phi(T)$. This suggests—but does *not* prove—that, in this case, Σ_α left-divides Δ_T, and Δ_T left-divides $\Sigma_\alpha \, \Delta_{T \bullet \Sigma_\alpha}$ in M_{LD}. The latter statements can actually be proved, that is, one can get a syntactic counterpart to Lemma 2.10 at the level of the monoid M_{LD}.

Lemma 3.20. [77, Lemmas VII.3.16 and VII.3.17] *If T is a term and $T \bullet \Sigma_\alpha$ is defined, then, in M_{LD}, we have*

$$\Sigma_\alpha \preccurlyeq \Delta_T \preccurlyeq \Sigma_\alpha \cdot \Delta_{T \bullet \Sigma_\alpha}. \tag{3.21}$$

Proof (*sketch*). This computation is the key for the existence of the Garside structure we are currently describing. The proof uses several inductive arguments both on the length of α and on the size of T. Here, we shall establish (3.21) in the base case, namely for $\alpha = \emptyset$ and $T = T_0{}^\wedge(T_1{}^\wedge x)$. Let $T' = T \bullet \Sigma_\emptyset = (T_0{}^\wedge T_1)^\wedge(T_0{}^\wedge x)$. Applying the definitions gives

$$\Delta_T = \text{sh}_0(\Delta_{T_0}) \cdot \text{sh}_{10}(\Delta_{T_1}) \cdot \Sigma_\emptyset \cdot \text{sh}_0(\delta_{\phi(T_1)}) \tag{3.22}$$

(3.23) $\Sigma_\emptyset{\cdot}\Delta_{T'} = \Sigma_\emptyset{\cdot}\mathrm{sh}_{00}(\Delta_{T_0}){\cdot}\mathrm{sh}_{01}(\Delta_{T_1}){\cdot}\mathrm{sh}_0(\delta_{\phi(T_1)}){\cdot}\mathrm{sh}_{10}(\Delta_{T_0}){\cdot}\Sigma_\emptyset{\cdot}\mathrm{sh}_0(\delta_{\phi(T_0)})$.

First, we observe that, in (3.22), the factor $\Sigma_\emptyset$ can be moved to the left using nested case relations of type 1 and 2, yielding

$$\Delta_T = \Sigma_\emptyset \cdot \mathrm{sh}_{00}(\Delta_{T_0}) \cdot \mathrm{sh}_{10}(\Delta_{T_0}) \cdot \mathrm{sh}_{01}(\Delta_{T_1}) \cdot \mathrm{sh}_0(\delta_{\phi(T_1)}) :$$

this shows that $\Sigma_\emptyset$ left-divides Δ_T. Similarly, in (3.23), we can move the factor $\mathrm{sh}_{10}(\Delta_{T_0})$ to the left and the factor $\Sigma_\emptyset$ to the right to obtain

$$\Sigma_\emptyset \cdot \Delta_{T'} = \mathrm{sh}_0(\Delta_{T_0}) \cdot \mathrm{sh}_{10}(\Delta_{T_1}) \cdot \Sigma_\emptyset \cdot \mathrm{sh}_0(\delta_{\phi(T_1)}) \cdot \Sigma_\emptyset \cdot \mathrm{sh}_0(\delta_{\phi(T_0)}),$$

which shows the equality $\Sigma_\emptyset\, \Delta_{T'} = \Delta_T\, \Sigma_\emptyset\, \mathrm{sh}_0(\delta_{\phi(T_0)})$. Hence Δ_T left-divides $\Sigma_\emptyset\, \Delta_{T'}$.

The general cases are similar. Full details appear in [77, Section VII.3]. □

Proposition 3.24 (right-Garside sequence). *The sequence $(\Delta_T)_{T\in\mathcal{T}}$ is a right-Garside sequence in the monoid M_{LD} with respect to its partial action on terms via self-distributivity.*

Proof. First, Lemma 3.17 guarantees that, for every term T, the term $T \bullet \Delta_T$ is defined, so the sequence $(\Delta_T)_{T\in\mathcal{T}}$ satisfies (1.11).

Next, by definition, the monoid M_{LD} is generated by the elements Σ_α. Now, for every address α, there exists a term T such that $T \bullet \Sigma_\alpha$ is defined. Then, by Lemma 3.20, Σ_α left-divides $\Delta(T)$, hence it belongs to $\mathcal{D}iv(\Delta)$. It follows that $\mathcal{D}iv(\Delta)$ generates $\mathcal{LD}$, and the sequence $(\Delta_T)_{T\in\mathcal{T}}$ satisfies (1.12).

Then, assume that $T \bullet \Sigma_\alpha$ is defined. According to (3.21), $\Delta_T \preccurlyeq \Sigma_\alpha\, \Delta_{T\bullet\Sigma_\alpha}$ holds in M_{LD}, which means that the sequence $(\Delta_T)_{T\in\mathcal{T}}$ satisfies (1.13).

As for the existence of a left-gcd for g and Δ_T when $T \bullet g$ is defined, it is automatic, as, by Proposition 3.10, any two elements of M_{LD} with the same source admit a left-gcd. So the sequence $(\Delta_T)_{T\in\mathcal{T}}$ satisfies (1.14).

Finally, assume that $T \bullet g$ and $T' \bullet g$ are defined and that $g \preccurlyeq \Delta_T$ holds. We claim that $g \preccurlyeq \Delta_{T'}$ must hold as well. Let T_0 be an injective term with the same skeleton as T. By Lemma 3.4(ii), the assumption that $T \bullet g$ is defined implies that $T_0 \bullet g$ is defined as well, and there exists a substitution σ satisfying $T_0 = (T_g^-)^\sigma$. As T_0 is injective and g left-divides Δ_T, which is also Δ_{T_0}, Proposition 2.18 implies that $T_0 \bullet g$ is semi-injective. Now, we have $T_0 \bullet g = (T_g^+)^\sigma$, and $T_0 \bullet g$ being semi-injective implies that T_g^+ must be semi-injective as well.

Now, let T_0' be an injective term with the same skeleton as T'. Then there exists a substitution σ' satisfying $T_0' = (T_g^-)^{\sigma'}$. One easily checks that the conjunction of $(T_g^-)^{\sigma'}$ being injective and T_g^+ being semi-injective implies that $(T_g^+)^{\sigma'}$ is semi-injective. So $T_0' \bullet g$ is semi-injective. By Proposition 2.18, this implies that g left-divides $\Delta_{T_0'}$, which is also $\Delta_{T'}$. Hence the sequence $(\Delta_T)_{T\in\mathcal{T}}$ satisfies (1.15), and it is a right-Garside sequence. □

Proposition 3.25 (Garside map). *For every term T, put $\Delta(T) = T \xrightarrow{\Delta_T} \phi(T)$. Then Δ is a target-injective right-Garside map in the category $\mathcal{LD}$.*

Proof. We apply Proposition 1.18 to the monoid M_{LD} and the right-Garside sequence $(\Delta_T)_{T\in\mathcal{T}}$. The property that Δ is target-injective, that is, ϕ is injective on terms, follows from Lemma 2.10(iii). □

Thus the category $\mathcal{LD}$ is an example of a left-cancellative category that admits a natural and nontrivial right-Garside map. As in the case of the category $\mathcal{LD}_0$, the left and the right sides do not play symmetric roles and the right-Garside map Δ is not a Garside map. Indeed, the map ϕ is not surjective on terms: for instance, no term T satisfies $\phi(T) = x^\wedge(x^\wedge x)$, and, therefore, $(x^\wedge(x^\wedge x), 1, x^\wedge(x^\wedge x))$ is an element of $\mathcal{D}iv(\Delta)$ that cannot lie in $\widetilde{\mathcal{D}iv}(\Delta)$. More generally, the proportion of terms of a given size that lie in the image of ϕ is exponentially small, so the lack of surjectivity of ϕ is by no means a marginal phenomenon.

Corollary 3.26 (right-lcm). (i) *The category $\mathcal{LD}$ admits right-lcms and left-gcds.*
(ii) *The monoid M_{LD} admits right-lcms and left-gcds.*

Proof. (i) Since $\mathcal{LD}$ admits a right-Garside map, it is eligible for Corollary V.1.48 (common right-multiple), so any two elements of $\mathcal{LD}$ with the same source admit a common right-multiple. On the other hand, by Proposition 3.12, any two elements of $\mathcal{LD}$ that admit a common right-multiple admit a right-lcm and a left-gcd. Hence, any two elements of $\mathcal{LD}$ with the same source admit a right-lcm and a left-gcd.

For (ii), we repeat the argument using Proposition 3.10 instead of Proposition 3.12 or, equivalently, using Lemma 1.9 and its counterpart involving right-lcms. □

3.4 The Embedding Conjecture

From the viewpoint of self-distributive algebra, the main benefit of the current approach might be that it leads to a natural program for possibly establishing the so-called Embedding Conjecture. This conjecture, at the moment the most puzzling open question involving free LD-systems, can be stated in several equivalent forms.

Proposition 3.27 (equivalence). [77, Section IX.6] *The following are equivalent:*
(i) *The monoid M_{LD} embeds in a group.*
(ii) *The monoid M_{LD} is right-cancellative.*
(iii) *The category $\mathcal{LD}$ is right-cancellative.*
(iv) *The functor ϕ_Δ is injective on $\mathcal{LD}$.*
(v) *The answer to Question 3.6 is positive.*
(vi) *The categories $\mathcal{LD}_0$ and $\mathcal{LD}$ are isomorphic.*

Moreover, if (i)–(vi) *are true, then so is*

(vii) *The answer to Question* 2.11 *is positive.*

Conjecture 3.28 (Embedding Conjecture). *The equivalent statements of Proposition* 3.27 *are true.*

Because of its importance, we shall give an almost complete proof of Proposition 3.27. First, we need a preliminary result about $\mathcal{R}_{\mathrm{LD}}$-equivalence of signed Σ-words, that is, words in the alphabet $\{\Sigma_\alpha | \alpha \in \mathbb{A}\} \cup \{\overline{\Sigma_\alpha} | \alpha \in \mathbb{A}\}$. We recall from Section II.3 that, for w a signed $\mathcal{S}$-word (or path), $\overline{w}$ denotes the signed word (or path) obtained from w by exchanging s and $\overline{s}$ everywhere and reversing the order of letters.

Lemma 3.29. *For T a term of $\mathcal{T}_1$, define signed Σ-words $\chi_T, \hat{\chi}_T$ by $\chi_T = \hat{\chi}_T = \varepsilon$ for $T = x_1$ and, for $T = T_0{}^\wedge T_1$,*

$$\chi_T = \chi_{T_0} | \mathrm{sh}_1(\chi_{T_1}) | \Sigma_\emptyset | \mathrm{sh}_1(\overline{\chi_{T_0}}) \quad \textit{and} \quad \hat{\chi}_T = \chi_{T_0} | \mathrm{sh}_1(\hat{\chi}_{T_1}).$$

Then, if w is a signed Σ-word and $T \bullet w$ is defined, $T' = T \bullet w$ implies $\hat{\chi}_{T'} \equiv_{\mathcal{R}_{\mathrm{LD}}} \hat{\chi}_T\, w$.

Proof (*sketch*). For an induction it is sufficient to establish the result when w has length one and is positive, say $w = \Sigma_\alpha$. Then the proof is a verification. □

Proof of Proposition 3.27. If the monoid M_{LD} embeds in a group, it must be cancellative. Conversely, by Proposition 3.10, the monoid M_{LD} is left-cancellative, and, by Proposition 1.21, which is valid as $(\Delta_T)_{T \in \mathcal{T}}$ is a right-Garside sequence on M_{LD}, any two elements of M_{LD} admit a common right-multiple. So, if M_{LD} is right-cancellative, it satisfies the right Ore conditions, and embeds in a group of right-fractions. We see that (i) and (ii) are equivalent.

Next, as $\mathcal{LD}$ is $\mathcal{C}(M_{\mathrm{LD}}, F)$, (ii) and (iii) are equivalent by (the right counterpart) of Lemma 1.8.

Then, by Proposition 3.25, $\mathcal{LD}$ is a left-cancellative category, $\mathcal{D}iv(\Delta)$ is a Garside family of $\mathcal{LD}$ that is right-bounded by Δ, and Δ is target-injective. Hence, by Proposition V.1.36 (right-cancellative I), $\mathcal{LD}$ is right-cancellative if and only if ϕ_Δ is injective on $\mathcal{LD}$. So (iii) and (iv) are equivalent.

Assume now that w, w' are Σ-words and we have $T \bullet w = T \bullet w'$ for some term T. Then, by Lemma 3.29, the signed Σ-words $\hat{\chi}_T\, w$ and $\hat{\chi}_T\, w'$ both are $\mathcal{R}_{\mathrm{LD}}$-equivalent to $\hat{\chi}_{T \bullet w}$. By left-cancelling $\hat{\chi}_T$, we deduce $w \equiv_{\mathcal{R}_{\mathrm{LD}}} w'$. Now, if (i) is true, the monoid M_{LD} embeds in its enveloping group, and, therefore, $w \equiv_{\mathcal{R}_{\mathrm{LD}}} w'$ implies $w \equiv^+_{\mathcal{R}_{\mathrm{LD}}} w'$, that is, w and w' represent the same element of M_{LD}. This means that the answer to Question 3.6 is positive. Hence (i) implies (v).

On the other hand, if the answer to Question 3.6 is positive, then the categories $\mathcal{LD}_0$ and $\mathcal{LD}$ are isomorphic, that is, (v) implies (vi). Indeed, in this case, for every pair (T, T')

in $\mathcal{LD}_0$, that is, every pair such that T' is an LD-expansion of T, there exists a unique element g of M_{LD} satisfying $T' = T \bullet g$ and, therefore, the projection of $\mathcal{LD}$ onto $\mathcal{LD}_0$ is injective.

And (vi) implies (iii) since, by Proposition 2.13, the category $\mathcal{LD}_0$ is, in every case, right-cancellative. So the equivalence of (i)–(vi) is established.

Finally, according to Corollary 3.26, any two elements of $\mathcal{LD}$ sharing the same source admit a right-lcm and a left-gcd. If (vi) is true, the same properties hold in the category $\mathcal{LD}_0$. This means, in particular, that, if T', T'' are two LD-expansions of some term T, the elements (T, T') and (T, T'') of $\mathcal{LD}_0$ admit a right-lcm and a left-gcd, that is, the terms T' and T'' admit a least upper bound and a greatest lower bound with respect to $\leqslant_{\text{LD}}$. So $(\mathcal{T}, \leqslant_{\text{LD}})$ is a lattice, and the answer to Question 2.11 is positive. □

At the moment, the Embedding Conjecture remains open, although a number of partial positive statements have been established. One of the reasons for emphasizing the Garside structure involved in the framework is to suggest a possible approach for establishing the conjecture. The method relies on the following technical result.

Lemma 3.30. *The functor ϕ_Δ is injective on $\mathcal{D}iv(\Delta)$.*

Proof (sketch). Assume that $T \xrightarrow{g} \cdot$ left-divides $\Delta(T)$ and that we have the equality $\phi_\Delta(T \xrightarrow{g} \cdot) = \phi_\Delta(T' \xrightarrow{g'} \cdot)$. Proposition 1.22 gives $\phi(T) = \phi(T')$ and $\phi(T \bullet g) = \phi(T' \bullet g')$, whence, by Lemma 2.10(ii), $T = T'$ and $T \bullet g = T' \bullet g'$, that is, $T \bullet g = T \bullet g'$. By [77, Proposition IX.6.6], the latter equality implies $g = g'$ provided g left-divides Δ_T in M_{LD}. □

Then we deduce

Lemma 3.31. *Assume that the functor ϕ_Δ preserves Δ-normality. Then the Embedding Conjecture is true.*

Proof. We apply Proposition V.1.59 (right-cancellative II). Indeed, $\mathcal{LD}$ is a left-cancellative category, and $\mathcal{D}iv(\Delta)$ is a Garside family in $\mathcal{LD}$ that is right-bounded by Δ. By Proposition 3.25, Δ is target-injective. Moreover, as the relation $\leqslant_{\text{LD}}$ is a partial ordering, the category $\mathcal{LD}$ contains no non-trivial invertible element. By Proposition V.1.59, the assumption that ϕ_Δ preserves Δ-normality together with the result of Lemma 3.30 that ϕ_Δ is injective on $\mathcal{D}iv(\Delta)$ imply that $\mathcal{LD}$ is right-cancellative. According to Proposition 3.27, this is one of the forms of the Embedding Conjecture. □

We are thus left with the question of proving that the functor ϕ_Δ preserves normality.

Lemma 3.32. *Assume that $\mathcal{C}$ is a left-cancellative category and Δ is a right-Garside map in $\mathcal{C}$. Then a sufficient condition for ϕ_Δ to preserve normality is that $\phi_\Delta(s)$ and $\phi_\Delta(t)$ are left-coprime whenever s and t are left-coprime elements of $\mathcal{D}iv(\Delta)$.*

Proof. Assume that $s_1|s_2$ is an $\mathcal{S}$-normal sequence. By Proposition V.1.53 (recognizing greedy), the elements $\partial_\Delta s_1$ and s_2 are left-coprime. Hence, by assumption, so are $\phi_\Delta(\partial_\Delta s_1)$ and $\phi_\Delta(s_2)$. By (V.1.30), we have $\phi_\Delta(\partial_\Delta s_1) = \partial_\Delta(\phi_\Delta(s_1))$, so $\partial_\Delta(\phi_\Delta s_1)$ and $\phi_\Delta(s_2)$ are left-coprime and, by Proposition V.1.53 again, we deduce that $\phi_\Delta(s_1)|\phi_\Delta(s_2)$ is $\mathcal{S}$-greedy, hence $\mathcal{S}$-normal. So ϕ_Δ preserves Δ-normality. □

In the case of the category $\mathcal{LD}$ and the right-Garside map Δ, the functor ϕ_Δ takes the specific form described in Proposition 1.22: for every term T, there exists a partial map ϕ_T on M_{LD} such that $\phi_T(g)$ is defined if and only if $T \bullet g$ is defined and, in this case, we have $\phi_\Delta(T \xrightarrow{g} T') = \phi(T) \xrightarrow{\phi_T(g)} \phi(T')$. Then the criterion of Lemma 3.32 directly leads to the following result:

Proposition 3.33 (Embedding Conjecture). *A sufficient condition for the Embedding Conjecture to be true is:*

(3.34) *If s, t are left-coprime simple elements of M_{LD} and T is a term such that $T \bullet s$ and $T \bullet t$ are defined, then $\phi_T(s)$ and $\phi_T(t)$ are left-coprime as well.*

Example 3.35 (Embedding Conjecture). Assume $s = \Sigma_\emptyset$, $t = \Sigma_1$, and let T be the term $x^\wedge(x^\wedge(x^\wedge x))$. Then $T \bullet s$ and $T \bullet t$ are defined, s and t are simple, and they are left-coprime. On the other hand, we have $\phi(T) = ((x^\wedge x)^\wedge(x^\wedge x))^\wedge((x^\wedge x)^\wedge(x^\wedge x))$. An easy computation gives $\phi_T(\Sigma_\emptyset) = \Sigma_0\Sigma_1$ and $\phi_T(\Sigma_1) = \Sigma_\emptyset$, see Figure 5. Then $\phi_T(s)$ and $\phi_T(t)$ are left-coprime, and the corresponding instance of (3.34) is true, see Figure 5.

Remark 3.36. Relation (3.34) would follow from the more general relation that ϕ_T preserves left-gcds, in the natural sense that, whenever defined, the left-gcd of $\phi_T(s)$ and $\phi_T(t)$ is the image under ϕ_T of the left-gcd of s and t. We have no counter-example to the latter statement, but we can note here that the counterpart involving right-lcms fails. Indeed, for $s = \Sigma_\emptyset$, $t = \Sigma_1$, and $T = x^\wedge(x^\wedge(x^\wedge x))$ as in Example 3.35, the right-lcm of $\phi_T(s)$ and $\phi_T(t)$ is a proper right-multiple of the ϕ_T-image of the right-lcm of s and t, namely the product of the latter by $\Sigma_0\Sigma_1$. This corresponds to the fact that the terms $\phi(T \bullet \Sigma_\emptyset)$ and $\phi(T \bullet \Sigma_1)$ admit a common LD-expansion that is smaller than $\phi_T(T \bullet \mathrm{lcm}(\Sigma_\emptyset, \Sigma_1))$, which turns out to be $\phi^2(T)$, see Figure 5 again.

4 Connection with braids

There exists a deep, multiform connection between left self-distributivity and braids. Among others, this connection includes orderability properties (using orderable LD-systems to color braids) and topology (knot invariants based on quandles), which are beyond the scope of this text. Here we shall concentrate on the algebraic side. The connection can then be summarized in the unique result that the Garside structure of braids as described in the category $\mathcal{C}(B_\infty^+, \mathbb{N})$ and the associated (right)-Garside map Δ is a projection of the Garside structure involving the category $\mathcal{LD}$ and the associated right-Garside map Δ. It follows in particular that a number of algebraic results involving braids can be deduced from the results about the LD law that were established or mentioned above.

The section is organized as follows. In Subsection 4.1, we explain the connection between the categories $\mathcal{LD}$ and $\mathcal{C}(B_\infty^+, \mathbb{N})$. Then, in Subsection 4.2, we show how the results

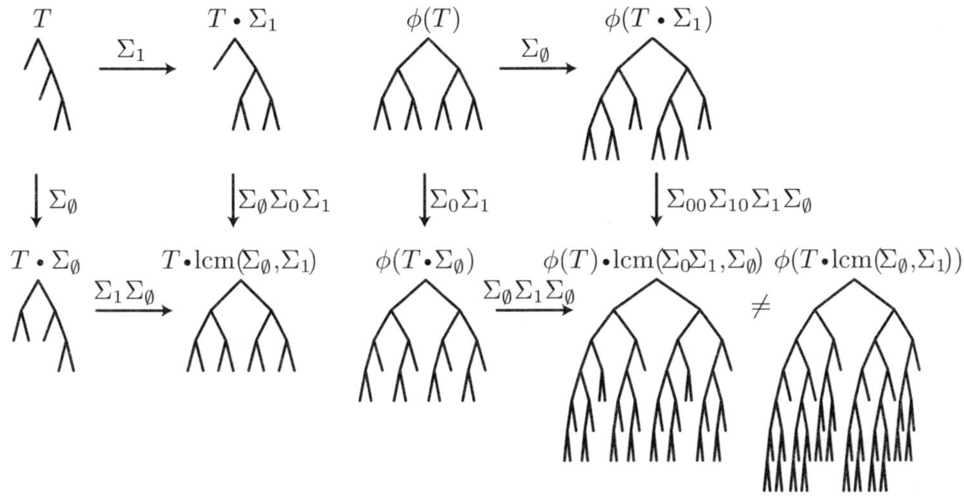

Figure 5. An instance of Relation (3.34): $\Delta_T = \Sigma_\emptyset\Sigma_1\Sigma_\emptyset$, $\Delta_{T\bullet\Sigma_1} = \Sigma_1\Sigma_\emptyset\Sigma_0\Sigma_{00}\Sigma_1\Sigma_{10}$, leading to $\phi_T(\Sigma_1) = \Sigma_\emptyset$ and $\phi_T(\Sigma_\emptyset) = \Sigma_0\Sigma_1$. Here $\phi_T(\Sigma_\emptyset)$ and $\phi_T(\Sigma_1)$ are left-coprime, so (3.34) is true. The right diagram shows that the counterpart involving right-lcms fails.

about the LD law can be used to (re)prove results about braids. Finally, in Section 4.3, we describe intermediate categories that naturally arise between $\mathcal{LD}$ and $\mathcal{C}(B_\infty^+, \mathbb{N})$ in connection with the Hurwitz action of braids on sequences from an LD-system.

4.1 The main projection

We wish to define a projection from the world of left self-distributivity onto the world of braids. The first step is to define a projection at the levels of the monoids M_{LD} and B_∞^+.

Lemma 4.1. *Define* $\pi : \{\Sigma_\alpha \mid \alpha \in \mathbb{A}\} \to \{\sigma_i \mid i \geqslant 1\} \cup \{1\}$ *by*

$$\pi(\Sigma_\alpha) = \begin{cases} \sigma_{i+1} & \text{if } \alpha \text{ is the address } 1^i, \text{ that is, } 11\cdots 1,\ i \text{ times } 1, \\ 1 & \text{otherwise.} \end{cases} \tag{4.2}$$

Then π *induces a surjective homomorphism of the monoid* M_{LD} *onto the monoid* B_∞^+.

Proof. The point is that every relation of $\mathcal{R}_{\mathrm{LD}}$ projects under π onto a braid equivalence. All relations involving addresses that contain at least one 0 collapse to equalities. The remaining relations are

$$\Sigma_{1^i}\Sigma_{1^j} = \Sigma_{1^j}\Sigma_{1^i} \quad \text{with } j \geqslant i+2,$$

which projects to the valid braid relation $\sigma_{i+1}\sigma_{j+1} = \sigma_{j+1}\sigma_{i+1}$, and

$$\Sigma_{1^i}\Sigma_{1^j}\Sigma_{1^i} = \Sigma_{1^j}\Sigma_{1^i}\Sigma_{1^j}\Sigma_{1^i0}, \quad \text{with } j = i+1,$$

which projects to the not less valid braid relation $\sigma_{i+1}\sigma_{j+1}\sigma_{i+1} = \sigma_{j+1}\sigma_{i+1}\sigma_{j+1}$. ☐

The second step is to go to the level of the categories $\mathcal{LD}$ and $\mathcal{C}(B_\infty^+, \mathbb{N})$, that is, to take into account the partial actions of the monoids M_{LD} and B_∞^+ on terms and on positive integers, respectively. The point is that these partial actions are compatible.

Definition 4.3 (compatible). If F and F' are partial actions of monoids M and M' on sets X and X', respectively, a morphism $\varphi: M \to M'$ is said to be *compatible* with a map $\psi: X \to X'$ with respect to F and F' if

$$\psi(x \bullet g) = \psi(x) \bullet \varphi(g) \tag{4.4}$$

holds whenever $x \bullet g$ is defined. Then, we denote by $\mathcal{C}(\varphi, \psi)$ the functor from $\mathcal{C}_F(M,X)$ to $\mathcal{C}(M', F')$ that coincides with ψ on objects and maps $x \xrightarrow{g} \cdot$ to $\psi(x) \xrightarrow{\varphi(g)} \cdot$.

Lemma 4.5. *The morphism* π *of* (4.2) *is compatible with the* right height *of terms, inductively defined by* $\mathrm{ht}_{\mathrm{R}}(x_i) = 0$ *and* $\mathrm{ht}_{\mathrm{R}}(T_0{}^\wedge T_1) = \mathrm{ht}_{\mathrm{R}}(T_1) + 1$, *and* $\mathcal{C}(\pi, \mathrm{ht}_{\mathrm{R}})$ *is a surjective functor from* $\mathcal{LD}$ *onto* $\mathcal{C}(B_\infty^+, \mathbb{N})$.

The number $\mathrm{ht}_{\mathrm{R}}(T)$ is the length of the rightmost branch in T viewed as a tree or, equivalently, the number of final closing parentheses in T viewed as a bracketed expression.

Proof. Assume that $T \xrightarrow{g} T'$ belongs to $\mathcal{LD}$. Put $n = \mathrm{ht}_{\mathrm{R}}(T)$. The LD law preserves the right height of terms, so we have $\mathrm{ht}_{\mathrm{R}}(T') = n$ as well. The assumption that $T \bullet g$ exists implies that the factors Σ_{1^i} that occur in some (hence in every) expression of g satisfy $i < n - 1$. Hence $\pi(g)$ is a braid of B_n^+, and $n \bullet \pi(g)$ is defined. Then the compatibility condition (4.4) is clear, and $\mathcal{C}(\pi, \mathrm{ht}_{\mathrm{R}})$ is a functor of $\mathcal{LD}$ to $\mathcal{C}(B_\infty^+, \mathbb{N})$.

Surjectivity is clear, as every braid σ_i belongs to the image of π. □

Finally, a simple relation connects the elements Δ_T of M_{LD} and the braids Δ_n.

Lemma 4.6. *We have* $\pi(\Delta_T) = \Delta_n$ *whenever* T *has right height* $n \geqslant 1$.

Proof. We first prove that $\mathrm{ht}_{\mathrm{R}}(T) = n$ implies

$$\pi(\delta_T) = \sigma_1 \sigma_2 \cdots \sigma_n \tag{4.7}$$

using induction on the size of T. If T is a variable, we have $\mathrm{ht}_{\mathrm{R}}(T) = 0$ and $\delta_T = 1$, so the equality is clear. Otherwise, write $T = T_0{}^\wedge T_1$. By definition, we have

$$\delta_T = \Sigma_\emptyset \cdot \mathrm{sh}_0(\delta_{T_0}) \cdot \mathrm{sh}_1(\delta_{T_1}).$$

Let sh denote the endomorphism of B_∞^+ that maps σ_i to σ_{i+1} for each i. Then π collapses every term in the image of sh_0, and $\pi(\mathrm{sh}_1(g)) = \mathrm{sh}(\pi(g))$ holds for each g in M_{LD}. Hence, using the induction hypothesis $\pi(\delta_{T_1}) = \sigma_1 \cdots \sigma_{n-1}$, we deduce

$$\pi(\delta_T) = \sigma_1 \cdot 1 \cdot \mathrm{sh}(\sigma_1 \cdots \sigma_{n-1}) = \sigma_1 \cdots \sigma_n,$$

which is (4.7).

Put $\Delta_0 = 1$ (= Δ_1). We prove that $\mathrm{ht}_{\mathrm{R}}(T) = n$ implies $\pi(\Delta_T) = \Delta_n$ for $n \geqslant 0$, using induction on the size of T again. If T is a variable, we find $n = 0$ and $\Delta_T = 1$, as expected. Otherwise, write $T = T_0{}^\wedge T_1$. We have $\Delta_T = \mathrm{sh}_0(\Delta_{T_0}) \cdot \mathrm{sh}_1(\Delta_{T_1}) \cdot \delta_{\phi(T_1)}$ by definition. As above, π collapses the term in the image of sh_0, and it transforms sh_1 into sh. Hence, using the induction hypothesis $\pi(\Delta_{T_1}) = \Delta_{n-1}$ and (4.7) for $\phi(T_1)$, whose right height is that of T_1, we finally obtain $\pi(\Delta_T) = 1 \cdot \mathrm{sh}(\Delta_{n-1}) \cdot \sigma_1\sigma_2\cdots\sigma_{n-1}$, whence $\pi(\Delta_T) = \Delta_n$. □

Summarizing the results, we obtain

Proposition 4.8 (projection). *The functor $\mathcal{C}(\pi, \mathrm{ht}_\mathrm{R})$ is a surjective functor from the category $\mathcal{LD}$ onto the category $\mathcal{C}(B_\infty^+, \mathbb{N})$, and it maps the right-Garside map Δ of $\mathcal{LD}$ to the right-Garside map Δ of $\mathcal{C}(B_\infty^+, \mathbb{N})$, in the sense that, for every term T, we have*

$$\mathcal{C}(\pi, \mathrm{ht}_\mathrm{R})(\Delta(T)) = \Delta(\pi(T)). \tag{4.9}$$

In other words, the Garside structure of braids is a projection of the Garside structure associated with the LD law.

4.2 Reproving braid properties

One benefit of the Garside structure on the category $\mathcal{LD}$ is that it allows for proving specific results about the LD law, such as solving the associated Word Problem or addressing the Embedding Conjecture. In addition, it provides new proofs of some algebraic results involving braids as a direct application of the projection of Subsection 4.1.

If $S, \underline{S}$ are alphabets and π is a map of S to the free monoid $\underline{S}^*$, we denote by π^* the alphabetical homomorphism of S^* to $\underline{S}^*$ that extends π, that is, the homomorphism defined by $\pi^*(\varepsilon) = \varepsilon$ and $\pi^*(s_1|\cdots|s_\ell) = \pi(s_1)|\cdots|\pi(s_\ell)$.

Lemma 4.10. *Assume that (S, R) and $(\underline{S}, \underline{R})$ are right-complemented presentations of two monoids $M, \underline{M}$, respectively associated with syntactic right-complements θ and $\underline{\theta}$, and π is a map from S to $\underline{S} \cup \{\varepsilon\}$ satisfying $\pi(S) \supseteq \underline{S}$ and*

$$\forall s, t {\in} S\ (\underline{\theta}(\pi(s), \pi(t)) = \pi(\theta(s, t))). \tag{4.11}$$

Assume moreover that M admits common right-multiples.

(i) *The map π^* induces a surjective homomorphism from M onto $\underline{M}$.*

(ii) *The monoid $\underline{M}$ admits common right-multiples.*

(iii) *If right-reversing is complete for (S, R), it is complete for $(\underline{S}, \underline{R})$ too.*

Proof. (i) By definition, the relations $s\theta(s, t) = t\theta(t, s)$ with s, t in S make a presentation of M. Now, for s, t in S, we have

$$\pi(s)\underline{\theta}(\pi(s), \pi(t)) = \pi(s\theta(s, t)) = \pi(t\theta(t, s)) = \pi(t)\theta(\pi(t), \pi(s))$$

in $\underline{M}$. It follows that, for all S-words u, v,

(4.12) $$u \equiv_R^+ v \quad \text{implies} \quad \pi^*(u) \equiv_{\underline{R}}^+ \pi^*(v),$$

and π^* induces a well-defined homomorphism of M to $\underline{M}$. This homomorphism, still denoted by π, is surjective since, by assumption, its image includes $\underline{S}$.

(ii) Let $\underline{f}, \underline{g}$ belong to $\underline{M}$. As π is surjective, there exist f, g in M satisfying $\pi(f) = \underline{f}$ and $\pi(g) = \underline{g}$. By assumption, f and g admit a common right-multiple, say h, in M. Then $\pi(h)$ is a common right-multiple of $\underline{f}$ and $\underline{g}$ in $\underline{M}$.

(iii) An easy induction shows that, if u, v are S-words and $\theta^*(u, v)$ exists, then $\underline{\theta}^*(\pi^*(u), \pi^*(v))$ exists as well and we have

(4.13) $$\underline{\theta}^*(\pi^*(u), \pi^*(v)) = \pi^*(\theta^*(u, v)).$$

According to Exercise 22 (complete *vs.* cube, right-complemented case), in order to show that right-reversing is complete for $(\underline{S}, \underline{R})$, it is sufficient to establish that the $\underline{\theta}$-cube condition is valid for every triple of $\underline{S}$-words. So assume that $\underline{u}, \underline{v}, \underline{w}$ are $\underline{S}$-words sharing the same source and $\underline{\theta}^*(\underline{u}, \underline{v}, \underline{w})$ is defined. The assumption that $\pi(S)$ includes $\underline{S}$ guarantees that there exist S-words u, v, w satisfying $\pi^*(u) = \underline{u}$, $\pi^*(v) = \underline{v}$, $\pi^*(w) = \underline{w}$. The assumption that any two elements of M admit a common right-multiple implies that θ^* is defined on every pair of S-words. Then the assumption that right-reversing is complete for (S, R) implies that the θ-cube condition is valid for u, v, w. By applying π^* and using (4.13), we obtain

$$\pi^*(\theta_3^*(u, v, w)) \equiv_{\underline{R}}^+ \pi^*(\theta_3^*(v, u, w)),$$

which, by (4.12), is $\underline{\theta}_3^*(\underline{u}, \underline{v}, \underline{w}) \equiv_{\underline{R}}^+ \theta_3^*(\underline{v}, \underline{u}, \underline{w})$. So the $\underline{\theta}$-cube condition is valid for $(\underline{S}, \underline{R})$, and we conclude that right-reversing is complete for $(\underline{S}, \underline{R})$. □

In our current framework, we immediately deduce

Proposition 4.14 (braids I). *For every positive n, the braid monoid B_n^+ is left-cancellative and admits right-lcms.*

Proof. The projection π defined in (4.2) satisfies (4.11): using θ and $\underline{\theta}$ for the syntactic right-complements involved in the presentations of M_{LD} and B_∞^+, respectively, we check, for all addresses α, β, the equality

(4.15) $$\pi(\theta(\Sigma_\alpha, \Sigma_\beta)) = \underline{\theta}(\pi(\Sigma_\alpha), \pi(\Sigma_\beta)).$$

For instance, we find

$$\pi(\theta(\Sigma_1, \Sigma_\emptyset)) = \pi(\Sigma_\emptyset | \Sigma_1 | \Sigma_0) = \sigma_1 | \sigma_2 = \underline{\theta}(\sigma_2, \sigma_1) = \underline{\theta}(\pi(\Sigma_1), \pi(\Sigma_\emptyset)),$$

and similar relations hold for all pairs of generators $\Sigma_\alpha, \Sigma_\beta$. Then, by Lemma 4.10, we deduce that (as already stated in Subsection 4.1) π induces a surjective homomorphism from M_{LD} onto B_∞^+, that any two elements of B_∞^+ admit a common right-multiple, and that right-reversing is complete for the braid presentation (I.1.5). Then, by Propositions II.4.44 (left-cancellativity) and II.4.46 (common right-multiple), this implies that B_∞ is left-cancellative and that any two elements of B_∞ admit a right-lcm. □

We now turn to right-Garside sequences, and, pretending that we ignore the properties of the fundamental braids Δ_n, deduce them from the properties of the elements Δ_T of M_{LD}. In general, there is no reason why the image of a right-Garside sequence under a surjective homomorphism should be a right-Garside sequence. However, we can easily state sufficient conditions for this to happen: the following result may look intricate, but it just says that, when natural compatibility conditions are met, the projection of a right-Garside family is a right-Garside family.

Lemma 4.16. *Assume that M is a left-cancellative monoid, F is a proper partial action of M on some set X, and $(\Delta_x)_{x\in X}$ is a right-Garside map on M with respect to F. Assume moreover that S is a family of atoms that generate M. Now assume that π : $M \to \underline{M}$ is a surjective homomorphism, $\pi_\bullet : X \to \underline{X}$ is a surjective map, and, for all x in X and g in M,*

(4.17) *The value of $\pi_\bullet(x \bullet g)$ only depends on $\pi_\bullet(x)$ and $\pi(g)$.*

(4.18) *The value of $\pi(\Delta_x)$ only depends on $\pi_\bullet(x)$.*

Assume finally that $\widetilde{\pi} : \underline{S} \to S$ is a section of π such that, for x in X, $\underline{s}$ in $\underline{S}$, and $\underline{w}$ in $\underline{S}^*$,

(4.19) *If $\pi_\bullet(x) \bullet \underline{s}$ is defined, then so is $x \bullet \widetilde{\pi}(\underline{s})$.*

(4.20) *The relation $[\underline{w}] \preccurlyeq \Delta_{\pi_\bullet(x)}$ implies $[\widetilde{\pi}^*(\underline{w})] \preccurlyeq \Delta_x$.*

Define a partial action $\underline{F}$ of $\underline{M}$ on $\underline{X}$ by $\pi_\bullet(x) \bullet \pi(g) = \pi_\bullet(x \bullet g)$, and, for $\underline{x}$ in $\underline{X}$, let $\Delta_{\underline{x}}$ be the common value of $\pi(\Delta_x)$ for x satisfying $\pi_\bullet(x) = \underline{x}$. Then $\underline{F}$ is a proper partial action of $\underline{M}$ on $\underline{X}$ and $(\Delta_{\underline{x}})_{\underline{x}\in\underline{X}}$ is a right-Garside sequence on $\underline{M}$.

The easy proof is left to the reader.

Proposition 4.21 (braids II). *The sequence $(\Delta_n)_{n\in\mathbb{N}}$ is a right-Garside sequence in the monoid B_∞^+.*

Proof. We check that the assumptions of Lemma 4.16 are satisfied for the monoids M_{LD} and B_∞^+, with $S = \{\Sigma_\alpha \mid \alpha \in \mathbb{A}\}$, $\underline{S} = \{\sigma_i \mid i \geqslant 1\}$ and π as defined in (4.2). We already established in Proposition 4.14 that the mapping π induces a surjective homomorphism of M_{LD} onto B_∞^+, and that B_∞^+ admits right-lcms and left-gcds. The property that every element Σ_α of M_{LD} is an atom directly follows from Lemma 3.9.

Now, for T in $\mathcal{T}$, let $\pi_\bullet(T)$ be the right height of T, that is, the length of the rightmost branch of T viewed as a tree. The right height is invariant under LD-equivalence, and, therefore, we have $\pi_\bullet(T \bullet g) = \pi_\bullet(T)$ whenever $T \bullet g$ is defined. Hence (4.17) is satisfied. On the other hand, the equality $\pi(\Delta_T) = \Delta_{\pi_\bullet(T)}$ was established in Lemma 4.6, so (4.18) is satisfied.

Next, define $\widetilde{\pi} : S \to \underline{S}$ by $\widetilde{\pi}(\sigma_i) = \Sigma_{1^{i-1}}$. Assume that T is a term and $\pi_\bullet(T) \bullet \sigma_i$ is defined. By definition, this implies $\pi_\bullet(T) \geqslant i + 1$, that is, the right height of T is at least $i + 1$. This guarantees that $T \bullet \Sigma_{1^{i-1}}$ is defined. Hence (4.19) is satisfied.

Finally, assume that w is a braid word that represents a left-divisor of Δ_n. A direct inspection shows that any two strands in the diagram associated with the recursive definition of Δ_n cross at most once, hence the same must be true for the braid diagram encoded in w. On the other hand, an easy induction shows that, if w is a word in the letters Σ_{1^i}

and T is an injective term $x_1{}^\wedge(x_2{}^\wedge(...))$, then $T \bullet w$ is semi-injective if and only if, in the braid diagram encoded in $\pi^*(w)$, any two strands cross at most once. It follows that (4.20) is satisfied.

We then apply Lemma 4.16. □

Corollary 4.22 (braids III). *For every n, the braid monoid B_n^+ is a Garside monoid.*

Proof. We apply Proposition 1.25 to the right-Garside sequence $(\Delta_n)_{n\in\mathbb{N}}$ in B_∞^+. Then every integer n is eligible, as $n \bullet g$ is defined if and only if g has at least one expression that involves no σ_i with $i \geqslant n$, which implies that $n \bullet gh$ is defined if and only if $n \bullet g$ and $n \bullet h$ are defined. □

So, as announced, the right-Garside structure of braids can be recovered from the right-Garside structure associated with the LD law. Since the right-Garside structure has no left counterpart, we cannot expect for more. However, by the symmetry of the braid relations, the existence of a right-Garside structure immediately implies the existence of a left-structure, and merging the left- and the right-structures completes the results.

4.3 Hurwitz action of braids on LD-systems

We conclude with a related but different topic. The projection of the self-distributivity category $\mathcal{LD}$ to the braid category $\mathcal{C}(B_\infty^+, \mathbb{N})$ described above is partly trivial, in that terms are involved through their right height only and the corresponding action of braids on integers is constant. One can consider alternative projections that correspond to less trivial braid actions and lead to factorizations

$$\mathcal{LD} \longrightarrow \mathcal{C}(B_\infty^+, ...) \longrightarrow \mathcal{C}(B_\infty^+, \mathbb{N})$$

for the projection of $\mathcal{LD}$ onto $\mathcal{C}(B_\infty^+, \mathbb{N})$, and, in good cases, to new examples of categories admitting non-trivial right-Garside maps. We shall describe two such examples.

The first example comes from the action of braids on sequences of integers provided by the associated permutations. For $(p_1, ..., p_n)$ a finite sequence of nonnegative integers and $i < n$, let us define

(4.23) $$(p_1, ..., p_n) \bullet \sigma_i = (p_1, ..., p_{i-1}, p_{i+1}, p_i, p_{i+2}, ..., p_n).$$

In this way we obtain a partial action of the braid monoid B_∞^+ on the set $\mathrm{Seq}(\mathbb{N})$ of all finite sequences of natural numbers. We deduce a new category, hereafter denoted by $\mathcal{C}(B_\infty^+, \mathrm{Seq}(\mathbb{N}))$. A typical element of $\mathcal{C}(B_\infty^+, \mathrm{Seq}(\mathbb{N}))$ is a triple of the form

$$(1, 2, 2) \xrightarrow{\sigma_1} (2, 1, 2).$$

Clearly, $\mathcal{C}(\mathrm{id}, \mathrm{lg})$ defines a surjective functor from $\mathcal{C}(B_\infty^+, \mathrm{Seq}(\mathbb{N}))$ onto $\mathcal{C}(B_\infty^+, \mathbb{N})$. Our aim is then to describe a natural surjective functor from $\mathcal{LD}$ onto this category. We recall that terms have been defined to be bracketed expressions constructed from a fixed sequence of variables $x_1, x_2, \ldots$ (or as binary trees with leaves labeled with variables x_p), and that, for T a term and α a binary address, $T_{/\alpha}$ denotes the subterm of T whose root, when T is viewed as a binary tree, has address α.

Proposition 4.24 (Hurwitz I). *For T a term with right height n, define*

$$\mathrm{var}_{\mathrm{R}}^*(T) = (\mathrm{var}_{\mathrm{R}}(T_{/0}), \mathrm{var}_{\mathrm{R}}(T_{/10}), \ldots, \mathrm{var}_{\mathrm{R}}(T_{/1^{n-1}0})), \tag{4.25}$$

where $\mathrm{var}_{\mathrm{R}}(T)$ denote the index of the rightmost variable occurring in T. Then the surjective morphism $\mathcal{C}(\pi, \mathrm{ht})$ from $\mathcal{LD}$ onto $\mathcal{C}(B_\infty^+, \mathbb{N})$ factors into

$$\mathcal{LD} \xrightarrow{\mathcal{C}(\pi, \mathrm{var}_{\mathrm{R}}^*)} \mathcal{C}(B_\infty^+, \mathrm{Seq}(\mathbb{N})) \xrightarrow{\mathcal{C}(\mathrm{id}, \mathrm{lg})} \mathcal{C}(B_\infty^+, \mathbb{N}).$$

Moreover, the map Δ defined by $\Delta((p_1, \ldots, p_n)) = (p_1, \ldots, p_n) \xrightarrow{\Delta_n} (p_n, \ldots, p_1)$ is a right-Garside map in $\mathcal{C}(B_\infty^+, \mathrm{Seq}(\mathbb{N}))$.

In the above context, terms are mapped to sequences of integers by

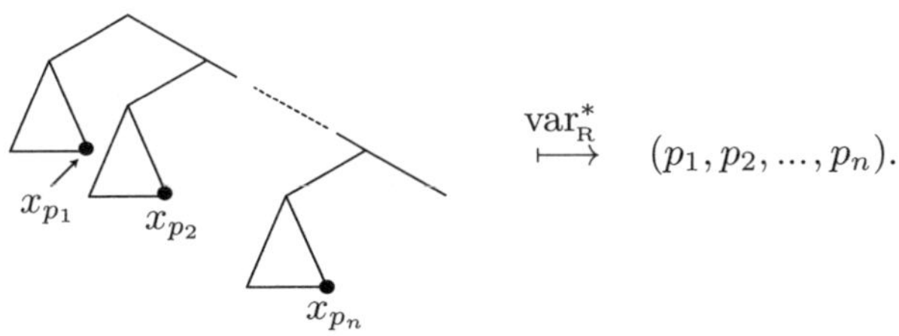

Proof of Proposition 4.24 (*sketch*). The point is to check that the action of the LD law on the indices of the right variables of the subterms with addresses 1^i0 is compatible with the action of braids on sequences of integers. It suffices to consider the basic case of $\Sigma_{1^{i-1}}$, and the expected relation is shown on the right. Details are easy. □

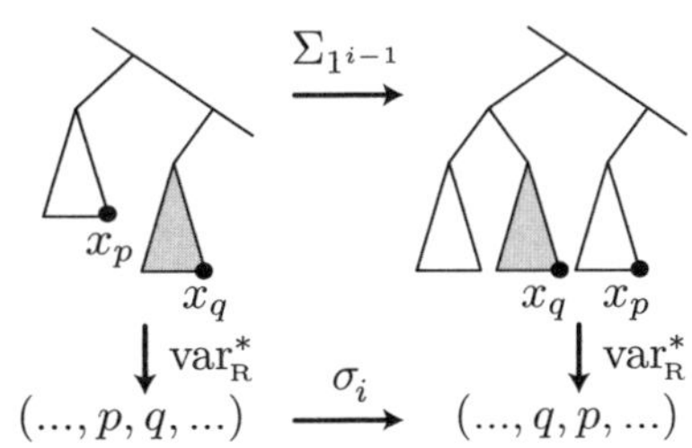

As can be expected, for symmetry reasons, the map Δ of Proposition 4.24 is not only a right-Garside map, but even a Garside map in the category $\mathcal{C}(B_\infty^+, \mathrm{Seq}(\mathbb{N}))$.

The action of positive braids on sequences of integers defined in (4.23) is just one example of a much more general situation, namely the action of positive braids on sequences of elements of any LD-system. It is well known—see, for instance, [35] or [77, Chapter I]—that, if $(G, {}^\wedge)$ is an LD-system, that is, ${}^\wedge$ is a binary operation on G that obeys the LD law, then setting for $i < n$

$$(a_1, \ldots, a_n) \bullet \sigma_i = (a_1, \ldots, a_{i-1}, a_i{}^\wedge a_{i+1}, a_i, a_{i+2}, \ldots, a_n) \tag{4.26}$$

provides a well-defined action of the monoid B_n^+ on the set G^n, and, from there, a partial action of B_∞^+ on the set $\mathrm{Seq}(G)$ of all finite sequences of elements of G. It is natural to call this action a *Hurwitz action*, since it is a direct extension of the Hurwitz action as considered in Example I.2.8 when conjugacy is replaced with an arbitrary left-selfdistributive operation $^\wedge$—the case of Proposition 4.24 then corresponding to the trivial operation $x^\wedge y = y$. For each choice of the LD-system $(G, {}^\wedge)$, we obtain an associated category that we shall simply denote by $\mathcal{C}(B_\infty^+, \mathrm{Seq}(G))$.

Proposition 4.27 (Hurwitz II). *Assume that $(G, {}^\wedge)$ is an* LD*-system and $\vec{c}$ is a sequence of elements of G. For T a term with right height n, define*

$$\mathrm{ev}_{\vec{c}}(T) = (T_{/0}(\vec{c}), \dots, T_{/1^{n-1}0}(\vec{c})),$$

where $T(\vec{c})$ denotes the evaluation of T in $(G, {}^\wedge)$ when x_i is given the value c_i for every i. Then the surjective morphism $\mathcal{C}(\pi, \mathrm{ht}_R)$ of $\mathcal{LD}$ onto $\mathcal{C}(B_\infty^+, \mathbb{N})$ factors into

$$\mathcal{LD} \xrightarrow{\mathcal{C}(\pi, \mathrm{ev}_{\vec{c}})} \mathcal{C}(B_\infty^+, \mathrm{Seq}(G)) \xrightarrow{\mathcal{C}(\mathrm{id}, \mathrm{lg})} \mathcal{C}(B_\infty^+, \mathbb{N}).$$

Moreover, the map Δ defined by

$$\Delta((a_1, \dots, a_n)) = (a_1, \dots, a_n) \xrightarrow{\Delta_n} (a_1{}^\wedge \cdots {}^\wedge a_n, a_1{}^\wedge \cdots {}^\wedge a_{n-1}, \dots, a_1{}^\wedge a_2, a_1),$$

where $x^\wedge y^\wedge z$ stands for $x^\wedge(y^\wedge z)$, is a right-Garside map in $\mathcal{C}(B_\infty^+, \mathrm{Seq}(G))$.

We skip the proof, which is an easy verification similar to that of Proposition 4.24. When $(G, {}^\wedge)$ is $\mathbb{N}$ equipped with $a^\wedge b = b$ and x_p is sent to p, the category $\mathcal{C}(B_\infty^+, \mathrm{Seq}(G))$ is the category $\mathcal{C}(B_\infty^+, \mathrm{Seq}(\mathbb{N}))$ of Proposition 4.24. In the latter case, the (partial) action of braids is not constant as in the case of $\mathcal{C}(B_\infty^+, \mathbb{N})$, but it factors through an action of the associated permutations, and, therefore, it is far from being free. By contrast, take for G the braid group B_∞ with $^\wedge$ defined by $a^\wedge b = a\,\mathrm{sh}(b)\,\sigma_1\,\mathrm{sh}(a)^{-1}$, where we recall sh is the shift endomorphism of B_∞ that maps σ_i to σ_{i+1} for each i, and map x_p to 1 (or to any other fixed braid) for each p. Then the corresponding action (4.26) of B_∞^+ on $\mathrm{Seq}(B_\infty)$ is free, in the sense that $g = g'$ holds whenever $(a_1, \dots, a_n) \bullet g = (a_1, \dots, a_n) \bullet g'$ holds for at least one sequence $(a_1, \dots, a_n)$ in $(B_\infty)^*$: this follows from Lemma III.1.10 of [100]. This suggests that the associated category $\mathcal{C}(B_\infty^+, \mathrm{Seq}(B_\infty))$ has a rich structure.

Exercises

Exercise 104 (skeleton). Say that a set of addresses A is an *antichain* if, for all α, β in A, neither α is a prefix of β, nor β is a prefix of α. Call an antichain *maximal* if it is properly included in no antichain. (i) Show that a finite maximal antichain is a family $\{\alpha_1, \dots, \alpha_n\}$ such that every long enough address admits as a prefix exactly one of the addresses α_i. (ii) Show that, for every Σ-word w, there exists a unique finite maximal antichain A_w such that $T \bullet w$ is defined if and only if the skeleton of T includes A_w.

Exercise 105 (preservation). Assume that M is a left-cancellative monoid and F is a partial action of M on a set X. (i) Show that, if M admits right lcms (*resp.* conditional right-lcms), then so does the category $\mathcal{C}_F(M,X)$. (ii) Show that, if M is right-Noetherian, then so is $\mathcal{C}_F(M,X)$.

Exercise 106 (right-Garside sequence). Assume that M is a left-cancellative monoid, α is a partial action of M on X and $(\Delta_x)_{x\in X}$ is a Garside sequence in M with respect to α. (i) Show that, for every x in X, any two elements of $\mathrm{Def}(x)$ admit a common right-multiple in M, and the latter lies in $\mathrm{Def}(x)$. (ii) Show that any two elements of M admit a common right-multiple. (iii) Show that, if $x \bullet g$ is defined, then $\Delta_x \preccurlyeq g\,\Delta_{x\bullet g}$ holds and that, if we define $\phi(x)$ by $\phi(x) = x \bullet \Delta_x$ and $\phi(g)$ by $g\,\Delta_{x\bullet g} = \Delta_x\,\phi(g)$, then the functor ϕ_Δ of $\mathcal{C}_F(M,X)$ associated with the right-Garside map Δ is given by $\phi_\Delta((x,g,y)) = (\phi(x),\phi(g),\phi(y))$.

Exercise 107 (Noetherianity). Assume that F is a proper partial action of a monoid M on a set X and there exists a map $\mu : X \to \mathbb{N}$ such that $\mu(x\bullet g) > \mu(x)$ holds whenever g is not invertible. Show that M is Noetherian and every element of M has a finite height.

Exercise 108 (relations $\mathcal{R}_{\mathrm{LD}}$). Show that, for every Σ-word w and every term T such that $T \bullet w$ is defined, one has $\delta_T \cdot w = \mathrm{sh}_1(w) \cdot \delta_{T\bullet w}$. [Hint: Show the result for w of length one and use induction on the size of T.]

Exercise 109 (common multiple). Assume that M is a left-cancellative monoid, F is a partial action of M on X, and $(\Delta_x)_{x\in X}$ is a right-Garside sequence in M with respect to F. Show that, for every x in X, any two elements of $\mathrm{Def}(x)$ admit a common right-multiple that lies in $\mathrm{Def}(x)$.

Exercise 110 (Embedding Conjecture). Show that (3.34) holds for $f = \Sigma_0$ and $g = \Sigma_1$ with $T = (x^\wedge(x^\wedge x))^\wedge(x^\wedge(x^\wedge x))$. [Hint: The values are $\phi_T(\Sigma_0) = \Sigma_{000}\Sigma_{010}\Sigma_{100}\Sigma_{110}$ and $\phi_T(\Sigma_1) = \Sigma_\emptyset$.]

Notes

Sources and comments. The content of this chapter is based on the approach to left self-distributivity developed in [77]. Although implicit from the early developments—introducing the operation ϕ on terms to prove the existence of a common LD-expansions is totally similar to introducing Garside's fundamental braid Δ_n to prove the existence of common right-multiples in the braid monoid B_n^+—the connection with Garside's theory of braids was discovered a posteriori, when it appeared that the subword reversing machinery specially developed to prove that the monoid M_{LD} is left-cancellative is also relevant for the braid monoids B_n^+, and is actually closely connected with the Higman–Garside strategy for proving that the braid monoids are left-cancellative. The current exposition in terms of a right-Garside map in the category $\mathcal{LD}$ or, equivalently, a right-Garside sequence in the monoid M_{LD}, appeared more recently, in [88].

The interest in the self-distributivity law was renewed by the discovery in the 1980's and the early 1990's of unexpected connections with low-dimensional topology in Joyce

[155], Matveev [179], Fenn–Rourke [121] and axiomatic set theory in [70] and in Laver [169]. The latter connection led to deriving several purely algebraic statements about the LD law from certain higher infinity axioms, but, at least at first, no proof of these statements in the standard framework of the Zermelo–Fraenkel system was known. This unusual situation provided a strong motivation for the quest of alternative elementary proofs, and a rather extensive theory of the LD law was developed in the decade 1985-95 by the first author in [77], with a partial success: some of the statements first derived from large cardinal axioms did receive elementary proofs, in particular the decidability of the Word Problem for the LD law, that is, the question of algorithmically recognizing whether two terms are $=_{\text{LD}}$-equivalent, but other questions remain open, in particular those involving Laver's tables, a sequence of finite LD-systems of sizes $1, 2, 4, 8, \ldots$ reminiscent of 2-adic integers.

The above mentioned theory of the LD law was centered on the study of the relation $=_{\text{LD}}$, and it was developed in two steps. The first one [71] consisted in investigating $=_{\text{LD}}$ directly, and it essentially corresponds to the results of Section 2 in this chapter. However, because Question 2.11 remained open, these results were not sufficient to establish the desired decidability of $=_{\text{LD}}$. Here came the second step, namely introducing the monoid M_{LD} (and the group G_{LD} specified by the same presentation) and using it to describe $=_{\text{LD}}$ more precisely. This approach, which includes the results of Section 3, turned out to be successful in [73, 75], and, in addition, it led to unexpected braid applications when the connection between M_{LD} and B_∞^+ became clear in [72].

By the way, we insist that the slogan "the Garside structure of braids is the projection of the Garside structure on the selfdistributivity monoid M_{LD}" is fully justified and it provides really new proofs for the algebraic properties of braids: the verifications of Subsection 4.2 to show that braids are eligible require no background result about braids: we do not have to assume half the results to reprove them!

Further questions. Many puzzling questions remain open in the domain of left-selfdistributive algebra, like the question of whether the results about periods in Laver's tables, currently established using large cardinal assumptions, can be established in the standard framework of the Zermelo–Fraenkel system, see [77, Chapter XIII], or the termination of the so-called Polish algorithm, see [79]. It is however unlikely that the Garside approach may be useful in solving these questions.

By contrast, another significant open problem is the already mentioned Embedding Conjecture, which we recall can be expressed as:

Question 32. Is the monoid M_{LD} right-cancellative?

This is indeed a structural question about the LD law since, by Proposition 3.27, a positive solution would in particular guarantee that any two LD-equivalent terms admit a unique least common LD-expansion.

Proposition 3.33 leads to a realistic program that would reduce the proof of the Embedding Conjecture to a (long) sequence of verifications. Indeed, it is shown in Proposition VIII.5.15 of [77] that every element of M_{LD} that left-divides an element Δ_T admits a unique expression of the form $\prod^{>}_{\alpha\in\mathbb{A}} \Sigma_{\alpha,e_\alpha}$, where $\Sigma_{\alpha,e}$ denotes $\Sigma_\alpha\Sigma_{\alpha1}\ldots\Sigma_{\alpha1^{e-1}}$ and $>$ refers to the unique linear ordering on binary addresses that satisfies $\alpha > \alpha0\beta > \alpha1\gamma$

for all α, β, γ. In this way, we associate with every simple element of M_{LD} a sequence of nonnegative integers $(e_\alpha)_{\alpha \in \mathbb{A}}$ that plays the role of coordinates for the considered element. Then it *should* be possible to express the coordinates of $\phi_T(s)$ in terms of those of s, and then express the coordinates of $\gcd(s, t)$ in terms of those of s and t. If this were done, proving (or disproving) the equalities (3.34) should be easy.

In another direction, the presentation of M_{LD}, which we saw is right-complemented, is also left-complemented, that is, for every pair of generators $\Sigma_\alpha, \Sigma_\beta$, there exists in $\mathcal{R}_{\text{LD}}$ at most one (actually exactly one) relation of the form $\cdots \Sigma_\alpha = \cdots \Sigma_\beta$. But the presentation fails to satisfy the counterpart of the cube condition, and it is extremely unlikely that one can prove that the monoid M_{LD} is possibly right-cancellative simply using the counterpart of Proposition II.4.16. Now the germ approach of Chapter VI provides another powerful criterion for establishing cancellativity: a second, independent program for proving the Embedding Conjecture would be to prove that the family of simple elements in M_{LD} forms a (right)-germ that satisfies, say, the conditions of Proposition VI.2.8 or VI.2.28 (recognizing Garside germ). As the considered germ is infinite, this is *not* an easy task but, clearly, only finitely many combinatorial situations may appear and, again, a (long) sequence of verifications could be sufficient.

So, in both approaches, it seems that Garside's theory will be crucial in a possible solution of the Embedding Conjecture. In any case, it is crucial to work with the structural monoid M_{LD} efficiently, and this is precisely what the Garside approach provides.

Other algebraic laws. The above approach of self-distributivity can be developed for other algebraic laws. However, at least from the viewpoint of Garside structures, the case of self-distributivity seems rather particular.

A typical case is associativity, that is, the law $x(yz) = (xy)z$. It is syntactically close to self-distributivity, the only difference being that the variable x is not duplicated in the right-hand side. Let us say that a term T' is an A-expansion of another term T if T' can be obtained from T by applying the associativity law in the left-to-right direction only, that is, by iteratively replacing subterms of the form $T_1{}^\wedge(T_2{}^\wedge T_3)$ by the corresponding term $(T_1{}^\wedge T_2){}^\wedge T_3$. Then the counterpart of Proposition 2.6 is true, that is, two terms T, T' are equivalent up to associativity if and only if they admit a common A-expansion, a trivial result since every size n term T admits as an A-expansion the term $\phi(T)$ obtained from T by pushing all brackets to the left.

Then, as in Section 2, we can introduce a category $\mathcal{A}_0$ whose objects are terms, and whose morphisms are pairs (T, T') with T' an A-expansion of T. Next, as in Section 3, we can take positions into account and, using A_α when associativity is applied at address α, introduce a monoid M_A that describes the connections between the generators A_α, see [84]. Here the relations of Lemma 3.5 are to be replaced by analogous new relations, among which the MacLane–Stasheff Pentagon relations $A_\alpha^2 = A_{\alpha 1} A_\alpha A_{\alpha 0}$ plays the critical role and replaces the relation $\Sigma_\emptyset \Sigma_1 \Sigma_\emptyset = \Sigma_1 \Sigma_\emptyset \Sigma_0 \Sigma_1$. The monoid M_A turns out to be a well-known object: indeed, its group of fractions is Richard Thompson's group F, see Cannon, Floyd, and Parry [48], or [90]. Finally, we can introduce the category $\mathcal{A}$, whose objects are terms, and whose morphisms are triples $T \xrightarrow{g} T'$ with g in M_A and $T \bullet g = T'$. Then, using $\widetilde{\phi}(T)$ for the term obtained from T by pushing all brackets to the right, one easily proves that the categories $\mathcal{A}_0$ and $\mathcal{A}$ are isomorphic—that is, the analog of the Embedding Conjecture is true—and that the map $T \mapsto (T, \phi(T))$ is a right-Garside map in $\mathcal{A}_0$, whereas $T \mapsto (T, \widetilde{\phi}(T))$ is, in the obvious sense, a left-Garside map. This seemingly promising result is *not* interesting. Indeed, the Garside structure(s) is trivial: the maps ϕ and $\widetilde{\phi}$ are constant on each orbit for the action of M_A on terms, and every element in the categories $\mathcal{A}_0$ and $\mathcal{A}$ is a divisor of Δ. In other words, the underlying Garside families are the whole categories and, therefore, nothing is to be expected from the derived normal decompositions.

By contrast, let us mention that the central duplication law $x(yz) = (xy)(yz)$ turns out to be similar to self-distributivity, with a non-trivial Garside structure, see [81]. As there is no known connection between this exotic law and other widely investigated objects like braids, we do not go into details.

Chapter XII
Ordered groups

A priori, there is no obvious connection between ordered groups and Garside structures. In particular, the definition of a Garside monoid includes Noetherianity assumptions that are never satisfied in those monoids that occur in connection with ordered groups. However, as already seen with the Klein bottle monoid (Reference Structure 5, page 17), the current, extended framework developed in this text allows for non-Noetherian structures. What we do in this chapter is to show that the above Klein bottle monoid is just the very first one in a rich series of monoids that are canonically associated with ordered groups and that contain a Garside element.

In recent years, a number of interesting questions arose about the topology of the space of all orderings of a given orderable group. We shall show, in particular, that an approach based on monoids and Garside elements is specially efficient for constructing examples of orderable groups with orderings that are isolated in the space of their orderings, a not so frequent situation of which few examples were known. Here Garside theory is mainly useful through the result that a Garside element Δ and the associated functor ϕ_Δ (here a morphism since we are in a monoid) provide an effective criterion for the existence of common multiples.

The chapter is divided in three sections. We begin in Section 1 with some background about ordered groups and the topology of their space of orderings (Proposition 1.14). Next, we consider in Section 2 the specific question of constructing isolated orderings on an orderable group and we explain how to use Garside elements for this task, in the specific case of what we call triangular presentations, along the scheme described in Proposition 2.7, leading to a practical criterion in Proposition 2.15. Finally, in Section 3, we point out some limits of the approach, showing both that some further examples can be constructed without using Garside element and that some examples remain inaccessible to triangular presentations, in particular in the case of braids with 4 strands and more and the associated Dubrovina–Dubrovin ordering.

1 Ordered groups and monoids of O-type

In this introductory section, we review basic results about orderable groups and their space of orderings. The only non-standard point is the introduction of the notion of a monoid of O-type, which is relevant for addressing orderability questions in a monoid framework.

We start with background on ordered groups in Subsection 1.1. Then Subsection 1.2 is devoted to the space of orderings of an orderable group, with the specific question of the existence of isolated orderings that is central in the rest of the chapter. Finally, in Subsection 1.3, we describe the action of automorphisms on orderings with a special emphasis on the case of Artin's braid groups.

1.1 Orderable and bi-orderable groups

In the sequel, when considering an ordering, we always use $<, \prec, ...$ for the strict (antireflexive) version and $\leqslant, \preccurlyeq, ...$ for the associated non-strict (reflexive) version: $g < h$ stands for $g \leqslant h$ with $g \neq h$, and $g \leqslant h$ stands for $g < h$ or $g = h$. An ordering is called *linear* (or *total*) if any two elements are comparable.

Definition 1.1 (orderable, bi-orderable). (i) A linear ordering $<$ on a group G is called *left-invariant* (*resp. right-invariant*, *resp. bi-invariant*) if $g < g'$ implies $fg < fg'$ (*resp.* $gh < g'h$, *resp.* both) for all f, g, g', h in G.

(ii) A group G is called *orderable* (*resp. bi-orderable*) if it admits at least one left-invariant (*resp.* bi-invariant) linear ordering.

Example 1.2 (orderable group). The standard ordering on $(\mathbb{Z}, +)$ is bi-invariant, so $(\mathbb{Z}, +)$ is a bi-orderable group. More generally, a lexicographic ordering on a free abelian group is bi-invariant, so every such group is bi-orderable. By contrast, it is clear that, in an orderable group, $1 < g$ implies $g^n < g^{n+1}$ for every n, so such a group admits no torsion element. Therefore, an abelian group if orderable if and only if it is bi-orderable if and only if it is torsion-free.

By contrast, Artin's 3-strand braid group B_3 (Reference Structure 2, page 5) is not bi-orderable: indeed, assume that $<$ is a linear ordering on B_3 with, say, $\sigma_1 < \sigma_2$. If $<$ is bi-invariant, then, for every g, we have $g^{-1}\sigma_1 g < g^{-1}\sigma_2 g$, that is, with the notation of Chapter VIII, $\sigma_1^g < \sigma_2^g$. Now, for $g = \Delta_3$, we have $\sigma_1^{\Delta_3} = \sigma_2$ and $\sigma_2^{\Delta_3} = \sigma_1$, so it is impossible that $\sigma_1 < \sigma_2$ and $\sigma_1^{\Delta_3} < \sigma_2^{\Delta_3}$ hold simultaneously. However, we shall see below that B_3 is left-orderable.

As can be expected, there is no notion of a left-orderable group because every group that admits a left-invariant ordering automatically admits a right-invariant ordering, and vice versa: a linear ordering $<$ is left-invariant if and only if the ordering $\tilde{<}$ defined by $g \,\tilde{<}\, h \Leftrightarrow h^{-1} < g^{-1}$ is right-invariant.

A left-invariant ordering on a group turns out to be entirely determined by what is called its *positive cone*.

Lemma 1.3. (i) *Assume that $<$ is a left-invariant ordering on a group G. Define the* positive cone $P_<^+$ *of $<$ to be $\{g \in G \mid g > 1\}$. Then $P_<^+$ is a subsemigroup of G such that $P_<^+$, $(P_<^+)^{-1}$, and $\{1\}$ make a partition of G.*

(ii) *Conversely, assume that P is a subsemigroup of a group G such that P, P^{-1}, and $\{1\}$ make a partition of G. Then the relation $g^{-1}h \in P$ defines a left-invariant ordering on G, and P is the associated positive cone.*

Proof. (i) Assume that g and h lie in $P_<^+$. Then we have $1 < h$, whence $g < gh$, since $<$ is left-invariant. We deduce the inequalities $1 < g < gh$, so gh lies in $P_<^+$, and the latter is a subsemigroup of G.

Next, using the invariance of $g < 1$ under left-multiplication by g^{-1} and that of $g^{-1} > 1$ under left-multiplication by g, we see that $g < 1$ is always equivalent to $g^{-1} > 1$.

Now let g belong to G. If g does not lie in $P_<^+ \cup \{1\}$, the relation $g \geqslant 1$ fails, so we have $g < 1$, whence $g^{-1} > 1$, and g lies in $(P_<^+)^{-1}$. So G is the union of $P_<^+$, $(P_<^+)^{-1}$, and $\{1\}$. Then the assumption that $<$ is strict implies that 1 lies neither in $P_<^+$ nor in $(P_<^+)^{-1}$. Finally, $g \in P_<^+ \cap (P_<^+)^{-1}$ would imply both $g > 1$ and $g < 1$, which is impossible. So $P_<^+$, $(P_<^+)^{-1}$, and $\{1\}$ make a partition of G.

(ii) Write $g < h$ for $g^{-1}h \in P$. First 1 does not lie in P, so $g < g$ never holds, that is, $<$ is antireflexive. Next, assume $f < g < h$. As P is a subsemigroup of G, we find $f^{-1}h = (f^{-1}g)(g^{-1}h) \in P^2 \subseteq P$, whence $f < h$. So $<$ is transitive, and it is a strict partial ordering on G. Assume $g \neq h \in G$. If $g < h$ fails, that is, if $g^{-1}h$ does not lie in P, then, by assumption, $(g^{-1}h)^{-1}$ lies in P. This means that $h^{-1}g$ lies in P, hence $h < g$ holds. So $<$ is a linear ordering on G. Then, as we have $(fg)^{-1}(fh) = g^{-1}h$, it is obvious that $g < h$ implies $fg < fh$ for every f. So $<$ is left-invariant. Finally, by definition of $<$, the set $\{g \in G \mid g > 1\}$ coincides with P. □

In the current text, monoids and the associated divisibility relations play an important rôle. It is then easy to adapt Lemma 1.3 to a monoid terminology. We recall that, if M is a left-cancellative monoid, the left-divisibility relation $\preccurlyeq$ is a partial ordering on M if and only if M admits no nontrivial invertible elements.

Definition 1.4 (monoid of O-type). A monoid M is said to be *of right-O-type* (*resp.* *left-O-type*) if M is left-cancellative (*resp.* right-cancellative) and left-divisibility (*resp.* right-divisibility) is a linear ordering on M. A monoid is *of O-type* if it is both of right- and left-O-type.

In other words, a monoid M is of right-O-type if it is left-cancellative, has no nontrivial invertible element, and, for all g, h in M, at least one of $g \preccurlyeq h$, $h \preccurlyeq g$ holds.

Example 1.5 (monoid of O-type). The free abelian monoid $(\mathbb{N}, +)$ is of O-type: the associated divisibility relation is the standard ordering of natural numbers, which is linear.

Now, let M be the monoid $\langle \mathtt{a}, \mathtt{b} \mid \mathtt{a} = \mathtt{ba} \rangle^+$. Mapping $\mathtt{a}$ to the ordinal ω and $\mathtt{b}$ to 1 defines an isomorphism from M to the monoid $(\omega^2, +)$, and one easily deduces that $(M, \preccurlyeq)$ is isomorphic to the ordinal ω^2 equipped with its standard ordering, which is linear. Hence M is of right-O-type. Note that M is left-Noetherian, since $(\omega^2, <)$ is a well-order, but not right-Noetherian, since $\mathtt{a} \mathrel{\tilde{\succneqq}} \mathtt{a}$ holds.

Next, let K^+ be the Klein bottle monoid $\langle \mathtt{a}, \mathtt{b} \mid \mathtt{a} = \mathtt{bab} \rangle^+$ (Reference Structure 5, page 17). We saw in Chapter I that K^+ is cancellative and that the left- and the right-divisibility relations of K^+ are linear orderings. Hence K^+ is a monoid of O-type.

By contrast, for $n \geqslant 2$, a free abelian monoid $\mathbb{N}^n$ is not of left-O-type: if $\mathtt{a}_1, \dots, \mathtt{a}_n$ is a basis, neither $\mathtt{a}_1 \preccurlyeq \mathtt{a}_2$, nor $\mathtt{a}_2 \preccurlyeq \mathtt{a}_1$ holds. Similarly, the braid monoid B_n^+ (Reference Structure 2, page 5) is not of right-O-type for $n \geqslant 3$, since neither $\sigma_1 \preccurlyeq \sigma_2$, nor $\sigma_2 \preccurlyeq \sigma_1$ holds.

Generalizing the above examples, we immediately see that a monoid that contains at least two atoms cannot be of left-O-type: if s_1, s_2 are distinct atoms, then, by the definition of an atom, both $s_1 \prec s_2$ and $s_2 \prec s_1$ are impossible. Owing to Proposition II.2.58 (atoms generate), we deduce

Proposition 1.6 (not O-type). *The only Noetherian monoid of right-O-type is* $(\mathbb{N}, +)$.

The connection between ordered groups and monoids of O-type is simple:

Lemma 1.7. *For M a submonoid of a group G, the following conditions are equivalent:*
(i) *The group G admits a left-invariant ordering whose positive cone is $M\backslash\{1\}$.*
(ii) *The monoid M is of O-type and generates G.*

Proof. Assume (i). Put $P = M\backslash\{1\}$. First, since M is included in a group, it must be cancellative. Next, assume that g is an invertible element of M, that is, there exists h in M satisfying $gh = 1$. If g belongs to P, then so does h and, therefore, g belongs to $P \cap P^{-1}$, contradicting the assumption that P is a positive cone. So 1 must be the only invertible element of M. Now, let g, h be distinct elements of M. Then one of $g^{-1}h$, $h^{-1}g$ belongs to P, hence to M: in the first case, $g \preccurlyeq h$ holds, in the second, $h \preccurlyeq g$ does. Symmetrically, one of gh^{-1}, hg^{-1} belongs to P, hence to M, now implying $g \succcurlyeq h$ or $h \succcurlyeq g$. So any two elements of M are comparable with respect to $\preccurlyeq$ and $\succcurlyeq$. Hence M is of O-type, and (i) implies (ii).

Conversely, assume that M is of O-type. Put $P = M\backslash\{1\}$ again. Then P is a subsemigroup of G. The assumption that 1 is the only invertible element in M implies $P \cap P^{-1} = \emptyset$. Next, the assumption that any two elements of M are comparable with respect to $\preccurlyeq$ implies a fortiori that any two of its elements admit a common right-multiple. By (the easy direction of) Proposition II.3.11 (Ore's theorem), this implies that G is a group of right-fractions for M. Let f be an element of G. There exist g, h in M satisfying $f = gh^{-1}$. By assumption, at least one of $g \succcurlyeq h$, $h \succcurlyeq g$ holds in M. This means that at least one of $f \in M$, $f \in M^{-1}$ holds. Therefore, we have $G = M \cup M^{-1}$, which is also $G = P \cup P^{-1} \cup \{1\}$. So P is a positive cone on G, and (ii) implies (i). □

It will be convenient to restate the orderability criterion of Lemma 1.7 in terms of presentations. We recall from Chapter II that a group presentation (S, R) is called *positive* if R a family of relations of the form $u = v$, where u, v are nonempty words in the alphabet S (no empty word, and no letter s^{-1}). We recall that a monoid admits a positive presentation if and only if 1 is the only invertible element. Also remember that, in general, the monoid $\langle S \mid R\rangle^+$ need not embed in the group $\langle S \mid R\rangle$.

Proposition 1.8 (orderability). *A group G is orderable if and only if*

(1.9) *G admits a positive presentation (S, R) such that the monoid $\langle S \mid R\rangle^+$ is of O-type.*

In this case, the subsemigroup of G generated by S is the positive cone of a left-invariant ordering on G.

Proof. Assume that G is an orderable group. Let P be the positive cone of a left-invariant ordering on G, and let $M = P \cup \{1\}$. By the implication (i)$\Rightarrow$(ii) of Lemma 1.7, the monoid M is of O-type. As 1 is the only invertible element in M, the latter admits a positive presentation (S, R). As $G = M \cup M^{-1}$ holds, G is a group of right-fractions for M. By Proposition II.3.11 (Ore's theorem), this implies that (S, R) is also a presentation of G.

Conversely, assume that G admits a positive presentation (S, R) such that the monoid $\langle S \mid R\rangle^+$ is of O-type. Let M be the submonoid of G generated by S, and let $P = M \setminus \{1\}$. As observed in the proof of Lemma 1.7, Ore's theorem implies that $\langle S \mid R\rangle^+$ embeds in a group of fractions, which admits the presentation (S, R), hence is isomorphic to G. Hence, the identity mapping on S induces an embedding ι of $\langle S \mid R\rangle^+$ into G. Therefore the image of ι, which is the submonoid of G generated by S, hence is M, admits the presentation (S, R). So the assumption implies that M is of O-type. Then, by the implication (ii) $\Rightarrow$ (i) of Lemma 1.7, P is the positive cone of a left-invariant ordering on G. □

Example 1.10 (orderability). The Klein bottle group K (Reference Structure 5, page 17) admits the presentation $\langle \mathtt{a}, \mathtt{b} \mid \mathtt{a} = \mathtt{bab}\rangle$, and we saw in Example 1.5 that the monoid $\langle \mathtt{a}, \mathtt{b} \mid \mathtt{a} = \mathtt{bab}\rangle^+$ is of O-type, so Proposition 1.8 implies that K is orderable, by a left-invariant ordering that admits $K^+ \setminus \{1\}$ as positive cone.

1.2 The spaces of orderings on a group

We now consider the family of all left-invariant (or bi-invariant) orderings on a given orderable (or bi-orderable) group. This family can be given a natural topology, making it a compact, totally disconnected space, and this will be our main subject of interest in the sequel.

If G is a group—actually, any set—we denote here by $\mathfrak{P}(G)$ the powerset of G, that is, the family of all subsets of G. Identifying a subset of G with its indicator function, we identify $\mathfrak{P}(G)$ with the set $\{0, 1\}^G$ of all functions from G to $\{0, 1\}$. Then, starting with the discrete topology on $\{0, 1\}$, we equip $\mathfrak{P}(G)$ with the product topology. A basic open set consists of all functions in $\{0, 1\}^G$ with specified values on a specified finite subset of G, that is, we choose two finite (possibly empty) subsets $\{g_1, ..., g_p\}$, $\{h_1, ..., h_q\}$ of G and define the corresponding open set to be

$$\{X \subseteq G \mid g_1 {\in} X, ..., g_p {\in} X,\ h_1 {\notin} X, ..., h_q {\notin} X\}. \tag{1.11}$$

Since $\{0, 1\}$ is a compact space, $\mathfrak{P}(G)$ is compact by Tychonoff's theorem. In addition, $\mathfrak{P}(G)$ is totally disconnected, that is, any two points lie in disjoint open sets whose union is the whole space. Indeed, if X_1 and X_2 are distinct subsets of G, there exists g satisfying $g \in X_1$ and $g \notin X_2$ (or vice versa), and then the two open sets $\{X \subseteq G \mid g \in X\}$ and $\{X \subseteq G \mid g \notin X\}$ separate X_1 and X_2 and their union is all of $\mathfrak{P}(G)$.

In the case of ordered groups, we observed in Lemma 1.3 that, if G is an orderable group, there exists a one-to-one correspondence between the left-invariant orderings of G and the positive cones of G. The latter are subsets of G, hence points in the powerset $\mathfrak{P}(G)$. Then, at the expense of identifying a left-invariant ordering with its positive cone, the topology of $\mathfrak{P}(G)$ naturally induces a topology on the family of all left-invariant orderings of G, and similarly for bi-invariant orderings.

Definition 1.12 (spaces $\mathrm{LO}(G)$ and $\mathrm{O}(G)$). For G a group, we define $\mathrm{LO}(G)$ (*resp.* $\mathrm{O}(G)$) to be the family of the positive cones of the left-invariant (*resp.* bi-invariant) orderings on G, equipped with the topology induced by that of $\mathfrak{P}(G)$.

Of course, the space $\mathrm{LO}(G)$ is nonempty if and only if G is orderable, whereas $\mathrm{O}(G)$ is nonempty if and only if G is bi-orderable.

Lemma 1.13. *If G is an orderable group, then the family $\{U_{g,h} \mid g, h \in G\}$ is a base of open sets for the topology of $\mathrm{LO}(G)$, where, for g, h in G, the set $U_{g,h}$ is defined to be $\{P \in \mathrm{LO}(G) \mid g^{-1}h \in P\}$.*

Proof. Every open set of the form (1.11) is an intersection of finitely many sets $U_{g,h}$, since we consider positive cones, so that, when g, h are disjoint, $g^{-1}h \notin P$ is equivalent to $h^{-1}g \in P$. □

Note that $U_{g,h}$ is the family of all left-invariant orderings $<$ satisfying $g < h$.

Proposition 1.14 (spaces $\mathrm{LO}(G)$ and $\mathrm{O}(G)$). (i) *For every group G, the spaces $\mathrm{LO}(G)$ and $\mathrm{O}(G)$ are closed subsets of $\{0,1\}^G$; they are compact, totally disconnected spaces.*

(ii) *For every countable group G, the space $\mathrm{LO}(G)$ (resp. $\mathrm{O}(G)$) is either empty, or nonempty finite, or homeomorphic to the Cantor set, or infinite homeomorphic to a closed subset of the Cantor set with isolated points.*

Proof. (i) As the space $\mathfrak{P}(G)$ is compact and totally disconnected, it suffices to see that $\mathrm{LO}(G)$ and $\mathrm{O}(G)$ are closed in $\mathfrak{P}(G)$, that is, that not being a positive cone is an open condition. Now, P is a positive cone if it satisfies the three conditions (i) $P^2 \subseteq P$, (ii) $P \cap P^{-1} = \emptyset$, (iii) $P \cup P^{-1} = G \setminus \{1\}$. Then P fails to satisfy (i) if there exist g, h in P such that gh does not belong to P, that is, if P belongs to the open set

$$\bigcup_{g,h \in G} \{X \mid g \in X\,,\ h \in X\,,\ gh \notin X\}.$$

Similarly, P fails to satisfy conditions (ii) or (iii) if it belongs to the open sets

$$\bigcup_{g \in G} \{X \mid g \in X\,,\ g^{-1} \in X\} \quad \text{or} \quad \bigcup_{g \in G \setminus \{1\}} \{X \mid g \notin X\,,\ g^{-1} \notin X\},$$

respectively. So $\mathfrak{P}(G) \setminus \mathrm{LO}(G)$ is the union of three open sets, and it follows that $\mathrm{LO}(G)$ is closed in $\mathfrak{P}(G)$. Similarly, the condition that the left-invariant ordering associated with P fails to be right-invariant means that there exist g, h in G satisfying $g \in P$ and $hgh^{-1} \notin P$, again an open condition.

(ii) Assume that G is countable. Fix an enumeration $g_1, g_2, \ldots$ of G and, for X_1, X_2 included in G, define

$$\mathrm{dist}(X_1, X_2) = \sum_{g_k \in (X_1 \setminus X_2) \cup (X_2 \setminus X_1)} 2^{-k}.$$

Then dist is a distance on $\mathfrak{P}(G)$. Hence $\mathfrak{P}(G)$ is a metric space, and so are its subspaces $\mathrm{LO}(G)$ and $\mathrm{O}(G)$. Now a classic result [149, Corollary 2.98] states that a nonempty compact metric space that is totally disconnected is homeomorphic to a subspace of the Cantor set, and it is homeomorphic to the Cantor set if and only if it has no isolated point.

Now the possibilities for $\mathrm{LO}(G)$, when it is nonempty, are that it is finite, countably infinite, homeomorphic to the Cantor space, or homeomorphic to a closed subspace of the Cantor space with isolated points. By a result of P. Linnell [175], $\mathrm{LO}(G)$ cannot be countably infinite. So we are left with the options of the statement. Results are similar for $\mathrm{O}(G)$. □

Example 1.15 (space $\mathrm{LO}(G)$). The monoid $(\mathbb{N}, +)$ admits only two left-invariant (hence bi-invariant) orderings, namely the standard ordering and the opposite ordering, so the corresponding space is finite. Another similar example is the Klein bottle group K of Example 1.5, which admits exactly four left-invariant orderings, namely the one associated with the monoid K^+ and the variants obtained by exchanging $\mathtt{a}$ and $\mathtt{a}^{-1}$ and/or $\mathtt{b}$ and $\mathtt{b}^{-1}$.

By contrast, for $n \geqslant 2$, if G is a countable free abelian group of rank at least two, the space $\mathrm{LO}(G)$ (which coincides with $\mathrm{O}(G)$) is infinite, homeomorphic to the Cantor set [213]. In the case of $\mathbb{Z}^2$, the space of orderings has a simple description: the positive cones correspond to half-planes containing $(0, 0)$, with the additional fact that an irrational value for the slope of the boundary of the half-plane gives rise to one cone, whereas a rational value gives rise to two cones due to the two options for which half-line the cone includes. It follows that $\mathrm{LO}(\mathbb{Z}^2)$ can be seen as the union of two circles $\mathbb{R}/\mathbb{Z}$ identified along the complement of $\mathbb{Q}/\mathbb{Z}$.

The case of non-abelian free groups is similar: for G free of rank n with $n \geqslant 2$, the group G is bi-orderable, and the spaces $\mathrm{LO}(G)$ and $\mathrm{O}(G)$ are homeomorphic to the Cantor set [180, 188].

Left-orderable groups having only finitely many left-invariant orderings have been fully classified by Tararin [223, 160]. So, in view of Proposition 1.14 and of the situations described in Example 1.15, we are left with

Question 1.16. *Does there exist a countable orderable group G such that the space $\mathrm{LO}(G)$ is infinite, but admits isolated points?*

This is the main problem addressed in the rest of this chapter. Coming back to the definition of the topology on the space $\mathrm{LO}(G)$, we directly obtain a characterization of the (positive cones of) left-invariant orderings that are isolated.

Lemma 1.17. *Assume that $<$ is a left-invariant ordering on a group G.*

(i) *(The positive cone of) $<$ is an isolated point in $\mathrm{LO}(G)$ if and only if there exists a finite subset $\{g_1, \ldots, g_p\}$ of G such that $<$ is the only left-invariant ordering of G satisfying $1 \leqslant g_i$ for $i = 1, \ldots, p$.*

(ii) *This situation occurs, in particular, if the positive cone of* $<$ *is a finitely generated semigroup.*

Proof. (i) By definition, the positive cone $P_<^+$ of $<$ is isolated in $\mathrm{LO}(G)$ if and only if the singleton $\{P_<^+\}$ is open, hence, by the construction of the topology and with the notation of Lemma 1.13, if there exists a finite family $g_1, \dots, g_p$ satisfying

$$\{P_<^+\} = U_{1,g_1} \cap \cdots \cap U_{1,g_p}. \tag{1.18}$$

Now (1.18) means that $P_<^+$ is the only positive cone in $\mathrm{LO}(G)$ that contains $g_1, \dots, g_p$.

(ii) Assume that $P_<^+$ is generated by $g_1, \dots, g_p$. Let P be any positive cone on G that contains $g_1, \dots, g_p$. Then P includes the subsemigroup of G generated by $g_1, \dots, g_p$, which is $P_<^+$. This implies $P = P_<^+$. Indeed, if we had $g \in P \setminus P_<^+$, we would deduce $g^{-1} \in P_<^+$, whence $g^{-1} \in P$, contradicting the assumption that P is a cone. □

So, in order to obtain a positive answer to Question 1.16, it is sufficient to construct orderable groups with infinitely many left-invariant orderings and such that at least one such ordering has a positive cone that is a finitely generated semigroup—we do not claim that Lemma 1.17(ii) is an equivalence, so the approach need not be the only possible one. In the language of monoids, we can summarize the situation in the following refinement of Proposition 1.8:

Proposition 1.19 (isolated point). *Let G be a group. Assume that*

(1.20) *G admits a positive presentation (S, R) such that S is finite and the monoid $\langle S \mid R\rangle^+$ is of O-type.*

Then the subsemigroup of G generated by S is the positive cone of a left-invariant ordering on G that is isolated in $\mathrm{LO}(G)$.

Proof. Condition (1.20) refines (1.9) and, therefore, Proposition 1.8 implies that the subsemigroup of G generated by S is the positive cone of a left-invariant ordering on G. Now, by assumption, S is finite, so the latter submonoid is finitely generated and Lemma 1.17 implies it is an isolated point in $\mathrm{LO}(G)$. □

Note that, in order to use a group G satisfying (1.20) to answer Question 1.16, one still has to verify that G admits infinitely many left-orderings: for instance, the presentation $\langle \mathtt{a}, \mathtt{b} \mid \mathtt{a} = \mathtt{bab}\rangle$ of the Klein bottle group K is eligible for (1.20) but, as was mentioned above, the group K admits exactly four left-invariant orderings, which are of course isolated in the discrete space $\mathrm{LO}(K)$. So this example does not answer Question 1.16.

1.3 Two examples

Automorphisms, in particular inner automorphisms, of the ambient group induce actions on (invariant) orderings. Such actions may be used in view of recognizing limit points in $\mathrm{LO}(G)$, as we shall see on two examples.

Lemma 1.21. *For G a group, ϕ an automorphism of G, and $<$ an ordering on G, define $<_\phi$ by $g <_\phi h \Leftrightarrow \phi^{-1}(g) < \phi^{-1}(h)$. Then $<_\phi$ is a left-invariant (resp. bi-invariant) ordering on G if and only if $<$ is, and defining $\phi \bullet < \, = \, <_\phi$ provides actions of* $\mathrm{Aut}(G)$ *on* $\mathrm{LO}(G)$ *and* $\mathrm{O}(G)$ *by homeomorphisms.*

Proof. The verifications are straightforward. For g, h in G and ϕ, ψ in $\mathrm{Aut}(G)$, we find

$$g <_{\phi\psi} h \Leftrightarrow (\phi\psi)^{-1}(g) < (\phi\psi)^{-1}(h) \Leftrightarrow \phi^{-1}(g) <_\psi \phi^{-1}(h) \Leftrightarrow g\ (<_\psi)_\phi\ h,$$

that is, $\phi\psi \bullet < \, = \phi \bullet (\psi \bullet <)$, as expected for a (left-)action. □

Restricting the above actions to inner automorphisms of G amounts to defining actions of G on $\mathrm{LO}(G)$ and $\mathrm{O}(G)$ by $f \bullet < \, = \, <_f$, with $g <_f h$ meaning $f^{-1}gf < f^{-1}hf$, that is, $g^f < h^f$. The ordering $<_f$ will be called the *conjugate* of $<$ by f. In terms of positive cones, if P is the positive cone of $<$, then the positive cone of $<_f$ is ${}^fP$, that is, fPf^{-1}.

Note that $\mathrm{O}(G)$ is the subspace of $\mathrm{LO}(G)$ made of all elements that are fixed under conjugacy: if $g < h$ is equivalent to $g^f < h^f$ for every f, then $g < h$ implies $f^{-1}gf < f^{-1}hf$, whence $gf < hf$ as $<$ is left-invariant, so $<$ is right-invariant; conversely, if $<$ is bi-invariant, then $g < h$ implies $f^{-1}gf < f^{-1}hf$, whence $<_f \, = \, <$ for every f.

Example 1.22 (no isolated point). Let F_∞ be a free group of countable rank. Then, using automorphisms, we can easily show that the space $\mathrm{LO}(F_\infty)$ has no isolated point and therefore is homeomorphic to the Cantor set. Indeed, fix a countable base $\{\mathtt{a}_1, \mathtt{a}_2, \ldots\}$ of F_∞, and assume that $<$ is a left-invariant ordering of F_∞. Let U be a neighborhood of $<$ in $\mathrm{LO}(F_\infty)$. By the definition of the topology, there exists a finite sequence $g_1, \ldots, g_p$ of elements of F_∞ such that U includes $U_{1,g_1} \cap \cdots \cap U_{1,g_p}$. Each element g_k involves only finitely many generators $\mathtt{a}_i$, so there exists n such that $\mathtt{a}_n$ does not occur in $g_1, \ldots, g_p$. Let ϕ be the automorphism of F_∞ that exchanges $\mathtt{a}_n$ and $\mathtt{a}_n^{-1}$ and preserves $\mathtt{a}_i$ for $i \neq n$. Then ϕ fixes $g_1, \ldots, g_p$ and, therefore, $g_k >_\phi 1$ holds for every k. This implies that the ordering $<_\phi$ belongs to $U_{1,g_1} \cap \cdots \cap U_{1,g_p}$, hence to U. Now $<$ and $<_\phi$ cannot coincide, since, by construction, $\mathtt{a}_n$ is larger than 1 in one ordering and smaller in the other. So the ordering $<$ is not isolated in $\mathrm{LO}(F_\infty)$.

We now describe another example involving Artin's braid groups B_n, which are also known to be orderable.

Example 1.23 (limit of conjugates). Consider Artin's braid group B_n (Reference Structure 2, page 5). Say that a braid word is *σ_k-positive* (*resp.* *σ_k-negative*) if it contains no letter $\sigma_i^{\pm 1}$ with $i < k$ and it contains no letter σ_k^{-1} (*resp.* no letter σ_k). Then say that an element g of B_n is *σ_k-positive* if, among the various expressions of g in terms of the generators σ_i, at least one is σ_k-positive. It turns out that the family of all braids in B_n that are σ_k-positive for at least one k is a positive cone in B_n. We denote by $<_{\mathrm{D}}$ the associated ordering. For instance, we have $\sigma_2 <_{\mathrm{D}} \sigma_1$, since the quotient-braid $\sigma_2^{-1}\sigma_1$ is represented by the σ_1-positive word $\sigma_2^{-1}|\sigma_1$, and $\sigma_2^{-1}\sigma_1 <_{\mathrm{D}} \sigma_1\sigma_2^{-1}$, since the quotient-braid $\sigma_1^{-1}\sigma_2\sigma_1\sigma_2^{-1}$ is represented (among others) by the σ_1-positive word $\sigma_2|\sigma_1|\sigma_2^{-1}|\sigma_2^{-1}$.

We claim that the ordering $<_{\mathrm{D}}$ is not isolated in $\mathrm{LO}(B_n)$ and, to this end, we will show that $<_{\mathrm{D}}$ is a limit of its conjugates. For simplicity, we consider the case $n = 3$ only. Let P be the positive cone of $<_{\mathrm{D}}$ in B_3. Proving that P is a limit of its conjugates means showing that, for every finite subset S of P, we can find f such that fPf^{-1} includes S, but is distinct from P.

As a preparatory result, we observe that, if g is a σ_1-positive braid, there exists f that is σ_1-positive, does not commute with σ_2, and satisfies $1 <_{\mathrm{D}} f \leqslant_{\mathrm{D}} g$: if g does not commute with σ_2, we may take $f = g$, otherwise, the explicit description of the commutator of σ_2 in B_3 shows that g can be expressed as $(\sigma_1\sigma_2\sigma_1)^{2p}\sigma_2^q$ with $p > 0$, in which case we may take $f = \sigma_1\sigma_2$ (see Exercise 111).

So, assume that S is a finite subset of P. Let g be the $<_{\mathrm{D}}$-smallest σ_1-positive element of $S \cup \{\sigma_1\}$. By the above observation, we find f that satisfies $1 <_{\mathrm{D}} f \leqslant_{\mathrm{D}} g$ and does not commute with σ_2. Let h belong to S. If h is a power of σ_2, then one can show (Property S, [100, page 28]) that $1 <_{\mathrm{D}} f^{-1}hf$ holds. Otherwise, by assumption, we have $f \leqslant_{\mathrm{D}} g \leqslant_{\mathrm{D}} h$, hence $1 \leqslant_{\mathrm{D}} f^{-1}h$ and, a fortiori, $1 <_{\mathrm{D}} f^{-1}hf$ since f is σ_1-positive. So we have $f^{-1}Sf \subseteq P$, that is, $P \subseteq fPf^{-1}$. On the other hand, σ_2 is the least element of P, so $f\sigma_2 f^{-1}$, which, by construction, is not σ_2, is the least positive element of fPf^{-1}. Hence fPf^{-1} does not coincide with P. We conclude that the order $<_{\mathrm{D}}$ is the limit of its conjugates in $\mathrm{LO}(B_3)$. See Exercise 112 for the extension of the result to B_n.

Owing to Lemma 1.17, the above result shows that the positive cone associated with the ordering $<_{\mathrm{D}}$ on B_n is not finitely generated as a semigroup. Among the subsequent results that can be established about the ordering $<_{\mathrm{D}}$ on B_n is the fact that the closure of the (countable) family of all conjugates of $<_{\mathrm{D}}$ in $\mathrm{LO}(B_n)$ is a Cantor set, see Exercise 113.

2 Construction of isolated orderings

We will now show how an approach based on Garside elements in a non-Noetherian context can lead to explicit examples of groups with isolated left-invariant orderings. Several families of such groups will be described, including torus knot groups and some of their amalgamated products. In doing so, we shall answer Question 1.16 in the positive.

The construction involves presentations of a particular type called triangular (Subsection 2.1), and then concentrates on proving the existence of common right-multiples (Subsection 2.2). Various examples are mentioned in Subsection 2.3. Finally, effectivity questions are discussed in Subsection 2.4.

2.1 Triangular presentations

Proposition 1.19 invites to look for finite positive presentations (or, at least, positive presentations with a finite set of generators) that define monoids of O-type. Owing to the symmetry of the definition, we shall concentrate on recognizing monoids of right-O-type and then appeal to the criteria for the opposite presentation. To this end, we shall focus on presentations of a certain syntactical type called triangular.

If a monoid M is of right-O-type and it is generated by some subset S, then, for all s, t in S, the elements s and t are comparable with respect to $\preccurlyeq$, that is, $t = sg$ holds for some g, or *vice versa*. In other words, some relation of the particular form $t = sw$ must be satisfied in M. We shall consider presentations in which all relations have this form (see Subsection 3.3 for the limits of this approach).

Definition 2.1 (triangular relation). A positive relation $u = v$ is called *triangular* if either u or v consists of a single letter.

So, a triangular relation has the generic form $t = sw$, where s, t belong to the reference alphabet. The problem we shall address now is whether, assuming that (S, R) is a presentation consisting of triangular relations, the associated monoid is necessarily of right-O-type. The question is ill-posed: the presentation $(\mathtt{a}, \mathtt{b}, \mathtt{c}, \mathtt{c} = \mathtt{ab}, \mathtt{c} = \mathtt{ba})$ consists of two triangular relations, but the associated monoid M is a rank 2 free abelian monoid based on $\mathtt{a}$ and $\mathtt{b}$, and neither of $\mathtt{a}, \mathtt{b}$ is a right-multiple of the other, so M is not of right-O-type. Clearly, the problem in the above example is the existence of several relations $\mathtt{c} = ...$ simultaneously. We are thus led to restricting to particular families of triangular relations, namely the right-triangular presentations already mentioned in Section II.4. Let us repeat the definition here.

Definition 2.2 (triangular presentation). A right-complemented monoid presentation (S, R) associated with a syntactic right-complement θ is called *right-triangular* if there exists a (finite or infinite) enumeration $\mathtt{a}_1, \mathtt{a}_2, ...$ of S such that $\theta(\mathtt{a}_i, \mathtt{a}_{i+1})$ is defined and empty for every i and, for $i + 1 < j$, either $\theta(\mathtt{a}_i, \mathtt{a}_j)$ is undefined, or we have

$$\theta(\mathtt{a}_i, \mathtt{a}_j) = \varepsilon \quad \text{and} \quad \theta(\mathtt{a}_j, \mathtt{a}_i) = \theta(\mathtt{a}_j, \mathtt{a}_{j-1})|\cdots|\theta(\mathtt{a}_{i+1}, \mathtt{a}_i) \quad \text{for } i < j; \tag{2.3}$$

Then (R, S) is called *maximal* (*resp. minimal*) if $\theta(\mathtt{a}_i, \mathtt{a}_j)$ is defined for all i, j (*resp.* for $|i - j| \leqslant 1$ only). A *left-triangular* presentation is defined symmetrically by relations $\mathtt{a}_i = w_{i+1}\mathtt{a}_{i+1}$. A presentation is *triangular* if it is both right- and left-triangular.

As already noted in Chapter II, if a right-triangular presentation is not minimal, the relations $\mathtt{a}_i... = \mathtt{a}_j...$ with $|i - j| \geqslant 2$ are redundant due to (2.3), so removing them does not change the associated monoid. On the other hand, there is always only one way to complete a right-triangular presentation to a maximal one.

Example 2.4 (triangular presentation). Let $S = \{\mathtt{a}, \mathtt{b}, \mathtt{c}\}$ and R consist of $\mathtt{a} = \mathtt{bac}$ and $\mathtt{b} = \mathtt{cba}$. Then (S, R) is a minimal right-triangular presentation with respect to the enumeration $(\mathtt{a}, \mathtt{b}, \mathtt{c})$ of S, as we have $R = \{\mathtt{a} = \mathtt{b} \cdot \mathtt{ac}, \mathtt{b} = \mathtt{c} \cdot \mathtt{ba}\}$; the associated maximal right-triangular presentation is $\widehat{R} = R \cup \{\mathtt{a} = \mathtt{cba}^2\mathtt{c}\}$. The presentation (S, R) is also left-triangular, but now with respect to the enumeration $(\mathtt{b}, \mathtt{a}, \mathtt{c})$, since we can write $R = \{\mathtt{b} = \mathtt{cb} \cdot \mathtt{a}, \mathtt{a} = \mathtt{ba} \cdot \mathtt{c}\}$.

The main technical result is the following criterion for recognizing which right-triangular presentations give rise to a monoid of right-O-type.

Lemma 2.5. *For every right-triangular presentation (S, R), the following are equivalent:*
(i) *The monoid $\langle S \mid R\rangle^+$ is of right-O-type.*
(ii) *Any two elements of $\langle S \mid R\rangle^+$ admit a common right-multiple.*

The proof of Lemma 2.5 relies on using the reversing technique of Section II.4 and Proposition II.4.51 (completeness), which states that right-reversing is complete for a maximal right-triangular presentation, implying in particular that, if u, v are $\mathcal{R}$-equivalent words, then the signed word $\overline{u}v$ right-reverses to the empty word. The (not very difficult) proof of Proposition II.4.51 is given in Appendix at the end of the book. Here comes the key observation for proving Lemma 2.5.

Lemma 2.6. *If (S, R) is a positive presentation consisting of triangular relations, and u, v, u', v' are S-words and $u^{-1}v$ is right-reversible to $v'u'^{-1}$, then at least one of u', v' is empty.*

The proof uses induction on the lengths of the words, and it relies on the fact that concatenating reversing diagrams in which at least one of the output edges is empty yields a diagram in which at least one of the output edge is empty, as illustrated in Figure 1.

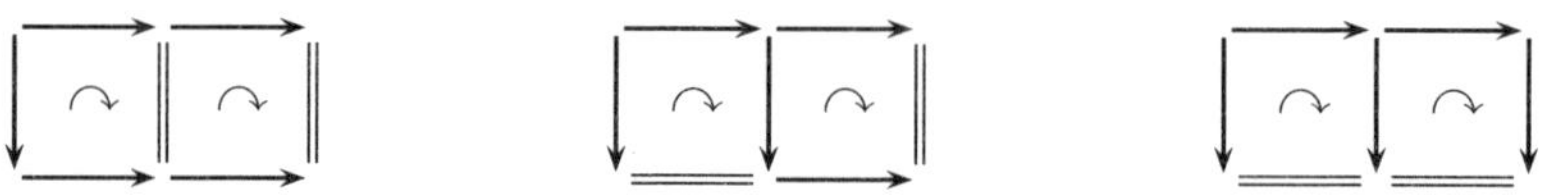

Figure 1. The three possible ways of concatenating two reversing diagrams in which one of the output words is empty: in each case, one of the final output words has to be empty.

It follows that, in the context of triangular relations, when reversing is terminating, it shows not only that the elements of the monoid represented by the initial words admit a common right-multiple, but also that these elements are comparable with respect to left-divisibility. It is then easy to establish Lemma 2.5.

Proof of Lemma 2.5. Put $M = \langle S \mid R\rangle^+$. If M is of right-O-type, any two elements of M are comparable with respect to left-divisibility, hence they certainly admit a common right-multiple, namely the larger of them. So (i) trivially implies (ii).

Conversely, assume that any two elements of M admit a common right-multiple. First, as M admits a right-triangular presentation, hence a positive presentation, 1 is the only invertible element in M.

Next, by Proposition II.4.44 (left-cancellativity), M must be left-cancellative since the presentation $(S, \widehat{R})$ is right-complemented, hence contains no relation of the form $su = sv$ with $u \neq v$.

Finally, let g, h belong to M. By Proposition II.4.46 (common right-multiple), which is relevant as right-reversing is complete for $(S, \widehat{R})$, there exist S-words u, v representing g and h and such that right-reversing $u^{-1}v$ terminates, that is, there exist S-words u', v' satisfying $u^{-1}v \curvearrowright_{\widehat{R}} v'u'^{-1}$. By construction, the family $\widehat{R}$ consists of triangular relations so, by Lemma 2.6, at least one of the words u', v' is empty. This means that at least one of $g \preccurlyeq h$ or $h \preccurlyeq g$ holds in M, that is, g and h are comparable with respect to left-divisibility. So M is a monoid of right-O-type, and (ii) implies (i). □

Once Lemma 2.5 is available, it is easy to merge the results and establish the following criterion for identifying orderable groups using the current approach:

Proposition 2.7 (triangular presentation). *If a group G admits a triangular presentation (S, R) such that any two elements of the monoid $\langle S \mid R\rangle^+$ have a common right-multiple and a common left-multiple, then G is orderable, and the subsemigroup of G generated by S is the positive cone of a left-invariant ordering on G. If S is finite, this ordering is an isolated point in the space $\mathrm{LO}(G)$.*

Proof. By Lemma 2.5, the monoid $\langle S \mid R\rangle^+$, which admits a right-triangular presentation is of right-O-type, and so is the opposite monoid. Hence $\langle S \mid R\rangle^+$ is also of left-O-type, and therefore it is of O-type. Then Propositions 1.8 and 1.19 imply that G is orderable, with the expected explicit ordering. □

2.2 Existence of common multiples

Owing to Lemma 2.5, to establish that a monoid specified by a right-triangular presentation is of right-O-type, one needs to prove that any two elements admit a common right-multiple. This is the point where Garside theory naturally appears. Indeed, in Chapter V, we derived the existence of common multiples from the existence of a right-Garside map, or even of a map that satisfies some, but not necessarily all of the conditions defining right-Garside maps. In our current monoidal context, the relevant result is Corollary V.1.48 (common right-multiple), which says a left-cancellative monoid M containing an element Δ such that the left-divisors of Δ generate M and include the right-divisors of Δ must admit common right-multiples. In a context of right-triangular presentations, the conditions can be restated in a more convenient way, leading to the following criterion.

Proposition 2.8 (right-Garside element). *If a monoid M admits a right-triangular presentation (S, R) and contains an element Δ satisfying $s \preccurlyeq \Delta \preccurlyeq s\Delta$ for every s in S, then M is a monoid of right-O-type and Δ is a right-Garside element in M.*

Proof. First, every element of S is a left-divisor of Δ, so $\mathrm{Div}(\Delta)$ generates M. Next, assume that g right-divides Δ, say $fg = \Delta$. As $\mathrm{Div}(\Delta)$ generates M, there exists a decomposition $f = s_1 \cdots s_p$ with $s_1, \ldots, s_p$ in $\mathrm{Div}(\Delta)$. By assumption, for every i, there exists h_i satisfying $\Delta h_i = s_i \Delta$. Then we find the equalities

$$fgh_1 \cdots h_p = \Delta h_1 \cdots h_p = s_1 \Delta h_2 \cdots h_p = \cdots = s_1 \cdots s_p \Delta = f\Delta,$$

whence $gh_1 \cdots h_p = \Delta$ by left-cancelling f. So g left-divides Δ and $\widetilde{\mathrm{Div}}(\Delta) \subseteq \mathrm{Div}(\Delta)$ holds. Then Corollary V.1.48 (common multiple) implies that any two elements of M admit a common right-multiple, and Lemma 2.5 then implies that M is of right-O-type.

Moreover, once we know that any two elements of M are comparable with respect to $\preccurlyeq$, we immediately deduce that any two elements of M admit a left-gcd, namely the smaller of them. Hence Δ is a right-Garside element in M. □

Corollary 2.9 (right-Garside element). *If a monoid M admits a right-triangular presentation (S, R) and contains a central element Δ satisfying $s \preccurlyeq \Delta$ for every s in S, then M is a monoid of right-O-type and Δ is a right-Garside element in M.*

Proof. If Δ is central, it commutes with every element, hence in particular with every element s of S, so $\Delta \preccurlyeq s\Delta$ automatically holds since $s\Delta$ is Δs. Then Proposition 2.8 applies. □

Example 2.10 (right-Garside element). For $p, q, r \geqslant 1$, let $M_{p,q,r}$ be the monoid $\langle \mathtt{a}, \mathtt{b} \mid \mathtt{a} = \mathtt{b}(\mathtt{a}^p\mathtt{b}^r)^q \rangle^+$. The given presentation is right-triangular, and we claim that the monoid $M_{p,q,r}$ is eligible for Proposition 2.8 with respect to $\Delta = \mathtt{a}^{p+1}$.

Indeed, we first find $\mathtt{b} \preccurlyeq \mathtt{a} \preccurlyeq \Delta$, and $\Delta \preccurlyeq \mathtt{a}\Delta$ trivially holds since Δ is a power of $\mathtt{a}$. So, it just remains to establish $\Delta \preccurlyeq \mathtt{b}\Delta$. Now, applying the defining relation, we first obtain

$$\mathtt{a} = \mathtt{b}(\mathtt{a}^p\mathtt{b}^r)^q = \mathtt{b} \cdot \mathtt{a} \cdot (\mathtt{a}^{p-1}\mathtt{b}^r)(\mathtt{a}^p\mathtt{b}^r)^{q-1},$$

whence, repeating the operation r times and moving the brackets,

$$\mathtt{a} = \mathtt{b}^r \cdot \mathtt{a} \cdot ((\mathtt{a}^{p-1}\mathtt{b}^r)(\mathtt{a}^p\mathtt{b}^r)^{q-1})^r = \mathtt{b}^r(\mathtt{a}^p\mathtt{b}^r)^q \cdot ((\mathtt{a}^{p-1}\mathtt{b}^r)(\mathtt{a}^p\mathtt{b}^r)^{q-1})^{r-1}.$$

Substituting the above value of $\mathtt{a}$ at the underlined position, moving the brackets, and applying the relation once in the contracting direction, we deduce the equalities

$$\begin{aligned}\mathtt{b} \cdot \Delta = \mathtt{b}\mathtt{a}^p\underline{\mathtt{a}} &= \mathtt{b}\mathtt{a}^p \cdot \mathtt{b}^r(\mathtt{a}^p\mathtt{b}^r)^q \cdot ((\mathtt{a}^{p-1}\mathtt{b}^r)(\mathtt{a}^p\mathtt{b}^r)^{q-1})^{r-1}\\ &= \mathtt{b}(\mathtt{a}^p\mathtt{b}^r)^q \cdot \mathtt{a}^p\mathtt{b}^r((\mathtt{a}^{p-1}\mathtt{b}^r)(\mathtt{a}^p\mathtt{b}^r)^{q-1})^{r-1}\\ &= \mathtt{a} \cdot \mathtt{a}^p \cdot \mathtt{b}^r((\mathtt{a}^{p-1}\mathtt{b}^r)(\mathtt{a}^p\mathtt{b}^r)^{q-1})^{r-1} = \Delta \cdot \mathtt{b}^r((\mathtt{a}^{p-1}\mathtt{b}^r)(\mathtt{a}^p\mathtt{b}^r)^{q-1})^{r-1},\end{aligned}$$

whence $\Delta \preccurlyeq \mathtt{b}\Delta$ as expected. So $M_{p,q,r}$ is eligible for Proposition 2.8, it is a monoid of right-O-type, and Δ is a right-Garside element in $M_{p,q,r}$. The above computations show that the associated endomorphism is given by

$$\phi_\Delta(\mathtt{a}) = \mathtt{a}, \qquad \phi_\Delta(\mathtt{b}) = \mathtt{b}^r((\mathtt{a}^{p-1}\mathtt{b}^r)(\mathtt{a}^p\mathtt{b}^r)^{q-1})^{r-1}. \tag{2.11}$$

For $r = 1$, the formula reduces to $\phi_\Delta(\mathtt{b}) = \mathtt{b}$, hence ϕ_Δ is the identity and Δ is a central element, and so a Garside element. On the other hand, for $r \geqslant 2$, the functor ϕ_Δ is not surjective: for instance, we have $\phi_\Delta(\mathtt{b}) = \mathtt{b}^4$ in $M_{1,1,2}$, and $\phi_\Delta(\mathtt{b}) = \mathtt{b}^2\mathtt{ab}^2$ in $M_{2,1,2}$. So, in such cases, Δ is a right-Garside element that is not a Garside element.

In order to obtain ordered groups, we need monoids of O-type, and not only of right-O-type. Now applying Proposition 2.8 to the opposite monoid $\widetilde{M}$ gives a straightforward criterion for establishing that a monoid M is possibly of left-O-type, leading to:

Proposition 2.12 (O-type). *Every monoid that admits a triangular presentation (S, R) and contains elements $\Delta, \widetilde{\Delta}$ satisfying $s \preccurlyeq \Delta \preccurlyeq s\Delta$ and $\widetilde{\Delta}s \succneqq \widetilde{\Delta} \succneqq s$ for every s in S is of O-type.*

Corollary 2.13 (O-type). *Every monoid that admits a triangular presentation (S, R) and contains a central element Δ satisfying $s \preccurlyeq \Delta$ and $\Delta \succneqq s$ for every s in S is of O-type.*

Example 2.14 (O-type). Consider the monoids $M_{p,q,r}$ of Example 2.10 with $r = 1$. In this case, the defining relation reduces to $\mathtt{a} = \mathtt{b}(\mathtt{a}^p\mathtt{b})^q$, a symmetric relation. As already noted, the element Δ, that is, $\mathtt{a}^{p+1}$, is central. We trivially have $\mathtt{b} \preccurlyeq \mathtt{a} \preccurlyeq \Delta$ and $\Delta \succneqq \mathtt{a} \succneqq \mathtt{b}$, so Corollary 2.13 says that $M_{p,q,1}$ is of O-type. As the Garside morphism ϕ_Δ is the identity, Δ is a Garside element.

By contrast, for $r \geqslant 2$, the monoid $M_{p,q,r}$ is not of left-O-type, and it embeds in a group of right-fractions that is not a group of left-fractions: one can show that $\mathtt{a}$ and $\mathtt{ab}$ have no common left-multiple in $M_{p,q,r}$ and the right-fraction $\mathtt{aba}^{-1}$ is a typical element of the enveloping group that cannot be expressed as a left-fraction. In this case, the semigroup $M_{p,q,r}\backslash\{1\}$ defines a partial left-invariant ordering on the enveloping group only: for instance, the elements $\mathtt{b}^{-1}\mathtt{a}^{-1}$ and $\mathtt{a}^{-1}$ are not comparable, as their quotient $\mathtt{aba}^{-1}$ belongs neither to $M_{p,q,r}$ nor to $M_{p,q,r}^{-1}$. Note that, for $p = q = 1$, the involved group $\langle \mathtt{a}, \mathtt{b} \mid \mathtt{a} = \mathtt{bab}^{r+1}\rangle$ is the Baumslag–Solitar group $\mathrm{BS}(r+1, -1)$, whereas the opposite group $\langle \mathtt{a}, \mathtt{b} \mid \mathtt{a} = \mathtt{b}^{r+1}\mathtt{ab}\rangle$ is $\mathrm{BS}(-1, r+1)$.

Returning to groups, we thus obtain a method for identifying orderable groups and, more specifically, isolated left-invariant orderings.

Proposition 2.15 (isolated orderings I). *Suppose the group G admits a triangular presentation (S, R) and, in the monoid $\langle S \mid R\rangle^+$, there exist elements $\Delta, \widetilde{\Delta}$ satisfying $s \preccurlyeq \Delta \preccurlyeq s\Delta$ and $\widetilde{\Delta}s \succneqq \widetilde{\Delta} \succneqq s$ for every s in S. Then G is orderable and the sub-semigroup of G generated by S is the positive cone of a left-invariant ordering on G. If S is finite, this ordering is an isolated point in the space $\mathrm{LO}(G)$.*

Proof. The monoid $\langle S \mid R\rangle^+$ is eligible for Proposition 2.12, so it is of O-type. Then we apply Proposition 2.7. □

If, instead of using Proposition 2.12, we appeal to Corollary 2.13, we obtain similarly:

Corollary 2.16 (isolated orderings I). *Suppose the group G admits a triangular presentation (S, R) and, in the monoid $\langle S \mid R\rangle^+$, there exists a central element Δ satisfying $s \preccurlyeq \Delta$ and $\widetilde{\Delta} \succneqq s$ for every s in S. Then G is orderable and the subsemigroup of G generated by S is the positive cone of a left-invariant ordering on G. If S is finite, this ordering is an isolated point in the space* $\mathrm{LO}(G)$.

Example 2.17 (isolated orderings I). For $p, q \geqslant 1$, let $T_{p,q}$ be the (p, q)-torus knot group of Example I.2.7, defined by the presentation $\langle \mathtt{x}, \mathtt{y} \mid \mathtt{x}^{p+1} = \mathtt{y}^{q+1}\rangle$. Put $\mathtt{a} = \mathtt{x}$ and $\mathtt{b} = \mathtt{x}^{-p}\mathtt{y}$. Then $\mathtt{a}$ and $\mathtt{b}$ generate $T_{p,q}$, and standard verifications show that, in terms of $\mathtt{a}$ and $\mathtt{b}$, the group $T_{p,q}$ admits the presentation $\langle \mathtt{a}, \mathtt{b} \mid \mathtt{a} = \mathtt{b}(\mathtt{a}^p\mathtt{b})^q\rangle$. But we observed in Example 2.14 that the monoid $\langle \mathtt{a}, \mathtt{b} \mid \mathtt{a} = \mathtt{b}(\mathtt{a}^p\mathtt{b})^q\rangle^+$ is eligible for Corollary 2.13 since, putting $\Delta = \mathtt{a}^{p+1}$, the element Δ is a left- and a right-multiple of both $\mathtt{a}$ and $\mathtt{b}$. Then Corollary 2.16 says that $T_{p,q}$ is orderable, and that the subsemigroup of $T_{p,q}$ generated by $\mathtt{x}$ and $\mathtt{x}^{-p}\mathtt{y}$ is the positive cone of a left-invariant ordering on G which is isolated in $\mathrm{LO}(G)$.

Note that, once observed that the group $\langle \mathtt{a}, \mathtt{b} \mid \mathtt{a} = \mathtt{b}(\mathtt{a}^p\mathtt{b})^q\rangle$ is isomorphic to the group $\langle \mathtt{x}, \mathtt{y} \mid \mathtt{x}^{p+1} = \mathtt{y}^{q+1}\rangle$, it is obvious that $\mathtt{x}^{p+1}$, that is, $\mathtt{a}^{p+1}$, is central in the group. However, this is *not* sufficient to deduce that $\mathtt{a}^{p+1}$ is central in the monoid $\langle \mathtt{a}, \mathtt{b} \mid \mathtt{a} = \mathtt{b}(\mathtt{a}^p\mathtt{b})^q\rangle^+$ as the latter is not a priori known to embed in the group: it is crucial to make all verifications inside the monoid, that is, without using inverses except possibly those provided by cancellativity.

Two special cases are worth mentioning. First, $T_{1,1}$, that is, $\langle \mathtt{x}, \mathtt{y} \mid \mathtt{x}^2 = \mathtt{y}^2\rangle^+$, is what we called the Klein bottle group $\langle \mathtt{a}, \mathtt{b} \mid \mathtt{a} = \mathtt{bab}\rangle$ (Reference Structure 5, page 17), which we already mentioned as an example of an orderable group with finitely many left-invariant orderings.

Less trivial is $T_{2,1}$, that is, $\langle \mathtt{x}, \mathtt{y} \mid \mathtt{x}^3 = \mathtt{y}^2\rangle^+$. As noted in Chapter IX, this group is the 3-strand braid group B_3. The submonoid of B_3 involved above is generated by $\mathtt{a} = \mathtt{x}$ and $\mathtt{b} = \mathtt{x}^{-2}\mathtt{y}$, that is, in terms of Artin's generators, $\mathtt{a} = \sigma_1\sigma_2$ and $\mathtt{b} = \sigma_2^{-1}$. Thus we established that the subsemigroup of B_3 generated by $\sigma_1\sigma_2$ and σ_2^{-1} is the positive cone of a left-invariant ordering $<_{\mathrm{DD}}$ on B_3 that is isolated in the space $\mathrm{LO}(B_3)$. In terms of the σ_k-positive braids introduced in Example 1.23, its definition implies that the positive cone of $<_{\mathrm{DD}}$ consists of all 3-strand braids that are either σ_1-positive or σ_2-negative, contrasting with the positive cone of the ordering $<_{\mathrm{D}}$ which consists of all 3-strand braids that are either σ_1-positive or σ_2-positive. The braid group B_3 is also obtained as $T_{1,2}$, corresponding to the presentation $\langle \mathtt{a}, \mathtt{b} \mid \mathtt{a} = \mathtt{babab}\rangle$, with $\mathtt{a}$ and $\mathtt{b}$ now realizable as $\sigma_1\sigma_2\sigma_1$ and σ_2^{-1} (see Exercise 115).

The above example shows that Artin's braid group B_3 admits an isolated ordering. So, owing to Example 1.23, in which we showed that there exist infinitely many left-invariant orderings on B_3, we obtain a positive answer to Question 1.16:

Proposition 2.18 (not Cantor set). *The space* $\mathrm{LO}(B_3)$ *is infinite, but not homeomorphic to the Cantor set.*

2.3 More examples

We describe further examples of groups for which the current method based on Garside elements is relevant. We begin with a fairly large family of amalgamated products of torus knot groups.

Proposition 2.19 (amalgamated torus groups). *Let G be the group*

$$\langle \mathtt{x}_1, \mathtt{x}_2, \ldots, \mathtt{x}_\ell \mid \mathtt{x}_1^{m_2+1} = \mathtt{x}_2^{n_2+1}, \mathtt{x}_2^{m_3+1} = \mathtt{x}_3^{n_3+1}, \ldots, \mathtt{x}_{\ell-1}^{m_\ell+1} = \mathtt{x}_\ell^{n_\ell+1} \rangle \tag{2.20}$$

with $\ell \geqslant 2$ and $m_2, n_2, m_3, n_3, \ldots, m_\ell, n_\ell \geqslant 1$. Then G is orderable and the subsemigroup of G generated by $\mathtt{x}_1$, $\mathtt{x}_1^{-m_2}\mathtt{x}_2$, ..., $\mathtt{x}_1^{-m_2} \cdots \mathtt{x}_{\ell-1}^{-m_\ell}\mathtt{x}_\ell$ is the positive cone of a left-invariant ordering on G which is isolated in $\mathrm{LO}(G)$.

Proof. Put $\mathtt{a}_1 := \mathtt{x}_1$ and $\mathtt{a}_i := \mathtt{x}_1^{-m_2} \cdots \mathtt{x}_{i-1}^{-m_i}\mathtt{x}_i$ in G for $2 \leqslant i \leqslant \ell$, and inductively define positive words $w_1, \ldots, w_\ell$ in the alphabet $\{\mathtt{a}_1, \ldots, \mathtt{a}_\ell\}$ by $w_1 := \mathtt{a}_1$ and $w_i := w_{i-1}^{m_i} \cdots w_2^{m_3} w_1^{m_2} \mathtt{a}_i$ for $i \geqslant 2$. Easy inductions on i give $[w_i] = \mathtt{x}_i$ for every i and $\mathtt{a}_{i-1} = \mathtt{a}_i[w_i]^{n_i}$ in G for $i \geqslant 2$. Hence G is a quotient of the group G' that admits the triangular presentation

$$\langle \mathtt{a}_1, \ldots, \mathtt{a}_\ell \mid \mathtt{a}_1 = \mathtt{a}_2 w_2^{n_2}, \ldots, \mathtt{a}_{\ell-1} = \mathtt{a}_\ell w_\ell^{n_\ell} \rangle. \tag{2.21}$$

Let M be the monoid defined by the (positive) presentation (2.21) and, for $1 \leqslant i \leqslant \ell$, let g_i be the element of M represented by w_i. Using the relation $\mathtt{a}_{i-1} = \mathtt{a}_i g_i^{n_i}$, we obtain

$$g_{i-1}^{m_i+1} = g_{i-1}^{m_i} g_{i-1} = g_{i-1}^{m_i} g_{i-2}^{m_{i-1}} \cdots g_1^{m_2} \mathtt{a}_{i-1} = g_{i-1}^{m_i} \cdots g_1^{m_2} \mathtt{a}_i g_i^{n_i} = g_i \cdot g_i^{n_i} = g_i^{n_i+1}$$

in M for $i \geqslant 2$. The same computation is a fortiori valid in the group G', yielding $[w_{i-1}]^{m_i+1} = [w_i]^{n_i+1}$ in this group: this implies that G and G' are isomorphic, that is, (2.21) is a presentation of G.

Next, put $\Delta := \mathtt{a}_1^{(m_2+1)\cdots(m_\ell+1)}$ in M. Using the relations $g_{i-1}^{m_i+1} = g_i^{n_i+1}$, we inductively obtain $\Delta = g_i^{e_i}$ with $e_i = (n_2+1)\cdots(n_i+1)(m_{i+1}+1)\cdots(m_{\ell+1}+1)$ for $i \geqslant 2$. It follows that Δ commutes with every element g_i in M. We inductively deduce that Δ commutes with every element $\mathtt{a}_i$. For $i = 1$, this follows from $\mathtt{a}_1 = g_1$. For $i \geqslant 2$, using the relation $\mathtt{a}_{i-1} = \mathtt{a}_i g_i$ and the induction hypothesis, we write

$$\Delta \mathtt{a}_i g_i = \Delta \mathtt{a}_{i-1} = \mathtt{a}_{i-1}\Delta = \mathtt{a}_i g_i \Delta = \mathtt{a}_i \Delta g_i,$$

whence $\Delta \mathtt{a}_i = \mathtt{a}_i \Delta$ by right-cancelling g_i, which is legitimate as M, which admits a left-triangular presentation (by definition, the word w_i finishes with the letter $\mathtt{a}_i$), must be right-cancellative. As $\mathtt{a}_1, \ldots, \mathtt{a}_\ell$ generate M, the element Δ is central in M. By Corollary 2.16, M is of O-type, and G is orderable with an ordering as expected. $\square$

Example 2.22 (amalgamated torus groups). For $m_2 = n_2 = \cdots = m_\ell = n_\ell = p$ in Proposition 2.19, G admits the presentation $\langle \mathtt{x}_1, \dots, \mathtt{x}_\ell \mid \mathtt{x}_1^{p+1} = \mathtt{x}_2^{p+1} = \cdots = \mathtt{x}_\ell^{p+1} \rangle$, and the result applies with $\Delta = \mathtt{x}_1^{p+1}$. The positive cone of the associated isolated ordering is defined by the presentation (2.21), whose relations, in the current case, take the form (we write $\mathtt{a}, \mathtt{b}, \dots$ for $\mathtt{a}_1, \mathtt{a}_2, \dots$) $\mathtt{a} = \mathtt{b}(\mathtt{a}^p\mathtt{b})^p$, $\mathtt{b} = \mathtt{c}((\mathtt{a}^p\mathtt{b})^p\mathtt{a}^p\mathtt{c})^p$, $\mathtt{c} = \mathtt{d}(((\mathtt{a}^p\mathtt{b})^p\mathtt{a}^p\mathtt{c})^p(\mathtt{a}^p\mathtt{b})^p\mathtt{a}^p\mathtt{d})^p$, *etc.*

Further examples of eligible groups for Proposition 2.15 are listed in Table 1.

1:	$\langle \mathtt{x}, \mathtt{y} \mid \mathtt{x}^{r+1} = (\mathtt{y}\mathtt{x}^2\mathtt{y})^{q+1} \rangle$ with $r = 0$ or $r = 1$	$\langle \mathtt{a}, \mathtt{b}, \mathtt{c} \mid \mathtt{a} = \mathtt{b}\mathtt{a}^2(\mathtt{b}^2\mathtt{a}^2)^q\mathtt{c}, \mathtt{b} = \mathtt{c}(\mathtt{b}\mathtt{a}^2)^r\mathtt{b}\mathtt{a} \rangle^+$ $\Delta = (\mathtt{b}\mathtt{a}^2)^{2q+r+3}$ Garside central
2:	$\langle \mathtt{x}, \mathtt{y} \mid \mathtt{x}^{q+2} = \mathtt{y}^2 \rangle$	$\langle \mathtt{a}, \mathtt{b} \mid \mathtt{a} = \mathtt{b}(\mathtt{a}\mathtt{b}^p)^q\mathtt{a}\mathtt{b} \rangle^+$ $\Delta = (\mathtt{a}\mathtt{b}^{p-1})^2$ Garside central
3:	$\langle \mathtt{x}, \mathtt{y} \mid \mathtt{x}^p = (\mathtt{y}\mathtt{x}^{r-p}\mathtt{y})^2 \rangle$ with $p + 1 \mid r$	$\langle \mathtt{a}, \mathtt{b} \mid \mathtt{a} = \mathtt{b}\mathtt{a}^r\mathtt{b}\mathtt{a}^p\mathtt{b}\mathtt{a}^r\mathtt{b} \rangle^+$ $\Delta = (\mathtt{a}^r\mathtt{b})^2$ right-Garside with $\phi_\Delta(\mathtt{a}) = \mathtt{a}^p(\mathtt{b}\mathtt{a}^r\mathtt{b})^2$, $\phi_\Delta(\mathtt{b}) = \mathtt{b}$
4:	$\langle \mathtt{x}, \mathtt{y}, \mathtt{z} \mid \mathtt{x}^{p+1} = \mathtt{y}^2, \mathtt{y} = \mathtt{z}\mathtt{y}\mathtt{x}^{r-p}\mathtt{z} \rangle$ with $p + 1 \mid r + 1$	$\langle \mathtt{a}, \mathtt{b}, \mathtt{c} \mid \mathtt{a} = \mathtt{b}\mathtt{a}^p\mathtt{b}, \mathtt{b} = \mathtt{c}\mathtt{b}\mathtt{a}^r\mathtt{c} \rangle^+$ $\Delta = \mathtt{a}^{p(r-p)+1}$ Garside central
5:	$\langle \mathtt{x}, \mathtt{y}, \mathtt{z} \mid \mathtt{x}^{p+1} = \mathtt{y}^2, \mathtt{y} = \mathtt{z}\mathtt{y}\mathtt{x}^{r-p}\mathtt{z} \rangle$ with $p + 1 \mid r$	$\langle \mathtt{a}, \mathtt{b}, \mathtt{c} \mid \mathtt{a} = \mathtt{b}\mathtt{a}^p\mathtt{b}, \mathtt{b} = \mathtt{c}\mathtt{b}\mathtt{a}^r\mathtt{c} \rangle^+$ $\Delta = \mathtt{a}^r\mathtt{b}^2$ Garside with $\phi_\Delta(\mathtt{b}) = \mathtt{b}$, $\phi_\Delta(\mathtt{a}) = \mathtt{a}^p\mathtt{b}\mathtt{a}^{p-1}\mathtt{b}^3$, $\phi_\Delta(\mathtt{c}) = \mathtt{c}$
6:	$\langle \mathtt{x}, \mathtt{y}, \mathtt{z} \mid \mathtt{x}^2 = (\mathtt{x}\mathtt{y})^{p+1}, \mathtt{z}\mathtt{y}\mathtt{z} = \mathtt{y}(\mathtt{x}^r\mathtt{y})^q \rangle$ with r odd	$\langle \mathtt{a}, \mathtt{b}, \mathtt{c} \mid \mathtt{a} = \mathtt{b}(\mathtt{a}\mathtt{b})^p, \mathtt{b} = \mathtt{c}\mathtt{b}(\mathtt{a}^r\mathtt{b})^p\mathtt{c} \rangle^+$ $\Delta = \mathtt{a}^{p(p+1)(r-1)+2}$ Garside central

Table 1. Some groups eligible for Proposition 2.15, hence ordered with an isolated point in the space of orderings: on the left, a "torus-type" presentation of the group, on the right, a triangular presentation such that the associated monoid is of O-type, together with a witnessing right-Garside element and the associated Garside morphism ϕ_Δ.

2.4 Effectivity questions

One benefit of the current approach is that it leads to effective procedures: not only do we obtain orderable groups, but also simple algorithms for solving related questions. Two different types of questions occur here naturally, namely solving the decision problem and the Word Problem once we know that a presentation (S, R) is eligible for Proposition 2.15 and deciding whether a candidate-presentation is eligible on the other hand. We begin with the first type. Here the answers are easy.

If (S, R) is a right-triangular presentation, we shall denote by $(S, \widehat{R})$ the maximal right-triangular presentation deduced from (S, R), that is, the equivalent presentation in which we add the "transitive closure" of the relations connecting adjacent generators. Similarly, if (S, R) is a left-triangular presentation, we shall denote by $(S, \check{R})$ the maximal left-triangular presentation obtained, with hopefully obvious notation, by adding the relations $\mathtt{a}_i = w_{i+1} \cdots w_{i+k}\mathtt{a}_{i+k}$.

Algorithm 2.23 (deciding order).

Context: A group G, a triangular presentation (S, R) of G, an element Δ of $\langle S \mid R\rangle^+$
Input: A signed S-path w
Output: The position of $[w]$ with respect to 1 in G

```
1: right-R̂-reverse w into v ū with u, v in S*;
2: left-Ř-reverse v ū into ū' v' with u', v' in S*;
3: if u' = ε then
4:     if v' = ε then
5:         return [w] = 1
6:     else return [w] > 1
7: else return [w] < 1.
```

Proposition 2.24 (deciding order). *If a group G admits a finite triangular presentation (S, R) eligible for Proposition* 2.15 *or Corollary* 2.16*, then Algorithm* 2.23 *decides the left-invariant ordering of G whose positive cone is the subsemigroup of G generated by S, and it solves the Word Problem for G.*

Proof. Let $(S, \widehat{R})$ be the maximal right-triangular presentation deduced from (S, R). By Proposition II.4.51 (completeness), right-reversing is complete for $(S, \widehat{R})$ and, by Proposition 2.12, any two elements in the monoid $\langle S \mid R\rangle^+$, which is also $\langle S \mid \widehat{R}\rangle^+$, admit a common right-multiple. It follows from Proposition II.4.46 (common multiple) that right-$\widehat{R}$-reversing is always terminating. For symmetric reasons, left-$\check{R}$-reversing is also always terminating. Hence Steps 1 and 2 in Algorithm 2.23 always terminate in finite time. Then, by construction, $[w] = [\overline{u'}v']$ holds in G and, by Lemma 2.6, at least one of the words u', v' is empty. If u' and v' are empty, $[w] = 1$ is obvious. If u' is empty and v' is not, $[w]$, which is $[v']$, belongs to the subsemigroup of G generated by S, that is, to the positive cone of the considered ordering. If u' is empty and v' is nonempty, $[w]$ is $[u']^{-1}$, and its inverse belongs to the positive cone.

A solution of the Word Problem immediately follows as both $[w] > 1$ and $[w] < 1$ imply $[w] \neq 1$. $\square$

Remark 2.25. Algorithm 2.23 actually gives more than just a "$=/>/<$" answer: for every initial signed S-word w, the method provides a positive S-word w' such that w is equivalent either to w' or or to w'^{-1}. In other words, it gives for every element g of the group G an explicit positive or negative decomposition of g in terms of the distinguished generators of the considered positive cone.

Example 2.26 (deciding order). Consider the braid ordering $<_{\mathrm{DD}}$ on B_3 mentioned at the end of Example 2.17 and, for instance, $g = \sigma_1^2\sigma_2^2$. Using the dictionary $\sigma_1 = \mathtt{ab}$,

$\sigma_2 = \mathtt{b}^{-1}$, we find $g = \mathtt{abab}^{-1}$. Running Algorithm 2.23 on the word $\mathtt{a}|\mathtt{b}|\mathtt{a}|\overline{\mathtt{b}}$ yields $\mathtt{a}|\mathtt{b}|\mathtt{b}|\mathtt{a}|\mathtt{a}$, a positive word. We deduce $1 <_{\mathrm{DD}} g$ and, more interestingly, the decomposition $\sigma_1\sigma_2|\sigma_2^{-1}|\sigma_2^{-1}|\sigma_1\sigma_2|\sigma_1\sigma_2$ of g in terms of $\sigma_1\sigma_2$ and σ_2^{-1}.

We observed in Chapter II that, when reversing is terminating and there exists a finite set of words (or paths) that is closed under reversing, then there exists a quadratic upper bound for the number of reversing steps in function of the length of the initial word (or path). In the current approach, except in degenerate cases, one is not in such a situation, and no uniform upper bound on the complexity of reversing can be expected. For instance, the presentation $(\mathtt{a}, \mathtt{b}, \mathtt{a} = \mathtt{bab}^{r+1})$ of the Baumslag–Solitar group $\mathrm{BS}(r+1, -1)$ corresponds to the monoid $M_{1,1,r}$ of Example 2.17. For every n, the signed word $\mathtt{a}^{-n}\mathtt{ba}^n$ reverses to the word $\mathtt{b}^{(r+1)^n}$, whose length is exponential in n. As every reversing step adds at most r letters, the number of steps needed to reverse the length-$(2n+1)$ word $\mathtt{a}^{-n}\mathtt{ba}^n$ must be exponential in n. However, the monoid $M_{1,1,r}$ is of right-O-type only, and such behaviors could not be found for monoids of O-type.

Question 2.27. *If a triangular presentation defines a monoid of O-type, does the associated reversing necessarily have a polynomial (quadratic?) complexity?*

Note that the existence of a right-Garside element that is not central need not imply an exponential complexity. For instance, for the presentation $(\mathtt{a}, \mathtt{b}, \mathtt{a} = \mathtt{ba}^2\mathtt{baba}^2\mathtt{b})$ with $\Delta = (\mathtt{a}^2\mathtt{b})^2$, we have $\phi_\Delta(\mathtt{a}) = \mathtt{a}(\mathtt{ba}^2\mathtt{b})^2$, and the shortest expression of $\phi_\Delta(\mathtt{a})$ is longer by 8 letters than that of $\mathtt{a}$. However, $\phi_\Delta(\mathtt{a}^2) = \mathtt{a}^2$ holds, and reversing $\Delta^{-n}\mathtt{a}\Delta^n$ leads to a word of length linear in n in a quadratic number of steps.

The monoid of Exercise 122 below is a good test-case for Question 2.27. It turns out that this monoid is of O-type (but it is not eligible for Proposition 2.15), and that reversing the length $2n$ word $\mathtt{a}^{-(n-1)}\mathtt{b}^{-1}\mathtt{a}^n$ leads to a word of length $(2n)^2$ in a number of steps that is (exactly) cubic in n: this is compatible with a positive answer to Question 2.27, but discards a uniform quadratic upper bound.

To conclude this subsection, let us briefly address the second question mentioned at the beginning, namely the question of effectively recognizing whether a group is eligible for Proposition 2.12 or Proposition 2.15. We can expect no method for finding a convenient presentation or a right-Garside element, so the only reasonable question is whether, given a candidate presentation (S, R) and a candidate element Δ, one can effectively recognize whether they are eligible. Recognizing whether a (finite) positive presentation is triangular is easy, and, therefore, the problem is to be able to check relations of the form $s \preccurlyeq \Delta$ or $\Delta \preccurlyeq s\Delta$. Then the situation is that of a typical semi-decidable problem: if Δ is indeed a right-Garside element, then $\langle S \mid R\rangle^+$ is a right-Ore monoid, hence right-reversing must be terminating, so, in this case, a positive answer can certainly be obtained in finite time. By contrast, if Δ is not a right-Garside element, then it may be that reversing $\overline{s}|\underline{\Delta}$ or $\overline{\underline{\Delta}}|s|\underline{\Delta}$ (where $\underline{\Delta}$ is a word representing Δ) never terminates, and we shall never know that $\langle S \mid R\rangle^+$ is not of O-type. However, it can be mentioned that one can identify specific patterns in triangular presentations that prevent reversing from being terminating (see Exercise 116), which enables one to a priori discard a number of presentations, see the remarks about the two generators case at the end of the chapter.

3 Further results

We thus explained how looking for monoids and possible Garside elements may lead to constructing interesting families of ordered groups. For completeness, we now briefly explain how to extend this approach and construct further examples by developing more general tools: this shows that Garside elements cannot be the end of this history in this topic but, on the other hand, no more general approach known so far enjoys the same effectivity properties as the Garside approach.

We begin in Subsection 3.1 with dominating elements, a proper extension of the notion of a right-Garside element. Then, in Subsection 3.2, we introduce the right-S-ceiling of a monoid, a general notion that enables one to somehow classify monoids of right-O-type. Finally, in Subsection 3.3, we concentrate on the braid group B_n, with additional results about the Dubrovina–Dubrovin ordering and the fact that the latter is not directly accessible to the Garside approach for $n \geqslant 4$.

3.1 Dominating elements

In view of Lemma 2.5, the central question when (S, R) is a right-triangular presentation is to establish that any two elements in the monoid $\langle S \mid R\rangle^+$ admit a common right-multiple. Right-Garside elements are relevant for this task: if Δ is a right-Garside element, then every element of the monoid left-divides every sufficiently large power of Δ. However, there may exist elements with this property that are not right-Garside elements.

Definition 3.1 (dominating). For S included in a monoid M, we say that an element δ of M *right-dominates* S if $\forall n{\geqslant}0\,(g\delta^n \preccurlyeq \delta^{n+1})$ holds for each g in S.

Lemma 3.2. *Assume that M is a left-cancellative monoid.*

(i) *If Δ is a right-Garside element in M, then Δ dominates* $\mathrm{Div}(\Delta)$.

(ii) *If δ is an element of M that dominates some generating family S of M, then any two elements of M admit a common right-multiple.*

Proof. (i) Assume that Δ is a right-Garside element in M. Let ϕ_Δ be the associated Garside morphism, and let g belong to $\mathrm{Div}(\Delta)$. For $n = 0$, we have $g \preccurlyeq \Delta$ by assumption. For $n \geqslant 1$, by Proposition V.1.28 (functor ϕ_Δ), we have $g\Delta^n = \Delta^n\phi_\Delta^n(g)$ and $\phi_\Delta^n(g) \in \mathrm{Div}(\Delta)$, whence $g\Delta^n \preccurlyeq \Delta^{n+1}$. So Δ right-dominates $\mathrm{Div}(\Delta)$.

(ii) Assume that δ right-dominates S. We prove, using induction on n, that $g \in S^n$ implies $g \preccurlyeq \delta^n$. For $n = 0$, that is, for $g = 1$, the property is obvious and, for $n = 1$, it directly follows from the assumption. Assume $n \geqslant 2$ and $g \in S^n$. Write $g = sg'$ with $s \in S$ and $g' \in S^{n-1}$. The induction hypothesis gives $g' \preccurlyeq \delta^{n-1}$, and we deduce $g = sg' \preccurlyeq s\delta^{n-1} \preccurlyeq \delta^n$, the last relation by the definition of right-domination. If S generates M, we deduce that every element of M left-divides δ^n for every n large enough and, from there, that any two elements of M admit a common right-multiple. □

Adapting the argument of Section 2, we obtain:

Proposition 3.3 (isolated orderings II). *Suppose the group G admits a triangular presentation (S, R) and the monoid $\langle S \mid R\rangle^+$ contains elements $\delta, \tilde{\delta}$ satisfying $\forall n{\geqslant}0\,(s\delta^n \preccurlyeq \delta^{n+1})$ and $\forall n{\geqslant}0\,(\tilde{\delta}^{n+1} \succneqq \tilde{\delta}^n s)$ for every s in S. Then G is orderable and the subsemigroup of G generated by S is the positive cone of a left-invariant ordering on G. If S is finite, this ordering is an isolated point in the space* $\mathrm{LO}(G)$.

Example 3.4 (isolated orderings II). For $p, q, \ell \geqslant 0$ and $m \geqslant p$, let G be the group $\langle \mathtt{x}, \mathtt{y} \mid \mathtt{x}^{p+1} = (\mathtt{y}(\mathtt{x}^{m-p}\mathtt{y})^{\ell})^{q+1}\rangle$. Considering the elements $\mathtt{a} = \mathtt{x}$, $\mathtt{c} = \mathtt{y}\mathtt{x}^{-p}$, and $\mathtt{b} = \mathtt{c}(\mathtt{a}^m\mathtt{c})^{\ell}$, one checks that G also admits the presentation

$$\langle \mathtt{a}, \mathtt{b}, \mathtt{c} \mid \mathtt{a} = \mathtt{b}(\mathtt{a}^p\mathtt{b})^q, \mathtt{b} = \mathtt{c}(\mathtt{a}^m\mathtt{c})^{\ell}\rangle. \tag{3.5}$$

Now the point is that $\mathtt{a}$ right-dominates $\{\mathtt{a}, \mathtt{b}, \mathtt{c}\}$ in the monoid M presented by (3.5). The relation $\mathtt{a}\mathtt{a}^n \preccurlyeq \mathtt{a}^{n+1}$ is trivial for every n, and it is easy to check that $\mathtt{a}^{p+1}$ commutes with $\mathtt{b}$ and then to deduce that $\mathtt{a}$ right-dominates $\mathtt{b}$ from the relation $\mathtt{b}\mathtt{a}^p \preccurlyeq \mathtt{a}$. The proof that $\mathtt{a}$ right-dominates $\mathtt{c}$ is (much) more delicate, see Exercise 118 and [92]. It follows that M is of right-O-type, hence, by symmetry, of O-type, and that the group G is orderable and the subsemigroup of G generated by $\mathtt{x}$ and $\mathtt{y}\mathtt{x}^{-p}$ is the positive cone of a left-invariant ordering on G which is isolated in $\mathrm{LO}(G)$.

Note that the monoid M in the Example 3.4 is generated by $\mathtt{a}$ and $\mathtt{c}$ alone and admits the corresponding (less readable) presentation

$$\langle \mathtt{a}, \mathtt{c} \mid \mathtt{a} = \mathtt{c}(\mathtt{a}^m\mathtt{c})^{\ell}(\mathtt{a}^p\mathtt{c}(\mathtt{a}^m\mathtt{c})^{\ell})^q\rangle^+.$$

The 4-parameter family of monoids defined by (3.5) contains in particular all monoids $\langle \mathtt{a}, \mathtt{b} \mid \mathtt{a} = \mathtt{b}\mathtt{a}^m\mathtt{b}\mathtt{a}^p\mathtt{b}\mathtt{a}^m\mathtt{b}\rangle^+$ and $\langle \mathtt{a}, \mathtt{b} \mid \mathtt{a} = (\mathtt{b}(\mathtt{a}^m\mathtt{b})^{\ell})^q\rangle^+$. It can be proved that, for $m \geqslant 2$ and $p+1$ not dividing $m+1$, these monoids admit no right-Garside element that is a power of $\mathtt{a}$: indeed, calling m the largest multiple of $p+1$ below r, one can show the relation $\mathtt{c}\mathtt{a}^{im+r} \preccurlyeq \mathtt{a}^{im+1}$ for every i, implying $\mathtt{c}\mathtt{a}^{(i+1)m} \preccurlyeq \mathtt{a}^{im+1}$ for every i, whence $\mathtt{c}\mathtt{a}^n \prec \mathtt{a}^n$ for every n. So $\mathtt{a}^n \preccurlyeq \mathtt{c}\mathtt{a}^n$ is always impossible, and $\mathtt{a}^n$ is not a right-Garside element for any n.

3.2 Right-ceiling

If a monoid M is generated by a finite set S of cardinality n and M is of right-O-type, then, for every ℓ, some (a priori not necessarily unique) word w of $S^{[\ell]}$ represents the $\preccurlyeq$-largest element of S^{ℓ}, that is, $[w']^+ \preccurlyeq [w]^+$ holds for every w' in $S^{[\ell]}$. It turns out that such maximal words are unique and, moreover, that the maximal words corresponding to different lengths are the final fragments of some well-defined left-infinite word.

Definition 3.6 (right-ceiling). If a monoid M is generated by a set S, a left-infinite S-word $...|s_2|s_1$ is said to be a *right-S-ceiling* for M if, for every length-ℓ word w in S^*, the relation $[w]^+ \preccurlyeq s_\ell \cdots s_1$ holds in M.

Example 3.7 (right-ceiling). Let K^+ be the Klein bottle monoid, that is, $\langle \mathtt{a}, \mathtt{b} \mid \mathtt{a} = \mathtt{bab} \rangle^+$ (Reference Structure 5, page 17). Then the left-infinite word ${}^\infty\mathtt{a}$ is a right-$\{\mathtt{a}, \mathtt{b}\}$-ceiling for K^+. Indeed, we have $\mathtt{b} \preccurlyeq \mathtt{a}$ and $\mathtt{b}^2 \preccurlyeq \mathtt{ba} \preccurlyeq \mathtt{ab} \preccurlyeq \mathtt{a}^2$. Next, $\mathtt{a}^2$ is central, and an easy induction gives $g \preccurlyeq \mathtt{a}^{2\ell}$ for every g in $\{\mathtt{a}, \mathtt{b}\}^{2\ell}$: for $\ell > 1$, writing $g = g'g''$ with g' in $\{\mathtt{a}, \mathtt{b}\}^{2\ell-2}$ and g'' in $\{\mathtt{a}, \mathtt{b}\}^2$, we obtain $g = g'g'' \preccurlyeq g'\mathtt{a}^2 = \mathtt{a}^2 g' \preccurlyeq \mathtt{a}^2\mathtt{a}^{2\ell-2} = \mathtt{a}^{2\ell}$.

Then one can show:

Proposition 3.8 (right-ceiling). (i) *Every monoid of right-O-type generated by a finite set S admits a unique right-S-ceiling.*

(ii) *Conversely, every monoid with a finite right-triangular presentation (S, R) that admits a right-S-ceiling is of right-O-type.*

We skip the proof. When it exists, (finite fragments of) the right-S-ceiling W can be computed inductively using the observation that, if $s_{\ell-1}|\cdots|s_1$ is a length-$(\ell-1)$ suffix of W, then the letter s_ℓ such that $s_\ell|\cdots|s_1$ is the length ℓ suffix of W is determined by the condition that $s_\ell s_{\ell-1} \cdots s_1$ is the $\preccurlyeq$-largest element of the set $\{s s_{\ell-1} \cdots s_1 \mid s \in S\}$: for S of size n, at most $n-1$ (and not $n^\ell - 1$) comparisons have to be performed. Thus experiments are easy—but, of course, no finite fragment of a ceiling is sufficient to show that the latter exists.

We shall not go very far in the investigation of right-ceilings, but only mention without proof a few results that show how complicated the situation can be. In particular, although, by definition, the first generator $\mathtt{a}_1$ in a triangular presentation is always the largest generator, it may happen that the right-ceiling is not the left-infinite power ${}^\infty\mathtt{a}_1$ and that neither $\mathtt{a}_1$ nor any power of $\mathtt{a}_1$ is dominating. In general, the right-ceiling is difficult to determine. However, the following connection between dominating elements and periodic right-ceiling can help to determine the right-ceiling and then deduce new results about (other) dominating elements.

Lemma 3.9. *For every cancellative monoid M with no nontrivial invertible element and every generating family S of M, the following are equivalent:*

(i) *The monoid M admits a right-S-ceiling that is periodic with period $s_\ell \cdots s_1$.*

(ii) *The element $s_\ell \cdots s_1$ right-dominates S^ℓ in M.*

Example 3.10 (periodic ceiling). Let M be defined by the presentation

$$\langle \mathtt{a}, \mathtt{b}, \mathtt{c} \mid \mathtt{a} = \mathtt{bac}, \mathtt{b} = \mathtt{cba} \rangle^+. \tag{3.11}$$

Put $\Delta = \mathtt{b}^2\mathtt{a}^2$. One easily checks that Δ is a right-Garside element in M, satisfying $\phi_\Delta(\mathtt{a}) = \mathtt{aba}^2\mathtt{cac}^3$, $\phi_\Delta(\mathtt{b}) = \mathtt{ba}^2\mathtt{c}^2$, and $\phi_\Delta(\mathtt{c}) = \mathtt{c}$. So M is of right-O-type. Next, one can check that $\mathtt{b}^2\mathtt{a}^2$ dominates $\{\mathtt{a}, \mathtt{b}, \mathtt{c}\}^4$. So, by Lemma 3.9, the right-ceiling is the periodic word ${}^\infty(\mathtt{b}^2\mathtt{a}^2)$. But then we deduce that $\mathtt{a}$ cannot dominate $\mathtt{b}$ and $\mathtt{c}$ in M. On the other hand, $\mathtt{ba}^2 = \mathtt{a}^3 \cdot \mathtt{ba}^2\mathtt{c}^3$ holds in M, and one finds $\mathtt{b}(\mathtt{a}^n) = (\mathtt{a}^n)^2 \cdot \mathtt{ba}^2\mathtt{c}^{2n}\mathtt{a}^{n-2}$ for $n \geqslant 2$, which shows that $\mathtt{a}^n$ does not dominate $\mathtt{b}$ for any n.

3.3 The specific case of braids

We conclude with a few additional observations about Artin's braid groups B_n and their orderings, in particular a negative result about the use of the Garside approach in the case of n-strand braids with $n \geqslant 4$.

In the case of 3-strand braids, we saw in Example 2.17 that the group B_3 admits a left-invariant ordering that is isolated in the space $\mathrm{LO}(B_3)$, namely the ordering $<_{\mathrm{DD}}$ whose positive cone is generated by $\sigma_1\sigma_2$ and σ_2^{-1}. The construction of $<_{\mathrm{DD}}$ can be extended to n-strand braids for every n.

Definition 3.12 (Dubrovina–Dubrovin monoid). For $n \geqslant 2$, the n-strand *Dubrovina–Dubrovin monoid* is the submonoid $B_n^\oplus$ of B_n generated by

$$s_1 = \sigma_1 \cdots \sigma_{n-1},\ s_2 = (\sigma_2 \cdots \sigma_{n-1})^{-1},\ s_3 = \sigma_3 \cdots \sigma_{n-1},\ \ldots,\ s_{n-1} = \sigma_{n-1}^{(-1)^n}.$$

Lemma 3.13. *For every n, the monoid $B_n^\oplus$ is of O-type; the positive cone of the associated ordering $<_{\mathrm{DD}}$ of B_n consists of the n-strand braids that are σ_k-positive for some odd k or σ_k-negative for some even k.*

Proof (sketch). Let P_{DD} be the set of all braids that are σ_k-positive for some odd k or σ_k-negative for some even k. It follows from the basic results about σ_k-positive braids that a braid can be σ_k-positive or σ_k-negative for at most one k, and that every nontrivial braid is σ_k-positive or σ_k-negative for at least one k (hence for exactly one k). This implies that P_{DD} (as every variation of the same style) is the positive cone of a left-invariant ordering on B_n. Thus it just remains to check by a rather simple direct computation that every σ_k-positive braid can be expressed as a finite product of the generators s_k^e with $e = +1$ for odd k and $e = -1$ for even k. □

Without appealing to the above result, we established in Section 2 that the monoid $B_3^\oplus$ is of O-type since it admits the triangular presentation $\langle \mathtt{a}, \mathtt{b} \mid \mathtt{a} = \mathtt{b}\mathtt{a}^2\mathtt{b}\rangle^+$, which is eligible for Proposition 2.15—and, in doing so, we reproved the above mentioned properties of σ_k-positive braids in the case of 3-strand braids. We shall now see that, for $n \geqslant 4$, the monoid $B_n^\oplus$ admits no triangular presentation based on $\{s_1, \ldots, s_{n-1}\}$. This is a consequence of the following general negative result, which we do not prove here (see [92, Proposition 9.5]):

Proposition 3.14 (no triangular presentation). *If a monoid of right-O-type M admits a generating subfamily S with $\#S \geqslant 3$ and there exists s in S satisfying $g \mathrel{\widetilde{\preccurlyeq}} hs$ for all g, h in the submonoid generated by $S \setminus \{s\}$, then M admits no right-triangular presentation based on S.*

Proposition 3.14 prevents a number of monoids of right-O-type from admitting a right-triangular presentation, see Exercise 123. In the case of braids, we obtain:

Corollary 3.15 (no triangular presentation). *For $n \geqslant 4$, the monoid $B_n^\oplus$ is of O-type, but it admits no right-triangular presentation based on $\{s_1, \ldots, s_{n-1}\}$.*

Proof. We apply Proposition 3.14 with the specific element s_1. If g and h belong to the submonoid of $B_n^\oplus$ generated by $s_2, \ldots, s_{n-1}$, then, by the properties of σ_k-positive braids recalled in the proof of Lemma 3.13, they may be neither σ_1-positive nor σ_1-negative. On the other hand, the braid s_1 is σ_1-positive, hence so is hs_1. It follows that $g <_{\mathrm{DD}} hs$ holds in B_n, hence that $g \prec hs_1$ holds in $B_n^\oplus$, the expected property. ☐

One can indeed convert the standard presentation of the braid group B_n into a presentation in terms of the generators $s_1, \ldots, s_{n-1}$ of Corollary 3.15. For instance, writing $\mathtt{a}, \mathtt{b}, \ldots$ for $s_1, s_2, \ldots$, one can check that $B_4^\oplus$ admits the presentation

$$(\mathtt{a}, \mathtt{b}, \mathtt{c}, \mathtt{a} = \mathtt{b}^2\mathtt{a}^2\mathtt{baba}^2\mathtt{b}^2, \mathtt{b} = \mathtt{cb}^2\mathtt{c}, \mathtt{abc} = \mathtt{cab}), \tag{3.16}$$

a triangular presentation augmented with a third, additional relation. But the triangular presentation made of the first two relations in (3.16) is not a presentation of $B_4^\oplus$, nor of any monoid of O-type either. On the other hand, Corollary 3.15 does not discard the possibility that $B_n^\oplus$ admits a triangular presentation based on other generators and, therefore, the question of whether the monoid $B_n^\oplus$ admit a (finite) triangular presentation for $n \geqslant 4$ arises immediately. It remains open, but leads to unexpected results.

Natural candidates arise in connection with the Birman–Ko–Lee band generators (Reference Structure 3, page 10). We recall that, for $1 \leqslant i < j \leqslant n$, one puts

$$a_{i,j} = \sigma_i \cdots \sigma_{j-2}\sigma_{j-1}\sigma_{j-2}^{-1} \cdots \sigma_i^{-1},$$

whence in particular $\sigma_i = a_{i,i+1}$. Then there exists a simple connection between the monoid $B_n^\oplus$ and the elements $a_{i,j}$, namely $B_n^\oplus$ is generated by the elements $a_{i,j}^{(-1)^{i+1}}$ (see Exercise 124). It is then natural to wonder whether $B_n^\oplus$ admits a triangular presentation in terms of the above generators or of related generators. As no complete (positive or negative) result is known so far, we skip the discussion, but we conclude with an amusing application, namely the existence, for $n = 3$ and $n = 4$, of a braid ordering on B_n that is isolated (contrary to the Dehornoy ordering $<_{\mathrm{D}}$) and, at the same time, includes the positive braid monoid B_n^+ (contrary to the Dubrovina–Dubrovina ordering $<_{\mathrm{DD}}$).

Proposition 3.17 (exotic braid orderings). *The subsemigroup generated by σ_1, σ_2, and $\sigma_1\sigma_2^{-1}\sigma_1^{-1}$ is the positive cone of an isolated left-invariant ordering on B_3. The subsemigroup generated by σ_1, σ_2, σ_3, $\sigma_1\sigma_2\sigma_1^{-1}$, $\sigma_2\sigma_3^{-1}\sigma_2^{-1}$, and $\sigma_1\sigma_2\sigma_3^{-1}\sigma_2^{-1}\sigma_1^{-1}$ is the positive cone of an isolated left-invariant ordering on B_4.*

Proof. Conjugating by $\sigma_1 \cdots \sigma_n$ defines an order-n automorphism ϕ_n of B_n that rotates the Birman–Ko–Lee generators, see Chapter IX. For $n = 3$, one has $\phi_3 : \sigma_1 \mapsto \sigma_2 \mapsto a_{1,3} \mapsto \sigma_1$. By the above remark, $B_3^{\oplus}$ is generated by σ_1, $a_{1,3}$, and σ_2^{-1}. Hence the monoid $\phi_3(B_3^{\oplus})$ is generated by σ_2, σ_1, and $a_{1,3}^{-1}$, and it is (when 1 is removed) the positive cone of a left-invariant ordering on B_3.

Similarly, for $n = 4$, we have $\phi_4 : \sigma_1 \mapsto \sigma_2 \mapsto \sigma_3 \mapsto a_{1,4} \mapsto \sigma_1$ and $a_{1,3} \leftrightarrow a_{2,4}$. By the above remark again, the monoid $B_4^{\oplus}$ is generated by σ_1, $a_{1,3}$, $a_{1,4}$, σ_2^{-1}, $a_{2,4}^{-1}$, and σ_3. Hence $\phi_4^2(B_4^{\oplus})$ is generated by σ_3, $a_{1,3}$, σ_2, σ_3, $a_{1,4}^{-1}$, $a_{2,4}^{-1}$, and σ_1, and it is (when 1 is removed) the positive cone of a left-invariant ordering on B_4. □

The construction does not extend to $n \geqslant 5$, because the negative entries σ_2^{-1} and σ_4^{-1} cannot be eliminated simultaneously.

Exercises

Exercise 111 (braid ordering). Show that the relations $1 <_{\mathrm{D}} \sigma_1\sigma_2 \leqslant_{\mathrm{D}} (\sigma_1\sigma_2\sigma_1)^{2p}\sigma_2^q$ hold in B_3 for $p > 0$.

Exercise 112 (limit of conjugates). Assuming that $<_{\mathrm{D}}$ is a limit of its conjugates in B_3, show the same result in B_n. [Hint: Use the subgroup of B_n generated by σ_{n-2} and σ_{n-1}, which is isomorphic to B_3.]

Exercise 113 (closure of conjugates). Let P_n be the positive cone of the ordering $<_{\mathrm{D}}$ on B_n considered in Example 1.23. Show that the closure of the conjugates of P_n in $\mathrm{LO}(B_n)$ is a Cantor set.

Exercise 114 (space $\mathrm{LO}(B_\infty)$). Show that every point in the space $\mathrm{LO}(B_\infty)$ is a limit of its conjugates and that $\mathrm{LO}(B_\infty)$ is homeomorphic to the Cantor set (contrary to the spaces $\mathrm{LO}(B_n)$ for finite n).

Exercise 115 (braids). Show that, with the notation of Example 2.14, the monoids $M_{2,1,1}$ and $M_{1,2,1}$ are isomorphic, and deduce that the associated orderings of the braid group B_3 coincide. [Hint: Show the relations $\mathtt{a} = \mathtt{a}'\mathtt{b}$ and $\mathtt{a}' = \mathtt{b}\mathtt{a}^2$ between the involved generators.]

Exercise 116 (non-terminating reversing). Assume that (S, R) is a triangular presentation. (i) Show that, if a relation of $\widehat{R}$ has the form $s = w$ with $\lg(w) > 1$ and w finishing with s, then the monoid $\langle S \mid R\rangle^+$ is not of right-O-type. (ii) Let $(S, \widehat{R})$ be the maximal right-triangular presentation deduced from (S, R). Show that, if a relation of $\widehat{R}$ has the form $s = w$ with w beginning with $(uv)^r us$ with $r \geqslant 1$, u nonempty, and v such that $v^{-1}s$ reverses to a word beginning with s, hence, in particular, if v is empty or it can be decomposed as $u_1, \dots, u_m$ where $u_k s$ is a prefix of w for every k, then $s^{-1}us$ cannot be terminating, and deduce that $\langle S \mid R\rangle^+$ is not of right-O-type. (iii) Show that a relation $\mathtt{a} = \mathtt{babab}^3\mathtt{a}^2\ldots$ is impossible in a right-triangular presentation for a monoid of right-O-type.

Exercise 117 (roots of Garside element). Assume that M is a left-cancellative monoid generated by a set S. (i) Show that, for δ, g in a left-cancellative monoid M, a necessary and sufficient condition for δ to right-dominate g is that there exist $m \geqslant 1$ satisfying $(*) \forall k{\geqslant}0\ (\,g\delta^{km+m-1} \preccurlyeq \delta^{km+1}\,)$. (ii) Assume that δ^m is a right-Garside element in M. Show that δ right-dominates every element g that satisfies $g\delta^{m-1} \preccurlyeq \delta$.

Exercise 118 (dominating element). Let M be the monoid defined by (3.5). (i) Show that, for $q = 0$, the element $\mathtt{a}^{r+1}$ is central and that M is of O-type. (ii) From now on, we assume $q \geqslant 1$. Prove $\mathtt{a}^{p+1}\mathtt{b} = \mathtt{b}\mathtt{a}^{p+1}$ and deduce that $\mathtt{a}$ right-dominates $\mathtt{b}$ [Hint: Use Exercise 117.] (iii) Write $r = m + p'$ with $p + 1 \,|\, m$ and $0 \leqslant p' \leqslant p$. Show that $\mathtt{a}^m$ commutes with $\mathtt{a}$ and $\mathtt{b}$. From here, we separate two cases. (iv) Assume that $p + 1$ does not divide $r + 1$, that is, $p' < p$ holds. Put $\mathtt{c}' = \mathtt{c}\mathtt{a}^p$, and let M' be the submonoid of M generated by $\mathtt{a}$, $\mathtt{b}$, and $\mathtt{c}'$. Let $\Delta = \mathtt{a}^m$. Prove that $\mathtt{a} \preccurlyeq \Delta \preccurlyeq \mathtt{a}\Delta$ and $\mathtt{b} \preccurlyeq \Delta \preccurlyeq \mathtt{b}\Delta$ hold in M'. (v) Prove $\mathtt{c}' \preccurlyeq \mathtt{a}$ in M', and then $\mathtt{c}'\Delta = \Delta\mathtt{b}\mathtt{a}^{p'}\mathtt{c}'(\mathtt{b}\mathtt{a}^q)^{q-1}\mathtt{b}\mathtt{a}^{p-p'-1}$. Deduce that $\mathtt{c}\mathtt{a}^{km+r} \preccurlyeq \mathtt{a}^{km+1}$ holds for every k in M, and conclude that $\mathtt{a}$ right-dominates $\mathtt{c}$. (vi) Conclude that M is of O-type. (vii) Repeat the argument when $p + 1$ divides $r + 1$ and prove $\mathtt{c}\mathtt{a}^{r+km} \preccurlyeq \mathtt{a}^{1+km}$ for $k = 0, ..., q$. (viii) Show that, for $r < p$ with $q \neq 0$, the monoid M is not of right-O-type [Hint: Use Exercise 116.]

Exercise 119 (right-ceiling). Assume that M is a cancellative monoid of right-O-type, and that $s_\ell \cdots s_1$ is a right-top S-word in M such that $[s_\ell \cdots s_1]^+$ is central in M. Show that $s_i = s_1$ must hold for every i, and deduce that ${}^\infty s_1$ is the right-S-ceiling in M.

Exercise 120 (power dominating). Let M be defined by $(\mathtt{a}, \mathtt{b}, \mathtt{c}, \mathtt{a} = \mathtt{bcb}, \mathtt{b} = \mathtt{cbabc})$. (i) Show that M is generated by $\mathtt{b}$ and $\mathtt{c}$, with the presentation $(\mathtt{b}, \mathtt{c}, \mathtt{b} = \mathtt{c}\mathtt{b}^2\mathtt{c}\mathtt{b}^2\mathtt{c})$. (ii) Show that $\mathtt{b}^3$ is a central Garside element in M and that M is of O-type. (iii) Check $\mathtt{a}^3 = \mathtt{b}^3$ and deduce that $\mathtt{a}^3$ dominates $\mathtt{b}$ and $\mathtt{c}$. (iv) Check the equality $\mathtt{ba} = \mathtt{a}^2 \cdot \mathtt{b}\mathtt{c}^2\mathtt{b}$, deduce $\mathtt{ba} \not\preccurlyeq \mathtt{a}^2$, and conclude that $\mathtt{a}$ does not dominate $\mathtt{b}$.

Exercise 121 (periodic ceiling). Let M_n be the monoid defined by the cycling presentation $(\mathtt{a}_1, ..., \mathtt{a}_n, \mathtt{a}_1 = \mathtt{a}_2 \cdots \mathtt{a}_n, \mathtt{a}_2 = \mathtt{a}_3 \cdots \mathtt{a}_n\mathtt{a}_1, ..., \mathtt{a}_{n-1} = \mathtt{a}_n\mathtt{a}_1 \cdots \mathtt{a}_{n-1})$. Show that $\mathtt{a}_1^2$ is central in M_n, and the right-ceiling is ${}^\infty(\mathtt{a}_{n-1} \cdots \mathtt{a}_1)$, hence it has period $n - 1$.

Exercise 122 (exotic dominating element). Let M be the monoid defined by $(\mathtt{a}, \mathtt{b}, \mathtt{a} = \mathtt{babab}^2\mathtt{ab}^2\mathtt{abab})$. (i) Show that M is also defined by $(\mathtt{a}, \mathtt{b}, \mathtt{c}, \mathtt{a} = \mathtt{bcacb}, \mathtt{b} = \mathtt{cacac})$. (ii) Put $\delta = \mathtt{b}^2$. Prove $\mathtt{a}\delta^n \cdot \mathtt{cacbc}(\mathtt{bc})^{2n} = \mathtt{b}\delta^n \cdot \mathtt{b} = \mathtt{c}\delta^n \cdot \mathtt{acacb} = \delta^{n+1}$ for every n. (iii) Deduce that δ dominates $\mathtt{a}$, $\mathtt{b}$, and $\mathtt{c}$ in M, and that M is of O-type. [It is conjectured that the right-ceiling is ${}^\infty\mathtt{a}$.]

Exercise 123 (no triangular presentation). Assume that M is a monoid of right-O-type that is generated by $\mathtt{a}, \mathtt{b}, \mathtt{c}$ with $\mathtt{a} \succ \mathtt{b} \succ \mathtt{c}$ and $\mathtt{b}, \mathtt{c}$ satisfying some relation $\mathtt{b} = \mathtt{c}v$ with no $\mathtt{a}$ in v. (i) Prove that, unless M is generated by $\mathtt{b}$ and $\mathtt{c}$, there is no way to complete $\mathtt{b} = \mathtt{c}v$ with a relation $\mathtt{a} = \mathtt{b}u$ so as to obtain a presentation of M. (ii) Deduce that no right-triangular presentation made of $\mathtt{b} = \mathtt{cbc}$ (Klein bottle relation) or $\mathtt{b} = \mathtt{c}\mathtt{b}^2\mathtt{c}$ (Dubrovina–Dubrovin braid relation) plus a relation of the form $\mathtt{a} = \mathtt{b}...$ may define a monoid of right-O-type.

Exercise 124 (Birman–Ko–Lee generators). Put $b_{i,j} = a_{i,j}^{(-1)^{i+1}}$ in the braid group B_n. Show that, for every n, the monoid $B_n^\oplus$ is generated by the elements $b_{i,j}$.

Notes

Sources and comments. The connections between orderable groups and topology recently became an active domain of research. On the one hand, deep relationships were discovered between properties of 3-manifolds and the possible orderability of their fundamental group involving in particular L-spaces and co-oriented taut foliations, see for instance Boyer, Rolfsen, and Wiest [29] or Boyer, Gordon, and Watson [28]. On the other hand, and this is the aspect that was discussed in this chapter, the idea of putting a topology on the family of all invariant orderings on a group has been considered for several years by such experts as E. Ghys, A. Sikora and others, and it led to a number of developments and unexpected applications, see Morita [185].

More specifically, the quest for isolated orderings was launched by A. Sikora in [213] and the result that, in the fundamental cases of a free abelian group and of a free group, no isolated ordering exists, appears in Navas [188]. The fact that a positive cone that is finitely generated as a semigroup gives an isolated ordering is mentioned in [213]. The converse implication is false: C. Rivas shows in [202] that a non-finitely generated semigroup may give rise to an isolated ordering

The orderability of braid groups was proved by P. Dehornoy in [72, 75], and the result that there exists an isolated ordering in the space $\mathrm{LO}(B_n)$ follows from the construction of the Dubrovina–Dubrovin ordering in [115] and the result by A. Navas [188] that the Dehornoy ordering is a limit of its conjugates, both in the case of 3-strand braids and in the general case using the argument of Exercise 112. Constructions of isolated orderings on torus knot groups and related groups appear in [92], Navas [189], and Ito [151], relying on various approaches that are not directly comparable. As already mentioned in the text, the classification of groups with finitely many left-invariant orderings is due to Tararin [223, 160].

The current exposition of the space $\mathrm{LO}(G)$ in Section 1 is based on Dehornoy *et al.* [100], itself based on Sikora [213]. In particular, Proposition 1.14 appears in [213]—an alternative argument can be found in Dabkovska *et al.* [69]. The rest of the chapter is directly inspired by [92], where monoids of O-type are introduced and a number of examples are mentioned. Monoids of O-type are connected with divisibility monoids [167], which essentially correspond to the case when divisibility relations are lattice orderings, but not necessarily linear orderings. Also, right-triangular presentations are those whose left-graph, in the sense of Adjan [1] and Remmers [201], is a chain. The letter O stands for "order", reminiscent of the monoids of I-type considered in Chapter XIII; it may seem strange that the notion connected with left-divisibility is called right-O-type, but this option is natural when one thinks in terms of multiples and Garside elements.

The definition of triangular presentations considered in this chapter is slightly more restricted than that of [92], but one can show that this does not really restrict the family of monoids of O-type that are eligible for the approach.

Further questions. At the moment, the range of the approach described in this chapter remains unknown: we know that some monoids of O-type admit no triangular presentations, and that some monoids of O-type admit no Garside element, at least of a certain form, but, on the other hand and although triangular presentations may seem to be extremely particular, a number of monoids of O-type with such presentations were found—much more than was expected first. So the main open question in this approach is to further explore its range and understand which ordered groups are eligible. For instance, it is natural to raise:

Question 33. Does every monoid of O-type that admits a triangular presentation contain a right-Garside element?

We saw in Example 3.4 a monoid of O-type that admits a right-triangular presentation, but in which no power of the leading generator is a right-Garside element. This however does not dismiss the possibility of the existence of a Garside element of another type. A positive answer to Question 33 seems unlikely but, on the other hand, it is uneasy to a priori discard the existence of exotic Garside elements. For instance, in the monoid $\langle \mathtt{ab} \mid \mathtt{a} = \mathtt{bab}^3\mathtt{ab}\rangle^+$, no power of $\mathtt{a}$ is a Garside element, but $(\mathtt{ab})^3$ is a central Garside element. In the context of Example 3.4, the first critical case for which we (weakly) conjecture that no right-Garside element exists is $\langle \mathtt{a}, \mathtt{b}, \mathtt{c} \mid \mathtt{a} = \mathtt{ba}^2\mathtt{b}, \mathtt{b} = \mathtt{ca}^4\mathtt{c}\rangle^+$.

As in the case of right-Garside elements, the above results say nothing about dominating elements that are not powers of the top generator and they leave the following natural questions open:

Question 34. Does every monoid of O-type that admits a triangular presentation based on a set S contain an element that dominates S? Is the right-ceiling necessarily periodic?

By Lemma 3.9, a positive answer to the second question implies the existence of an element of S^n that dominates all of S^n for some $n \geqslant 1$, hence a fortiori S, so it implies a positive answer to the first question. Thinking of the examples known so far, it seems reasonable to conjecture a positive answer to both questions, but so far no clue toward a proof is in view. See Exercises 120–122 for more examples witnessing various behaviors of the ceiling.

Returning to the general question of the range of the current "Garside-type" approach, an exhaustive investigation is certainly out of reach in the general case, the particular case of two-generator monoids seems more accessible, and several natural questions arise. Typically, we mentioned in Proposition 3.14 that some monoids of O-type admit no triangular presentation, but the argument of the proof requires the existence of at least three generators.

Question 35. Does every two-generator monoid of right-O-type admit a right-triangular presentation?

On the other hand, if we start with a two-generator triangular presentation, that is, if we consider monoids of the form $\langle \mathtt{a}, \mathtt{b} \mid \mathtt{a} = \mathtt{b}w\rangle^+$, we can ask

Question 36. Is every two-generator triangular presentation that defines a monoid of right-O-type eligible for Proposition 3.3?

At the moment, we have no counter-example: more precisely, all presentations that involve a word w of length at most 10 and are not discarded by the syntactic property of Exercise 116 turn out to contain a right-Garside element or, at least, a dominating element—a rather unexpected result. Also, experiments show that all two-generator triangular presentations defining a monoid of O-type are palindromic, that is, the relation is invariant under reversing the order of letters, and that the associated right-ceiling is equal to $^{\infty}\mathtt{a}$: so the question is, are these general facts?

In a different direction, Propositions 2.15 and 3.3 are valid in the case of an infinite presentation, thus leading to orderable groups with an explicit positive cone. But the argument showing that the involved ordering is isolated in its space of orderings is valid only when the presentation is finite. However, as mentioned above, a non-finitely generated monoid may give rise to an isolated ordering, so it makes sense to raise

Question 37. If (S, R) is an infinite triangular presentation that defines a monoid of O-type, may the associated ordering be isolated in the space $\mathrm{LO}(\langle S \mid R\rangle)$?

In the direction of a positive answer, it is natural to address the above question in the context of a direct limit of finitely generated monoids. The properties of subword reversing make this situation easy to analyze. Indeed, if (S, R) is an infinite triangular presentation such that all (or at least unboundedly many) finite approximations (S_n, R_n) define monoids of O-type, one can show that the monoid $\langle S \mid R\rangle^+$ is a direct limit of the monoids $\langle S_n \mid R_n\rangle^+$ and it is of O-type. A typical test-case for the above question is the torus-type group $G = \langle \mathtt{x}_1, \mathtt{x}_2 ... \mid \mathtt{x}_1^2 = \mathtt{x}_2^q, \mathtt{x}_2^2 = \mathtt{x}_3^q, ...\rangle$. For $q = 2$, the element $\Delta = \mathtt{x}_1^2$ is a central Garside element in $\langle S \mid R\rangle^+$, but, for odd q, the element $\Delta_n = \mathtt{x}_1^{2^{n-2}}$ is central in the finite approximation $\langle S_n \mid R_n\rangle^+$, but no power of $\mathtt{x}_1$ may be central in G, and it would be interesting to know whether the associated ordering is isolated in $\mathrm{LO}(G)$.

Chapter XIII

Set-theoretic solutions of Yang–Baxter equation

The Yang–Baxter equation (YBE) is a fundamental equation occurring in integrable models in statistical mechanics and quantum field theory. Some of its many solutions called set-theoretic turn out to be directly connected with a particular family of monoids, the monoids of I-type, and the latter turn out to be Garside monoids of a simple type. In this chapter, we describe the correspondences between the above mentioned objects and show how using what can be called a Garside approach leads to a conceptually simple and technically efficient exposition of the main results.

The chapter is organized as follows. In Section 1, we introduce set-theoretic solutions of the Yang–Baxter equation and describe various equivalent algebraic structures, in particular biracks, and, on the other hand, (bijective) right-cyclic quasigroups, or RC-quasigroups (Proposition 1.34).

Then, in Section 2, we associate with every (convenient) set-theoretic solution of YBE—or, equivalently, with every bijective RC-quasigroup—a monoid called the structure monoid, and, after developing some elementary results about the RC-law, we characterize the structure monoids of solutions of YBE as those Garside monoids that admit a presentation of a certain syntactic form (Proposition 2.34).

Further results about structure monoids (and their groups of fractions) are established in Section 3, in particular the important two-way connection with monoids of I-type (Propositions 3.5 and 3.6). As an application, we deduce in Proposition 3.23 the existence, for every finitely generated group of I-type, of a short exact sequence similar to the one that connects a pure braid group, a braid group, and the corresponding Coxeter group; here the role of pure braids is played by a free abelian group, whereas the role of the Coxeter group is played by some finite group for which we give an explicit presentation.

1 Several equivalent frameworks

We introduce particular solutions of YBE called the (involutive nondegenerate) set-theoretic solutions (Subsection 1.1). Then we show that such structures can be equivalently described as involutive biracks, which are systems consisting of a set equipped with two binary operations obeying certain algebraic laws (Subsection 1.2). Finally—this is more interesting—we show in Subsection 1.3 that another equivalent framework is provided by bijective RC-quasigroups, other algebraic systems consisting of a set equipped with a binary operation obeying a certain algebraic law that, in some sense, is the inverse of the laws defining biracks.

1.1 Set-theoretic solutions of YBE

We consider the (non-parametric form of) the (quantum) Yang–Baxter equation.

Definition 1.1 (solution of YBE). If V is a vector space, an element R of $\mathrm{GL}(V \otimes V)$ is called a *solution of the Yang–Baxter equation* (YBE) if we have

$$(R \otimes \mathrm{id})(\mathrm{id} \otimes R)(R \otimes \mathrm{id}) = (\mathrm{id} \otimes R)(R \otimes \mathrm{id})(\mathrm{id} \otimes R). \tag{1.2}$$

Example 1.3 (solution of YBE). Let $A = \mathbb{C}[q, q^{-1}]$ and $V = A \times A$ with the standard basis (e_1, e_2). Then one can check that the automorphism of $V \otimes V$ defined in the basis $(e_1 \otimes e_1,\ e_1 \otimes e_2,\ e_2 \otimes e_1,\ e_2 \otimes e_2)$ by the matrix $q^{-1/2}\begin{pmatrix} 1 & 0 & 0 & 0 \\ 0 & 0 & 1 & 0 \\ 0 & 1 & q - q^{-1} & 0 \\ 0 & 0 & 0 & 1 \end{pmatrix}$ satisfies (1.2), that is, it is a solution of YBE. This solution is connected with the basic representation of the quantum group $U_q(\mathfrak{sl}(2))$ and the Jones polynomial [156].

In the current context, it is customary to denote by R^{ij} the automorphism of $V^{\otimes 3}$ that corresponds to R acting on the ith and jth coordinates, so, for instance, R^{12} stands for $R \otimes \mathrm{id}$. Then the Yang–Baxter equation takes the simpler form

$$R^{12} R^{23} R^{12} = R^{23} R^{12} R^{23}; \tag{1.4}$$

note the similarity with the Coxeter relation of $\mathfrak{S}_3$ and the braid relation of B_3 (Reference Structure 2, page 5 and Chapter IX).

Among the (many) solutions of the Yang–Baxter equation, we consider here those that preserve some fixed basis of the considered vector space.

Lemma 1.5. *Assume that V is a vector space and X is a basis of V.*

(i) *If R is a solution of* YBE *that maps $X^{\otimes 2}$ into itself, the restriction of R to $X \otimes X$ yields a bijection ρ of $X \times X$ to itself that satisfies the relation*

$$\rho^{12} \rho^{23} \rho^{12} = \rho^{23} \rho^{12} \rho^{23}. \tag{1.6}$$

(ii) *Conversely, if ρ is a bijection of $X \times X$ into itself that satisfies* (1.6), *then ρ induces a solution of* YBE *that maps $X^{\otimes 2}$ into itself.*

Of course, in (1.6), ρ^{12} means the bijection of $X \times X \times X$ to itself consisting in applying ρ in the first two positions. The verifications are straightforward. In particular, the map ρ must be bijective, since as a solution of YBE is demanded to be an automorphism of the ambient vector space.

Example 1.7 (preservation of basis). The solution of YBE given in Example 1.3 does not enter the framework of Lemma 1.5: unless q is specialized at $q = 1$, the (operator associated with the) matrix does not preserve the canonical basis, nor any other basis of V. By contrast, for $q = 1$, the matrix becomes a permutation matrix, and enters the framework of Lemma 1.5 with an associated bijection ρ of $X \times X$ given by $\rho((e_1, e_1)) = (e_1, e_1)$, $\rho((e_1, e_2)) = (e_2, e_1)$, $\rho((e_2, e_1)) = (e_1, e_2)$, $\rho((e_2, e_2)) = (e_2, e_2)$.

From now on, we shall concentrate on solutions of YBE of the type considered in Lemma 1.5. Such solutions are called *set-theoretic* because they are entirely determined by their action on the basis, with no linearity condition. So we can forget about vector spaces and restart from the following definition.

Definition 1.8 (set-theoretic solution of YBE). (i) A *set-theoretic solution of YBE* is a pair (X, ρ), where X is a set and ρ is a bijection of $X \times X$ into itself that satisfies (1.6). In this case, we denote by $\rho_1(s,t)$ and $\rho_2(s,t)$ the first and second entries of $\rho(s,t)$.

(ii) A set-theoretic solution (X, ρ) of YBE is called *nondegenerate* if, for every s in X, the left translation $y \mapsto \rho_1(s, y)$ is one-to-one and, for every t in X, the right translation $x \mapsto \rho_2(x,t)$ is one-to-one.

(iii) A set-theoretic solution (X, ρ) of YBE is called *involutive* if $\rho \circ \rho$ is the identity of $X \times X$.

Example 1.9 (set-theoretic solution of YBE). Assume $X = \{\mathtt{a}, \mathtt{b}\}$. Then there are 4!, that is, 24, bijections of $X \times X$, among which six are set-theoretic solutions of YBE. Among the latter, two are degenerate, namely (we always adopt the convention that, in a table, the first argument corresponds to rows and the second one to columns)

	a	b
a	(a, a)	(a, b)
b	(a, a)	(b, b)

and

	a	b
a	(b, b)	(b, a)
b	(a, b)	(a, a)

with constant left translations in the left table and constant right translations in the right table. Next, two are non-involutive, having order 4 and not 2:

	a	b
a	(a, b)	(b, b)
b	(a, a)	(b, a)

and

	a	b
a	(b, a)	(a, a)
b	(b, b)	(a, b)

Finally, two are nondegenerate and involutive:

	a	b
a	(a, a)	(b, a)
b	(a, b)	(b, b)

and

	a	b
a	(b, b)	(a, b)
b	(b, a)	(a, a)

Remark 1.10. The reason why one is interested in nondegenerate set-theoretic solutions of YBE will become clear soon, as it is a necessary condition for the duality described in Subsection 1.3 to be possible. One of the reasons for dealing with involutive set-theoretic solutions is that assuming $\rho \circ \rho = \mathrm{id}$ is the simplest way to guarantee that ρ is bijective.

Note that there is no connection between the above two conditions, namely having bijective translations and being globally bijective. The table displayed on the right defines a degenerate bijection of $X \times X$: here the left translation by $\mathtt{a}$ is not one-to-one, since we have $\rho_1(\mathtt{a}, \mathtt{a}) = \mathtt{a} = \rho_1(\mathtt{a}, \mathtt{b})$.

	a	b
a	(a, a)	(a, b)
b	(b, a)	(b, b)

On the other hand, this next table on the right defines a nondegenerate map from $X \times X$ to itself that is not bijective: here left and right translations are bijective, but the global map ρ is not one-to-one.

	a	b
a	(a, b)	(b, a)
b	(b, a)	(a, b)

1.2 Involutive biracks

As already noted in Definition 1.8, a bijection, more generally a map, from $X \times X$ to itself is a pair of maps from $X \times X$ to X, hence it can be viewed as a pair of binary operations on X. This framework will provide slightly shorter notation. The first step is to translate the defining properties of an (involutive, nondegenerate) set-theoretic solution of YBE.

Lemma 1.11. (i) *If (X, ρ) is a set-theoretic solution of* YBE, *then the binary operations $\rceil$ and $\lceil$ defined on X by*

$$a \rceil b = \rho_1(a, b) \quad \text{and} \quad a \lceil b = \rho_2(a, b), \tag{1.12}$$

respectively, obey the laws

$$(a \rceil b) \rceil ((a \lceil b) \rceil c) = a \rceil (b \rceil c), \tag{1.13}$$

$$(a \rceil b) \lceil ((a \lceil b) \rceil c) = (a \lceil (b \rceil c)) \rceil (b \lceil c), \tag{1.14}$$

$$(a \lceil b) \lceil c = (a \lceil (b \rceil c)) \lceil (b \lceil c). \tag{1.15}$$

Moreover, if (X, ρ) is involutive, then $\rceil$ and $\lceil$ obey the laws

$$(a \rceil b) \rceil (a \lceil b) = a \quad \text{and} \quad (a \rceil b) \lceil (a \lceil b) = b. \tag{1.16}$$

(ii) *Conversely, if $\rceil$ and $\lceil$ are binary operations on X that satisfy* (1.13)–(1.15) *and* (1.16), *and if ρ is defined by*

$$\rho(a, b) = (a \rceil b, a \lceil b), \tag{1.17}$$

then (X, ρ) is an involutive set-theoretic solution of YBE.

(iii) *In the above context, (X, ρ) is nondegenerate if and only if $(X, \rceil)$ is a left-quasigroup and $(X, \lceil)$ is a right-quasigroup, that is, the left translations of $\rceil$ and the right translations of $\lceil$ are one-to-one.*

Proof. (i) A simple computation: ρ^{12} maps (a, b, c) to $(a \rceil b, a \lceil b, c)$, then ρ^{23} maps the latter to $((a \rceil b, (a \lceil b) \rceil c, (a \lceil b) \lceil c)$, and ρ^{12} maps the above triple to the triple consisting of the left-hand terms in (1.13)–(1.15). Starting from (a, b, c) and applying $\rho^{23}\rho^{12}\rho^{23}$, one similarly obtains the triple consisting of the right-hand terms in (1.13)–(1.15). Then the assumption that (X, ρ) is a set-theoretic solution of YBE implies that the entries of the above triples pairwise match.

Then $\rho\circ\rho$ maps (a, b) to $((a\rceil b)\rceil(a\lceil b), (a\rceil b)\lceil(a\lceil b))$, whence (1.16) whenever (S, ρ) is involutive.

(ii) For every pair (a, b) in $X \times X$, the above computation and the assumption that $(X, \rceil, \lceil)$ satisfies (1.16) implies that $\rho \circ \rho$ maps (a, b) to itself. Hence ρ is an involution of $X \times X$, and a fortiori it is a bijection of $X \times X$. Next, for every triple (a, b, c) in $X \times X \times X$, the above computation and the assumption that $(X, \rceil, \lceil)$ satisfies (1.13)–(1.15) imply that $\rho^{12}\rho^{23}\rho^{12}$ and $\rho^{23}\rho^{12}\rho^{23}$ map (a, b, c) to the same triple. So (X, ρ) satisfies (1.6), and it is a set-theoretic solution of YBE, which is moreover involutive.

(iii) Owing to (1.12) and (1.17), the equivalence is obvious, as, for a in X, the left translation $y \mapsto \rho_1(a, y)$ coincides with the left translation $y \mapsto a\rceil y$ and, symmetrically, the right translation $x \mapsto \rho_2(x, b)$ coincides with the right translation $x \mapsto x\lceil b$. □

Structures satisfying the laws of Lemma 1.11 have already been considered and they are usually called biracks:

Definition 1.18 (birack). A *birack* is a triple $(X, \rceil, \lceil)$, where X is a set and $\rceil, \lceil$ are binary operations on X satisfying (1.13)–(1.15) and such that the left translations of $\rceil$ and the right translations of $\lceil$ are one-to-one. A birack is called *involutive* if it satisfies (1.16).

Example 1.19 (birack). By Lemma 1.11, every involutive nondegenerate set-theoretic solution of YBE gives a birack. So, for instance, the involutive birack associated with the two involutive nondegenerate set-theoretic solutions of YBE described in Example 1.9 correspond respectively to

$\rceil$	a	b
a	a	b
b	a	b

$\lceil$	a	b
a	a	a
b	b	b

and

$\rceil$	a	b
a	b	a
b	b	a

$\lceil$	a	b
a	b	b
b	a	a

Here is an example of a different flavor. Assume that $*$ is a binary operation on X, and let $*_0$ be the trivial binary operation defined by $s *_0 t = s$. Replacing $x\lceil y$ with x in (1.13)–(1.15), one sees that the latter relations are satisfied in $(X, *, *_0)$ if and only if $*$ satisfies the left-selfdistributivity law LD $x * (y * z) = (x * y) * (x * z)$ investigated in Chapter XI. The right translations of $*_0$ are the identity, hence one-to-one. So $(X, *, *_0)$ is a birack if and only if $(X, *)$ is an LD-system whose left translations are one-to-one: such a structure is called a *rack* in literature. Thus completing a rack with a trivial second operation gives a birack. Except in trivial cases, the latter birack is not involutive: indeed, $(X, *, *_0)$ is involutive if and only if $s * t = t$ holds for all s and t.

We can then restate Lemma 1.11 as

Proposition 1.20 (set-theoretic solution to birack). (i) *If* (X, ρ) *is an involutive nondegenerate set-theoretic solution of* YBE, *and* $\rceil$ *and* $\lceil$ *are defined by* (1.12), *then* $(X, \rceil, \lceil)$ *is an involutive birack.*

(ii) *Conversely, if* $(X, \rceil, \lceil)$ *is an involutive birack, and* ρ *is defined by* (1.17), *then* (X, ρ) *is an involutive nondegenerate set-theoretic solution of* YBE.

Thus, investigating involutive nondegenerate set-theoretic solutions of YBE and investigating involutive biracks are equivalent tasks.

1.3 RC- and RLC-quasigroups

We now make a second step and move to a new framework. Here the transformation is less trivial, and it amounts to replacing a multiplication with the corresponding division. If $(X, *)$ is a left-quasigroup, that is, $*$ is a binary operation on X whose left translations are one-to-one, we can consider the left-inverse operation $\overline{*}$ such that $a \overline{*} b$ is the unique c satisfying $a * c = b$. Here we will consider the left- and right-inverses of the operations in a birack, and see that these new operations are characterized by simple laws, actually simpler that the laws that define biracks. For our current purpose, the point will be that the new laws are closely connected with the cube conditions of Section II.4. We introduce two versions, one involving a single operation, one involving two operations which, we shall see below, are essentially equivalent.

Definition 1.21 (RC-system, RC-quasigroup). A *right-cyclic system*, or RC-*system*, is a pair $(X, \star)$ where X is a set and $\star$ is a binary operation on X that obeys the *right-cyclic law* RC

(1.22) $$(x \star y) \star (x \star z) = (y \star x) \star (y \star z).$$

An RC-*quasigroup* is an RC-system whose left translations are one-to-one, that is, for every s in X, the map $t \mapsto s \star t$ is one-to-one. An RC-system is called *bijective* if the map $(s, t) \mapsto (s \star t, t \star s)$ is a bijection of $X \times X$ to itself.

Example 1.23 (RC-system). Let X be any set, and let σ is a permutation of X. Then defining $s \star t = \sigma(t)$ provides a (semi-trivial) bijective RC-quasigroup. Indeed, all left translations coincide with σ, hence they are one-to-one by assumption, and, for all r, s, t in X, we have $(s \star t) \star (s \star t) = \sigma^2(t) = (s \star r) \star (s \star t)$, so the RC-law is obeyed. Finally, the map $(s, t) \mapsto (s \star t, t \star s)$ is $(s, t) \mapsto (\sigma(t), \sigma(s))$, a bijection, so $(X, \star)$ is bijective.

Assume now that M is a left-cancellative monoid that admits unique right-lcms. We recall from Definition II.2.11 that, for f, g in M, the unique element g' such that fg' is the right-lcm of f and g is denoted by $f \backslash g$ and called the *right-complement* of f in g. Then, as observed in Proposition II.2.15 (triple lcm), $(M, \backslash)$ is an RC-system. More generally, if S is a subset of M that is closed under right-complement, then $(S, \backslash)$ is also an RC-system: this happens in particular when S is a Garside family of M that includes 1. In general, the RC-systems of the type above are not RC-quasigroups: if f does not left-divide g, the right-lcm h of f and g is distinct from g but $f\backslash g = f\backslash h$ holds. Nor are they bijective either: for all f, g, we have $f\backslash f = g\backslash g = 1$.

It will also be useful to consider a two-sided version of RC-quasigroups.

Definition 1.24 (**RLC-system, RLC-quasigroup**). An RLC-*system* is a triple $(X, \star, \tilde{\star})$ where $(X, \star)$ is an RC-system, $\tilde{\star}$ is a binary operation on X obeying the *left-cyclic law* LC

(1.25) $$(z \,\tilde{\star}\, x) \,\tilde{\star}\, (y \,\tilde{\star}\, x) = (z \,\tilde{\star}\, y) \,\tilde{\star}\, (x \,\tilde{\star}\, y),$$

and the two operations are connected by

(1.26) $$(y \star x) \,\tilde{\star}\, (x \star y) = x = (y \,\tilde{\star}\, x) \star (x \,\tilde{\star}\, y).$$

An RLC-*quasigroup* is an RLC-system $(X, \star, \tilde{\star})$ such that the left translations of $\star$ and the right translations of $\tilde{\star}$ are one-to-one.

Example 1.27 (**RLC-system**). On the shape of Example 1.23, let X be any set and define $s \star t = \sigma(t)$ and, symmetrically, $s \,\tilde{\star}\, t = \tau(s)$ where σ, τ map X to itself. Then the RC and LC laws are automatically satisfied, and (1.26) amount to $\sigma \circ \tau$ and $\tau \circ \sigma$ being the identity map, which is possible only if σ and τ are bijective. So, for every bijection σ from X to itself, we obtain an RLC-quasigroup by defining $s \star t = \sigma(t)$ and $s \,\tilde{\star}\, t = \sigma^{-1}(s)$.

By contrast, if M is a monoid that admits both unique right- and left-lcms, $(M, \backslash)$ is an RC-quasigroup and, in an obvious sense, $(M, /)$ is an LC-quasigroup, but $(M, \backslash, /)$ is not an RLC-system in general, as (1.26) is not satisfied: for all f, g in M, we have $f = h \cdot ((g\backslash f)/(f\backslash g))$ and $g = h \cdot ((f\backslash g)/(g\backslash f))$, where h is the left-gcd of f and g, which, for instance, is not 1 for $f \neq g \neq 1$.

In an RLC-quasigroup, the operations determine one another and, as a consequence, RLC-quasigroups and bijective RC-quasigroups are equivalent structures.

Lemma 1.28. *For all binary operations $\star, \tilde{\star}$ on X, the following are equivalent:*

(i) *The system $(X, \star, \tilde{\star})$ obeys the involutivity laws* (1.26).

(ii) *The map $\Psi : (s, t) \mapsto (s \star t, t \star s)$ is a bijection of $X \times X$ to itself and the operation $\tilde{\star}$ is the unique operation on X such that the map $(s, t) \mapsto (s \,\tilde{\star}\, t, t \,\tilde{\star}\, s)$ is the inverse of Ψ.*

Proof. (i) Assume that $(X, \star, \tilde{\star})$ satisfies (1.26). Let (s', t') belong to $X \times X$. Then put $s = t' \,\tilde{\star}\, s'$ and $t = s' \,\tilde{\star}\, t'$. The right-hand equality in (1.26) gives $s \star t = s'$ and $t \star s = t'$, whence $\Psi(s, t) = (s', t')$. So Ψ is surjective. Conversely, assume $\Psi(s, t) = (s', t')$. Then the left-hand equality in (1.26) gives $s = t' \star s'$ and $t = s' \star t'$. So Ψ is injective. Moreover, the equalities show that the map $(s, t) \mapsto (s \,\tilde{\star}\, t, t \,\tilde{\star}\, s)$ is Ψ^{-1}.

(ii) Assume that Ψ is a bijection from $X \times X$ to itself. Then there exists a unique operation $\tilde{\star}$ on X such that the map $(s, t) \mapsto (s \tilde{\star} t, t \tilde{\star} s)$ is Ψ^{-1}, namely the one given by

(1.29) $$s' \,\tilde{\star}\, t' = \text{ the unique } t \text{ such that } s \star t = s' \text{ and } t \star s = t' \text{ hold for some } s.$$

Then (1.26) is satisfied by definition. □

Building on Lemma 1.28, we shall mainly consider bijective RC-quasigroups in the sequel, and appeal to RLC-quasigroups only when necessary. To follow the subsequent computations involving $\rceil$ and $\lceil$, it can be useful to associate with every pair of equalities $a' = a \rceil b, b' = a \lceil b$ the square diagram $\begin{array}{ccc} & \xrightarrow{a'} & \\ a\big\downarrow & & \big\downarrow b' \\ & \xrightarrow[b]{} & \end{array}$, that is, $\begin{array}{ccc} & \xrightarrow{a \rceil b} & \\ a\big\downarrow & & \big\downarrow a \lceil b \\ & \xrightarrow[b]{} & \end{array}$. Similarly, for $\star$ and $\tilde{\star}$, we draw diagrams $\begin{array}{ccc} & \xrightarrow{t} & \\ s\big\downarrow & & \big\downarrow t \star s \\ & \xrightarrow[s \star t]{} & \end{array}$ and $\begin{array}{ccc} & \xrightarrow{t' \tilde{\star} s'} & \\ s' \tilde{\star} t'\big\downarrow & & \big\downarrow s' \\ & \xrightarrow[t']{} & \end{array}$. For the moment—things will be different in Section 2 when structure monoids are introduced—the diagrams are not meant to represent any multiplication, but just serve as a mnemonic help: for instance, comparing the above diagrams helps recording that, if (1.26) holds, the conjunction of $s' = s \star t$ and $t' = t \star s$ is equivalent to the conjunction of $s = t' \tilde{\star} s'$ and $t = s' \tilde{\star} t'$.

We now come back to the connection between biracks and RC-systems. To formalize it, we introduce a general algebraic transformation on binary operations whose left translations are bijective.

Definition 1.30 (left-inverse operation). If X is a set and $*$ is a binary operation on X whose left translations are one-to-one, that is, for every a in X the map $x \mapsto a * x$ is one-to-one, the *left-inverse* of $*$ is the binary operation $\overline{*}$ on X such that $a \overline{*} b$ is the unique element c satisfying $a * c = b$.

By definition, left-inverting an operation is an involutive transformation: for every operation $*$ with bijective left translations, the left translations of $\overline{*}$ are bijective as well and the left-inverse of $\overline{*}$ is $*$. Here comes the main observation:

Proposition 1.31 (birack to RC-quasigroup). *If $(X, \rceil, \lceil)$ is an involutive birack, and $\star$ is the left-inverse of $\rceil$, then $(X, \star)$ is a bijective* RC*-quasigroup.*

Proof. By the definition of a birack, the left translations of $\rceil$ are bijective, which guarantees the existence of $\star$, and the fact that the left translations of $\star$ are one-to-one. The task is to show that the assumption that $(X, \rceil, \lceil)$ satisfies the laws (1.13)–(1.16) implies that $\star$ obeys the RC-law (1.25).

Claim. *For all x, y, z, the relation $y = x \rceil z$ is equivalent to $x \star y = z$ and it implies $y \star x = x \lceil z$.*

Proof of the claim. Assume $y = x \rceil z$. First, by definition of $\star$, this relation is equivalent to $x \star y = z$. Next, (1.16), implies $(x \rceil z) \rceil (x \lceil z) = x$, so we obtain $y \rceil (x \lceil z) = x$. By the definition of $\star$, the latter relation is equivalent to $y \star x = x \lceil z$, see on the right. □

$$\begin{array}{ccc} & \xrightarrow{y = x \rceil z} & \\ x\big\downarrow & & \big\downarrow y \star x = x \lceil z \\ & \xrightarrow[x \star y = z]{} & \end{array}$$

Now let r, s, t lie in X. Put $a = t$, $b = t \star s$, and $c = (t \star s) \star (t \star r)$. We shall step by step compute the expressions $(r \star s) \star (r \star t)$ and $(s \star r) \star (s \star t)$ in terms of a, b, and c and, using (1.13)–(1.15), establish that these expressions are equal. The proof consists in repeatedly using the claim for various relations $z = x \star y$. The corresponding diagrams are displayed in Figure 1. The latter shows that we are actually completing a cube and it should make the order of the verifications clear.

Applying the claim to the definition $b = t{\star}s$ with $t = a$ gives $s = a \rceil b$ and $s{\star}t = a \lceil b$.

Next, applying the claim to $c = (t \star s) \star (t \star r)$ with $t \star s = b$ gives $t \star r = b \rceil c$ and $(r \star t) \star (r \star s) = b \lceil c$.

Then, applying the claim to $b \rceil c = t \star r$ with $t = a$ gives $r = a \rceil (b \rceil c)$, hence also $r = (a \rceil b) \rceil ((a \lceil b) \rceil c)$ by (1.13), and $r \star t = a \lceil (b \rceil c)$.

Next, the relations $s = a \rceil b$ and $r = (a \rceil b) \rceil ((a \lceil b) \rceil c)$ imply $s{\star}r = (a \lceil b) \rceil c$, and the claim implies $r{\star}s = s \lceil (s{\star}r) = (a \rceil b) \lceil ((a \lceil b) \rceil c))$, hence also $r{\star}s = (a \lceil (b \rceil c)) \rceil (b \lceil c)$ by (1.14).

Next, the relations $r \star t = a \lceil (b \lceil c)$ and $r \star s = r \star s = (a \lceil (b \rceil c)) \rceil (b \lceil c)$ imply $(r \star t) \star (r \star s) = b \lceil c$, and the claim implies $(r \star s) \star (r \star t) = (a \lceil (b \rceil c)) \lceil (b \lceil c)$, hence $(r \star s) \star (r \star t) = (a \lceil b) \lceil c$ by (1.15).

Finally, the relations $s \star t = a \lceil b$ and $s \star r = (a \lceil b) \rceil c$ imply $(s \star t) \star (s \star r) = c$, and the claim implies $(s \star r) \star (s \star t) = (a \lceil b) \lceil c$.

We thus established the three equalities $(r \star t) \star (r \star s) = b \lceil c = (t \star r) \star (t \star s)$, $(t{\star}s){\star}(t{\star}r) = c = (s{\star}t){\star}(s{\star}r)$, and $(r{\star}s){\star}(r{\star}t) = (a \lceil b) \lceil c = (s{\star}r){\star}(s{\star}t)$. Hence $(X, \star)$ is an RC-quasigroup (one equality would be sufficient as r, s, t are arbitrary).

Now, consider the binary operation $\tilde{\star}$ on X such that $x \,\tilde{\star}\, y = z$ is equivalent to $z \lceil x = y$, that is, in an obvious sense, the right-inverse of $\lceil$, which makes sense since, by assumption, the right translations of $\lceil$ are one-to-one. Then an entirely symmetric verification shows that $\tilde{\star}$ obeys the LC-law.

Finally, we consider (1.26). Let r, s belong to X. Put $a = s$ and $b = s \star r$. Then the definition of $\star$ gives $r = a \rceil b$, and the claim then implies $r \star s = b \lceil a$. Now, owing to the relations $s \star r = b$ and $r \star s = b \lceil a$, the definition of $\tilde{\star}$ gives $(r \star s) \,\tilde{\star}\, (s \star r) = a$, and the symmetric counterpart of the claim then implies $(s \star r) \,\tilde{\star}\, (r \star s) = a \rceil b$. We deduce $(r \star s) \,\tilde{\star}\, (s \star r) = s$ and $(s \star r) \,\tilde{\star}\, (r \star s) = r$. So (1.26) is obeyed, $(X, \star, \tilde{\star})$ is an RLC-quasigroup, and, by Lemma 1.28, $(X, \star)$ is a bijective RC-quasigroup. □

We now establish a converse for Proposition 1.31, thus defining an involutive birack starting from a bijective RC-quasigroup.

Proposition 1.32 (**RC-quasigroup to birack).** *Assume that* $(X, \star)$ *is a bijective* RC-*quasigroup. Let* $\rceil$ *be the left-inverse of* $\star$ *and* $\lceil$ *be the right-inverse of the operation* $\tilde{\star}$ *defined by* (1.29). *Then* $(X, \rceil, \lceil)$ *is an involutive birack.*

Proof. We argue as for Proposition 1.31, now using the following result:

Claim. *For all* x, y, z, *the relation* $z = x \star y$ *is equivalent to* $x \rceil z = y$ *and it implies* $x \lceil z = y \star x$.

Proof of the claim. Assume $z = x \star y$. By the definition of $\rceil$, this relation is equivalent to $x \rceil z = y$. Then, by definition of Ψ, we have $\Psi(x, y) = (z, y \star x)$, hence $(x, y) = \Psi^{-1}(z, y \star x)$. By the definition of $\tilde{\star}$, this implies $z \,\tilde{\star}\, (y \star x) = x$, whence $z \lceil x = y \star x$ by definition of the operation $\lceil$. □

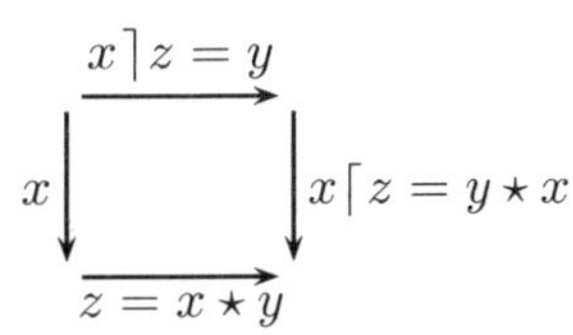

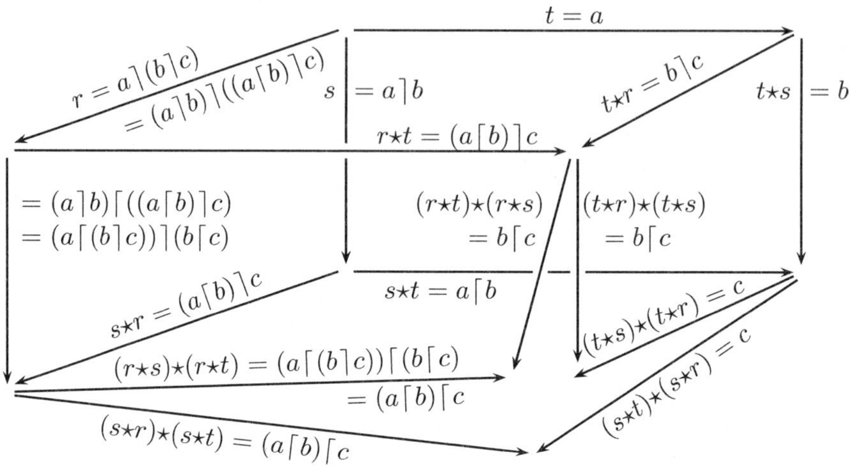

Figure 1. Proof of Proposition 1.31(i): one successively evaluates the edges of the cube in terms of a, b, c and the relations (1.13)–(1.15) guarantee that the cube closes. The same diagram can be used to follow the proof of Proposition 1.32 below, except that one starts with a closed cube and evaluates some edges in two different ways to establish (1.13)–(1.15).

Note that the diagram corresponding to the current claim is the same as the one for the claim for Proposition 1.31, but the meaning is different as, here, we assume $\star$-relations and are interested in deducing $\rceil$, $\lceil$-relations whereas, in the previous case, we went the other way.

Now, let a, b, c belong to X. Put $r = a\rceil(b\rceil c)$, $s = a\rceil b$, and $t = a$. We shall now compute the expressions involved in (1.13)–(1.15) in terms of r, s, t, and establish the expected equalities.

First, applying the claim to $s = a\rceil b$ with $t = a$ gives $b = t \star s$ and $a\lceil b = s \star t$.

Next, applying it to $r = a\rceil(b\rceil c)$ with $t = a$ gives $b\rceil c = t \star r$ and $a\lceil(b\rceil c) = r \star t$.

Then, applying the claim to $t \star r = b\rceil c$ with $t \star s = b$ gives $c = (t \star s) \star (t \star r)$, hence also $c = (s\star t)\star(s\star r)$ by (1.22), and $b\lceil c = (t\star r)\star(t\star s)$, hence also $b\lceil c = (r\star t)\star(r\star s)$, again by (1.22).

Next, the relation $(s \star t) \star (s \star r) = c$ with $(s \star t = a\lceil b$ implies $(a\lceil b)\rceil c = s \star r$, and the claim then implies $(a\lceil b)\lceil c = (s \star r) \star (s \star t)$, whence also $(a\lceil b)\lceil c = (r \star s) \star (r \star t)$ by (1.22).

Then, the relation $(a\lceil b)\rceil c = s \star r$ with $s = a\rceil b$ implies $(a\rceil b)\rceil((a\lceil b)\rceil c) = r$, which, together with the previously established relation $r = a\rceil(b\rceil c)$, gives (1.13). By the claim, we deduce $(a\rceil b)\lceil((a\lceil b)\rceil c) = r \star s$.

Now, $b\lceil c = (r \star t) \star (r \star s)$ with $r \star t = a\lceil(b\rceil c)$ implies $r \star s = (a\lceil(b\rceil c))\rceil(b\lceil c)$ which, together the previously established relation $(a\rceil b)\lceil((a\lceil b)\rceil c) = r \star s$ gives (1.14). Moreover, the claim then implies $(r \star s) \star (r \star t) = (a\lceil(b\rceil c))\lceil(b\lceil c)$ which, together the previously established relation $(a\lceil b)\lceil c = (r \star s) \star (r \star t)$ gives (1.15). Hence $(X, \rceil, \lceil)$ is a birack.

We conclude with involutivity. Let a, b belong to X. Put $r = a$ and $s = a\rceil b$. By the claim, we have $r \star s = b$ and $s \star r = a\lceil b$, that is, $(a\rceil b) \star a = a\lceil b$. By the claim again, the latter is equivalent to $(a\rceil b)\rceil(a\lceil b) = a$. The argument for $(a\rceil b)\lceil(a\lceil b) = b$ is symmetric. □

Thus, involutive biracks and bijective RC-quasigroups are equivalent frameworks. By the way, we also observe that the conjunction of Propositions 1.31 and 1.32 implies

Corollary 1.33 (bijective RC implies LC). *If* $(X, \star)$ *is a bijective* RC-*quasigroup, then the operation* $\tilde{\star}$ *provided by* (1.29) *satisfies the* LC-*law.*

Proof. Proposition 1.32 provides from $\star$ two operations $\rceil, \lceil$ such that $(X, \rceil, \lceil)$ is an involutive birack. Then Proposition 1.31 provides from $\rceil$ and $\star$ an operation $\tilde{\star}$ such that $(X, \star, \tilde{\star})$ is an RLC-quasigroup. So, in particular, the operation $\tilde{\star}$ satisfies the LC-law. By uniqueness, the operation $\tilde{\star}$ is the one provided from $\star$ by Lemma 1.28. □

Whether a simple direct argument exists for Corollary 1.33 is not clear: introducing the auxiliary operations $\rceil$ and $\lceil$ can be avoided but, in any case, evaluating all expressions occurring in the cube of Figure 1 is probably necessary.

Summarizing the results, we can state:

Proposition 1.34 (equivalence). *The following three (actually four) frameworks are equivalent:*

(i) *nondegenerate involutive set-theoretic solutions of the Yang–Baxter equation,*

(ii) *involutive biracks, and*

(iii) *bijective* RC-*quasigroups (or, equivalently,* RLC-*quasigroups),*

with an explicit way of going from one to the other.

To conclude this section, we mention without proof a beautiful result of W. Rump [204, Theorem 2] that will enable us to skip bijectivity assumptions in the sequel when dealing with finite RC-systems.

Proposition 1.35 (bijectivity). *Every finite* RC-*quasigroup is bijective.*

2 Structure monoids and groups

Here comes the key step from our current point of view, namely associating with every involutive nondegenerate set-theoretic solution of YBE a certain group called its structure group. In this section, we shall show how the general methods developed in this text enable one to investigate structure groups very easily and, in particular, to prove that they are Garside groups.

The section is organized as follows. In Subsection 2.1, we introduce the structure monoids and groups and state the main algebraic results that will be established subsequently. In Subsection 2.2, we develop a sort of polynomial calculus for structures satisfying the RC-law. Then, in Subsection 2.3, we complete the proof of the results announced in Subsection 2.1, in particular the result that the structure group of an involutive nondegenerate set-theoretic solution of YBE is a Garside group. Finally, in Subsection 2.4, we show that, conversely, every Garside group that admits a presentation of a certain type is the structure group of a set-theoretic solution of YBE.

2.1 Structure monoids and groups

We first restart from involutive nondegenerate set-theoretic solutions of the Yang–Baxter equation. Then there exists a standard way of associating a monoid and a group.

Definition 2.1 (structure group I). If (X, ρ) is an involutive nondegenerate set-theoretic solution of YBE, the *structure group* (*resp. monoid*) of (X, ρ) is the group (*resp.* the monoid) defined by the presentation

(2.2) $$\langle X, \{ab = a'b' \mid a, b, a', b' \in X \text{ satisfying } \rho(a, b) = (a', b')\}\rangle.$$

The presentation (2.2) is redundant and contains trivial relations: as ρ is bijective, most relations occur twice and $\rho_1(a, b) = a$ implies $\rho_2(a, b) = b$ and, in this case, we obtain the trivial relation $ab = ab$. Note that saying the relations of (2.2) are satisfied exactly means that the diagram $\begin{array}{ccc} & \xrightarrow{\rho_1(a,b)} & \\ a\big\downarrow & & \big\downarrow\rho_2(a,b) \\ & \xrightarrow[b]{} & \end{array}$ is commutative when interpreted in the structure monoid of (X, ρ).

Example 2.3 (structure group I). Consider the two involutive nondegenerate solutions of Example 1.9. For the left-hand table, (2.2) gives the four relations $\mathtt{aa} = \mathtt{aa}$, $\mathtt{ab} = \mathtt{ba}$, $\mathtt{ba} = \mathtt{ab}$, and $\mathtt{bb} = \mathtt{bb}$. Erasing the trivial and redundant relations, we obtain that the structure group is $\langle \mathtt{a}, \mathtt{b} \mid \mathtt{ab} = \mathtt{ba}\rangle$, a free abelian group of rank 2, and the structure monoid is a free abelian monoid of rank 2.

Similarly, the right-hand table gives the relations $\mathtt{a}^2 = \mathtt{b}^2$, $\mathtt{ab} = \mathtt{ab}$, $\mathtt{ba} = \mathtt{ba}$, and $\mathtt{b}^2 = \mathtt{a}^2$. Erasing the trivial and redundant relations, we see that the structure group is $\langle \mathtt{a}, \mathtt{b} \mid \mathtt{a}^2 = \mathtt{b}^2\rangle$ and the structure monoid is $\langle \mathtt{a}, \mathtt{b} \mid \mathtt{a}^2 = \mathtt{b}^2\rangle^+$.

The next statement summarizes the results about structure groups and monoids we shall establish below. We recall that, in a monoid that admits unique right-lcms, $f\backslash g$ denotes the unique element g' such that fg' is the right-lcm of f and g.

Proposition 2.4 (structure group I). *Assume that (X, ρ) is an involutive nondegenerate set-theoretic solutions of* YBE *and M, G are the associated structure monoid and group.*

(i) *The monoid M contains no nontrivial invertible element and is Noetherian.*

(ii) *The monoid M is an Ore monoid, it admits unique left- and right-lcms and left- and right-gcds, G is a group of left- and right-fractions for M; this group is torsion-free.*

(iii) *The solution (X, ρ) can be retrieved from M: the set X is the atom set of M and, for s, t in X, the value of $\rho(a, b)$ is determined by $\rho(a, b) = (a', a'\backslash a)$ if there exists a' in M satisfying $a\backslash a' = b$, and $\rho(a, b) = (a, b)$ otherwise.*

(iv) *The right-lcm Δ_I of a size-n subset I of X belongs to X^n, it is the left-lcm of (another) size-n subset of X, the map $I \mapsto \Delta_I$ is injective, and its image is the smallest Garside family containing 1 in M.*

(v) *If X is finite with n elements and Δ is the right-lcm of X, then (M, Δ) is a Garside monoid, G is a Garside group, and the above Garside family has 2^n elements and is bounded by Δ, which is also the left-lcm of X.*

The proof of Proposition 2.4 will be completed at the end of Subsection 2.3 only. Our goal here will be to show that mixing an approach based on RC-quasigroups with the general tools previously developed in this text make that proof easy and quick. The initial remark is that, as seen in Section 1, the framework of (involutive nondegenerate) set-theoretic solutions of YBE can be replaced with a framework of bijective RC-quasigroups or, equivalently, RLC-quasigroups. This rephrasing is specially useful in view of applying the reversing method of Section II.4.

Definition 2.5 (structure group II). If $(X, \star)$ is an RC-quasigroup, the *structure group* (*resp. monoid*) of $(X, \star)$—or, simply, the group and monoid *associated with* $(X, \star)$—is the group (*resp.* the monoid) defined by the presentation

$$(X, \{s(s \star t) = t(t \star s) \mid s \neq t \in X\}). \tag{2.6}$$

For instance, it turns out that the RC-quasigroups associated with the biracks of Example 1.19 admit the same table as the corresponding operation $\rceil$, and, therefore, the associated structure monoids are $\langle \mathtt{a}, \mathtt{b} \mid \mathtt{ab} = \mathtt{ba} \rangle^+$ and $\langle \mathtt{a}, \mathtt{b} \mid \mathtt{a}^2 = \mathtt{b}^2 \rangle^+$, respectively—hence, as can be expected, the same monoids as in Example 2.3. Note that, if X is finite with n elements, then there are exactly $\binom{n}{2}$ relations in (2.6), and that they correspond to diagrams $\begin{array}{ccc} & \xrightarrow{\;t\;} & \\ s\big\downarrow & & \big\downarrow t \star s \\ & \xrightarrow[s \star t]{} & \end{array}$ similar to those considered in Subsection 1.3. We immediately observe that Definitions 2.1 and 2.5 are compatible.

Lemma 2.7. *If (X, ρ) is an involutive nondegenerate set-theoretic solution of YBE and $(X, \star)$ is the associated* RC-*quasigroup as described in Proposition* 1.31, *then the structure monoid of (X, ρ) coincides with the monoid associated with $(X, \star)$, and so do the corresponding groups.*

Proof. Assume that $ab = a'b'$ is a relation of (2.2). Then, by the definition of $\star$ from ρ, we have $b = a \star a'$ and $b' = a' \star a$. If a and a' coincide, the assumption that ρ is nondegenerate implies that b and b' coincide as well, and the relation $ab = a'b'$ is trivial. Otherwise, the relation rewrites as $a(a \star a') = a'(a' \star a)$, and it is a relation of (2.6).

Conversely, consider a relation $s(s\star t) = t(t\star s)$ of (2.6). Put $a = s$, $a' = t$, $b = s\star t$, and $b' = t' \star s'$. Then, by the claim in the proof of Proposition 1.32, we have $a' = a\rceil b$ and $t' = a\lceil b$ in the language of biracks, that is, $(a', b') = \rho(a, b)$ in the language of set-theoretic solutions of YBE. So the relation $s(s \star t) = t(t \star s)$, which is $ab = a'b'$, is a relation of (2.2). □

Thus establishing results for the structure monoids of (involutive nondegenerate) set-theoretic solutions of YBE and for the structure monoids of bijective RC-quasigroups are equivalent tasks. We shall see now that the second framework is specially convenient. Here is the result we shall actually establish:

Proposition 2.8 (structure monoid II). *If $(X, \star)$ is a bijective* RC*-quasigroup and M, G are the associated monoid and group, then M and G satisfy all properties listed in Proposition* 2.4, *with the only difference that* (iii) *now says that $s\star t$ is the right-complement $s\backslash t$ for $s \neq t$, and is the unique element of $X \setminus \{s\backslash t \mid t \neq s \in X\}$ otherwise.*

The specific point that makes the RC-quasigroup approach specially efficient here is that, by definition, the presentation (2.6) is right-complemented (Definition II.2.11): it contains no relation of the form $s... = s...$ and, for $s \neq t$, it contains at most one, actually exactly one, relation of the form $s... = t...$. With the formalism of Section II.4, this presentation is associated with the syntactic right-complement θ defined by

(2.9)
$$\theta(s,t) = \begin{cases} s\star t & \text{for } s \neq t, \\ \varepsilon & \text{for } s = t. \end{cases}$$

Moreover, the syntactic right-complement θ is short, that is, the length of every word $\theta(s,t)$ is at most one. It follows that the monoid associated with $(X, \star)$ is eligible for the results of Section II.4.

Lemma 2.10. *Assume that $(X, \star)$ is an* RC*-quasigroup and M is the associated monoid.*

(i) *The monoid M has no nontrivial invertible element and is Noetherian.*

(ii) *The monoid M is left-cancellative, and it admits unique right-lcms and left-gcds.*

(iii) *The system $(X, \star)$ can be retrieved from M: the set X is the atom set of M, for $s \neq t$, the value of $s \star t$ is the right-complement $s\backslash t$ in M and the value of $s \star s$ is the unique element of $X \setminus \{s\backslash t \mid t \neq s \in X\}$.*

Proof. (i) First, the relations of the presentation (2.6) preserve the length, so, by Proposition II.2.32 (homogeneous), the monoid M is (strongly) Noetherian. As there is no ε-relation in (2.6), M contains no nontrivial invertible element.

(ii) The assumption that the operation $\star$ satisfies the RC-law implies that the presentation (2.6) satisfies the sharp θ-cube condition for every triple of pairwise distinct elements of X. Indeed, if r, s, t are pairwise distinct, we obtain, using (2.9),

$$\theta_3^*(r,s,t) = \theta(\theta(r,s), \theta(r,t)) = \theta(r \star s, r \star t) = (r \star s) \star (r \star t),$$

the third equality because the assumption $s \neq t$ implies $r \star s \neq r \star t$ since the left translations of $\star$ are injective. We similarly find $\theta_3^*(s,r,t) = (s \star r) \star (s \star t)$ and the assumption that $\star$ satisfies the RC-law gives $\theta_3^*(r,s,t) = \theta_3^*(s,r,t)$, as expected.

Then Proposition II.4.16 (right-complemented) states that M is left-cancellative and that any two elements of M that admit a common right-multiple admit a right-lcm. Moreover, as the syntactic right-complement θ defined by (2.9) is defined for every pair of letters and it is short, right-reversing is always terminating, that is, $\theta^*(u,v)$ is defined for all X-words u, v. So, by Proposition II.4.16 again, any two elements of M admit a common right-multiple, and therefore a right-lcm. Finally, as is M is (right)-Noetherian, Lemma II.2.37 then implies that M admits left-gcds.

(iii) As there is no relation involving a word of length one in (2.6), the elements of X are atoms, and every element not lying in $X \cup \{1\}$ is not an atom. So X is the atom set of M.

Next, for distinct s, t in X, the right-lcm of s and t is $s(s \star t)$, so, by definition, $s\backslash t$ is equal to $s \star t$. Hence, all nondiagonal values $s \star t$ can be retrieved from M.

Finally, as all left translations of $(X, \star)$ are one-to-one, $s \star s$ must be the unique element of $X \setminus \{s \star t \mid s, t \in X, s \neq t\}$, that is, of $X \setminus \{s\backslash t \mid t \neq s \in X\}$. □

We recall that the proof of Proposition II.4.16 (right-complemented) relies on the completeness of right-reversing for the presentation (2.6), that is, on the result that two X-words u, v represent the same element in the monoid M if and only if the signed word $\overline{u}v$ is right-reversible to the empty word. This result requires an induction, which is relatively delicate in general, but is easy in the current case because all relations involve words of length two, making termination trivial.

Remark 2.11. Definition 2.5 makes sense for every RC-system, and not only for an RC-quasigroup. However, if $(X, \star)$ is not an RC-quasigroup, then the RC-law need not guarantee that the presentation (2.6) satisfies the cube condition, because of possible equalities $r \star s = r \star t$ with $s \neq t$, implying $\theta(r \star s, r \star t) = \varepsilon$ instead of $\theta(r \star s, r \star t) = (r \star s) \star (r \star t)$. In this case, our current methods say nothing, and another approach is needed—see Notes.

2.2 RC-calculus

Before proceeding, we develop a sort of polynomial calculus for expressions involving the binary operators $\star$ or $\tilde{\star}$ and multiplication. This framework will allow us to easily perform computations that, otherwise, would require tedious verifications.

Definition 2.12 (monomials Ω_n and $\widetilde{\Omega}_n$). For $n \geqslant 1$, we inductively define formal expressions $\Omega_n(x_1, \ldots, x_n)$ and $\widetilde{\Omega}_n(x_1, \ldots, x_n)$ by $\Omega_1(x_1) = \widetilde{\Omega}_1(x_1) = x_1$ and

(2.13) $$\Omega_n(x_1, \ldots, x_n) = \Omega_{n-1}(x_1, \ldots, x_{n-1}) \star \Omega_{n-1}(x_1, \ldots, x_{n-2}, x_n),$$

(2.14) $$\widetilde{\Omega}_n(x_1, \ldots, x_n) = \widetilde{\Omega}_{n-1}(x_1, x_3, \ldots, x_n) \,\tilde{\star}\, \widetilde{\Omega}_{n-1}(x_2, \ldots, x_n).$$

Example 2.15 (monomial Ω_n). We find $\Omega_2(x_1, x_2) = x_1 \star x_2$, then $\Omega_3(x_1, x_2, x_3) = (x_1 \star x_2) \star (x_1 \star x_3)$, etc. It should be clear that 2^{n-1} variables x_i occur in $\Omega_n(x_1, \dots, x_n)$, with brackets corresponding to a balanced binary tree. For instance, for $n = 4$, the variables occur in the order 12131214 and, for $n = 5$, in the order 1213121412131215.

The expression $\Omega_n(x_1, \dots, x_n)$—a *term* in the language of model theory—is a sort of n-variable monomial involving the binary operator $\star$ in place of the standard multiplication, and similarly for $\widetilde{\Omega}_n(x_n, \dots, x_1)$ with $\tilde{\star}$. Diagrams have been associated in Section 1 with expressions involving $\star$ and $\tilde{\star}$. In particular, each instance of the RC-law gives rise to the cubic diagram of Figure 1 and, similarly, an n-simplex comes associated with every size-n family of elements. Then the monomials $\Omega_i(x_1, \dots, x_i)$ arise in the labels of the ith level in such n-simplices, see Figure 2.

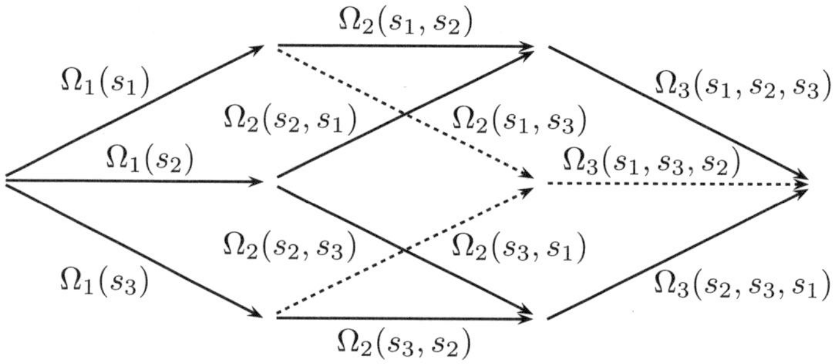

Figure 2. The monomials Ω_i occur at the ith level in the n-simplex associated with the RC-law, here the cube corresponding to $n = 3$. Of course, $\widetilde{\Omega}_i(x_1, \dots, x_i)$ symmetrically appears in connection with the LC-law.

The next result is an iterated version of the RC-law, which, in terms of the expressions Ω_i, is $\Omega_3(x, y, z) = \Omega_3(y, x, z)$.

Lemma 2.16. *If $(X, \star)$ is an* RC*-system, then, for all $s_1, \dots, s_n$ in X and π in $\mathfrak{S}_{n-1}$, we have*

(2.17) $$\Omega_n(s_{\pi(1)}, \dots, s_{\pi(n-1)}, s_n) = \Omega_n(s_1, \dots, s_n).$$

Proof. An induction on n. For $n = 1$ and $n = 2$, there is nothing to prove. For $n = 3$, the equality $\Omega_3(s_1, s_2, s_3) = \Omega_3(s_2, s_1, s_3)$ is the RC-law. Assume $n \geqslant 4$. As transpositions of adjacent entries generate the symmetric group $\mathfrak{S}_n$, it is sufficient to prove the result when π is a transposition $(i, i+1)$. For $i < n-2$, the definition plus the induction hypothesis give

$$\begin{aligned}
&\Omega_n(s_1, \dots, s_i, s_{i+1}, \dots, s_n)\\
&\quad = \Omega_{n-1}(s_1, \dots, s_i, s_{i+1}, \dots, s_{n-1}) \star \Omega_{n-1}(s_1, \dots, s_i, s_{i+1}, \dots, s_{n-2}, s_n)\\
&\quad = \Omega_{n-1}(s_1, \dots, s_{i+1}, s_i, \dots, s_{n-1}) \star \Omega_{n-1}(s_1, \dots, s_{i+1}, s_i, \dots, s_{n-2}, s_n)\\
&\quad = \Omega_n(s_1, \dots, s_{i+1}, s_i, \dots, s_n).
\end{aligned}$$

For $i = n-2$, writing $\vec{s}$ for $s_1, \dots, s_{n-3}$, the definition plus the RC-law give

$$\Omega_n(s_1, \dots, s_n) = \Omega_n(\vec{s}, s_{n-2}, s_{n-1}, s_n)$$

$$
\begin{aligned}
&= \Omega_{n-1}(\vec{s}, s_{n-2}, s_{n-1}) \star \Omega_{n-1}(\vec{s}, s_{n-2}, s_n) \\
&= (\Omega_{n-2}(\vec{s}, s_{n-2}) \star \Omega_{n-2}(\vec{s}, s_{n-1})) \star (\Omega_{n-2}(\vec{s}, s_{n-2}) \star \Omega_{n-2}(\vec{s}, s_n)) \\
&= (\Omega_{n-2}(\vec{s}, s_{n-1}) \star \Omega_{n-2}(\vec{s}, s_{n-2})) \star (\Omega_{n-2}(\vec{s}, s_{n-1}) \star \Omega_{n-2}(\vec{s}, s_n)) \\
&= \Omega_{n-1}(\vec{s}, s_{n-1}, s_{n-2}) \star \Omega_{n-1}(\vec{s}, s_{n-1}, s_n) = \Omega_n(\vec{s}, s_{n-2}, s_{n-1}, s_n). \qquad \square
\end{aligned}
$$

Of course, the counterpart of (2.17) involving $\widetilde{\Omega}_n$ is valid when $\tilde{\star}$ satisfies the LC-law (1.25).

Definition 2.18 (polynomial Π_n and $\widetilde{\Pi}_n$). For $n \geqslant 1$, we define the formal expressions $\Pi_n(x_1, \dots, x_n)$ and $\widetilde{\Pi}_n(x_1, \dots, x_n)$ by

(2.19) $$\Pi_n(x_1, \dots, x_n) = \Omega_1(x_1) \cdot \Omega_2(x_1, x_2) \cdot \dots \cdot \Omega_n(x_1, \dots, x_n),$$

(2.20) $$\widetilde{\Pi}_n(x_1, \dots, x_n) = \widetilde{\Omega}_n(x_1, \dots, x_n) \cdot \widetilde{\Omega}_{n-1}(x_2, \dots, x_n) \cdot \dots \cdot \widetilde{\Omega}_1(x_n).$$

Note that (2.19) implies $\Pi_n(x_1, \dots, x_n) = \Pi_{n-1}(x_1, \dots, x_{n-1}) \cdot \Omega_n(x_1, \dots, x_n)$ for $n \geqslant 2$. The defining relation $r(r \star s) = s(s \star r)$ of the monoid associated with $(X, \star)$ takes the form $\Pi_2(r, s) = \Pi_2(s, r)$. Here is an iterated version.

Lemma 2.21. *If $(X, \star)$ is an* RC*-system and M is the associated monoid, the evaluation of Π_n in M is a symmetric function: for all $s_1, \dots, s_n$ in X and π in $\mathfrak{S}_n$, we have*

(2.22) $$\Pi_n(s_{\pi(1)}, \dots, s_{\pi(n)}) = \Pi_n(s_1, \dots, s_n).$$

Proof. We use induction on n. For $n = 1$, there is nothing to prove. For $n = 2$, (2.22) is the equality $s_1(s_1 \star s_2) = s_2(s_2 \star s_1)$, which is valid in M. Assume $n \geqslant 3$. As in Lemma 2.16, it is sufficient to consider transpositions $(i, i+1)$, that is, to compare $\Pi_n(s_1, \dots, s_n)$ and $\Pi_n(s_1, \dots, s_{i+1}, s_i, \dots, s_n)$. By definition, $\Pi_n(s_1, \dots, s_n)$ is the product of the values $\Omega_j(s_1, \dots, s_j)$ for j increasing from 1 to n; on the other hand, $\Pi_n(s_1, \dots, s_{i+1}, s_i, \dots, s_n)$ is a similar product of $\Omega_j(s'_1, \dots, s'_j)$ with $s'_i = s_{i+1}, s'_{i+1} = s_i$, and $s'_k = s_k$ for $k \neq i, i+1$. For $j < i$, the entries s_i and s_{i+1} do not occur in $\Omega_j(s_1, \dots, s_j)$ and $\Omega_j(s'_1, \dots, s'_j)$, which are therefore equal. For $j > i+1$, the expressions $\Omega_j(s_1, \dots, s_j)$ and $\Omega_j(s'_1, \dots, s'_j)$ differ by the permutation of two non-final entries, so they are equal by Lemma 2.16. There remains to compare the central entries

$$t = \Omega_i(s_1, \dots, s_i) \cdot \Omega_{i+1}(s_1, \dots, s_{i+1}) \quad \text{and} \quad t' = \Omega_i(s'_1, \dots, s'_i) \cdot \Omega_{i+1}(s'_1, \dots, s'_{i+1}).$$

Now put $r = \Omega_i(s_1, \dots, s_i)$ and $r' = \Omega_i(s_1, \dots, s_{i-1}, s_{i+1})$. By the definition of s'_k, we have also $r = \Omega_i(s'_1, \dots, s'_{i-1}, s'_{i+1})$ and $r' = \Omega_i(s'_1, \dots, s'_i)$. Then, by the definition of Ω_i and Ω_{i+1}, we have $t = r(r \star r')$ and $t' = r'(r' \star r)$, whence $t = t'$ in M. $\square$

So far, we considered arbitrary RC-systems. Further results appear when we restrict to RC-quasigroups, that is, when left translations are one-to-one.

Lemma 2.23. *Assume that $(X,\star)$ is an* RC-*quasigroup and $s_1, \dots, s_n$ lie in X.*

(i) *The map $s \mapsto \Omega_{n+1}(s_1, \dots, s_n, s)$ is a bijection of X into itself.*

(ii) *There exist $r_1, \dots, r_n$ in X satisfying $\Omega_i(r_1, \dots, r_i) = s_i$ for $1 \leqslant i \leqslant n$.*

(iii) *Put $\tilde{s}_i = \Omega_n(s_1, \dots, \widehat{s_i}, \dots, s_n, s_i)$ for $1 \leqslant i \leqslant n$. Then, for all i, j, the relations $s_i = s_j$ and $\tilde{s}_i = \tilde{s}_j$ are equivalent.*

Proof. (i) We use induction on n. For $n = 1$, the considered map is the left translation $s \mapsto s_1 \star s$, a bijection of X into itself by assumption. Assume $n \geqslant 2$. The definition of Ω_{n+1} gives $\Omega_{n+1}(s_1, \dots, s_n, s) = t \star \Omega_n(s_1, \dots, s_{n-1}, s)$, with $t = \Omega_n(s_1, \dots, s_{n-1})$. By the induction hypothesis, the map $s \mapsto \Omega_n(s_1, \dots, s_{n-1}, s)$ is bijective. Hence composing it with the left translation by t yields a bijection.

(ii) Use once more induction on n. For $n = 1$, take $t_1 = s_1$. Assume $n \geqslant 2$. By the induction hypothesis, there exist $r_1, \dots, r_{n-1}$ satisfying $\Omega_i(r_1, \dots, r_i) = s_i$ for $1 \leqslant i \leqslant n-1$. Then the definition of Ω_n and the equality $\Omega_{n-1}(r_1, \dots, r_{n-1}) = s_{n-1}$ imply $\Omega_n(r_1, \dots, r_{n-1}, x) = s_{n-1} \star \Omega_{n-1}(r_1, \dots, r_{n-2}, x)$. As the left translation by s_{n-1} is surjective, there exists s satisfying $s_{n-1} \star s = s_n$. Then, by (i), there exists r_n satisfying $\Omega_{n-1}(r_1, \dots, r_{n-2}, r_n) = s$, whence $\Omega_n(r_1, \dots, r_n) = s_n$.

(iii) Again an induction on n. For $n = 1$ there is nothing to prove. For $n = 2$, we find $\tilde{s}_1 = s_2 \star s_1$ and $\tilde{s}_2 = s_1 \star s_2$. It is clear that $s_1 = s_2$ implies $\tilde{s}_1 = \tilde{s}_2$. Conversely, assume $s_1 \star s_2 = s_2 \star s_1$. Using the assumption, the RC-law, and the assumption again, we obtain

$$(s_1 \star s_2) \star (s_2 \star s_2) = (s_2 \star s_1) \star (s_2 \star s_2) = (s_1 \star s_2) \star (s_1 \star s_2) = (s_1 \star s_2) \star (s_2 \star s_1).$$

As the left translations by $s_1 \star s_2$ and s_2 are injective, we first deduce $s_2 \star s_2 = s_2 \star s_1$, and then $s_2 = s_1$. Assume now $n \geqslant 3$. Fix i, j, write $\vec{s}$ for $s_1, \dots, \widehat{s_i}, \dots, \widehat{s_j}, \dots, s_n$ and put $t_k = \Omega_{n-1}(\vec{s}, s_k)$. Then, by Lemma 2.16 and by definition, we have

$$\tilde{s}_i = \Omega_n(\vec{s}, s_j, s_i) = \Omega_{n-1}(\vec{s}, s_j) \star \Omega_{n-1}(\vec{s}, s_i) = t_j \star t_i,$$

and, similarly, $\tilde{s}_j = t_i \star t_j$. If $s_i = s_j$ holds, we have $t_i = t_j$, whence $\tilde{s}_i = \tilde{s}_j$. Conversely, assume $\tilde{s}_i = \tilde{s}_j$, that is, $t_j \star t_i = t_i \star t_j$. By the result for $n = 2$, we deduce $t_i = t_j$, that is, $\Omega_{n-1}(\vec{s}, s_i) = \Omega_{n-1}(\vec{s}, s_j)$, which is an equality of the form $r_1 \star (\dots \star (r_{n-2} \star s_i)\dots) = r_1 \star (\dots \star (r_{n-2} \star s_j)\dots)$. By applying $n-2$ times the assumption that the left translations of $(X,\star)$ are injective, we deduce $s_i = s_j$. □

Let us now consider the involutivity laws (1.26). In the language of Ω_1 and Ω_2, the latter say that, if we put $\tilde{s}_1 = \Omega_2(s_1, s_2)$ and $\tilde{s}_2 = \Omega_2(s_2, s_1)$, we have $s_1 = \widetilde{\Omega}_2(\tilde{s}_1, \tilde{s}_2)$ and $s_2 = \widetilde{\Omega}_2(\tilde{s}_2, \tilde{s}_1)$: two elements can be retrieved from their Ω_2 images using the polynomial $\widetilde{\Omega}_2$. Here is an n-variable version of this result.

Lemma 2.24. *If $(X, \star, \tilde{\star})$ is an involutive* RLC-*system, $s_1, \dots, s_n$ belong to X, and, for $1 \leqslant i \leqslant n$, we put $\tilde{s}_i = \Omega_n(s_1, \dots, \widehat{s_i}, , \dots, s_n, s_i)$, then, for $1 \leqslant i \leqslant n$, and for every permutation π in $\mathfrak{S}_n$, we have*

$$\Omega_i(s_{\pi(1)}, \dots, s_{\pi(i)}) = \widetilde{\Omega}_{n+1-i}(\tilde{s}_{\pi(i)}, \dots, \tilde{s}_{\pi(n)}), \tag{2.25}$$

$$\Pi_n(s_1, \dots, s_n) = \widetilde{\Pi}_n(\tilde{s}_1, \dots, \tilde{s}_n). \tag{2.26}$$

Proof. For $n = 1$, (2.25) reduces to the tautology $s_{\pi(1)} = s_{\pi(1)}$. Now we fix $n \geqslant 2$ and use induction on i decreasing from n to 1. Assume first $i = n$. Then (2.25) is $\Omega_n(s_{\pi(1)}, \dots, s_{\pi(i)}) = \widetilde{\Omega}_1(\tilde{s}_{\pi(i)})$. By Lemma 2.16, the value of the left-hand term does not depend on the order of the $n-1$ first entries, so it is $\Omega_n(s_1, \dots, \widehat{s_{\pi(i)}}, \dots, s_{n-1}, s_{\pi(i)})$, which, by definition, is $\tilde{s}_{\pi(i)}$, hence $\widetilde{\Omega}_1(\tilde{s}_{\pi(i)})$, so (2.25) is satisfied.

Assume now $i < n$. Put

$$s = \Omega_i(s_{\pi(1)}, \dots, s_{\pi(i)}), \qquad s' = \Omega_i(s_{\pi(1)}, \dots, s_{\pi(i-1)}, s_{\pi(i+1)}),$$
$$t = \Omega_{i+1}(s_{\pi(1)}, \dots, s_{\pi(i)}, s_{\pi(i+1)}), \qquad t' = \Omega_{i+1}(s_{\pi(1)}, \dots, s_{\pi(i-1)}, s_{\pi(i+1)}, s_{\pi(i)}).$$

Using the definition of Ω_{i+1} from Ω_i, we find $t = s \star s'$ and $t' = s' \star s$, whence $s = t' \,\tilde{\star}\, t$ and $s' = t \,\tilde{\star}\, t'$ by the involutivity law. Now the induction hypothesis gives

$$t = \widetilde{\Omega}_{n-i}(\tilde{s}_{\pi(i+1)}, \dots, \tilde{s}_{\pi(n)}), \qquad t' = \widetilde{\Omega}_{n-i}(\tilde{s}_{\pi(i)}, \tilde{s}_{\pi(i+2)}, \dots, \tilde{s}_{\pi(n)}).$$

Using the definition of Π_{n+1-i} from Π_{n-i}, we find $s = t' \,\tilde{\star}\, t = \widetilde{\Omega}_{n+1-i}(\tilde{s}_{\pi(i)}, \dots, \tilde{s}_{\pi(n)})$ (and $s' = t \,\tilde{\star}\, t = \widetilde{\Omega}_{n+1-i}(\tilde{s}_{\pi(i+1)}, \tilde{s}_{\pi(i)}, \tilde{s}_{\pi(i+2)}, \dots, \tilde{s}_{\pi(n)})$), which is (2.25).

Then, using (2.25) and the definitions of Π_n and $\widetilde{\Pi}_n$, we obtain

$$\begin{aligned}\Pi_n(s_1, \dots, s_n) &= \Omega_1(s_1) \cdot \Omega_2(s_1, s_2) \cdot \dots \cdot \Omega_n(s_1, \dots, s_n)\\ &= \widetilde{\Omega}_n(\tilde{s}_1, \dots, \tilde{s}_n) \cdot \widetilde{\Omega}_{n-1}(\tilde{s}_2, \dots, \tilde{s}_n) \cdot \dots \cdot \widetilde{\Omega}_1(\tilde{s}_n) = \widetilde{\Pi}_n(\tilde{s}_1, \dots, \tilde{s}_n). \quad \square\end{aligned}$$

Lemma 2.24 says, in particular, that when we start with n elements $s_1, \dots, s_n$ and construct the n-simplex of Figure 2 from the left, starting from $s_1, \dots, s_n$, then this n-simplex can be (re)-constructed from the right starting from $\tilde{s}_1, \dots, \tilde{s}_n$.

Lemma 2.27. *If $(X, \star)$ is an RC-quasigroup and M is the associated monoid, then, for all $s_1, \dots, s_n$ in X, the following conditions are equivalent:*

(i) *The elements $s_1, \dots, s_n$ are pairwise distinct;*

(ii) *The element $\Pi_n(s_1, \dots, s_n)$ is the right-lcm of $s_1, \dots, s_n$ in M.*

If the above relations hold and, in addition, $(X, \star)$ is bijective, then $\Pi_n(s_1, \dots, s_n)$ is also the left-lcm of the elements $\tilde{s}_1, \dots, \tilde{s}_n$ defined by $\tilde{s}_i = \Omega_n(s_1, \dots, \widehat{s_i}, \dots, s_n, s_i)$.

Proof. Assume first that $s_1, \dots, s_n$ are pairwise distinct in X. Let Ω'_n be the counterpart of Ω_n where $\backslash$ replaces $\star$. We prove, using induction on i, the equality

$$\Omega_i(s_{\pi(1)}, \dots, s_{\pi(i)}) = \Omega'_i(s_{\pi(1)}, \dots, s_{\pi(i)})$$

for every i and every permutation π in $\mathfrak{S}_i$. For $i = 1$, we have $\Omega_1(s_{\pi(1)}) = s_{\pi(1)} = \Omega'_1(s_{\pi(1)})$, and the result is straightforward. Assume $n \geqslant 2$. Put

$$s = \Omega_i(s_{\pi(1)}, \dots, s_{\pi(i)}) \quad \text{and} \quad s' = \Omega_i(s_{\pi(1)}, \dots, s_{\pi(i-2)}, s_{\pi(i)}, s_{\pi(i-1)}),$$
$$t = \Omega_i(s_{\pi(1)}, \dots, s_{\pi(i)}) \quad \text{and} \quad t' = \Omega_{i-1}(s_{\pi(1)}, \dots, s_{\pi(i-2)}, s_{\pi(i)}).$$

By the definition of Ω_i from Ω_{i-1}, we have $s = t \star t'$ and $s' = t' \star t$. By Lemma 2.23 applied to $(s_{\pi(1)}, \dots, s_{\pi(i)})$, the assumption $s_{\pi(i-1)} \neq s_{\pi(i)}$ implies $s \neq s'$, which implies

$t \star t' \neq t' \star t$. By Lemma 2.10, the latter relation implies $t \star t' = t\backslash t'$ and $t' \star t = t'\backslash t$ in M. The induction hypothesis implies

$$t = \Omega'_i(s_{\pi(1)}, ..., s_{\pi(i)}) \quad \text{and} \quad t' = \Omega'_{i-1}(s_{\pi(1)}, ..., s_{\pi(i-2)}, s_{\pi(i)}),$$

so we deduce $s = t\backslash t' = \Omega'_i(s_{\pi(1)}, ..., s_{\pi(i)})\backslash\Omega'_{i-1}(s_{\pi(1)}, ..., s_{\pi(i-2)}, s_{\pi(i)})$, that is, $s = \Omega'_i(s_{\pi(1)}, ..., s_{\pi(i)})$. We thus established that every face in the n-simplex of Figure 2 corresponds to forming a right-lcm. By the formula for an iterated right-lcm, $\Pi_n(s_1, ..., s_n)$ is the right-lcm of $s_1, ..., s_n$ in M. So (i) implies (ii).

Now, let n' be the cardinal of $\{s_1, ..., s_n\}$. The above argument shows that, if I is a size-n' subset of X, then the right-lcm Δ_I of I has length n' in M. So, if $n' < n$ holds, the right-lcm of $\{s_1, ..., s_n\}$ is an element of M that has length n', and it cannot be $\Pi_n(s_1, ..., s_n)$ which, by definition, has length n. So (ii) implies (i).

Assume now that $(X, \star)$ is bijective and (i)–(ii) are satisfied. Let $\tilde{\star}$ be the second operation provided by Lemma 1.28. Then $(X, \star, \tilde{\star})$ is an RLC-quasigroup. By Lemma 2.23, the assumption that $s_1, ..., s_n$ are pairwise distinct implies that $\tilde{s}_1, ..., \tilde{s}_n$ are pairwise distinct. Then $(X, \tilde{\star})$ is an LC-quasigroup, so the counterpart of the above results implies that $\widetilde{\Pi}_n(\tilde{s}_1, ..., \tilde{s}_n)$ is a left-lcm of $\tilde{s}_1, ..., \tilde{s}_n$ in M. Now, by (2.26), $\widetilde{\Pi}_n(\tilde{s}_1, ..., \tilde{s}_n)$ is equal to $\Pi_n(s_1, ..., s_n)$. □

2.3 Every structure monoid is a Garside monoid

With Lemma 2.10 and the results of Subsection 2.2 at hand, we can now say more about lcms in the structure monoid of an RC-quasigroup. Note that, by definition, every element g in the monoid associated with an RC-quasigroup $(X, \star)$ belongs to a unique family X^n; with our general definitions, the parameter n is the height of g but, in this specific context, it is more natural to call it the *length* of g, as it is the common length of all X-words that represent g.

The first result is a complete description of the smallest Garside family in the monoid associated with an RC-quasigroup.

Lemma 2.28. *If $(X, \star)$ is an RC-quasigroup and M is the associated monoid, then there exists a smallest Garside family containing 1 in M, namely the family S of all right-lcms of finite subsets of X. Mapping a finite subset of X to its right-lcm defines a bijection from the set $\mathfrak{P}_{\text{fin}}(X)$ of finite subsets of X to S.*

Proof. By Corollary IV.2.41 (smallest Garside), there exists a smallest Garside family containing 1 in M, namely the closure S of X under the right-lcm and right-complement operations. We claim that S coincides with the closure S' of X under the sole right-lcm operation. By definition, S' is included in S, and the point is to prove that S' is closed under the right-complement operation.

Now this follows from Proposition II.2.15 (triple lcm): indeed, in a monoid that admits unique right-lcms, (II.2.16) gives $f\backslash\mathrm{lcm}(g, h) = \mathrm{lcm}(f\backslash g, f\backslash h)$. An immediate induction then implies

$$f\backslash\mathrm{lcm}(g_1, ..., g_n) = \mathrm{lcm}(f\backslash g_1, ..., f\backslash g_n) \tag{2.29}$$

for every finite family $g_1, \ldots, g_n$. Assume that g belongs to S', that is, g is a right-lcm of elements $t_1, \ldots, t_n$ of X. If f lies in X, then, for every i, the element $f\backslash t_i$ belongs to $X \cup \{1\}$ since it is either $f \star t_i$ (if f and t_i are distinct) or 1 (if f and t_i coincide). Then (2.29) shows that $f\backslash g$ belongs to S' for every f in X. Using induction on the length of f, we deduce a similar result for every f in M from the formula (II.2.14) for an iterated right-complement, namely $(f_1f_2)\backslash g = f_2\backslash(f_1\backslash g)$. So S' is closed under $\backslash$, it coincides with S, and it is the smallest Garside family containing 1 in M.

For I a finite subset of X, write Δ_I for the right-lcm of I. Lemma 2.27 implies that, if I has p elements, say $s_1, \ldots, s_p$, then Δ_I is equal to $\Pi_p(s_1, \ldots, s_p)$. So, in particular, Δ_I has length p. Now, assume that I, J are finite subsets of X and $\Delta_I = \Delta_J$ holds. Then every element of $I \cup J$ left-divides Δ_I, so we must have $\Delta_{I\cup J} = \Delta_I = \Delta_J$. It follows that $I \cup J$ has the same cardinal as I and J, implying $I = I \cup J = J$. So the map $I \mapsto \Delta_I$ is a bijection of $\mathfrak{P}_{\mathrm{fin}}(X)$ to S. □

When we add the assumption that the RC-quasigroup is finite, the Garside family of Lemma 2.28 is finite and we obtain more complete results.

Proposition 2.30 (Garside monoid). *If $(X, \star)$ is a finite* RC-*quasigroup of size n and M is the associated monoid, then the right-lcm Δ of X is a Garside element in M and (M, Δ) is a Garside monoid. Moreover, Δ admits 2^n divisors in M, and M admits a presentation in terms of X consisting of $\binom{n}{2}$ relations $u = v$ with u, v of length two and such that every length-two X-word appears in at most one relation.*

Proof. For $I \subseteq X$, let Δ_I be the right-lcm of I in M, write Δ for Δ_X, and let S be the family of all elements Δ_I when I ranges over the subsets of X. By Lemma 2.28, S is the smallest Garside family containing 1 in M, and it has 2^n elements. By definition, Δ_I left-divides Δ_X for every I. So, as Δ_X itself lies in S, the Garside family S is right-bounded by Δ, and Δ is a right-Garside element in M.

Now, by Proposition 1.35, the assumption that X is finite implies that the RC-quasigroup $(X, \star)$ is bijective. Hence, by Lemma 1.28, there exists a second operation $\tilde{\star}$ such that $(S, *, \tilde{\star})$ is an RLC-quasigroup, and M is then also associated (in the obvious symmetric sense) with $(S, \tilde{\star})$. The counterparts of the above results are therefore satisfied and, in particular, M is right-cancellative. As S is finite, Proposition V.2.6 (finite bounded) says that S is not only right-bounded, but even bounded by Δ, so Δ is a Garside element in M, and that the Garside endomorphism ϕ_Δ must be a finite-order automorphism of M. Moreover, Proposition V.2.35 (bounded implies gcd) implies that M admits left-lcms and right-gcds. So (M, Δ) is a Garside monoid.

The last properties in the statement have already been established: the (left- or right-) divisors of Δ make the smallest Garside family containing 1 in M and, by Lemma 2.28, they are 2^n in number. On the other hand, by definition, the presentation (2.6) of M contains $\binom{n}{2}$ relations involving length-two words. It is impossible that a word occurs in two relations at the same time, for this would mean that there exist s, t, t' with $t \neq t'$ and $s \star t \neq s \star t'$, contradicting the assumption that the left translations of $\star$ are injective. □

Example 2.31 (Garside monoid). Let $X = \{\mathtt{a}, \mathtt{b}, \mathtt{c}\}$, and let $\star$ be determined by $x \star y = \sigma(y)$, where σ is the cycle $a \mapsto b \mapsto c \mapsto a$. Then, as seen in Example 1.23, $(X, \star)$ is a (bijective) RC-quasigroup, and it is eligible for the above results. The associated monoid admits the presentation

$$\langle \mathtt{a}, \mathtt{b}, \mathtt{c} \mid \mathtt{ac} = \mathtt{b}^2, \mathtt{ba} = \mathtt{c}^2, \mathtt{cb} = \mathtt{a}^2 \rangle^{+}.$$

The right-lcm Δ of X is then $\mathtt{a}^3$, which is also $\mathtt{b}^3$ and $\mathtt{c}^3$, and the lattice of the 8 divisors of Δ is shown on the right.

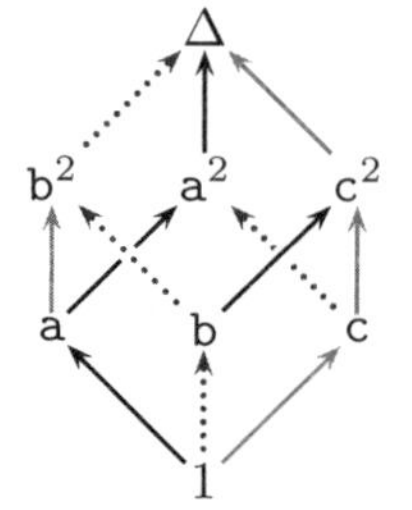

We can now complete the proof of Proposition 2.8—and, therefore, of Proposition 2.4.

Proof of Proposition 2.8. By Lemma 2.10, the monoid M has no nontrivial invertible element and it is Noetherian, which completes the proof of (i).

By Lemma 2.10 again, M is left-cancellative, and it admits unique right-lcms and left-gcds. Now, let $\tilde{\star}$ be the operation on X derived from $\star$ by Lemma 1.28. Then $(X, \star, \tilde{\star})$ is an RLC-quasigroup, and $(X, \tilde{\star})$ is a bijective LC-quasigroup. We observe that the presentation

(2.32) $$\langle X \mid \{(s \tilde{\star} t)t = (t \tilde{\star} s)s \mid s \neq t \in X\}\rangle^{+},$$

coincides with the one of (2.6), and therefore is a presentation of M. Indeed, let $(s \tilde{\star} t)t = (t \tilde{\star} s)s$ be a relation of (2.32). Put $s' = s \tilde{\star} t$ and $t' = t \tilde{\star} s$. As $(X, \star, \tilde{\star})$ is involutive, we obtain $s' \star t' = (s \tilde{\star} t) \star (t \tilde{\star} s) = t$ and $t' \star s' = (t \tilde{\star} s) \star (s \tilde{\star} t) = s$ by (1.26), so the above relation is the relation $s'(s' \star t') = t'(t' \star s')$ of (2.6). A symmetric argument shows that every relation of (2.6) is a relation of (2.32). Then, by the counterpart of Lemma 2.10—or by Lemma 2.10 applied to the opposed monoid M^{opp} and to the RC-quasigroup $(X, \tilde{\star}^{\mathrm{opp}})$—$M$ must be right-cancellative and admit left-lcms and right-gcds. Hence M is an Ore monoid, and, by Ore's theorem (Proposition II.3.11), its enveloping group G is a group of left- and right-fractions for M. By Corollary II.3.24 (torsion-free), G is torsion-free since it is the group of fractions of a torsion-free monoid. This completes the proof of (ii).

As for (iii), it follows from Lemma 2.10(iii).

Most results in item (iv) have already been established in Lemma 2.28, with the exception of the property involving left-lcms. Now the latter follows from the last sentence in Lemma 2.27: if $s_1, \ldots, s_n$ are pairwise distinct elements of X, then $\Pi_n(s_1, \ldots, s_n)$ is the right-lcm of $s_1, \ldots, s_n$ and the latter is also the left-lcm of the elements $\tilde{s}_1, \ldots, \tilde{s}_n$ defined by $\tilde{s}_i = \Omega_n(s_1, \ldots, \widehat{s_i}, \ldots, s_n, s_i)$.

Finally, (v) follows from Proposition 2.30 directly. □

Then Proposition 2.4 follows, as, owing to Lemma 2.7, the latter is a restatement of Proposition 2.8 using the language of set-theoretic solutions of YBE instead of the language of RC-quasigroups.

Remark 2.33. The notion of a Garside family is *not* symmetric: it involves the left-normal form based on a largest left-divisor, and it need not coincide with its symmetric counterpart. So, for instance, the uniqueness of the smallest Garside family in M cannot be invoked in Proposition 2.8 to establish that the right-lcm of X is the left-lcm of X.

2.4 A converse connection

According to Proposition 2.30, if a monoid M is associated with a finite RC-quasigroup $(X, \star)$, then M (equipped with the right-lcm of X) is a Garside monoid that admits in terms of X a presentation of a certain type. It turns out that the latter syntactic property characterizes the Garside monoids that arise in this context:

Proposition 2.34 (characterization). *If M is a monoid with atom set X of size n, the following are equivalent:*

(i) *There exists a map ρ such that (X, ρ) is an involutive nondegenerate set-theoretic solution of* YBE *and M is isomorphic to the structure monoid of (X, ρ).*

(ii) *There exist two operations $\star, \tilde{\star}$ such that $(X, \star, \tilde{\star})$ is an* RLC*-quasigroup and M is the monoid associated with $(X, \star)$.*

(iii) *There exists an operation $\star$ such that $(X, \star)$ is a bijective* RC*-quasigroup and M is the monoid associated with $(X, \star)$.*

(iv) *There exists Δ such that (M, Δ) is a Garside monoid and M admits a presentation in terms of X consisting of $\binom{n}{2}$ relations $u = v$ with u, v of length-two such that every length two X-word appears in at most one relation.*

Proof. We already proved that (i) and (ii) are equivalent, and that so are (ii) and (iii). By Proposition 2.30, (iii) implies (iv). Hence, in order to complete the proof, it is sufficient to show for instance that (iv) implies (ii). So assume that M is a Garside monoid satisfying (iv) and R is the list of relations involved in the considered presentation. Starting from the right- and left-complement operations associated with the Garside structure on M, we shall construct operations $\star$ and $\tilde{\star}$ that make X into an RLC-quasigroup whose associated monoid is M. The only technical point in the argument is to take care of the exceptional values where the operations $\backslash$ and $\star$ do not coincide.

First, we observe that the presentation (X, R) is right-complemented and, more precisely, that it contains exactly one relation of the form $s... = t...$ for all distinct s, t in X. Indeed, assume that s, t are distinct elements of S and R contains at least two relations $s... = t...$, say $st' = ts'$ and $st'' = ts''$ with $(s', t') \neq (s'', t'')$. As M is cancellative, we have $st' \neq st''$, so st' and st'' are two common right-multiples of s and t of length-two: this contradicts the existence of a right-lcm for s and t, as the latter can have neither length one, nor length two. Hence R contains at most one relation $s... = t...$ for all distinct s, t in X. On the other hand, R contains no relation $s... = s...$, since M is left-cancellative and $st = st'$ would imply $t = t'$. As there are $\binom{n}{2}$ pairs of distinct elements of X, we deduce that R contains exactly one relation of the form $s... = t...$ for all distinct s, t in X. By symmetric arguments using left-lcms and right-cancellativity (or by applying the previous result to the opposite monoid), we see that (X, R) contains exactly one relation of the form $...s = ...t$ for all distinct s, t in X.

We now define a binary operation $\star$ on X. First, we put $s \star t = s\backslash t$ for $s \neq t$, that is, we define $s \star t$ to be the unique element t' such that st' is the right-lcm of s and t. Then $t \neq t'$ implies $s \star t \neq s \star t'$, since otherwise there would be two relations of the form $s(s \star t) = ...$ in R. So the map $x \mapsto s \star x$ is injective on $S \setminus \{s\}$ and, therefore, the complement of $\{s{\star}x \mid x \neq s\}$ in X consists of a unique element: we define $s{\star}s$ to be that element. In this way, we obtained a binary operation $\star$ whose left translations are one-to-one. Of course, we define the operation $\tilde{\star}$ symmetrically using the left-complement operation $/$, and its right translations are one-to-one.

We claim that $\star$ and $\tilde{\star}$ satisfy the involutivity laws (1.26). First, assume $s \neq t$. Then $s(s{\star}t) = t(t{\star}s)$ is a relation of R, hence we must have $s{\star}t \neq t{\star}s$. Next, by the definition of $\star$ and $\tilde{\star}$, the element $s(s{\star}t)$ is the right-lcm of s and t, and, as $s{\star}t$ and $t{\star}s$ are distinct, $s(s \star t)$ is also the left-lcm of $s \star t$ and $t \star s$. This exactly means that $(s \star t) \,\tilde{\star}\, (t \star s) = t$ holds in this case. Next, assume $t \neq s$, and put $s' = s \star s$ and let $r = s' \,\tilde{\star}\, (s \star t)$. We have $s \star t \neq s'$, whence $r = s'/(s\backslash t)$. Then $r(s\backslash t) = ((s\backslash t)/s')s'$ is a relation of R, which implies $(s\backslash t)/s' \neq s$ since, by assumption, R contains no relation $ss' = \ldots$. Since $(s\backslash t)/s' \neq s$ holds for every t distinct of s, we deduce $s' \,\tilde{\star}\, s' = s$ since, by definition, $s' \,\tilde{\star}\, s'$ is the only element of X that is not of the form $(s\backslash t) \,\tilde{\star}\, s'$ with $t \neq s$. In other words, $(s \star s) \,\tilde{\star}\, (s \star s) = s$ holds, and the first involutivity law is satisfied in $(X, \star, \tilde{\star})$. By a symmetric argument, the second involutivity law is satisfied as well.

Next, we claim that $(X, \star)$ satisfies the RC-law. Let r, s, t belong to X. Assume first that r, s, t are pairwise distinct. Then we have $r \star s \neq r \star t$ and $s \star r \neq s \star t$, whence

$$(r \star s) \star (r \star t) = (r\backslash s)\backslash(r\backslash t) = (s\backslash r)\backslash(s\backslash t) = (s \star r) \star (s \star t),$$

the second equality because, as observed in Example 1.23, the right-complement operation $\backslash$ always satisfies the RC-law. Assume now that r and s coincide. Then the RC-law tautologically holds. So there only remain the cases when $r \neq s$ and t is either r or s, we would like to establish the equalities

$$(r \star s) \star (r \star s) = (s \star r) \star (s \star s) \text{ and } (s \star r) \star (s \star r) = (r \star s) \star (r \star r),$$

that is, owing to $r \neq s$,

$$(r\backslash s) \star (r\backslash s) = (s\backslash r)\backslash(s \star s) \text{ and } (s\backslash r) \star (s\backslash r) = (r\backslash s)\backslash(r \star r). \tag{2.35}$$

Assume $z \neq r, s$ and put $z' = (r\backslash s)\backslash(r\backslash z)$, which is also $z' = (s\backslash r)\backslash(s\backslash z)$, since $\backslash$ satisfies the RC-law. Then we have $r\backslash z \neq r\backslash s$, whence $z' \neq (r\backslash s) \star (r\backslash s)$. Also, we have $s\backslash z \neq s \star s$, whence $z' \neq (s\backslash r)\backslash(s \star s)$. Arguing similarly with r and s exchanged, we find $z' \neq (s\backslash r) \star (s\backslash r)$ and $z' \neq (r\backslash s)\backslash(r \star r)$. It follows that z' is distinct from the four expressions occurring in (2.35) and, therefore, the only possible values for the latter are the two elements of X that are not of the form $(r\backslash s)\backslash(r\backslash z)$ with $z \neq r, s$. Now, as left translations of $\star$ are injective, we must have $(r\backslash s) \star (r\backslash s) \neq (r\backslash s)\backslash(r \star r)$ and $(s\backslash r)\backslash(s \star s) \neq (s\backslash r) \star (s\backslash r)$. So, in order to conclude that (2.35) is true, it is sufficient to show that $(r\backslash s) \star (r\backslash s) = (s\backslash r) \star (s\backslash r)$ is impossible. Now $r \neq s$ implies $r{\star}s \neq s{\star}s$, so it is enough to prove that $x \neq y$ implies $x \star x \neq y \star y$: this follows from the above established involutivity relation $(x \star x) \,\tilde{\star}\, (x \star x) = x$.

We are done: $(X, \star)$ is an RC-quasigroup, by a symmetric argument $(X, \tilde{\star})$ is an LC-quasigroup, and $(X, \star, \tilde{\star})$ is an RLC-quasigroup. Now, by construction, M admits the presentation (X, R), so it is (isomorphic to) the monoid associated with $(X, \star)$. □

3 I-structure

In this final section, we establish further results involving structure monoids of involutive nondegenerate set-theoretic solutions of YBE (or, equivalently, with bijective RC-quasigroups), centered on the existence of an I-structure, an algebraic way of expressing the fact that the Cayley graph of the considered monoid is an n-dimensional Euclidean lattice and, therefore, this monoid resembles a free abelian monoid.

There are three subsections. In Subsection 3.1, we show that every structure monoid admits an I-structure whereas, in Subsection 3.2, we explain how to establish that, conversely, every monoid that admits an I-structure is a structure monoid. Finally, in Subsection 3.3, we construct, as an application of the I-structure, finite groups that play for our current groups the role played by Coxeter groups for the spherical Artin–Tits groups of Chapter IX.

3.1 From RC-quasigroups to I-structures

We saw in Proposition 2.8 that, if a monoid M is associated with an RC-quasigroup $(X, \star)$, it admits right-lcms and left-gcds, so that M equipped with the left-divisibility relation $\preccurlyeq$ is a lattice. We shall see now that the particular form of the defining relations implies that this lattice is very simple, namely it is isomorphic to the grid $\mathbb{N}^{(X)}$. In other words, the Cayley graph of the monoid M is a grid, and we can attribute integer coordinates to the elements of M.

Example 3.1 (Cayley graph). Let M be the monoid $\langle \mathtt{a}, \mathtt{b} \mid \mathtt{a}^2 = \mathtt{b}^2 \rangle^+$ associated with the bijective RC-groupoid

$\star$	a	b
a	b	a
b	b	a

. The Cayley graph of M is the rectangular grid displayed in Figure 3, and we can attribute to each vertex two integer coordinates, thus attributing $(0,0)$ to 1, $(1,0)$ to $\mathtt{a}$, $(0,1)$ to $\mathtt{b}$, $(2,1)$ to a^3, etc. In other words, putting $\mathtt{a} = (1,0)$, $\mathtt{b} = (0,1)$, and considering $\mathbb{N}^2$ as a free abelian monoid based on $\mathtt{a}$ and $\mathtt{b}$, we have a bijection ν of $\mathbb{N}^2$ to M satisfying $\nu(1) = 1$, $\nu(\mathtt{a}) = \mathtt{a}$, $\nu(\mathtt{b}) = \mathtt{b}$, $\nu(\mathtt{a}^2\mathtt{b}) = \mathtt{a}^3$, etc. By construction, there exist two arrows respectively labeled $\mathtt{a}$ and $\mathtt{b}$ starting from every node, and, therefore, for every a in $\mathbb{N}^2$, we have

$$\{\nu(a\mathtt{a}), \nu(a\mathtt{b})\} = \{\nu(a)\mathtt{a}, \nu(a)\mathtt{b}\}.$$

We cannot say more: for instance, we have $\nu(\mathtt{a} \cdot \mathtt{a}) = \mathtt{ab} = \nu(\mathtt{a})\mathtt{b}$, but $\nu(\mathtt{a}^2 \cdot \mathtt{b}) = \mathtt{a}^3 = \nu(\mathtt{a}^2)\mathtt{a}$, witnessing that right-multiplying by $\mathtt{a}$ may correspond either to a step right, or to a step up in the figure. This corresponds to the fact that the elementary tiles of the Cayley graph of M may have different orientations: the latter can be specified by attaching with every vertex a a permutation $\psi(a)$ of $\{\mathtt{a}, \mathtt{b}\}$ that prescribes which edges correspond to the horizontal and vertical moves. In the current case, we see that $\psi(a)$ is the identity when a has even length, and is the transposition $(\mathtt{a}, \mathtt{b})$ otherwise.

An indexation of the Cayley graph as in the above example is called a (*right*) *I-structure* for the monoid M. We shall see in this subsection that every structure monoid

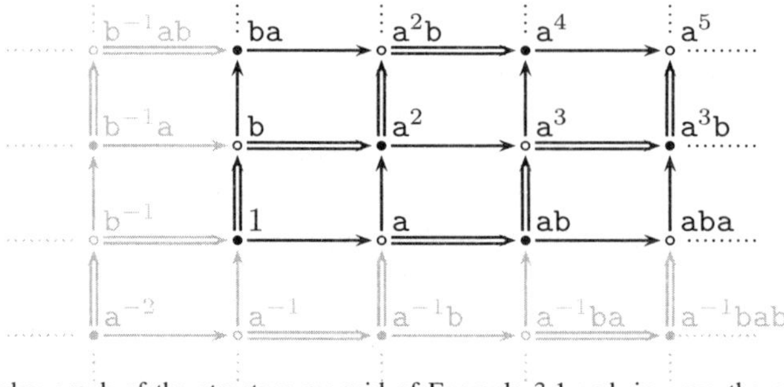

Figure 3. The Cayley graph of the structure monoid of Example 3.1 and, in grey, the one of its group of fractions; simple (*resp.* double) arrows correspond to right-multiplication by a (*resp.* by b). The graph consists of matching squares, all of the same type, but that can have two orientations: the vertices where the orientation is the same as in 1, that is, a is horizontal and b is vertical, (*resp.* is opposed) are marked with solid (*resp.* empty) disks.

of an involutive nondegenerate set-theoretic solution of YBE admits such a structure and that, conversely, every monoid that admits an I-structure is the structure monoid of an involutive nondegenerate set-theoretic solution of the YBE. Hereafter, we shall simultaneously consider a monoid M that we wish to investigate and a free abelian monoid $\mathbb{N}^{(X)}$ used as a sort of template. To avoid ambiguity, we shall (as usual) write $g, h, ...$ for the elements of M, and (as already done in Example 3.1 above) reserve $a, b, ...$ for the elements of the reference free abelian monoid.

Definition 3.2 (I-structure, monoid of I-type). If M is a monoid and X is included in M, a *right-I-structure based on X* for M is a bijective map $\nu : \mathbb{N}^{(X)} \to M$ satisfying $\nu(1) = 1$, $\nu(s) = s$ for s in X, and, for every a in $\mathbb{N}^{(X)}$,

$$\{\nu(as) \mid s \in X\} = \{\nu(a)s \mid s \in X\}. \tag{3.3}$$

A monoid is said to be *of right-I-type* if it admits a right I-structure.

Note that (3.3) is equivalent to the existence, for every a in $\mathbb{N}^{(X)}$, of a permutation $\psi(a)$ of X such that, for every s in X, one has

$$\nu(as) = \nu(a) \cdot \psi(a)(s). \tag{3.4}$$

It immediately follows from the definition that every element in the image of ν is a product of elements of X and, therefore, the existence of a right-I-structure based on X for M implies that X generates M. The condition $\nu(s) = s$ for s in X, which amounts to $\psi(1)$ being the identity of X, can be ensured by precomposing ν with the automorphism induced by $\psi(1)^{-1}$ and, therefore, it could be removed from the definition without changing the range of the latter.

What we saw in Example 3.1 is that the monoid $\langle a, b \mid a^2 = b^2\rangle^+$ is a monoid of right-I-type. We now extend this observation to all monoids associated with RC-quasigroups.

Proposition 3.5 (**RC-quasigroup to I-structure**). *For every* RC-*quasigroup* $(X, \star)$, *the monoid M associated with $(X, \star)$ is of right-I-type: the map ν defined from $\star$ by $\nu(s_1 \cdots s_n) = \Pi_n(s_1, \ldots, s_n)$ is a right I-structure on M.*

Proof. Appealing to the symmetric polynomials Π_n of Definition 2.18, we first define a map ν_* from the free monoid X^* to M by

$$\nu_*(\varepsilon) = 1 \quad \text{and} \quad \nu_*(s_1|\cdots|s_n) = \Pi_n(s_1, \ldots, s_n) \quad \text{for } n \geqslant 1.$$

By Lemma 2.21, the value of $\nu_*(s_1|\cdots|s_n)$ does not depend on the order of the entries, so ν_* induces a well-defined map ν from the free abelian monoid $\mathbb{N}^{(X)}$ to M. We claim that the latter provides the expected right-I-structure on M. First, the equalities $\nu(1) = 1$ and $\nu(s) = s$ for s in X are obvious. Next, let a belong to $\mathbb{N}^{(X)}$, say $a = s_1 \cdots s_n$. Then the definition of ν gives $\nu(as) = \Pi_{n+1}(s_1, \ldots, s_n, s)$, whence

$$\nu(as) = \Pi_n(s_1, \ldots, s_n) \cdot \Omega_{n+1}(s_1, \ldots, s_n, s) = \nu(a) \cdot \Omega_{n+1}(s_1, \ldots, s_n, s).$$

By Lemma 2.23(i), the map $s \mapsto \Omega_{n+1}(s_1, \ldots, s_n, s)$ is a bijection of X into itself, hence (3.3) holds.

It remains to show that ν is a bijection from $\mathbb{N}^{(X)}$ to M. Let g be an arbitrary element of M, say $g = s_1 \cdots s_n$ with $s_1, \ldots, s_n$ in X. By Lemma 2.23(ii), there exist $r_1, \ldots, r_n$ in X satisfying $\Omega_i(r_1, \ldots, r_i) = s_i$ for $1 \leqslant i \leqslant n$, whence $\Pi_n(r_1, \ldots, r_n) = s_1 \cdots s_n = g$. By definition, this means that we have $\nu(r_1 \cdots r_n) = g$ holds, and so ν is surjective.

Finally, assume that a, a' belong to $\mathbb{N}^{(X)}$ and $\nu(a) = \nu(a')$ holds. As the elements of M have a well-defined length, the length of a and a' must be the same. Write $a = r_1 \cdots r_n$, $a' = r'_1 \cdots r'_n$ with $r_1, \ldots, r'_n$ in X. Define $s_i = \Omega_i(r_1, \ldots, r_i)$ and $s'_i = \Omega_i(r'_1, \ldots, r'_i)$. By definition, $\nu(a)$ is the class of the X-word $s_1|\cdots|s_n$ in M, whereas $\nu(a')$ is the class of $s'_1|\cdots|s'_n$. The assumption $\nu(a) = \nu(a')$ means that these X-words are connected by a finite sequence of defining relations of M. By Lemma 2.23, the map $(x_1, \ldots, x_n) \mapsto (\Omega_1(x_1), \ldots, \Omega_n(x_1, \ldots, x_n))$ of $X^{[n]}$ to itself is surjective, so we can assume without loss of generality that $s_1|\cdots|s_n$ and $s'_1|\cdots|s'_n$ are connected by one relation exactly, that is, there exist i satisfying

$$s_{i+1} = s_i \star s'_i, \quad s'_{i+1} = s'_i \star s_i, \quad \text{and} \quad s'_k = s_k \text{ for } k \neq i, i+1.$$

The relations $s'_k = s_k$ inductively imply $r'_k = r_k$ for $k < i$. Next, writing $\vec{r}$ for $r_1, \ldots, r_{i-1}$, we have $s_i = \Omega_i(\vec{r}, r_i)$ and $s'_i = \Omega_i(\vec{r}, r'_i)$, whence

$$\Omega_i(\vec{r}, r_i) \star \Omega_i(\vec{r}, r_{i+1}) = \Omega_{i+1}(\vec{r}, r_i, r_{i+1}) = s_{i+1} = s_i \star s'_i = \Omega_i(\vec{r}, r_i) \star \Omega_i(\vec{r}, r'_i).$$

As the left translation by $\Omega_i(\vec{r}, r_i)$ is injective, we deduce $\Omega_i(\vec{r}, r_{i+1}) = \Omega_i(\vec{r}, r'_i)$, whence $r_{i+1} = r'_i$ by Lemma 2.23(i). A symmetric argument gives $r'_{i+1} = r_i$. From there, everything is easy and, for $k > i + 1$, the relations $s'_k = s_k$ inductively imply $r'_k = r_k$. Indeed, we have

$$\Omega_k(\vec{r}, r_i, r'_i, r_{i+2}, \ldots, r_k) = s_k = s'_k = \Omega_k(\vec{r}, r'_i, r_i, r_{i+2}, \ldots, r'_k),$$

and, by Lemma 2.16, we know that switching the non-final entries r_i and r'_i in Ω_k changes nothing, and $r'_k = r_k$ follows by Lemma 2.23. We thus proved that the words $r_1|\cdots|r_n$ and $r'_1|\cdots|r'_n$ are obtained by switching two (adjacent) entries, hence they represent the same element in the free abelian monoid $\mathbb{N}^{(X)}$. It follows that ν is injective, hence bijective, and it provides the expected right-I-structure on M. □

3.2 From I-structures to RC-quasigroups

We now go in the other direction, and show that every finitely generated monoid of I-type is the structure monoid of some RC-quasigroup.

Proposition 3.6 (I-structure to RC-quasigroup). *Assume that M is a finitely generated monoid of right-I-type.*

(i) *There exists a unique finite* RC*-quasigroup $(X, \star)$ with whom M is associated: the set X is the atom set of M and $\star$ is determined by $s \star t = s\backslash t$ for $s \neq t$ and $\{s \star s\} = X \setminus \{s\backslash t \mid t \neq s\}$.*

(ii) *The right-I-structure on M is unique: it is defined from the operation $\star$ of* (i) *by $\nu(s_1 \cdots s_n) = \Pi_n(s_1, ..., s_n)$.*

Some of the formulas we shall establish below for proving Proposition 3.6 will be used in a slightly different context in Subsection 3.3 and, to avoid repeating the computations, we shall extend the notion of an I-structure so as to be able to replace the reference monoid $\mathbb{N}^{(X)}$ with another monoid, typically a quotient $(\mathbb{Z}/d\mathbb{Z})^{(X)}$.

Definition 3.7 (right-IM_0-structure). If M_0 and M are monoids and X is a subset of M that generates M_0, a *right-IM_0-structure based on X* for M is a bijection $\nu : M_0 \to M$ satisfying $\nu(1) = 1$, $\nu(s) = s$ for s in X and, for every a in M_0,

$$\{\nu(as) \mid s \in X\} = \{\nu(a)s \mid s \in X\}. \tag{3.8}$$

So, a right-I-structure as in Definition 3.2 is a right-$I\mathbb{N}^{(X)}$-structure in the current extended sense. As above, we shall in general omit the base set X when mentioning an IM_0-structure.

As above, (3.8) is equivalent to the existence, for every a in M_0, of a permutation $\psi(a)$ of X such that, for every s in X, one has

$$\nu(as) = \nu(a) \cdot \psi(a)(s). \tag{3.9}$$

Note that the assumption $\psi(1) = \mathrm{id}_X$ is innocent only if every permutation of X induces an automorphism of the monoid M_0, as is the case for $\mathbb{N}^{(X)}$ or $(\mathbb{Z}/d\mathbb{Z})^{(X)}$.

Lemma 3.10. *Assume that M is a left-cancellative monoid and ν is a right-IM_0-structure based on X for M, where M_0 is abelian and satisfies*

$$\forall s, t, s', t' \in X\ ((s \neq t \text{ and } st' = ts') \Longrightarrow (s' = s \text{ and } t' = t)). \tag{3.11}$$

Put $s \star t = \psi(s)(t)$ for s, t in X. Then $(X, \star)$ is an RC*-quasigroup.*

Proof. We first claim that, for all s, t, s', t' in X with $s \neq t$, the only equality $st' = ts'$ holding in M is $s(s \star t) = t(t \star s)$. Indeed, by assumption, M_0 is abelian and, in M, using (3.9), we obtain

$$s(s \star t) = s\psi(s)(t) = \nu(st) = \nu(ts) = t\psi(t)(s) = t(t \star s).$$

On the other hand, assume $st' = ts'$ in M. Let $t_0 = \psi(s)^{-1}(t')$ and $s_0 = \psi(t)^{-1}(s')$. Always using (3.9), we find $\nu(st_0) = st = ts' = \nu(ts_0)$ in M, whence $st_0 = ts_0$ in M_0 since ν is bijective. The assumption on M_0 implies $s_0 = s$ and $t_0 = t$, whence $s' = \psi(t)(s) = t \star s$ and $t' = \psi(s)(t) = s \star t$. This establishes the claim.

Now, let $a \in M_0$ and $s, t \in X$. Using (3.9), we find

$$\nu(ast) = \nu(as) \cdot \psi(as)(t) = \nu(a) \cdot \psi(a)(s) \cdot \psi(as)(t),$$

and, similarly, $\nu(ats) = \nu(a) \cdot \psi(a)(t) \cdot \psi(at)(s)$. By assumption, M_0 is abelian, so we have $ast = ats$, whence $\nu(ast) = \nu(ats)$ and, merging the above expressions and left-cancelling $\nu(a)$, which is legal as M is assumed to be left-cancellative, we find

$$\psi(a)(s) \cdot \psi(as)(t) = \psi(a)(t) \cdot \psi(at)(s). \tag{3.12}$$

Assume first $s \neq t$. The elements $\psi(a)(s)$, $\psi(as)(t)$, $\psi(a)(t)$, and $\psi(at)(s)$ lie in X, so, by the claim above, (3.12) implies

$$\psi(as)(t) = \psi(a)(s) \star \psi(a)(t) \quad \text{and} \quad \psi(at)(s) = \psi(a)(t) \star \psi(a)(s). \tag{3.13}$$

When t ranges over $X \setminus \{s\}$, the element $\psi(a)(t)$ ranges over $X \setminus \{\psi(a)(s)\}$, and $\psi(a)(s) \star \psi(a)(t)$ ranges over $X \setminus \{\psi(a)(s) \star \psi(a)(s)\}$. As $\psi(as)$ is a bijection of X, the only possibility is therefore $\psi(as)(s) = \psi(a)(s) \star \psi(a)(s)$. In other words, (3.13) is valid in X for all a in M_0 and s, t in X.

Now, let r lie in X. Taking $a = r$ in (3.13) and applying the definition of $\star$ gives $\psi(rs)(t) = (r \star s) \star (r \star t)$ and, similarly, $\psi(sr)(t) = (s \star r) \star (s \star t)$. Now, in M_0, we have $rs = sr$, whence $\psi(rs)(t) = \psi(sr)(t)$, and this gives $(r \star s) \star (r \star t) = (s \star r) \star (s \star t)$, the RC-law. So $(X, \star)$ is an RC-system. Moreover, by definition, $\psi(s)$ belongs to $\mathfrak{S}_X$, so the left translations of $\star$ are one-to-one, and $(X, \star)$ is an RC-quasigroup. □

Lemma 3.14. *Assume that M is a left-cancellative monoid and ν is a right-IM_0-structure based on X for M, where M_0 is abelian and satisfies* (3.11). *Then, for every $p \geqslant 1$ and for all $s_1, \dots, s_p$ in X, we have*

$$\psi(s_1 \cdots s_{p-1})(s_p) = \Omega_p(s_1, \dots, s_p) \quad \textit{and} \quad \nu(s_1 \cdots s_p) = \Pi_p(s_1, \dots, s_p), \tag{3.15}$$

where values are taken in M. Moreover, for all a, b in M_0, we have

$$\nu(ab) = \nu(a)\nu(\psi(a)[b]) \quad \textit{and} \quad \psi(ab) = \psi(\psi(a)[b]) \circ \psi(a) \tag{3.16}$$

where $\psi(a)[b]$ is the result of applying $\psi(a)$ to b componentwise.

Proof. We begin with the left equality in (3.15), which is proved using induction on p. For $p = 1$, we have $\psi(1)(s_1) = s_1 = \Omega_1(s_1)$ and, for $p = 2$, we have $\psi(s_1)(s_2) = s_1 \star s_2 = \Omega_2(s_1, s_2)$. For $p \geqslant 3$, using (3.13), the induction hypothesis, and the inductive definition of the monomials Ω_p, we find

$$\begin{aligned}\psi(s_1 \cdots s_{p-1})(s_p) &= \psi(s_1 \cdots s_{p-2})(s_{p-1}) \star \psi(s_1 \cdots s_{p-2})(s_p)\\ &= \Omega_{p-1}(s_1, \dots, s_{p-1}) \star \Omega_{p-1}(s_1, \dots, s_{p-2}, s_p) = \Omega_p(s_1, \dots, s_p).\end{aligned}$$

The right equality in (3.15) then follows using a similar induction on p. For $p = 1$, we have $\nu(s_1) = s_1 = \Pi(s_1)$. For $p \geqslant 2$, using (3.9), the left equality in (3.15), and the inductive definition of Π_p, we obtain

$$\begin{aligned}\nu(s_1 \cdots s_p) &= \nu(s_1 \cdots s_{p-1}) \cdot \psi(s_1 \cdots s_{p-1})(s_p)\\ &= \Pi_{p-1}(s_1, \dots, s_{p-1}) \cdot \Omega_p(s_1, \dots, s_p) = \Pi_p(s_1, \dots, s_p).\end{aligned}$$

The definition of Ω_n implies, for $p, q \geqslant 1$, the formal equality

$$\Omega_{p+q}(\vec{x}, y_1, \dots, y_q) = \Omega_q(\Omega_{p+1}(\vec{x}, y_1), \dots, \Omega_{p+1}(\vec{x}, y_q)),$$

where $\vec{x}$ stands for $x_1, \dots, x_p$. This is a formal identity, not using the RC-law or any specific relation; for instance, it says that $\Omega_3(x, y_1, y_2)$, that is, $(x \star y_1) \star (x \star y_2)$, is also $\Omega_2(\Omega_2(x, y_1), \Omega_2(x, y_2))$. With the same convention, one deduces

$$\Pi_{p+q}(\vec{x}, y_1, \dots, y_q) = \Pi_p(\vec{x}) \cdot \Pi_q(\Omega_{p+1}(\vec{x}, y_1), \dots, \Omega_{p+1}(\vec{x}, y_q)). \tag{3.17}$$

Now, assume that a, b lie in M_0. Write $a = s_1 \cdots s_p$ and $b = t_1 \cdots t_q$ with $s_1, \dots, t_q$ in X. The right equality in (3.15) gives $\nu(ab) = \Pi_{p+q}(s_1, \dots, s_p, t_1, \dots, t_q)$. On the other hand, we have $\nu(a) = \Pi_p(s_1, \dots, s_p)$ and the left equality in (3.15) implies $\psi(a)(t) = \Omega_{p+1}(s_1, \dots, s_p, t)$ for every t. Owing to $b = t_1 \cdots t_q$, we deduce the equalities

$$\begin{aligned}\nu(\psi(a)[b]) &= \Pi_q(\psi(a)(t_1), \dots, \psi(a)(t_q))\\ &= \Pi_q(\Omega_{p+1}(s_1, \dots, s_p, t_1), \dots, \Omega_{p+1}(s_1, \dots, s_p, t_q)),\end{aligned}$$

whence, by (3.17),

$$\begin{aligned}\nu(a)\nu(\psi(a)[b]) &= \Omega_p(s_1, \dots, s_p) \cdot \Pi_q(\Omega_{p+1}(s_1, \dots, s_p, t_1), \dots, \Omega_{p+1}(s_1, \dots, s_p, t_q))\\ &= \Omega_{p+q}(s_1, \dots, s_p, t_1, \dots, t_q)).\end{aligned}$$

By definition, the latter expression is $\nu(ab)$, so the left formula in (3.16) is established.

Finally, assume $s \in X$. On the one hand, (3.9) gives $\nu(abs) = \nu(ab)\psi(ab)(s)$. On the other hand, the left formula in (3.16) gives

$$\begin{aligned}\nu(abs) &= \nu(a) \cdot \nu(\psi(a)[bs]) = \nu(a) \cdot \nu(\psi(a)[b] \cdot \psi(a)(s))\\ &= \nu(a) \cdot \nu(\psi(a)[b]) \cdot \psi(\psi(a)[b])(\psi(a)(s)) = \nu(ab) \cdot \psi(\psi(a)[b])(\psi(a)(s)).\end{aligned}$$

Merging the two expressions and using the assumption that M is left-cancellative, we deduce $\psi(ab)(s) = \psi(\psi(a)[b])(\psi(a)(s))$, which is the right equality in (3.16). □

Before completing the proof of Proposition 3.6, we establish a few more preparatory results, this time under the assumption that there exists a right-$I\mathbb{N}^{(X)}$-structure, that is, a genuine right-I-structure.

Lemma 3.18. *Assume that ν is a right-I-structure based on X for a monoid M.*

(i) *There exists an additive length function on M and X is the atom set in M.*

(ii) *The map ν is compatible with left-division in the sense that, for all a, b in $\mathbb{N}^{(X)}$, we have $a \preccurlyeq b$ in $\mathbb{N}^{(X)}$ if and only if $\nu(a) \preccurlyeq \nu(b)$ holds in M.*

(iii) *The monoid M admits right-lcms; for $s \neq t$ in X, the element $s\backslash t$ lies in X.*

(iv) *If, moreover, X is finite, then M is left-cancellative, and it admits the presentation $\langle X \mid \{s(s\backslash t) = t(t\backslash s) \mid s \neq t \in X\}\rangle^+$.*

Proof. (i) Defining $\lambda(g)$ to be the length of $\nu^{-1}(g)$ provides a function from M to $\mathbb{N}$ that satisfies $\lambda(1) = 0$, $\lambda(gh) = \lambda(g) + \lambda(h)$, and $\lambda(s) = 1$ for every s in X. It follows that M contains no nontrivial invertible element, that M is Noetherian, and that X is the atom set of M.

(ii) Assume $a \preccurlyeq b$ in $\mathbb{N}^{(X)}$. For an induction on length, we may assume $b = as$ with s in X. Now, by (3.9), we have $\nu(b) = \nu(a)\psi(a)(s)$, whence $\nu(a) \preccurlyeq \nu(b)$ in M. Conversely, assume $\nu(a) \preccurlyeq \nu(b)$. Again, it is enough to consider the case $\nu(b) = \nu(a)s$ with s in X. Now, as $\psi(a)$ is bijective, there exists a unique r in X satisfying $\psi(a)(r) = s$, and, by (3.9), we have then $\nu(ar) = \nu(a)\psi(a)(r) = \nu(a)s = \nu(b)$, whence $b = ar$ since ν is injective, and $a \preccurlyeq b$ in $\mathbb{N}^{(X)}$.

(iii) The monoid $\mathbb{N}^{(X)}$ admits right-lcms, and (ii) enables us to transfer the result to M. Indeed, let g, h belong to M. Put $a = \nu^{-1}(g)$ and $b = \nu^{-1}(h)$. Let ab' be the right-lcm of a and b in $\mathbb{N}^{(X)}$. By (ii), $\nu(ab')$ is a common right-multiple of g and h in M. Now, assume that f is a common right-multiple of g and h in M. By (ii) again, we have $a \preccurlyeq \nu^{-1}(f)$ and $b \preccurlyeq \nu^{-1}(f)$ in $\mathbb{N}^{(X)}$, whence $ab' \preccurlyeq \nu^{-1}(f)$. By (ii) once more, this implies $\nu(ab') \preccurlyeq f$ in M. So $\nu(ab')$ is a right-lcm of g and h in M, and M admits right-lcms. In the case when g and h are distinct elements of X, then $\nu^{-1}(g)$ and $\nu^{-1}(f)$ also are distinct elements of X, so their right-lcm in $\mathbb{N}^{(X)}$ has length two, and so does the right-lcm of g and h in M. In other words, the right-complement $g\backslash h$ lies in X.

(iv) We assume now that X is finite. Fix g in M, and put $a = \nu^{-1}(g)$. For every b in $\mathbb{N}^{(X)}$, we have $g \preccurlyeq g\nu(b)$ in M, whence, by (ii), $a \preccurlyeq \nu^{-1}(g\nu(b))$ in $\mathbb{N}^{(X)}$. So, as $\mathbb{N}^{(X)}$ is left-cancellative, there exists a well-defined map ψ from $\mathbb{N}^{(X)}$ to itself such that, for every b in $\mathbb{N}^{(X)}$, we have $\nu^{-1}(g\nu(b)) = a\psi(b)$, that is, equivalently, $g\nu(b) = \nu(a\psi(b))$. Put $\mathbb{N}^{(X)}_{(\ell)} = \{b \in \mathbb{N}^{(X)} \mid \|b\| = \ell\}$. The additivity of length implies $\|\psi(b)\| = \|b\|$, so, for every ℓ, the restriction ψ_ℓ of ψ to $\mathbb{N}^{(X)}_{(\ell)}$ maps $\mathbb{N}^{(X)}_{(\ell)}$ to itself. The map ψ_ℓ is surjective. Indeed, let b' belong to $\mathbb{N}^{(X)}_{(\ell)}$. Then we have $a \preccurlyeq ab'$ in $\mathbb{N}^{(X)}$, whence, by (ii), $g \preccurlyeq \nu(ab')$ in M. So some element $\nu(b)$ of M satisfies $g\nu(b) = \nu(ab')$, whence $b' = \psi_\ell(b)$ since b must be of length ℓ. As $\mathbb{N}^{(X)}_{(\ell)}$ is finite, ψ_ℓ must be injective for every ℓ, and so is ψ. Now, assume $gh = gh'$ in M. Put $b = \nu^{-1}(h)$ and $b' = \nu^{-1}(h')$. By the definition of ψ, we have $\nu(a\psi(b)) = \nu(a\psi(b'))$, whence $a\psi(b) = a\psi(b')$ in $\mathbb{N}^{(X)}$, since ν is injective. We deduce $\psi(b) = \psi(b')$, since $\mathbb{N}^{(X)}$ is left-cancellative, then $b = b'$, since ψ is injective and, finally, $h = h'$. So M is left-cancellative.

Finally, as M is Noetherian, left-cancellative, and admits right-lcms, and as X is the atom set of M, we know by Proposition IV.3.21 (lcm-witness) that the list of all relations $s(s\backslash t) = t(t\backslash s)$ with $s \neq t \in X$ make a presentation of M. □

We can now easily complete the proof of Proposition 3.6.

Proof of Proposition 3.6. Assume that ν is a right-I-structure for M, based on a set X. By Lemma 3.18(i), X must be the atom set of M, and the assumption that M is finitely generated implies that X is finite.

Next, the free abelian monoid $\mathbb{N}^{(X)}$ satisfies the assumptions of Lemma 3.10 and, therefore, the latter applies. Hence, if we define $s \star t = \psi(s)(t)$, then $(X, \star)$ is an RC-quasigroup.

Assume $s \neq t \in X$. By (3.9), we have $s(s \star t) = \nu(st) = \nu(ts) = t(t \star s)$, whereas, by Lemma 3.18(i), (iii) and (iv), the monoid M admits unique right-lcms and is left-cancellative. It follows that $s(s \star t)$ is necessarily the right-lcm of s and t, and we must have $s \star t = s\backslash t$. Then Lemma 3.18(iv) implies that M admits the presentation $\langle X \mid \{s(s \star t) = t(t \star s) \mid s \neq t \in X\}\rangle^{+}$, so M is the structure monoid of $(X, \star)$.

Finally, the connection between ν and the polynomials Π is given by the right formula in (3.15). The uniqueness of ν follows, as X is the atom set of M, and $\star$ is the only possible extension of the right-complement operation outside the diagonal that admits bijective left translations, so they only depend on M. □

Relation (3.9) is reminiscent of a semi-direct product. We recall from [152] and [53] that, once the equalities (3.16) are established, one deduces the following connection, here in the slightly extended context of Lemma 3.10:

Proposition 3.19 (I-structure to twisted free abelian). *If M is a left-cancellative monoid, N is either $\mathbb{N}$ or $\mathbb{Z}/d\mathbb{Z}$ for some d, and ν is a map from $N^{(X)}$ to M, then the following conditions are equivalent:*

(i) *The map ν is a right-$IN^{(X)}$-structure based on X for M.*

(ii) *The map ν is bijective and there exists a map ψ from $N^{(X)}$ to $\mathfrak{S}_X$ such that $g \mapsto (\nu^{-1}(g), \psi(\nu^{-1}(g))^{-1})$ defines an injective homomorphism of M to the wreath product $N \wr \mathfrak{S}_X$.*

Proof. Assume that ν is a right $IN^{(X)}$-structure for M. As usual, we write ψ for the map from $N^{(X)}$ to $\mathfrak{S}_X$ specified by (3.9). We consider the wreath product $N \wr \mathfrak{S}_X$, that is, the semidirect product $N^{(X)} \rtimes \mathfrak{S}_X$, where $\mathfrak{S}_X$ acts on $N^{(X)}$ by permuting positions. Define $\iota : M \to N \wr \mathfrak{S}_X$ by $\iota(g) = (\nu^{-1}(g), \psi(\nu^{-1}(g))^{-1})$.

The map ν is bijective by definition, hence ι is injective, and the point is to check that ι is a homomorphism. Let g, h belong to M. Putting $a = \nu^{-1}(g)$, $b = \nu^{-1}(h)$, $\sigma = \psi(a)$, and $\tau = \psi(b)$, we find $\iota(g) = (a, \sigma^{-1})$ and $\iota(h) = (b, \tau^{-1})$, whence, in $N \wr \mathfrak{S}_X$,

$$\iota(g)\iota(h) = (a\sigma^{-1}[b], \sigma^{-1} \circ \tau^{-1}). \tag{3.20}$$

On the other hand, the monoid $N^{(X)}$ satisfies (3.11), so Lemma 3.14 applies and, using the left formula in (3.16), we find

$$\nu(a\sigma^{-1}[b]) = \nu(a)\nu(\psi(a)(\sigma^{-1}[b])) = g\nu(\sigma[\sigma^{-1}[b]]) = g\nu(b) = gh,$$

whence $\nu^{-1}(gh) = a\sigma^{-1}[b]$. Using the right formula in (3.16) similarly, we find

$$\psi(a\sigma^{-1}[b]) = \psi(\psi(a)[\sigma^{-1}[b]]) \circ \psi(a) = \psi(\sigma[\sigma^{-1}[b]]) \circ \sigma = \psi[b] \circ \sigma = \tau \circ \sigma.$$

We deduce $\iota(gh) = (a\sigma^{-1}[b], \sigma^{-1} \circ \tau^{-1})$. So, owing to (3.20), ι is a homomorphism, and (i) implies (ii).

Conversely, assume that ν and ψ ensure that ι is an embedding. Then the above computation gives, for all a, b in $N^{(X)}$, the equality $\nu(ab) = \nu(a)\nu(\psi(a)[b])$, whence in particular $\nu(as) = \nu(a)\psi(a)(s)$ for a in $N^{(X)}$ and s in X. So ν is a right-$IN^{(X)}$-structure on M, and (ii) implies (i) □

3.3 Coxeter-like groups

In the case of Artin's braid group B_n, the seminal example of a Garside group, the Garside structure (B_n^+, Δ_n) is directly connected with the exact sequence $1 \to PB_n \to B_n \to \mathfrak{S}_n \to 1$, where PB_n is the pure braid group and $\mathfrak{S}_n$ is the symmetric group on n objects: the lattice $\mathrm{Div}(\Delta_n)$ is isomorphic to $\mathfrak{S}_n$ equipped with the weak order and a presentation of $\mathfrak{S}_n$ is obtained by adding $n-1$ relations $\sigma_i^2 = 1$ to the standard presentation of B_n. Moreover, we saw in Chapter VI that $\mathfrak{S}_n$ leads to a Garside germ for B_n^+, meaning that the whole structure of B_n can be recovered from $\mathfrak{S}_n$; in particular, the Cayley graph of the germ derived from $\mathfrak{S}_n$ coincides with the Hasse diagram of $\mathrm{Div}(\Delta_n)$. We saw in Chapter IX that these results extend to other types in the Cartan classification, thus providing similar connections between all spherical Artin–Tits groups and the associated finite Coxeter groups.

It is natural to wonder whether some finite groups could play a similar role for other families of Garside groups. The aim of this subsection is to answer this in the affirmative in the case of finitely generated groups of I-type—that is, equivalently, in the case of structure groups of finite, involutive, nondegenerate set-theoretic solutions of YBE or, equivalently again, in the case of groups associated with finite RC-quasigroups. To this end, we first attach with every finite RC-quasigroup a parameter called its *class*.

Definition 3.21 (class). An RC-quasigroup $(X, \star)$ is said to be *of class* d if $\Omega_{d+1}(s, ..., s, t) = t$ holds for all s, t in X.

So an RC-quasigroup is of class 1 if $s \star t = t$ holds for all s, t, and it is of class 2 if $(s \star s) \star (s \star t) = t$ holds for all s, t.

Lemma 3.22. *Every finite* RC*-quasigroup of cardinal n is of class d for some number d satisfying $d < (n^2)!$.*

Proof. Let $(X,\star)$ be a finite RC-quasigroup with cardinal n. By Proposition 1.35, the map $\Psi : (s,t) \mapsto (s{\star}t, t{\star}s)$ is bijective on X^2. Consider the map $\Phi : (s,t) \mapsto (s{\star}s, s{\star}t)$ on X^2. Assume $(s,t) \neq (s',t')$. If s and s' are distinct, we have $\Psi(s,s) \neq \Psi(s',s')$, hence $s \star s \neq s' \star s'$, and $\Phi(s,t) \neq \Phi(s',t')$. If s and s' coincide, we must have $t \neq t'$, whence $s \star t \neq s \star t'$ and, again, $\Phi(s,t) \neq \Phi(s',t')$ since left translations of $\star$ are injective. So Φ is injective, hence bijective on the finite set X^2. As X^2 has cardinal n^2, then the order of Φ in $\mathfrak{S}_{X^2}$ is at most $(n^2)!$. So there exists $d < (n^2)!$ such that Φ^{d+1} is the identity. Now, an easy induction gives $\Phi^m(s,t) = (\Omega_m(s,...,s,s), \Omega_m(s,...,s,t))$ for every m. So $\Phi^{d+1} = \mathrm{id}$ implies $\Omega_{d+1}(s,...,s,t) = t$ for all s,t in X, meaning that $(X,\star)$ is of class d. □

There exist finite RC-quasigroups with an arbitrarily high minimal class, see Exercise 127. Here is now the main result:

Proposition 3.23 (Coxeter-like group). *If $(X,\star)$ is an RC-quasigroup of size n and class d and G (resp. M) is the associated group (resp. monoid), there exist a Garside element Δ in M and a finite group W of order d^n such that W enters a short exact sequence*

$$1 \longrightarrow \mathbb{Z}^X \longrightarrow G \longrightarrow W \longrightarrow 1$$

and M is generated by the germ $\underline{W}$ derived from (W,X). The Cayley graph of $\underline{W}$ coincides with the Hasse diagram of $\mathrm{Div}(\Delta)$ in M. A presentation of W is obtained by adding to (2.6) the relations $s^{[d]} = 1$ for s in X, where $s^{[d]}$ stands for $\Pi_d(s,...,s)$.

Definition 3.24 (Coxeter-like group). In the framework of Proposition 3.23, the finite group W will be called the *Coxeter-like group associated with $(X,\star)$ and d.*

The proof of Proposition 3.23 consists in using the I-structure on M, which exists by Proposition 3.5, to carry the results from the (trivial) case of $\mathbb{Z}^n$ to the case of M. As above, the I-structure on M will be denoted by ν, and the associated map from $\mathbb{N}^X$ to $\mathfrak{S}_X$ as defined in (3.4) will be denoted by ψ. It can easily be shown that ν and ψ, respectively, extend to a bijection from $\mathbb{Z}^X$ to G and a map from $\mathbb{Z}^X$ to $\mathfrak{S}_X$ satisfying (3.3) and (3.4), but we shall not use the result here.

Lemma 3.25. *Assume that $(X,\star)$ is an RC-quasigroup of class d and M is the associated monoid. For s in X and $q \geqslant 0$, let $s^{[q]} = \Pi_q(s,...,s)$. Then*

$$\nu(s^d a) = s^{[d]}\nu(a) \tag{3.26}$$

holds for all s in X and a in $\mathbb{N}^{(X)}$. The permutation $\psi(s^d)$ is the identity and, for all s,t in X, the elements $s^{[d]}$ and $t^{[d]}$ commute in M.

Proof. Let $t_1 \cdots t_q$ be a decomposition of a in terms of elements of X. By Proposition 3.5, we have

$$
\begin{aligned}
(3.27)\qquad \nu(s^d a) &= \Pi_{d+q}(s, ..., s, t_1, ..., t_q)\\
&= \Pi_d(s, ..., s)\Pi_q(\Omega_{d+1}(s, ..., s, t_1), ..., \Omega_{d+1}(s, , ..., s, t_q))\\
&= \Pi_d(s, ..., s)\Pi_q(t_1, ..., t_q) = \nu(s^d)\nu(t_1 \cdots t_q) = s^{[d]}\nu(a),
\end{aligned}
$$

in which the second equality comes from expanding the terms and the third one from the assumption that M is of class d. Applying (3.27) with $a = t$ in X and merging with $\nu(s^d t) = \nu(s^d)\,\psi(s^d)(t)$, we deduce that $\psi(s^d)$ is the identity. On the other hand, applying (3.27) with $a = t^{[d]}$, we find $s^{[d]}t^{[d]} = \nu(s^d t^d) = \nu(t^d s^d) = t^{[d]}s^{[d]}$. □

Lemma 3.28. (i) *Assume that $(X, \star)$ is a finite* RC*-quasigroup, M is the associated monoid, and $d \geqslant 2$ holds. Let $\Delta_0 = \prod_{s \in X} s$ in $\mathbb{N}^X$ and $\Delta_d = \nu(\Delta_0^{d-1})$. Then we have $\Delta_d = \Delta^{d-1}$, where Δ is the right-lcm of X, and Δ_d is a Garside element in M.*

(ii) *If, moreover, $(X, \star)$ is of class d, then Δ^d and $(\Delta_d)^d$ lie in the center of M.*

Proof. (i) By Lemma 2.27, we have $\Delta = \Pi_n(s_1, ..., s_n) = \nu(\Delta_0)$, where $(s_1, ..., s_n)$ is any enumeration of X. In other words, we have $\Delta = \Delta_2$. Now, we observe that $f[\Delta_0] = \Delta_0$ holds in $\mathbb{N}^X$ for every f in $\mathfrak{S}_X$ since every element of X occurs once in the definition of Δ_0. By (3.16), we deduce

$$
(3.29)\qquad \nu(a\Delta_0) = \nu(a)\nu(\psi(a)[\Delta_0]) = \nu(a)\nu(\Delta_0),
$$

whence $\nu(\Delta_0^k) = \nu(\Delta_0)^k$ for every k and, in particular, $\Delta_d = \nu(\Delta_0)^{d-1} = \Delta^{d-1}$. By Proposition 2.30, Δ is a Garside element in M and, by Proposition V.2.32(iv) (Garside map), so is every power of Δ, hence, in particular Δ_d.

(ii) Assume now that $(X, \star)$ is of class d. Let t belong to X. Then, by (3.29), we obtain $\nu(t\Delta_0^d) = \nu(t)\nu(\Delta_0)^d = t\Delta^d$. On the other hand, (3.29) and (3.26) give

$$
(3.30)\qquad \Delta^d = \nu(\Delta_0^d) = \prod_{s \in X} s^{[d]} \text{ and } \nu(\Delta_0^d t) = \prod_{s \in X} s^{[d]}t = \Delta^d t.
$$

Merging the values of $\nu(t\Delta_0^d)$ and $\nu(\Delta_0^d t)$, we obtain $t\Delta^d = \Delta^d t$, so that Δ^d, hence its power $(\Delta_d)^d$ as well, lie in the center of M. □

We are now ready to introduce the equivalence relation on $\mathbb{Z}^X$ that, when carried to G, will induce the expected quotient of G (and M). We recall that, for a in a free abelian group $\mathbb{Z}^{(X)}$, we denote by $a(s)$ the (well-defined) algebraic number of s in any X-decomposition of a.

Lemma 3.31. *Assume that $(X, \star)$ is an* RC*-quasigroup of class d and M and G are the associated monoid and group. For a, a' in $\mathbb{Z}^X$, write $a \equiv_0 a'$ if $a(s) = a'(s) \pmod{d}$ holds for every s in X.*

(i) *For g, g' in M, declare $g \equiv g'$ for $\nu^{-1}(g) \equiv_0 \nu^{-1}(g')$. Then $\equiv$ is an equivalence relation on M that is compatible with left- and right-multiplication. The class of 1 is the abelian submonoid M_1 of M generated by the elements $s^{[d]}$ with s in X.*

(ii) *For g, g' in G, declare that $g \equiv g'$ holds if there exist h, h' in M and r, r' in $\mathbb{Z}$ satisfying $g = \Delta^{dr}h$, $g' = \Delta^{dr'}h'$, and $h \equiv h'$. Then $\equiv$ is a congruence on G, and the kernel of the projection of G to $G/{\equiv}$ is the group of fractions of M_1.*

Proof. (i) As ν is bijective, carrying the equivalence relation $\equiv_0$ of $\mathbb{N}^X$ to M yields an equivalence relation on M. Assume $g \equiv g'$. Let $a = \nu^{-1}(g)$ and $a' = \nu^{-1}(g')$. Without loss of generality, we may assume $a' = as^d = s^d a$ for some s in X. Applying (3.16) and Lemma 3.25, we obtain $\psi(a') = \psi(\psi(s^d)[a]) \circ \psi(s^d) = \psi(a)$. Let t belong to X. Using (3.16) again, we deduce

$$\begin{aligned} g \cdot \psi(a)(t) &= \nu(a) \cdot \psi(a)(t) = \nu(at) \\ &\equiv \nu(a't) = \nu(a') \cdot \psi(a')(t) = \nu(a') \cdot \psi(a)(t) = g' \cdot \psi(a)(t). \end{aligned}$$

As $\psi(a)(t)$ takes every value in X when t ranges over X, we deduce that $\equiv$ is compatible with right-multiplication. On the other hand, $a \equiv_0 a'$ implies $f[a] \equiv_0 f[a']$ for every permutation f of X. Let t belong to X. Always by (3.16), we obtain

$$t \cdot g = t \cdot \nu(a) = \nu(t \cdot \psi(t)^{-1}[a]) \equiv \nu(t \cdot \psi(t)^{-1}[a']) = t \cdot \nu(a') = t \cdot g',$$

and $\equiv$ is compatible with left-multiplication by X.

The $\equiv_0$-class of 1 in $\mathbb{N}^X$ is the free abelian submonoid generated by the elements s^d with s in X. The $\equiv$-class of 1 in M consists of the image under ν of the products of such elements s^d. By Lemma 3.25, the latter are the products of elements $s^{[d]}$.

(ii) As Δ^d is a Garside element in M, every element of G admits an expression $\Delta^{dr}h$ with r in $\mathbb{Z}$ and h in M. This expression is not unique, but, if $g = \Delta^{dr}h = \Delta^{dr_1}h_1$ holds with, say, $r_1 < r$, then, as M is left-cancellative, we must have $h_1 = \Delta^{d(r-r_1)}h$, whence $h_1 \equiv h$ by (3.30). So, for every h' in M, the relations $h \equiv h'$ and $h_1 \equiv h'$ are equivalent. It is then easy to deduce that $\equiv$ is a equivalence relation on G and that it extends the relation $\equiv$ on M: in particular, the reflexivity of $\equiv$ follows from the fact that every element of G admits an expression $\Delta^{dr}h$ with h in M. Then the compatibility of $\equiv$ with multiplication on G follows from the compatibility on M and the fact that Δ^d lies in the center of G.

Finally, the $\equiv$-class of 1 in G consists of all elements $\Delta^{dr}h$ with h in M_1. As Δ^d belongs to M_1, this is the group of fractions of M_1 in G, hence the free abelian subgroup of G generated by the elements $s^{[d]}$ with s in X. □

We can now conclude.

Proof of Proposition 3.23. Let W be the quotient-group $G/{\equiv}$. By Lemma 3.31, the kernel of the projection of G onto W is the free abelian group based on $\{s^d \mid s \in X\}$, hence it is isomorphic to $\mathbb{Z}^X$, so we have an exact sequence $1 \to \mathbb{Z}^X \to G \to W \to 1$. The cardinality of W is the number of $\equiv$-classes in G. As every element of G is $\equiv$-equivalent to an element of M, this number is also the number of $\equiv$-classes in M, hence the number d^n of $\equiv_0$-classes in $\mathbb{N}^X$, and we have $W = M/{\equiv}$ as well.

By definition, $s^{[d]} \equiv 1$ holds for every s in X. Conversely, the congruence $\equiv_0$ on $\mathbb{Z}^X$ is generated by the pairs $(s^d, 1)$ with s in X, hence the congruence $\equiv$ on G is generated by the pairs $(s^{[d]}, 1)$ with s in X. Hence a presentation of W is obtained by adding to the presentation (2.6) of G and of M the n relations $s^{[d]} = 1$ with s in X.

By construction, the bijection ν is compatible with the congruences $\equiv_0$ on $\mathbb{N}^X$ and $\equiv$ on M, so it induces a bijection $\overline{\nu}$ of $\mathbb{Z}^X/{\equiv_0}$, which is $(\mathbb{Z}/d\mathbb{Z})^X$, onto $G/{\equiv}$, which is W, providing a commutative diagram

(3.32)
$$\begin{array}{ccc} \mathbb{N}^X & \xrightarrow{\;\nu\;} & M \\ {\scriptstyle \pi_0}\big\downarrow & & \big\downarrow{\scriptstyle \pi} \\ (\mathbb{Z}/d\mathbb{Z})^X & \xrightarrow[\;\overline{\nu}\;]{} & W\,. \end{array}$$

Now, let $\underline{W}$ be the germ derived from (W, X), and let σ_0 be the section of π_0 from $(\mathbb{Z}/d\mathbb{Z})^X$ to $\mathbb{N}^X$ that maps every $\equiv_0$-class to the unique n-tuple of $\{0, \dots, d-1\}^X$ that lies in that class. Let $\sigma : W \to M$ be defined by $\sigma(g) = \nu(\sigma_0(\overline{\nu}^{-1}(g)))$. Then, for every g in W, we obtain

$$\pi(\sigma(g)) = \pi(\nu(\sigma_0(\overline{\nu}^{-1}(g)))) = \overline{\nu}(\pi_0(\sigma_0(\overline{\nu}^{-1}(g)))) = g,$$

since σ_0 is a section of π_0. Hence σ is a section of π. Next, by construction, the image of W under σ is the image under ν of $\{0, \dots, d-1\}^X$, hence the image under ν of the family of all left-divisors of Δ_0^{d-1} in $\mathbb{N}^X$, hence the family of all left-divisors of Δ^{d-1}, that is, of Δ_d, in M. Finally, $\sigma(f)\sigma(g) = \sigma(fg)$ holds in M if and only if the relation $\sigma_0(\overline{\nu}(f))\sigma_0(\overline{\nu}(g)) = \sigma_0(\overline{\nu}(fg))$ holds in $\mathbb{N}^X$, hence if and only if, for every i, the sum of the ith coordinates of $\overline{\nu}(f)$ and $\overline{\nu}(g)$ does not exceed $d-1$. This happens if and only if $\|\overline{\nu}(f)\|_X + \|\overline{\nu}(g)\|_X = \|\overline{\nu}(fg)\|_X$ holds in $(\mathbb{Z}/d\mathbb{Z})^X$, hence if and only if $\|f\|_{\overline{X}} + \|g\|_{\overline{X}} = \|fg\|_{\overline{X}}$ holds in W. By construction, the family X is included in the image of σ, and all length-two relations of (2.6) belong to the previous list of relations, hence the latter make a presentation of M. In other words, M is (isomorphic to) the monoid $\mathcal{M}on(\underline{W})$ generated by the germ $\underline{W}$. □

Example 3.33 (Coxeter-like group). For an RC-quasigroup of class 1, that is, satisfying $s \star t = t$ for all s, t, we find $\Delta_1 = \Delta^0 = 1$. The group G is a free abelian group, the group W is trivial, and Proposition 3.23 here reduces to the isomorphism $\mathbb{Z}^X \cong G$.

For class 2, that is, when $(s \star s) \star (s \star t) = t$ holds for all s, t, we find $\Delta_2 = \Delta^1 = \Delta$, where Δ is the right-lcm of X. The element Δ_2 has 2^n divisors which are the right-lcms of subsets of X, and the group W is the order-2^n quotient of G obtained by adding the relations $s(s \star s) = 1$. For instance, in the case of $\{\mathtt{a}, \mathtt{b}\}$ with $s \star t = f(t)$, $f : \mathtt{a} \mapsto \mathtt{b} \mapsto \mathtt{a}$, the group G has the presentation $\langle \mathtt{a}, \mathtt{b} \mid \mathtt{a}^2 = \mathtt{b}^2 \rangle$, the relations $\mathtt{a}^{[2]} = \mathtt{b}^{[2]} = 1$ both amount to $\mathtt{ab} = 1$, and the associated Coxeter-like group W is a cyclic group of order 4.

For class 3, let us consider as in Example 2.31 the RC-quasigroup $\{\mathtt{a}, \mathtt{b}, \mathtt{c}\}$ with $s \star t = f(t)$ and $f : \mathtt{a} \mapsto \mathtt{b} \mapsto \mathtt{c} \mapsto \mathtt{a}$. The presentation of the associated group G is $\langle \mathtt{a}, \mathtt{b}, \mathtt{c} \mid \mathtt{ac} = \mathtt{b}^2, \mathtt{ba} = \mathtt{c}^2, \mathtt{cb} = \mathtt{a}^2 \rangle$. With the same notation as above, the smallest Garside element Δ is $\mathtt{a}^3$. As the class of $(X, \star)$ is 3, we consider here $\Delta_3 = \Delta^2 = \mathtt{a}^6$. The lattice $\mathrm{Div}(\Delta_3)$ has 27 elements, its Hasse diagram is the cube shown in Figure 4. The latter is also the Cayley graph of the germ derived from $(W, \{\mathtt{a}, \mathtt{b}, \mathtt{c}\})$, that is, the restriction of the Cayley graph of W to the partial product of the germ. Adding to the above presentation the relations $s^{[3]} = 1$, that is, $s(s \star s)((s \star s) \star (s \star s)) = 1$, namely $\mathtt{abc} = \mathtt{bca} = \mathtt{cab} = 1$, here reducing to $\mathtt{abc} = 1$, yields for W the presentation $\langle \mathtt{a}, \mathtt{b}, \mathtt{c} \mid \mathtt{ac} = \mathtt{b}^2, \mathtt{cb} = \mathtt{a}^2, \mathtt{ba} = \mathtt{c}^2, \mathtt{abc} = 1 \rangle$. One can check that alternative presentations of W are $\langle \mathtt{a}, \mathtt{b} \mid \mathtt{a} = \mathtt{b}^2\mathtt{ab}, \mathtt{b} = \mathtt{aba}^2 \rangle$ and $\langle \mathtt{a}, \mathtt{b} \mid \mathtt{a} = \mathtt{b}^2\mathtt{ab}, \mathtt{a}^3 = \mathtt{b}^3 \rangle$.

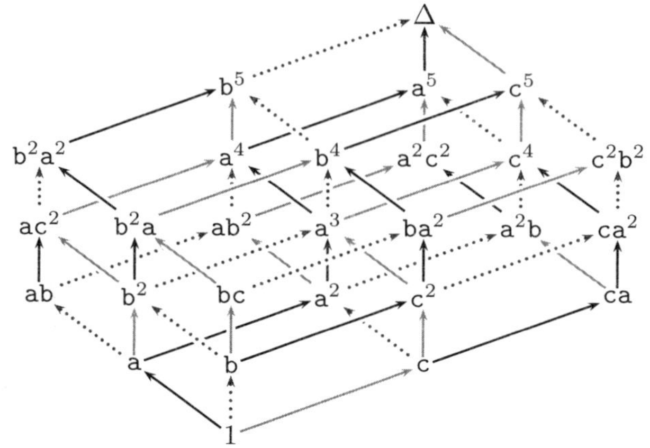

Figure 4. The Coxeter-like group associated with the RC-quasigroup of Example 2.31. The lattice of divisors of $\mathtt{a}^6$ in the associated monoid M, the Hasse diagram of the weak order on the finite group W with respect to the generators a, b, c, and the Cayley graph of the germ derived from W with respect to the previous generators all three correspond to the 27-vertex cube shown above.

The question naturally arises of characterizing Coxeter-like groups associated with finite RC-quasigroups (hence, equivalently, with solutions of YBE) as described in Proposition 3.23. We shall establish now that, exactly as structure groups of solutions of YBE are those groups that admit an I-structure, their Coxeter-like quotients are those finite groups that admit the counterpart of an I-structure where some cyclic group $\mathbb{Z}/d\mathbb{Z}$ replaces $\mathbb{Z}$, that is, what was called an $I(\mathbb{Z}/d\mathbb{Z})^X$-structure in Definition 3.2.

Proposition 3.34 (characterization). *For every finite group W, the following conditions are equivalent:*

(i) *There exists a finite* RC-*quasigroup $(X, \star)$ of class d such that W is the Coxeter-like group associated with $(X, \star)$ and d.*

(ii) *The group W admits a right-$I(\mathbb{Z}/d\mathbb{Z})^X$-structure based on X.*

Proof. Assume that $(X, \star)$ is a finite RC-quasigroup and G and M are the associated group and monoid. Then M admits an I-structure ν. With the notation of Lemma 3.31, the congruences $\equiv_0$ on $\mathbb{Z}^X$ and $\equiv$ on G are compatible and ν induces a well-defined map $\overline{\nu}$ of $(\mathbb{Z}/d\mathbb{Z})^X$ to $G/{\equiv}$ that makes the diagram of (3.32) commutative. Then $\overline{\nu}$ is bijective by construction, and projecting the relation (3.4) for ν gives its counterpart for $\overline{\nu}$. So $\overline{\nu}$ is a right-$I(\mathbb{Z}/d\mathbb{Z})^X$-structure for $G/{\equiv}$, and (i) implies (ii).

Conversely, assume that W is a finite group admitting a right-$I(\mathbb{Z}/d\mathbb{Z})^X$-structure based on X. As $(\mathbb{Z}/d\mathbb{Z})^X$ satisfies (3.11), Lemmas 3.10 and 3.14 apply. Thus putting $s \star t = \psi(s)(t)$ yields an RC-quasigroup and the formulas (3.15) are valid. Let s and t belong to X. In $(\mathbb{Z}/d\mathbb{Z})^X$, we have $s^d = 1$, whence $\psi(s^d) = \psi(1) = \mathrm{id}_X$. Applying (3.15) (left), we deduce $\Omega_{d+1}(s, \ldots, s, t) = \psi(s^d)(t) = t$. Hence $(X, \star)$ is of class d.

Now, let M be the monoid associated with $(X, \star)$, and W' be the associated finite quotient as provided by Proposition 3.23. The monoid M is generated by X and it admits a presentation consisting of all relations $s(s \star t) = t(t \star s)$ with s, t in X. By assumption, the group W is generated by X, and the relations $s(s \star t) = t(t \star s)$ with s, t in X are satisfied in W since they are equivalent to $\nu(st) = \nu(ts)$ and $(\mathbb{Z}/d\mathbb{Z})^X$ is abelian. Hence, there exists a surjective homomorphism θ from M to W that is the identity on X:

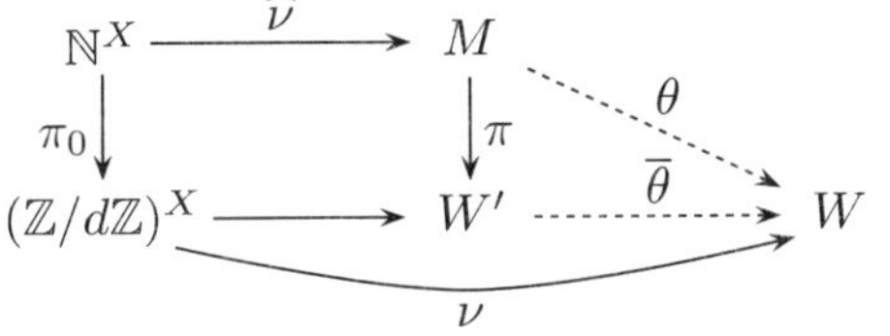

Next, W' is the quotient of M (and of its group of fractions) obtained by adding the relations $s^{[d]} = 1$, that is, $\Pi_d(s, \ldots, s) = 1$ for s in X. Now, in W, we have $\Pi_d(s, \ldots, s) = \nu(s^d)$ by (3.15); as $s^d = 1$ holds in $(\mathbb{Z}/d\mathbb{Z})^X$, we deduce from (3.15) that, in W, we have $\Pi_d(s, \ldots, s) = 1$ for every s in X. Thus the surjective homomorphism θ factorizes through W, yielding a surjective homomorphism $\overline{\theta}$ from W' to W. As both W' and W have cardinality d^n, the homomorphism $\overline{\theta}$ must be an isomorphism. Hence W is the Coxeter-like quotient of the group associated with $(X, \star)$ and d. So (ii) implies (i). □

Using Proposition 3.19, we deduce

Corollary 3.35 (wreath product). *Every finite group that is the Coxeter-like group associated with an* RC*-quasigroup of size n and class d embeds into the wreath product $(\mathbb{Z}/d\mathbb{Z}) \wr \mathfrak{S}_n$ so that the first component is a bijection.*

Example 3.36 (wreath product). For the last group W of Example 3.33, owing to the fact that the permutations of $\{1, 2, 3\}$ associated with $\mathtt{a}, \mathtt{b}, \mathtt{c}$ all are the cycle $f : 1 \mapsto 2 \mapsto 3 \mapsto 1$, one obtains a description as the family of the 27 tuples $(p, q, r; f^{p+q+r})$ with p, q, r in $\mathbb{Z}/3\mathbb{Z}$, the product of triples being twisted by the action of f^{p+q+r} on positions.

We conclude with observations about linear representations of the groups G and W associated with a finite RC-quasigroup. Here again, we start from the trivial case of a free abelian group and use the I-structure to carry the results.

Proposition 3.37 (linear representation). *Assume that $(X, \star)$ is an* RC*-quasigroup of cardinal n and class d. Let G be the associated group. For s the ith element of X (in some fixed enumeration), define*

$$\Theta(s) = \Theta_0(s) P_{\psi(s)}, \tag{3.38}$$

where $\Theta_0(s)$ is the diagonal $n \times n$-matrix with diagonal entries $(1, \ldots, 1, q, 1, \ldots, 1)$, q at position i and P_f is the permutation matrix associated with a permutation f of $\{1, \ldots, n\}$. Then Θ provides a faithful representation of G into $\mathrm{GL}(n, \mathbb{Q}[q, q^{-1}])$; specializing at $q = \exp(2i\pi/d)$ gives a faithful representation of the associated Coxeter-like group W.

Proof. First, Θ_0, which is initially defined on X, extends multiplicatively to a well-defined faithful representation of $\mathbb{Z}^X$ into $\mathrm{GL}(n, \mathbb{Q}[q, q^{-1}])$, since $\Theta_0(\prod \mathsf{s}_i^{e_i})$ is the diagonal matrix with diagonal $(q^{e_1}, \dots, q^{e_n})$. Specializing at $q = \exp(2i\pi/d)$ gives a faithful representation of $(\mathbb{Z}/d\mathbb{Z})^X$.

We now carry the results to G. First, let Θ^* be the multiplicative extension of Θ to X^*, and let s, t lie in X. By construction, we have $s \star t = \psi(s)(t)$, and we find

$$\begin{aligned}
\Theta^*(s|s \star t) &= \Theta(s)\,\Theta(\psi(s)(t)) \\
&= \Theta_0(s)\,P_{\psi(s)}\,\Theta_0(\psi(s)(t))\,P_{\psi(\psi(s)(t))} && \text{by definition} \\
&= \Theta_0(s)\,\Theta_0(t)\,P_{\psi(s)}\,P_{\psi(\psi(s)(t))} \\
&\qquad \text{by conjugating a diagonal matrix by a permutation matrix} \\
&= \Theta_0(st)\,P_{\psi(\psi(s)(t))\circ\psi(s)} && \text{by definition} \\
&= \Theta_0(st)\,P_{\psi(st)}. && \text{by (3.16)}
\end{aligned}$$

As $st = ts$ holds in $\mathbb{N}^{(X)}$, we deduce $\Theta^*(s|s \star t) = \Theta^*(t|t \star s)$. As M is presented by the relations $s(s \star t) = t(t \star s)$ with s, t in X, the map Θ^* induces a well-defined homomorphism Θ of M to $\mathrm{GL}(n, \mathbb{Q}[q, q^{-1}])$. As we are working with invertible matrices, the latter homomorphism extends to the group of fractions G of M.

Further, the relation (3.38) extends to

$$\Theta(\nu(a)) = \Theta_0(a) P_{\psi(a)}, \tag{3.39}$$

which holds for every a in $\mathbb{N}^{(X)}$, and then in $\mathbb{Z}^{(X)}$: for an induction, the point is to pass from a to as for a in $\mathbb{N}^{(X)}$ and s in X, and the verification is exactly similar to the above computation, using the general form of (3.16), namely $\psi(as) = \psi(\psi(a)(s)) \circ \psi(a)$.

As for faithfulness, (3.39) implies that $\Theta_0(\nu^{-1}(g))$ is the unique diagonal matrix obtained from $\Theta(g)$ by right-multiplication by a permutation matrix, so $\Theta(g)$ determines $\nu^{-1}(g)$, hence g.

Finally, specializing at a dth root of unity induces a well-defined faithful representation of the finite group W since, by definition, g and g' represent the same element of W if and only if $\nu^{-1}(g)$ and $\nu^{-1}(g')$ are $\equiv_0$-equivalent, hence if and only if the matrices $\Theta_0(\nu^{-1}(g))_{q=\exp(2i\pi/d)}$ and $\Theta_0(\nu^{-1}(g'))_{q=\exp(2i\pi/d)}$ are equal. □

Example 3.40 (linear representation). Coming back to the last case in Example 3.33 with the enumeration $(\mathsf{a}, \mathsf{b}, \mathsf{c})$, the permutations $\psi(\mathsf{a})$, $\psi(\mathsf{b})$, and $\psi(\mathsf{c})$ all are the 3-cycle $(1, 2, 3)$, and we find the explicit representation

$$\Theta(\mathsf{a}) = \begin{pmatrix} 0 & q & 0 \\ 0 & 0 & 1 \\ 1 & 0 & 0 \end{pmatrix}, \quad \Theta(\mathsf{b}) = \begin{pmatrix} 0 & 1 & 0 \\ 0 & 0 & q \\ 1 & 0 & 0 \end{pmatrix}, \quad \Theta(\mathsf{c}) = \begin{pmatrix} 0 & 1 & 0 \\ 0 & 0 & 1 \\ q & 0 & 0 \end{pmatrix}.$$

Specializing at $q = \exp(2i\pi/3)$ gives a faithful unitary representation of the associated 27-element group W. Using the latter, it is easy to check for instance that W has exponent 9: $\mathsf{a}, \mathsf{b}, \mathsf{c}$ have order 9, and all elements have order 1, 3, or 9.

Corollary 3.41 (group of isometries). *Every finite group that is the Coxeter-like group associated with a size-n RC-quasigroup can be realized as a group of isometries in an n-dimensional Hermitian space.*

Proof. The matrices $\Theta_0(s)_{q=\exp(2i\pi/d)}$ correspond to order-d complex reflections, whereas permutation matrices are finite products of hyperplane symmetries. $\square$

Exercises

Exercise 125 (bijective RC-quasigroup). Assume that $(X, \star)$ is an RC-quasigroup. Let $\psi : X \to X$ and $\Psi : X \times X \to X \times X$ be defined by $\psi(a) = a \star a$ and $\Psi(a, b) = (a \star b, b \star a)$. Show that Ψ is injective (*resp.* bijective) if and only if ψ is.

Exercise 126 (right-complement). Assume that $(X, \star)$ is an RC-quasigroup and M is the associated structure monoid. (i) Show that, for every element f in $M \cap X^p$, the map from X to $X \cup \{1\}$ that sends t to $f\backslash t$ takes pairwise distinct values in X and takes at most p times the value 1. (ii) Deduce that, for I a finite subset of X with cardinal n, the right-lcm Δ_I of I lies in X^n.

Exercise 127 (class). For $X = \{\mathtt{a}_1, ..., \mathtt{a}_n\}$, define $s{\star}t = f(t)$, where f is the cyclic permutation that maps $\mathtt{a}_i$ to $\mathtt{a}_{i+1 (\mathrm{mod}\, n)}$ for every i. Show that, for all p, i, and $s_1, ..., s_p$ in X, we have $\Omega_{p+1}(s_1, ..., s_p, \mathtt{a}_i) = \mathtt{a}_{i+p\, (\mathrm{mod}\, n)}$. Deduce that the minimal class of $(X, \star)$ is n.

Exercise 128 (I-structure). Assume that $(X, \star)$ is a bijective RC-quasigroup. (i) Show (by a direct argument) that the map $\underline{\star}$ from $X^* \times X$ to X defined by $1\underline{\star}t = t$, $s\underline{\star}t = s \star t$ for s in X, and $(u|v)\underline{\star}t = v\underline{\star}(u\underline{\star}t)$ induces a well-defined map from $M \times X$ to X. (ii) Show that the map ν from X^* to M defined by $\nu(1) = 1$, $\nu(s) = s$, and $\nu(ws) = \nu(w) \cdot (\nu(w)\underline{\star}s)$ for s in X induces a well-defined map from $\mathbb{N}^{(X)}$ to M.

Exercise 129 (parabolic submonoid). (i) Assume that (X, ρ) is a finite involutive nondegenerate set-theoretic solution of YBE and M is the associated structure monoid. Show that a submonoid M_1 of M is parabolic if and only if there exists a (unique) subset I of X satisfying $\rho(I \times I) = I \times I$ such that M_1 is the submonoid of M generated by I. (ii) Assume that $(X, \star)$ is a finite RC-system and M is the associated structure monoid. Then a submonoid M_1 of M is parabolic if and only if there exists a (unique) subset I of X such that I is closed under $\star$ and M_1 is the submonoid generated by I. (iii) Show that, if $(X, \star)$ is an infinite RC-quasigroup, there may exist subsets I of X that are closed under $\star$, but the induced RC-system $(I, \star)$ is not an RC-quasigroup.

Notes

Sources and comments. We refer to Jimbo [154] for an introduction to the Yang–Baxter equation. The latter takes its name from independent work of C. N. Yang from 1968, and

R. J. Baxter from 1971. The original form is

$$(\star) \qquad R^{12}(a)R^{13}(a+b)R^{23}(b) = R^{23}(b)R^{13}(a+b)R^{12}(a),$$

where R is a function of a complex parameter a taking values in $\mathrm{GL}(V \otimes V)$. When one restricts to solutions that do not depend on the parameter, the relation becomes $R^{12}R^{13}R^{23} = R^{23}R^{13}R^{12}$. From there one goes to the form (1.2) considered in Section 1 by composing with the switch operator P defined by $P(x \otimes y) = y \otimes x$. Then R satisfies $(\star)$ if and only if PR satisfies (1.4).

The investigation of set-theoretic solutions of YBE goes back to V. Drinfeld in 1992 [114]—where a connection with the self-distributivity law of Chapter XI is mentioned. A pair (X, ρ), where ρ is a bijection of $X \times X$, is sometimes called a *quadratic set*, and a quadratic set is said to be *braided* if it satisfies (1.6). So a set-theoretic solution of the Yang–Baxter equation is also a braided quadratic set.

Biracks and, previously, racks and quandles, have been introduced in low-dimensional topology as an algebraic counterpart of Reidemeister moves, see Joyce [155], Fenn and Rourke [121], and [77]: the basic principle is that, when one colors the strands of a braid diagram and pushes the colors through crossing according to the rule

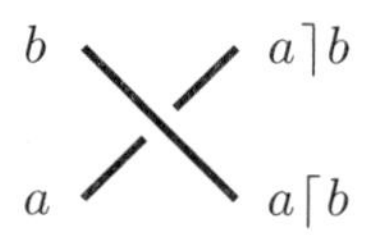

that is, when b overcrosses a, the colors become $a \lceil b$ and $a \rceil b$, then the birack laws naturally appear as the condition for an invariance of the output colours under braid relations (Reidemeister move III), and the assumption that translations are bijective appears as the condition for an invariance under Reidemeister move II. The connection between set-theoretic solutions of YBE and biracks can then be considered as folklore.

The connection between set-theoretic solutions of YBE and the right-cyclic law was described by W. Rump in [204]; the latter paper also contains (among others) the result that every finite RC-quasigroup is bijective.

The introduction of the structure group associated with a set-theoretic solution of YBE goes back to P. Etingof, T. Schedler, and A. Soloviev in [120]. These groups then received much attention. The results about the I-structure go back to T. Gateva-Ivanova and M. van den Bergh [127], building on earlier results of J. Tate and M. van den Bergh about Sklyanin algebras and what they called algebras of I-type [224]. Further results about monoids of I-type, in particular the connection with a wreathed abelian monoid, appear in Jespers and Oknińsky [152] and in Cedó, Jespers, and Okniński [53]; a comprehensive exposition is available as Chapter 8 of the book [153] by E. Jespers and J. Okniński.

The connection between set-theoretic solution of YBE and Garside monoids, in particular the characterization of Proposition 2.34, was established by F. Chouraqui in [59]; further results appear in Chouraqui and Godelle [60] and [61]. See also Gateva–Ivanova [126], in which *square-free* solutions of YBE are considered, namely those satisfying $\rho(a, a) = (a, a)$ for every a in the reference set. The results of Subsection 3.3 are inspired by those of Chouraqui–Godelle [61], which correspond to the special case of class 2 RC-quasigroups; the current results have been announced in [91]. Corollary 3.35 implies that the Coxeter-like groups W associated with finite RC-quasigroups are IG-monoids in the sense of [141]. It follows that they inherit all properties of such monoids established there, in particular in terms of the derived algebras $K[W]$ and their prime ideals. Concerning the existence of a finite quotient whose associated kernel is free abelian, one can show, using general results of Gromov, that every finitely generated group G whose Cayley graph is (quasi)-isometric to that of $\mathbb{Z}^n$ must be virtually $\mathbb{Z}^n$, that is, there exists an exact sequence $1 \to \mathbb{Z}^n \to G \to W \to 1$ with W finite, see [33]. By definition, an I-structure is an isometry as above, and, therefore, the existence of a finite quotient W as in Proposition 3.23 can be seen as a concrete instance of the above (abstract) result.

The current exposition is original in that it emphasizes the crucial role of RC-quasigroups. Section 1 is reminiscent of Rump's approach in [204], with more explicit details. All results in Section 2 appear in literature, but the current exposition, based on general results involving complemented presentations and on developing the polynomial calculus of Subsection 2.2, was not used so far, but it is close to the work of W. Rump in [207]. Note the efficiency of this approach, which allows for short inductive proofs for formulas like (2.22) or (2.25) which, when fully expanded, would be awful. Up to our knowledge, the argument for Proposition 2.34 is new. Similarly, the exposition in Subsections 3.1 and 3.2 does not follow the one of [153] but rather puts the emphasis on the connection between the I-structure and the RC-law.

Further results and questions. W. Rump's result that every finite RC-quasigroup is bijective (Proposition 1.35) is used in the proof of Proposition 2.30 only once, in order to guarantee that the monoid associated with a finite RC-quasigroup is right-cancellative, that is, equivalently, that the functor ϕ_Δ is injective. A natural question is whether either of these results can be proved without appealing to Proposition 1.35 (for instance, by showing that ϕ_Δ is surjective on atoms), in which case one would obtain an alternative proof of the latter based on a Garside approach.

At the moment, no exhaustive classification of the finite (involutive, nondegenerate) set-theoretic solutions of the Yang–Baxter equation is in view. By Proposition 3.19, the monoids associated with solutions based on S embed in the wreath product $\mathbb{N} \wr \mathfrak{S}_S$ and it was suggested in [53] to use the second projections of the images, called involutive Yang–Baxter (IYB) groups, for classifying the solutions of YBE. The Coxeter-like groups of Proposition 3.23 appear as other natural candidates for a classification. As every such group characterizes the associated RC-quasigroup, classifying all Coxeter-like groups is a priori not easier than classifying all solutions of YBE, a presumably (very) difficult task. On the other hand, the analogy with Coxeter groups suggests to further investigate their geometrical properties.

So far, Proposition 2.34 is the only known global characterization of a relatively large family of Garside groups: together with the results of Section 3, it identifies the Garside groups that admit a presentation of a certain form with those that admit an I-structure, hence resemble a free abelian group $\mathbb{Z}^n$ in some precise sense. One might think of replacing free abelian groups with other reference (Garside) groups Γ and consider those groups that admit a "$I\Gamma$-structure" in the sense of Definition 3.7, that is, their Cayley graph is that of Γ up to relabeling the edges. Should the above approach make sense, a natural problem would be

Question 38. Can one characterize the Garside groups that admit an $I\Gamma$-structure?

By the results of Section 3, the Garside groups that admit an $I\mathbb{Z}^n$-structure are those associated with RC-quasigroups, and Question 38 asks for a similar characterization of those admitting for instance an IB_n-structure, whatever it means, where B_n is Artin's n-strand braid group. From there, one can imagine various conjectures in the direction of a classification of (all) Garside groups. Addressing the most general case is just a dream at the moment, but considering particular classes, such as the class of modular Garside groups of [203], might lead to more accessible problems.

Returning to Earth, we refer to Section XIV.2 in the next chapter for further results involving the monoids associated with RC-systems that are not necessarily RC-quasigroups, that is, structures $(X, \star)$ where $\star$ satisfies the RC-law $(x \star y) \star (x \star z) = (y \star x) \star (x \star z)$, but the left translations $t \mapsto s \star t$ need not be bijective.

Chapter XIV
More examples

This final chapter is an introduction to four unrelated situations where interesting Garside structures arise. The first two are general schemes for deriving (new) Garside structures from previously given structures, whereas the last two sections describe particular families of Garside structures appearing in a topological context.

In Section 1, we show how to associate with every category equipped with a distinguished Garside family new categories whose objects are decompositions of (certain) elements of the category. As an application, we give the algebraic part of the proof by D. Bessis in [11] that periodic elements of the same period are conjugate in the braid group of a well-generated complex reflection group (Proposition 1.8), as well as a new proof of (a variant of) Deligne's Theorem stated as Proposition X.3.5.

Next, we show in Section 2 how to attach a category equipped with a distinguished Garside family to every (weak) RC-system, a general construction that extends the one of Chapter XIII where a Garside monoid is associated with every RC-quasigroup, see Proposition 2.27. The interest of the extension is to potentially capture a number of new examples.

In Section 3, we describe what is called the braid group of $\mathbb{Z}^n$ in Krammer [164]. This group is a sort of counterpart of the braid group B_n in which the symmetric group $\mathfrak{S}_n$ and its action on the orderings of $\{1, ..., n\}$ are replaced by the group $\mathrm{GL}(n, \mathbb{Z})$ and its action on the left-invariant orderings of $\mathbb{Z}^n$. It is a seminal example for a Garside germ derived from a lattice (Proposition 3.11).

Finally, Section 4 contains an introduction to the family of groupoids arising from decompositions of a punctured disk investigated in Krammer [165] (Proposition 4.21). These groupoids can be seen as a generalization of the braid group B_n, here viewed as the mapping class group of an n-punctured disk. They involve unexpected combinatorial objects, like the MacLane–Stasheff associahedra, opening fascinating questions.

1 Divided and decomposition categories

Starting with a fixed Garside family $\mathcal{S}$ in a cancellative category $\mathcal{C}$, we define and investigate new categories derived from the pair $(\mathcal{C}, \mathcal{S})$, whose objects are $\mathcal{S}$-decompositions of (some) elements of $\mathcal{C}$. To this end, we shall define several germs, prove that they are Garside germs, and describe their elements precisely. As mentioned in the introduction, this approach will enable us to reprove results by D. Bessis about periodic elements and Deligne's Theorem about the action of braid groups on bicategories.

There are two subsections. In Subsection 1.1, we investigate divided categories, a construction developed by D. Bessis in [7], which corresponds to the specific case when the initial Garside family is bounded by a map Δ and one considers the decompositions of the elements $\Delta(x)$, and contains the application to periodic elements. In Subsection 1.2, we develop an extended approach in which, instead of specifically considering the roots of a Garside map Δ, we consider decompositions of arbitrary elements and establish general versions of the common basic results, leading to the new proof of Deligne's Theorem.

In this section, we apply Convention V.3.7 (omitting source), thus writing Δ instead of $\Delta(x)$ when there is no need to specify x explicitly.

1.1 Divided categories

The construction of *divided categories* comes from D. Bessis in [7]. The construction of [7] is based on a topological intuition, whereas the current exposition will be exclusively algebraic. As an application, we shall define a context where periodic elements in a category become Garside maps in an associated divided category. As the current construction is a particular case of the constructions of Subsection 4.2, several proofs will be deferred to the latter.

Here we start with a Garside map Δ in a cancellative category $\mathcal{C}$, and aim at adding nth roots of Δ. For a very coarse intuition, we may think of extending, say, the field $\mathbb{Q}$ to a larger field $\mathbb{Q}[\sqrt[n]{\alpha}]$ by adding new elements. In the current case, the natural idea is to add new objects to our initial category so as to create new length-n paths for every element (that is, length-one path) of which we wish to add an nth root.

We recall that, by Proposition V.2.32 (Garside map II), every Garside map Δ in a cancellative category $\mathcal{C}$ gives rise to an automorphism of $\mathcal{C}$ denoted by ϕ_Δ.

Convention 1.1 (sequences). Hereafter in this section, when g is a sequence of elements of a category (or a germ), g_i refers to the ith entry of g. This applies, in particular, to paths, which are particular sequences (namely those in which the target of every element coincides with the source of the next one). On the other hand, as in Chapters VIII, X, and XI, a triple (u, f, v) will be denoted by $u \xrightarrow{f} v$.

Definition 1.2 (divided category). Assume that $\mathcal{C}$ is a cancellative category that is Noetherian and admits no nontrivial invertible element, and Δ is a Garside map in $\mathcal{C}$. Write $\mathcal{S}$ for $\mathcal{D}iv(\Delta)$. For $n \geqslant 1$, we define a family $\mathcal{S}_n$ and a category $\mathcal{C}_n$, called the *nth divided category* of $\mathcal{C}$, as follows.

(i) An object of $\mathcal{C}_n$ is a length n decomposition of some element $\Delta(\cdot)$.

(ii) An element of $\mathcal{S}_n$ is a triple $u \xrightarrow{f} v$ with u, v in $\mathcal{O}bj(\mathcal{C}_n)$ and f in $\mathcal{S}^n$ such that there exists f' in $\mathcal{S}^n$ satisfying $u_i = f_i f'_i$ and $f'_i f_{i+1} = v_i$ for every i, with $f_{n+1} = \phi_\Delta(f_1)$.

(iii) An element of $\mathcal{C}_n$ is a finite product of elements of $\mathcal{S}_n$, the product being defined by $(u \xrightarrow{f} v)(v \xrightarrow{g} w) = u \xrightarrow{h} w$ with $h_i = f_i g_i$ for each i.

Thus every element $u \xrightarrow{f} v$ of $\mathcal{S}_n$ gives rise to a commutative diagram

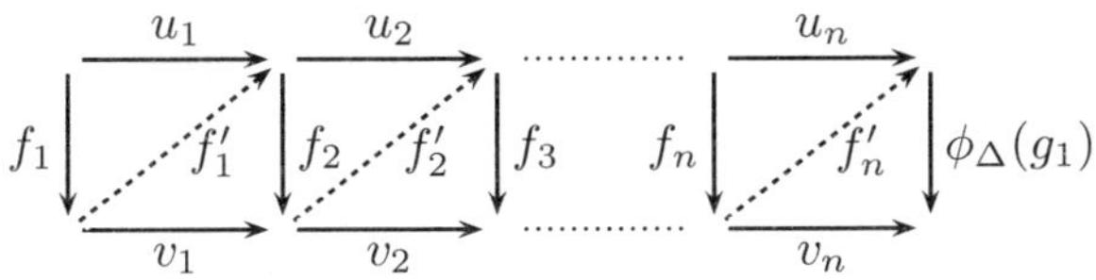

Notation 1.3. For an object u of $\mathcal{C}_n$, we denote by $[u]$ the element of $\mathcal{C}$ that is equal to the product of the successive entries of u.

Example 1.4 (divided category). Mapping every object x of $\mathcal{C}$ to the (length-one) path $(\Delta(x))$, and every element s of $\mathcal{S}(x, y)$ to the element $\Delta(x) \xrightarrow{s} \Delta(y)$, that is, to the square shown on the right, defines an isomorphism from $\mathcal{C}$ to $\mathcal{C}_1$.

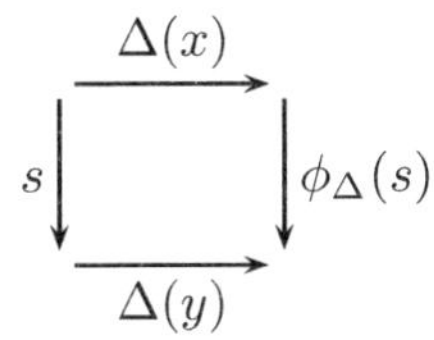

By definition, an element of $\mathcal{C}_n$ corresponds to a rectangular grid, which is a vertical juxtaposition of elementary diagrams corresponding to elements of $\mathcal{S}_n$. We will show in the next subsection (Corollary 1.18) that such an element is indeed determined by its source u and by the sequence $(f_1, \ldots, f_n)$, where f_i is the product of the elements that occur in the ith column of a grid as above. This makes the notation $u \xrightarrow{f} \cdot$ unambiguous for such an element, and the same applies to the elements of the enveloping groupoid of $\mathcal{C}_n$.

On the model of Example 1.4, it should be clear that mapping x to $1_x|\cdots|1_x|\Delta(x)$ and, for s in $\mathcal{D}iv(\Delta)$, mapping s to

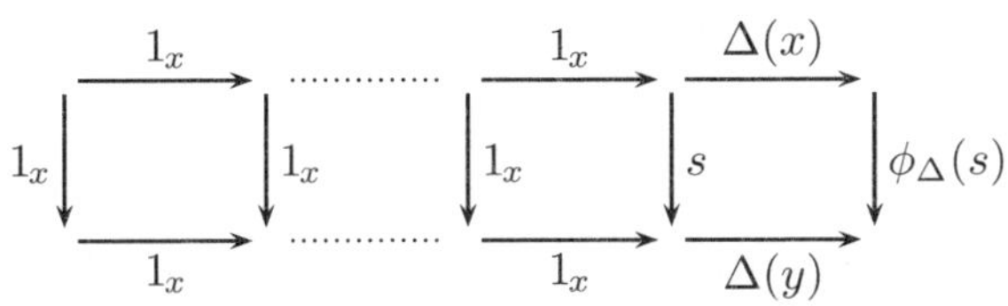

yields an injective functor from $\mathcal{C}$ to $\mathcal{C}_n$.

Proposition 1.5 (Garside map Δ_n). *Assume that $\mathcal{C}$ is a cancellative category that is Noetherian, and admits no nontrivial invertible element, and that Δ is a Garside map. Then, for every n, the category $\mathcal{C}_n$ is cancellative, Noetherian, admits no nontrivial invertible element, and $\mathcal{S}_n$ is a Garside family in $\mathcal{C}_n$ that is bounded by the map Δ_n that sends every object u of $\mathcal{C}_n$ to the element*

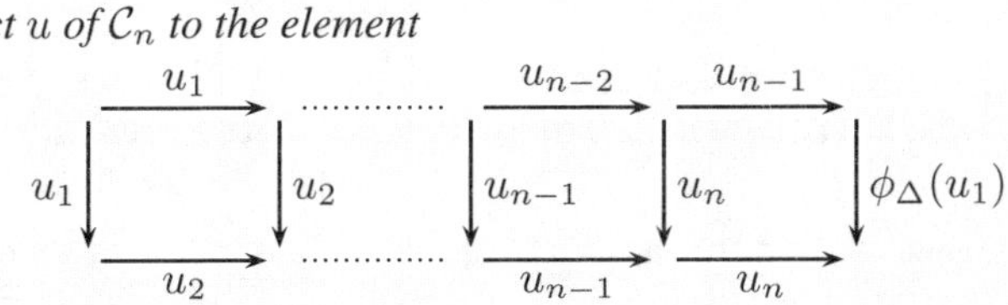

A direct verification is easy, but we skip it because, with the notation of Subsection 4.2, the category $\mathcal{C}_n$ identifies with the full subcategory of $\mathcal{C}_n(\phi_\Delta)$ whose objects are decompositions of Δ, and Proposition 1.5 immediately follows from Proposition 1.13.

We now compare the enveloping groupoids of a category $\mathcal{C}$ and of its nth divided category $\mathcal{C}_n$. The (slightly surprising) result is that, whereas $\mathcal{C}_n$ is intuitively larger than $\mathcal{C}$, neverthess the enveloping groupoids $\mathcal{E}nv(\mathcal{C})$ and $\mathcal{E}nv(\mathcal{C}_n)$ are equivalent. The result is due to D. Bessis in [7, 9.4]; we shall establish it here in a more explicit way than in [7], replacing the topological arguments in *loc. cit.* by an explicit natural transformation.

We recall from [177] that, if $\mathcal{C}, \mathcal{C}'$ are categories, an *equivalence of categories* between $\mathcal{C}$ and $\mathcal{C}'$ consists of a pair of functors $\phi : \mathcal{C} \to \mathcal{C}'$, $\phi' : \mathcal{C}' \to \mathcal{C}$, plus natural isomorphisms from $\phi' \circ \phi$ to $\mathrm{id}_{\mathcal{C}}$ and from $\phi \circ \phi'$ to $\mathrm{id}_{\mathcal{C}'}$.

Proposition 1.6 (equivalence of groupoids). *Assume that $\mathcal{C}$ is a cancellative category that is Noetherian, that admits no nontrivial invertible element, and that Δ is a Garside map. For x in $\mathcal{O}bj(\mathcal{C})$ and for s in $\mathcal{S}(x, y)$, put*

$$\rho_n(x) = 1_x|1_x|\cdots|1_x|\Delta(x) \quad \textit{and} \quad \rho_n(s) = \rho_n(x) \xrightarrow{(s,\ldots,s)} \rho_n(y).$$

On the other hand, for u in $\mathcal{O}bj(\mathcal{C}_n)$ with source x and $u \xrightarrow{f} v$ in $\mathcal{C}_n$, put

$$\pi_n(u) = x \quad \textit{and} \quad \pi_n(u \xrightarrow{f} v) = f_1.$$

Then ρ_n and π_n provide well-defined functors from $\mathcal{C}$ to $\mathcal{C}_n$ and from $\mathcal{C}_n$ to $\mathcal{C}$, respectively, and induce inverse equivalences of categories between $\mathcal{E}nv(\mathcal{G})$ and $\mathcal{E}nv(\mathcal{G}_n)$.

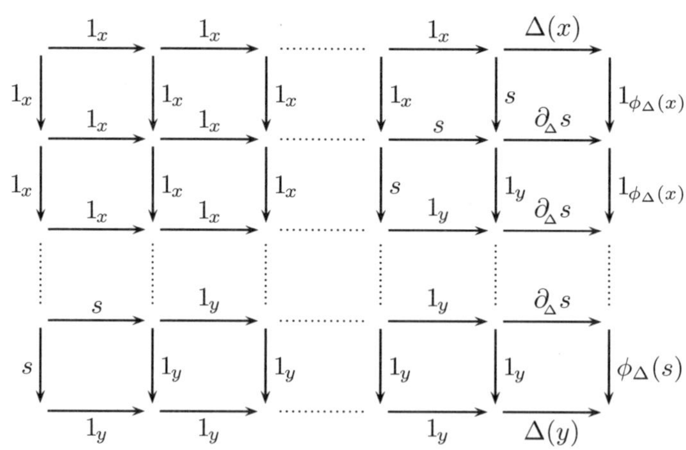

Figure 1. Decomposition of $\rho_n(s)$ into the product of n elements of $\mathcal{S}_n$; we recall that $\partial_\Delta s$ is the unique element satisfying $s\,\partial_\Delta s = \Delta(\cdot)$.

Proof. First, the diagram of Figure 1 shows that, for s in $\mathcal{S}$, the element $\rho_n(s)$ does lie in $\mathcal{C}_n$, that is, it admits a decomposition into the product of finitely many elements of $\mathcal{S}_n$.

Next, the composition $\pi_n \circ \rho_n$ is the identity of $\mathcal{C}$. Conversely, we define a natural transformation Dil_n from $\rho_n \circ \pi_n$ to the identity functor. For an object u of $\mathcal{C}_n$, say $u = s_1|\cdots|s_n$ of $\mathcal{C}_n$ with s_1 in $\mathcal{S}(x,\cdot)$, we define $\mathrm{Dil}_n(u)$ to be the dilatation

$$1_x|\cdots|1_x|\Delta(x) \xrightarrow{(1_x, s_1, s_1s_2, \ldots, s_1\cdots s_{n-1})} s_1|\cdots|s_n,$$

which is indeed an element of $\mathcal{C}_n$ with source $\rho_n(\pi_n(u))$, that is, $\rho_n(x)$, and target u as witnesses the diagram of Figure 2.

Let $u \xrightarrow{f} v$ with f_1 in $\mathcal{C}(x,y)$. Then we have $\pi_n(u \xrightarrow{f} v) = f_1$, and

$$(\rho_n \circ \pi_n)(u \xrightarrow{f} v) = \rho_n(x) \xrightarrow{(f_1,\ldots,f_1)} \rho_n(y).$$

Now, in $\mathcal{C}$, we have $f_1[v] = [u]\phi_\Delta(f_1)$, where $[u]$ is as in Notation 1.3. In the current case, this reads $g_1\Delta(y) = \Delta(x)\phi_\Delta(f_1)$. From there, one obtains a commutative diagram

$$\begin{array}{ccc} \rho_n(x) & \xrightarrow{(\rho_n \circ \pi_n)(u \xrightarrow{f} v)} & \rho_n(y) \\ {\scriptstyle \mathrm{Dil}_n(u)}\downarrow & & \downarrow{\scriptstyle \mathrm{Dil}_n(v)} \\ u & \xrightarrow[u \xrightarrow{f} v]{} & v \end{array} \tag{1.7}$$

witnessing that Dil_n is a natural transformation, as expected. Now the point is that, since $\mathrm{Dil}_n(u)$ becomes invertible in the groupoid $\mathcal{G}_n$, the natural transformation Dil_n induces in $\mathcal{G}_n$ a natural equivalence, whence the proposition. □

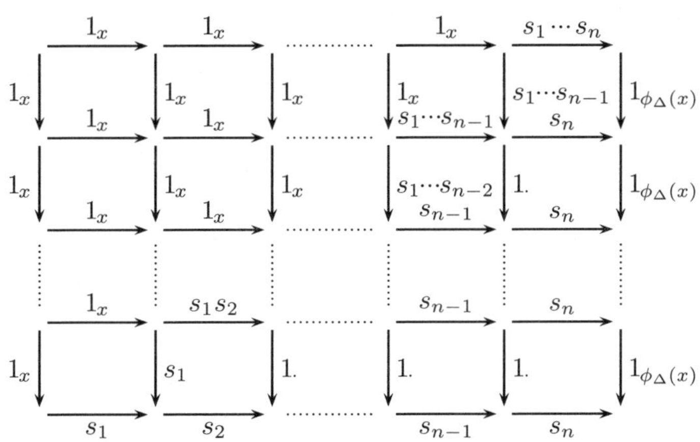

Figure 2. Decomposition of $\mathrm{Dil}_n(s_1|\cdots|s_n)$ into the product of $n-1$ elements of $\mathcal{S}_n$.

Application to periodic elements. We shall now apply the above construction to the study of periodic elements as developed in Subsection VIII.3. Keeping the above notation, assume that $s^n = \Delta$ holds, that is, s is an $(n,1)$-periodic element of $\mathcal{S}$. Then $s|\cdots|s$ is an object of $\mathcal{C}_n$, and we have $\pi_n(\Delta_n(s|\cdots|s)) = s$ in $\mathcal{C}$. We recall that Δ_n is the distinguished Garside map in $\mathcal{C}_n$ specified in Proposition 1.5. Then Equation 1.7 gives

$$\mathrm{Dil}_n(s|\cdots|s)\Delta_n(s|\cdots|s) = \rho_n(s)\mathrm{Dil}_n(s|\cdots|s),$$

which shows that, in $\mathcal{C}_n$, the element $\rho_n(s)$ is conjugate to $\Delta_n(w)$ for $w = s|\cdots|s$. Extending the argument, we shall now reprove [7, Theorem10.1]; here again we replace the topological proof of Bessis with a direct argument.

Proposition 1.8 (periodic elements). *In the context of Definition 1.2, assume in addition that the order of ϕ_Δ is finite. Then, if d is a (p,q)-periodic element of $\mathcal{C}$, there exists an object w of $\mathcal{C}_p$ such that $\rho_p(d)$ is conjugate to $\Delta_p^{[q]}(w)$.*

Proof. By Proposition VIII.3.34 (periodic elements), there exist p', q' satisfying $pq' - p'q = 1$ and a conjugate e of d that left-divides $\Delta^{[q']}$, such that the element s in S defined by $e^{p'}s = \Delta^{[q']}$ satisfies the equalities $e = s\phi_\Delta^{q'}(s)\phi_\Delta^{2q'}(s)\cdots\phi_\Delta^{(q-1)q'}(s)$ and $s\phi_\Delta^{q'}(s)\phi_\Delta^{2q'}(s)\cdots\phi_\Delta^{(p-1)q'}(s) = \Delta$. Note that, since we assume that there are no non-trivial invertible element in $\mathcal{C}$, the element e, because it commutes to its power $\Delta^{[p]}$, is fixed under $\phi_\Delta^{[q]}$. Hence the equality $e^{p'}s = \Delta^{[q']}$ implies that s is also fixed under $\phi_\Delta^{[q]}$. It follows that the path w defined by $w = s|\phi_\Delta^{q'}(s)|\phi_\Delta^{2q'}(s)|\cdots|\phi_\Delta^{(p-1)q'}(s)$ is an object of $\mathcal{C}_p$ such that, in $\mathcal{C}$, the element defined by $g = \pi_p(\Delta_p^{[q]}(w)) = s\phi_\Delta^{q'}(s)\phi_\Delta^{2q'}(s)\cdots\phi_\Delta^{(q-1)q'}(s)$ is conjugate to d. Now, by (1.7), we have $\mathrm{Dil}_p(w)\Delta_p^{[q]}(w) = \rho_p(g)\mathrm{Dil}_p(w)$. □

Note that not all paths $\Delta_p^{[q]}(w)$ are periodic; the condition for such an element to be periodic is that $\Delta_p^{[q]}(w)$ belong to $\mathcal{G}^{\circlearrowleft}$, which happens if and only if $\phi_{\Delta_p}^q(w) = w$ holds; this is the case for the element s involved in Proposition 1.8. The image under ρ_p of the centralizer of d in $\mathcal{C}_p$ consists of the elements fixed by the automorphism $\phi_{\Delta_p}^q$, hence it is eligible for Proposition VII.4.2 (fixed points).

Proposition 1.8 does not prove that all (p,q)-periodic elements are pairwise conjugate; this statement would be equivalent to the fact that the decompositions w and w' corresponding to two different (p,q)-periodic elements d and d' are in the same connected component of the category of fixed points $\mathcal{C}_p^{\phi_{\Delta_p}^q}$. In the cases described in Bessis [7] where the result holds, a topological proof of the connectedness of this category is given, interpreted as corresponding to a simplicial structure on the Garside nerve of a category with a Garside family.

One can show that the category $\mathcal{C}_p^{\phi_{\Delta_p}^q}$ is equivalent to the category $(\mathcal{C}yc\mathcal{C})^{\phi_\Delta^q}$ of Proposition VIII.3.37(iii) (periodic elements, twisted case).

Remark 1.9. When $\mathcal{C}$ is the dual monoid of a well-generated complex reflection group (see Section IX.3) with reflection degrees $d_1, ..., d_n$ and $h = d_n$ is the Coxeter number, Bessis–Reiner's conjecture [14, 6.5] predicts that when p is the order of a regular element the number of objects in the category $\mathcal{C}_p^{\Phi_{\Delta_p}}$ is

$$\prod_i \frac{[(p-1)h + d_i]_{\zeta_p}}{[d_i]_{\zeta_p}},$$

where we put $[n]_\zeta = [1 + \zeta + \cdots + \zeta^{n-1}]$ and $\zeta_p = e^{2i\pi/p}$.

1.2 Decomposition categories

We now explain a construction that is slightly more general than the one of Subsection 1.1 in that we consider the decompositions of arbitrary elements rather than just elements $\Delta(x)$, where Δ is a Garside map. So we now start from a cancellative category $\mathcal{C}$ with a distinguished Garside family $\mathcal{S}$ and an automorphism (possibly the identity), and construct for each n a new category whose objects are $\mathcal{S}$-decompositions of length n of the elements of $\mathcal{C}$. In order to establish that the extended categories have the desired properties, it will be convenient to use some criteria from Chapter VI and, accordingly, we shall adopt a germ framework. We recall from Proposition VI.1.11 (germ from Garside) that, whenever $\mathcal{C}$ is a (left)-cancellative category and $\mathcal{S}$ is a Garside family in $\mathcal{C}$, then $\mathcal{S}$ equipped with the partial product induced by the product of $\mathcal{C}$ is a Garside germ, denoted by $\underline{\mathcal{S}}$, and that $\mathcal{C}$ can then be recovered as the associated category $\mathcal{C}at(\underline{\mathcal{S}})$. So, in practice, we shall start with a Garside germ and an automorphism of this germ, and produce a new Garside germ consisting of sequences of given length of the given germ. Some of the results proved in this subsection were used in of Subsection 1.1 in the particular case of divided categories.

Definition 1.10 (germ $\underline{\mathcal{S}}_n(\phi)$). Assume that $\mathcal{C}$ is a cancellative category that is right-Noetherian and admits conditional right-lcms, $\mathcal{S}$ is a Garside family of $\mathcal{C}$ that is closed under left- and right-divisor, and ϕ is an automorphism of $\mathcal{C}$. For every positive integer n, we define a germ $\underline{\mathcal{S}}_n(\phi) = (\mathcal{S}_n, 1_{\mathcal{S}_n}, \bullet)$ in the following way.

(i) The objects of $\mathcal{S}_n(\phi)$ are the $\mathcal{S}$-paths u of length n in $\mathcal{S}$ such that the target of u_n is the image of the source of u_1 under ϕ.

(ii) An element of $\mathcal{S}_n(\phi)$ is a triple $u \xrightarrow{f} v$ with u, v in $\mathcal{O}bj(\mathcal{S}_n(\phi))$ and f in $\mathcal{S}^n$ such that there exists f' in $\mathcal{S}^n$ satisfying $u_i = f_i f'_i$ and $f'_i f_{i+1} = v_i$ for every i, where we put $g_{n+1} = \phi(g_1)$.

(iii) The product of two elements $u \xrightarrow{f} v$ and $v \xrightarrow{g} w$ of $\mathcal{S}_n(\phi)$ is defined when $f_i g_i \preccurlyeq u_i$ holds for every i, and it is given by $(u \xrightarrow{f} v) \bullet (v \xrightarrow{g} w) = u \xrightarrow{h} w$, with h defined by $h_i = f_i \bullet g_i$ for each i.

(iv) For u in $\mathcal{O}bj(\mathcal{S}_n(\phi))$, the identity element 1_u is $u \xrightarrow{(1_{u_1}, \dots, 1_{u_n})} u$.

It should be obvious that Definition 1.10 extends Definition 1.2, which corresponds to the case when ϕ is ϕ_Δ and the objects are restricted to decompositions of Δ. Item (ii) in Definition 1.10 is illustrated by the following diagram, which is entirely similar to the diagram after Definition 1.2:

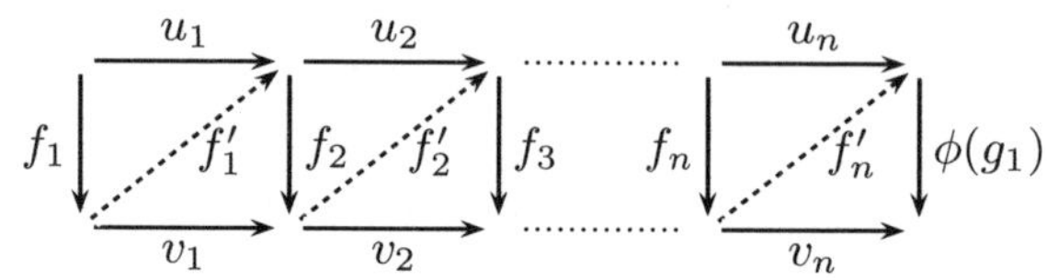

Lemma 1.11. *Under the assumptions of Definition* 1.10, $\underline{S}_n(\phi)$ *is a well-defined germ.*

Proof. The point to see is that (iii) defines a partial product on $\underline{S}_n(\phi)$. We have to show that under the condition of (iii) $u \xrightarrow{fg} w$ satisfies the conditions for belonging to $\underline{S}_n(\phi)$. Now, if we define h_i by $u_i = f_i g_i h_i$, the equality to prove is $w_i = h_i f_{i+1} g_{i+1}$. This equality is equivalent by left-cancellation to the equality $f_i g_i w_i = u_i f_{i+1} g_{i+1}$, which is true since, if we write $g_i g'_i = v_i$, we have $f_i g_i w_i = f_i g_i g'_i g_{i+1} = f_i v_i g_{i+1} = f_i f'_i f_{i+1} g_{i+1} = u_i f_{i+1} g_{i+1}$. □

Lemma 1.12. *Under the assumptions of Definition* 1.10, *let* $u \xrightarrow{f} v$ *and* $u \xrightarrow{g} w$ *be two elements of* $\underline{S}_n(\phi)$. *Then* f *left-divides* g *in* $\underline{S}_n(\phi)$ *if and only if, for every* i, *the element* f_i *left-divides* g_i *in* $\mathcal{C}$. *Then the associated right-quotient* h *is unique, given by* $f_i h_i = g_i$.

Proof. By the description of the product, it is clear that if h satisfies $f \bullet h = g$ in $\underline{S}_n(\phi)$, then we have $f_i h_i = g_i$ for every i. The unicity of h comes from the assumption that $\underline{S}$ is left-cancellative. Let us see that, conversely, the above condition implies that h is an element of $S_n(\phi)$. Indeed, from the equality $g_i w_i = u_i g_{i+1}$, we deduce $f_i h_i w_i = u_i f_{i+1} h_{i+1} = f_i v_i h_{i+1}$, which by left-cancellation gives $h_i w_i = v_i h_{i+1}$. Now let us define g'_i by $w_i = g'_i g_{i+1}$. Then the equality $h_i w_i = v_i h_{i+1}$ is equivalent to $h_i g'_i g_{i+1} = v_i h_{i+1}$, that is, $h_i g'_i f_{i+1} h_{i+1} = v_i h_{i+1}$, which implies by right-cancellativity that h_i left-divides v_i. So, if we define h'_i by $v_i = h_i h'_i$, we obtain $w_i = h'_i h_{i+1}$, as required. □

Proposition 1.13 (Garside germ). *Assume that* $\mathcal{C}$ *is a cancellative category that is right-Noetherian and admits conditional right-lcms,* $\mathcal{S}$ *is a Garside family of* $\mathcal{C}$ *that is closed under left- and right-divisor, and* ϕ *is an automorphism of* $\mathcal{C}$. *Then* $\underline{S}_n(\phi)$ *is a bounded Garside germ.*

Proof. We will apply Corollary VI.2.45 (recognizing Garside germ, conditional right-lcm case) to $\underline{S}_n(\phi)$. We first prove that $\underline{S}_n(\phi)$ is associative. To this end, we consider a path $u \xrightarrow{f} v | v \xrightarrow{g} w | w \xrightarrow{h} \cdot$ in $S_n(\phi)$. The conditions for $(u \xrightarrow{f} v \bullet v \xrightarrow{g} w) \bullet w \xrightarrow{h} \cdot$ and for $u \xrightarrow{f} v \bullet (v \xrightarrow{g} w \bullet w \xrightarrow{h} \cdot)$ to be defined in $\underline{S}_n(\phi)$ are the same, namely that $f_i g_i h_i$ left-divides u_i for every i. This gives left- and right-associativity. The germ $\underline{S}_n(\phi)$ is clearly left-cancellative. Right-Noetherianity comes from the right-Noetherianity of $\mathcal{C}$ since, by Lemma 1.12, the ith entries of the elements of a decreasing sequence for right-divisibility in $\underline{\mathcal{S}}_n$ form a decreasing sequence in $\mathcal{S}$.

We show now that $\underline{\mathcal{S}}_n(\phi)$ admits conditional right-lcms. If two elements $u \xrightarrow{f} \cdot$ and $u \xrightarrow{g} \cdot$ of $\underline{\mathcal{S}}_n(\phi)$ have a common right-multiple $u \xrightarrow{h} \cdot$, then, by Lemma 1.12, for every i the element h_i is a common right-multiple of f_i and g_i, so that f_i and g_i have a right-lcm, say h'_i in $\underline{\mathcal{S}}$. Since f_i and g_i left-divide u_i, we deduce that h'_i left-divides u_i. Then $u \xrightarrow{h'} \cdot$ is a well-defined element of $\underline{\mathcal{S}}_n(\phi)$, which is clearly a right-lcm of $u \xrightarrow{f} \cdot$ and $u \xrightarrow{g} \cdot$.

The same kind of argument shows that Relation (VI.2.46) holds.

We now show that the germ $\underline{\mathcal{S}}_n(\phi)$ is bounded. For every object u of $\underline{\mathcal{S}}_n(\phi)$, there exists an element of the form $u \xrightarrow{u} \cdot$ in $\underline{\mathcal{S}}_n(\phi)$, namely $u \xrightarrow{u} v$, with $v_i = u_{i+1}$ for $i < n$ and $v_n = \phi(u_1)$. We define $\Delta_n(u)$ to be this element. By definition, every element of $\underline{\mathcal{S}}_n(\phi)$ with source u left-divides $\Delta_n(u)$, so that Δ_n is a right-bound for $\underline{\mathcal{S}}_n(\phi)$. Finally, the functor Φ_{Δ_n} maps an object u to the shifted path $(u_2, u_3, \dots, u_n, \phi(u_1))$ and an element $u \xrightarrow{f} v$ to $\Phi_{\Delta_n}(u) \xrightarrow{f'} \Phi_{\Delta_n}(v)$ where f' is $(f_2, f_3, \dots, f_n, \phi(f_1))$. Since ϕ is an automorphism of $\mathcal{C}$, the functor Φ_{Δ_n} is bijective, hence the germ $\underline{\mathcal{S}}_n(\phi)$ is not only right-bounded, but even bounded by Δ_n. □

Notation 1.14 (category $\mathcal{C}_n(\phi)$). Under the assumptions of Proposition 1.13, we shall denote by $\mathcal{C}_n(\phi)$ the category $\mathcal{C}at(\underline{\mathcal{S}}_n(\phi))$. Hereafter we shall not distinguish the products in the germ $\underline{\mathcal{S}}_n(\phi)$ from the product in $\mathcal{C}_n(\phi)$.

If f and g are elements of $\underline{\mathcal{S}}_n(\phi)$ whose product belongs to $\underline{\mathcal{S}}_n(\phi)$, their ith entries satisfy $f_i g_i = (fg)_i$, hence we can extend the notation f_i to all elements of $\mathcal{C}_n(\phi)$. Note that f_i is the product of the entries in the ith column of a diagram associated with a decomposition $f = f^1 \cdots f^p$ of the form

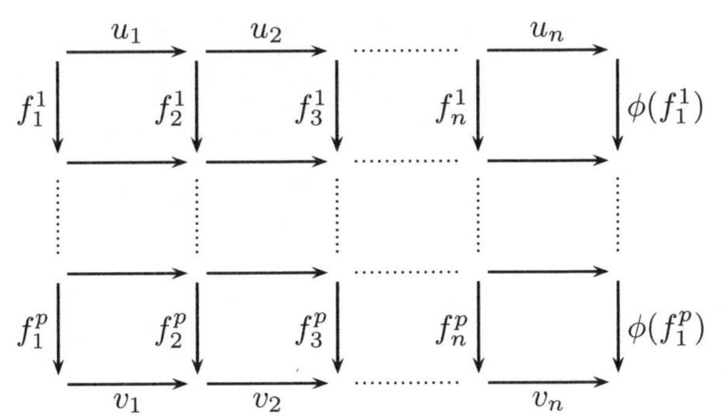

We shall use the notation $u \xrightarrow{\boldsymbol{f}} v$ to refer to an element of $\mathcal{C}_n(\phi)$ with source u, target v, and corresponding grid $\boldsymbol{f}$: here $\boldsymbol{f}$ represents the entries in the columns of the grid, hence the product in $\mathcal{C}_n$ of a sequence of the form $(f^1, \dots, f^p)$ with $f^1, \dots, f^p$ in $\mathcal{S}_n(\phi)^n$. For $u \xrightarrow{\boldsymbol{f}} v$ as above, we denote by $\boldsymbol{f}_i$ the product $f^1_i \cdots f^p_i$.

We now want to compute the head of an element of $\mathcal{C}_n(\phi)$. We begin with a preliminary result.

Lemma 1.15. *Under the assumptions of Definition* 1.10, *assume that* $u \xrightarrow{f} v$ *and* $v \xrightarrow{g} \cdot$ *lie in* $\mathcal{S}_n(\phi)$, *and* $u \xrightarrow{\boldsymbol{h}} \cdot$ *is an element of* $\mathcal{C}_n(\phi)$ *such that* h_i *is a left-gcd of* $f_i g_i$ *and* u_i *for every* i. *Then* $u \xrightarrow{\boldsymbol{h}} \cdot$ *is an* $\mathcal{S}_n(\phi)$*-head of* $(u \xrightarrow{f} v)(v \xrightarrow{g} \cdot)$.

Proof. First, h_i lies in $\mathcal{S}$, since it left-divides u_i and $\mathcal{S}$ is assumed to be closed under left-divisor. Since f_i left-divides u_i for every i, it left-divides the left-gcd h_i of u_i and $f_i g_i$. So there exist $\hat{f}_i$ and $\hat{g}_i$ in $\mathcal{S}$ satisfying $h_i = f_i \hat{f}_i$ and $g_i = \hat{f}_i \hat{g}_i$. Next, there exists h'_i in $\mathcal{S}$ satisfying $h_i h'_i = f_i \hat{f}_i h'_i = u_i$. Let $g''_i = g'_i \hat{f}_{i+1}$, where f_{n+1} and $\hat{f}_{n+1}$ denote, respectively, $\phi(f_1)$ and $\phi(\hat{f}_1)$. We have the following commutative diagram.

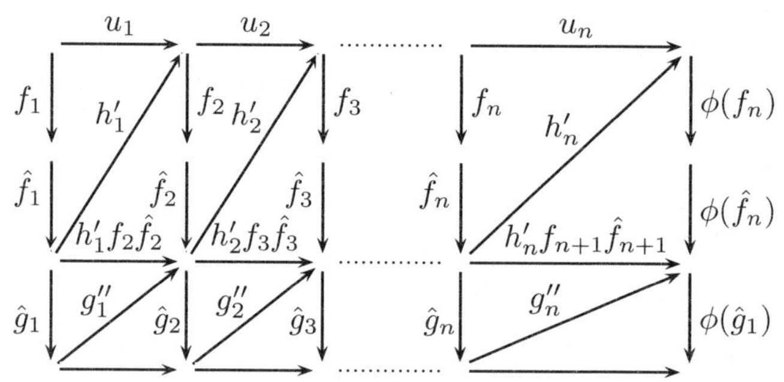

Proving that $u \xrightarrow{\boldsymbol{h}} \cdot$ defines an element of $\mathcal{S}_n(\phi)$ amounts to proving that $h'_i f_{i+1} \hat{f}_{i+1}$ belongs to $\mathcal{S}$ for every i. We claim that $h'_i f_{i+1} \hat{f}_{i+1}$ is the left-lcm of $f_{i+1} \hat{f}_{i+1}$ and g''_i for $i = 1, ..., n$. Indeed, it is a common left-multiple of these elements and, if their left-lcm was a strict left-divisor, then $\hat{g}_i$ and h'_i would have a nontrivial common left-divisor e_i, which would give a common left-divisor $f_i \hat{f}_i e_i$ of u_i and $f_i g_i$, a strict right-multiple of their left-gcd $f_i \hat{f}_i$. We conclude as $\mathcal{S}$ is closed under left-lcm.

So $u \xrightarrow{\boldsymbol{h}} \cdot$ is a legitimate element of $\mathcal{S}_n(\phi)$ that left-divides $(u \xrightarrow{f} v)(v \xrightarrow{g} \cdot)$. If $u \xrightarrow{e} \cdot$ is any left-divisor of $(u \xrightarrow{f} v)(v \xrightarrow{g} \cdot)$, then e_i left-divides u_i and $f_i g_i$ for every i, and we deduce that $u \xrightarrow{\boldsymbol{h}} \cdot$ is an $\mathcal{S}_n(\phi)$-head of $(u \xrightarrow{f} v)(v \xrightarrow{g} \cdot)$. □

Proposition 1.16 (head). *Assume that* $\mathcal{C}$ *is a cancellative category that is right-Noetherian and admits conditional right-lcms,* $\mathcal{S}$ *is a Garside family of* $\mathcal{C}$ *that is closed under left- and right-divisor, and* ϕ *is an automorphism of* $\mathcal{C}$. *Assume that* $u \xrightarrow{\boldsymbol{f}} \cdot$ *is an element of* $\mathcal{C}_n(\phi)$, *and* $u \xrightarrow{\boldsymbol{g}} \cdot$ *is such that* $\boldsymbol{g}_i$ *is a left-gcd of* u_i *and* $\boldsymbol{f}_i$ *for each* i. *Then* $u \xrightarrow{\boldsymbol{g}} \cdot$ *is an* $\mathcal{S}_n(\phi)$*-head of* $u \xrightarrow{\boldsymbol{f}} \cdot$.

Proof. We write $u \xrightarrow{\boldsymbol{f}} \cdot = (u \xrightarrow{f^1} u^1) \cdots (u^{p-1} \xrightarrow{f^p} u^p)$ with $u^{j-1} \xrightarrow{f^j} u^j$ in $\mathcal{S}_n(\phi)$. The proof uses induction on p. Lemma 1.15 gives the result for $p = 2$. By Proposition IV.1.50 (recognizing Garside III), there exists a sharp head-function H on $\mathcal{C}_n(\phi)$. Then we have $H(\boldsymbol{f}) = H(f^1 H(f^2 \cdots f^p))$. By the induction hypothesis applied to $f^2 \cdots f^p$, the following diagram, in which v stands for u^1, represents $f^1 H(f^2 \cdots f^p)$

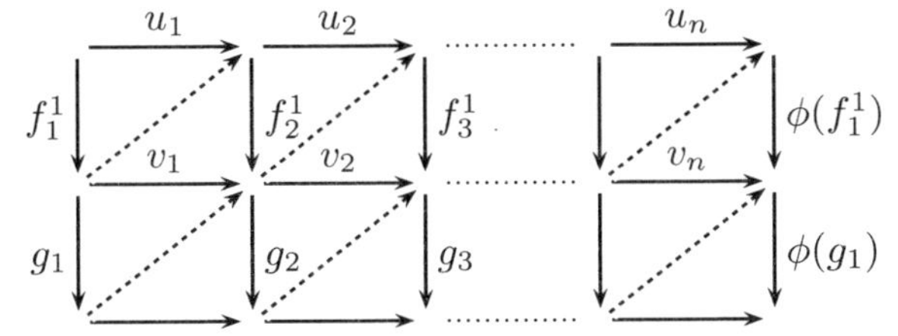

where g_i is a left-gcd of v_i and $(f^2 \cdots f^p)_i$ for every i. We apply now Lemma 1.15 to the two-term product of $u \xrightarrow{f^1} v$ and $v \xrightarrow{H(f^2 \cdots f^p)} \cdot$. We will be done if a left-gcd of $f_i^1 g_i$ and u_i is also a left-gcd of $\boldsymbol{f}_i$ and u_i, that is, if a left-gcd of the product of f_i^1 by a left-gcd of v_i and $(f^2 \cdots f^p)_i$ on the one hand, and of u_i on the other hand, is also a left-gcd of $\boldsymbol{f}_i$ and u_i for $i \leqslant n$. This is true as the former is a multiple of f_i^1, so has to be the product of f_i^1 by a common divisor of $(f^2 \cdots f^p)_i$ and v_i. □

Corollary 1.17 (left-division). *Under the assumptions of Proposition* 1.16, *let* $u \xrightarrow{\boldsymbol{f}} \cdot$ *and* $u \xrightarrow{\boldsymbol{g}} \cdot$ *be elements of* $\mathcal{C}_n(\phi)$. *Then* $u \xrightarrow{\boldsymbol{f}} \cdot$ *left-divides* $u \xrightarrow{\boldsymbol{g}} \cdot$ *in* $\mathcal{C}_n(\phi)$ *if and only if* $\boldsymbol{f}_i$ *left-divides* $\boldsymbol{g}_i$ *in* $\mathcal{C}$ *for every* i.

Proof. Assume that $\boldsymbol{f}_i$ left-divides $\boldsymbol{g}_i$ for every i. If $u \xrightarrow{\boldsymbol{f}} \cdot$ and $u \xrightarrow{\boldsymbol{g}} \cdot$ belong to $\mathcal{S}_n$, then we are done by Lemma 1.12. In general, we prove the result using induction on the length of the normal form of $u \xrightarrow{\boldsymbol{f}} \cdot$. Using, as in the above proof, a fixed sharp head-function H, and referring to the resulting head-function in $\mathcal{C}_n(\phi)$, we first show that the head of $u \xrightarrow{\boldsymbol{f}} \cdot$ left-divides the head of $u \xrightarrow{\boldsymbol{g}} \cdot$. Indeed, by Lemma 1.15, $H(u \xrightarrow{\boldsymbol{f}} \cdot)_i$ is a left-gcd of $\boldsymbol{f}_i$ and u_i, hence it left-divides $H(u \xrightarrow{\boldsymbol{g}} \cdot)_i$, which is a left-gcd of $\boldsymbol{g}_i$ and u_i for $i \leqslant n$, whence the result as $H(u \xrightarrow{\boldsymbol{f}} \cdot)$ and $H(u \xrightarrow{\boldsymbol{g}} \cdot)$ are two elements of $\mathcal{S}_n$. After left-cancelling $H(u \xrightarrow{\boldsymbol{f}} \cdot)$, we can apply the induction hypothesis, which gives that, if $\cdot \xrightarrow{\boldsymbol{f}'} \cdot$ is defined by $(u \xrightarrow{\boldsymbol{f}} \cdot) = H(u \xrightarrow{\boldsymbol{f}} \cdot)(\cdot \xrightarrow{\boldsymbol{f}'} \cdot)$, then $(\cdot \xrightarrow{\boldsymbol{f}'} \cdot)$ left-divides $H(u \xrightarrow{\boldsymbol{f}} \cdot)^{-1}(u \xrightarrow{\boldsymbol{g}} \cdot)$, so that $u \xrightarrow{\boldsymbol{f}} \cdot$ left-divides $u \xrightarrow{\boldsymbol{g}} \cdot$. The converse is clear. □

Corollary 1.18 (characterization of elements). *Under the assumptions of Proposition* 1.16, *and assuming in addition that* $\mathcal{C}$ *has no nontrivial invertible elements, an element* $u \xrightarrow{\boldsymbol{f}} \cdot$ *of* $\mathcal{C}_n(\phi)$ *is determined uniquely by its source* u *and its entries* $\boldsymbol{f}_i$ *in* $\mathcal{C}$.

Proof. Note first that, under our assumptions, $\mathcal{C}_n(\phi)$ has no nontrivial invertible elements either. If two elements $u \xrightarrow{\boldsymbol{f}} \cdot$ and $u \xrightarrow{\boldsymbol{g}} \cdot$ satisfy the equalities $\boldsymbol{f}_i = \boldsymbol{g}_i$ for all i, then Corollary 1.17 implies that they left-divide each other, hence they are equal. □

We conclude with a characterization of $\mathcal{C}_n(\phi)$ inside its enveloping groupoid. We naturally use for the elements of the latter a similar notation as for elements of $\mathcal{C}_n(\phi)$.

Proposition 1.19 (enveloping groupoid). *Under the assumptions of Proposition* 1.16, *let $\mathcal{G}_n(\phi)$ be the enveloping groupoid of $\mathcal{C}_n(\phi)$. Then an element $u \xrightarrow{\boldsymbol{f}} \cdot$ of $\mathcal{G}_n(\phi)$ belongs to $\mathcal{C}_n(\phi)$ if and only if all elements $\boldsymbol{f}_i$ belong to $\mathcal{C}_n(\phi)$.*

Proof. Suppose that Δ_n is a Garside map in $\mathcal{C}_n(\phi)$. An element of $\mathcal{G}_n$ can be written $\Delta_n^{[-m]} \cdot u \xrightarrow{\boldsymbol{f}} \cdot$ with $u \xrightarrow{\boldsymbol{f}} \cdot$ in $\mathcal{C}_n(\phi)$. We claim that, if $u \xrightarrow{\boldsymbol{f}} \cdot$ belongs to $\mathcal{C}_n(\phi)$ and all the entries of $\Delta_n^{[-m]} \cdot u \xrightarrow{\boldsymbol{f}} \cdot$ belong to $\mathcal{C}$, then $\Delta_n^{[-1]} \cdot u \xrightarrow{\boldsymbol{f}} \cdot$ belongs to $\mathcal{C}_n(\phi)$, which gives the result using induction on m. The ith entry of $\Delta_n^{[-m]} \cdot u \xrightarrow{\boldsymbol{f}} \cdot$ is $(v_i v_{i+1} \cdots v_{i+m-1})^{-1} \boldsymbol{f}_i$ where, for every j, we denote by v_j the jth term of the sequence

$$(u_1, \dots, u_n, \phi_{\Delta_n}(u_1), \dots, \phi_{\Delta_n}(u_n), \phi^2_{\Delta_n}(u_1), \dots).$$

So, in particular, $u_i^{-1} \boldsymbol{f}_i$ belongs to $\mathcal{C}$. By Corollary 1.17, the n-tuple $(u_1^{-1} \boldsymbol{f}_1, \dots, u_n^{-1} \boldsymbol{f}_n)$ defines an element $\cdot \xrightarrow{\boldsymbol{g}} \cdot$ of $\mathcal{C}_n$ that satisfies $u \xrightarrow{\boldsymbol{f}} \cdot = \Delta_n \cdot \cdot \xrightarrow{\boldsymbol{g}} \cdot$, whence our claim. □

Decomposition categories. We shall now briefly explain how to use a slightly extended version of the previous constructions to obtain in Proposition 1.24 below a new proof of Deligne's Theorem (Proposition X.3.5).

As above, our framework will be a cancellative category that is Noetherian and admits conditional right-lcms. To simplify the exposition in this part, we assume that it has no nontrivial invertible element. We will construct a category whose object family consists of *all* $\mathcal{S}$-decompositions of elements of the ambient category, where $\mathcal{S}$ is some fixed (solid) Garside family. The category will be defined by means of a germ which can be seen as a limit of subgerms of the germ $\underline{\mathcal{S}}_n(\mathrm{id})$. This germ is very rigid, in that it has at most one element with given source and target.

Definition 1.20 (germ $\underline{\mathcal{S}}^a_n$). Assume that $\mathcal{C}$ is a cancellative category that is Noetherian and admits unique conditional right-lcms. For every element a of $\mathcal{C}$, we define $\underline{\mathcal{S}}^a_n$ to be the subgerm of $\underline{\mathcal{S}}_n(\mathrm{id})$ whose objects u are decompositions of a, that is, satisfy $u_1 \cdots u_n = a$, and whose elements $u \xrightarrow{\boldsymbol{f}} \cdot$ satisfy $\boldsymbol{f}_1 \in \mathbf{1}_\mathcal{C}$. We embed $\mathcal{S}^a_n$ into $\mathcal{S}^a_{n+1}$ by mapping an object u to $u|1.$ and an element $u \xrightarrow{\boldsymbol{f}} \cdot$ to the element $u|1. \xrightarrow{\boldsymbol{f},1.} \cdot$. The germ $\underline{\mathcal{S}}^a$ is then defined as the union (direct limit) of the germs $\underline{\mathcal{S}}^a_n$ when n varies.

Thus an object of $\mathcal{S}^a$ is a right-infinite $\mathcal{S}$-path $(u_i)_{i \geqslant 1}$ that is eventually constant with identity-entries and whose evaluation in $\mathcal{C}$ is a. It is easy to check that every germ $\mathcal{S}^a$ is left- and right-associative and, therefore, embeds in the associated category $\mathcal{C}at(\mathcal{S}^a)$. The latter has a natural grading: the degree of an object u is the least integer n such that $u_i \in \mathbf{1}_\mathcal{C}$ holds for $i > n$.

It is clear that $\mathcal{S}^a$ is a bounded Garside germ, since each property to check is inherited from an appropriate $\underline{\mathcal{S}}^a_n$. In particular, the germ $\mathcal{S}^a$ is bounded by the map Δ^a defined by $\Delta^a(u) = \Delta^a_n(u \restriction n)$, where n is the degree of u and $u \restriction n$ means the length-n prefix of u.

Lemma 1.21. *Under the assumptions of Definition* 1.20*, the germ* $\underline{\mathcal{S}}_n^a$ *contains at most one element with given source and target. If* $u \xrightarrow{\boldsymbol{g}} v$ *is an element of* $\mathcal{S}^a$*, the degree of* v *is smaller than or equal to the degree of* u*, and, if* u *has degree* n*, then* $\boldsymbol{g}_i = 1$ *holds for* $i \geqslant n$.

Proof. Let $u \xrightarrow{\boldsymbol{g}} v$ be an element of $\mathcal{S}^a$. Starting from the assumption $\boldsymbol{g}_1 \in \mathbf{1}_\mathcal{C}$ and using induction on i, we see that $\boldsymbol{g}_{i+1}$ is determined by the existence of g'_i satisfying $\boldsymbol{g}_i g'_i = u_i$ and $g'_i \boldsymbol{g}_{i+1} = v_i$.

The assumption $\boldsymbol{g}_i \preccurlyeq u_i$ implies immediately the remark on the degree. □

There is a distinguished object $\mathrm{NF}(a)$ of $\mathcal{S}^a$ corresponding to the (strict) $\mathcal{S}$-normal decomposition $(s_1, ..., s_p)$ of a. The degree of $\mathrm{NF}(a)$ is $\|x\|_\mathcal{S}$. This object is well-defined, since we assume that there is no nontrivial invertible element in $\mathcal{C}$.

Lemma 1.22. *Under the assumptions of Definition* 1.20*, for every decomposition* u *of* a*, there exists a unique element of* $\mathcal{C}at(\mathcal{S}^a)$ *with source* u *and target* $\mathrm{NF}(a)$.

Proof. Using Algorithm III.1.48 (left-multiplication) inductively, we can construct at least one element in $\mathcal{C}at(\mathcal{S}^a)$ with source u and target $\mathrm{NF}(a)$.

For uniqueness, we first show that $\mathrm{NF}(a)$ is a final object in $\mathcal{C}at(\mathcal{S}^a)$, that is, the identity is the only element with source $\mathrm{NF}(a)$. Indeed, assume that $\mathrm{NF}(a) \xrightarrow{\boldsymbol{g}} v$ is an element of $\mathcal{S}^a$. The assumption $\boldsymbol{g}_1 \in \mathbf{1}_\mathcal{C}$, and the existence of g'_i satisfying $\boldsymbol{g}_i g'_i = \mathrm{NF}(a)_i$ and $g'_i \boldsymbol{g}_{i+1} = v_i$ for each i inductively imply $\boldsymbol{g}_{i+1} \in \mathbf{1}_\mathcal{C}$. Indeed, if $\boldsymbol{g}_i$ is an identity element then g'_is is equal to $\mathrm{NF}(a)_i$, so has no proper right-multiple $g'_i \boldsymbol{g}_{i+1}$ in $\mathcal{S}$ left-dividing $\mathrm{NF}(a)_i \mathrm{NF}(a)_{i+1}$.

We now show that every element of $\mathcal{C}at(\underline{\mathcal{S}}^a)$ with source u and target $\mathrm{NF}(a)$ is left-divisible by $\Delta^a(u)$. Let $u \xrightarrow{\boldsymbol{f}} \mathrm{NF}(a)$ be such an element. Express it as a product of elements $u \xrightarrow{f^1} u_1, ..., u_{m-1} \xrightarrow{f^m} \cdot$ of $\mathcal{S}^a$. By definition of $\Delta^a(u)$, we have $u \xrightarrow{f^1} u_1 \preccurlyeq \Delta^a(u)$. Using left-cancellativity, we define $u_1 \xrightarrow{g^1} \cdot$ by $\Delta^a(u) = (u \xrightarrow{f^1} u_1)(u_1 \xrightarrow{g^1} \cdot)$. Since $u_1 \xrightarrow{f^2} \cdot$ and $u_1 \xrightarrow{g^1} \cdot$ both left-divide $\Delta^a(u_1)$, they have a right-lcm of the form $(u_1 \xrightarrow{f^2} \cdot)(\cdot \xrightarrow{g^2} \cdot) = (u_1 \xrightarrow{g^1} \cdot)(\cdot \xrightarrow{h^2} \cdot)$. Using induction on i, we can repeat the process and obtain elements $\cdot \xrightarrow{g^i} \cdot$ and $\cdot \xrightarrow{h^i} \cdot$ of $\mathcal{S}^a$ that satisfy

$$(\cdot \xrightarrow{f^i} \cdot)(\cdot \xrightarrow{g^i} \cdot) = (\cdot \xrightarrow{g^{i-1}} \cdot)(\cdot \xrightarrow{h^i} \cdot).$$

As $\mathrm{NF}(a)$ is a final object, $\cdot \xrightarrow{g^{m+1}} \cdot$ must be an identity element of $\mathcal{S}^a$, and we obtain $(\cdot \xrightarrow{f^1} \cdot) \cdots (\cdot \xrightarrow{f^m} \cdot) = \Delta^a(u)(\cdot \xrightarrow{h^2} \cdot) \cdots (\cdot \xrightarrow{h^m} \cdot)$, which shows that $\Delta^a(u)$ left-divides $u \xrightarrow{\boldsymbol{f}} \mathrm{NF}(a)$.

From there, using induction, and because the Noetherianity assumption for $\mathcal{S}$, hence for $\mathcal{C}at(\mathcal{S}^a)$, implies the termination, we can express every element with target $\mathrm{NF}(a)$ as a (necessarily unique) finite product of elements $\Delta^a(\cdot)$, whence the result. □

We can then extend the result of Lemma 1.22.

Lemma 1.23. *Under the assumptions of Definition* 1.20, *there exists at most one element with given source and target in the category* $\mathcal{C}at(\underline{\mathcal{S}}^a)$.

Proof. Suppose that $u \xrightarrow{f} v$ and $u \xrightarrow{g} v$ belong to $\mathcal{C}at(\underline{\mathcal{S}}^a)$. By composing these elements with the canonical element of the form $v \xrightarrow{h} \mathrm{NF}(a)$, we obtain elements with source u and target $\mathrm{NF}(a)$, and Lemma 1.22 implies $(u \xrightarrow{f} v)(v \xrightarrow{h} \mathrm{NF}(a)) = (u \xrightarrow{g} v)(v \xrightarrow{h} \mathrm{NF}(a))$. By assumption, $\mathcal{C}$ is cancellative, so we deduce $f = g$. □

We can now apply the above results to establish another version of Deligne's Theorem X.3.5. The proof is more natural than the one of Chapter X, but requires stronger assumptions.

Proposition 1.24 (Deligne's Theorem II). *Assume that* $\mathcal{C}$ *is a cancellative category that is Noetherian and admits unique conditional right-lcms, that* a *is an element of* $\mathcal{C}$, *and that* $\mathcal{O}$ *is a functor from* $\mathcal{C}at(\mathcal{S}^a)$ *to a groupoid. Let us call* elementary isomorphism *a map of the form* $\mathcal{O}(u \xrightarrow{f} v)$ *or* $\mathcal{O}(u \xrightarrow{f} v)^{-1}$, *where* v *has the form* $u_1|\cdots|u_{i-1}|u_iu_{i+1}|u_{i+2}|\cdots$. *Then all the compositions of elementary isomorphisms between two objects in the image of* $\mathcal{O}$ *are equal.*

Proof. Let E_u be the image under $\mathcal{O}$ of the unique element in $\mathcal{C}at(\mathcal{S}^a)(u, \mathrm{NF}(a))$ whose existence is granted by Lemma 1.22. The elementary isomorphism $\mathcal{O}(u \xrightarrow{f} v)$ from $\mathcal{O}(u)$ to $\mathcal{O}(v)$ is equal to $E_u E_v^{-1}$. It follows that, for n larger than the degree of all the objects involved, we find by composing the above formula along a path of elementary isomorphisms, that any composition of elementary isomorphisms between $\mathcal{O}(u)$ and $\mathcal{O}(v)$ is equal to $E_u E_v^{-1}$. Thus all such compositions are equal. □

2 Cyclic systems

In Chapter XIII we investigated RC-quasigroups, which are algebraic systems $(S, \star)$, where $\star$ is a binary operation on S that satisfies the right-cyclic law $(x \star y) \star (x \star z) = (y \star x) \star (x \star z)$ and whose left-translations are bijective. We saw, in particular, that such structures are often connected with set-theoretic solutions of the Yang–Baxter equation, and with Garside monoids admitting presentations of a simple syntactic type. We shall now investigate more general algebraic structures, namely structures that again satisfy the RC law, but whose left-translations are not assumed to be bijective.

2.1 Weak RC-systems

We restart from the context of Section XIII.1 and the notion of an RC-system as introduced in Definition XIII.1.21: an RC-*system* is a structure $(S, \star)$ where $\star$ is a binary operation on S that satisfies the RC *law* :

(2.1) $$(x \star y) \star (x \star z) = (y \star x) \star (y \star z).$$

Example 2.2 (RC**-system).** As already noted in Example XIII.1.23, if S is any nonempty set and σ is any map of S to itself, the operation defined by $x \star y = \sigma(y)$ makes S into an RC-system.

On the other hand, if M is any left-cancellative monoid that admits unique right-lcms and $\backslash$ is the associated right-complement operation, that is, $g\backslash h$ is the (unique) element h' such that gh' is the right-lcm of g and h, then $(M, \backslash)$ is an RC-system, which, in general, is not an RC-quasigroup. More generally, if S is a subset of M that is closed under right-complement, then $(S, \backslash)$ is also an RC-system: this happens, in particular, when S is a Garside family of M that includes 1, or when S is the closure of the atoms under $\backslash$.

For instance, starting from the braid monoid B_3^+ (Reference Structure 2, page 5), the closure of the atom set $\{\sigma_1, \sigma_2\}$ under the operation $\backslash$ has 5 elements 1, σ_1, σ_2, $\sigma_1\sigma_2$, $\sigma_2\sigma_1$, giving rise to the 5-element RC-system shown on the right (in which non-unit elements have been renamed).

	1	a	b	c	d
1	1	a	b	c	d
a	1	1	d	b	d
b	1	c	1	c	a
c	1	1	a	1	a
d	1	b	1	b	1

Similarly, starting from the dual braid monoid B_3^{+*} (Reference Structure 3, page 10), the closure of the atom set $\{a_{1,2}, a_{2,3}, a_{1,3}\}$ under the operation $\backslash$ has 4 elements, leading to the RC-system whose table is shown on the right. Note that neither of these RC-systems is an RC-quasigroup: some values appear twice or more in the same row.

	1	a	b	c
1	1	a	b	c
a	1	1	b	b
b	1	c	1	c
c	1	a	a	1

Also note that the operations defined by $p \star q = \max(p, q) - p$ on $\mathbb{N}$ and $\mathbb{Z}$ (and their counterparts in any ordered group) enter the previous framework.

See more examples at the end of the section. Before proceeding, we shall extend our framework in two directions. The first extension consists in replacing sets with categories. This small change just means that we attach to each element a source and a target that put constraints on the existence of $s \star t$. In practice, we start with a precategory, that is, a family $\mathcal{S}$ plus two maps, source and target, from $\mathcal{S}$ to (another) family $\mathcal{O}bj(\mathcal{S})$, and we add the restriction that, for s, t in $\mathcal{S}$, the element $s \star t$ is defined only when s and t share the same source and, then, the source of $s \star t$ is the target of s and the target of $s \star t$ is that of $t \star s$. In this way, it is still true that, whenever $s \star t$ and $t \star s$ are defined, they naturally enter a square diagram as on the right. The initial case of Definition XIII.1.21 corresponds to the case of one object only.

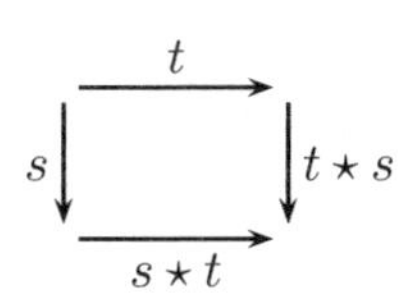

The second extension is then natural. Going to a category framework induces constraints for the existence of $s \star t$ and, in general, makes the operation partial. It is then coherent to allow the operation to be really partial, this meaning that $s \star t$ may be undefined even if s and t share the same source. In this way, we arrive at the following weakening of Definition XIII.1.21.

Definition 2.3 (weak RC-system). A *weak* RC-*system* is a precategory $\mathcal{S}$ equipped with a partial binary operation $\star$ satisfying

(2.4) If $s \star t$ exists, the sources of s and t coincide, and the source of $s \star t$ is the target of s.

(2.5) If $s \star t$ exists, so does $t \star s$, and they share the same target.

(2.6) If $(r \star s) \star (r \star t)$ exists, so does $(s \star r) \star (s \star t)$, and they are equal.

Condition (2.6) will be still called the RC *law*. Note that, in the context of (2.6), the assumption that $(r \star s) \star (r \star t)$ exists implies that $r \star s$ and $r \star t$ are defined as well, hence so do $s \star r$ and $(s \star r) \star (s \star t)$ when (2.6) is satisfied. Moreover, according to (2.5), if $(s \star r) \star (s \star t)$ exists, so does $(s \star t) \star (s \star r)$, and a new instance of (2.6) guarantees that $(t \star s) \star (t \star r)$ exists too and is equal to the latter. By the same argument, we deduce that $(r \star t) \star (r \star s)$ exists as well and is equal to $(t \star r) \star (t \star s)$. Thus each triple (r, s, t) that is eligible for (2.6) gives rise to three equalities, as illustrated in the cubic diagram on the right: starting from three arrows sharing the same source, if three faces of the cube exist, then the other three faces also exist and the last three arrows are pairwise equal.

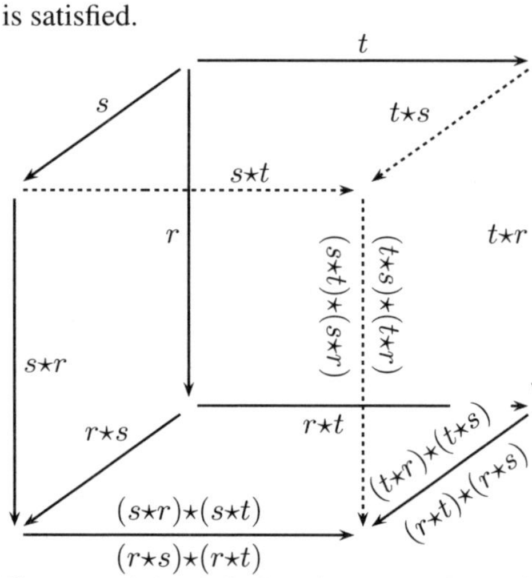

In the sequel, the operation of a weak RC-system is by default written $\star$, and we often say "the RC-system $\mathcal{S}$" for "the RC-system $(\mathcal{S}, \star)$".

Definition 2.7 (unit, unital). If $\mathcal{S}$ is a weak RC-system, a *unit family* in $\mathcal{S}$ is a subfamily $(\epsilon_x)_{x \in \mathcal{O}bj(\mathcal{S})}$ of $\mathcal{S}$ satisfying $\epsilon_x \in \mathcal{S}(x, x)$ for every x and, for every s in $\mathcal{S}(x, y)$,

(2.8) $$s \star \epsilon_x = \epsilon_y, \quad \epsilon_x \star s = s, \quad \text{and} \quad s \star s = \epsilon_y.$$

We say that $\mathcal{S}$ is *unital* if it has a unit family $(\epsilon_x)_{x \in \mathcal{O}bj(\mathcal{S})}$ and, for all s, t in $\mathcal{S}(\cdot, y)$,

(2.9) the conjunction of $s \star t = \epsilon_y$ and $t \star s = \epsilon_y$ implies $s = t$.

We shall write $\boldsymbol{\epsilon}_{\mathcal{S}}$ for a family of the form $(\epsilon_x)_{x \in \mathcal{O}bj(\mathcal{S})}$. Saying that $\boldsymbol{\epsilon}_{\mathcal{S}}$ is a unit family in $\mathcal{S}$ means that, for every s in $\mathcal{S}(x, y)$, the relations corresponding to the diagrams

$\begin{array}{ccc} \cdot & \xrightarrow{s} & \cdot \\ {\scriptstyle \epsilon_x}\big\downarrow & & \big\downarrow{\scriptstyle \epsilon_y} \\ \cdot & \xrightarrow[s]{} & \cdot \end{array}$ and $\begin{array}{ccc} \cdot & \xrightarrow{s} & \cdot \\ {\scriptstyle s}\big\downarrow & & \big\downarrow{\scriptstyle \epsilon_y} \\ \cdot & \xrightarrow[\epsilon_y]{} & \cdot \end{array}$ are satisfied. Note that units are unique when they exist, since, if ϵ_x and ϵ'_x are two units with source x, (2.8) implies $\epsilon_x = \epsilon_x \star \epsilon_x = \epsilon'_x$. So, when we speak of a unital weak RC-system, there is no ambiguity on the involved units.

According to the already established properties of the right-complement operation in a category that admits unique right-lcms, we can immediately state:

Proposition 2.10 (right-complement). *If $\mathcal{C}$ is a left-cancellative category that admits unique conditional right-lcms and $\mathcal{S}$ is a subfamily of $\mathcal{C}$ that is closed under $\backslash$, then $(\mathcal{S}, \backslash)$ is a unital weak* RC*-system with unit family $\mathbf{1}_{\mathcal{S}}$.*

Proof. Conditions (2.4) and (2.5) about source and target are trivial; that the RC law (2.6) is valid follows from Proposition II.2.15 (triple lcm). For g in $\mathcal{C}(x, y)$, the right-lcm of g and 1_x is g, implying $g\backslash 1_x = 1_y$, $1_x\backslash g = g$, and $g\backslash g = 1_y$, so (2.8) is satisfied. Finally, if both $g\backslash h$ and $h\backslash g$ are identity elements, then the right-lcm of g and h is both g and h, implying $g = h$. Hence, $(\mathcal{C}, \backslash)$ is a unital weak RC-system and so is $(\mathcal{S}, \backslash)$ for every nonempty subfamily $\mathcal{S}$ of $\mathcal{C}$ that is closed under right-complement. □

Corollary 2.11 (right-complement). *If $\mathcal{C}$ is a left-cancellative category that admits unique conditional right-lcms and $\mathcal{S}$ is a Garside family of $\mathcal{C}$, then $(\mathcal{S} \cup \{\mathbf{1}_{\mathcal{C}}\}, \backslash)$ is a unital weak* RC*-system.*

Proof. If $\mathcal{S}$ is a Garside family in $\mathcal{C}$, then $\widehat{\mathcal{S}}$, which is $\mathcal{S} \cup \mathbf{1}_{\mathcal{C}}$ in the current context, is closed under right-complement, hence eligible for Proposition 2.10. □

So, for instance, the last two tables in Example 2.2 resort to Corollary 2.11, and they are indeed unital weak RC-systems (with unit 1).

What we shall do now is establish a converse for Proposition 2.10 and Corollary 2.11, that is, starting from a unital weak RC-system $\mathcal{S}$, construct a category $\mathcal{C}(\mathcal{S})$ that admits unique (conditional) right-lcms and such that $\star$ (essentially) is the right-complement operation on some generating subfamily of $\mathcal{C}(\mathcal{S})$. The result is similar to those involving RC-quasigroups in Chapter XIII but, on the other hand, as left-translations are not assumed to be bijective, the context is quite different and we shall heve to use a different construction.

2.2 Units and ideals

Controlling units turns out to be technically important, and it requires to develop specific tools. As the following result shows, starting with an arbitrary weak RC-system, it is easy to add units and obtain a unital system.

Lemma 2.12. *Assume that $\mathcal{S}$ is a weak* RC*-system and $\mathcal{S} \subseteq \mathcal{S}'$ holds. For every object x of $\mathcal{S}$, define $\mathcal{S}^\sharp(x,x)$ to be $\mathcal{S}(x,x) \cup \{\epsilon_x\}$, where ϵ_x is a distinguished element of $\mathcal{S}'(x,x)$ (belonging to $\mathcal{S}$ or not), and define $\star^\sharp$ on $\mathcal{S}^\sharp$ by*

$$s \star^\sharp t = \begin{cases} t & \text{for } s = \epsilon_x \text{ and } t \text{ in } \mathcal{S}^\sharp(x,\cdot), \\ \epsilon_y & \text{for } s \text{ in } \mathcal{S}(x,y) \setminus \{\epsilon_x\} \text{ and } t = \epsilon_x \text{ or } t = s, \\ s \star t & \text{otherwise, if } s \star t \text{ is defined.} \end{cases}$$

Then $(\mathcal{S}^\sharp, \star^\sharp)$ is a unital weak RC*-system with unit family $\boldsymbol{\epsilon}_{\mathcal{S}}$.*

Proof. That $(\mathcal{S}^\sharp, \star^\sharp)$ satisfies (2.4), (2.5), and that $\boldsymbol{\epsilon}_{\mathcal{S}}$ satisfies (2.8) follows from the definitions. So one need to check that the RC law (2.6) is still satisfied although the initial operation has been changed or extended. Assume that r, s, t belong to $\mathcal{S}^\sharp$ and $(r \star^\sharp s) \star^\sharp (r \star^\sharp t)$ exists.

For $t \in \boldsymbol{\epsilon}_{\mathcal{S}}$, say $t = \epsilon_x$, we directly obtain

$$(r \star^\sharp s) \star^\sharp (r \star^\sharp \epsilon_x) = \epsilon_z = (s \star^\sharp r) \star^\sharp (s \star^\sharp \epsilon_x),$$

where z is the target of $r \star^\sharp s$. For $t \notin \boldsymbol{\epsilon}_{\mathcal{S}}$ and $s \in \boldsymbol{\epsilon}_{\mathcal{S}}$, say $s = \epsilon_x$, we obtain

$$(r \star^\sharp \epsilon_x) \star^\sharp (r \star^\sharp t) = \epsilon_y \star^\sharp (r \star^\sharp t) = r \star^\sharp t = (\epsilon_x \star^\sharp r) \star^\sharp (\epsilon_x \star^\sharp \epsilon_x),$$

where y is the target of r. The argument is symmetric for $r \in \boldsymbol{\epsilon}_{\mathcal{S}}$. So assume now $r, s, t \notin \boldsymbol{\epsilon}_{\mathcal{S}}$. For $r = s$, the expressions $(r \star^\sharp s) \star^\sharp (r \star^\sharp t)$ and $(s \star^\sharp r) \star^\sharp (t \star^\sharp r)$ formally coincide. Next, for $r \neq s = t$, we find

$$(r \star^\sharp s) \star^\sharp (r \star^\sharp t) = (r \star^\sharp s) \star^\sharp (r \star^\sharp s) = \epsilon_z = (s \star^\sharp r) \star^\sharp \epsilon_y = (s \star^\sharp r) \star^\sharp (s \star^\sharp t),$$

where y is the target of s and z the one of $r \star s$. The result is similar for $t = r$. Finally, for r, s, t pairwise distinct, we have $(s \star^\sharp t) \star^\sharp (s \star^\sharp r) = (s \star t) \star (s \star r)$ and $(t \star^\sharp s) \star^\sharp (t \star^\sharp r) = (t \star s) \star (t \star r)$, and these expressions are equal since $(\mathcal{S}, \star)$ satisfies (2.6). So, in all cases, $(\mathcal{S}^\sharp, \star^\sharp)$ satisfies (2.6).

Finally $(\mathcal{S}^\sharp, \star^\sharp)$ is unital because, by definition, $s \star^\sharp t$ may lie in $\boldsymbol{\epsilon}_{\mathcal{S}}$ only if s does, or if s and t coincide. So, for $s \neq t$, it is impossible to have $s \star^\sharp t \in \boldsymbol{\epsilon}_{\mathcal{S}}$ and $t \star s \in \boldsymbol{\epsilon}_{\mathcal{S}}$ simultaneously. □

What is (slightly) surprising in Lemma 2.12 is that we still obtain a weak RC-system when we randomly choose units among the already present elements. If we start from a weak RC-system $\mathcal{S}$ that admits units and we choose these units as new units, then the new operation coincides with the initial operation of $\mathcal{S}$ but, if we choose elements that were not units, the new units supersede the old ones. In the sequel, we shall use the construction of Lemma 2.12 to add new elements, in which case there is only one way to extend the system.

Definition 2.13 (completion). The *completion* $\hat{\mathcal{S}}$ of a weak RC-system $\mathcal{S}$ is defined to be the unital weak RC-system obtained from $\mathcal{S}$ by adding, for every object x of $\mathcal{S}$, a unit ϵ_x that does not belong to $\mathcal{S}$.

Example 2.14 (completion). Unless it has only one element, an RC-quasigroup S as considered in Chapter XIII contains no unit, since, for $s \neq \epsilon$, the conjunction of $s \star \epsilon = \epsilon$ and $s \star s = \epsilon$ would contradict the injectivity of left-translations.
Going to the completion $\widehat{S}$ means adding one new element ϵ to S and extend the operation from S to $S \cup \{\epsilon\}$ by putting $s \star \epsilon = s \star s = \epsilon$ and $\epsilon \star s = s$ for every s. In the extended system $\widehat{S}$, the element ϵ is a unit, and $\widehat{S}$ is unital. The table on the right displays the completion of the two-element RC-quasigroup of Example XIII.1.23.

	ϵ	a	b
ϵ	ϵ	a	b
a	ϵ	ϵ	a
b	ϵ	b	ϵ

In a different direction, when we consider an arbitrary weak RC-system that admits units, we can obtain a unital weak RC-system by going to a convenient quotient.

Definition 2.15 (ideal, relations $\preccurlyeq_{\mathcal{I}}$, $\equiv_{\mathcal{I}}$). If $\mathcal{S}$ is a weak RC-system that admits a unit family, an *ideal* of $\mathcal{S}$ is a subfamily $\mathcal{I}$ of $\mathcal{S}$ such that $\mathcal{I}$ includes the unit family, and, for every s in $\mathcal{I}(x, \cdot)$, the target of s is x and, for every t in $\mathcal{S}(x, \cdot)$, we have $s \star t = t$ and $t \star s \in \mathcal{I}$ (this meaning, in particular, that $s \star t$ and $t \star s$ are defined). If $\mathcal{I}$ is an ideal, we write $s \preccurlyeq_{\mathcal{I}} t$ if $t \star s$ exists and lies in $\mathcal{I}$, and $s \equiv_{\mathcal{I}} t$ for the conjunction of $s \preccurlyeq_{\mathcal{I}} t$ and $t \preccurlyeq_{\mathcal{I}} s$, that is, of $s \star t \in \mathcal{I}$ and $t \star s \in \mathcal{I}$.

The definition directly implies that, if $\mathcal{S}$ is a weak RC-system that admits units, then the latter form an ideal of $\mathcal{S}$ and this ideal is the smallest ideal in $\mathcal{S}$.

Lemma 2.16. *Assume that $\mathcal{S}$ is a weak RC-system admitting a unit family and $\mathcal{I}$ is an ideal of $\mathcal{S}$.*

(i) *The relation $\preccurlyeq_{\mathcal{I}}$ is a preordering on $\mathcal{S}$ that is compatible with $\star$ on the left in the sense that, if $s \preccurlyeq_{\mathcal{I}} t$ holds and $r \star t$ exists, then $r \star s$ also exists and we have $r \star s \preccurlyeq_{\mathcal{I}} r \star t$.*

(ii) *The relation $\equiv_{\mathcal{I}}$ is an equivalence relation on $\mathcal{S}$ that is compatible with $\star$ on both sides, in the sense that if $s \equiv_{\mathcal{I}} t$ holds, then $r \star s$ exists if and only if $r \star t$ does, in which case we have $r \star s \equiv_{\mathcal{I}} r \star t$, and similarly $s \star r$ exists if and only if $t \star r$ does, in which case we have $s \star r \equiv_{\mathcal{I}} t \star r$ (and even $s \star r = t \star r$).*

(iii) *For each object x, the family $\mathcal{I}(x, x)$ is the $\equiv_{\mathcal{I}}$-class of each of its elements.*

Proof. (i) Let r, s, t belong to $\mathcal{S}(x, \cdot)$. First, by (2.8), we have $s \star s = \epsilon_y$ where y is the target of s and ϵ_y is the corresponding unit. We deduce $s \star s \in \mathcal{I}$, whence $s \preccurlyeq_{\mathcal{I}} s$. So the relation $\preccurlyeq_{\mathcal{I}}$ is reflexive. Then assume $s \preccurlyeq_{\mathcal{I}} t \preccurlyeq_{\mathcal{I}} r$, that is, $t \star s \in \mathcal{I}$ and $r \star t \in \mathcal{I}$. Then we find

$$\begin{aligned} r \star s &= (r \star t) \star (r \star s) && \text{by definition, as } r \star t \text{ lies in } \mathcal{I} \\ &= (t \star r) \star (t \star s) && \text{by the RC law} \\ &\in \mathcal{I} && \text{by definition, as } t \star s \text{ lies in } \mathcal{I}, \end{aligned}$$

so $s \preccurlyeq_{\mathcal{I}} r$ is satisfied. Hence the relation $\preccurlyeq_{\mathcal{I}}$ is transitive, and it is a preordering.

Assume now $s \preccurlyeq_{\mathcal{I}} t$, and $r \star t$ exists. The assumption that $t \star s$ belongs to $\mathcal{I}$ implies that $(t \star r) \star (t \star s)$ is defined, and the RC law then implies that $s \star r$, $r \star s$, and $(r \star t) \star (r \star s)$ exist, and that we have $(r \star t) \star (r \star s) = (t \star r) \star (t \star s) \in \mathcal{I}$, whence $r \star s \preccurlyeq_{\mathcal{I}} r \star t$.

(ii) By definition, $\equiv_{\mathcal{I}}$ is the equivalence relation associated with the preordering $\preccurlyeq_{\mathcal{I}}$. The compatibility result with $\star$ on the left established for $\preccurlyeq_{\mathcal{I}}$ directly implies a similar result for $\equiv_{\mathcal{I}}$ and, as $\equiv_{\mathcal{I}}$ is symmetric, it takes the form of the lemma. As for the other side, assume $s \equiv_{\mathcal{I}} t$ and $s \star r$ exists. Then we obtain

$$\begin{aligned} s \star r &= (s \star t) \star (s \star r) && \text{by definition, as } s \star t \text{ lies in } \mathcal{I} \\ &= (t \star s) \star (t \star r) && \text{by the RC law} \\ &= t \star r && \text{by definition, as } t \star s \text{ lies in } \mathcal{I} \\ &\equiv_{\mathcal{I}} t \star r && \text{by the reflexivity of } \equiv_{\mathcal{I}}. \end{aligned}$$

Hence $\equiv_{\mathcal{I}}$ is compatible with operation $\star$ on both sides.

(iii) Assume that s, t lie in $\mathcal{I}(x, x)$. By definition, we have $s \star t = t \in \mathcal{I}$ and, similarly, $t \star s = s \in \mathcal{I}$, whence $s \equiv_{\mathcal{I}} t$. Conversely, assume $s \in \mathcal{I}(x, x)$ and $t \equiv_{\mathcal{I}} s$. First the source of t must be x. Then, by definition, we have $s \preccurlyeq_{\mathcal{I}} t$, whence $s \star t \in \mathcal{I}$. On the other hand, as s belongs to $\mathcal{I}$, we have $s \star t = t$. So we deduce that t lies in $\mathcal{I}$, that is, the $\equiv_{\mathcal{I}}$-class of s is exactly $\mathcal{I}(x, x)$. □

Notation 2.17 (quotient). When $\mathcal{I}$ is an ideal on a weak RC-system $\mathcal{S}$ and s lies in $\mathcal{S}$, we shall write $\mathcal{S}/\mathcal{I}$ for $\mathcal{S}/{\equiv_{\mathcal{I}}}$, and $[s]_{\mathcal{I}}$, or simply $[s]$, for the $\equiv_{\mathcal{I}}$-class of s.

Proposition 2.18 (unital quotient). *If $\mathcal{S}$ is a weak* RC*-system admitting a unit family and $\mathcal{I}$ is an ideal of $\mathcal{S}$, the operation of $\mathcal{S}$ induces a well-defined operation on $\mathcal{S}/\mathcal{I}$, and the quotient-structure is a unital weak* RC*-system.*

Proof. Lemma 2.16(ii) implies that $\star$ induces a well-defined operation on $\mathcal{S}/\mathcal{I}$. The relation $s \preccurlyeq_{\mathcal{I}} t$ requires that $t \star s$ be defined, hence that s and t share the same source. Therefore, $s \equiv_{\mathcal{I}} t$ implies that s and t have the same source and the same target. So the objects of $\mathcal{S}/\mathcal{I}$ coincide with those of $\mathcal{S}$, and the rules (2.4) and (2.5) are obeyed in $\mathcal{S}/\mathcal{I}$. As (2.6) and (2.8) are algebraic laws, they go to the quotient and, therefore, $\mathcal{S}/\mathcal{I}$ is a weak RC-system.

Let $\epsilon_{\mathcal{S}}$ be the unit family in $\mathcal{S}$. Then the $\mathcal{I}$-classes of the elements of $\epsilon_{\mathcal{S}}$ make a unit family in $\mathcal{S}/\mathcal{I}$. Moreover, the latter is unital: indeed, if we have $[s] \star [t] = [t] \star [s] = [\epsilon_y]$, then $s \star t$ and $t \star s$ lie in $\mathcal{I}$ by the definition of the operation on $\mathcal{S}/\mathcal{I}$ and, therefore, the $\mathcal{I}$-classes of s and t are equal. □

Corollary 2.19 (maximal unital quotient). *Every weak* RC*-system $\mathcal{S}$ admitting a unit family $\epsilon_{\mathcal{S}}$ admits a maximal unital quotient, namely $\mathcal{S}/\epsilon_{\mathcal{S}}$.*

Proof. As $\epsilon_{\mathcal{S}}$ is an ideal in $\mathcal{S}$, the quotient $\mathcal{S}/\epsilon_{\mathcal{S}}$ is a unital weak RC-system. In the other direction, assume that $\mathcal{S}^\sharp$ is a unital weak RC-system and ϕ is a functor from $\mathcal{S}$ to $\mathcal{S}^\sharp$. Let s, t be $\epsilon_{\mathcal{S}}$-equivalent elements of $\mathcal{S}$. By assumption, $s \star t = t \star s = \epsilon_y$ holds for some object y, whence $\phi(s) \star \phi(t) = \phi(t) \star \phi(s) = \epsilon_{\phi(x)}$ in $\mathcal{S}^\sharp$. As $\mathcal{S}^\sharp$ is unital, this implies $\phi(s) = \phi(t)$, that is, ϕ factors through $\mathcal{S}/\epsilon_{\mathcal{S}}$. □

Corollary 2.19 implies, in particular, that, if $\mathcal{S}$ is unital with unit family $\epsilon_{\mathcal{S}}$, then $\mathcal{S}/\epsilon_{\mathcal{S}}$ is isomorphic to $\mathcal{S}$.

2.3 The structure category of a weak RC-system

In Chapter XIII, we associated with every RC-*quasigroup* a structure monoid. Mimicking the definition in the current context would naturally lead to considering the category $\langle \mathcal{S} \mid \{s(s \star t) = t(t \star s) \mid s \neq t \in \mathcal{S}\}\rangle^+$ for every weak RC-system $\mathcal{S}$. This however is not convenient: typically, the units of $\mathcal{S}$ cannot coincide with the identity elements of the category and there is no natural embedding of $\mathcal{S}$ in the category. We shall therefore take another route, first extending the operation $\star$ from $\mathcal{S}$ to the free category $\mathcal{S}^*$ and then considering an appropriate quotient of this category. We recall that, if $\mathcal{S}$ is any precategory, $\mathcal{S}^*$ is the free category based on $\mathcal{S}$, realized as the category of all $\mathcal{S}$-paths equipped with concatenation (Definition II.1.28). Also, for x in $\mathit{Obj}(\mathcal{C})$, the empty path with source and target x is denoted by ε_x, and we write $\varepsilon_\mathcal{S}$ for $(\varepsilon_x)_{x \in \mathit{Obj}(\mathcal{S})}$.

Definition 2.20 (grid). Assume that $\mathcal{S}$ is a weak RC-system. For u, v in $\mathcal{S}^*$ nonempty and sharing the same source, say $u = s_1|\cdots|s_p$, $v = t_1|\cdots|t_q$, the (u, v)-*grid* is (when it exists) the double $\mathcal{S}$-sequence $(s_{i,j})_{1 \leqslant i \leqslant p, 0 \leqslant j \leqslant q}$, $(t_{i,j})_{0 \leqslant i \leqslant p, 1 \leqslant j \leqslant q}$ satisfying the initial conditions $s_{i,0} = s_i$ and $t_{0,j} = t_j$, and the inductive relations

$$s_{i,j} = t_{i-1,j} \star s_{i,j-1} \quad \text{and} \quad t_{i,j} = s_{i,j-1} \star t_{i-1,j}. \tag{2.21}$$

for $1 \leqslant i \leqslant p$ and $1 \leqslant j \leqslant q$. We then put $u \star v = t_{p,1}|\cdots|t_{p,q}$. For u in $\mathcal{S}^*$ with source x and target y, the (u, ε_x)-grid and the (ε_x, u)-grid are defined to be u, and we put $\varepsilon_x \star u = u$, $v \star \varepsilon_x = \varepsilon_y$, and $\varepsilon_x \star \varepsilon_x = \varepsilon_x$.

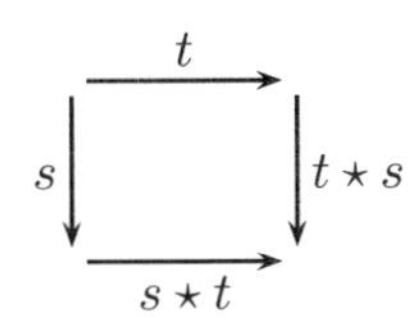

The current grids are reminiscent of the many grids already considered in this text, in particular in Chapters II and IV. The (u, v)-grid is the rectangular diagram obtained by starting with a vertical sequence of arrows labeled u and a horizontal sequence of arrows labeled v and using the operation $\star$ to construct square tiles as on the right.

By construction, the length of $u \star v$ is always that of v, and the (v, u)-grid is the image of the (u, v)-grid under a diagonal symmetry: if $(s_{i,j})_{1 \leqslant i \leqslant p, 0 \leqslant j \leqslant q}$, $(t_{i,j})_{0 \leqslant i \leqslant p, 1 \leqslant j \leqslant q}$ is the (u, v)-grid, then $v \star u$ is the path $s_{1,q}|\cdots|s_{p,q}$.

Lemma 2.22. *If $\mathcal{S}$ is a weak* RC*-system, then $\mathcal{S}^*$ is a weak* RC*-system with unit family $\varepsilon_\mathcal{S}$, and the following mixed relations hold:*

$$u \star (v_1|v_2) = u \star v_1|(v_1 \star u) \star v_2 \quad \text{and} \quad (v_1|v_2) \star u = v_2 \star (v_1 \star u). \tag{2.23}$$

Proof. The $(u, v_1|v_2)$-grid is a concatenation of the (u, v_1)-grid and the $(v_1 \star u, v_2)$-grid as shown in the diagram

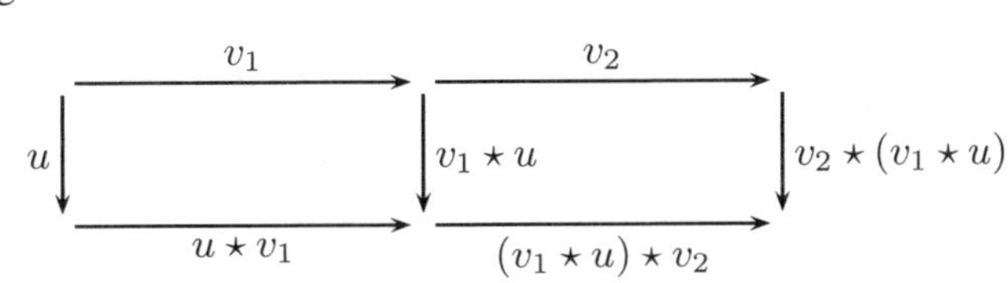

Then the relations of (2.23) directly follow.

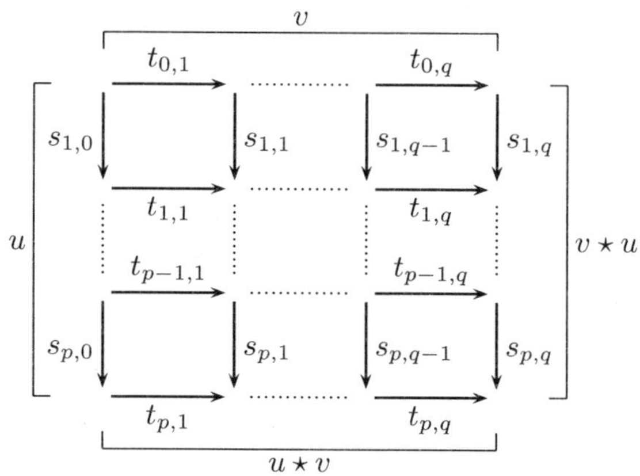

Figure 3. The (u, v)-grid: we start from u and v, the left and top sides of the diagram, and fill it in pq steps using $\star$-tiles, that is, complete each open pattern s, t with $s \star t$ and $t \star s$.

As for the right-cyclic law RC, the construction is exactly same as in the proof of Lemma II.4.60. By assumption, we have small cubes with $\mathcal{S}$-labeled edges, and we want to construct a large $\mathcal{S}^*$-labeled cube as the one after Definition 2.3. So, starting from the common source of the three labels u, v, w, we stack $\mathcal{S}$-cubes one after the other. If the lengths of u, v, w are p, q, r, exactly pqr cubes are used. □

The weak RC-system $\mathcal{S}^*$ is never unital: (2.8) is not satisfied since, for every s in $\mathcal{S}(x, \cdot)$, we have $s \star s = \epsilon_x$, and the length one path ϵ_x is *not* equal to the length zero path ε_x. This however is easily repaired at the expense of going to a convenient quotient.

Lemma 2.24. *If $\mathcal{S}$ is a weak* RC*-system admitting a unit family and $\mathcal{I}$ is an ideal of $\mathcal{S}$, the family $\mathcal{I}^*$ of all $\mathcal{I}$-paths is an ideal of $\mathcal{S}^*$, the relation $\equiv_{\mathcal{I}^*}$ is compatible with concatenation on both sides, and mapping $[s]_{\mathcal{I}}$ to $[s]_{\mathcal{I}^*}$ defines an injection from $\mathcal{S}/\mathcal{I}$ into $\mathcal{S}^*/\mathcal{I}^*$.*

Proof. First, every empty path ε_x belongs to $\mathcal{I}^*$ by definition. Assume $u \in \mathcal{S}^p$ and $v \in \mathcal{I}^q$. Let $((s_{i,j})_{i,j}, (t_{i,j})_{i,j})$ be the (u, v)-grid. As $\mathcal{I}$ is an ideal, a straightforward induction gives $s_{i,j} = s_i$ for $0 \leqslant j \leqslant q$, for each i, and $t_{i,j} \in \mathcal{I}$ for $0 \leqslant i \leqslant p$, for each j. So we conclude that $u \star v$, which is $t_{p,1}|\cdots|t_{p,q}$, belongs to $\mathcal{I}^*$, and that $v \star u$, which is $s_{1,q}|\cdots|s_{p,q}$, is equal to u. So $\mathcal{I}^*$ is an ideal of $\mathcal{S}^*$.

Assume now $u' \equiv_{\mathcal{I}} u$ and $v' \equiv_{\mathcal{I}} v$. Then the diagram on the right shows that $u'|v' \equiv_{\mathcal{I}} u|v$ holds as well. Finally, assume $s \equiv_{\mathcal{I}^*} t$ with s, t in $\mathcal{I}$. By definition, this means that the length-one paths $s\star t$ and $t\star s$ belong to $\mathcal{I}^*$. By construction, the latter belong to $\mathcal{I}^*$ if and only if $s \star t$ and $t \star \mathcal{I}$ belong to $\mathcal{I}$, hence if and only if $s \equiv_{\mathcal{I}} t$ holds. □

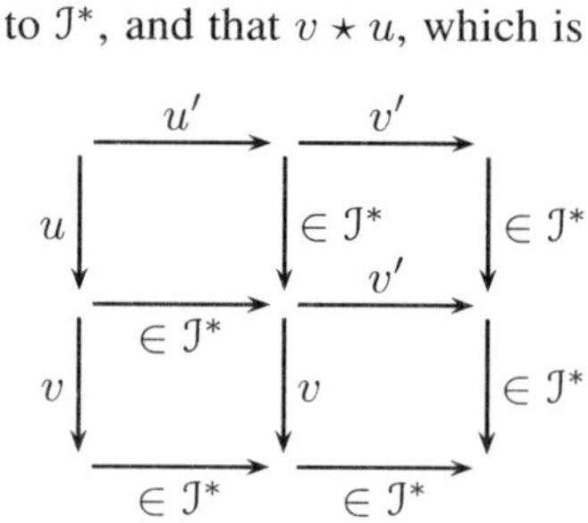

We are now ready to introduce the expected structure category.

Definition 2.25 (structure category). If $\mathcal{S}$ is a weak RC-system admitting a unit family $\boldsymbol{\epsilon}_{\mathcal{S}}$, the *structure category* $\mathcal{C}(\mathcal{S})$ of $\mathcal{S}$ is defined to be the category $\mathcal{S}^*/\boldsymbol{\epsilon}_{\mathcal{S}}^*$.

Of course, if there is only one object, we rather speak of a *structure monoid.* Definition 2.25 is legal since, by Proposition 2.18, the structure $\mathcal{S}^*/\boldsymbol{\epsilon}_{\mathcal{S}}^*$ is a unital weak RC-system, and, by Lemma 2.24, it is equipped with a well-defined partial associative multiplication. In the sequel, we write $\equiv$ for $\equiv_{\boldsymbol{\epsilon}_{\mathcal{S}}^*}$, and $[u]$ for the $\equiv$-class of an $\mathcal{S}$-path u. Saying that $u \equiv v$ holds means that the (u, v)-grid is defined and finishes with units everywhere. Note that $\epsilon_x \equiv \varepsilon_x$ holds for every object x: indeed, by definition, we have $\epsilon_x \star \varepsilon_x = \varepsilon_x$ and $\varepsilon_x \star \epsilon_x = \epsilon_x$.

Before investigating the properties of structure categories, we begin with an easy preliminary remark. In Definition 2.25, we did not require that the RC-system $\mathcal{S}$ be unital. However, the structure category of $\mathcal{S}$ only depends on the maximal unital quotient of $\mathcal{S}$:

Lemma 2.26. *Assume that $\mathcal{S}$ is a weak* RC*-system with unit family $\boldsymbol{\epsilon}_{\mathcal{S}}$. Then the categories $\mathcal{C}(\mathcal{S})$ and $\mathcal{C}(\mathcal{S}/\boldsymbol{\epsilon}_{\mathcal{S}})$ are isomorphic.*

Proof. Let π be the canonical projection from $\mathcal{S}$ to $\mathcal{S}/\boldsymbol{\epsilon}_{\mathcal{S}}$ and π^* be the componentwise extension of π to $\mathcal{S}$-paths. If u, v are $\mathcal{S}$-paths and $u \equiv v$ holds, then we have $u \star v \in \boldsymbol{\epsilon}_{\mathcal{S}}^*$ and $v \star u \in \boldsymbol{\epsilon}_{\mathcal{S}}^*$, whence $\pi^*(u) \star \pi^*(v) \in \pi^*(\boldsymbol{\epsilon}_{\mathcal{S}}^*)$ and $\pi^*(v) \star \pi^*(u) \in \pi^*(\boldsymbol{\epsilon}_{\mathcal{S}}^*)$, and $\pi^*(u) \equiv \pi^*(v)$. Hence π^* induces a functor from $\mathcal{C}(\mathcal{S})$ to $\mathcal{C}(\mathcal{S}/\boldsymbol{\epsilon}_{\mathcal{S}})$. This functor is obviously surjective. It is also injective. Indeed, assume $\pi^*(u) \equiv \pi^*(v)$. Applying the definition, we deduce $\pi^*(u \star v) \in \pi^*(\boldsymbol{\epsilon}_{\mathcal{S}}^*)$ and $\pi^*(v \star u) \in \pi^*(\boldsymbol{\epsilon}_{\mathcal{S}}^*)$. By Lemma 2.16(iii), $\pi(\epsilon_x)$ consists of ϵ_x alone, so the previous relations imply $u \star v \in \boldsymbol{\epsilon}_{\mathcal{S}}^*$ and $v \star u \in \boldsymbol{\epsilon}_{\mathcal{S}}^*$, whence $u \equiv v$. □

By Corollary 2.19, a weak RC-system of the form $\mathcal{S}/\boldsymbol{\epsilon}_{\mathcal{S}}$ is unital, so Lemma 2.26 shows that we lose no generality in restricting to unital weak RC-systems when introducing structure categories. The following statement summarizes some properties of these categories.

Proposition 2.27 (structure category). *Assume that $\mathcal{S}$ is a unital weak* RC*-system with unit family $\boldsymbol{\epsilon}_{\mathcal{S}}$ and $\mathcal{C}$ is the associated structure category.*

(i) *The map $s \mapsto [s]$ is an injection from $\mathcal{S}$ to $\mathcal{C}$, and $\mathcal{C}$ admits the presentation*

$$\langle \mathcal{S} \mid \{s(s \star t) = t(t \star s) \mid s \neq t \in \mathcal{S} \backslash \boldsymbol{\epsilon}_{\mathcal{S}} \text{ and } s \star t \text{ is defined}\} \cup \{\epsilon_x = 1_x \mid x \in \mathit{Obj}(\mathcal{S})\}\rangle^+, \tag{2.28}$$

and satisfies a quadratic isoperimetric inequality with respect to this presentation.

(ii) The category $\mathcal{C}$ has no nontrivial invertible element, it is left-cancellative, and it admits unique conditional right-lcms. Moreover, $g\backslash h = g \star h$ holds for all g, h in $\mathcal{C}$ that admit a common right-multiple.

(iii) If, in addition, $\mathcal{S}$ satisfies the condition

(2.29) $$\text{for } s \notin \epsilon_{\mathcal{S}}, \text{ the element } s \star t \text{ lies in } \epsilon_{\mathcal{S}} \text{ only for } t = s,$$

the category $\mathcal{C}$ admits the presentation

(2.30) $$\langle \mathcal{S}\backslash\epsilon_{\mathcal{S}} \mid \{s(s \star t) = t(t \star s) \mid s \neq t \in \mathcal{S}\backslash\epsilon_{\mathcal{S}} \text{ and } s \star t \text{ is defined}\}\rangle^{+},$$

it is strongly Noetherian, it admits left-gcds, the atoms of $\mathcal{C}$ are the elements of $\mathcal{S} \setminus 1_{\mathcal{S}}$, and $\mathcal{C}$ admits a smallest Garside family including $1_{\mathcal{C}}$, namely the family $\widehat{\mathcal{S}}$ of all right-lcms of finite subfamilies of $\mathcal{S}$.

Proof. (i) Assume $[s] = [t]$ in $\mathcal{C}$. By definition, this means that the (s, t)-grid exists and finishes with units. Thus $s \star t$ and $t \star s$ must be defined, and we have $s \star t = t \star s = \epsilon_y$ for some object y. As $\mathcal{S}$ is unital, (2.9) implies $s = t$. Building on this, we shall identity every element of $\mathcal{S}$ with its image in $\mathcal{C}$, that is, consider $\mathcal{S}$ as a subfamily of $\mathcal{C}$.

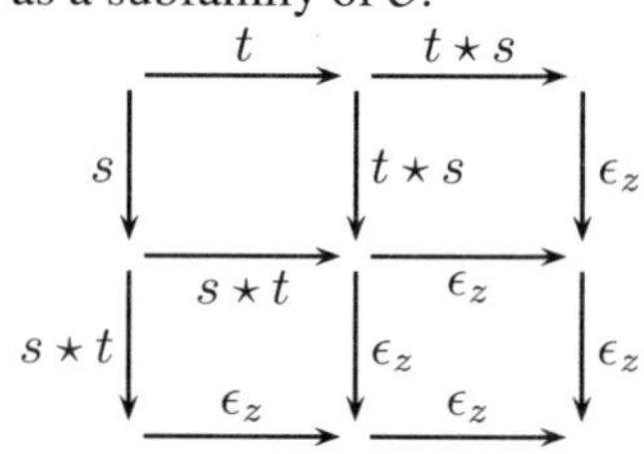

By construction, $\mathcal{S}$ generates the free category $\mathcal{S}^*$, hence (the image of) $\mathcal{S}$ generates $\mathcal{C}$. Assume that s, t lie in $\mathcal{S}$ and $s \star t$ is defined. Then, by (2.5), $t \star s$ is defined as well and, by (2.4), $s|s \star t$ and $t|t \star s$ are $\mathcal{S}$-paths with the same target, say z. Then the diagram on the right is well-defined and it gives the equivalence $s|s \star t \equiv t|t \star s$. So the relations of (2.28) are valid in $\mathcal{C}$.

Assume now that u, v are $\equiv$-equivalent $\mathcal{S}$-paths. By definition, this means that the (u, v)-grid is defined and finishes with unit elements. By construction, this grid consists of elementary squares, each of which correspond to a relation of one of the following four types: $s(s \star t) = t(t \star s)$, $s\epsilon_y = s\epsilon_y$, $s\epsilon_y = \epsilon_x s$, or $\epsilon_x \epsilon_x = \epsilon_x \epsilon_x$. The relations of the first type belong to the list (2.28), those of the second and fourth types are tautologies, and those of the third type follow from the relations $\epsilon_x = 1_x$. Writing z for the target of u and p, q for the length of u and v, this shows that the equivalence of $u|(\epsilon_z)^q$ and $v|(\epsilon_z)^p$ can be established using at most pq relations from (2.28). As we then consider $\epsilon_{\mathcal{S}}^*$-classes, all arrows labeled ϵ_z collapse, and we have a similar result for u and v. This shows that (2.28) is a presentation of $\mathcal{C}$ and that $\mathcal{C}$ satisfies a quadratic isoperimetric inequality with respect to this presentation.

(ii) Once we have a presentation of $\mathcal{C}$, the proof goes on exactly as for Proposition XIII.2.8. First, the presentation (2.28) is right-complemented, associated with the short syntactic right-complement θ defined by

(2.31) $$\theta(s,t) = \begin{cases} s \star t & \text{for } s \neq t \text{ and } s \star t \text{ defined,} \\ \varepsilon_x & \text{for } s = t \text{ with source } x, \end{cases}$$

and the assumption that $(\mathcal{S}, \star)$ satisfies the RC law (2.6) implies that θ satisfies the sharp right-cube condition for every triple of pairwise distinct elements of $\mathcal{S}$. Then Proposition II.4.16 (right-complemented) implies that $\mathcal{C}$ has no nontrivial invertible element, is left-cancellative, and admits unique conditional right-lcms. For $s \neq t$ in $\mathcal{S}$, the right-lcm of s and t is represented by $s\theta(s,t)$, which, by (2.31), is $s(s \star t)$. This means that $s \backslash t$ equals $s \star t$ in this case.

(iii) Assume that (2.29) is satisfied. Then no element ϵ_x appears in the first list of relations in (2.28), and, therefore, we can simply remove the generators $\epsilon_{\mathcal{S}}$, which are to be collapsed, thus obtaining for $\mathcal{C}$ the presentation is (2.30). Now, every relation in the latter involves two paths of length two, so the presentation is homogeneous. Hence, by Proposition II.2.32 (homogeneous), $\mathcal{C}$ is strongly Noetherian, and the height of an element of $\mathcal{C}$ is the common length of all paths that represent it and contain no entry in $\epsilon_{\mathcal{S}}$. In particular, the atoms of $\mathcal{C}$ are the elements of height one, namely the elements of $\mathcal{S} \setminus \epsilon_{\mathcal{S}}$.

Then, by Corollary IV.2.41 (smallest Garside), $\mathcal{C}$ admits a smallest Garside family that includes $\mathbf{1}_{\mathcal{C}}$, namely the closure of the atoms under right-lcm and right-complement. Now, as in the case of Lemma XIII.2.28, the closure $\widehat{\mathcal{S}}$ of $\mathcal{S} \setminus \epsilon_{\mathcal{S}}$ under right-lcm is closed under right-lcm by definition, and the assumption that $\mathcal{S}$ is closed under right-complement implies that $\widehat{\mathcal{S}}$ is closed under right-complement too. Hence, $\mathcal{S}$ is the smallest Garside family including $\mathbf{1}_{\mathcal{C}}$ in $\mathcal{C}$. □

Before discussing examples, let us immediately explain the connection between the current construction and those of Chapter XIII.

Corollary 2.32 (coherence). *If S is an RC-quasigroup, the structure monoid of S as introduced in Definition XIII.2.5 coincides with the structure monoid of its completion $\widehat{S}$ as defined in Definition* 2.25.

Proof. In the case of $\widehat{S}$, (2.29) is satisfied by definition, so (2.30) is a presentation of the associated structure in the sense of Definition 2.25. Now, (2.30) is also the defining presentation involved in Definition XIII.2.5. □

Example 2.33 (structure monoid). Consider the two RC-systems of Example 2.2. In both cases, we recover the expected monoid. In the first case, (2.28) gives $\mathtt{ab} = \mathtt{c1}$, whence $\mathtt{ab} = \mathtt{c}$ when 1 is collapsed, and, similarly, $\mathtt{ba} = \mathtt{d}$, so, at the end, the presentation is equivalent to $\langle \mathtt{a}, \mathtt{b} \mid \mathtt{aba} = \mathtt{bab} \rangle^+$, a presentation of B_3^+. In the second case, we similarly retrieve the presentation $\langle \mathtt{a}, \mathtt{b}, \mathtt{c} \mid \mathtt{ab} = \mathtt{bc} = \mathtt{ca} \rangle^+$of B_3^{+*}. Then both monoids are eligible for Proposition 2.27, which states in particular that they are left-cancellative and admit unique conditional right-lcms.

Now, in the first table, the unit 1 appears elsewhere than in the first column and the diagonal ($\mathtt{c} \star \mathtt{a} = 1$ holds), so (2.29) fails, and we cannot assert that the properties of Proposition 2.27(iii) are true—although, in this case, they are.

By contrast, in the second table, (2.29) holds and, therefore, Proposition 2.27 guarantees in addition that the monoid is Noetherian and admits a smallest Garside family containing 1, namely the closure of the involved set S under right-lcm: in the current case, this family is $\{1, \mathtt{a}, \mathtt{b}, \mathtt{c}\} \cup \{\Delta\}$, where Δ is the common value of $\mathtt{ab}$, $\mathtt{bc}$, and $\mathtt{ca}$. Note that the Garside family $\mathrm{Div}(\Delta)$ has 5 elements: owing to the results of Chapter XIII, this shows that there is no way of removing 1 from the table and completing it into an RC-quasigroup since, otherwise, the cardinality of the Garside family should be a power of 2.

When (2.29) is not satisfied, the results of Proposition 2.27 may seem frustrating and, in particular, it is natural to wonder how to recognize Noetherianity from a (weak) RC-system. We have no complete answer, but some conditions are clearly sufficient.

Proposition 2.34 (Noetherianity). *Assume that $\mathcal{S}$ is a weak* RC*-system that admits a unit family $\epsilon_{\mathcal{S}}$ and there exists a map $\lambda : \mathcal{S} \to \mathbb{N}$ such that $\lambda(s) = 0$ holds if and only if s lies in $\epsilon_{\mathcal{S}}$ and, for all s, t in $\mathcal{S}$ such that $s\star t$ is defined, we have $\lambda(s)+\lambda(s\star t) = \lambda(t)+\lambda(t\star s)$. Then the structure category $\mathcal{C}$ of $\mathcal{S}$ is strongly Noetherian, it admits left-gcds, the atoms of $\mathcal{C}$ are the elements of $\mathcal{S} \setminus 1_{\mathcal{S}}$, and $\mathcal{C}$ admits a smallest Garside family including $1_{\mathcal{C}}$, namely the family of all right-lcms of finite subfamilies of $\mathcal{S}$.*

Proof. For u an $\mathcal{S}$-path, say $u = s_1|\cdots|s_p$, define $\lambda^*(u) = \lambda(s_1) + \cdots + \lambda(s_p)$. Then λ^* induces a well-defined map on $\mathcal{C}$, which by definition is a strong Noetherianity witness. Hence $\mathcal{C}$ is (strongly) Noetherian and, from there, we repeat the proof of Proposition 2.27(iii). ☐

Example 2.35 (Noetherianity). Consider the first table of Examples 2.2 and 2.33. Define λ by $\lambda(1) = 0$, $\lambda(\mathtt{a}) = \lambda(\mathtt{b}) = 1$, and $\lambda(\mathtt{c}) = \lambda(\mathtt{d}) = 2$. Then the assumptions of Proposition 2.34 are satisfied, and, this time, we can indeed deduce that the smallest Garside family containing 1 in B_3^+ is the 6-element family $\{1, \mathtt{a}, \mathtt{b}, \mathtt{c}, \mathtt{d}\} \cup \{\Delta\}$ with $\Delta = \mathtt{ad} = \mathtt{bc} = \mathtt{ca} = \mathtt{db}$.

Of course, the assumptions of Proposition 2.34 need not be always satisfied, even in a finite RC-system. For instance, the structure monoid associated with the RC-system on the right admits the presentation $\langle \mathtt{a}, \mathtt{b} \mid \mathtt{ab} = \mathtt{b}\rangle^+$, and it is neither Noetherian nor right-cancellative.

	1	a	b
1	1	a	b
a	1	1	b
b	1	1	1

In the case when the considered RC-system is finite and the operation $\star$ is defined everywhere, the Garside family of Proposition 2.34 is right-bounded by the right-lcm Δ of $\mathcal{S}$. From there, the results of Chapter V possibly apply: for instance, by Proposition V.1.36 (right-cancellativity I), right-cancellativity reduces to the injectivity of the functor ϕ_Δ, and therefore to that of the map $g \mapsto g \star \Delta$. We shall not give more details in this direction here.

3 The braid group of $\mathbb{Z}^n$

In a completely different direction, we now describe a group that can be seen as an analog of the braid group B_n in which the role of the symmetric group $\mathfrak{S}_n$ is played by a linear group $\mathrm{GL}(n, \mathbb{Z})$. The group was introduced and investigated by D. Krammer in [164]. The construction relies on the scheme of Subsection VI.3.3 and consists in defining a germ using a lattice ordering.

3.1 Ordering orders

Restarting from Example VI.3.18 (germ from lattice), we consider the weak order on the symmetric group $\mathfrak{S}_n$. It will be useful to consider here $\mathfrak{S}_n$ as the quotient of the braid group B_n (Reference Structure 2, page 5) by the subgroup PB_n made of all *pure* braids, defined as those braids whose associated permutation is the identity. We shall now use a similar scheme with $\mathfrak{S}_n$ replaced by $\mathrm{GL}(n, \mathbb{Z})$ and PB_n replaced by the subgroup H of $\mathrm{GL}(n, \mathbb{Z})$ consisting of those elements that preserve the standard lexicographic order on $\mathbb{Z}^n$. A lattice ordering on $\mathrm{GL}(n, \mathbb{Z})/H$ is then defined using a lattice structure on the linear orders of $\mathbb{Z}^n$ that arise from the standard lexicographic order via the action of some element of $\mathrm{GL}(n, \mathbb{Z})$. Thus the general idea is to replace the action of the symmetric group $\mathfrak{S}_n$ on $\{1, ..., n\}$ with the action of $\mathrm{GL}(n, \mathbb{Z})$ on $\mathbb{Z}^n$.

The first step of the construction is to define a lattice ordering on the family of all linear orders of a set S. In the sequel, as in Chapter XII, an order on S is viewed as a set of pairs, hence a subset of S^2.

Definition 3.1 (order on orders). Let $\mathcal{O}(S)$ be the family of all linear orders on a set S. For O_*, O in $\mathcal{O}(S)$, we put

$$\mathrm{Inv}_{O_*}(O) = \{(x, y) \in S^2 \mid (x, y) \in O_* \text{ and } (y, x) \in O\},$$

and define $O_1 \preccurlyeq_{O_*} O_2$ to mean $\mathrm{Inv}_{O_*}(O_1) \subseteq \mathrm{Inv}_{O_*}(O_2)$.

Thus $\mathrm{Inv}_{O_*}(O)$ is the family of all inversions of O with respect to the reference order O_*. We say that O_1 is smaller than O_2 if all inversions of O_1 are inversions of O_2. If S is $\{1, ..., n\}$ and O_* is the standard order, at the expense of identifying a permutation of $\{1, ..., n\}$ with the linear order it induces, we retrieve the usual weak order on $\mathfrak{S}_n$.

Lemma 3.2. *For every set S and every linear order O_* on S, the relation $\preccurlyeq_{O_*}$ is a lattice ordering on $\mathcal{O}(S)$.*

Proof (*sketch*). That $\preccurlyeq_{O_*}$ is a partial ordering is straightforward, and the point is to show that least upper bounds exist. Now one observes that a set $\mathrm{Inv}_{O_*}(O)$ is always transitive (meaning that, if it contains (x, y) and (y, z), then it also contains (x, z)) and co-transitive (meaning that its complement in S^2 is transitive) and, conversely, every subset of S^2 that is transitive and co-transitive is the set $\mathrm{Inv}_{O_*}(O)$ for some linear order O. Then one easily checks that the least upper bound of two orders O_1, O_2 is the linear order O such that $\mathrm{Inv}_{O_*}(O)$ is the smallest transitive co-transitive subset of S^2 that includes $\mathrm{Inv}_{O_*}(O_1) \cup \mathrm{Inv}_{O_*}(O_2)$. The argument works not only for two orders, but also for an arbitrary family of orders. Greatest lower bounds can then be obtained as least upper bounds of the family of all common lower bounds. □

Note that the smallest element in the lattice $(\mathcal{O}(S), \preccurlyeq_{O_*})$ is O_*, since $\mathrm{Inv}_{O_*}(O_*)$ is empty, whereas the largest element is the reversed order O_*^{rev}, since $\mathrm{Inv}_{O_*}(O_*^{\mathrm{rev}})$ is all of S^2.

3.2 Lexicographic orders of $\mathbb{Z}^n$

The second step of the construction consists in fixing $S = \mathbb{Z}^n$ and considering a particular family of linear orders called lexicographic. Hereafter we use $x, y, ...$ for the elements of $\mathbb{Z}^n$. The latter are viewed as $\mathbb{Z}$-valued sequences of length n, and we write x_i for the ith entry in x. We let $\mathrm{GL}(n, \mathbb{Z})$ act on $\mathbb{Z}^n$ on the right, thus writing xg for the image of x under the linear transformation g; accordingly, we assume that the product in $\mathrm{GL}(n, \mathbb{Z})$ is reversed composition.

Definition 3.3 (lexicographic). Let the *standard lexicographic order* on $\mathbb{Z}^n$ be

$$O_{\text{lex}} = \{\ (x, y) \in (\mathbb{Z}^n)^2 \mid \exists i \leqslant n\ (x_i < y_i \text{ and } \forall j < i\ (x_j = y_j))\ \}.$$

We put $\mathcal{O}_{\text{lex}}(\mathbb{Z}^n) = \{O_{\text{lex}} \bullet g \mid g \in \mathrm{GL}(n, \mathbb{Z})\}$, with $O \bullet g = \{(xg, yg) \mid (x, y) \in O\}$. The elements of $\mathcal{O}_{\text{lex}}(\mathbb{Z}^n)$ are called *lexicographic* orders.

Example 3.4 (lexicographic). We mentioned in Example XII.1.15 that, for $n = 2$, the positive cones of the invariant orders on $\mathbb{Z}^n$ are the intersections of $\mathbb{Z}^2$ with open half-planes completed with one of the two possible half-lines of the border. The lexicographic orders are those for which the slope of the border line is rational. See Figure 4.

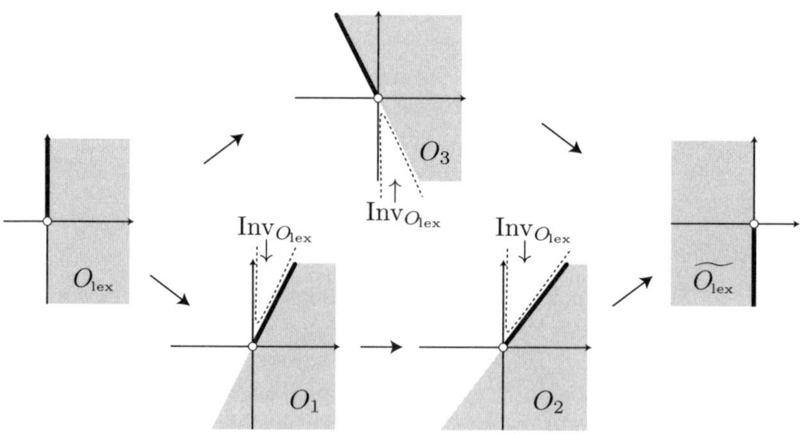

Figure 4. Ordering the left-invariant orderings on $\mathbb{Z}^2$ with respect to the lexicographical ordering O_{lex}: one compares the "inversions" with respect to O_{lex} (the zone delimited in dashed lines), and an ordering is smaller than another if the set of inversions of the former is included in the set of inversions of the latter, for instance, O_1 is smaller than O_2. The greatest lower bound of O_1 and O_3 is O_{lex}, because $\mathrm{Inv}(O_1)$ and O_3 are disjoint; similarly, the least upper bound of O_1 and O_3 is the maximal element, namely the reversing lexicographical ordering.

The group $\mathrm{GL}(n, \mathbb{Z})$ acts on $\mathcal{O}_{\text{lex}}(\mathbb{Z}^n)$ on the right: by construction, for all g, h in $\mathrm{GL}(n, \mathbb{Z})$ and O in $\mathcal{O}_{\text{lex}}(\mathbb{Z}^n)$, we have $(O \bullet g) \bullet h = O \bullet gh$. Beware that, with the above definition, if we write O and $O \bullet g$ as binary relations, say $<_O$ and $<_{O \bullet g}$, then $x <_{O \bullet g} y$ is equivalent to $xg^{-1} <_O y^{-1}$.

Lemma 3.5. *For every lexicographic order O_* on $\mathbb{Z}^n$, the restriction of $\preccurlyeq_{O_*}$ to $\mathcal{O}_{\text{lex}}(\mathbb{Z}^n)$ is a lattice ordering on $\mathcal{O}_{\text{lex}}(\mathbb{Z}^n)$.*

Proof (sketch). The point is to show that, if O_1 and O_2 are lexicographic, then their sup and their inf as determined in Lemma 3.2 still are lexicographic. This will follow from a characterization of lexicographic orders that we describe now.

The lexicographic order on $\mathbb{Z}^n$ is (left- and right-) invariant with respect to the additive group structure in the sense of Chapter XII, that is, $(x, y) \in O_{\text{lex}}$ is equivalent to $(x + z, y + z) \in O_{\text{lex}}$ for every z. As the elements of $\mathrm{GL}(n, \mathbb{Z})$ preserve addition, every lexicographic order on $\mathbb{Z}^n$ is invariant too.

On the other hand, if O is an invariant linear order on $\mathbb{Z}^n$, then $(x, y) \in O$ is equivalent to $(qx, qy) \in O$ for every q. It follows that O admits a unique well-defined extension $\widehat{O}$ to a linear order on $\mathbb{Q}^n$: say that (x, y) lies in $\widehat{O}$ if (qx, qy) lies in O for q large enough to ensure $qx \in \mathbb{Z}^n$ and $qy \in \mathbb{Z}^n$. Then the positive cone of $\widehat{O_{\text{lex}}}$ is piecewise linear in $\mathbb{Q}^n$, this meaning that it is a finite intersection of half-spaces $\{x \mid x \bullet g > 0\}$ or $\{x \mid x \bullet g \geqslant 0\}$ with g in $\mathrm{GL}(n, \mathbb{Q})$. This immediately implies that, for every lexicographic order O, the positive cone of $\widehat{O}$ is piecewise linear.

Now, using an explicit description of all invariant orders on $\mathbb{Q}^n$, one can show that, conversely, every linear order O on $\mathbb{Z}^n$ that is invariant and such that the positive cone of $\widehat{O}$ is piecewise linear must be lexicographic. It is then easy to check that, if O is the least upper bound of two lexicographic orders with respect to $\preccurlyeq_{O_*}$ and O_* is lexicographic, then O is invariant and the positive cone of $\widehat{O}$ is piecewise linear. Hence, this least upper bound is lexicographic. The result for greatest lower bounds follows using the duality that stems from reversing the orders. □

3.3 A lattice ordering on $\mathrm{GL}(n, \mathbb{Z})$

The third step of the construction is to define a lattice ordering on a certain quotient of $\mathrm{GL}(n, \mathbb{Z})$. Hereafter, we use $<_{\text{lex}}$ for the lexicographic order on $\mathbb{Z}^n$: so $x <_{\text{lex}} y$ is just the same as $(x, y) \in O_{\text{lex}}$. Also, we fix O_{lex} as the reference order and write $\preccurlyeq$ for $\preccurlyeq_{O_{\text{lex}}}$ on $\mathcal{O}_{\text{lex}}(\mathbb{Z}^n)$. We write $\underline{0}$ for $(0, \dots, 0)$.

Definition 3.6 (relation $\lesssim$ on $\mathrm{GL}(n, \mathbb{Z})$). For g, h in $\mathrm{GL}(n, \mathbb{Z})$, we declare that $g \lesssim h$ is true if we have

$$\forall z >_{\text{lex}} \underline{0} \; (zg <_{\text{lex}} \underline{0} \;\Rightarrow\; zh <_{\text{lex}} \underline{0}). \tag{3.7}$$

Lemma 3.8. (i) *For g, h in $\mathrm{GL}(n,\mathbb{Z})$, the relation $g \lesssim h$ is equivalent to $O_{\text{lex}} \bullet g \preccurlyeq O_{\text{lex}} \bullet h$.*

(ii) *The relation $\lesssim$ is reflexive and transitive. Write $g \sim h$ if both $g \lesssim h$ and $h \lesssim g$ hold. Then $g \sim h$ is equivalent to $g^{-1}h \in H$, where H is the subgroup of $\mathrm{GL}(n,\mathbb{Z})$ consisting of all linear transformations that fix O_{lex}.*

(iii) *The relation $\lesssim$ induces on $\mathrm{GL}(n,\mathbb{Z})/H$ a partial ordering $\leqslant$ that is a lattice with least element H and greatest element ΔH, where Δ is defined by $x\Delta = -x$ for every x. For all g, h, the relations $gH \leqslant hH$ and $\Delta g \Delta^{-1} H \leqslant \Delta h \Delta^{-1} H$ are equivalent.*

(iv) *The relations* (VI.3.16) *and* (VI.3.17) *are satisfied by $\mathrm{GL}(n,\mathbb{Z})$, H, and $\leqslant$.*

Proof. (i) By definition, $O_{\text{lex}} \bullet g \preccurlyeq O_{\text{lex}} \bullet h$ means that $\mathrm{Inv}_{O_{\text{lex}}}(O_{\text{lex}} \bullet g)$ is included in $\mathrm{Inv}_{O_{\text{lex}}}(O_{\text{lex}} \bullet h)$, hence that every pair (x, y) in $(\mathbb{Z}^n)^2$ satisfying $x <_{\text{lex}} y$ and $xg >_{\text{lex}} yg$ satisfies $xh >_{\text{lex}} yh$. For $x = \underline{0}$ and $z = y$, we obtain (3.7). Conversely, if $g \lesssim h$ holds, every pair (x, y) satisfying $y - x >_{\text{lex}} \underline{0}$ and $(y-x)g <_{\text{lex}} \underline{0}$ satisfies $(y-x)h <_{\text{lex}} \underline{0}$. As $<_{\text{lex}}$ is invariant and g and h are linear, this implies that, for all x, y satisfying $x <_{\text{lex}} y$ and $xg >_{\text{lex}} yg$, we have $xh >_{\text{lex}} yh$. This is the definition of $O_{\text{lex}} \bullet g \preccurlyeq O_{\text{lex}} \bullet h$.

(ii) Reflexivity and transitivity are obvious from the definition. Assume $g \lesssim h$ and $h \lesssim g$. By assumption, for every z satisfying $z >_{\text{lex}} \underline{0}$, we have $zg <_{\text{lex}} \underline{0} \Leftrightarrow zh <_{\text{lex}} \underline{0}$ and therefore $zg >_{\text{lex}} \underline{0} \Leftrightarrow zh >_{\text{lex}} \underline{0}$ as well. Assume $z <_{\text{lex}} \underline{0}$. Then $-z >_{\text{lex}} \underline{0}$ holds, so the assumption implies $(-z)g <_{\text{lex}} \underline{0} \Leftrightarrow (-z)h <_{\text{lex}} \underline{0}$, whence $zg >_{\text{lex}} \underline{0} \Leftrightarrow zh >_{\text{lex}} \underline{0}$ in this case also. So, for every z in $\mathbb{Z}^n$, we have $zg >_{\text{lex}} \underline{0} \Leftrightarrow zh >_{\text{lex}} \underline{0}$. As g is surjective, we equivalently obtain that, for every x in $\mathbb{Z}^n$, we have $x >_{\text{lex}} \underline{0} \Leftrightarrow xg^{-1}h >_{\text{lex}} \underline{0}$, meaning that $g^{-1}h$ fixes O_{lex}.

(iii) By Lemma 3.5, the ordering $\preccurlyeq$ is a lattice ordering on $\mathcal{O}_{\text{lex}}(\mathbb{Z}^n)$. Then the map $g \mapsto O_{\text{lex}} \bullet g$ induces an isomorphism from $(\mathrm{GL}(n,\mathbb{Z})/H, \leqslant)$ to $(\mathcal{O}_{\text{lex}}(\mathbb{Z}^n), \preccurlyeq)$. The image of H is O_{lex}, which is the smallest element in $(\mathcal{O}_{\text{lex}}(\mathbb{Z}^n), \preccurlyeq)$, so H is the smallest element in $(\mathrm{GL}(n,\mathbb{Z})/H, \leqslant)$; the image of ΔH is the reversed order $O_{\text{lex}}^{\text{rev}}$, which is the largest element in $(\mathcal{O}_{\text{lex}}(\mathbb{Z}^n), \preccurlyeq)$, so ΔH is the largest element in $(\mathrm{GL}(n,\mathbb{Z})/H, \leqslant)$. Finally, we have $x\Delta g \Delta^{-1} = -((-x)g) = xg$ for every x in $\mathbb{Z}^n$, so the last equivalence is clear.

(iv) The argument is similar to the one of Lemma IX.1.7 for the symmetric group or, more generally, an arbitrary Coxeter system, and it consists in using the map N, here defined by

$$N(g) = \{x \in \mathbb{Z}^n \mid (x >_{\text{lex}} \underline{0} \text{ and } xg <_{\text{lex}} \underline{0}) \text{ or } (x <_{\text{lex}} \underline{0} \text{ and } xg >_{\text{lex}} \underline{0})\},$$

which will play the role of the set of reflections in the Coxeter case. Then one easily obtains the equality $N(gh) = N(g) \,\triangle\, N(h)g^{-1}$, where $\triangle$ is the symmetric difference, and deduces that $g \lesssim h$ is equivalent to $N(gh)$ being the disjoint union of $N(g)$ and $N(h)g^{-1}$, and also to $N(g) \cap N(h)g^{-1}$ being empty. From there, the relations (VI.3.16) and (VI.3.17) follow easily. □

Example 3.9 (case $n = 2$). In the (trivial) case $n = 2$, the subgroup H is isomorphic to $(\mathbb{Z}, +)$ as it identifies with the group of matrices $\begin{pmatrix} 1 & 0 \\ e & 1 \end{pmatrix}$ and the relation $gH = hH$ reduces to $\begin{pmatrix} a & b \\ c & d \end{pmatrix} \sim \begin{pmatrix} a+eb & b \\ c+ed & d \end{pmatrix}$.

Lemma 3.8 precisely shows that $(\mathrm{GL}(n, \mathbb{Z}), H, \leqslant)$ is eligible for Proposition VI.3.15 (germ from lattice), making it natural to introduce the corresponding monoid:

Definition 3.10 (braid monoid of $\mathbb{Z}^n$). For $n \geqslant 2$, we let $B^+(\mathbb{Z}^n)$ to be the monoid generated by the germ defined in Proposition VI.3.15 from $(\mathrm{GL}(n, \mathbb{Z}), H, \leqslant)$.

Proposition 3.11 (braid monoid of $\mathbb{Z}^n$). *For every n, the monoid $B^+(\mathbb{Z}^n)$ is an Ore monoid, it admits left-gcds and right-lcms, the element Δ is central, it is a Garside element in $B^+(\mathbb{Z}^n)$, and its divisors make a lattice isomorphic to* $\mathrm{GL}(n, \mathbb{Z})/H$; *the invertible elements of $B^+(\mathbb{Z}^n)$ are the elements of H.*

Proof. Apply Proposition VI.3.15. □

Of course, one can then define the *braid group of* $\mathbb{Z}^n$ to be the group of fractions $B(\mathbb{Z}^n)$ of $B^+(\mathbb{Z}^n)$. Then the results of Chapters III and V apply to $B(\mathbb{Z}^n)$ and, in particular, every element of $B(\mathbb{Z}^n)$ admits a Δ-normal decomposition that is unique up to deformation by elements of H.

4 Cell decompositions of a punctured disk

We now give an introduction to another geometric construction, developed by D. Krammer in [165], which consists in defining a category equipped with a distinguished Garside map Δ for every finite sequence of integers larger than 1. In every case, the family $\mathcal{D}iv(\Delta)$ is finite and it is a lattice. In the case of a length-n sequence $(2, ..., 2)$, the category is the braid monoid B_n^+ and Δ coincides with the standard Garside element Δ_n of B_n^+ (Reference Structure 2, page 5), whereas, in the case of a length-n sequence $(3, ..., 3)$, the category is closely connected with the Tamari lattice on the set of vertices of the nth associahedron, and, in the case of a sequence with several entries, the category has more than one object.

4.1 Braid groups as mapping class groups

Hereafter it will be convenient to look at the braid group B_n (Reference Structure 2, page 5) in one more different way, namely as a *mapping class group* of a punctured disk, and we begin with a brief description of this viewpoint, which consists in looking at braids from one end rather than from the side.

Definition 4.1 (mapping class group). If Σ is an oriented compact surface, possibly with boundary $\partial\Sigma$, and $\boldsymbol{P}$ is a finite set of interior points of Σ, the *mapping class group* $\mathit{MCG}(\Sigma, \boldsymbol{P})$ of Σ relative to $\boldsymbol{P}$ is the group of all isotopy classes of orientation-preserving self-homeomorphisms of Σ that fix $\partial\Sigma$ pointwise and preserve $\boldsymbol{P}$ globally.

So a homeomorphism ϕ from Σ to itself taking punctures to punctures represents an element of $\mathit{MCG}(\Sigma, \boldsymbol{P})$ whenever it acts as the identity on the boundary of Σ. Note that the punctures may be permuted by ϕ. Two homeomorphisms represent the same element if and only if they are isotopic through a family of boundary-fixing homeomorphisms that also fix $\boldsymbol{P}$. Such homeomorphisms necessarily induce the same permutation of the punctures. It can be noticed that, in Definition 4.1, the word "isotopic" can be replaced by "homotopic": by a theorem of Epstein [118], two homeomorphisms of a compact surface are homotopic if and only if they are isotopic.

Mapping class groups, also known as modular groups, play a prominent role in the study of the topology and geometry of surfaces, as well as in 3-dimensional topology. They admit finite presentations, but establishing it requires deep arguments, see Hatcher and Thurston [144] and Wajnryb [228].

From now on, we denote by D the unit disk of the real plane, identified for instance with $\{z \in \mathbb{C} \mid |z| \leqslant 1\}$. The boundary of D is denoted by ∂D.

Proposition 4.2 (braid group as mapping class group). *If $\boldsymbol{P}$ is a set of n elements, then the mapping class group $\mathit{MCG}(D, \boldsymbol{P})$ is isomorphic to the braid group B_n.*

Proof (sketch). (See Figure 5; for a complete argument, see for instance [17].) Let β be an n-strand geometric braid, sitting in the cylinder $[0, 1] \times D$, whose n strands are starting at the puncture points of $\{0\} \times D$ and ending at the puncture points of $\{1\} \times D$. Then β may be considered to be the graph of the motion, as time goes from 0 to 1, of n points moving in the disk, starting and ending at the puncture points. This motion can be extended to a continuous family of homeomorphisms of the disk, starting with the identity and fixed on the boundary at all times: think of the cylinder as filled with a viscous fluid. The end map of this isotopy is the corresponding homeomorphism from D to itself, which is is well-defined up to isotopy, globally fixes the punctures, and pointwise fixes the boundary.

Conversely, assume that ϕ is a homeomorphism of D to itself that represents some element of the mapping class group. By a trick of Alexander, every homeomorphism of a disk that fixes the boundary is isotopic to the identity, through homeomorphisms fixing the boundary. The corresponding braid is then the graph of the restriction of such an isotopy to the puncture points. □

In this description, if the punctures are enumerated $P_1, \ldots, P_n$, the braid σ_i corresponds to the class of a homeomorphism that performs in a neighborhood of $\{P_i, P_{i+1}\}$ a clockwise half-turn centered at the middle of P_iP_{i+1}.

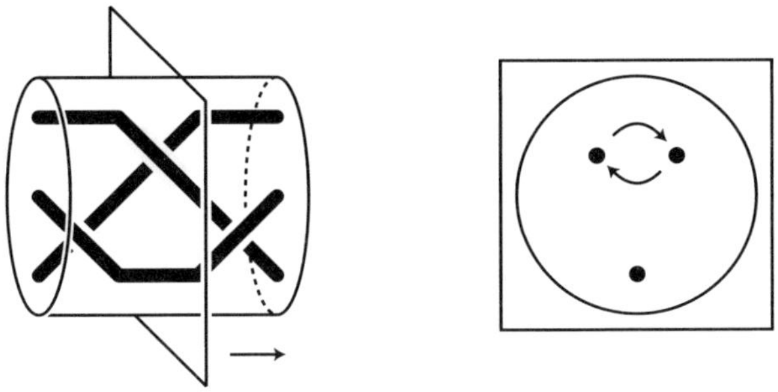

Figure 5. Viewing an n-strand geometric braid drawn in a cylinder as the motion (the dance) of n points in a disk when time goes from 0 to 1: cut the cylinder using a mobile vertical plane and look at the motion of the punctures.

4.2 Cell decompositions

The notion of a triangulation of a polygon is standard: it is a partition of the polygon into (disjoint) triangles. A triangulation of an N-gon involves $N - 2$ triangles. Here we shall consider similar decompositions involving pieces that are not necessarily triangles, but can be arbitrary polygons, including bigons—the viewpoint is topological, not geometrical, and our polygons may have curved edges.

Definition 4.3 (decomposition type, cell decomposition). A *decomposition type* is a finite sequence of positive integers at least equal to 2. For $\vec{\ell} = (\ell_1, \dots, \ell_n)$, a *cell decomposition* of type $\vec{\ell}$ is a partition of an N-gon Π into n polygonal cells with respectively $\ell_1, \dots, \ell_n$ edges, with $N = \sum_i (\ell_i - 2) + 2$.

Two cell decompositions are isotopic if there exists a continuous deformation that keeps the boundary of the polygon pointwise fixed and maps one decomposition to the other: at the expense of considering that the vertices of the polygon lie on the circle ∂D, the isotopy class of a cell decomposition is its orbit under the action of the mapping class group $MCG(D, \emptyset)$. So, for instance, there is only one isotopy type of cell decomposition of type $(2, \dots, 2)$, 2 repeated n times, corresponding to the action of orientation-preserving self-homeomorphisms of D on the $n - 1$ edges that separate adjacent bigons, see Figure 6(i). A triangulation of an N-gon is a cell decomposition of type $(3, \dots, 3)$, 3 repeated $N - 2$ times, and there are $\frac{1}{N+1}\binom{2N}{N}$ (the Nth Catalan number) isotopy classes of such triangulations, see Figure 6(ii).

When (as in Figure 6) the base polygon is represented as N points $Q_1, \dots, Q_N$ drawn on ∂D, a cell decomposition of type $\vec{\ell}$ can be described as a collection of $n - 1$ noncrossing arcs $\gamma_1, \dots, \gamma_{n-1}$ in $\mathring{D}$ separating the adjacent cells of the decomposition; by construction, the endpoints of these arcs belong to $\{Q_1, \dots, Q_N\}$.

(i) 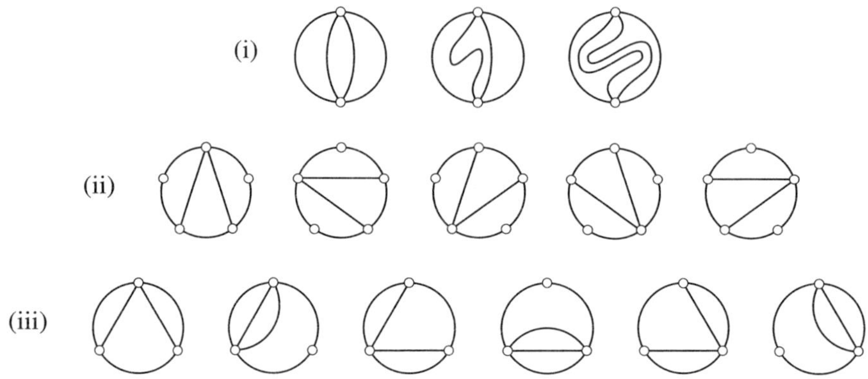

Figure 6. Cell decompositions of a polygon: (i) Three isotopic cell decompositions of type $(2, 2, 2)$, that is, three decompositions of a bigon ($N = 2$) into three bigons; (ii) The five isotopy classes of cell decompositions of type $(3, 3, 3)$, that is, the decompositions of a pentagon($N = 5$) into three triangles; (iii) The six isotopy classes of cell decompositions of type $(2, 2, 3)$, that is, the decompositions of a triangle ($N = 3$) into two bigons and one triangle.

4.3 The group B_ℓ and the category $\mathcal{B}_\ell$

For each decomposition type $\vec{\ell}$ of integers larger than 1, we shall now define a category whose objects are (certain) isotopy classes of cell decompositions of type $\vec{\ell}$. The elements of the categories are (certain) elements of the mapping class group $MCG(D, \boldsymbol{P})$ where the cardinality of $\boldsymbol{P}$ is the length n of $\vec{\ell}$—hence, by Proposition 4.2, they are (certain) n strand braids.

Hereafter, "homeomorphism of D" always means "orientation-preserving homeomorphism of D that fixes the boundary ∂D pointwise". When there is no puncture, homeomorphisms act transitively on cell decompositions of a given type, and there is not much to say. Things become interesting when punctures are posted in each of the n regions of a cell decomposition.

From now on, the n-strand braid group B_n is identified with the mapping class group of an n-punctured disk, that is, with the family of isotopy classes of orientation-preserving homeomorphisms of D that fix the boundary ∂D pointwise and leave the set of punctures globally invariant. The family of punctures will always be written as $\underline{P} = (P_1, \ldots, P_n)$.

Definition 4.4 (group $B_{\vec{\ell}}$). For $\vec{\ell}$ a decomposition type of length n, we denote by $B_{\vec{\ell}}$ the subgroup of B_n consisting of those braids that map every puncture P_i to a puncture P_j satisfying $\ell_j = \ell_i$.

If all entries in $\vec{\ell}$ are equal, the group $B_{\vec{\ell}}$ is B_n, whereas, if the entries of $\vec{\ell}$ are pairwise distinct, $B_{\vec{\ell}}$ is the pure braid group PB_n, that is, the index $n!$ subgroup of B_n made by the braids that leave the punctures pointwise fixed.

Our aim is to let the group $B_{\vec{\ell}}$ act on cell decompositions of type $\vec{\ell}$. To this end, we shall restrict to cell decompositions that are compatible with the punctures in that every cell contains one puncture exactly, with an additional requirement on the number of edges.

Definition 4.5 (admissible cell decomposition). (See Figure 7.) For $n \geqslant 1$ and $N \geqslant 2$, we denote by $D_{n,N}$ the disk D with N distinguished points $Q_1, \ldots, Q_N$ in ∂D and n distinguished points $P_1, \ldots, P_n$ in $\mathring{D}$. A type $\vec{\ell}$ cell decomposition $\boldsymbol{c}$ of $D_{n,N}$ is called *admissible* if, for every i, the vertices of the cell c_i belong to $\{Q_1, \ldots, Q_N\}$ and $\mathring{c}_i$ contains exactly one point of $\{P_1, \ldots, P_n\}$, say $P_{f(i)}$, and $\ell_{f(i)} = \ell_i$ holds. We write $\mathcal{CD}_{\vec{\ell}}$ for the family of all admissible cell decompositions of type $\vec{\ell}$.

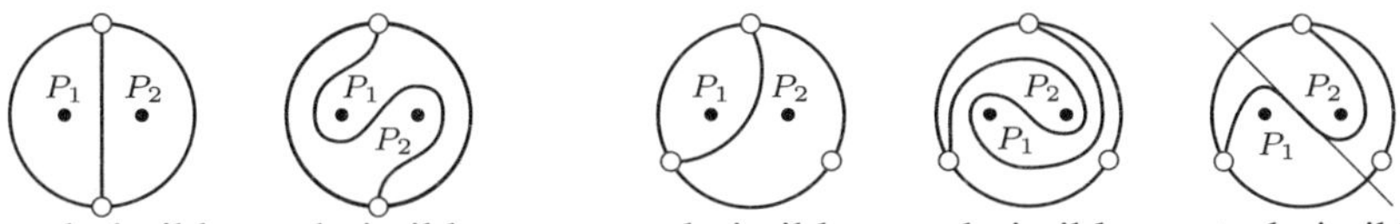

Figure 7. Admissible cell decomposition: on the left, two admissible cell decompositions of type $(2, 2)$, on the right, two admissible cell decompositions of type $(2, 3)$, and one that is not admissible: each region contains one puncture, but the region containing P_1 is a triangle, whereas ℓ_1 is 2, not 3.

The definition of the group $B_{\vec{\ell}}$ immediately implies:

Lemma 4.6. *For every decomposition type $\vec{\ell}$, every braid in $B_{\vec{\ell}}$ maps every admissible decomposition of type $\vec{\ell}$ to an admissible decomposition of type $\vec{\ell}$.*

Therefore, we can put

Notation 4.7 (class $[\boldsymbol{c}]$, family $[\mathcal{CD}_{\vec{\ell}}]$). For $\boldsymbol{c}$ an admissible cell decomposition of type $\vec{\ell}$, we write $[\boldsymbol{c}]$ for the $B_{\vec{\ell}}$-class of $\boldsymbol{c}$. We write $[\mathcal{CD}_{\vec{\ell}}]$ for the family of all classes $[\boldsymbol{c}]$ with $\boldsymbol{c}$ an admissible cell decomposition of type $\vec{\ell}$.

Note that, as $B_{\vec{\ell}}$ acts transitively on the punctures with a given label, we can always assume that the cells of a decomposition are enumerated so that the cell c_i contains the puncture P_i.

We are now ready to introduce the category $\mathcal{B}_{\vec{\ell}}$ that is our main subject of interest. It consists of transformations on the family of (classes of) cell decompositions, properly extending braids (viewed as homeomorphisms of the punctured disk) in the general case. However, we shall not define the involved transformations starting from geometric transformations of the disk, but rather start from a global definition of the expected category and then show how its elements can be generated using convenient geometrically defined transformations.

Definition 4.8 (transformation $\tau_{\boldsymbol{c}_1,\boldsymbol{c}_2}$, category $\mathcal{B}_{\vec{\ell}}$). (i) If $\boldsymbol{c}_1, \boldsymbol{c}_2$ are cell decompositions of type $\vec{\ell}$, then $\tau_{\boldsymbol{c}_1,\boldsymbol{c}_2}$ is the mapping from $[\boldsymbol{c}_1]$ to $[\boldsymbol{c}_2]$ that maps $\beta(\boldsymbol{c}_1)$ to $\beta(\boldsymbol{c}_2)$ for every braid β in $B_{\vec{\ell}}$.

(ii) We define $\mathcal{B}_{\vec{\ell}}$ to be the category made of the transformations $\tau_{\boldsymbol{c}_1,\boldsymbol{c}_2}$ with $\boldsymbol{c}_1, \boldsymbol{c}_2$ cell decompositions of type $\vec{\ell}$, source, target, and composition being those of functions acting on the right, that is, gh means "g then h".

Thus, by definition, the object family of the category $\mathcal{B}_{\vec{\ell}}$ is the family $[\mathcal{CD}_{\vec{\ell}}]$ of all classes of admissible cell decompositions of type $\vec{\ell}$. Before giving examples, we note a few easy properties of the category $\mathcal{B}_{\vec{\ell}}$ that follow almost directly from the definition.

Lemma 4.9. (i) *Assume that $\boldsymbol{c}_1, \boldsymbol{c}_2, \boldsymbol{c}'_2, \boldsymbol{c}'_3$ are cell decompositions of type $\vec{\ell}$, and the $B_{\vec{\ell}}$-classes of $\boldsymbol{c}_2$ and $\boldsymbol{c}'_2$ coincide. Then, in $\mathcal{B}_{\vec{\ell}}$, we have*

(4.10)
$$\tau_{\boldsymbol{c}_1,\boldsymbol{c}_2} \cdot \tau_{\boldsymbol{c}'_2,\boldsymbol{c}'_3} = \tau_{\boldsymbol{c}_1,\beta(\boldsymbol{c}'_3)},$$

where β is the unique braid of $B_{\vec{\ell}}$ satisfying $\beta(\boldsymbol{c}'_2) = \boldsymbol{c}_2$

(ii) *For every x in $[\mathcal{CD}_{\vec{\ell}}]$, the group $\mathcal{B}_{\vec{\ell}}(x,x)$ is isomorphic to the braid group $B_{\vec{\ell}}$.*

Proof. Assume that the target of $[\boldsymbol{c}_1, \boldsymbol{c}_2]$ is the source of $[\boldsymbol{c}'_2, \boldsymbol{c}'_3]$. By definition, this means that the $B_{\vec{\ell}}$-classes of $\boldsymbol{c}_2$ and $\boldsymbol{c}'_2$ coincide, that is, there exists a braid β in $B_{\vec{\ell}}$ mapping $\boldsymbol{c}'_2$ to $\boldsymbol{c}_2$. Then $(\boldsymbol{c}_2, \beta(\boldsymbol{c}'_3))$ is another representative of $[\boldsymbol{c}'_2, \boldsymbol{c}'_3]$, and the product of $[\boldsymbol{c}_1, \boldsymbol{c}_2]$ and $[\boldsymbol{c}'_2, \boldsymbol{c}'_3]$ is the well-defined class of $(\boldsymbol{c}_1, \beta(\boldsymbol{c}'_3))$.

Assume that x belongs to $\mathcal{CD}_{\vec{\ell}}$. Choose a cell decomposition $\boldsymbol{c}_*$ with $B_{\vec{\ell}}$-class x. Define $\Phi_{\boldsymbol{c}_*} : B_{\vec{\ell}} \to \mathcal{B}_{\vec{\ell}}(x,x)$ by $\Phi_{\boldsymbol{c}_*}(\beta) = \tau_{\boldsymbol{c}_*,\beta(\boldsymbol{c}_*)}$. First, by definition, $\beta(\boldsymbol{c}_*)$ lies in the class x, so $\Phi_{\boldsymbol{c}_*}(\beta)$ does belong to $\mathcal{B}_{\vec{\ell}}(x,x)$. Next, $B_{\vec{\ell}}$ acts transitively on its classes, so every decomposition c whose class is x is the image of $\boldsymbol{c}_*$ under some braid. So $\Phi_{\boldsymbol{c}_*}$ is surjective. On the other hand, $\Phi_{\boldsymbol{c}_*}(\beta) = \Phi_{\boldsymbol{c}_*}(\beta')$ means that there exists a braid β_0 satisfying $\boldsymbol{c}_* = \beta_0(\boldsymbol{c}_*)$ and $\beta(\boldsymbol{c}_*) = \beta_0(\beta'(\boldsymbol{c}_*))$. The group $B_{\vec{\ell}}$ acts freely on its classes, so the first equality implies $\beta_0 = 1$, whence $\beta' = \beta$, so $\Phi_{\boldsymbol{c}_*}$ is injective. Finally, we have

$$\begin{aligned}\Phi_{\boldsymbol{c}_*}(\beta_1)\Phi_{\boldsymbol{c}_*}(\beta_2) &= \tau_{\boldsymbol{c}_*,\beta_1(\boldsymbol{c}_*)}\tau_{\boldsymbol{c}_*,\beta_2(\boldsymbol{c}_*)} \\ &= \tau_{\boldsymbol{c}_*,\beta_1(\boldsymbol{c}_*)}\tau_{\beta_1(\boldsymbol{c}_*),\beta_1(\beta_2(\boldsymbol{c}_*))} = \tau_{\boldsymbol{c}_*,\beta_1(\beta_2(\boldsymbol{c}_*))} = \Phi_{\boldsymbol{c}_*}(\beta_1\beta_2),\end{aligned}$$

so $\Phi_{\boldsymbol{c}_*}$ is a homomorphism, hence an isomorphism. □

Note that, viewed as a set of pairs, the transformation $\tau_{\boldsymbol{c}_1,\boldsymbol{c}_2}$ is the family of all pairs $(\beta(\boldsymbol{c}_1), \beta(\boldsymbol{c}_2))$ with β in $B_{\vec{\ell}}$, hence it is simply the $B_{\vec{\ell}}$-class of $(\boldsymbol{c}_1, \boldsymbol{c}_2)$ when the action of $B_{\vec{\ell}}$ is extended from $\mathcal{CD}_{\vec{\ell}}$ to $\mathcal{CD}_{\vec{\ell}} \times \mathcal{CD}_{\vec{\ell}}$ componentwise and, therefore, it could be denoted by $[\boldsymbol{c}_1, \boldsymbol{c}_2]$. However, owing to the case of braids or to that of flips below, it seems more intuitive to view this class as a transformation on $\mathcal{CD}_{\vec{\ell}}$. Speaking of braids, it should be noted that the isomorphism $\Phi_{\boldsymbol{c}_*}$ is by no means canonical, because the geometric action of braids on cell decompositions is *not* invariant under $B_{\vec{\ell}}$-equivalence: for instance, in Figure 8 below, the braid σ_2 maps the top left decomposition $\boldsymbol{c}_1$ to the top right decomposition $\boldsymbol{c}_2$, but the σ_1-image of $\boldsymbol{c}_1$ is mapped to the σ_1-image of $\boldsymbol{c}_2$ by $\sigma_1^{-1}\sigma_2\sigma_1$, not by σ_2.

Example 4.11 (category $\mathcal{B}_{\vec{\ell}}$). Write $2^{[n]}$ for $(2, ..., 2)$ with 2 repeated n times. Then we have $N = 2$ and, as seen in Figure 6(i), there is only one class of (admissible) cell decompositions, namely a sequence of n adjacent bigons $c_1, ..., c_n$, each containing one puncture. So $[\mathcal{CD}_{2^{[n]}}]$ has only one element, and the category $\mathcal{B}_{2^{[n]}}$ identifies with the braid group B_n. So, in this case, $\mathcal{B}_{\vec{\ell}}$ contains only braids.

By contrast, for all other decomposition types $\vec{\ell}$, there is more than one element in $\mathcal{CD}_{\vec{\ell}}$, hence more than one object in the category $\mathcal{B}_{\vec{\ell}}$, giving rise to elements that are not braids. For instance, $\mathcal{CD}_{(2,3)}$ has three elements, and the (classes of the) following pairs of cell decompositions are typical elements of $\mathcal{B}_{(2,3)}$ that are not braids, since they can be realized by no homeomorphism of the disk D pointwise preserving the boundary ∂D:

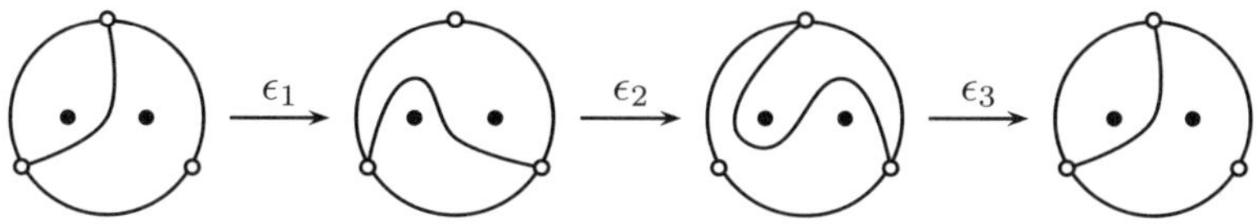

Note that, if x is the class of the left decomposition, the product $\epsilon_1\epsilon_2\epsilon_3$ is the identity element 1_x.

4.4 Flips

We will now identify a family of generators in the category $\mathcal{B}_{\vec{\ell}}$. To this end, we introduce a (mild) generalization of the notion of a *flip*, which is standard for triangulations.

Definition 4.12 (flip). (See Figure 8). Assume that $\mathbf{c}$ is an admissible cell decomposition in which the cells containing P_i and P_j are adjacent, that is, they are separated by an arc γ not reduced to a point. Then we denote by $f_{i,j}(\mathbf{c})$ the (admissible) cell decomposition obtained from $\mathbf{c}$ by rotating γ by one step, this meaning that each endpoint of γ is moved in the positive direction to the next endpoint Q_k of the region $c_i \cup c_j$, the rest of γ being continuously deformed so as to avoid the (two) punctures of the region.

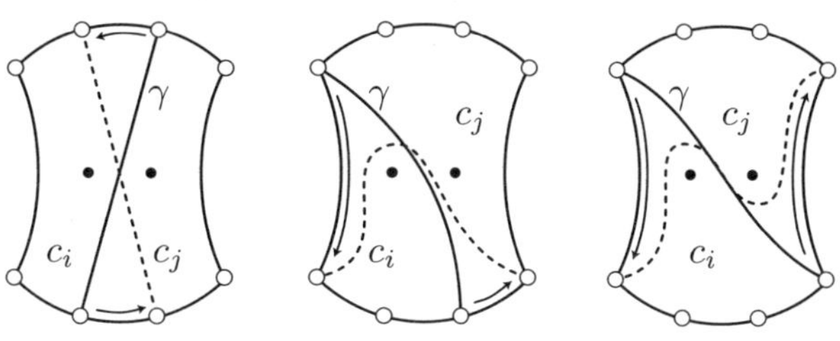

Figure 8. Flipping the arc that separates two adjacent cells c_i and c_j: one moves each endpoint to the next one in the polygon $c_i \cup c_j$, skirting around the punctures, obtaining the new dotted arc. Note that the number of edges around each puncture does not change, so the image of an admissible decomposition is an admissible decomposition.

Lemma 4.13. *Assume that $f_{i,j}(\mathbf{c})$ is defined and $[\mathbf{c}'] = [\mathbf{c}]$ holds. Then there exist a unique pair $\{i', j'\}$ such that $f_{i',j'}(\mathbf{c}')$ is defined, and then $[f_{i',j'}(\mathbf{c}')] = [f_{i,j}(\mathbf{c})]$ holds.*

Proof. By definition, a braid β (viewed as a self-homeomorphism of the disk) maps adjacent cells to adjacent cells. By assumption, the endpoints Q_k are fixed under β, and β preserves the orientation, hence applying β and flipping commute. The numbers i' and j' are the images of i and j under β. $\square$

Remark 4.14. Except for decomposition types with pairwise distinct entries, there is no canonical indexation of flips: according to Lemma 4.13, what makes sense is a $B_{\vec{\ell}}$-class to be the image of another one under *some* flip, but not under a flip involving specific cells (or punctures), see Figure 9.

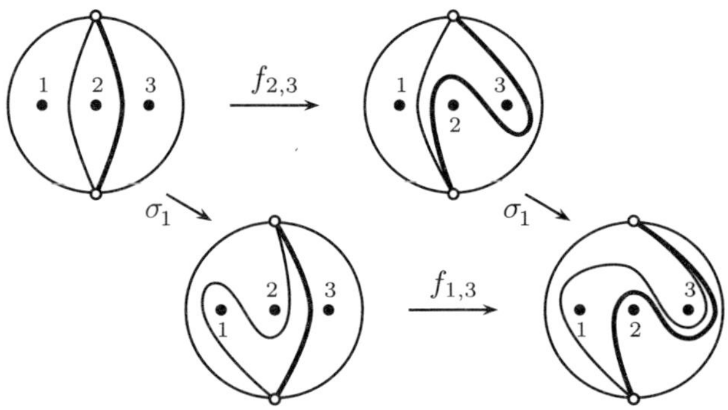

Figure 9. Flip *vs.* isotopy: the top pair is obtained by flipping the thick arc separating 2 and 3 in the left cell decomposition of type $(2, 2)$; when an eligible isotopy is applied, here the braid σ_1, we still obtain a pair of decompositions in which the right decomposition is obtained by flipping the thick arc in the left decomposition, but now this corresponds to flipping the arc separating 1 and 3.

Definition 4.15 (flip, category $\mathcal{B}^+_{\vec{\ell}}$). An element τ_{c_1,c_2} of $\mathcal{B}_{\vec{\ell}}$ is called a *flip* if we have $c_2 = f_{i,j}(c_1)$ for some i, j. The subcategory of $\mathcal{B}_{\vec{\ell}}$ generated by all flips is denoted by $\mathcal{B}^+_{\vec{\ell}}$.

Example 4.16 (flip). Let us consider again the decomposition types of Example 4.11. In the case of type 2^n, the following picture shows that flipping the arc that separates the punctures P_i and P_{i+1} amounts to applying the braid σ_i:

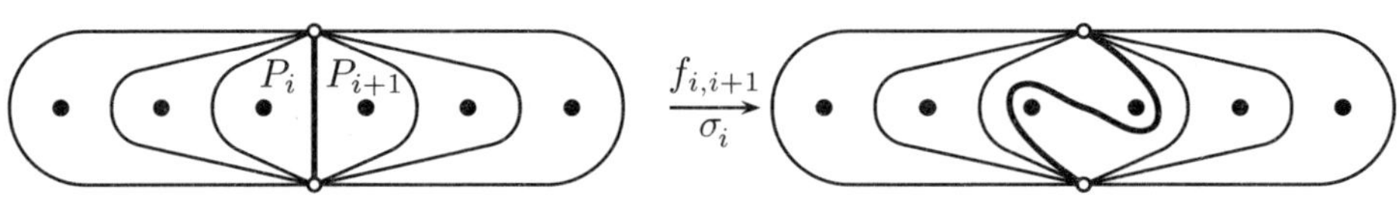

Thus, every braid σ_i is a flip and, therefore, when a base decomposition c is fixed, the subcategory $\mathcal{B}^+_{2(n)}$ is the image of the submonoid of B^+_n under the isomorphism Φ_{c_*}.

Consider now the type $(2, 3)$. Each cell decomposition has only one interior arc, so it is eligible for one flip. Starting as in Example 4.11, we find now:

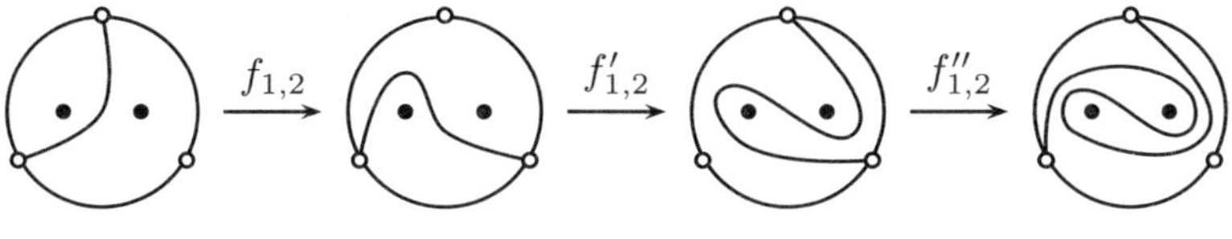

Here the product of three successive flips (which are distinct since their sources are) is the braid σ_1^2. We deduce that the submonoid $\mathcal{B}^+_{(2,3)}(x, x)$ is an isomorphic image of the submonoid $B^+_{(2,3)}$ of $B_{(2,3)}$ generated by σ_1^2. Observe that the braid σ_1 does not appear here—but, as noticed in Figure 7, this braid does *not* correspond to a pair of admissible cell decompositions.

Here comes the first result.

Proposition 4.17 (presentation). *Let $\vec{\ell}$ be a decomposition type. Write $\mathcal{F}_{\vec{\ell}}$ for the family of all flips of type $\vec{\ell}$. Then the groupoid $\mathcal{B}_{\vec{\ell}}$ admits the presentation $\langle \mathcal{F}_{\vec{\ell}} \mid \mathcal{R}_{\vec{\ell}} \rangle$, where $\mathcal{R}_{\vec{\ell}}$ is the family of all relations displayed in Figure* 10.

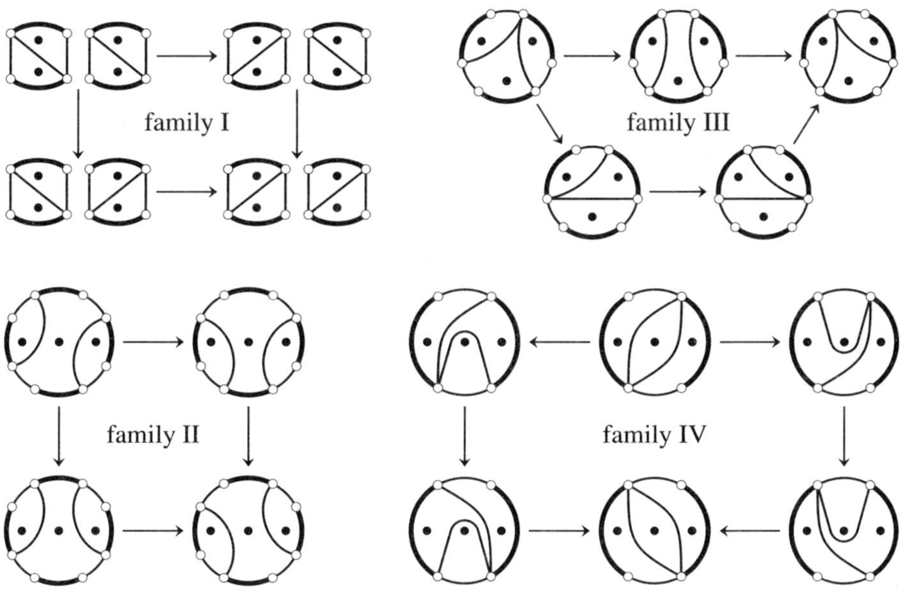

Figure 10. Four generic families of relations connecting flips: in this schematic representation, thick arcs represent portions of ∂D that may contain any number of points Q_k.

Proof (sketch). It is easy to check that the relations of $\mathcal{R}_{\vec{\ell}}$ are valid in $\mathcal{B}_{\vec{\ell}}$. In the other direction, as we know that the groups $\mathcal{B}_{\vec{\ell}}(x, x)$ are braid groups and that the latter admit the usual presentation in terms of the braids σ_i (or σ_i^2 and its conjugates) and the braid relations of (I.1.5), the main point is to prove that σ_i can be recovered from flips and that the braid relations follow from the relations of $\mathcal{R}_{\vec{\ell}}$. An example is displayed in Figure 11 for the case of the decomposition type $(3, 3, 3)$, that is, the triangulations of a pentagon. □

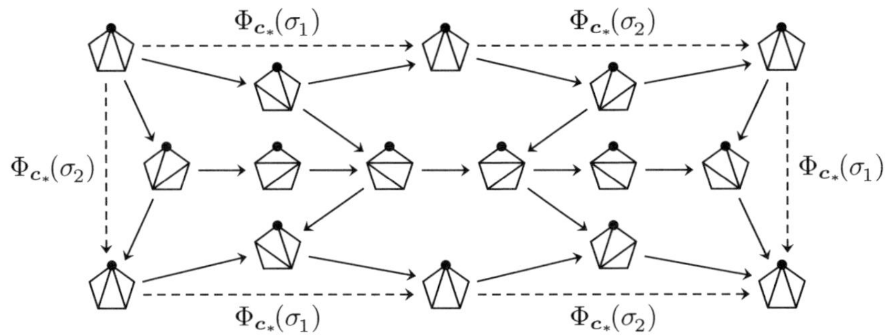

Figure 11. Derivation of the braid relation $\sigma_1\sigma_2\sigma_1 = \sigma_2\sigma_1\sigma_2$ from six family III flip relations; here the punctures are omitted, so what remains is just the $B_{\vec{\ell}}$-class of the decompositions; do not forget that braids occur under their $\Phi_{\mathbf{c}_*}$-image.

One easily checks that all relations of $\mathcal{R}_{\vec{\ell}}$ are positive, that is, involve only positively oriented flips. So these relations take place in the category $\mathcal{B}_{\vec{\ell}}^+$, and, according to a scheme that is usual here, we can wonder whether they make a presentation. As in the case of braids, the method of Chapter II proves to be relevant.

Lemma 4.18. *The presentation $(\mathcal{F}_{\vec{\ell}}, \mathcal{R}_{\vec{\ell}})$ is right-complemented, Noetherian, and the cube condition is satisfied for every triple of flips.*

As can be expected, the verification is elementary, but gathering the cases into a not too large family requires combinatorially intricate definitions. Applying Proposition II.4.16 (right-complemented) and its symmetric counterpart (the presentation $(\mathcal{F}_{\vec{\ell}}, \mathcal{R}_{\vec{\ell}})$ is symmetric), we deduce

Proposition 4.19 (left-Ore category). *For every decomposition type $\vec{\ell}$, the category $\mathcal{B}_{\vec{\ell}}^+$ admits the presentation $\langle \mathcal{F}_{\vec{\ell}} \mid \mathcal{R}_{\vec{\ell}} \rangle^+$ and the groupoid $\mathcal{B}_{\vec{\ell}}$ is its groupoid of fractions.*

At this point, we retrieved for every decomposition type the situation that prevails for the braid group B_n and the braid monoid B_n^+, which, we recall, correspond to the particular type $(2, ..., 2)$, n times 2.

4.5 A bounded Garside family

The obvious task now is now to define a (bounded) Garside family in the category $\mathcal{B}^+_{\vec{\ell}}$ that would coincide with the usual divisors of the fundamental braid Δ_n in the case of the monoid B^+_n, that is, in for type $(2, ..., 2)$. This can be done by explicitly defining a family $\mathcal{S}_{\vec{\ell}}$ of simple elements in $\mathcal{B}^+_{\vec{\ell}}$ using a geometric construction reminiscent of what is known as the Kreweras complement in the language of non-crossing partitions.

Definition 4.20 (simple element, family $\mathcal{S}_{\vec{\ell}}$). (See Figures 12 and 13.) For $\vec{\ell}$ a decomposition type, we say that a transformation $\tau_{\boldsymbol{c}_1,\boldsymbol{c}_2}$ is *simple* if there exists an isotopy ϕ of D leaving ∂D and $\boldsymbol{P}$ pointwise invariant and $0 < \theta < 2\pi/N$ such that every arc in $\mathrm{rot}_\theta(\phi(\boldsymbol{c}_1))$ intersects every arc of $\boldsymbol{c}_2$ at most once. The family of all simple elements in $\mathcal{B}_{\vec{\ell}}$ is denoted by $\mathcal{S}_{\vec{\ell}}$.

Of course, rot_θ stands for the angle-θ rotation centered at the center of D. So the criterion is that a copy of $\boldsymbol{c}_1$ slightly rotated counterclockwise can be deformed so that each arc of the copy intersects each arc of $\boldsymbol{c}_2$ at most once.

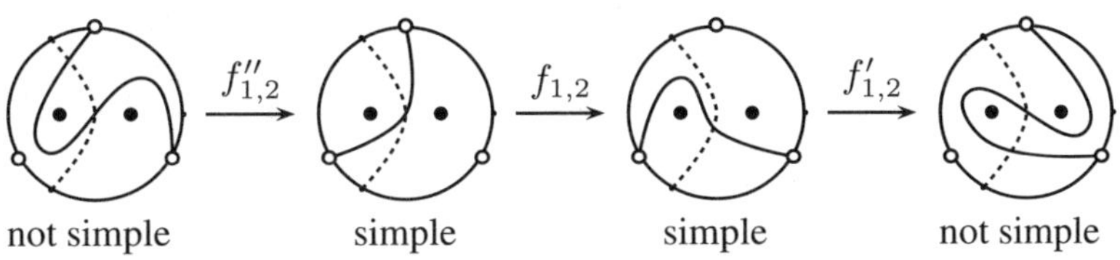

Figure 12. Simple elements in $\mathcal{B}_{(2,3)}$: only two elements, the identity and the flip, are simple, whereas the inverse of a flip and the product of two flips are not simple.

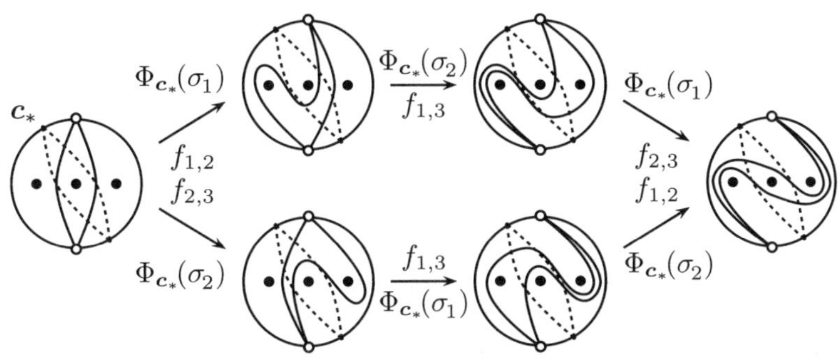

Figure 13. Simple elements in $\mathcal{B}_{(2,2,2)}$: we retrieve the expected six divisors of Δ_3 in B_3^+. Note that the action of $\Phi_{\boldsymbol{c}_*}(\sigma_i)$ need not be that of the braid σ_i.

Proposition 4.21 (Garside family). *For every decomposition type $\vec{\ell}$, the family $\mathcal{S}_{\vec{\ell}}$ is a bounded Garside family in the category $\mathcal{B}^+_{\vec{\ell}}$.*

Proof (sketch). The key point is the closure of $\mathcal{S}_{\vec{\ell}}$ under right-lcm. As $\mathcal{S}_{\vec{\ell}}$ contains all flips and is closed under right-divisors, this implies that $\mathcal{S}_{\vec{\ell}}$ is a Garside family. The existence of a bound then stems from the fact that $\mathcal{S}_{\vec{\ell}}$ has to be finite. Note that, starting with a cell decomposition $\boldsymbol{c}$, an image of $\boldsymbol{c}$ lies in $\mathcal{S}_{\vec{\ell}}$ if it is compatible with $\boldsymbol{c}$ in the sense that, up to a deformation, all triangles in $\boldsymbol{c}$ and in its image intersect in three points, in the obvious sense. Of course, the image of $\boldsymbol{c}$ under Δ is then the largest compatible image. □

We shall not go into details, but just give, in addition to the types $(2,3)$ and $(2,2,2)$ addressed in Figures 12 and 13, two other examples in Figures 14 and 15, respectively, involving the "non-classical" type $(2,2,3)$ and the triangulation type $(3,3,3)$. We refer to the original article [165] for a more complete insight.

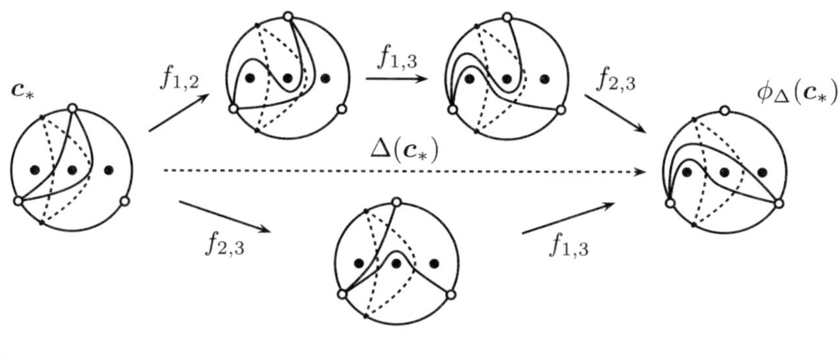

Figure 14. Simple elements in $\mathcal{B}_{(2,2,3)}$ and the associated value of the Garside map. In this case, the category has three objects, and it is not a monoid. We see on the diagram that ϕ_Δ has order 3.

Exercise

Exercise 130 (RC-**systems).** (i) Assume that $(B, \vee, \wedge, *)$ is a Boolean algebra. Show that the operation $\star$ defined by $s \star t = (s \vee t) \wedge s^*$ obeys the RC law. (ii) Assume that V is a Euclidean vector space. Show that the operation $\star$ defined by $E \star F = (F + F) \cap E^\perp$ for E, F subspaces of V obeys the RC law. (iii) Assume that $(S, *)$ is an LD-system (Subsection XI.2.1) and $<$ is a linear ordering on S with the property that $r < s < t$ implies $r < s * t$ and $r * s < r * t$. Show that defining $r \star s$ to be $r * s$ for $r < s$, to be ϵ for $r = s$, and to be s for $r > s$ defines an operation on $S \cup \{\epsilon\}$ that obeys the RC law.

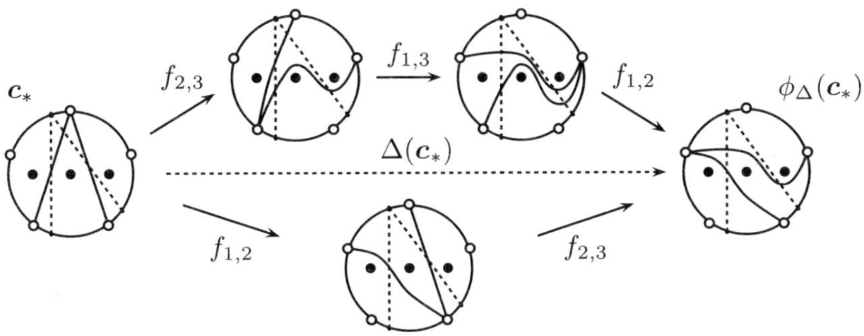

Figure 15. Simple elements of $\mathcal{B}_{(3,3,3)}$ starting from a cell decomposition c and the associated value of the Garside map. Here, the category has five objects, corresponding to the five triangulations of a pentagon, and ϕ_Δ has order 5. The lattice so obtained is the MacLane–Stasheff pentagon, which is the skeleton of the associahedron K_3.

Notes

Divided and decomposition categories

As already mentioned, the construction of divided categories and their application to the study of periodic elements appears in Bessis [7]. Our current exposition is more algebraic than the original topological approach of [7], which in particular allows for more explicit versions. Factorizations analogous to the current category $\mathcal{C}_n(\mathrm{id})$ appear in Krammer [164], see Section 3. The fact that such decompositions could lead to Garside maps was suggested by D. Bessis following a discussion with three of the authors of this book (FD, DK, JM).

Weak RC-systems

The results of Subsection 2 seem to be new. They are mild extensions of the results in Chapter XIII, and their main interest is to show that the restriction to RC-quasigroups is by no means necessary for obtaining a Garside structure on the associated monoid or category. There exists an obvious connection between the RC law and the θ-cube condition of Section II.4: what the sharp θ-cube condition says is that the partial binary operation defined by θ satisfies the (weak) RC law. It is therefore not surprising that a connection exists with the Garside approach and the current developments are just another variation on the basic principle that the cube condition is what is needed to make grid diagrams useful.

The situation with RC-systems, that is, structures involving a binary relation obeying the right-cyclic law $(xy)(xz) = (yx)(yz)$—which could also be called the *cube law*—is similar to the situation with LD-systems, which are structures involving a binary relation obeying the left-selfdistributive law $x(yz) = (xy)(xz)$: some families satisfying restrictive additional conditions received some attention, but the most general structures (typically free ones) are much less understood, yet potentially richer. In the case of LD-systems, quandles, which are very special LD-systems, are easily usable in topology and therefore received much attention, whereas other examples, like the Laver tables of [77, Chapter X], might prove even richer. Similarly, as already mentioned, in the case of RC-systems, RC-quasigroups (*alias* right-cyclic sets, and the related braces and radical rings), because of their connection with set-theoretic solutions of the Yang–Baxter equation, received much attention, see for instance Rump [206, 205] or Cedó, Jespers, and Okniński [53]. However, here again, examples of a different kind might enjoy interesting properties. In particular, let us raise

Question 39. In the context of Exercise 130, are the conditions of (iii) satisfiable (that is, does there exist a linear ordering with the expected properties) when S is a free group and $*$ is the conjugacy operation $s * t = s^{-1}ts$? Can one retrieve in this way the Garside structure connected with the Hurwitz action on a free group described in Example I.2.8? Same question when S is the infinite braid group B_∞ and $*$ is the shifted conjugacy operation $s * t = s\,\mathrm{sh}(t)\,\sigma_1\,\mathrm{sh}(s^{-1})$.

Let us mention the recent reference [203] by W. Rump in which further connections between Garside structures, lattice groups, and various algebraic laws reminiscent of the RC law are explored.

The braid group of $\mathbb{Z}^n$

The braid group of $\mathbb{Z}^n$ was introduced and investigated by D. Krammer in [164], as a sort of baby-example in the direction of a Garside structure on general mapping class groups. The current exposition follows that of [164] with only minor differences. The monoid $B^+(\mathbb{Z}^n)$ and the group $B(\mathbb{Z}^n)$ are very large and they remain incompletely understood. For instance, no explicit presentation (except the one directly stemming from the germ of Subsection 3.3) is known so far. In [164], it is shown that a presentation of $B(\mathbb{Z}^n)$ can be obtained from a presentation of $B(\mathbb{Z}^3)$, but no explicit presentation is known in the case $n = 3$. Note that the braid group of $\mathbb{Z}^n$ extends the classical braid group B_n, as one obtains an embedding of B_n into $B(\mathbb{Z}^n)$ by representing simple braids with permutation matrices. However, this embedding does *not* preserve the lattice structures: the right-lcm of the (images of) two permutation matrices need not be (the image of) a permutation matrix in $B(\mathbb{Z}^n)$.

Cell decompositions

Cell decompositions were investigated by D. Krammer in [165]. They can be seen as a geometric solution for forcing the existence of roots in braid groups, and are somehow reminiscent of the divided categories of Section 1. The Garside structure described in Proposition 4.21 provides in particular for every cell decomposition $\boldsymbol{c}$ of type $\vec{\ell}$ a lattice structure whose minimal element is $\boldsymbol{c}$. In the particular case of triangulations, that is, for types of the form $(3, ..., 3)$, this means providing for every initial triangulation $\boldsymbol{c}$ on an

n-gon a lattice structure on the family of all triangulations of the n-gon whose smallest element is $\boldsymbol{c}$ (see Figure 15). When $\boldsymbol{c}$ is a fan, meaning that there exists a vertex that belongs to all triangles, the lattice so obtained turns out to be the $(n-2)$nd Tamari lattice [222], a central object in finite combinatorics that admits a number of simple characterizations, in particular in connection with the MacLane–Stasheff associahedron K_{n-2}, and with the action of R. Thompson's group F on size $n-1$ dyadic decompositions of the interval $[0,1]$, see for instance [186].

Question 40. Does there exist, for every triangulation $\boldsymbol{c}$ of a polygon Π, a simple combinatorial description for the (unique) lattice on the triangulations of Π, as provided by Proposition 4.21, that admits $\boldsymbol{c}$ as its smallest element?

Raised in [165], the question seems to have received no answer so far.

Appendix

Here we gather the missing proofs of some results of Chapter II. The involved results are Ore's theorem (Proposition II.3.11) and Proposition II.3.18 in Section II.3 and, mainly, the technical results of Section II.4 about reversing, in particular the completeness criteria of Proposition II.4.51.

1 Groupoids of fractions

Ore's theorem

The context is: we start with a category $\mathcal{C}$ and look for conditions guaranteeing that $\mathcal{C}$ embeds in its enveloping groupoid and the latter is a groupoid of left-fractions of $\mathcal{C}$, that is, every element of the groupoid can be expressed as a left-fraction whose numerator and denominator belong to $\mathcal{C}$. Ore's theorem provides a necessary and sufficient condition involving cancellativity and existence of common multiples in $\mathcal{C}$.

Proposition II.3.11 (**Ore's theorem**). *For every category $\mathcal{C}$, the following are equivalent:*

(i) *There exists an injective functor ι from $\mathcal{C}$ to $\mathcal{E}nv(\mathcal{C})$ and every element of $\mathcal{E}nv(\mathcal{C})$ has the form $\iota(f)^{-1}\,\iota(g)$ for some f, g in $\mathcal{C}$.*

(ii) *The category $\mathcal{C}$ is a left-Ore category.*

When (i) *and* (ii) *are met, every element of $\mathcal{E}nv(\mathcal{C})$ is represented by a negative–positive path $\overline{f}|g$ with f, g in $\mathcal{C}$, and two such paths $\overline{f}|g$, $\overline{f'}|g'$ represent the same element of $\mathcal{E}nv(\mathcal{C})$ if and only if they satisfy the relation*

$$\text{(II.3.12)} \qquad \exists h, h' \in \mathcal{C}\ (hf' = h'f \quad \text{and} \quad hg' = h'g),$$

hereafter denoted by $(f, g) \bowtie (f', g')$. Moreover, every presentation of $\mathcal{C}$ as a category is a presentation of $\mathcal{E}nv(\mathcal{C})$ as a groupoid.

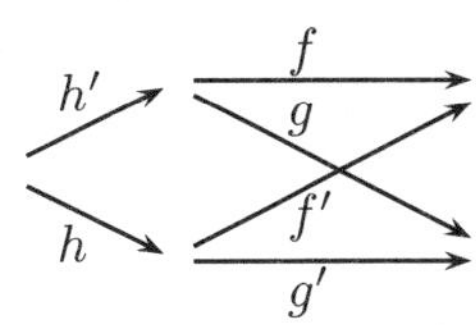

Figure 1. The relation $\bowtie$: two commutative diagrams. Note that $(f, g) \bowtie (f', g')$ implies that f and g share the same source, and so do f' and g'.

The proof of Ore's theorem is not difficult provided one starts with the correct definition of the relation $\bowtie$, see Figure 1. We split the construction into a series of claims. We begin with necessary conditions, that is, with the proof that (i) implies (ii). For further use, we state a result that involves an arbitrary groupoid rather than necessarily $\mathcal{E}nv(\mathcal{C})$.

Claim 1. *Assume that ϕ is an injective functor from a category $\mathcal{C}$ into a groupoid $\mathcal{G}$ and every element of $\mathcal{G}$ belongs to $\phi(\mathcal{C})^{-1}\phi(\mathcal{C})$. Then $\mathcal{C}$ is a left-Ore category, every element of $\mathcal{G}$ is represented by a negative–positive path $\overline{f}|g$, and two such paths $\overline{f}|g$, $\overline{f'}|g'$ represent the same element of $\mathcal{G}$ if and only if $(f,g) \bowtie (f',g')$ holds.*

Proof. First, Lemma II.3.8 implies that $\mathcal{C}$ must be cancellative. Next, assume that f,g are elements of $\mathcal{C}$ with the same target. Then $\phi(f)$ and $\phi(g)$ are elements of $\mathcal{G}$ with the same target. Therefore, $\phi(g)\phi(f)^{-1}$ exists in $\mathcal{G}$. By assumption, there exist f',g' in $\mathcal{C}$ such that, in $\mathcal{G}$, we have $\phi(g)\phi(f)^{-1} = \phi(f')^{-1}\phi(g')$. We deduce

$$\phi(f'g) = \phi(f')\phi(g) = \phi(g')\phi(f) = \phi(g'f),$$

whence $f'g = g'f$, since ϕ is injective. So f and g admit a common left-multiple in $\mathcal{C}$.

Next, for f,g in $\mathcal{C}$, an element of $\mathcal{G}$ equals $\phi(f)^{-1}\phi(g)$ if and only if it is the equivalence class of the signed path $\overline{f}|g$, so, by assumption, every element of $\mathcal{G}$ is represented by such a path.

Assume $\phi(f)^{-1}\phi(g) = \phi(f')^{-1}\phi(g')$ in $\mathcal{G}$. Let x be the target of f, hence the source of $\phi(f)^{-1}$. The above equality implies that x is also the source of $\phi(f')^{-1}$, hence it is the target of f'. As $\mathcal{C}$ is a left-Ore category, and f, f' share the same target, there exist h, h' in $\mathcal{C}$ satisfying $hf' = h'f$. As ϕ is a functor, this implies $\phi(h)\phi(f') = \phi(h')\phi(f)$ in $\mathcal{G}$, whence

$$\phi(f')\phi(f)^{-1} = \phi(h)^{-1}\phi(h'). \tag{1.1}$$

Now the assumption $\phi(f)^{-1}\phi(g) = \phi(f')^{-1}\phi(g')$ implies $\phi(f')\phi(f)^{-1} = \phi(g')\phi(g)^{-1}$. Merging with (1.1), we deduce $\phi(g')\phi(g)^{-1} = \phi(h)^{-1}\phi(h')$, which implies

$$\phi(hg') = \phi(h)\phi(g') = \phi(h')\phi(g) = \phi(h'g),$$

and $hg' = h'g$ as ϕ is injective. So (h,h') witnesses for $(f,g) \bowtie (f',g')$.

Conversely, assume $(f,g) \bowtie (f',g')$, say $hf' = h'f$ and $hg' = h'g$. As ϕ is a functor, we deduce $\phi(h)\phi(f') = \phi(h')\phi(f)$ and $\phi(hg') = \phi(h)\phi(g') = \phi(h')\phi(g) = \phi(h'g)$ in $\mathcal{G}$, whence $\phi(f)^{-1}\phi(g) = \phi(f')^{-1}\phi(g')$, using the computation rules of a groupoid. □

Applying Claim 1 to the groupoid $\mathcal{E}nv(\mathcal{C})$ and to the functor ι shows that (i) implies (ii) in Proposition II.3.11. It remains to show that (ii) implies (i). From now on, we assume that $\mathcal{C}$ is a left-Ore category. We shall construct a groupoid of fractions $[\mathcal{E}]$ into which $\mathcal{C}$ embeds, and then show that $[\mathcal{E}]$ has the universality property that characterizes $\mathcal{E}nv(\mathcal{C})$. To this end, we investigate the relation $\bowtie$ more closely. Put

$$\mathcal{E} = \{(f,g) \in \mathcal{C} \times \mathcal{C} \mid f \text{ and } g \text{ share the same source}\}.$$

Claim 2. *For (f,g), (f',g') in $\mathcal{E}$, the relation $(f,g) \bowtie (f',g')$ is satisfied if and only if the targets of f and f' coincide, so do those of g and g', and we have*

$$\forall h, h' \; (hf' = h'f \;\Leftrightarrow\; hg' = h'g). \tag{1.2}$$

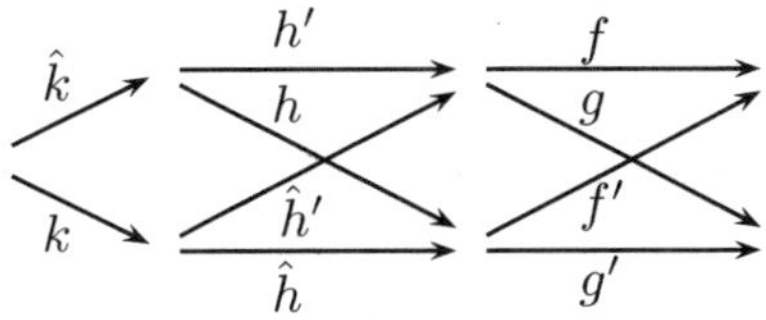

Figure 2. Proof of Claim 2

Proof. Assume $(f, g) \bowtie (f', g')$. By definition, there exist two elements, say h and h', satisfying $hf' = h'f$ and $hg' = h'g$, see Figure 2. The first equality implies that the targets of f and f' coincide, and the second equality implies that the targets of g and g' coincide. Assume now that $\hat{h}$ and $\hat{h}'$ are any elements satisfying $\hat{h}f' = \hat{h}'f$. By construction, the targets of $\hat{h}$ and h both coincide with the source of f', so there exist $k, \hat{k}$ satisfying $\hat{k}h = k\hat{h}$. Then we find $\hat{k}h'f = \hat{k}hf' = k\hat{h}f' = k\hat{h}'f$, hence $\hat{k}h' = k\hat{h}'$ by right-cancelling f. But then we obtain $k\hat{h}'g = k\hat{h}'g = k\hat{h}g' = k\hat{h}g'$, hence $\hat{h}'g = \hat{h}g'$. So (1.2) holds.

Conversely, assume that the targets of f and f' coincide and (1.2) holds. The assumption about the targets of f and f' implies the existence of $\hat{h}, \hat{h}'$ satisfying $\hat{h}f' = \hat{h}'f$. Then (1.2) implies $\hat{h}g' = \hat{h}'g$ and, therefore, we have $(f, g) \bowtie (f', g')$. □

Claim 3. $\bowtie$ *is an equivalence relation.*

Proof. Reflexivity and symmetry are obvious, so it remains to show that $\bowtie$ is transitive. So assume $(f, g) \bowtie (f', g') \bowtie (f'', g'')$. Then f, f', f'' share the same target. An obvious induction on the cardinal of the family shows that each finite family of elements of $\mathcal{C}$ sharing the same target admits a common left-multiple. Hence, there exist h, h', h'' satisfying $hf'' = h'f' = h''f$. By Claim 2, the first equality implies $hg'' = h'g'$ since $(f', g') \bowtie (f'', g'')$ is true. Similarly, the second equality implies $h'g' = h''g$, since $(f, g) \bowtie (f', g')$ is true. We deduce $hg'' = h''g$, whence $(f, g) \bowtie (f'', g'')$. □

For (f, g) in $\mathcal{E}$, we denote by $[f, g]$ the $\bowtie$-class of (f, g) and, for x, y in $\mathcal{O}bj(\mathcal{C})$, put

$$[\mathcal{E}](x, y) = \{[f, g] \mid f \text{ has target } x \text{ and } g \text{ has target } y\}.$$

So $[\mathcal{E}](x, y)$ consists of the equivalence classes of pairs (f, g) such that f and g share the same source and their targets are x and y, respectively. We shall now define a category $[\mathcal{E}]$ whose objects are those of $\mathcal{C}$ and such that the family of elements with source x and target y is precisely $[\mathcal{E}](x, y)$. The question is to define a composition map from $[\mathcal{E}](x, y) \times [\mathcal{E}](y, z)$ to $[\mathcal{E}](x, z)$. There is only one possible definition. Indeed, assume that $[f_1, g_1]$ and $[f_2, g_2]$ are two classes such that the targets of g_1 and f_2 coincide. Then there exist h_1, h_2 satisfying $h_1 g_1 = h_2 f_2$ (see the figure below) and, in the groupoid $\mathcal{E}nv(\mathcal{C})$, we must have

$$\begin{aligned}\iota(f_1)^{-1}\iota(g_1)\iota(f_2)^{-1}\iota(g_2) &= \iota(f_1)^{-1}\iota(h_1)^{-1}\iota(h_1)\iota(g_1)\iota(f_2)^{-1}\iota(h_2)^{-1}\iota(h_2)\iota(g_2)\\ &= \iota(h_1 f_1)^{-1}\iota(h_1 g_1)\iota(h_2 f_2)^{-1}\iota(h_2 g_2) = \iota(h_1 f_1)^{-1}\iota(h_2 g_2).\end{aligned}$$

Thus, if ι is injective, the product of $[f_1, g_1]$ and $[f_2, g_2]$ must be $[h_1 f_1, h_2 g_2]$, where h_1, h_2 satisfy $h_1 g_1 = h_2 f_2$.

So, we are led to defining, for (f_1, g_1) and (f_2, g_2) in $\mathcal{E}$ such that the targets of g_1 and f_2 coincide,

$$(f_1, g_1) * (f_2, g_2) = [h_1 f_1, h_2 g_2]$$

for h_1, h_2 satisfying $h_1 g_1 = h_2 f_2$, with the hope that the operation $*$ is well defined and induces the desired composition map. This is what happens.

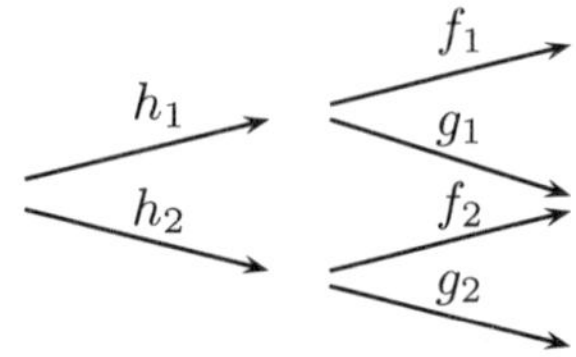

Claim 4. *The operation $*$ is well defined, that is, the result does not depend on the choice of h_1 and h_2, and it is compatible with $\bowtie$, that is, the value of $(f_1, g_1) * (f_2, g_2)$ only depends on the $\bowtie$-classes of (f_1, g_1) and (f_2, g_2).*

Proof. Assume $h_1 g_1 = h_2 f_2$ and $h_1' g_1 = h_2' f_2$. Then h_1 and h_1' share the same target, namely the common source of f_1 and g_1, so there exist k, k' satisfying $k h_1' = k' h_1$. Applying the assumptions, we find

$$k h_2' f_2 = k h_1' g_1 = k' h_1 g_1 = k' h_2 f_2,$$

whence $k h_2' = k' h_2$ by right-cancelling f_2. So we have $k(h_1' f_1) = k'(h_1 f_1)$ on the one hand and, on the other hand, $k(h_2' g_2) = k'(h_2 g_2)$. Hence (k, k') witnesses that $(h_1 f_1, h_2 g_2)$ is $\bowtie$-equivalent to $(h_1' f_1, h_2' g_2)$.

Assume now $(f_1, g_1) \bowtie (f_1', g_1')$ and $(f_2, g_2) \bowtie (f_2', g_2')$. By assumption, the four elements f_2, f_2', g_1, g_1' share the same target. Hence they admit a common left-multiple, that is, there exist h_1, h_1', h_2, h_2' satisfying $h_1 g_1 = h_1' g_1' = h_2 f_2 = h_2' f_2'$. By Claim 2, as we have $(f_1, g_1) \bowtie (f_1', g_1')$, the first equality implies $h_1 f_1 = h_1' f_1'$. Similarly, as we have $(f_2, g_2) \bowtie (f_2', g_2')$, the last equality implies $h_2 g_2 = h_2' g_2'$. Consequently, $(h_1 f_1, h_2 g_2) = (h_1' f_1', h_2' g_2')$ holds, whence, a fortiori, $(h_1 f_1, h_2 g_2) \bowtie (h_1' f_1', h_2' g_2')$. □

So, for all objects x, y, z, we obtain a well defined map $*$ from $[\mathcal{E}](x, y) \times [\mathcal{E}](y, z)$ to $[\mathcal{E}](x, z)$ by defining $[f_1, g_1] * [f_2, g_2]$ to be $[h_1 f_1, h_2 g_2]$, where h_1, h_2 satisfy $h_1 g_1 = h_2 f_2$. It remains to see that this map satisfies the axioms of the composition in a category.

Claim 5. *The map $*$ is associative and, for each class $[f, g]$ in $[\mathcal{E}](x, y)$, we have $[1_x, 1_x] * [f, g] = [f, g] * [1_y, 1_y] = [f, g]$.*

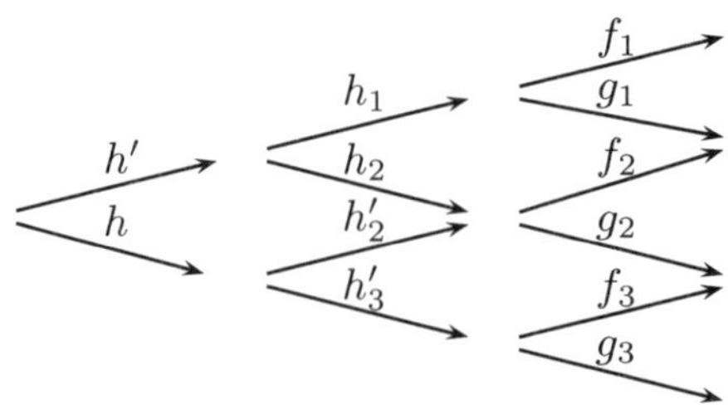

Figure 3. Associativity of composition

Proof. (See Figure 3.) Assume that $(f_1, g_1) * (f_2, g_2)$ and $(f_2, g_2) * (f_3, g_3)$ are defined. Choose h_1, h_2 satisfying $h_1 g_1 = h_2 f_2$, then h'_2, h'_3 satisfying $h'_2 g_2 = h'_3 f_3$, and, finally, h, h' satisfying $hh'_2 = h'h_2$. Then, by construction, the following equalities are satisfied:

$$[f_1, g_1] * ([f_2, g_2] * [f_3, g_3]) = [f_1, g_1] * [h'_2 f_2, h'_3 g_3] = [(h'h_1)f_1, h(h'_3 g_3)],$$
$$([f_1, g_1] * [f_2, g_2]) * [f_3, g_3] = [h_1 f_1, h_2 g_2] * [f_3, g_3] = [h'(h_1 f_1), (hh'_3)g_3],$$

and the associativity of composition in $\mathcal{C}$ gives the equality.

Assume now that $[f, g]$ belongs to $[\mathcal{E}](x, y)$. Then, calling z the common source of f and g, we have $1_z f = f 1_x$, whence $[1_x, 1_x] * [f, g] = [f 1_x, 1_z g] = [f, g]$, and, similarly, $1_z g = g 1_y$, whence $[f, g] * [1_y, 1_y] = [1_z f, g 1_y] = [f, g]$. □

We can then define the category $[\mathcal{E}]$. Its objects are those of $\mathcal{C}$, the family of elements with source x and target y is defined to be $[\mathcal{E}](x, y)$, and composition is defined to be $*$. Finally, the identity functor at x is defined to be $[1_x, 1_x]$. Then Claim 5 states that $[\mathcal{E}]$ satisfies the axioms of categories.

Claim 6. *The category $[\mathcal{E}]$ is a groupoid. The map ϕ which is the identity on $\mathcal{O}bj(\mathcal{C})$ and is defined by $\phi(f) = [1_x, f]$ for f in $\mathcal{C}(x, y)$ is an injective functor from $\mathcal{C}$ to $[\mathcal{E}]$. Moreover, $[\mathcal{E}]$ is a groupoid of left fractions for $\mathcal{C}$.*

Proof. Assume that $[f, g]$ belongs to $[\mathcal{E}](x, y)$. Let z be the common source of f and g. Then we have $1_z g = 1_z g$, whence

$$[f, g] * [g, f] = [1_z f, 1_z f] = [1_x, 1_x].$$

Exchanging f and g gives $[g, f] * [f, g] = [1_y, 1_y]$. Hence, every element of $[\mathcal{E}]$ has an inverse, and $[\mathcal{E}]$ is a groupoid.

As for ϕ, first, $(1_x, f)$ belongs to $\mathcal{E}$ since 1_x and f share the same source, so $[1_x, f]$ is defined. Assume $[1_x, f] = [1_x, f']$. By definition, this means that $(1_x, f) \bowtie (1_x, f')$ is true, hence that there exist h, h' satisfying $h 1_x = h' 1_x$ and $hf' = h'f$. The first equality implies $h = h'$, and then the second one implies $f = f'$, since $\mathcal{C}$ is left-cancellative. So ϕ is injective.

Next, assume $f \in \mathcal{C}(x, y)$ and $g \in \mathcal{C}(y, z)$. Then we have $1_x f = f 1_y$, whence

$$[1_x, f] * [1_y, g] = [1_x 1_x, fg] = [1_x, fg],$$

that is, $\phi(f) * \phi(g) = \phi(fg)$, and ϕ is a functor.

Finally, assume $[f, g] \in [\mathcal{E}](x, y)$. For z the common source of f and g, we find

$$\phi(f)^{-1} * \phi(g) = [1_z, f]^{-1} * [1_z, g] = [f, 1_z] * [1_z, g] = [f, g],$$

so every element of $[\mathcal{E}]$ is a left fraction with respect to the image of $\mathcal{C}$ under ϕ. □

It remains to prove that $[\mathcal{E}]$ is (isomorphic to) the enveloping groupoid $\mathcal{E}nv(\mathcal{C})$.

Claim 7. *The groupoid* $[\mathcal{E}]$ *admits the presentation* (II.3.4).

Proof. By construction, $[\mathcal{E}]$ is generated (as a groupoid) by $\phi(\mathcal{C})$. Next, the (image under ϕ of the) relations of $Rel(\mathcal{C})$ are satisfied in $[\mathcal{E}]$ since ϕ is a functor: every equality $h = fg$ in $\mathcal{C}$ implies $\phi(h) = \phi(f)\phi(g)$ in $[\mathcal{E}]$. Finally, the (image under ϕ of the) relations of $Free(\mathcal{C})$ are satisfied in $[\mathcal{E}]$, since $[\mathcal{E}]$ is a groupoid. So $[\mathcal{E}]$ is a homomorphic image of $\mathcal{E}nv(\mathcal{C})$ (as is obvious from the universality property of the latter).

So the point is to prove that any relation satisfied in $[\mathcal{E}]$ follows from the relations of (II.3.4). As every element of $[\mathcal{E}]$ is a fraction of the form $\phi(f)^{-1}\phi(g)$, hence is represented in terms of the generating family $\phi(\mathcal{C})$ by a path of the form $\overline{\phi(f)}|\phi(g)$, it is sufficient to show that, if two such paths represent the same element of $[\mathcal{E}]$, then they are equivalent with respect to the relations of (II.3.4). So assume that $\overline{\phi(f)}|\phi(g)$ and $\overline{\phi(f')}|\phi(g')$ represent the same element of $[\mathcal{E}]$. Applying Claim 1 to $[\mathcal{E}]$ and ϕ, we know that $(f,g) \bowtie (f',g')$ holds, that is, there exist h, h' in $\mathcal{C}$ satisfying $hf' = h'f$ and $hg' = h'g$. Consequently, we obtain

$$\begin{aligned}\overline{\phi(f)}|\phi(g) &\equiv^{\pm}_{\mathcal{C}} \overline{\phi(f)}|\overline{\phi(h')}|\phi(h')|\phi(g) \equiv^{\pm}_{\mathcal{C}} \overline{\phi(h'f)}|\phi(h'g)\\ &= \overline{\phi(hf')}|\phi(hg') \equiv^{\pm}_{\mathcal{C}} \overline{\phi(f')}|\overline{\phi(h)}|\phi(h)|\phi(g') \equiv^{\pm}_{\mathcal{C}} \overline{\phi(f')}|\phi(g'). \qquad \square\end{aligned}$$

This completes the proof of Ore's theorem (Proposition II.3.11).

Ore subcategories

Proposition II.3.18 (**left-Ore subcategory**). *Assume that* $\mathcal{C}_1$ *is a left-Ore subcategory of a left-Ore category* $\mathcal{C}$.

(i) *The inclusion of* $\mathcal{C}_1$ *in* $\mathcal{C}$ *extends to an embedding of* $\mathcal{E}nv(\mathcal{C}_1)$ *into* $\mathcal{E}nv(\mathcal{C})$.

(ii) *One has* $\mathcal{C}_1 = \mathcal{E}nv(\mathcal{C}_1) \cap \mathcal{C}$ *if and only if* $\mathcal{C}_1$ *is closed under right-quotient in* $\mathcal{C}$.

Proof. (i) We recall from the proof of Proposition II.3.11 (Ore's theorem) that $\mathcal{E}nv(\mathcal{C})$ consists of the equivalence classes of pairs (f, g) of elements of $\mathcal{C}$ with the same source under the relation $\bowtie$ such that $(f,g) \bowtie (f',g')$ holds if and only if there exist h, h' satisfying $hf' = h'f$ and $hg' = h'g$. In the current context, two equivalence relations arise for pairs of elements of $\mathcal{C}_1$, namely one for $\mathcal{E}nv(\mathcal{C})$, where h and h' are supposed to lie in $\mathcal{C}$, that will be denoted by $\bowtie_{\mathcal{C}}$, and one for $\mathcal{E}nv(\mathcal{C}_1)$, where h and h' lie in $\mathcal{C}_1$, denoted by $\bowtie_{\mathcal{C}_1}$. As $\mathcal{C}_1$ is a subcategory of $\mathcal{C}$, an equality $hf' = h'f$ in $\mathcal{C}_1$ implies the same equality in $\mathcal{C}$ and, therefore, $(f,g) \bowtie_{\mathcal{C}_1} (f',g')$ implies $(f,g) \bowtie_{\mathcal{C}} (f',g')$. Therefore, we obtain a well-defined map ι from $\mathcal{E}nv(\mathcal{C}_1)$ to $\mathcal{E}nv(\mathcal{C})$ by mapping, for all f, g in $\mathcal{C}_1$ with the same source, the $\bowtie_{\mathcal{C}_1}$-class of (f, g) to its $\bowtie_{\mathcal{C}}$-class.

Next, for g in $\mathcal{C}_1$ with source x, the image under ι of g, identified with the $\bowtie_{\mathcal{C}_1}$-class of $(1_x, g)$, is the $\bowtie_{\mathcal{C}}$-class of $(1_x, g)$, itself identified with g: in other words, ι extends the inclusion of $\mathcal{C}_1$ in $\mathcal{C}$.

Then, with the notation of Claim 4 in the proof of Proposition II.3.11, if we have $[f, g] = (f_1, g_1) * (f_2, g_2)$ in $\mathcal{C}_1$, then there exist h_1, h_2 in $\mathcal{C}_1$ satisfying $h_1g_1 = h_2, f_2$

and $(f, g) \bowtie_{\mathcal{C}_1} (h_1 f_1, h_2 g_2)$. Now h_1 and h_2 belong to $\mathcal{C}$ as well, and similar relations hold in $\mathcal{C}$. It follows that ι preserves the products of $\mathcal{E}nv(\mathcal{C}_1)$ and $\mathcal{E}nv(\mathcal{C})$ and, therefore, it is a functor from $\mathcal{E}nv(\mathcal{C}_1)$ to $\mathcal{E}nv(\mathcal{C})$.

It remains to establish that ι is injective. This follows from the assumption that $\mathcal{C}_1$ is a left-Ore subcategory of $\mathcal{C}$. Indeed, assume $f, g, f', g' \in \mathcal{C}_1$ with $(f, g) \bowtie_{\mathcal{C}} (f', g')$. So there exist h, h' in $\mathcal{C}$ satisfying $hf' = h'f$ and $hg' = h'g$. Necessarily, f and f' have the same target, so they admit a common left-multiple in $\mathcal{C}_1$: there exist h_1, h_1' in $\mathcal{C}_1$ satisfying $h_1 f' = h_1' f$. Now, by Claim 2 in the proof of Proposition II.3.11, as we have $(f, g) \bowtie (f', g')$ in $\mathcal{C}$, the equality $h_1 f' = h_1' f$ implies $h_1 g' = h_1' g$ and, therefore, h_1 and h_1' witness that $(f, g) \bowtie (f', g')$ holds in $\mathcal{C}_1$. So, for f, g, f', g' in $\mathcal{C}_1$, the relations $(f, g) \bowtie_{\mathcal{C}} (f', g')$ and $(f, g) \bowtie_{\mathcal{C}_1} (f', g')$ are equivalent, and ι is injective.

(ii) Assume that $\mathcal{C}_1 = \mathcal{E}nv(\mathcal{C}_1) \cap \mathcal{C}$ holds and that, in $\mathcal{C}$, we have $g = fg'$ with f, g in $\mathcal{C}_1$. In $\mathcal{E}nv(\mathcal{C}_1)$, there exists an element $f^{-1}g$ satisfying $f \cdot f^{-1}g = g$. So, in $\mathcal{C}$, we have $f \cdot f^{-1}g = g = f \cdot g'$, whence $f^{-1}g = g' \in \mathcal{E}nv(\mathcal{C}_1) \cap \mathcal{C}$. The assumption $\mathcal{C}_1 = \mathcal{E}nv(\mathcal{C}_1) \cap \mathcal{C}$ implies $g' \in \mathcal{C}_1$. So $\mathcal{C}_1$ must be closed under right-quotient in $\mathcal{C}$.

Conversely, assume that $\mathcal{C}_1$ is closed under right-quotient in $\mathcal{C}$, and let h be an element of $\mathcal{E}nv(\mathcal{C}_1) \cap \mathcal{C}$. By definition, we have $h = f^{-1}g$ with f, g in $\mathcal{C}_1$, so $fh = g$ holds in $\mathcal{C}$. As f and g lie in $\mathcal{C}_1$, the assumption implies $h \in \mathcal{C}_1$, and we deduce $\mathcal{E}nv(\mathcal{C}_1) \cap \mathcal{C} \subseteq \mathcal{C}_1$. The inclusion $\mathcal{C}_1 \subseteq \mathcal{E}nv(\mathcal{C}_1) \cap \mathcal{C}$ is always true, so we have $\mathcal{C}_1 = \mathcal{E}nv(\mathcal{C}_1) \cap \mathcal{C}$. □

2 Working with presented categories

Right-reversing: termination

Lemma II.4.23. *Assume that $(\mathcal{S}, \mathcal{R})$ is a category presentation, w, w' are signed $\mathcal{S}$-paths, and $w \curvearrowright_{\mathcal{R}}^{n} w'$ holds. For each decomposition $w = w_1 w_2$, there exist a decomposition $w' = w_1' w_0' w_2'$ and two $\mathcal{S}$-paths u', v' satisfying*

$$w_1 \curvearrowright_{\mathcal{R}}^{n_1} w_1' \overline{u'}, \quad \overline{u'} v' \curvearrowright_{\mathcal{R}}^{n_0} w_0', \quad \text{and} \quad w_2 \curvearrowright_{\mathcal{R}}^{n_2} v' w_2'$$

with $n_1 + n_0 + n_2 = n$.

Proof. We use induction on n. For $n = 0$, we have $w' = w$, and the result is obvious with $w_1' = w_1$, $w_2' = w_2$, $u' = v' = \varepsilon_x$, where x is the target of w_1, and $n_1 = n_2 = n_0 = 0$.

Assume now $n \geqslant 1$, and let w'' be the second signed path in a length-n reversing sequence from w to w', so that we have $w \curvearrowright_{\mathcal{R}}^{1} w'' \curvearrowright_{\mathcal{R}}^{n-1} w'$. By definition, w'' is obtained from w by reversing some pattern $\overline{s}t$ into $v\overline{u}$ such that $sv = tu$ is a relation of $\mathcal{R}$, or $t = s$ holds and u, v are empty. Three cases may occur according to the position of the considered pattern $\overline{s}t$ inside w.

Assume first that the pattern is entirely included in w_1, say $w_1 = \hat{w}_1 \overline{s} t \hat{w}_{1,1}$. Then we have $w'' = w_1'' w_2''$ with $w_1'' = \hat{w}_1 v \overline{u} \hat{w}_{1,1}$ and $w_2'' = w_2$. Applying the induction hypothesis to $w'' \curvearrowright_{\mathcal{R}}^{n-1} w'$ and $w'' = w_1'' w_2''$ yields paths w_1', w_0', w_2', u', v' satisfying

$w_1'' \curvearrowright_{\mathcal{R}}^{n_1'} w_1'\overline{u'}$, $\overline{u'}v' \curvearrowright_{\mathcal{R}}^{n_0} w_0'$, and $w_2'' \curvearrowright_{\mathcal{R}}^{n_2} v'w_2'$ with $n_1' + n_0 + n_2 = n - 1$. We deduce $w_1 \curvearrowright_{\mathcal{R}}^{n_1'+1} w_1'\overline{u'}$, and the claimed result is established.

The argument is similar when the reversed pattern is entirely included in w_2.

So we are left with the nontrivial case, namely when the reversed pattern is split between w_1 and w_2. In this case, there exist signed paths $\hat{w}_1, \hat{w}_2$ such that w_1 is $\hat{w}_1\overline{s}$ and w_2 is $t\hat{w}_2$, in which case we have $w'' = \hat{w}_1 v\overline{u}\hat{w}_2$, see Figure 4. Applying the induction hypothesis to the relation $(\hat{w}_1 v)(\overline{u}\hat{w}_2) \curvearrowright_{\mathcal{R}}^{n-1} w'$, we obtain a decomposition $w' = w_1''w_0''w_2''$ and $\mathcal{S}$-paths u'', v'' satisfying $\hat{w}_1 v \curvearrowright_{\mathcal{R}}^{n_1''} w_1''\overline{u''}$, $\overline{u''}v'' \curvearrowright_{\mathcal{R}}^{n_0''} w_0''$, and $\overline{u}\hat{w}_2 \curvearrowright_{\mathcal{R}}^{n_2''} v''w_2''$ with $n_1'' + n_0'' + n_2'' = n - 1$. Now, as $n_1'' < n$ holds, we can apply the induction hypothesis to the relation $\hat{w}_1 v \curvearrowright_{\mathcal{R}}^{n_1''} w_1''\overline{u''}$ and the decomposition of $\hat{w}_1 v$ into $\hat{w}_1$ and v, obtaining a decomposition $w_1'' = w_1' w_{0,1}$ and an $\mathcal{S}$-path $\hat{u}$ satisfying $\hat{w}_1 \curvearrowright_{\mathcal{R}}^{n_1} w_1'\overline{\hat{u}}$ and $\overline{\hat{u}}v \curvearrowright_{\mathcal{R}}^{n_{0,1}} w_{0,1}\overline{u''}$ (there is no third piece in the final decomposition because the second piece of the initial decomposition, here v, is an $\mathcal{S}$-path) with $n_1 + n_{0,1} = n_1''$. Symmetrically, we obtain a decomposition $w_2'' = w_{0,2}w_2'$ and $\hat{v}$ in $\mathcal{S}^*$ satisfying $\overline{u}\hat{v} \curvearrowright_{\mathcal{R}}^{n_{0,2}} v''w_{0,2}$ and $\hat{w}_2 \curvearrowright_{\mathcal{R}}^{n_2} \hat{v}w_2'$ with $n_{0,2} + n_2 = n_2''$. Put $u' = s\hat{u}$, $v' = t\hat{v}$, $w_0' = w_{0,1}w_0''w_{0,2}$. Then we have $w_1 = \hat{w}_1\overline{s} \curvearrowright_{\mathcal{R}}^{n_1} w_1'\overline{\hat{u}}\overline{s} = w_1'\overline{u'}$, $\overline{u'}v' \curvearrowright_{\mathcal{R}}^{n_0} w_0'$ with $n_0 = n_{0,1} + n_0'' + n_{0,2} + 1$, and $w_2 = t\hat{w}_2 \curvearrowright_{\mathcal{R}}^{n_2} t\hat{v}w_2' = v'w_2'$. As we have $n_1 + n_0 + n_2 = (n_1 + n_{0,1}) + n_0'' + (n_{0,2} + n_2) + 1 = n$, we are done. □

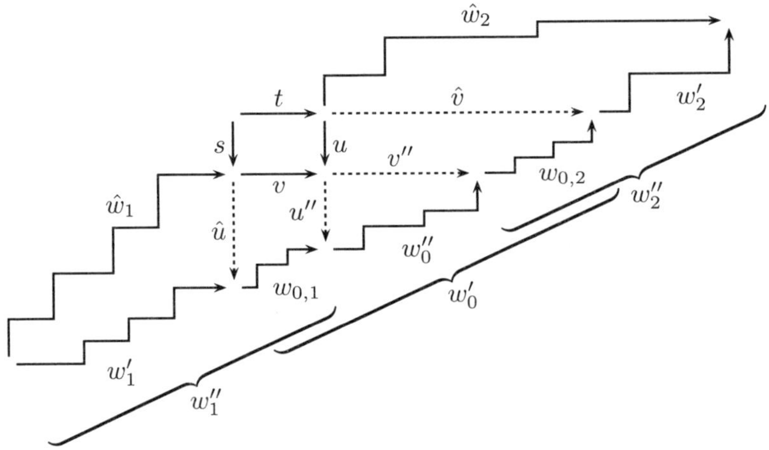

Figure 4. Inductive proof of Lemma II.4.23

Right-reversing: completeness

Lemma II.4.60. *If $(\mathcal{S}, \mathcal{R})$ is a short category presentation and the sharp cube condition is true on $\mathcal{S}$, then the sharp cube condition is true on $\mathcal{S}^*$.*

Proof. First, if at least one of the paths u, v, w is empty, then the sharp cube condition is automatically true for (u, v, w).

Then we prove, using induction on ℓ, that the sharp cube condition holds for all (u, v, w) satisfying $\lg(u)+\lg(v)+\lg(w) = \ell$. For $\ell \leqslant 2$, at least one of u, v, w is empty, and the result follows from the remark above. For $\ell = 3$, either at least one of u, v, w is empty, and the result follows again from the remark above, or we have $\lg(u) = \lg(v) = \lg(w) = 1$, that is, u, v, w belong to $\mathcal{S}$, and the result follows from the assumption that the sharp cube condition is true on $\mathcal{S}$.

Assume now $\ell \geqslant 4$ and consider u, v, w satisfying $\lg(u) + \lg(v) + \lg(w) = \ell$ and

$$\overline{u}w \curvearrowright_{\mathcal{R}} v_1\overline{u_0}, \quad \overline{w}v \curvearrowright_{\mathcal{R}} v_0\overline{u_1}, \quad \overline{u_0}v_0 \curvearrowright_{\mathcal{R}} v_2\overline{u_2}.$$

At least one of u, v, w has length at least 2. Assume $\lg(v) \geqslant 2$ and write $v = v^1v^2$ with both v^1 and v^2 nonempty. By Lemma II.4.24, there exist $u_2^1, u_1^1, v_0^1, v_0^2, v_2^1, v_2^2$ in $\mathcal{S}^*$ satisfying $v_0 = v_0^1v_0^2$, $v_2 = v_2^1v_2^2$ and

$$\overline{w}v^1 \curvearrowright_{\mathcal{R}} v_0^1\overline{u_1^1}, \quad \overline{u_1^1}v^2 \curvearrowright_{\mathcal{R}} v_0^2\overline{u_1}, \quad \overline{u_0}v_0^1 \curvearrowright_{\mathcal{R}} v_2^1\overline{u_2^1} \quad \overline{u_2^1}v_0^2 \curvearrowright_{\mathcal{R}} v_2^2\overline{u_2},$$

see Figure 5. The inequality $\lg(u) + \lg(v^1) + \lg(w) < \ell$ implies that the sharp cube condition holds for (u, v^1, w) and we obtain the existence of u'^1, v'^1, w'^1 satisfying

$$\overline{u}v^1 \curvearrowright_{\mathcal{R}} u'^1\overline{v'^1}, \quad \overline{u'^1}u_1^1 \curvearrowright_{\mathcal{R}} w'^1\overline{u_2^1}, \quad \overline{v_1}v'^1 \curvearrowright_{\mathcal{R}} v_2^1\overline{w'^1}.$$

As $(\mathcal{S}, \mathcal{R})$ is short, we must have $\lg(u'^1) \leqslant \lg(u)$ and $\lg(u_1^1) \leqslant \lg(w)$, hence $\lg(u'^1) + \lg(v^2) + \lg(u_1^1) < \ell$. The induction hypothesis implies that the sharp cube condition is true for (u'^1, v^2, u_1^1) and we obtain the existence of u', v'^2, w' satisfying

$$\overline{u'^1}v^2 \curvearrowright_{\mathcal{R}} v'^2\overline{u'}, \quad \overline{u'}u_1 \curvearrowright_{\mathcal{R}} w'\overline{u_2}, \quad \overline{w'^1}v'^2 \curvearrowright_{\mathcal{R}} v_2^2\overline{w'}.$$

Put $v' = v'^1v'^2$. Then we have

$$\overline{u}v \curvearrowright_{\mathcal{R}} v'\overline{u'}, \quad \overline{u'}u_1 \curvearrowright_{\mathcal{R}} w'\overline{u_2}, \quad \overline{v_1}v' \curvearrowright_{\mathcal{R}} v_2\overline{w'},$$

and we conclude that the sharp cube condition is true for (u, v, w).

The argument for $\lg(u) \geqslant 2$ is identical, as u and v play symmetric roles. The argument for $\lg(w) \geqslant 2$ is similar, writing $w = w^1w^2$ and beginning this time with w^1. □

Lemma II.4.61. *If $(\mathcal{S}, \mathcal{R})$ is a category presentation that contains no ε-relation and the cube condition is true on $\mathcal{S}^*$, then right-reversing is complete for $(\mathcal{S}, \mathcal{R})$.*

Proof. Write $u \cong_{\mathcal{R}}^+ v$ for $\overline{u}v \curvearrowright_{\mathcal{R}} \varepsilon_y$, with y the target of u. It follows from the definition of right-reversing that $\cong_{\mathcal{R}}^+$ is reflexive, symmetric, and compatible with concatenation on the left and on the right. We claim that the cube condition implies that $\cong_{\mathcal{R}}^+$ is transitive. Indeed, assume $u \cong_{\mathcal{R}}^+ w \cong_{\mathcal{R}}^+ v$. By assumption, we have $\overline{u}w \curvearrowright_{\mathcal{R}} \varepsilon_y$ and $\overline{w}v \curvearrowright_{\mathcal{R}} \varepsilon_y$, where y is the (necessarily common) target of u, v, w. Then $\overline{\varepsilon_y}\varepsilon_y \curvearrowright_{\mathcal{R}} \varepsilon_y\overline{\varepsilon_y}$ also holds, and the cube condition for u, v, w implies the existence of u', v', w' satisfying

$$\overline{u}v \curvearrowright_{\mathcal{R}} v'\overline{u'}, \quad \varepsilon_y \equiv_{\mathcal{R}}^+ u'w', \quad \varepsilon_y \equiv_{\mathcal{R}}^+ v'w'.$$

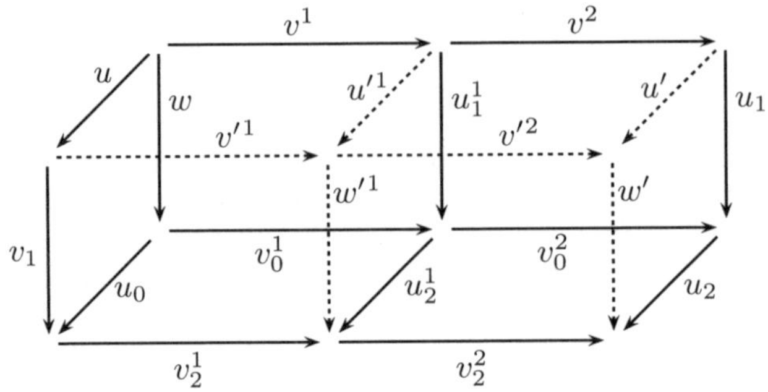

Figure 5. Induction for the sharp cube condition.

The assumption that $(\mathcal{S}, \mathcal{R})$ contains no ε-relation then implies $u' = v' = w' = \varepsilon_y$, and $u \cong^+_{\mathcal{R}} v$ holds.

Moreover, assume that $u = v$ is a relation of $\mathcal{R}$. Then, by definition, the relation $\overline{u}v \curvearrowright_{\mathcal{R}} \varepsilon_y$ is satisfied, where y is the target of u and v, and, therefore, $u \cong^+_{\mathcal{R}} v$ is true. So $\cong^+_{\mathcal{R}}$ is a congruence on $\mathcal{S}^*$ that includes all pairs $\{u, v\}$ such that $u = v$ belongs to $\mathcal{R}$. As $\equiv^+_{\mathcal{R}}$ is, by definition, the smallest congruence with these properties, we deduce that $\equiv^+_{\mathcal{R}}$ is included in $\cong^+_{\mathcal{R}}$, that is, $u \equiv^+_{\mathcal{R}} v$ implies $u \cong^+_{\mathcal{R}} v$.

On the other hand, by Proposition II.4.34 (reversing implies equivalence), $u \cong^+_{\mathcal{R}} v$ implies $u \equiv^+_{\mathcal{R}} v$, so, finally, $u \equiv^+_{\mathcal{R}} v$ is equivalent to $u \cong^+_{\mathcal{R}} v$, that is, to $\overline{u}v \curvearrowright_{\mathcal{R}} \varepsilon.$. By Lemma II.4.42, which is relevant since $(\mathcal{S}, \mathcal{R})$ contains no ε-relation, this implies that right-reversing is complete for $(\mathcal{S}, \mathcal{R})$. □

Lemma II.4.62. *If $(\mathcal{S}, \mathcal{R})$ is a category presentation that is right-Noetherian, contains no ε-relation, and the cube condition is true on $\mathcal{S}$, then right-reversing is complete for $(\mathcal{S}, \mathcal{R})$.*

Proof. We fix a right-Noetherianity witness λ^* for $(\mathcal{S}, \mathcal{R})$ (Definition II.2.31). For every s in $\mathcal{S}$, the assumption that $(\mathcal{S}, \mathcal{R})$ contains no ε-relation implies that (the class of) s is not invertible in $\langle \mathcal{S} \mid \mathcal{R} \rangle^+$ and, therefore, we have a strict inequality $\lambda^*(sw) > \lambda^*(w)$ for every w in $\mathcal{S}^*$ such that sw is defined. For α an ordinal, let $\mathcal{E}_\alpha$ be the statement

$(\mathcal{E}_\alpha)$ Every quadruple $(u, v, \hat{u}, \hat{v})$ satisfying $u\hat{v} \equiv^+_{\mathcal{R}} v\hat{u}$ and $\lambda^*(u\hat{v}) \leqslant \alpha$ is $\curvearrowright$-factorable.

We shall establish that $\mathcal{E}_\alpha$ is true for every α using induction on α, that is, we shall prove that $\mathcal{E}_\alpha$ is true whenever $\mathcal{E}_\beta$ is true for every $\beta < \alpha$. This will imply that right-reversing is complete for $(\mathcal{S}, \mathcal{R})$.

The proof of $\mathcal{E}_\alpha$ uses a nested induction. For ℓ a natural number, let $\mathcal{E}_{\alpha,\ell}$ be the statement

$(\mathcal{E}_{\alpha,\ell})$ Every quadruple $(u, v, \hat{u}, \hat{v})$ satisfying $u\hat{v} \equiv^+_{\mathcal{R}} v\hat{u}$, $\lambda^*(u\hat{v}) \leqslant \alpha$, and $\lg(u) + \lg(v) \leqslant \ell$ is $\curvearrowright$-factorable.

First, $\mathcal{E}_{\alpha,1}$ is true for every α. Indeed, $\lg(u) + \lg(v) \leqslant 1$ implies that u or v is empty. Assume for instance that u is empty. Then the choice $u' = \varepsilon_y$ (y the target of v), $v' = v$, $w' = \hat{u}$ witnesses that $(u, v, \hat{u}, \hat{v})$ is $\curvearrowright$-factorable. The argument is similar for v empty.

We turn to $\mathcal{E}_{\alpha,2}$. If u or v is empty, the above argument shows that $(u, v, \hat{u}, \hat{v})$ is $\curvearrowright$-factorable. So we are left with the case when u and v have length one, that is, belong to $\mathcal{S}$. We treat this case using a third, nested induction, on the combinatorial distance $\mathrm{dist}_{\mathcal{R}}(u\hat{v}, v\hat{u})$ between $u\hat{v}$ and $v\hat{u}$, that is, the minimal number of relations of $\mathcal{R}$ needed to transform $u\hat{v}$ into $v\hat{u}$. So let $\mathcal{E}_{\alpha,2,d}$ be the statement

$(\mathcal{E}_{\alpha,2,d})$ Every quadruple $(u, v, \hat{u}, \hat{v})$ satisfying $u\hat{v} \equiv^+_{\mathcal{R}} v\hat{u}$, $\lambda^*(u\hat{v}) \leqslant \alpha$, $\lg(u) = \lg(v) = 1$, and $\mathrm{dist}_{\mathcal{R}}(u\hat{v}, v\hat{u}) \leqslant d$ is $\curvearrowright$-factorable.

For $d = 0$, we have $u = v$ and $\hat{u} = \hat{v}$, and $u' = v' = \varepsilon_y$ (y the target of s), $w' = \hat{u}$ witness that $(u, v, \hat{u}, \hat{v})$ is $\curvearrowright$-factorable, so $\mathcal{E}_{\alpha,2,0}$ is true.

Assume now $d = 1$, that is, $v\hat{u}$ is obtained from $u\hat{v}$ by applying one relation of $\mathcal{R}$. Two cases may arise according to the position where the relation is applied. If the position does not involve the first generator, we have $u = v$ and $\hat{u} \equiv^+_{\mathcal{R}} \hat{v}$, and the choice $u' = v' = \varepsilon_y$ (y the target of u), $w' = \hat{u}$ witness that $(u, v, \hat{u}, \hat{v})$ is $\curvearrowright$-factorable. Otherwise, there exists u', v', and w' such that $uv' = vu'$ is a relation of $\mathcal{R}$ and we have $\hat{u} = u'w'$ and $\hat{v} = v'w'$. In this case, $\overline{u}v \curvearrowright_{\mathcal{R}} v'\overline{u'}$ holds by definition and, again, u', v', w' witness that $(u, v, \hat{u}, \hat{v})$ is $\curvearrowright$-factorable. So, in every case, $\mathcal{E}_{\alpha,2,1}$ is true.

Next, assume $d \geqslant 2$. Let $w\hat{w}$ be an intermediate path in an $\mathcal{R}$-derivation from $u\hat{v}$ to $v\hat{u}$. Then we have $u\hat{v} \equiv^+_{\mathcal{R}} w\hat{w}$ and $v\hat{u} \equiv^+_{\mathcal{R}} w\hat{w}$, with $\mathrm{dist}_{\mathcal{R}}(u\hat{v}, w\hat{w}) < d$ and $\mathrm{dist}_{\mathcal{R}}(v\hat{u}, w\hat{w}) < d$. By the induction hypothesis, $(u, w, \hat{w}, \hat{v})$ and $(w, v, \hat{u}, \hat{w})$ are $\curvearrowright$-factorable. This means that there exist $u_0, v_0, u_1, v_1, \hat{u}_0, \hat{v}_0$ in $\mathcal{S}^*$ satisfying

$$\overline{u}w \curvearrowright_{\mathcal{R}} v_1\overline{u_0}, \quad \hat{v} \equiv^+_{\mathcal{R}} v_1\hat{v}_0, \quad \hat{w} \equiv^+_{\mathcal{R}} u_0\hat{v}_0, \tag{2.1}$$

$$\overline{w}v \curvearrowright_{\mathcal{R}} v_0\overline{u_1}, \quad \hat{u} \equiv^+_{\mathcal{R}} u_1\hat{u}_0, \quad \hat{w} \equiv^+_{\mathcal{R}} v_0\hat{u}_0, \tag{2.2}$$

see Figure 6. So we have $u_0\hat{v}_0 \equiv^+_{\mathcal{R}} v_0\hat{u}_0$. Let $\beta = \lambda^*(u_0\hat{v}_0)$. By construction, we have $\beta = \lambda^*(\hat{w}) < \lambda^*(w\hat{w}) \leqslant \alpha$. Hence, by induction hypothesis, $\mathcal{E}_\beta$ is true. Therefore, $(u_0, v_0, \hat{u}_0, \hat{v}_0)$ is $\curvearrowright$-factorable, that is, there exist u_0', v_0', w_0' in $\mathcal{S}^*$ satisfying

$$\overline{u_0}v_0 \curvearrowright_{\mathcal{R}} v_0'\overline{u_0'}, \quad \hat{u}_0 \equiv^+_{\mathcal{R}} u_0'w_0', \quad \hat{v}_0 \equiv^+_{\mathcal{R}} v_0'w_0'. \tag{2.3}$$

The first relations in (2.1), (2.2), and (2.3) imply $\overline{u}w\overline{w}v \curvearrowright_{\mathcal{R}} v_1v_0'\overline{u_0'}\,\overline{u_1}$. As the cube condition is true for (u, v, w), we deduce the existence of u', v', w_1 satisfying

$$\overline{u}v \curvearrowright_{\mathcal{R}} v'\overline{u'}, \quad u_1u_0' \equiv^+_{\mathcal{R}} u'w_1, \quad v_1v_0' \equiv^+_{\mathcal{R}} v'w_1.$$

Putting $w' = w_1w_0'$, we obtain

$$\hat{u} \equiv^+_{\mathcal{R}} u_1u_0'w_0' \equiv^+_{\mathcal{R}} u'w_1w_0' = u'w', \quad \hat{v} \equiv^+_{\mathcal{R}} v_1v_0'w_0' \equiv^+_{\mathcal{R}} v'w_1w_0' = v'w',$$

and u', v', w' witness that $(u, v, \hat{u}, \hat{v})$ is $\curvearrowright$-factorable. So $\mathcal{E}_{\alpha,2,d}$ holds for every d, and therefore so does $\mathcal{E}_{\alpha,2}$.

Finally, assume $\ell \geqslant 3$. If $\lg(u) + \lg(v) \leqslant 2$ holds, then $\mathcal{E}_{\alpha,2}$ gives the result. Otherwise, at least one of u, v has length at least two. Assume that v does. Write $v = v_1v_2$ with v_1, v_2 nonempty. Let $\ell_1 = \lg(u) + \lg(v_1)$. Then we have $u\hat{v} \equiv^+_{\mathcal{R}} v_1(v_2\hat{u})$ with $\lg(u) + \lg(v_1) = \ell_1 < \ell$. By the induction hypothesis, $\mathcal{E}_{\alpha,\ell_1}$ is true, so there exist u_1', v_1', w_1' satisfying

$$\overline{u}v_1 \curvearrowright_{\mathcal{R}} v_1'\overline{u_1'}, \quad v_2\hat{u} \equiv^+_{\mathcal{R}} u_1'w_1', \quad \hat{v} \equiv^+_{\mathcal{R}} v_1'w_1', \tag{2.4}$$

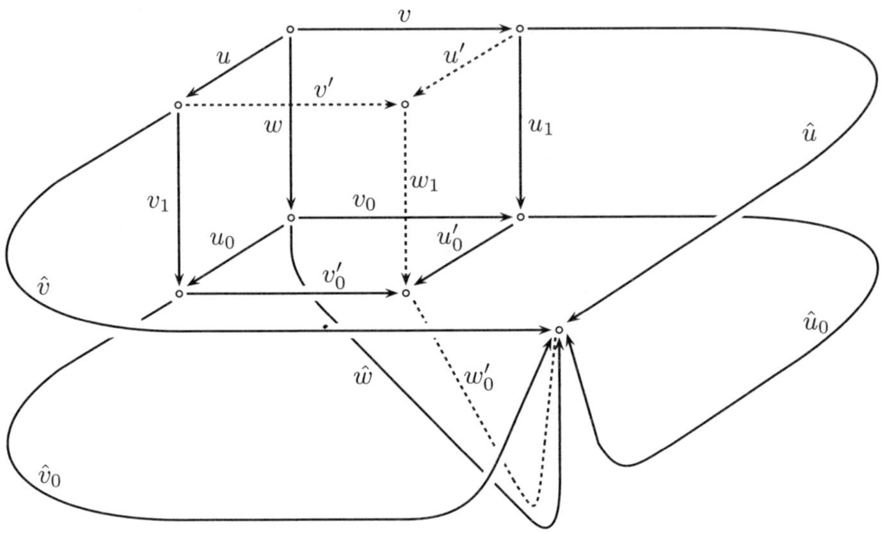

Figure 6. Inductive argument for $\mathcal{E}_{\alpha,2,d}$ with $d \geqslant 2$.

see Figure 7. Let $\beta = \lambda^*(v_2\hat{u})$. Then we have $\beta = \lambda^*(v_2\hat{u}) < \lambda^*(v_1v_2\hat{u}) \leqslant \alpha$. Hence, by the induction hypothesis, $\mathcal{E}_\beta$ is true. Therefore $u'_1w'_1 \equiv^+_{\mathcal{R}} v_2\hat{u}_0$ implies that $(u'_1, v_2, \hat{u}, w'_1)$ is $\curvearrowright$-factorable, that is, there exist u', v'_2, w' satisfying

$$\overline{u'_1}v_2 \curvearrowright_{\mathcal{R}} v'_2\overline{u'}, \quad \hat{u} \equiv^+_{\mathcal{R}} u'w', \quad w'_1 \equiv^+_{\mathcal{R}} v'_2w'. \tag{2.5}$$

Put $v' = v'_1v'_2$. Then we have $\overline{u}v \curvearrowright_{\mathcal{R}} v'\overline{u'}$, and $\hat{v} \equiv^+_{\mathcal{R}} v'_1w'_1 \equiv^+_{\mathcal{R}} v'_1v'_2w' = v'w'$. So u', v', w' witness that $(u, v, \hat{u}, \hat{v})$ is $\curvearrowright$-factorable, and $\mathcal{E}_\alpha$ is true. □

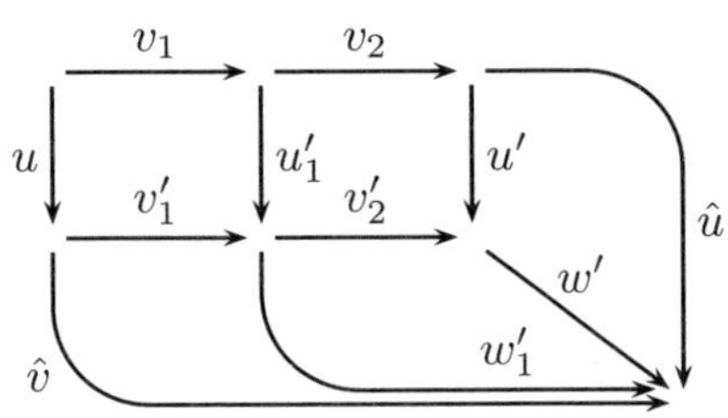

Figure 7. Inductive argument for $\mathcal{E}_{\alpha,\ell}$ with $\ell \geqslant 3$.

See Exercise 131 for a (partial) extension of Lemma II.4.62 to the case when ε-relations may exist.

Lemma II.4.63. *If $(\mathcal{S}, \mathcal{R})$ is a maximal right-triangular presentation, then right-reversing is complete for $(\mathcal{S}, \mathcal{R})$.*

Proof. We denote by $s_1, s_2, \ldots$ the enumeration of $\mathcal{S}$ involved in the right-triangular structure, and by θ the associated syntactic right-complement. The assumption that $(\mathcal{S}, \mathcal{R})$ is maximal means that $\theta(s_i, s_j)$ is defined for all i, j. As in the previous proof, we write $\mathrm{dist}_{\mathcal{R}}(u, v)$ for the combinatorial distance between two $\mathcal{R}$-equivalent paths, that is, the length of a shortest $\mathcal{R}$-derivation from u to v. We shall prove, using induction on d, that $\mathrm{dist}_{\mathcal{R}}(u, v) \leqslant d$ implies $\overline{u}\, v \curvearrowright_{\mathcal{R}} \varepsilon_y$ (with y the common target of u and v): by Lemma II.4.42, this is enough to conclude that right-reversing is complete for $(\mathcal{S}, \mathcal{R})$.

Assume that w is a nonempty $\mathcal{S}$-path beginning with s_i, say $w = s_i w'$. We then define $I(w) = s_i$, $T(w) = w'$ (like "tail"), and $E(w) = s_1\theta(s_1, s_i)w'$. By assumption, $s_1\theta(s_1, s_i) = s_i$ is a relation of $\mathcal{R}$, so $E(w) \equiv^+_{\mathcal{R}} w$ always holds.

Claim 1. *Assume that u, v are nonempty $\mathcal{S}$-paths satisfying $\mathrm{dist}_{\mathcal{R}}(u, v) = 1$. Then we have $\mathrm{dist}_{\mathcal{R}}(E(u), E(v)) \leqslant 1$, and exactly one of the following:*

(i) $I(u) = I(v)$ *and* $\mathrm{dist}_{\mathcal{R}}(T(u), T(v)) = 1$.

(ii) $I(u) \neq I(v)$ *and* $E(u) = E(v)$.

Proof. The assumption $\mathrm{dist}_{\mathcal{R}}(u, v) = 1$ means that there exist a number $p \geqslant 1$ and a relation of $\mathcal{R}$ such that v is obtained from u by applying that relation to its subpath starting at position p. For $p \geqslant 2$, the initial letter is not changed, that is, we have $I(u) = I(v)$, whereas $T(v)$ is obtained from $T(u)$ by applying a relation of $\mathcal{R}$ at position $p - 1$, and we have $\mathrm{dist}_{\mathcal{R}}(T(u), T(v)) = 1$.

For $p = 1$, there exist $i < j$ and u', v' satisfying $u = s_i u'$ and $v = s_j v'$ with $u' = \theta(s_i, s_j)v'$, or vice versa. Then we find $I(u) = s \neq N(s) = I(v)$ and the definition gives $E(u) = s_1\theta(s_1, s_i)u' = s_1\theta(s_1, s_i)\theta(s_i, s_j)v' = s_1\theta(s_1, s_j)v' = E(v)$. □

Claim 2. *Assume that u, v are nonempty $\mathcal{S}$-paths satisfying $\mathrm{dist}_{\mathcal{R}}(u, v) = d$. Then we have at least one of the following:*

(i) $I(u) = I(v)$ *and* $\mathrm{dist}_{\mathcal{R}}(T(u), T(v)) = d$.

(ii) $\mathrm{dist}_{\mathcal{R}}(E(u), E(v)) < d$.

Proof. Let $(w_0, \ldots, w_d)$ be an $\mathcal{R}$-derivation from u to v. Two cases are possible. Assume first that the initial letter never changes in the considered derivation, that is, $I(w_k) = I(u)$ holds for every k. Then all one-step derivations (w_k, w_{k+1}) correspond to case (i) in Claim 1. The latter implies $I(w_k) = I(w_{k+1})$ and $\mathrm{dist}_{\mathcal{R}}(T(w_k), T(w_{k+1})) = 1$ for every k, whence $I(u) = I(v)$ and $\mathrm{dist}_{\mathcal{R}}(T(u), T(v)) = d$.

Assume now that the initial letter changes at least once in $(w_0, \ldots, w_d)$, say we have $I(w_i) \neq I(w_{i+1})$. Then (w_i, w_{i+1}) corresponds to case (ii) in Claim 1. So, we have $E(w_i) = E(w_{i+1})$. On the other hand, Claim 1 implies $\mathrm{dist}_{\mathcal{R}}(E(w_k), E(w_{k+1})) \leqslant 1$ for $k \neq i$. Summing up, we obtain $\mathrm{dist}_{\mathcal{R}}(E(u), E(v)) < d$. □

We now show, using induction on $d \geqslant 0$, and, for each value of d, on $\max(\lg(u), \lg(v))$, that $\mathrm{dist}_{\mathcal{R}}(u, v) = d$ implies $\overline{u}\, v \curvearrowright_{\mathcal{R}} \varepsilon_y$. Assume first $d = 0$. Then the assumption implies $u = v$, in which case $\overline{u}\, v$ reverses to the empty path by $\lg(u)$ successive deletions of patterns $\overline{s}|s$.

Now assume $d \geqslant 1$. Then u and v must be nonempty. Assume first that $I(u) = I(v)$ and $\mathrm{dist}_{\mathcal{R}}(T(u), T(v)) = d$ hold.

Then, by definition, we have

$$\max(\lg(T(u)), \lg(T(v))) = \max(\lg(u), \lg(v)) - 1,$$

so the induction hypothesis implies $\overline{T(u)}\, T(v) \curvearrowright_{\mathcal{R}} \varepsilon_y$. On the other hand, as $I(u)$ and $I(v)$ are equal, we have

$$\overline{u}\, v = \overline{T(u)}\, \overline{I(u)}\, I(v)\, T(v) \curvearrowright_{\mathcal{R}} \overline{T(u)}\, T(v).$$

By transitivity of reversing, we deduce that $\overline{u}\, v$ right-reverses to ε_y.

Assume now that at least one of $I(u) = I(v)$ and $\mathrm{dist}_{\mathcal{R}}(T(u), T(v)) = d$ does not hold. Then, by Claim 2, we must have $\mathrm{dist}_{\mathcal{R}}(E(u), E(v)) = d'$ for some $d' < d$. Then the induction hypothesis implies $\overline{E(u)}\, E(v) \curvearrowright_{\mathcal{R}} \varepsilon_y$. Assume $I(u) = s_i$ and $I(v) = s_j$ with, say, $i \leqslant j$. By definition, we have

$$E(u) = s_1\, \theta(s_1, s_i)\, T(u), \quad E(v) = s_1\, \theta(s_1, s_j)\, T(v) = s_1\, \theta(s_1, s_i)\, \theta(s_i, s_j)\, T(v),$$

so that, if ℓ is the length of $s_1\, \theta(s_1, s_i)$, the first ℓ steps in any reversing sequence starting from $\overline{E(u)}\, E(v)$ must be

$$\begin{aligned}\overline{E(u)}\, E(v) = \overline{T(u)}\; \overline{\theta(s_1, s_i)}\; \overline{s_1}\, s_1\, \theta(s_1, s_i)\, \theta(s_i, s_j)\, T(v) \\ \curvearrowright_{\mathcal{R}} \overline{T(u)}\, \theta(s_i, s_j)\, T(v).\end{aligned}$$

It follows that the relation $\overline{E(u)}\, E(v) \curvearrowright_{\mathcal{R}} \varepsilon_y$ deduced above from the induction hypothesis implies

$$\overline{T(u)}\, \theta(s_i, s_j)\, T(v) \curvearrowright_{\mathcal{R}} \varepsilon_y. \tag{2.6}$$

Now, let us consider the right-reversing of $\overline{u}\, v$, that is, of $\overline{T(u)}\, \overline{s_i}\, s_j\, T(v)$. By definition, the only relation of $\mathcal{R}$ of the form $s_i... = s_j...$ is $s_i\theta(s_i, s_j) = s_j$, so the first step in the reversing must be $\overline{T(u)}\, \overline{s_i}\, s_j\, T(v) \curvearrowright_{\mathcal{R}} \overline{T(u)}\, \theta(s_i, s_j)\, T(v)$. Concatenating this with (2.6), we deduce $\overline{u}\, v \curvearrowright_{\mathcal{R}} \varepsilon_y$, which completes the induction. □

Exercises

Exercise 131 (invertible elements). Say that a category presentation $(\mathcal{S}, \mathcal{R})$ has *explicitly invertible generators only* if there exists $\mathcal{S}^{\times} \subseteq \mathcal{S}$ such that, for every s in $\mathcal{S}^{\times}$, there exists $\widetilde{s}$ in $\mathcal{S}^{\times}$ such that the relations $s\widetilde{s} = 1_x$ and $\widetilde{s}s = 1_y$ belong to $\mathcal{R}$ (x the source of s and y its target), these are the only ε-relations in $\mathcal{R}$ and, for all s in $\mathcal{S}^{\times}$ and t in $\mathcal{S} \setminus \mathcal{S}^{\times}$, there exist s' in $\mathcal{S}^{\times}$ and t' in in $\mathcal{S} \setminus \mathcal{S}^{\times}$ such that $st' = ts'$ belongs to $\mathcal{R}$. Show that, if $(\mathcal{S}, \mathcal{R})$ has explicitly invertible generators only, then it is eligible for Lemma II.4.62. [Hint: In the statement $\mathcal{E}_{\alpha,\ell}$ replace $\lg(u)$ and $\lg(v)$ with the number of generators of $\mathcal{S} \setminus \mathcal{S}^{\times}$ in u and v.]

Bibliography

[1] S.I. Adyan, *Defining relations and algorithmic problems for groups and semigroups*, Proc. Steklov Inst. Math., vol. 85, Amer. Math. Soc. (English translation), 1966.

[2] ______, *Fragments of the word Delta in a braid group*, Mat. Zam. Acad. Sci. SSSR **36** (1984), no. 1, 25–34, (Russian); English translation in *Math. Notes of the Acad. Sci. USSR* **36** (1984), no. 1, p. 505-510.

[3] I. Anshel, M. Anshel, and D. Goldfeld, *An algebraic method for public-key cryptography*, Math. Research Letters **6** (1999), 287–291.

[4] E. Artin, *Theorie der Zöpfe*, Abh. Math. Sem. Univ. Hamburg **4** (1925), 47–72.

[5] M. Autord, *Comparing Gröbner bases and word reversing*, Southeast Asian Bull. Math. **33** (2009), 639–663.

[6] E. Bannai, *Fundamental groups of the spaces of regular orbits of the finite unitary reflection groups of dimension* 2, J. Math. Soc. Japan **28** (1976), 447–454.

[7] D. Bessis, *Garside categories, periodic loops and cyclic sets*, arXiv:0610778.

[8] ______, *Zariski theorems and diagrams for braid groups*, Invent. Math. **145** (2001), 487–507.

[9] ______, *The dual braid monoid*, Ann. Sci. École Norm. Sup. **36** (2003), 647–683.

[10] ______, *A dual braid monoid for the free group*, J. Algebra **302** (2006), 55–69.

[11] ______, *Finite complex reflection arrangements are* $K(\pi, 1)$, Ann. of Math. **181** (2015), 809–904.

[12] D. Bessis and R. Corran, *Non-crossing partitions of type* (e, e, r), Adv. in Math. **202** (2006), 1–49.

[13] D. Bessis, F. Digne, and J. Michel, *Springer theory in braid groups and the Birman-Ko-Lee monoid*, Pacific J. Math. **205** (2002), 287–309.

[14] D. Bessis and V. Reiner, *Cyclic sieving of noncrossing partitions for complex reflection groups*, Ann. of Combin. **15** (2011), no. 2, 197–222.

[15] M. Bestvina, *Non-positively curved aspects of Artin groups of finite type*, Geom. Topol. **3** (1999), 269–302.

[16] P. Biane and P. Dehornoy, *Dual Garside structure of braids and free cumulants of products*, Sém. Lothar. Combin. **72** (2014), Article B72b, 15 pp.

[17] J. Birman, *Braids, Links, and Mapping Class Groups*, Annals of Math. Studies, vol. 82, Princeton Univ. Press, 1974.

[18] J. Birman, V. Gebhardt, and J. González-Meneses, *Conjugacy in Garside groups I: cyclings, powers and rigidity*, Groups Geom. Dyn. **1** (2007), 221–279.

[19] ______, *Conjugacy in Garside groups III: periodic braids*, J. Algebra **316** (2007), 746–776.

[20] ______, *Conjugacy in Garside groups II: structure of the ultra summit set*, Groups Geom. Dyn. **2** (2008), 16–31.

[21] J. Birman, K.H. Ko, and S.J. Lee, *A new approach to the word problem in the braid groups*, Advances in Math. **139** (1998), no. 2, 322–353.

[22] ______, *The infimum, supremum and geodesic length of a braid conjugacy class*, Advances in Math. **164** (2001), 41–56.

[23] A. Bjorner and F. Brenti, *Combinatorics of Coxeter Groups*, Graduate Texts in Mathematics, no. 231, Springer Verlag, 2005.

[24] C.F. Bödigheimer and B. Visy, *Factorable groups and their homology*, 10.41717OWR/2010/32, 2010.

[25] A. Borel, *Algebraic Groups*, Graduate Texts in Mathematics, no. 126, Springer Verlag, 1991.

[26] S. Bouc, *Homologie de certains ensembles de 2-sous-groupes des groupes symétriques*, Journal of Algebra **150** (1992), 158–186.

[27] N. Bourbaki, *Groupes et Algèbres de Lie, chapitres IV, V, VI*, Hermann, Paris, 1968.

[28] S. Boyer, C. McA. Gordon, and L. Watson, *On L-spaces and left-orderable fundamental groups*, Math. Annalen **356** (2013), 1213–1245.

[29] S. Boyer, D. Rolfsen, and B. Wiest, *Orderable 3-manifold groups*, Ann. Inst. Fourier **55** (2005), 243–288.

[30] T. Brady and C. Watt, *A partial order on the orthogonal group*, Comm. Algebra **30** (2002), 3749–3754.

[31] ______, *Non-crossing partition lattices in finite real reflection groups*, Trans. Amer. Math. Soc. **360** (2008), 1983–2005.

[32] C. Brav and H. Thomas, *Braids groups and Kleinian singularities*, Math. Ann. **351** (2011), 1005–1017.

[33] M.R. Bridson and S.M. Gersten, *The optimal isoperimetric inequality for torus bundles over the circle*, Quart. J. Math. Oxford **47** (1996), 1–23.

[34] E. Brieskorn, *Die Fundamentalgruppe des Raumes der regulären Orbits einer endlichen komplexen Spiegelungsgruppe*, Invent. Math. **12** (1971), 57–61.

[35] ______, *Automorphic sets and braids and singularities*, Braids, Contemporary Mathematics, vol. 78, American Mathematical Society, 1988, pp. 45–117.

[36] E. Brieskorn and K. Saito, *Artin-Gruppen und Coxeter-Gruppen*, Invent. Math. **17** (1972), 245–271.

[37] B. Brink and R. B. Howlett, *Normalizers of parabolic subgroups in Coxeter groups*, Inventiones Math. **136** (1999), 323–351.

[38] M. Broué, *Isométries parfaites, types de blocs, catégories dérivées*, Astérisque **181-182** (1990), 61–92.

[39] ______, *Introduction to Complex Reflection Groups and their Braid Groups*, Lectures Notes in Math., no. 1988, Springer Verlag, 2010.

[40] M. Broué and G. Malle, *Théorèmes de Sylow génériques pour les groupes réductifs sur les corps finis*, Mathematische Annalen **292** (1992), 241–262.

[41] M. Broué, G. Malle, and R. Rouquier, *Complex reflection groups, braid groups, Hecke algebras*, J. Reine Angew. Mathematik **500** (1998), 127–190.

[42] M. Broué and J. Michel, *Sur certains éléments réguliers des groupes de Weyl et les variétés de Deligne-Lusztig associées*, Progress in Math., vol. 141, pp. 73–139, Birkhauser, 1996.

[43] K.S. Brown, *Cohomology of Groups*, Springer Verlag, 1982.

[44] S. Burckel, *The well-ordering on positive braids*, J. Pure Appl. Algebra **120** (1997), no. 1, 1–17.

[45] F. Callegaro and I. Marin, *Homology computations for complex braid groups*, J. European Math. Soc. **16** (2014), 103–164.

[46] M. Calvez, *Fast Nielsen-Thurston classification of braids*, Algebr. Geom. Topol. **14** (2014), no. 3, 1745–1758.

[47] M. Calvez and B. Wiest, *A fast solution to the conjugacy problem in the 4-strand braid group*, J. Group Th. **17** (2014), no. 5, 757–780.

[48] J.W. Cannon, W.J. Floyd, and W.R. Parry, *Introductory notes on Richard Thompson's groups*, Enseign. Math. **42** (1996), 215–257.

[49] O. Caramello and A.C. Russo, *The Morita-equivalence between MV-algebras and abelian ℓ-groups with strong unit*, arXiv:1312.1272.

[50] L. Carlucci, P. Dehornoy, and A. Weiermann, *Unprovability statements involving braids*, Proc. London Math. Soc. **102** (2011), no. 1, 159–192.

[51] H. Cartan and S. Eilenberg, *Homological Algebra*, Princeton University Press, Princeton, 1956.

[52] R. W. Carter, *Finite Groups of Lie type: Conjugacy Classes and Complex Characters*, Wiley, 1985.

[53] F. Cedó, E. Jespers, and J. Okniński, *Braces and the Yang–Baxter equation*, Comm. Math. Physics **327** (2014), 101–116, arXiv:1205.3587.

[54] R. Charney, *Artin groups of finite type are biautomatic*, Math. Ann. **292** (1992), no. 4, 671–683.

[55] ———, *Geodesic automation and growth functions for Artin groups of finite type*, Math. Ann. **301** (1995), no. 2, 307–324.

[56] R. Charney and J. Meier, *The language of geodesics for Garside groups*, Math. Zeitschr. **248** (2004), 495–509.

[57] R. Charney, J. Meier, and K. Whittlesey, *Bestvina's normal form complex and the homology of Garside groups*, Geom. Dedicata **105** (2004), 171–188.

[58] R. Charney and D. Peifer, *The $K(\pi, 1)$-conjecture for the affine braid groups*, Comm. Math. Helv. **78** (2003), 584–600.

[59] F. Chouraqui, *Garside groups and Yang-Baxter equations*, Comm. Algebra **38** (2010), no. 12, 4441–4460.

[60] F. Chouraqui and E. Godelle, *Folding of set-theoretical solutions of the Yang-Baxter equation*, Algebra Represent. Th. **15** (2012), 1277–1290.

[61] ———, *Finite quotients of groups of I-type*, Adv. in Math. **258** (2014), 46–68.

[62] A.H. Clifford and G.B. Preston, *The Algebraic Theory of Semigroups, volume 1*, Amer. Math. Soc. Surveys, vol. 7, Amer. Math. Soc., 1961.

[63] A.M. Cohen and D.B. Wales, *Linearity of Artin groups of finite type*, Israel J. Math. **131** (2002), 101–123.

[64] R. Corran, *A normal form for a class of monoids containing the singular braid monoids*, J. Algebra **223** (2000), 256–282.

[65] ———, *On monoids related to braid groups*, PhD. Thesis, University of Sydney, 2000.

[66] R. Corran and M. Picantin, *A new Garside structure for braids groups of type (e,e,r)*, J. London Math. Soc. **84** (2011), no. 3, 689–711.

[67] J. Crisp, *Symmetric subgroups of Artin groups*, Adv. Math. **152** (2000), 159–177.

[68] J. Crisp and L. Paris, *Representations of the braid group by automorphisms of groups, invariants of links, and Garside groups*, Pac. J. Maths **221** (2005), 1–27.

[69] M. Dabkovska, M. Dabkowski, V. Harizanov, J. Przytycki, and M. Veve, *Compactness of the space of left orders*, J. Knot Th. and Ramifications **16** (2007), 267–256.

[70] P. Dehornoy, Π^1_1-*complete families of elementary sequences*, Ann. P. Appl. Logic **38** (1988), 257–287.

[71] ______, *Free distributive groupoids*, J. Pure Appl. Algebra **61** (1989), 123–146.

[72] ______, *Deux propriétés des groupes de tresses*, C. R. Acad. Sci. Paris Sér. I Math. **315** (1992), 633–638.

[73] ______, *Preuve de la conjecture d'irréflexivité pour les structures distributives libres*, C. R. Acad. Sci. Paris Sér. I Math. **314** (1992), 333–336.

[74] ______, *Structural monoids associated to equational varieties*, Proc. Amer. Math. Soc. **117** (1993), no. 2, 293–304.

[75] ______, *Braid groups and left distributive operations*, Trans. Amer. Math. Soc. **345** (1994), no. 1, 115–151.

[76] ______, *Groups with a complemented presentation*, J. Pure Appl. Algebra **116** (1997), 115–137.

[77] ______, *Braids and Self-Distributivity*, Progress in Math., vol. 192, Birkhäuser, 2000.

[78] ______, *On completeness of word reversing*, Discrete Math. **225** (2000), 93–119.

[79] ______, *The fine structure of LD-equivalence*, Advances in Math. **155** (2000), 264–316.

[80] ______, *Groupes de Garside*, Ann. scient. Éc. Norm. Sup. 4e série **35** (2002), 267–306.

[81] ______, *Study of an identity*, Algebra Universalis **48** (2002), 223–248.

[82] ______, *Complete positive group presentations*, J. of Algebra **268** (2003), 156–197.

[83] ______, *The group of fractions of a torsion free lcm monoid is torsion free*, J. of Algebra **281** (2004), 303–305;.

[84] ______, *Geometric presentations of Thompson's groups*, J. Pure Appl. Algebra **203** (2005), 1–44.

[85] ______, *Combinatorics of normal sequences of braids*, J. Combinatorial Th. Series A **114** (2007), 389–409.

[86] ______, *Alternating normal forms for braids and locally Garside monoids*, J. Pure Appl. Algebra **212** (2008), no. 11, 2416–2439.

[87] ______, *Efficient solutions to the braid isotopy problem*, Disc. Appl. Math. **156** (2008), 3094–3112.

[88] ———, *Left-Garside categories, self-distributivity, and braids*, Ann. Math. Blaise Pascal **16** (2009), 189–244.

[89] ———, *The word reversing method*, Intern. J. Alg. and Comput. **21** (2011), 71–118.

[90] ———, *Tamari Lattices and the symmetric Thompson monoid*, Associahedra, Tamari lattices, and Related Structures (F.Mueller-Hoissen, J.Pallo, J.Stasheff, and H.O.Walther, eds.), Trends in Mathematics, Birkhäuser Verlag, 2012.

[91] ———, *Coxeter-like groups for groups of set-theoretic solutions of the Yang–Baxter equation*, Comptes Rendus Math. **351** (2013), 419–424.

[92] ———, *Monoids of O-type, subword reversing, and ordered groups*, J. of Group Theory **17** (2014), no. 3, 465–524.

[93] P. Dehornoy, F. Digne, and J. Michel, *Garside families and Garside germs*, J. of Algebra **380** (2013), 109–145.

[94] P. Dehornoy, M. Dyer, and C. Hohlweg, *Garside families in Artin–Tits monoids and low elements in Coxeter groups*, Comptes Rendus Math. **353** (2015), 403–408.

[95] P. Dehornoy and V. Gebhardt, *Algorithms for Garside calculus*, J. of Symbol. Comput. **63** (2014), 68–116.

[96] P. Dehornoy and E. Godelle, *A conjecture about Artin–Tits groups*, J. Pure Appl. Algebra **217** (2013), 741–756.

[97] P. Dehornoy and Y. Guiraud, *Quadratic normalisation in monoids*, arXiv:1504.02717.

[98] P. Dehornoy and Y. Lafont, *Homology of Gaussian groups*, Ann. Inst. Fourier (Grenoble) **53** (2003), no. 2, 1001–1052.

[99] P. Dehornoy and L. Paris, *Gaussian groups and Garside groups, two generalisations of Artin groups*, Proc. London Math. Soc. **79** (1999), no. 3, 569–604.

[100] P. Dehornoy, with I. Dynnikov, D. Rolfsen, and B. Wiest, *Ordering Braids*, Math. Surveys and Monographs vol. 148, Amer. Math. Soc., 2008.

[101] P. Deligne, *Les immeubles des groupes de tresses généralisés*, Invent. Math. **17** (1972), 273–302.

[102] ———, *Action du groupe des tresses sur une catégorie*, Invent. Math. **128** (1997), 159–175.

[103] P. Deligne and G. Lusztig, *Representations of reductive groups over finite fields*, Ann. of Math. **103** (1976), 103–161.

[104] F. Deloup, *Palindromes and orderings in Artin groups*, J. Knot Theory Ramifications **19** (2010), no. 2, 145–162.

[105] F. Digne, *On the linearity of Artin braid groups*, J. Algebra **268** (2003), no. A, 39–57.

[106] ______, *Présentations duales pour les groupes de tresses de type affine $\widetilde{A}$*, Comm. Math. Helvetici **8** (2008), 23–47.

[107] ______, *A Garside presentation for Artin-Tits groups of type $\tilde{C}_n$*, Ann. Inst. Fourier **62** (2012), no. 2, 641–666.

[108] F. Digne, I. Marin, and J. Michel, *The center of pure complex braid groups*, J. Algebra **347** (2011), 206–213.

[109] F. Digne and J. Michel, *Garside and locally Garside categories*, arXiv:math.GR/0612652.

[110] ______, *Representations of Finite Groups of Lie Type*, London Math. Soc. Student Texts, no. 21, Cambridge University Press, 1991.

[111] ______, *Endomorphisms of Deligne-Lusztig varieties*, Nagoya Math. J. **183** (2006), 35–103.

[112] ______, *Parabolic Deligne-Lusztig varieties*, Adv. Math. **257** (2014), 136–218.

[113] F. Digne, J. Michel, and R. Rouquier, *Cohomologie des variétés de Deligne-Lusztig*, Advances in Math. **209** (2007), 749–822.

[114] V. Drinfeld, *On some unsolved problems in quantum group theory*, Quantum Groups, Lecture Notes in Math., vol. 1510, Springer Verlag, 1992, pp. 1–8.

[115] T. Dubrovina and N. Dubrovin, *On braid groups*, Sbornik Math. **192** (2001), 693–703.

[116] M. Dyer and C. Hohlweg, *Small roots, low elements, and the weak order in Coxeter groups*, arXiv:1505.02058.

[117] E.A. El-Rifai and H.R. Morton, *Algorithms for positive braids*, Quart. J. Math. Oxford Ser. **45** (1994), no. 2, 479–497.

[118] D. Epstein, *Curves on 2-manifolds and isotopies*, Acta Math. **115** (1966), 83–107.

[119] D. Epstein, with J.W. Cannon, D.F. Holt, S.V.F. Levy, M.S. Paterson, and W.P. Thurston, *Word Processing in Groups*, Jones and Bartlett Publ., 1992.

[120] P. Etingof, T. Schedler, and A. Soloviev, *Set-theoretical solutions to the quantum Yang-Baxter equation*, Duke Math. J. **100** (1999), no. 2, 169–209.

[121] R. Fenn and C.P. Rourke, *Racks and links in codimension 2*, J. Knot Th. and its Ramifications **1** (1992), 343–406.

[122] N. Franco and J. González-Meneses, *Conjugacy problem for braid groups and Garside groups*, J. Algebra **266** (2003), 112–132.

[123] J. Fromentin, *Every braid admits a short sigma-definite expression*, J. Europ. Math. Soc. **13** (2011), 1591–1631.

[124] F.A. Garside, *The theory of knots and associated problems*, PhD. Thesis, Oxford University, 1965.

[125] F.A. Garside, *The braid group and other groups*, Quart. J. Math. Oxford Ser. **20** (1969), 235–254.

[126] T. Gateva-Ivanova, *Garside structures on monoids with quadratic square-free relations*, Algebra Represent. Th. **14** (2011), 779–802.

[127] T. Gateva-Ivanova and M. Van den Bergh, *Semigroups of I-type*, J. of Algebra **206** (1998), 97–112.

[128] S. Gaussent, Y. Guiraud, and P. Malbos, *Coherent presentations of Artin monoids*, Compos. Math., to appear; arXiv:1203.5358.

[129] V. Gebhardt, *A new approach to the conjugacy problem in Garside groups*, J. Algebra **292** (2005), 282–302.

[130] ______, *Computing growth functions of braid monoids and counting vertex-labelled bipartite graphs*, J. Combinat. Th., Series A **120** (2013), no. 1, 232–244.

[131] V. Gebhardt and J. González-Meneses, *The cyclic sliding operation in Garside groups*, Math Z. **265** (2010), 85–114.

[132] ______, *Solving the conjugacy problem in Garside groups by cyclic sliding*, J. Symb. Computation **45** (2010), 629–656.

[133] E. Godelle, *Garside subgroups of Garside groups*, unpublished notes.

[134] ______, *Normalisateurs et centralisateurs des sous-groupes paraboliques dans les groupes d'Artin–Tits*, PhD. Thesis, Université d'Amiens, 2001.

[135] ______, *Normalisateurs et groupes d'Artin-Tits de type sphérique*, J. Algebra **269** (2003), 263–274.

[136] ______, *Parabolic subgroups of Artin groups of type FC*, Pacific J. Math. **208** (2003), 243–254.

[137] ______, *Quasi-centraliser of LCM-homomorphisms*, Comm. in Algebra **34** (2006), 3167–3181.

[138] ______, *Parabolic subgroups of Garside groups*, J. Algebra **317** (2007), 1–16.

[139] ______, *Parabolic subgroups of Garside groups II: ribbons*, J. Pure Appl. Algebra **214** (2010), 2044–2062.

[140] E. Godelle and L. Paris, *PreGarside monoids and groups, parabolicity, amalgamation, and FC property*, Int. J. Alg. Comput. **23** (2013), 1431–1467.

[141] I. Goffa and E. Jespers, *Monoids of IG-type and maximal orders*, J. Algebra **308** (2007), 44–62.

[142] J. González-Meneses, *The nth root of a braid is unique up to conjugacy*, Alg. and Geom. Topology **3** (2003), 1103–1118.

[143] J. González-Meneses and B. Wiest, *On the structure of the centralizer of a braid*, Ann. Sci. Ec. Norm. Sup **37** (2004), no. 5, 729–757.

[144] A. Hatcher and W. Thurston, *A presentation for the mapping class group of a closed orientable surface*, Topology **19** (1980), no. 3, 221–237.

[145] X. He and S. Nie, *Minimal length elements of finite coxeter groups*, Duke Math. J. **161** (2012), no. 15, 2945–2967.

[146] A. Hess, *Factorable monoids: resolutions and homology via discrete Morse theory*, PhD. Thesis, University of Bonn, 2012, http://hss.ulb.uni-bonn.de/2012/2932/2932.pdf.

[147] A. Hess and V. Ozornova, *Factorability, string rewriting and discrete Morse theory*, arXiv:1412.3025.

[148] F. Hivert, J.-C. Novelli, and J.-Y. Thibon, *Sur une conjecture de Dehornoy*, Comptes-Rendus Math. **346** (2008), 375–378.

[149] J.G. Hocking and G.S. Young, *Topology*, Addison-Wesley, Reading MA, 1961.

[150] J.E. Humphreys, *Linear Algebraic Groups*, Graduate Texts in Mathematics, no. 21, Springer Verlag, 1975.

[151] T. Ito, *Construction of isolated left-orderings via partially central cyclic amalgamation*, arXiv:1107.0545.

[152] E. Jespers and J. Okninski, *Monoids and groups if I-type*, Algebra and Represent. Th. **8** (2005), 709–729.

[153] ———, *Noetherian Semigroup Algebras*, Algebra and Applications, vol. 7, Springer-Verlag, 2007.

[154] M. Jimbo, *Introduction to the Yang–Baxter equation*, Int. J. of Modern Physics A **4** (1989), no. 15, 3759–3777.

[155] D. Joyce, *A classifying invariant of knots: the knot quandle*, J. Pure Appl. Algebra **23** (1982), 37–65.

[156] C. Kassel, *Quantum Groups*, Grad. Texts in Math., Springer Verlag, 1994.

[157] C. Kassel and V. Turaev, *Braid Groups*, Grad. Texts in Math., Springer Verlag, 2008.

[158] K.H. Ko, S. Lee, J.H. Cheon, J.W. Han, J. Kang, and C. Park, *New public-key cryptosystem using braid groups*, Proc. Crypto 2000, Lecture Notes in Comput. Sci., vol. 1880, Springer Verlag, 2000, pp. 166–184.

[159] Y. Kobayashi, *Complete rewriting systems and homology of monoid algebras*, J. Pure Appl. Algebra **65** (1990), 263–275.

[160] A.I. Kokorin, V.M. Kopyutov, and N.Ya. Medvedev, *Right-Ordered Groups*, Plenum Publishing Corporation, 1996.

[161] D. Krammer, *Garside theory*, homepages.warwick.ac.uk/~masbal/index_files/gt080128.pdf.

[162] ———, *The braid group B_4 is linear*, Invent. Math. **142** (2000), 451–486.

[163] ———, *Braid groups are linear*, Ann. of Math. **155** (2002), no. 1, 131–156.

[164] ———, *The braid group of $\mathbb{Z}^n$*, J. Lond. Math. Soc. **76** (2007), 293–312.

[165] ———, *A class of Garside groupoid structures on the pure braid group*, Trans. Amer. Math. Soc. **360** (2008), 4029–4061.

[166] ———, *An asymmetric generalisation of Artin monoids*, Groups, Complexity Cryptology **5** (2013), no. 1, 141–167.

[167] D. Kuske, *Divisibility monoids: presentation, word problem, and rational languages*, FCT01, Lecture Notes in Computer Science, vol. 2138, Springer Verlag, 2001, pp. 227–239.

[168] Y. Lafont, *A new finiteness condition for monoids presented by complete rewriting systems (after Craig C. Squier)*, J. Pure Appl. Algebra **98** (1995), 229–244.

[169] R. Laver, *The left distributive law and the freeness of an algebra of elementary embeddings*, Adv. in Math. **91** (1992), no. 2, 209–231.

[170] E.K. Lee and S.J. Lee, *A Garside-theoretic approach to the reducibility problem in braid groups*, J. Algebra **320** (2008), 783–820.

[171] S.J. Lee, *Garside groups are strongly translation discrete*, J. Algebra **309** (2007), 594–609.

[172] G.I. Lehrer and J. Michel, *Invariant theory and eigenspaces for unitary reflection groups*, Comptes-Rendus Acad. Sci. Paris **336** (2003), 795–800.

[173] G.I. Lehrer and T. A. Springer, *Reflection subquotients of unitary reflection groups*, Canad. J. Math. **51** (1999), 1175–1193.

[174] A. Levy, *Basic Set Theory*, Springer Verlag, 1979.

[175] P. Linnell, *The space of left orders of a group is either finite or uncountable*, Bull. Lond. Math. Soc. **43** (2011), no. 1, 200–202.

[176] G. Lusztig, *On certain varieties attached to a Weyl group element*, Bull.Inst.Math.Acad.Sinica **6** (2011), 377–414.

[177] S. MacLane, *Categories for the Working Mathematician*, Grad. Texts in Math., Springer Verlag, 1998.

[178] J. Mairesse and F. Matheus, *Growth series for Artin groups of dihedral type*, Internat. J. Algebra Comput. **16** (2006), 1087–1107.

[179] S.V. Matveev, *Distributive groupoids in knot theory*, Math. Sbornik **119** (1982), no. 1-2, 78–88.

[180] J. McCammond, *An introduction to Garside structures* , http://www.math.ucsb.edu/~mccammon/papers/intro-garside.pdf.

[181] ______, *Dual Euclidean Artin groups and the failure of the lattice property*, http://www.math.ucsb.edu/ jon.mccammond/papers/lattice-failure.pdf; Talk given in Cap Hornu Garside meeting, May 2012.

[182] J. Michel, *Sigma-divisibilité*, Talk at Journées Tresses, Marne-la-Vallée, November 16-18, 2000.

[183] ______, *The development version of the CHEVIE package of GAP3*, arXiv:1310.7905 [math.RT], to appear in J. of Algebra.

[184] ______, *A note on words in braid monoids*, J. Algebra **215** (1999), 366–377.

[185] D. Morris, *Amenable groups that act on the line*, Algebr. Geom. Topol. **6** (2006), 2509–2518.

[186] F. Mueller-Hoissen, J.Pallo, J.Stasheff, and H.O.Walther (eds), *Associahedra, Tamari Lattices, and Related Structures*, Trends in mathematics, Birkhäuser Verlag, 2012.

[187] D. Mundici, *Mapping abelian ℓ-groups with strong unit one-one into MV-algebras*, J. Algebra **98** (1986), 76–81.

[188] A. Navas, *On the dynamics of (left) orderable groups*, Ann. Inst. Fourier **60** (2010), 1685–1740.

[189] ______, *A remarkable family of left-ordered groups: central extensions of Hecke groups*, J. Algebra **328** (2011), 31–42.

[190] S. Orevkov, *Quasipositivity Problem for 3-Braids*, Turk. J. Math. **28** (2004), 89–93.

[191] V. Ozornova, *Factorability, Discrete Morse Theory and a Reformulation of the $K(\pi,1)$-Conjecture*, PhD Thesis, Universität Bonn, 2013, http://hss.ulb.uni-bonn.de/2013/3117/3117.pdf.

[192] L. Paris, *Centralizers of parabolic subgroups of Artin groups of type A_l, B_l, and D_l*, J. Algebra **196** (1997), no. 2, 400–435.

[193] ———, *Parabolic subgroups of Artin groups*, J. Algebra **196** (1997), no. 2, 369–399.

[194] ———, *Artin monoids embed in their groups*, Comment. Math. Helv. **77** (2002), no. 3, 609–637.

[195] M. Picantin, *Tree products of cyclic groups and HNN extensions*, arXiv:1306.5724.

[196] ———, *Petits groupes gaussiens*, PhD Thesis, Université de Caen, 2000.

[197] ———, *The center of thin Gaussian groups*, J. Algebra **245** (2001), no. 1, 92–122.

[198] ———, *The conjugacy problem in small Gaussian groups*, Comm. Algebra **29** (2001), no. 3, 1021–1038.

[199] ———, *Automatic structures for torus link groups*, J. Knot Th. Ramifications **12** (2003), no. 6, 833–866.

[200] ———, *Garside monoids vs divisibility monoids*, Math. Struct. Comp. Sci. **15** (2005), no. 2, 231–242.

[201] J.H. Remmers, *On the geometry of positive presentations*, Advances in Math. **36** (1980), 283–296.

[202] C. Rivas, *Left-orderings on free products of groups*, J. Algebra **350** (2012), 318–329.

[203] W. Rump, *Right l-groups, geometric Garside groups, and solutions of the quantum Yang–Baxter equation*, J. Algebra, to appear.

[204] ———, *A decomposition theorem for square-free unitary solutions of the quantum Yang-Baxter equation*, Adv. Math. **193** (2005), 40–55.

[205] ———, *Braces, radical rings, and the quantum Yang-Baxter equation*, J. of Algebra **307** (2007), 153–170.

[206] ———, *Classification of cyclic braces*, J. Pure Appl. Algebra **209** (2007), 671–685.

[207] ———, *Generalized radical rings, unknotted biquandles, and quantum groups*, Colloq. Math. **109** (2007), 85–100.

[208] M. Salvetti, *Topology of the complement of real hyperplanes in* $\mathbf{C}^N$, Invent. Math. **88** (1987), no. 3, 603–618.

[209] V. Sergiescu, *Graphes planaires et présentations des groupes de tresses*, Math. Zeitschr. **214** (1993), 477–490.

[210] G. Shephard and J. Todd, *Finite unitary reflection groups*, Canad. J. Math. **6** (1954), 274–304.

[211] H. Sibert, *Extraction of roots in Garside groups*, Comm. Algebra **30** (2002), no. 6, 2915–2927.

[212] ———, *Tame Garside monoids*, J. Algebra **281** (2004), 487–501.

[213] A.S. Sikora, *Topology on the spaces of orderings of groups*, Bull. London Math. Soc. **36** (2004), 519–526.

[214] L. Solomon, *A Mackey formula in the group ring of a Coxeter group*, J. Algebra **41** (1976), 255–268.

[215] T. A. Springer, *Regular elements of finite reflection groups*, Invent. Math. **25** (1974), 159–198.

[216] ———, *Algebraic Groups, 2nd Edition*, Progress in Mathematics, no. 9, Birkhaüser, 1998.

[217] C. Squier, *A finiteness condition for rewriting systems, revision by F. Otto and Y. Kobayashi*, Theoret. Compt. Sci. **131** (1994), 271–294.

[218] ———, *The homological algebra of Artin groups*, Math. Scand. **75** (1994), 5–43.

[219] R. Steinberg, *Differential equations invariant under finite reflection groups*, Trans. Amer. Math. Soc. **112** (1964), 392–400.

[220] V.B. Stychnev, *Izvlechenie kornya v grupe koc*, Seria Matematichevskaya **42** (1978), no. 5, 1120–1131.

[221] J. Szép, *On the structure of groups which can be represented as the product of two subgroups*, Acta Sci. Math. Szeged **12** (1950), 57–61.

[222] D. Tamari, *The algebra of bracketings and their enumeration*, Nieuw Archief voor Wiskunde **10** (1962), 131–146.

[223] V. Tararin, *On the theory of right orderable groups*, Matem. Zametki **54** (1993), no. 2, 96–98, Translation to english in Math. Notes 54 (1994), 833-834.

[224] J. Tate and M. Van den Bergh, *Homological properties of Sklyanin algebras*, Invent. Math. **124** (1996), 619–647.

[225] K. Tatsuoka, *An isoperimetric inequality for Artin groups of finite type*, Trans. Amer. Math. Soc. **339** (1993), no. 2, 537–551.

[226] W. Thurston, *Finite state algorithms for the braid group*, Circulated notes, 1988.

[227] B. Visy, *Factorable groups and their homology*, PhD Thesis, Universität Bonn, 2010, http://hss.ulb.uni-bonn.de/2011/2559/2559.pdf.

[228] B. Wajnryb, *An elementary approach to the mapping class group of a surface*, Geometry and Topology **3** (1999), 405–466.

[229] B. Wiest, *How to read the length of a braid from its curve diagram*, Groups, Geometry and Dynamics **5** (2011), no. 3, 673–681.

[230] G. Zappa, *Sulla costruzione dei gruppi prodotto di due dati sottogruppi permutabili traloro*, Atti Secondo Congresso Un. Mat. Ital., Edizioni Cremonense, Rome, 1942, pp. 119–125.

Index